Amino Acids in Higher Plants

Amino Acids in Higher Plants

Edited by

J.P.F. D'Mello

Formerly of SAC, University of Edinburgh King's Buildings Campus, Edinburgh, UK

www.cabi.org

CABI is a trading name of CAB International

CABI
Nosworthy Way
Wallingford
Oxfordshire OX10 8DE
UK

Tel: +44 (0)1491 832111
Fax: +44 (0)1491 833508
E-mail: info@cabi.org
Website: www.cabi.org

CABI
38 Chauncy Street
Suite 1002
Boston, MA 02111
USA

Tel: +1 800 552 3083 (toll free)
E-mail: cabi-nao@cabi.org

A catalogue record for this book is available from the British Library, London, UK.

Library of Congress Cataloging-in-Publication Data

Amino acids in higher plants / edited by J.P.F. D'Mello.
pages cm
Includes bibliographical references and index.
ISBN 978-1-78064-263-5 (alk. paper)
1. Amino acids. 2. Plants--Metabolism. I. D'Mello, J.P. Felix.

QK898.A5A56 2015
572′.65--dc23

2014033212

ISBN-13: 978 1 78064 263 5
Commissioning editor: Rachel Cutts
Assistant editor: Alexandra Lainsbury
Production editor: James Bishop

Typeset by SPi, Pondicherry, India
Printed and bound in the UK by CPI Group (UK) Ltd, Croydon, CR0 4YY

Contents

Contributors

Arnao, M.B., Department of Plant Physiology, Faculty of Biology, University of Murcia, 30100-Murcia, Spain. E-mail: marino@um.es

Azevedo, R.A., Departamento de Genética, Escola Superior de Agricultura "Luiz de Queiroz", Universidade de São Paulo, Piracicaba, Brazil. E-mail: raa@usp.br

Barroso, J.B., Departamento de Bioquímica, Biología Celular y Molecular de Plantas, Estacíon Experimental del Zaídin, Consejo Superior de Investigaciones Científicas (CSIC), Apartado 419, E-18080 Granada, Spain. E-mail: jbarroso@ujaen.es

Bell, T.L., Faculty of Agriculture and Environment, University of Sydney, 1 Central Avenue, Eveleigh, NSW 2015, Australia. E-mail: tina.bell@sydney.edu.au

Bleeke, M.S., Regulatory Product Safety Center, Monsanto, 800 North Lindberg Boulevard, St Louis, MO 63167, USA. E-mail: marian.s.bleeke@monsanto.com

Brown, P.D., Departments of Chemistry, Biology and Environmental Studies, Trinity Western University, 7600 Glover Road, Langley, BC V2Y 1Y1, Canada. E-mail: paul.brown@twu.ca

Chandler, J.W., Institute of Developmental Biology, Cologne Biocenter, Cologne University, Zulpicher Strasse 47b, D-50674 Cologne, Germany. E-mail: john.chandler@uni-koeln.de

Corpas, F.J., Departamento de Bioquímica, Biología Celular y Molecular de Plantas, Estacíon Experimental del Zaídin, Consejo Superior de Investigaciones Científicas (CSIC), Apartado 419, E-18080 Granada, Spain. E-mail: Javier.corpas@eez.csic.es

Dayan, F.E., USDA Agricultural Research Service, Natural Products Utilization Research Unit, Oxford, MS 38677, USA. E-mail: franck.dayan@ars.usda.gov

del Rio, L.A., Departamento de Bioquímica, Biología Celular y Molecular de Plantas, Estacíon Experimental del Zaídin, Consejo Superior de Investigaciones Científicas (CSIC), Apartado 419, E-18080 Granada, Spain. E-mail: luisalfonso.delrio@eez.csic.es

D'Mello, J.P.F., Formerly of SAC (Scottish Agricultural College), University of Edinburgh King's Buildings Campus, West Mains Road, Edinburgh, EH9 3JG, UK. E-mail: jpfdmello@hotmail.co.uk

Duff, S.M.G., Monsanto, 700 Chesterfield Parkway West, Chesterfield, MO 63017, USA. E-mail: Stephen.m.duff@monsanto.com

Duke, S.O., USDA Agricultural Research Service, Natural Products Utilization Research Unit, Oxford, MS 38677, USA. E-mail: stephen.duke@ars.usda.gov

Farmer, D.R., Regulatory Product Safety Center, Monsanto, 800 North Lindberg Boulevard, St Louis, MO 63167, USA. E-mail: donna.r.farmer@monsanto.com

Fatima, T., Sustainable Agricultural Systems Laboratory, USDA Agricultural Research Service, Beltsville Agricultural Research Center, Beltsville, MD 20705-2350, USA. Present address:

Department of Physiology and Pharmacology, University of Western Ontario, London, ON N6A 5C1, Canada. E-mail: tahirakazmi@yahoo.com

Gallardo, F., Departamento de Biología Molecular y Bioquímica, Facultad de Ciencias, Universidad de Málaga, E-29071 Malaga, Spain. E-mail: fgallardo@uma.es

Garcia, I., Instituto de Bioquímica Vegetal y Fotosíntesis, Consejo Superior de Investigaciones Científicas (CSIC) and Universidad de Sevilla, Sevilla, Spain. E-mail: Irene.garcia@ibvf.csic.es

Garrido-Aranda, A., Departamento de Biología Molecular y Bioquímica, Facultad de Ciencias, Universidad de Malága, E-29071 Malaga, Spain. E-mail: Aliciagarrido@uma.es

Gaziola, S.A., Departamento de Genética, Escola Superior de Agricultura "Luiz de Queiroz", Universidade de São Paulo, Piracicaba, Brazil. E-mail: sagaziol@usp.br

Gessler, A., Swiss Federal Institute for Forest, Snow and Landscape Research WSL, Zurcherstrasse 111, 8903 Birmensdorf, Switzerland. E-mail: arthur.gessler@wsl.ch

Gotor, C., Instituto de Bioquímica Vegetal y Fotosíntesis, Consejo Superior de Investigaciones Científicas (CSIC) and Universidad de Sevilla, Sevilla, Spain. E-mail: gotor@ibvf.csic.es

Handa, A.K., Department of Horticulture and Landscape Architecture, Purdue University, West Lafayette, IN 47907, USA. E-mail: ahanda@purdue.edu

He, Z-H., Department of Biology, San Franscisco State University, 1600 Holloway Avenue, San Franscisco, CA 94132, USA. E-mail: zhe@sfsu.edu

Hernandez-Ruiz, J., Department of Plant Physiology, Faculty of Biology, University of Murcia, 30100-Murcia, Spain. E-mail: jhruiz@um.es

Hoefgen, R., Max Planck Institute of Molecular Plant Physiology, 14424 Potsdam-Golm, Germany. E-mail: hoefgen@mpimp-golm.mpg.de

Hubberten, H-M., Max Planck Institute of Molecular Plant Physiology, 14424 Potsdam-Golm, Germany. E-mail: hmhubberten@gmail.com

Hudson, A.O., Thomas H. Gosnell School of Life Sciences, Rochester Institute of Technology, 85 Lomb Memorial Drive, Rochester, NY 14623, USA. E-mail: aohsbi@rit.edu

Ingle, R.A., Department of Molecular and Cell Biology, University of Cape Town, Private Bag, Rondebosch 7701, South Africa. E-mail: Robert.Ingle@uct.ac.za

Ishihara, A., Faculty of Agriculture, Tottori University, Koyama, Tottori 680-8553, Japan. E-mail: aishihara@muses.tottori-u.ac.jp

Leasure, C.D., Department of Biology, San Franscisco State University, 1600 Holloway Avenue, San Franscisco, CA 94132, USA. E-mail: cleasure@sfsu.edu

Levesque, C.L., Department of Animal Science, South Dakota State University, Box 2170, Brookings, SD 57007, USA. E-mail: crystal.levesque@sdstate.edu

Limami, A.M., Research Institute of Horticulture and Seeds, University of Angers, 2 Bd Lavoisier, F-49045 Angers, France. E-mail: anis.limami@univ-angers.fr

Madu, W.C., Imo State Polytechnic, Owerri, Nigeria. E-mail: frmadu2000@aol.com

Majumdar, R., USDA Agricultural Research Service, 308 Sturtevant Hall, Geneva, NY 14456, USA. E-mail: rajtilak.majumdar@gmail.com

Matsukawa, T., Faculty of Biology-Oriented Science and Technology, Kinki University, Kinokawa 649-6493, Japan. E-mail: tmatsu@waka.kindai.ac.jp

Mattoo, A.K., Sustainable Agricultural Systems Laboratory, USDA Agricultural Research Service, Beltsville Agricultural Research Center, Beltsville, MD 20705-2350, USA. E-mail: autar.mattoo@ars.usda.gov

Medici, L.O., Departamento de Ciências Fisiológicas, Universidade Federal Rural do Rio de Janeiro, Seropédica CEP 23890-000, Brazil. E-mail: lomedici@ufrrj.br

Mehrsheikh, A., Regulatory Product Safety Center, Monsanto, 800 North Lindberg Boulevard, St Louis, MO 63167, USA. E-mail: akbar.mehrsheikh@monsanto.com

Minocha, R., USDA Forest Service, Northern Research Station, Durham, NH 03824, USA. E-mail: rminocha@fs.fed.us

Minocha, S.C., Department of Biological Sciences, University of New Hampshire, Durham, NH 03824, USA. E-mail: sminocha@unh.edu

Molina-Rueda, J.J., Departamento de Biología Molecular y Bioquímica, Facultad de Ciencias, Universidad de Malága, E-29071 Malaga, Spain. E-mail: juanjmolina@uma.es

Nazareno, A.C., Departamento de Genética, Escola Superior de Agricultura "Luiz de Queiroz", Universidade de São Paulo, Piracicaba, Brazil. E-mail: liriocirqueira@yahoo.com.br

Nomura, T., Biotechnology Research Center, Toyama Prefectural University, 5180 Kurokawa, Imizu, Toyama 939-0398, Japan. E-mail: tnomura@pu-toyama.ac.jp

Oikawa, A., Faculty of Agriculture, Yamagata University, Tsuruoka 997-8555, Japan. E-mail: oikawa@tds1.tr.yamagata-u.ac.jp

Okazaki, Y., Metabolomics Research Group, RIKEN Center for Sustainable Resource Science, Yokohama 230-0045, Japan. E-mail: yozo.okazaki@riken.jp

Okumoto, S., Department of Plant Pathology, Physiology and Weed Science, Virginia Tech, 512 Latham Hall, Blacksburg, VA 24061, USA. E-mail: sokumoto@vt.edu

O'Leary, B., Department of Plant Sciences, University of Oxford, South Parks Road, Oxford OX1 3RB, UK. E-mail: brendan.oleary@plants.ox.ac.uk

Ortega, J.L., Department of Plant and Environmental Sciences, New Mexico State University, Las Cruces, New Mexico 88003, USA. E-mail: jortegac@nmsu.edu

Osuji, G.O., College of Agriculture and Human Sciences, Prairie View A&M University, Prairie View, TX 77446, USA. E-mail: goosuji@pvamu.edu

Palma, J.M., Departamento de Bioquímica, Biología Celular y Molecular de Plantas, Estación Experimental del Zaídin, Consejo Superior de Investigaciones Científicas (CSIC), Apartado 419, E-18080 Granada, Spain. E-mail: josemanuel.palma@eez.csic.es

Pfautsch, S., Hawkesbury Institute of the Environment, University of Western Sydney, Locked Bag 1797, Penrith, NSW 2751, Australia. E-mail: s.pfautsch@uws.edu.au

Planchet, E., Research Institute of Horticulture and Seeds, University of Angers, 2 Bd Lavoisier, F-49045 Angers, France. E-mail: elisabeth.planchet@univ-angers.fr

Plaxton, W.C., Department of Biology, Queen's University, Kingston, Ontario K7L 3N6, Canada. E-mail: plaxton@queensu.ca

Price, M.B., Department of Plant Pathology, Physiology and Weed Science, Virginia Tech, 512 Latham Hall, Blacksburg, VA 24061, USA. E-mail: mprice@plant-ss.com

Raychaudhuri, A., Monsanto, 700 Chesterfield Parkway West, Chesterfield, MO 63011, USA. E-mail: aarayc@monsanto.com

Romero, L.C., Instituto de Bioquímica Vegetal y Fotosíntesis, Consejo Superior de Investigaciones Científicas (CSIC) and Universidad de Sevilla, Sevilla, Spain. E-mail: lromero@ibvf.csic.es

Rutherfurd, S.M., Riddet Institute, Massey University, Private Bag 11222, Palmerston North, New Zealand. E-mail: S.M.Rutherfurd@massey.ac.nz

Saito, K., Metabolomics Research Group, RIKEN Center for Sustainable Resource Science, Yokohama 230-0045, Japan. E-mail: kazuki.saito@riken.jp

Saltmiras, D.A., Regulatory Product Safety Center, Monsanto, 800 North Lindberg Boulevard, St Louis, MO 63167, USA. E-mail: david.a.saltmiras@monsanto.com

Schmidt, D., Departamento de Genética, Escola Superior de Agricultura "Luiz de Queiroz", Universidade de São Paulo, Piracicaba, Brazil. E-mail: dschmidt@usp.br

Sengupta-Gopalan, C., Department of Plant and Environmental Sciences, New Mexico State University, Las Cruces, New Mexico 88003, USA. E-mail: csgopala@nmsu.edu

Stotz, H.U., School of Life and Medical Science, University of Hertfordshire, Hatfield, AL10 9AB, UK. E-mail: h.stotz@herts.ac.uk

Sue, M., Department of Applied Biology and Chemistry, Tokyo University of Agriculture, Setagaya, Tokyo 156-8502, Japan. E-mail: sue@nodai.ac.jp

Tebayashi, S., Faculty of Agriculture, Kochi University, Nangoku 783-8502, Japan. E-mail: tebayasi@kochi-u.ac.jp

Tokuhisa, J., Department of Horticulture, Virginia Tech., Blacksburg, VA 24061, USA. E-mail: tokuhisa@vt.edu

Upadhyay, R.K., Sustainable Agricultural Systems Laboratory, USDA Agricultural Research Service, Beltsville Agricultural Research Center, Beltsville, MD 20705-2350, USA. Present address: Pennsylvania State University Harrisburg, Middletown, PA 17057, USA. E-mail: rakesh.upadhyay@ars.usda.gov

Watanabe, M., Max Planck Institute of Molecular Plant Physiology, 14424 Potsdam-Golm, Germany. E-mail: watanabe@mpimp-golm.mpg.de

Preface

Rationale

As a long-standing member of the academic community in Edinburgh, I have been immensely enthused by the very concept of the Higgs boson particle and the profound implications of recent discoveries highlighted in the media. The ensuing exuberance is still palpable and continues to cause reverberations throughout the scientific world, even as I compose this Preface. Biochemists, of course, have their own 'God Molecules' in the form of amino acids, the fundamental 'particles' of all living organisms. In 'Astrobiology: seeds of life?' Shock (2002) referred to the occurrence of amino acids in the interstellar medium, and speculation continues to this day as to the primordial processes preceding the formation of these indispensable molecules. It is salutary to note the extreme conditions permitting synthesis of amino acids in ongoing research investigations (Neish *et al.*, 2010; Parker *et al.*, 2011).

The consistent and ubiquitous distribution of precisely the same 20 amino acids in diverse living organisms is widely recognized. The preordained sequence and configurational prerequisites of amino acids in cellular protein structure and activity are features of profound biological significance. As a consequence, the term 'canonical' amino acids is frequently invoked to exemplify the fundamental requirements for biosynthetic processes in living organisms. Nevertheless, there are efforts now being directed at expanding the genetic lexicon to circumvent the above-mentioned sequence and structural constraints in order to facilitate the incorporation of xenobiotic amino acids into proteins (Young and Schultz, 2010).

In addition, specific amino acids have also been selected for special attention. For example, glutamate has been accorded with the title of an amino acid of 'particular distinction' in the context of mammalian metabolism (Young and Ajami, 2000). Equally, the diversity of functions ensures an important position for glutamate in the biochemistry of higher plants. However, the statement by Young and Ajami (2000) inevitably invites contradiction as there is much evidence in the literature to support the contention that other amino acids are similarly endowed with unique functions in higher plants, as in mammals. In all living organisms, the transport characteristics of amino acids are determined by structural and chemical orientation. The metabolic significance in human and livestock nutrition is further enhanced by classification of amino acids into essential (indispensable) and non-essential (dispensable) nutrients and glucogenic and ketogenic precursors. Thus, there has long been a rich diversity of interdisciplinary issues which, over the years, I have exploited in the publication of three titles: *Amino Acids in Farm Animal Nutrition* (D'Mello, 1994), *Amino Acids in Animal Nutrition* (D'Mello, 2003) and *Amino Acids in Human Nutrition and Health* (D'Mello, 2012).

The publication of *Amino Acids in Higher Plants* reflects my continuing commitment to the cause of amino acids both at the fundamental and practical levels, following the model adopted in my previous books (D'Mello, 1994, 2003, 2012). *Amino Acids in Higher Plants* has been designed specifically for academic, research and corporate institutions worldwide, particularly in Europe, the USA, Canada, Japan, Australia and New Zealand, but generally in all countries where English is a primary medium for instruction. This book should appeal to final year undergraduate and graduate students as well as to teaching and research staff. It is anticipated that it will be recommended reading for courses in the biological sciences, including botany, biochemistry, genetics and agronomy, but the multidisciplinary approach adopted in this volume should serve to attract a wider readership. *Amino Acids in Higher Plants* is also designed with the commercial sector in mind, particularly companies with progressive R&D agendas.

Overview

The chapters in *Amino Acids in Higher Plants* are arranged within a thematic structure as outlined below. The nature of the subjects under consideration and the need for continuity necessarily involves a certain amount of overlap. This should not be perceived as a detraction, as individual chapters are self-contained as a result, thereby reducing the need for cross-referencing to other sections of the book. This approach has also allowed authors increased flexibility in terms of emphasis and interpretation.

Part I Enzymes and Metabolism

As in my most recent volume (D'Mello, 2012), this section pursues the theme of amino acid metabolism through the driving actions of the principal enzymes, emphasizing recent advances particularly with reference to localization, biophysical characterization and regulation. Indeed, several of the enzymes and their associated pathways are common to both plants and animals. A key feature concerns the expression of enzymes and genetic manipulation. Other subdivisions in these chapters are designed to address issues such as cellular and sub-cellular functions, substrate requirements and availability, cofactor constraints and enzyme kinetics. A number of enzymes under review in this section catalyse rate-limiting steps in important metabolic pathways leading to the synthesis of physiologically active intermediates and end-products, including phytoalexins. In terms of regulation, both transcriptional and translational mechanisms are considered here for several enzymes and pathways. The changing facets of amino acid biochemistry in plants is another pervading theme allowing the consideration of comparative issues and integration of nitrogen with carbon metabolism. Part I has also been designed with the aim of underpinning issues such as signalling as well as immunity and environmental responses in plants.

Part II Dynamics

This section includes two chapters designed to explore transport dynamics at the cellular and whole-plant levels and the relationship to external factors. Transport and fluxes at the whole-organism level have been reviewed in *Amino Acids in Human Nutrition and Health* (D'Mello, 2012). The extent to which comparative issues will emerge in the two chapters in Part II is a cogent theme for current thinking and future research. Not unexpectedly, therefore, there is a need to consider the uptake of amino acids, transporter superfamilies and associated proteins as well as exchanges between cellular compartments. Specifically, metabolic transactions between the xylem and phloem, regulatory mechanisms in export dynamics and interactions with the rhizosphere will be pertinent components of these two chapters. Thus, the extensive transport and metabolic networking within plants and their interactions with the rhizosphere are significant issues for consideration in this section.

Part III Chemical Ecology

This series of chapters is designed to encapsulate the theme of biochemical diversity with reference to the impact of amino acids and related derivatives on metabolic pathways, ecology and the environment. Although other sections of this volume contain references to unique features, Part III has been designed to show biochemical diversity in a more comprehensive and deliberate manner.

Among the physiologically important compounds to be considered are those hitherto associated with defined roles within the animal domain but which are now attributed with actual or potential activity in higher plants. Although much remains to be elucidated regarding precise functions, this volume should contribute to the ongoing debate.

The issue of secondary metabolism has recently taken on renewed significance as a major component of plant defence strategy and stress management. There is no doubt that current perceptions of secondary metabolism are evolving at an unprecedented pace and several chapters in this section should reflect these changes.

Although there is an underlying theme of ecology in this series of chapters, other features should also emerge as part of the individual submissions. For example, the concept of modulation should be relevant in the topics covered here. Furthermore, the general principle of interactions will be developed more fully in Part III, with examples drawn from the effects of a variety of other endogenous bioactive molecules, including phytohormones. The role of abiotic factors may well impinge on these relationships and authors were encouraged to develop relevant concepts to underline biochemical diversity.

Part IV Plant Products: Quality and Safety

In this section, three chapters are presented to indicate the potential and actual value of plant products as sources of amino acids for the nutrition of humans and other vertebrates. The amino acid composition of dietary proteins is an important indicator of nutritional quality, and a chapter in this section is devoted to recent advances in methodologies concerning acid and alkaline hydrolysis and chromatographic analysis. In a rapidly changing field, this chapter is designed to update a previous review by Fontaine (2003) published in *Amino Acids in Animal Nutrition*. The determination of amino acid composition is only a first step in the evaluation of protein quality of foods. The actual value for humans and other animals is influenced by digestibility and the ultimate availability of the constituent amino acids at tissue level. These aspects are addressed in a dedicated chapter in Part IV. The toxicology of plant non-protein amino acids is relevant in this section as it is an important determinant of food and feed safety. In addition, however, there are potential applications for herbicidal and insecticidal activities of certain non-protein amino acids that deserve particular attention.

Part V Conclusions

The primary objective in the final section is to collate the main findings in the foregoing chapters within an integrated overview. Inevitably, there will be scope for additional information to incorporate developments emanating from a wide and fast-moving field of research. In this respect, issues such as comparative metabolism, signal transduction and molecular interactions may be considered within a more comprehensive perspective. Similarly, the effects of biotic and environmental pressures on amino acid metabolism and defence mechanisms may be developed into an integrated model of plant responses to a changing environment.

Acknowledgements

I have selected an international team of contributing authors who share my ideals and are actively publishing in peer-reviewed journals. I am indebted to them for sacrificing valuable time to prepare and submit authoritative and interesting chapters which should serve our readers well for the next decade or so. I am convinced that this volume will inspire many to enquire further and to develop new hypotheses for the future.

It was important to me that authors should be given freedom of expression in their unique areas of expertise, particularly as the terminology and technology becomes increasingly more complex. This approach has served me well in the past and I trust will succeed again with this volume.

Disclaimer

As might be expected, this edition *of Amino Acids in Higher Plants* contains references to and descriptors of commercial products. Authors were asked to refrain from excessive usage of trade names unless there were compelling reasons for doing so. It should be emphasized here that no endorsement of these products is implied or should be attributed to the Editor or to CAB International, the publishers of this volume. We confirm our absolute impartiality in the choice of chapter titles and authors and there has been no conflict of interests during this process.

The information set out within *Amino Acids in Higher Plants* is presented in good faith and in accordance with the principles of 'best practice' and 'due diligence'. Although every effort has been made to verify the facts and figures, neither the Editor nor CAB International can accept responsibility for the data presented in individual chapters or for any consequences of their use.

The popular press, and particularly internet sites, contain numerous articles extolling the virtues of a variety of amino acids. However, publication of this volume should not be interpreted as a recommendation for individuals to use these or any other amino acids for whatever purpose. *Amino Acids in Higher Plants* is intended exclusively for use as a text in education and research.

References

D'Mello, J.P.F. (ed.) (1994) *Amino Acids in Farm Animal Nutrition*. CAB International, Wallingford, UK.

D'Mello, J.P.F. (ed.) (2003) *Amino Acids in Animal Nutrition*. CAB International, Wallingford, UK.

D'Mello, J.P.F. (ed.) (2012) *Amino Acids in Human Nutrition and Health*. CAB International, Wallingford, UK.

Fontaine, J. (2003) Amino acid analysis of feeds. In: D'Mello, J.P.F. (ed.) *Amino Acids in Animal Nutrition*. CAB International, Wallingford, UK, pp. 15–40.

Neish, C.D., Somogyi, A. and Smith, M.A. (2010) Titan's primordial soup: formation of amino acids via low-temperature hydrolysis of tholins. *Astrobiology* 10, 337–347.

Parker, E.T., Cleaves, H.J., Dworkin, J.P., Callahan, M., Aubrey, A., Lazcano, A. and Bada, J.L. (2011) Primordial synthesis of amino acids in a 1958 Miller H_2S-rich spark discharge experiment. *Proceedings of the National Academy of Sciences of the United States of America* 108, 5526–5531.

Shock, E.L. (2002) Astrobiology: seeds of life? *Nature* 416, 380–381.

Young, T.S. and Schultz, P.G. (2010) Beyond the canonical 20 amino acids: expanding the genetic lexicon. *The Journal of Biological Chemistry* 285, 11039–11044.

Young, V.R and Ajami, A.M (2000) Glutamate: an amino acid of particular distinction. *Journal of Nutrition* 130, 892S–900S.

J.P.F. D'Mello
Editor

Glossary

Introduction

It is now virtually impossible to effectively consider the functional metabolism of amino acids in living organisms without at least some knowledge of the specific nomenclature and technical descriptors associated with this very diverse subject. Although many of the terms and acronyms appearing in *Amino Acids in Higher Plants* are already in the public domain, it was deemed important to provide as comprehensive a guide as possible to assist those readers who are new to this field. Further definitions are available in a wide range of scientific dictionaries, including, for example, the compilations of Thain and Hickman (2004), Allaby (2006), Parish *et al.* (2006), Singleton and Sainsbury (2006), Martin and Hine (2008), Lawrence (2011) and Lackie (2013). Readers will be aware of several free dictionaries available and accessible online. Glossaries and other relevant information also appear in specialist monographs and handbooks such as those edited by D'Mello *et al.* (1991), D'Mello (1997) and D'Mello (2012). In addition, standard textbooks including works by Alberts *et al.* (2008), Nelson and Cox (2008), Voet and Voet (2011) and Klug *et al.* (2014) are recommended as sources of in-depth information on different aspects of the biological sciences. Considerable emphasis is placed in this volume on metabolic and signal transduction pathways, and it is assumed that readers will be conversant with these aspects, at the minimum, to the level covered by Michal and Schomburg (2012). Research in the life sciences is profoundly dependent upon the application of cell and molecular interpretation, and authoritative and up-to-date information on molecular cell biology is available in the work of Lodish *et al.* (2013).

Definition of Terms and Acronyms

The important terms and acronyms used in *Amino Acids in Higher Plants* are listed and defined in Table 1. This compilation includes standard conventions as well as unique chapter-specific terms. Where appropriate, cross-referencing to individual chapters in this volume is included in order to facilitate a greater appreciation of the context of usage of selected terms.

Table 1. Explanation of relevant terms and acronyms used in *Amino Acids in Higher Plants*.

Abbreviation or Term	Definition
AA	Amino acid
AAA	Aromatic amino acid: phenylalanine, tyrosine and tryptophan (Chapters 21 and 28)
AAAH	Aromatic amino acid hydroxylase
AADC	Aromatic amino acid decarboxylase (Chapter 21)
AAP	Amino acid permease (Chapter 18)
AAT	Aspartate aminotransferase (Chapters 1 and 16), see also AspAT
ABA	Abscisic acid (Chapters 15 and 28)
ACC	1-Amino-cyclopropane-1-carboxylic acid (Chapters 10 and 28)
ACCase	Acetyl coenzyme A carboxylase (Chapter 1)
Acetyl-CoA	Acetyl coenzyme A (Chapter 11)
ACS	1-Amino-cyclopropane-1-carboxylic acid synthase (Chapter 7)
ACT	Amino acid/choline transporter (Chapter 17)
Acute toxicity	Severe adverse effects occurring within a relatively short period of exposure to a potentially harmful substance
ADC	Arginine decarboxylase (Chapters 9 and 10)
ADH	Alcohol dehydrogenase (Chapter 15)
ADI	Acceptable daily intake (Chapter 24)
ADim	Agmatine deiminase (Chapter 9)
ADME	Absorption, distribution, metabolism and excretion (Chapter 24)
ADP	Adenosine diphosphate (Chapter 5)
AEC	*S*-Aminoethylcysteine (Chapter 13)
AFLP	Amplified fragment length polymorphism (Chapter 11)
AFMK	N^1-Acetyl-N^2-formyl-5-methoxykynuramine (Chapter 21)
AGE	Advanced glycation end-products (Chapter 28)
AGT	Alanine:glyoxylate aminotransferase (Chapter 2)
AHAS	Acetohydroxyacid synthase (Chapter 13)
AICAR	5′-Phosphoribosyl-4-carboxamide-5-aminoimidazole (Chapter 14)
AIH	Agmatine iminohydrolase (Chapter 10)
AK	Aspartate kinase (Chapter 13)
AL	Argininosuccinate lyase
β-Ala	*β*-Alanine (Chapter 2)
AlaAT	Alanine aminotransferase (Chapters 2 and 15)
Allosteric	Multi-site enzyme modulation of structure and activity
ALS	Acetolactate synthase (Chapter 23)
ALS/PDC	Amyotrophic lateral sclerosis/Parkinsonism dementia complex (Chapter 27)
AMK	N^1-Acetyl-5-methoxykynuramine (Chapter 21)
AMP	Adenosine monophosphate (Chapter 21)
AMPA	Aminomethylphosphonic acid (Chapter 24)
AnS	Anthranilate synthase (Chapter 20)
ANT 1	Aromatic and neutral transporter (Chapter 17)
Antagonist	A compound acting as a metabolic inhibitor by virtue of structural analogy or functional competition
APA	Amino acid permease (Chapter 18)
APC	Amino acid-polyamine-organocation
APR	Adenosine 5′-phosphosulfate reductase (Chapters 11 and 12)
APS	Adenosine 5′-phosphosulfate (Chapter 12)
aRfD	Acute reference dose (Chapter 24)
Arg	Arginine
AS	Asparagine synthetase (Chapters 15, 18 and 23), see also AsnS
ASA	Aspartate semialdehyde (Chapter 13)
ASADH	Aspartate semialdehyde dehydrogenase (Chapter 13)
ASMT	Acetylserotonin methyltransferase (Chapter 21)
Asn	Asparagine (Chapter 6)

Continued

Table1. Continued.

Abbreviation or Term	Definition
Asnase	Asparaginase (Chapter 6)
AsnS	Asparagine synthetase (Chapter 6), see also AS
Asp	Aspartate
ASP	Asparaginase
AspAT	Aspartate aminotransferase (Chapters 2, 3, 6 and 18), see also AAT
ASS	Argininosuccinate synthetase (Chapter 28)
ATP	Adenosine triphosphate (Chapters 2, 5, 6, 12, 15 and 16)
ATP-PFK	ATP-dependent-phosphofructokinase (Chapter 16)
Avs	Avenanthramides (Chapter 20)
AVT	Amino acid vacuolar transporter (Chapter 17)
BABA	β-Aminobutyrate (Chapter 27)
BAT	Bidirectional amino acid transporter (Chapter 17)
BCAA	Branched-chain amino acid (Chapter 15 and 22)
BCAT	Branched-chain aminotransferase (Chapter 13, 15 and 22)
BCKA	Branched-chain keto acid
BCNA	β-Cyanoalanine (Chapter 27)
BH_4	Tetrahydrobiopterin (Chapter 8)
BMAA	β-*N*-Methylamino-L-alanine (Chapter 27)
BNF	Biological nitrogen fixation (Chapter 18)
BOAA	β-*N*-Oxalylamino-L-alanine (Chapter 27)
BR	Brassinosteroid (Chapter 28)
BTPC	Bacterial-type PEPC (Chapter 16)
Bxs	Benzoxazinones (Chapter 20)
CaM	Calmodulin (Chapter 7)
CaMV	*Cauliflower mosaic virus* (Chapter 9)
Carcinogenic	Causing cancer
CAS	β-Cyanoalanine synthase (Chapter 12)
CAT	Cationic amino acid transporter (Chapter 17)
CBL	Cystathionine-β-lyase (Chapter 13)
CCCP	Cyanide-*m*-chlorophenylhydrazone (Chapter 17)
cDNA	Complementary DNA (Chapters 1, 4, 13 and 22)
CDPK	Calcium-responsive protein kinase (Chapter 15)
cGMP	Cyclic guanosine monophosphate
CGS	Cystathionine-γ-synthase (Chapter 13)
Chronic toxicity	Adverse effects resulting from prolonged and repeated exposure to relatively small quantities of a potentially harmful substance
CID	Collision-induced dissociation (Chapter 25)
CIPK	Calcium-binding protein interacting kinase (Chapter 15)
Cit	Citrulline
CNS	Central nervous system (of animals)
CoA	Coenzyme A
CPA	*N*-Carbamoylputrescine amidohydrolase
CPS	Carbamoyl phosphate synthetase (Chapter 9)
CSA	CMP sialate:CMP antiporter (Chapter 17)
CSC	Cysteine synthase complex (Chapter 11 and 12)
CuAO	Copper-containing amine oxidase (Chapter 10)
Cys	Cysteine
DAA	Dispensable amino acid(s) (for animals; see also NEAA)
DABA	α,γ-Diaminobutyric acid (Chapter 27)
DAO	Diamine oxidase (Chapter 9)
DAPG	2,4-Diacetyl phloroglucinol (Chapter 17)
DART	Developmental and reproductive toxicity (Chapter 24)
DCAM	Decarboxylated 5-adenosyl-methionine
DES	Desulfhydrase (Chapter 12)
DHAD	Dihydroxy acid dehydratase (Chapter 13)

Continued

Table1. Continued.

Abbreviation or Term	Definition
DHAP	Dihydroxyacetone phosphate (Chapter 16)
DHDPS	Dihydrodipicolinate synthase (Chapter 13 and 23)
3,4-DHP	3-Hydroxy-4(1*H*)-pyridone (Chapter 27)
DIAAS	Digestible indispensable amino acid score (Chapter 25)
DIBOA	2,4-Dihydroxy-1,4-benzoxazin-3-one (Chapter 20)
DIBOA-Glc	2-*O*-β-D-glucopyranoside derivative of DIBOA (Chapter 20)
DIMBOA	7-Methoxy derivative of DIBOA (Chapter 20)
DIMBOA-Glc	2-*O*-β-D-glucopyranoside derivative of DIMBOA (Chapter 20)
DMT	Drug/metabolite transporter (Chapter 17)
DNA	Deoxyribonucleic acid (Chapter 21 and 24)
DOPA	3,4-Dihydroxyphenylalanine (Chapter 21 and 27)
DW	Dry weight
DXP	Deoxyxylulose phosphate (Chapter 3)
EAA	Essential amino acid(s) (for animals)
ECD	Electron capture dissociation (Chapter 25)
EDSP	Endocrine disruption screening program (Chapter 24)
EIN 3	ET insensitive 3 (Chapter 19)
EPSPS	5-Enolpyruvylshikimate-3-phosphate synthase (Chapter 23 and 24)
ET	Ethylene (ethene) (Chapter 28)
EU	European Union
FA	Fatty acid(s)
FAD	Flavin adenine dinucleotide (Chapter 8 and 19)
FAO	Food and Agriculture Organization (of the United Nations) (Chapter 24 and 25)
Fd_{ox}	Ferredoxin (oxidized)
Fd_{red}	Ferredoxin (reduced)
FDA	Food and Drug Administration (USA)
FDNB	Fluorodinitrobenzene (Chapter 25)
FMN	Flavin mononucleotide (Chapter 8)
FMO	Flavin monoxygenase (Chapter 19)
FMOC	9-Fluorenylmethyl chloroformate (Chapter 25)
FW	Fresh weight (Chapter 9)
GA	Gibberellin (Chapter 5)
GABA	γ-Aminobutyrate (Chapters 2, 5, 7, 9, 15, 16 and 17)
GABAP	GABA permease (Chapter 17)
GABA-T	GABA transaminase (Chapter 2 and 15)
GAD	Glutamic acid (glutamate) decarboxylase (Chapter 7, 9 and 15)
GAP	Good agricultural practice(s) (Chapter 24)
GAR	Glycinamide ribonucleotide (Chapter 1)
GARS	Glycinamide ribonucleotide synthase (Chapter 1)
GART	Glycinamide ribonucleotide transformylase (Chapter 1)
GB	Glycine betaine (Chapter 28)
GBSS	Granule-bound starch synthase (Chapter 1)
GCD	Glycine decarboxylase (Chapter 2)
GC-MS	Gas chromatography–mass spectroscopy (Chapter 19)
GCOS	Gene chip operating software (Chapter 11)
GDH	Glutamate dehydrogenase (Chapters 1, 6 and 18)
GDU 1	Glutamine dumper 1 (Chapter 17)
GGAT	Glutamate:glyoxylate aminotransferase (Chapter 2)
Gln	Glutamine (Chapter 5)
Glu	Glutamate
Gly	Glycine
GLR	Glutamate-like receptor (Chapter 28)
GMA	GDP-mannose:GMP antiporter (Chapter 17)

Continued

Table1. Continued.

Abbreviation or Term	Definition
GMO	Genetically modified organism (Chapter 24)
GMP	Guanosine monophosphate (Chapter 23)
GNMT	Glycine *N*-methyltransferase (Chapter 10)
GOGAT	Glutamate synthase (Chapters 1, 5, 6, 15, 16 and 18)
GOT	Glutamic oxaloacetic transaminase (Chapter 3)
GOX	Glyphosate oxidoreductase (Chapter 24)
GPAT	Glutamate:pyruvate aminotransferase (alternative for AlaAT) (Chapter 2)
GS	Glutamine synthetase (Chapters 1, 5, 6, 8, 16, 18 and 23)
GSA	Glutamate-semialdehyde (Chapter 15)
GSH	Glutathione (reduced) (Chapter 8 and 11)
GSL	Glucosinolate (Chapter 22)
GSNO	*S*-Nitrosoglutathione (Chapter 8)
GT	Glycosyltransferase (Chapter 20)
GTP	Guanosine triphosphate
GTR	Glucosinolate transporter (Chapter 22)
HATS	High-affinity transport system (Chapter 18)
Hcy	Homocysteine
HDC	Histidine decarboxylase
Hepatotoxic	Toxic to the liver
HIOMT	Hydroxyindole *O*-methyltransferase (Chapter 21)
His	Histidine
HISN1	ATP-phosphoribosyl transferase (Chapter 14)
HISN6	Histidinol-phosphate aminotransferase (Chapter 14)
HK	Homoserine kinase (Chapter 13)
HPLC	High-performance liquid chromatography (Chapter 25)
HR	Hypersensitive response (Chapter 12)
HSDH	Homoserine dehydrogenase (Chapter 13)
5-HT	5-Hydroxytryptamine (serotonin)
5-HTP	5-Hydroxytryptophan (Chapter 21)
IAA	Indole acetic acid (Chapters 5, 19, 21 and 28)
IAAld	Indole-3-acetaldehyde (Chapter 19)
4-Cl-IAA	4-Chloroindole-3-acetic acid (Chapter 19)
IAAO	Indicator amino acid oxidation (Chapter 26)
IAM	Indole-3-acetamide (Chapter 19)
IAN	Indole-3-acetonitrile (Chapter 19)
IAOX or IAO_x	Indole-3-acetaldoxime (Chapter 19 and 22)
IAP	Imidazole acetol-phosphate (Chapter 14)
IBA	Indole-3-butyric acid (Chapter 19)
ICP-MS	Inductively coupled plasma mass spectrometry (Chapter 11)
IDO	Indoleamine 2,3-dioxygenase (Chapter 21)
IG	Indole glucosinolate (Chapter 19)
iGluR	Ionotropic glutamate receptor (Chapter 28)
IGP	Indole-3-glycerol phosphate (Chapter 19)
Ile	Isoleucine
ImGP	Imidazole glycerol phosphate (Chapter 14)
IMP	Inosine monophosphate (inosinate)
IPA	Indole-3-pyruvic acid (Chapter 19)
IPMI	Isopropylmalate isomerase (Chapter 22)
ITC	Isothiocyanate (Chapter 22)
JA	Jasmonic acid (jasmonate) (Chapter 28)
JA-Ile	Jasmonate–isoleucine conjugate (Chapter 28)
α-KG	α-Ketoglutarate (Chapter 1, 5 and 18)
LAT	System-L amino acid transporter (Chapter 17)

Continued

Table1. Continued.

Abbreviation or Term	Definition
LATS	Low-affinity transport system (Chapter 18)
LCMT	Leucine carboxyl methyltransferase
LD_{50}	The dose that causes lethality in 50% of a population of test organisms (Chapter 24 and 27)
LDH	Lactate dehydrogenase (Chapter 15)
Leu	Leucine
Leucine-zipper proteins	Regulatory proteins with leucine residues facilitating interactions with DNA (Chapter 28)
LKR	Lysine α-ketoglutarate reductase (Chapter 13)
LNAA	Large neutral amino acids
LOR	Lysine 2-oxoglutarate reductase (Chapter 13)
LRR-RK	Leucine-rich repeat receptor kinase (Chapter 28)
Lys	Lysine
MA	Metabolic availability (Chapter 26)
MAM	Methylthioalkylmalate (Chapter 22)
MAO	Monoamine oxidase
MAPK	Mitogen-activated protein kinases (Chapter 15)
MAT	Methionine adenosyltransferase
MDH	Malate dehydrogenase
ME	Malic enzyme
MeJA	Methyl jasmonate (Chapter 9 and 28)
Met	Methionine
5-Me Trp	5-Methyl tryptophan (Chapter 17)
mGluR	Metabotropic G protein-coupled glutamate receptor
MRL	Maximum residue level (Chapter 24)
MRM	Multiple-reaction-monitoring (Chapter 25)
mRNA	Messenger RNA
MS	Methionine synthase (Chapter 13)
MSG	Methylsulfinylalkyl-GSL (Chapter 22)
MT	Methyltransferases
MTA	*S*-Methyl-5′-thioadenosine (Chapter 10)
MTG	Methylthioalkyl-GSL (Chapter 22)
Mutagenic	Causing mutations
N	Nitrogen
NAA	Neutral amino acids
NAD^+	Nicotinamide adenine dinucleotide (oxidized)
NADH	Nicotinamide adenine dinucleotide (reduced)
$NADP^+$	Nicotinamide adenine dinucleotide phosphate (oxidized)
NADPH	Nicotinamide adenine dinucleotide phosphate (reduced)
NAGK	*N*-Acetyl-Glu kinase (Chapter 9)
NAGPR	*N*-Acetyl-Glu-5-P reductase (Chapter 9)
NAGS	*N*-Acetyl-Glu synthase (Chapter 9)
NAOAT	N^2-Acetyl-Orn aminotransferase (Chapter 9)
NAOD	N^2-Acetyl-Orn deacetylase (Chapter 9)
NAOGAcT	N^2-Acetyl-Orn-Glu acetyltransferase (Chapter 9)
NAS	*N*-Acetylserine (Chapter 11)
NBUT	Nucleobase uptake transporter (Chapter 17)
NEAA	Non-essential amino acid(s)
NEM	*N*-Ethyl maleimide (Chapter 17)
Nephrotoxic	Toxic to the kidney
NGO	Non-governmental organization (Chapter 24)
NIH	National Institutes of Health (USA)
NiR	Nitrite reductase (Chapter 1)

Continued

Table1. Continued.

Abbreviation or Term	Definition
NMDA	*N*-Methyl-D-aspartate (Chapter 28)
NMR	Nuclear magnetic resonance (Chapter 25)
NO	Nitric oxide (Chapter 8, 9 and 12)
NOAEL	No observable adverse effect level (Chapter 24)
NOS	Nitric oxide synthase (Chapter 8)
NR	Nitrate reductase (Chapter 1)
NUE	Nitrogen use efficiency (Chapter 2)
OAA	Oxaloacetate (Chapter 16)
OAS	*O*-Acetylserine (Chapter 11)
OASTL	*O*-Acetylserine (thiol) lyase (Chapter 11 and 12)
OAT	Ornithine aminotransferase (Chapter 9, 15 and 28)
OCD	Ornithine cyclodeaminase (Chapter 9)
OCTase	Ornithine transcarbamoylase (Chapter 23)
ODC	Ornithine decarboxylase (Chapter 9 and 10)
OECD	Organisation for Economic Co-operation and Development (Chapter 24)
2-OG	2-Oxoglutarate (Chapter 2 and 16)
OPA	*O*-Phthalaldehyde (Chapter 25)
OPH	*O*-Phosphohomoserine (Chapter 13)
Orn	Ornithine
OTC	Ornithine transcarbamoylase (Chapter 9)
2-Oxoglu	2-Oxoglutarate (α-ketoglutarate) (Chapter 2)
PA	Polyamine (Chapter 9)
PAA	Phenylacetic acid (Chapter 19)
PAGE	Polyacrylamide gel electrophoresis
PAH	Phenylalanine hydroxylase
PAL	Phenylalanine ammonia-lyase (Chapter 20, 27 and 28)
PAO	Polyamine oxidase (Chapter 10)
PAPS	3′-Phosphoadenosine 5′-phosphosulfate (Chapter 11)
PBM	Peribacteroid membrane (Chapter 18)
P5C	Δ^1-Pyrroline-5-carboxylate (Chapter 15)
PCD	Programmed cell death (Chapter 7 and 12)
P5CDH	Δ^1-Pyrroline-5-carboxylate dehydrogenase (Chapter 15)
P5CS	Δ^1-Pyrroline-5-carboxylate synthetase (Chapter 15)
PCMBS	*P*-Chloromercuribenzene sulfonic acid (Chapter 17)
PCR	Polymerase chain reaction (Chapter 9 and 11)
P5CR	Δ^1-Pyrroline-5-carboxylate reductase
P5CS	Δ^1-Pyrroline-5-carboxylate synthetase (Chapter 15)
PDC	Pyruvate decarboxylase (Chapter 15)
PDC_m/PDC_p	Mitochondrial/plastidial pyruvate dehydrogenase complexes (Chapter 16)
PDME	Plant drug/metabolite exporter (Chapter 17)
PEP	Phosphoenolpyruvate (Chapter 2 and 16)
PEPC or PEPCase	Phosphoenolpyruvate carboxylase (Chapter 1 and 16)
PEPCK	Phosphoenolpyruvate carboxykinase (Chapter 3)
PFK 2	6-Phosphofructo-2-kinase/ fructose-2,6-bisphosphatase (Chapter 16)
3-PGA	3-Phosphoglycerate
PGM	Phosphoglucomutase (Chapter 1)
Phe	Phenyalanine
PHS	Polyamine H^+-symporter (Chapter 17)
PHYB	Phytochrome B (Chapter 19)
PITC	Phenylisothiocyanate (Chapter 25)
PK	Pyridoxal kinase (Chapter 3)
PK_c	Cytosolic pyruvate kinase (Chapter 16)
PKp	Plastidial pyruvate kinase (Chapter 16)

Continued

Table1. Continued.

Abbreviation or Term	Definition
PKU	Phenylketonuria (Chapter 28)
PLP	Pyridoxal phosphate (Chapters 2, 3, 4 and 12)
PMP	Pyridoxamine phosphate (Chapter 3)
PMT	Putrescine methyltransferase
PNP	Pyridoxine phosphate (Chapter 3)
POD	Peroxidase (Chapter 27)
POEA	Polyethoxylated tallow amine (Chapter 24)
POP	Plant organocation permease (Chapter 17)
POX	Pyridoxine oxidase (Chapter 3)
PPA	3-Phenylpropionic acid (Chapter 19)
PRFAR	*N*′-[(5′-phosphoribulosyl) formimino]-5-aminoimidazole-4-carboxamide (Chapter 14)
Pro	Proline
ProDH	Proline dehydrogenase (Chapter 15)
ProFAR	*N*′-[(5′-phosphoribosyl) formimino]-5-aminoimidazole-4-carboxamide ribonucleotide (Chapter 14)
PRP	Proline-rich proteins (Chapter 28)
PRPP	5-Phosphoribosyl-3-pyrophosphate (Chapter 14)
PTPC	Plant-type PEPC (Chapter 16)
Put	Putrescine (Chapter 9 and 10)
Pyr	Pyruvate
QPM	Quality protein maize (Chapter 13)
RAC	Raw agricultural commodities (Chapter 24)
R&D	Research and development
RIA	Radioimmunoassay (Chapter 21)
RNA	Ribonucleic acid (Chapter 22 and 23)
RNS	Reactive nitrogen species (Chapter 21)
ROS	Reactive oxygen species (Chapters 7, 11, 12, 15 and 21)
SA	Salicylic acid (Chapter 12 and 28)
SAA	Sulfur amino acid
SA-Asp	Salicylic acid-aspartate conjugate (Chapter 28)
SAG	Senescence-associated gene (Chapter 12)
SAH	*S*-Adenosylhomocysteine
SAM	*S*-Adenosylmethionine (Chapter 12 and 13)
SAMDC	*S*-Adenosylmethionine decarboxylase (Chapter 10)
SAR	Systemic acquired resistance (Chapter 12 and 28)
SAT	Serine acetyltransferase (Chapter 12)
SAV	Senescence-associated vacuole (Chapter 12)
SAV 3	Shade avoidance 3 (Chapter 19)
SBP	Solute-binding protein (Chapter 17)
SCN^-	Ionic thiocyanate (Chapter 22)
SDH	Saccharopine dehydrogenase (Chapter 13 and 15)
Ser	Serine
SERAT	Serine acetyltransferase (Chapter 11)
SGAT	Serine:glyoxylate aminotransferase (Chapter 2)
SHM	Serine hydroxylmethyltransferase gene (Chapter 11)
SLE	Systemic lupus erythematosus (Chapter 27)
SMCO	*S*-Methylcysteine sulphoxide (Chapter 27)
SNAT	Serotonin *N*-acetyltransferase (Chapter 21)
SNF	Symbiotic N fixation (Chapter 17)
SOD	Superoxide dismutase
Spd	Spermidine (Chapter 9 and 10)
SPDS	Spermidine synthase (Chapter 9 and 10)
Spm	Spermine (Chapter 9 and 10)

Continued

Table1. Continued.

Abbreviation or Term	Definition
SPMS	Spermine synthase (Chapter 9 and 10)
SSADH	Succinic semialdehyde dehydrogenase (Chapter 7 and 15)
TAA1	TRYPTOPHAN AMINOTRANSFERASE OF *ARABIDOPSIS* 1 gene (Chapter 19)
TAM	Tryptamine (Chapter 19)
TAR 1–4	TRYPTOPHAN AMINOTRANSFERASE-RELATED 1 to 4 genes (Chapter 19)
Tau	Taurine
TCA	Tricarboxylic acid (with reference to the TCA cycle) (Chapters 2, 6, 7, 9, 15 and 16)
TD	Threonine deaminase (Chapter 13 and 28)
TDC	Tryptophan decarboxylase (Chapter 19, 20 and 21)
TDI	Tolerable daily intake
Teratogenic	Causing birth defects
TH	Tyrosine hydroxylase
T5H	Tryptophan 5-hydroxylase (Chapter 21)
THF	Tetrahydrofolate
Thr	Threonine
TIR 2	Transport inhibitor response 2 (Chapter 19)
TMV	Tobacco mosaic virus (Chapter 8 and 27)
TPH	Tryptophan hydroxylase
TPPT	Thiamine pyrophosphate transporter (Chapter 17)
TPT	Triose phosphate transporter (Chapter 17)
tRNA	Transfer RNA
Trp	Tryptophan
TRR	Total radioactive residues (Chapter 24)
TS	Threonine synthase (Chapter 13)
TSA	Tryptophan α subunit (Chapter 20)
TSB	Tryptophan β subunit (Chapter 20)
T-Spm	Thermospermine (Chapter 10)
Tyr	Tyrosine
UDP	Uridine diphosphate (Chapter 1)
UGT	UDP-glycosyltransferase (Chapter 20)
UTR	Untranslated sequences (Chapter 11)
Val	Valine
VBA	Vacuolar basic amino acid transporter (Chapter 17)
WEI 8	Weak ethylene insensitive 8
WHO	World Health Organization (United Nations) (Chapter 13 and 24)

References

Alberts, B., Bray, D., Hopkin, K., Johnson, A., Lewis, J., Raff, M., Roberts, K. and Walter, P. (2014) *Essential Cell Biology*, 4th edn. Garland Science (Taylor & Francis Group), New York and Abingdon, UK.

Allaby, M. (ed.) (2006) *A Dictionary of Plant Sciences*, 2nd edn. Oxford University Press, Oxford, UK and New York.

D'Mello, J.P.F. (ed.) (1997) *Handbook of Plant and Fungal Toxicants*, 1st edn. CRC Press, Boca Raton, Florida.

D'Mello, J.P.F. (ed.) (2012) *Amino Acids in Human Nutrition and Health*, 1st edn. CAB International, Wallingford, UK.

D'Mello, J.P.F., Duffus, C.M. and Duffus, J.H. (eds) (1991) *Toxic Substances in Crop Plants*, 1st edn. The Royal Society of Chemistry, Cambridge, UK.

Klug, W.S., Cummings, M.R., Spencer, C. and Palladino, M. (2014) *Concepts of Genetics*, 10th edn. Pearson Education Limited, Harlow, UK.

Lackie, J. (ed.) (2013) *The Dictionary of Cell and Molecular Biology*, 5th edn. Academic Press, New York and London.

Lawrence, E. (ed.) (2011) *Henderson's Dictionary of Biology*, 15th edn. Pearson Education Limited, Harlow, UK.

Lodish, H., Berk, A., Kaiser, C.A., Krieger, M., Bretscher, A., Ploegh, H., Amon, A. and Scott, M.P. (2013) *Molecular Cell Biology*, 7th edn. W.H. Freeman and Company, New York.

Martin, E. and Hine, R.S. (eds) (2008) *A Dictionary of Biology*, 6th edn. Oxford University Press, Oxford, UK and New York.

Michal, G. and Schomburg, D. (eds) (2012) *Biochemical Pathways: An Atlas of Biochemistry and Molecular Biology*, 2nd edn. Wiley, Hoboken, New Jersey.

Nelson, D.C. and Cox, M.M. (2008) *Lehninger: Principles of Biochemistry*, 5th edn. W.H. Freeman and Company, New York.

Parish, H., Smith, T., Stirling, J. and Vella, F. (eds) (2006) *Oxford Dictionary of Biochemistry and Molecular Biology*, 2nd edn. Oxford University Press, Oxford, UK and New York.

Singleton, P. and Sainsbury, D. (2006) *Dictionary of Microbiology and Molecular Biology*, 3rd edn. Wiley-Blackwell, Hoboken, New Jersey.

Thain, M. and Hickman, M. (2004) *The Penguin Dictionary of Biology*, 11th edn. Penguin Books, London and New York.

Voet, D. and Voet, J.G. (2011) Biochemistry, 4th edn. Wiley, Hoboken.

1 Glutamate Dehydrogenase

G.O. Osuji[1]* and W.C. Madu[2]
[1]Prairie View A&M University, Prairie View, Texas, USA; [2]Imo State Polytechnic, Owerri, Nigeria

1.1 Abstract

Glutamate dehydrogenase (GDH; EC.1.4.1.2) aminating activity unleashes a metabolic propulsion in crop plants that optimizes, maximizes and doubles the amino acid, protein, fatty acid, carbohydrate and dry matter yields per hectare. It is possible that the molecular biology of the enzyme may hold clues for increasing the yields of food crops for feeding a burgeoning world population without cultivating more land, without applying more fertilizer and without increasing man-hour input. GDH is historically known to catalyse the reversible reductive amination of α-ketoglutarate (α-KG) in the synthesis of L-glutamate in higher plants. It also catalyses the template-independent polymerization of nucleoside triphosphate(s) to RNA. The discovery of the aminating activity of the glutamine synthetase (GS; EC6.3.1.2)–glutamate synthase (GOGAT; EC.2.6.1.53) cycle in chloroplasts triggered new conversations and extensive research on the role of GDH in the synthesis of glutamate. The K_m value for the NH_4^+ ion is much lower in GS than in GDH, which fact has increased understanding of the differential mechanisms of NH_4^+ ion assimilation and glutamate synthesis during deficit and surplus supplies of α-KG. It might appear from some repetitive experimentation on low-activity preparations of GDH that the deaminating direction is emerging as the consensus role of the enzyme. But many lines of new experimental evidence from systems biology illuminate the uninterruptible flow of α-KG via the citric acid cycle; and the nature of the aminating cassette of GDH isoenzymes reiterates and supports the amination role. Specifically, the kinetic property of GDH to discriminate and integrate the plant's environmental signals by synthesizing RNAs that regulate the mRNAs encoding many enzymes (phosphate translocator, granule-bound starch synthase, phosphoglucomutase, nitrate reductase, glycinamide ribonucleotide transformylase, etc.) of primary metabolism is the collateral chemical scaffolding on which the aminating natural role stands as it synchronizes carbon and nitrogen metabolic pathways, and assimilates exponential multiples of kilogram quantities of NH_4^+ ion as assimilated via the GS-GOGAT cycle during normal to maximum photosynthetic supplies of C_3, C_5, etc. compounds. GDH possibly deaminates vanishing fractions of glutamate as it synthesizes RNA that regulates the mRNAs so that glutamate biosynthesis is not shut down during low to normal flow of C_3, C_5, etc. compounds of photosynthesis. Ammonium ion assimilation and glutamate synthesis by GDH are regulated in higher plants by the supply of NH_4^+ ions, and the flow of α-KG from the citric acid cycle to GDH, not by the reverse-flow of α-KG from glutamate.

*E-mail address: goosuji@pvamu.edu

1.2 Introduction

L-Glutamic acid occupies the central position in amino acid metabolism because other amino acids (except lysine) can be derived from it when it transfers its amino group to their α-keto acids (D'Mello, 2003). Glutamate dehydrogenase (L-glutamic acid:GDH NADH oxidoreductase; EC.1.4.1.2) is the pre-eminent enzyme (Strecker, 1953; Pahlich *et al.*, 1980; Magalhaes, 1991) known to catalyse the reversible reductive amination of α-ketoglutarate (α-KG) in the synthesis of glutamate in all higher plants studied. It also catalyses template-independent polymerization of nucleoside triphosphate(s) to RNA (Osuji *et al.*, 2003a). The discovery of the aminating activity of glutamine synthetase (GS; EC6.3.1.2)-glutamate synthase (GOGAT; EC.2.6.1.53) cycle in chloroplasts (Lea and Miflin, 1974) triggered new conversations and intensive research on the role and regulation of GDH in the assimilation of NH_4^+ ions and synthesis of glutamate. The K_m value for the NH_4^+ ion is much lower in GS than in GDH, which fact has increased the understanding of the differential mechanisms of NH_4^+ ion assimilation and glutamate biosynthesis during normal and surplus supplies of α-KG (Robinson *et al.*, 1992; Osuji and Madu, 2012). This review discusses recent experimental evidence on the interwoven natural roles of GDH in glutamate synthesis, the coordination of nitrogen and carbon metabolic pathways, and glutamate deamination. GDH is multi-isoenzymatic and isomerizes in response to a wide assortment of genetic, biotic and abiotic environmental factors. Several experiments have focused narrowly and repeatedly on a handful of low activity GDH preparations while neglecting the natural aminating role of the preponderantly high activity GDH preparations. It might appear from such repetitive narratives on the handful of the same experimental approach that the consensus role of GDH is moving towards the deamination of glutamate (Robinson *et al.*, 1991; Glevarec *et al.*, 2004; Lehmann and Ratajczak, 2008; Labboun *et al.*, 2009). The approach of this review is experimentally different, being focused on the molecular biology of GDH amination and deamination reactions (Yamaya *et al.*, 1984; Oaks, 1994; Melo-Oliveira *et al.*, 1996; Pahlich, 1996; Syntichaki *et al.*, 1996; Madu, 1997; Turano *et al.*, 1997; Osuji and Braithwaite, 1999; Ameziane *et al.*, 2000; Pavesi *et al.*, 2000), especially those that explain the biochemistry of the isomerization reaction in crops treated with pesticides, mineral fertilizer, *N*-carboxymethylchitosan and plant hormones; and synchronizing the primary nitrogen and carbon metabolic pathways of citric acid cycle, glycolysis, photosynthesis and nitrogen assimilation, leading to maximum yields of glutamate and other amino acids, proteins, carbohydrates, fatty acids, etc. (Osuji and Madu, 2012). Some of the incomplete arguments embedded in earlier conversations on the roles of the enzyme will be highlighted. The review discusses many lines of new experimental evidence from systems biology (metabolism) that illuminate the uninterruptible flow of C_3, C_4, C_5, etc. compounds; the nature of the aminating cassette of GDH isoenzymes; the large kilogram quantities of amino acids and proteins that result from GDH amination activity; and the application of the GDH amination role to increase the yields of maize, roots/tubers and leguminous crops for combating kwashiorkor (protein energy malnutrition) and food insecurity in some arid regions.

1.3 Glutamate Dehydrogenase Structure and Localization

GDH has a hexameric structure of three subunit polypeptides controlled by two non-allelic genes GDH_1 and GDH_2, responsible for the synthesis of subunits 'a' and 'b' respectively; with gene GDH_1 synthesizing subunits 'a' and 'α', two codominant allelic forms; and GDH_2 synthesizing the 'b' subunit (Cammaerts and Jacobs, 1983). The binomial combination of the three subunits into hexamers gives the ~28 isoenzyme distribution typically obtained in 2-D electrophoretic resolution of the enzyme extracted from soybean (Osuji *et al.*, 2003b), peanut (Osuji and Braithwaite, 1999), sweet potato (Osuji and Madu, 1997), maize (Osuji and Madu, 1995), yam tuber (Osuji *et al.*, 1991) and many other higher plants. Since isomerization is the committed step in the oxidoreductase catalytic mechanism of GDH (Osuji and Braithwaite, 1999), the activities of the enzyme changing with the alterations in the isoenzyme pattern (Cammaerts and Jacobs, 1985), it is necessary in all experimentation with the enzyme to demonstrate the qualitative alterations in the isoenzyme pattern (Osuji and Madu, 1996; Watanabe *et al.*, 2011).

Presentation of the GDH isomerization profiles inspires confidence in the quality assurance evaluation, in the baseline for validating the experimental biological tissues, and authenticates the quality control measures practised during the instrumental analytical data generation. In this regard, the free solution (Rotofor) isoelectric focusing (IEF) of GDH, followed by native PAGE of the fractions, introduced a considerable advancement in the analytical chemistry of the enzyme (Osuji and Braithwaite, 1999; Osuji *et al.*, 1999) because the isoenzymes become fully separated one from the other, being elegantly projected in three dimensions in some cases. It is noteworthy, therefore, that all the conversations that withheld the aminating GDH isoenzyme profiles have invariably supported the deaminating function of the enzyme (Miflin and Lea, 1976; Robinson *et al.*, 1991, 1992; Stewart *et al.*, 1995; Sadunishvili *et al.*, 1996; Aubert *et al.*, 2001; Miflin and Habash, 2002; Glevarec *et al.*, 2004; Masclaux-Daubresse *et al.*, 2006; Purnell and Botella, 2007; Skopelitis *et al.*, 2007; Miyashita and Good, 2008; Labboun *et al.*, 2009).

Nucleotide sequences of GDH cDNA and deduced amino acid sequences of GDH polypeptides (Syntichaki *et al.*, 1996; Pavesi *et al.*, 2000) have been studied as well as the protein structure of hexameric GDH (Britton *et al.*, 1992). The highest level of sequence similarities in GDHs were found within the N-terminal half of the polypeptides, the region thought to be involved in oxoglutarate/glutamate binding and catalytic activity. Analyses of the conserved residues in 3-D projection attest to a remarkable similarity in the layout of the active site of GDHs. Specifically, in the hexamer, each subunit polypeptide is folded into two domains separated by a deep cleft which harbours the NADH binding site. Adjacent to the domain of the NADH is a strategic cluster of lysine residues, one of which forms the Schiff base intermediate and facilitates the transfer of the hydrogen of the nicotinamide (Brown *et al.*, 1973) to α-iminoglutarate in the catalytic reduction reaction (Osuji *et al.*, 1999). These structural details support the non-redundant role of the enzyme in the reversible amination of α-KG (Melo-Oliveira *et al.*, 1996).

GDH is easily found in mitochondria (Yamaya *et al.*, 1984; Calle *et al.*, 1986; Bravo and Maeso, 1987). GDHs have also been identified in chloroplasts (Turano *et al.*, 1996) and, earlier, in the cytosol (Chou and Splittstoesser, 1972). These strategic localizations, especially in the cytosol and mitochondria of phloem companion cells (Tercé-Laforgue *et al.*, 2004a), enable the enzyme to play a vigorous role, as supported by mechanisms that export α-KG from the citric acid cycle, in the assimilation of ammonium ion inside and outside the mitochondria in naturally photosynthesizing plants that are challenged by biotic and abiotic stress environments (Osuji *et al.*, 1997; Debouba *et al.*, 2006; Lehmann *et al.*, 2010), including antisense and mutational modifications of plants (Ameziane *et al.*, 2000; Fontaine *et al.*, 2006; Lehmann and Ratajczak, 2008).

1.4 Control Plants and Control Glutamate Dehydrogenase

GDH isomerizes in response to genetic, biotic and abiotic factors in the plant's environment (Osuji and Madu, 1996; Osuji *et al.*, 1998; Ameziane *et al.*, 2000; Fontaine *et al.*, 2006; Labboun *et al.*, 2009). Therefore, project design needs to take steps to normalize not only the experimental plant material but also the GDH isoenzymes. Many project designs did not distinguish between the control crop and the control GDH; they simply regarded the control crop tissue as being synonymous with the control GDH. This kind of confusion has led to exaggerated conclusions being made about the role of the enzyme. Similarly, projects in which plant materials are harvested at lapsed time intervals from a common pool, and analysed for GDH activities, need to demonstrate the zero time control GDH isoenzyme pattern followed by the lapsed-time changes in GDH isoenzyme patterns that corroborate the activity changes. GDH is a complex mix of isoenzymes. With so many types of environmental stimuli affecting the enzyme, it can be overwhelming to arbitrarily decide which responses have been overridden and excluded from an experiment. A single treatment of plant materials with a nutrient solution is not sufficient for deducing the natural role of the enzyme because the control plant cannot be biochemically distinguished from the control GDH. Therefore, a system approach in the design of GDH research is to adopt a broad-based interlocking multiple factorial experimentation. Many GDH research projects that were biologically broad-based and comprehensive in incorporating control GDHs as distinct from control crop materials produced results that confirmed the aminating role of the enzyme (Yamaya *et al.*, 1984; Oaks, 1994; Melo-Oliveira *et al.*, 1996; Pahlich, 1996; Turano *et al.*, 1997; Osuji and Braithwaite, 1999; Ameziane *et al.*, 2000; Pavesi *et al.*, 2000; Dubois *et al.*, 2003; Restivo 2004; Tercé-Laforgue *et al.*, 2004a,b; Jaspard, 2006; Osuji and Madu, 2012). The control crop materials are the baseline metabolic signalling environments, whereas the control GDHs are the baseline for rationally judging the natural role of the enzyme in the experimental crop materials.

Melo-Oliveira *et al.* (1996) deployed a battery of environmental conditions encompassing controls in the absence of nitrate and NH_4^+ ions, and with three concentrations of NH_4^+ ions without nitrate; a control without sucrose as carbon source, 3% sucrose; continuous darkness, continuous light, dark-adapted, transitions from darkness to light and vice versa; and a stratified timescale from zero to 16 h to study GDH isomerization, GDH mRNA accumulation, and the glutamine synthetase mRNA accumulation in the roots, leaves, and flowers of mutant and wild type *Arabidopsis thaliana*. It was a biologically comprehensive research plan because it made for compelling and incontrovertible interpretation of results. For each genetic, biochemical and physiological factorial experiment, there were specific progressive, hierarchical layers of controls in time and space for the GDH and plant tissues, leading logically and scientifically to the experimental GDH and plant tissue of focus. That was how the non-redundant role of GDH was proved. The research project of Turano *et al.* (1996) was also detailed in biological comprehensiveness, encompassing GDH-encoded mRNAs, GDH isoenzyme patterns, differentiating the control *Arabidopsis* from the control GDH subunit polypeptides, and the deployment of a sophisticated complexity of mineral nutrients combined with daylight and dark treatments, all of which converged to show that NADH-GDH activity was controlled by the transcript levels, an important landmark that supports the natural role of amination.

Evidence that the cationic isoenzymes of GDH control the protein yield and, therefore, amination in peanut seeds came from a criss-cross research plan involving three aromatic pesticides (fungicide, herbicide, insecticide), each at three rates of treatment superimposed on zero and 25 mM NH_4^+ ion fertilization, and two controls with or without NH_4^+ ion fertilization (Osuji and Braithwaite, 1999). With the cascade of external and internal control peanuts, and resultant arrays of GDH isoenzyme profiles, the interpretation of quantitative data easily and visually related the cationic isoenzymes to the combined pesticide and NH_4^+ ion treatments with resultant double protein yields. Doubling of crop protein yield through the enhancement of the cationic isoenzymes of GDH has since then become one of the litmus tests for the aminating function of GDH. But increasing the carbohydrate yield through the enhancement of the anionic isoenzymes is yet to be experimentally demonstrated in support for the deaminating function of GDH.

Where multiple, interlocking and differentiating experimental conditions are not adopted, systems analysis (Pahlich, 1996), in which GDH activities were integrated into a metabolic chain of enzymes that determine function, has been adopted (Oaks, 1994). In the project on primary nitrogen assimilation and its regulation in maize seedlings, a battery of enzymes, viz. nitrate reductase (NR), nitrite reductase (NiR), glutamine synthetase (GS), Fd (ferredoxin)-glutamate synthase (Fd-GOGAT), NADH-glutamate synthase (NADH-GOGAT), phosphoenolpyruvate carboxylase (PEPCase), and aspartate aminotransferase (AAT) were assayed side by side with GDH in maize shoots and roots, with and/or without NO_2^- and NH_4^+ ions in the growth media. All the enzymes were subjected to immunochemical analyses in order to assure biological uniformity. There were, therefore, many layers of maize tissue controls, external nutrient controls and internal GDH controls that added up in the metabolic pathway integrating NR, NiR, GS, GOGAT, AAT, and PEPCase to validate the unequivocal conclusion that GDH is aminating in function (Oaks, 1994).

1.5 Availability of Ammonium Ions

1.5.1 Ammonium ion contents of experimental tissues and plants

At the centre of GDH project designs is the availability of NH_4^+ ions, either metabolically derived or externally added (Lea and Miflin, 1974). Some earlier research projects on the deamination function of GDH tried to induce NH_4^+ ion accumulation in suspension-cultured carrot cells that were deprived of a carbon nutrient source (Robinson *et al.*, 1991, 1992). Such conditions were presumed to encourage carrot protein catabolism to amino acids that would be deaminated to release NH_4^+ ions. In addition to the conversations contributed by several authorities, the choice of carrot as the experimental organism was slightly inappropriate because it possesses $<1.0\%$ protein; catabolism of which storage protein, or was it catabolism of the carrot's housekeeping enzymes? In those research projects, including that on germinating mutant maize seeds (Stewart *et al.*, 1995), there were no assays for proteolytic enzyme (pepsin, trypsin, elastase, proteinases, carboxypeptidase, aminopeptidase, dipeptidase, etc.) activities at any growth stages of the tissues/cells to prove and buttress the protein catabolism dogma. In particular, the carrot cell suspension project did not determine the concentrations of the insoluble proteins. Therefore, the tissues/cells may not have suffered protein catabolism. Metabolic release of NH_4^+ ion from free glutamate is not the exclusive activity of GDH, as has been emphasized (Oaks, 1994). At the stagnant phases of carrot cell or maize shoot growth, some purine and pyrimidine molecules could be subject to degradation, releasing more NH_4^+ ions than a glutamate or other amino acid molecule (Osuji and Ory, 1986). So it is not clear that the NH_4^+ ion detected in the stagnant carrot cells or mutant maize was released by GDH and aminotransferases. The carrot cell and maize shoot protein catabolism projects also did not present the GDH isoenzyme profiles at any steps of the experimentation to corroborate the GDH deamination activities. The GDH isoenzyme profile of carrot is similar to those of (other) root and tuber crops, especially sweet potato and the yam tuber (Osuji *et al.*, 1991), in being rich in the cationic (aminating) isoenzymes. Therefore, in moving from its natural aminating profile to the presumed deaminating profile, the GDH isoenzymes would have undergone dramatic and visible alterations in the 14 days of carrot cell suspension incubation (Robinson *et al.*, 1991, 1992).

1.5.2 Glutamate deamination in mitochondria

The use of mitochondrial culture also needed experimental clarification of the injection of pyridoxal phosphate into the GDH assay cocktail (Aubert *et al.*, 2001), because the authors asserted that in the absence of aminotransferases, the pyridoxal phosphate-mediated deamination of glutamate can be attributed entirely to GDH. It is novel to use mitochondria to perform GDH experimentation because the procedure carries the additional requirement to determine the level of depletion in the mitochondria of α-KG (substrate of amination reaction) prior to the deamination assays (Hartmann and Ehmke, 1980). Although pyridoxal phosphate is not *in vivo* substrate of GDH, it inhibits the amination reaction by mimicking α-KG *in vitro* (Osuji *et al.*, 1999). Therefore, it is a useful analogue for demonstrating the principle of the Schiff base reaction of GDH (Brown *et al.*, 1973), besides the aminotransferase reaction. Glutamate, the substrate of GDH in the deamination direction, does not form a Schiff base with either GDH or α-KG. The biochemical logic for the induction of the amination molecular mechanism (Schiff base reaction) in a protocol (Aubert *et al.*, 2001) designed to prove GDH deamination activity needed detailed and sophisticated justification. Was pyridoxal phosphate injected into the reaction cocktail to supplement α-KG or the deactivated GDH, or to prove that aminotransferases were absent? Similarly, mitochondria were subjected to >11 h experimentation without any explanation of the reaction temperature and the reasons for their rugged thermostability; however, mitochondria do not function as free-living organisms.

1.6 Glutamate Dehydrogenase-Linked Schiff Base Amination Complex

1.6.1 Pesticide treatment and ammonium ion fertilization

The foregoing discussion on the chemistry of Schiff base formation leads to a conversation on more litmus tests that have confirmed the GDH amination role. The first reaction product of GDH with α-KG, its natural substrate in the amination direction, is the enzyme-linked Schiff base complex (Brown *et al.*, 1973; Smith *et al.*, 1975; Fersht, 1985). GDH does not form GDH-linked Schiff base with glutamate, the natural substrate of the enzyme in the deamination direction (Osuji *et al.*, 1999). After stabilization, the GDH-linked Schiff base reacts with NH_4^+ ion to liberate the GDH and to produce α-iminoglutarate, the reduction of which by NADH produces glutamate. Deamination (reversal) by GDH may kick in after the α-iminoglutarate has been produced. Therefore, the chemical mechanism supports the amination role of the enzyme (Osuji and Madu, 2012).

To qualitatively freeze capture the GDH-linked Schiff base amination complex, visually demonstrate its physical existence, and thereby prove the aminating role of GDH, it was made to react *in vitro* with Pb^{++}, or *in vivo* with aromatic pesticides, all of which are strong nucleophiles (Osuji *et al.*, 1999, 2001). The GDH-linked Schiff base was in each case immunochemically detected with peanut GDH antibodies. The protein bands immunochemically detected on the electrophoresed denaturing polyacrylamide gels were the degradation products of the GDH subunit polypeptides. Those results also proved the biochemical basis of the enzyme's isomerization. Besides the NH_4^+ ion, several nucleophiles (nucleosides, pesticides, monosaccharides, metabolic intermediates, amino acids, mineral nutrients, etc.) may attack the GDH-linked Schiff base to produce substituted imine complexes that are stable dead-end complexes. Such modified GDH subunits are degraded because inactivated enzymes are removed by degradation (Davies, 1987). In the cases of GDH-linked Pb^{++}- and GDH-linked pesticide-substituted imine complexes (Osuji *et al.*, 1999, 2001), the resultant degraded GDH fragments were massive. It is noteworthy that when glutamate, including other amino acids, attack the GDH-linked Schiff base, they are not necessarily deaminated in the reaction; rather they could form dead-end GDH-linked amino acid substituted imine complexes. GDH-linked Schiff base formation and GDH isomerization are therefore some of the litmus tests for the aminating role of GDH.

1.6.2 Pesticide treatment, ammonium ion fertilization and protein contents

Demonstration that GDH isomerization-linked degradation was synonymous with the amination role propelled the search for the specific isoenzyme subunits that increase the amination activity, with concomitant increases in protein yield. Earlier studies had demonstrated the increase of storage protein contents by inducing GDH isomerization in cultured yam and sweet potato cells with auxins, chitosan and NH_4^+ ions (Madu, 1997). Other studies had applied GDH amination kinetic properties (Osuji and Braithwaite, 1999) to explain the relationship between the signal integrating and discriminating function, storage protein and the total RNA contents (Osuji and Cuero, 1991, 1992a,b; Osuji *et al.*, 2003c). Pesticides were chosen as the xenobiotic inducers of the isomerization of peanut GDH because peanut is protected in the field and in storage with many different pesticides, having developed an arsenal of metabolic reactions for detoxifying them (Osuji, 1997; Osuji *et al.*, 2008). Pesticide treatment of peanut unequivocally demonstrated the pivotal role of the Schiff base intermediate in the amination reaction mechanism of GDH, and the consequent degradation of the enzyme subunits (Osuji *et al.*, 1999). Combining the pesticides with NH_4^+ ion fertilization doubled the seed protein yields (Osuji and Braithwaite, 1999). Seed and seed protein yields for NH_4Cl-fertilized and not-fertilized peanuts were about the same; and, in addition, the peanut without pesticide treatment but with NH_4Cl fertilization harboured the control GDH. Therefore, in the project, the control peanut was different from the control GDH. Immunochemical assay of the peanut GDH following denaturing PAGE revealed that wherever the peanut protein yields doubled, the b-, a-, and α-subunits of the enzyme were present, and the peanuts were those treated with NH_4Cl together with pesticides (Fig. 1.1). These results pronouncedly projected the aminating role of GDH. On the other hand, contrast, wherever the peanut protein yields were normal, the α-subunit of GDH was present together with the a-subunit, but the b-subunit was absent, and the peanuts were those treated with pesticides without NH_4Cl (Osuji *et al.*, 1999). Therefore, when all the three subunits of GDH are present, amination that

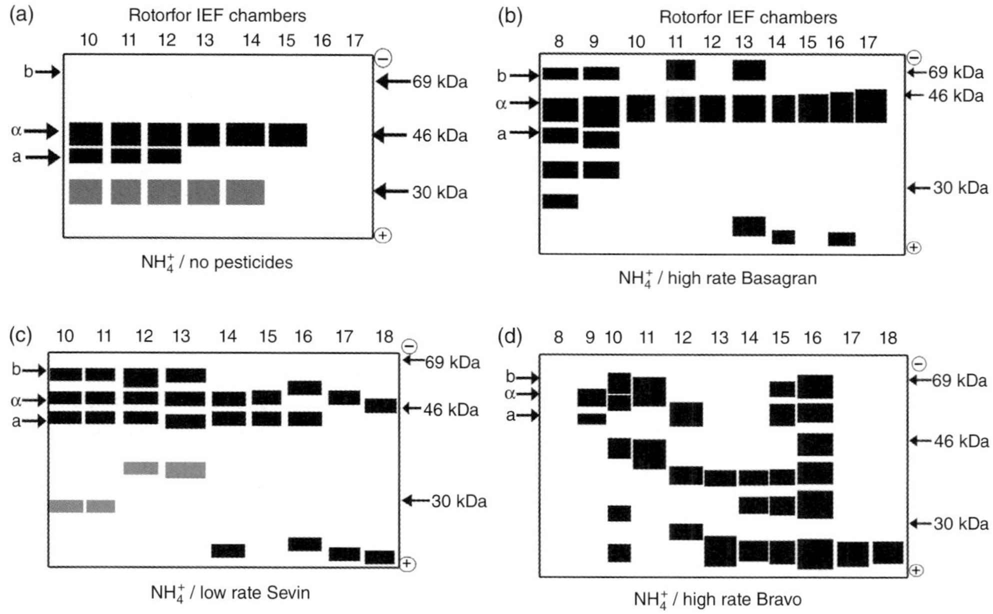

Fig. 1.1. Analysis of the degradation of the glutamate dehydrogenase (GDH) of NH_4^+ ion/pesticide-treated peanut. Each GDH preparation was subjected to Rotofor isoelectric focusing (IEF) to charge its isomers, which were then fractionated to their GDH subunit composition by sodium dodecyl sulfate (SDS)-10% polyacrylamide gel electrophoresis (PAGE). For the PAGE, 50μl of each Rotofor fraction of the control (a), 30 μl of each Rotofor fraction of Basagran-treated (b), 30 μl of each Rotofor fraction of Sevin-treated (c), and 30 μl of each Rotofor fraction of Bravo-treated (d) peanut were used. After Western transfer to nitrocellulose membrane, the GDH subunits and degradation polypeptides were immunodetected with anti-GDH (Osuji *et al.*, 1999).

leads to the doubling of peanut protein yield occurs; but in the absence of b-subunit, amination that produces normal protein yield occurs. These results have encouraged important conversations that have assigned deaminating and/or aminating functions to specific GDH subunits (Skopelitis *et al.*, 2006, 2007; Purnell and Botella, 2007; Watanabe *et al.*, 2011). However, control peanut without pesticide and NH_4Cl treatments produced merely normal protein yield, although all the three subunits of the enzyme were present (Osuji *et al.*, 1999), thus suggesting that additional factors control the amination activity of the enzyme. With and without pesticide or NH_4^+ ion treatment, the peanut's GDH subunits suffered the usual isomerization-related degradation (Osuji *et al.*, 1999). Other than the subunits of GDH, the supply of C_3, C_4, C_5, etc. compounds (Osuji *et al.*, 2011, 2012a) is the major propulsion driving the amination role forwards. The K_m values for α-KG of GDH and GOGAT are in the low mM range (Osuji *et al.*, 1991), thereby positioning GDH as an efficient contender in the synthesis of glutamate under the mM concentration range of α-KG in mitochondria (Bowman *et al.*, 1976).

1.7 Protect the Glutamine Synthetase-Glutamate Synthase Cycle in Glutamate Dehydrogenase Research

Following the research projects of Oaks (1994), Melo-Oliveira *et al.* (1996), Pahlich (1996) and Turano *et al.* (1996), there was need to biologically widen the scope of GDH experimentation in order to make for biochemical comprehensiveness in the analyses of and conversation on its natural function. As the GS-GOGAT cycle cannot be overlooked when GDH function is being discussed, it became obvious that experimental conditions that excluded the inhibition of the cycle with amino acid analogues (azaserine, methionine sulfoximine, phosphinothricine, etc.) (Lacuesta *et al.*, 1989; Glevarec *et al.*, 2004; Masclaux-Daubresse *et al.*, 2006) needed to be developed in order to understand the regulation of the flow of α-KG to GDH and GOGAT under normal plant growth conditions. Methionine sulfoximine and related inhibitors of amino acid metabolism are strong nucleophiles that also attack the GDH-linked Schiff base, thereby converting it to dead-end complex (Osuji and Madu, 1996; Osuji *et al.*, 1999) that has aberrant Michaelis-Menten kinetic properties (Stone *et al.*, 1980). Oaks (1994) and Melo-Oliveira *et al.* (1996) had demonstrated that the GS-GOGAT cycle was neither inhibited nor activated by mineral nutrients in the forms of NO_3^- and/or NH_4^+ ions. Therefore, an experimental protocol was developed for the treatment of peanut with mineral nutrients (Osuji *et al.*, 2011) that allowed the GS-GOGAT cycle to function normally (Osuji and Madu, 2012), while targeting and altering the activities of GDH and regulating primary carbon and nitrogen metabolism (Osuji *et al.*, 2012a) in a holistic biological systems approach, as advocated by Pahlich (1996). Plant GDH is the target site of mineral salt action (Osuji and Madu, 1995; Osuji *et al.*, 1998).

At this juncture, it is necessary to distinguish between the function of mineral salts as inducers of GDH isomerization and mineral salts as plant nutrients (Marschner, 1990; Rengel, 1999). As inducers of GDH isomerization, mineral ions are applied in stoichiometric (molar) ratios and miniscule (μM) quantities to target the subunits of GDH. But as fertilizers, mineral ions are applied in kilogram weight ratios to target and saturate almost all the macromolecular components of the plant. Mineral ions in stoichiometric ratios function as electromagnets in their sensitization and induction of GDH isomerization and the resultant resolution of the treated crop to biochemical (metabolic) variants (Osuji *et al.*, 2011). Stoichiometric nutrients-based differentiation of crop plants to their arrays of metabolic variants elevates the biochemical comprehensiveness of the GDH research approach and enables the application of systems biological computation in the analyses of data.

In order to protect the GS-GOGAT cycle under GDH experimental procedures, stoichiometric ratios of mineral nutrients were developed (Osuji and Madu, 2012). Furthermore, the wide variations in the concentration and composition of the interlocking stoichiometric combinations made for internal repeats in the mineral nutrient solutions, thus limiting stochastic variability in the plot treatments; imposed firm control on the number of experimental repeats; and consolidated the biochemical comprehensiveness of the project design by serving as the base line validation for the synchronous synthesis of RNA by GDH and silencing of homologous mRNAs by the GDH-synthesized RNAs (Osuji *et al.*, 2012b).

1.8 Molecular Biology of Glutamate Dehydrogenase

1.8.1 The supply of α-ketoglutarate from the citric acid cycle to glutamate dehydrogenase and glutamate synthase

This section opens a limited conversation on the chemistry of carbon limitation/protein catabolism in the context of the synchronized supply of NH_4^+ ion and C_3, C_4, C_5, etc. compounds to GDH and GOGAT for amino acid metabolism. It is also necessary to integrate all the reactions of GDH into the plant's primary metabolic pathways in order to begin to unravel the natural role of the enzyme (Pahlich, 1996). Most publications have focused exclusively on the supply of NH_4^+ ions to GDH, while ignoring the flow of α-KG from citric acid cycle to the enzyme. Peanut GDH is active *in vitro* in the main reactions, which are reductive amination of α-KG, reversal (deamination) of glutamate to give α-KG and RNA synthesis. Peanut is ideal for the study of glutamate metabolism because all the pathways are present, peanut seeds being very rich in amino acids, fatty acids, carbohydrates, proteins, nucleic acids, antioxidants and vitamins.

Although the metabolic pathways elegantly show that plant metabolism is controlled by numerous enzymes, they do not show the pathway integration and discrimination that produce differential yield (proteins, carbohydrate, fats and vitamins) deposits from one soil environment to another. In addition to feedback inhibition regulation at the enzyme level, metabolic pathways are coordinately silenced, permuted and regulated at the mRNA level by cross-talks catalysed by GDH-synthesized RNAs, which function has been in existence since evolutionary time (Osuji *et al.*, 2012a,b). GDH-synthesized RNAs silence mRNAs that are homologous to them using antisense and siRNA mechanisms. The cross-talks between two or more mRNAs arise because GDH-synthesized RNAs, being isomeric in their primary structure, share extensive plus/plus, and/or plus/minus sequence homologies among themselves and with mRNAs that encode different enzymes in the metabolic pathways (Osuji *et al.*, 2010, 2011). The isomeric primary structure of the RNAs is the evidence for the repeatability of the synthetic reaction although GDH is a template-independent RNA polymerase; and for the fidelity of the silencing reaction although the RNA is non-genomic being metabolic in origin. In the peanut, the sequence homologies between GDH-synthesized RNAs and many mRNAs encoding different enzymes demonstrate a spider web-like molecular scaffolding between photosynthesis, glycolysis and the Krebs cycle, biosyntheses of cellulose, purine nucleotides, amino acids, fatty acids and flavonoids, and nitrogen assimilation. This connectivity describes the molecular biology of GDH as not visualized before (Figs 1.2 and 1.3), and the metabolic pathway integration (systems metabolism) at the mRNA level for the propulsive torrent of C_3, C_5, C_6, etc. compounds from photosynthesis through glycolysis to carbohydrates and the citric acid cycle to energy, and to nucleotide to amino acid, etc. biosynthetic pathways at the enzyme level. The induction of carbon starvation and protein catabolism in relation to GDH functions demands some detailed discussion on the integration of carbon and nitrogen metabolism.

The synchronization of photosynthesis, glycolysis and the citric acid cycle with fatty acid, glutamate, etc. biosyntheses at the peanut's mRNA level involves many mRNAs, including those encoding the phosphate (Pi) translocator, granule-bound starch synthase (GBSS), phosphoglucomutase (PGM), acetyl coenzyme A carboxylase (ACCase), NR, glycinamide ribonucleotide (GAR) synthase (GARS)/GAR transformylase (GART), NADH-GOGAT and glucosyltransferase (Osuji *et al.*, 2011). GDH reversible oxidoreductase activity links the regulation of primary metabolism at the mRNA level with the regulation at the enzyme level (Osuji and Madu, 2012). The mRNAs encoding the regulatory enzymes share sequence homologies with GDH-synthesized RNA. The GDH-synthesized RNA that is homologous to the mRNA encoding Pi translocator (photosynthesis) shared plus/plus sequence similarity with that homologous to the mRNA encoding glucosyltransferase (cellulose biosynthesis). Peanut photosynthetic reduction of CO_2 produces triose phosphates, the key compounds used in starch synthesis in the chloroplast or, after leaving the chloroplasts to the cytosol, in sucrose synthesis in cytosol (Cakmak and Engels, 1999). The Pi translocator exports phosphorylated C_3, C_5, C_6 compounds from the chloroplast to the cytosol in exchange for Pi import from the cytosol into the chloroplast (Stitt and Quick, 1989). Glucosyltransferases catalyse the formation of glucosidic linkages between monosaccharide residues from UDP (uridine diphosphate)-glucose and specific acceptors (Keegstra and Raikhel, 2001) in the initiation of cellulose biosynthesis (Saxena and Brown, 1997). The GDH-synthesized RNA that is homologous to the mRNA encoding GBSS shared twofold plus/plus sequence similarity with that homologous to the mRNA encoding ACCase in lipid biosynthesis. The plus/plus sequence similarities between GDH-synthesized RNAs ensured an ordered variation of the abundances of their homologous mRNAs. Granule-bound starch synthase I (GBSSI) is responsible for the synthesis of linear glucan (amylase) in the starch granule (Smith *et al.*, 1997), whereas another isoform, GBSSII, is important in the synthesis of the branched (amylopectin) component (Edwards *et al.*, 1996). In the peanut chloroplast, the export of triose phosphates to the cytosol by the Pi translocator competed with their utilization in the chloroplast for starch biosynthesis. ACCase catalyses the first (committed) reaction in lipid biosynthesis with the production of malonyl-CoA. The regulation of the enzyme is an important control point in lipid biosynthesis (Page *et al.*, 1994). Plants contain ACCases in plastids and the cytosol (Ke *et al.*, 2000). The two ACCases generate physically separated pools of malonyl-CoA. The plastidic malonyl-CoA pool is used for *de novo* fatty acid biosynthesis to produce 16 and 18 carbon fatty acids. The cytosolic malonyl-CoA pool is used for the elongation of fatty acids to 20 carbons and longer, and for biosynthesis of flavonoid, stilbenoid and malonyl derivatives (Nikolau *et al.*, 1984). ACCase controls the production of fats in peanut (Osuji *et al.*, 2010),

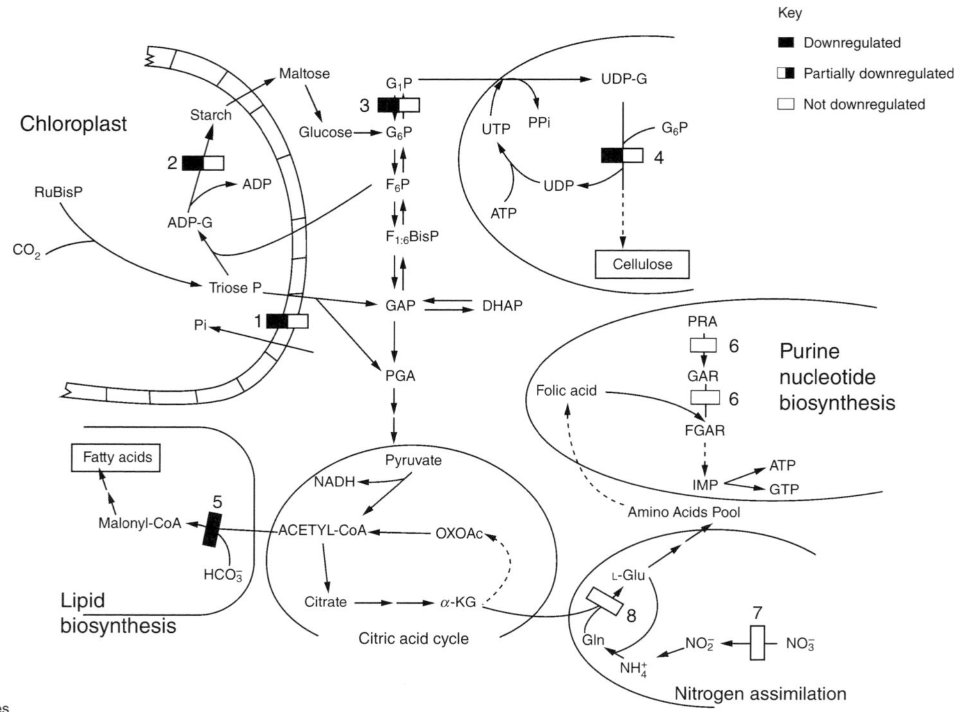

Notes
1, Pi translocator; 2, GBSS; 3, PGM; 4, Glucosyltransferase;
5, ACCase; 6, GARS/GART; 7, NR; 8, NADH-GOGAT.

Fig. 1.2. P + N metabolic variant of peanut. Sketch of simplified primary carbon and nitrogen metabolic pathways highlighting the enzymes whose encoding mRNAs were synchronized by the RNAs synthesized by the GDH of the P + N-treated peanut. Only the chloroplastic GBSS is shown. Only the cytosolic PGM is shown. Only the plastidial ACCase is shown. Because under the P + N mineral ion concentrations, the mRNA (1) encoding the Pi translocator was partially downregulated, there was limited Pi translocator activity to import Pi into chloroplasts; mRNA (2) encoding GBSS was partially downregulated, phosphorylated C_3, C_5 or C_6 compounds were partly expended in starch synthesis, thereby minimizing the quantity of triose phosphates available to power glycolysis and citric acid operations. Also the mRNA encoding PGM (3) was partially downregulated, thereby only incompletely blocking utilization of G_6P for cellulose biosynthesis via glucosyltransferase (4). The mRNA (5) encoding ACCase was downregulated, thereby minimizing fatty acid synthesis. Nitrogen assimilation is important in doubling peanut yield, therefore P + N mineral ion composition ensured that the mRNAs (7) and (8) encoding NR and NADH-GOGAT, respectively, and also the mRNA (6) that regulates purine nucleotide synthesis via GARS/GART were not simultaneously silenced as ACCase by the GDH-synthesized RNAs during the GDH amination-based doubling of dry matter yield. Abbreviations: ACCase, acetyl-CoA carboxylase; DHAP, dihydroxyacetone phosphate; FGAR, formyl-glycinamide ribonucleotide; F_6P, fructose-6-phosphate; $F_{1,6}$ BisP, fructose-1,6-bisphosphate; GAP, glyceraldehyde 3-phosphate; GAR, glycinamide ribonucleotide; GARS, GAR synthase; GART, GAR transformylase; GBSS, granule-bound starch synthase; GDH, glutamate dehydrogenase; Gln, glutamine; L-Glu, glutamate; GOGAT, glutamate synthase; G_1P, glucose-1-phosphate; G_6P, glucose-6-phosphate; IMP, inosine monophosphate; α-KG, α-ketoglutarate; NR, nitrate reductase; OXOAc, oxaloacetate; PGA, phosphoglyceric acid; PGM, phosphoglucomutase; Pi, phosphate; PPi, pyrophosphate; PRA, phosphoribosylamine; RuBisP, ribulose 1,5-bisphosphate.

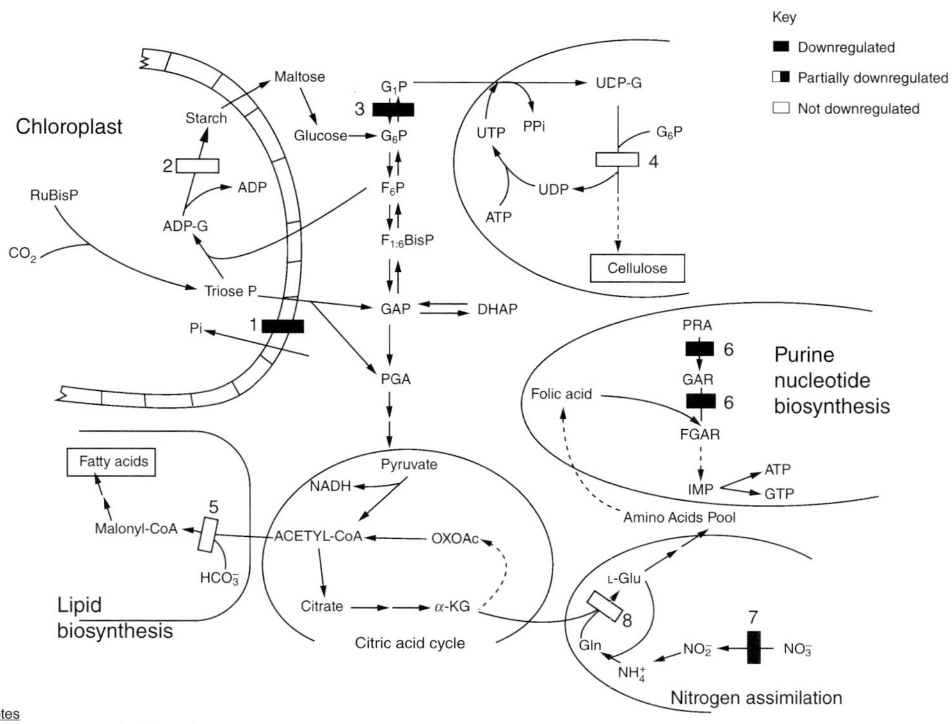

Notes
1, Pi translocator; 2, GBSS; 3, PGM; 4, Glucosyltransferase;
5, ACCase; 6, GARS/GART; 7, NR; 8, NADH-GOGAT.

Fig. 1.3. Control metabolic variant of peanut. Sketch of simplified primary carbon and nitrogen metabolic pathways highlighting the enzymes whose encoding mRNAs were synchronized by the RNAs synthesized by the GDH of the control untreated peanut. Only the chloroplastic GBSS is shown. Only the cytosolic PGM is shown. Only the plastidial ACCase is shown. Because under the untreated control peanut, the mRNA (1) encoding the Pi (phosphate) translocator was downregulated, there was minimum Pi translocator activity to import Pi into chloroplasts; mRNA (2) encoding GBSS was not downregulated, phosphorylated C_3, C_5 or C_6 compounds were maximally expended in starch synthesis, thereby minimizing the quantity of triose phosphates available to power glycolysis and citric acid operations. However, the mRNA encoding PGM (3) was downregulated, thereby blocking free utilization of G_6P (glucose-6-phosphate) for cellulose biosynthesis (4). The mRNAs (5) and (8) encoding ACCase and NADH-GOGAT were not downregulated, therefore respective citric acid intermediates were readily distributed to fatty acid, glutamate, etc. biosynthetic pathways. Nitrogen assimilation is important in GDH amination-based doubling of crop yield, but the control peanut in its inefficiency permitted the mRNAs (6) and (7) encoding the enzymes of purine nucleotide synthesis and NR, respectively, to be silenced simultaneously in contrast with the more efficient P + N metabolic variant. For abbreviations, see Fig. 1.2 caption.

and similarly in soybean (Reverdatto *et al.*, 1999). The GDH-synthesized RNA that is homologous to the mRNA encoding GBSS also shared sixfold plus/minus sequence similarities with the GDH-synthesized RNA that is homologous to the mRNAs encoding GARS/GART (purine nucleotide biosynthesis); twofold plus/plus sequence similarities with that homologous to the mRNA encoding NR (nitrogen assimilation); and twofold plus/minus sequence similarities with that homologous to the mRNA encoding NADH-GOGAT (nitrogen assimilation). This is a reciprocal regulation of the mRNAs that are homologous to the same GDH-synthesized RNA, so that at least one in the group of mRNAs could not be knocked out with the other three. Several metabolic pathways co-operated at the mRNA level to assimilate nitrogen in peanut. They included purine biosynthesis, nitrate reduction and the GS-GOGAT cycle. The purine metabolic pathway is also employed in tropical legumes to assimilate and detoxify ammonia (Smith and Atkins, 2002). GARS catalyses the second step, and GART catalyses the third step in the purine biosynthetic pathway. GDH-synthesized RNA was found to be homologous to the twin mRNAs encoding GARS and GART (Osuji *et al.*, 2010). This is in agreement with the finding that the two enzymes were encoded by the same genetic locus (Caperelli, 1989). NR catalyses the reduction of nitrate to nitrite, which is then reduced by nitrite reductase to an NH_4^+ ion (Kleinhofs *et al.*, 1989), followed by incorporation into amino acids by GS-GOGAT cycle (Vance *et al.*, 1995), mainly in the control peanuts. Reciprocal regulation of glycolysis and nitrogen assimilation was further enforced by the GDH-synthesized RNA that is homologous to the mRNAs encoding PGM, GARS/GART and NADH-GOGAT. The GDH-synthesized RNA shared tenfold plus/minus sequence similarities with the GDH-synthesized RNA that is homologous to the mRNAs encoding GARS/GART; and twofold plus/minus sequence similarities with that homologous to the mRNA encoding NADH-GOGAT. These sequence similarities created structural loopholes for incomplete silencing, meaning that at least one of the mRNAs could not suffer a demise with the other two. Phosphoglucomutase generates glucose-1-phosphate (Glc-1-P) from glucose-6-phosphate (Glc-6-P). It is localized both in the plastids and the cytosol (Streb *et al.*, 2009). The plastidic PGM is essential for starch synthesis to store photosynthates in leaves during the day, and also in the degradation of starch (Hattenbach and Heineke, 1999). The cytosolic PGM is involved in sucrose catabolism to provide intermediates for glycolysis, etc. (Manjunath *et al.*, 1998). Deficiency in PGM activity resulted in 'starchless' and wrinkled seed phenotypes accompanied by significantly increased seed lipid contents (Periappuram *et al.*, 2000). Therefore, PGM distributes Glc-6-P between the glycolysis, starch, citric acid, amino acid and fatty acid pathways. The GDH-synthesized RNA that is homologous to the mRNA encoding ACCase (fatty acid biosynthesis) shared 22-fold plus/plus sequence similarity with the GDH-synthesized RNA that is homologous to the mRNAs encoding GARS/GART, but sevenfold plus/minus sequence similarities with those homologous to the mRNAs encoding NR and NADH-GOGAT. This spider web-like coordination by the GDH-synthesized RNAs that are homologous to the mRNAs encoding NADH-GOGAT, GARS/GART and NR provide all-around metabolic pathway synchronization so that all three of them could not be totally silenced with the mRNAs encoding ACCase (fatty acid biosynthesis), PGM (glycolysis) and NH_4^+ ion assimilation under any applied carbon limitation and/or mineral ion regimen. The GDH-synthesized RNA that is homologous to the mRNAs encoding the GARS/GART shared sixfold plus/plus sequence similarities with that homologous to the mRNAs encoding NR, and 16-fold plus/plus similarity with that homologous to the mRNA encoding NADH-GOGAT. This structural integration is the molecular description of the kinetic signal integration and discrimination function of GDH (Osuji and Braithwaite, 1999), and it meant that all the molecularly connected mRNAs were under orderly control of their relative abundance. These (Table 1.1) were some of the important molecular provisions at the mRNA level that made for propulsive flow and prevented carbon limitation (starvation) or the total shut down/reversal of the pathways for amino acid syntheses, especially of glutamate, while C_3, C_4, C_5, C_6, etc. compounds were preferentially channelled towards saccharide and/or fatty acid metabolism. They explain the robust capability of peanut to produce photosynthates and to coordinately regulate at the mRNA level their ordered transportation/distribution (Figs 1.2 and 1.3) through downstream saccharide (carbohydrate biosynthesis) and glycolytic transactions to acetyl-CoA (for fatty acid synthesis), α-KG (for glutamate synthesis), oxaloacetate and all citric acid cycle intermediates, no matter what the inadequacy of the environment (Wright, 2008).

Ammonium ion assimilation is regulated by the propulsion of α-KG from the citric acid cycle, not by the reverse flow of α-KG from glutamate. Therefore, metabolic pathways at the enzyme level do not narrate the whole story about intermediary metabolic synchronization. The reactions catalysed by GDH-synthesized RNA at the mRNA level narrate another part of the events because they are as molecularly sophisticated and intertwined as those catalysed by enzymes. The flow of C_3, C_5, C_6 compounds from photosynthesis provides an abundant supply of α-KG, both in the mitochondria and the chloroplasts (Chen and Gadal, 1990), which physiological circumstance

Table 1.1. Synchronization of carbon and nitrogen biochemical pathways (photosynthesis, glycolysis, saccharide metabolism, the citric acid cycle; and purine nucleotide, fatty acid, nitrogen assimilation, glutamate, etc. biosyntheses) by glutamate dehydrogenase (GDH)-synthesized RNA in response to stoichiometric mineral ion treatment of peanut.[a]

	Stoichiometric mineral ion combinations									
Regulatory enzyme[b]	Control	NPKS	NS	Pi	K	S	PS	PN	N	PK
Pi translocator	– –	+ –	++	++	+ –	+ –	– –	+ –	+ –	++
GBSS	++	– –	++	++	++	++	+ –	+ –	+ –	– –
PGM	– –	– –	– –	+ –	++	– –	++	+ –	– –	– –
Glucosyltransferase	++	– –	+ –	+ –	– –	++	– –	+ –	– –	– –
ACCase	++	++	– –	+ –	++	++	++	– –	++	++
GARS/GART	– –	+ –	++	– –	++	– –	– –	++	++	+ –
NR	– –	– –	+ –	++	– –	– –	– –	++	++	– –
NADH-GOGAT	++	++	++	– –	++	++	+ –	++	++	++
Dry matter (kg ha^{-1})[c]	27057.7	19428.7	27057.7	39170.5	21628.7	27057.6	31996.9	27057.7	32380.6	41323.2

[a]Key: ++ not silenced mRNA was >90% digitalized mRNA band presence; +– partially silenced mRNA ranged from ~40-60% digitalized mRNA band presence; – – fully silenced mRNA was <10% digitalized mRNA band presence (Osuji *et al.*, 2011, 2012a).

[b]Abbreviations: Pi translocator, phosphate translocator; GBSS, granule-bound starch synthase; PGM, phosphoglucomutase; ACCase, acetyl coenzyme A carboxylase; GARS/GART, glycinamide ribonucleotide (GAR) synthase/GAR transformylase; NR, nitrate reductase; NADH-GOGAT, glutamate synthase.

[c]Osuji *et al.*, 2012a.

creates no irresistible metabolic need to deaminate glutamate. At the enzyme level, the carbon and nitrogen metabolic pathways are integrated with the carbon and nitrogen regulatory pathways at the mRNA level by GDH oxidoreductase activity (Fig. 1.4a); the two levels of metabolism cooperate to assure the propulsion of C_3, C_5, C_6, etc. compounds from photosynthesis through glycolysis/gluconeogenesis, citric acid/glyoxylate cycles to energy metabolism,

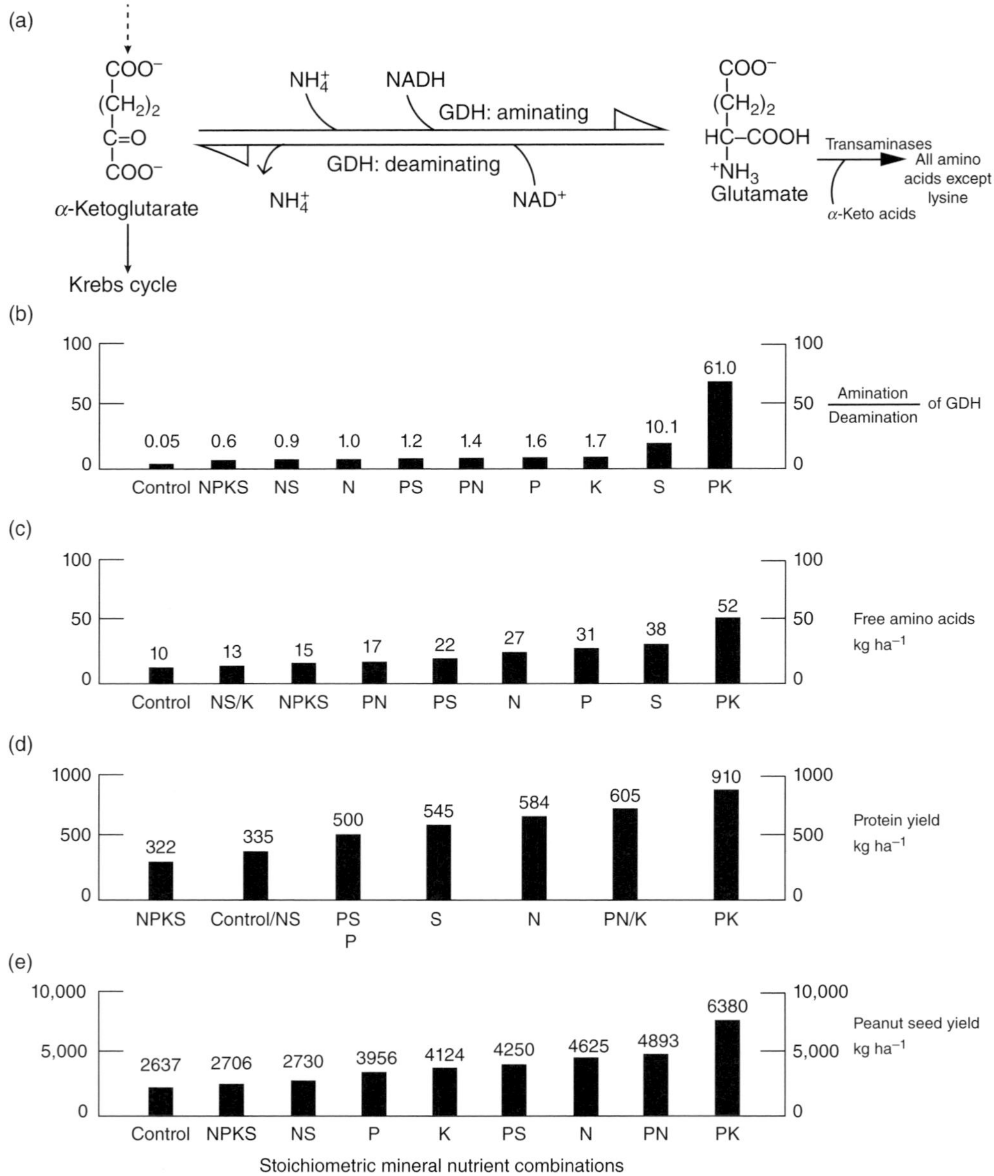

Fig. 1.4. Amination–deamination ratios of GDH (glutamate dehydrogenase) in relation to free amino acids, total proteins and yields of peanut treated with mineral nutrients. (a) Sketch of GDH reversible amination activity unleashing metabolic propulsion in peanut that: (b) maximizes the amination–deamination ratios of the enzyme; optimizes (c) free amino acid and (d) total protein contents; and (e) doubles the seed yields of interlocking stoichiometric mineral fertilizer-treated plots relative to the controls.

glutamate and other amino acids, fatty acids, flavonoids, etc.

GDH-synthesized RNAs are statistical in nature, their primary structure being dependent on the binomial subunit compositions of the GDH isoenzymes (Osuji *et al.*, 2009). The cobweb-like cross-connections between metabolic pathways at the mRNA level and at the enzyme level introduce the concept of probability or likelihood for metabolites to flow in one or alternative directions in peanuts subjected to interlocking environmental conditions. Many probabilities arise under a prevailing environmental condition for the GDH-synthesized RNA to reprogramme the mRNA abundances and consequently C_3, C_4, C_5, etc. metabolic propulsion and the rates of yield production. The reprogramming of carbon metabolic flow in response to the prevailing environment in order to maximize crop yield was best approached by permuting the yield-specific pathways guided by the enzyme steps that were not silenced, partially silenced and fully silenced (Table 1.1, Figs 1.2 and 1.3) by GDH-synthesized RNAs per environmental condition, and limited to the primary metabolic pathways that synchronize photosynthesis, glycolysis, the citric acid cycle, and saccharide, glutamate, fatty acid metabolism at the mRNA level. There is a multitude of enzymes whose encoding mRNAs share short sequence homologies with GDH-synthesized RNA, but only eight of such enzymes were selected to simplify this systems biology flow model (see Table 1.1, Figs 1.2 and 1.3). The responses of the eight mRNAs were different from one treated peanut to the other. The number of permutations varied widely: control peanut produced 70 permutations; N + P + K + S-treated peanut produced 420; N + S-treated peanut produced 1680; Pi-treated peanut produced 560; P + N-, P + K- and NH_4Cl-treated peanuts each produced 280; Na_2SO_4-, P + S- and KCl-treated peanuts each produced 168. There were many probable rearrangements of peanut carbon and nitrogen metabolic pathways in response to any mineral nutrient regimen, each rearrangement being a metabolic variant of peanut (Figs 1.2 and 1.3), as different from genetic variants. This is a sophisticated biochemical comprehensiveness that GDH experimentation demands in a conversation on NH_4^+ ion availability and carbon starvation. The permutation of the metabolic pathways is the explanation for the contradictions in the literature surrounding the responses of peanut yields to applied fertilizer (Henning *et al.*, 1982; Kidder, 1994; Nasr-Alla *et al.*, 1998; Jamal *et al.*, 2010). A frequency distribution plot of the metabolic pathways versus the number of permutations gave a positively skewed graph reminiscent of the GDH binomial isoenzyme population (Fig. 1.5) distribution pattern (Osuji *et al.*, 2010). This is proof that the stoichiometric combinations of mineral nutrients targeted the GDH isoenzyme subunit polypeptides, and of the multidimensional biological scope of the project design. The mode (224) of the frequency distribution graph had a frequency of four, with the metabolic pathways of the S-, K-, N-, P + S-, P + N-, and P + K-treated peanuts clustered around it. Therefore, when peanuts were treated with electromagnetic mineral nutrients, their carbon and nitrogen metabolism (photosynthesis, glycolysis, citric acid cycle, saccharide, nitrogen assimilation etc) gravitated towards increased efficiency of the flow of C_3, C_4, C_5, C_6 compounds and of the synthesis of glutamate compared with the untreated control peanut (Fig. 1.3), and the control GDHs from inadequately fertilized (N + P + K + S-, Pi- and N + S-treated) peanuts (Osuji *et al.*, 2012a). Specifically, the unidirectional flow of α-KG from the citric acid cycle to GDH and GOGAT increased in efficiency in the S-, K-, N-, P + S-, P + N- (Fig. 1.2) and P + K-treated peanuts and they recorded higher yields of amino acids and protein compared with the untreated control (Fig. 1.3). The frequency distribution graph also showed that the GDH from the untreated control peanut was different from, not synonymous with, the control GDHs from inappropriately fertilized peanuts. Table 1.1 and Figs 1.2–1.4 illustrate the signal integration, discrimination and synchronization functions of GDH in intermediary metabolism.

The responses (Table 1.1) of the target mRNAs to the mineral nutrients showed that the P + K-treated peanut recorded the highest yields of fatty acids, amino acids, proteins, carbohydrates, etc. (Fig. 1.4) because the mRNA encoding Pi translocator was not silenced, but the mRNAs encoding GBSS and PGM were fully silenced. This meant that the C_3, C_5 and C_6 compounds in the P + K-treated peanut enjoyed propulsive metabolism from the chloroplast into glycolysis and downstream to the citric acid cycle where the intermediates, including acetyl-CoA and α-KG, were channelled to fatty acid, glutamate, etc. biosynthetic pathways, leading to the record doubling of the yields of fats, amino acids, storage protein and seeds (Fig. 1.4). Partial or full silencing of the mRNA encoding Pi translocator in the NH_4^+ ion-, P + N-, P + S- and K-treated peanuts inhibited photosynthesis by impeding the exit of the products of photosynthesis from the chloroplasts, with consequent decreases in fatty acid, amino acid, protein, etc. yields of the peanuts (Fig. 1.4), compared with the P + K-treated peanut. In the controls (untreated, N + P + K + S- and N + S-treated) the mRNA encoding Pi translocator was partially, fully or not silenced in tandem with the mRNAs encoding GBSS and PGM that were not silenced or fully silenced (Table 1.1), thereby causing reduced export of photosynthates to glycolysis, with consequent lower fatty acid, amino acid, carbohydrate and protein yields compared with the P + N-, N-, S- and P + S-treated peanuts (Fig. 1.4). However, the controls did not suffer protein catabolism because their seed, protein and fatty acid yields were normal, being similar to peanut farmers' yields (Pease *et al.*, 2012), their limited photosynthetic supply of C_3, C_5, etc. compounds being adequately redistributed by the permutation of their metabolic pathways to make for normal growth.

In claiming carbon limitation, protein catabolism, amino acid deamination and glutamate degradation, the changes in the primary metabolic flow of C_3, C_4, C_5 and C_6 compounds in response to the environment and carbon nutrients, including all other external conditions applied, need to be demonstrated as done by Oaks (1994) and Melo-Oliveira *et al.* (1996). It is incomplete to assume that the textbook metabolic pathways function in all plants as is and under all experimental conditions. GDH is a broad-spectrum enzyme; its activity modifies the efficiency of many other enzymes. Therefore, experimentations on it needs to be equally broad-based to assure that the conclusions reached are balanced and are exhaustively justifiable in science.

1.8.2 Aminating and deaminating activities

Besides the flow of α-KG *in vivo* to GDH, the perceived natural role of the enzyme also hinges on the accuracy of the *in vitro* assays of its activities in the amination direction as separate from the deamination direction. It is important to differentiate completely the deaminating reaction from the aminating reaction. In this regard, the reversibility of the oxidoreductase activity embodies a considerable level of non-comparative incompleteness of all assays that seek to quantify NADH, glutamate, NH_4^+ ion, and/or α-KG resulting from the reactions; furthermore, it is not possible for the reactions to be stopped prior to analysis of the products and/or the remaining substrates. Therefore, monitoring of the NADH/NAD^+ and glutamate/NH_4^+ equilibria in the GDH oxidoreductase reaction is more of a qualitative exercise because it aims at capturing the products 'in flight' in their reversible transition to the substrates. When the GDH is not very active, the monitoring of the NADH/NAD^+ and glutamate/NH_4^+ equilibria gives many non-reproducible results. It is similarly novel to monitor the levels of the transcripts encoding the GDH polypeptides and to calculate the aminating activity as different from the deaminating activity from the transcript levels, where possible. On the other hand, assay of the RNA synthetic activity embodies the advantage that the GDH does not hydrolyse the product RNA and, very importantly, the reaction is arrested and/or instantly stoppable at will prior to product analysis (Osuji *et al.*, 2003a). Therefore, RNA synthesis gives reproducible, quantitative and more complete assays of GDH activity.

Stoichiometric formulations of mineral nutrients did not affect the GS-GOGAT cycle activity (Osuji and Madu, 2012). This has allowed increased dissection of the deaminating activity of GDH from its aminating activity. Molecular analyses of GDH aminating as different from deaminating activity showed that the array of RNA synthesized in the deamination direction was different from that synthesized in the amination direction. Then, there were non-reciprocal relationships in the quantities of RNA synthesized in the two reaction directions. The GDHs from the Pi- and NH_4Cl-treated peanuts were super active in the amination direction and also very active in the deamination direction. The GDH from P + K-treated peanut was super active in the amination direction but was minimally active in the deamination direction. The GDH from N + P + K + S-treated peanut was very active, whereas the GDH from the untreated peanut was only minimally active in the amination direction, but both GDHs were very active in the deamination direction. However, no GDH was virtually inactive in both deamination and amination directions, thus confirming the results of Melo-Oliveira *et al.* (1996) that GDH is not redundant in higher plants.

Based on the RNA synthetic activity, GDH amination–deamination ratios (assumed to approximate to the ratio of amination–deamination equilibrium constants) varied widely. Compared with the control, treatment of peanut with P + K increased the amination–deamination ratio (aminating capacity) by >1000-fold (Osuji and Madu, 2012), thus showing that the stoichiometric mineral ions functioned as proposed to displace the equilibrium position of the GDH oxidoreductase activity, thereby permitting the aminating activity to function with minimal interference from the deaminating activity and vice versa per hexameric GDH molecule and per field plot landscape of peanut. Spectral methods that monitor $NADH/NAD^+$ and glutamate/NH_4^+ equilibrium positions have been unable to detect such a complete separation of aminating from deaminating activities of the enzyme (Stone *et al.*, 1980; Robinson *et al.*, 1991; Osuji and Braithwaite, 1999; Labboun *et al.*, 2009).

The GDH amination–deamination ratios (Fig. 1.4b) and the free amino acid yields (Fig. 1.4c) enabled calculation of the quantities of NH_4^+ ion assimilated and of glutamate deaminated by GDH. The normal GOGAT activity (Osuji and Madu, 2012), taken together with the high deamination reaction mode of the GDH of the untreated control peanut, suggested that GDH did not assimilate a considerable quantity of NH_4^+ ion, while the GS-GOGAT cycle assimilated considerable quantities of NH_4^+ ion and α-KG, the product glutamate being available for synthesis of most amino acids (Figs 1.3, 1.4a and 1.4c). Using the free amino acid yield as the comparative index for NH_4^+ ion assimilated, the free amino acid yield of 9.8 kg ha^{-1} for the control peanut was assumed to be due to the activity of the GS-GOGAT cycle (Fig. 1.3), as the GDH aminating activity was almost silent in the treatment and could not make a substantial contribution to the production of glutamate. Furthermore, the high deaminating activity of the GDH of the control peanut could have eroded some of the glutamate by deaminating it to α-KG, although AAT and glutamate decarboxylase also degrade glutamate even more efficiently (Oaks, 1994). Therefore, the adjusted total free amino acid should be slightly higher than 9.8; but for the purpose of estimating the quantity of amination it was assumed that GS-GOGAT accounted for the NH_4^+ ion assimilation of the control peanut, which represented the baseline biological nitrogen fixation.

Mineral treatment of the peanut increased the GDH aminating activity but decreased the deaminating activity, with corresponding increases in the free amino acid yields (Fig. 1.4c). As GOGAT activity remained normal and constant throughout the peanuts, the mineral nutrient-induced increases in free amino acid yields were attributed to GDH aminating activity: the higher the GDH amination–deamination ratio the lower the possible quantity of glutamate lost via the GDH deamination function. Therefore, in the N + P + K + S-treated peanut with 14.7 kg ha^{-1} of free amino acids (Fig. 1.4c), the quantity of amino acids resulting from amination by GDH was 4.9 kg ha^{-1} by subtraction of the free amino acids (9.8 kg ha^{-1}) due from amination by GS-GOGAT cycle. The increased flow of C_3, C_5 and C_6 compounds from photosynthesis (mRNA encoding Pi translocator was only partially silenced), and consequently that of α-KG from the citric acid cycle in the N + P + K + S-treated compared with the untreated control (Table 1.1), accounted in part for the increased aminating activity of GDH. The amination capacity (0.6) of the GDH of the N + K + P + S-treated peanut permitted the calculation of the possible quantity of free amino acids (8.17 kg ha^{-1}) lost as NH_4^+ ion due to the possible deamination function of GDH.

In the NH_4Cl-treated peanut with 27.12 kg ha^{-1} of free amino acids (Fig. 1.4c), the quantity of amino acids resulting from amination by GDH was 17.32 kg ha^{-1} by subtraction of the free amino acids (9.8 kg ha^{-1}) due from amination by the GS-GOGAT cycle. The increased flow of C_3, C_5 and C_6 compounds from photosynthesis (mRNAs encoding the Pi translocator and GBSS were only partially silenced), and consequently of α-KG from the citric acid cycle (Table 1.1), in the NH_4Cl-treated compared with the untreated control accounted in part for the increased aminating activity of GDH. The amination–deamination ratio (1.1) of the GDH of the

NH_4Cl-treated peanut permitted the calculation of the possible quantity of free amino acids (~15.7 kg ha^{-1}) lost as NH_4^+ ions due to the deamination function of GDH. It was expected (Lettgen *et al.*, 1989) that the added NH_4Cl would have induced more deamination. Instead, it induced both increased amination and deamination almost equally. The added NH_4^+ ion did not change the position of equilibrium between the amination and deamination reactions. Therefore, increased NH_4^+ ion concentrations are not necessarily the perfect inducer of amination because in relation to carbon limitation the consolidated supply of α-KG through the permutation of metabolic pathways should not be overlooked.

In the Pi-treated peanut with 30.81 kg ha^{-1} of free amino acids (Fig. 1.4c), the quantity of amino acids resulting from amination by GDH was 21.01 kg ha^{-1} by subtraction of the free amino acids (9.8 kg ha^{-1}) due from amination by the GS-GOGAT cycle. The elevated flow of C_3, C_5 and C_6 compounds from photosynthesis (the mRNA encoding GBSS was not silenced), and consequently of α-KG from the citric acid cycle, in the Pi-treated compared with the untreated control accounted in part for the increased aminating activity of GDH. The amination–deamination ratio (1.6) of the GDH of the Pi-treated peanut permitted the calculation of the possible quantity of free amino acids (~13.1 kg ha^{-1}) lost as NH_4^+ ions due to the deamination function of GDH, etc. It was expected that the added Pi would have maximally increased the export of photosynthates (C_3, C_5, C_6 compounds) from chloroplasts to the cytosol for increased saccharide metabolism, glycolysis, citric acid, fatty acid, amino acid, biosyntheses. Therefore, increased Pi concentrations are not necessarily the perfect inducer of GDH deamination because in relation to the differential photosynthetic response, the increased supply of α-KG should not be overlooked.

Amination by the GDHs of the other mineral-treated peanuts were similarly calculated. In the P + K-treated peanut with 52.39 kg ha^{-1} of free amino acid yield (Fig. 1.4c), the quantity of amino acids resulting from amination by GDH was 42.59 kg ha^{-1} by subtraction of the free amino acids (9.8 kg ha^{-1}) due from amination by the GS-GOGAT cycle. The optimized supply of C_3, C_5 and C_6 compounds from the chloroplast (the mRNA encoding Pi translocator was not silenced, but the mRNA encoding GBSS was fully silenced), uninterrupted transactions (the mRNA encoding PGM was fully silenced) to the citric acid cycle, and consequently of α-KG from the citric acid cycle to GDH in the P + K-treated compared with the untreated control (Table 1.1), accounted for the record high aminating activity of GDH. Although there was no added NH_4^+ ion, the P + K mix induced the highest amination, synthesis of glutamate and other amino acids, and accumulation of proteins (Fig. 1.4d). The controlling factor was that under the P + K electromagnetic mix, the peanut's mRNA encoding Pi translocator was not silenced (Table 1.1). The amination–deamination ratio (61) of the GDH of the P + K-treated peanut permitted the calculation of the quantity of amino acids (0.75 kg ha^{-1}) lost as NH_4^+ ions due to the possible deamination function of GDH. The electromagnetic property of the P + K mineral salt mix displaced the position (Fig. 1.4a) of equilibrium between amination and deamination to the far right in favour of glutamate synthesis, thereby unleashing the peanut's metabolic potential that optimizes, maximizes and doubles amino acid, protein, carbohydrate, fatty acid, etc. and crop yields per hectare.

In the literature, estimates of the GDH deamination role were expressed grossly as activity per unit weight of protein, which was doubly inappropriate because of the dependence of proximate compositions on the plant's environment (Osuji *et al.*, 2012a) and on the activities of proteinases. The RNA synthetic activity of GDH was different because it advanced far beyond the proximate level to permit the estimation of the quantities of amino acids possibly eroded by the enzyme's deaminating activity per unit area of land. The results show that the natural role of the GDH aminating activity is to assimilate exponential multiples of the quantities of NH_4^+ ion as assimilated via the GS-GOGAT cycle, whereas the possible natural role of the GDH deaminating activity could be to erode a vanishing fraction of the NH_4^+ ions assimilated by the enzyme (Fig. 1.4). The stability of GOGAT over wide variations of mineral nutrient concentrations and compositions (Osuji and Madu, 2012) is evidence for the role of the GS-GOGAT cycle as the housekeeping device for the assimilation of low to normal concentrations of α-KG and NH_4^+ ions.

On the other hand, GDH aminating activity is the responsive mechanism for large-scale assimilation of NH_4^+ ions and α-KG under normal to maximum photosynthetic yields of C_3, C_5, etc. compounds.

1.8.3 Amination-based crop yield doubling biotechnology

Following the confirmation that GDH is aminating (Oaks, 1994; Melo-Oliveira *et al.*, 1996; Pahlich, 1996; Osuji and Braithwaite, 1999), there arose the need to interject a game changer into the biochemical conversation and to establish the technological foundation for translating the amination reaction to achieve economic benefit (Ameziane *et al.*, 2000) in increased and maximized protein and crop yields without additional fertilizer input, without additional irrigation and without cultivation of more land. Pernicious food insecurity (FAO, 2006) is a pending global calamity based on human population that is projected to rise above 9 billion in the near future, when feminine energy will power most of the peasant labour in the developing countries (FAO, 2000). The coincidence in concomitance of the increasing trends of GDH amination–deamination ratios, glutamate and other amino acids, total protein, fatty acids and cellulose yields culminating in a doubling of crop dry matter yields (Osuji *et al.*, 2011) without the cultivation of more land constituted the technological bedrock for the GDH amination-based crop yield doubling biotechnology (Figs 1.1–1.4; Table 1.1). Crop yield doubling biotechnology (Osuji *et al.*, 2010, 2011) is the collateral chemical scaffolding (Figs 1.2 and 1.3) on which GDH stands as it performs its natural amination role, which unleashes the crop's potential to double the dry matter yield as it is coupled to the synchronization of carbon and nitrogen metabolic pathways. The cobweb-like chemical scaffolding (Figs 1.2 and 1.3) is different from one treated peanut to another, each treated peanut being a metabolic variant, as different from genetic variants. The cobweb-like synchronization of carbon and nitrogen metabolic pathways by GDH also seems to disallow considerable glutamate deamination in photosynthesizing crops. These confirm that the GDH amination–deamination ratio is the biochemical molecular hinge linking primary metabolism at the enzyme level with metabolism at the mRNA level. Labelling of the NH_4^+ ion or glutamate, and computational simulations (Rhodes *et al.*, 1989; Morgan and Rhodes, 2002) of amino acid flux, did not illuminate the natural role of GDH probably because among other kinetic considerations, the differential regulation of C_3, C_4, C_5, etc. compounds at the mRNA level was not considered. The assays of activities at the RNA level instead are an analytically exact science because the GDH-synthesized RNA is stable, and not hydrolysed by GDH. GDH amination-based crop yield doubling biotechnology complements and strengthens the results of research projects on the genetic (Reynolds *et al.*, 2009; Xing and Zhang 2010), plant breeding and crop protection (Hinchee *et al.*, 1993; Gould, 1998; Horton, 2000; Century *et al.*, 2008) bases of crop yield improvement.

1.8.4 The aminating cassette of glutamate dehydrogenase isoenzymes

Only miniscule quantities of α-KG resulted from possible deaminating activity of GDH in all of the peanuts (Osuji and Madu, 2012). Judging from the continuous torrent of C_3, C_4, C_5, etc. compounds in the peanut's glycolysis and citric acid cycles, the large kilogram yields of especially those amino acids (proline, glutamate, glutamine, histidine, arginine, aspartate, asparagine, etc.) that originate directly from citric acid cycle intermediates, and the record high kilogram yields of storage proteins (Fig. 1.4d), the supply of α-KG for citric acid operation was not from the deamination of glutamate. The peanuts that produced the record high yields of free amino acids and storage proteins also produced record high yields of cellulose, fatty acids and seeds, thus confirming the absence of GDH deamination of glutamate. Had GDH deaminating activity made any contribution to α-KG supply, the dry matter and carbohydrate yields of the untreated control, N + P + K + S- and N + S-treated peanuts would have been higher than in those peanuts with high aminating activities of GDH, thereby creating a potentially dangerous counter metabolic flux. Also, had the GS-GOGAT cycle been more efficient than GDH in the synthesis of glutamate, the yields of amino acids and protein by the untreated control, N + P + K + S- and N + S-treated peanuts would have been higher than in the P + N- and P + K-treated peanuts (Figs 1.4c and 1.4d). These results suggest that GDH deaminating activity might be primarily a kinetic metabolic state that enables the enzyme to synthesize RNA for silencing and coordinating the mRNAs encoding Pi translocator, GBSS, PGM, pyruvate decarboxylase, glyceraldehyde-3-phosphate dehydrogenase, phosphoglycerate mutase, phosphoenolpyruvate carboxylase, malate dehydrogenase, GARS/GART, NR, etc. to assure that carbon and nitrogen metabolism are synchronized in same direction at all times, even when the crop is growing in deficit environments.

GDH synthesizes RNA, some of which remains tightly bound to it even after exhaustive dialysis of the enzyme. Ribonuclease A treatment of GDH, followed by dialysis, successfully removed the bound RNA (Fig. 1.5). Digestion of the GDH with proteinase K, followed by anion exchange chromatography of the solution, gave a single strong peak of nucleosides which represented the residue at the growing end of the RNA (Osuji *et al.*, 2004). Therefore, some RNA is tightly bound to GDH in the natural state. In order to visually demonstrate the natural state of GDH, and distinguish that state from the normalized state, GDH was fractionated to isoenzymes before and then after RNase treatment (Fig. 1.5). The staining intensities of the isoenzymes were digitally quantified, and the results showed there were twofold to fourfold more isoenzymes per band after the RNase treatment (normalized state) than before the treatment (natural state). These results clearly showed that the GDH that was complexed to RNA displayed no oxidoreductase activity before the RNase treatment (Fig. 1.5a), but displayed oxidoreductase activity only after the removal of the RNA (Fig. 1.5b). The oxidoreductive isoenzymes in the natural state (Fig. 1.5a) were overwhelmingly aminating because they were cationic in the Rotofor direction and in the PAGE direction, and constituted 20–30% of the total GDH isoenzymes (Fig. 1.5b). Deaminating GDHs are predominantly the anionic isoenzymes (Osuji and Madu, 2012), and they manifested themselves (Fig. 1.5b) prominently after the bound RNA had been removed. Figure 1.5a represents the aminating cassette of the GDH isoenzyme; Fig. 1.5b represents the full (normalized) profile of all the isoenzymes. When steps were taken during GDH purification to minimize proteolysis by adding phenylmethylsulfonyl fluoride, and to protect RNA by adding ribonuclease inhibitor to the homogenization buffer cocktail, the GDH isoenzyme profile was overwhelmingly the aminating cationic isoenzymes cassette. The tenacious binding of RNA to GDH isoenzymes differentiated the two molecular states. The aminating cassette (the molecular kinetic state for oxidoreductase), dominated by the aminating (cationic) isoenzymes, agrees with the predominant natural aminating role of GDH (Fig. 1.4); the deaminating activity remained as a minor role manifested in the synthesis of RNA by the enzyme and when crops are cultivated in harsh environments similar to those of the control peanuts (Fig. 1.3).

Figure 1.5 explains why spectral methods of GDH assay have failed to unravel the natural aminating role of the enzyme. The kinetic molecular state for RNA synthesis obstructs the kinetic molecular state for oxidoreductase. The tenacious binding of RNA to GDH isoenzymes (Fig. 1.5) suggests that ribonucleases need to be added to the homogenization buffer cocktail (Osuji *et al.*, 2004) so as to ensure removal of scientific bias from *in vitro* assays of the enzyme activities. Experimental strategies also need to be developed to ensure that the RNA bound to the enzyme is removed before *in vivo* assays (Labboun *et al.*, 2009), because a situation where more than 50% of the isoenzymes are excluded from investigation is likely to produce occluded pictures of the role of the enzyme.

1.9 Food Security

A litmus test for the natural aminating role of GDH is the doubling of dry matter, amino acid, protein, carbohydrate, fatty acid, etc. yields when photosynthetic production of C_3, C_5 and C_6 compounds is above normal (Fig. 1.4). GDH isomerization has been applied to determine mineral nutrient sufficiency during crop growth and development in maize, soybeans, sweet potato, yam tuber, etc. (Osuji and Cuero, 1992b; Osuji and Madu, 1997; Osuji *et al.*, 1998, 2003b). Enzyme kinetics studies permitted

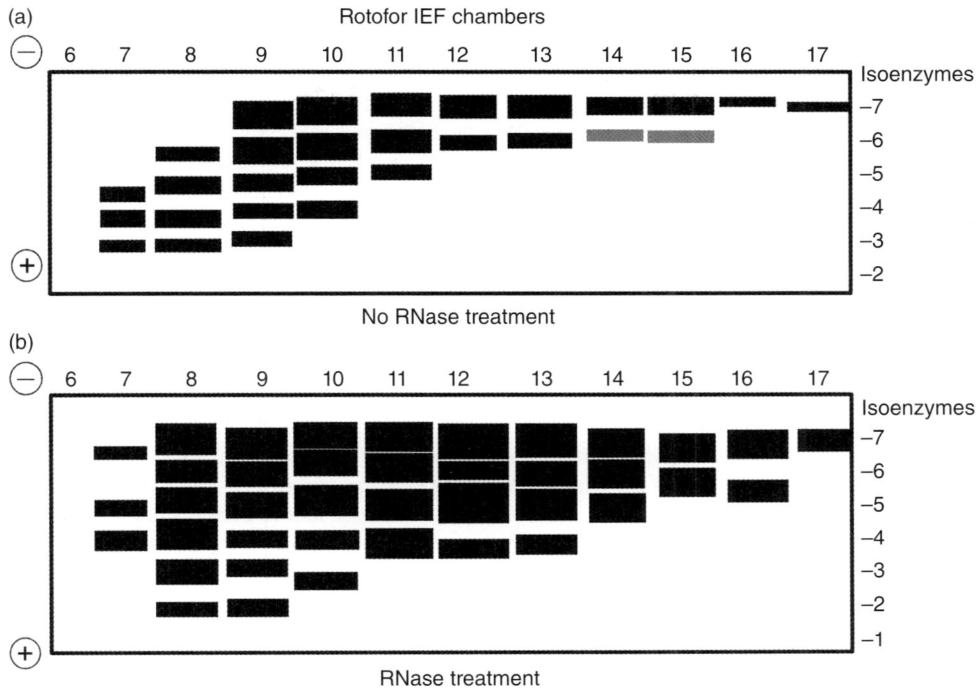

Fig. 1.5. Glutamate dehydrogenase (GDH) isoenzyme profiles before and after RNase treatment. GDH was extracted from equal weights of peanut seed using buffer cocktail (a) without RNase A, and (b) containing RNase A. In each case, an equal volume of the GDH was subjected to Rotofor isoelectric focusing (IEF) to charge its isomers; and equal volumes of Rotofor fractions were then fractionated to isoenzymes by native polyacrylamide gel electrophoresis (PAGE). The GDH isoenzyme profile was visualized by activity staining of the electrophoresed gel with tetrazolium blue reagent (Cammaerts and Jacobs, 1983). Figure (a) is the natural (cationic) aminating cassette of GDH isoenzymes; Figure (b) is the normalized GDH isoenzyme profile.

the explanation of nutrient insufficiency in terms of the signal integration and discrimination function of the enzyme (Osuji and Braithwaite, 1999). The discovery of the RNA synthetic activity of GDH (Osuji *et al.*, 2008) permitted incisive explanation of the signal integration/discrimination function in molecular terms as the silencing and permutation of the mRNAs encoding the regulatory enzymes in the metabolic pathways (Osuji *et al.*, 2010). GDH amination-dependent crop yield doubling biotechnology without the cultivation of more land and without additional man-hour input (Osuji *et al.*, 2011) could be put to practical advantage to produce more food crops (groundnuts, beans, root/tuber crops, etc.) for the poor (Chrispeels, 2000), especially in arid zones where inadequacy of protein-rich diets is causing kwashiorkor among infants and the elderly, and in expectant mothers (Waterlow, 1994). Although cowpeas, millet, sorghum and groundnuts are cultivated in arid zones, agricultural extension agents have taught the farmers to cultivate the crops under GDH deamination conditions that minimize photosynthetic production of C_3, C_4, C_5 and C_6 compounds, and limit the assimilation of nitrogen to biological nitrogen fixation and the GS-GOGAT cycle, as these are inefficient mechanisms reminiscent of the control peanut situation (Table 1.1 and Fig. 1.3), and with consequent insufficient crop yields that herald food insecurity. Cowpea yields are lowest in the African Sahel and South Asia compared with other primary production zones (IITA, 2006). Such minimal yields of cowpeas, groundnuts, yam tubers and other food crops of the poor are insufficient to feed the population throughout the hot dry season, which extends over 7–12 months yearly. This is the scientific basis of the protein–energy malnutrition (kwashiorkor) disease afflicting children and elderly mothers in many parts of the arid world. Therefore, the prevalence of kwashiorkor in the arid zones is in part a consequence of the misinterpretation of the natural aminating role of GDH. Doctors Without Borders/Médicines San Frontières (2012) is prescribing/distributing peanut butter to the severely malnourished. The kwashiorkor population could be assisted further by training the farmers to cultivate their crops in the GDH aminating direction so as to double the harvest to last throughout the hot, dry and lean season.

Similarly, sustainable and subsistence agricultural practices in many parts of the world that are devastated by climate change, land degradation, extreme temperatures, strained water resources, etc. are in part responsible for the food insufficiency ravaging many nations. Subsistence agriculture (Crouch, 1995) that depends mainly on biological nitrogen fixation and/or the application of inadequate mineral nutrient mixes mimics the control and N + P + K + S-treated peanuts (Fig. 1.3), and the control, NH_4^+- and K + N + N-treated cowpeas (Table 1.2) that lead to insufficient yields because the Pi translocator, GBSS, PGM, glyceraldehyde-3-phosphate dehydrogenase, etc. are unable to export the products of photosynthesis to glycolysis, the citric acid cycle, etc. for the maximum synthesis of carbohydrates, amino acids, proteins, fatty acids, etc. GDH amination-based crop yield doubling biotechnology has been applied to cowpeas with success (Table 1.2), to double the amino acid, dry matter and seed yields without application of more mineral fertilizer, water and pesticides,

Table 1.2. Glutamate dehydrogenase (GDH) amination activity unleashing cowpea (*Vigna unguiculata*, cream 40 variety) metabolic propulsion that optimizes and doubles the free amino acid, total protein and seed yields of control and stoichiometric mineral salt-treated plots.

	Mineral salt treatment						
Parameter	NH_4Cl	K + N + N	K + K + N	N + N	Control	KCl	K + K
GDH[a]	<1.0	1.39	3.21	3.19	3.78	12.67	13.51
Total protein[b]	387.0	322.0	376.0	381.0	348.0	584.0	561.0
Seed yield[c]	1167.2	1296.7	1552.4	1490.0	1419.5	2363.0	2297.4
Asp[d]	34.31	19.85	16.21	14.33	34.12	22.60	15.99
Glu[d]	265.25	261.67	271.12	234.31	263.77	265.05	265.09
Gly[d]	33.33	27.90	33.18	32.04	34.73	35.48	33.26
His[d]	105.21	111.93	101.16	117.11	124.95	118.10	130.57
Arg[d]	275.92	255.10	246.60	289.34	316.33	324.32	323.26
Thr[d]	<0.2	<0.2	<0.2	<0.2	<0.2	<0.2	<0.2
Ala[d]	30.53	31.02	25.78	27.65	29.54	35.46	33.49
Pro[d]	17.78	23.05	24.53	21.42	21.61	19.44	25.25
Tyr[d]	0.30	0.40	<0.3	0.46	0.42	0.42	0.47
Lys[d]	<0.2	<0.2	<0.2	<.02	<0.2	<0.2	<0.2
Val[d]	15.75	28.42	24.23	33.54	15.19	17.88	26.50
Met[d]	5.27	<0.2	0.49	0.82	7.35	1.44	0.32
Ile[d]	12.40	18.59	14.63	16.65	13.11	12.39	15.27
Leu[d]	34.16	40.63	31.85	38.32	33.79	31.30	36.43
Phe[d]	29.20	39.30	32.27	31.79	33.17	27.74	37.46
Try[d]	6.49	7.30	7.05	6.78	8.53	7.46	7.67
AA(kg ha^{-1})[e]	11.62	13.01	14.83	14.81	15.40	24.29	25.14

[a]GDH, (kg ha^{-1}) amino acid equivalent synthesized as glutamate via GDH.

[b]kg ha^{-1} seed protein.

[c]kg ha^{-1} seed.

[d]Amino acids as mg 100 g^{-1} seeds.

[e]Total amino acids (kg ha^{-1}).

and without cultivating more land, etc. Dry bean yield recorded from KCl-fertilized cream 40 cowpeas was 2363 kg ha^{-1} as compared with a yield of <1500 kg ha^{-1} in NPK-fertilized cowpeas in the tropical USA (Duke, 1983). It is noteworthy that in the cowpea, as in the peanut, the control cowpea did not harbour the control GDH; instead, the NH_4Cl-treated cowpea harboured the control GDH. Food crop insufficiencies are, therefore, in part the consequence of the misinterpretation of the natural aminating role of GDH. Many biology texts still erroneously project GDH as deaminating in its natural role. Stoichiometric and interlocking mixes of electromagnetic mineral nutrients could supplant commercial fertilizers for the primary production of peanut, cowpea, millet, etc. in torrid zones. Sufficient food crops could be produced by educating the farmers on the practice of GDH amination-based crop yield doubling biotechnology.

1.10 Conclusions

GDH is historically known to catalyse the reversible reductive amination of α-KG in the synthesis of L-glutamate in all higher plants studied. It also catalyses the template-independent polymerization of nucleoside triphosphate(s) to RNA. The discovery of the aminating activity of the glutamine synthetase (GS; EC.6.3.1.2)-glutamate synthase (GOGAT; EC.2.6.1.53) cycle reopened conversations and extensive research on the role of GDH in the synthesis of glutamate. But many lines of new experimental evidence: from the isomerization kinetics of the hexameric subunits in response to nucleophiles (pesticides, fertilizer mineral ions, auxins, xenobiotics); the nature of the aminating cassette of GDH isoenzymes; the normal kinetic properties of glutamate synthase; the ability of GDH-synthesized RNAs to form cobweb-like molecular scaffolding that regulates metabolism at the mRNA level, thereby linking metabolism at the enzyme level with metabolism at the mRNA level, and the resultant permutation of biochemical pathways by GDH-synthesized RNA; systems metabolism and molecular biology experimental approaches in which many primary biochemical pathways are integrated; that illuminate the uninterruptible propulsion of phosphorylated C_3, C_5 and C_6 compounds from the chloroplasts to the cytosol, and of α-KG via the citric acid cycle to GDH, when research protocols carefully delineate the control experimental tissues from the control GDHs, reiterating and supporting the amination role of GDH. When the control GDH is not identified and delineated from the control plant tissues, experimental results are obtained that demean the role of GDH. Specifically, the kinetic property of GDH to discriminate and integrate the plant's environmental signals by synthesizing RNAs that regulate the mRNAs encoding many enzymes (phosphate translocator, granule-bound starch synthase, acetyl coenzyme A carboxylase, phosphoglucomutase, nitrate reductase, glycinamide ribonucleotide transformylase, etc.) of primary metabolism is the collateral chemical scaffolding on which the aminating natural role stands as it synchronizes carbon and nitrogen metabolic pathways, and assimilates exponential multiples of kilogram NH_4^+ ion as assimilated via the GS-GOGAT cycle during normal to maximum photosynthetic supplies of phosphorylated C_3, C_5, etc. compounds. Only miniscule quantities of α-KG resulted from the possible deaminating activity of GDH in all the peanuts and cowpeas studied. Judging from the continuous torrent of C_3, C_4, C_5, etc. compounds via glycolysis and the citric acid cycle, the large kilogram yields of especially those amino acids (proline, glutamate, glutamine, histidine, arginine, aspartate, asparagine, etc.) that originate directly from citric acid cycle intermediates, and the record high yields in kilograms of storage proteins, the supply of α-KG for citric acid operation was not from the deamination of glutamate. The crops that produced the record high yields of free amino acids and storage proteins also produced the record high yields of cellulose, fatty acids and seeds, thus confirming the absence of GDH deamination of glutamate. Had GDH deaminating activity made any contribution to the α-KG supply, the dry matter and carbohydrate yields of the untreated control, N + P + K + S- and N + S-treated crops would have been higher than in those peanuts with high aminating activities of GDH, thereby creating a potentially dangerous counter metabolic flux.

GDH possibly deaminates diminishingly small fractions of glutamate as it synthesizes RNA that regulates the mRNAs, so that glutamate biosynthesis is not shut down during low to normal flow of C_3, C_5, etc. compounds of photosynthesis. Ammonium ion assimilation and glutamate synthesis by GDH are regulated in higher plants by the supply of NH_4^+ ions and the flow of α-KG from citric acid cycle to GDH, but not by the reverse flow of α-KG from glutamate.

Acknowledgements

The research on glutamate dehydrogenase was funded in parts by: Prairie View A&M University; the Capacity Building Research Grant Program of the USDA; Evans Allen Fund of the USDA; and Texas Agricultural Experiment Station funding made to Prairie View A&M University. W.C. Madu was supported by the Graduate Assistantship Program of Prairie View A&M University.

References

Ameziane, R., Bernhard, K. and Lightfoot, D. (2000) Expression of the bacterial *gdhA* gene encoding a NADPH glutamate dehydrogenase in tobacco affects plant growth and development. *Plant and Soil* 221, 47–57.

Aubert, S., Bligny, R., Douce, R., Gout, E., Ratcliffe, R.G. and Roberts, J.K.M. (2001) Contribution of glutamate dehydrogenase to mitochondrial glutamate metabolism studied by ^{13}C and ^{31}P nuclear magnetic resonance. *Journal of Experimental Botany* 52, 37–45.

Bowman, E.J., Ikuma, H. and Stein, H.J. (1976) Citric acid cycle activity in mitochondria isolated from mung bean hypocotyls. *Plant Physiology* 58, 426–432.

Bravo, G. and Maeso, M. (1987) Effect of nitrate and ammonium on NADH-dependent glutamate dehydrogenase activity on detached roots and cotyledons of *Citrullus vulgaris* L. *Journal of Plant Physiology* 127, 331–337.

Britton, K.L., Baker, P.J., Rice, D.W. and Stillman, T.J. (1992) Structural relationships between the hexameric and tetrameric family of glutamate dehydrogenase. *European Journal of Biochemistry* 209, 851–859.

Brown, A., Culver, J.M. and Fisher, H.F. (1973) Mechanism of inactivation of L-glutamate dehydrogenase by pyridoxal and pyridoxal phosphate. *Biochemistry* 12, 4367–4374.

Cakmak, I. and Engels, C. (1999) Role of mineral nutrients in photosynthesis and yield formation. In: Rengal, Z. (ed.) *Mineral Nutrition of Crops: Fundamental Mechanisms and Implications*. Food Products Press (imprint of The Haworth Press), Binghampton, New York, pp. 144–168.

Calle, F., Martín, M. and Sabater, B. (1986) Cytoplasmic and mitochondrial localization of the glutamate dehydrogenase induced by senescence in barley (*Hordeum vulgare*). *Physiologia Plantarum* 66, 451–456.

Cammaerts, D. and Jacobs, M. (1983) A study of the polymorphism and the genetic control of glutamate dehydrogenase isoenzymes in *Arabidopsis thaliana*. *Plant Science Letters* 31, 65–73.

Cammaerts, D. and Jacobs, M. (1985) A study of the role of glutamate dehydrogenase in the nitrogen metabolism of *Arabidopsis thaliana*. *Planta* 163, 517–526.

Caperelli, C.A. (1989) Mammalian glycinamide ribonucleotide transformylase. *The Journal of Biological Chemistry* 264, 5053–5057.

Century, K., Reuber, T.L. and Ratcliffe, O.J. (2008) Regulating the regulators: the future prospects of transcription factor-based agricultural biotechnology products. *Plant Physiology* 147, 20–29.

Chen, R. and Gadal, P. (1990) Do the mitochondria provide the 2-oxoglutarate needed for glutamate synthesis in higher plant chloroplasts? *Plant Physiology and Biochemistry* 28, 141–145.

Chou, K. and Splittstoesser, W.E. (1972) Glutamate dehydrogenase from pumpkin cotyledons. *Plant Physiology* 49, 550–554.

Chrispeels, M.J. (2000) Biotechnology and the poor. *Plant Physiology* 124, 3–6.

Crouch, M.L. (1995) Biotechnology is not compatible with sustainable agriculture. *Journal of Agricultural and Environmental Ethics* 8, 98–111.

Davies, K.J.A. (1987) Protein damage and degradation by oxygen radicals. I. General aspects. *The Journal of Biological Chemistry* 262, 9914–9920.

Debouba, M., Gouia, H., Suzuki, A. and Ghorbel, M.H. (2006) NaCl stress effects on enzymes involved in nitrogen assimilation pathway in tomato '*Lycopersicon esculentum*' seedlings. *Journal of Plant Physiology* 163, 1247–1258.

D'Mello, J.P.F. (2003) An outline of pathways in amino acid metabolism. In: D'Mello, J.P.F. (ed.) *Amino Acids in Animal Nutrition*, 2nd edn. CAB International, Wallingford, UK, pp. 71–86.

Doctors Without Borders/Médicines San Frontières (2012) Available at: http://www.doctorswithoutborders.org/news/issue.cfm?id=2396 (accessed 27 November 2012).

Dubois, F., Tercé-Laforgue, T., Gonzalez-Moro, M., Estavillo, J.-M., Sangwan, R., Gallais, A. and Hirel, B. (2003) Glutamate dehydrogenase in plants: is there a new story for an old enzyme? *Plant Physiology and Biochemistry* 41, 565–576.

Duke, J.A. (1983) *Vigna unguiculata (L.) Walp. spp. unguiculata*. Available at: http://www.hort.purdue.edu/newcrop/duke_energy/Vigna_unguiculata.html (accessed 17 February 2010).

Edwards, A., Marshall, J., Denyer, K., Sidebottom, C., Visser, R.G.F., Martin, C. and Smith, A.M. (1996) Evidence that a 77-kilodalton protein from the starch of pea embryos is an isoform of starch synthase that is both soluble and granule-bound. *Plant Physiology* 112, 89–97.

FAO (2000) *The State of Food and Agriculture 2002*. Food and Agriculture Organization of the United Nations, Available at: http://www.fao.org/docrep/004/y6000e/y6000e00.htm (accessed 23 July 2014).

FAO (2006) *The State of Food Insecurity in the World 2006*. Food and Agriculture Organization of the United Nations, Rome. Available at: http://www.fao.org/docrep/009/a0750e/a0750e00.htm (accessed 23 July 2014).

Fersht, A. (1985) *Enzyme Structure and Mechanism*, 2nd edn. Freeman, New York.

Fontaine, J-X., Saladino, F., Agrimonti, C., Bedu, M., Tercé-Laforgue, T., Tétu, T., Hirel, B., Restivo, F.M. and Dubois, F. (2006) Control of the synthesis and subcellular targeting of the two genes products in leaves and stem of *Nicotiana plumbaginifolia* and *Arabidopsis thaliana*. *Plant and Cell Physiology* 47, 410–418.

Glevarec, G., Bouton, S., Jaspard, E., Riou, M.T., Cliquet, J.B. and Suzuki, A. (2004) Respective roles of the glutamine synthetase/glutamate synthase cycle and glutamate dehydrogenase in ammonium and amino acid metabolism during germination and post-germinative growth in model legume *Medicago truncatula*. *Planta* 219, 286–297.

Gould, F. (1998) Sustainability of transgenic insecticidal cultivars: integrating pest genetics and ecology. *Annual Review of Entomology* 43, 701–726.

Hartmann, T. and Ehmke, A. (1980) Role of mitochondrial glutamate dehydrogenase in the reassimilation of ammonia produced by glycine serine transformation. *Planta* 149, 207–208.

Hattenbach, A. and Heineke, D. (1999) On the role of chloroplastic phosphoglucomutase in the regulation of starch turn over. *Planta* 207, 527–532.

Henning, R.J., Allison, A.H. and Tripp, L.D. (1982) Cultural practices. In: Pattee, H.E. and Young, C.T. (eds) *Peanut Science and Technology*. American Peanut Research and Education Society, Inc., Yoakum, Texas, pp. 126–137.

Hinchee, M.A.W., Padgette, S.R., Kishmore, G.M., Delannay, X. and Fraley, R.T. (1993) Herbicide-tolerant crops. In: Kung, S. and Wu, R. (eds) *Transgenic Crops. Volume 1: Engineering and Utilization*. Academic Press, San Diego, California, pp. 243–363.

Horton, P. (2000) Prospects for crop improvement through the genetic manipulation of photosynthesis: morphological and biochemical aspects of light capture. *Journal of Experimental Botany* 51, 475–585.

IITA (2006) *Cowpea (Vigna unguiculata)*. Available at: http://www.iita.org/cowpea (accessed 23 July 2014).

Jamal, A., Moon, Y.S. and Abdin, M.Z. (2010) Enzyme activity assessment of peanut (*Arachis hypogea* L.) under slow-release sulphur fertilization. *Australian Journal of Crop Science* 4, 169–174.

Jaspard, E. (2006) A computational analysis of the three isoforms of glutamate dehydrogenase reveals structural features of the isoform EC 1.4.1.4 supporting a key role in ammonium assimilation by plants. *Biology Direct* 1, 38.

Ke, J., Wen, T., Nicolau, B.J. and Wurtele, E.S. (2000) Coordinate regulation of the nuclear and plastidic genes coding for the subunits of the heteromeric acetyl-CoA carboxylase. *Plant Physiology* 122, 1657–1071.

Keegstra, K. and Raikhel, N. (2001) Plant glycosyltransferases. *Current Opinion in Plant Biology* 4, 219–224.

Kidder, G. (1994) Nitrogen and sulfur. Chapter 3 in: Mitchell, C.C. and Stribling, L.H. (eds) *Research-based Soil Testing Interpretation and Fertilizer Recommendations for Peanuts on Coastal Plain Soils*. Southern Cooperative Series Bulletin 380, Southern Extension and Research Activity Information Exchange Group-6 (SERA-IEG-6), Auburn University, Alabama. Available at: http://www.aaes.auburn.edu/comm/pubs/SCSB/380site/chapterthree.htm (accessed 23 July 2014).

Kleinhofs, A., Warner, R.L. and Melzer, J.M. 1989) Genetics and molecular. biology of higher plant nitrate reductase. In: Poulton, J.E., Romeo, J.T. and Conn, E.E. (eds) *Plant Nitrogen Metabolism*. Plenum, New York, pp. 117–155.

Labboun, S., Tercé-Laforgue, T., Roscher, A. Bedu, M., Restivo, F.M., Velanis, C.N., Skopelitis, S.S., Moshou, P.N., Roubelakis-Angelakis, K.A., Suzuki, A. and Hirel, B. (2009) Resolving the role of plant glutamate dehydrogenase. I. *In vivo* real time nuclear magnetic resonance spectroscopy experiments. *Plant and Cell Physiology* 50, 1761–1773.

Lacuesta, M., Gonzalez-Moro, B., Gonzalez-Murua, C., Aparicio-Tejo, P. and Munoz-Rueda, A. (1989) Effect of phosphinothricine (glufosinate) on activities of glutamine synthetase and glutamate dehydrogenase in *Medicago sativa* L. *Journal of Plant Physiology* 134, 304–307.

Lea, P.J. and Miflin, B.J. (1974) Alternative route for nitrogen assimilation in higher plants. *Nature* 251, 614–616.

Lehmann, T. and Ratajczak, L. (2008) The pivotal role of glutamate dehydrogenase (GDH) in the mobilization of N and C from storage material to asparagines in germinating seeds of yellow lupine. *Journal of Plant Physiology* 165, 149–158.

Lehmann, T., Skrok, T. and Dabert, M. (2010) Stress-induced changes in glutamate dehydrogenase activity imply its role in adaptation to C and N metabolism in lupin embryos. *Physiologia Plantarum* 138, 35–47.

Lettgen, W., Britsch, L. and Kasemir, H. (1989) The effect of light and exogenously supplied ammonium ions on glutamate dehydrogenase activity and isoforms in young mustard (*Sinapis alba* L.) seedlings. *Botanica Acta* 102, 189–195.

Madu, W.C. (1997) The effects of elicitors on growth and nitrogen metabolism of cell cultures of yam (*Dioscorea rotundata* and *Dioscorea alata*) and sweet potato (*Ipomoea batatas*). PhD thesis, Abia State University, Uturu, Nigeria.

Magalhaes, J.R. (1991) Kinetics of $^{15}NH_4^+$ assimilation in tomato plants: evidence for $^{15}NH_4^+$ assimilation via GDH in tomato roots. *Journal of Plant Nutrition* 14, 1341–1353.

Manjunath, S., Lee, C.H.K., VanWinkle, P. and Bailey-Serres, J. (1998) Molecular and biochemical characterization of cytosolic phosphoglucomutase in maize expression during development and in response to oxygen deprivation. *Plant Physiology* 117, 997–1006.

Marschner, H. (1990) *Mineral Nutrition of Higher Plants*. Academic Press, London.

Masclaux-Daubresse, C., Reisdorf-Cren, M., Pageau, K., Lelandais, M., Grandjean, O., Kronenberger, J., Valadier, M.H., Feraud, M., Jouglet, T. and Suzuki, A. (2006) Glutamine synthetase-glutamate synthase pathway and glutamate dehydrogenase play distinct roles in the sink–source nitrogen cycle in tobacco. *Plant Physiology* 140, 444–456.

Melo-Oliveira, R., Oliveira, I.C. and Coruzzi, G.M. (1996) *Arabidopsis* mutant analysis and gene regulation define a nonredundant role for glutamate dehydrogenase in nitrogen assimilation. *Proceedings of the National Academy of Sciences of the United States of America* 93, 4718–4723.

Miflin, B.J. and Habash, D.Z. (2002) The role of glutamine synthetase and glutamate dehydrogenase in nitrogen assimilation and possibilities for improvement in the nitrogen utilization of crops. *Journal of Experimental Botany* 53, 979–987.

Miflin, B.J. and Lea, P.J. (1976) The pathway of nitrogen assimilation in plants. *Phytochemistry* 15, 873–885.

Miyashita, Y. and Good, A.G. (2008) Glutamate deamination by glutamate dehydrogenase plays a central role in amino acid catabolism in plants. *Plant Signaling and Behavior* 3, 842–843.

Morgan, J.A. and Rhodes, D. (2002) Mathematical modeling of plant metabolic pathways. *Metabolic Engineering* 4, 80–89.

Nasr-Alla, A.E., Osman Fatma, A.A. and Soliman, K.G. (1998) Effects of increased phosphorus and potassium or sulfur application in their different combinations on yield, yield components, and chemical composition of peanut in a newly reclaimed sand soil. *Zagazig Journal of Agricultural Research* 25, 557–579.

Nikolau, B.J., Wurtele, E.S. and Stumpf, P.K. (1984) Tissue distribution of acetyl-coenzyme A carboxylase in leaves. *Plant Physiology* 75, 895–901.

Oaks, A. (1994) Primary nitrogen assimilation in higher plants and its regulation. *Canadian Journal of Botany* 72, 739–750.

Osuji, G.O. (1997) Peanut glutamate dehydrogenase: a target site of herbicide action. In: Ando, T., Fujita, K., Mae, T., Matsumoto, H., Mori, S. and Sekiya, J. (eds) *Plant Nutrition for Sustainable Food Production and Environment*. Kluwer Academic Publishers, Dordrecht, The Netherlands, pp. 845–850.

Osuji, G.O. and Braithwaite, C. (1999) Signaling by glutamate dehydrogenase in response to pesticide treatment and nitrogen fertilization of peanut (*Arachis hypogaea* L.). *Journal of Agricultural and Food Chemistry* 47, 3332–3344.

Osuji, G.O. and Cuero, R.G. (1991) A rapid method for enhancing the storage protein yields of yam tuber, sweetpotato, and maize through *N*-carboxymethylchitosan treatment of the crops. *Mededelingen Faculteit Landbouwwetetnschappen Rigksuniversiteit Gent* 56, 1661–1672.

Osuji, G.O. and Cuero, R.G. (1992a) *N*-carboxymethylchitosan enhancement of the storage protein contents of maize seeds (*Zea mays* L.). *Food Biotechnology* 6, 105–126.
Osuji, G.O. and Cuero, R.G. (1992b) The biochemical basis of *N*-carboxymethylchitosan-mediated enhancement of the storage protein content of sweetpotato. In: Hill, W.A., Bonsi, C.K. and Loretan, P.A. (eds) *Sweetpotato Technology for the 21st Century*. Tuskegee University, Tuskegee, Alabama, pp. 78–86.
Osuji, G.O. and Madu, W.C. (1995) Ammonium-ion dependent isomerization of glutamate dehydrogenase in relation to glutamate synthesis in maize. *Phytochemistry* 39, 495–503.
Osuji, G.O. and Madu, W.C. (1996) Ammonium ion salvage by glutamate dehydrogenase during defense response in maize. *Phytochemistry* 42, 1491–1498.
Osuji, G.O. and Madu, W.C. (1997) Regulation of sweetpotato growth and differentiation by glutamate dehydrogenase. *Canadian Journal of Botany* 75, 1070–1078.
Osuji, G.O. and Madu, W.C. (2012) Using the RNA synthetic activity of glutamate dehydrogenase to illuminate the natural role of the enzyme. *Advances in Biological Chemistry* 2, 379–389.
Osuji, G.O. and Ory, R.L. (1986) Purine degradative pathway of the yam and sweetpotato. *Journal of Agricultural and Food Chemistry* 34, 599–602.
Osuji, G.O., Cuero, R.G. and Washington, A.C. (1991) Effects of α-ketoglutarate on the activities of the glutamate synthase, glutamate dehydrogenase, and aspartate transaminase of sweetpotato, yam tuber, and cream pea. *Journal of Agricultural and Food Chemistry* 39, 1590–1596.
Osuji, G.O., Haby, V.A., Beyene, A., Madu, W.C. and Mangaroo, A.S. (1997) The Isomerization of glutamate dehydrogenase in response to lead toxicity in maize. *Biologia Plantarum* 40, 389–398.
Osuji, G.O., Reyes, J.C. and Mangaroo, A.S. (1998) Glutamate dehydrogenase isomerization: a simple method for diagnosing nitrogen, phosphorus, and potassium sufficiency in maize (*Zea mays* L.). *Journal of Agricultural and Food Chemistry* 46, 2395–2401.
Osuji, G.O., Braithwaite, C., Pointer, R. and Reyes, J. (1999) Pesticide inactivation of peanut glutamate dehydrogenase: biochemical basis of the enzyme's isomerization. *Journal of Agricultural and Food Chemistry* 47, 3345–3351.
Osuji, G.O., Mangaroo, A.S. and Roberts, P.S. (2001) *In vitro* isomerization of glutamate dehydrogenase in relation to phytosequestration of lead. *Southern Association of Agricultural Scientists Bulletin: Biochemistry and Biotechnology* 14, 60–72.
Osuji, G.O., Braithwaite, C., Fordjour, K., Madu, W.C., Beyene, A., Roberts, P.S. and Wright, V. (2003a) Purification of glutamate dehydrogenase isoenzymes and characterization of their substrate specificities. *Preparative Biochemistry and Biotechnology* 33, 13–28.
Osuji, G.O., Mangaroo, A.S., Reyes, J., Bulgin, A. and Wright, V. (2003b) Biomass enhancement in maize and soybean in response to glutamate dehydrogenase isomerization. *Biologia Plantarum* 47, 45–52.
Osuji, G.O., Madu, W.C., Braithwaite, C., Beyene, A., Roberts, P.S., Bulgin, A. and Wright, V. (2003c) Nucleotide-dependent isomerization of glutamate dehydrogenase in relation to total RNA contents of peanut. *Biologia Plantarum* 47, 195–202.
Osuji, G.O., Konan, J. and M'Mbijjewe, G. (2004) RNA synthetic activity of glutamate dehydrogenase. *Applied Biochemistry and Biotechnology* 119, 209–228.
Osuji, G.O., Brown, T.K. and South, S.M. (2008) Discovery of the RNA synthetic activity of glutamate dehydrogenase and its application in drug metabolism research. *The Open Drug Metabolism Journal* 2, 1–13.
Osuji, G.O., Brown, T.K. and South, S.M. (2009) Nucleotide-dependent reprogramming of mRNAs encoding acetyl coenzyme A carboxylase and lipoxygenase in relation to fat contents of peanut. *Journal of Botany* (2009), Article ID 278324. Available at: http://dx.doi.org/10.1155/2009/278324 (accessed 23 July 2014).
Osuji, G.O., Brown, T.K. and South, S.M. (2010) Optimized fat and cellulosic biomass accumulation in peanut through biotechnology. *International Journal of Biochemistry and Biotechnology* 3, 455–476.
Osuji, G.O., Brown, T.K., South, S.M., Duncan, J.C. and Johnson, D. (2011) Doubling of crop yield through permutation of metabolic pathways. *Advances in Bioscience and Biotechnology* 2, 364–379.
Osuji, G.O., Brown, T.K., South, S.M., Duncan, J.C., Johnson, D. and Hyllam, S. (2012a) Molecular adaptation of peanut metabolic pathways to wide variations of mineral ion composition and concentration. *American Journal of Plant Sciences* 3, 33–50.
Osuji, G.O., Brown, T.K., South, S.M., Johnson, D. and Hyllam, S. (2012b) Molecular modeling of metabolism for allergen-free low linoleic acid peanuts. *Applied Biochemistry and Biotechnology* 168, 805–823.
Page, R.A., Okada, S. and Harwood, J.L. (1994) Acetyl-CoA carboxylase exerts a strong flux control over lipid synthesis in plants. *Biochimica et Biophysica Acta* 1210, 369–372.
Pahlich, E. (1996) Remarks concerning the dispute related to the function of plant glutamate dehydrogenase: Commentary. *Canadian Journal of Botany* 74, 512–515.

Pahlich, E., Ott, W. and Schad, B. (1980) Immunochemical investigations with highly purified glutamate dehydrogenase from pea seeds by means of the Ouchterlony test. *Journal of Experimental Botany* 31, 419–423.

Pavesi, A., Ficarelli, A., Tassi, F. and Restivo, F.M. (2000) Cloning of two glutamate dehydrogenase cDNAs from *Asparagus officinalis:* sequence analysis and evolutionary implications. *Genome* 4, 306–316.

Pease, J., Bullen, G., Roberts, M. and Shokes, F. (2012) *Economic Impacts of the 2002 Farm Bill on Peanut Farms in Six States*. National Peanut Board, Atlanta, Georgia.

Periappuram, C., Steinhauer, L., Barton, D.L., Taylor, D.C., Chatson, B. and Zou, J. (2000) The plastidic phosphoglucomutase from *Arabidopsis*. A reversible enzyme reaction with an important role in metabolic control. *Plant Physiology* 122, 1193–1199.

Purnell, M.P. and Botella, J.M. (2007) Tobacco isoenzyme 1 of NAD(H)-glutamate dehydrogenase catabolizes glutamate *in vivo*. *Plant Physiology* 143, 530–539.

Rengel, Z. (ed.) (1999) *Mineral Nutrition of Crops:* Fundamental Mechanisms and Implications. Food Products Press (imprint of The Haworth Press), Binghampton, New York.

Restivo, F.M. (2004) Molecular cloning of glutamate dehydrogenase genes of *Nicotiana plumbaginifolia:* structure analysis and regulation of their expression by physiological and stress conditions. *Plant Science* 166, 971–982.

Reverdatto, S., Beilinson, V. and Nielsen, N.C. (1999) A multisubunit acetyl coenzyme A carboxylase from soybean. *Plant Physiology* 119, 961–978.

Reynolds, M., Foulkes, M.J., Slafer, G.A., Berry, P., Parry, M.A.J., Snape, J.W. and Angus, W.J. (2009) Raising yield potential in wheat. *Journal of Experimental Botany* 60, 1899–1918.

Rhodes, D., Brunk, D.G. and Magalhaes, J.R. (1989) Assimilation of ammonia by glutamate dehydrogenase? In: Poulton, J.E., Romeo, J.T. and Conn, E.E. (eds) *Recent Advances in Phytochemistry. Volume 23: Plant Nitrogen Metabolism*. Plenum, New York, pp. 191–226.

Robinson, S.A., Slade, A.P., Fox, G.G., Phillips, R., Ratcliffe, R.G. and Stewart, G.R. (1991) The role of glutamate dehydrogenase in plant nitrogen metabolism. *Plant Physiology* 95, 509–516.

Robinson, S.A., Stewart, G.R. and Phillips, R. (1992) Regulation of glutamate dehydrogenase activity in relation to carbon limitation and protein catabolism in carrot cell suspension cultures. *Plant Physiology* 98, 1190–1195.

Sadunishvili, T., Gvarliani, N., Nutsubidze, N. and Kvesitadze, G. (1996) Effect of methionine sulfoximine on nitrogen metabolism and externally supplied ammonium assimilation in kidney bean. *Ecotoxicology and Environmental Safety* 34, 70–75.

Saxena, I.M. and Brown, R.M. (1997) Identification of cellulose synthases in higher plants: sequence analysis of processive β-glycosyltransferases with the common motif 'D, D, D35Q(R,Q)XRW'. *Cellulose* 4, 33–49.

Skopelitis, D.S., Paranychianakis, N.V., Paschalidis, K.A., Pliakonis, E.D., Delis, I.D., Yakoumakis, D.I., Kouvarakis, A., Papadakis, A.K., Stephanou, E.G. and Roubelakis-Angelakis, K.A. (2006) Abiotic stress generates ROS that signal expression of anionic glutamate dehydrogenases to form glutamate for proline synthesis in tobacco and grapevine. *The Plant Cell* 18, 2767–2781.

Skopelitis, D.S., Paranychiankis, N.V., Kouvarakis, A., Spyros, A., Stephanou, E.G. and Roubelakis-Angelakis, K.A. (2007) The isoenzyme 7 of tobacco NAD(H)-glutamate dehydrogenase exhibits high deaminating and low aminating activity. *Plant Physiology* 145, 1726–1734.

Smith, A.M., Denyer, K. and Martin, C. (1997) The synthesis of the starch granule. *Annual Review of Plant Physiology and Plant Molecular Biology* 48, 67–87.

Smith, E.L., Austen, B.M., Blumenthal, K.M. and Nyc, J.F. (1975) Glutamate dehydrogenase. In: Boyer, P.D. (ed.) *Enzymes: Volume II*, 3rd edn. Academic Press, New York, pp. 293–367.

Smith, P.M.C. and Atkins, C.A. (2002) Purine biosynthesis. Big in cell division, even bigger in nitrogen assimilation. *Plant Physiology* 128, 793–802.

Stewart, G.R., Shatilov, V.R., Turnbull, M.H., Robinson, S.A. and Goodall, R. (1995) Evidence that glutamate dehydrogenase plays a role in oxidative deamination of glutamate in seedlings of *Zea mays*. *Australian Journal Plant Physiology* 22, 805–809.

Stitt, M. and Quick, W.P. (1989) Photosynthetic carbon partitioning: its regulation and possibilities for manipulation. *Physiologia Plantarum* 77, 633–641.

Stone, S.R., Copeland, L. and Heyde, E. (1980) Glutamate dehydrogenase of lupin nodules: kinetics of the deamination reaction. *Archives of Biochemistry and Biophysics* 199, 550–559.

Streb, S., Elgi, B., Eicke, S. and Zeeman, S. (2009) The debate on the pathway of starch synthesis: a closer look on low-starch mutants lacking plastidial phosphoglucomutase supports the chloroplast-localized pathway. *Plant Physiology* 151, 1769–1777.

Strecker, H.J. (1953) Glutamate dehydrogenase. *Archives of Biochemistry and Biophysics* 46, 128–140.

Syntichaki, K.M., Loulakakis, K.A. and Roubelakis-Angelakis, K.A. (1996) The amino acid sequence similarity of plant glutamate dehydrogenase to extremophilic archaeal enzyme conforms to its stress-related function. *Gene* 168, 87–92.

Tercé-Laforgue, T., Mack, G. and Hirel, B. (2004a) New insights towards the function of glutamate dehydrogenase revealed during source–sink transition of tobacco (*Nicotiana tabacum* L.) plants grown under different nitrogen regimes. *Physiologia Plantarum* 120, 220–228.

Tercé-Laforgue, T., Dubois, F., Ferrario-Mery, S., Pou de Crecenzo, M., Sangwan, R. and Hirel, B. (2004b) Glutamate dehydrogenase of tobacco is mainly induced in the cytosol of phloem companion cells when ammonia is provided either externally or released during photorespiration. *Plant Physiology* 136, 4308–4417.

Turano, F.J., Dashner, R., Upadhyaya, A. and Caldwell, C.R. (1996) Purification of mitochondrial glutamate dehydrogenase from dark-grown soybean seedlings. *Plant Physiology* 112, 87–92.

Turano, F.J., Thakkar, S.S., Fang, T. and Weisemann, J.M. (1997) Characterization and expression of NAD(H)-dependent glutamate dehydrogenase genes in *Arabidopsis*. *Plant Physiology* 113, 1329–1341.

Vance, C.P., Miller, S.S., Gregerson, R.G., Samac, D.A., Robinson, D.L. and Gantt, J.S. (1995) Alfalfa NADH-dependent glutamate synthase: structure of the gene and importance in symbiotic N_2 fixation. *The Plant Journal* 8, 345–358.

Watanabe, M., Yumi, O., Itoh, Y., Yasuda, K., Kamachi, K. and Ratcliffe, G. (2011) Deamination role of inducible glutamate dehydrogenase isoenzyme 7 in *Brassica napus* leaf protoplasts. *Phytochemistry* 72, 587–593.

Waterlow, J.C. (1994) Childhood malnutrition in developing nations: looking back and looking forward. In: Olson, P.E., Bier, D.M. and McCormick, D.B. (eds) *Annual Review of Nutrition* 14, 1–19.

Wright, D.L. (2008) *Production of Biofuel Crops in Florida: Peanut*. Publication SS AGR 295, Agronomy Department, Florida Cooperative Extension Service, Institute of Food and Agricultural Sciences, University of Florida, Gainseville, Florida. Available as revised edition (2010) at: http://ufdc.ufl.edu/IR00003731/00001 (accessed 23 July 2014).

Xing, Y. and Zhang, Q. (2010) Genetic and molecular bases of rice yield. *Annual Review of Plant Biology* 61, 421–442.

Yamaya, T., Oaks, A. and Matsumoto, H. (1984) Characteristics of glutamate dehydrogenase in mitochondria prepared from corn shoots. *Plant Physiology* 76, 1009–1013.

2 Alanine Aminotransferase: Amino Acid Metabolism in Higher Plants

A. Raychaudhuri*
Monsanto, Chesterfield, Missouri, USA

2.1 Abstract

Alanine aminotransferase (AlaAT, EC 2.6.1.2) catalyses the reversible transfer of an amino group from alanine (Ala) to 2-oxoglutarate to form pyruvate and glutamate. AlaAT is a pyridoxal phosphate-dependent enzyme that serves as the main route of alanine biosynthesis and degradation. This chapter provides a comprehensive review of the biochemical characteristics of AlaAT and provides insight into the physiological and metabolic functions of the enzyme as reported from several elegant studies across the world. The diversity in cellular and sub-cellular localizations of multiple AlaAT homologues indicates their involvement in multifaceted metabolic processes, including amino acid biosynthesis and primary carbon metabolism. The enzyme is fairly ubiquitous, occurring across different phyla – fungi, algae, mycorrhizae and higher plants, and existing as various isoforms. Detailed kinetic characterization of the enzyme from several laboratories provides interesting details about its plethora of biochemical functions. The recent elucidation of the crystal structure of plant AlaAT paves opportunities for expanding our understanding of the AlaAT reaction mechanism and catalysis. Additionally, the roles of AlaAT in hypoxia, photorespiration and nitrogen use efficiency are reviewed in terms of stress amelioration and overall plant physiology.

2.2 Introduction

Alanine aminotransferase (AlaAT, EC 2.6.1.2) catalyses the reversible transfer of an amino group from alanine (Ala) to 2-oxoglutarate to form pyruvate and glutamate (Arnon, 1949). AlaAT is a type of aminotransferase enzyme that shares significant similarity in protein structure and features with other aminotransferases such as glutamate:glyoxylate aminotransferases (GGATs). AlaAT has been reported to be ubiquitous across bacteria, yeast, plants and mammals. In plants, the enzyme activity is detected in various tissues as well as in several sub-cellular compartments, such as mitochondria and peroxisomes. This diversity in localization has been found to be associated with a spectrum of biological functions, making AlaAT one of the most interesting enzymes of nitrogen metabolism.

AlaAT is a pyridoxal phosphate-dependent enzyme that serves as the main route of Ala biosynthesis and degradation. The enzyme is a major link between primary carbon metabolism and amino acid biosynthesis, and plays a critical role in the intercellular carbon shuttle associated with NAD-malic enzyme and PEP carboxykinase-type C_4 photosynthesis (Givan, 1980; Furbank, 1998) and in anaerobic glycolysis (Good and Crosby,

*E-mail address: aarayc@monsanto.com

1989; Miyashita *et al.*, 2007). AlaAT exists in various isoforms, and detailed kinetic characterization of the enzyme from several laboratories throughout the world has provided interesting details about the plethora of biochemical functions of the enzyme. Due to its importance in abiotic and biotic stress tolerance, AlaAT has been the subject of intense research over decades, and the transgenic expression of AlaAT in plants indicates its role in nitrogen use efficiency and yield gain in rice (Shrawat *et al.*, 2008) and canola (Good *et al.*, 2007a,b).

We will begin this chapter by introducing the chemical properties of Ala and its biosynthesis catalysed by AlaAT. The chapter then reviews the diversity of occurrence and phylogeny of AlaAT, followed by discussion of the biochemical and physiological importance of the enzyme in terms of its diversity and localization in cellular and sub-cellular compartments. We have attempted to capture the extensive biochemical work done on AlaAT, reviewing the range of studies on enzyme purification and enzyme characterization, with an emphasis on kinetics. The chapter also provides a glimpse of the recently solved crystal structure of AlaAT. A major section of the chapter is dedicated to highlighting the functional diversity of AlaAT and its critical role in plant metabolism, physiology and stress tolerance.

2.3 Structure and Functions of Alanine

2.3.1 Structure of alanine

Ala was first isolated in 1879 from silk fibroin, which is a rich source of L-alanine. Ala is an aliphatic amino acid in which the α-carbon atom is bound with a methyl group (-CH_3) as its side chain; its chemical formula is $CH_3CH(NH_2)COOH$ and it has a molecular weight of 89.09318. Ala is an ambivalent amino acid, found in or outside the protein molecule. It is a non-essential amino acid; in animals, it is synthesized by the transamination of pyruvic acid and is therefore not required through dietary sources.

In nature, Ala exists in two forms: L-alanine or alpha-alanine (α-alanine) and D-alanine or beta-alanine (β-alanine). L-Alanine is a constituent of proteins encoded by the genetic codons GCU, GCC, GCA and GCG, and its α-carbon is optically active in protein molecules. Ala constitutes one of the most frequently occurring amino acid in proteins (Doolittle, 1989). D-Ala, or beta-Ala (β-Ala), is not found in proteins but occurs naturally in mammalian muscle, bacterial cell walls and in some peptide antibiotics; it functions as an important constituent of the B vitamin pantothenic acid. The comparative chemical structure of L-Ala and D-Ala is illustrated in Fig. 2.1; the biosynthetic paths of the D and L forms are distinct. This chapter highlights only the metabolism and functions of L-alanine, and the biosynthesis of D-Ala is not discussed.

2.3.2 Functions of alanine

Although Ala is not considered as the prime amino acid for nitrogen uptake or signalling, its roles in plant metabolism and stress responses have been stellar. Ala has been reported to be the major form of nitrogen transport from nitrogen-fixing bacteria into the plant root (Waters *et al.*, 1998) and may account for up to 94% of the total nitrogen assimilated by the alga *Selenastrum minutum* under anaerobic conditions, probably utilizing pyruvate – the major available carbon source under anaerobic conditions – as the sole substrate (Vanlerberghe and Turpin, 1990). In addition to its role in both the transport and storage of nitrogen, Ala acts as an important connector of carbon and nitrogen metabolism and balance (Miyashita *et al.*, 2007; Rocha *et al.*, 2010a). Several reports indicate

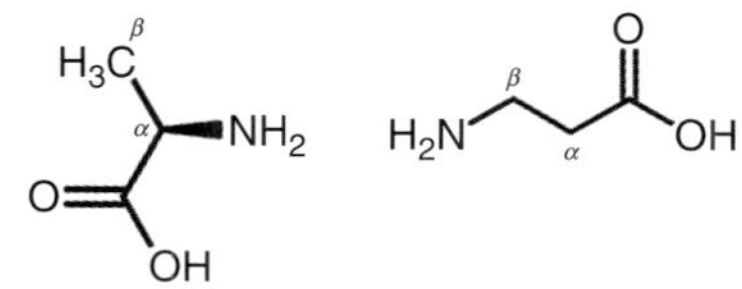

Fig. 2.1. L-Alanine and D-alanine are enantiomers. As shown in the figure, the two structures are non-superimposable. In nature, only the L-form is used to make proteins, while the D-form is present mostly in the cell walls of bacteria, but not in their proteins.

that plants increase their levels of Ala to mitigate various adverse environmental conditions, such as hypoxia, drought, unfavourable temperatures and biotic stress (Drossopoulos *et al.*, 1985; Puiatti and Sodek, 1999; de Sousa and Sodek, 2003; Mazzucotelli *et al.*, 2006; Miyashita *et al.*, 2007; Limami *et al.*, 2008; Rocha *et al.*, 2010a,b). For more references on this topic, please see Section 2.9.2. Recently, it has been proposed that Ala may save the C_3 skeleton by competing with the ethanolic fermentation of pyruvate, consequently limiting the accumulation of toxic acetaldehydes (Ricoult *et al.*, 2006; Rocha *et al.*, 2010a).

2.4 Alanine Metabolism

The metabolism of Ala is mediated through aminotransferase or transaminase reactions. Ala is an α-amino acid analogue of the α-keto acid pyruvate and is interchangeable by transamination reactions in both sugar and amino acid metabolic pathways. It should be noted that although transamination reactions are freely reversible, irreversible transamination reactions are not unprecedented (Givan, 1980). As most of the plant amino acid metabolic pathways are L-enantiomer stereospecific (Hatch, 1987), this chapter illustrates the pathways and enzymes that are only L-specific. These classes of enzymes catalyse the transfer of an amino group, together with a proton and an electron pair, from an amino donor to the carbonyl position of an amino acceptor. Generally speaking, an acidic amino acid, such as glutamate or aspartate, acts as an amino donor, while a 2-keto acid acts as the amino acceptor, with the exception of certain aldehydes serving as amino acceptor and amines as donors (Givan, 1980; Stewart and Larher, 1980; Ireland and Lea, 1999).

Aminotransferases are a large and diverse group of pyridoxal phosphate-dependent enzymes that are present in all living organisms, catalysing transamination between different pairs of amino acid:2-oxo acid substrates. AlaAT is one of the most thoroughly characterized plant aminotransferases (Rech and Crouzet, 1974; Otter *et al.*, 1992; Fujii and Yoshii, 1995; Igarashi *et al.*, 2003).

There are several transaminases involved in Ala synthesis and metabolism, as illustrated in Fig. 2.2. However, the intent of this present chapter is to review only the metabolism and functions of AlaAT. Several metabolic reactions of Ala are outlined below:

$$\text{Glutamate} + \text{Pyruvate} \leftrightarrow \text{2-Oxoglutarate} + \text{Ala (EC 2.6.1.2)} \quad (2.1)$$

$$\text{Ala} + \text{Glyoxylate} \leftrightarrow \text{Pyruvate} + \text{Glycine (EC 2.6.1.44)} \quad (2.2)$$

$$\text{Pyruvate} + \text{GABA } (\gamma\text{-Aminobutyric acid}) \leftrightarrow \text{Ala} + \text{Succinic semialdehyde (similar to EC 2.6.1.19)} \quad (2.3)$$

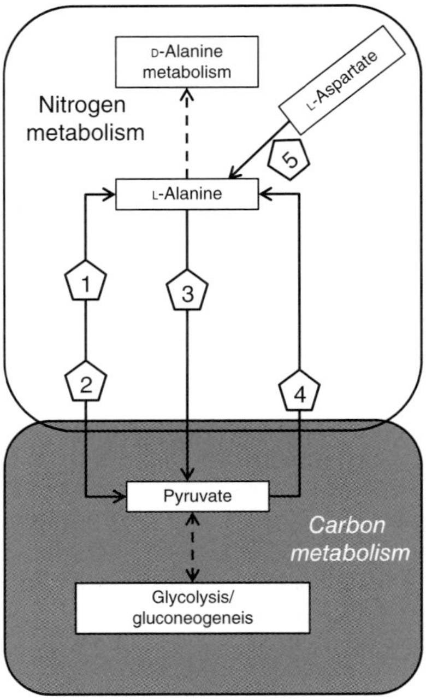

Fig. 2.2. Alanine metabolism serves a major link between carbon and nitrogen metabolism. This figure depicts alanine metabolic enzymes catalysing different sets of reactions (numbered 1–5), using different substrates (such as alanine, pyruvate, aspartate, etc.) and performing important roles in energy and amino acid metabolism. The reactions are: (1) alanine aminotransaminase/AlaAT (EC: 2.6.1.2); (2) L-alanine:2-oxo-acid aminotransferase (EC:2.6.1.12); (3): L-alanine:glyoxylate aminotransferase (EC:2.6.1.44); (4) L-alanine:NAD$^+$ oxidoreductase (EC:1.4.1.1); and (5) L-aspartate 4-carboxy-lyase (EC:4.1.1.12). This chapter reviews in detail the function of AlaAT, which catalyses the reversible transfer of an amino group from alanine to 2-oxoglutarate to form pyruvate and glutamate.

It is also interesting to note that several homologues of alanine aminotransferases with GGAT activity have been reported in plants; these will be discussed in more details in Sections 2.5 and 2.6.

2.4.1 Alanine metabolism by alanine aminotransferase

AlaAT, also known as glutamate-pyruvate aminotransferase (GPAT or GPT), is a pyridoxal-5′-phosphate (PLP) cofactor-requiring enzyme that catalyses the reversible transfer of an amino group from Ala to 2-oxoglutarate to form pyruvate and glutamate. The enzyme serves as major route of Ala biosynthesis and degradation in plants, and plays a critical role in carbon and nitrogen metabolism (Givan, 1980; Furbank, 1998). AlaAT catalyses a reversible reaction with an equilibrium constant of near one, as noted in Box 2.1. This is significant in terms of plant metabolism, as it indicates that the reaction has no specific preference towards either glutamate or Ala formation, and therefore the reaction may occur in either direction based on substrate availability and *in planta* physiological conditions.

AlaAT shares significant similarity in protein structure and features with other subgroup I aminotransferases, such as GGATs, aspartate aminotransferase (AspAT) and tyrosine aminotransferase (Arnon 1949; McAllister *et al.*, 2012). Both AlaAT and GGAT have eleven invariant residues essential for binding to pyridoxal phosphate for stabilization of the enzyme:substrate transition state (Arnon 1949; Duff *et al.*, 2012). The following sections describe in detail the diversity of the locations and functions of AlaAT.

2.5 Specific Cellular and Sub-Cellular Functions of Alanine Aminotransferase

The diversity in cellular localization of multiple AlaAT homologues indicates their involvement in multifaceted processes that include: amino acid biosynthesis and metabolism (Sechley *et al.*, 1992; Kikuchi *et al.*, 1999); pyruvate transfer across cells in C_4 plants (Son *et al.*, 1991; Son and Sugiyama, 1992); several stress responses, such as hypoxia and plant defence responses (de Sousa and Sodek, 2003; Taler *et al.*, 2004; Kim *et al.*, 2005; Miyashita *et al.*, 2007); and the regulation of photorespiratory metabolism (Fukao *et al.*, 2002; Igarashi *et al.*, 2003; Liepman and Olsen, 2003; Niessen *et al.*, 2012). Multiple tissue- and organelle- specific AlaAT activities appear to be associated with specific metabolic roles, as discussed in the section below. This involvement in a variety of biochemical reactions during the whole life cycle of the plant is not surprising because AlaAT catalyses a near equilibrium reaction in the reversible conversion of Ala and 2-oxoglutarate into pyruvate and glutamate. Sub-cellular fractionation analysis and a range of biochemical studies indicate

Box 2.1. Reaction catalysed by AlaAT (E.C. 2.6.1.2).

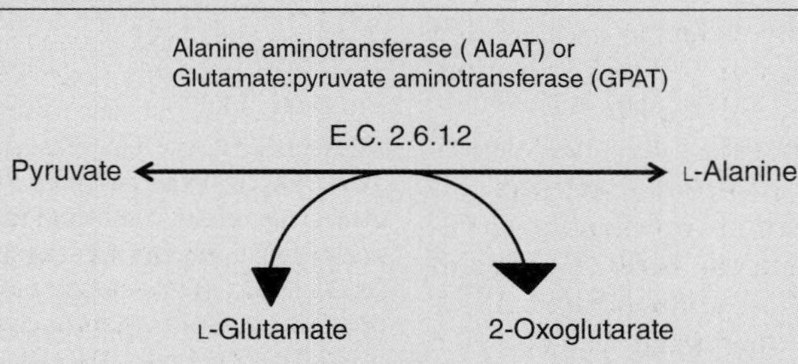

Alanine aminotransferase (AlaAT), also known as glutamate pyruvate aminotransferase (GPAT), is a pyridoxal 5-phosphate (PLP) cofactor-requiring enzyme that catalyses a reversible transfer of an amino group from alanine to 2-oxoglutarate to form pyruvate and glutamate with an equilibrium constant of near one. This enzyme serves as the major route of alanine biosynthesis and degradation in plants.

that AlaAT or similar activities are located in the cytosol (Liepman and Olsen, 2003) and the mitochondria (Igarashi *et al.*, 2003; Liepman and Olsen, 2003). The enzyme is also located in peroxisomes, suggesting important roles in photorespiration and amino acid metabolism (Liepman and Olsen, 2003; Liepman and Olsen, 2004; Shrawat *et al.*, 2008).

2.5.1 Homologues and tissue localization

AlaAT-like activities and significant pools of Ala have been detected in several genera across the evolutionary tree: in fungi, algae, mycorrhizae and higher plants (Genetet *et al.*, 1984; Lain-Guelbenzu *et al.*, 1991; Chen *et al.*, 1996; Bhadauria *et al.*, 2012). The enzyme exists as different homologues that perform specific functions, as illustrated below. In higher plants, the enzyme has been reported to be present in various species, such as *Arabidopsis thaliana* (Igarashi *et al.*, 2003; Liepman and Olsen 2003; Miyashita *et al.*, 2007), *Glycine max* (Rocha *et al.*, 2010b), *Medicago truncatula* (Ricoult *et al.*, 2006) and rice (Niessen *et al.*, 2012). It has also been reported in various tissues, such as the leaves, roots and floral organs of *A. thaliana* (Igarashi *et al.*, 2003), (the tomato) fruit (Rech and Crouzet, 1974), bundle sheath cells of maize and broomcorn millet (*Panicum miliaceum*), and ear shoots of maize (Valle and Heldt, 1991; Son and Sugiyama, 1992; Seebauer *et al.*, 2004), endosperm of developing seeds in *Oryza sativa* (Kikuchi *et al.*, 1999) and the seed coat and cotyledons in peas (Murray and Kennedy, 1980).

AlaAT is encoded by multiple genes: *AlaAT1* and *AlaAT2* encode Ala:glyoxylate aminotransferase activities (EC 2.6.1.2), while *AtGGAT1* and *AtGGAT* encode GGAT or glutamate: glyoxylate aminotransferase activities (GGAT; EC 2.6.1.4) (Igarashi *et al.*, 2003; Liepman and Olsen, 2003; Niessen *et al.*, 2012). Several AlaAT isoforms are observed in barley (Good and Muench, 1992), mustard (Penther, 1991) and *Arabidopsis* (Igarashi *et al.*, 2003; Wiśniewski *et al.*, 2006). In cereals such as millet, three AlaAT homologues (Son *et al.*, 1991) are present, while six AlaAT homologues are present in maize (Orzechowski *et al.*, 1999b). In the C_4 plant *P. miliaceum*, cytosolic AlaAT was found in both mesophyll and bundle sheath cells and is likely to be involved in the transfer of pyruvate between these cell types (Son and Sugiyama, 1992). Three Ala aminotransferases, two minor (AlaAT1, AlaAT3) and one major (AlaAT2), were detected in leaf extracts from *P. miliaceum*, and the activity of AlaAT2 was found to increase markedly during leaf greening and is suggested to function in the C_4 photosynthetic cycle (Son *et al.*,1991). A putative AlaAT (EC 2.6.1.2) in *Capsicum anuum* exhibited an organ-specific expression pattern and there is abundant accumulation of transcript in ripe fruit tissues and in older leaves during senescence. Miyashita *et al.* (2007) observed the induction of transcript levels as well as enzyme activities in *Arabidopsis* AlaAT1 and AlaAT2 in hypoxically treated roots. Tissue-specific expression of promoter:-GUS fusions of AlaAT1 and AlaAT2 indicated that both AlaAT genes were expressed predominantly in vascular tissues. Additionally, the AlaAT1 knockout mutant (alaat1-1) showed a dramatic reduction in AlaAT activity, suggesting that AlaAT1 is the major AlaAT isoenzyme in *Arabidopsis* (Miyashita *et al.*, 2007).

AlaAT has been reported from several crop plants, including rice, soybean, maize and wheat. Seebauer *et al.* (2004) reported that Ala is one of the major amino acids in maize cobs that accumulated over time in unpollinated spikelets and cobs. They found that the cob performs significant enzymatic interconversions among Gln (glutamine), Ala, Asp (aspartate) and Asn (asparagine) during early reproductive development, which may precondition the nitrogen assimilate supply for sustained kernel growth. In soybean, AlaAT activity was detected in all leaves, pods, roots and nodules, with the lowest activity in roots under normal growth conditions (Rocha *et al.*, 2010b). However, waterlogging-induced hypoxia increased the activity of AlaAT, with a concomitant accumulation of Ala. Interestingly, during re-oxygenation, the AlaAT activity remained high, but the transcript level and the Ala content decreased, indicating the role for AlaAT in the catabolism of Ala during the initial period of re-oxygenation following hypoxia (Rocha *et al.*, 2010a,b). In young wheat seedlings, both AlaAT and GGAT are downregulated by nitrogen deficiency, whereas AlaAT1 is upregulated by hypoxia and GGAT1 by light (Kendziorek *et al.*, 2012b).

These examples signify the diverse roles of AlaAT across different tissues and across the life cycles of the plants.

2.5.2 Sub-cellular localization

A significant diversity of Ala metabolism has been reported in mitochondria, peroxisomes and chloroplasts, as illustrated below. *In silico* predictions indicate that AlaAT1 is located in the cytosol, whereas AlaAT2 is a mitochondrial enzyme (Liepman and Olsen, 2003; Ricoult *et al.*, 2006; Niessen *et al.*, 2012). The presence of two AlaATs in mitochondria is most likely involved in linking mitochondrial glycolate oxidation to major photorespiratory pathways in green algae, rice and *Arabidopsis*. Peroxisomal AlaAT exhibiting GGAT activity has been implicated in the regulation of photorespiration (Fukao *et al.*, 2002; Igarashi *et al.*, 2003; Liepman and Olsen, 2003; Niessen *et al.*, 2012).

Depending on the plant species, two to six isoforms of AlaAT are located in different sub-cellular compartments, such as the cytoplasm, mitochondria and peroxisomes (Liepman and Olsen, 2001). There are four homologous genes encoding AlaAT-like functions (Igarashi *et al.*, 2003); the product AlaAT1 is located in the cytosol while AlaAT2 is a mitochondrial enzyme (NCBI protein accessions AAF82782 and AAF82781, respectively) (Liepman and Olsen, 2003). The other two AlaAT homologues have a peroxisomal targeting signal (PTS1) in their C termini, characteristic of peroxisomal GGAT. These homologues were identified as GGAT1 and GGAT2 (NCBI protein accessions AAK25905 and AAL34156, respectively) that can catalyse the transfer of an α-amino group between L-glutamate (L-Glu) or L-Ala and glyoxylate (Liepman and Olsen, 2003). Although GGAT activity has been observed in several sub-cellular locations in higher plants and algae, including the cytoplasm and mitochondria, sub-cellular fractionation and biochemical analyses indicate that the activity was mainly located within peroxisomes (Liepman and Olsen, 2003). The glyoxylate aminotransferase reactions in peroxisomes appear to be catalysed by at least two distinct types of aminotransferases: an AGT1 (alanine:glyoxylate aminotransferase) homologue with serine:glyoxylate aminotransferase activity (Liepman and Olson, 2001), and a pair of closely related potentially redundant AlaAT homologues with GGAT activity (Liepman and Olsen, 2001).

Transgenic expression studies of four AlaATs in *M. truncatula* by Ricoult *et al.* (2006) provide significant information on the organellar and tissue specific functions of different forms of Ala aminotransferase. Two genes encoding mitochondrial isoforms of AlaAT (m-AlaAT, EC 2.6.1.2) and Ala-glyoxylate aminotransferase (AGT, EC 2.6.1.44) were expressed in roots, stems and leaves. A gene encoding a cytosolic AlaAT isoform (c-AlaAT) catalysing the same reaction as m-AlaAT, was expressed specifically in leaves, while a gene encoding an isoform involved in branched chain amino acid metabolism was expressed in stems and roots. In young seedlings, only m-AlaAT and AGT were expressed in embryo axes. In hypoxic embryo axes, the amounts of transcript and putative protein of m-AlaAT (EC 2.6.1.2) increased. Isotope labelling studies indicated that the *in vivo* enzyme activities also changed concomitantly. Under hypoxia, m-AlaAT catalysed only Ala synthesis, while glutamate synthesis using Ala as amino donor was inhibited. As a result, in hypoxic seedlings, Ala accumulated as the major amino acid instead of asparagines. These results reiterate the involvement of the fermentative AlaAT pathway in hypoxia tolerance. Regulation of m-AlaAT at both the transcriptional and post-translational levels allowed for an increase in gene expression and orientation of the activity of the product of its transcription towards Ala synthesis under hypoxia. Additionally, labelling experiments showed that glycine synthesis occurred at the expense of either Ala or glutamate as amino donor, indicating that a glutamate-glyoxylate aminotransferase was operating together with AGT in *M. truncatula* seedlings. Both enzymes seemed to be inhibited by hypoxia, resulting in a very low amount of glycine in hypoxic seedlings (Ricoult *et al.*, 2006).

2.6 A Phylogenetic Analysis of Alanine Aminotransferase

AlaAT is fairly ubiquitous across several phyla in bacteria, yeast, plants and mammals (Rech and Crouzet, 1974; Otter *et al.*, 1992; Mehta *et al.*, 1993; Fujii and Yoshii, 1995; Igarashi *et al.*, 2003). The gene encoding AlaAT has a broad expression profile and distribution in several plant tissues and organelles, and this diversity indicates the biochemical reaction that it catalyses is essential in plant metabolism, as discussed in the preceding sections. Depending on the plant species, two to six isoforms of this enzyme are present in different cellular compartments, such as the cytoplasm, mitochondria and peroxisomes (Liepman and Olsen, 2001, 2003). AlaAT homologues from plants can be divided into two distinct groups based on substrate specificity, sub-cellular localization and their similarities in gene sequence. Functional analysis of four aminotransferases from *Arabidopsis* revealed at least two classes of sequences encoding for glutamate: pyruvate aminotransferases/ AlaAT/E.C.2.6.1.2 enzymes, whereas two homologues that cluster within the same gene subfamily as the AlaAT genes were shown to act as glutamate:glyoxylate aminotransferase/GGAT/ E.C.2.6.1.4 (Igarashi *et al.*, 2003).

Comparison of amino acid sequences showed 93% of identity between GGAT1 and GGAT2 and 85% of identity between AlaAT1 and AlaAT2, whereas the sequence identity between these two groups of aminotransferases from *Arabidopsis* is about 44% (Rocha *et al.*, 2010b). The AlaATs and GGATs differ mostly in their subunit composition and molecular mass, as described in Section 2.8. The family exhibits a division into two major subfamilies, referred to as A and B (Rocha *et al.*, 2010b).

AlaAT homologues are dimers with a molecular weight of about 100 kDa (Lain-Guelbenzu *et al.*, 1991; Son *et al.*, 1991; Good and Muench, 1992; Orzechowski *et al.*, 1999b), unable to utilize glyoxylate as an amino acceptor as in GGATs. The GGATs are monomeric with a molecular mass of about 60 kDa (Paszkowski and Niedzielska, 1989; Orzechowski *et al.*, 1999a). A phylogenetic tree is shown in Fig. 2.3, which was constructed based on the available sequences (Rocha *et al.*, 2010b). Phylogenetic analysis of the subgroup I aminotransferases (Ward *et al.*, 2000) shows that they are characterized by 11 invariant residues in the primary sequence of the protein. Sequence analysis (McAllister *et al.*, 2012) indicates that the protein sequences of AtGGAT1 and AtGGAT2 are more similar to plant AlaAT enzymes than are mammalian AlaAT enzymes to plant homologues, while the non-eukaryotic sequence as exemplified in the archaean *Pyrococcus furiosus* (PfAlaAT) also seems to be kinetically the most divergent.

2.7 Purification of Alanine Aminotransferase

This section describes the wide range of protein purification techniques attempted by various laboratories to obtain near homogeneous preparations of AlaAT for further *in vitro* biochemical characterization of the enzyme. To date, AlaATs have been purified from a variety of organisms, including bacteria, fungi, algae and higher plants, as summarized in Table 2.1. As illustrated in the table, there is a wide range of specific activities expressed in terms of μmol min^{-1} mg^{-1}, with the lowest (18.66) noted in *Chlamydomonas* (Lain-Guelbenzu *et al.*, 1991) and the highest (2231) in barley (Good and Muench, 1992), while the specific activities reported from the fungus *Candida maltosa* (Umemura *et al.*, 1994) and other plant species, such as *Solanum lycopersicum* (Ricoult *et al.*, 2006) and *Zea mays* (Orzechowski *et al.*, 1999a) are comparable to one other. As shown in Table 2.1, it appears that a combination of several chromatographic techniques, including gel filtration and ion chromatography, was used to achieve the purification of AlaAT. These reports provide a comprehensive view of the purification of AlaAT and act as a good source of knowledge for future purification and biochemical work on AlaAT.

2.8 Protein Characterization of Alanine Aminotransferase

2.8.1 Subunits and substrate specificities

Table 2.2 provides a summary of the AlaAT enzyme characteristics as reported from several laboratories. This section will highlight the molecular weight and subunit structures of AlaAT, and the variety of substrates used by the enzyme. A more detailed discussion on enzyme kinetics will be discussed in Section 2.8.2.

Most enzymes with AlaAT activities that have been characterized to date are reported to

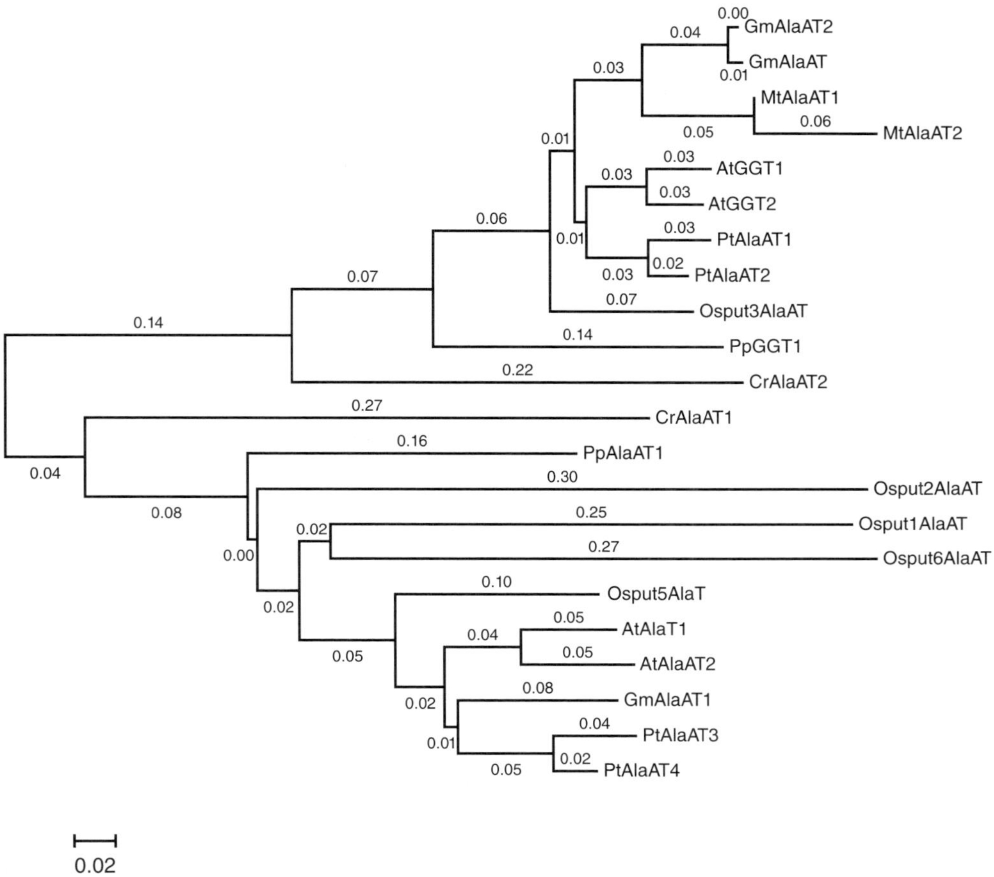

Fig. 2.3. The phylogenetic tree of the alanine aminotransaminase (AlaAT) enzyme. The tree is created based on the full coding sequenced of AlaAT genes in the alga *Chlamydomonas reinhardtii* and the bryophyte *Physcomitrella patens*, and several (higher) plant species including both monocotyledons and dicotyledons.

be nearly 100 kDa, with homodimers comprising two identical subunits (Noguchi and Hayashi, 1980; Noguchi and Fujiwara, 1982; Lain-Guelbenzu *et al.*, 1991; Son *et al.*, 1991; Good and Muench, 1992; Umemura *et al.*, 1994; Orzechowski *et al.*, 1999a; Ricoult *et al.*, 2006; Kameya *et al.*, 2010). Monomeric structure is rare among aminotransferases, although some of them are able to retain activity after dissociation at a lower pH (Truszkiewicz and Paszkowski, 2005). Several monomeric structures were reported for AlaAT homologues from *Arabidopsis* leaves (Wiśniewski *et al.*, 2006), and for GGATs from rye leaves (Paszkowski and Niedzielska, 1989), *Chlorella fusca* (Meish *et al.*, 1983) and the phototrophic bacteria *Rhodopseudomonas palustris* (Yamaguchi *et al.*, 2003). It is interesting to note that the molecular weight of 55 kDa of *Arabidopsis* AlaAT homologues with glutamate:glyoxylate activity was lower than that of the 62 kDa homologue with no GGAT activity (Wiśniewski *et al.*, 2006). These results are similar to the those reported from wheat seedlings, in which each of the two GGAT homologues was 50kDa (Kendziorek *et al.*, 2012a).

It is also interesting to note the diversity of substrates and products catalysed by AlaAT, as noted in Table 2.1 and the references therein. A more detailed discussion of this will follow in the next section, which elucidates the kinetic characteristics of AlaAT from several species, and indicates the range of substrate specificities and diversity in catalytic functions. In contrast with most bacterial aminotransferases (Hwang *et al.*, 2005), including those annotated as AlaAT or AspAT, the few eukaryotic AlaATs examined to date seemed to have a narrow substrate range and specificities (Lain-Guelbenzu *et al.*, 1991; Vedavanthi *et al.*, 2004, 2006).

2.8.2 Kinetics and reaction mechanism

Detailed kinetic comparisons of AlaAT enzymes from a variety of organisms have been obtained from a number of studies over decades (Noguchi and Hayashi, 1980; Noguchi and Fujiwara, 1982; Lain-Guelbenzu *et al.*, 1991; Son *et al.*, 1991; Good and Muench, 1992; Umemura, *et al.*, 1994; Orzechowski *et al.*, 1999a,b; Ward *et al.*, 2000; Marangoni, 2003; Liepman and Olsen, 2004; Ricoult *et al.*, 2006; Wiśniewski *et al.*, 2006; Kameya *et al.*, 2010; Duff *et al.*, 2012; McAllister *et al.*, 2013). However, comparative analyses of the enzymatic properties of AlaAT are difficult owing to the fact that most of the biochemical studies on AlaAT have been organism- or tissue-specific, focusing on a single enzyme or isoenzymes. Additionally, differences in enzyme source and purification procedures, compounded by disparate assay conditions and expression differences, could have led to altered kinetic values (Noguchi and Hayashi, 1981; Noguchi and Fujiwara, 1982; Lain-Guelbenzu *et al.*, 1991; Son *et al.*, 1991; Good and Muench, 1992; Umemura, *et al.*, 1994; Orzechowski *et al.*, 1999a,b; Ward *et al.*, 2000; Marangoni, 2003; Ricoult *et al.*, 2006; Kameya *et al.*, 2010). Nevertheless, a direct kinetic comparison has

Table 2.1. Protein purification of alanine aminotransferase (AlaAT) from a bacterium, fungus, alga and several higher plants. The different analytical techniques used to obtain purified AlaAT are shown.

Organism	Purification method	Specific activity (μmol min^{-1} mg^{-1})[a]	Reference
Hydogenobacter thermophilus	DE52 column, butyl-Toyopearl column, DEAE-Toyopearl column, CHT ceramic hydroxyapatite column and Mono Q column	NA	Kameya *et al.*, 2010
Candida maltosa	Ni^{2+}-NTA agarose resin chromatography and Poly-Prep mini-column chromatography	156 Reverse reaction, 594 Forward reaction	Umemura *et al.*, 1994
Chlamydomonas reinhardtii	Ammonium sulfate precipitation, gel filtration, hydroxyapatite chromatography and anion exchange	18.66	Lain-Guelbenzu *et al.*, 1991
Hordeum vulgare	Ammonium sulfate, gel filtration, Q-Sepharose, Mono P and Mono Q columns	2231	Good and Muench, 1992
Medicago truncatula	Dowex 50WX8 200-400 mesh chromatography	NA	Ricoult *et al.*, 2006
Panicum miliaceum	Ammonium sulfate, DEAE-cellulose, Sephacryl S-200 and phenyl-Sepharose	387.9	Son *et al.*, 1991
Solanum lycopersicum	Ammonium sulfate, DEAE-Sephadex and Sephadex G-200	120	Ricoult *et al.*, 2006
Triticum aestivum	Saline precipitation, gel filtration, preparative electrophoresis and anion exchange chromatography on protein-Pak Q 8HR column attached to HPLC	NA	Kendziorek *et al.*, 2012a
Zea mays	Ammonium sulfate, hydroxyapatite and preparative electrophoresis	127.5	Orzechowski *et al.*, 1999a

[a]NA, not available.

Table 2.2. Protein characterization of AlaAT from a bacterium, fungus, alga and several higher plants. The K_m values for different substrates is specified. For each set of enzymatic reactions, the products are represented in parentheses.

Organism	Molecular weight (MW)	Subunits	K_m – Substrate	pH Optimum range	Substrates (products)	Reference
Hydogenobacter thermophilus	69,000	NA[a]	0.9 – Glyoxylate 8.1 – L-Alanine	7.9–8.4	L-Alanine + glyoxylate (pyruvate + glycine)	Kameya *et al.*, 2010
Candida maltosa	99,000	NA	0.16 – Pyruvate 0.37 – 2-Oxoglutarate 2.6 – L-Alanine 6.3 – L-Glutamate	6–7.5	L-Alanine + 2-oxobutyrate (pyruvate + 2-aminobutanoate) L-Alanine + 2-oxoglutarate (pyruvate + L-glutamate) L-Glutamate + 2-oxobutyrate (maltose + 2-oxoglutarate + 2-aminobutanoate)	Umemura *et al.*, 1994
Chlamydomonas reinhardtii	105,000	Dimer, 45,000	0.05 – 2-Oxoglutarate 0.24 – Pyruvate 0.52 – L-Glutamate 2.7 – L-Alanine	7.3	Glyoxylate + alanine (glycine + pyruvate) L-Alanine + 2-oxoglutarate (pyruvate + L-glutamate)	Lain-Guelbenzu *et al.*, 1991
Hordeum vulgare	97,000	Dimer, 50,000	0.06 – Pyruvate 1 – L-Glutamate 5 – 2-Oxoglutarate 17 – L-Alanine	8-8.3	L-Alanine + 2-oxoglutarate: highly specific for alanine and 2-oxoglutarate Pyruvate + L-glutamate (L-alanine + 2-oxoglutarate): cannot use glyoxylate as keto donor	Good and Muench, 1992
Panicum miliaceum	102,000	Dimer, 50,000	0.15 – 2-Oxoglutarate 0.33 – Pyruvate 5 – L-Glutamate 6.67 – L-Alanine	6.5–7.5	L-Alanine + 2-oxoglutarate (pyruvate + L-glutamate)	Son *et al.*, 1991
Solanum lycopersicum	100,000	NA	0.09 – Pyruvate 0.28 – 2-Oxoglutarate 2.3 – L-Glutamate 2.8 – L-Alanine	7.3–7.5	L-Alanine + 2-oxoglutarate: rate of the forward reaction is 4 times higher than the rate of the reverse reaction, no activity with aspartate, glycine, threonine, leucine, methionine, arginine, phenylalanine, tyrosine, tryptophan, glyoxylate and 2-oxo-3-methylvalerate	Ricoult *et al.*, 2006

Continued

Table 2.2. Continued.

Organism	Molecular weight (MW)	Subunits	K_m – Substrate	pH Optimum range	Substrates (products)	Reference
Triticum aestivum	61,300	Monomer, 65,000	0.19 – 2-Oxoglutarate 0.22 – Pyruvate (AlaAT1) 0.29 – Pyruvate (AlaAT2) 0.35 – 2-Oxoglutarate (AlaAT2) 6.05 – L-Alanine (AlaAT1) 6.25 – L-Glutamate (AlaAT1) 6.66 – L-Glutamate (AlaAT2) 6.9 – L-Alanine (AlaAT2)	7.3	L-Alanine + 2-oxoglutarate (pyruvate + L-glutamate)	Kendziorek *et al.*, 2012a
Zea mays	95,000	Dimer, 50,000	NA	NA	L-Alanine + 2-oxoglutarate: highly specific for alanine and 2-oxoglutarate (pyruvate + L-glutamate)	Orzechowski *et al.*, 1999a

[a]NA, not available.

been provided by several recent studies (Duff *et al.*, 2012; Kendziorek *et al.*, 2012a; and McAllister *et al.*, 2013). Interestingly, the studies by McAllister *et al.* (2013) placed emphasis on the differences in K_m values between homologous enzymes, in contrast to the enzyme-specific activities that were reported by Duff *et al.* (2012).

Despite the problems noted above, kinetic studies of AlaAT enzymes with differential substrate specificities and kinetic properties, as well as the distinct cellular roles of the isoenzymes discussed in the preceding sections, provide some interesting implications for the diversity of physiological functions mediated by AlaAT (Duff *et al.*, 2012; Kendziorek *et al.*, 2012a,b; McAllister *et al.*, 2013). Table 2.2 has provided a summary of the basic kinetics, substrate and product specificities of AlaAT from a diverse group of organisms. This section will now describe in detail the recent kinetic characterization that has been done by Duff *et al.* (2012), Kendziorek *et al.* (2012b) and McAllister *et al.* (2013).

The work by McAllister *et al.* (2013) provided a direct K_m comparison of several glutamate: pyruvate aminotransferases/AlaATs and glutamate:glyoxylate aminotransferases/GGATs, as shown in Table 2.3. Although the K_m values for AlaAT analysed by this group have been reported previously, the current study (McAllister *et al.*, 2013) enables a direct kinetic comparison of the enzymes from different sources using a single assay system. McAllister *et al.*'s study shows that the AlaAT homologues and two glutamate:glyoxylate aminotransferase (GGAT) enzymes that have secondary glutamate:pyruvate aminotransferase activities have different K_m values for co-substrates, indicating that the *in vivo* rate and direction of the reaction catalysed by each enzyme may differ dramatically, even under similar substrate concentrations, thus reaffirming the results obtained by Duff *et al.* (2012).

Based on the results from the kinetic assays by McAllister *et al.* (2013), as shown in Table 2.3, it appears that *A. thaliana* AlaAT1 (AtAlaAT1) is most similar to barley (*Hordeum vulgare*) AlaAT1 (HvAlaAT). Both AtAlaAT enzymes have higher K_m values for glutamate and Ala, and lower K_m values for 2-oxoglutarate and pyruvate, when compared with HvAlaAT. In spite of the functional similarity of the GGAT enzymes in catalysing both glutamate:glyoxylate and glutamate:pyruvate aminotransferase reactions (Liepman and Olsen, 2003), the kinetic differences identified by McAllister *et al.* (2013) are most likely caused by non-conserved residues (Ward *et al.*, 2000) in the primary sequences of the two types of enzymes, leading to changes in enzyme folding, substrate binding affinity and catalytic rate. For more details of the kinetic comparison of the different AlaAT, please refer to McAllister *et al.* (2013).

Kinetic analysis of the AlaAT enzymes by Duff *et al.* (2012) also provides a detailed determination of the substrate specificity of plant, bacterial and yeast AlaATs. The authors report that the barley AlaAT catalyses the Ala-forming reaction with a K_{cat} of 25.6 s^{-1}, while the reverse reaction has a K_{cat} of 12.1 s^{-1}, and an equilibrium constant of ~0.5. The enzyme also utilizes aspartate and oxaloacetate with ~10% efficiency compared with the native substrates, which makes it much more specific than related bacterial/archaeal enzymes with lower K_m values. Bacterial enzymes tested in various studies have a fairly broad substrate specificity compared with plant enzymes (Ward *et al.*, 2000; Duff *et al.*, 2012; and references in Table 2.1). However, the *P. furiosus* AlaAT seems to have a more specific substrate choice than other bacterial forms, most likely supporting its function as an electron sink in the archaeon, along with glutamate dehydrogenase (Ward *et al.*, 2000). Duff *et al.* (2012) also suggests that the bacterial enzymes had lower K_m values, especially for pyruvate, which might be advantageous under conditions in which pyruvate is limiting. The K_m values reported by McAllister *et al.* (2013) and Duff *et al.* (2012) for HvAlaAT and *Saccharomyces cerevisiae* AlaAT1 (ScAlaAT1) are fairly similar, with the largest difference in K_m values reported for ScAlaAT1 with pyruvate as substrate: 11.0 by McAllister *et al.* (2013), compared with values of 0.4 previously reported by Duff *et al.* (2012), which is nearly a 28-fold difference (see Table 2.3). The K_m values reported for HvAlaAT did not show a significant differences for the various substrates, with 6.7-, 1.1-, 9.1- and 2.1-fold differences for Ala, 2-oxoglutarate, glutamate and pyruvate, respectively (McAllister *et al.*, 2013). All other K_m values for Ala, 2-oxoglutarate and glutamate showed lower fold differences when K_m values for ScAlaAT were compared – 12, 2.5 and 2.9, respectively, reiterating the importance of obtaining enzymatic data from a single source for a meaningful comparative analysis (McAllister *et al.*, 2013).

Table 2.3. Summary of the kinetics of alanine aminotransferase (AlaAT) homologues. The K_m values are from both glutamate:pyruvate aminotransferase (AlaAT) and glutamate:glyoxalate aminotransferase (GGAT) from various species.[a] Standard errors given in parentheses.

	K_m (mM)				
Enzyme[a]	L-Alanine	2-Oxoglutarate	L-Glutamate	Pyruvate	Reference
AtAlaAT1	2.4 (0.5)	0.1	2.5 (0.4)	0.1	McAllister *et al.*, 2013
AtAlaAT2	10.4 (1.2)	1.0 (0.1)	4.9 (1.1	5.1 (0.9)	
HvAlaAT	5.6 (0.6)	0.2	4.9 (0.1)	0.1	
MtAlaAT1	1.0 (0.2)	0.2	0.1	18.6 (2.7)	
MtAlaAT2	1.5 (0.2)	0.3	0.3 (0.1)	18.0 (1.9)	
PfAlaAT	4.0 (0.5)	0.02	0.9 (0.1)	16.5 (2.2)	
ScAlaAT1	0.3 (0.1)	0.5 (0.1)	0.7 (0.1)	11.0 (1.7)	
ScAlaAT	3.6 (0.07)	0.2 (0.02)	0.24 (0.01)	0.4 (0.01)	Duff *et al.*, 2012
HvAlaAT	0.83 (0.01)	0.18 (0.02)	0.54 (0.01)	0.21 (0.03)	
HvAlaAT (Tag free)	3.79 (0.08)	0.28 (0.02)	0.79 (0.02)	0.17 (0.02)	
OsAlaAT	0.4 (0.02)	0.25	0.68 (0.01)	0.11 (0.01)	
ZmAlaAT1	0.62 (0.01)	0.37 (0.01)	0.56 (0.01)	0.42 (0.05)	
ZmAlaAT2	3.06 (0.06)	0.2 (0.01)	0.55 (0.01)	0.54 (0.03)	
TaAlaAT1	6.0	0.19	6.25	0.22	Kendziorek *et al.*, 2012a
TaAlaAT2	6.9	0.35	6.66	0.29	
TaGGAT1	8.33	0.21	9.52	0.35	
TaGGAT2	8.69	0.19	9.09	0.33	
AtGGAT1	1.9 (0.5)	0.2 (0.1)	0.2 (0.1)	0.6 (0.1)	McAllister *et al.*, 2013
AtGGAT2	1.2 (0.1)	8.8 (1.8)	0.2 (0.1)	18.5 (2.7)	

[a]Abbreviations of the enzyme sources by species origin are as follows: At, *Arabidopsis thaliana*; Hv, *Hordeum vulgare*; Mt, *Medicago truncatula*; Os, *Oryza sativa*; Pf, *Pyrococcus furiosus*; Sc, *Saccharomyces cerevisiae*; Ta, *Triticum aestivum*; and Zm, *Zea mays*.

Kendziorek's (2012a,b) studies on wheat seedlings provide additional insight into the kinetics of the different AlaATs. Kinetic studies of four homologues of alanine aminotransferases from the shoots of wheat seedlings, e.g. AlaAT1, AlaAT2, GGAT1 GGAT2, showed insignificant differences in K_m values using Ala as substrate and in catalytic efficiency for L-Ala in reaction with 2-oxoglutarate, indicating comparable kinetic constants for AlaAT1 and AlaAT2. Similar kinetic constants for L-Ala in reaction with 2-oxoglutarate, and for L-glutamate in reaction with pyruvate, for all four homologues suggest an equally efficient reaction in both forward and reverse directions (Table 2.3). GGAT1 and GGAT2 were able to catalyse transamination between L-glutamate and glyoxylate, L-Ala and glyoxylate, and the reverse reactions between glycine and 2-oxoglutarate or pyruvate. The authors noted that the estimated K_m values for GGAT1 (3.22 M) and GGAT2 (1.27 M) using L-glutamate as substrate was lower in transamination with glyoxylate than with pyruvate (9.52 and 9.09 mM, respectively), see in Table 2.3.

Several studies (Lain- Gulebenzu *et al.*, 1991; Rech and Crouzet, 1974) in higher plants have indicated that AlaAT exhibits a bi-bi ping-pong reaction mechanism, as observed for other aspartate aminotransferases (Ward *et al.*, 2000). Inhibition and protection studies strongly support the involvement of PLP in enzyme catalysis through a Schiff's base formation (Lain-Gulebenzu *et al.*, 1991). Recently, Duff *et al.* (2012) observed a similar reaction mechanism for AlaAT, indicating that the AlaAT reaction mechanism is of a bi-bi ping-pong type and that each substrate binds sequentially and competitively to the enzyme, as observed for other Ala and aspartate aminotransferases (Ward *et al.*, 2000).

2.8.3 Inhibitors of the enzyme

Several studies (Goil, 1978; Lain-Guelbenzu *et al.*, 1991; Good and Muench, 1992, 1993; Umemura *et al.*, 1994; Orzechowski *et al.*, 1999a; Duff *et al.*, 2012; Kendziorek *et al.*, 2012a,b) have been dedicated to determining the impact of different inhibitors of AlaAT, as shown in Table 2.4. Kendziorek *et al.* (2012a) reported significant inhibition of AlaAT and GGAT activities by *p*-hydroxymercuribenzoate in wheat, indicating the presence of essential sulfhydryl groups at the active sites of AlaAT homologues in a manner similar to that reported from other plants (Goil, 1978; Orzechowski *et al.*, 1999b).

Recently, Duff *et al.* (2012) investigated the effects of known aminotransferase inhibitors on the activity of this enzyme, including *O*-carboxymethyl hydroxylamine hemihydrochloride, DL-vinylglycine, DL-propargylglycine, 4-chloromercuribenzoic acid, cycloserine and penicillamine, and found significant inhibitions, especially by *O*-carboxymethyl hydroxylamine hemihydrochloride and cycloserine. Both *O*-carboxymethyl hydroxylamine hemihydrochloride and penicillamine are known to be general inhibitors of PLP-containing enzymes, because they react with the pyridoxal moiety to form an oxide (Cornell *et al.*, 1984; Markovic-Housley *et al.*, 1996). Cycloserine, in contrast, is a structural analogue of Ala and is known to be a more effective inhibitor of AlaAT activity than of other transaminases (Cornell *et al.*, 1984). The inactivation by cycloserine is irreversible because it reacts with pyridoxal-linked aminotransferases to form a highly reactive acylating agent that attacks other active-site residues to form a stable complex (Rando, 1975). The reaction mechanism based inhibition of PLP-containing enzymes is also known for DL-vinylglycine and DL-propargylglycine (Abeles and Maycock, 1976; Burnett *et al.*, 1980). 4-Chloromercuribenzoic acid is a thiol-modifying reagent that reacts with free catalytically important cysteine, leading to the inhibition of enzyme activity (Tsubota and Hoshino, 1969). Table 2.4 highlights the range of inhibitors of AlaAT activity.

Table 2.4. Inhibition of alanine aminotransferase (AlaAT) activities from a fungus, alga and various higher plants by a selection of metals, biochemicals and metabolites.

Organism	Inhibitors	Reference/s
Candida maltosa	Aminooxyacetate 2 mM, complete inhibition Hg^{2+} 0.1 mM, 81% inhibition Hydroxylamine 2 mM, 73% inhibition *p*-Hydroxymercuribenzoate 1 mM, 87% inhibition	Umemura *et al.*, 1994
Chlamydomonas reinhardtii	1,10-Phenanthroline 2 mM, 51% inhibition Aminooxyacetate 1 mM, complete inhibition Hydroxylamine 2 mM, 96% inhibition L-Ascorbic acid 2 mM, 25% inhibition *p*-Hydroxymercuribenzoate 1 mM, complete inhibition	Lain-Guelbenzu *et al.*, 1991
Hordeum vulgare	Aminooxyacetate 1 mM, 98% inhibition Glutarate 10 mM, 94% inhibition Semicarbazide 10 mM, 97% inhibition *O*-Carboxymethyl hydroxylamine hemihydrochloride 6.9 μM, 50% inhibition Cycloserine 119 μM , 50% inhibition	Good and Muench, 1992 Duff *et al.*, 2012
Triticum aestivum	4-Hydroxymercuribenzoate	Kendziorek *et al.*, 2012a
Zea mays	3-Chloro-L-alanine 0.1 mM, 78% inhibition Cd^{2+} 0.1 mM, complete inhibition Cycloserine 0.1 mM, 98% inhibition *p*-Hydroxymercuribenzoate 0.1 mM, 72% inhibition Zn^{2+} 1 mM, 75% inhibition	Orzechowski *et al.*, 1999a

2.8.4 Crystal structure

The crystal structure of barley AlaAT was solved as a complex with PLP and L-cycloserine at 2.7Å resolution (Duff *et al.*, 2012; Protein Data Bank (PDB) entry 3TCM). The structure represents the first of a plant AlaAT enzyme to be solved by crystallography, and confirms that barley AlaAT is a dimer and requires PLP. The structure of molecule A of the barley AlaAT dimer is displayed in Fig. 2.4, and the active site of molecule A is shown in Fig. 2.5. The crystal structure shows that both molecules of the dimer are ordered from residues 3–481 and contain a bound L-cycloserine-PLP adduct. Previous reports also determined covalent adduct formation between D-cycloserine and PLP in the crystal structure of D-amino acid aminotransferase (PDB entry 2DAA) (Peisach *et al.*, 1998). Barley AlaAT possesses a 'PLP-dependent transferase' fold similar to that of AspAT TM1255 from *Thermotoga maritima* (PDB entry 1O4S) (Schwarzenbacher *et al.*, 2004). This aminotransferase fold is also found in the structures of AlaAT from *T. maritima*, *P. furiosus* and humans, which highlights the similarity in enzyme structure across phyla. The ribbon-style figure of molecule A of barley AlaAT (see Fig. 2.4) demarcates the key structural elements of the canonical aminotransferase fold – these being the N-terminal arm (residues 2–23), the small alpha/beta domain (residues 24–64 and 356–481) and the large alpha-beta-alpha

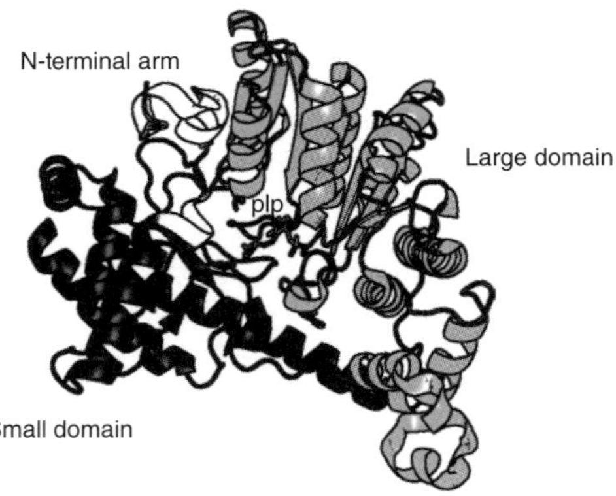

Fig. 2.4. Ribbons diagram of barley alanine aminotransaminase (AlaAT) molecule A (i.e. molecule A of Protein Data Bank (PDB) entry 3TCM). The different components of the structure are distinguished by differences in grey shading and are labelled: N-terminal arm (residues 2–23; white); small alpha/beta domain (residues 24–64, 356–481; medium grey); large domain (residues 65–355; black). The pyridoxal-5′-phosphate (PLP) molecule is located in the centre of the image between the large and small domains.

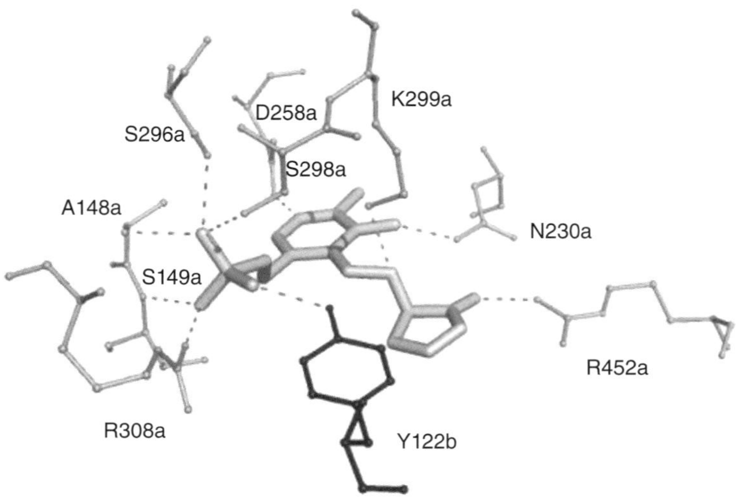

Fig. 2.5 The binding interactions of the pyridoxal-5′-phosphate (PLP)-L-cycloserine adduct (501a) in molecule A of the barley alanine aminotransaminase (AlaAT) structure. Hydrogen-bonded interactions are rendered as dashed lines. The side chain atoms of molecule that provide H-bond interactions are in grey, and the side chain of molecule B (Y122b) which provides an H-bond interaction, is rendered in black. The PLP-L-cycloserine adduct lies at the centre, rendered in medium grey.

domain (residues 65–355), with the PLP bound in the centre between the two domains.

The numerous protein-ligand hydrogen-bonded interactions between the PLP-cycloserine adduct in molecule A are displayed in Fig. 2.5. The catalytic lysine, K299a, forms a Schiff base with the aldehyde group of the PLP cofactor via its ε-amino group (Schwarzenbacher *et al.*, 2004) in the transamination reactions of the enzyme. In Fig. 2.5, lysine 299a lies above the adduct and the ε-amino nitrogen makes a polar contact of 3.2 Å with the nitrogen atom that bridges L-cycloserine with PLP. The figure also shows that many H-bond interactions stabilize the binding of the adduct in the barley AlaAT active site. This structure opens new avenues for our understanding of the AlaAT reaction mechanism and of the specifics of the catalysis performed by this protein across different phyla (Duff *et al.*, 2012).

2.9 Diverse Roles of Alanine Aminotransferase in Plants

A large number of scientific projects have been dedicated to unravelling the physiological, metabolic and biochemical functions of AlaAT and its role in plant stress. AlaAT seems to be the primary anabolic enzyme for biosynthesis of Ala, shuttling fixed carbon molecules between the mesophyll and bundle sheath cells in C_4 plants (Hatch and Mau, 1972; Hatch, 1987). Its role in connecting carbon and nitrogen metabolism is undisputed. Additionally, the roles of AlaAT in hypoxia, alanine transport, increasing biomass, nitrogen use efficiency and yield (Good and Crosby, 1989; Ricoult *et al.*, 2006; Good *et al.*, 2007a,b; Miyashita *et al.*, 2007; Shrawat *et al.*, 2008, Good and Beatty, 2011) have also been fairly well established. Interestingly, as discussed in Section 2.5, different homologues and the diverse localization of multiple forms of AlaAT across tissues and organelles makes AlaAT one of the most interesting enzymes, with a wide spectrum of functions important for plant growth and survival. The following sections delve in more detail into the diverse functional roles of AlaAT in plants.

2.9.1 Roles in metabolism

As described in Section 2.5, multiple AlaAT homologues with diverse cellular and sub-cellular localization perform different metabolic and physiological functions. AlaAT is primarily involved in general amino acid metabolism to modulate nitrogen acquisition (Sechley *et al.*, 1992) and utilization (Kikuchi *et al.*, 1999). The enzyme serves as a critical link between carbon and nitrogen metabolism, and is known to transfer pyruvate from the mesophyll to the bundle sheath cells in C_4 plants (Son *et al.*, 1991; Son and Sugiyama, 1992). Figure 2.6 illustrates the metabolic function of AlaAT in the transport of pyruvate and its role in serving as a link between carbon and nitrogen metabolism. As the literature suggests, and as described in the following sections, several critical roles have been assigned to AlaAT: response to abiotic and biotic stress, such as oxygen deficiency (de Sousa and Sodek, 2003 and Miyashita *et al.*, 2007), plant defence responses triggered by bacterial and virus infections or various other pathogen attacks (Taler *et al.*, 2004; Kim *et al.*, 2005). Several peroxisomal and mitochondrial AlaAT homologues function in the regulation of photorespiratory metabolism (Fukao *et al.*, 2002; Igarashi *et al.*, 2003; Liepman and Olsen, 2003; Niessen *et al.*, 2012). The following sections review in detail the role of AlaAT in carbon and nitrogen metabolism (Sections 2.9.1.1–2.9.1.3) and stress responses (Section 2.9.2 and its subsections) in plants.

2.9.1.1 *Roles in carbon metabolism*

Because AlaAT catalyses a reversible reaction with an equilibrium constant of near one (as shown in Box 2.1), which indicates no specific preference towards the formation of glutamate or Ala, it serves as a major route for the biosynthesis of either Ala or pyruvate, based on the metabolic demand of the cell. This provides a greater opportunity for an effective cross-talk between carbon and nitrogen metabolic pathways, thus achieving an overall efficacy and success of plant metabolism, as has been reiterated through various studies (Streeter and Thompson, 1972; Givan, 1980; Scheible *et al.*, 1997; Furbank, 1998; Stitt *et al.*, 2002; Rocha *et al.*, 2010a; Duff *et al.*, 2012). AlaAT is suggested to be involved in the transfer of pyruvate between different cell types (Son and Sugiyama, 1992), as has been demonstrated in the C_4 plant, *P. miliaceum*, in which cytosolic AlaAT was found in both the mesophyll and bundle sheath cells, as illustrated in Fig. 2.6. The critical role of AlaAT in carbon metabolism has been well illustrated by the recent work of Rocha *et al.* (2010a) on *Lotus japonicum*. Biochemical analyses of AlaAT using transgenic plants with silenced nodular leghaemoglobin have provided important clues of the role of AlaAT in hypoxia and its significance in carbon metabolism (Rocha *et al.*, 2010a,b). These authors showed that the production of Ala was critical to hypoxic pathways and that it provides support for carbon metabolism and energy homeostasis by linking glycolysis with the tricarboxylic acid (TCA) cycle. They also demonstrated that Ala is the only amino acid whose biosynthesis is not inhibited by nitrogen deficiency (Rocha *et al.*, 2010a,b). The authors also propose an interesting hypothesis that the metabolic activation of Ala, in combination with the modular operation of a 'split' TCA pathway, resulting in an increased accumulation of both Ala and succinate, may provide the extra energy under hypoxia. Additionally, the authors emphasized the importance of Ala in regulating the level of pyruvate, which otherwise could have a detrimental effect on energy homeostasis and the carbon pool during hypoxia.

The details of the role of AlaAT metabolism in regulating this carbon and energy homeostasis are further explained below. When the oxygen availability decreases to such a level that oxidative phosphorylation becomes limiting, the plant seeks alternative metabolic pathways to produce ATP, such as the glycolytic pathway, which produces two ATP and two pyruvate molecules per unit of hexose, while concomitantly reducing NAD^+ to NADH. In addition, in order to alleviate an increased demand for carbon and energy, several plant species respond by depleting their protein and carbohydrate contents (Thorne, 1982; Geigenberger *et al.*, 2000; Gibon *et al.*, 2002; Rolletschek *et al.*, 2003; van Dongen *et al.*, 2003, 2004; Rocha *et al.*, 2010a). However, in order to maintain glycolysis under anoxic conditions, NAD^+ must be continuously regenerated from NADH via fermentative reactions. Using pyruvate as substrate, fermentative metabolism either produces lactate, which is toxic, or produces ethanol, which may easily diffuse out of cells, leading to a considerable loss of carbon during hypoxia (Kendziorek *et al.*, 2012a).

In addition to lactate and ethanol fermentation, many plant species accumulate Ala under

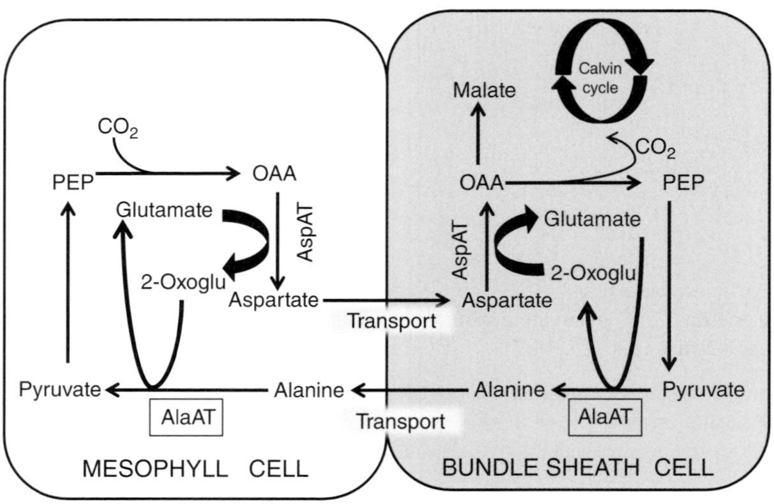

Fig. 2.6. The role of alanine aminotransaminase (AlaAT) in carbon and amino acid shuttling and metabolism. AlaAT serves as a shuttle linking carbon and nitrogen metabolism, transporting amino acids between mesophyll and bundle sheath cells in selected C_4 leaves. Note that the reversibility of the AlaAT reaction provides the flexibility of either using pyruvate as a substrate or forming it as a product. Abbreviations: AspAT, aspartate aminotransferase; OAA, oxaloacetate; 2-Oxoglu, 2-oxoglutarate; PEP, phosphoenolpyruvate. Figure adapted from Kelly *et al.*, 1976.

anoxic conditions (de Sousa and Sodek, 2003; Miyashita *et al.*, 2007) without any detrimental effects by acting as a compatible solute or regulating the pH balance in an anoxic cells (Reggiani *et al.*, 1988). However, the NAD^+ limitation may lead to an ultimate decrease in the glycolytic flux, as Ala production proceeds without producing NAD^+ (de Sousa and Sodek, 2002). For a detailed explanation of the role of AlaAT metabolism in regulating carbon and energy homeostasis, please see Section 2.9.2. Rocha *et al.* (2010a) argued that the biosynthesis of Ala using pyruvate as a substrate has several advantages. It enables the conservation of carbon skeletons that otherwise would be lost via the ethanolic fermentation pathway (Good and Muench 1992, 1993; Muench and Good, 1994). Pyruvate is not only a known activator of the alternative oxidase (AOX, part of the electron transport chain in mitochondria) (Vanlerberghe *et al.*, 1999), but has also been shown to interfere with the hypoxia-induced inhibition of respiration (Gupta *et al.*, 2009; Zabalza *et al.*, 2009). Therefore, the depletion of pyruvate may prove advantageous in reducing the oxygen demand during hypoxia. Additionally, AlaAT plays a significant role in readjusting the cells' carbon metabolism during post-hypoxia. The enhanced activity of AlaAT, but the rapid decrease in Ala, in plants recovering from hypoxia suggests the role of AlaAT in the degradation of Ala to replenish the carbon skeleton (de Sousa and Sodek, 2003; Rocha *et al.*, 2010a). This view seems to be supported by the *Arabidopsis* AlaAT1 knockout mutant (alaat1-1), which is able to accumulate more Ala during hypoxia than wild type plants, and shows a delay in the decline of accumulated Ala during the recovery period, suggesting that the elevated levels of AlaAT activity enable the conversion of the accumulated Ala back into pyruvate during the recovery phase (Miyashita *et al.*, 2007).

2.9.1.2 Roles in photorespiration

In higher plants, the major photorespiratory pathway is distributed across chloroplasts, mitochondria and peroxisomes. The diverse roles of Ala metabolism in photorespiration are reported from several detailed biochemical studies (Betsche and Eisling, 1986; Liepman and Olsen, 2001, 2003; Igarashi *et al.*, 2003, 2006; Niessen *et al.*, 2012).

Betsche and Eisling (1986) used ^{15}N-labelled glutamate and Ala to examine the nitrogen flux to glycine during photorespiratory nitrogen metabolism in oat leaves. Their results showed that the nitrogen flux from Ala to glycine was estimated to be three times higher than that from glutamate, indicating that Ala is a direct and important amino donor for photorespiratory glycine formation in oat leaves. It has been established through several studies that the photorespiratory nitrogen cycle contributes to the metabolism of certain amino acids, such as Gln, Glu (glutamine), Ser (serine) and Gly (glycine). In photorespiration, peroxisomal glutamate:glyoxylate aminotransferase (GGAT) catalyses the reaction of glutamate and glyoxylate to produce 2-oxoglutarate and glycine, thus playing a key role in the photorespiratory pathway. The reactions in this pathway are catalysed by glutamate: glyoxylate aminotransferase and serine:glyoxylate aminotransferase (SGAT). Several studies have demonstrated that an Ala aminotransferase-like protein functions as a photorespiratory GGAT and SGAT (Liepman and Olsen, 2001, 2003; Igarashi *et al.*, 2003). Transgenic studies using 42 GGAT1 overexpressing lines confirmed that the photorespiratory aminotransferase reactions catalysed by GGAT and SGAT are important regulators of amino acid content, implicating the importance of Ala metabolism in plants (Igarashi *et al.*, 2006). As discussed in Section 2.5, the sub-cellular localization of AlaAT in the peroxisomes and mitochondria has been implicated in the regulation of photorespiratory metabolism (Fukao *et al.*, 2002; Igarashi *et al.*, 2003; Liepman and Olsen, 2003; Niessen *et al.*, 2012).

The involvement of peroxisomal GGAT1 in photorespiration was confirmed by studies of mutants that are deficient in GGAT1 (Igarashi *et al.*, 2003), which showed that the transcript levels for the gene encoding GGAT1 were undetectable and that AlaAT and GGAT activities were significantly lower than in wild type plants. It is also interesting to note that the AlaAT may regulate the consumption of respiratory oxygen via the activation of AOX in mitochondria by increased concentrations of pyruvate (Gupta *et al.*, 2009). Niessen *et al.* (2012) identified two *Arabidopsis* mitochondrial alanine:glyoxylate aminotransferases, which they termed ALAATs, that link glycolate oxidation to glycine formation.

The authors describe the role of alanine aminotransferases by saying that the mitochondrial side pathway produces glycine from glyoxylate, acting as a precursor for glycine decarboxylase (GCD). RNA interference (RNAi) suppression of mitochondrial ALAAT RNAi lines showed reduced photorespiratory CO_2 release and a lower CO_2 compensation point. Mitochondria isolated from RNAi lines are incapable of converting glycolate to CO_2, whereas simultaneous overexpression of glycolate dehydrogenase (GlcDH) and ALAATs in transiently transformed tobacco leaves enhances glycolate conversion. The conservation of such biochemical reactions in monocotyledons was also confirmed by studies in rice. The authors suggest that one of the functions of ALAATs could be to convert excess glyoxylate that might diffuse from peroxisomes to the mitochondria, as previously suggested for formate resulting from the decarboxylation of glyoxylate (Wingler *et al.*, 1999; Cousins *et al.*, 2008). Alternatively, a mitochondrial GlcDH might convert photorespiratory glycolate to glyoxylate and provide substrate for ALAATs. Niessen *et al.* (2012) suggested that the mitochondrial GlcDH and ALAATs are conserved from chlorophytes that use homologous enzymes for photorespiration (Nakamura *et al.*, 2005; Stabenau and Winkler, 2005). Therefore, it is interesting that such catalytic reactions seem to be ubiquitous in algae and higher plants.

2.9.1.3 Role in nitrogen use efficiency

Apart from its direct role in alanine metabolism, AlaAT has been implicated in increasing nitrogen use efficiency (NUE) in plants. A series of transgenic studies from Good's laboratory indicate the importance of AlaAT in increasing NUE and yield in plants. In an attempt to develop nitrogen-efficient plants, rice (*O. sativa*) and canola (*Brassica napus*) were genetically engineered by introducing a barley AlaAT cDNA driven by a rice tissue-specific promoter (OsAnt1) and a root epidermal-specific promoter (Good *et al.*, 2007a,b; Shrawat *et al.*, 2008). Recent studies carried out on transgenic canola (Good *et al.*, 2007a,b; Good and Beatty, 2011) and rice (Shrawat *et al.*, 2008) with an introduced barley AlaAT gene were characterized by a much higher efficiency of nitrogen assimilation than non-transgenic plants. The overexpression of AlaAT in transgenic plants has been shown to increase yield in both rice and canola (Good *et al.*, 2007a,b; Shrawat *et al.*, 2008). In canola, this increase has been linked to improvements in NUE, as evident by increased nitrate influx, and increased biomass and seed yield under low nitrogen, but not under high nitrogen (Good *et al.*, 2007a,b; Shrawat *et al.*, 2008). In transgenic rice plants, an increase in AlaAT activity in both the shoot and the root resulted in significant increases of major transportable amino acids such as Gln, Glu and Asn, and of the total nitrogen content. Interestingly, an increase in the AlaAT activity in roots resulted in an increased flow of nitrogen into Gln, Glu and Asp, as indicated by significant increases in these key amino acids (Shrawat *et al.*, 2008). The authors proposed that the observed growth may result from a direct increase in available nitrogen in the form of amino acids and other nitrogenous compounds, or may be an effect of the increase in Gln acting as a signalling molecule (Forde and Lea, 2007). The same conjecture of increased nitrogen metabolism resulting from enhanced AlaAT activity as a means of increasing biomass and yield in canola and other dicotyledons was also mentioned earlier by Good *et al.* (2007a) in their discussion of the possibilities for using the method to increase the NUE of monocotyledons. Differences in NUE in AlaAT overexpressing plants were correlated with increased Ala levels and mobilization as well as increased uptake of nitrate in roots during vegetative growth (Good *et al.*, 2007a,b; Good and Beatty, 2011). Interestingly, because transgenic manipulations of key nitrogen metabolic genes, such as NR (nitrate reductase), NiR (nitrite reductase), GS (glutamine synthase) and GOGAT (glutamate synthase), which were hypothesized to affect NUE, did not result in consistently nitrogen use efficient phenotypes, Good *et al.* (2007a,b) and Shrawat *et al.* (2008) indicate that AlaAT may significantly modulate NUE in plants to increase plant growth and yield (McAllister *et al.*, 2012).

2.9.2 Role in stress biology

An intriguing aspect of plant response to unfavourable environmental conditions is the accumulation of Ala and the regulation of AlaAT activity, as reported by work from the various laboratories referenced below. Because Ala can function as a compatible solute, its accumulation in high concentrations will most likely cause no detrimental effects. In plants, the intracellular levels of Ala are generally low but have been found to increase several fold under several suboptimal conditions, such as under flooding (Limami *et al.*, 2008) in roots, hypoxia and anoxia (de Sousa and Sodek, 2003; Miyashita *et al.*, 2007; Rocha *et al.*, 2010a,b), nitrogen deficit (Puiatti and Sodek, 1999), water deficit (Drossopoulos *et al.*, 1985), low temperatures (Mazzucotelli *et al.*, 2006), etc. The exact physiological significance of AlaAT induction with the concomitant increased production of Ala under these suboptimal conditions is debatable. However, this section illustrates the possible physiological roles of AlaAT in ameliorating stress. As mentioned above, Ala may function as a compatible solute and contribute to the osmotic adjustment of cells by stabilizing the quaternary structure of proteins and membranes (Yancey, 2005). It is suggested that the production of Ala may help to regulate the pH balance within an anoxic cell (Reggiani *et al.*, 1988). Additionally, recent findings by Rocha *et al.* (2010a) indicate that the primarily function of Ala accumulation is regulation of the pyruvate level and, therefore, control of the rate of respiration and energy homeostasis, which probably contributes to reprogramming the plants to recover from stress. As AlaAT catalyses an equilibrium reaction, it is likely to play a flexible role in both the biosynthesis and the breakdown of Ala, based on *in planta* substrate availability and physiological requirements. There is evidence of this hypothesis from *Medicago*, in which Miyashita *et al.* (2007) described the rapid conversion of Ala during the post-hypoxia recovery period in roots. The nearest homologue of the *Arabidopsis* AlaAT enzyme in *M. truncatula* was shown to be induced in the embryo axis of young seedlings under hypoxia and to catalyse the synthesis of Ala instead of using it as an amino donor. However, the same enzyme catalyses glutamate synthesis from Ala and oxoglutarate under non-hypoxic conditions (Ricoult *et al.*, 2006), so demonstrating the biochemical flexibility of the enzyme.

2.9.2.1 Roles in hypoxia

Oxygen deprivation or hypoxia caused by flooding or waterlogging of the soil is one of the major environmental factors that are detrimental to plant growth and development. Plants respond to oxygen deprivation by a combination of several adaptive responses, together known as low oxygen escape syndrome (Bailey-Serres and Voesenek, 2008). During hypoxia, the glycolysis and fermentation pathways are stimulated, resulting in the production of ethanol and lactate (Good and Muench, 1993). In addition to lactate and ethanol fermentation, many plant species accumulate Ala during anoxic conditions (de Sousa and Sodek, 2003; Miyashita *et al.*, 2007). Several reports (de Sousa and Sodek 2003; Ricoult *et al.*, 2006; Miyashita *et al.*, 2007; Rocha *et al.*, 2010a,b) indicate the importance of Ala metabolism in mitigating some of the biochemical damages caused by hypoxia. Refer also to the previous (main) Section 2.9.2 for plausible stress-ameliorating roles of Ala.

Flooding induced increase in the production of Ala has been explained as a metabolic outcome of the rapid induction of the expression of the gene for AlaAT, with a concomitant increase in enzyme activity during flooding (Good and Crosby, 1989; Good and Muench, 1992; Muench and Good, 1994). Hypoxia-induced AlaAT transcript levels or activity have been reported in barley, maize and soybean roots (Good and Muench, 1992, 1993; Muench *et al.*, 1998; de Sousa and Sodek 2003) and in seedlings of *M. truncatula* (Ricoult *et al.*, 2006). Interestingly, Ala accumulation was also observed in T-DNA knockout plants where no AlaAT activity was detected (Miyashita *et al.*, 2007). A significant increase in the concentration of two amino acids, Ala and GABA, has been observed in roots of oxygen-deficient plants (Roberts *et al.*, 1992; Reggiani *et al.*, 2000). Therefore, it seems likely that Ala production does not depend solely on the activity of AlaAT but also on an alternative reaction, such as that catalysed by γ-aminobutyric acid transaminase (GABA-T) using pyruvate as co-substrate. Like the AlaAT knockout plants, GABA-T null mutants accumulated only slightly less Ala upon hypoxia compared with wild type plants (Miyashita and Good, 2008), suggesting the partial redundancy of AlaAT and GABA-T under hypoxic conditions (Rocha *et al.*, 2010a,b). The enhanced activity of AlaAT seems to be equally important for plants recovering from hypoxia. In roots of soybean, AlaAT activity continued to increase during hypoxia (de Sousa and Sodek, 2003), although the level of Ala decreases rapidly to its pre-hypoxic level, suggesting that AlaAT is involved in the degradation of Ala during recovery (Rocha *et al.*, 2010b). Miyashita *et al.* (2007) showed that the *Arabidopsis* AlaAT1 knockout mutant (alaat1-1) was able to accumulate more Ala during hypoxia than wild type plants, and a decline in accumulated Ala was delayed during recovery period. Therefore, it appears that elevated levels of AlaAT activity enable conversion of the accumulated Ala back into pyruvate during the recovery phase (Miyashita *et al.*, 2007).

Rocha *et al.* (2010b) observed pronounced accumulation of succinate and Ala, independent of the nitrogen status of the plant, whereas all other amino acids were strongly influenced by the nitrogen status during flooding (Rocha *et al.*, 2010b). Based on this observation, the authors proposed a metabolic model that explains the role of Ala accumulation during waterlogging via reorganization of the TCA cycle in a modular manner in which ATP is produced from an oxidizing pathway from oxoglutarate to succinate, and a parallel reducing branch that oxidizes NADH during hypoxia. Additionally, the model suggests that the metabolism of Ala will prevent pyruvate accumulation, thereby facilitating the continued operation of glycolysis during waterlogging. The biochemical flexibility of AlaAT is further exemplified in *Arabidopsis* and *Medicago*. AlaAT has been shown to contribute to the rapid conversion of Ala during the post-hypoxia recovery period in *Arabidopsis* roots (Miyashita *et al.*, 2007). In *M. truncatula*, AlaAT has been shown to be induced in the embryo axis of young seedlings under hypoxia and to catalyse synthesis of Ala instead of using it as an amino donor. However, the same enzyme catalyses glutamate synthesis from Ala and oxoglutarate under non-hypoxic conditions (Ricoult *et al.*, 2006). This indicates the flexible roles of AlaAT as demanded by various environmental conditions. Biochemical analyses of AlaAT using transgenic plants with silenced nodular leghaemoglobin provide important clues of the role of the enzyme in hypoxia and its significance in carbon metabolism (Rocha *et al.*, 2010a), as explained in details in Section 2.9.1.1.

2.9.2.2 Other abiotic and biotic stresses

Several reports indicate that biotic stress increases the levels of intracellular Ala in plants as in flooding (Limami *et al.*, 2008), hypoxia and anoxia (de Sousa and Sodek, 2003; Miyashita *et al.*, 2007; Rocha *et al.*, 2010a,b), nitrogen deficit (Puiatti and Sodek, 1999), water deficit (Drossopoulos *et al.*, 1985), low temperature (Mazzucotelli *et al.*, 2006), etc. AlaAT has been implicated in the maintenance of the functional integrity of proteins as intracellular pH conditions change (Reggiani *et al.*, 1988), and may also be involved in maintaining energy homeostasis and increased carbon use efficiency in response to biotic stress conditions (Rocha *et al.*, 2010a). Temperature-induced changes in Ala metabolism were reported from several plant species and have been implicated in temperature-induced stress tolerance in cowpea, barley and soybean (Wallace *et al.*, 1984; Mayer *et al.*, 1990; Mazzucotelli *et al.*, 2006).

Kim *et al.* (2005) reported a *Capsicum annuum* AlaAT (CaAlaAT1) encoding a putative Ala aminotransferase. The gene exhibited an organ-specific expression pattern and the transcript accumulated abundantly in ripe fruit tissues and senescing leaves. There was a marked increase in the transcript abundance in response to infection by *Xanthomonas campestris*. The expression of the *CaAlaAT1* gene was triggered by salicylic acid and ethylene, but not by methyl jasmonate. Therefore, AlaAT may be one of the components in the plant's defence signalling pathway against pathogens (Kim *et al.*, 2005). Taler *et al.* (2004) showed that transgenic melon plants overexpressing either of the enzymatic resistance (*eR*) genes, *At1* and *At2*, displayed enhanced activity of glyoxylate aminotransferases and remarkable resistance against *Pseudoperonospora cubensis*, indicating a new resource for developing downy mildew-resistant melon varieties.

2.10 Conclusions

As discussed in the preceding section, alanine aminotransferase (AlaAT, EC 2.6.1.2) is one of the most interesting and significant enzymes of nitrogen metabolism. The enzyme catalyses the reversible transfer of an amino group from alanine (Ala) to 2-oxoglutarate to form pyruvate and glutamate, and the reaction proceeds with an equilibrium constant of nearly one, providing an interesting metabolic flexibility for the biosynthesis as well as catabolism of alanine. This is significant in terms of plant metabolism, as evident from several studies, especially the role of AlaAT in hypoxia in conserving the C_3 carbon skeleton and maintaining energy homeostasis (Niessen *et al.*, 2012; Rocha *et al.*, 2010a). Additionally, the role of AlaAT in executing nitrogen economy in plants, as evident from several transgenic studies, provides significant insight into its role in modulating nitrogen metabolism and consequently having an impact on plants growth and development (Shrawat *et al.*, 2008; Good and Beatty, 2011). As evident from several studies, AlaAT is an ubiquitous enzyme, conserved across different phyla. The occurrence of multiple homologues, in different tissues and sub-cellular compartments, signifies the undisputed roles of the enzyme in affecting different physiological processes across the life cycle of plants. This chapter provides a comprehensive summary of the wide spectrum of biochemical and metabolic functions of AlaAT, portraying insightful details into its molecular functions, not only in carbon and nitrogen metabolism, but also in ameliorating stress in plants (see Fig. 2.7). It is interesting to note that at least 44 gene sequences encoding putative aminotransferases have been found in the *Arabidopsis* genome, although most of the aminotransferase-like sequences remain uncharacterized so far. The recent advances in kinetic studies and the availability of a crystal structure of AlaAT (Duff *et al.*, 2012; McAllister *et al.*, 2013) pave the path for further elucidation of the functionalities and role of the enzyme in higher plants.

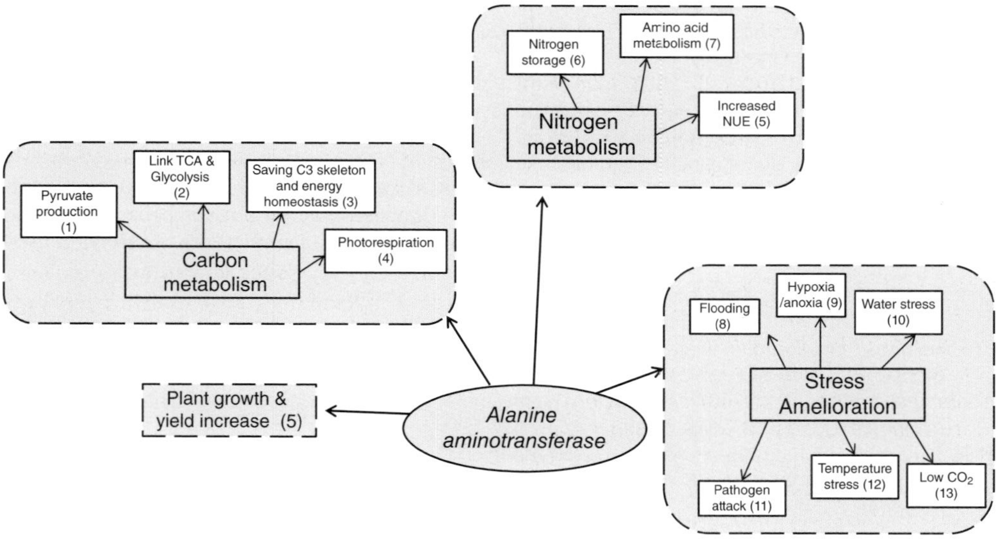

Fig. 2.7. The diversity of functions of alanine aminotransaminase (AlaAT) in plants. Dashed boxes represent the major areas of AlaAT functions. Several subgroups in each of these categories are represented in white boxes with a solid outline. The numbers in parenthesis represent the most recent reference(s). Please also refer to the text for more references in Section 2.9. I1) Givan (1980); (2) Miyashita *et al.* (2007), Rocha *et al.* (2010a); (3) Ricoult *et al.* (2006), deSousa and Sodek (2003), Rocha *et al.* (2010a); (4) Liepman and Olsen (2003), Niessen *et al.* (2012); (5) Shrawat *et al.* (2008), Good and Beatty (2011); (6) Givan (1980), Waters *et al.* (1998); (7) Good *et al*, (2007a,b); (8) Limami *et al.* (2008); (9) Muench and Good (1994), deSousa and Sodek (2003), Miyashita *et al.* (2007); (10) Drosssopoulos *et al.* (1985); (11) Kim *et al.* (2005), Taler *et al.* (2004); (12) Wallace *et al.* (1984), Mayer *et al.* (1990), Mazzucotelli *e al.* (2006); and (13) Chen *et al.* (1996). Abbreviations: NUE, nitrogen use efficiency; TCA, tricarboxylic acid cycle.

References

Abeles, R.H. and Haycock, A.L. (1976) Suicide enzyme inactivators. *Accounts of Chemical Research* 9, 313–319.

Arnon, D.I. (1949) Copper enzymes in isolated chloroplasts; polyphenolase in *Beta vulgaris*. *Plant Physiology* 24, 1–15.

Bailey-Serres, J. and Voesenek, L.A. (2008) Flooding stress: acclimations and genetic diversity. *Annual Review of Plant Biology* 59, 313–339.

Betsche, T. and Eisling, R. (1986) Refixation of photorespiratory ammonia and the role of alanine in photorespiration: studies with ^{15}N. *Plant and Soil* 91, 367–371.

Bhadauria, V., Banniza, S., Vandenberg, A., Selvaraj, G. and Wei Y. (2012) Peroxisomal alanine:glyoxylate aminotransferase AGT1 is indispensable for appressorium function of the rice blast pathogen, *Magnaporthe oryzae*. *PLoS One* 7(4):e36266.

Burnett, G., Marcotte, P. and Walsh, C. (1980) Mechanism-based inactivation of pig heart L-alanine transaminase by L-propargylglycine. Half-site reactivity. *The Journal of Biological Chemistry* 255, 3487–3491.

Chen, Z.Y., Burow, M.D., Mason, C.B. and Moroney, J.V. (1996) A low-CO_2-inducible gene encoding an alanine:alpha-ketoglutarate aminotransferase in *Chlamydomonas reinhardtii*. *Plant Physiology* 112, 677–684.

Cornell, N.W., Zurendonk, P.F., Kerich, M.J. and Straight, C.B. (1984) Selective inhibition of alanine aminotransferase and aspartate aminotransferase in rat hepatocytes. *Biochemical Journal* 220, 707–716.

Cousins, A.B., Pracharoenwattana, I., Zhou, W., Smith, S.M. and Badger, M.R. (2008) Peroxisomal malate dehydrogenase is not essential for photorespiration in *Arabidopsis* but its absence causes an increase in the stoichiometry of photorespiratory CO_2 release. *Plant Physiology* 148, 786–795.

de Sousa, C.A.F. and Sodek, L. (2002) The metabolic response of plants to oxygen deficiency. *Brazilian Journal of Plant Physiology* 14, 83–94.

de Sousa, C.A.F. and Sodek, L. (2003) Alanine metabolism and alanine aminotransferase activity in soybean (*Glycine max*) during hypoxia of the root system and subsequent return to normoxia. *Environmental and Experimental Botany* 50, 1–8.

Doolittle, R.F. (1989) Redundancies in protein sequences. In: Fasman, G.D. (ed.) *Prediction of Protein Structures and the Principles of Protein Conformation*. Plenum, New York, pp. 599–623.

Drossopoulos, J.B., Karamanos, A.J. and Niavis, A. (1985) Changes in amino acid compounds during the development of two wheat cultivars subjected to different degrees of water stress. *Annals of Botany* 56, 291–305.

Duff, S.M.G., Rydel, T.J., McClerren, A.L., Zhang, W., Li, J.Y., Sturman, E.J., Halls, C., Chen, S., Zeng, J., Peng, J., Kretzler, C.N. and Evdokimov, A. (2012) The enzymology of alanine aminotransferase (AlaAT) isoforms from *Hordeum vulgare* and other organisms, and the HvAlaAT crystal structure. *Archives of Biochemistry and Biophysics* 528, 90–101.

Forde, B.G. and Lea, P.J. (2007) Glutamate in plants: metabolism, regulation, and signaling. *Journal of Experimental Botany* 58, 2339–2358.

Fujii, K. and Yoshii, I. (1995) Purification and properties of L-alanine aminotransferase from germinating castor bean endosperms. *Bulletin of the University of Osaka Prefecture, Ser. B, Agriculture and Life Sciences* 47, 95–104

Fukao, Y., Hayashi, M. and Nishimura, M. (2002) Proteomic analysis of leaf peroxisomal proteins in greening cotyledons of *Arabidopsis thaliana*. *Plant Cell and Physiology* 43, 689–696.

Furbank, R.T. (1998) C_4 pathway. In: Raghavendra, A.S. (ed.) *Photosynthesis. A Comprehensive Treatise*. Cambridge University Press, Cambridge, UK, pp. 123–135.

Geigenberger, P., Fernie, A.R., Gibon, Y., Christ, M. and Stitt, M. (2000) Metabolic activity decreases as an adaptive response to low internal oxygen in growing potato tubers. *Biological Chemistry* 381, 723–740.

Genetet, I., Martin, F. and Stewart, G.R. (1984) Nitrogen assimilation in mycorrhizas. Ammonium assimilation in the starved ectomycorrhizal fungus *Cenococcum graniforme*. *Plant Physiology* 76, 395–399.

Gibon, Y., Vigeolas, H., Tiessen, A., Geigenberger, P. and Stitt, M. (2002) Sensitive and high throughput metabolite assays for inorganic pyrophosphate, ADPGlc, nucleotide phosphates, and glycolytic intermediates based on a novel enzymatic cycling system. *The Plant Journal* 30, 221–235.

Givan, C.V. (1980) Aminotransferases in higher plants. In: Miflin B.J. (ed.) *The Biochemistry of Plants, Vol. 5*. Academic Press, New York, pp. 329–357.

Goil, M.M. (1978) Effect of some -SH and other reagents on aspartate aminotransferase and L-alanine aminotransferase of *Paramphistomum explanatum* Fischoeder, 1901. *Zeitschrift für Parasitenkunde* 55, 59–62.

Good, A. and Beatty, P. (2011) Biotechnological approaches to improving nitrogen use efficiency in plants: alanine aminotransferase as a case study. In: Hawkesford, M.J. and Barraclough, P. (eds) *The Molecular and Physiological Basis of Nutrient Use Efficiency in Crops*. Wiley, Chichester, UK, pp. 165–192.

Good, A.G. and Crosby, W.L. (1989) Anaerobic induction of alanine aminotransferase in barley root tissue. *Plant Physiology* 90,1305–1309.

Good, A.G. and Muench, D.G. (1992) Purification and characterization of an anaerobically induced alanine aminotransferase from barley roots. *Plant Physiology* 99, 1520–1525.

Good, A.G. and Muench, D.G. (1993) Long term anaerobic metabolism in root tissue: metabolic products of pyruvate metabolism. *Plant Physiology* 101, 1163–1168.

Good, A.G., DePauw, M., Kridl, J.C., Theodoris, G. and Shrawat, A.K. (2007a) Nitrogen-efficient monocot plants. US [Patent] 20070162995. Available at: http://www.google.com/patents/US20070162995 (accessed 25 July 2014).

Good, A.G., Johnson, S.J., DePauw, M., Carrol, R.T., Savidov, N., Vidmar, J., Lu, Z., Taylor, G. and Stroeher, V. (2007b) Engineering nitrogen use efficiency with alanine aminotransferase. *Canadian Journal of Botany* 85, 252–262.

Gupta, K.J., Zabalza, A. and van Dongen, J.T. (2009) Regulation of respiration when the oxygen availability changes. *Physiologia Plantarum* 137, 383–391.

Hatch, M.D. (1987) C_4 photosynthesis: a unique blend of modified biochemistry, anatomy and ultrastructure. *Biochimica et Biophysica Acta* 895, 81–106.

Hatch, M.D. and Mau, S.L. (1972) Activity, location and role of aspartate aminotransferase isoenzymes in leaves with C_4 pathway biosynthesis. *Archives of Biochemistry and Biophysics* 156, 195–206.

Hwang, B.-Y., Cho, B.-K., Yun, H., Koteshwar, K. and Kim, B.-G. (2005) Revisit of aminotransferase in the genomic era and its application to biocatalysis. *Journal of Molecular Catalysis B: Enzymatic* 37, 47–55.

Igarashi, D., Miwa, T., Seki, M., Kobayashi, M., Kato, T., Tabata, S., Shinozaki, K. and Ohsumi, C. (2003) Identification of photorespiratory glutamate:glyoxylate aminotransferase (GGAT) gene in *Arabidopsis*. *The Plant Journal* 33, 975–987.

Igarashi, D., Tsuchida, H., Miyao, M. and Ohsumi, C. (2006) Glutamate:glyoxylate aminotransferase modulates amino acid content during photorespiration. *Plant Physiology* 142, 901–910.

Ireland, R.J. and Lea, P.J. (1999) The enzymes of glutamine, glutamate, asparagine, and aspartate metabolism. In: Singh, B.K. (ed.) *Plant Amino Acids: Biochemistry and Biotechnology*. Marcel Dekker, New York, pp. 49–109.

Kameya, M., Arai, H., Ishii, M. and Igarashi, Y. (2010) Purification of three aminotransferases from *Hydrogenobacter thermophilus* TK-6 – novel types of alanine or glycine aminotransferase. *FEBS Journal* 277, 1876–1885.

Kelly, G.L., Latzko, E. and Gibbs, M. (1976) Regulatory aspects of photosynthetic carbon metabolism. *Annual Review of Plant Physiology* 27, 181–205.

Kendziorek, M., Paszkowski, A. and Zagdańska, B. (2012a) Biochemical characterization and kinetic properties of alanine aminotransferase homologues partially purified from wheat (*Triticum aestivum* L.) *Phytochemistry* 82, 7–14.

Kendziorek, M., Paszkowski, B. and Zagdańska, B. (2012b) Differential regulation of alanine aminotransferase homologues by abiotic stresses in wheat (*Triticum aestivum* L.) seedlings. *Plant Cell Reports* 3, 1105–1117.

Kikuchi, H., Hirose, S., Toki, S., Akama, K. and Takaiwa, F. (1999) Molecular characterization of a gene for alanine aminotransferase from rice (*Oryza sativa*). *Plant Molecular Biology* 39, 149–159.

Kim, K.J., Park, C.J., An, J.M., Ham, B.K., Lee, B.J. and Paek, K.H. (2005) CaAlaAT1 catalyses the alanine:2-oxoglutarate aminotransferase reaction during the resistance response against *Tobacco mosaic virus* in hot pepper. *Planta* 221, 857–867.

Lain-Guelbenzu, B., Cardenas, J. and Munoz-Blanco, J. (1991) Purification and properties of L-alanine aminotransferase from *Chlamydomonas reinhardtii*. *European Journal of Biochemistry* 202, 881–887.

Liepman, A.H. and Olsen, L.J. (2001) Peroxisomal alanine:glyoxylate aminotransferase (AGT1) is a photorespiratory enzyme with multiple substrates in *Arabidopsis thaliana*. *The Plant Journal* 25, 487–498.

Liepman, A.H. and Olsen, L.J. (2003) Alanine aminotransferase homologs catalyze the glutamate: glyoxylate aminotransferase reaction in peroxisomes of *Arabidopsis*. *Plant Physiology* 131, 215–227.

Liepman, A.H. and Olsen, L.J. (2004) Genomic analysis of aminotransferases in Arabidopsis thaliana. *Critical Reviews in Plant Sciences* 23, 73–89.
Limami, A.M., Glévarec, G., Ricoult, C., Cliquet, J.B. and Planchet, E. (2008) Concerted modulation of alanine and glutamate metabolism in young *Medicago truncatula* seedlings under hypoxic stress. *Journal of Experimental Botany* 59, 2325–2335.
Marangoni, B. (2003) *Enzyme Kinetics: A Modern Approach*. Wiley, Hoboken, New Jersey.
Markovic-Housley, Z., Schirmer, T., Hohenester, E., Khomutov A.R., Khomutov, R.M. Karpeisky, M.Y., Sandmeier, E., Christen, P. and Jansonius, J.N. (1996) Crystal structures and solution studies of oxime adducts of mitochondrial aspartate aminotransferase. *European Journal of Biochemistry* 236, 1025–1032.
Mayer, R.R., Cherry, J.H. and Rhodes, D. (1990) Effects of heat shock on amino acid metabolism of cowpea cells. *Plant Physiology* 94, 796–810.
Mazzucotelli, E., Tartari, A., Cattivelli, L. and Forlani, G. (2006) Metabolism of g-aminobutyric acid during cold acclimation and freezing and its relationship to frost tolerance in barley and wheat. *Journal of Experimental Botany* 57, 3755–3766.
McAllister, C.H., Beatty, P.H. and Good, A.G. (2012) Engineering nitrogen use efficient crop plants; the current status. *Plant Biotechnology Journal* 10, 1011–1025.
McAllister, C.H., Facette, M., Holt, A. and Good, A.G. (2013) Analysis of the enzymatic properties of a broad family of alanine aminotransferases. *PLoS One* 8(2):e55032.
Mehta, P.K., Hale, T.I. and Christen, P. (1993) Aminotransferases: demonstration of homology and division into evolutionary subgroups. *FEBS Journal* 214, 549–561.
Meish, H.U., Hoffman, H. and Reinle, W. (1983) Biosynthesis of chlorophyll precursors in green algae. Purification and characterisation of L-glutamate:4,5-dioxovaleric acid aminotransferase from *Chlorella fusca*. *Biochimica et Biophysica Acta* 743, 281–289.
Miyashita, Y. and Good, A.G. (2008) NAD(H)-dependent glutamate dehydrogenase is essential for the survival of *Arabidopsis thaliana* during dark-induced carbon starvation. *Journal of Experimental Botany* 59, 667–680.
Miyashita, Y., Dolferus, R., Ismind, K.P. and Good, A.G. (2007) Alanine aminotransferase catalyses the breakdown of alanine after hypoxia in *Arabidopsis thaliana*. *The Plant Journal* 49, 1108–112.
Muench, D.G. and Good, A.G. (1994) Hypoxically inducible barley alanine aminotransferase: cDNA cloning and expression analysis. *Plant Molecular Biology* 24, 417–427.
Muench, D., Christopher, M.E. and Good, A.G. (1998) Cloning and expression of a hypoxic and nitrogen inducible maize alanine aminotransferase gene. *Physiologia Plantarum* 103, 503–512.
Murray, D. R. and Kennedy, I. R. (1980) Changes in activities of enzymes of nitrogen metabolism in seed coats and cotyledons during embryo development in pea seeds. *Plant Physiology* 66, 782–786.
Nakamura, Y., Kanakagiri, S., Van, K., He, W. and Spalding, M. (2005) Disruption of the glycolate dehydrogenase gene in the high-CO_2-requiring mutant HCR89 of *Chlamydomonas reinhardtii*. *Canadian Journal of Botany* 83, 820–833.
Niessen, M., Krause, K., Horst, I., Staebler, N., Klaus, S., Gaertner, S., Kebeish, R., Araujo, W. L., Fernie, A. R. and Peterhansel, C. (2012) Two alanine aminotransferases link mitochondrial glycolate oxidation to the major photorespiratory pathway in *Arabidopsis* and rice. *Journal of Experimental Botany* 63, 2705–2716.
Noguchi, T. and Fujiwara, S. (1982) Development of glutamate:glyoxylate aminotransferase in the cotyledons of cucumber (*Cucumis sativus*) seedlings. *Biochemical Journal* 201, 209–214.
Noguchi, T. and Hayashi, S. (1980) Peroxisomal localization and properties of tryptophan aminotransferase in plant leaves. *The Journal of Biological Chemistry* 255, 2267–2269.
Noguchi, T. and Hayashi, S. (1981) Plant leaf alanine:2-oxoglutarateaminotransferase. Peroxisomal localization and identity with glutatmate glyoxalate aminotransferase. *Biochemical Journal* 195, 235–239.
Orzechowski, S., Socha-Hanc, J. and Paszkowski, A. (1999a) Purification and properties of alanine aminotransferase from maize (*Zea mays* L.) leaves. *Acta Physiologiae Plantarum* 21, 323–330.
Orzechowski, S., Socha-Hanc, J. and Paszkowski, A. (1999b) Alanine aminotransferase and glycine aminotransferase from maize (*Zea mays* L.) leaves. *Acta Biochimica Polonica* 46, 447–457.
Otter, T., Penther, J.M. and Mohr, H. (1992) Control of the appearance of alanine aminotransferase in the Scots pine (*Pinus silvestris* L.) seedling. *Planta* 188, 376–383.
Paszkowski, A. and Niedzielska, A. (1989) Glutamate:glyoxylate aminotransferase from the seedlings of rye (*Secale cereale* L.). *Acta Biochimica Polonica* 36, 17–29.
Peisach, D., Chipman, D.M., Van Ophem, P.W., Manning, J.M. and Ringe, D.J. (1998) D-Cycloserine inactivation of D-amino acid aminotransferase leads to a stable noncovalent protein complex with an aromatic cycloserine-PLP derivative. *Journal of the American Chemical Society* 120, 2268–2274.

Penther, J.M. (1991) Analysis of alanine and aspartate aminotransferase isoforms in mustard (*Sinapis alba* L.) cotyledons. *Journal of Chromatography* 587, 101–108.

Puiatti, M. and Sodek, L. (1999) Water logging affects nitrogen transport in the xylem of soybean. *Plant Physiology and Biochemistry* 37, 767–773.

Rando R. R. (1975) Mechanisms of action of naturally occurring irreversible enzyme inhibitors. *Accounts of Chemical Research* 8, 281–288.

Rech, J. and Crouzet, J. (1974) Partial purification and initial studies of the tomato L-alanine:2-oxoglutarate aminotransferase. *Biochimica et Biophysica Acta* 350, 392–399.

Reggiani, R., Cantú, C.A., Brambilla, I. and Bertani, A. (1988) Accumulation and interconversion of amino acids in rice roots under anoxia. *Plant Cell Physiology* 29, 981–987.

Reggiani, R., Nebuloni, M., Mattana, M. and Brambilla, I. (2000) Anaerobic accumulation of amino acids in rice roots: role of the glutamine synthetase/glutamate synthase cycle. *Amino Acids* 18, 207–217.

Ricoult, C., Echeverria, L.O., Cliquet, J.-B. and Limami, A.M. (2006) Characterization of alanine aminotransferase (AlaAT) multigene family and hypoxic response in young seedlings of the model legume *Medicago truncatula*. *Journal of Experimental Botany* 57, 3079–3089.

Roberts, J.K.M., Hooks, M., Miaullis, A.P., Edwards, S. and Webster, C (1992) Contribution of malate and amino acid metabolism to cytoplasmic pH regulation in hypoxic maize root tips studied using nuclear magnetic resonance spectroscopy. *Plant Physiology* 98, 480–487.

Rocha, M., Licausi, F., Araújo, W.L., Nunes-Nesi, A., Sodek, L., Fernie, A.R. and van Dongen, J.T. (2010a) Glycolysis and the tricarboxylic acid cycle are linked by alanine aminotransferase during hypoxia induced by waterlogging of *Lotus japonicus*. *Plant Physiology* 152, 1501–1513.

Rocha, M., Sodek, L., Licausi, F., Hameed, M.W., Dornelas, M.C. and van Dongen, J.T. (2010b) Analysis of alanine aminotransferase in various organs of soybean (*Glycine max*) and independence of different nitrogen fertilisers during hypoxic stress. *Amino Acids* 39, 1043–1053.

Rolletschek, H., Weber, H. and Borisjuk, L. (2003) Energy status and its control on embryogenesis of legumes. Embryo photosynthesis contributes to oxygen supply and is coupled to biosynthetic fluxes. *Plant Physiology* 132, 1196–1206.

Scheible, W.R., Gonzalez-Fontes, A., Lauerer, M., Muller-Rober, B., Caboche, M. and Stitt, M. (1997) Nitrate acts as a signal to induce organic acid metabolism and repress starch metabolism in tobacco. *The Plant Cell* 9, 783–798.

Schwarzenbacher, R., Jaroszewski, L., von Delft, F., Abdubek, P., Ambling, E. and Biorac, T. (2004) Crystal structure of an aspartate aminotransferase (TM1255) from *Thermotoga maritima* at 1.90 Å resolution. *Proteins: Structure, Function, and Bioinformatics* 55, 759–763.

Sechley, K.A , Yamaya, T. and Oaks, A. (1992) Compartmentation of nitrogen assimilation in higher plants. *International Review of Cytology* 136, 85–163.

Seebauer, J.R., Moose, S.P., Fabbri, B.J., Crossland, L.D. and Below, F.E. (2004) Amino acid metabolism in maize earshoots. Implications for assimilate preconditioning and nitrogen signaling. *Plant Physiology* 136, 4326–4334.

Shrawat, A.K., Carroll, R.T., DePauw, M., Taylor, G.J. and Good, A.G. (2008) Genetic engineering of improved nitrogen use efficiency in rice by the tissue-specific expression of alanine aminotransferase. *Plant Biotechnology Journal* 6, 722–732.

Son, D. and Sugiyama,Y. (1992) Molecular cloning of an alanine aminotransferase from NAD-malic enzyme type C_4 plant *Paniculum miliaceum*. *Plant Molecular Biology* 20, 705–713.

Son, D., Jo, J. and Sugiyama, T. (1991) Purification and characterization of alanine aminotransferase from *Panicum miliaceum* leaves. *Archives of Biochemistry and Biophysics* 289, 262–266.

Stabenau, H. and Winkler, U. (2005) Glycolate metabolism in green algae. *Physiologia Plantarum* 123, 235–245.

Stewart, G.R. and Larher, F. (1980) Accumulation of amino acids and related compounds in relation to environmental stress. In: Miflin B.J. (ed.) *The Biochemistry of Plants, Vol. 5*. Academic Press, New York, pp. 609–635.

Stitt, M., Müller, C., Matt, P., Gibon, Y., Carillo, P., Morcuende, R., Scheible, W.R. and Krapp, A. (2002) Steps towards an integrated view of nitrogen metabolism. *Journal of Experimental Botany* 53, 959–970.

Streeter, J.G. and Thompson, J.F. (1972) Anaerobic accumulation of gamma-aminobutyric acid and alanine in radish leaves (*Raphanus sativus* L.). *Plant Physiology* 49, 572–578.

Taler, D., Galperin, M., Benjamin, I., Cohen, Y. and Kenigsbuch, D. (2004) Plant *eR* genes that encode photorespiratory enzymes confer resistance against disease. *The Plant Cell* 16, 172–184.

Thorne, J.H. (1982) Temperature and oxygen effects on C_{14}-photosynthate unloading and accumulation in developing soybean seeds. *Plant Physiology* 69, 48–53.

Truszkiewicz, W. and Paszkowski. A. (2005) Some structural properties of plant serine:glyoxylate aminotransferase. *Acta Biochimica Polonica* 52, 527–534.

Tsubota, H. and Hoshino, S. (1969) Transaminase activity in sheep rumen content. *Journal of Dairy Science* 52, 2024–2028.

Umemura, I., Yanagiya, K., Komatsubara, S., Sato, T. and Tosa, T. (1994) Purification and some properties of alanine aminotransferase from *Candida maltos*. *Bioscience, Biotechnology and Biochemistry* 58, 283–287.

Valle, E.M. and Heldt, H.W. (1991) Alanine synthesis by bundle sheath cells of maize. *Plant Physiology* 95, 839–845.

van Dongen, J.T., Schurr, U., Pfister, M. and Geigenberger, P. (2003) Phloem metabolism and function have to cope with low internal oxygen. *Plant Physiology* 131, 1529–1543.

van Dongen, J.T., Roeb, G.W., Dautzenberg, M., Froehlich, A., Vigeolas, H., Minchin, P.E. and Geigenberger, P. (2004) Phloem import and storage metabolism are highly coordinated by the low oxygen concentrations within developing wheat seeds. *Plant Physiology* 135, 1809–1821.

Vanlerberghe, G.C. and Turpin, D.H. (1990) Anaerobic metabolism in the N-limited green algae *Selenastrum minutum*. II. Assimilation of ammonium by anaerobic cells. *Plant Physiology.* 94, 1124–1130.

Vanlerberghe, G.C., Yip, J,Y. and Parsons, H.L. (1999) *In organello* and *in vivo* evidence of the importance of the regulatory sulfhydryl/disulfide system and pyruvate for alternative oxidase activity in tobacco. *Plant Physiology* 121, 793–803.

Vedavathi, M., Girish, K.S. and Kumar, M.K. (2004) Isolation and characterization of cytosolic alanine aminotransferase isoforms from starved rat liver. *Molecular and Cellular Biochemistry* 267, 13–23.

Vedavathi, M., Girish, K.S. and Karuna, M.K. (2006) A novel low molecular weight alanine aminotransferase from fasted rat liver. *Biochemistry* 71, 105–112.

Wallace, W., Secor, J. and Schrader, L.E. (1984) Rapid accumulation of γ-aminobutyric acid and alanine in soybean leaves in response to an abrupt transfer to lower temperature, darkness, or mechanical manipulation. *Plant Physiology* 75, 170–175.

Ward, D.E., Kengen, S.W., van Der Oost, J. and de Vos, W.M. (2000) Purification and characterization of the alanine aminotransferase from the hyperthermophilic archaeon *Pyrococcus furiosus* and its role in alanine production. *Journal of Bacteriology* 182, 2559–2566.

Waters, J.K., Hughes, B.L. II, Purcell, L.C., Gerhardt, K.O., Mawhinney, T.P. and Emerich, D.W. (1998) Alanine, not ammonium, is excreted from N_2-fixing soybean nodule bacteroids. *Proceedings of the National Academy of Sciences of the United States of America* 95, 12038–12042.

Wingler, A., Lea, P.J. and Leegood, R.C. (1999) Photorespiratory metabolism of glyoxylate and formate in glycine-accumulating mutants of barley and *Amaranthus edulis*. *Planta* 207, 518–526.

Wiśniewski, P.J., Szklarczyk, J.M. and Paszkowski, M.A. (2006) L-Alanine:2-oxoglutarate aminotransferase isoenzymes from *Arabidopsis thaliana* leaves. *Acta Physiologiae Plantarum* 28, 577–588.

Yamaguchi, H., Ohtani, M., Amachi, S., Shinoyama, H. and Huji, T. (2003) Some properties of glycine aminotransferase from *Rhodopseudomonas palustris* No. 7 concerning extracellular porphyrin production. *Bioscience, Biotechnology and Biochemistry* 67, 783–789.

Yancey, P.H. (2005) Organic osmolytes as compatible, metabolic and cytoprotectants in high osmolarity and other stresses. *Journal of Experimental Biology* 208, 2819–2830.

Zabalza, A., van Dongen, J.T., Froehlich, A., Oliver, S.N., Faix, B., Gupta, K.J., Schmäzlin, E., Igal, M., Orcaray, L. and Royuela, M. (2009) Regulation of respiration and fermentation to control the plant internal oxygen concentration. *Plant Physiology* 149, 1087–1098.

3 Aspartate Aminotransferase

C.D. Leasure and Z.-H. He*
San Francisco State University, San Francisco, California, USA

3.1 Abstract

Aspartate aminotransferase (EC 2.6.1.1; AspAT) is present in all cellular life on the earth. It catalyses the removal of an amino group from an amino acid, followed by the subsequent amination of a keto acid, to create glutamate from aspartate or vice versa. AspAT is a vitamin B_6-dependent enzyme that plays a pivotal role in carbon and nitrogen metabolism. While bacteria have only one AspAT isoenzyme, eukaryotic organisms have multiple isoenzymes that are located in different sub-cellular compartments. The structure of the AspAT enzyme was determined decades ago, and has since been refined. Most of this work was done using either the chicken (*Gallus gallus*), pig (*Sus scrofa*) or *Escherichia coli* versions of the enzyme. Many residues of the enzyme are 100% conserved and, as a convention, to make things comparable between species, they are numbered based on the pig version of the enzyme.

AspAT functions as a homodimer, with each monomer binding to one molecule of pyridoxal-5′-phosphate (PLP), the active form of vitamin B_6. The binding of the PLP cofactor to the AspAT monomer is required for both formation of the dimer and the enzymatic activity of AspAT. PLP is a cofactor for more than 100 enzymes in biology, and acts as an excellent electron sink for modifying the substrates of the PLP-dependent enzymes in many biochemical reactions, including decarboxylation, transamination, racemization and elimination. Recent studies suggest that PLP-dependent AspAT may play a role in PLP homeostatic sensing. The transgenic overexpression of various AspAT isoforms results in altered nitrogen metabolism and amino acid content in rice and *Arabidopsis*. Additional roles of AspATs associated with other physiological functions are also discussed in the chapter.

3.2 Introduction

Aspartate aminotransferases (AspAT) play a central role in metabolism in all organisms. In all known examples in plants, there exist multiple isoforms of AspAT enzymes. These generally are differentiated by their sub-cellular localization. As in animals, there is a mitochondrial and cytoplasmic isoform, but in plants there are also plastidic and peroxisomal/glyoxysomal isoforms.

The amino acid aspartate plays a critical role in plant metabolism. These roles include ammonium fixation, amino acid biosynthesis and roles in C_4 metabolism, in addition, of course, to its role as a building block in proteins. Aspartate is one of the first amino acids formed during ammonium fixation and in many species

*E-mail address: zhe@sfsu.edu

 Amino Acids in Higher Plants (ed. J.P.F. D'Mello)

is present in high concentrations in the phloem. AspATs convert the low C:N (4:1) aspartate (Asp) into the higher C:N (5:1) glutamate (Glu), and vice versa. Thus, AspATs, Asp and Glu are intimately involved in the balance of C:N in plants under various growth conditions (Schultz *et al.*, 1998).

The AspAT enzyme utilizes the cofactor pyridoxal-5′-phosphate (PLP) to carry catalyse the conversion of Asp and α-ketoglutarate to Glu and oxaloacetate (OAA): Asp + α-KG ↔ Glu + OAA. This reaction, and its mechanism, are well understood, both biochemically and physically within the enzyme, and the enzymatic mechanisms of AspAT will be discussed in some detail in this chapter. PLP is the enzymatically active form of vitamin B_6, and PLP and other vitamin B_6 vitamers (i.e. different forms of vitamin B_6) have a role in quenching reactive oxygen species that develop during photosynthesis. Recent genetic results suggesting a possible role for AspATs in the regulation of the homeostasis of vitamin B_6 will also be discussed.

3.3 The Vitamin B_6 Cofactor

Perhaps the best place to start with AspAT is with the enzyme cofactor: vitamin B6. Of course, the term 'vitamin' is technically incorrect when referring to plants, because they are perfectly capable of synthesizing vitamin B_6 themselves and do not need it from their environment. Vitamin B_6 exists in several different forms, or 'vitamers', within plants (and animals as well). PLP is the vitamer form that is used enzymatically, and it binds first to apoproteins and then to substrates in the reactions that are catalysed. In addition to PLP, the vitamin has the forms pyridoxine-5′-phosphate (PNP) and pyridoxamine-5′-phosphate (PMP), and also non-phosphorylated versions of all three: PL, PN and PM (Tambasco-Studart *et al.*, 2005).

The so-called 'salvage' pathway of enzymes exists in plants to convert different forms of vitamin B_6 into the enzymatically relevant form of PLP (Herrero *et al.*, 2011). The key enzymes in the salvage pathway are pyridoxine oxidase (POX) and pyridoxal kinase (PK). POX can convert PN or PM to PL, as well as PNP or PMP to PLP. PK can phosphorylate any of the non-phosphorylated vitamers. There is some evidence linking PK to the actual delivery of PLP to the enzymes that require it, although this evidence comes from animal studies and has not been investigated in plants.

In plants, PLP is actually the first vitamer produced. The first biosynthetic pathway discovered for vitamin B_6 was in bacteria, and this utilized deoxyxylulose-5-phosphate (DXP) at an intermediate step of the synthesis. Hence, the pathway became known as the DXP-dependent pathway. By contrast, the PLP synthesis pathway that was discovered later was named the DXP-independent pathway. Plants utilize this DXP-independent pathway, which combines ribulose-5-phosphate, glyceraldehyde-3-phosphate and ammonia to directly synthesize the enzymatically active PLP (Tambasco-Studart *et al.*, 2005).

The enzyme that is involved in the biosynthesis of PLP in plants is pyridoxal synthase. This is a dodecamer, i.e. it has 12 subunits, which consist of six molecules of PDX1 and six molecules of PDX2. PDX1 combines the three substrates (ribulose-5-phosphate, glyceraldehyde-3-phosphate and ammonia) to form PLP, while PDX2 is critical for the removal and delivery of ammonia from glutamine (Gln) to PDX1. In *Arabidopsis*, there are three *PDX1* genes (*PDX1.1*, *PDX1.2*, *PDX1.3*) and a single *PDX2* gene. Loss-of-function *pdx2* mutants are lethal, and *pdx1* mutants display altered levels of vitamin B_6 and of the expression of related genes (Tambasco-Studart *et al.*, 2007).

Vitamin B_6 vitamers are linked to a role in stress responses, specifically as antioxidants important for photo-oxidative stress (Havaux *et al.*, 2009; Raschke *et al.*, 2011; Vanderschuren *et al.*, 2013). The primary role of vitamin B_6 in photoprotection has been suggested to be a result of the ability of vitamin B_6 to both limit the accumulation of and prevent damage from singlet oxygen ($^1O_2$) (Havaux *et al.*, 2009). This is especially relevant in the chloroplasts during high light conditions that may damage the photosynthetic machinery (Havaux *et al.*, 2009).

3.4 Enzyme Function

AspAT is a Fold Type I PLP-dependent enzyme (Schneider *et al.*, 2000). AspATs were the first and most studied enzymes of the PLP-dependent enzyme group. They arguably have the most extensive and well-understood structure and function of any enzyme. In addition to multiple wild type AspAT crystal structures having been solved, many single amino acid-altered mutant protein structures have also been determined. The physical structures that have been defined, in combination with the effects of point mutations on function, have provided incredible insight into the function of AspATs.

The PLP cofactor seems particularly ideal for performing many reactions with amino acids, including transamination, elimination, decarboxylation and racemization reactions (Eliot and Kirsch, 2004). In a sense, the major role of the various PLP-dependent enzyme apoproteins is to choose the substrates and direct which of the reactions that PLP is going to perform on them. This section deals primarily with the enzymatic reaction. The specific roles of AspAT protein residues will be discussed in Section 3.5.

The primary reaction catalysed by AspAT is the transfer of the α-amino group from an amino acid (Asp or Glu) to a keto acid (α-KG or OAA), resulting in a new amino acid and a new keto acid being formed. The AspAT transamination reaction occurs through a ping–pong bi–bi mechanism, in which the enzyme performs two half reactions, with the enzyme itself existing in a modified form in the intervening stage between the two half reactions (Fig. 3.1). For AspAT, this intermediate form consists of the AspAT enzyme having PMP as a cofactor instead of PLP. During the intermediate state, the cofactor is holding the amino group to be used for the subsequent transamination reaction of a keto acid receiver.

The overall reaction that AspAT catalyses is the reversible transamination reaction:

$$\text{Aspartate} + \alpha\text{-Ketoglutarate} \leftrightarrow \text{Oxaloacetate} + \text{Glutamate} \quad (3.1)$$

This reaction is better thought of as two separate half reactions:

$$\text{AspAT-PLP} + \text{Aspartate} \leftrightarrow \text{AspAT-PMP} + \text{oxaloacetate} \quad (3.2)$$

$$\text{AspAT-PMP} + \alpha\text{-Ketoglutarate} \leftrightarrow \text{AspAT-PLP} + \text{Glutamate} \quad (3.3)$$

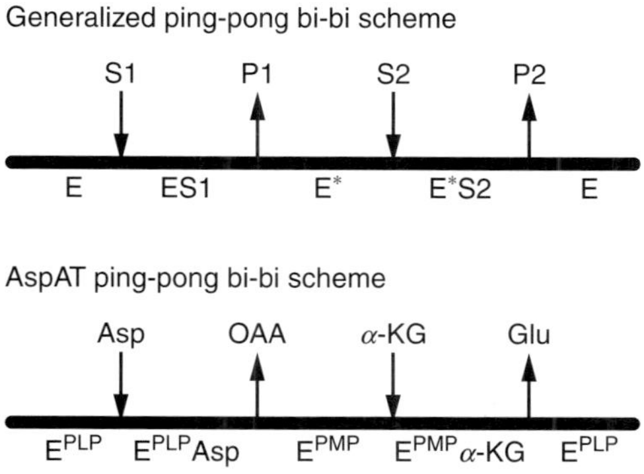

Fig. 3.1. Mechanism of the aspartate transaminase (AspAT) transamination reaction. The coenzyme pyridoxal-5′-phosphate (PLP) of the enzyme (E^{PLP}) accepts an amino group from aspartate (Asp) and becomes aminated to pyridoxamine-5′-phosphate (E^{PMP}). The keto acid, oxaloacetate (OAA), thus produced is released and the coenzyme then undergoes a reversal of the process, giving up its newly acquired amino group to the other keto acid, α-ketoglutarate (α-KG), to produce the new amino acid product glutamate (Glu).

AspATs can perform this reaction in either direction, suggesting that substrate concentrations are a major determinant in reaction directionality. As discussed later (Section 3.4.2), the K_m values for the four substrates vary, but not hugely so. There is no input of ATP energy required for AspATs to function. Upon removal of the amino group from the amino acid donor, the vitamin B_6 cofactor is turned into the B_6 vitamer PMP. PMP contains the amino group that has been removed from the amino acid donor and can then be transferred to a keto acid receiver to make a new amino acid. PMP is not covalently linked to the protein component of the enzyme, but is instead held in place through hydrogen and ionic bonding.

All known examples of AspATs are homodimers, with two active sites per holoenzyme (Rhee *et al.*, 1997; Jeffery *et al.*, 1998; Deu *et al.*, 2009). The mechanism, which is discussed below, occurs at each active site. There is no evidence that the active sites are distinct (Eliot and Kirsch, 2004), which means that the activity at one site of the enzyme neither interferes with nor enhances the activity of the other site of the enzyme (Eliot and Kirsch 2004). Each active site exists at the subunit interface, so that each active site is made up of residues from each subunit.

3.4.1 The reaction mechanism

The initial step of the AspAT reaction can be thought of as the step in which the cofactor binds to the apoprotein. There is limited evidence that PLP is delivered to apoprotein through an interaction between the AspAT apoprotein (encoded by *ASP1* to *ASP5* in *Arabidopsis*) and the PK enzyme (encoded by *SOS4* in *Arabidopsis*) (Kim *et al.*, 1988). This interaction may exist between diverse PLP-dependent enzymes and PK (Cheung *et al.*, 2003). However, these data come from studies in animals, and plants may use a different mechanism for PLP delivery (Cheung *et al.*, 2003; Kim *et al.*, 1988).

PK phosphorylates pyridoxal using ATP as a phosphate donor. If PK does indeed interact with PLP-dependent enzymes, the interaction of PK with those enzymes is likely to be important for limiting the amount of free PLP present in the cytosol or organelle matrices. PLP is less stable than the non-phosphorylated forms of vitamin B_6, and this mechanism would represent an efficient means by which a cell could deliver the cofactor to PLP-dependent enzymes, while limiting the time that it spends freely roaming in the cell.

The AspAT apoprotein and PLP combine to form the mature holoenzyme. As already noted, the mature AspAT functions as a homodimer with two equivalent active sites per complete enzyme dimer. Each monomer contains a single PLP-binding site, where the PLP is held in place by interaction with several residues in the binding pocket. A covalent linkage is made between the aldehyde group of the 4′-C of PLP and the ε-amino group of Lys258 (K258) of the apoprotein (Malashkevich *et al.*, 1995a). This initial bonding of PLP and K258, followed by dehydration, yields a Schiff base (aldimine) linkage that is critical for the proper function of the enzyme (Malashkevich *et al.*, 1995a). This first AspAT-PLP structure is referred to as the *internal* aldimine, as the cofactor is bound 'internally' to the AspAT protein through an aldimine linkage.

The PLP-enzyme takes on a slightly 'open' conformation that can bind to its substrates (Asp or Glu) to form a Michaelis complex (Jager *et al.*, 1994; Malashkevich *et al.*, 1995b). Once the substrate is bound to the enzyme, the amino group of the substrate becomes de-protonated and can attack the double bond of the PLP-Lys258 Schiff base. After passing through intermediate forms, including a tetrahedral arrangement, a new aldimine linkage is created between the PLP cofactor and the amino group of the amino acid substrate. Lys258 is no longer covalently attached to the PLP cofactor at this point, although it does still play a role in later steps of the AspAT reaction. At this stage, the cofactor is now bound to the substrate, rather than the protein. This step in the reaction is termed the *external* aldimine, because the cofactor is now bound 'externally' to the substrate.

The external aldimine intermediate then passes through a quinonoid form before reaching the ketimine intermediate. This process requires the formation of a carbanion in the substrate, which occurs when the α-C of the amino acid substrate is deprotonated (Jager *et al.*, 1994; Griswold *et al.*, 2012). This important reaction is facilitated by the hyperconjugation of the sigma orbital of the α-C proton with the pi bond of the external aldimine linkage (Griswold *et al.*, 2012). The quinonoid form is next rearranged to form a ketimine complex, which is a double bond formed between the amino group of the substrate and its own α-C.

The removal of the proton from the α-C of the amino acid substrate is critical for the reaction to proceed. In fact, this deprotonation reaction appears to be the major role of PLP in most PLP-dependent enzymes, and illustrates the role of PLP as an electron sink. The removal of this proton is of major importance, because it allows for the ketimine complex to form. This proton is essentially irremovable from Asp or Glu in solution, having a pK_a of ~30. K258 is thought to play a role as proton acceptor during this 1,3-prototropic shift part of the reaction (Malashkevich *et al.*, 1995a). Also, Tyr225 is important for hydrogen bonding with the 3′-O of the coenzyme at this point and stabilizing the intermediates prior to the formation of the ketimine (Birolo *et al.*, 1995).

The ketimine intermediate form is next readily hydrolysed to form the end product of the half reaction: the conjugate keto acid of the amino acid substrate. This product is OAA or α-KG, depending on whether Asp or Glu, respectively, was the initial substrate. The vitamin B_6 cofactor itself now contains the amino group and is still bound to the protein, although non-covalently. At this point, the form of vitamin B_6 is PMP, which will then be utilized as the amino donor for the reverse reaction that will create a new amino acid.

3.4.2 Enzyme properties

In plants, multiple measurements of the K_m values for the four AspAT substrates (OAA, α-KG, Asp, Glu) have yielded similar values for AspATs from different species. The strength of the affinity of AspATs for their four substrates is invariably in the order OAA > α-KG > Asp > Glu. A glance at reported K_m values yields values such as: ~2 mM for Asp; ~0.3 mM for α-KG; ~12 mM for Glu; and ~0.05 mM for OAA (Martins *et al.*, 2002). Higher reported values may come from enzyme preparations that were less pure. These values are very similar to those reported for the *E. coli* AspAT, suggesting that billions of years of evolution have not altered the function of this enzyme in the least.

3.5 Enzyme Structure

The AspAT enzyme is one of the most heavily studied enzymes in terms of both its enzyme function and its protein structure. The crystal structure has been solved at multiple resolutions, and at multiple intermediate steps in the enzymatic reaction. AspAT crystals have been formed in complex with many different substrates, including inhibitory substrates such as maleate, or PLP analogues such as deaza-PLP (Jager *et al.*, 1994; Griswold *et al.*, 2011). These inhibitory substrates act to 'lock' the enzyme into one of the intermediate steps of the reaction, thus allowing a uniform crystal to be made for analysis.

The amino acid numbering used for inter-species comparisons is based on the sequence of the porcine cytosolic AspAT. The apoprotein comprises two domains: a large one and a small one. The large domain corresponds to amino acid residues 48–325, and the small domain is made up of residues 15–47 and 326–410 (Danishefsky *et al.*, 1991). As already mentioned, AspATs function as homodimers, and the complete holozyme contains two active sites, each with its own PLP cofactor. The active enzyme is an α_2-dimer, containing two active sites that are each made up of residues from both individual proteins. The enzyme has both closed and open positions. Studies indicate that the two active sites are equivalent in terms of function. The active sites are made up of residues from each monomer.

When a substrate binds, the enzyme closes slightly around it. The dicarboxlyate nature of the substrates is held in place by the positive charges of two arginine (Arg) residues, R386 (from the same monomer that bound the PLP) and R292* (from the other monomer). The substrate is also bound to the PLP cofactor through a Schiff base formation. The PLP cofactor is bound to the enzyme through a Schiff base formation (removable), and hydrogen bonds and salt bridges.

3.5.1 K258

K258 is the lysine (Lys) to which the cofactor, PLP, covalently binds in the holozyme. K258 forms a Schiff base covalent linkage to the PLP cofactor, termed the internal aldimine. This linkage is broken during the initial steps of the reaction. K258 is thought to play an additional role as a proton acceptor/donor in later steps of the reaction. In the PMP-enzyme intermediate, the PMP is not covalently bound to K258, or to any other residue.

3.5.2 R292*

The R292* (Arg) binds ionically to the side-chain carboyxlate group of the AspAT substrates. This residue is provided to the active site by the other protein in the dimer (i.e. not the monomer that is bound to the PLP). R292*D mutants actually convert the enzyme into a transaminase that works on basic amino acids like R (Arg) and K (Lys).

3.5.3 R386

The R386 (Arg) binds ionically to the α-carboxylate group of the AspAT substrates. R386 is part of the small domain of the protein and is also a part of the same subunit that contains the K (lysine) bound to PLP. In the R386Y and R386F mutants the enzymatic activity is virtually eliminated, being five orders of magnitude lower, when using Asp or Glu as a substrate (Danishefsky *et al.*, 1991).

3.5.4 D222

D222 (Asp) binds to the N(1) position of the PLP through strong ionic interaction. Asp222 acts to keep the cofactor in a protonated state at position N(1) (Yano *et al.*, 1992; Griswold *et al.*, 2011). This greatly increases the ability of the cofactor to withdraw the α-proton of the amino acid substrates, which is a critical step in the removal of the amino group of the substrate. The replacement of D222 with alanine (Ala) or asparagine (Asn) results in a decrease of enzyme activity of about four orders of magnitude. The replacement of D222 with Glu (to give E222) yields a somewhat active enzyme, but it is interesting to note that D222 appears to be invariant in nature; that is, no known sequences have E222 naturally (Yano *et al.*, 1992).

3.5.5 Y225

Y225 (tyrosine, or Tyr) is importantly located near the PLP so that the -OH group of the phenol side chain of Tyr225 can hydrogen bond with the deprotonated 3′-O of the PLP (Inoue *et al.*, 1991). When this residue is mutated to Arg or phenylalanine (Phe), the enzymatic activity is largely abolished.

3.6 Enzyme Genetics

Before discussing the genetics of AspATs in plants, it should be noted that there is apparently no consistent naming scheme for the genes or gene products that are involved between species. For example, the gene encoding mitochondrial AspAT in *Arabidopsis* is *ASP1*, but the orthologous gene in rice is called *AAT3* (Zhou *et al.*, 2009). Not to be outdone, the enzymes are referred to in wheat as AAT-1 (mitochondrial), AAT-2 (plastid) and AAT-3 (cytoplasmic) (Maciaga and Paszkowski, 2004). The enzyme itself has been referred to as AspAT throughout this chapter, but is abbreviated as AAT in other sources, or even named for the reverse activity in other organisms, i.e. glutamic oxaloacetic transaminase (GOT). Wherever possible, we will clarify the function/location of the gene product.

The *Arabidopsis thaliana* genome encodes for five 'eukaryotic' AAT (i.e. AspAT) isoenzymes and one 'prokaryotic' AAT isoenzymes (Schultz and Coruzzi, 1995; Wilkie and Warren, 1998; de la Torre *et al.*, 2006). The prokaryotic type is thought to have different functions from the AspAT reaction, leaving only five true AspAT isoenzymes in *Arabidopsis*. The ASP1 and ASP2 proteins are orthologues of the mitochondrial and cytoplasmic versions of the enzyme, respectively. ASP1 is more similar to mitochondrial AATs from animals than it is to ASP2. ASP3 and ASP5 are potential chloroplastic isoforms, although there does not seem to be too much data to confirm this beyond predictive models.

D-Asp exists in plants, as it does in animals. At least one enzyme seems to have broad substrate specificity that includes D-Asp. The role of D-amino acids is not that clear in plants, although it is well understood in animals.

Schultz *et al.* (1998) isolated several mutant versions of the AAT2 and AAT3 isoforms in *Arabidopsis*. They utilized an in-gel assay for the activity of the enzyme, and found several mutants that lacked either the AAT2 or AAT3 activity on the gel. They mapped the AAT2 function to the *ASP2* gene in the genome and the AAT3 function to the *ASP5* gene in the genome. Interestingly, apart from one notable exception (*asp2-2*), none of these alleles had any large impact on the phenotype of the plant. Plants with *asp2-2* had shorter roots than wild type (WT) plants, and transported less Asp in the phloem than WT plants. This is interesting because all of their alleles for *asp2* mutants seemed to completely lack activity in their assay, so it is not clear why they should be different in terms of phenotype.

The mutations in *ASP2* and *ASP5*, which seemingly greatly reduce the available levels of AspAT activity in the plant, do not cause lethality, or even a very severe phenotype. This is consistent with other alleles for *ASP2* isolated by Leasure *et al.* (2011). These authors identified several new point mutations that lack detectable AAT2 activity with an in-gel assay, and an insertion that deletes most of the gene and has undetectable ASP2 protein by western blot.

None of these mutations has a noticeable phenotype under growth chamber conditions. However, the mutants are slightly chlorotic under higher light conditions (personal communication).

Studies with two *Arabidopsis root uv-b sensitive* (*rus1* and *rus2*) mutants suggested that AspAT may play a role in vitamin B_6 homeostatic sensing. Both *rus1* and *rus2* mutants are developmentally arrested when exposed to light conditions containing a trace amount of ultraviolet-B (UV-B) light, and their UV-B hypersensitive phenotypes can be partially rescued by either eliminating UV-B or by exogenously added vitamin B_6 (Tong *et al.*, 2008; Leasure *et al.*, 2011). Further analyses show that these two rescue conditions are related, as Vitamin B_6 can be photolytically destroyed by UV-B (Leasure *et al.*, 2011). RUS1 and RUS2 are homologous proteins functionally working together in a complex, and are members of a large protein family containing DUF647 (domain of unknown function 647), a domain found in almost all multicellular organisms (Leasure *et al.*, 2009). Interestingly, although *rus1* and *rus2* mutants can be partially rescued by exogenous vitamin B_6, the actual endogenous vitamin B_6 levels in both mutants are comparable with those in the WT (Leasure *et al.*, 2009). Mutagenesis screens for second-site suppressors of the *rus* mutant phenotypes have identified many *sor* (*suppressor of rus*) individuals, which were mapped to five genetic complementation groups. One of the *sor* groups (sor1) carried mutations in the aspartate aminotransferase 2 gene (*ASP2*) that encodes for cytosolic AspAT. Genetic analyses of six *asp2* mutant alleles have shown that only specific amino acid substitutions in ASP2 override the phenotypes of *rus1* or *rus2* single mutants and a *rus1 rus2* double mutant, but a complete knockout of the *ASP2* gene does not. The authors reported that all of these substitutions reside at specific positions in the vitamin B_6-binding pocket, suggesting the binding of vitamin B_6 to AspAT may play a key role in the vitamin B_6 homeostatic response in *Arabidopsis* (Leasure *et al.*, 2011).

3.7 The Enzyme during Plant Development

Liu and Huang (1977) found that there were as many as six isoforms of AspAT present in developing cucumber seedlings. Through sub-cellular isolation, they determined that the AspATs were present in distinct sub-cellular compartments, including the glyoxysomes, plastids and cytoplasm. The overall activity of the enzyme increased dramatically (~17-fold) during the first 2 days of plant development, but following this, the levels of enzyme gradually declined. The addition of light at day 3 increased the rate of decline. Interestingly, the levels of the glyoxysomal isoforms decreased essentially to zero after 10 days (when the experiment ended). The cytosolic and proplastid/chloroplastic isoforms decreased slightly, but seemed to be maintained at about 25 and 60% of their peak levels, respectively. This information suggests that AspATs peak during the mobilization of stored resources that occurs during development. They then decrease and level out during the transition to normal growth and development.

In rice, there are three genes that encode AspATs: *OsAAT1* (the plastid isoform), *OsAAT2* (the cytosolic isoform) and *OsAAT3* (the mitochondrial isoform) (Zhou *et al.*, 2009). qRT-PCR (quantitative reverse transcriptase PCR) analyses revealed that the cytosolic isoform was generally the most highly expressed isoform, especially during early development and in roots. The plastid isoform had its highest expression, not surprisingly, in leaves. The expression of the mitochondrial isoform was very low in early development, but increased in more mature tissues ((Zhou *et al.*, 2009).

3.8 The Role of Aspartate in Plants

Asp is synthesized in plants by the AspAT enzyme utilizing Glu and OAA as substrates. The synthesis of Glu is catalysed by glutamate synthase (GOGAT) and OAA derives from the TCA (tricarboxylic acid) cycle. Asp represents the third amino acid into which ammonia is assimilated as part of the ammonia assimilation pathway in plants, the other two being Glu (synthesized via GOGAT) and Gln, synthesized via glutamine synthetase (GS) as part of the GS/GOGAT pathway. As such, Asp has a central role in nitrogen metabolism in plants, and is present in concentrations much higher than those needed for protein synthesis. It is a precursor for four other amino acids and is also utilized in certain situations as an electron carrier.

3.8.1 C_4 metabolism

In C_4 plants that utilize the NAD-malic enzyme (ME) in the mitochondria of bundle sheath cells (BSC), Asp is commonly the molecule (or one of the molecules), that is transported between the mesophyll and BSCs. OAA, the product of phosphoenolpyruvate carboxylase (PEPC/PEPCase), is aminated into Asp by AspAT in the cytosol (cAspAT) of NAD-ME C_4 plants. Asp then moves to the BSCs, where AspATs in either the cytosol or the mitochondria (mAspAT) create OAA from Asp. In the NAD-ME type of C_4 metabolism, OAA is reduced to form malate, which is subsequently oxidized to form pyruvate and CO_2 for use in the Calvin Cycle. In the phosphoenolpyruvate carboxykinase (PEPCK) type of C_4 metabolism, OAA is directly converted to PEP and CO_2 for the Calvin Cycle. Alanine is generally transported back into mesophyll cells.

An important regulatory mechanism for AspATs appears to exist in these C_4 plants. This is that the genes for cAspAT and mAspAT are highly expressed in either the mesophyll cells or the BSCs, respectively. Additionally, both genes are light regulated, which is key for their role in photosynthesis.

3.9 Other Roles of Aspartate Aminotransferase

3.9.1 Moonlighting

The chicken mitochondrial isoform of AspAT was shown to not only be important for its enzymatic (i.e. aspartate aminotransferase) activity but also for long chain fatty acid uptake in cells (Bradbury *et al.*, 2011) – hence could be described as 'moonlighting'. A hydrophobic cleft is found in the crystal structure of this protein, between amino acid positions 166 to 270. R201 (Arg) was identified as the likely candidate for binding to the carboxylate group of the fatty acid. This protein was identified genetically as being important for fatty acid uptake and biochemically as localized to the plasma membrane. Interestingly, the cytoplasmic isoform has threonine (Thr) at position 201, which most likely reduces its ability to bind to fatty acids. More interestingly, all of the isoforms in *Arabidopsis* have Thr at position 201. This includes the mitochondrial isoform, ASP1, which is clearly more closely related to the animal mitochondrial isoform than to any of the other plant isoforms. The same holds true in moss and rice, where all of the AspAT enzymes show either Thr or serine (Ser) at this position (see NCBI Database: http://www.ncbi.nlm.nih.gov/protein/?term=Aspartate%20 Aminotransferase). Bradbury *et al.* (2011) suggest that Thr removes all of the specific binding of oleic acid to this cleft. Thus, the plant AspATs seem to have lost this ability early in evolution, or never gained it to begin with.

3.9.2 Genetic engineering with aspartate aminotransferases

The central role of AspATs in plant metabolism and, more specifically, in amino acid metabolism, make them an attractive target for genetic manipulation aimed at altering the amino acid content of plants for human consumption.

Zhou *et al.* (2009) investigated the overexpression of four genes for AspAT – three native rice genes and a gene from *E. coli* – by introducing them into rice. While the overexpression of the *OsAAT1* (the rice plastid isoform) or the *OsAAT2* (the rice cytosolic isoform) genes led to increased amino acid content in rice seeds, the overexpression of the *OsAAT3* (the rice mitochondrial isoform) gene did not affect the seed amino acid content. The overexpression of *EcAAT* (from *E. coli*) had a similar effect to that of the native rice *OsAAT1* and *OsAAT2* genes. The overexpression of *OsAAT1*, *OsAAT2* or *EcAAT* also increased the total measurable AspAT activity in mature leaves, but *OsAAT3* increased neither AspAT activity nor amino acid content in leaves.

In *Arabidopsis*, it was shown that overexpression of a chloroplastic AspAT gene from soybean (called *GmAAT5*) led to increases in many of the amino acids in dry seeds (Murooka *et al.*, 2002). However, Asp and Glu levels were actually lower in lines overexpressing this AspAT gene. The biggest increases were seen in glycine (Gly), Ala, Asn and Gln.

3.10 Future Research

Future research could cover the following topics. At least five AspAT isoenzymes in *Arabidopsis* and three isoenzymes in rice have been documented. A prokaryotic-type AspAT has also been identified in plants. It is not clear whether and how these different isoenzymes play different metabolic functions in plant growth and development. Knockout mutants for individual *Arabidopsis* AspATs in that plant give few to modest numbers of phenotypes, suggesting that AspAT may share functional redundancy. The reactions catalysed by AspATs are reversible and it is not clear what major factors determine the directions of the reactions catalysed by the various isoenzymes located in different sub-cellular compartments.

3.11 Conclusions

AspATs play an essential role in carbon and nitrogen metabolism. Multiple AspAT isoenzymes are localized in different sub-cellular compartments. While AspAT protein structures are well characterized, how an individual AspAT isoenzyme plays a role in plant morphogenesis and development is not fully understood. Molecular and genetic studies suggest some AspAT isoenzymes may play redundant roles in amino acid metabolism. There is also increasing evidence that suggests additional functional roles for AspATs in signal transductions.

References

Birolo, L., Sandmeier, E., Christen, P. and John, R.A. (1995) The roles of Tyr70 and Tyr225 in aspartate aminotransferase assessed by analysing the effects of mutations on the multiple reactions of the substrate analogue serine o-sulphate. *European Journal of Biochemistry* [now *The FEBS Journal*] 232, 859–864.

Bradbury, M.W., Stump, D., Guarnieri, F. and Berk, P.D. (2011) Molecular modeling and functional confirmation of a predicted fatty acid binding site of mitochondrial aspartate aminotransferase. *Journal of Molecular Biology* 412, 412–422.

Cheung, P.-Y., Fong, C.-C., Ng, K.-T., Lam, W.-C., Leung, Y.-C., Tsang, C.-W., Yang, M. and Wong, M.-S. (2003) Interaction between pyridoxal kinase and pyridoxal-5-phosphate-dependent enzymes. *The Journal of Biochemistry* 134, 731–738.

Danishefsky, A.T., Onnufer, J.J., Petsko, G.A. and Ringe, D. (1991) Activity and structure of the active-site mutants R386Y and R386F of *Escherichia coli* aspartate aminotransferase. *Biochemistry* 30, 1980–1985.

de la Torre, F., De Santis, L., Suarez, M.F., Crespillo, R. and Canovas, F.M. (2006) Identification and functional analysis of a prokaryotic-type aspartate aminotransferase: implications for plant amino acid metabolism. *The Plant Journal* 46, 414–425.

Deu, E., Dhoot, J. and Kirsch, J.F. (2009) The partially folded homodimeric intermediate of *Escherichia* coli aspartate aminotransferase contains a "molten interface" structure. *Biochemistry* 48, 433–441.

Eliot, A.C. and Kirsch, J.F. (2004) Pyridoxal phosphate enzymes: mechanistic, structural, and evolutionary considerations. *Annual Review of Biochemistry* 73, 383–415.

Griswold, W.R., Fisher, A.J. and Toney, M.D. (2011) Crystal structures of aspartate aminotransferase reconstituted with 1-deazapyridoxal 5′-phosphate: internal aldimine and stable L-aspartate external aldimine. *Biochemistry* 50, 5918–5924.

Griswold, W.R., Castro, J.N., Fisher, A.J. and Toney, M.D. (2012) Ground-state electronic destabilization via hyperconjugation in aspartate aminotransferase. *Journal of the American Chemical Society* 134, 8436–8438.

Havaux, M., Ksas, B., Szewczyk, A., Rumeau, D., Franck, F., Caffarri, S. and Triantaphylidès, C. (2009) Vitamin B_6 deficient plants display increased sensitivity to high light and photo-oxidative stress. *BMC Plant Biology* 9, 130.

Herrero, S., González, E., Gillikin, J.W., Vélëz, H. and Daub, M.E. (2011) Identification and characterization of a pyridoxal reductase involved in the vitamin B_6 salvage pathway in *Arabidopsis*. *Plant Molecular Biology* 76, 157–169.

Inoue, K., Kuramitsu, S., Okamoto, A., Hirotsu, K., Higuchi, T., Morino, Y. and Kagamiyama, H. (1991) Tyr225 in aspartate aminotransferase: contribution of the hydrogen bond between Tyr225 and coenzyme to the catalytic reaction. *The Journal of Biochemistry* 109, 570–576.

Jager, J., Moser, M., Sauder U. and Jansonius, J.N. (1994) Crystal structures of *Escherichia coli* aspartate aminotransferase in two conformations. Comparison of an unliganded open and two liganded closed forms. *Journal of Molecular Biology* 239, 285–305.

Jeffery, C.J., Barry, T., Doonan, S., Petsko, G.A. and Ringe, D. (1998) Crystal structure of *Saccharomyces cerevisiae* cytosolic aspartate aminotransferase. *Protein Science* 7, 1380–1387.

Kim, Y.T., Kwok, F. and Churchich, J.E. (1988) Interactions of pyridoxal kinase and aspartate aminotransferase emission anisotropy and compartmentation studies. *The Journal of Biological Chemistry* 263, 13712–13717.

Leasure, C.D., Tong, H., Yuen, G., Hou, X., Sun, X., He, Z.-H. (2009) ROOT UV-B SENSITIVE2 acts with ROOT UV-B SENSITIVE1 in a root ultraviolet B-sensing pathway. *Plant Physiology* 150, 1902–1915.

Leasure, C.D., Tong, H.-Y., Hou, X.-W., Shelton, A., Minton, M., Esquerra, R., Roje, S., Hellmann, H. and He, Z.-H.. (2011) *root uv-b sensitive* Mutants are suppressed by specific mutations in *ASPARTATE AMINOTRANSFERASE2* and by exogenous vitamin B_6. *Molecular Plant* 4, 759–770.

Liu, K.D.F. and Huang, A.H.C. (1977) Subcellular localization and developmental changes of aspartate-alpha-Ketoglutarate transaminase isozymes in the cotyledons of cucumber seedlings. *Plant Physiology* 59, 777–782.

Maciaga, M. and Paszkowski, A. (2004) Genetic control of aspartate aminotransferase isoenzymes in *Aegilops* and *Triticum* species. *Journal of Applied Genetics* 45, 411–417.

Malashkevich, V.N., Jager, J., Ziak, M., Sauder, U., Gehring, H., Christen, P. and Jansonius, J.N. (1995a) Structural basis for the catalytic activity of aspartate aminotransferase K258H lacking the pyridoxal 5′-phosphate-binding lysine residue. *Biochemistry* 34, 405–414.

Malashkevich, V.N., Strokopytov, B.V., Borisov, V.V., Dauter, Z., Wilson, K.S. and Torchinsky, Y.M. (1995b) Crystal structure of the closed form of chicken cytosolic aspartate aminotransferase at 1.9 Å resolution. *Journal of Molecular Biology* 247, 111–124.

Martins, M.L.L., Mourato, M.P. de, F.B. and de Varennes e Mendonça, A.P.P. (2002) Characterization of aspartate aminotransferase isoenzymes from leaves of *Lupinus albus* L. cv Estoril. *Journal of Biochemistry and Molecular Biology* 35, 220–227.

Murooka, Y., Mori, Y. and Hayashi, M. (2002) Variation of the amino acid content of *Arabidopsis* seeds by expressing soybean aspartate aminotransferase gene. *Journal of Bioscience and Bioengineering* 94, 225–230.

Raschke, M., Boycheva, S., Crèvecoeur, M., Nunes-Nesi, A., Witt, S., Fernie, A.R., Amrhein, N. and Fitzpatrick, T.B. (2011) Enhanced levels of vitamin B_6 increase aerial organ size and positively affect stress tolerance in *Arabidopsis*. *The Plant Journal* 66, 414–432.

Rhee, S., Silva, M.M., Hyde, C.C., Rogers, P.H., Metzler, C.M., Metzler, D.E. and Arnone, A. (1997) Refinement and comparisons of the crystal structures of pig cytosolic aspartate aminotransferase and its complex with 2-methylaspartate. *The Journal of Biological Chemistry* 272, 17293–17302.

Schneider, G., Kack, H. and Lindqvist, Y. (2000) The manifold of vitamin B_6 dependent enzymes. *Structure* 8, R1–R6.

Schultz, C.J. and Coruzzi, G.M. (1995) The aspartate aminotransferase gene family of *Arabidopsis* encodes isoenzymes localized to three distinct subcellular compartments. *The Plant Journal* 7, 61–75.

Schultz, C.J., Hsu, M., Miesak, B. and Coruzzi, G.M. (1998) *Arabidopsis* mutants define an *in vivo* role for isoenzymes of aspartate aminotransferase in plant nitrogen assimilation. *Genetics* 149, 491–499.

Tambasco-Studart, M., Tews, I., Amrhein, N. and Fitzpatrick, T.B. (2007) Functional analysis of PDX2 from *Arabidopsis*, a glutaminase involved in vitamin B_6 biosynthesis. *Plant Physiology* 144, 915–925.

Tambasco-Studart, M., Titiz, O., Raschle, T., Forster, G., Amrhein, N. and Fitzpatrick, T.B. (2005) Vitamin B_6 biosynthesis in higher plants. *Proceedings of the National Academy of Sciences of the United States of America* 102, 13687–13692.

Tong, H., Leasure, C.D., Hou, X., Yuen, G., Briggs, W. and He, Z.-H. (2008) Role of root UV-B sensing in *Arabidopsis* early seedling development. *Proceedings of the National Academy of Sciences of the United States of America* 105, 21039–21044.

Vanderschuren, H., Boycheva, S., Li, K.-T., Szydlowski, N., Gruissem, W. and Fitzpatrick, T.B. (2013) Strategies for vitamin B_6 biofortification of plants: a dual role as a micronutrient and a stress protectant. *Frontiers in Plant Science* 4, 00143.

Wilkie, S.E. and Warren, M.J. (1998) Recombinant expression, purification, and characterization of three isoenzymes of aspartate aminotransferase from *Arabidopsis thaliana*. *Protein Expression and Purification* 12, 381–389.

Yano, T., Kuramitsu, S., Tanase, S., Morino, Y. and Kagamiyama, H. (1992) Role of Asp222 in the catalytic mechanism of *Escherichia coli* aspartate aminotransferase: the amino acid residue which enhances the function of the enzyme-bound coenzyme pyridoxal 5′-phosphate. *Biochemistry* 31, 5878–5887.
Zhou, Y., Cai, H., Xiao, J., Li, X., Zhang, Q. and Lian, X. (2009) Over-expression of aspartate aminotransferase genes in rice resulted in altered nitrogen metabolism and increased amino acid content in seeds. *Theoretical and Applied Genetics* 118, 1381–1390.

4 Tyrosine Aminotransferase

A.O. Hudson*
Rochester Institute of Technology, Rochester, New York, USA

4.1 Abstract

In plants, the pyridoxal-5′-phosphate (PLP)-dependent enzyme tyrosine aminotransferase (EC 2.6.15) is involved in the interconversion of tyrosine and 4-hydoxyphenylpyruvate, and phenylalanine and phenylpyruvate. Tyrosine and phenylalanine are amino acids that are ubiquitously involved in protein synthesis and are deemed proteogenic. In plants, these amino acids are also involved in reactions that are instrumental in the synthesis of compounds that are involved in plant secondary metabolism. Specifically, these two amino acids are catabolized in plants to provide precursor intermediates to facilitate the anabolism of phenylpropanoid compounds that are involved in plant responses to biotic and abiotic stresses.

The model organism *Arabidopsis thaliana* provides an excellent tool to study aminotransferases because the genome of the organism has been sequenced and completely annotated. Recently, several tyrosine aminotransferases from *Arabidopsis* and several other plant species have been characterized using *in vitro* and *in vivo* biochemical analyses. The results from these studies show that, unlike the bacterial enzyme, which is primarily involved in the synthesis of tyrosine and phenylalanine to facilitate protein synthesis, the major role of plant the plant enzyme is in the biosynthesis of compounds that are involved in secondary metabolism facilitated through the catabolism of tyrosine and phenylalanine. Recent studies have confirmed that tyrosine aminotransferase is involved in many aspects of secondary metabolism; these include the biosynthesis of compounds such as rosmarinic acid in *Salvia miltiorrhiza* and tocopherol in *Arabidopsis thaliana*. The goal of this chapter is to present a synthesis of the recent work on tyrosine aminotransferases in plants by highlighting the kinetic and physical properties of the enzyme in addition to its overarching role(s) in plants.

4.2 Introduction

4.2.1 Aminotransferases: a brief introduction

Aminotransferases, also known as transaminases (EC 2.6.1.X), are a class of enzyme that is found in all three domains of life (Bacteria, Archaea and Eukaryota). Aminotransferases transfer an amino group, plus a proton and an electron pair, from a donor (usually an amino acid) to the carbonyl position of an amino acceptor (usually a keto acid) (Braunstein, 1973) (see Fig. 4.1). The reactions catalysed by aminotransferases are mostly reversible, with the exception of unidirectional transamination reactions that result from instability of the product or from product channelling, or from the affinity of the enzyme

*E-mail address: aohsbi@rit.edu

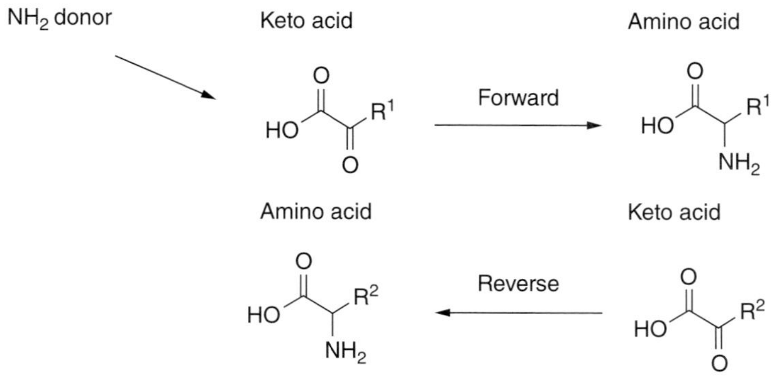

Fig. 4.1. The aminotransferase reaction between an amino acid and a keto acid. The enzyme facilitates the interconversion of the amino (NH_2) and the keto (=O) groups to facilitate the synthesis of a new keto acid or a new amino acid.

for a product that is formed but which does not serve as a substrate for the enzyme to facilitate catalysis in the reverse direction (Givan, 1980).

4.2.2 A brief history of aminotransferase activity in plants

Early work by Braustein and Kritzmann (1937) in the 1930s was instrumental in developing our current understanding of aminotransferases. The early biochemical experiments were performed with crude enzyme extracts and demonstrated that 14 amino acids served as amino donors to either of the keto acids 2-ketoglutarate and oxaloacetic acid (Kritzmann, 1939). The first reported aminotransferase activity from plants was observed using crude extracts from pea (*Pisum sativum*) seedlings (Virtanen and Laine, 1938; Cedrangolo and Carandante, 1940). Leonard and Burris (1947) reported the first comprehensive study of aminotransferases in plants when they surveyed the occurrence of aspartate/glutamate aminotransferase in 22 plant species at various stages of development.

Aminotransferases are important in metabolism and play an integral role in many metabolic pathways in which the interconversions of nitrogen-containing molecules are involved. Nitrogen is acquired either from the environment or through endosymbiotic relationships with nitrogen-fixing bacteria. The majority of this nitrogen is then assimilated into glutamate or glutamine, two proteogenic amino acids that normally serve as the amino donors for transamination reactions (Stumpf and Conn, 1980). In primary metabolism, aminotransferases are responsible for the anabolism of the amino acids that are involved in protein synthesis. Examples of the involvement of aminotransferase in amino acid anabolism include the synthesis of alanine from pyruvate, aspartate from oxaloacetate, glycine from glyoxylate, valine from oxoisovalerate, leucine from oxoisocaproate and isoleucine from oxo-3-methylvalerate. Recently, an aminotransferase from plants and photosynthetic cohorts was identified and characterized that defined a novel variant of the lysine biosynthesis pathways (Hudson *et al*., 2006). In addition, aminotransferases are also involved in the synthesis of many secondary metabolites, such as aliphatic amines (Hasse and Schmid, 1963) and alkaloids (Roberts, 1977), and also the porphyrin precursor aminolevulinic acid, which is involved in chlorophyll anabolism (Beale and Castelfranco, 1974).

4.2.3 Oligomeric state, cofactor requirement and mechanism of action of aminotransferases

Aminotransferases are dimeric in nature, consisting of two identical monomeric units. With respect to molecular weight, most aminotransferases fall within a range from 100 kDa for aspartate aminotransferase from *Glycine max* (Ryan and Fottrell, 1972) to 48 kDa for ornithine 2-oxoacid aminotransferase from *Solanum lycopersicum* (Lu and Mazelis, 1975). It should be noted that there are exceptions to the dimeric nature of aminotransferases; for instance, an aspartate aminotransferase from the bacterium *Methanobacterium thermoformicium* was reported to be active as a tetramer (Tanaka *et al.*, 1994), and there are examples of aminotransferases from *Secale cereale* and *Zea mays* that are active as monomers (Paszkowski and Niedzielska, 1989; Orzechowski *et al.*, 1999).

Most aminotransferases require the cofactor pyridoxal-5'-phosphate (PLP) for activity. The purpose of PLP is to serve as an electron sink, which accepts electrons from substrates (John, 1995). The involvement of PLP in amino acid metabolism is not unique to transamination reactions as it is involved in many reactions pertaining to decarboxylation, racemization, aldol cleavage and β- and γ-substitutions (Leipman and Olsen, 2004). In enzymes that require PLP as a cofactor for enzyme activity, the PLP is covalently linked to a conserved lysine residue through an imine linkage to the ε-amino group. The conserved lysine is part of the active site of the enzyme (Mehta and Christen, 2000).

Aminotransferase reactions proceed through two steps facilitated by a bimolecular ping-pong mechanism (Velick and Vavra, 1962). In the first step, an amino acid substrate is converted into its cognate 2-oxo acid. This first step occurs in three distinct stages. The first stage is a Schiff base formation facilitated by the PLP cofactor that is covalently linked to a lysine residue in the active site of the enzyme. The second stage is tautomerization, and this is followed by the third and final stage of the first step, hydrolysis. In the second step of the mechanism, the 2-oxo acid is converted to its amino acid cognate by transfer of the amino group (NH_2) from the amino donor to the exposed keto group (=O) of the amino acceptor (Leipman and Olsen, 2004) (Fig. 4.2).

4.3 Aminotransferases from the Model Organism *Arabidopsis thaliana*

The genome of *Arabidopsis thaliana*, a model organism, was the first plant genome to be elucidated (The Arabidopsis Genome Initiative, 2000). As such, *Arabidopsis* provides an excellent tool to study many enzymes from plants, including

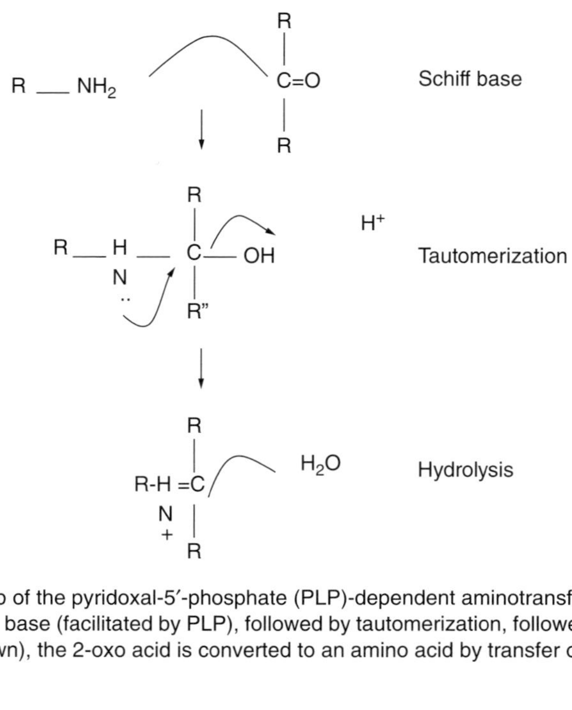

Fig. 4.2. The first step of the pyridoxal-5′-phosphate (PLP)-dependent aminotransferase reaction: formation of the Schiff base (facilitated by PLP), followed by tautomerization, followed by hydrolysis. In the second step (not shown), the 2-oxo acid is converted to an amino acid by transfer of the amino group from the donor.

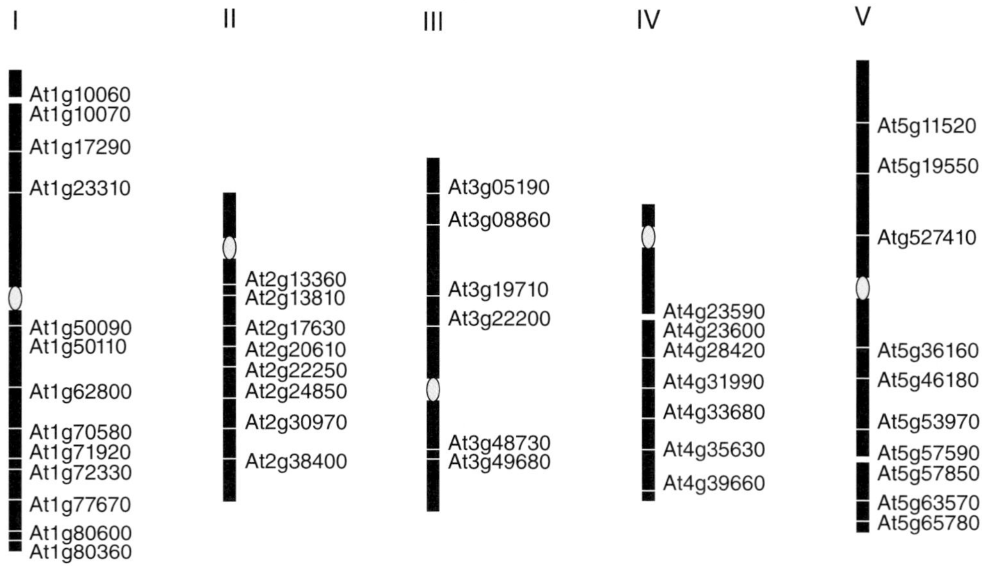

Fig. 4.3. The relative chromosomal map positions of aminotransferase and aminotransferase-like genes on the five chromosomes (I–V) of *Arabidopsis thaliana*. (Adapted from Liepman and Olsen, 2004.)

aminotransferases. Genomic mining by Liepman and Olsen (2004) demonstrated that the plant contains 44 annotated aminotransferase genes or aminotransferase-like genes that are distributed on the five chromosomes of the plant (see Fig. 4.3). Of the 44 aminotransferase or aminotransferase-like genes that are present, seven are annotated as tyrosine aminotransferases (Prabhu and Hudson, 2010) (Table 4.1). To date, five of the seven annotated tyrosine aminotransferases have been confirmed to be authentic by biochemical experimentation.

4.4 The Anabolism of Tyrosine and Phenylalanine in Plants and Bacteria

The proteogenic aromatic amino acids – tyrosine, phenylalanine and tryptophan – are all synthesized from the common intermediate chorismate. Chorismate is synthesized via the shikimate pathway, which is found in plants, bacteria and fungi. This pathway is the central metabolic pathway for the biosynthesis of aromatic compounds that are carbocyclic in nature. In addition to being involved in the synthesis of the three aromatic amino acids (tyrosine, phenylalanine and tryptophan), chorismate is also involved in the synthesis of vitamins K and E, folate, plastoquinones and coenzyme Q, among other compounds.

4.4.1 The anabolism of tyrosine and phenylalanine in bacteria

In bacteria, tyrosine, phenylalanine and tryptophan are all synthesized from chorismate. Starting with chorismate, the first step in the synthesis is catalysed by the bifunctional enzyme, chorismate mutase (EC 5.4.99.5)/prephenate dehydratase (EC 4.2.1.51), which is encoded by the *pheA* gene. This enzyme (Phe) catalyses the conversion of chorismate to prephenate and of prephenate to phenylpyruvate. For the synthesis of tyrosine, the bifunctional enzyme chorismate mutase/prephenate dehydrogenase (EC 5.4.99.5) encoded by the *tyrA* gene, Tyr A, catalyses the conversion of chorismate to 4-hydroxyphenylpyruvate (rather than phenylpyruvate). The final reaction in the synthesis of both tyrosine and phenylalanine is facilitated by the enzyme tyrosine aminotransferase, TyrB, encoded by the *tyrB* gene, which is

Table 4.1. List of the seven annotated tyrosine aminotransferases and tyrosine aminotransferase-like genes from the model plant organism *Arabidopsis thaliana.* The accession numbers and predicted function of each orthologue are from GenBank™ (http://www.ncbi.nlm.nih.gov/genbank/).

Accession no.	Locus tag	Annotation in GenBank™	Reference
AAG37061	At2g20610	Tyrosine aminotransferase	Mikkelsen *et al.*, 2004
AAN31921	At5g53970	Tyrosine aminotransferase	Riewe *et al.*, 2012
AAP31937	At2g24850	Tyrosine aminotransferase	Höllander-Czytko *et al.*, 2005
CAB79314	AT4g23590	Putative tyrosine aminotransferase	–
CAB79315	At4g23600	Tyrosine aminotransferase	Lopukhina *et al.*, 2001
CAB79644	AT4g28420	Putative tyrosine aminotransferase	–
NP_198465	At5g36160	Tyrosine aminotransferase	Prabhu and Hudson, 2010

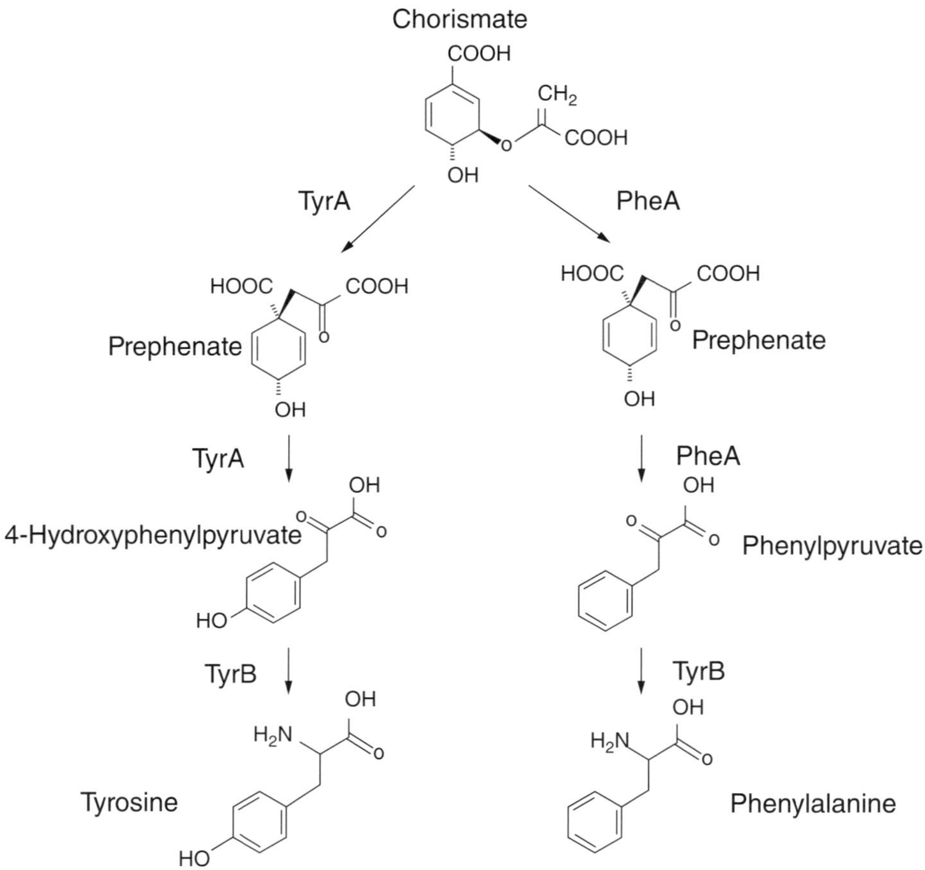

Fig. 4.4. The biosynthesis of tyrosine and phenylalanine from chorismate in *Escherichia coli*. PheA, chorismate mutase/prephenate dehydratase, encoded by the *pheA* gene; TyrA, chorismate mutase/prephenate dehydratase, encoded by the *tyrA* gene; TyrB, tyrosine aminotransferase, encoded by the *tyrB* gene. (Prabhu and Hudson, 2010.)

responsible for the conversion of 4-hydroxyphenylpyruvate to tyrosine and of phenylpyruvate to phenylalanine (Fig. 4.4). It should be noted that aspartate aminotransferase (AspC) and the branched-chain aminotransferase (IlvE) are also involved in tyrosine and phenylalanine anabolism, as these enzymes are promiscuous in their substrate specificity and have overlapping activities with TyrB (Mavrides and Orr, 1975; Gelfand and Steinberg, 1977; Whitaker *et al.*, 1982).

4.4.2 A second pathway for the synthesis of tyrosine and phenylalanine in plants

In plants, the synthesis of tyrosine and phenylalanine also proceeds through the intermediate chorismate (Bender, 2012), but the pathway is distinct from that in fungi and bacteria. The major difference is that plants have the aminotransferase step at the penultimate rather than the ultimate step. In plants, the prephenate is synthesized from chorismate by the enzyme chorismate mutase. The next step in the pathway is facilitated by a glutamate-dependent aminotransferase reaction of prephenate by prephenate aminotransferase (EC 2.6.1.79) to produce arogenate. Arogenate, the result of the aminotransferase reaction from the previous step, is then either oxidized by the oxidoreductase enzyme arogenate dehydrogenase (EC 1.3.1.43) to form tyrosine or converted to phenylalanine by the lyase enzyme arogenate dehydratase (EC 4.2.1.91) (Buchanan *et al.*, 2000) (Fig. 4.5).

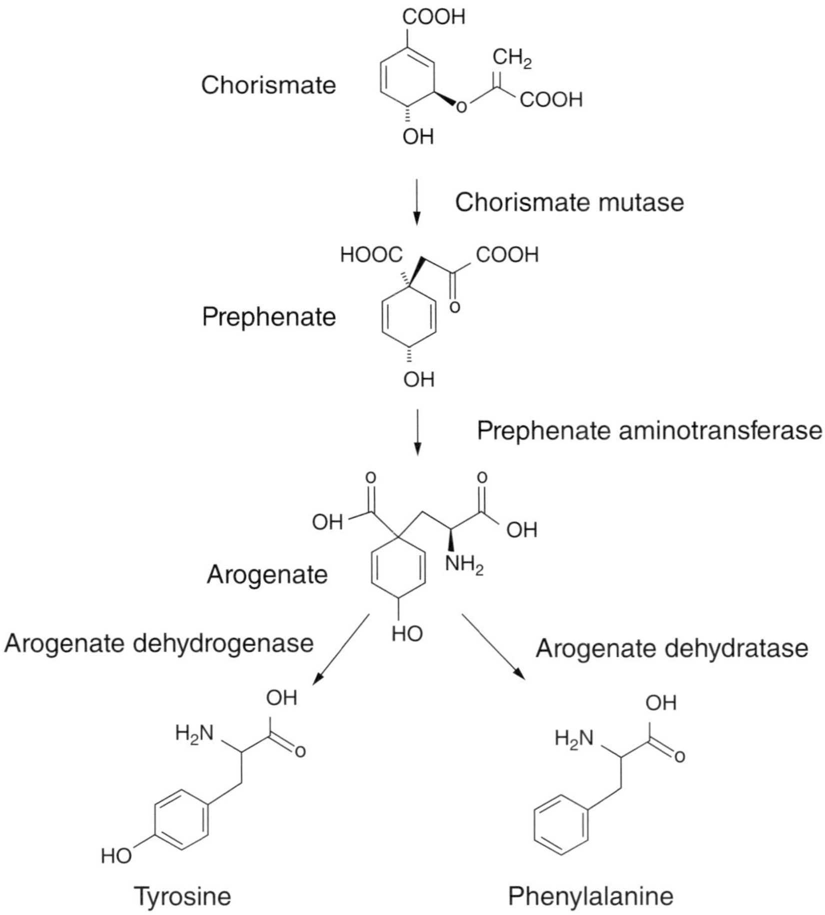

Fig. 4.5. The biosynthesis of tyrosine and phenylalanine from chorismate in plants. In plants, the prephenate is synthesized from chorismate by the enzyme chorismate mutase. The major difference from the pathway in bacteria (see Fig. 4.4) is that plants have the aminotransferase step at the penultimate rather than the ultimate step. (Prabhu and Hudson, 2010.)

4.5 Properties of Tyrosine Aminotransferase Annotated by the Locus Tag At5g36160 from *Arabidopsis thaliana*

As already noted, the enzymatic activities of five of the seven tyrosine aminotransferases from *A. thaliana* have been elucidated (see Table 4.1). In these studies, a series of *in vitro* and/or *in vivo* experiments was carried out to demonstrate the activity of the enzyme.

For example, the gene annotated by the locus tag At5g36160 from *A. thaliana* was shown to encode tyrosine aminotransferase (Prabhu and Hudson, 2010). To demonstrate the activity of this enzyme, the full-length cDNA of At5g36160 was amplified by PCR. For cloning, the PCR fragment was double digested and ligated into the *Escherichia coli* plasmid pET30a to produce a plasmid pET30a + At5g36160 construct. The resultant recombinant enzyme was purified using affinity chromatography facilitated by the hexahistidine S-TAG epitope from the plasmid at the amino terminus of the recombinant protein (Fig. 4.6). The purified recombinant enzyme was then characterized using *in vitro* experiments (described in Sections 4.5.1 and 4.5.2) and also by an *in vivo* analysis of functional complementation (described in Section 4.5.3). Some data on other tyrosine aminotransferases are also briefly outlined in these sections.

4.5.1 Kinetic and physical properties

The purified recombinant tyrosine aminotransferase whose production is described above should be able to convert tyrosine to 4-hydroxyphenylpyruvate and phenylalanine to phenylpyruvate. As most aminotransferases are able to catalyse reversible reactions, the enzyme should also be capable of synthesizing tyrosine from 4-hydroxyphenylpruvate and phenylalanine from phenylpyruvate (Fig. 4.7). To demonstrate enzymatic activity in the forward direction, tyrosine and phenylalanine serve as the two amino donors and 2-ketoglutarate as the amino acceptor. The ability of the enzyme to catalyse the reaction in reverse is demonstrated using glutamate as the amino donor and 4-hydroxphenylpyruvate and phenylpyruvate as the amino acceptors.

The synthesis of 4-hydroxyphenylpyruvate and phenylpyruvate resulting from the transamination of tyrosine and phenylalanine was demonstrated using increasing amounts of the purified recombinant enzyme (Fig. 4.8a). In the

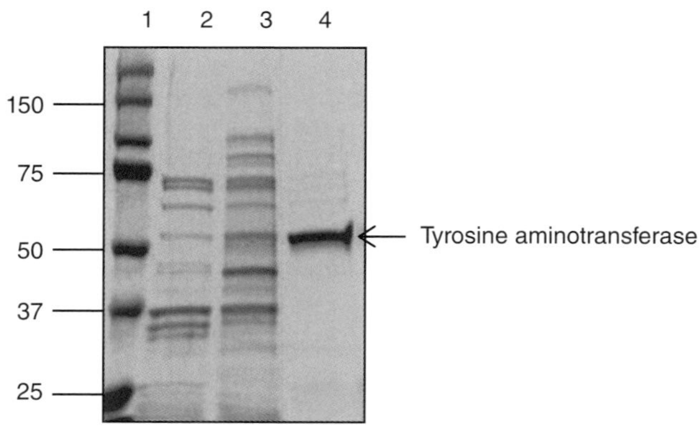

Fig. 4.6. Expression and purification of recombinant tyrosine aminotransferase annotated by the locus At5g36160 from *Arabidopsis thaliana* using *Escherichia coli*. Lane 1, molecular weight markers in kDa; lane 2, 10 µg of soluble proteins extracted from uninduced *E. coli*; lane 3, 10 µg of soluble proteins extracted from induced *E. coli*; lane 4, 1.0 g of the purified recombinant enzyme. The proteins were resolved on an SDS-PAGE (sodium dodecyl sulfate polyacrylamide gel electrophoresis) gel containing 10% (w/v) acrylamide, and the gel was stained using Coomassie Blue. (Prabhu and Hudson, 2010.)

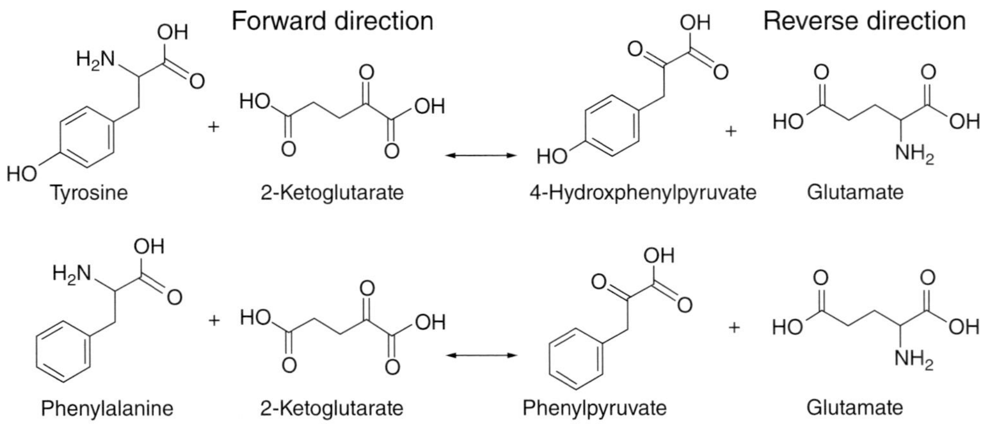

Fig. 4.7. Reactions catalysed by tyrosine aminotransferase as measured using enzyme annotated by the locus tag At5g36160 from *Arabidopsis thaliana*. In the forward direction, tyrosine and phenylalanine serve as the amino donors, while 2-ketoglutarate serves as the acceptor. In the reverse direction, glutamate serves as the amino donor, while 4-hydroxyphenylpyruvate and phenylpyruvate serve as the acceptors. (Prabhu and Hudson, 2010.)

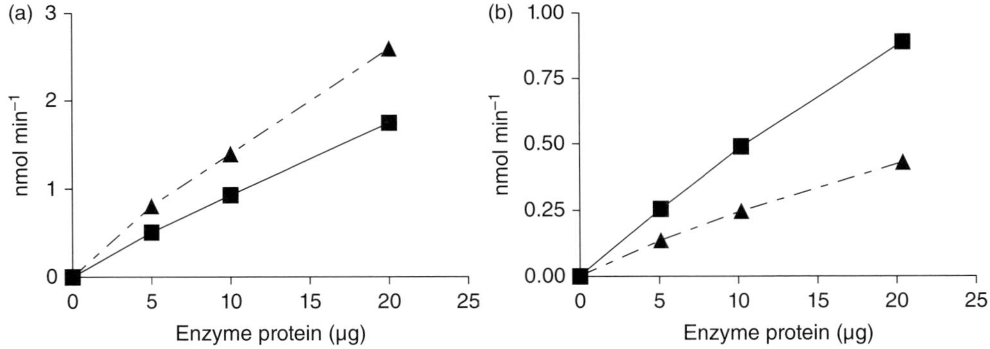

Fig. 4.8. Measurements of the activity of tyrosine aminotransferase annotated by the locus tag At5g36160 from *Arabidopsis thaliana* in the forward and reverse directions. Both graphs show the correlation between reaction rate and the amount of enzyme protein added to the reaction. (a) shows the synthesis of 4-hydroxyphenylpyruvate (squares) and phenylpyruvate (triangles) in the forward direction, while (b) shows the synthesis of 2-ketoglutarate from 4-hyrdoxyphenylpyruvate (squares) and phenylpyruvate (triangles) in the reverse direction. (Prabhu and Hudson, 2010.)

reverse direction, the synthesis of 2-ketoglutarate, which is the result of the transamination reaction using glutamate as the amino donor and 4-hydroxyphenylpyruvate and phenylpyruvate as the amino acceptors, was also observed (Fig. 4.8b) (Prabhu and Hudson, 2010).

The kinetic properties of the tyrosine aminotransferase annotated by At5g36160 from *A. thaliana* are outlined in Table 4.2. The kinetic analysis assessed the activity of the enzyme at several concentrations of one substrate and at the saturation levels of other substrates. The reciprocal plots for the six possible substrates were linear and are consistent with traditional Michaelis–Menten kinetics. The enzyme has a maximum velocity (V_{max}) of approximately 5.0 μmol min^{-1} mg^{-1} in the forward direction and 23.5 μmol min^{-1} mg^{-1} in the reverse direction. The apparent K_m values for the substrates were 0.19 mM for tyrosine, 0.84 mM for phenylalanine, 1.2 mM for

Table 4.2. Kinetic properties of recombinant tyrosine aminotransferase annotated by locus tag At5g36160 from *Arabidopsis thaliana*. V_{max} is represented in μmol min⁻¹ mg⁻¹ of protein; K_m values of the various substrates are given as millimolar (mM) (Prabhu and Hudson, 2010).

Assay	V_{max}	Substrate	K_m
Forward direction	5 ± 0.3	Tyrosine	0.19 ± 0.16
		Phenylalanine	0.84 ± 0.2
		2-Ketoglutarate	1.2 ± 0.27
Reverse direction	23.5 ± 1.0	Glutamate	1.4 ± 0.16
		4-Hydroxyphenylpyruvate	0.22 ± 0.05
		Phenylpyruvate	0.13 ± 0.06

2-ketoglutarate, 1.4 mM for glutamate, 0.22 mM for 4-hydroxyphenylpyruvate and 0.13 mM for phenylpyruvate (Table 4.2) (Prabhu and Hudson, 2010). The tyrosine aminotransferase from *Papaver somniferum* was shown to have K_m values of a similar order of magnitude as those of the recombinant enzyme annotated by At5g36160 from *A. thaliana*: 1.8 mM for tyrosine and 6.33 mM for phenylalanine (Lee and Facchini, 2011).

With respect to pH and temperature optima, the recombinant enzyme annotated by At5g36160 has a pH optimum of 8.2 and a temperature optima range between 30 °C and 40 °C (Prabhu and Hudson, 2010). This is comparable to the tyrosine aminotransferase from *P. somniferum*, which has a pH optimum of 8.5 (Lee and Facchini, 2011).

4.5.2 Substrate specificity

Conceptually, all aminotransferases catalyse a conical reaction by transferring an amino group from a donor to an acceptor. As such, they can be promiscuous, i.e. they are able to use more than one amino donor and/or amino acceptor compound. The At5g36160 orthologue was tested with five different acceptor compounds in the forward assay using tyrosine as the amino donor and either 2-ketoglutarate, pyruvate, oxaloacetate, oxovalerate or prephenate as the amino acceptors. The results from this analysis demonstrated that the enzyme is promiscuous for the amino acceptors tested, although 2-ketoglutarate is the preferred substrate (Table 4.3) (Prabhu and Hudson, 2010).

A tyrosine aminotransferase from *P. somniferum* was also shown by Lee and Facchini (2011) to be promiscuous, in demonstrating the ability to use the three aromatic amino acids tyrosine, phenylalanine and tryptophan as amino donor substrates, in addition to three oxo acids (2-ketoglutarate, pyruvate and oxaloacetate) as amino acceptors. Three isoforms of tyrosine aminotransferase from cell cultures of *Althaea officinialis* (marshmallow) were isolated and characterized by De-Ekanmkul and Ellis (1987). Studies of these gave contrasting results, showing that one of the isoform was very specific for tyrosine as the amino donor and 2-ketoglutrate as the amino acceptor, with another isoform showing higher activity with aspartate as the amino donor and 2-ketoglutarate as the amino acceptor.

Table 4.3. Relative substrate specificity of tyrosine aminotransferase annotated by locus tag At5g36160 from *Arabidopsis thaliana* with tyrosine as the amino donor and various amino acceptors.

Amino donor	Amino acceptor	Relative activity (%)
Tyrosine	2-Ketoglutarate	100
Tyrosine	Oxaloacetate	75
Tyrosine	Pyruvate	40
Tyrosine	Prephenate	23
Tyrosine	Oxovalerate	3

4.5.3 *In vivo* analysis of tyrosine aminotransferase

If the purified recombinant enzyme is able to synthesize tyrosine and phenylalanine as demonstrated by the *in vitro* assays, it should be possible to complement an *E. coli* auxotrophic mutant that is dependent on tyrosine and phenylalanine for viability. A functional complementation assay was employed to assess this using the *E. coli* mutant DL39. DL39 is a triple mutant that has mutations in the aminotransferase genes *aspC* (aspartate aminotransferase), *tyrB* (tyrosine aminotransferase), and *ilvE* (branched-chain aminotransferase) (LeMaster and Richards, 1988). These three aminotransferases have been shown to have overlapping activities in *E. coli* (Gu *et al.*, 1998). As a result of the three mutations, DL39 is auxotrophic for the amino acids aspartate, tyrosine and phenylalanine, and also for the branched-chain amino acids valine, leucine and isoleucine. See Section 4.4.1 for additional information on the overlapping functions of TyrB, AspC and IlvE.

E. coli DL39 was transformed with either an empty plasmid vector (pBAD33) or a vector expressing the tyrosine aminotransferase annotated by the *Arabidopsis* locus tag At5g36160 (pBAD33::At5g36160), which had been produced from the pET30a+At5g36160 construct described in Section 4.5. The analysis demonstrated that while DL39 is able to grow on minimal (M9) medium supplemented with tyrosine and phenylalanine, it is able to grow on minimal medium lacking tyrosine and phenylalanine only when transformed into expressing the plant enzyme. This growth on tyrosine- and phenylalanine-free medium demonstrates the ability of the plant enzyme to functionally complement the *tyrB* mutation present in the *E. coli* DL39 mutant (Fig. 4.9). (Prabhu and Hudson, 2010).

4.6 The Role of Tyrosine Aminotransferase in Plants

In plants, the phenylpropanoid pathway is involved in the biosynthesis of a variety of compounds that are generated from the shikimate pathway. As discussed earlier in the chapter, tyrosine and phenylalanine are anabolized from the shikimate pathway. The enzyme phenylalanine ammonia lyase (PAL) (EC 4.3.1.24) converts phenylalanine to cinnamate. Cinnamate is a precursor of *p*-coumaroyl CoA, the central metabolite in the phenylpropanoid pathway (Fig. 4.10).

The sub-cellular synthesis of tyrosine and phenylalanine has been confirmed to be localized in the chloroplasts of plants (Gaines *et al.*, 1982; Benesova and Bode, 1992), while the sub-cellular localization of tyrosine aminotransferase annotated by At5g36160 is predicted to be in the cytoplasm using various sub-cellular localization prediction algorithms (Prabhu and Hudson, 2010). This leads to the idea that tyrosine aminotransferases from plants are likely to be involved in the synthesis of secondary metabolites, given that the enzyme is not necessary for the synthesis of tyrosine and phenylalanine and also the fact that the enzyme is not localized in a cellular compartment where tyrosine and phenylalanine are known to occur. The products that result from the catabolism of tyrosine and phenylalanine are shuttled into the phenylpropanoid pathway, which is responsible for the synthesis of many compounds, including coumarins,

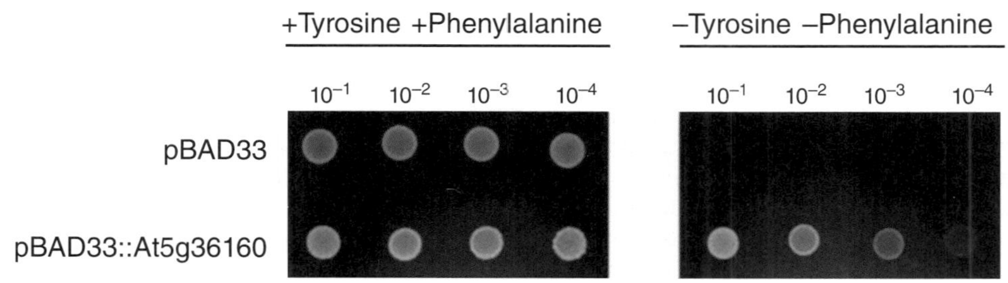

Fig. 4.9. Functional complementation assay using the *Escherichia coli* triple mutant DL39 which is auxotrophic for several amino acids, including tyrosine and phenylalanine. DL39 was transformed with either an empty plasmid vector (pBAD33) or a vector expressing the tyrosine aminotransferase annotated by the *Arabidopsis* locus tag At5g36160 (pBAD33::At5g36160). While DL39 is able to grow on minimal (M9) medium supplemented with tyrosine and phenylalanine (left), it is able to grow on minimal medium lacking tyrosine and phenylalanine only when transformed into expressing the plant enzyme (right bottom). The figures along the top indicate bacterial dilutions. (Prabhu and Hudson, 2010.)

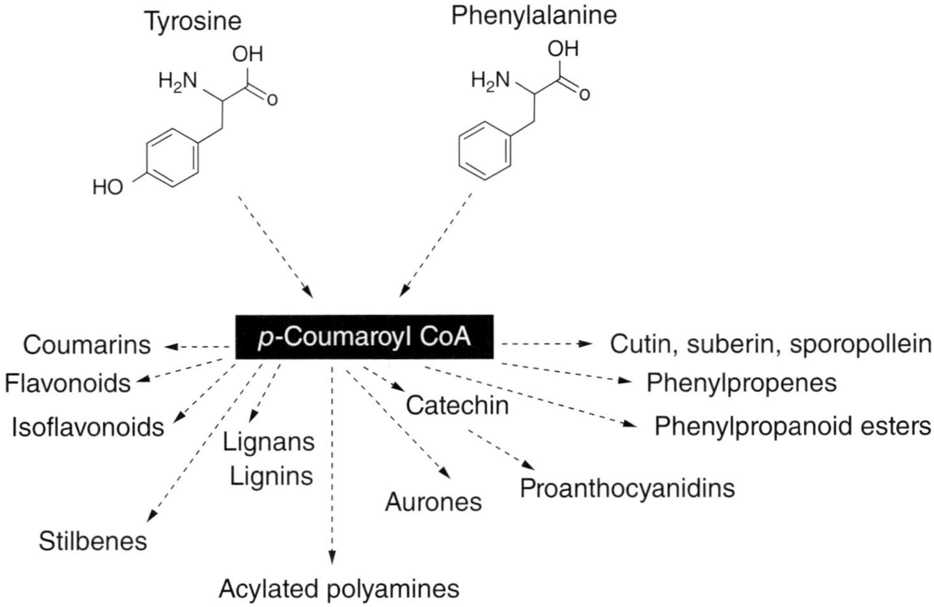

Fig. 4.10. The phenylpropanoid pathway derived from the catabolism of tyrosine and phenylalanine in plants.

flavonoids, isoflavonoids, stilbenes, aurones, cutin, suberin, sporopolleins, catechins, proanthocyanidins, lignans, lignins, phenylpropenes, acylated polyamines and many other alkaloid derivatives (Fig. 4.10) (Fraser and Chapple, 2011).

In plants, the secondary metabolites that are derived from the phenylpropanoid pathway are directly involved in the response of plants to biotic and abiotic stresses. These compounds have integral roles in providing mechanical strength to plant cells, and in pollen viability, pest deterrence, protection against UV, disease resistance, drought resistance and insect resistance, among many other roles (Nair *et al.*, 2004).

An example of secondary metabolite synthesis in response to biotic stress has been studied in *A. thaliana*. The plant bacterial pathogen *Pseudomonas syringae* produces the phytotoxin coronatine, which belongs to the polyketide class of secondary metabolites and is an analogue of octadecanoids (jasmonic acid or 12-oxo-phytodienoic acid). Plants that are exposed to coronatine typically express a chlorotic phenotype (Gnanamanickam *et al.*, 1982). In response to this phytotoxin, plants produce and accumulate anthocyanins, which are directly derived from the phenylpropanoid pathway (Bent *et al.*, 1992; Kong *et al.*, 2003). The production of flavonoids via the phenylpropanoid pathway is important because these compounds are known to possess antioxidant properties that are involved in protection from free radical compounds that are harmful to the plant cell. Lopukhina *et al.* (2001) showed that the tyrosine aminotransferase annotated by the locus tag At4g23600 from *A. thaliana* was regulated by coronatine.

Tyrosine aminotransferase has also been implicated in tocopherol anabolism in plants (Lopukhina *et al.*, 2001; Höllander-Czytko *et al.*, 2005). In addition, the enzyme was also shown to be involved in the synthesis of rosmarinic acid in whole plants and in hairy root cell cultures (De-Eknamkul and Ellis, 1987; Xiao *et al.*, 2009). Tocopherols and rosmarinic acid are involved in the scavenging of free radicals and help to protect the plant from various stresses (Xiao *et al.*, 2009). Recently, a tyrosine aminotransferase from *P. somniferum* was implicated in the synthesis of the alkaloid benzylisoquinoline, which is involved in the synthesis of many alkaloid compounds, including papaverine, noscapine, codeine, morphine, apomorphine, berberine, protopine, tubocurarine and sanquinarine (Lee and Facchini, 2011).

4.7 Conclusions

In *A. thaliana*, aminotransferases comprise a significant subset of the approximately 25,000 genes of the plant (The Arabidopsis Genome Initiative, 2000). Many of these genes have been shown to be involved integral pathways for facilitating proper growth, development and physiological responses to biotic and abiotic stresses. Tyrosine aminotransferase is an interesting enzyme given that it is not necessary for the anabolism of tyrosine and phenylalanine in plants, as it is in bacteria, but is primarily involved in the catabolism of these two amino acids as a means of facilitating the shunting of precursor intermediates into the phenylpropanoid pathway. Numerous studies outlined in this chapter have implicated tyrosine aminotransferase in the metabolism of compounds involved in the synthesis of secondary metabolites that affect the plant response(s) to both biotic and abiotic stresses. The enzymatic activity of the enzyme has also been elucidated from a number of plant species, although there is a lack of structural data available in the literature. The protein structure of the tyrosine aminotransferase orthologue from *E. coli* has been elucidated using X-ray crystallography (Ko *et al.*, 1999). Interestingly, the orthologue annotated by At5g36160 is only 7.6% identical to its *E. coli* counterpart, yet it was still able to functionally complement an *E. coli* mutant that is auxotrophic for tyrosine and phenylalanine (Prabhu and Hudson, 2010). It would be interesting to elucidate the structure of the plant enzyme to garner some insights into its structure. In addition, it would be interesting to study plant-based systems expressing varying amount of the enzyme. As the enzyme has been shown to be involved in the synthesis of compounds that are involved in plant defence, one wonders whether plants that overexpress tyrosine aminotransferase will be more resistant or tolerant to biotic and or abiotic stresses? Another important question is whether the overexpression of the enzyme will lead to adverse effects resulting from the over accumulation of toxic precursor(s). It would also be of interest to ascertain the phenotype of plants that lack tyrosine aminotransferase activity; will this plant be more susceptible and less tolerant to biotic and abiotic stresses?

Acknowledgement

This work was funded by a National Science Foundation (NSF) MCB-1120541 award to André O. Hudson.

References

Beale, S.I. and Castelfranco, P.A. (1974) ^{14}C incorporation from exogenous compounds into aminolevulinic acid by greening cucumber cotyledons. *Biochemical and Biophysical Research Communications* 52, 143–149.

Bender, D.A. (2012) The aromatic amino acids: phenylalanine, tyrosine, and tryptophan. *Amino Acid Metabolism*, 3rd edn. Wiley, Chichester, UK.

Benesova, M. and Bode, R. (1992) Chorismate mutase isoforms from seeds and seedlings of *Papaver somniferum*. *Phytochemistry* 31, 2983–2987.

Bent, A.F., Innes, R.W., Ecker, J.R. and Staskawicz, B.J. (1992) Disease development in ethylene-insensitive *Arabidopsis thaliana* infected with virulent and avirulent *Pseudomonas* and *Xanthomonas* pathogens. *Molecular Plant–Microbe Interactions* 5, 372–378.

Braunstein, A.E. (1973) Amino group transfer. In: Boyer, P.D. (ed.) *The Enzymes. Vol. IX: Group Transfer, Part B.* Academic Press, New York, pp. 379–471.

Braunstein, A.E. and Kritzmann, M.G. (1937) Formation and breakdown of amino acids by intermolecular transfer of the amino group. *Nature* 140, 503–504.

Buchanan, B., Gruissem, W. and Jones, R.L. (eds) (2000) *Biochemistry and Molecular Biology of Plants.* American Society of Plant Biologists, Rockville, Maryland.

Cedrangolo, F. and Carandante, G. (1940) Transamination in higher plants. *Bollettino della Società Italiana di Biologia Sperimentale* 15, 482–484.

De-Eknamkul, W., Ellis, B.E. (1987) Purification and characterization of tyrosine aminotransferase activities from *Anchusa officinalis* cell cultures. *Archives of Biochemistry and Biophysics* 257, 430–438.
Fraser, C.M. and Chappel, C. (2011) The phenylpropanoid pathway in *Arabidopsis*. *The Arabidopsis Book* 9:e0152.
Gaines, C.G., Byng, G.S., Whitaker, R.J. and Jensen, R.A. (1982) l-tyrosine regulation and biosynthesis via arogenate dehydrogenase in suspension-cultured cells of *Nicotiana silvestris* Speg. et Comes. *Planta* 156, 233–240.
Gelfand, D.H. and Steinberg, R.A. (1977) *Escherichia coli* mutants deficient in the aspartate and aromatic amino acid aminotransferases. *Journal of Bacteriology* 130, 429–440.
Givan, C.V. (1980) Aminotransferases in higher plants. In: Milfin, B.J. (ed.) *The Biochemistry of Plants, Vol. 5: Amino Acids and Derivatives*. Academic Press, New York, pp. 329–357.
Gnanamanickam, S.S., Starratt, A.N. and Ward, E.W.B. (1982) Coronatine production *in vitro* and *in vivo* and its relation to symptom development in bacterial blight of soybean. *Canadian Journal of Botany* 60, 645–650.
Gu, W., Song, J., Bonner, C.A., Xie, G. and Jensen, R.A. (1998) PhhC is an essential aminotransferase for aromatic amino acid catabolism in *Pseudomonas aeruginosa*. *Microbiology* 144, 3127–3134.
Hasse, K. and Schmid, G. (1963) [Synthesis and degradation of biogenic amines by enzymatic transamination]. *Biochemische Zeitschrift* 337, 69–79.
Höllander-Czytko, H., Grabowski, J., Sandorf, I., Weckermann, K. and Weiler, E.W. (2005) Tocopherol content and activities of tyrosine aminotransferase and cysteine lysase in *Arabidopsis* under stress conditions. *Plant Physiology* 162, 767–770.
Hudson, A.O., Singh, B.K., Leustek, T. and Gilvarg, C. (2006) An L,L-diaminopimelate aminotransferase defines a novel variant of the lysine biosynthesis pathway in plants. *Plant Physiology* 140, 292–301.
John, R.A. (1995) Pyridoxal phosphate dependent enzymes. *Biochimica et Biophysica Acta* 1248, 81–96.
Ko, T.-P., Wu, S.-P., Yang, W.-Z., Tsai, H. and Yuan, H.S. (1999) Crystallization and preliminary crystallographic analysis of the *Escherichia coli* tyrosine aminotransferase. *Acta Crystallographica Section D: Biological Crystallography* 55, 1474–1477.
Kong, J.M., Chia, L.S., Goh, N.K., Chia, T.F. and Brouillard, R. (2003) Analysis and biological activities of anthocyanins. *Phytochemistry* 64, 923–933.
Kritzmann, M.G. (1939) The isolation and properties of the enzyme effecting transamination of aspartic acid (aspartic aminopherase). Communication of the formation and breakdown of amino acids by intermolecular transfer of amino-groups. *Biokhimiya* 4, 691.
Lee, E. and Facchini, P.J. (2011) Tyrosine aminotransferase contributes to benzylisoquinoline alkaloid biosynthesis in opium poppy. *Plant Physiology* 157, 1067–1078.
Leipmann, A.H. and Olsen, L. (2004) Genomic analysis of aminotransferases in *Arabidopsis thaliana*. *Critical Reviews in Plant Sciences* 23, 73–89.
LeMaster, D.M. and Richards, F.M. (1988) NMR sequential assignment of *Escherichia coli* thioredoxin utilizing random fractional deuteriation. *Biochemistry* 27, 142–150.
Leonard, M.J.K. and Burris, R.H. (1947) A survey of transaminases in plants. *The Journal of Biological Chemistry* 170, 701–709.
Lopukhina, A., Dettenberg, M., Weiler, E.W. and Höllander-Czytko, H. (2001) Cloning and characterization of a coronatine-regulated tyrosine aminotransferase from *Arabidopsis*. *Plant Physiology* 126, 1678–1687.
Lu, T.S. and Mazelis, M. (1975) L-ornithine 2-oxoacid aminotransferase from squash cotyledons. *Plant Physiology* 55, 502–506.
Mavrides, C. and Orr, W. (1975) Multispecific asparate and aromatic amino acid aminotransferases in *Escherichia coli*. *The Journal of Biological Chemistry* 250, 4128–4133.
Mehta, P.K. and Christen, O. (2000) The molecular evolution of pyridoxal-5′-phosphate dependent enzymes. *Advances in Enzymology and Related Areas of Molecular Biology* 74, 129–184.
Mikkelsen, M.D., Naur, P. and Halkier, B.A. (2004) *Arabidopsis* mutants in the C-S lyase of glucosinolate biosynthesis establish a critical role for indole-3-acetaldoxime in auxin homeostasis. *The Plant Journal* 37, 770–777.
Nair, R.B., Bastress, K.L., Ruegger, M.O., Denault, J.W., Chapple, C., Smulevich, G. and Welinder, K.G. (2004) The *Arabidopsis thaliana REDUCED EPIDERMAL FLUORESCENCE1* gene encodes an aldehyde dehydrogenase involved in ferulic acid and sinapic acid biosynthesis. *The Plant Cell* 16, 544–554.
Orzechowski, S., Socha-Hanc, J. and Paszkowski, A.(1999) Alanine aminotransferase and glycine aminotransferase from maize (*Zea mays L.*) leaves. *Acta Biochimica Polonica* 46, 447–457.

Paszkowski, A. and Niedzielska, A. (1989) Glutamate:glyoxylate aminotransferase from the seedlings of rye (*Secale cereal L.*). *Acta Biochimica Polonica* 36, 17–29.

Prabhu, P.R. and Hudson, A.O. (2010) Identification and partial characterization of an L-tyrosine aminotransferase (TAT) from *Arabidopsis thaliana*. *Biochemistry Research International (2010)* Article ID 549572.

Riewe, D., Koohi, M., Lisec, J., Pfeiffer, M., Lippmann, R., Schmeichel, J., Willmitzer, L. and Altmann, T. (2012) A tyrosine aminotransferase involved in tocopherol synthesis in *Arabidopsis*. *The Plant Journal* 71, 850–859.

Roberts, M.F. (1977) Purification and properties of L-alanine: 5-ketooctanal aminotransferase from *Conium macualatum*. *Phytochemistry* 16, 1381–1386.

Ryan, E. and Fottrell, P.F. (1972) Interactions between an unusual aspartate aminotransferase from *Rhizobium japonicum* and pyridoxal-5′-phosphate studied by affinity chromatography. *FEBS Letters* 23, 73–76.

Stumpf, P.K. and Conn, E.E. (1980) *The Biochemistry of Plants: A Comprehensive Treatise*. Academic Press, New York.

Tanaka, T., Yamamoto, S., Moriya, T., Taniguchi, M., Hayahi, H. and Kagamiyama, H. (1994) Aspartate aminotransferase from a thermophilic formate-utilizing methanogen, *Methanobacterium thermoformicium* strain SF-4: relation to serine and phophoserine aminotransferase, but not to aspartate aminotransferase family. *Journal of Biochemistry (Tokyo)* 115, 309–317.

The Arabidopsis Genome Initiative (2000) Analysis of the genome sequence in the flowering plant *Arabidopsis thaliana*. *Nature* 408, 796–815.

Velick, S.F. and Vavra, J. (1962) A kinetic and equilibrium analysis of the glutamic oxaloacetate transaminase mechanism. *The Journal of Biological Chemistry* 237, 2109–2122.

Virtanen, A.I. and Laine, T. (1938) Biological synthesis of amino acids from atmospheric nitrogen. *Nature* 141, 748–749.

Whitaker, R.J., Gaines, C,G. and Jensen, R.A. (1982) A multispecific quintet of aromatic aminotransferases that overlap different biochemical pathways in *Pseudomonas aeruginosa*. *The Journal of Biological Chemistry* 257, 13550–13556.

Xiao, Y., Gao, S., Di, P., Chen, J., Chen, W. and Zhang, L. (2009) Methyl jasmonate dramatically enhances the accumulation of phenolic acids in *Salvia miltiorrhiza* hairy root cultures. *Physiologia Plantarum* 137, 1–9.

5 An Insight into the Role and Regulation of Glutamine Synthetase in Plants

C. Sengupta-Gopalan* and J.L. Ortega
New Mexico State University, Las Cruces, New Mexico, USA

5.1 Abstract

Nitrogen is the most essential plant macronutrient and the major limiting factor for plant growth and productivity. It is assimilated into glutamine by the activity of the enzyme glutamine synthetase (GS), and the activity of this enzyme plays a central role in the assimilation and re-assimilation of ammonia in higher plants. GS catalyses the ATP-dependent condensation of ammonia with glutamate to yield glutamine. There are two major isoforms of GS in plants, GS_1, which is cytosolic, and GS_2, which is located in the chloroplast), and these two isoforms play non-redundant, non-overlapping metabolic roles. GS_2 is the main isoform that is responsible for the assimilation of ammonia produced by nitrate reduction and photorespiration, but its function in non-photosynthetic tissues is not known. GS_1 assimilates ammonia produced by all other physiological processes. While the norm is for there being one GS_2 gene member, GS_1 in all plants is encoded by small gene families, with each member of the family performing different functions.

With the goal of improving plant performance, several attempts have been made to modulate the levels of GS enzyme in different plants using genetic engineering tools. However, outcomes have been rather inconsistent. The inconsistency points to the complexity of the regulatory mechanisms underlying the GS genes, specifically GS_1. The indispensability of GS, and its role in maintaining homeostasis between the prevention of ammonia build-up and depletion of carbon skeletons, requires that it be regulated very precisely. In plants, GS is regulated at multiple steps, which include transcription, through the stability of mRNA, translation rates and post-translational modification. The stability of the transcript is regulated via the 3′UTR (untranslated sequence) and is in response to plant nitrogen status, while translation can be regulated via the TOR (target of rapamycin) signalling pathway. The highest level of regulation appears to be at the post-translational level, and involves oxidative modification, phosphorylation, tyrosine (Tyr) nitration, ubiquitination and association with other phosphoproteins. The available data further suggest that GS genes in plants may be regulated at different steps in the flow of gene expression, and by C and N metabolites or the C/N ratio.

5.2 Introduction

Nitrogen is an essential macronutrient for all plants and is therefore indispensable for plant growth (Hirel *et al.*, 2007; Liu *et al.*, 2011). Quantitatively, nitrogen is the most essential nutrient for plants and the major limiting factor in plant growth and productivity (Tabuchi *et al.*, 2005). In plants, nitrogen is assimilated into glutamine by the activity of the enzyme glutamine

*E-mail address: csgopala@nmsu.edu

synthetase, which plays a central role in the assimilation and re-assimilation of ammonia in higher plants. Glutamine synthetase (GS; EC 6.3.1.2) catalyses the ATP-dependent condensation of ammonia with glutamate (Glu) to yield glutamine (Gln). This reaction is followed by the reductive transfer of the amide amino group of glutamine to α-ketoglutarate (α-KG) to yield two glutamates, this reaction being catalysed by glutamate synthase (GOGAT). Together the two enzymes form the GS-GOGAT cycle (Forde and Lea, 2007; Lea and Miflin, 2011). Hence, GS is the key enzyme in nitrogen metabolism in plants, as it catalyses the first step in the conversion of an inorganic form of nitrogen, ammonia, to an organic form, Gln, which is the starting point for the synthesis of all other N-containing molecules that make up a cell (Lea and Miflin, 2011; Swarbreck *et al.*, 2011). Because of its central role in N metabolism, GS is highly regulated, and this regulation takes place at multiple levels. Understanding the regulatory mechanism of GS expression and the signals involved, is crucial in our efforts to improve nitrogen use efficiency.

5.3 Classification of Glutamine Synthetase

Based on structure and sequence, three different types of glutamine synthetases have been described in living organisms: GS-I, GS-II and GS-III (Betti *et al.*, 2012). GS type I (GS-I) is mostly found in prokaryotes, and is by far the best characterized of all GS types. It has a dodecameric quaternary structure and subunit molecular masses ranging from 44 to 60 kDa. The other GS found in prokaryotes is GS-III. This is composed of larger subunits of 730 amino acids in average (75–83 kDa), with a dodecameric organization, in which the two hexameric rings associate on the opposite surface relative to GS types I and II (van Rooyen *et al.*, 2011). GS type II (GS-II) is found in eukaryotes and in some species of soil bacteria. This is an octameric or decameric enzyme with smaller protein subunits in the range of 35–50 kDa (Kumada *et al.*, 1993; Woods and Reid, 1993; Betti *et al.*, 2012; Seabra *et al.*, 2013). GSI-encoding genes have been identified in some plants (Mathis *et al.*, 2000), but their function is not known.

The three-dimensional structure of the cytosolic GS1a from maize (*Zea mays*) has been determined by crystallography. The crystal structure of the monocotyledonous ZmGS1a reveals a decamer composed of two face-to-face pentameric rings with a total of ten active sites located between adjacent monomers within each pentameric ring (Unno *et al.*, 2006). The ADP-binding sites are located near the openings of the ten catalytic sites. Preliminary crystallographic characterization of the quaternary structure of the cytosolic GS1a from the dicotyledonous plant *Medicago truncatula* shows that MtGS1a is arranged as two superposed, face-to-face, pentameric rings similar to the maize GS1a (Seabra *et al.*, 2013). However, structural and biochemical results indicate that GS type II from bean (*Phaseolus vulgaris*) is an octameric enzyme built by two tetramers placed back to back and rotated 90° with respect to each other. The GS type II enzyme from bean, GS1α, has four active sites, each one located in the inter-monomer region of a dimer (Llorca *et al.*, 2006). The findings of an octameric organization of the cytosolic GS1α from bean suggests that the quaternary structure of the GS type II enzymes may differ between species. Binding experiments reveal that each subunit of the octameric GSII type enzyme from *P. vulgaris* binds one ATP molecule, which indicates that each dimer maintains two ATP-binding sites, one of them in the active site and one free ATP-binding site in each of the four side faces of the octamer, which might have a regulatory role (Llorca *et al.*, 2006).

5.4 Glutamine Synthetase in Plants

In plants, GS occurs as different isoforms within the cytosol (GS_1) and the chloroplast (GS_2). These display multiple physiological functions controlled by a complex system of developmental, organ, cell and specific sub-cellular factors (Miflin and Habash, 2002). Phylogenetic studies have shown that the GS_1 and GS_2 genes have evolved through gene duplication, and that the duplication occurred before the divergence of monocotyledons and dicotyledons (García-Gutiérrez *et al.*, 1998) and long before the divergence of angiosperms and gymnosperms, as pine has only two cytosolic GSs (GS_1), but no GS_2 (Sáez *et al.*, 2000).

GS activity plays a central role in the assimilation and re-assimilation of ammonia in higher plants. Glutamine, asparagine, aspartate and glutamate all play an important role in N storage and distribution and, in addition, they act as precursors for other amino acids (Brauc *et al.*, 2012). The expression of the GS genes reveals that the different GS isoforms play non-redundant, non-overlapping metabolic roles (Edwards *et al.*, 2004; Cánovas *et al.*, 2007), and the relative abundance of GS isoforms varies within different plant species and even within different organs in a plant.

5.4.1 Chloroplastic glutamine synthetase

Chloroplastic glutamine synthetase (GS_2) is encoded by one or two nuclear genes (Seabra *et al.*, 2010) and is translated as a precursor with a chloroplast-targeting sequence. The chloroplast transit peptide directs the protein into the chloroplast where it functions to assimilate NH_4^+ produced from NO_3^- reduction and photorespiration. However, the same GS_2 precursor in *Arabidopsis* has been shown to be targeted to both the chloroplast and the mitochondria, and it has been proposed that mitochondrial GS, like the chloroplastic GS, functions in the re-assimilation of photorespiratory ammonia (Taira *et al.*, 2004). GS_2 is found primarily in the mesophyll cells of the leaves, which play a crucial role in plant growth as the NH_4^+ released by photorespiration exceeds primary N assimilation by tenfold. Additionally, GS_2 is found in the roots of legumes, and has been suggested to play an important role in the assimilation of NH_4^+ from NO_3^- reduction (Zozaya-Hinchliffe *et al.*, 2005). GS_2 has also been found at significant levels in the root nodules of lucerne (*Medicago sativa*; also known as alfalfa) and *M. truncatula*, but its function in the root nodules is not known. It is interesting to note that about 40% of the GS_2 in the nodules is in the form of the precursor (Melo *et al.*, 2003; Zozaya-Hinchliffe *et al.*, 2005). However, this precursor is only present in the wild type (WT) nodules and not in non-fixing (Fix^-) nodules, implying that GS_2 import into the nodules may have a regulatory function (Melo *et al.*, 2003).

The function of GS_2 in the leaves has a major impact on nodule function as seen in the *Lotus japonicus–Rhizobium* symbiosis. Under non-photorespiratory conditions, mutant plants lacking GS_2 are able to nodulate and fix N_2 like WT plants, but not under photorespiratory conditions. Sucrose is the major photosynthate exported to the nodules, where it functions to provide both energy and C skeletons for the purpose of N_2 fixation and ammonia assimilation. In GS_2 mutants, the relative levels of sucrose are considerably lower in the nodules under photorespiratory than under non-photorespiratory conditions, which negatively affects nodule function. Thus, it can be proposed that GS_2 has a role in the C/N balance of *L. japonicus* plants, and that this is independent of photorespiratory activity (García-Calderón *et al.*, 2012). GS_2 has also been implicated as having a role in photoxidation and salt stress tolerance. More recently, it has been shown that GS_2 also has a role in drought tolerance as well, and this is attributed to the role that it plays in proline synthesis and metabolism (Díaz *et al.*, 2010).

Plants typically contain a single gene that encodes for chloroplastic GS_2, although a second GS_2 gene was recently found in *M. truncatula* (Seabra *et al.*, 2010; Swarbreck *et al.*, 2011). The enzyme encoded by this GS_2 gene member in *M. truncatula* has unique kinetic properties and is expressed only in developing seeds, specifically during seed filling. The authors infer that the second plastid-localized GS enzyme found in *M. truncatula* may function in N remobilization from amide amino acids to other amino acids that are required for the synthesis of storage protein (Seabra *et al.*, 2010). Whether this phenomenon of a second GS_2 gene with a unique function is specific to all legumes is not known at this stage.

5.4.2 Cytosolic glutamine synthetase

Cytosolic glutamine synthetase (GS_1) genes are part of a complex multi-gene family, with a range of at least two to five homologues, depending on the plant species (Lea and Miflin, 2011; Swarbreck *et al.*, 2011). In soybean, three different classes of GS_1 genes, α, β and γ, have been identified, each with two members, which can be accounted for by the allotetraploid nature of soybean (Morey *et al.*, 2002). Only two isoforms of GS_1 have been identified in lucerne (Temple *et al.*, 1995) and in the related plant, *M. truncatula* (Seabra *et al.*, 2013). In *Arabidopsis* and maize, five different GS_1 gene members have been identified (Unno, 2006; Lothier *et al.*, 2011), with three in rice (*Oryza sativa*) (Zhao and Shi, 2005). In *Populus*, there are six different genes for GS_1, organized in groups of duplicated genes (Castro-Rodríguez *et al.*, 2011). Ten GS sequences, including the GS_2 genes, were identified in wheat (Bernard and Habash, 2009). Two different isoforms of cytosolic GS were characterized in pine, GS1a and GS1b, with GS1a being expressed in photosynthetic parenchyma while GS1b is located in the vasculature. Functionally, GS1a is similar to GS_2 in angiosperms, and the two GS_1 genes in pine are evolutionarily divergent (Avila *et al.*, 2001). It is clear that there is no consensus as to how many GS_1 gene members are present in any particular plant species.

The different GS_1 genes within any particular plant exhibit a differential expression pattern between tissue types, and based on their location, they also perform different functions (Bernard and Habash, 2009). In lucerne, both *GS1b* and *GS1a* are expressed constitutively; however, *GS1a* exhibits nodule-enhanced expression (Temple *et al.*, 1995). In soybean, the α forms are the major isoforms in the cotyledons and young roots. The β forms, although constitutive, are ammonia inducible in the N_2-fixing nodules. The two members of the γ class shows differences in the expression pattern, in that the $\gamma 1$ is nodule-specific, while $\gamma 2$ is expressed not only in the nodules but also in the cotyledons and flowers (Roche *et al.*, 1993; Morey *et al.*, 2002).

Cytosolic GS_1, localized in the vasculature elements, is the predominant isoform found in non-photosynthetic tissues, and its role is more complex as demonstrated to the numerous isoforms that are found in plants (Bernard and Habash, 2009; Lea and Miflin, 2011). It has recently been shown that GS_1 is located in the intermediary cells, where nitrogen is exchanged between different cell types, for instance the core linking the mesophyll cells to the vascular bundles in sugarcane (Kichey *et al.*, 2005; Nogueira *et al.*, 2005). In roots, GS_1 assimilates ammonia derived from NO_3^- reduction or directly from the soil (White *et al.*, 2007; Bernard and Habash, 2009; Lothier *et al.*, 2011); in the root nodules, the primary function of the GS_1 is the rapid assimilation of ammonia excreted into the plant cytosol of infected cells by the N_2-fixing bacteroids (Oldroyd and Downie, 2008). The GS_1 in leaves and stems is located in the phloem and functions to generate glutamine for nitrogen transport (Bernard and Habash, 2009; Lea and Miflin, 2011), and it also plays a major role in the synthesis of proline, which functions as a nitrogen source and a key metabolite synthesized in response to water stress (Brugiere *et al.*, 1999). GS_1 also performs other roles as an assimilation enzyme involved in the capture of ammonia released by phenylalanine ammonia lyase (PAL) during the synthesis of secondary products. Phenylpropanoid metabolism during lignification of the vascular cells releases considerable amounts of ammonia that has to be re-assimilated by the GS_1 in these cells (Gómez-Maldonado *et al.*, 2004a). Overlapping expression of GS_1 and PAL has been reported during lignin deposition in maturing leaf blades (Sakurai *et al.*, 2001).

Gene members from the GS_1 multi-gene families might have a role in plant adaptation to low and high nitrate environments, as seen in *Arabidopsis*, where one of the GS_1 gene member (GLN1:2), is expressed at very high levels in the roots and is essential for nitrogen assimilation under high nitrate supply and also for ammonium detoxification (Lothier *et al.*, 2011). GS_1 is commonly believed to be responsible for ammonia re-assimilation during natural and stress-induced senescence and during pathogen attack (Mattsson and Schjoerring, 2003; Kichey *et al.*, 2005; Pageau *et al.*, 2006; Seifi *et al.*, 2013). Overactivation of the GS_1 during the plant defence response indicates a concerted interplay between defence response and primary metabolism, suggesting that GS_1 is not merely involved in scavenging ammonia – the by-product of a defensive pathway, but could also be an influential part of an activated anti-cell death mechanism of an infected plant (Seifi *et al.*, 2013).

While GS_1 activity is key in the control of ammonia and glutamate pools, it is also involved in the control of the synthesis of γ-amino butyric acid (GABA) (Chapter 7). Both glutamate and GABA are considered to have a signalling role in plants (Forde and Lea, 2007; Roberts, 2007; Seifi *et al.*, 2013).

There is evidence showing that GS_1 genes may play a role in grain filling in rice and maize (Tabuchi *et al.*, 2005; Martin *et al.*, 2006). It would thus follow that GS_1 plays a key role both in the assimilation and recycling of nitrogen. Moreover, functional genomics and quantitative trait locus (QTL) approaches, mainly in rice and maize, have emphasized the importance of GS_1

in N management, growth rate, yield and grain filling (Bernard and Habash, 2009; Masclaux-Daubresse *et al.*, 2010; Lea and Miflin, 2011). In fact, both genetic and molecular approaches have shown that GS_1 may be a key component of efficiency of nitrogen use and yield (Hirel *et al.*, 2001; Man *et al.*, 2005; Tabuchi *et al.*, 2005; Martin *et al.*, 2006). In addition to its central role in plant productivity, GS integrates a global N-metabolic network in plants (Kusano *et al.*, 2011).

5.5 Modulation of Glutamine Synthetase Expression in Transgenic Plants

Several attempts have been made to modulate the levels of GS enzyme in different plants using genetic engineering tools, with the goal of improving plant performance, but the outcomes have been rather inconsistent (Temple *et al.*, 1993, 1994, 1998; Fuentes *et al.*, 2001; Ortega *et al.*, 2001, 2004; Carvalho *et al.*, 2003; Fu *et al.*, 2003; Harrison *et al.*, 2003). This inconsistency can be attributed to the differences in the transgene being used and the plant in question. Nevertheless, analysis of these transformants has added to our understanding of how GS is regulated. Tobacco transformants with different GS_1 gene constructs driven by the CaMV 35S promoter showed significantly higher levels of GS enzyme activity over control plants. Under N-sufficient conditions, the transformants showed no physiological differences from the control plants. However, the same transformants under N-deficient conditions exhibited enhancement in all the physiological traits that were measured: shoot and root biomass, leaf area, photosynthetic rates and plant growth (Fuentes *et al.*, 2001). In a similar study by Oliveira *et al.* (2002), transgenic tobacco plants ectopically expressing GS_1 showed an increase in fresh and dry weight and leaf soluble proteins when grown in the light under both N-limited and N-unlimited conditions. These transformants showed a huge decrease in the level of free ammonium in the leaves, which is attributed to the assimilation by the GS_1 expressed in the mesophyll cells of the transformants of ammonia produced during photorespiration. These transformants also exhibited an increased ratio of serine (Ser) to glycine (Gly), an indicator of increased photorespiratory activity. The authors concluded that the improved growth phenotype in the GS_1 transgenic tobacco and other C_3 plants may be a consequence of increased photosynthesis/photorespiration, combined with enhanced nitrogen use efficiency.

Poplar is one of the more well-studied plants regarding the overexpression of GS_1. Poplar plants overexpressing the pine cytosolic *GS1a* gene displayed higher photosynthetic rates and greater vegetative growth than those that were not (Gallardo *et al.*, 1999; El-Khatib *et al.*, 2004; Man *et al.*, 2005; Pascual *et al.*, 2008). In search of the possible link between plant growth and increased GS activity, Man *et al.* (2011) investigated the other roles, besides ammonia assimilation, that GS may play in transgenic poplar. GS is directly involved in the pathway leading to *de novo* purine biosynthesis (Smith and Atkins, 2002), and one of the intermediates in this pathway is the signalling molecule cAMP. Furthermore, glutamine also plays a role in the synthesis of indole-3-acetic acid (IAA) by transferring its α-amino group to chorismate to produce anthranilate, in a reaction catalysed by the enzyme anthranilate synthase (ASA). An increase in both the transcript and the protein corresponding to ASA is seen in these transformants, resulting from an increase in the glutamine pools. Increased vegetative growth of the poplar transformants can thus be attributed to an increased level of IAA resulting from increased ASA (Gallardo *et al.*, 1999; Fu *et al.*, 2003; Man *et al.*, 2005).

Our efforts to express a lucerne (*MsGS1b*) or soybean GS_1 (Gm*gln*β_1) gene in lucerne did not show any increase in GS_1 polypeptide and enzyme activity (Ortega *et al.*, 2001). However, when the soybean GS_1 gene, devoid of its 3′UTR, was used as a transgene, the levels of accumulation of the transgene transcript and of the corresponding protein were many fold higher than the levels obtained in the transformants produced with the gene with its 3′UTR. Nuclear run-off analysis showed that transgene transcription was not a factor in the reduction of the steady state level of the transcript; rather, it was a post-transcriptional event. The transformants with the Gm*gln*β_1 gene with its 3′UTR showed transcripts of higher and lower molecular weight homologous to the Gm*gln*β_1 gene, suggesting a mechanism involving the generation of a double-stranded (ds)RNA and subsequent cleavage of that dsRNA (unpublished data).

The post-transcriptional silencing (PTGS) of soybean GS_1 expression does not involve co-suppression of an endogenous gene, and seems to depend on the 3′UTR, as the silencing effect is suppressed by replacing the GS_1 3′UTR with the NOS terminator (a stop signal for gene transcription), suggesting that neither the GS_1 coding region nor the GS activity have a role in the silencing process. In addition, a GUS reporter gene exhibits post-transcriptional silencing if the GS_1 3′UTR replaces the terminator in the GUS gene construct (Ortega *et al.*, 2006).

The lucerne transformants with the soybean GS_1 gene devoid of its 3′UTR, as in *Lotus corniculatus* and *L. japonicus* overexpressing GS_1 (Vincent *et al.*, 1997; Ortega *et al.*, 2004), showed increased growth and premature flowering and also an increase in proline and asparagine levels (unpublished data). Tobacco plants transformed with a soybean GS_1 gene without its 3′UTR, showed threefold to fourfold increase in GS activity and growth enhancement under both N-sufficient and N-deficient conditions, a phenomenon not seen in any other study. Moreover, these transformants showed delayed flowering (Seger, 2012). These results suggest that the 3′UTR of the GS_1 gene from soybean could have a role in regulating GS activity in order to maintain a balance between preventing a build-up of ammonia to toxic levels and draining C skeletons at the cost of other essential functions.

While in most of the transgenic studies, the CaMV 35S promoter has been used to drive the transgene, there have been some studies done using organ-specific promoters. Pea (*Pisum sativum*) plants transformed with a cytoplasmic GS gene (Gm*glnβ*$_1$) driven by organ-specific promoters showed increased GS_1 levels in all of the expected organs, but other than in the roots of transformants, with the gene driven by the rolD promoter, none of the other transformants showed any increase in GS activity. None of the different classes of transformants, including the transformants with the rolD-GS_1 construct, showed any consistent outcome in any increase in biomass and N content (Fei *et al.*, 2003, 2006).

Lucerne plants expressing a soybean cytosolic gene (without its 3′UTR) driven either by the constitutive CaMV 35S promoter or by the photosynthetic cell-specific Rubisco (ribulose-1, 5-bisphosphate carboxylase/oxygenase) promoter, were compared (Seger *et al.*, 2009). Both sets of transformed plants performed better than control plants, but while the transformants with the CaMV 35S promoter driving the transgene showed a threefold to fivefold higher level of transgene transcript and GS_1 polypeptide and holoprotein than the transformants with the Rubisco promoter, very little difference was seen between them at the enzyme activity level. Tobacco transformants with the same gene constructs showed a similar trend to the lucerne transformants. This discrepancy between transcript levels and enzyme activity can be attributed to the nature of the holoprotein made in the different cell types. The CaMV 35S promoter, being constitutive, is functional both in the mesophyll cells and the vasculature, and the vasculature is where endogenous GS_1 is normally expressed, but the Rubisco promoter is active only in the mesophyll cells. The holoprotein formed in the transformants with the CaMV 35S promoter-GS_1 construct is a heteromer made up of the endogenous GS_1 subunits and the transgene subunits, while the transformants with the RUB-GS_1 constructs form a homomer made up of only the transgene products. These results would strongly suggest that there is regulation beyond the assembly of the holoprotein, probably at the level of enzyme activity (Seger *et al.*, 2009).

Upregulation of GS activity in the roots of *L. japonicus* showed a decrease in plant biomass (Limami *et al.*, 1999). This is in keeping with previous findings that selective inhibition of root GS activity in both cereals and temperate legumes results in enhanced growth (Knight and Langston-Unkefer, 1988). Limami *et al.* (1999) proposed that the higher GS activity results in an increased level of amino acids, which leads to a decreased nitrate uptake and assimilation by feedback inhibition. Analysis of *M. truncatula* transformed with the nodule-specific leghaemoglobin promoter driving *MtGS1* genes in both sense and antisense directions showed an inverse correlation between GS_1 and asparagine synthetase at the transcript and protein levels (Carvalho *et al.*, 2003). However, no physiological analysis was done on these transformants. Nodule-specific downregulation of GS in *L. japonicus*, using an antisense *GS1a* gene construct driven by the leghaemoglobin promoter, resulted in a twofold increase in fresh and dry weight of the nodules (Harrison *et al.*, 2003), although there were no other significant changes in shoot and root weight, flowering time and N_2 fixation. The level of asparagine synthetase transcript and protein, along with the level of free asparagine, was enhanced in the nodules of the transformed plants. The data suggest that there are alternate pathways to assimilate ammonium when GS is limiting, but whether these have a role in controlling biomass is not known. Complete sterility was seen in *L. japonicus* transformed with a lucerne cytosolic GS gene driven by a *Sesbania* leghaemoglobin promoter (Suárez *et al.*, 2003). The sterility was attributed to an increase in glutamine content in the flowers resulting from the activity of the *Sesbania* leghaemoglobin promoter in the ovules and pollen.

There are very few studies reported on the overexpression of GS_2. Tobacco transformants overexpressing GS_2 in a leaf specific manner showed a 15–18-fold increase in the transcript level, but only about a twofold increase in GS_2 activity, suggesting the possibility that GS_2, like GS_1, may be subject to post-transcriptional regulation (Migge *et al.*, 2000). Similarly, lucerne transformed with the lucerne GS_2 gene driven by the CaMV 35S promoter showed transgene transcript accumulation in all the organs, but the corresponding protein was seen only in the leaves and not in the other organs, thus further confirming the possibility of GS_2 being under post-transcriptional regulation (Zozaya-Hinchliffe *et al.*, 2005). In both cases, however, the transformants showed improved performance compared with control plants.

5.6 Regulation of Glutamine Synthetase Gene Expression in Plants

The indispensability of GS and its role in maintaining homeostasis between the prevention of ammonia build-up and the depletion of carbon skeletons requires that it be regulated very precisely. The presence of multiple GS_1 gene members in any particular plant that exhibit organ specificity and physiological differences would imply that the enzymes differ in their ammonia assimilation efficiency.

In bacteria, GS is regulated by feedback inhibition by a number of nitrogenous metabolites (Liaw *et al.*, 1993). In addition, the enzyme is regulated by reversible covalent adenylation (Stadtman, 1990). Further, the structural gene for GS, *glnA*, is subject to transcriptional regulation. Altogether, these regulatory mechanisms ensure that GS activity is precisely controlled and can be rapidly adjusted in response to environmental stimuli. The stimuli controlling GS activity in bacteria include carbon and nitrogen signals, which are antagonistic (Stadtman, 1990; Jiang *et al.*, 1998). The key regulatory stimuli for GS are the intracellular concentrations of α-KG and Gln, as these play a key role in controlling both the adenylation status of GS and the transcription of N-regulated genes. Although normally stable, bacterial GS is turned over when cells are starved of nitrogen (Fulks and Stadtman, 1985), suggesting that the intracellular level of GS in bacterial cells is also regulated by proteolysis. So it is not surprising that GS in plants is regulated at multiple steps, which include transcription, through the stability of mRNA, translation rates and post-translational modification.

5.6.1 Transcriptional regulation

The GS_1 and GS_2 genes are regulated differently within specific cell types and organs, and in response to different developmental, metabolic and environmental cues. This specialization ensures a rapid re-assimilation of ammonium derived from multiple sources (Miflin and Habash, 2002). Overall, these two major GS isoforms play primarily non-overlapping roles in plant N metabolism. Earlier studies focused mainly on the regulation of GS_2, which has been shown to be light regulated and *cis*-element specific for light-regulated genes have been identified in the GS_2 promoters (Tjaden *et al.*, 1994). However, as GS_2 is also expressed in non-photosynthetic tissues, such as the roots and root nodules (Zozaya-Hinchliffe *et al.*, 2005), it must also have other *cis* elements other than the light-regulated elements. Light affects GS_2 gene expression by acting either directly via phytochrome activation or indirectly via changes in the levels of carbon metabolites, i.e. the activation of photosynthesis that leads to an increase in the biosynthesis of sucrose (Oliveira *et al.*, 2002). Feeding experiments in lucerne, using different combinations of C and N metabolites, showed the highest induction of GS_2 in both leaves and roots when fed with both nitrate and sucrose/α-KG simultaneously, suggesting a requirement for both C and N metabolites in the regulation of GS_2 expression (Zozaya-Hinchliffe *et al.*, 2005).

Studies of the regulatory mechanisms that control GS_1 expression have centred mostly at the transcriptional level. Transcription profiling has shown that a large proportion of genes regulated transcriptionally have metabolic functions (Nicolaï *et al.*, 2006). Transcriptome analysis has revealed that sugars effectively regulate genes associated with nitrogen metabolism, including GS, GOGAT, glutamate dehydrogenase (GDH) and glutamine-dependent asparagine synthetase (AS) (Price *et al.*, 2004). Nitrogen-metabolizing genes are also transcriptionally regulated by Glu or Glu-derived metabolites. Systems analysis of expression data has uncovered the mechanisms underlying gene regulation in response to sensing Glu or Glu-derived metabolites. As the master regulators of this network are an MYB family transcription factor, the central clock gene (*CCA1*) and a golden 2-related transcription factor (*GLK1*). Both *CCA1* and *GLK1* positively affect the expression of a gene for GS, while repressing the expression of AS and GDH by binding to the promoters of a leucine zipper transcription factor, bZIP1. Thus, the circadian clock regulates N nutrition and in turn, N status feedback to the circadian clock in part through its effect on *CCA1* regulation (Gutiérrez *et al.*, 2008; Gutiérrez, 2012).

Analyses of the pine *GS1b* promoter have identified regulatory elements that are also found in the promoters of genes encoding enzymes involved in phenylpropanoid metabolism. Transcription factors of the MYB family bind to these regulatory elements and coordinate the expression of GS_1 with the expression of genes involved in the phenylpropanoid pathway (Gómez-Maldonado *et al*, 2004b). Computer analysis of putative regulatory *cis* elements in the promoter of pine *GS1b* revealed the presence of an AT box and an AC motif similar to those found in other genes expressed in the vascular tissue, and a gibberellin (GA)-responsive element (Gómez-Maldonado *et al.*, 2004b). The *GS1b* promoter may be the only GS gene promoter in the plant world that is controlled by GA. Further analysis of the promoters of the *GS1a* and *GS1b* genes of pine have identified putative *cis* elements that could interact with Dof factors; Dof (DNA binding with one finger) protein is a class of transcription factors that play diverse roles in regulating gene expression (Yanagisawa *et al.*, 2004). *Arabidopsis* plants expressing a *Dof1* gene from maize showed improved nitrogen assimilation and growth under low nitrogen conditions, suggesting that *Dof1* could be a key factor in coordinating gene expression involved in carbon skeleton production for the assimilation of nitrogen in plants (Yanagisawa *et al.*, 2004). Overexpression of *Dof* has been shown to increase the activity of GS and promote plant growth (Yanagisawa *et al.*, 2004). Dof factors have also been considered to play a role in the regulation of lignin production (Rogers *et al.*, 2005). A Dof protein was identified that could differentially trans-regulate *GS1a* and *GS1b* genes, acting as an activator for *GS1b* and a repressor for *GS1b* (Rueda-López *et al.*, 2008). Promoter regions of the GS_1 genes in pine and GS_1 gene promoters in angiosperms contain Dof-binding sites, although it is not known as to whether the Dof factors play a role in regulating GS gene expression.

5.6.2 Post-transcriptional regulation

While it has already been established that GS genes are regulated transcriptionally, we now have evidence that GS_1 genes are also subject to regulation at the level of RNA stability. Just as a word of caution, in many cases where we have assumed regulation to be transcriptional based on the steady-state level of transcript, we cannot rule out the possibility that the regulation is at the level of transcript turnover. In bacteria, there are many reports of regulation at the post-translational level, but not at the level of transcript turnover, though there is a recent report of a GS being regulated by ammonium in the filamentous cyanobacterium *Anabaena*, which is subject to post-transcriptional regulation via a protein factor (Galmozzi *et al.*, 2010).

Lucerne transformants having a soybean GS_1 gene (Gm*gln*β_1) with its 5′ and 3′UTR driven by the CaMV 35S promoter showed a low level of transgene transcript accumulation with little to no increase in GS_1 polypeptide or GS activity (Ortega *et al.*, 2001). However, removal of the 3′UTR from the Gm*gln*β_1 transgene abolishes the regulatory constraints in the expression of the transgene in lucerne and the transformants show a transgene GS_1 transcript level which is many fold higher than that seen with the GS_1 transgene with its 3′UTR (Ortega *et al.*, 2006). Gene constructs consisting of the GUS gene fused with the 3′UTRs of the soybean Gm*gln*β_1 and the lucerne *MsGS1b* and *MsGS1a*, and driven by the CaMV 35S promoter, were introduced into lucerne, with the goal of checking whether or not the 3′UTR-mediated transcript destabilization is limited to just the Gm*gln*β_1 gene. The results showed that the 3′UTR of *MsGS1b* is as effective as that of Gm*gln*β_1 in transcript turnover, while the 3′UTR of the *MsGS1a* gene is not (Simon and Sengupta-Gopalan, 2010). It is important to point out that the *GS1b* gene from *M. truncatula* driven by the leghaemoglobin promoter, when introduced into *M. truncatula*, did not show accumulation of the transcript for the transgene, while the transformant with the *MtGS1a* showed significant accumulation of the transcript (Carvalho *et al.*, 2003). These results would seem to imply that, like *MsGS1b* and Gm*gln*β_1, *MtGS1b* is probably also regulated at the level of transcript turnover. Furthermore, to check whether or not the 3′UTR mediated post-transcriptional regulation is unique to lucerne, the gene constructs were introduced into tobacco. The tobacco transformants showed the same trend as those of lucerne, suggesting that the 3′UTR-mediated transcript destabilization is not specific to lucerne, but is more universal.

While we still do not know how the 3′UTR functions in transcript destabilization, glutamine and nitrate seem to play a large role in this activity (Ortega *et al.*, 2001, 2006; Simon and Sengupta-Gopalan, 2010). It would, therefore, follow that when glutamine levels are adequate, the transcript turns over to limit GS enzymatic activity and avoid draining the glutamate pools. The 3′UTR of the Gm*gln*β_1 gene had the same destabilizing effect on the transgene transcript in the nodules as it did in the leaves, but not in the roots (Ortega *et al.*, 2001, 2006). The transcript corresponding to the Gm*gln*β_1 with the 3′UTR showed a many fold higher level of accumulation in the nodules formed by a Fix^- strain of *Sinorhizobium meliloti* than in nodules formed by WT *S. meliloti*. These results lead us to postulate that ammonia, or a metabolite resulting from it, mediates transcript turnover via its 3′UTR. Ammonium produced either by nitrate reduction (in the leaves) or by symbiotic N_2-fixation (WT nodules) promotes RNA turnover via its assimilation product, glutamine. Nitrate by itself could also be the trigger for transcript turnover. Under N_2-limited conditions in the leaves, or in Fix^- nodules, the absence of ammonia or glutamine prevents transcript turnover.

Analysis of the 3′UTR of the soybean Gm*gln*β_1 and the lucerne *MsGS1b* genes have revealed not only regions that are conserved in the two genes, but also the involvement of these sequences in the formation of stem-loop structures. Furthermore, the conserved regions found in the 3′UTR of the GS_1 genes showed the presence of sequences with homology to sequences found in domains I and III of the rice α-amylase gene, which have been shown to play a role in the turnover of the α-amylase mRNA (Chan and Yu, 1998). These results suggest that sequences in the 3′UTR in a class of GS_1 genes play a central role in the post-transcriptional regulation of GS_1 gene expression. We further propose that a product of N assimilation acts directly or indirectly in regulating the level of GS_1 transcript and that the 3′UTR of the GS_1 mRNA may be the target of this regulatory step. Whether this regulation involves an RNA-binding protein or a miRNA is under investigation, as sequences found in the 3′UTRs of the different GS_1 genes could be targets for miRNAs (Zhang *et al.*, 2005; Wang and Xu, 2010). Nutrient-responsive miRNAs have been discovered by next-generation sequencing, thus providing evidence that small RNAs play a role in signalling during the adaptation of plants to nutrient deprivation (Vidal *et al.*, 2010). Numerous studies have implicated miRNAs as key regulators of post-transcriptional and translational gene expression. Recent investigations have examined the changes in expression levels of miRNAs in response to nitrogen starvation in maize, *Arabidopsis* and bean. Upregulated miRNAs include miR156, miR160 and miR447. Shared downregulated miRNAs include miR169, miR397, miR399 and miR408 (Fischer *et al.*, 2013). Repression of miR169 in *M. truncatula* by N limitation point to a mechanistic link between N status and nodule development (Pant *et al.*, 2008). Interestingly, a recent report observed that a moderate expression of miR156 improved N uptake and increased biomass (Fu *et al.*, 2012).

5.6.3 Translational regulation

As described later, plant GS has been shown to be subject to post-translational regulation in diverse ways, but recent data have suggested that a GS_1 gene from soybean (Gm*gln*β_1) is regulated at the level of translation via its 5′UTR (Ortega *et al.*, 2012). The 5′UTR of the Gm*gln*β_1 gene not only acts as a translational enhancer for Gm*gln*β_1 mRNA, but also for a reporter gene. At this stage, it has not been established if the 5′UTR mediated translation enhancement is a regulatory step, but sequence analysis of the 5′UTR of the Gm*gln*β_1 mRNA showed a track of oligopyrimidines at 48–33 nucleotides upstream of the translation start codon (Ortega *et al.*, 2012). 5′-Terminal oligopyrimidine tracts (5′-TOP) at the transcriptional start site are translational enhancers found in many mRNAs in animals and plants (Hamilton *et al.*, 2006; Tzeng *et al.*, 2009).

One important mechanism for translational control is mediated through TOR, and the effect of TOR regulation is the translational control of mRNAs containing a 5′-TOP sequence in their mRNA transcripts. The ribosomal S6 protein (RPS6) in eukaryotes is one of the key elements in the control of the translation rates of mRNAs coding for proteins involved in protein synthesis. Many of these transcripts contain a *cis* element, an oligopyrimidine tract that is recognized by the RPS6 protein and leads to higher translation rates under stress conditions. The oligopyrimidine tract present in the mRNA of the Gm*gln*β_1 mRNA may be an RPS6 protein binding site or a site for interaction with another ribosomal protein or a translation initiation factor.

Translational control has been observed for genes involved in many functions with responses to light, sugars and hormones, programmed cell death, protein synthesis and metabolic functions among the most important (Kawaguchi and Bailey-Serres, 2002; Nicolaï *et al.*, 2006). The mammalian target of rapamycin (mTOR) is a multi-domain protein that is important in the regulation of the translation machinery in response to hormones, such as insulin, and amino acids. mTOR signalling is fundamental in regulating physiological processes, such as the regulation of cell size, and in integrating amino acid- and energy-sensing pathways; it also plays a role in pathological conditions, such as cancer (Ren *et al.*, 2012). The TOR regulation of energy and nutrient availability is known to be mediated by alterations in protein and ribosome production (Wullschleger *et al.*, 2006). TOR suppression in *Arabidopsis* significantly extended lifespan and had a negative effect in biomass; the genes for both RPS6A and RPS6B in *Arabidopsis* function as downstream mediator components of TOR signalling and are required for regulating growth processes (Ren *et al.*, 2012). TOR and RPS6 overexpression contributed to accelerated senescence and flowering. TOR requires both light and nutrition signals to regulate the translation machinery. The TOR pathway is a highly conserved signalling pathway in other eukaryotes as well, and has a central role in sensing nutrient and energy status. It has a critical role in regulating cell metabolism and anabolic functions through alterations in protein synthesis and ribosome biogenesis. The essential role of TOR in plants became evident because mutations in TOR in *Arabidopsis* caused embryo lethality. In plants, TOR is a key element in the control of metabolic processes associated with cell growth and phase transition (Ren *et al.*, 2012). It is tempting to speculate that the 5′UTR of Gm*gln*β_1 (and other GS) genes acts through the TOR pathway as an enhancer of translation under conditions when nitrogen needs to be remobilized, such as during stress or senescence or phase transition.

5.6.4 Post-translational regulation

While in the past many the studies have focused on the transcriptional regulation of GS, attention is now being diverted towards steps downstream of transcription. In bacteria, the regulation of GS by cumulative feedback inhibition, adenylation/deadenylation, and repression/derepression is well documented (Stadtman, 1990). In plants, it appears that there are multiple ways to regulate GS levels at the post-translational level.

Oxidative modification has been implicated as the first step in the turnover of GS in bacteria (Rivett and Levine, 1990). In plants, GS is subject to a two-step mechanism involving oxidative modifications and turnover. There are two modes of turnover (Ortega *et al.*, 1999): (i) conformational or hydrophobicity changes of the holoprotein, followed by proteolysis; and (ii) disassembly followed by proteolysis. In the presence of the substrate, the active site of the enzyme is less accessible to oxidative modification or the production of oxygen radicals is enhanced under low nitrogen conditions. In the absence of the substrate, the enzyme is biologically inactive and, therefore, its selective degradation can have little effect on its biological functions. However, by degradation, it can yield amino acids needed for the synthesis of other proteins. In the same context, it can be argued that in plants, the GS enzyme is always available to metabolize any ammonia that may be produced, thereby avoiding the toxic build-up of ammonia. It would then follow that in plants, the process of GS turnover and synthesis is a continuous process: when the substrate is limiting, the enzyme is turned over and when the substrate is available, the enzyme is stabilized. Earlier work from our laboratory had shown instability of a GS holoprotein in the nodules of soybean in the absence of N_2 fixation, suggesting that in the absence of ammonia, the holoprotein turns over (Temple *et al.*, 1996).

GS – both GS_2 and GS_1 – are regulated post-translationally by reversible phosphorylation catalysed by protein kinases and protein phosphatases (Finnemann and Schjoerring, 2000). Reversible phosphorylation has been shown for other key enzymes in C and N metabolism too, such as nitrate reductase and sucrose phosphate synthase (SPS) (Sugden *et al.*, 1999). GS_2 is regulated by phosphorylation and the subsequent binding of 14-3-3 proteins (a family of conserved regulatory molecules that are expressed in all eukaryotes), which then induces selective proteolysis of the GS_2 protein (Riedel *et al.*, 2001; Lima *et al.*, 2006a). This manner of regulation is reminiscent of how nitrate reductase is regulated (Mackintosh and Meek, 2001). GS_1, like GS_2, is also subject to phosphorylation, but unlike GS_2 does not interact with 14-3-3 proteins (Lima *et al.*, 2006b). However, GS_1 in *M. truncatula* has been shown to associate with some phosphopolypeptides. These putative GS-binding phosphoproteins were found to be differentially phosphorylated under different physiological conditions, with higher level of phosphorylation in Fix^- nodules than in Fix^+ nodules. The authors propose that these GS-binding phosphopolypeptides could be involved in regulating GS activity in the root nodules of *M. truncatula* in response to the nitrogen status (Lima *et al.*, 2006a,b).

Tyr nitration has been implicated in the regulation of GS in animals (Gorg *et al.*, 2007). Protein Tyr nitration is a nitric oxide (NO)-mediated post-translational modification and involves the addition of a nitro group to the aromatic ring of Tyr residues, which can alter the conformation and structure of proteins, as well as the catalytic properties of enzymes and their susceptibility to proteolysis (Abello *et al.*, 2009). Tyr nitration represents an important NO-mediated post-translational modification mechanism in plants (Lozano-Juste *et al.*, 2011). Even though NO inhibits nitrogenase, it appears that NO production is required for nodule development and function. In functional nodules of *M. truncatula*, NO production coincides with the location of GS in the nitrogen fixation zone. In root nodules of *M. truncatula*, it has been shown that GS is inactivated by Tyr nitration *in planta* and that GS nitration status is positively correlated with nitrogen fixation (Melo *et al.*, 2011). Silva and Carvalho (2013) have put forward a model for the involvement of GS in the NO signalling pathway in the root nodules. The inhibition of GS activity by Tyr nitration is directly related to NO-mediated nitrogenase inhibition, and inhibition of GS activity results in increased levels of glutamate, which is both a substrate for GS activity and a precursor of glutathione (GSH). GSH is a major player in antioxidant defence and so one could propose that GS is involved in the signalling pathway, functioning both as a sensor of increased levels of NO inside the nodules and to divert primary N assimilation to the synthesis of GSH.

A detailed analysis of the expression of GS gene family in *M. truncatula* has shown multiple molecular weight GS polypeptides representing either post-translational modification, or GS associations with other proteins in protein complexes in different organs (Seabra *et al.*, 2013). A higher molecular weight GS_1 polypeptide was specifically detected in nodules, whereas a higher molecular weight GS_2 polypeptide was detected in nodules, flowers, pods and seeds. Furthermore, native gel electrophoresis showed other bands in addition to the expected bands for GS_1 and GS_2 holoproteins, and the pattern appeared to be organ specific. A particular higher molecular weight GS_1 polypeptide was specifically found in the nodules and its accumulation pattern appeared to be associated with the nodule nitrogen fixation zone. This could imply a nodule-specific post-translational regulation related to N_2-fixation. The higher molecular weight GS_2

polypeptide was, however, found in all of the tissues analysed. These higher molecular weight polypeptides could result from monoubiquitination of GS protein in an organ-specific manner (monoubiquitination is a reversible post-translational modification involved in the regulation of diverse cellular processes; Hicke, 2001). Also, GS_2 has been identified as a potential substrate for small ubiquitin-like modifier (SUMO) by a two-hybrid system in *Arabidopsis* (Elrouby and Coupland, 2010).

GS interactions with other proteins also contribute to the particular metabolic roles that this enzyme performs. Soybean nodulin-26 (nod26) is a multifunctional aquaglyceroporin in the symbiosome membrane of root nodules that functions in facilitating the transport of fixed ammonia from the symbiosome to the cytosol for assimilation. Affinity chromatography experiments show that the nod26 C-terminal domain has a site for association to cytosolic GS. The major interacting 40 kDa protein is a mixture of GS_1 β and γ isoforms, suggesting that the interaction of nodule GS_1 with nod26 could serve to localize GS_1 to the surface of the symbiosome (Masalkar *et al.*, 2010).

5.7 Concluding Remarks

In plants, nitrogen assimilation is integrated with photosynthesis, photorespiration and respiration in different tissues and organs and at both intracellular and intercellular levels. Current knowledge indicates that plants possess an intricate regulatory machinery that coordinates their N and C metabolism with nutrient availability, environmental factors and the demands for growth and development (Nunes-Nesi *et al.*, 2010). As GS plays such a central role in primary metabolism, it is not surprising that it is so heavily regulated. The review of the literature on GS that has been presented here indicates that GS genes in plants may be regulated at different steps in the flow of gene expression by C and N metabolites. Preliminary work from our laboratory has shown that SPS-overexpressing nodulated lucerne plants have higher GS activity in the nodules and higher elemental N content in the plants (Seger, 2012). We ascribe this to the increased importation of sucrose into the nodules, which then provides both the energy and C skeletons for the GS-mediated assimilation of ammonia. This would lead to the proposition that a common pathway leading to C:N sensing may exist in plants, as it does in bacteria (Coruzzi and Bush, 2001). In bacteria, the regulation of GS activity involves post-translational modification of GS by reversible covalent adenylation and transcriptional control of the *GlnA*. In *Escherichia coli*, these regulatory steps are linked to a signalling process monitoring N and C supply in which the metabolic signals are Gln as an indicator of N status and α-KG as an indicator of C status. In this process, the protein allosteric effector PII interacts with other regulatory proteins to modulate GS at the transcriptional and post-translational level in response to the C/N ratio (Jiang and Ninfa, 2007). A homologue of the bacterial PII has been isolated from many plants, including *Arabidopsis* (Hsieh *et al.*, 1998; Smith *et al.*, 2003) and lucerne (Ibilcieta-Garcia and Sengupta-Gopalan, unpublished results), and it is tempting to speculate that PII also has a role in plants in regulating GS at different levels.

The control of GS gene expression by energy availability is an area of research that needs to be explored because the GS reaction involves the hydrolysis of ATP. In bacteria, changes in energy charge are monitored by adenylation of the GS enzyme. In plants, adverse conditions are decoded as energy deficiency signals. Energy deficit triggers metabolic and structural modifications that result from global changes in gene expression, affecting transcription, but also polysome loading, translation rates and cell growth (Baena-González and Sheen, 2008). Energy charge may also be sensed at the post-translational level as changes in protein phosphorylation.

Based on all the studies that are reported here, we propose that GS_1 is subject to multiple levels of regulation. All of the steps may not be applicable to all the gene members and it would be good to figure out what differentiates the different members in the regulation process. GS_2 has not been as well studied from the aspect of regulation past transcription, but there is evidence for the post-transcriptional regulation of GS_2 genes (Zozaya-Hinchliffe *et al.*, 2005) and GS_2 is also known to be subject to post-translational modification (Lima *et al.*, 2006a).

Acknowledgements

We thank Dr Suman Bagga for her careful reading of the manuscript and helpful comments. This work was supported by the National Institutes of Health (Grant number S06 GM08136-32), by the US Department of Agriculture (Grant number 2007-03596), and by the Agricultural Experimental Station at New Mexico State University.

References

Abello, N., Kerstjens, H.A.M., Postma, D.S. and Bischoff, R. (2009) Protein tyrosine:selectivity, physicochemical and biological consequences, denitration and proteomics methods for the identification of tyrosine-nitrated protein. *Journal of Proteome Research* 8, 3222–3238.

Avila, C., Suárez, M.F., Gómez-Maldonado, J. and Cánovas, F.M. (2001) Spatial and temporal expression of two cytosolic glutamine synthetase genes in Scots pine: functional implications on nitrogen metabolism during early stages of conifer development. *The Plant Journal* 25, 93–102.

Baena-González, E. and Sheen, J. (2008) Convergent energy and stress signaling. *Trends in Plant Sciences* 13, 474–482.

Bernard, S.M. and Habash, D.Z. (2009) The importance of cytosolic glutamine synthetase in nitrogen assimilation and recycling. *New Phytologist* 182, 608–620.

Betti, M., García-Calderón, M., Pérez-Delgado, C.M., Credali, A., Estivill, G., Galván, F., Vega, J.M. and Marquez, A.J. (2012) Glutamine synthetase in legumes: recent advances in enzyme structure and functional genomics. *International Journal of Molecular Sciences* 13, 7994–8024.

Brauc, S., De Vooght, E., Claeys, M., Geuns, J.M.C., Hofte, M. and Angenon, G. (2012) Overexpression of arginase in *Arabidopsis thaliana* influences defence responses against *Botrytis cinerea*. *Plant Biology* 14, 39–45.

Brugiere, N., Dubois, F., Limami, A.M., Lelandais, M., Roux, Y., Sangwani, R.S. and Hirel, B. (1999) Glutamine synthetase in the phloem plays a major role in controlling proline production. *The Plant Cell* 11, 1995–2012.

Cánovas, F.M., Avila, C., Cantón, F.R., Cañas, R.A. and de la Torre, F. (2007) Ammonium assimilation and amino acid metabolism in conifers. *Journal of Experimental Botany* 58, 2307–2318.

Carvalho, H.G., Lopes-Cardoso, I.A., Lima, L.M., Melo, P.M. and Cullimore, J.V. (2003) Nodule-specific modulation of glutamine synthetase in transgenic *Medicago truncatula* leads to inverse alterations in asparagine synthetase expression. *Plant Physiology* 133, 243–252.

Castro-Rodríguez, V., García-Gutiérrez, A., Canales, J., Avila, C., Kirby, E.G. and Cánovas, F.M. (2011) The glutamine synthetase gene family in *Populus*. *BMC Plant Biology* 11, 119.

Chan, M.T. and Yu, S.M. (1998) The 3′ untranslated region of a rice a-amylase gene functions as a sugar-dependent mRNA stability determinant. *Proceedings of the National Academy of Sciences of the United States of America* 95, 6543–6547.

Coruzzi, G.M. and Bush, D.R. (2001) Nitrogen and carbon nutrient and metabolite signaling in plants. *Plant Physiology* 125, 61–64.

Díaz, P., Betti, M., Sánchez, D.H., Udvardi, M.K., Monza, J. and Márquez, A.J. (2010) Deficiency in plastidic glutamine synthetase alters proline metabolism and transcriptome response in *Lotus japonicus* under drought stress. *New Phytologist* 188, 1001–1013.

Edwards, J.W., Walker, E.L. and Coruzzi, G.M. (1990) Cell-specific expression in transgenic plants reveals nonoverlapping roles for chloroplast and cytosolic glutamine synthetase. *Proceedings of the National Academy of Sciences of the United States of America* 87, 3459–3463.

El-Khatib, R.T., Hamerlynck, E.P., Gallardo F. and Kirby, E.G. (2004) Transgenic poplar characterized by ectopic expression of a pine glutamine synthetase gene exhibits enhanced tolerance to water stress. *Tree Physiology* 24, 729–736.

Elrouby, N. and Coupland, G. (2010) Proteome-wide screens for small ubiquitin-like modifier (SUMO) substrates identify *Arabidopsis* proteins implicated in diverse biological processes. *Proceedings of the National Academy of Sciences of the United States of America* 107, 17415–17420.

Fei, H., Chaillou, S., Hirel, B., Mahon, J.D. and Vessey, K.J. (2003) Overexpression of a soybean cytosolic glutamine synthetase gene linked to organ-specific promoters in pea plants grown in different concentrations of nitrate. *Planta* 216, 467–474.

Fei, H., Chaillow, S., Hirel, B., Polowick, P., Mahon, J.D. and Vessey, J.K. (2006) Effects of overexpression of a soybean cytosolic glutamine synthetase gene (*GS15*) linked to organ-specific promoters on growth and nitrogen accumulation of pea plants supplied with ammonium. *Plant Physiology and Biochemistry* 44, 543–550.

Finnemann, J. and Schjoerring, J.K. (2000) Post-translational regulation of cytosolic glutamine synthetase by reversible phosphorylation and 14-3-3 protein interaction. *The Plant Journal* 24, 171–181.

Fischer, J.J., Beatty, P.H., Good, A.G. and Muench, D.G. (2013) Manipulation of microRNA expression to improve nitrogen use efficiency. *Plant Science* 210, 70–81.

Forde, B.G. and Lea, P.J. (2007) Glutamate in plants: metabolism, regulation, and signaling. *Journal of Experimental Botany* 58, 2339–2358.

Fu, C., Sunkar, R. and Wang, Z.-Y. (2012) Overexpression of miR156 in switchgrass (*Panicum virgatum* L.) results in various morphological alterations and leads to improved biomass production. *Plant Biotechnology Journal* 10, 443–452.

Fu, J., Sampalo, R., Gallardo, F., Cánovas, F.M. and Kirby, A.G. (2003) Assembly of a cytosolic pine glutamine synthetase holoenzyme in leaves of transgenic poplar leads to enhanced vegetative growth in young leaves. *Plant, Cell and Environment* 26, 411–418.

Fuentes, S.I., Allen, D.J., Ortiz-Lopez, A. and Hernández, G. (2001) Over-expression of cytosolic glutamine synthetase increases photosynthesis and growth at low nitrogen concentrations. *Journal of Experimental Botany* 52, 1071–1081.

Fulks, R.M. and Stadtman, E.R. (1985) Regulation of glutamine synthetase, aspartokinase and total protein turnover in *Klebsiella arogenes*. *Biochimica et Biophysica Acta* 843, 214–229.

Gallardo, F., Fu, J., Cantón, F.R., García-Gutiérrez, A., Cánovas, F.M. and Kirby, E.G. (1999) Expression of a conifer glutamine synthetase gene in transgenic poplar. *Planta* 210, 19–26.

Galmozzi, C.V., Saelices, L., Florencio, F.J. and Muro-Pastor, M.I. (2010) Posttranscriptional regulation of glutamine synthetase in the filamentous cyanobacterium *Anabaena* sp. PCC 7120: differential expression between vegetative cells and heterocysts. *Journal of Bacteriology* 192, 4701–4711.

García-Calderón, M., Chiurazzi, M., Espuny, M.R. and Márquez, A.J. (2012) Photorespiratory metabolism and nodule function: behavior of *Lotus japonicus* mutant deficient in plastid glutamine synthetase. *Molecular Plant–Microbe Interactions* 25, 211–219.

García-Gutiérrez, A., Dubois, F., Cantón, F.R., Gallardo, F., Sangawan, R.S. and Cánovas, F.M. (1998) Two different modes of early development and nitrogen assimilation in gymnosperm seedlings. *The Plant Journal* 13, 187–199.

Gómez-Maldonado, J., Avila, C., de la Torre, F., Cañas, R, Cánovas, F.M. and Campbell, M.A. (2004a) Functional interactions between a glutamine synthetase promoter and MYB proteins. *The Plant Journal* 39, 513–526.

Gómez-Maldonado, J., Cánovas, F.M. and Avila, C. (2004b) Molecular analysis of the 5′ upstream region of a gibberellin-inducible cytosolic glutamine synthetase gene (*GS1b*) expressed in the pine vascular tissue. *Planta* 218, 1036–1045.

Gorg, B., Ovartshkava, N., Voss, P., Grune, T., Haussinger, D. and Schliess, F. (2007) Reversible inhibition of mammalian glutamine synthetase by tyrosine nitration. *FEBS Letters* 581, 84–90.

Gutiérrez, R.A. (2012) Systems biology for enhanced plant nitrogen nutrition. *Science* 336, 1673–1675.

Gutiérrez, R.A., Stokes, T.L., Thum, K., Xu, X., Obertello, M., Katari, M.S., Tanurdzic, M., Dean, A., Nero, D.C., McClung, C.R. and Coruzzi, G.M. (2008) Systems approach identifies an organic nitrogen-responsive gene network that is regulated by the master clock control gene *CCA1*. *Proceedings of the National Academy of Sciences of the United States of America* 105, 4939–4944.

Hamilton, T.L., Stoneley, M., Spriggs, K.A. and Bushell, M. (2006) TOPs and their regulation. *Biochemical Society Transactions* 34, 12–16.

Harrison, J., de Crescenzo, M.P., Sene, O. and Hirel, B. (2003) Does lowering glutamine synthetase activity in nodules modify nitrogen metabolism and growth of *Lotus japonicus*? *Plant Physiology* 133, 253–262.

Hicke, L. (2001) Protein regulation by monoubiquitination. *Nature Reviews Molecular and Cell Biology* 2, 195–201.

Hirel, B., Bertin, P., Quilleré, I., Bourdoncle, W., Attagant, C., Dellay, C., Gouy, A., Cadiou, S., Retailliau, C., Falque, M. and Gallais, A. (2001) Towards a better understanding of the genetic and physiological basis for nitrogen use efficiency in maize. *Plant Physiology* 125, 1258–1270.

Hirel, B., Le Gouis, J., Ney, B. and Gallais, A. (2007) The challenge of improving nitrogen use efficiency in crop plants: towards a more central role for genetic variability and quantitative genetics within integrated approaches. *Journal of Experimental Botany* 58, 2369–2387.

Hsieh, M.H., Lam, H.M., van de Loo, F.J. and Coruzzi, G (1998) A PII-like protein in *Arabidopsis*: putative role in nitrogen sensing. *Proceedings of the National Academy of Sciences of the United States of America* 95, 13965–13970.

Jiang, P. and Ninfa, A.J. (2007) *Escherichia coli* PII signal transduction protein controlling nitrogen assimilation acts as a sensor of adenylate energy change *in vitro*. *Biochemistry* 46, 12979–12996.

Jiang P., Peliska, J.A. and Ninfa, A.J. (1998) The regulation of *Escherichia coli* glutamine synthetase revisited: role of 2-ketoglutarate in the regulation of glutamine synthetase adenylation state. *Biochemistry* 37, 12802–12810.

Kawaguchi, R. and Bailey-Serres, J. (2002) Regulation of translational initiation in plants. *Current Opinions in Biology* 5, 460–465.

Kichey, T., Le Gouis, J., Sangwan, B., Hirel, B. and Dubois, F. (2005) Changes in the cellular and subcellular localization of glutamine synthetase and glutamate dehydrogenase during flag leaf senescence in wheat (*Triticum aestivum*, L.). *Plant and Cell Physiology* 33, 964–974.

Knight, T.J. and Langston-Unkeferr, P.J. (1988) Enhancement of symbiotic dinitrogen fixation by a toxin-releasing plant pathogen. *Science* 241, 951–954.

Kumada, Y., Benson, D.R., Hillemann, D., Hosted T.J., Rochefort, D.A., Thompson, C.J., Wohlleben, W. and Tateno, Y. (1993) Evolution of the glutamine synthetase gene, one of the oldest existing and functioning genes. *Proceedings of the National Academy of Sciences of the United States of America* 90, 3009–3013.

Kusano, M., Tabuchi, M., Fukushima, A., Funayama, K., Diaz, C., Kobayashi, M., Hayashi, N., Tsuchiya, Y.N., Takahashi, H., Kamata, A., Yamaya, T. and Saito, K. (2011) Metabolomics data reveal a crucial role of cytosolic glutamine synthetase1;1 in coordinating metabolic balance in rice. *The Plant Journal* 66, 456–466.

Lea, P.J. and Miflin, B.J. (2011) Nitrogen metabolism in plants in the post-genomic era. In: Foyer, C.H. and Zhang, H. (eds) *Nitrogen Assimilation and its Relevance to Crop Improvement. Annual Plant Reviews, Vol. 42*. Wiley-Blackwell, Chichester, UK, pp. 1–40.

Liaw, S.H., Pan, C. and Eisenberg, D. (1993) Feedback inhibition of fully unadenylylated glutamine-synthetase from *Salmonella typhimurium* by glycine, alanine, and serine. *Proceedings of the National Academy of Sciences of the United States of America* 90, 4996–5000.

Lima, L., Saebra, A., Melo, P. and Cullimore, J. (2006a) Phosphorylation and subsequent interaction with 14-3-3 proteins regulate plastid glutamine synthetase in *Medicago truncatula*. *Planta* 223, 558–567.

Lima, L., Saebra, A., Melo, P., Cullimore, J. and Carvalho, H. (2006b) Post-translational regulation of cytosolic glutamine synthetase of *Medicago truncatula*. *Journal of Experimental Botany* 57, 2751–2761.

Limami, A., Phillipson, B., Ameziane, R., Pernollet, N., Jiang, Q., Roy, R., Deleens, E., Chaumont-Bonnet, M., Gresshoff, P.M. and Hirel, B. (1999) Does root glutamine synthetase control plant biomass production in *Lotus japonicus* L.? *Planta* 209, 495–502.

Liu, H., Liu, H. and Song Y. (2011) Absorption and translocation of nitrogen in rhizomes of *Leymus chinensis*. *Rapid Communications in Mass Spectrometry* 25, 665–671.

Llorca, O., Betti, M., Gonzalez, J.M., Valencia, A., Márquez, A.J. and Valpuesta, J.M. (2006) The three-dimensional structure of an eukaryotic glutamine synthetase: functional implications of its oligomeric structure. *Journal of Structural Biology* 156, 469–479.

Lothier, J., Gaufichon, L., Sormani, R., Lemaître, T., Azzopardi, M., Morin, H., Chardon, F., Reisdorf-Cren, M., Avice, J.C. and Masclaux-Daubresse, C. (2011) The cytosolic glutamine synthetase GLN1;2 plays a role in the control of plant growth and ammonium homeostasis in *Arabidopsis* rosettes when nitrate supply is not limiting. *Journal of Experimental Botany* 62, 1375–1390.

Lozano-Juste, J., Colom-Moreno, R. and León, J. (2011) *In vivo* protein tyrosine nitration in *Arabidopsis thaliana*. *Journal of Experimental Botany* 62, 3501–3517.

Mackintosh, C. and Meek, S.E. (2001) Regulation of plant NR activity by reversible phosphorylation, 14-3-3 proteins and proteolysis. *Cell and Molecular Life Sciences* 58, 205–214.

Man, H.M., Boriel, R., El Khatib, R. and Kirby, E.G. (2005) Characterization of transgenic poplar with ectopic expression of pine cytosolic glutamine synthetase under conditions of varying nitrogen availability. *New Phytologist* 167, 31–39.

Man, H.M., Pollman, S., Weiler, E.W. and Kirby, E.G. (2011) Increased glutamine in leaves of poplar transgenic with pine GS1a caused greater anthranilate synthetase α-subunit (*ASA1*) transcript and protein abundances: an auxin-related mechanism for enhanced growth in GS transgenic? *Journal of Experimental Botany* 62, 4423–4431.

Martin, A., Lee, J., Kichey, T., Gerentes, D., Zivy, M., Tatout, C., Dubois, F., Balliau, T., Valot, B., Davanture, M., Tercé-Laforgue, T., Quilleré, I. *et al.* (2006) Two cytosolic glutamine synthetase isoforms of maize are specifically involved in the control of grain production. *The Plant Cell* 18, 3252–3274.

Masalkar, P., Wallace, I.S., Hwang, J.H. and Roberts, D.M. (2010) Interaction of cytosolic glutamine synthetase of soybean root nodules with the C-terminal domain of the symbiosome membrane nodulin 26 aquaglyceroporin. *The Journal of Biological Chemistry* 285, 23880–23888.

Masclaux-Daubresse, C., Daniel-Vedele, A., Dechorgnat, J., Chardon, F., Gaufichon, L. and Suzuki, A. (2010) Nitrogen uptake, assimilation and remobilization in plants: challenges for sustainable and productive agriculture. *Annals of Botany* 105, 1141–1157.

Mathis, R., Gamas, P., Meyer, Y. and Cullimore, J.V. (2000) The presence of GSI-like genes in higher plants: support for the paralogous evolution of GSI and GSII genes. *Journal of Molecular Evolution* 50, 116–122.

Mattsson, M. and Schjoerring, J.K. (2003) Senescence-induced changes in apoplastic and bulk tissue ammonia concentrations of ryegrass leaves. *New Phytologist* 160, 489–499.

Melo, P.M., Lima, L.M., Santos, I.M., Carvalho, H.G. and Cullimore, J.V. (2003) Expression of plastid-located glutamine synthetase of *Medicago truncatula*. Accumulation of precursor in root nodules reveals an *in vivo* control at the level of protein import. *Plant Physiology* 132, 390–399.

Melo, P.M., Silva, L.S., Ribeiro, L., Seabra, A.R. and Carvalho, H.G. (2011) Glutamine synthetase is a molecular target of nitric oxide in root nodules of *Medicago truncatula* and is regulated by tyrosine nitration. *Plant Physiology* 157, 1505–1517.

Miflin, B.J. and Habash, D.Z. (2002) The role of glutamine synthetase and glutamate dehydrogenase in nitrogen assimilation and possibilities for improvement in the nitrogen utilization of crops. *Journal of Experimental Botany* 53, 979–987.

Migge, A., Carrayol, E., Hirel, B. and Becker, T.W. (2000) Leaf-specific overexpression of plastidic glutamine synthetase stimulates the growth of transgenic tobacco seedlings. *Planta* 210, 252–260.

Morey, K., Ortega, J.L. and Sengupta-Gopalan, C. (2002) Cytosolic glutamine synthetase in soybean is encoded by a multigene family and the members are regulated in an organ-specific and developmental manner. *Plant Physiology* 128, 182–193.

Nicolaï, M., Roncato, M.A., Canoy, A.S., Roquié, D., Sarda, X., Freyssinet, G., and Robaglia, C. (2006) Large-scale analysis of mRNA translation states during sucrose starvation in *Arabidopsis* cells identifies cell proliferation and chromatin structure as targets of translational control. *Plant Physiology* 141, 663–673.

Nogueira, E.D., Olivares, F.L., Japiassu, J.C., Vilar, C., Vinagre, F., Baldani, J.I. and Hemerly, A.S. (2005) Characterization of glutamine synthetase genes in sugarcane genotypes with different rates of biological nitrogen fixation. *Plant Science* 169, 819–832.

Nunes-Nesi, A., Fernie, A.R. and Stitt, M. (2010) Metabolic and signaling aspects underpinning the regulation of plant carbon nitrogen interactions. *Molecular Plant* 3, 973–996.

Oldroyd, G.E. and Downie, J.A. (2008) Coordinating nodule morphogenesis with rhizobial infection in legumes. *Annual Reviews in Plant Biology* 59, 519–546.

Oliveira, I.C., Brears, T., Knight, T.J., Clark, A and Coruzzi, G.M. (2002) Overexpression of cytosolic glutamine synthetase. Relation to nitrogen, light, and photorespiration. *Plant Physiology* 129, 1170–1180.

Ortega, J.L., Roche, D. and Sengupta-Gopalan, C. (1999) Oxidative turnover of soybean root glutamine synthetase. *In vitro* and *in vivo* studies. *Plant Physiology* 119, 1483–1495.

Ortega, J. L., Temple, S.J. and Sengupta-Gopalan, C. (2001) Constitutive overexpression of cytosolic glutamine synthetase (GS_1) gene in transgenic alfalfa demonstrates that GS_1 may be regulated at the level of RNA stability and protein turnover. *Plant Physiology* 126, 109–121.

Ortega, J.L., Temple, S., Bagga, S., Ghoshroy, S. and Sengupta-Gopalan, C. (2004) Biochemical and molecular characterization of transgenic *Lotus japonicus* plants constitutively over-expressing a cytosolic glutamine synthetase gene. *Planta* 219, 807–818.

Ortega, J.L., Moguel-Esponda, S., Potenza, C., Conklin, C., Quintana, A. and Sengupta-Gopalan, C. (2006) The 3′ untranslated region of a soybean cytosolic glutamine synthetase (GS_1) affects transcript stability and protein accumulation in transgenic alfalfa. *The Plant Journal* 45, 832–846.

Ortega, J.L., Wilson, O.L. and Sengupta-Gopalan, C. (2012) The 5′ untranslated region of the soybean cytosolic glutamine synthetase β_1 gene contains prokaryotic translation initiation signals and acts as a translational enhancer in plants. *Molecular Genetics and Genomics* 287, 881–893.

Pageau, K., Reisdorf-Cren, M., Morot-Gaudry, J.F. and Masclaux-Daubresse, C. (2006) The two senescence-related markers, GS1 (cytosolic glutamine synthetase) and GDH (glutamate dehydrogenase), involved in nitrogen mobilization, are differentially regulated during pathogen attack and by stress hormone and reactive oxygen species in *Nicotiana tabaccum* L. leaves. *Journal of Experimental Biology* 57, 547–557.

Pant, B.D., Buhtz, A., Kehr, J. and Scheible, W.R. (2008) MicroRNA399 is a long-distance signal for the regulation of plant phosphate homeostasis. *The Plant Journal* 53, 731–738.

Pascual, M.B., Jing, Z.P., Kirby, E.G., Cánovas, F.M. and Gallardo, F. (2008) Response of transgenic poplar overexpressing cytosolic glutamine synthetase to phosphinothricin. *Phytochemistry* 69, 382–389.

Price, J., Laxmi, A., St Martin, S.K. and Jang, J-C. (2004) Global transcription profiling reveals multiple sugar signal transduction mechanisms in *Arabidopsis*. *Plant Cell* 16, 2128–2150.

Ren, M., Venglat, P., Qiu, S., Feng, L., Cao, Y., Wang, E., Xiang, D., Wang, J., Alexander, D., Chalivendra, S., Logan, D., Matoo, A. *et al.* (2012) Target of rapamycin regulates metabolism, growth and life span in *Arabidopsis*. *Plant Cell* 24, 4850–4874.

Riedel, J., Tischner, R. and Mäck, G. (2001) The chloroplastic glutamine synthetase (GS-2) of tobacco is phosphorylated and associated with 14-3-3 proteins inside the chloroplast. *Planta* 213, 396–401.

Rivett A.J. and Levine, R.L. (1990) Metal-catalyzed oxidation of *Escherichia coli* glutamine synthetase: correlation of structural and functional changes. *Archives of Biochemistry and Biophysics* 278, 26–34.

Roberts, M.R. (2007) Does GABA act as a signal in plants? *Plant Signaling and Behavior* 2, 408–409.

Roche, D., Temple, S. J. and Sengupta-Gopalan, C. (1993) Two classes of differentially regulated glutamine synthetase genes are expressed in the soybean nodule: a nodule-specific class and a constitutively expressed class. *Plant Molecular Biology* 22, 971–983.

Rogers, L.A., Dubos, C., Surman, C., Willment, J., Cullis, I.F., Mansfield, S.D. and Campbell, M.M. (2005) Comparison of lignin deposition in three ectopic lignification mutants. *New Phytologist* 168, 123–140.

Rueda-López, M., Crespillo, R., Cánovas, F.M. and Avila, C. (2008) Differential regulation of two glutamine synthetase genes by a single Dof transcription factor. *The Plant Journal* 56, 73–85.

Sáez, C.Á., Muñoz-Chapuli, R., Plomion, C., Frigerio, J.M. and Cánovas, F.M. (2000) Two genes encoding distinct cytosolic glutamine synthetases are closely linked in the pine genome. *FEBS Letters* 477, 237–243.

Sakurai, N., Katayama, Y. and Yamaya, T. (2001) Overlapping expression of cytosolic glutamine synthetase and phenylalanine ammonia-lyase in immature leaf blades of rice. *Physiologia Plantarum* 113, 400–408.

Seabra, A.R., Viera, C.P., Cullimore, C.V. and Carvalho, H.G. (2010) *Medicago truncatula* contains a second gene encoding a plastid located glutamine synthetase exclusively expressed in developing seeds. *BMC Plant Biology* 10:183.

Seabra, A.R., Silva, L.S. and Carvalho, H.G. (2013) Novel aspects of glutamine synthetase (GS) regulation revealed by a detailed expression analysis of the entire GS gene family of *Medicago truncatula* under different physiological conditions. *BMC Plant Biology* 13, 137.

Seger, M., Ortega, J.L., Bagga, S. and Sengupta-Gopalan, C. (2009) Repercussion of mesophyll-specific overexpression of a soybean cytosolic glutamine synthetase gene in alfalfa (*Medicago sativa* L.) and tobacco (*Nicotiana tabaccum* L.) *Plant Science* 176, 119–129.

Seger, M. (2012) The biochemical and physiological repercussions of the constitutive expression of a maize sucrose phosphate synthase gene in alfalfa and tobacco plants. PhD thesis, New Mexico State University, Las Cruces, New Mexico.

Seifi, H.S., Curvers, K., De Vleesschauwer, D., Delaere, I., Aziz, A. and Hofte, M. (2013) Concurrent overactivation of the cytosolic glutamine synthetase and the GABA shunt in the ABA-deficient *sitiens* mutant of tomato leads to resistance against *Botrytis cinerea*. *New Phytologist* 199, 490–504.

Silva, L. and Carvalho, H. (2013) Possible role of glutamine synthetase in the NO signaling response in root nodules by contributing to the antioxidant defense. *Frontiers in Plant Science* 4, 372.

Simon, B. and Sengupta-Gopalan, C. (2010) The 3′ untranslated region of the two cytosolic glutamine synthetase (GS_1) genes in alfalfa (*Medicago sativa*) regulates transcript stability in response to glutamine. *Planta* 232, 1151–1162.

Smith, C.A., Weljie, A.M. and Moorhead, G.B.G. (2003) Molecular properties of the putative nitrogen sensor PII from *Arabidopsis thaliana*. *The Plant Journal* 33, 353–360.

Smith, P.M.C. and Atkins, C.A. (2002) Purine biosynthesis. Big in cell division, even bigger in nitrogen assimilation. *Plant Physiology* 128, 793–802.

Stadtman, E.R. (1990) Discovery of glutamine synthetase cascade. *Methods in Enzymology* 182, 793–809.

Suárez, R., Márquez, J., Shishkova, S. and Hernández, G. (2003) Overexpression of alfalfa cytosolic glutamine synthetase in nodules and flowers of transgenic *Lotus japonicus* plants. *Physiologia Plantarum* 117, 326–336.

Sugden, C., Donaghy, P.G., Halford, N.G. and Hardie, D.G. (1999) Two SNF1-related protein kinases from spinach leaf phosphorylate and inactivate 3-hydroxy-3-methylglutaryl-coenzyme A reductase, nitrate reductase, and sucrose phosphate synthase *in vitro*. *Plant Physiology* 120, 257–274.

Swarbreck, S.M., Defoin-Platel, M., Hindle, M., Saqi, M. and Habash, D.Z. (2011) New perspectives on glutamine synthetase in grasses. *Journal of Experimental Botany* 62, 1511–1522.

Tabuchi, M., Sugiyama, K., Ishiyama, K., Inoue, E., Sato, T., Takahashi, H. and Yamaya, T. (2005) Severe reduction in growth rate and grain filling of rice mutants lacking OsGS1;1, a cytosolic glutamine synthetase 1;1. *The Plant Journal* 42, 641–651.
Taira, M., Valtersson, U., Burkhardt, B. and Ludwig, R.A. (2004) *Arabidopsis thaliana* GLN2-encoded glutamine synthetase is dual targeted to leaf mitochondria and chloroplasts. *Plant Cell* 16, 2048–2058.
Temple, S.J., Knight, T.J., Unkefer, P.J. and Sengupta-Gopalan, C. (1993) Modulation of glutamine synthetase gene expression in tobacco by the introduction of an alfalfa glutamine synthetase gene in sense and antisense orientations: molecular and biochemical analysis. *Molecular and General Genetics* 236, 315–325.
Temple, S.J., Bagga, S. and Sengupta-Gopalan, C. (1994) Can glutamine synthetase activity levels be modulated in transgenic plants by the use of recombinant DNA technology? *Biochemical Society Transactions* 22, 915–920.
Temple, S. J., Heard, J., Ganter, G., Dunn, K. and Sengupta-Gopalan, C. (1995) Characterization of a nodule-enhanced glutamine synthetase from alfalfa: nucleotide sequence, *in situ* localization and transcript analysis. *Molecular Plant–Microbe Interactions* 8, 218–227.
Temple, S.J., Kunjibettu, S., Roche, D. and Sengupta-Gopalan, C. (1996) Total glutamine synthetase activity during soybean nodule development is controlled at the level of transcription and holoprotein turnover. *Plant Physiology* 112, 1723–1733.
Temple, S.J., Bagga, S. and Sengupta-Gopalan, C. (1998) Down regulation of specific members of the glutamine synthetase gene family in alfalfa by antisense RNA technology. *Plant Molecular Biology* 37, 535–547.
Tjaden, G., Edwards, J.W. and Coruzzi, G.M. (1995) *cis* Elements and *trans*-acting factors affecting regulation of a nonphotosynthetic light-regulated gene for chloroplast glutamine synthetase. *Plant Physiology* 108, 1109–1117.
Tzeng, T-Y., Kong, L-R., Chen, C-H., Shaw, C-C. and Yang, C-H. (2009) Overexpression of the Lily $p70^{s6k}$ gene in *Arabidopsis* affects elongation of flower organs and indicates TOR-dependent regulation of *AP3*, *PI* and *SUP* translation. *Plant and Cell Physiology* 50, 1695–1709.
Unno, H., Uchida, T., Sugawara, H., Kurisu, G., Sugiyama, T., Yamaya, T., Sakakibara, H., Hase, T. and Kusunoki, M. (2006) Atomic structure of plant glutamine synthetase – a key enzyme for plant productivity. *The Journal of Biological Chemistry* 281, 29287–29296.
van Rooyen, J.M., Abratt, V.R., Belrhall, H. and Sewell, T. (2011) Crystal structure of Type III glutamine synthetase: surprising reversal of the inter-ring interface. *Structure* 19, 471–483.
Vidal, E.A., Araus, V., Lu, C., Parry, G., Green, P.J., Coruzzi, G.M. and Gutiérrez, R.A. (2010) Nitrate-responsive miR393/*AFB3* regulatory module controls root system architecture in *Arabidopsis thaliana*. *Proceedings of the National Academy of Sciences of the United States of America* 107, 4477–4482.
Vincent, R., Fraiser, V., Chaillou, S., Limami, M.A., Deleens, E., Phillipson, B., Douat, C., Boutin, J.P. and Hirel, B. (1997) Overexpression of a soybean gene encoding glutamine synthetase in shoots of transgenic *Lotus corniculatus* L. plants triggers changes in ammonium assimilation and plant development. *Planta* 201, 424–433.
Wang, C.T. and Xu, Y.N. (2010) The 5′ untranslated region of the FAD3 mRNA is required for its translational enhancement at low temperature in *Arabidopsis* roots. *Plant Science* 179, 234–240.
White, J., Prell, J., James, E.K. and Poole, P. (2007) Nutrient sharing between symbionts. *Plant Physiology* 144, 604–614.
Woods, D.R. and Reid, S.J. (1993) Recent developments on the regulation and structure of glutamine synthetase enzymes from selected bacterial groups. *FEMS Microbiology Reviews* 11, 273–284.
Wullschleger, S., Loewith, R. and Hall, M.N. (2006) TOR signaling in growth and metabolism. *Cell* 124, 471–484.
Yanagisawa, S., Akiyama, A., Kisaka, H., Uchimiya, H. and Miwas, T. (2004) Metabolic engineering with Dof1 transcription factor in plants: improved nitrogen assimilation and growth under low-nitrogen conditions. *Proceedings of the National Academy of Sciences of the United States of America* 101, 7833–7838.
Zhang, B.H., Pan, X.P., Wang, Q.L., Cobb, G.P. and Anderson, T.A. (2005) Identification and characterization of new plant microRNAs using EST analysis. *Cell Research* 15, 336–360.
Zhao, X-Q. and Shi, W-M. (2005) Expression analysis of the glutamine synthetase and glutamate synthase gene families in young rice (*Oryza sativa*) seedlings. *Plant Science* 170, 748–754.
Zozaya-Hinchliffe, M., Potenza, C., Ortega, J.L. and Sengupta-Gopalan, C. (2005) Nitrogen and metabolic regulation of the expression of plastidic glutamine synthetase in alfalfa (*Medicago sativa*). *Plant Science* 168, 1041–1052.

6 Asparagine Synthetase

S.M.G. Duff*
Monsanto, Chesterfield, Missouri, USA

6.1 Abstract

Asparagine (Asn) is an amino acid which serves as a carrier and storage form of nitrogen. In this chapter, the history, chemical properties and general role of Asn will be examined, followed by an account of the history of asparagine synthetase (AsnS) research in humans, microbes and plants. The occurrence of AsnS across plant phyla will be then surveyed, and its expression and functions in plants will be examined and compared with those in other organisms. After this, the phylogeny and enzymology (including the reaction mechanism and three-dimensional structure of *Escherichia coli* asparagine synthetase type B, AsnSB) of AsnS will be looked at in detail. This will be followed by a brief investigation of other possible routes of Asn synthesis in plants, as well as Asn catabolism. Next, the relationship between asparagine and seed storage protein will be investigated, as will the applications of Asn in crop plants, including Illinois high and low protein maize. Finally, recent transgenic studies using AsnS in plants, which open up several possibilities for crop metabolic engineering, will be reviewed. Throughout the chapter, our emphasis will be on the role of asparagine in transport, storage and nitrogen signalling, including the importance of the glutamine:asparagine ratio.

6.2 Introduction: the Role of Asparagine and Asparagine Synthetase in Nitrogen Metabolism

Nitrogen storage and remobilization are important metabolic processes that occur continually throughout a plant's life cycle. As the seedling develops post germination, the supply of nitrogen shifts from the diminishing seed storage protein pool to the absorption of nitrate and/or ammonia from the soil by the root. The reduction and assimilation of nitrate typically occur in both the leaf and the root, and nitrate is then transported throughout the plant (Oaks and Hirel, 1985; Becker *et al.*, 2000). Although different molecules can be used to transport nitrogen, many plants rely mainly on the amides asparagine (Asn) and glutamine (Glu) (Lea and Miflin, 1980; Sieciechowicz *et al.*, 1988; Fischer *et al.*, 1998). Asn is more chemically and biologically inert and has a higher N: C ratio than most other amino acids, making it an ideal nitrogen transport molecule. The level of Asn varies greatly between different species. It is the major nitrogenous compound found in the phloem of several legumes (up to 30 mM) (Atkins *et al.*, 1975), whereas in maize, asparagine has been reported to represent only about 2% (approximately 250 μM) of the phloem free amino acid pool (Ohshima *et al.*, 1990).

*E-mail address: Stephen.m.duff@monsanto.com

Asparagine synthetase (AsnS; EC 6.3.5.4), the topic of this chapter, catalyses the ATP-dependent transfer of an amino group from ammonia or Gln to aspartate to generate Asn. Asn is a major amino acid in many plants and because AsnS is the primary means by which Asn is synthesized in all plants, it plays a key role in nitrogen metabolism (Sieciechowicz *et al.*, 1988; Lea *et al.*, 1990; Richards and Schuster, 1992; Lohaus *et al.*, 1998; Lea *et al.*, 2007; Todd *et al.*, 2008; Gaufichon *et al.*, 2010). AsnS preferentially uses Gln as a nitrogen donor in plants, and AsnS activity is often positively associated with Asn content (Chevalier *et al.*, 1996). Ammonia is also utilized as an AsnS substrate, particularly in maize roots (Oaks and Ross, 1984). AsnS may act in ammonia detoxification when plants encounter high concentrations of ammonia (Givan, 1979; Sieciechowicz *et al.*, 1988). All plant forms of AsnS studied to date have shown a strong preference for Gln as an amino donor and are phylogenetically and kinetically much more closely related to *Escherichia coli* AsnSB (*E. coli* asparagine synthetase type B; glutamine-preferred) than *E. coli* AsnSA (*E. coli* asparagine synthetase type A; NH_4^+-dependent) (Ohshima *et al.*, 1990; Gaufichon *et al.*, 2010). On the basis of amino acid sequence analyses, AsnSB belongs to a larger family of enzymes referred to as the glutamine amidotransferases, enzymes for which Gln serves as the preferred source of nitrogen whereby an acceptor molecule is subsequently aminated (Zalkin, 1993; Zalkin and Smith, 1998). In plants, Gln-preferred AsnS is thought to be the major route of Asn biosynthesis (Sieciechowicz *et al.*, 1988), although other routes may exist and will be covered later in this chapter (Stulen and Oaks, 1977; Joy, 1988).

We will begin the chapter by reviewing the history, chemical properties and general role of Asn. Next, we will follow the history of AsnS research since its discovery in humans, microbes and plants, and then move into a discussion of its occurrence across the phyla. We will then take a detailed look at how AsnS is expressed and how it functions in plants and compare this with how AsnS functions in other organisms. In the sections that follow, we will study the phylogeny and enzymology (including the reaction mechanism and three-dimensional structure of *E. coli* AsnSB) in detail, and then take a brief look at other possible routes of Asn synthesis in plants, as well as Asn catabolism. We will investigate the relationship between Asn and seed storage protein and look at applications of Asn in crop plants, including Illinois high and low protein maize. Finally, we will discuss some recent transgenic studies of AsnS in plants which open up several possibilities for crop metabolic engineering. Throughout the chapter, our emphasis will be on the role of Asn in transport, storage and N signalling, including the importance of the Gln:Asn ratio.

However, as the function of AsnS is inarguably to produce Asn, we will begin our discussion by investigating the history and role of its product, Asn, in plants.

6.3 Asparagine: History, Chemical Properties and Role in Plants

The metabolism of Asn in plants was reviewed in 1975 (Lea and Fowden, 1975a), 1980 (Lea and Miflin, 1980), 1982 (Miflin and Lea, 1982), 1988 (Sieciechowicz *et al.*, 1988), 2007 (Lea *et al.*, 2007) and then again in 2010 (Gaufichon *et al.*, 2010). Because of its importance in plant nitrogen metabolism, it has also been frequently covered more peripherally in many other reviews that focus on related aspects of plant metabolism (e.g. Xu *et al.*, 2012).

Asn has the dual distinction of being both first identified in its natural form from a plant and being the first amino acid to be discovered. In 1802, Delaville (1802) observed cubic crystals in concentrated solutions of sap from asparagus plants. Four years later, Asn was isolated and identified from the sap (Vauquelin and Robiquet, 1806; Lea *et al.*, 2007). Asn is an amino acid amide with a molecular mass of 132.12 and an isoelectric point of 5.4 (Lea *et al.*, 2007). Its chemical structure is shown in Fig. 6.1. It was found to be a constituent of proteins when it was discovered in the Brazil nut protein edestin in

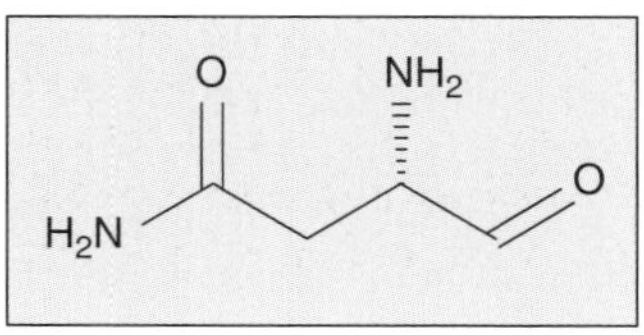

Fig. 6.1. The chemical structure of asparagine. Note that with two nitrogen atoms, asparagine has a higher N:C ratio than most other amino acids.

1932 (Damodaran, 1932). It is a moderately polar molecule which often plays a role in the active site of enzymes (Mansfield *et al.*, 2006).

As one of the few amino acids containing multiple nitrogen atoms, asparagine has one of the higher nitrogen to carbon ratios among the standard twenty such acids. It is relatively neutral in charge, is metabolically stable and is a substrate for only a small number of highly specific enzymes, such as asparaginase (Joy and Ireland, 1990; Xu *et al.*, 2012). This makes it an ideal candidate for the transport and temporary storage of nitrogen, to the extent that each of these is dependent upon the plant nitrogen status and species. For instance, it has been suggested that when the N:C ratio is optimal in maize, AsnS may function to redirect the flow of nitrogen into Asn for storage and/or long-distance transport (Chevalier *et al.*, 1996; Herrera-Rodríguez *et al.*, 2002).

As would be expected of a high nitrogen-containing amino acid, Asn is also a major component of many storage proteins (Ratajczak, 1986). Asn and Gln are the most common molecules used by plants for the short-term storage and transport of nitrogen (Lea and Miflin, 1980; Ireland and Lea, 1999; Gaufichon *et al.*, 2010). Analysis of xylem sap from a number of different species confirms that Asn plays a major role in nitrogen transport (Bollard, 1960; Pate *et al.*, 1980; Schmidt and Stewart, 1998; do Amarante *et al.*, 2006). Asn is also the primary assimilation product of N_2 fixation in many temperate legumes, including lupins (Pate *et al.*, 1980, 1984), *Medicago* (Snapp and Vance, 1986; Shi *et al.*, 1997), *Lotus japonicus* (Waterhouse *et al.*, 1996) and *Pisum* (Scharff *et al.*, 2003).

In addition to its primary role in nitrogen transport and storage, Asn is also thought to be involved in signalling. Of particular importance as a signal, perhaps, is the Asn:Gln ratio (Seebauer *et al.*, 2004), which will be discussed in more detail in a later section.

The initial substrates for Asn biosynthesis in plants are oxaloacetate (OAA)and glutamate (Glu). Starting with these compounds, as shown in Fig. 6.2 (Sieciechowicz *et al.*, 1988; Todd *et al.*, 2008) Asn is synthesized in plants by the combined action of four enzymes, the final step being catalysed by AsnS, which catalyses the ATP-dependent transfer of an amino group from ammonia or Gln to aspartate (Asp) to generate Glu and Asn.

We can see from Fig. 6.2 that Asp is synthesized by aspartate aminotransferase in a reversible reaction using Glu and OAA. Glutamine is then synthesized by the combined action of glutamine synthetase (GS), in an irreversible ATP-dependent reaction, along with glutamate synthetase (GOGAT), which together make up the GS/GOGAT nitrogen assimilation cycle. Finally, Asn is synthesized from Asp and Gln by the glutamine-preferring plant AsnS in an irreversible reaction.

6.4 Asparagine Synthetase: an Early History of Research in Humans, Microbes and Plants

Whereas Asn was first discovered in plants, the discovery and subsequent study of AsnS occurred first in mammals, where the importance of the enzyme was realized from studies of leukaemia and other cancers. In 1957, Leon Levintow provided the initial evidence that Gln was a precursor to Asn in human tissue culture (Levintow, 1957), and more than a decade later AsnS was partially purified from embryo chicken liver (Arfin, 1967) and Novikoff hepatoma (Patterson and Orr, 1968). The enzyme was later purified from leukaemia cells (Horowitz and Meister, 1972). It quickly became clear that leukaemia cells were susceptible to drug therapy in a near inverse correlation to their capacity for intracellular Asn biosynthesis (Cooney and Handschumacher, 1970). For this reason, L-asparaginase (Asnase) is used broadly in chemotherapeutic protocols for treating childhood acute lymphoblastic leukaemia (Amylon *et al.*, 1999; Pui *et al.*, 2002). It is believed that normal and malignant lymphocytes depend upon the uptake of Asn from circulating plasma for their growth, and it is likely that Asnase functions by depleting Asn in the blood, followed by transport of cytoplasmic Asn from the leukaemic blasts (Amylon *et al.*, 1999; Graham, 2003). However, the treatment of leukaemia using Asnase leads to side effects (Hersh, 1971; Terebello *et al.*, 1986), and 10–12% of patients frequently relapse with new tumours (Lobel *et al.*, 1979; Terebello *et al.*, 1986; Kiriyama *et al.*, 1989). The treatment also enhances the growth and increases the metastatic activity of resistant

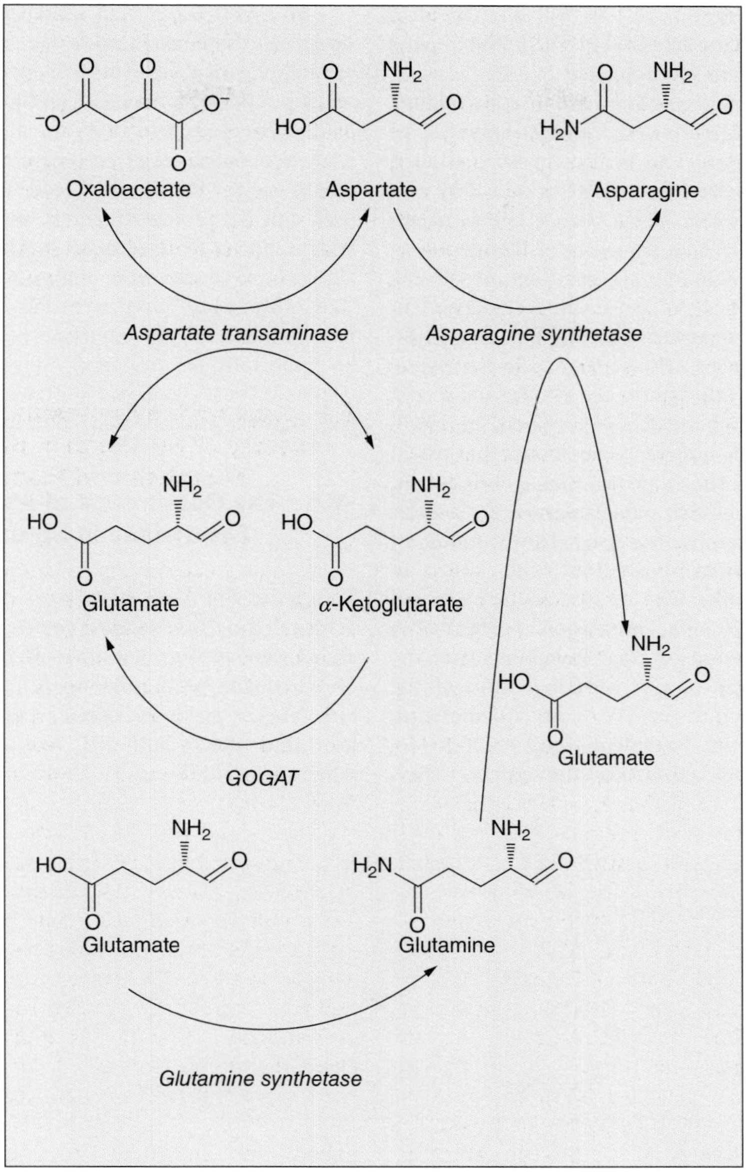

Fig. 6.2. The biochemical pathway to the synthesis of asparagine. GOGAT, glutamate synthetase.

tumours owing to the increased expression of AsnS (Tallal *et al.*, 1970; Capizzi *et al.*, 1971; Chakrabarti *et al.*, 1989; Aslanian *et al.*, 2001). Increased AsnS expression appears to be a component of solid tumour adaptation to nutrient deprivation and/or hypoxia (Balasubramanian *et al.*, 2013). For this reason, human AsnS is now considered a specific target for therapies directed at Asnase-resistant tumours and leukaemia (Aslanian *et al.*, 2001; Richards and Kilberg, 2006; Balasubramanian *et al.*, 2013). Human AsnS has only recently been expressed in Rat Sarcoma cells (Andrulis *et al.*, 1987; Van Heeke and Schuster, 1989) and it was later purified for kinetic studies and antibody production from yeast cells (Shend *et al.*, 1992). The most highly studied eukaryotic AsnS is probably the form isolated and characterized in 1985 from bovine pancreas (Luehr and Schuster, 1985).

It has long been known that many plants, like mammalian tumours, can accumulate large amounts of Asn. This is in contrast to what is generally observed in bacteria and healthy animals (Meiss, 1952; Tallen *et al.*, 1954; Kittredge *et al.*, 1962). A half decade after the identification of AsnS in mammals, a synthetase converting aspartic acid to Asn in an ATP-dependent manner was reported in several bacteria: *Lactobaccillus arabinosus* (Ravel *et al.*, 1962); *Streptococcus bovis* (Burchall *et al.*, 1964); and *E. coli* (Cedar and Schwartz, 1969). Then, in 1968, using ^{14}C radio tracers, it was shown that wheat seedlings and shoots grown under normal conditions produce Asn independently of proteolysis; they accumulate free Asn as the predominant amino acid by synthesis (Lees *et al.*, 1968). In 1980, ^{15}N amide Gln fed to detached soybean nodules led to the accumulation of ^{15}N label in the amide group of Asn, indicating that the amide nitrogen of Gln was the source of the amide nitrogen in Asn (i.e. a direct demonstration of the Gln-dependent AsnS reaction described above) (Fujihara and Yamaguchi, 1980).

The synthesis of Asn generally occurs at a slower rate than that of serine (Ser) and Ala, but is virtually identical to the rate at which many other major amino acids, such as Glu, Gln and Asp are made (Lees *et al.*, 1968). As the authors of this study were not able to detect the hydrolysis of Asn, they suggested that it was a dead-end metabolite used only for protein synthesis, and they could find no support for the concept that it was specifically involved in carbon or nitrogen translocation (Lees *et al.*, 1968). However, by this time, Asn had been found in the phloem tissue. Long known to be important for transport in several plants (Chibnall, 1930; McKee, 1962), its role in transport, signalling and storage is now well accepted.

6.5 The Occurrence of Asparagine Synthetase in Nature

The genes for AsnS have been demonstrated across all three taxonomic domains. As discussed above, some prokaryotes have two distinct forms of asparagine synthetase; an A form that uses only NH_4^+ as an amide donor (AsnSA) and a B form that utilizes both NH_4^+ and Gln as amide donors (AsnSB) (Fig. 6.3). These two forms have

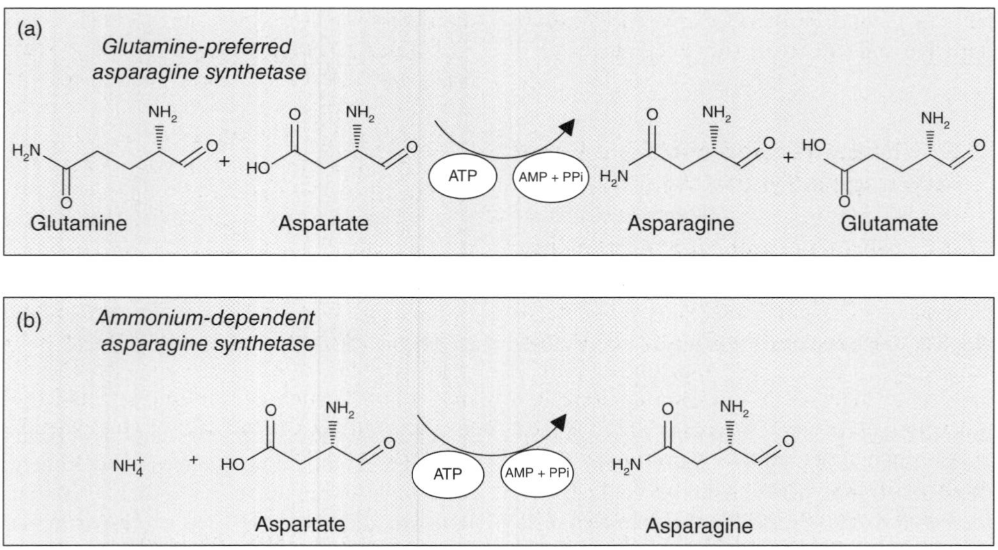

Fig. 6.3. The two basic forms of asparagine synthetase (AsnS). *Escherichia coli* AsnSB is a glutamine-preferred asparagine synthetase (a), whereas *E. coli* AsnSA is an ammonium-dependent asparagine synthetase (b). PPi, pyrophosphate.

very low homology; for example AsnSA and AsnSB from *E. coli* are only 16% identical and are likely to have evolved independently (Scofield *et al.*, 1990), the B form being the ancestral form of all eukaryotic AsnS genes. In fact, AsnSA has more similarity to aspartic aspartyl-tRNA synthetase (EC 6.1.1.12) in *E. coli* than it does to AsnSB. AsnSA and aspartic aspartyl-tRNA synthetase use two common substrates (Mg^{2+}ATP and aspartic acid), share most of their catalytic residues and form an aminoacyl-adenylate intermediate (Nakatsu *et al.*, 1998). Therefore, AsnSA and aspartyl-tRNA synthetase appear to have evolved from a common ancestor unrelated to AsnSB and eukaryotic asparagine synthetases (Nakatsu *et al.*, 1998). Many other bacteria and all eukaryotes appear to have only single or multiple AsnS genes of similar phylogeny and structure to their ancestral *E. coli* B form (Gaufichon *et al.*, 2010).

In eukaryotes, AsnS appears to be ubiquitous, having been identified in all kingdoms: yeast, algae, fungi, mammals and a wide variety of higher plants (Coruzzi, 2003; US National Center for Biotechnology Information at http://www.ncbi.nlm.nih.gov/). Asn is a necessary component in most proteins and the main route in nature for its synthesis is AsnS. Therefore, AsnS, in one or both of its forms, is probably present in all living organisms.

We will survey several common examples of how plants use AsnS before considering its function and structure in greater detail.

6.6 The Expression and Function of Asparagine Synthetase in Plants

6.6.1 Nutritional and mineral deficiency

Several environmental conditions are known to affect AsnS expression (Beatoa *et al.*, 2010; Herrera-Rodríguez, 2002), in particular, conditions that lead to an increase in ammonia by proteolysis (Stewart and Larher, 1980; Rabe, 1990; Vierstra, 1993), including several forms of nutritional deficiency. In these cases, ammonium may replace Gln as the primary amide donor of the AsnS reaction (Sivasankar and Oaks, 1996). In addition, gene expression profiling experiments also suggest a link between ammonium metabolism and Asn (Canales *et al.*, 2012).

Nutrient solutions deficient in many minerals can lead to increased Asn concentrations, especially if the nitrogen supply is high (Sivasankar and Oaks, 1996; Camacho-Cristóbal and González-Fontes, 2007; Huang, 2008; Beatoa *et al.*, 2010). As many of these deficiencies are known to lead to intracellular proteolysis, resulting in the production of ammonia, it is probable that AsnS functions to complement GOGAT and glutamate dehydrogenase (GDH) to prevent ammonium from reaching toxic levels. In some studies, such as a 20 day phosphorus deficiency in soybean, Asn accumulated in the roots and stems, while arginine (Arg) accumulated in the leaves (Rufty *et al.*, 1990). In the case of boron deficiency, an increase in AsnS transcripts was reported to coincide with the increase of Asn, thus directly implicating AsnS in plant response to this stress (Beatoa *et al.*, 2010).

With a plentiful supply of nitrogen, deficiencies in other mineral ions and micronutrients can also stimulate the accumulation of Asn (Possingham, 1956; Stewart and Larher, 1980). Sulfur deficiency in developing maize seeds can alter the types of proteins formed and result in a substantial increase in soluble Asn (Baudet *et al.*, 1986); however, surprisingly, a ^{15}N tracer study showed that the accumulation of Asn following sulfur starvation in maize suspension cells is due to *de novo* synthesis, presumably by AsnS, rather than by protein hydrolysis (Amancio *et al.*, 1997). This appears to be a general response to sulfur depletion in plants because it has also been observed in certain wheat varieties (Zhao *et al.*, 1996) and ryegrass (Mortensen *et al.*, 1993); potentially, it is in response to protein degradation leading to an increase in ammonia resulting from the stress. This hypothesis is supported by the observation that in plant cell cultures, sulfur deficiency leads to a parallel reduction of protein content with increased free amino acids (Klapheck *et al.*, 1982). However, that this increase is mostly due to the primary amino acid assimilation products such as Gln and Asn and their *de novo* synthesis can be shown by tracer studies (Amancio *et al.*, 1997).

6.6.2 Seed germination

Seeds must mobilize resources for growth before, during and after germination, especially nitrogen and carbon resources (Gómez-Cadenas *et al.*, 2001). These reserves are metabolized in the endosperm and destined to supply the embryo to facilitate growth (Lea *et al.*, 2007). Asn is, in many cases, the primary nitrogen carrier (Lea *et al.*, 2007). Upon germination, long-lived embryo RNA molecules must be quickly translated into proteins and this requires the immediate availability of amino acids (Koorneef *et al.*, 2002). Moreover, mRNA from the AsnS gene (*ASN1*) has been observed as one of the long-lived mRNAs detected in quiescent dry seeds of *Arabidopsis* (Nakabayashi *et al.*, 2005). In this way, AsnS mRNA can serve as a pre-existing template for the rapid synthesis of AsnS protein, thus facilitating mobilization of the other nitrogen reserves into Asn. Additional evidence for the role of AsnS in germination has been described in mung beans, where AsnS activity is barely detectable in the cotyledons of dry seeds, but appears very early during germination, from the second day. However, this process can be prevented by protein synthesis inhibitors (Dilworth and Dure, 1978; Kern and Chrispeels, 1978), suggesting that AsnS was translated from a long-lived mRNA rather than being present in the dry mature seed.

In *Pisum sativum*, two AsnS mRNAs, named AS1 and AS2, also accumulate early to high levels in the cotyledons of germinating seedlings, as well as later on in the mature plant in nitrogen-fixing root nodules. The expression of the AsnS genes correlates well with the physiological role of Asn as a nitrogen carrier amino acid throughout plant development (Tsai and Coruzzi, 1990).

In other cases, such as sunflower, AsnS mRNA is not detected in dry seeds but accumulates very rapidly after imbibition (Herrera-Rodríguez *et al.*, 2002). In pine seed embryos, the initial mobilization is from the gametophyte via Arg, and Asn accumulates later by conversion from Arg (King and Gifford, 1997). In all of these examples, however, Asn is used for nitrogen transport, either very early or slightly later on during plant development.

6.6.3 Light signalling

Light serves as a source of energy for plants and acts as a signal by means of its quality, quantity and duration. Consequently, many genes in plants are ultimately regulated by light, including those involved in nitrogen metabolism and carbon fixation (Gaufichon *et al.*, 2010 and references therein). As nitrogen uptake and metabolism is coordinated with photosynthesis, several nitrogen metabolism genes are also upregulated or downregulated by light. One of these genes is AsnS, a form of which has been shown to be upregulated in the dark in *Arabidopsis* (Koorneef *et al.*, 2002). This light regulation may possibly be related to low carbon reserves and the advantage that an amino acid with a high N:C ratio confers under these conditions.

In *P. sativum*, dark treatment was shown to induce high-level accumulation of pea AS1 mRNA (PsAS1) in leaves, while light treatment represses this effect as much as 30-fold. The dark-induced accumulation of PsAS1 mRNA is a phytochrome-mediated response. In contrast, white light represses the transcription of AS1 within 6 h in isolated pea nuclei (Tsai and Coruzzi, 1990) and so light and darkness are involved in the reciprocal regulation of the AS1 gene.

In *Arabidopsis*, AsnS is encoded by a small gene family – *ASN1*, *ASN2* and *ASN3*. *ASN2* is expressed at low levels in the dark, and its expression is relatively higher in light-treated plants (Lam *et al.*, 1994, 1995, 1998). Levels of ASN1 and ASN2 mRNA in *Arabidopsis* are reciprocally regulated by light and by organic carbon and nitrogen (Lam *et al.*, 1998). Asparagine levels track with the expression of *ASN1*, but not *ASN2*. The light-induced increase in asparagine levels in *Arabidopsis* is ammonium dependent (Wong *et al.*, 2004). These data suggest that ASN1 and ASN2 are housekeeping enzymes, regulated reciprocally by light and amino acids. ASN1 regulates asparagine synthesis under carbon-limiting conditions, whereas ASN2 regulates asparagine synthesis in the light (Lam *et al.*, 1998).

6.6.4 Developmental stage and tissue specificity

The expression of the maize (*Zea mays*) AsnS gene (*ZmASN*) was originally described as root specific and was not detected in other organs of 6-day-old maize seedlings (Chevalier *et al.*, 1996). However, it was later shown that this maize AsnS gene is probably a chimera of *AsnS1* and AsnS3. The expression of the latter (AsnS3) is mostly restricted to the root (Todd *et al.*, 2008), whereas the former is expressed across many tissues and developmental stages. When nitrogen: carbon ratios are high, one or both of these genes may redirect the flow of nitrogen into asparagine for storage and/or long-distance transport of nitrogen. (Chevalier *et al.*, 1996). In addition, the later study (by Todd *et al.*, 2008) also identified two other asparagine synthetase genes, *AsnS2*, expressed in roots, cob and the stem ear node at R3, and *AsnS4*, which is very similar in both sequence and expression pattern to *AsnS3* (Todd *et al.*, 2008).

A similar expression pattern was observed in barley (*Hordeum vulgare*), where the expression of two AsnS genes has been studied (Moller *et al.*, 2003). *HvAS1* expression is found in the roots, stems and leaves of barley plants in both light-grown and dark-adapted seedlings at various developmental stages (similar to *AsnS2* or *AsnS3* in maize). *HvAS2* expression is clearly detected in all tissue types except roots, and probably acts as a housekeeping gene in the leaves under normal growth conditions (similar to AsnS1 in maize), while the *HvAS1* gene (similar to *AsnS2* in maize) seems to be more important in the roots, where nitrogen is assimilated into Asn for long-distance transport within the plant (Moller *et al.*, 2003). Therefore, in both monocotyledons, a non-root housekeeping gene exists as well as an important form in roots that probably functions in making AsnS for nitrogen transport.

In rice, the expression of a single gene has been described, although other AsnS genes are known. Rice (*Oryza sativa*) AsnS (OsAS) is localized in vascular tissues of the leaf sheath and in grains at the middle stage of ripening. OsAS protein content increases during the early stage of spikelet ripening, reaches maximum levels at 21 days after flowering and rapidly declines thereafter. The AsnS protein is localized in the companion cells of vascular bundles of leaf sheath and phloem-parenchyma cells, the nucellar projection and the nucellar epidermis of dorsal vascular bundles of grains (Nakano *et al.*, 2000). Like the root-expressed versions in maize and barley, it probably functions to synthesize Asn for long distance transport.

In another study in rice, Kawachi *et al.* (2002) showed that OsAS (a class I AsnS) also accumulates in the leaf sheath of fully expanded leaves and is not detectable in mature leaf blades. Asn, which is very low in rice leaves and roots under nitrogen starvation, increases significantly after 1 mM NH_4^+ is supplied to the nutrient solution, thus supporting the notion that one of the functions of AsnS is to detoxify and assimilate ammonium. OsAS activity or OsAS protein could be detected in leaves and roots in plants undergoing nitrogen starvation; this activity increased in roots but not in leaves within 12 h of the supply of 1 mM NH_4^+ or 10 mM Gln. The NH_4^+-induced accumulation of AsnS protein was inhibited by methionine sulfoximine (a Gln analogue), but methionine sulfoximine did not affect the Gln-induced accumulation of the enzyme, suggesting that Gln or Gln-derived metabolites are signals for the regulation of AsnS expression in rice roots. Other nitrogen compounds, such as nitrate, Glu, Asp and Asn had no effect on the expression of AsnS (Kawachi *et al.*, 2002).

Similarly, in soybean (*Glycine max*) *GmAS2* is expressed highly in root and stem of light-grown tissue and in the leaves of dark-treated soy seedlings; and light downregulates *GmAS2* expression in the leaves of dark-grown seedlings (Hughes *et al.*, 1997). In all of these cases, the tissue and development-specific expression of AsnS is consistent with either a housekeeping role or a role in the transportation of nitrogen throughout the plant as Asn.

6.6.5 Environmental stress and carbohydrate depletion

Apart from nutritional deficiency, which has been covered above, a variety of stresses in plants can lead either directly or indirectly to carbon starvation. These include darkness, drought, salinity, wounding and pathogen infection. The ultimate substrates for Asn synthesis (Fig. 6.2) include Glu and OAA, which are key metabolites in both carbon and nitrogen metabolism. As already mentioned, Asn has a high N:C ratio and can serve as a temporary storage form of nitrogen. As would be expected, the expression of AsnS is very sensitive to changes in carbon/nitrogen status. Accordingly, transcriptome analysis shows that dark stress induces ASN1, cytosolic aspartate aminotransferase (At1g62800) and other genes for pyruvate metabolism within a day in *Arabidopsis* leaves (Lin and Wu, 2004).

However, in *Zea mays*, AsnS genes are upregulated under glucose starvation, salt and heavy metal stress. In fact, *ZmASN* expression reaches maximum levels between 18 and 24 h after carbohydrate deprivation (Chevalier *et al.*, 1996).

ASN1 from *Arabidopsis* was induced in response to several stresses that limit photosynthesis and/or deplete sugar. These include darkness, herbicide treatment, hypoxia and bacterial pathogenesis (Baena-González *et al.*, 2007). In plants, sugars are sensed by two systems; a hexokinase-dependent system and a hexokinase-independent system which detects hexose and regulates genes such as *Arabidopsis ASN1* and *ASN2* (Smeekens, 2000; Rolland *et al.*, 2002). Because a number of dark-induced genes like *ASN* are also upregulated under diverse stresses, a convergent signal perception and signal transduction pathway has been proposed as a response to stresses that impose carbon depletion (Fujiki *et al.*, 2001). Recently, an AsnS gene was found to be among those upregulated under sucrose depletion in cowpea (Kaneko *et al.*, 2013). Therefore, it appears AsnS has a major role in the responses of many plants to carbon stress, potentially as a way to store nitrogen in an inert form using less carbon per unit nitrogen.

6.6.6 Senescence and nitrogen remobilization

During senescence, proteins, lipids, polysaccharides and nucleic acids are degraded in the source organs (Hörtensteiner and Feller, 2002). Endogenous nitrogen from the leaf is supplied first to the flowers and then later on to the seeds. Proteins must be broken down in the organelles and cytoplasm and in different cell types and tissues. Multiple types of proteases, especially including cysteine proteases, are induced by senescence in plants (Donnison *et al.*, 2007; Diaz *et al.*, 2008; De Michele *et al.*, 2009), leading to phloem loading of amino acids. However, as described above, conditions favouring proteolysis also lead to the production of ammonium, which is toxic to the plant. Given the hypothesized role of AsnS in detoxifying ammonium *and* nitrogen transport, it would be expected that AsnS might be involved in the recycling and remobilization of ammonium to prevent nitrogen loss. Accordingly, transcriptome studies using leaf extracts show that the genes for AsnS, as well as for several transaminases, are induced during senescence (Guo *et al.*, 2004).

It is also known that rice OsAS is expressed mainly in the mature and senescing leaf sheath, supporting the view that its main role is in nitrogen remobilization (Guo *et al.*, 2004). In *Arabidopsis*, this gene is also called *DIN2*, a dark-inducible gene (it is also called *ASN1*), and it is associated with senescence and repressed by sugars (Fujiki *et al.*, 2001). The dark-induced expression of the gene may provide asparagine as a nitrogen transport and/or storage compound, especially when carbon supplies are limited (Nozawa *et al.*, 1999).

In maize, leaves below and above the ear act as the main sources for nitrogen reallocation after pollination (Crawford *et al.*, 1982). Most of the leaf nitrogen reduction occurs in the mesophyll cells, but nitrogen assimilation via the GS/GOGAT cycle can occur in both mesophyll and bundle sheath cells. During grain development and nitrogen remobilization, an increasing amount of ammonium is produced in the ear leaf by proteolysis compared with nitrogen reduction, and this needs to be remobilized to the grain (Lohaus *et al.*, 1998; Valadier *et al.*, 2008). Asn and Gln levels increase in the grain and are higher at about 14 days after pollination (Canas *et al.*, 2009). However, if the nitrogen is transported by Gln and/or Asn at the same time remobilization is occurring in the leaves, transamination and asparaginase reactions must also occur in the grain to allow nitrogen to be deposited in grain proteins. Indeed, this type of cycle has been observed (Seebauer *et al.*, 2004). Therefore, this data is consistent with a role for AsnS during senescence, in both nitrogen assimilation and ammonium detoxification.

6.6.7 Seed maturation

Selected amino acids are transported via the phloem or xylem from the senescing leaves into the grain. Asn is a relatively inert amino acid, so it must be converted back into other amino acids to facilitate protein production during seed fill (Credali *et al.*, 2013). For example, Asn content decreases in mature *Arabidopsis* leaves at 35 days after germination (Igarashi *et al.*, 2009), indicating that the Asn has been incorporated into seed storage proteins. In *Arabidopsis*, seed maturation is linked to a general decrease in free amino acids and sugars in leaves, starting around 14 days following pollination (Fait *et al.*, 2006). Asn, Gln and lysine (Lys) decrease to a much greater extent than other amino acids. Tricarboxylic acid (TCA) cycle intermediates, such as 2-oxoglutarate and malate (Fait *et al.*, 2006), which are carbon skeletons for amino acid biosynthesis, also decrease, indicating that Asn, Asp, Glu and Gln are extensively converted into other amino acids for incorporation into storage proteins. Taken together, these studies indicate that Asn is crucially important in nitrogen remobilization. The synthesis of Asn is induced by proteolysis and occurs via AsnS in the leaf; the Asn is subsequently transaminated and hydrolysed in the maturing seed.

6.6.8 Photorespiration

Photorespiration is one of the main factors limiting the growth and yield of C_3 plants (Leegood, 2013). Higher temperatures and water stress in C_3 plants contribute to photorespiration and are in fact initiating factors in the evolution of C_4 and CAM (Crassulacean acid metabolism) metabolism (Keeley and Rundel, 2003; Sage *et al.*, 2012; Leegood, 2013). Photorespiration results in the release of ammonium during methylene tetrahydrofolate synthesis from glycine (Gly) (Rachmilevitch *et al.*, 2004; Keys, 2006), and this ammonium is recycled by glutamine synthetase. As AsnS is involved under other conditions for the mediation of ammonium toxicity, Asn has been proposed to play a role in photorespiration (Ta *et al.*, 1984; Bernard *et al.*, 2008). Other studies have shown that in addition to Glu and Ser, alanine (Ala) and Asn provide amino nitrogen for the synthesis of photorespiratory glycine (Wingler *et al.*, 2000). The main photorespiratory cycle is well known to involve GS (Lam *et al.*, 1996). However, studies with ^{15}N-labelled amino acid demonstrated that the application of an inhibitor of the conversion of Gly to Ser in pea leaves undergoing photorespiration caused a decrease in the transfer of the Asn amino nitrogen to Ser, ammonium and Gln, but a concomitant increase in the label to Gly, indicating a possible role for Asn in photorespiration (Ta *et al.*, 1985).

6.6.9 Nitrogen signalling and the glutamine:asparagine ratio

In a study on maize cob tissue, it was found that Gln prevailed as the major amino acid until about 8 days after pollination (Seebauer *et al.*, 2004). During grain fill, however, levels of both Gln and GS declined, while Asn levels increased over time and to a greater extent when ear shoots were supplied with higher levels of nitrogen. The authors suggested that the coordinate activity of GS, aspartic acid aminotransferase and asparagine synthetase converted incoming Gln into Asn to provide a more metabolically stable amino acid with a higher nitrogen to carbon ratio for transfer into the pedicel and developing sink. They inferred the existence of a maize cob AsnS, which would be induced by N supply and play an important role in amino acid metabolism, although they were not able to measure it. Later, it was shown by Todd *et al.* (2008) that two AsnS genes (*AsnS1* and *AsnS2*) are expressed in the cob. For one (*AsnS1*), cob expression of the gene was higher than in any other tissue tested, and this may be the previously inferred maize cob AsnS. These authors also suggested that the conversion of Gln to Asn occurred to a greater extent under high N supply and that the Asn:Gln ratio could act as a signal to indicate high N status. This conclusion was supported by a previous *in vitro* study which demonstrated that maize ovules grew faster and accumulated more total protein when they were grown on media containing a greater Asn:Gln ratio (Rafalski and Dembinski, 1990). Also, several studies have shown that the Asn:Gln ratio increases after defoliation (Bigot *et al.*, 1991; Thornton and MacDuff, 2002), and that this change is accounted for mainly by the decrease in Gln levels. It has been suggested many times that both Gln and Asn rapidly turn over in xylem sap and that this is advantageous because it allows for the shoot–root cycling of amino acids to act as a signal for the regulation of N uptake (Cooper and Clarkson, 1989; Muller *et al.*, 1995). Therefore, a high Asn:Gln ratio indicates a high nitrogen status and a high Gln:Asn ratio indicates nitrogen deficiency. All in all, it appears likely that the Asn:Gln ratio may be a major signal indicating nitrogen status in many plants.

6.6.10 Asparagine: a nitrogen carrier, storage compound, detoxification mechanism and signal

AsnS (of either the A or B type) is the primary mechanism for the generation of Asn in all organisms that have been studied and therefore must always be an important enzyme in nitrogen metabolism. From the proceeding sections, we can see that plants use Asn in multiple ways that include the following. First, it is used as a nitrogen transport molecule, either in a general housekeeping fashion or under conditions when a lot of nitrogen has to be mobilized quickly. This means that Asn is a primary means of moving nitrogen around the plant to facilitate the remobilization of nitrogen, either from the endosperm to the growing seedling, or from the leaf where it is made by either *de novo* synthesis or the degradation of proteins to a sink tissue (Pate, 1980). Secondly, it is used as a storage form of nitrogen, either as a major constituent of storage proteins or as a free amino acid. The former occur in many seeds, the latter occurs particularly in woody plants where nitrogen, stored as Asn over the winter, is rapidly remobilized in the spring (Millard, 1996). Thirdly, it is used as a detoxification mechanism to remove ammonium under conditions when high levels of this would be generated, such as under stresses that enhance photorespiration or the degradation of proteins (Lea *et al.*, 2007). Finally, Asn acts as a signal for N status, possibly manifesting as the Asn:Gln ratio in plant tissue (Seebauer *et al.*, 2004).

6.7 Phylogeny, Subunit Structure and Enzymatic Activity of Asparagine Synthetase

6.7.1 Phylogeny

The number of available sequences of asparagine synthetases has expanded dramatically in recent years. In order to examine the phylogenetic differences and evolution of the enzyme, plant AsnS protein sequences were chosen, along with a few non-plant sequences, and used to generate an un-rooted phylogenetic tree (Fig. 6.4). All plant AsnS genes are related to *E. coli* AsnSB, which is also included in this tree.

Several previous studies have identified two major classes of AsnS (I and II) in plants (Gaufichon *et al.*, 2010). Many plants for which multiple sequences of AsnS have been identified (e.g. maize, *Arabidopsis*, sorghum, soybean) appear to have representatives from both classes (Gaufichon *et al.*, 2010; see Fig. 6.4) although the number of representatives from each class varies, suggesting that additional gene duplication or deletion has occurred in recent history (Gaufichon *et al.*, 2010).

A previous phylogenetic analysis in our laboratory predicted the existence of a third class of plant asparagine synthetases (Todd *et al.*, 2008), found only in monocotyledons, suggesting that gene duplication of an AsnS ancestral to AsnS2, AsnS3 and AsnS4 may have occurred immediately before the divergence of monocotyledons and dicotyledons, but disappeared very rapidly in dicotyledons after the split. Conversely, a later analysis using additional and more recently identified protein sequences contradicted this finding. The authors identified only the usual two classes and, more importantly, put ZmAsnS2 in the same clade, closely related to AsnS3 and AsnS4 (class I) (Gaufichon *et al.*, 2010). In our analysis, AsnS3 and AsnS4 were placed in a separate third clade (class III) (Todd *et al.*, 2008). In both studies, ZmAsnS1 is in a separate clade (class II). The phylogenetic tree shown in Fig. 6.4 extends these earlier analyses, and includes additional newly identified protein sequences. In this analysis, the class III proteins identified in our earlier study (Todd *et al.*, 2008) appear to form a distinct clade. As described in both earlier studies (Todd *et al.*, 2008; Gaufichon *et al.*, 2010), classes I and II have representatives from both monocotyledons and dicotyledons, and the pattern of distribution suggests that the classes evolved before the divergence of monocotyledons and dicotyledons, but then either gene duplication and/or gene loss in both classes has occurred relatively recently and unequally between species. For instance, *Helianthus* has two class I genes and a single class II gene (Herrera-Rodríguez *et al.*, 2002), while *Arabidopsis* has a single class I and two class II genes (Lam *et al.*, 1997). Non-plant AsnS proteins are more distantly related, as would be expected. This new phylogenetic analysis supports and extends those previously published (Lam *et al.*, 1998; Osuna *et al.*, 1999; Moller *et al.*, 2003; Todd *et al.*, 2008, Gaufichon *et al.*, 2010).

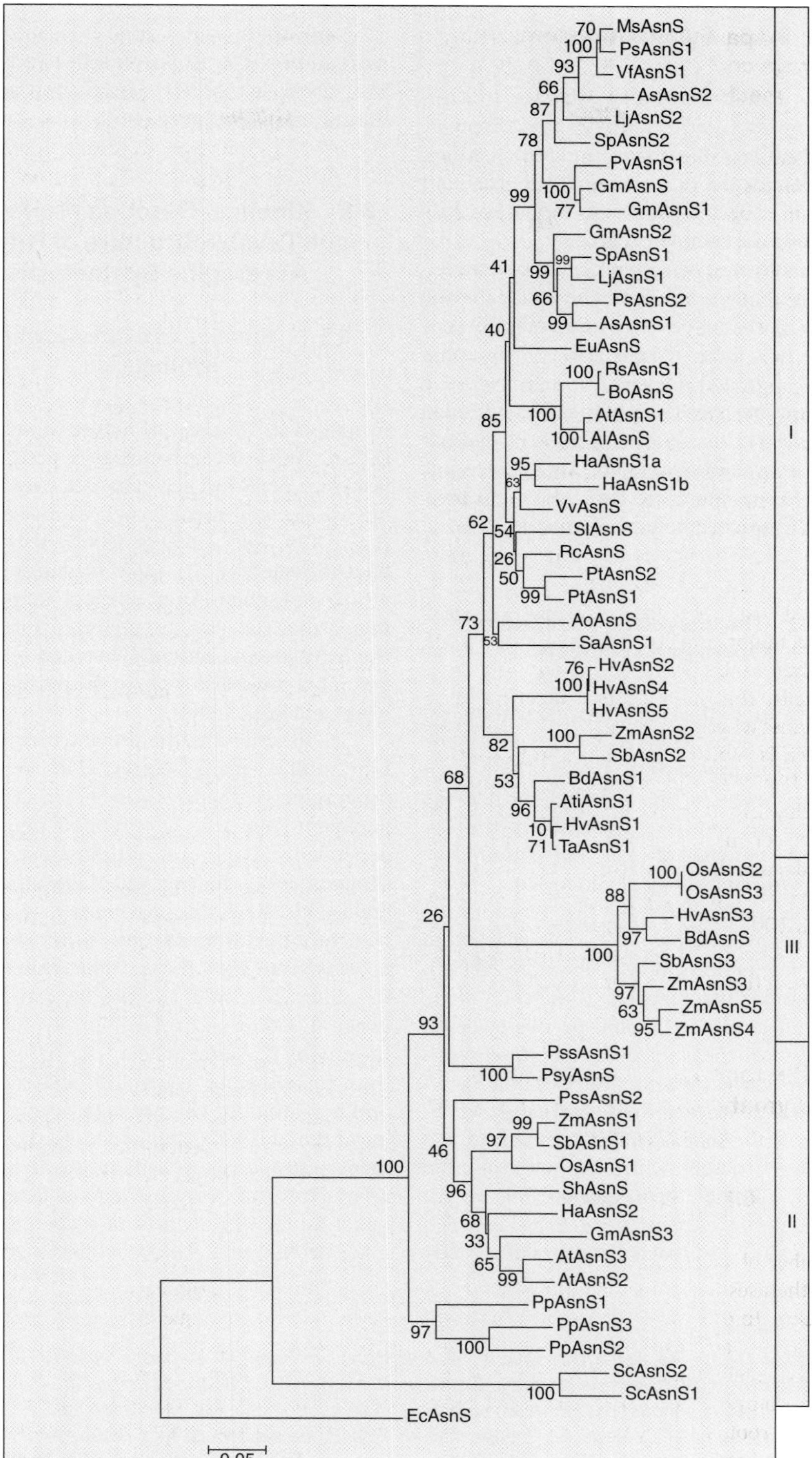

Fig. 6.4. Un-rooted phylogenetic tree of plant asparagine synthetase (AsnS) proteins. Class I, class II and class III are indicated on the figure as I, II and III, respectively. Species abbreviations: Ao, *Asparagus officinalis*; *Continued*

6.7.2 Subunit structure

In *E. coli*, AsnSB forms a homodimer. In *Klebsiella aerogenes*, AsnSA forms a homodimer of 80 kDa and AsnSB exists as a tetramer, which can be dissociated to a monomer (Reitzer and Magasanik, 1982); this AsnSB monomer does not catalyse the complete reaction.

In eukaryotes, AsnS may commonly form either a dimer or tetramer. The highly studied bovine pancreatic AsnS was observed under different conditions to be a dimer (Pfeiffer *et al.*, 1986) and a tetramer (Luehr and Schuster, 1985), whereas the rat liver enzyme appears to be a tetramer (Hongo and Sato, 1981). In a study on glutamine-dependent AsnS from *Lupinus luteus*, the AsnS protein was observed to be a dimer but tetramized in the presence of ATP and magnesium (Rognes, 1975).

6.7.3 The enzymatic activities of asparagine synthesis

In the test tube, three different enzymatic reactions have been observed for the eukaryotic AsnS and the prokaryotic AsnSB. These reactions are shown in Fig. 6.5 and include the ammonium-dependent synthesis of Asn, the Gln-dependent synthesis of Asn and glutaminase activity in the conversion of Gln to Glu.

In eukaryotic AsnS, the first reaction, the Gln-dependent synthesis of Asn, is at least an order of magnitude catalytically more efficient than the other two reactions (Gaufichon *et al.*, 2010; Duff *et al.*, 2011) and by this evidence must be deemed the main function of the enzyme. In fact, it is unclear whether the glutaminase reaction has a biological purpose or is simply a consequence of the main reaction mechanism and the enzyme structure, which has not evolved to completely eliminate it. In Section 6.6, we encountered several incidences (e.g. stress) in which the synthesis of Asn probably occurs using NH_4^+ as a substrate, and several isoforms of plant AsnS utilize NH_4^+ with different efficiencies. Therefore, it is likely that NH_4^+ assimilation occurs by the glutamine-preferred AsnS.

6.8 Kinetics, Reaction Mechanism and Crystal Structure of B-type Asparagine Synthetases

6.8.1 Kinetics of plant asparagine synthetase

In spite of its recalcitrant nature (Romagni and Dayan, 2000), the enzymology of partially purified plant AsnS has been characterized for various species and tissues, such as maize (Stulen *et al.*, 1979; Oaks and Ross, 1984), soybean seedlings (Streeter, 1973), lupin seedlings (Rognes, 1970) and others. Each of these studies determined that Gln was the preferred substrate of the AsnS enzyme, although in maize root tissue, ammonia could also serve as the nitrogen donor (Oaks and Ross, 1984).

As described in the previous section, certain studies clearly suggest that some plant asparagine synthetases and bacteria type B asparagine synthetases have residual ammonium-dependent AsnS and glutaminase activity (Rognes, 1975; Reitzer and Magasanik, 1982; Todd *et al.*, 2008). Approximate K_m values can be determined from partially purified extracts. It is apparent from these studies that substantial differences in K_m values for Gln exist between different plant AsnS forms (Table 6.1). In general, K_m values of AsnS for ATP are lower than for either of its amino acid substrates, and K_m values for Gln are always considerably lower than for NH_4^+. This appears to be a fundamental property of asparagine synthetases related to *E. coli* AsnSB. The AsnS from *L. albus* has a K_m for Gln of 0.04 mM compared with a K_m for NH_4^+ of 2.1 mM, while *E. coli* AsnSB

Fig. 6.4. continued

Al, *Arabidopsis lyrata;* As, *Astragalus sinicus*; At, *Arabidopsis thaliana*; Ati, *Aegilops tauschii*; Bd, *Brachypodium distachyon*; Bo, *Brassica oleracea*; Ec, *Escherichia coli*; Eu, *Elaeagnus umbellata*; Gm, *Glycine max*; Ha, *Helianthus annuus*; Hv, *Hordeum vulgare*; Lj, *Lotus japonicus*; Ms, *Medicago sativa*; Os, *Oryza sativa*; Pp, *Physcomitrella patens*; Ps, *Pisum sativum*; Pss, *Picea sitchensis*; Psy, *Pinus sylvestris*; Pt, *Populus tremula*; Pv, *Populus vulgaris*; Rc, *Ricinus communis*; Rs, *Raphanus sativus*; Sa, *Sandersonia aurantiaca*; Sb, *Sorghum bicolor*; Sc, *Saccharomyces cerevisiae*; Sh, *Striga hermonthica*; Sl, *Solanum lycopersicum*; Sp, *Securigera parviflora*; Ta, *Triticum aestivum*; Vf, *Vicia faba*; Vv, *Vitis vinifera*; Zm, *Zea mays.*

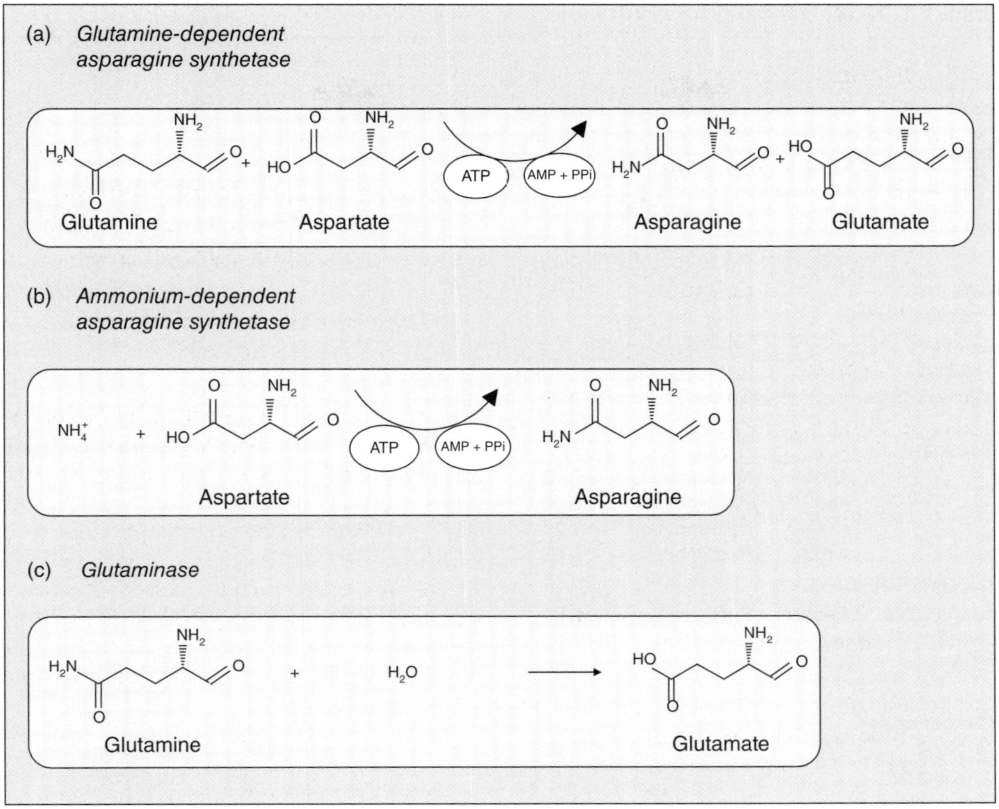

Fig. 6.5. The reactions of asparagine synthetase (AsnS). (After Lea and Fowden, 1975a; Gaufichon *et al.*, 2010). Cys, cysteine; E, enzyme (AsnS); PPi, pyrophosphate.

has K_m values for Gln and NH_4^+ of 0.66 and 17 mM, respectively. (Lea and Fowden, 1975b; Boehlein *et al.*, 1994). For the different plant asparagine synthetases, K_m (Gln) values as high as 1 mM (Stulen *et al.*, 1979) and as low as 40 μM (Oaks and Ross, 1984) have been reported, providing support for functional differences between the various plant enzymes. Interestingly, the variation in other kinetic parameters, such as the K_m values for Asp and ATP were much less than for Gln, which may mean that in many tissues Gln is the primary limiting substrate for AsnS and that the enzyme has evolved in different tissues to make use of differing levels of Gln.

Owing to difficulties in purifying the soluble enzyme, studies of AsnS have historically examined only crude extracts from eukaryotic species that have multiple isoforms. Therefore, it is difficult to get a handle on how the classes of AsnS may differ kinetically. For this reason, in our laboratory, we compared recombinant maize isoforms expressed in different tissues (Table 6.1). The binding constants for Asp were within the range of 800–1200 μM for the different maize AsnS forms, consistent with what is observed in the literature (Duff *et al.*, 2011). However, the K_m (Gln) values of other plant forms varied. Interestingly, ZmAsnS2 has a significantly lower K_m (Gln) than any other tested maize AsnS and it also had a higher V_{max}, suggesting that this isoform of AsnS has a distinct function. As might be expected, the kinetic values for ZmAsnS3 and ZmAsnS4 were very similar, the largest difference being a slightly less than twofold difference in K_m (Gln). GmAsnS had a low V_{max} but was otherwise kinetically similar to ZmAsnS1, although it was less related phylogenetically to this enzyme than to the other maize forms.

What possible functional significance might the two main classes (I and II) of asparagine

Table 6.1. Kinetic comparison of various asparagine synthetases.

Species/ cell type	Enzyme name	Preparation	K_m (mM)				Reference
			Aspartate	MgATP	Glutamine	NH_4^+	
Plant enzymes							
Glycine max	–	Cotyledons	2.2	0.066	0.12	3	Streeter, 1973
	–	Root nodules	1.24	0.076	0.16	nd[a]	Huber and Streeter, 1985
	AS1	Recombinant	0.85	0.1	0.25	nd	Duff *et al.*, 2011
Lupinus albus	–	Cotyledons	0.8	0.15	0.04	2.1	Lea and Fowden, 1975b
Lupinus luteus	–	Shoots			0.5		Rognes, 1975
Vigna radiata	–	Crude	1.3	0.53	0.19		Pike and Beevers, 1982
Zea mays	–	Root extract			0.6	2.6	Oaks and Ross, 1984
	AsnS1	Recombinant	0.98	0.11	0.54	9.9	Duff *et al.*, 2011
	AsnS2	Recombinant	0.91	0.13	0.11	0.75	
	AsnS3	Recombinant	1.2	0.1	0.42	8.4	
	AsnS4	Recombinant	0.93	0.13	0.23	9	
Bacterial enzymes							
Escherichia coli	ASB	Pure	0.85	0.26	0.66	17	Boehlein *et al.*, 1994
Klebsiella	–	Pure	0.5	0.24	0.55	nd	Reitzer and Magasanik, 1982
Cancer cells							
Novikoff heptoma	–	–	0.58	0.11	1.1	120	Patterson and Orr, 1968
Rat sarcoma lines	–	Pure	0.82	nd	0.9	nd	Sugiyama *et al.*, 1983

[a]nd, no data.

synthetases have? In maize, the kinetic parameters for multiple AsnS enzymes have been determined (Duff *et al.*, 2011). ZmAsnS1 is a class II protein while AsnS2, AsnS3 and AsnS4 were originally all class I, with AsnS3 and AsnS4 now possibly relegated to a third clade, class III (Fig. 6.4). Interestingly, ZmAsnS1 (class II) has broad expression and a considerably higher K_m (Gln) and lower V_{max} than the three class I maize enzymes, which are typical characteristics of a housekeeping enzyme. In contrast, the class I AsnS enzymes, and especially ZmAsnS2, have more specific expression patterns, are more active and bind Gln tighter, all of these characteristics suggesting a more specific function. More detailed side-by-side comparisons of class I and class II AsnS enzymes will be required to determine whether the separate clades of AsnS in plants have distinct functions.

6.8.2 The crystal structure and reaction mechanism of asparagine synthetase

What can the structure of AsnS tell us about the function of AsnS? As this chapter is written, no eukaryotic crystal structures have been solved for the enzyme, probably because of its recalcitrant nature, which was noted previously. The only B-type asparagine structure so far described is that of AsnSB from *E. coli* (Larsen *et al.*, 1999). As no other B class AsnS prokaryotic structures are available and no eukaryotic AsnS structures exist, the inferences we can make for plant AsnS enzymes are limited.

AsnSB and the related eukaryotic types are members of the Ntn (formerly referred to as the *PurF* or class II) group of glutamine amidotransferases (Zalkin, 1993; Zalkin and Smith, 1998), which are defined by an active-site cysteine (Cys), invariably located at the N-terminus.

This Ntn family of enzymes contains an approximately 190 amino acid, glutamine-amide transfer domain at the N-terminus, including the active site cysteine (Boehlein *et al.*, 1994), as their name implies. This family also includes NADPH-GOGAT, Fd (ferredoxin)-GOGAT, amidophosphoribosyltransferase, glutamine:fructose 6-phosphate transaminase (Vanoni and Curti, 2008) and glucosamine 6-phosphate synthase (Isupov *et al.*, 1996).

Not surprisingly, the three-dimensional architecture of the N-terminal domain of AsnSB is similar to that of glutamine phosphoribosylpyrophosphate amidotransferase, an Ntn class glutamine amidotransferase enzyme with a similar reaction mechanism and function to AsnSB (Larsen *et al.*, 1999). AsnSB has two catalytic domains, the glutamine-amide transfer domain described above and the C-terminal synthetase domain that forms Asn from ATP, Asp and magnesium (Fig. 6.4) (Larsen *et al.*, 1999). The two subunits of the AsnSB dimer fold into two distinct N-terminal and C-terminal domains (Chakrabarti *et al.*, 1989). Analysis of the structure of AsnSB (Larsen *et al.*, 1999) has defined the spatial relationship between the domains as well as the binding of Gln. It is probable that a molecular tunnel connecting these two domains exists that facilitates the reaction mechanism (Larsen *et al.*, 1999).

The available data indicate that the reaction proceeds according to the scheme shown in Fig. 6.6 (Boehlein *et al.*, 1998; Larsen *et al.*, 1999). Glutamine binds to the Cys2 nucleophile in the N-terminal domain to form a covalent cysteinyl-glutamine tetrahedral intermediate, which removes the NH_3, making a γ-glutamyl thioester intermediate which is then hydrolysed to form Glu. Asp and Mg^{2+}-ATP bind to a C-terminal synthetase domain, activating the β-aspartyl side of the Asp molecule by adenylation to form a β-aspartyl intermediate plus pyrophosphate (Mei and Zalkin, 1989). The ammonia moves through a 19 Å tunnel (Boehlein *et al.*, 1998; Mei and Zalkin, 1989) producing a nucleophilic assault on the intermediate, and hydrolysing it to produce Asn (Fig. 6.6).

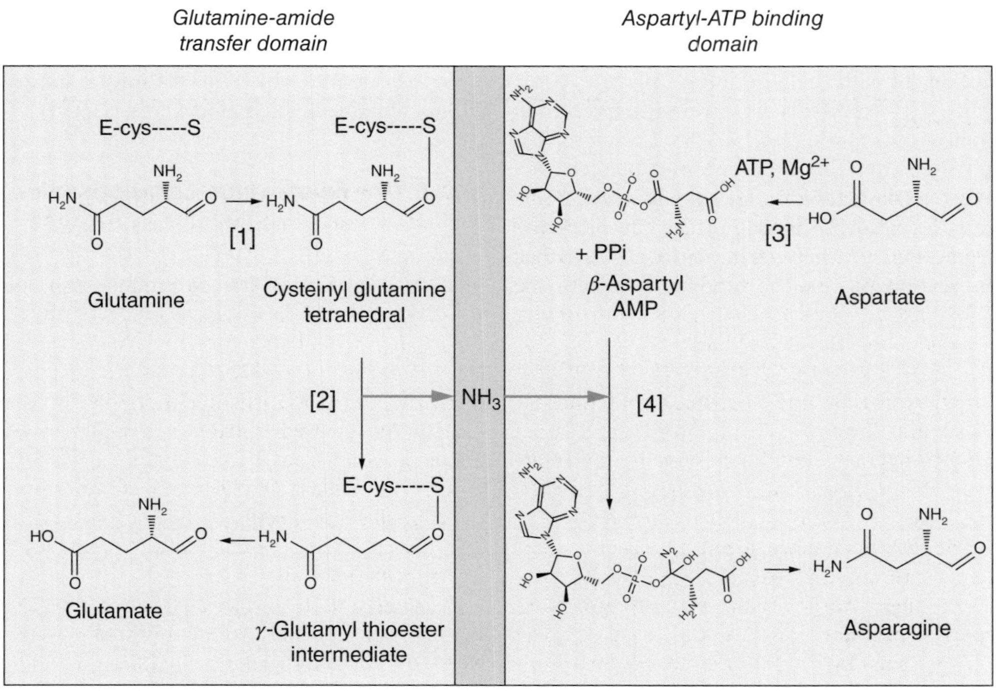

Fig. 6.6. The reaction mechanism of *Escherichia coli* asparagine synthetase (AsnSB). See text for details of the various steps. PPi, pyrophosphate.

Interestingly, monoclonal antibodies to bovine AsnS can inhibit the Gln-dependent synthesis of Asn but enhance the ammonium-dependent activity. This indicates that the ammonium-dependent activity might proceed independently, but less efficiently, in the synthetase domain when the ammonia is supplied endogenously rather than by molecular tunnelling from the glutamine-amide transfer domain (Markin *et al.*, 1981; Pfeiffer *et al.*, 1986; Mehlhaff and Schuster, 1991). The reaction mechanism of bovine AsnS is consistent with that of the enzyme from *E. coli* (Pfeiffer *et al.*, 1987). A quick glance at Table 6.1 reveals that the K_m (ammonium) for the prokaryotic AsnSB and the eukaryotic type AsnS is on average 40 times higher than the K_m (glutamine).

Evidence does exist for *in vivo* ammonium-dependent synthesis of Asn in plants. For example, AsnS activity from maize roots is not inhibited by a Gln analogue (Stulen *et al.*, 1979) and selective inhibition of tobacco GS/GOGAT leads to direct incorporation of ^{15}N-ammonium into Asn in the dark (Masclaux-Daubresse *et al.*, 2006). From Table 6.1, it is apparent that eukaryotic asparagine synthetases vary considerably in the efficiency of their ammonium-dependent reaction. In particular, ZmAsnS2, a type I AsnS, has a relatively low K_m (ammonium) of 0.75 mM compared with the other maize enzymes (Duff *et al.*, 2011) and it is also strongly expressed in maize roots (Todd *et al.*, 2008). Therefore, it is possible that this ammonium-dependent activity in dark tobacco leaves and maize roots may be catalysed by one of the Gln-preferred asparagine synthetases in plants. As described in Section 6.6 on the function of AsnS, the ammonium-dependent activity may function under certain types of stress as an ammonium detoxification mechanism by directly incorporating ammonium into Asn (Lea *et al.*, 2007), or indirectly by converting Asp to Asn for temporary storage and recycling of Gln to Glu to anaplerotically replenish the GS/GOGAT cycle.

6.9 Other Routes of Asparagine Synthesis in Plants

In addition to the two types of asparagine synthetase reactions already described, other pathways for Asn synthesis also occur in plants. An additional known pathway for Asn synthesis that is dependent upon cyanide and cysteine as initial substrates has been demonstrated (Joy, 1988) (Fig. 6.7). Cyanide is a by-product of ethylene biosynthesis (Miller and Conn, 1980). Since the availability of both cyanide and cysteine in plants is limited, the overall relevance of this pathway to plants is hard to assess (Joy, 1988).

A fourth possible route for the manufacture of Asn in plants is via the transamination of an amino nitrogen group on to 2-oxosuccinamic acid (the corresponding keto acid for asparagine) by the enzyme asparagine-oxo-acid transaminase. However, the extent to which this pathway is utilized *in vivo* is unknown (Meister and Fraser, 1954).

6.10 Asparagine Catabolism

Asparaginase (EC 3.5.1.1) catalyses the irreversible hydrolysis of Asn to produce Asp and

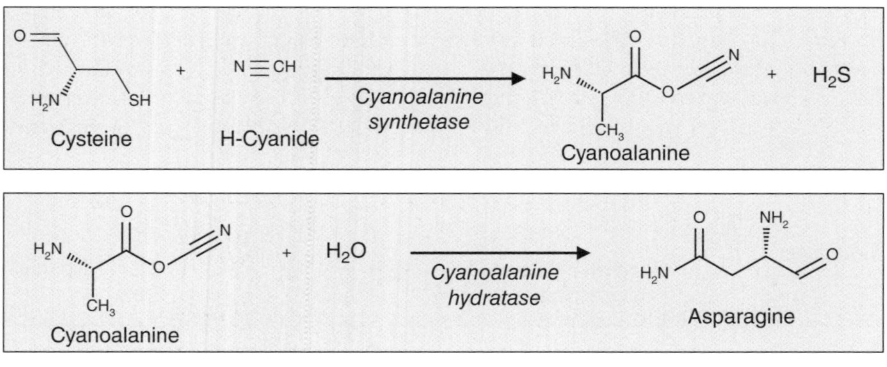

Fig. 6.7. The synthesis of asparagine from cysteine and cyanide.

ammonia, as shown in Fig. 6.8. The ammonia can be subsequently reassimilated by glutamine synthetase.

The first clear indication of asparaginase activity in the synthesis of Asn came in 1975 from an experiment in which dual-labelled ^{14}C and ^{15}N-Asn was fed to excised shoots of *L. albus* (Atkins *et al.*, 1975). The ^{15}N label was recovered as ammonia or amino acids in the endosperm fluid, whereas most of the ^{14}C label was not found in amino acids; the separation of the two labels demonstrated that hydrolysis of Asn had occurred. Subsequently, asparaginase activity was confirmed in legume seeds (Lea *et al.*, 1978). Later, a K^+-dependent asparaginase was discovered in the testa and maturing cotyledons of pea seeds (Sodek *et al.*, 1980). The activity of a K^+-dependent asparaginase in pea leaves was shown to increase during the light, and decrease during the dark (Sieciechowicz *et al.*, 1985), the regulation being due to enzyme synthesis and proteolytic degradation. The NH_4^+ produced by hydrolysis would need to be reassimilated immediately to prevent toxicity, and so light-dependent activity would ensure that the breakdown of Asn coincides with the presence of ATP to fuel reassimilation by GS (Sieciechowicz *et al.*, 1988).

Arabidopsis has two genes encoding a K^+-dependent asparaginase as well as a K^+-independent asparaginase (Bruneau *et al.*, 2006). The proteins encoded by these two genes belong to two distinct evolutionary subfamilies, but have very similar structures as well as cleavage sites, and probably a similar enzymatic mechanism. In *Arabidopsis*, the K^+-dependent form of asparaginase has a much higher (about 80-fold) catalytic efficiency than the K^+-independent form (Bruneau *et al.*, 2006). Both genes are expressed mainly in sink tissues and they often overlap, indicating redundant functions (Bruneau *et al.*, 2006), although this is not yet a settled question. In plants, Asn residues in proteins are frequently converted to isoaspartyl residues. When these proteins are degraded, they produce β-aspartyl-containing N-terminal isoaspartyl residues that require specialized hydrolytic enzymes (Borek *et al.*, 2004). Therefore, it is possible that the isoaspartyl peptidase activity of the K^+-independent asparaginase could be particularly important in reducing the frequency of occurrence of isoaspartyl peptidases forming in seeds that have to maintain their viability for long periods of time (Borek *et al.*, 2004). In any case, the specificity of the expression of these two asparaginase genes is crucial for the plant's ability to remobilize nitrogen from Asn after nitrogen transport and temporary storage.

6.11 Asparagine Synthetase and Agriculture

6.11.1 Seed protein content and crop yield

Although more basic research is necessary to integrate what we know about AsnS to help improve crop nitrogen use efficiency, it is clear that as an important plant transport and storage molecule, asparagine, and by implication AsnS, have critical functions in crops. While there are relatively few studies on free or protein-bound Asn content in seeds, it is clearly a significant constituent, especially in storage proteins. Conversely, free amino acids account for a small proportion of cereal nitrogen content and Asn generally constitutes a small proportion of this fraction (Dembinski and Bany, 1991).

Recently, a positive correlation between Asn metabolism and protein content in soybean seeds was described (Wan *et al.*, 2006; Pandurangan *et al.*, 2012). Cultivars with higher protein in the seeds were confirmed to have higher levels of asparaginase transcript, enzyme and activity levels, and also of AS1 transcript

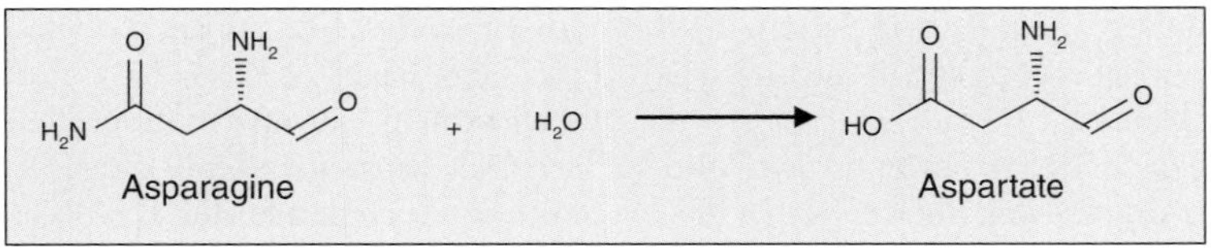

Fig. 6.8. The reaction catalysed by asparaginase.

(Wan *et al.*, 2006). The authors suggested that the level of *AS1* expression may be a potential supplementary marker to screen for high protein soybean cultivars in breeding programmes. This direct link will aid elucidation of the function of the genes determining protein concentration, a key trait for soybean genetic improvement.

Illinois high and low protein inbred maize strains were developed by continuous selection in the direction of high and low kernel protein, respectively, to investigate the metabolic processes governing protein content (Uribelarrea *et al.*, 2004). The two strains differ in many ways, and studies have demonstrated that grain protein and starch composition in maize hybrids is under genetic control and can be modulated by the positive effects of nitrogen on reproductive sink capacity and seed protein synthesis.

An earlier study comparing Illinois high protein (IHP) and Illinois low protein (ILP) lines (Lohaus *et al.*, 1998) demonstrated that the IHP lines had much higher nitrate reductase activity in their roots, a higher root-to-shoot biomass ratio, and higher amino nitrogen in their leaves, roots, phloem and grains. The enhanced amino nitrogen was mainly due to an increase in Asn (Lohaus *et al.*, 1998), which was 13-fold greater in the IHP lines than in the ILP lines. The authors concluded that IHP maize differs from ILP maize in that it has a much higher capacity to deliver nitrogen in the form of Asn from the roots and also has higher nitrogen assimilation in the leaves (Lohaus *et al.*, 1998).

6.11.2 The impact of plant nutrition

Winkler and Schon (1980) grew barley plants in pots under different levels of nitrogen and demonstrated that while the total free amino acids and total Asn increased with grain nitrogen, the proportion of Asn remained constant at about 15% of the fraction. Thus, increasing seed nitrogen does not necessarily disproportionally affect Asn accumulation (Winkler and Schon, 1980) differing from what is observed in the maize IHP lines described above.

Later, in a study of barley grown under varying levels of nitrogen, Shewry *et al.* (1987) determined that while nitrogen alone did not seem to affect Asn levels in the endosperm disproportionately to those of the other amino acids, the S:N ratio was important. They grew plants under sulfur-deficient conditions, which restricted their ability to synthesize the major hordein (prolamin) storage proteins that contain cysteine and methionine. Under these conditions, total hordein decreased from 51.0 to 27.0% of total seed nitrogen in one cultivar, and from 46.7 to 26.9% in another, while non-protein nitrogen increased from 6.8 to 29.1% and from 7.3 to 32.6%, respectively. Analysis of the total grain demonstrated increases in the proportion of Asp plus Asn relative to the total amino acids from 5.7 to 19.2% for one cultivar and from 5.3 to 18.5% for the other. This study indicates that these changes resulted from the accumulation of free Asn that acted as an alternative nitrogen store when the ability to synthesize hordein was compromised by insufficient sulfur.

Studies such as this are important because high levels of Asn are not desirable in food due to the connection between Asn levels and acrylamide formed during cooking (Mottram *et al.*, 2002). Recently, many areas of Western Europe have become at risk for cereal sulfur deficiency as a result of decreased use of sulfur-containing fertilizers and reduced atmospheric deposition (Zhao *et al.*, 1999). Therefore, decreased sulfur in the soil may raise Asn levels in food, thus increasing the risk of acrylamide poisoning. Fortunately, a follow-up study on wheat demonstrated that similar high levels of Asn accumulation occurred in grain grown under normal conditions in the field and in the highly sulfur-deficient grain grown in glasshouse experiments (Muttucumaru *et al.*, 2006). So it is unlikely that sulfur deficiency will lead to higher levels of Asn in food products.

6.11.3 Metabolic engineering and transgenic studies

A number of transgenic studies on plants expressing various forms of AsnS are available. In one such study, transgenic *Arabidopsis* plants mis-expressing *AtASN1* had elevated soluble seed protein, elevated total protein and higher seedling tolerance to low nitrogen, leading the authors to conclude that the overexpression of the gene had enhanced the nitrogen content in their seeds (Lam *et al.*, 2003).

In another study from the same laboratory, *Arabidopsis* plants overexpressing *AtASN2* were grown on MS medium containing 50 mM ammonium, and were shown to accumulate less endogenous ammonium than the wild type Colombia-0 and ASN2 underexpressers (Wong *et al.*, 2004). Under high light, ammonium levels are high, and under these conditions, transgenic plants with reduced AtASN2 accumulated more endogenous ammonium than the wild type Colombia-0 and AtASN2 overexpressers. These results provide more support for the notion that AtASN2 is closely correlated to ammonium metabolism in higher plants. (Wong *et al.*, 2004).

Transgenic tobacco plants expressing the pea AsnS (AS1) under a 35S promoter demonstrated increased vegetative growth and accumulated 10–100 fold higher free Asn levels than the control plants (Brears *et al.*, 1993). These plants also had lower levels of the AsnS substrates Gln and Asp. Pea AS1 was truncated to remove the Gln-binding domain and enhance the ammonia-dependent activity. When this truncated pea AsnS was expressed in tobacco, the transgenic plants had 3–19 fold higher Asn levels, but also displayed an 'off-type' with poorer growth (Brears *et al.*, 1993). Recently, simultaneous downregulation of two AsnS genes in potato was shown to limit Asn formation and reduce the acrylamide-forming potential of tubers (Chawla *et al.*, 2012). Interestingly, grafting experiments described in the study suggested that Asn formation in potato plants occurs mainly in tubers, and the increase in Gln suggests that in potatoes the main form of nitrogen transported is Gln. The Gln is formed in the leaf and transported to the tuber, where it is converted into Asn. Therefore, modifying the amount of Asn by changing the levels of AsnS can affect nitrogen status and growth of plants, although the phenotypes seem very case specific.

Transgenic oilseed rape (*Brassica napus*) expressing the *E. coli* AsnA (ammonium-dependent AsnS) gene under the control of the CaMV-35S promoter had substantially higher AsnS activity than wild type plants (Seiffert *et al.*, 2004). However, despite higher levels of enzyme activity, no changes in the amino acid composition in the leaves or in the phloem of transgenic plants were detected. In a pot experiment, two transgenic lines expressing the AsnA gene clearly performed worse than control plants under limiting nitrogen. Although the seed nitrogen content was higher, the seed yield and the seed nitrogen yield were lower, which was interpreted to mean that increased nitrate assimilation led, at limiting N supply, to a reduced seed yield and seed nitrogen yield. Under high nitrogen, differences were less pronounced for one transgenic line, whereas the other showed a higher seed nitrogen yield and an improved nitrogen harvest index. These results show that the expression of the *E. coli* AsnA gene in oilseed rape could be of advantage at high but not under limiting nitrogen (Seiffert *et al.*, 2004).

More recently, transgenic lettuce plants expressing *E. coli* AsnA displayed reduced levels of γ-aminobutric acid (GABA) and Glu but relatively similar levels of other amino acids (Sobolev *et al.*, 2007). However, a much greater change in carbon metabolism was observed in the transgenic plants, as the contents of sucrose, glucose and fructose decreased, whereas that of inulin increased by up to 30 times, accompanied by the alteration of most TCA cycle organic acids and higher tartaric acid compared with the non-transformed control plants. The authors concluded that as inulins are beneficial for human health and are extracted from plants, lettuce genotypes with a high content of foliar short-chain inulin represent useful materials for breeding strategies and a potential source for low molecular weight inulin.

Nitrogen mobilization is promoted by pathogen infection, a strategy that deprives the pathogen of nutrients. Nutritional and metabolic changes also occur in the defence response against pathogens (Hammond-Kosack and Parker, 2003). As already mentioned, in many higher plants, Asn and Gln represent central intermediates in nitrogen metabolism that also contribute to nitrogen transport. High levels of Asn are observed in infected leaves (Zurbriggen *et al.*, 2009). When pepper was infected by *Xanthomonas campestris* pv. *vesicatoria*, strong and early expression of *CaAS1* occurred in pepper leaves. Moreover, silencing this gene resulted in enhanced susceptibility to *Xanthomonas* infection (Hwang *et al.*, 2011). In *CaAS1*-silenced pepper and/or *CaAS1*-overexpressing *Arabidopsis*, CaAS1-dependent decreases or increases in Asn levels correlated with increased susceptibility or defence responses to microbial pathogens, respectively. Linking transcriptional and metabolite studies, this study suggests that the CaAS1-dependent synthesis of Asn is required for proper disease resistance in plants.

6.12 Conclusions

Despite the fact that asparagine was the first amino acid to be discovered in nature and was identified first from plants (Sieciechowicz *et al.*, 1988), less is known about its metabolism and biosynthesis than most other amino acids. For example, while the relationship between light and asparagine was first observed in the mid-19th century, it took over a 100 years for this process to begin to be elucidated (Cedar and Schwartz, 1969), and much still remains to be discovered.

There are many reasons why asparagine research in plants has lagged behind that of other amino acids. Glutamine and asparagine are integrated in a highly complicated fashion, though glutamine is readily catabolized and asparagine only selectively so. This discriminating catabolism of asparagine, however, increases its value as a transport and storage molecule. Further complicating asparagine research, most eukaryotic asparagine synthetases are extremely tricky to study. They are difficult to detect, assay and express in a soluble form in bacteria. In fact, only recently has plant asparagine synthetase been expressed in *E. coli* (Duff *et al.*, 2011), and even now no crystal structures of eukaryotic asparagine synthetase have been solved.

What is known is that asparagine, and by extension asparagine synthetase, play a major role in nitrogen metabolism, storage and transport in plants. This role is facilitated by the chemical properties of asparagine. As we have seen in this chapter, asparagine accumulates under stress, especially those types of stress under which the plant cannot support normal levels of protein synthesis or under conditions where proteins undergo abnormally high levels of proteolysis. In these cases, asparagine can serve as a temporary storage form for nitrogen as well as a mechanism for detoxifying ammonium. Due to the centrality of asparagine in nitrogen storage and transport, asparagine synthetase is involved in many stress responses and also in phenomena such as nutritional deficiency, seed germination, light signalling, carbohydrate depletion, nitrogen remobilization, seed maturation, mineral deficiencies and nitrogen signalling.

Further studies, especially those involving transgenes, in which the pathways involved in asparagine synthesis can be easily manipulated, are likely to turn up more surprises for researchers investigating asparagine synthetase; without doubt, there is much more to be discovered.

Acknowledgements

I would like to thank Drs James Crowley and Charles Armstrong for helpful discussions and editing the manuscript.

References

Amancio, S., Clarkson, D.T., Diogo, E., Lewis, M. and Santos, H. (1997) Assimilation of nitrate and ammonium by sulphur deficient *Zea mays* cells. *Plant Physiology and Biochemistry* 35, 41–48.

Amylon, M.D., Shuster, J., Pullen, J., Bernard, C. and Link, M.P. (1999) Intensive high-dose asparaginase consolidation improves survival for pediatric patients with T cell acute lymphoblastic leukemia and advanced stage lymphoblastic lymphoma: a Pediatric Oncology Group Study. *Leukemia* 13, 335–342.

Andrulis, I.L., Chen, J. and Ray, P.N. (1987) Isolation of human cDNAs for asparagine synthetase and expression in Jensen rat sarcoma cells. *Molecular and Cell Biology* 7, 2435–2443.

Arfin, S.M. (1967) Asparagine synthesis in the chick embryo liver. *Biochimica et Biophysica Acta* 136, 233–244.

Aslanian, A.M., Fletcher, B.S. and Kilberg, M.S. (2001) Asparagine synthetase expression alone is sufficient to induce L-asparaginase resistance in MOLT-4 human leukaemia cells. *Biochemical Journal* 357, 321–328.

Atkins, C.A., Pate, J.S. and Sharkey, P.J. (1975) Asparagine metabolism – key to the nitrogen nutrition of developing legume seeds. *Plant Physiology* 56, 807–812.

Baena-González, E., Rolland, F., Thevelein, J. and Sheen, J. (2007) A central integrator of transcription networks in plant stress and energy signaling, *Nature* 448, 938–942.

Balasubramanian, M.N., Butterworth, E.A. and Kilberg, M.S. (2013) Asparagine synthetase: regulation by cell stress and involvement in tumor biology. *American Journal of Physiology – Endocrinology and Metabolism* 304, 789–799.

Baudet, J., Huet, J.-C., Jolivet, E., Lesaint, C., Mossé, J. and Pernollet, J.-C. (1986) Changes in accumulation of seed nitrogen compounds in maize under conditions of sulphur deficiency. *Physiologia Plantarum* 68, 608–614.

Beatoa, V.M., Rexach, J., Navarro-Gochicoa, M.T., Camacho-Cristóbal, J.J., Herrera-Rodríguez, M., Maldonado, J.M. and González-Fontes, A. (2010) A tobacco asparagine synthetase gene responds to carbon and nitrogen status and its root expression is affected under boron stress. *Plant Science* 178, 289–298.

Becker, T.W., Carrayol, E. and Hirel, B. (2000) Glutamine synthetase and glutamate dehydrogenase isoforms in maize leaves: localization, relative proportion and their role in ammonium assimilation or nitrogen transport. *Planta* 211, 800–806.

Bernard, S.M., Laurell, A., Møller, B., Dionisio, G., Kichey, T., Jahn, T.P., Dubois, F., Baudo, M., Lopes, M.S., Tercé-Laforgue, T., Foyer, C.H., Parry, M.A.J. *et al.* (2008) Gene expression, cellular localization and function of glutamine synthetase isozymes in wheat (*Triticum aestivum* L.). *Plant Molecular Biology* 67, 89–105.

Bigot, J., Lefevre, J. and Boucard, J. (1991) Changes in the amide and amino acid composition of xylem exudate from perennial ryegrass (*Lolium perenne* L.) during regrowth after defoliation. *Plant and Soil* 136, 59–64.

Boehlein, S.K., Richards, N.G.J. and Schuster, S.M. (1994) Glutamine-dependent nitrogen transfer in *Escherichia coli* asparagine synthetase B. *The Journal of Biological Chemistry* 269, 7450–7457.

Boehlein, S.K., Stewart, J.D., Walworth, E.S., Thirumoorthy, R., Richards, N.G.J. and Schuster S.M. (1998) Kinetic mechanism of *Escherichia coli* asparagine synthetase B. *Biochemistry* 37, 13230–13238.

Bollard, E.G. (1960) Transport in the xylem. *Annual Review of Plant Physiology* 11, 141–166.

Borek, D., Michalska, K., Brzezinski, K., Kisiel, A., Podkowinski, J., Bonthron, D.T., Krowarsch, D., Otlewski, J. and Jaskolski, M. (2004) Expression, purification and catalytic activity of *Lupinus luteus* asparagine β-amidohydrolase and its *Escherichia coli* homolog. *European Journal of Biochemistry* 271, 3215–3226.

Brears, T., Liu, C., Knight, T.J. and Coruzzi, G.M. (1993) Ectopic over expression of asparagine synthetase in transgenic tobacco. *Plant Physiology* 103, 1285–1290.

Bruneau, L., Chapman, R. and Marsolais, F. (2006) Co-occurrence of both L-asparaginase subtypes in *Arabidopsis*: At3g16150 encodes a K+-dependent L-asparaginase. *Planta* 224, 668–679.

Burchall, J.J., Reichelt, E.C. and Wolin, M.J. (1964) Purification and properties of the asparagine synthetase of *Streptococcus bovis*. *The Journal of Biological Chemistry* 239, 1794–1798.

Camacho-Cristóbal, J.J. and González-Fontes, A. (2007) Boron deficiency decreases plasmalemma H+-ATPase expression and nitrate uptake, and promotes ammonium assimilation into asparagine in tobacco roots. *Planta* 226, 443–451.

Canales, J., Ávila, C., Cantón, F.R., Pacheco-Villalobos, D., Díaz-Moreno, S., Ariza, D., Molina-Rueda, J.J., Navarro-Cerrillo, R.M., Claros, M.G. and Cánovas, F.M. (2012) Gene expression profiling in the stem of young maritime pine trees: detection of ammonium stress-responsive genes in the apex. *Trees* 26, 609–619.

Canas, R., Quilleré, I., Christ, A. and Hirel, B. (2009) Nitrogen metabolism in the developing ear of maize (*Zea mays*): analysis of two lines contrasting in their mode of nitrogen management. *New Phytology* 184, 340–352.

Capizzi, R.L., Bertino, J.R., Skeel, R.T., Creasey, W.A., Zanes, R., Olayon, C., Peterson, R.G. and Handschumacher, R.E. (1971) L-asparaginase: clinical, biochemical, pharmacological, and immunological studies. *Annals of Internal Medicine* 74, 893–901.

Cedar, H. and Schwartz, J.H. (1969) The asparagine synthetase of *Escherichia. coli*. Biosynthetic role of the enzyme, purification and characterization of the reaction products. *The Journal of Biological Chemistry* 244, 1794–1798.

Chakrabarti, R., Wylie, D.E. and Schuster, S.M. (1989) Transfer of monoclonal antibodies into mammalian cells by electroporation. *The Journal of Biological Chemistry* 264, 15494–15500.

Chawla, R., Shakya, R. and Rommens, C.M. (2012) Tuber-specific silencing of asparagine synthetase-1 reduces the acrylamide-forming potential of potatoes grown in the field without affecting tuber shape and yield. *Plant Biotechnology Journal* 10, 913–924.

Chevalier, C., Bourgeois, E., Just D. and Raymond P. (1996) Metabolic regulation of asparagine synthetase gene expression in maize (*Zea mays* L.) root tips. *The Plant Journal* 9, 1–11.

Chibnall, A.C. (1930) *Protein Metabolism in the Plant*. Yale University Press, New Haven, Connecticut.

Cooney, D.A. and Handschumacher, R.E. (1970) L-Asparaginase and L-asparagine metabolism. *Annual Review of Pharmacology* 10, 421–440.

Cooper, H.D. and Clarkson, D.T. (1989) Cycling of amino-nitrogen and other nutrients between shoots and roots in cereals – a possible mechanism integrating the shoot and root in the regulation of nutrient uptake. *Journal of Experimental Botany* 40, 753–762.

Coruzzi, G. (2003) Primary N-assimilation into amino acids In: Somerville, C.M.E. (ed.) *The Arabidopsis Book*. American Society of Plant Biologists, Rockville, Maryland, pp. 1–17.

Crawford, J.T., Rending, V. and Broadbent, F. (1982) Sources, fluxes, and sinks of nitrogen during early reproductive growth of maize (*Zea mays* L.). *Plant Physiology* 70, 1654–1660.

Credali, A., García-Calderón, M., Dam, S., Perry, J., Díaz-Quintana, A., Parniske, M., Wang, T.L., Stougaard, J., Vega, J.M. and Márquez, A.J. (2013) The K^+-dependent asparaginase, NSE1, is crucial for plant growth and seed production in *Lotus japonicus*. *Plant and Cell Physiology* 54, 107–118.

Damodaran, M. (1932) The isolation of asparagine from an enzymic digest of edestin. *Biochemical Journal* 26, 235–247.

Delaville, M. (1802) Sur les seves d'asperges et de choux. *Annales de Chimie* 41, 298.

De Michele, R., Formentin, E., Todesco, M., Toppo, S., Carimi, F., Zottini, M., Barizza, E., Ferrarini, A., Delledonne, M., Fontana, P. and Lo Schiavo, F. (2009) Transcriptome analysis of *Medicago truncatula* leaf senescence: similarities and differences in metabolic and transcriptional regulations as compared with *Arabidopsis*, nodules senescence and nitric oxide signaling, *New Phytologist* 181, 563–575.

Dembinski, E. and Bany, S. (1991) The amino acid pool of high and low protein rye inbred lines (*Secale cereale* L.). *Journal of Plant Physiology* 138, 494–496.

Diaz, C., Lemaître, T., Christ, A., Azzopardi, M., Kato, Y., Sato, F., Morot-Gaudry, J.F., Le Dily, F. and Masclaux-Daubresse, C. (2008) Nitrogen recycling and remobilization are differentially controlled by leaf senescence and development stage in *Arabidopsis* under low nitrogen nutrition. *Plant Physiology* 147, 1437–1449.

Dilworth, M. and Dure, L. (1978) Developmental biochemistry of cotton seed embryogenesis and germination: X. Nitrogen flow from arginine to asparagine in germination. *Plant Physiology* 61, 698–702.

do Amarante, L., Lima, L.D. and Sodek, L. (2006) Growth and stress conditions cause similar changes in xylem amino acids for different legume species. *Environmental and Experimental Botany* 58, 123–129.

Donnison, I.S., Gay, A.P., Thomas, H., Edwards. K.J., Edwards, D., James, C.L., Thomas, A.M. and Ougham, H.J. (2007) Modification of nitrogen remobilization, grain filling and leaf senescence in maize (*Zea mays*) by transposon insertional mutagenesis in a protease gene. *New Phytologist* 173, 481–494.

Duff, S.M.G., Qi, Q., Reich, T., Wu, X., Brown, T., Crowley, J.H. and Fabbri, B. (2011) A kinetic comparison of asparagine synthetase isozymes from higher plants. *Plant Physiology and Biochemistry* 49, 251–256.

Fait, A., Angelovici, R., Less, H., Ohad, I., Urbanczyk-Wochniak, E., Fernie, A.R. and Galili, G. (2006) *Arabidopsis* seed development and germination is associated with temporally distinct metabolic switches. *Plant Physiology* 142, 839–854.

Fischer W.-N., André, B., Rentsch, D., Krolkiewicz, S., Tegeder, M., Breitkreuz, B. and Frommer, W.B. (1998) Amino acid transport in plants. *Trends in Plant Science* 3, 188–195.

Fujihara, S. and Yamaguchi, M. (1980) Asparagine formation in soybean nodules. *Plant Physiology* 66, 139–141.

Fujiki, Y., Yoshikawa, Y., Sato, T., Inada, N., Ito, M., Nishida, I. and Watanabe, A. (2001) Dark-inducible genes from *Arabidopsis thaliana* are associated with leaf senescence and repressed by sugars, *Physiologia Plantarum* 111, 345–352.

Gaufichon, L., Reisdorf-Cren, M., Rothstein, S.J., Chardon, F. and Suzuki, A. (2010) Biological functions of asparagine synthetase in plants. *Plant Science* 179, 141–153.

Givan, C.V. (1979) Metabolic detoxification of ammonia in tissues of higher plants. *Phytochemistry* 18, 375–382.

Gómez-Cadenas, A., Zentella, R., Walker-Simmons, M.K. and Ho, T.H. (2001) Gibberellin/abscisic acid antagonism in barley aleurone cells: site of action of the protein kinase PKABA1 in relation to gibberellin signaling molecules. *The Plant Cell* 13, 667–679.

Graham, M.L. (2003) Pegaspargase: a review of clinical studies. *Advanced Drug Delivery Reviews* 55, 1293–1302.

Guo, Y., Cai, Z. and Gan, S. (2004) Transcriptome of *Arabidopsis* leaf senescence. *Plant, Cell and Environment* 27, 521–549.

Hammond-Kosack, K. and Parker, J. (2003) Deciphering plant–pathogen communication: fresh perspectives for molecular resistance breeding. *Current Opinions in Plant Biology* 14, 177–193.

Herrera-Rodríguez, M.B., Carrasco-Ballesteros, S., Maldonado, J.M., Pineda, M., Aguilar, M. and Pérez-Vicente, R. (2002) Three genes showing distinct regulatory patterns encode the asparagine synthetase of sunflower (*Helianthus annuus*). *New Phytologist* 155, 33–45.

Hersh, E.M. (1971) Immunosuppression by L-asparaginase and related enzymes: a review. *Transplantation* 12, 368–376.

Hongo, S. and Sato, T. (1981) Purification of rat liver asparagine synthetase by affinity chromatography on reactive blue 2-agarose. *Analytical Biochemistry* 114, 163–166.

Horowitz, B. and Meister, A. (1972) Glutamine-dependent asparagine synthetase from leukemia cells. *The Journal of Biological Chemistry* 247, 6708–6719.

Hörtensteiner, S. and Feller, U. (2002) Nitrogen metabolism and remobilization during senescence, *Journal of Experimental Botany* 53, 927–937.

Huang, C.Y., Roessner, U., Eickmeier, I., Genc, Y., Callahan, D.L., Shirley, N., Langridge, P. and Bacic, A. (2008) Metabolite profiling reveals distinct changes in carbon and nitrogen metabolism in phosphate-deficient barley plants (*Hordeum vulgare* L.). *Plant and Cell Physiology* 49, 691–703.

Huber, T.A. and Streeter, J.G. (1985) Purification and properties of asparagine synthetase from soybean root nodules. *Plant Science* 42, 9–17.

Hughes, C.A., Beard, H.S. and Matthews, B.F. (1997) Molecular cloning and expression of two cDNAs encoding asparagine synthetase in soybean. *Plant Molecular Biology* 33, 301–311.

Hwang, I.S., An S.H. and Hwang B.K. (2011) Pepper asparagine synthetase 1 (CaAS1) is required for plant nitrogen assimilation and defense responses to microbial pathogens. *The Plant Journal* 67, 749–762.

Igarashi, D., Ishizaki, T., Totsuka, K. and Ohsumi, C. (2009) ASN2 is a key enzyme in asparagine biosynthesis under ammonium sufficient conditions. *Plant Biotechnology* 26, 153–159.

Ireland, R.J. and Lea, P.J. (1999) The enzymes of glutamine, glutamate, asparagine and aspartate metabolism. In: Singh, B.K. (ed.) *Plants, Amino Acids, Biochemistry, and Biotechnology*. Marcel Dekker, New York, pp. 49–109.

Isupov, M.N., Obmolova, G., Butterworth, S., Badet-Denisot, M.A., Badet, B., Polikarpov, I., Littlechild, J.A. and Teplyakov, A. (1996) Substrate binding is required for assembly of the active conformation of the catalytic site in Ntn amidotransferases: evidence from the 1.8 A crystal structure of the glutaminase domain of glucosamine 6-phosphate synthase *Structure* 4, 801–810.

Joy, K.W. (1988) Ammonia, glutamine and asparagine: a carbon–nitrogen interface. *Canadian Journal of Botany* 66, 2103–2109.

Joy, K.W. and Ireland, R.J. (1990) Enzymes of asparagine metabolism. In: Lea, P.J. (ed.) *Enzymes of Primary Metabolism. Methods in Plant Biochemistry, Vol. 3*. Academic Press, San Diego, pp 287–296.

Kaneko, A., Noguchi, E., Ishibashi, Y., Yuasa, T. and Iwaya-Inoue, M. (2013) The sucrose starvation signal mediates induction of autophagy- and amino acid catabolism-related genes in cowpea seedling. *American Journal of Plant Science* 4, 647–653.

Kawachi, T., Sueyoshi, K., Nakajima, A., Yamagata. H., Sugimoto, T. and Oji, Y. (2002) Expression of asparagine synthetase in rice (*Oryza sativa*) roots in response to nitrogen. *Physiologia Plantarum* 114, 41–46.

Keeley, J.E. and Rundel, P.W (2003) Evolution of CAM and C_4 carbon concentrating mechanisms *International Journal of Plant Sciences* 164(suppl.), S55–S77.

Kern, R. and Chrispeels, M. (1978) Influence of the axis in the enzymes of protein and amide metabolism in cotyledons of mung bean seedlings. *Plant Physiology* 62, 815–819.

Keys, A.J. (2006) The re-assimilation of ammonia produced by photorespiration and the nitrogen economy of C_3 higher plants. *Photosynthesis Research* 87, 165–175.

King, J.E. and Gifford, D.J. (1997) Amino acid utilization in seeds of loblolly pine during germination and early seedling growth (I. Arginine and arginase activity). *Plant Physiology* 113, 1125–1135.

Kiriyama, Y., Kubota, M., Takimoto, T., Kitoh, J., Tanizawa, A., Akiyama, Y.H. and Mikawa, H. (1989) Biochemical characterization of U937 cells resistant to L-asparaginase: the role of asparagine synthetase. *Leukemia* 3, 294–297.

Kittredge, J.S., Simonson, D.G, Roberts, E. and Jelinek, B. (1962) Free amino acids of marine invertebrates. In: Holden, J.T. (ed.) *Amino Acid Pools: Distribution, Formation and Function of Free Amino Acids*. Elsevier, Amsterdam, pp. 176–186.

Klapheck, S., Grosse, W. and Bergmann, L. (1982) Effect of sulphur deficiency on protein synthesis and amino acid accumulation in cell suspensions of *Nicotiania tabacum*. *Zeitschrift für Pflanzenphysiologie* 108, 235–245.

Koorneef, M., Bentsink, L. and Hilhost, H. (2002) Seed dormancy and germination. *Current Opinions in Plant Biology* 5, 33–36

Lam, H.-M., Peng, S. and Coruzzi, G. (1994) Metabolic control of asparagine synthetase gene expression in *Arabidopsis thaliana*. *Plant Physiology* 106, 1347–1357.

Lam, H.-M., Coschigano, K., Schultz, C., Melo-Oliveira, R., Tjaden, G., Oliveira, I., Ngai, N., Hsieh, M-H. and Coruzzi, G. (1995) Use of *Arabidopsis* mutants and genes to study amide amino acid biosynthesis. *The Plant Cell* 7, 887–898.

Lam, H.-M., Coschigano, K., Oliveira, I.C., Melo-Oliveira, R. and Coruzzi, G. (1996) The molecular-genetics of nitrogen assimilation into amino acids in higher plants. *Annual Reviews in Plant Physiology and Plant Molecular Biology* 47, 569–593.

Lam, H.-M., Hsieh, M-H. and Coruzzi, G. (1998) Reciprocal regulation of distinct asparagine synthetase genes by light and metabolites in *Arabidopsis thaliana*. *The Plant Journal* 16, 345–353.

Lam, H.-M., Wong, P., Chan, H-K., Yam, K-M., Chen, L., Chow, C-M. and Coruzzi, G.M. (2003) Overexpression of the *ASN1* gene enhances nitrogen status in seeds of *Arabidopsis*. *Plant Physiology* 132, 926–935.

Larsen, T.M., Boehlein, S.K., Schuster, S.M., Richards, N.G.J., Thoden, J.B., Holden, H.M. and Rayment, I. (1999) Three-dimensional structure of *Escherichia coli* asparagine synthetase B: a short journey from substrate to product. *Biochemistry* 38, 16146–16157.

Lea, P.J. and Fowden, L. (1975a) Asparagine metabolism in higher plants. *Biochemie und Physiologie der Pflanzen* 168, 3–14.

Lea, P.J. and Fowden, L. (1975b) The purification and properties of glutamine-dependent asparagine synthetase isolated from *Lupinus albus*, *Proceedings of the Royal Society of London Series B, Biological Sciences* 192, 13–26.

Lea, P.J. and Miflin, B.J. (1980) Transport and metabolism of asparagine and other nitrogen compounds within the plant. In: Miflin, B.J. (ed.) *The Biochemistry of Plants*. Academic Press, New York, pp. 569–607.

Lea, P.J., Fowden, L. and Miflin, B.J. (1978) The purification and properties of asparaginase from *Lupinus* species. *Phytochemistry* 17, 217–222.

Lea, P.J., Robinson, S.A. and Stewart, G.R. (1990) The enzymology and metabolism of glutamine, glutamate, and asparagine. In Miflin, B.J. and Lea, P.J. (eds) *The Biochemistry of Plants*. Academic Press, New York, pp. 151–159.

Lea, P.J., Sodek, M.A., Shewry, P.R. and Halford, N.G. (2007) Asparagine in plants. *Annals of Applied Biology* 150, 1–26.

Leegood, R.C. (2013) Strategies for engineering C_4 photosynthesis. *Journal of Plant Physiology* 170, 378–388.

Lees, E.M., Farnden, K.J.F. and Elliot, W.H. (1968) Studies on asparagine synthesis and utilization in seedlings. *Archives of Biochemistry and Biophysics* 126, 539–546.

Levintow, L. (1957) Evidence that glutamine is a precursor of asparagine in a human cell in tissue culture. *Science* 126, 611–612.

Lin, J. and Wu, S. (2004) Molecular events in senescing *Arabidopsis* leaves. *The Plant Journal* 39, 612–628.

Lobel, J.S., O'Brien, R.T., McIntosh, S., Aspnes, G.T. and Capizzi, R.L. (1979) Methotrexate and asparaginase combination chemotherapy in refractory acute lymphoblastic leukemia of childhood. *Cancer* 43, 1089–1094.

Lohaus, G., Buker, M., HuBmann, M., Soave C. and Heldt, H.-W. (1998) Transport of amino acids with special emphasis on the synthesis and transport of asparagine in the Illinois Low Protein and Illinois High Protein strains of maize. *Planta* 205, 181–188.

Luehr, A.C. and Schuster, S.M. (1985) Purification and characterization of beef pancreatic asparagine synthetase. *Archives of Biochemistry and Biophysics* 237, 335–346.

Mansfield, J., Gebauer, S., Dathe, K. and Ulbrich-Hofmann, R. (2006) Secretory phospholipase A2 from *Arabidopsis thaliana*: insights into the three-dimensional structure and the amino acids involved in catalysis. *Biochemistry* 45, 5687–5694.

Markin, R.S., Luehr, C.A. and Schuster S.M. (1981) Kinetic mechanism of beef pancreatic L-asparagine synthetase. *Biochemistry* 20, 7226–7232.

Masclaux-Daubresse, C., Reisdorf-Cren, M., Pageau, K., Lelandais, M., Grandjean, O., Kronenberger, J., Valadier, M.H., Feraud, M., Jouglet, T. and Suzuki, A. (2006) Glutamine synthetase-glutamate synthase pathway and glutamate dehydrogenase play distinct roles in the sink–source nitrogen cycle in tobacco. *Plant Physiology* 140, 444–456.

McKee, H.S. (1962) *Nitrogen Metabolism in Plants*. Clarendon Press, Oxford, UK.

Mehlhaff, P.M. and Schuster S.M. (1991) Bovine pancreatic asparagine synthetase explored with substrate analogs and specific monoclonal antibodies. *Archives of Biochemistry and Biophysics* 284, 143–150.

Mei, B. and Zalkin, H. (1989) A cysteine-histidine-aspartate catalytic triad is involved in glutamine amide transfer function in pur-F-type glutamine amidotransferases. *The Journal of Biological Chemistry* 264, 16613–16619.

Meiss, A.N, (1952) *The Formation of Asparagine in Etiolated Seedlings of Lupinus albus L*. Bulletin 553, The Connecticut Agricultural Experiment Station, New Haven, Connecticut.

Meister, A. and Fraser, P.E. (1954) The enzymatic formation of L-asparagine by transamination. *The Journal of Biological Chemistry* 210, 37–43.

Miflin, B.J. and Lea, P.J. (1982) Ammonia assimilation and amino acid metabolism. In: Boulter, D. and Parthier, B. (eds) *Encyclopedia of Plant Physiology, Vol. 14a, Nucleic Acids and Proteins in Plants I*. Springer, Berlin, pp. 5–64.

Millard, P. (1996) Ecophysiology of the internal cycling of nitrogen for tree growth. *Journal of Plant Nutrition and Soil Science* 159, 1–10.

Miller, J. and Conn, E. (1980) Metabolism of hydrogen cyanide by higher plants. *Plant Physiology* 65, 1199–1202.

Moller, M.G., Taylor, C., Rasmussen, S.K. and Holm, P.B. (2003) Molecular cloning and characterization of two genes encoding asparagine synthetase in barley (*Hordeum vulgare* L.). *Biochimica et Biophysica Acta* 1628, 123–132.

Mortensen, J., Eriksen, J. and Nielsen, J.D. (1993) Sulphur deficiency and amino acid composition in seeds and grass. *Phyton – Annales Rei Botanicae* 32, 85–90.

Mottram, D.S., Wedzicha, B.L. and Dodson A.T. (2002) Acrylamide is formed in the Maillard reaction. *Nature* 419, 448–449.

Muller, B., Tillard, P. and Touraine, B. (1995) Nitrate fluxes in soybean seedling roots and their response to amino acids: an approach using 15N. *Plant, Cell and Environment* 18, 1267–1279.

Muttucumaru, N., Halford, N.G., Elmore, J.S., Dodson, A.T., Parry, M.A.J., Shewry, P.R. and Mottram, D.S. (2006) The formation of high levels of acrylamide during the processing of flour derived from sulfate-deprived wheat. *Journal of Agricultural and Food Chemistry* 54, 8951–8955.

Nakabayashi, K., Okamoto, M., Koshiba, T., Kamiya, Y. and Nambara, E. (2005) Genome wide profiling of stored mRNA in *Arabidopsis thaliana* seed germination: epigenetic and genetic regulation of transcription in seed. *The Plant Journal* 41, 697–709.

Nakano, K., Suzuki, T., Hayakawa, T. and Yamaya, T. (2000) Organ and cellular localization of asparagine synthetase in rice plants. *Plant and Cell Physiology* 41, 874–880.

Nakatsu, T., Kato, H. and Oda, J. (1998) Crystal structure of asparagine synthetase reveals a close evolutionary relationship to class II aminoacyl-tRNA synthetase. *Nature Structural and Molecular Biology* 5, 15–19.

Nozawa, A., Ito, M., Hayashi, H. and Watanabe A (1999) Dark-induced expression of genes for asparagine synthetase and cytosolic glutamine synthetase in radish cotyledons is dependent on the growth stage. *Plant and Cell Physiology* 40, 942–948.

Oaks, A. and Hirel, B. (1985) Nitrogen metabolism in roots. *Annual Review of Plant Physiology* 36, 345–365.

Oaks, A. and Ross, D.W. (1984) Asparagine synthetase in *Zea mays*. *Canadian Journal of Botany* 62, 68–73.

Ohshima, T., Hayashi, H. and Chino, M. (1990) Collection and chemical composition of pure phloem sap from *Zea mays* L. *Plant and Cell Physiology* 31, 735–737.

Osuna, D.G., Galvez, G., Pineda, M. and Aguilar, M. (1999) RT-PCR cloning, characterization and mRNA expression analysis of a cDNA encoding a type II asparagine synthetase in common bean. *Biochimica et Biophysica Acta* 1445, 75–85.

Pandurangan, S., Pajak, A., Molnar, S.J., Cober, E.R., Dhaubhadel, S., Hernandez-Sebastia, C., Kaiser, W.M., Nelson, R.L., Huber, S.C. and Marsolais, F. (2012) Relationship between asparagine metabolism and protein concentration in soybean seed. *Journal of Experimental Botany* 63, 3173–3184.

Pate, J.S. (1980) Transport and partitioning of nitrogenous solutes. *Annual Review of Plant Physiology* 31, 313–340.

Pate, J.S., Atkins, C.A., White, S.T., Rainbird R.M. and Woo, K.C. (1980) Nitrogen nutrition and xylem transport of nitrogen in ureide-producing grain legumes. *Plant Physiology* 65, 961–965.

Pate, J.S., Atkins, C.A., Layzell, D.B. and Shelp, B.J. (1984) Effect of N_2 deficiency on transport and partitioning of C and N in a nodulated lupin (*Lupinus albus* L.). *Plant Physiology* 76, 59–64.

Patterson, M.K. and Orr, G.R. (1968) Asparagine biosynthesis by the Novikoff hepatoma. *The Journal of Biological Chemistry* 243, 376–380.

Pfeiffer, N.E., Mehlhaff, P.M., Wylie, D.E. and Schuster, S.M. (1986) Monoclonal antibodies specific for bovine pancreatic asparagine synthetase. *The Journal of Biological Chemistry* 261, 1914–1919.

Pfeiffer, N.E., Mehlhaff, P.M., Wylie, D.E. and Schuster, S.M. (1987) Topographical separation of the catalytic sites of asparagine synthetase explored with monoclonal antibodies. *The Journal of Biological Chemistry* 262, 11565–11570.

Pike, D.C. and Beevers, L. (1982) Kinetic studies on asparagine synthetase from mung beans (*Vigna radiata* L. Wilczek) with a phosphonate analogue with a proposed reaction intermediate. *Biochimica et Biophysica Acta* 708, 203–209.

Possingham, J.V. (1956) The effect of mineral nutrition on the content of free amino acids and amides in tomato plants. I. A comparison of the effects of deficiencies of copper, zinc, manganese, iron and molybdenum. *Australian Journal of Biological Sciences* 9, 539–551.

Pui, C.H., Relling, M.V., Campana, D. and Evans, W.E. (2002) Childhood acute lymphoblastic leukemia. *Reviews in Clinical and Experimental Hematology* 6, 161–180.

Rabe, E. (1990) Stress physiology: the functional significance of the accumulation of nitrogen-containing compounds. *Journal of Horticultural Science* 65, 231–243.

Rachmilevitch, S., Cousins, A.B. and Bloom, A.J. (2004) Nitrate assimilation in plant shoots depends on photorespiration. *Proceedings of the National Academy of Sciences of the United States of America* 101, 11506–11510.

Rafalski, A. and Dembinski, E. (1990) Genotypic difference in ammonia metabolism of the developing maize kernel. *Acta Physiologiae Plantarum* 12, 331–339.

Ratajczak, W. (1986) Asparagine metabolism in developing seeds of *Lupinus leuteus* L. *Biochemie und Physiologie der Pflanzen* 181, 17–22.

Ravel, J.M., Norton, S.J., Humphreys, J. and Shive, W. (1962) Asparagine biosynthesis in *Lactobacillus arabinosus* and its control by asparagine through enzyme inhibition and repression. *The Journal of Biological Chemistry* 237, 2845–2849.

Reitzer, L.J. and Magasanik, B. (1982) Asparagine synthetases of *Klebsiella aerogenes*: properties and regulation of synthesis. *Journal of Bacteriology* 151, 1299–1313.

Richards, N.G.J. and Kilberg, M.S. (2006) Asparagine synthetase chemotherapy. *Annual Review of Biochemistry* 75, 629–654.

Richards, N.G.J. and Schuster, S.M. (1992) An alternative mechanism for the nitrogen transfer reaction in asparagine synthetase. *FEBS Letters* 313, 98–102.

Rognes, S.E. (1970) A glutamine-dependent asparagine synthetase from yellow lupine seedlings, *FEBS Letters* 10, 62–66.

Rognes, S.E. (1975) Glutamine-dependent asparagine synthetase from *Lupinus luteus*. *Phytochemistry* 14, 1975–1982.

Rolland, F., Moore, B. and Sheen, J. (2002) Sugar sensing and signaling in plants. *The Plant Cell* 14(suppl 1), S185–S205.

Romagni, J.G. and Dayan, F.E. (2000) Measuring asparagine synthetase activity in crude plant extracts. *Journal of Agricultural and Food Chemistry* 48, 1692–1696.

Rufty, T.W., MacKown, C.T. and Israel, D.W. (1990) Phosphorus stress effects on the assimilation of nitrate. *Plant Physiology* 94, 328–333.

Sage, R.F., Sage, T.L. and Kocacinar, F. (2012) Photorespiration and the evolution of C_4 photosynthesis. *Annual Review of Plant Biology* 63, 19–47.

Scharff, A.M., Eggaard, H., Hansen, P.E. and Rosendahl, L. (2003) Exploring symbiotic nitrogen fixation and assimilation in pea root nodules by *in vivo* 15N nuclear magnetic resonance spectroscopy and liquid chromatography–mass spectrometry. *Plant Physiology* 131, 367–378.

Schmidt, S. and Stewart, G.R. (1998) Transport, storage and mobilization of nitrogen by trees and shrubs in the wet/dry tropics of northern Australia. *Tree Physiology* 18, 403–410.

Scofield, M., Lewis, W. and Schuster, S. (1990) Nucleotide sequence of *Escherichia coli asnB* and deduced amino acid sequence of asparagine synthetase B. *The Journal of Biological Chemistry* 265, 12895–12902.

Seebauer, J.R., Moose, S.P., Fabbri, B.J., Crossland, L.D. and Below, F.E. (2004) Amino acid metabolism in maize earshoots. implications for assimilate preconditioning and nitrogen signaling. *Plant Physiology* 136, 4326–4334.

Seiffert, B., Zhou, Z.W. and Wallbraun, M. (2004) Expression of a bacterial asparagine synthase gene in oilseed rape (*Brassica napus*) and its effect on traits related to nitrogen efficiency. *Physiologia Plantarum* 121, 656–665.

Shend, S., Moraga, D.A., Van Heeke, G. and Schuster, S.M. (1992) High-level expression of human asparagine synthetase and production of monoclonal antibodies for enzyme purification. *Protein Expression and Purification* 3, 337–346.

Shewry, P.R., Williamson, M.S. and Kreis, M. (1987) Effects of mutant genes on the synthesis of storage components in developing barley endosperms. In: Thomas, H. and Grierson, D. (eds) *Mutant Genes that Affect Plant Development*. Cambridge University Press, Cambridge, UK, pp. 95–118.

Shi, L., Twary, S.N., Yoshiokay, H., Gregerson, R.G., Miller, S.S., Samacyay, D.A., Gantt, J.S., Unkeferyc, P.J. and Vancea, C.P. (1997) Nitrogen assimilation in alfalfa: isolation and characterization of an asparagine synthetase gene showing enhanced expression in root nodules and dark-adapted leaves. *The Plant Cell* 9, 1339–1356.

Sieciechowicz, K.A., Ireland, R.J. and Joy, K.W. (1985) Diurnal variation of asparaginase in developing pea leaves. *Plant Physiology* 77, 506–508.

Sieciechowicz, K.A., Joy, K.W. and Ireland, R.J. (1988) The metabolism of asparagine in plants. *Phytochemistry* 27, 663–671.

Sivasankar, S. and Oaks, A. (1996) Nitrate assimilation in higher plants: the effect of metabolites and light. *Plant Physiology and Biochemistry* 34, 609–620.

Smeekens, S. (2000) Sugar-induced signal transduction in plants. *Annual Review of Plant Physiology and Plant Molecular Biology* 51, 49–81.

Snapp, S.S. and Vance, C.P. (1986) Asparagine biosynthesis in alfalfa (*Medicago sativa* L.) root nodules. *Plant Physiology* 82, 390–395.

Sobolev, A.P., Segre, A.L., Giannino, D., Mariotti, D., Nicolodi, C., Brosio, E. and Amato, M.E. (2007) Strong increase of foliar inulin occurs in transgenic lettuce plants (*Lactuca sativa* L.) over-expressing the *Asparagine Synthetase A* gene from *Escherichia coli*. *Journal of Agricultural and Food Chemistry* 55, 10827–10831.

Sodek, L., Lea, P.J. and Miflin, B.J. (1980) Distribution and properties of a potassium-dependent asparaginase isolated from developing seeds of *Pisum sativum* and other plants. *Plant Physiology* 65, 22–26.

Stewart, G.R. and Larher, F. (1980) Accumulation of amino acids and related compounds in relation to environmental stress. In: Miflin B.J. (ed.), *The Biochemistry of Plants*. Academic Press, New York, pp. 609–635.

Streeter, J.G. (1973) *In vivo* and *in vitro* studies on asparagine biosynthesis in soybean seedlings. *Archives of Biochemistry and Biophysics* 157, 613–624.

Stulen, I. and Oaks A., (1977) Asparagine synthetase in corn roots. *Plant Physiology* 60, 680–683.

Stulen, I., Israelstam, G.F. and Oaks, A. (1979) Enzymes of asparagine synthesis in maize roots, *Planta* 146, 237–241.

Sugiyama, R.H., Arfin, S.M. and Harris, M. (1983) Properties of asparagine synthetase in asparagine-independent variants of Jensen rat sarcoma cells induced by 5-azacytidine. *Molecular and Cellular Biology* 3, 1937–1942.

Ta, T.C., Joy, K.W. and Ireland, R.J. (1984) Amino acid metabolism in pea leaves. *Plant Physiology* 74, 822–826.

Ta, T.C., Joy, K.W. and Ireland, R.J. (1985) Role of asparagine in photorespiratory nitrogen metabolism in pea leaves. *Plant Physiology* 78, 334–337.

Tallal, L., Tan, C., Oettgen, H.,Wollner, N., McCarthy, M., Helson, L., Burchenal, J., Karnofsky, D. and Murphy, M.L. (1970) *E. coli* L-asparaginase in the treatment of leukemia and solid tumors in 131 children. *Cancer* 25, 306–320.

Tallen, H.H., Moore, S. and Stein, W.H. (1954) Studies on the free amino acids and related compounds in the tissues of the cat. *The Journal of Biological Chemistry* 211, 927–939.

Terebello, H.R., Anderson, K., Wiernik, P.H., Cuttner, J., Cooper, R.M., Faso, L. and Berenberg, J.L. (1986) Therapy of refractory adult acute lymphoblastic leukemia with vincristine and prednisone plus tandem methotrexate and L-asparaginase. *Results of a Cancer and Leukemia Group B Study. American Journal of Clinical Oncology* 9, 411–415.

Thornton, B. and MacDuff, J.H. (2002) Short-term changes in xylem N compounds in *Lolium perenne* following defoliation. *Annals of Botany* 89, 715–722.

Todd, J., Screen, S., Crowley, J., Peng, J., Andersen, S., Brown, T., Qi, Q., Fabbri, B. and Duff, S.M.G. (2008) Identification and characterization of four distinct asparagine synthetase (*AsnS*) genes in maize (*Zea mays* L.). *Plant Science* 175, 799–804.

Tsai, F.Y. and Coruzzi, G.M. (1990) Dark-induced and organ-specific expression of two asparagine synthetase genes in *Pisum sativum*. *EMBO Journal* 9, 323–332.

Uribelarrea, M., Below, F.E. and Moose, S.P. (2004) Grain composition and productivity of maize hybrids derived from the Illinois protein strains in response to variable nitrogen supply. *Crop Science* 44, 1593–1600.

Valadier, M.H., Yoshida, A., Grandjean, O., Morin, H., Kronenberger, J., Boutet, S., Raballand, A., Hase, T., Yoneyama, T. and Suzuki, A. (2008) Implication of the glutamine synthetase-glutamate synthase pathway in conditioning the amino acid metabolism in bundle sheath and mesophyll cells of maize leaves. *FEBS Journal* 275, 3193–3206.

Van Heeke, G. and Schuster, S.M. (1989) Expression of human asparagine synthetase in *Escherichia coli*. *The Journal of Biological Chemistry* 264, 5503–5509.

Vanoni, M. and Curti, B. (2008) Structure–function studies of glutamate synthase: a class of self-regulated iron-sulfur flavoenzymes essential for nitrogen assimilation. *IUBMB Life* 60, 287–300.

Vauquelin, L.N. and Robiquet, P.J. (1806) The discovery of a new plant principle in *Asparagus sativus*. *Annales de Chimie* 57, 88–93.

Vierstra, R.D. (1993) Protein degradation in plants, *Annual Review of Plant Physiology and Plant Molecular Biology* 44, 385–410.

Wan, T.F., Shao, G.H., Shan, X.C., Zeng, N.Y. and Lam, H.M. (2006) Correlation between *AS1* gene expression and seed protein contents in different soybean (*Glycine max* [L.] Merr.) cultivars. *Plant Biology* 8, 271–276.

Waterhouse, R.N., Smyth, A.J., Massonneau, A., Prosser, I.M. and Clarkson, D.T. (1996) Molecular cloning and characterization of asparagine synthetase from *Lotus japonicus:* dynamics of asparagine synthesis in N-sufficient conditions. *Plant Molecular Biology* 30, 883–897.

Winkler, U. and Schon, W.J. (1980) Amino acid composition of the kernel proteins in barley resulting from nitrogen fertilization at different stages of development. *Journal of Agronomy and Crop Science* 149, 503–512.

Wingler, A., Lea, P.J., Quick, W.P. and Leegood, R.C. (2000) Photorespiration: metabolic pathways and their role in stress protection. *Philosophical Transactions of the Royal Society of London B, Biological Sciences* 355, 1517–1529.

Wong, H.-K., Chan, H.-K., Coruzzi, G. and Lam, H.-M. (2004) Correlation of *ASN2* gene expression with ammonium metabolism in *Arabidopsis*. *Plant Physiology* 134, 332–338.

Xu., G., Fan, X. and Miller, A.J. (2012) Plant nitrogen assimilation and use efficiency. *Annual Review of Plant Biology* 3, 153–182.

Zalkin, H. (1993) The amidotransferases. *Advances in Enzymology and Related Areas of Molecular Biology* 66, 203–309.

Zalkin, H. and Smith, J. L. (1998) Enzymes using glutamine as an amide donor. *Advances in Enzymology and Related Areas of Molecular Biology* 72, 87–144.

Zhao, F.J., Hawkesford, M.J., Warrilow, A.G.S., McGrath, S.P. and Clarkson, D.T. (1996) Responses of two wheat varieties to sulphur addition and diagnosis of sulphur deficiency. *Plant and Soil* 181, 317–327.

Zhao, F.J., Hawkesford, M.J. and McGrath, S.P. (1999) Sulphur assimilation and effects on yield and quality of wheat. *Journal of Cereal Science* 30, 1–17.

Zurbriggen, M.D., Carrillo, N., Tognetti, V.B., Melzer, M., Martin, P., Hause, B. and Hajirezaei, M.R. (2009) Chloroplast-generated reactive oxygen species play a major role in localized cell death during the non-host interaction between tobacco and *Xanthomonas campestris pv. vesicatoria*. *The Plant Journal* 60, 962–973.

7 Glutamate Decarboxylase

J.J. Molina-Rueda, A. Garrido-Aranda and F. Gallardo*
Universidad de Málaga, Málaga, Spain

7.1 Abstract

Glutamate decarboxylase (GAD) catalyses the decarboxylation of glutamate to produce γ-aminobutyric acid (GABA). In plants, its activity is modulated by pH and calcium. The regulation of GAD activity by calcium is through a calmodulin binding domain that seems to be specific for the plant kingdom, having resulted from an ancient event during evolution. Plant GAD is usually encoded by several genes and these show different expression patterns, including tissue specificity and response to stress, which suggest that there are specialization phenomena within this gene family. Although the role of GABA as a neurotransmitter is well known in animals, several roles for GABA and its metabolism have been proposed in plants over the last few decades. The roles for GABA in plants include acting as a biochemical pH-stat, in temporary nitrogen storage, as a compatible osmolyte, in defence against biotic stress, in the control of reactive oxygen species (ROS) and in the metabolism of photorespiratory intermediates. In addition, GABA is considered to be a signal controlling the expression of regulatory genes. However, the available information supporting the involvement of GAD and/or GABA in many processes is very limited, as is the contribution of polyamine metabolism to the cellular levels of GABA. Studies on the regulation of GABA levels are required for a better understanding of the involvement of GAD and GABA in plant development.

7.2 Introduction

Glutamate decarboxylase (GAD) has been known to decarboxylase glutamate to form γ-aminobutyric acid (GABA) for over 50 years. Surprisingly, however, although the role of GAD as a neurotransmitter is well known in animals, information on plant GAD is limited and very little is known about the regulation of its enzymatic activity by calcium, or about the expression of the GAD gene family. GABA is a metabolite that is traditionally associated with stress, but its roles in plant processes are also under revision. During the last two decades, experimental results on GAD and GABA in plants, including investigations of the changed phenotypes shown by transgenic plants with altered expression of GAD, and changes caused by GABA treatments, have generated intriguing questions about the role of GABA in plant cell signalling.

This chapter reviews: (i) the characteristics of the GAD domains in plants, including the calmodulin-binding domain and its acquisition during evolution; (ii) GAD gene families from different plant species and GAD regulation; (iii) the metabolism of GABA and its metabolic

*E-mail address: fgallardo@uma.es

connections and transport; (iv) the roles of GABA in plants as a metabolite, considering its physicochemical properties and its putative role as a signalling molecule. A possible role for GAD and GABA in vascular development in woody plants is also discussed.

7.3 Characteristics of Glutamate Decarboxylase in Plants

GAD is a cytosolic enzyme catalysing the decarboxylation of glutamate to produce GABA; this is a proton-consuming reaction that has been considered relevant to the control of the cytosolic pH. GABA has traditionally been associated with stress in plants and therefore the proton-consuming nature of the reaction that produces it has been viewed as part of the cellular mechanism for coping with stress. As a decarboxylase, GAD requires pyridoxal-5′-phosphate (PLP) as a cofactor. GAD has been studied only in a small number of plant species, mainly angiosperms, but also in pine (a gymnosperm). Plant GAD activity is regulated by pH and by Ca^{2+} and calmodulin (CaM) binding (Crawford *et al.*, 1994; Arazi *et al.*, 1995; Bouché and Fromm, 2004). The enzyme is more active at an acidic pH (e.g. 5.8) than at neutral pH, whereas its regulation by Ca^{2+}/CaM binding seems to be relevant at neutral pH. These features indicate that GAD activity and GABA production can be adjusted in response to physiological and environmental signals as many signalling transduction pathways affect both cytosolic pH and Ca^{2+} levels.

The GAD protein is encoded by a small family of nuclear genes (Akama *et al.*, 2001). In contrast to animals, which have two different GADs of molecular weight 65 and 67 kDa (Ueno *et al.*, 2012), only one form of GAD has so far been described in higher plants. The size of the plant GAD polypeptide is 43–62 kDa (Shelp *et al.*, 2012a). All plant GAD sequences in the US National Institutes of Health GenBank share a central domain, spanning from the N-terminal end, that it is involved in catalysis and interaction with the cofactor PLP. Thus it is a very well-conserved domain and is typical in decarboxylases that include a conserved SSEA (serine-serine-glutamic acid-alanine) sequence involved in the interaction with PLP, and an aspartate residue and a conserved sequence SGHK (serine-glycine-histidine-lysine) at the active site (Sukhareva and Mamaeva, 2002). In addition, most plant GAD sequences also contain a CaM-binding domain in the carboxyl terminal of the protein that is absent in animal sequences (Molina-Rueda *et al.*, 2010; Fig. 7.1). The carboxyl moiety of the protein is much less conserved than the central domain involved in catalysis. The CaM binding domain is a region of about 28 residues, with some conserved residues, including one flanking lysine at each edge of the domain and a tryptophan residue in a WKK (tryptophan-lysine-lysine) sequence close to the centre that is typically considered a key residue

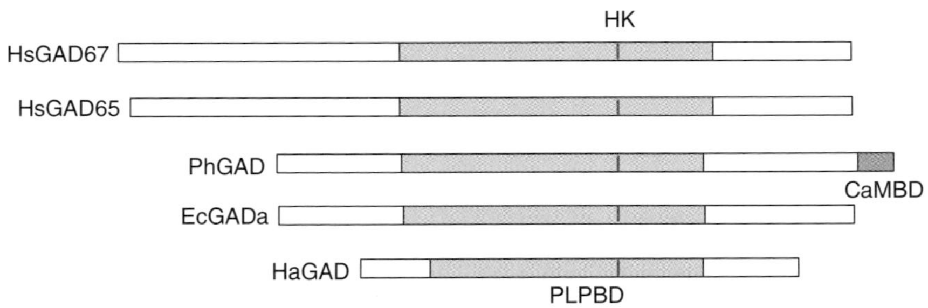

Fig. 7.1. Comparison of glutamate decarboxylase (GAD) polypeptides from different kingdoms. The pyridoxal-phosphate binding domain (PLPBD), calmodulin-binding domain (CaMBD) and conserved HK (histidine-lysine-lysine) sequence in PLPBD are shown. Animals: *Homo sapiens* HsGAD67 (accession numbers: Q99259.1; 594 amino acids (aa), 66.9 kDa) and HsGAD65 (Q05329.1; 585 aa, 65.4 kDa); Plants: *Petunia hybrida*, PhGAD (Q07346.1; 500 aa, 56.7 kDa); Bacteria: *Escherichia coli*, EcGADa (P69908.1; 466 aa, 52.7 kDa); Archaea: *Halobacterium sp.* HaGAD (Q07346.1; 355 aa, 37.4 kDa). Data for domain definition and localization were taken from Sukhareva and Mamaeva (2002).

in the binding to CaM (Arazi *et al.*, 1995; Baum *et al.*, 1996; Molina-Rueda *et al.*, 2010). The CaM-binding domain can be identified after prediction of its secondary structure (O'Neil and DeGrado, 1990), and corresponds to an amphipathic alpha helix in which the conserved tryptophan residue is located in the hydrophobic part and most charged residues are in the hydrophilic part. This domain acts as an auto-inhibitory domain for the enzymatic activity and its release after Ca^{2+}/CaM binding activates the enzyme. It is also interesting to note that not all plant GADs are regulated by Ca^{2+}/CaM binding. A few GAD sequences have been reported to lack the CaM-binding capability, including some rice sequences that have a non-functional CaM-binding domain and the *Bryophyta*, in which GADs lack of the typical sequence found in angiosperms and show an even lower degree of conservation of the carboxyl moiety of the protein. The evolution of these sequences will be commented upon below.

The interaction of GAD with Ca^{2+}/CaM has been also been described as a possible mechanism for assembling the holoenzyme, but to date only a few studies have considered the characterization of the holoenzyme, and the results are not conclusive. Native GAD holoenzymes of 550 and 340 kDa have been reported (Zik *et al.*, 2006; Shelp *et al.*, 2012a), but the classical approach of protein purification and use of specific antibodies to determine the molecular characteristics of the holoenzyme has not been conclusive. Related inconclusive studies have been also reported in animals, where shorter and larger alternative gene products have been observed, possibly as result of multi-splicing processes (Ueno *et al.*, 2012).

7.4 Glutamate Decarboxylase Gene Family

Although GABA is ubiquitously present in prokaryotes and eukaryotes, GAD is encoded by just a few genes, usually one or two, in species from most of kingdoms, with the exception of plants, in which the number of genes encoding for GAD is more variable (Sukhareva and Mamaeva, 2002; Table 7.1), varying from just one, found in the *Aquilegia coerulea* genome, to ten in the *Glycine max* genome (Phytozome v9.0: http://www.phytozome.net/). The *Arabidopsis thaliana* genome has also suffered a reduction in the number of GAD genes during evolution (Yang *et al.*, 2013), although it still has five genes encoding for GAD in its genome. As for *A. thaliana*, several genes encode for GAD in most plant species, which suggests relevant roles and a possible specialization phenomena for GAD in plants. Gene duplication affecting plant GAD genes before and after specialization probably occurred during the evolutionary process. This suggestion is supported by a phylogenetic tree corresponding to GAD proteins from species of the principal plant taxonomic groups, which showed putative orthologous GAD genes (Fig. 7.2: Pt1, At5, Eg3, Gm1, Gm10, Sl1 and Zm4), and also branches with several GADs of the same species (Fig. 7.2: Gm5, Gm6, Gm7 and Gm8; Pt2 and Pt; and Pt5 and Pt6). Additionally, the number of exons across plant species has remain relatively constant (from five to seven in most of those in Table 7.1), but intron loss and shortening has been observed in GAD genes of the same species.

As indicated above, the CaM binding domain is present not just in the GADs of dicotyledonous plants, but also in monocotyledonous (*Oryza sativa* and *Zea mays*), and gymnosperm GADs (*Pinus pinaster*) (Fig. 7.3). We have found this domain conserved in a GAD from *Selaginella moellindorffii*, which belongs to the *Lycopodiophyta* division, an ancient group of vascular plants whose origins can be traced back as far as 400 million years ago. This fact proves that the CaM binding domain in plant GADs and, concomitantly, its regulation by calcium, were very ancient acquisitions during plant evolution.

7.5 Expression of Glutamate Decarboxylase Genes

For years, the expression of GAD has been studied in several species of higher plants, and as it may change in response to external stimuli, it has received special attention because of the implications that its product, GABA, may be involved in stress responses.

In general, and under normal growth conditions, the different GAD genes may have different expression patterns. Some GAD genes show

Table 7.1. Glutamate decarboxylase genes found in various types of plant genomes sequenced. (Phytozome V9.0: http://www.phytozome.net/.)

Taxonomic group	Plant species	Gene name	Transcript ID	No. exons
Eudicots				
Fabidae (Rosids)	*Glycine max*	*Gm1*	*Glyma16g34450*	6
		Gm2	*Glyma08g09650*	6
		Gm3	*Glyma08g09670*	6
		Gm4	*Glyma05g26660*	6
		Gm5	*Glyma14g39170*	7
		Gm6	*Glyma02g40840*	7
		Gm7	*Glyma18g04940*	7
		Gm8	*Glyma11g33280*	5
		Gm9	*Glyma08g09660*	6
		Gm10	*Glyma09g29900*	6
	Populus trichocarpa	*Pt1*	*Potri.012G039000*	6
		Pt2	*Potri.004G075200*	7
		Pt3	*Potri.T059200*	7
		Pt4	*Potri.004G075300*	7
		Pt5	*Potri.008G141100*	7
		Pt6	*Potri.010G100500*	7
Malvidae (Rosids)	*Arabidopsis thaliana*	*At1*	*At5g17330*	8
		At2	*At1g65960*	6
		At3	*At2g02000*	6
		At4	*At2g02010*	6
		At5	*At3g17760*	6
	Eucalyptus grandis	*Eg1*	*Eucgr.B02632*	8
		Eg2	*Eucgr.F02900*	7
		Eg3	*Eucgr.H00595*	6
Asterids	*Solanum lycopersicum*	*Sl1*	*Solyc03g098240*	5
		Sl2	*Solyc05g054050*	7
		Sl3	*Solyc11g011920*	7
		Sl4	*Solyc04g025530*	7
		Sl5	*Solyc01g005000*	7
Ranunculales	*Aquilegia coerulea*	*Ac1*	*Aquca_002_01342*	6
Monocots				
Grasses	*Oryza sativa*	*Os1*	*LOC_Os03g13300*	5
		Os2	*LOC_Os04g37500*	4
		Os3	*LOC_Os08g36320*	6
	Zea mays	*Zm1*	*GRMZM2G098875*	5
		Zm2	*GRMZM2G017110*	6
		Zm3	*GRMZM2G355906_T01*	5
		Zm4	*GRMZM2G101069_T01*	7
		Zm5	*GRMZM5G826838_T01*	7
Tracheophytes	*Selaginella moellindorffii*	*Sm1*	*164618*	7
Embryophytes	*Physcomitrella patens*	*Pp1*	*Pp1s68_116V6*	5
		Pp2	*Pp1s34_308V6*	5
		Pp3	*Pp1s8_163V6*	6
		Pp4	*Pp1s10_153V6*	7

high transcripts levels in the whole plant, while others have tissue-specific expression or show nearly undetectable transcript levels. An example is the *A. thaliana* GAD gene family (*AtGAD*), which is composed of five members. One of them, *AtGAD2*, is the most highly expressed in all organs of the plant, but the other genes have different expression patterns. Thus, *AtGAD1* is tissue specific showing higher levels of transcripts in roots, whereas *AtGAD4* and *AtGAD5* are expressed just in flowers, and *AtGAD3* seems to have no detectable expression under regular growth conditions (Miyashita and Good, 2008). In the case of *O. sativa*, no tissue specificity was

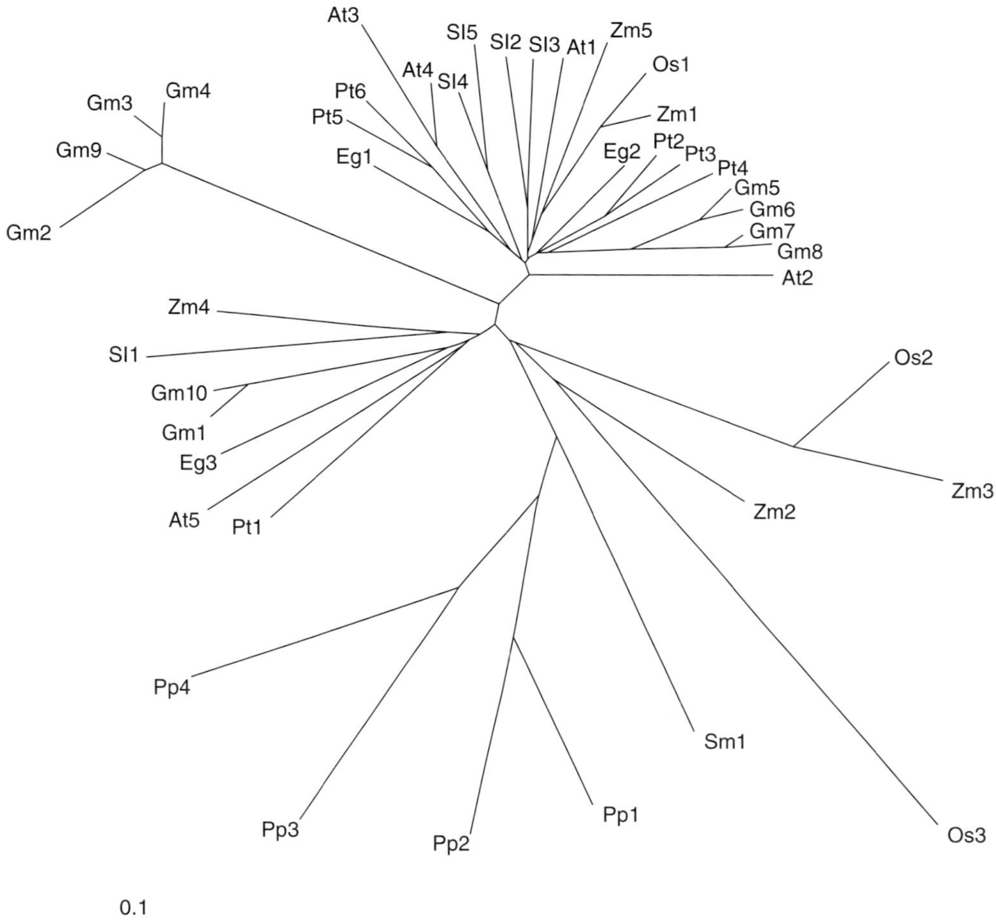

Fig. 7.2. Phylogenetic analysis of plant glutamate decarboxylases (GADs). Amino acids sequences of GADs from organisms representing the main plant and related taxonomic groups (Phytozome v9: http://www.phytozome.net/) (see also Table 7.1) were aligned using Clustal X (v2.1) and visualized by TreeView (v1.6.6). At, *Arabidopsis thaliana*; Eg, *Eucalyptus grandis*; Gm, *Glycine max*; Os, *Oryza sativa*; Pp, *Physcomitrella patens*; Pt, *Populus trichocarpa*; Sl, *Solanum lycopersicum*; Sm, *Selaginella moellindorffii*; Zm, *Zea mays*.

found for its GAD gene family (*OsGAD*), and the two genes, *OsGAD1* and *OsGAD2*, are expressed in all tissues (Akama *et al.*, 2001). In contrast, most of GAD gene family of *G. max* (*GmGAD*) show tissue-specific expression, except for *GmGAD2*, which is expressed in the whole plant (Hyun *et al.*, 2013). In the woody plant *Populus trichocarpa*, some GAD genes also show tissue-specific expression associated with xylogenesis (Sjödin *et al.*, 2009).

The expression of GAD has been also analysed during plant development and fruit ripening in different species. In *G. max*, *GmGADs* show an increased expression during germination (Hyun *et al.*, 2013). The GAD genes of *Solanum lycopersicum* (*SlGAD*) are differentially expressed during fruit development. mRNAs corresponding to two *SlGADs* increase during the fruit development process to the breaker stage (the first sign of colour change in the fruit), while transcripts for just one of the two genes are found from the breaker stage up to the end of the development (Akihiro *et al.*, 2008). In the case of woody plants, only one study has been conducted on GAD expression in trees to date. This described the expression of a pine GAD gene during hypocotyl

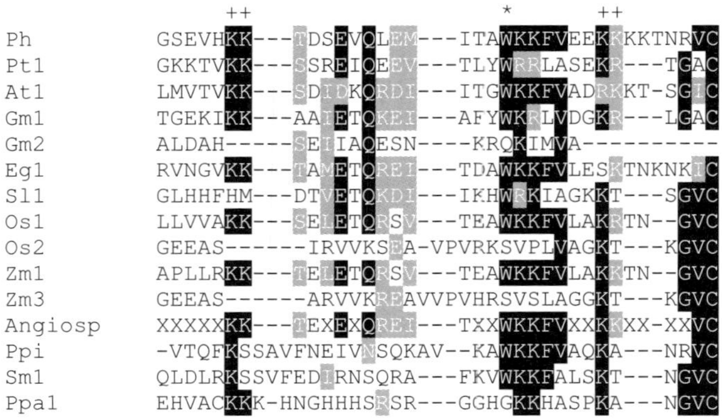

Fig. 7.3. Alignment of the C-terminal end of plant glutamate decarboxylase (GAD) proteins showing the calmodulin (CaM)-binding domain. Flanking lysines (K; ++) and the conserved tryptophan residue (W; *) are indicated at the top. Identical and similar residues are shown in black or grey, respectively. Residues in the consensus angiosperm sequence indicated by an X correspond to positions in which four or more different residues are found in different species (Molina-Rueda *et al.*, 2010). A dash represents a gap introduced to optimize the alignment. Os2 and Ppa1 lack the CaM-binding domain. Abbreviations: At, *Arabidopsis thaliana*; Eg, *Eucalyptus grandis*; Gm, *Glycine max*; Os, *Oryza sativa*; Ph, *Petunia × hybrid*; Ppa, *Physcomitrella patens* subsp. *patens*; Ppi, *Pinus pinaster*; Pt, *Populus trichocarpa*; Sl, *Solanum lycopersium*; Sm, *Selaginella moellindorffii*; Zm, *Zea mays*. See Table 7.1 for accession numbers (transcript IDs) of (most of) the sequences shown.

development, and also a differential expression during the production of reaction wood (Molina-Rueda *et al.*, 2010).

The expression pattern of GAD genes can change in response to various biotic and abiotic stresses, so that GAD genes with low transcript levels under normal growth conditions might present with high levels under environmental stress; the timing of the response differs depending on the plant species. The expression of *AtGAD4* is an example; it is expressed in low levels in flowers under normal growth conditions, but transcript levels increase in roots under low oxygen conditions, showing a peak of expression at 8 h (Miyashita and Good, 2008; Table 7.2), whereas slower responses ranging from 8 h to several days have been observed in other species (Table 7.2). In addition, different timing in the expression of GAD genes also depends upon the type of stress. Thus, in *Panax ginseng*, GAD transcripts show a maximum induction after just 1 h of anoxia or NaCl treatments, and a peak of expression is found at 8 h if the plant is exposed to cold or salt stress, or even later in the case of wounding (Table 7.2). These studies clearly show that GAD and GABA are associated with stress, but it is still difficult to understand the role of each GAD gene in the different responses.

The possible roles of plant GADs have been studied by altering their expression, and this has been done by both overexpression and silencing. The overexpression of a truncated petunia GAD in transgenic tobacco, which lacked 27 amino acids from the C-terminus corresponding to the CaM-binding domain, caused abnormal plant development, including inhibition of the elongation of stem cortex parenchyma cells (Baum *et al.*, 1996). The lack of the auto-inhibitory domain involved in Ca^{2+}/CaM regulation in this truncated GAD provoked an increase in GAD activity, which affected glutamate and GABA levels in the plant; this could be triggering the observed changes in plant development.

Additionally, overexpression of GAD may have biotechnological applications. For example, the overproduction of *OsGAD2* and knock-down of GABA-T in rice plants allows GABA accumulation in the grain for a long period, so that rice grains with a high content of GABA are obtained (Shimajiri *et al.*, 2013); this may prevent moderate hypertension in adults (Tsuchida *et al.*, 2003). The alteration of GABA synthesis may also alter the organoleptic properties of food. For instance, transgenic tomato plants with GAD antisense present higher levels of glutamate, an approach that may be used for the modification of taste

Table 7.2. Timing of expression of the glutamate dehydrogenase gene (GAD) in various species under stress treatments.

Species	Stress treatment	GAD gene and time to induce maximum peak of expression	Reference
Arabidopsis thaliana L.	Hypoxia	GAD4 8 h	Miyashita and Good, 2008
Glycine max L.	Stress (biotic) (*Phytophthora sojae*)	GAD1 6 h GAD4 6 h GAD5 6 h	Hyun *et al.*, 2013
Nicotiana sylvestris	NaCl (100 mM)	8 h	Akçay *et al.*, 2012
Panax ginseng C.A. Meyer		GAD consensus[a]	Lee *et al.*, 2010
	Anoxia	1 h	
	Cold (4 ºC)	8 h	
	Heat (37 ºC)	1 h	
	NaCl (100 mM)	8 h	
	Wounding	24 h	
	H_2O_2	Inhibited	
Solanum lycopersicum L.	Hypoxia treated with CO_2	GAD1 7 d GAD2 5 d GAD3 7 d	Mae *et al.*, 2012
	Hypoxia non-treated with CO_2	GAD1 5 d GAD2 3 d GAD3 7 d	

[a]A general expression of all GAD genes.

during the ripening of the fruit (Kisaka *et al.*, 2006). The alteration of gene expression seems as if it could be a reliable approach to understanding the physiological role of redundant GAD genes in plant species.

7.6 γ-Aminobutyric Acid Synthesis and its Metabolic Context

7.6.1 The γ-aminobutyric acid shunt pathway and stress

GABA shunt is a metabolic pathway that has been described as a bypass of the Krebs (tricarboxylic or TCA) cycle (Fig. 7.4). Three enzymes are involved in the GABA shunt, and GAD carries out the first reaction. GABA transaminase (GABA-T) catalyses the conversion of GABA to succinic semialdehyde using either pyruvate or glyoxylate as amino acceptor producing alanine or glycine respectively, additionally GABA-T activity using 2-oxoglutarate to produce glutamate has also been detected. Succinic semialdehyde dehydrogenase (SSADH) is the third enzyme of the bypass and it oxidizes the succinic semialdehyde to succinate, which might be incorporated to the Krebs cycle. GAD is localized in the cytosol and the other two, GABA-T and SSADH, are localized in the mitochondrion. GABA shunt connects nitrogen and carbon metabolism, considering that this bypass is fed by glutamate, and that GS/GOGAT cycle is described as the principal source of glutamate from inorganic nitrogen assimilation. GABA shunt can be regulated by pH, calcium, ATP and NADH. Low pH (≈5.8) and calcium increase GAD activity, high pH (8–10) positively regulates GABA-T and SSADH, and ATP and NADH inhibit SSADH activity. A correct functioning of the bypass has also been described to be required to restrict levels of reactive oxygen intermediates (ROI), considering that SSADH-deficiency in plants results in the accumulation of ROI, necrotic lesions, dwarfism, and hypersensitivity to environmental stresses (Fait *et al.*, 2008).

Under reducing intracellular conditions when NAD(P)H increases, like during hypoxia or drought and other stresses, succinic semialdehyde is not oxidized to succinate, but reduced by a group of proteins called the hydroxyisobutyrate dehydrogenase superfamily. This family includes enzymes that reduce succinic

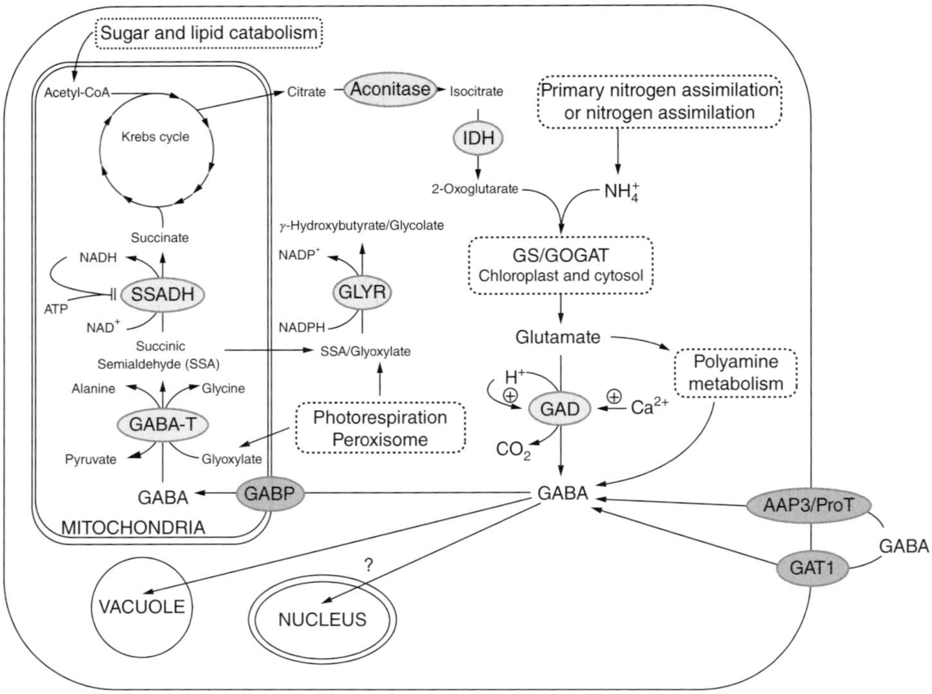

Fig. 7.4. The γ-aminobutyric acid (GABA) shunt and related metabolic pathways. Three enzymes are involved in the GABA shunt: cytosolic glutamate decarboxylase (GAD), mitochondrial GABA transaminase (GABA-T) and succinic semialdehyde dehydrogenase (SSADH). The GABA shunt is fed by glutamate. The glutamine synthetase/glutamate synthase (GS/GOGAT) cycle is described as the principal source of glutamate from inorganic nitrogen assimilation. The GABA shunt can be regulated by pH, calcium, ATP and NADH, as low pH (≈5.8) and calcium increase GAD activity, high pH (8–10) positively regulates GABA-T and succinic semialdehyde dehydrogenase (SSADH), and ATP and NADH inhibit SSADH activity. Under reducing intracellular conditions, when NAD(P)H increases, as during stress, succinic semialdehyde (SSA) is not oxidized to succinate, but reduced to γ-hydroxybutyrate by a group of proteins called the hydroxyisobutyrate dehydrogenase superfamily. In Fig. 7.4, only cytosolic SSA/glyoxylate reductases (GLYR) are shown. SSA and glyoxylate are typically generated from GABA and from the photorespiratory pathway (Allan *et al*., 2009). Putrescine metabolism can also be a source for GABA production. GABA levels can be also modulated by transport between cells and plant organs. The AAP3, ProT1 and ProT2 transporters have a low affinity (mM range) for GABA, and they can also transport other stress-related compounds. High affinity (μM range) transporters for GABA have also been discovered, including GAT1, which can transfer other related compounds across the plasma membrane as well. A member of the *Arabidopsis* amino acid/polyamine/organocation transporter family (GABP, GABA permease) is involved in the transfer from cytosol to mitochondria. GABA is a connecting point between nitrogen and carbon metabolism, but its metabolism also seems relevant in signalling the redox status of the cell, and pH and Ca^{2+} homeostasis. Other abbreviations: GLYR, glyoxylate reductase; IDH, isocitrate dehydrogenase.

semialdehyde and glyoxylate to γ-hydroxybutyrate and glycolate respectively. NADPH-dependent cytosolic and plastidial isoforms of succinic semialdehyde/glyoxylate reductases (GLYR) have been identified in *Arabidopsis*. Succinic semialdehyde and glyoxylate are typically generated from GABA and glycolate metabolisms respectively, which are frequently associated to stress conditions in general, including photorespiration especially for glycolate production (Allan *et al.*, 2009). The role of these reactions is not clear, although it might be related to the consumption of reducing power to alleviate stress situations.

7.6.2 Alternative sources of γ-aminobutyric acid in plant tissues and transport

Although the reaction catalysed by GAD is assumed to be the principal source of GABA, putrescine catabolism can also be a source for GABA production. As for GABA, various stresses, such as drought, salinity and cold, modulate polyamine levels, and high polyamine levels have been positively correlated with stress tolerance (Krasensky and Jonak, 2012). For example, stresses can contribute to GABA production both through the activation of GAD as indicated above, and also via the degradation of polyamines after modulation of their levels in the cell (Fig. 7.4). In addition to this latter source of GABA, the amino acid level in plant cells can be also modulated by transport between cells and plant organs. At the whole plant level, a number of transporters, including AAP3, ProT1 and ProT2, have been shown to have a relatively low affinity (mM range) for GABA. These transporters can also contribute to the transport of other stress-related compounds such as glycine, betaine or proline, and usually with higher efficiency (Grallath *et al.*, 2005). More recently, high affinity transporters for GABA have been discovered. A member of the *Arabidopsis* amino acid transporter family (AtGAT1) has found to act as a proton-driven transporter with a high affinity for GABA (μM range) although it can also transport other related compounds across the plasma membrane (Meyer *et al.*, 2006). Additionally, a member of the *Arabidopsis* amino acid/polyamine/organocation transporter family (GABA permease or GABP) localized in the mitochondrial membrane was found to complement GABA, but not proline transport, in a GABA-transport-deficient yeast mutant (Michaeli *et al.*, 2011; Shelp, *et al.*, 2012c).

The existence of several transporters for GABA indicates the relevance that the metabolite may have in signalling. GABA is seen as a connecting point between nitrogen and carbon metabolism, but its synthesis and metabolism also seem to be relevant in signalling the redox status of the cell, and pH and Ca^{2+} homeostasis (Shelp, *et al.*, 2012b).

7.7 Classical and Recent Evidence Supporting the Functions of Glutamate Decarboxylase and γ-Aminobutyric Acid

For a relatively long time, several functions have been attributed to GAD and GABA. A number of experiments have proved that GAD activity can significantly contribute to buffer acidosis effects in the cytosol in bacteria and plants. This role as a biochemical pH-stat is not surprising considering that the GAD reaction is proton consuming and it is activated by low pH, as mentioned before (reviewed in Shelp *et al.*, 1999; Kinnersley and Turano, 2000). GABA can be also considered to be a form of temporary nitrogen storage. It accumulates after ammonium inputs or when glutamine synthetase (GS) is inhibited or protein degradation is enhanced, and about 50% of the GABA content in a cell can be found in the vacuole. GABA has also been considered to be a compatible osmolyte in stress situations. As a zwitterion, GABA can reach high concentrations (mM range) in a cell but cause no apparent toxic effects as do other stress-related compounds such as proline and glycine betaine. It has been also reported that at high concentrations GABA stabilizes and protects isolated thylakoids against freezing damage in the presence of salt, exceeding the cryo-protective properties of proline and glycine betaine at the same concentrations. Additionally, the scavenging ability of GABA for the superoxide anion, peroxide hydrogen and singlet oxygen has been shown to be significantly higher than that of proline, which indicates its putative role in reducing the production of reactive oxygen species (ROS) under water stress (Liu *et al.*, 2011).

GABA has been also described as an element of the plant defence response to pathogen attack. For example, the incorporation of GABA into the synthetic diet of phytophagous larvae of *Choristoneura rosaceana* reduced their rates of growth, development and survival. In nature, these kinds of insects destroy vacuolar compartmentation as they feed from the leaves, increasing H^+ levels in the cytosol and stimulating the synthesis of GAD and GABA in attacked areas. GABA is an inhibitory neurotransmitter in animals; it hyper-polarizes the neural membrane by stimulating Cl^- influx through $GABA_A$-gated Cl^- channels. Therefore, increasing GABA levels in leaves as result of insect attack might cause inhibition of the normal neuromuscular activity of the attacking insects after feeding.

Variation in GABA levels has been observed during plant development as well. Thus, high GABA levels in *Arabidopsis* have been correlated with reduced apical dominance, leading to a higher degree of branching, and siliques (fruits) of reduced size, resulting in the production of significantly fewer seeds (Mirabella *et al.*, 2008). In tobacco, *Stellaria*, rice and *Arabidopsis*, high concentration of GABA or a high GABA/glutamate ratio have been correlated with the inhibition of stem elongation or dwarfism (Baum *et al.*, 1996; Kathiresan *et al.*, 1998; Akama and Takaiwa, 2007; Renault *et al.*, 2011).

Recently, a number of studies have been focused on other putative roles of GABA in plants, including ion transport and signalling. The fact that GABA regulates ion transport in animals prompts the hypothesis that GABA can also play a similar role in plants. The growth of *Lemna minor* was increased twofold to threefold by including 5 mM of GABA in the culture media *in vitro*. This growth was associated with an increase in the mineral content of treated plants in a dose-dependent manner, and inhibited by bicuculline and picrotoxin, which are, respectively, competitive and non-competitive antagonists of GABA receptors in the central nervous system (Kinnersley and Lin, 2000). In *Arabidopsis* seedlings cultivated *in vitro*, GABA promoted NO_3^- uptake at low NO_3^-, while GABA inhibited NO_3^- uptake at high NO_3^- (Barbosa *et al.*, 2010). These findings support the role of GABA as a potential modulator of ion transport in plants. There are also indications to support the role of GABA as a signal in plants, such as the existence of GABA receptors or glutamate receptors that may use GABA as a modulatory ligand (Lacombe *et al.*, 2001; Yu *et al.*, 2006). The production of a gradient of GABA has been shown to be involved in pollen tube growth and guidance in *Arabidopsis* (Palanivelu *et al.*, 2003), and Beuvé *et al.* (2004) showed that GABA might upregulate nitrate uptake via a long-distance signalling pathway in *Brassica napus*. A relationship between GABA and ethylene was observed in various studies. Kathiresan *et al.* (1997) showed that GABA causes increases in transcript levels for 1-aminocyclopropane-1-carboxylic acid synthase (ACS) and oxidase (ACO). The studies of Reggiani (2006) indicate that ethylene synthesis and signalling are required to induce GABA accumulation in rice. Additionally, Shi *et al.* (2010) showed that GABA increased transcript level for ACS and ACO in *Caragana* roots under salt stress, and that GABA treatment also altered the transcript level for genes related to signal transduction, transcriptional regulation, ROS and polyamine metabolism.

GABA is possibly involved in the storage and/or transport of nitrogen too, because *Arabidopsis* grows efficiently on medium containing GABA as a sole source of nitrogen. As mentioned before, the GABA shunt is also a way to translocate the ammonium assimilated into glutamate by the GS/GOGAT cycle in other amino acids and to generate carbon flux that enters the TCA cycle (Bouché and Fromm, 2004). More recently, Lancien and Roberts (2006) have shown that GABA downregulates the expression of a large subset of 14-3-3 gene family members in *Arabidopsis* seedlings in a calcium, ethylene and abscisic acid-dependent manner. The 14-3-3 proteins are ubiquitous regulatory proteins that function by interacting directly with a wide range of target proteins, usually in a phosphorylation-dependent manner. Several key enzymes in nitrogen and carbon metabolism are 14-3-3 targets, including nitrate reductase, GS, starch synthase III and glyceraldehyde-3-phosphate dehydrogenase, suggesting that GABA may act as part of the control of the carbon/nitrogen balance in the plant.

GAD expression and GABA levels are associated with hypocotyl development in pine (Molina-Rueda *et al.*, 2010). Mechanical stress that induces the production of reaction wood causes alteration not just in the transcript levels of GAD in pine and poplar, but also in the genes related with ethylene synthesis (Molina-Rueda *et al.*, 2010 and unpublished data). The localization of GABA has revealed its presence in vascular tissues, including differentiating xylem. Xylogenesis involves progressive programmed cell death (PCD), and this implies influx of Ca^{2+} and vacuolar collapse, resulting in decreases of cytosolic pH during the differentiation of tracheids. Both of these factors (Ca^{2+} influx and low pH) have also been shown to increase the activity of GAD. Thus, GABA production from GAD activity might be involved in the regulation of processes related to PCD.

7.8 Future Research

Besides the apparent relevance of GAD and GABA in many plant processes, the information that is currently available on their roles is very limited, and an effort is necessary in many areas to further elucidate their roles in development, stress responses and signalling. The molecular characterization of the GAD holoenzyme that is needed to obtain a better knowledge of its regulation by Ca^{2+}/CaM is still pending. It is also not clear why plant genomes have different genes encoding GAD. Additional basic studies of their expression in different organs and different physiological contexts are required to better define their roles in development and stress responses. The regulation of GAD gene expression, the contribution of polyamine synthesis and degradation to GABA levels, the localization of GABA in plant tissues, and GABA metabolism and its implications in photorespiration and ROS production are issues that need more attention. The fine and gross regulation of GABA levels has to be addressed too in order to understand the involvement of GAD and GABA in many processes in plants.

Acknowledgement

This work was supported by the grant AGL2009-11404 from the Ministerio de Economía y Competitividad, Spain.

References

Akama, K. and Takaiwa, F. (2007) C-terminal extension of rice glutamate decarboxylase (OsGAD2) functions as an autoinhibitory domain and overexpression of a truncated mutant results in the accumulation of extremely high levels of GABA in plant cells. *Journal of Experimental Botany* 58, 2699–2707.

Akama, K., Akihiro, T., Kitagawa, M. and Takaiwa, F. (2001) Rice (*Oryza sativa*) contains a novel isoform of glutamate decarboxylase that lacks an authentic calmodulin-binding domain at the C-terminus. *Biochimica et Biophysica Acta* 1522, 143–150.

Akçay, N., Bor, M., Karabudak, T., Özdemir, F. and Türkan, I. (2012) Contribution of gamma amino butyric acid (GABA) to salt stress responses of *Nicotiana sylvestris* CMSII mutant and wild type plants. *Journal of Plant Physiology* 169, 452–458.

Akihiro, T., Koike, S., Tani, R., Tominaga, T., Watanabe, S., Iijima, Y., Aoki, K., Shibata, D., Ashihara, H., Matsukura, C., Akama, K., Fujimura, T. and Ezura, H. (2008) Biochemical mechanism on GABA accumulation during fruit development in tomato. *Plant and Cell Physiology* 49, 1378–1389.

Allan, W.L., Clark, S.M., Hoover, G.J. and Shelp, B.J. (2009) Role of plant glyoxylate reductases during stress: a hypothesis. *Biochemical Journal* 423, 15–22.

Arazi, T., Baum, G., Snedden, W.A., Shelp, B.J. and Fromm, H. (1995) Molecular and biochemical analysis of calmodulin interactions with the calmodulin-binding domain of plant glutamate decarboxylase. *Plant Physiology* 108, 551–561.

Barbosa, J.M., Singh, N.K., Cherry, J.H. and Locy, R.D. (2010) Nitrate uptake and utilization is modulated by exogenous γ-aminobutyric acid in *Arabidopsis thaliana* seedlings. *Plant Physiology and Biochemistry* 48, 443–450.

Baum, G., Lev-Yadun, S., Fridmann, Y., Arazi, T., Katsnelson, H., Zik, M. and Fromm, H. (1996) Calmodulin binding to glutamate decarboxylase is required for regulation of glutamate and GABA metabolism and normal development in plants. *EMBO Journal* 15, 2988–2996.

Beuvé, N., Rispail, N., Laine, P., Cliquet, J.-B., Ourry, A. and Deunff, F. (2004) Putative role of γ-aminobutyric acid as a long-distance signal in up-regulation of nitrate uptake in *Brassica napus* L. *Plant, Cell and Environment* 27, 1035–1046.

Bouché, N. and Fromm, H. (2004) GABA in plants: just a metabolite? *Trends in Plant Science* 9, 110–115.

Crawford, L.A., Bown, A.W., Breitkreuz, K.E. and Guinel, F.C. (1994) The synthesis of [gamma]-aminobutyric acid in response to treatments reducing cytosolic pH. *Plant Physiology* 104, 865–871.

Fait, A., Fromm, H., Walter, D., Galili, G. and Fernie, A.R. (2008) Highway or byway: the metabolic role of the GABA shunt in plants. *Trends in Plant Science* 13, 14–19.

Grallath, S., Weimar, T., Meyer, A., Gumy, C., Suter-Grotemeyer, M., Neuhaus, J.M. and Rentsch, D. (2005) The AtProT family. Compatible solute transporters with similar substrate specificity but differential expression patterns. *Plant Physiology* 137, 117–126.

Hyun, T.K., Eom, S.H., Jeun, Y.C., Han, H.S. and Kim, J.-S. (2013) Identification of glutamate decarboxylases as a γ-aminobutyric acid (GABA) biosynthetic enzyme in soybean. *Industrial Crops and Products* 49, 864–870.

Kathiresan, A., Tung, P., Chinnappa, C.C. and Reid, D.M. (1997) [gamma]-Aminobutyric acid stimulates ethylene biosynthesis in sunflower. *Plant Physiology* 115, 129–135.

Kathiresan, A., Miranda, J., Chinnappa, C.C. and Reid, D.M. (1998) γ-Aminobutyric acid promotes stem elongation in *Stellaria longipes*. *Plant Growth Regulation* 26, 131–137.

Kinnersley, A.M. and Lin, F. (2000) Receptor modifiers indicate that 4-aminobutyric acid (GABA) is a potential modulator of ion transport in plants. *Plant Growth Regulation* 32, 65–76.

Kinnersley, A.M. and Turano, F.J. (2000) Gamma aminobutyric acid (GABA) and plant responses to stress. *Critical Reviews in Plant Science* 19, 479–509.

Kisaka, H., Kida, T. and Miwa, T. (2006) Antisense suppression of glutamate decarboxylase in tomato (*Lycopersicon esculentum* L.) results in accumulation of glutamate in transgenic tomato fruits. *Plant Biotechnology* 23, 267–274.

Krasensky, J. and Jonak, C. (2012) Drought, salt, and temperature stress-induced metabolic rearrangements and regulatory networks. *Journal of Experimental Botany* 63, 1593–1608.

Lacombe, B., Becker, D., Hedrich, R., DeSalle, R., Hollmann, M., Kwak, J.M., Schroeder, J.I., Le Novere, N., Nam, H.G., Spalding, E.P., Tester, M., Turano, F.J., Chiu, J. and Coruzzi, G. (2001) The identity of plant glutamate receptors. *Science* 292, 1486–1487.

Lancien, M. and Roberts, M.R. (2006) Regulation of *Arabidopsis thaliana* 14-3-3 gene expression by γ-aminobutyric acid. *Plant, Cell and Environment* 29, 1430–1436.

Lee, J.H., Kim, Y.J., Jeong, D.Y., Sathiyaraj, G., Pulla, R.K., Shim, J.S., In, J.G. and Yang, D.C. (2010) Isolation and characterization of a glutamate decarboxylase (GAD) gene and their differential expression in response to abiotic stresses from *Panax ginseng* C.A. Meyer. *Molecular Biology Reports* 37, 3455–3463.

Liu, C., Zhao, L. and Yu, G. (2011) The dominant glutamic acid metabolic flux to produce gamma-amino butyric acid over proline in *Nicotiana tabacum* leaves under water stress relates to its significant role in antioxidant activity. *Journal of Integrative Plant Biology* 53, 608–618.

Mae, N., Makino, Y., Oshita, S., Kawagoe, Y., Tanaka, A., Aoki, K., Kurabayashi, A., Akihiro, T., Akama, K., Koike, S., Takayama, M., Matsukura, C. and Ezura, H. (2012) Accumulation mechanism of γ-aminobutyric acid in tomatoes (*Solanum lycopersicum* L.) under low O_2 with and without CO_2. *Journal of Agricultural and Food Chemistry* 60, 1013–1019.

Meyer, A., Eskandari, S., Grallath, S. and Rentsch, D. (2006) AtGAT1, a high affinity transporter for gamma-aminobutyric acid in *Arabidopsis thaliana*. *The Journal of Biological Chemistry* 281, 7197–7204.

Michaeli, S., Fait, A., Lagor, K., Nunes-Nesi, A., Grillich, N., Yellin, A., Bar, D., Khan, M., Fernie, A.R., Turano, F.J. and Fromm, H. (2011) A mitochondrial GABA permease connects the GABA shunt and the TCA cycle, and is essential for normal carbon metabolism. *The Plant Journal* 67, 485–498.

Mirabella, R., Rauwerda, H., Struys, E.A., Jakobs, C., Triantaphylides, C., Haring, M.A. and Schuurink, R.C. (2008) The *Arabidopsis her1* mutant implicates GABA in *E*-2-hexenal responsiveness. *The Plant Journal* 53, 197–213.

Miyashita, Y. and Good, A.G. (2008) Contribution of the GABA shunt to hypoxia-induced alanine accumulation in roots of *Arabidopsis thaliana*. *Plant and Cell Physiology* 49, 92–102.

Molina-Rueda, J.J., Pascual, M.B., Cánovas, F.M. and Gallardo, F. (2010) Characterization and developmental expression of a glutamate decarboxylase from maritime pine. *Planta* 232, 1471–1483.

O'Neil, K.T. and DeGrado, W.F. (1990) How calmodulin binds its targets: sequence independent recognition of amphiphilic alpha-helices. *Trends in Biochemical Sciences* 15, 59–64.

Palanivelu, R., Brass, L., Edlund, A.F. and Preuss, D.P. (2003) Pollen tube growth and guidance is regulated by *POP2*, an *Arabidopsis* gene that controls GABA levels. *Cell* 114, 47–59.

Reggiani (2006) A role for ethylene in low-oxygen signaling in rice roots. *Amino Acids* 30, 299–301.

Renault, H., El Amrani, A., Palanivelu, R., Updegraff, E.P., Yu, A., Renou, J.P., Preuss, D., Bouchereau, A. and Deleu, C. (2011) GABA accumulation causes cell elongation defects and a decrease in expression of genes encoding secreted and cell wall-related proteins in *Arabidopsis thaliana*. *Plant and Cell Physiology* 52, 894–908.

Shelp, B.J., Bown, A.W. and McLean, M.D. (1999) Metabolism and functions of gamma-aminobutyric acid. *Trends in Plant Science* 4, 446–452.

Shelp, B.J., Bozzo, G.G., Trobacher, C.P., Chiu, G. and Bajwa, V.S. (2012a) Strategies and tools for studying the metabolism and function of γ-aminobutyrate in plants. I. Pathway structure. *Botany* 90, 651–668.

Shelp, B.J., Bozzo, G.G., Trobacher, C.P., Zarei, A., Deyman, K.L. and Brikis, C.J. (2012b) Hypothesis/review: contribution of putrescine to 4-aminobutyrate (GABA) production in response to abiotic stress. *Plant Science* 193–194, 130–135.

Shelp, B.J., Mullen, R.T. and Waller, J.C. (2012c) Compartmentation of GABA metabolism raises intriguing questions. *Trends in Plant Science* 17, 57–59.

Shi, S.Q., Shi, Z., Jiang, Z.P., Qi, L.W., Sun, X.M., Li, C.X., Liu, J.F., Xiao, W.F. and Zhang, S.G. (2010) Effects of exogenous GABA on gene expression of *Caragana intermedia* roots under NaCl stress: regulatory roles for H_2O_2 and ethylene production. *Plant, Cell and Environment* 33, 149–162.

Shimajiri, Y., Oonishi, T., Ozaki, K., Kainou, K. and Akama, K. (2013) Genetic manipulation of the γ-aminobutyric acid (GABA) shunt in rice: overexpression of truncated glutamate decarboxylase (GAD2) and knockdown of γ-aminobutyric acid transaminase (GABA-T) lead to sustained and high levels of GABA accumulation in rice kernels. *Plant Biotechnology Journal* 11, 594–604.

Sjödin, A., Street, N.R., Sandberg, G., Gustafsson, P. and Jansson, S. (2009) The Populus Genome Integrative Explorer (PopGenIE): a new resource for exploring the *Populus* genome. *New Phytologist* 182, 1013–1025.

Sukhareva, B.S. and Mamaeva, O.K. (2002) Glutamate decarboxylase: computer studies of enzyme evolution. *Biochemistry (Moscow)* 67, 1180–1188.

Tsuchida, T., Mashiko, K., Yamada, K., Hiratsuka, H., Shimada, T., Itagaki, Y., Fujinuma, H., Samejima, K., Nakamura, T., Hasegawa, T. and Matsubayashi, T. (2003) Clinical study of γ-aminobutyric acid-rich *Chlorella* for subjects with high-normal blood pressure and mild hypertension. *Journal of Japan Society of Nutrition and Food Science* 56, 97–102.

Ueno, H., Inoue, Y., Matsukawa, S. and Nakamura, Y. (2012) Glutamate decarboxylase. In: D'Mello, J.P.F. (ed.) *Amino Acids in Human Nutrition and Health*. CAB International, Wallingford, UK, pp 103–120.

Yang, Y.F., Zhu, T. and Niu, D.K. (2013) Association of intron loss with high mutation rate in *Arabidopsis*: implications for genome size evolution. *Genome Biology and Evolution* 5, 723–733.

Yu, G., Liang, J., He, Z. and Sun, M. (2006) Quantum dot-mediated detection of γ-aminobutyric acid binding sites on the surface of living pollen protoplasts in tobacco. *Chemistry and Biology* 13, 723–731.

Zik, M., Fridmann-Sirkis, Y. and Fromm, H. (2006) C-terminal residues of plant glutamate decarboxylase are required for oligomerization of a high-molecular weight complex and for activation by calcium/calmodulin. *Biochimica et Biophysica Acta* 1764, 872–876.

8 L-Arginine-Dependent Nitric Oxide Synthase Activity

F.J. Corpas,* L.A. del Río, J.M. Palma and J.B. Barroso
Estación Experimental del Zaidín, Consejo Superior de Investigaciones Científicas (CSIC), Granada, Spain

8.1 Abstract

Nitric oxide (NO) has long been recognized as a signalling molecule influencing diverse physiological processes in animals. More recently, it has emerged as an innovative topic of research in plant physiology and biochemistry. During the last decade, the involvement in a plethora of functions has been attributed to NO, including seed germination, plant development and senescence, as well as the mechanisms of response to abiotic and biotic stresses. Nevertheless, the enzymatic source of NO generated in plant cells from L-arginine is still an open question. In this chapter, the current knowledge and evidence on the presence of nitric oxide synthase (NOS) activity in plants will be reviewed.

8.2 Introduction

Nitric oxide (NO) is a free radical that diffuses readily through biomembranes and has been demonstrated to be involved, directly or indirectly, in almost all plant physiological and stress situations (Besson-Bard *et al.*, 2008; Corpas *et al.*, 2011). Thus, NO participates directly in certain processes by reaction with specific targets that affect its function or can act as a signal molecule (Corpas *et al.*, 2008c). During the last decade, one of the main goals in NO research in plants has been the identification and characterization of the potential enzymatic source(s) of this molecule because, considering the relevance of NO to a wide spectrum of functions, its synthesis must be carefully regulated. Therefore, a source of an enzymatic nature appears to be a very plausible mechanism. However, there are reports showing the non-enzymatic generation of NO from different substrates (del Río *et al.*, 2004) including, among others, nitrite in the presence of reducing compounds such as ascorbate or glutathione (Yamasaki, 2000). NO can also be produced during apoplastic acidification (pH 3–4) (Bethke *et al.*, 2004a) and by the oxidation of hydroxylamines (Rumer *et al.*, 2009).

Various potential enzymatic sources of NO have been reported so far; these mainly include L-arginine-dependent nitric oxide synthase (Barroso *et al.*, 1999; Corpas *et al.*, 2004, 2009), nitrate reductase (Dean and Harper, 1986; Yamasaki and Sakihama, 2000), a plasma membrane-bound enzyme (Stöhr *et al.*, 2001; Stöhr and Stremlau, 2006), mitochondrial electron transport (Gupta *et al.*, 2005; Planchet *et al.*, 2005) and polyamines (Tun *et al.*, 2006).

*E-mail address: javier.corpas@eez.csic.es

In this chapter, information on one of these enzymatic sources of NO will be reviewed: specifically, the activity that uses L-arginine to generate NO using the same cofactors as the animal nitric oxide synthase (NOS). In addition, a general overview of other pathways involved in plant L-arginine catabolism, such as the urea and the polyamine pathways, will also be provided (see Fig. 8.1).

8.3 Arginine Catabolism in Plants: Urea, Polyamines and Nitric Oxide

L-Arginine (L-Arg) is one of the 20 most common amino acids found in natural proteins. In plants, it is an important amino acid because it is used as a nitrogen storage and transport form. Thus, L-Arg could represent 50% of the nitrogen in seed protein, and up to 90% of the free nitrogen in vegetative tissues (Polacco and Holland, 1993;

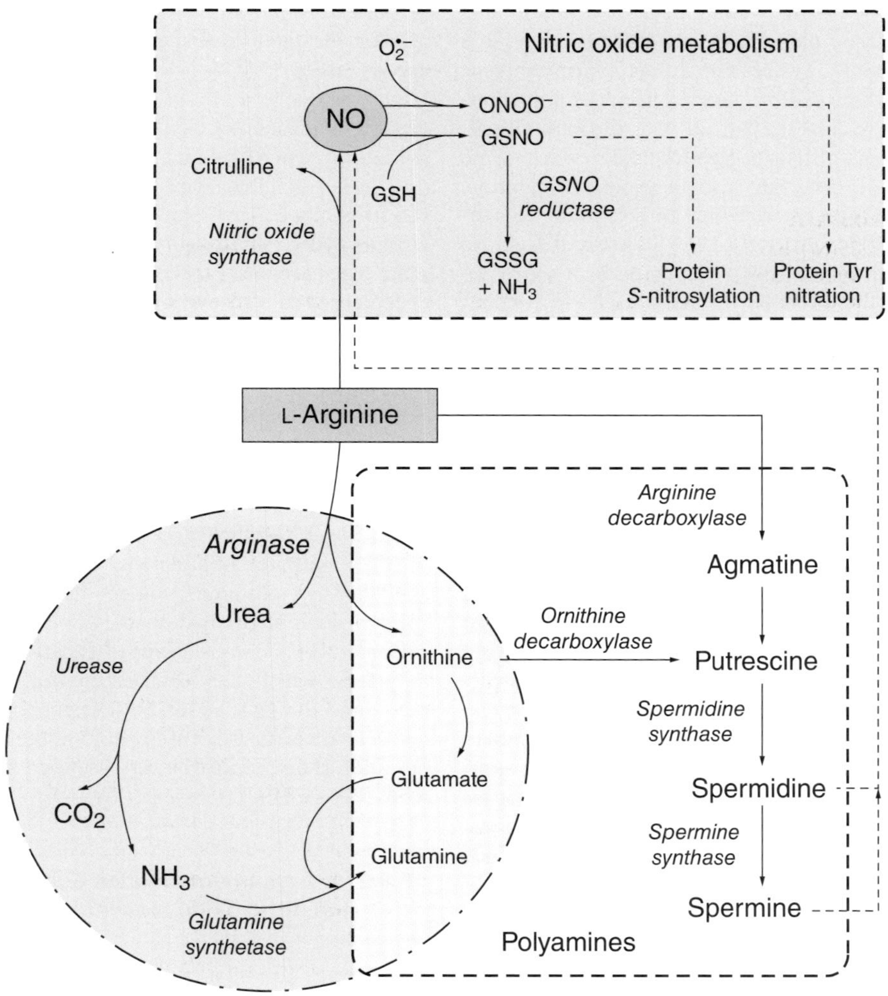

Fig. 8.1. L-Arginine catabolism in higher plants involves the biosynthesis of urea, polyamines and nitric oxide (NO). The names of the key enzymes involved are indicated in italics. NO generated by L-arginine-dependent nitric oxide synthase (NOS) activity can react with reduced glutathione (GSH) to form *S*-nitrosoglutathione (GSNO) which, in turn, can mediate *S*-nitrosylation processes. GSNO can be metabolized to oxidized glutathione (GSSG) and NH_3 by the enzyme GSNO reductase. NO can also react with superoxide radicals ($O_2^{\bullet-}$) to generate a powerful nitrating molecule, peroxynitrite ($ONOO^-$), which can mediate the tyrosine (Tyr) nitration process.

Nordin and Naesholm, 1997). Additionally, L-Arg is also a precursor for the synthesis of polyamines and NO, and an intermediary in the metabolism of other protein amino acids such as glutamine and proline (Forde and Lea, 2007).

In certain plant organs, such as seeds, the degradation of L-Arg provides nitrogen to support the growth of the seedlings after germination. Whereas L-Arg biosynthesis takes place predominantly in the plastids, with the participation of some cytosolic enzymes, L-Arg catabolism is localized in the mitochondria, where arginase produces the non-protein amino acid L-ornithine plus urea. This latter compound is decomposed by urease in the cytoplasm, whereas L-ornithine can be exported from the mitochondria to re-enter the pathway for L-Arg biosynthesis. L-Arg can also be decarboxylated by the enzyme arginine decarboxylase to produce agmatine, which is an intermediate in polyamine biosynthesis (Slocum *et al.*, 1989). Finally, it has also been proposed that L-Arg could be a source of NO via the activity of an L-arginine-dependent NOS. A simple model of L-arginine catabolism in higher plants involving the urea, polyamines and nitric oxide metabolism is shown in Fig. 8.1.

Interestingly, in some species of the *Fabaceae* family, L-arginine has a non-protein amino acid analogue called L-canavanine (Chapter 27), whose single difference is the replacement of a methylene bridge (-CH_2) in L-Arg by an oxygen group in L-canavanine. This amino acid is mainly accumulated in seeds such as *Canavalia ensiformis* (jack bean), where it serves both as a source of nitrogen and as a defensive compound against herbivores (Rosenthal, 2001; Huang *et al.*, 2011). The structural homology of L-Arg with L-canavanine can allow L-canavanine to act as a competitive inhibitor of the activities of some enzymes, such as arginase and NOS. As a result of this, L-canavanine is now being developed as a potential anticancer drug against human cancer cells, to which it is toxic; this is because the deprivation of L-Arg – which is presently being trialled as a novel anticancer therapy – provokes pro-apoptotic effects in human cancer cells, a process that is augmented by L-canavanine (Vynnytska *et al.*, 2011).

8.3.1 Urea metabolism

Urea is an intermediate in plant L-Arg catabolism that is involved in nitrogen remobilization from source tissues such as the leaves of old plants and germinating seeds (Mérigout *et al.*, 2008). However, it is also a primary nitrogen source, and is widely used as a fertilizer in agriculture because it is mobilized by plants from the soil (Witte, 2011). There are three relevant enzymes that are involved in plant urea metabolism: arginase, urease and glutamine synthetase (GS).

The first, arginase, also designated as L-arginine amidinohydrolase, hydrolyses arginine to urea and L-ornithine, the latter being a precursor for polyamine biosynthesis. This enzyme is decameric in structure, with subunits of approximately 35 kDa each. The second enzyme, plant urease, is a nickel metalloenzyme and exists as a hexamer of identical 90 kDa subunits containing two nickel ions per subunit; it is the only nickel metalloenzyme that has been identified in plants so far (Polacco *et al.*, 2013). The enzyme hydrolyses urea to ammonia and carbamate. The carbamate is then non-enzymatically decomposed to a second molecule of ammonia and carbon dioxide (Follmer, 2008; Witte, 2011). Urease is found in all plant tissues, and at the sub-cellular level, it is apparently located in the cytosol, although it predominates in seeds, where, in coordination with arginase, it plays an essential function in mobilizing protein reserves during germination, and consequently in seedling nitrogen metabolism (Polacco and Holland, 1993). Finally, the urea-derived ammonia is assimilated into glutamine via the action of the third enzyme mentioned above, GS (Miflin and Habash, 2002). Two isoforms of GS are present in plants, which are localized in the cytosol (GS1) and the plastids (GS2). It seems that the cytosolic GS is the isoform involved in the remobilization of ammonia from senescing leaves and seeds (Bernard and Habash, 2009).

8.3.2 L-Arginine modulates polyamine and nitric oxide biosynthesis

L-Arg serves also as a precursor of the biosynthesis of polyamines through the action of arginine decarboxylase or by a combination of the actions of arginase and ornithine decarboxylase (Fig. 8.1). These reactions allow the generation of a family of polyamines, including the diamine putrescine (Put), the triamine spermidine (Spd) and the tetraamine spermine (Spm). Put is the smallest and the most ubiquitous biological amine, being the primary substrate for Spd and Spm biosynthesis via spermidine synthase and spermine synthase, respectively (Fig. 8.1). Although these molecules have been known for many years, their physiological relevance is still increasing because it has been shown that they are involved in a wide spectrum of plant processes, including organogenesis, embryogenesis, leaf senescence, fruit development and ripening, and abiotic and biotic plant stress responses (Galston and Sawhney, 1990; Alcázar *et al.*, 2010; Fuell *et al.*, 2010). Nevertheless, new data are starting to show novel perspectives, as polyamine catabolism and/or its interconversion generate reactive oxygen species (ROS), particularly H_2O_2. This H_2O_2 is produced by two classes of amine oxidases: copper amine oxidases (CuAOs) and flavin-containing polyamine oxidases (Cona *et al.*, 2006), and its production has been correlated with different functions, such as cell wall maturation and lignification during development, wound healing and cell wall reinforcement during pathogen invasion (Angelini *et al.*, 2010). Moreover, there is new evidence showing that some polyamines can modulate the biosynthesis of NO. Thus, the exogenous addition of spermidine and spermine induced a rapid increase in the endogenous production of NO in *Arabidopsis* seedlings, in particular in the elongation zone of the *Arabidopsis* root tip and in primary leaves, and especially in the veins and trichomes (Tun *et al.*, 2006; Wimalasekera *et al.*, 2011) (Fig. 8.1). Moreover, it has been also reported that NO can regulate polyamine metabolism (Filippou *et al.*, 2013).

Therefore, all of these data indicate a clear interplay among molecules that are derived either directly or indirectly from L-Arg.

8.3.3 Arginine and nitric oxide synthesis in higher plants

In animal cells, NO is produced through the conversion of L-Arg by oxygen and NADPH, into citrulline, $NADP^+$ and NO (Fig. 8.2a) in a reaction catalysed by the homodimer haem-flavoprotein enzyme NOS (EC 1.14.13.3) (Groves and Wang, 2000). This enzyme is complex because each NOS monomer encloses an oxidase domain at its amino-terminal end and a reductase domain at its carboxy-terminal end. The NOS reductase domain contains binding sites for the redox cofactors NADPH, FMN (flavin mononucleotide), FAD (flavin adenine dinucleotide) and calmodulin. Two reducing equivalents are transferred from NADPH, via FMN and FAD, to the oxidase domain throughout the Ca^{2+}/calmodulin complex. In addition, the oxidase domain tightly binds one equivalent of tetrahydrobiopterin (BH_4) (Alderton *et al.*, 2001). In animals, there are three NOS isoforms: neuronal NOS (nNOS), endothelial NOS (eNOS) and inducible (iNOS). nNOS and eNOS are classified as constitutive NOSs because they are rapidly activated by an increase in calcium levels. They produce NO at low rate that is correlated with signalling functions. By contrast, the iNOS constitutively binds calcium and calmodulin and makes large amounts of NO as a cytotoxic agent in immune responses. iNOS and nNOS are soluble and mainly localized in the cytosol, whereas eNOS is membrane associated (Moncada *et al.*, 1991; Alderton *et al.*, 2001; Ghosh and Salerno, 2003).

In higher plants, there are several potential sources of NO, including both enzymatic and non-enzymatic systems (Wojtaszek, 2000; del Río *et al.*, 2004). However, the specific enzymatic source of NO is still controversial because among the different sequenced plant genomes, no gene similar to any of the NOS isoenzymes present in animal organisms has been identified. There are, though, abundant data supporting the existence of L-Arg-dependent NOS activity in plant cells (Corpas *et al.*, 2009), and the biochemical characterization of NOS-like enzymes by different methods has been reported in diverse plant species (see Table 8.1). For example, in pea leaf peroxisomes, a NOS activity was determined by monitoring the conversion of L-[^{3}H]Arg into L-[^{3}H]citrulline; this confirmed that NOS activity was strictly dependent on L-Arg, NADPH, calcium, calmodulin, FAD, FMN and BH_4, i.e. the same cofactors as are necessary for NOS activity in animals. Likewise, this NOS activity was sensitive to the characteristic inhibitors of the three NOS isoforms found in animals (Barroso *et al.*, 1999). However, because of the existing controversy on plant NOS activity and the existence of different pathways that can metabolize L-Arg, including arginase and arginine decarboxylase, which can mimic NOS activity, an alternative procedure to assay this enzyme was set up.

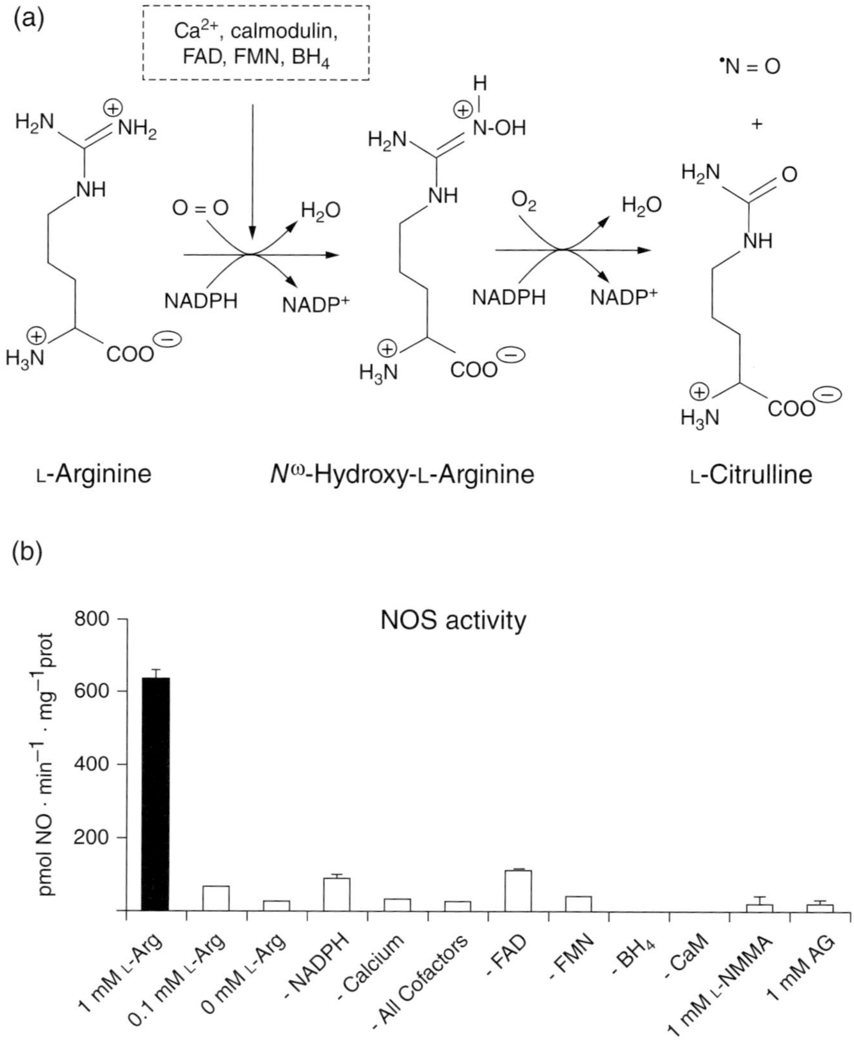

Fig. 8.2. (a) The nitric oxide synthase (NOS) reaction. (b) Biochemical characterization of NOS activity in hypocotyls of *Helianthus annuus* L. Reaction mixtures containing hypocotyl samples were incubated in the absence and presence of L-arginine (L-Arg, 1 or 0.1 mM), NADPH (1 mM), ethylene glycol tetraacetic acid (EGTA, 0.5 mM), the three cofactors (flavin adenine dinucleotide – FAD, 10 µM; flavin mononucleotide – FMN, 10 µM; and tetrahydrobiopterin – BH_4, 10 µM), 5 mg ml^{-1} calmodulin (CaM), 1 mM L-N^G-monomethyl arginine citrate (L-NMMA; a competitive and irreversible inhibitor of all three animal NOS isoforms), and 1 mM aminoguanidine (AG; an animal NOS inhibitor). The NOS activity was quantified from the NO produced, which was determined by the ozone chemiluminescence method (Corpas *et al*., 2008a). The production of NO was strictly dependent on the presence of calcium. (Reproduced, with some modifications, from Chaki *et al*., 2011a.)

This was the ozone chemiluminescence method, which specifically and unequivocally provides data on NO production (Corpas *et al*., 2008a). The method is based in the following reactions:

$$NO + O_3 \rightarrow NO_2^* + O_2 \tag{8.1}$$

and

$$NO_2^* \rightarrow NO_2 + h\nu \tag{8.2}$$

where NO_2^* denotes the NO_2 molecule in its excited state and $h\nu$ represents an emitted photon. Chemiluminescence resulting from these reactions is detected by a photomultiplier tube. The biochemical characterization of NOS activity in pea leaves based on this new method showed again that the activity required the same cofactors as determined by the previous assay in which the conversion of L-[^{3}H]Arg to L-[^{3}H] citrulline was monitored (Corpas *et al.*, 2004). A similar biochemical characterization based on the ozone chemiluminescence method has been carried out in other plant species as well, including olive (Valderrama *et al.*, 2007) and sunflower (Chaki *et al.*, 2011a), thus corroborating the existence of L-Arg-dependent NOS activity in plants. An example of the biochemical characterization of NOS activity in sunflower hypocotyls is shown in Fig. 8.2b.

In addition to the aforementioned techniques, other relatively simpler approaches have been used to detect NOS activity in plants, such as the oxyhaemoglobin method or different commercial NOS assay kits based on colorimetric or fluorimetric assays. In all of these methods, it is very important to use the pertinent controls to avoid false positive results (Ludidi, 2013). Accordingly, it is usual to complement with pharmacological analysis with inhibitors analogous to L-Arg, such as L-NAME (N^G-nitro-L-arginine methyl ester), L-NNA (N^G-nitro-L-arginine) or L-NMMA (L-N^G-monomethyl-arginine monoacetate), which can support the existence of an L-Arg-dependent NOS activity in plants (Corpas *et al.*, 2009). Together, all of these techniques have allowed extension of the detection of NOS activity to a significant number of plant species. Table 8.1 summarizes the detection of NOS activity, based on various techniques, in a series of plant organs and species. However, in many cases, the NOS activities that are detected are difficult to compare among different species owing to the varying techniques that have been used in each study. Altogether though, the data clearly demonstrate the existence of an L-Arg-dependent NOS activity in higher plants (Corpas *et al.*, 2009).

Recently, it has been reported that the unicellular marine green alga, *Ostreocccus tauri*, has a NOS activity which is encoded by a gene whose full-length sequence showed a similarity of 42, 43 and 34% with respect to eNOS, iNOS and nNOS, respectively (Foresi *et al.*, 2010). This putative NOS protein is a dimer with a subunit of 119 kDa, which is close to the molecular mass of the animal NOS subunits. Therefore, this finding further supports the presence in higher plants of a NOS-like protein that generates NO from L-Arg (Correa-Aragunde *et al.*, 2013).

8.4 Modulation of L-Arginine-Dependent Nitric Oxide Synthase Activity During Plant Development and Under Stress Conditions

In addition to the detection of L-Arg-dependent NOS activity in plants, there are also different reports on the behaviour of this activity during plant development and under adverse conditions (biotic and abiotic stresses) that provide evidence of the relevance of this activity in the physiology of plants.

8.4.1 Nitric oxide synthase activity during plant development

It is well established that NO is involved in many physiological plant processes, such as seed germination, the formation of root architecture and the plant actin cytoskeleton, plant development and senescence, and flower development (Bethke *et al.*, 2004b; Corpas *et al.*, 2004; Correa-Aragunde *et al.*, 2004; Simontacchi *et al.*, 2004; Kasprowicz *et al.*, 2009; Zafra *et al.*, 2010; Jin *et al.*, 2011; Chen and Kao, 2012). NOS activity has been described to be required in these physiological processes and to be modulated in different ways.

Roots and leaves were the first plant organs in which the presence of NOS activity was reported, in species such as *Lupinus albus* (Cueto *et al.*, 1996), maize (Ribeiro *et al.*, 1999) and pea (Barroso *et al.*, 1999). Later, it was demonstrated how the NO produced by NOS activity could affect the development of these organs, for example in root architecture. For instance, the inhibition of root elongation in maize was the result of a reduction in the endogenous NOS-dependent NO levels of root apical cells (Zhao *et al.*, 2007). Similarly, adventitious root formation in cucumber was blocked by treatment with two inhibitors of NOS activity (L-NAME and L-NNA)

Table 8.1. Summary of L-arginine-dependent nitric oxide synthase (NOS) activity reported in different plant species.

Species	Tissue or cell type	NOS activity (pmol min^{-1} mg^{-1} protein)	Reference
Arabidopsis thaliana	Leaves	4.5[a]	Zhao *et al.*., 2007
		Relative NOS activity[b]	Shi *et al.*, 2012
Brassica juncea	Seedling	Expressed as percentage[c]	Talwar *et al.*, 2012
Chorispora bungeana	Suspension cultured cells	0.2 U mg^{-1}protein[c]	Liu *et al.*, 2010
Festuca arundinacea (tall fescue)	Leaves	6 × 10^{3} [c]	Xu *et al.*, 2010
Glycine max (soybean)	Cotyledons	7.7[a]	Modolo *et al.*, 2002
	Leaves	760[d]	Jasid *et al.*, 2006
Helianthus annuus (sunflower)	Hypocotyls	280[e,f] 634[e,g]	Chaki *et al.*, 2009
	Roots	Relative NOS activity expressed as percentage[b]	David *et al.*, 2010
Hibiscus moscheutos	Roots	0.7[a]	Tian *et al.*, 2007
Lupinus albus (white lupin)	Roots and nodules	296[a]	Cueto *et al.*, 1996
Mucuna hassjoo (cowitch or velvet bean)		2.7[a]	Ninnemann and Maier, 1996
Nicotiana tabacum (TMV-infected tobacco)	Leaves	6[a]	Durner *et al.*, 1998
Olea europaea (olive)	Leaves	294[e]	Valderrama *et al.*, 2007
Oryza sativa (rice)	Seedlings	Relative NOS activity expressed as percentage[b]	Xiong *et al.*, 2009 Liu *et al.*, 2011
Pinus thunbergii	Leaves	2.0 × 10^{3} [h]	Yu *et al.*, 2012
Pisum sativum (pea)	Leaves	5.0 × 10^{3} [a,e]	Barroso *et al.*, 1999; Corpas *et al.*, 2004
	Roots, stems, leaves	240, 630, 120[e]	Corpas *et al.*, 2006
Rosa hybrida (rose)	Flowers	4.5[c]	Liao *et al.*, 2013
Solanum lycopersicum (tomato)	Leaves	19[a]	Negi *et al.*, 2010
	Roots	29[a]	
	Fruits	4.5 U mg^{-1} protein[h]	Zhang *et al.*, 2011
	Roots	Expressed as percentage[b]	Wang *et al.*, 2013
Sorghum bicolor (sorghum)	Seed embryonic axes	2.2[d]	Simontacchi *et al.*, 2004
Zea mays (maize)	Root tips and young leaves	0.18[a]	Ribeiro *et al.*, 1999
	Seedlings	410[i]	Hao *et al.*, 2008

[a]Arginine-citrulline assay; [b]NOS assay kit based on fluorescence of DAF-FM DA (Beyotime, Haimen, China) or the Fluorimetric Nitric Oxide Synthase Detection System (Sigma-Aldrich, USA); [c]Oxyhaemoglobin; [d]Spin trapping EPR (electron paramagnetic resonance) assay; [e]Ozone chemiluminiscence assay; [f]Sunflower cultivar X55 resistant to *Plasmopara halstedii* infection; [g]Sunflower cultivar HA89 (Peredovick variety) susceptible to *P. halstedii* infection; [h]NOS colorimetric assay kit based on nitroblue tetrazolium (NBT) (Nanjing Jiancheng Bioengineering Institute, Nanjing, China); [i]BIOXYTECHO® Nitric Oxide Synthase Assay Kit (OXIS International, Inc., Portland, Oregon, USA).

(Xuan *et al.*, 2012). It has also been shown that in *Arabidopsis* roots, salicylic acid induced NO synthesis through the participation of a NOS activity in which calcium and casein kinase 2 activity were essential components of the signalling cascade (Zottini *et al.*, 2007).

A more systematic analysis of NOS activity in the main plant organs (roots, stems and leaves) during early pea plant development provided a very complete picture of the relative level of L-Arg-dependent NOS activity in the those organs and its modulation. In 5-day-old pea seedlings, when the epicotyls and primary leaves expand, analysis in the main organs (roots, stems and leaves) of the L-Arg-dependent NOS activity by the ozone chemiluminescence method showed a clear correlation with the age of plants. In the roots, the highest NOS activity was found in 7-day-old seedlings, while the stem showed the highest NOS activity at the 11th day of growth and was the organ that displayed the highest activity during the period analysed (5 to 15 days). In contrast, the leaves had the lowest NOS activity of the three plant organs, and that activity did not show significant differences in leaves at different ages (Corpas *et al.*, 2006). There are also some reports showing that NOS activity is required for the development of functional nodules in soybean (Leach *et al.*, 2010).

Although it had been reported earlier that exogenous NO application could retard flower and plant senescence, as well as fruit ripening, in different plant species (Leshem, 2000), the involvement of NOS activity was only demonstrated later. For example, NOS activity associated with leaf peroxisomes was drastically downregulated in the leaves of senescent pea plants (Corpas *et al.*, 2004), a finding that is in good agreement with reports indicating that a high NO content can delay senescence. More recently, similar data collected using pharmacological approaches support the participation of the NO derived from NOS activity in the regulation of the senescence of cut rose flowers (Liao *et al.*, 2013). Overall, these examples are clear proofs of the involvement of NOS activity in plant development.

8.4.2 Nitric oxide synthase activity in plants under stress conditions

As it has been mentioned above, the involvement of NO in the mechanism of response of plants to biotic or abiotic stresses, either directly or through modifications of specific targets, has been clearly demonstrated. Thus, the first report showing the potential involvement of a NOS activity in the mechanism of response against pathogens was from the leguminous plant *Mucuna hassjoo* from a study that used specific inhibitors of NOS activity, such as L-NNA or L-NAME (Ninnemann and Maier, 1996). Later, using the Arg-citrulline assay, it was shown that the infection of tobacco plants by *Tobacco mosaic virus* (TMV) resulted in an enhancement of NOS activity (Durner *et al.*, 1998). Indeed, the number of publications showing the involvement of L-Arg-dependent NOS activity in plant stress episodes is still growing. In olive leaves of plants exposed to 200 mM NaCl, it was reported that salt stress caused an increase in L-Arg-dependent NOS activity which was accompanied by a process of nitro-oxidative stress (Valderrama *et al.*, 2007). In 4-day-old sunflower seedlings subjected to NaCl stress, a transient increase in endogenous NO content, primarily contributed by putative NOS activity, was accompanied by a relative reduction in the Na^+/K^+ ratio after 4 days, suggesting that NO was derived from the NOS activity regulated Na^+ accumulation, probably by affecting the associated transporter proteins (David *et al.*, 2010). In other cases, NOS activity could be either induced or not depending on the type and/or intensity of the environmental stress. Thus, in leaves of pea plants subjected to different abiotic stress conditions, L-Arg-dependent NOS activity assayed by the ozone-chemiluminescence method showed that whereas low temperature, wounding and high light intensity provoked an increase, other environmental conditions, such as high temperature, continuous light or continuous dark, did not produce any effect (Corpas *et al.*, 2008b). A similar situation was also described in sunflower seedlings, in which NOS activity did not seem to be directly involved because after mechanical wounding and under high temperature conditions, NOS activity was diminished, though this was accompanied by an increase in the *S*-nitrosothiol content which participated in the observed nitrosative stress (Chaki *et al.*, 2011a,b).

A high concentration of atmospheric carbon dioxide (CO_2) is recognized as an important pollutant. However, elevated levels of CO_2 also enhance photosynthesis and inhibit mitochondrial respiration, thereby promoting sucrose accumulation and plant growth. Under these conditions, an increase of lateral root formation has been described as consequence of a rise of NO generated by a NOS activity (Wang *et al.*, 2013). In ultraviolet-B-irradiated soybean plants, pharmacological evidence has shown that the NO generated by a NOS-like activity upregulates the enzyme haem oxygenase (Santa-Cruz *et al.*, 2010).

A role for NOS activity under biotic stress conditions has also been demonstrated. In *Arabidopsis* plants subjected to hypersensitive response (HR) the presence of NOS activity was demonstrated by pharmacological approaches (Delledonne *et al.*, 1998; Zhang *et al.*, 2003). The involvement of NOS activity in the mechanism of response to biotic stress has been also shown and an induction or inhibition of its activity depending on the plant species and type of pathogen has been reported. In tobacco (*Nicotiana tabacum*) leaves infected with the *Tobacco mosaic virus* (TMV), it was observed that infection of the resistant cultivar, but not of the susceptible cultivar, resulted in an enhancement of NOS activity (Durner *et al.*, 1988). Similarly, analysis of L-Arg-dependent NOS activity in two sunflower (*Helianthus annuus* L.) cultivars with different sensitivity to infection by the pathogen *Plasmopara halstedii* revealed differential behaviour between them. Measurement of NOS activity in the uninfected hypocotyls of susceptible and resistant cultivars revealed that the susceptible cultivar had a NOS activity twofold higher than that of the resistant cultivar (Table 8.1). In contrast, after infection, NOS activity was reduced in the susceptible cultivar but was induced in the resistant cultivar (Chaki *et al.*, 2009). This indicates that the cultivars displayed opposite behaviour against fungal infection. In other plant species under different biotic stresses, NOS activity has been shown to be augmented. For example, in soybean cotyledons, NOS activity-mediated phytoalexin accumulation in response to the pathogen *Diaporthe phaseolorum* (Modolo *et al.*, 2002). In *Arabidopsis*, the application of lipopolysaccharides from different bacteria activated NOS activity and induced defence genes (Zeidler *et al.*, 2004). In *Arabidopsis* leaves, after inoculation with *Pseudomonas syringae* pv. *maculiocola* (Psm) or *tomato* (Pst), NOS activity was also induced by fourfold to fivefold (Feechan *et al.*, 2005; Modolo *et al.*, 2005). More recently, it has been shown in tobacco suspension cells that NO produced by a NOS-like activity is rapidly accumulated during the induction of programmed cell death (PCD) by fusaric acid, a non-host-selective toxin mainly produced by *Fusarium oxysporum* (Jiao *et al.*, 2013).

8.5 A Genetic Engineering Approach to Study of the Relevance of Nitric Oxide Synthase Activity in Plants

Recently, researchers have started to use genetic approaches to obtain deeper insights into the significance of NOS activity in plants. As a matter of fact, several transgenic plants that overexpress mammalian NOS cDNA have been generated. In an *Arabidopsis thaliana* mutant expressing a rat neuronal NOS, a higher NO content was observed which resulted in an enhanced tolerance to abiotic stresses such as salinity and drought; the increased NO content was more significant in the guard cells under these stress conditions. In addition, these transgenic plants showed an increased resistance to biotic stress against *Pseudomonas syringae* pv. tomato (Shi *et al.*, 2012). A similar approach has been followed in tobacco plants expressing a constitutive mammalian nNOS, and it was also observed that this provided disease resistance to pathogens (Chun *et al.*, 2012). Furthermore, *nNOS*-transgenic tobacco plants revealed some special characteristics, in that they accumulated high amounts of H_2O_2 and salicylic acid (SA), and they induced an array of SA-, jasmonic acid-, and/or ethylene-related genes. As a result, these transgenic plants showed an enhanced resistance to a spectrum of pathogens, including bacteria, fungi, and viruses.

8.6 Conclusions

Although investigations of the generation of NO in plants and its physiological and biochemical implications in cell metabolism have provided significant advances, the same cannot be said for the identity of the enzymatic source of NO in plants, which is an essential part in the metabolism of NO in plant cells. Therefore, the search for the protein and gene involved in plant L-Arg-dependent NOS still continues. In the meantime, the amount of evidence demonstrating the presence of L-Arg-dependent NOS activity in plant cells keeps on growing.

Acknowledgements

Work in our laboratories was supported by a European Regional Development Fund (ERDF)-co-financed grant from the Ministry of Science and Innovation (project BIO2012-33904), and by the *Junta de Andalucía* (groups BIO-192 and BIO-286), Spain.

References

Alcázar, R., Altabella, T., Marco, F., Bortolotti, C., Reymond, M., Koncz, C., Carrasco, P. and Tiburcio, A.F. (2010) Polyamines: molecules with regulatory functions in plant abiotic stress tolerance. *Planta* 231, 1237–1249.

Alderton, W.K., Cooper, C.E. and Knowles, R.G. (2001) Nitric oxide synthases: structure, function and inhibition. *Biochemical Journal* 357, 593–615.

Angelini, R., Cona, A., Federico, R., Fincato, P., Tavladoraki, P. and Tisi, A. (2010) Plant amine oxidases "on the move": an update. *Plant Physiology and Biochemistry* 48, 560–564.

Barroso, J.B., Corpas, F.J., Carreras, A., Sandalio, L.M., Valderrama, R., Palma, J.M., Lupiáñez, J.A. and del Río, L.A. (1999) Localization of nitric-oxide synthase in plant peroxisomes. *The Journal of Biological Chemistry* 274, 36729–36733.

Bernard, S.M. and Habash, D.Z. (2009) The importance of cytosolic glutamine synthetase in nitrogen assimilation and recycling. *New Phytologist* 182, 608–620.

Besson-Bard, A., Pugin, A. and Wendehenne, D. (2008) New insights into nitric oxide signaling in plants. *Annual Review of Plant Biology* 59, 21–39.

Bethke, P.C., Badger, M.R. and Jones, R.L. (2004a). Apoplastic synthesis of nitric oxide by plant tissues. *The Plant Cell* 16, 332–341.

Bethke, P.C., Gubler, F., Jacobsen, J.V. and Jones, R.L. (2004b) Dormancy of *Arabidopsis* seeds and barley grains can be broken by nitric oxide. *Planta* 219, 847–855.

Chaki, M., Fernández-Ocaña, A.M., Valderrama, R., Carreras, A., Esteban, F.J., Luque, F., Gómez-Rodríguez, M.V., Begara-Morales, J.C., Corpas, F.J. and Barroso, J.B. (2009) Involvement of reactive nitrogen and oxygen species (RNS and ROS) in sunflower–mildew interaction. *Plant and Cell Physiology* 50, 265–79.

Chaki, M., Valderrama, R., Fernández-Ocaña, A.M., Carreras, A., Gómez-Rodríguez, M.V., Pedrajas, J.R., Begara-Morales, J.C., Sánchez-Calvo, B., Luque, F., Leterrier, M., Corpas, F.J. and Barroso, J.B. (2011a) Mechanical wounding induces a nitrosative stress by down-regulation of GSNO reductase and an increase in *S*-nitrosothiols in sunflower (*Helianthus annuus*) seedlings. *Journal of Experimental Botany* 62, 1803–1813.

Chaki, M., Valderrama, R., Fernández-Ocaña, A.M., Carreras, A., Gómez-Rodríguez, M.V., López-Jaramillo, J., Begara-Morales, J.C., Sánchez-Calvo, B., Luque, F., Leterrier, M., Corpas, F.J. and Barroso, J.B. (2011b) High temperature triggers the metabolism of *S*-nitrosothiols in sunflower mediating a process of nitrosative stress which provokes the inhibition of ferredoxin-NADP reductase by tyrosine nitration. *Plant, Cell and Environment* 34, 803–818.

Chen, Y.H. and Kao, C.H. (2012) Calcium is involved in nitric oxide- and auxin-induced lateral root formation in rice. *Protoplasma* 249,187–195.

Chun, H.J., Park, H.C., Koo, S.C., Lee, J.H., Park, C.Y., Choi, M.S., Kang, C.H., Baek, D., Cheong, Y.H., Yun, D.J., Chung, W.S., Cho, M.J. and Kim, M.C. (2012) Constitutive expression of mammalian nitric oxide synthase in tobacco plants triggers disease resistance to pathogens. *Molecules and Cells* 34, 463–471.

Cona, A., Rea, G., Angelini, R, Federico, R. and Tavladoraki, P. (2006) Functions of amine oxidases in plant development and defence. *Trends in Plant Science* 11, 80–88.

Corpas, F.J., Barroso, J.B., Carreras, A., Quirós, M., León, A.M., Romero-Puertas, M.C., Esteban, F.J., Valderrama, R., Palma, J.M., Sandalio, L.M., Gómez, M. and del Río, L.A. (2004) Cellular and subcellular localization of endogenous nitric oxide in young and senescent pea plants. *Plant Physiology* 136, 2722–2733.

Corpas, F.J., Barroso, J.B., Carreras, A., Valderrama, R., Palma, J.M., León, A.M., Sandalio, L.M. and del Río, L.A. (2006) Constitutive arginine-dependent nitric oxide synthase activity in different organs of pea seedlings during plant development. *Planta* 224, 246–254.

Corpas, F.J., Carreras, A., Esteban, F.J., Chaki, M., Valderrama, R., del Río, L.A. and Barroso, J.B. (2008a) Localization of *S*-nitrosothiols and assay of nitric oxide synthase and *S*-nitrosoglutathione reductase activity in plants. *Methods in Enzymology* 437, 561–574.

Corpas, F.J., Chaki, M., Fernández-Ocaña, A., Valderrama, R., Palma, J.M., Carreras, A., Begara-Morales, J.C., Airaki, M., del Río, L.A. and Barroso, J.B. (2008b) Metabolism of reactive nitrogen species in pea plants under abiotic stress conditions. *Plant and Cell Physiology* 49, 1711–1722.

Corpas, F.J., del Río, L.A. and Barroso, J.B. (2008c) Post-translational modifications mediated by reactive nitrogen species: nitrosative stress responses or components of signal transduction pathways? *Plant Signaling and Behaviour* 3, 301–303.

Corpas, F.J., Palma, J.M., del Río, L.A. and Barroso, J.B. (2009) Evidence supporting the existence of L-arginine-dependent nitric oxide synthase activity in plants. *New Phytologist* 184, 9–14.

Corpas, F.J., Leterrier, M., Valderrama, R., Airaki, M., Chaki, M., Palma, J.M. and Barroso, J.B. (2011) Nitric oxide imbalance provokes a nitrosative response in plants under abiotic stress. *Plant Science* 181, 604–611.

Correa-Aragunde, N., Graziano, M. and Lamattina, L. (2004) Nitric oxide plays a central role in determining lateral root development in tomato. *Planta* 218, 900–905.

Correa-Aragunde, N., Foresia N. and Lamattina L. (2013) Structure diversity of nitric oxide synthases (NOS): the emergence of new forms in photosynthetic organisms. *Frontiers in Plant Science* 4, 232.

Cueto, M., Hernández-Perea, O., Martín, R., Ventura, M.L., Rodrigo, J., Lamas, S. and Golvano, M.P. (1996) Presence of nitric oxide synthase activity in roots and nodules of *Lupinus albus*. *FEBS Letters* 398, 159–164.

David, A., Yadav, S. and Bhatla, S.C. (2010) Sodium chloride stress induces nitric oxide accumulation in root tips and oil body surface accompanying slower oleosin degradation in sunflower seedlings. *Physiologia Plantarum* 140, 342–354.

Dean, J.V. and Harper, J.E. (1986) Nitric oxide and nitrous oxide production by soybean and winged bean during the *in vivo* nitrate reductase assay. *Plant Physiology* 82, 718–723.

Delledonne, M., Xia, Y., Dixon, R.A. and Lamb, C. (1998) Nitric oxide functions as a signal in plant disease resistance. *Nature* 394, 585–588.

del Río, L.A., Corpas, F.J. and Barroso, J.B. (2004) Nitric oxide and nitric oxide synthase activity in plants. *Phytochemistry* 65, 783–792.

Durner, J., Wendehenne, D. and Klessig, D.F. (1998) Defense gene induction in tobacco by nitric oxide, cyclic GMP, and cyclic ADP-ribose. *Proceedings of the National Academy of Sciences of the United States of America* 95, 10328–10333.

Feechan, A., Kwon, E., Yun, B.W., Wang, Y., Pallas, J.A. and Loake, G.J. (2005) A central role for *S*-nitrosothiols in plant disease resistance. *Proceedings of the National Academy of Sciences of the United States of America* 102, 8054–8059.

Filippou, P., Antoniou, C. and Fotopoulos, V. (2013) The nitric oxide donor sodium nitroprusside regulates polyamine and proline metabolism in leaves of *Medicago truncatula* plants. *Free Radical Biology and Medicine* 56, 172–183.

Follmer, C. (2008) Insights into the role and structure of plant ureases. *Phytochemistry* 69, 18–28.

Forde, B.G. and Lea, P.J. (2007) Glutamate in plants: metabolism, regulation, and signalling. *Journal of Experimental Botany* 58, 2339–2358.

Foresi, N., Correa-Aragunde, N., Parisi, G., Caló, G., Salerno, G. and Lamattina, L. (2010) Characterization of a nitric oxide synthase from the plant kingdom: NO generation from the green alga *Ostreococcus tauri* is light irradiance and growth phase dependent. *The Plant Cell* 22, 3816–3830.

Fuell, C., Elliott, K.A., Hanfrey, C.C., Franceschetti, M. and Michael, A.J. (2010) Polyamine biosynthetic diversity in plants and algae. *Plant Physiology and Biochemistry* 48, 513–520.

Galston, A.W. and Sawhney, R.K. (1990) Polyamines in plant physiology. *Plant Physiology* 94, 406–410.

Ghosh, D.K. and Salerno, J.C. (2003) Nitric oxide synthases: domain structure and alignment in enzyme function and control. *Frontiers in Bioscience* 8, d193–209.

Groves, H.T. and Wang, C.C.Y. (2000) Nitric oxide synthase: models and mechanisms. *Current Opinion in Chemical Biology* 4, 687–695.

Gupta, K.J., Stoimenova, M. and Kaiser, W.M. (2005) In higher plants, only root mitochondria, but not leaf mitochondria reduce nitrite to NO, *in vitro* and *in situ*. *Journal of Experimental Botany* 56, 2601–2609.

Hao, G.P., Xing, Y. and Zhang, J.H. (2008) Role of nitric oxide dependence on nitric oxide synthase-like activity in the water stress signaling of maize seedling. *Journal of Integrative Plant Biology* 50, 435–442.

Huang, T., Jander, G. and de Vos, M. (2011) Non-protein amino acids in plant defense against insect herbivores: representative cases and opportunities for further functional analysis. *Phytochemistry* 72, 1531–1537.
Jasid, S., Simontacchi, M., Bartoli, C.G. and Puntarulo, S. (2006). Chloroplasts as a nitric oxide cellular source. Effect of reactive nitrogen species on chloroplastic lipids and proteins. *Plant Physiology* 142, 1246–1255.
Jiao, J., Zhou, B., Zhu, X., Gao, Z. and Liang, Y. (2013) Fusaric acid induction of programmed cell death modulated through nitric oxide signalling in tobacco suspension cells. *Planta* 238, 727–737.
Jin, C.W., Du, S.T., Shamsi, I.H., Luo, B.F. and Lin, X.Y. (2011) NO synthase-generated NO acts downstream of auxin in regulating Fe-deficiency-induced root branching that enhances Fe-deficiency tolerance in tomato plants. *Journal Experimental Botany* 62, 3875–3884.
Kasprowicz, A., Szuba, A., Volkmann, D., Baluska, F. and Wojtaszek, P. (2009) Nitric oxide modulates dynamic actin cytoskeleton and vesicle trafficking in a cell type-specific manner in root apices. *Journal of Experimental Botany* 60, 1605–1617.
Leach, J., Keyster, M., Du Plessis, M. and Ludidi, N. (2010) Nitric oxide synthase activity is required for development of functional nodules in soybean. *Journal of Plant Physiology* 167, 1584–1591.
Leshem, Y.Y. (2000) *Nitric Oxide in Plants: Occurrence, Function and Use*. Kluwer Academic Publishers, Dordrecht, The Netherlands.
Liao, W.-B., Zhang, M.-L. and Yu, J.-H. (2013) Role of nitric oxide in delaying senescence of cut rose flowers and its interaction with ethylene. *Scientia Horticulturae* 155, 30–38.
Liu, Y., Jiang, H., Zhao, Z. and An, L. (2010) Nitric oxide synthase like activity-dependent nitric oxide production protects against chilling-induced oxidative damage in *Chorispora bungeana* suspension cultured cells. *Plant Physiology and Biochemistry* 48, 936–944.
Liu, Y., He, J., Jiang, L., Wu, H., Xiao, Y., Liu, Y., Li, G., Du, Y., Liu, C. and Wan, J. (2011) Nitric oxide production is associated with response to brown plant hopper infestation in rice. *Journal of Plant Physiology* 168, 739–745.
Ludidi, N. (2013) Measurement of nitric oxide in plant tissue using difluorofluorescein and oxyhemoglobin. *Methods in Molecular Biology* 1016, 253–259.
Mérigout, P., Lelandais, M., Bitton, F., Renou, J.P., Briand, X., Meyer, C. and Daniel-Vedele, F. (2008) Physiological and transcriptomic aspects of urea uptake and assimilation in *Arabidopsis* plants. *Plant Physiology* 147, 1225–1238.
Miflin, B.J. and Habash, D.Z. (2002) The role of glutamine synthetase and glutamate dehydrogenase in nitrogen assimilation and possibilities for improvement in the nitrogen utilization of crops. *Journal of Experimental Botany* 53, 979–987.
Modolo, L.V., Cunha, F.Q., Braga, M.R. and Salgado, I. (2002) Nitric oxide synthase mediated phytoalexin accumulation in soybean cotyledons in response to the *Diaporthe phaseolorum* f. sp. *meridionalis* elicitor. *Plant Physiology* 130, 1288–1297.
Modolo, L.V., Augusto, O., Almeida, I.M., Magalhaes, J.R. and Salgado, I. (2005) Nitrite as the major source of nitric oxide production by *Arabidopsis thaliana* in response to *Pseudomonas syringae*. *FEBS Letters* 579, 3814–3820.
Moncada, S., Palmer, R.M.J. and Higgs, E.A. (1991) Nitric oxide: physiology, pathophysiology and pharmacology. *Pharmacological Reviews* 43, 109–142.
Negi, S., Santisree, P., Kharshiing, E.V. and Sharma, R. (2010) Inhibition of the ubiquitin-proteasome pathway alters cellular levels of nitric oxide in tomato seedlings. *Molecular Plant* 3, 854–869.
Ninnemann, H. and Maier, J. (1996) Indications for the occurrence of nitric oxide synthases in fungi and plants and the involvement in photoconidiation of *Neurospora crassa*. *Photochemistry and Photobiology* 64, 393–398.
Nordin, A., and Naesholm, T. (1997) Nitrogen storage forms in nine boreal understorey plant species. *Oecologia* 110, 487–492.
Planchet, E., Gupta, J.K., Sonoda, M. and Kaiser, W.M. (2005) Nitric oxide emission from tobacco leaves and cell suspensions: rate-limiting factors and evidence for the involvement of mitochondrial electron transport. *The Plant Journal* 41, 732–743.
Polacco, J.C. and Holland, M.A. (1993) Roles of urease in plant cells. *International Review of Cytology* 145, 65–103.
Polacco, J.C., Mazzafera, P. and Tezotto T. (2013) Opinion: Nickel and urease in plants: still many knowledge gaps. *Plant Science* 199–200, 79–90.
Ribeiro, E.A., Cunha, F.Q., Tamashiro, W.M.S.C. and Martins, I.S. (1999) Growth phase-dependent subcellular localization of nitric oxide synthase in maize cells. *FEBS Letters* 445, 283–286.
Rosenthal, G.A., (2001) L-Canavanine: a higher plant insecticidal allelochemical. *Amino Acids* 21, 319–330.

Rumer, S., Gupta, K.J. and Kaiser, W.M. (2009) Plant cells oxidize hydroxylamines to NO. *Journal of Experimental Botany* 60, 2065–2072.

Santa-Cruz, D.M., Pacienza, N.A., Polizio, A.H., Balestrasse, K.B., Tomaro, M.L. and Yannarelli, G.G. (2010) Nitric oxide synthase-like dependent NO production enhances heme oxygenase up-regulation in ultraviolet-B-irradiated soybean plants. *Phytochemistry* 71, 1700–1707.

Shi, H.T., Li, R.J., Cai, W., Liu, W., Wang, C.L. and Lu, Y.T. (2012) Increasing nitric oxide content in *Arabidopsis thaliana* by expressing rat neuronal nitric oxide synthase resulted in enhanced stress tolerance. *Plant and Cell Physiology* 53, 344–357.

Simontacchi, M., Jasid, S. and Puntarulo, S. (2004) Nitric oxide generation during early germination of sorghum seeds. *Plant Science* 167, 839–847.

Slocum, R.D., Flores, H.E., Galston, A.W. and Weinstein, L.H. (1989) Improved method for HPLC analysis of polyamines, agmatine and aromatic monoamines in plant tissue. *Plant Physiology* 89, 512–517.

Stöhr, C. and Stremlau, S. (2006) Formation and possible roles of nitric oxide in plant roots. *Journal of Experimental Botany* 57, 463–470.

Stöhr, C., Strube, F., Marx, G., Ullrich, W.R. and Rockel, P. (2001) A plasma membrane-bound enzyme of tobacco roots catalyses the formation of nitric oxide from nitrate. *Planta* 212, 835–841.

Talwar, P.S., Gupta, R., Maurya, A.K. and Deswal, R. (2012) *Brassica juncea* nitric oxide synthase like activity is stimulated by PKC activators and calcium suggesting modulation by PKC-like kinase. *Plant Physiology and Biochemistry* 60, 157–164.

Tian, Q.Y., Sun, D.H., Zhao, M.G. and Zhang, W.H. (2007) Inhibition of nitric oxide synthase (NOS) underlies aluminum-induced inhibition of root elongation in *Hibiscus moscheutos*. *New Phytologist* 174, 322–331.

Tun, N.N., Santa-Catarina, C., Begum, T., Silveira, V., Handro, W., Floh, E.I. and Scherer, G.F. (2006) Polyamines induce rapid biosynthesis of nitric oxide (NO) in *Arabidopsis thaliana* seedlings. *Plant and Cell Physiology* 47, 346–354.

Valderrama, R., Corpas, F.J., Carreras, A., Fernández-Ocaña A., Chaki, M., Luque, F., Gómez-Rodríguez, M.V., Colmenero-Varea, P., del Río, L.A. and Barroso, J.B. (2007) Nitrosative stress in plants. *FEBS Letters* 581, 453–461.

Vynnytska, B.O., Mayevska, O.M., Kurlishchuk, Y.V., Bobak, Y.P. and Stasyk, O.V. (2011) Canavanine augments proapoptotic effects of arginine deprivation in cultured human cancer cells. *Anticancer Drugs* 22, 148–157.

Wang, H., Xiao, W., Niu, Y., Jin, C., Chai, R., Tang, C. and Zhang, Y. (2013) Nitric oxide enhances development of lateral roots in tomato (*Solanum lycopersicum* L.) under elevated carbon dioxide. *Planta* 237, 137–144.

Wimalasekera, R., Tebartz, F. and Scherer, G.F. (2011) Polyamines, polyamine oxidases and nitric oxide in development, abiotic and biotic stresses. *Plant Science* 181, 593–603.

Witte, C.P. (2011) Urea metabolism in plants. *Plant Science* 180, 431–438.

Wojtaszek, P. (2000) Nitric oxide in plants. *To NO or not to NO. Phytochemistry* 54, 1–4.

Xiong, J., Lu, H., Lu, K., Duan, Y., An, L. and Zhu, C. (2009) Cadmium decreases crown root number by decreasing endogenous nitric oxide, which is indispensable for crown root primordia initiation in rice seedlings. *Planta* 230, 599–610.

Xu, Y., Sun, X., Jin, J. and Zhou, H. (2010) Protective effect of nitric oxide on light-induced oxidative damage in leaves of tall fescue. *Journal of Plant Physiology* 167, 512–518.

Xuan, W., Xu, S., Li, M., Han, B., Zhang, B., Zhang, J., Lin, Y., Huang, J, Shen, W. and Cui, J. (2012) Nitric oxide is involved in hemin-induced cucumber adventitious rooting process. *Journal of Plant Physiology* 169, 1032–1039.

Yamasaki, H. (2000) Nitrite-dependent nitric oxide production pathway: implications for involvement of active nitrogen species in photoinhibition *in vivo*. *Philosophical Transactions of the Royal Society of London, Series B: Biological Sciences* 355, 1477–1488.

Yamasaki, H. and Sakihama, Y. (2000) Simultaneous production of nitric oxide and peroxynitrite by plant nitrate reductase: *in vitro* evidence for the NR-dependent formation of active nitrogen species. *FEBS Letters* 468, 89–92.

Yu, L.Z., Wu, X.Q., Ye, J.R., Zhang, S.N. and Wang, C. (2012) NOS-like-mediated nitric oxide is involved in *Pinus thunbergii* response to the invasion of *Bursaphelenchus xylophilus*. *Plant Cell Reports* 31, 1813–1821.

Zafra, A., Rodríguez-García, M.I. and Alché, J.D. (2010) Cellular localization of ROS and NO in olive reproductive tissues during flower development. *BMC Plant Biology* 10, 36.

Zeidler, D., Zahringer, U., Gerber, I., Dubery, I., Hartung, T., Bors, W., Hutzler P. and Durner, J. (2004) Innate immunity in *Arabidopsis thaliana:* lipopolysaccharides activate nitric oxide synthase (NOS) and induce defense genes. *Proceedings of the National Academy of Sciences of the United States of America* 101, 15811–15816.

Zhang, C., Czymmek, K.J. and Shapiro, A.D. (2003) Nitric oxide does not trigger early programmed cell death events but may contribute to cell-to-cell signaling governing progression of the *Arabidopsis* hypersensitive response. *Molecular Plant–Microbe Interactions* 16, 962–972.

Zhang, X., Shen, L., Li, F., Meng, D. and Sheng, J. (2011) Methyl salicylate-induced arginine catabolism is associated with up-regulation of polyamine and nitric oxide levels and improves chilling tolerance in cherry tomato fruit. *Journal of Agricultural and Food Chemistry* 59, 9351–9357.

Zhao, D.Y., Tian, Q.Y., Li, L.H. and Zhang, W.H. (2007) Nitric oxide is involved in nitrate-induced inhibition of root elongation in *Zea mays*. *Annals of Botany* 100, 497–503.

Zottini, M., Costa, A., de Michele, R., Ruzzene, M., Carimi, F. and Lo Schiavo, F. (2007) Salicylic acid activates nitric oxide synthesis in *Arabidopsis*. *Journal of Experimental Botany* 58, 1397–1405.

9 Ornithine: At the Crossroads of Multiple Paths to Amino Acids and Polyamines

R. Majumdar,[1] R. Minocha[2] and S.C. Minocha[3]*

[1]USDA Agricultural Research Service, Geneva, New York, USA; [2]USDA Forest Service, Northern Research Station, Durham, New Hampshire, USA; [3]University of New Hampshire, Durham, New Hampshire, USA

9.1 Abstract

After the 20 amino acids that make up the proteins in all living organisms, ornithine perhaps occupies the most critical position among the non-protein amino acids. It sits at the crossroads of interconversions of glutamate and arginine on the one hand and the production of proline, polyamines and several alkaloids on the other: all products of tremendous importance to living cells. While ornithine is typically not an abundant amino acid, the metabolic flux of nitrogen through this amino acid is presumably quite rapid because of the cellular contents of the products for which it serves as the substrate. Our current knowledge of the regulation of ornithine biosynthesis is rather limited and mostly dependent upon the research targeted at understanding arginine and proline metabolism, and to some extent polyamine and alkaloid biosynthesis. Whereas most of the ornithine biosynthesis in animals occurs on the way to the production of glutamate, proline and putrescine from dietary arginine, in plants the pathway works in the reverse order, i.e. glutamate, the first product of nitrogen assimilation is the primary source of arginine, proline and putrescine. An understanding of the regulation of ornithine metabolism could help us in the genetic manipulation of plants for stress tolerance (via manipulation of proline and γ-aminobutyric acid, and the polyamine pathway), as well as in nutritional improvement (the cellular contents of important amino acids – arginine and citrulline among others). Based upon the recent observations of a strong connection between polyamine biosynthesis, ornithine metabolism and glutamic acid metabolism, it can be proposed that ornithine may act as a controlling metabolite to enhance nitrogen assimilation and, consequently, increased carbon assimilation.

9.2 Introduction

In addition to the 20 amino acids employed by nature for protein synthesis, numerous other amino acids are produced in plants, either as intermediates of metabolic pathways or for specific physiological roles in the life of a plant (Wagner and Musso, 1983; Bell, 2003). Ornithine (Orn) is such an amino acid that is not involved in protein synthesis, is often present in relatively small quantities in plants and animals compared with its related amino acids (e.g. glutamate, Glu; arginine, Arg; and proline, Pro), but plays a critical role in overall amino acid metabolism. While its importance as a key intermediate in the biosynthesis of Arg, Pro, polyamines (PAs), Glu (particularly in animals) and several alkaloids is well established in plants, no direct physiological

*E-mail address: sminocha@unh.edu

functions have been assigned to Orn (Bell, 2003; Slocum, 2005). In animals, Orn is synthesized predominantly from Arg obtained through dietary sources (Morris, 2006, 2007), whereas in plants it is mainly synthesized from Glu, the entry point of N assimilation, via a series of intermediates (Fig. 9.1). Although the enzymology of Orn biosynthesis from Glu (in plants) and from Arg (in animals) has been well characterized, the regulation of its biosynthesis and its steady-state

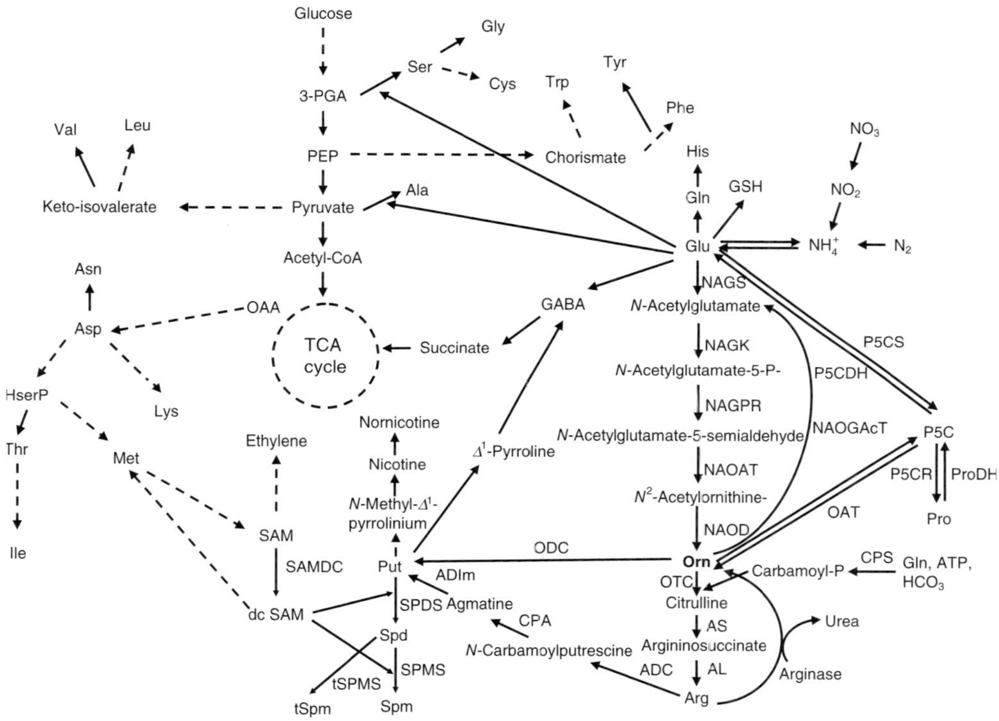

Fig. 9.1. Abbreviated pathway for the biosynthesis and metabolism of ornithine (Orn) in plants in connection to polyamines, amino acids, tricraboxylic acid (TCA) cycle metabolites and alkaloids. Abbreviations of enzymes, with EC numbers: ADC, arginine decarboxylase (EC 4.1.1.19); ADIm, agmatine deiminase (EC 3.5.3.12); AL, argininosuccinate lyase (EC 4.3.2.1); arginase (EC 3.5.3.1); AS, argininosuccinate synthase (EC 6.3.4.5); CPA, *N*-carbamoylputrescine amidohydrolase (EC 3.5.1.53); CPS, carbamoyl phosphate synthetase (EC 6.3.5.5); NAGK, *N*-acetylglutamate kinase (EC 2.7.2.8); NAGPR, *N*-acetylglutamate-5-phosphate reductase (EC 1.2.1.38); NAGS, *N*-acetylglutamate synthase (EC 2.3.1.1); NAOAT, N^2-acetylornithine aminotransferase (EC 2.6.1.11); NAOD, N^2-acetylornithine deacetylase (EC 3.5.1.16); NAOGAcT, acetyl-Orn-Glu acetyltransferase (EC 2.3.1.35); OAT, ornithine δ-aminotransferase (EC 2.6.1.13); ODC, ornithine decarboxylase (EC 4.1.1.17); OTC, ornithine transcarbamoylase (EC 2.1.3.3); P5CDH, Δ^1-pyrroline-5-carboxylate dehydrogenase (EC 1.5.1.12); P5CR, Δ^1-pyrroline-5-carboxylate reductase (EC 1.5.1.2); P5CS, Δ^1-pyrroline-5-carboxylate synthetase (EC 2.7.2.11/1.2.1.41); ProDH, proline dehydrogenase (EC 1.5.99.8); SAMDC, *S*-adenosylmethionine decarboxylase (EC 4.1.1.50); SPDS, spermidine synthase (EC 2.5.1.16); SPMS, spermine synthase (EC 2.5.1.22); tSPMS, thermospermine synthase (EC 2.5.1.79).
Other abbreviations: Ala, alanine; Asn, asparagine; Asp, aspartate; Cys, cysteine; dcSAM, decarboxylated *S*-adenosylmethionine; GABA, γ-aminobutryic acid; Gln, glutamine; Glu, glutamate; Gly, glycine; GSH, glutathione; His, histidine; HserP, homoserine phosphate; Ile, isoleucine; Leu, leucine; Lys, lysine; Met, methionine; OAA, oxaloacetic acid; P5C, Δ^1-pyrroline-5-carboxylate; PEP, phosphoenolpyruvate; 3-PGA, 3-phosphoglyceric acid; phe, phenylalanine; Pro, proline; Put, putrescine; SAM, *S*-adenosylmethionine; Ser, serine; Spd, spermidine; Spm, spermine; The, threonine; Trp, tryptophan; tSPM, thermospermine; Tyr, tyrosine; Val, valine. (Modified from Mohapatra *et al.*, 2010b.)

cellular content have not been well understood. For animals, several reviews have indicated that cellular Orn levels respond to various physiological and clinical conditions of the organism; however, in plants such studies have been rather limited. Several recent studies (Roosens *et al.*, 2002; Mohapatra *et al.*, 2010b; DeBoer *et al.*, 2011, 2013; Majumdar *et al.*, 2013; Gholami *et al.*, 2014) have implicated Orn as a regulatory molecule in plants for the biosynthesis of several amino acids and PAs, including its own biosynthesis and accumulation. Even though under normal conditions Orn concentrations in plants are often quite low, it can be argued that cells are capable of synthesizing it in large quantities in order to meet the demand for the biosynthesis of its products (i.e. Glu, Arg, Pro, PAs and alkaloids), which frequently are present in several-fold higher concentrations than Orn itself. This chapter summarizes different aspects of Orn metabolism in plants, including its biosynthesis and utilization, cellular contents, genetic manipulation of its biosynthesis through mutants and genetic engineering, and some of its regulatory roles. Limited comparisons of Orn metabolism in animal or microbial systems are made where appropriate. Finally, a model is proposed for the regulation of the metabolism of Orn and the associated flux through the Glu-Orn-Arg/Pro/PA pathways.

9.3 Ornithine Biosynthesis and Utilization

The biosynthesis of Arg has been well studied in bacteria and fungi (Cunin *et al.*, 1986; Davis, 1986) with only a limited focus on Orn (Slocum, 2005). In plants (and bacteria), Orn is synthesized mostly from Glu (Fig. 9.1); the first enzyme believed to regulate the entire pathway is *N*-acetyl-Glu synthase (NAGS; EC 2.3.1.1). *N*-acetyl-Glu kinase (NAGK; EC 2.7.2.8) then converts acetyl-Glu into *N*-acetyl-Glu-5-phosphate which, via the intermediates *N*-acetyl-Glu-5-semialdehyde and N^2-acetyl-Orn, is converted into Orn. The enzymes involved in these steps are *N*-acetyl-Glu-5-P reductase (NAGPR; EC 1.2.1.38), N^2-acetylOrn aminotransferase (NAOAT; EC 2.6.1.11) and N^2-Acetyl-Orn deacetylase (NAOD; EC 3.5.1.16), respectively (reviewed by Slocum, 2005; Kalamaki *et al.*, 2009a,b) (Fig. 9.1). Alternatively, N^2-acetyl-Orn is converted to Orn by acetyl-Orn-Glu acetyltransferase (NAOGAcT; EC 2.3.1.35), which conserves the acetyl group to generate *N*-acetyl-Glu from Glu. Under special circumstances (e.g. protein hydrolysis during seed germination), Orn can be synthesized from Arg via arginase, the by-product being urea; this pathway is analogous to the production of Orn in animals (Morris, 2006, 2007). In support of the earlier studies showing Arg production from Glu in plants, Bhatnagar *et al.* (2002) and Majumdar *et al.* (2013) have demonstrated that in cell cultures of poplar (*Populus nigra* × *maximowiczii*) and seedlings of *Arabidopsis thaliana*, respectively, Arg → Orn conversion does not occur even under conditions of Orn depletion in the cells.

Among the different enzymatic reactions required for Orn biosynthesis from Glu, only the reaction catalysed by the enzyme NAGK was considered to be the rate-limiting and the key regulatory step for Orn production in plants (Kalamaki *et al.*, 2009a,b). This enzyme was shown to be feedback inhibited by cellular Arg concentration in the context of the overall N status of cells (Burillo *et al.*, 2004; Sugiyama *et al.*, 2004). Presumably, this inhibition involved a group of ubiquitous proteins called P_{II} proteins (reviewed in Chellamuthu *et al.*, 2013; Huergo *et al.*, 2013; see also Section 9.5) through the formation of a complex with NAGK to activate it (Beez *et al.*, 2009); P_{II} proteins also help to overcome the negative effect of Arg-mediated feedback inhibition of NAGK (Beez *et al.*, 2009). Kalamaki *et al.* (2009a) later demonstrated that transgenic overexpression of a tomato (*Solanum lycopersicum* Mill. cv. Ailsa Craig) *SlNAGS1* gene in *A. thaliana* caused a twofold to ninefold increase in cellular Orn in the rosette leaves of pre-bolting plants. This increase in Orn was accompanied by a small increase in citrulline (Cit) and a significant decrease in Arg content in several of the transgenic lines. The transgenic plants with higher Orn content showed better germination, increased seed viability and tolerance to salt (NaCl) or drought stresses. In contrast, Brauc *et al.* (2012) have recently shown that transgenic expression of the homologous *Arginase 2* (*AtARGAH2*) gene under the control of a *Cauliflower mosaic virus* (CaMV) *35S* promoter surprisingly exhibited an approximately threefold decrease in Orn accompanied by a ~1.3-fold increase in putrescine (Put) and Pro contents in the leaves of transgenic *Arabidopsis* plants. The other amino acids that were affected in the *ARGAH2* overexpresser lines were: alanine (Ala), tyrosine (Tyr), valine (Val), isoleucine (Ile), leucine (Leu), phenylalanine (Phe), and lysine (Lys), which increased; and asparagine (Asn), glutamine (Gln), glycine (Gly) and histidine (His), whose concentration was lowered. The transgenic plants showed greater tolerance to the pathogen *Botrytis cinerea*. The authors suggested a possible role for urea accumulation or other stress-related metabolites in the *ARGAH2* overexpresser lines in enhancing basal resistance of the plants to disease.

Surprisingly, silencing (through T-DNA insertion) of *ARGAH2* had similar effects on PAs and amino acids as observed in the overexpresser lines, with the exception of Tyr and methionine (Met), both of which decreased in the silenced lines compared with the wild type (WT) plants. In another study, overexpression (*35S* promoter) of *AtARGAH1* and *AtARGAH2* (again in *Arabidopsis*) resulted in increased sensitivity to abiotic stresses in the transgenic plants, suggesting possible complex and differential regulation of Arg in connection with Orn responding to biotic and abiotic stresses (Shi *et al.*, 2013). As Orn is an intermediate metabolite of the PA and the amino acid biosynthetic pathways rather than a terminal product, the biosynthesis of Orn is likely to be regulated by its increased demand for use in the production of Put, Arg or Pro, depending on environmental or intracellular conditions. Increased biosynthesis of Orn is often observed by its rapid consumption into the allied pathways rather than it being accumulated in the cells (Mohapatra *et al.*, 2010b; Majumdar *et al.*, 2013). Furthermore, as indicated by its low cellular content, it can be surmised that the Orn titre in the cells is tightly monitored as well as regulated (Mohapatra *et al.*, 2010b; Majumdar *et al.*, 2013). More details on this aspect of Orn metabolism are covered later (see Section 9.9).

Arginine is also a source of nitric oxide (NO), which plays an important role in germination, defence responses, hormonal signalling, root growth and flowering (Crawford, 2006; Grün *et al.*, 2006); its role in plants has recently been reviewed by Mur *et al.* (2013). Taken together, Arg and Orn metabolism are of central importance in plant biology, but the regulation of gene expression and the enzymes of Glu/Orn/Arg biosynthesis are poorly understood at present.

9.4 Cellular Contents

Numerous reports on the cellular contents of Orn in a variety of plant(s)/tissues have been published. The indications are that Orn contents vary widely in different species and are subject to regulation by a multiplicity of growth conditions. Only in rare instances is Orn among the abundant amino acids (Tables 9.1 and 9.2). In our studies with several species (Table 9.2), Orn contents were rarely above 1–2% of the total amino acid pool in the tissue (versus the dominant related amino acids e.g. Arg, Pro or Glu, whose contents are several fold higher than those of Orn). Cyr and Bewley (1989) (Table 9.1) reported seasonal variation of Orn content in the roots of leafy spurge (*Euphorbia esula*); during the summer, for example, Orn content was ~3 μmol g^{-1} dry weight (DW) (~8% of the soluble amino acid pool), whereas during winter a significant increase in Orn content (~18 μM g^{-1} DW) was observed, when it became the predominant amino acid (~20% of the total soluble amino acid pool). The data presented in Table 9.2 support the results of Cyr and Bewley (1989) in that significant variations in different amino acids are common in relation to environmental conditions and/or in different tissues within the same species.

In a detached leaf assay with rice (*Oryza sativa*), Yang *et al.* (1999) reported that Orn content was higher under dark treatment (~10 nmol g^{-1} fresh weight (FW); 0.03% of total soluble amino acid pool) compared with light treatment (3 nmol g^{-1} FW; 0.01% of total soluble amino acid pool) (see Table 9.1). In 4-year-old saplings of white spruce (*Picea glauca*), cellular contents of Orn in the root, stem and buds were about 0.4% of the total soluble amino acids (Durzan, 2010). In poplar (*P. nigra* × *maximowiczii*) cell cultures, Orn was found to be ~1–4 nmol g^{-1} FW, constituting <0.1% of the total soluble amino acids (Mohapatra *et al.*, 2010b) (see Table 9.1). In cashew (*Anacardium occidentale*) leaves, Orn was about ~2 μmol g^{-1} DW, contributing to ~3% of the total soluble amino acids (da Rocha *et al.*, 2012; see Table 9.1). Likewise, seedless grapes obtained from different rootstocks showed Orn contents of 7 to 28 μg g^{-1} DW, constituting ~2–4% of the total soluble amino acids; the contents varied with the variety and the time of analysis (Jogaiah *et al.*, 2013). Cellular concentrations of Orn also varied among different tissues/organs of *Erythroxylum coca* plants;

Table 9.1. Cellular contents of glutamine (Glu), arginine (Arg) and ornithine (Orn) in different plant species and parts under different conditions.

Plant species/ conditions	Amino acid concentration (variable units, fresh weight, FW, or dry weight, DW)			Tissue	Reference
	Glu	Arg	Orn		
Anacardium occidentale	~15 (μmol g^{-1} DW)	~2 (μmol g^{-1} DW)	~2 (μmol g^{-1} DW)	Leaves	da Rocha *et al.*, 2012
Arabidopsis thaliana	~7.5 (nmol mg^{-1} FW)	~0.5 (nmol mg^{-1} FW)	~3 (nmol mg^{-1} FW)	Rosette leaves	Adio *et al.*, 2011
Arabidopsis thaliana	3.48 (μmol g^{-1} FW)	5.25 (μmol g^{-1} FW)	0.7 (μmol g^{-1} FW)	Whole seedlings	Majumdar *et al.*, 2013
Erythroxylum coca	47.35 (nmol mg^{-1} DW)	11.53 (nmol mg^{-1} DW)	0.11 (nmol mg^{-1} DW)	Leaves	Docimo *et al.*, 2012
	46.32 (nmol mg^{-1} DW)	15.3 (nmol mg^{-1} DW)	0.17 (nmol mg^{-1} DW)	Buds	
	47.65 (nmol mg^{-1} DW)	107.19 (nmol mg^{-1} DW)	0.56 (nmol mg^{-1} DW)	Stems	
	15.07 (nmol mg^{-1} DW)	9.02 (nmol mg^{-1} DW)	0.34 (nmol mg^{-1} DW)	Roots	
Euphorbia esula summer season	~6 (μmol g^{-1} FW)	~3 (μmol g^{-1} FW)	~2.5 (μmol g^{-1} FW)	Roots	Cyr and Bewley, 1989
winter season	~15 (μmol g^{-1} FW)	~5 (μmol g^{-1} FW)	~17.5 (μmol g^{-1} FW)	Roots	
Oryza sativa light	5589.9 (nmol g^{-1} FW)	191.2 (nmol g^{-1} FW)	2.9 (nmol g^{-1} FW)	Leaves	Yang *et al.*, 1999
dark	6070.3 (nmol g^{-1} FW)	717.7 (nmol g^{-1} FW)	10.3 (nmol g^{-1} FW)	Leaves	
Populus nigra × *maximowiczii*	818 (nmol g^{-1} FW)	522 (nmol g^{-1} FW)	3 (nmol g^{-1} FW)	Cells	Mohapatra *et al.*, 2010b

stems had the highest levels of Orn (~0.6 μmol g^{-1} DW), followed by roots (~0.3 μmol g^{-1} DW), buds (~0.2 μmol g^{-1} DW) and leaves (~0.1 μmol g^{-1} DW) (see Table 9.1). Overall, Orn constituted about 0.1–0.3% of the total soluble amino acids among different tissues (Docimo *et al.*, 2012). In 12-d-old seedlings of *Arabidopsis*, Orn content was only 0.7 μmol g^{-1} FW (≤0.5% of the total soluble amino acids (Majumdar *et al.*, 2013; see Table 9.1), whereas Kalamaki *et al.* (2009a,b) had reported Orn contents to be as high as ~50 μg 100 mg^{-1} FW (about ~4 μmol g^{-1}) in *Arabidopsis* leaves transformed with tomato *Sl-NAGS* versus about 0.4 μmol g^{-1} in the WT plants.

9.5 Mutants of Ornithine Biosynthesis

Whereas mutants of several genes in the Glu-Orn-Arg/Pro pathway have been studied in several plants, the studies do not cover all of the genes involved in the pathway. This is in contrast to a significant body of knowledge about inherited mutations of this pathway in animals and humans (Section 9.7). A few of the mutants that have been reported in detail are described here briefly.

A group of highly conserved proteins called P_{II} proteins are found in all living organisms and are believed to act as signalling molecules to activate NAGK in plants and cyanobacteria by forming a P_{II}-NAGK complex and thus participating in the control of Glu → Arg pathway (reviewed by Chellamuthu *et al.*, 2013; Huergo *et al.*, 2013). In *Arabidopsis*, P_{II}-deficient mutants P_{II}v1 and P_{II}s2 (T-DNA insertion in the P_{II} homologue *GLB1* gene) when grown in 1 mM nitrate medium (i.e. subjected to N deficiency) followed by resupply of different forms of N, showed a significant decrease (up to 70%) in Orn, Cit and Arg

Table 9.2. Cellular contents of glutamate (Glu), arginine + threonine (Arg + Thr), proline (Pro), γ-aminobutyric acid (GABA) and ornithine (Orn) in the leaves of different forest tree species at different sites.

	Mean ± SE amino acids (nmol g^{-1} fresh weight)					
Site/plant type and species (*n*)[a]	Glu	Arg + Thr	Pro	GABA	Orn	Reference
HBEF (Hubbard Brook Experimental Forest), New Hampshire, USA						
RS-CY (20)	230 ± 19	132 ± 30	31 ± 4	352 ± 65	11 ± 4	Minocha *et al.*, unpublished
RS-PY (20)	214 ± 16	81 ± 18	37 ± 8	237 ± 36	11 ± 3	Minocha *et al.*, unpublished
SM (30)	277 ± 25	292 ± 21	68 ± 5	283 ± 22	5 ± 1	Minocha *et al.*, 2010
YB (30)	797 ± 72	536 ± 111	28 ± 3	298 ± 21	22 ± 4	Minocha *et al.*, 2010
AB (30)	243 ± 47	196 ± 46	34 ± 3	1039 ± 80	10 ± 4	Minocha *et al.*, 2010
Teakettle, California, USA						
JP (30)	768 ± 54	1095 ± 461	138 ± 13	287 ± 31	15.78 ± 5.62	Minocha *et al.*, 2013
SP (30)	771 ± 42	680 ± 246	156 ± 13	163 ± 13	9.09 ± 1.96	
WF (30)	153 ± 11	172 ± 39	24.4 ± 2.2	504 ± 26	5.25 ± 1.04	
Harvard Forest						
RP-control (3)	174 ± 11	NA[b]	81 ± 6	109 ± 6	1 ± 1	Bauer *et al.*, 2004
RP-chronic-N (6)	773 ± 60	4072 ± 1233	174 ± 48	753 ± 95	55 ± 14	
Monagahela National Forest, West Virginia, USA						
QR (10)	718 ± 84	249 ± 44	270 ± 54	830 ± 95	NA	Minocha, R. *et al.*, unpublished
BC (10)	3883 ± 357	1014 ± 149	1323 ± 227	703 ± 85	NA	
TP (10)	2006 ± 173	304 ± 81	77 ± 7	1098 ± 152	1 ± 0	
SM (10)	1613 ± 94	469 ± 65	299 ± 74	1014 ± 100	3 ± 1	
Czech Republic						
NS-CY (15)	176 ± 13	166 ± 16	22 ± 1	284 ± 20	NA	Minocha, R. *et al.*, unpublished
NS-PY (15)	206 ± 13	32 ± 32	30 ± 3	315 ± 21	NA	
Juneau, Alaska, USA						
SS-CY (4)	85 ± 16	7 ± 4	52 ± 6	59 ± 11	7 ± 1	Minocha, R. *et al.*, unpublished
WH-CY (6)	117 ± 4	0 ± 0	10 ± 2	103 ± 11	1 ± 1	

Continued

Table9.2. Continued.

Site/plant type and species (n)[a]	Mean ± SE amino acids (nmol g^{-1} fresh weight)					Reference
	Glu	Arg + Thr	Pro	GABA	Orn	
Invasive plants						
MV (10)	1001 ± 43	237 ± 22	55 ± 5	729 ± 24	11 ± 2	Minocha, R. *et al.*, unpublished
AA (10)	2242 ± 83	1566 ± 139	339 ± 25	910 ± 38	3 ± 0	
GM (10)	1732 ± 36	483 ± 34	2474 ± 233	300 ± 17	1 ± 0	
Maine, USA						
BF-CY (15)	148 ± 16	104 ± 9	17 ± 1	495 ± 41	NA[b]	Minocha, R. *et al.*, unpublished

[a]Abbreviations used by site/plant type and in order of occurrence: n = number of replicates.

HEBF: RS-CY = red spruce (*Picea rubens*) – current year foliage; RS-PY = red spruce (*Picea rubens*) – 1-year-old year foliage; SM = sugar maple (*Acer saccharum*); YB = yellow birch (*Betula alleghaniensis*); AB = American beech (*Fagus grandifolia*).

Teakettle: JP = Jeffrey pine (*Pinus jeffreyi*); SP = sugar pine (*Pinus lambertiana*); WF = white fir (*Abies concolor*).

Harvard Forest: RP = red pine (*Pinus resinosa*).

Monagahela National Forest: QR = red oak (*Quercus rubra*): BC = black cherry (*Prunus serotina*); TP = tulip poplar (*Liriodendron tulipifera*); SM = sugar maple (*Acer saccharum*).

Czech Republic: NS-CY = Norway spruce (*Picea abies*), current year foliage; NS-PY = Norway spruce (*Picea abies*), 1-year-old foliage.

Juneau, Alaska: SS-CY = Sitka spruce (*Picea sitchensis*), current year foliage; WH-CY = western hemlock (*Tsuga heterophylla*), current year foliage.

Invasive plants: MV = Japanese stiltgrass (*Microstegium vimineum*); AA = tree of heaven (*Ailanthus altissima*); GM = garlic mustard (*Alliaria petiolata*).

Maine: BF-CY= balsam fir (*Abies balsamea*), current year foliage.

[b]NA, not available.

compared with the WT plants. Other major amino acids, such as Gln, Glu, aspartate (Asp) and Asn were also reduced, albeit to a lesser extent (15–50%) in response to ammonium resupply (Ferrario-Méry *et al.*, 2005, 2006). Microarray analysis of Arg metabolism genes, in contrast, did not show any differential expression in the P_{II} mutants compared to the WT under the same conditions. These results together suggest that Orn/Arg biosynthesis is significantly and specifically affected under excess N conditions, and is perhaps not subject to transcriptional regulation of the pathway genes. These results are consistent with our studies on poplar cells that showed a several-fold increase in Glu → Orn flux in response to increased demand for Orn production, but no change in the expression (analysed by microarrays and quantitative PCR) of most genes in the Glu-Orn pathway was seen (Page *et al.*, 2007, 2012).

In *Arabidopsis*, mutants of the *TUMOR PRONE5* (*TUP5*) gene (*tup5-1*) encoding the enzyme NAOAT had significantly lower (31–85%) Arg content in 11-d-old seedlings when grown under long-day (LD) conditions but not in the dark; the WT plants did not show this response (Frémont *et al.*, 2013). Lower cellular Arg in the mutant was accompanied by a significant increase (as % of WT) in several amino acids including, Gly (by ~800%), Gln (by ~450%); Asn, Val, tryptophan (Trp) and Phe by ~300%; Asp, threonine (Thr), Ala, Tyr and Ile, by ~200%; and Met and Lys by ~150%. Moreover, this overall increase of amino acids in the mutant varied with developmental stage in that only Gln and Asn increased in the 17-d-old seedlings. The authors suggested that the observed increase in Gln and Asn could be a compensation for cellular Arg (as an N sink) as these two amino acids are important in N storage and transport in plants. It was further proposed that the overall increase in amino acids in the mutant was perhaps a consequence of reduced protein synthesis, again due to Arg becoming a rate-limiting factor. Besides, the *tup5-1* mutant plants showed a blue light-dependent short-root phenotype (that could be restored by exogenous supply of Arg or its precursors) resulting from arrested cell division at the meristem region, which appeared to lack an organized structure and had no starch granules in the columella cells. The null mutants showed a decrease in gamete viability, accompanied by embryo lethality. Whereas overexpression of the *TUP5* gene in the *tup5-1* background was able to restore Arg content to the levels in the WT plants, no significant increase in Arg content was observed in the transgenic plants overexpressing this gene in the WT background. It can thus be argued that this step might not be rate limiting for Orn/Arg biosynthesis in *Arabidopsis*. The cellular Orn content in neither the mutant nor the overexpresser plants was reported.

A T-DNA insertion mutant of the N^{δ}-acetylOrn transferase gene (*nata1-1*) in *Arabidopsis* showed the absence of N^{δ}-acetylOrn in all organs but with no visual phenotype and no significant effect on PAs or amino acids in the mutant plants (versus the WT) under normal conditions of growth (Adio *et al.*, 2011). Methyl jasmonate (MeJa), a phytohormone that has been shown to induce the expression of N^{δ}-acetylOrn transferase in plants (Zimmermann *et al.*, 2004; Yan *et al.*, 2007), significantly increased N^{δ}-acetylOrn and caused a decrease in Put and spermidine (Spd) contents in the WT plants. There was also a concomitant increase in Put and agmatine (Agm) and a decrease in Spd contents with no N^{δ}-acetylOrn detected in the *nata1-1* mutant compared with its corresponding control. The amino acids that increased significantly upon MeJa treatment in the *nata1-1* mutant plants were Asp, Glu, Gly, His, Ala, N^{δ}-acetylOrn, Pro, Ile and Leu. In contrast, constitutive (*35S* promoter) expression of *NATA1* in the *nata1-1* background significantly decreased Pro content (by fivefold). There was a concomitant decrease in Orn and a significant increase in Thr, Val, Ile and Phe in the overexpressers, but without a significant change in total amino acid contents. Different biotic stresses, e.g. feeding on plants by *Myzus persicae* (the green peach aphid) or infection with *Pseudomonas syringae* increased the expression of *NATA1* accompanied by accumulation of N^{δ}-acetylOrn in the infected plants. Transient expression of *NATA1* or incorporation N^{δ}-acetylOrn in the aphid diet had a negative impact on insect reproduction, suggesting a role for N^{δ}-acetylOrn in tolerance to biotic stress.

The conversion of L-Arg to L-Orn is catalysed by the enzyme arginase (L-Arg amidinohydrolase; EC 3.5.3.1), which produces urea as a by-product; the urea is then rapidly recycled by urease to produce ammonia (Polacco and Holland, 1993).

During seed germination, Arg (a storage form of N) is catabolized by arginase to provide nourishment to the developing embryo. In *Arabidopsis*, mutations of either of the two arginase genes (*argah1-1* and *argah2-1*) show an increase in the formation of lateral and adventitious roots in the seedlings, accompanied by increased accumulation of NO (Flores *et al.*, 2008). The mutants exhibit enhanced accumulation of NO upon exposure to the synthetic auxin naphthaleneacetic acid (NAA) as well. Also, increased activity of β-glucuronidase (GUS) was observed in the root tips, lateral roots and hypocotyl under the control of an auxin-responsive promoter *DR5* when expressed in the *argah* background. NO in plants has been shown to be positively correlated with root formation (Hu *et al.*, 2005; Lombardo *et al.*, 2006) and is inducible by auxin. These observations, together with Arg-induced NO accumulation and increase in adventitious root formation in the mutants (versus WT plants), suggest that Arg is a potential source of NO regulating root development in concert with auxin (Flores *et al.*, 2008).

Besides *ARGAH*, dysfunction of an *OTC* (*Orn transcarbamoylase*) gene in *Arabidopsis* (a T-DNA insertion mutation) showed increased sensitivity (reflected in reduced germination and FW) of the plants to high Orn concentration in the medium, in contrast to Arg or Cit (Quesada *et al.*, 1999); once again, no information was available on the cellular concentration of Orn or other related compounds in this mutant line. Two venosa (*Ven*) mutants, *ven3* and *ven6*, that carry mutations in the large and small subunits of the enzyme carbamoyl phosphate synthetase (CPS; EC6.3.5.5; involved in Cit biosynthesis), respectively, show leaf reticulation and a decrease in mesophyll cell numbers in the leaves (Mollá-Morales *et al.*, 2011). The metabolites that were significantly affected in the mutant plants were increase in *N*-acetylOrn (35-fold), L-Orn (elevenfold), and decrease (<50% of WT) in Cit/Arg contents; this was accompanied by increased sensitivity to exogenous Orn compared with the WT plants. Both the defects in the mesophyll cells and the reticulated phenotype were rescued by exogenous application of Cit, yet this amino acid was unable to restore smaller leaf size in the *ven3/ven3 ven6/ven6* double mutant plants. These results together support a regulatory role for Cit in determining leaf morphology in plants. Changes in other metabolites observed in the *ven* mutants were associated with amino acid biosynthesis and the TCA (tricarboxylic acid) cycle (Mollá-Morales *et al.*, 2011).

9.6 Genetic Manipulation of Ornithine Metabolism and its Impact on Amino Acids and Other Related Compounds

The steady state cellular content of a metabolite is the sum total of its biosynthesis and utilization in different pathways. Of course, its transport out of the cells as well as its conjugation with or incorporation into macromolecules (e.g. for the 20 amino acids involved in protein biosynthesis) would also significantly affect its cellular content at any time. Orn is neither significantly transported from one organ to the other nor is it incorporated into proteins or any other macromolecule. For Orn, while its biosynthesis in plants occurs largely from Glu, its utilization involves multiple pathways, the three major ones being the biosynthesis of Arg (for use in protein synthesis, with some for PA biosynthesis via arginine decarboxylase, ADC), Pro (also for protein biosynthesis and for storage as an osmolyte) and the PAs via Orn decarboxylase (ODC) for their numerous physiological functions. Under conditions when Orn is produced from Arg instead of Glu, the biosynthesis of Glu and Pro are important consumers of Orn. Thus, the cellular content of Orn is subject to manipulation both through its increased biosynthesis (upregulation) and/or its increased utilization (depletion). As described above, upregulation of Orn biosynthesis via overexpression of *NAGS* gene has apparently been attempted with some success (Kalamaki *et al.*, 2009a,b); however, other genes of the pathway have not been successful in upregulating Orn biosynthesis. Enhancement of Orn biosynthesis via increased utilization into the PA or the Arg biosynthetic pathways have been quite effective in modulating its rate of biosynthesis, but without increased accumulation. Several laboratories, including ours, have used this approach, with the primary target of modulation of PAs in plants via the overexpression of both the *ADC* and *ODC* genes.

The alteration of Orn metabolism through the genetic manipulation of Orn aminotransferase

(OAT; EC 2.6.1.13) has been reported in plants as Orn serves as a major precursor for Pro biosynthesis as well (see Verslues and Sharma, 2010, and references therein). Overexpression of an *Arabidopsis* OAT in *Nicotiana plumbaginifolia* increased Pro content by threefold in the transgenic plants (Roosens *et al.*, 2002). The transgenic plants showed overall increased tolerance to osmotic and NaCl stresses with higher seed germination and increased biomass production under the influence of these stressors compared with the WT plants. The authors did not report the overall effect of OAT manipulation on the contents of Orn, PAs or other amino acids in the transgenic plants. It should be pointed out though that among the different fates of Orn into Put, Arg and Pro, the reaction carried out by ODC to make Put from Orn is the only known rate-limiting step in which the substrate Orn becomes limiting and not the enzyme ODC (Pegg and McCann, 1982; Hickok *et al.*, 1986; Majumdar *et al.*, 2013).

In another study, overexpression of an *O. sativa* OAT cDNA (*OsOAT*) under a constitutive maize ubiquitin promoter in Japonica rice showed significant increase in Pro content in both normal and drought conditions (You *et al.*, 2012). The other amino acids that increased significantly in the transgenic plants were Glu, serine (Ser), Lys and Met, concurrently with a significant decrease in Arg content. Transgenic plants with higher Pro and glutathione (GSH) contents showed significant tolerance to drought, osmotic, and oxidative stresses, mainly through enhanced reactive oxygen scavenging capacity.

Transgenic manipulation of Orn metabolism through overexpression of a mouse *ODC* (m*ODC*) in poplar (*P. nigra* × *maximowiczii*) cell cultures under a constitutive (CaMV 2x*35S*) promoter resulted in the alteration of a majority of the proteinaceous amino acids in the transgenic cells, especially those that are synthesized from Glu (Mohapatra *et al.*, 2010b). The transgenic cells showed threefold to sevenfold increase in Put content, with only small increases in Spd and spermine (Spm). Increased biosynthesis of Put was accompanied by a parallel increase in its catabolism, thus leading to increased accumulation of γ-aminobutyric acid (GABA; see Chapter 7); this must have created a significant increase in the metabolic flux of Orn consumption as well as its production. While the Orn content of cells was quite low to begin with (less than 0.1% of the total soluble amino acid pool), and it became even lower in the transgenic cells, its increased utilization as a substrate by the transgenic mODC had little effect on its utilization for the biosynthesis and accumulation of Arg. The increased flux of Orn into Put via ODC, while still not meeting the demand of transgenic mODC, resulted in elevated cellular Arg (albeit small). The observed change in cellular Arg accumulation was apparently the direct result of its increased biosynthesis from Orn, and not due to its reduced utilization in either protein biosynthesis or Put production via the ADC pathway. If anything, both these activities were higher in the transgenic cells, where a deficit of Orn was created by its overconsumption by mODC. Besides, there was no back conversion of Arg into Orn, which is biochemically possible (see above). Significant changes in the contents of Pro (also a part of the Glu-Orn-Arg-Pro-Put pathways) were also observed concurrently. Accompanying this increased flux of Orn consumption (and production), there was a significant increase in total soluble protein, as well as in total cellular N and C contents, in the transgenic cells overproducing Put via the ODC pathway (Bhatnagar *et al.*, 2001, 2002; Mohapatra *et al.*, 2010a). The increase in GABA in this case could be attributed to its overproduction via enhanced Put catabolism (Bhatnagar *et al.*, 2002; Shelp *et al.*, 2012; Shao *et al.*, 2014).

The above observations in cell cultures of poplar have been replicated in whole plants of *A. thaliana* using both a constitutive and an inducible (oestradiol mediated) system to regulate m*ODC* expression. The results again showed significant changes in the major amino acids of the Glu-Arg-Pro pathway in transgenic plants under both situations (Fig. 9.2), indicating a rapid metabolic flux adjustment in Orn production in response to its rapid utilization (Majumdar 2011; Majumdar *et al.*, 2013). Here, it should be pointed out that *Arabidopsis* is an exception in the plant kingdom in that it does not have a native *ODC* gene and Put is exclusively produced via ADC from Arg (Hanfrey *et al.*, 2001), which itself is entirely produced from Orn (Slocum, 2005). Once again, neither the production of Put from Arg nor the protein content in young seedlings was affected in spite of a 20- to 40-fold increase in Orn biosynthesis for increased Put production via the transgenic mODC step.

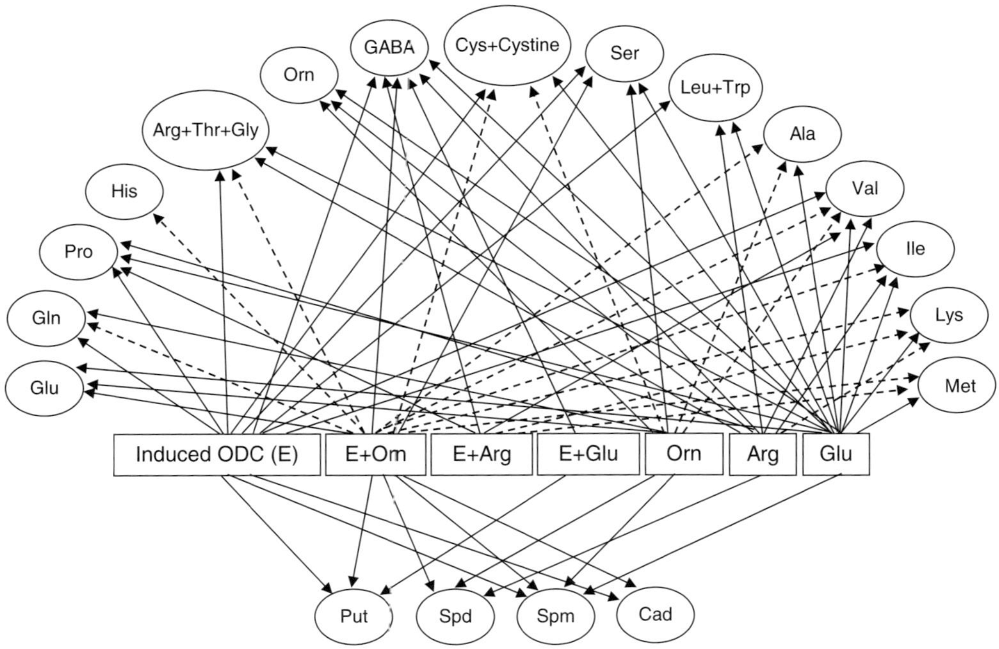

Fig. 9.2. Effects of exogenous Orn, Arg and Glu in uninduced and induced mouse ornithine decarboxylase (m*ODC*) transgenic *Arabidopsis thaliana* seedlings on amino acids and polyamines. Solid lines indicate increase and dotted lines indicate decrease (data are based on nmol g^{-1} FW basis; $P \leq 0.05$). Abbreviations: Ala, alanine; Arg, arginine; Cad, cadaverine; Cys, cysteine; E, induced ODC; GABA, γ-aminobutyric acid; Gln, glutamine; Glu, glutamate; Gly, glycine; His, histidine; Ile, isoleucine; Leu, leucine; Lys, lysine; Met, methionine; Orn, ornithine; Pro, proline; Put, putrescine; Ser, serine; Spd, spermidine; Spm, spermine; The, threonine; Trp, tryptophan; Val, valine. (Modified from Majumdar *et al.*, 2013.)

As with poplar cells, total N and C in the transgenic plants were also higher concomitant with the production of high Put.

The experiments with inducible mODC expression revealed how rapidly the metabolic flux from Glu to Orn (and then Put) could be induced. On short-term (24 h) induction of m*ODC* in 12-d-old *Arabidopsis* seedlings, significant increases in Gln, Pro, Arg, GABA, cysteine (Cys), Ser, Leu + Trp and Val were observed, whereas constitutive ODC activity in the same aged seedlings resulted in increase of GABA, Met, Val and Thr (Majumdar *et al.*, 2013). These differences in the response of inducible versus constitutive expression of the transgene are perhaps due to the long-term homeostatic adjustment of cellular metabolism in the constitutive transgenics. In both instances (i.e. under inducible or constitutive expression of m*ODC*), the transgenic seedlings showed up to a 40-fold increase in Put with only small changes in Spd and Spm contents; these results reinforce the argument that there is a tighter homeostatic regulation of higher PAs (i.e. Spd and Spm) than Put in plants. Increased Put production (inducible or constitutive) in the seedlings also showed increased catabolism of Put (Shao *et al.*, 2014), a result similar to that seen in poplar cells (Bhatnagar *et al.*, 2001, 2002). In mature plants, the amino acids whose contents increased in 2x*35S* mODC versus the WT plants in different organs included Glu, Gln, Leu, Val and His in the buds, and Pro and Ile in the rosette leaves. Conversely, Glu and Thr in the leaves, Ala in the buds, Orn in the flowers, and Arg, Orn, and His in the siliques were somewhat higher in the WT than in the transgenic plants. Notably, Glu content was significantly higher in the buds of the transgenic plants versus the buds of WT plants, but it was significantly lower (than WT) in rosette leaves and siliques (fruits) of the transgenic plants. Also, Orn was almost undetectable in many organs except the siliques and flowers. In general, total soluble amino acids were higher in the reproductive organs, among which flowers recorded the highest amount, compared with the rosette leaves in both the transgenic and the WT plants. All mature organs that were tested contained significant amounts of cadaverine (Cad, derived from Lys by the activity of mODC), which was actually even higher than Put in many organs. Supplementation of the growth medium with exogenous Orn further enhanced Put accumulation, indicating that in spite of its increased production, the demand for mODC was still not fully met (Majumdar, 2011; Majumdar *et al.*, 2013).

Our studies have led us to conclude that most of the changes in amino acids in the *ODC* overexpressing cells/plants are perhaps a compensation for the rapid consumption of Orn by the ODC enzyme, thereby creating a positive metabolic pull (a vacuum) to convert more Glu into PAs; this, we suggested, causes more N assimilation into Glu (Mohapatra *et al.*, 2010b; Majumdar 2011; Majumdar *et al.*, 2013). We further believe that this increase in N assimilation leads to increased C assimilation, thus providing an effective approach to increasing plant biomass through manipulation of the PA pathway.

In addition to overexpression, the inhibition of Put production via the mutation of *ADC* and other PA biosynthetic genes has been reported in plants, but no true mutants of the *ODC* gene are available (Watson *et al.*, 1998; Hanzawa *et al.*, 2000; Imai *et al.*, 2004a,b; Urano *et al.*, 2004, 2005; Kakehi *et al.*, 2008). Other approaches to reducing PA production in plants have included RNAi technology and chemical inhibitors. As Orn is both a direct and an indirect (via Arg) precursor of PA biosynthesis, alteration of PA biosynthesis would be expected to directly affect Orn metabolism. In tobacco (*N. tabacum*), inhibition of ODC activity via the RNAi approach showed downregulation of both *ODC* transcripts (by fourfold) and enzyme activity (~threefold) in transgenic hairy root cultures (DeBoer *et al.*, 2011). A slight increase of *ADC* transcripts was observed with a concomitant increase in ADC activity in response to ODC downregulation. Decreased activity of ODC altered alkaloid composition in the transgenic plants, in which the major alkaloid, nicotine, was decreased, but the second major alkaloid, anatabine, was increased, both in intact plants and in hairy root cultures. MeJa, which typically stimulates nicotine biosynthesis in tobacco, did not have any effect in the *ODC* RNAi plants; rather, the anatabine content was increased upon treatment with MeJa. Another study conducted by the same group (DeBoer *et al.*, 2013) showed that downregulation of *ODC* expression, again through constitutive RNAi, prevented the increase in anabasine in *N. glauca* hairy root cultures in response to wounding (another effect of MeJa) when compared with the vector-only controls, suggesting a possible role of ODC in elevating anabasine content in response to wound stress in this species. Besides affecting amino acid biosynthesis, the silencing of ODC was shown to affect several TCA cycle metabolites that are directly or indirectly derived from the diamine Put (DeBoer *et al.*, 2011, 2013). Arguably, Orn flux in these plants must have been altered because it is the primary precursor for the biosynthesis of both alkaloids (both via the ADC or ODC pathways) and several amino acids; however, no direct measurements of this or other amino acids were reported in these studies.

In plants, the conversion of Orn to Pro occurs via two steps, i.e. Orn → Δ^1-pyrroline-5-carboxylate (P5C) → Pro, while in some bacteria (e.g. *Agrobacterium* spp.) Orn can be directly converted to Pro by Orn cyclodeaminase (OCD; Trovato *et al.*, 2001). The expression of *AtOCD* was enhanced by Pro rather than its substrate Orn, and induced expression of *AtOCD* was especially observed under low water potential (Sharma *et al.*, 2013). In *Arabidopsis*, a T-DNA insertion mutant (*Atocd*) of the *AtOCD*-like gene, as well as *AtOCD*-RNAi transgenic plants, were shown to have increased (by ~15%) Pro accumulation, without any change in the expression of Pro catabolic genes, under low water potential conditions. Both the mutant and the overexpresser lines of *AtOCD* (*35S* promoter) showed a significant increase in Orn and a significant decrease in cellular oxalate contents only under low water potential compared with the WT plants, suggesting an indirect effect of *AtOCD* expression on redox status in plants under water stress conditions (Sharma *et al.*, 2013).

In nature, excess N absorbed from atmospheric acidic deposition, soil fertilization or constitutive overexpression of a gene in the N pathway can cause accumulation of N in the form of free amino acids, including Orn and PAs, thus causing a shift in the overall C and N metabolite ratios. Nutrient imbalances (Ca:Al and Ca:Mn ratios) and leaching of Ca and Mg from the soil is often observed under these conditions as a result of lowering of soil pH. If the conditions persist, a decline in growth is observed (Minocha et al., 2000; Bauer et al., 2004; Magill et al., 2004). These conditions may be reversed by the addition of Ca or by removal of the source of excess N (Wargo et al., 2002; Minocha et al., 2010). Hence, the accumulation of certain amino acids (including Orn) is indicative of their role in ammonia (excess N) detoxification. It was suggested by Kalamaki et al. (2009a,b) that engineering plants to constitutively accumulate a non-toxic metabolite such as Orn, that can readily be converted to osmo-protective molecules (e.g. Pro or PAs) upon stress induction, might present an alternative way of increasing abiotic stress tolerance in plants. These authors directly demonstrated improvement in stress tolerance by the accumulation of Orn via overexpression of the tomato *SlNAGS1* gene in *Arabidopsis*. They also proposed an interesting indirect role of Orn accumulation before the exposure of the plants to abiotic stress, during which Pro could be quickly produced on demand, thereby endowing the plant with future protection from abiotic stress. Relatively high levels of Pro in flowers and other reproductive structures (~60-fold higher than those in other organs) in several plants has been suggested to play a role in the protection of these parts from potential osmotic stress (Mattioli et al., 2009). It was further suggested that such high levels of Pro were synthesized from Orn by Δ^1-pyrroline-5-carboxylate synthetase (P5CS; EC 2.7.2.11/1.2.1.41) with the help of Pro transporter; under such conditions, a rapid flux of Orn → P5C pathway (see Fig. 9.1) will be essential.

9.7 Ornithine Biosynthesis and Functions in Animals

Like plants, in animals also, the cellular concentration of Orn is much lower than that of Arg or Glu. However, in contrast to plants, Orn biosynthesis in animals occurs solely from Arg by cleavage carried out by the enzyme arginase (Morris, 2006, 2007). Further metabolism of Orn is carried out by two enzymes, namely OAT to make Cit and Orn transcarbamoylase (OTC; EC 2.1.3.3) to make Pro. High activity of ODC and reduced activities of OAT and OTC have been correlated with rapid growth of hepatomas and associated tumorigenesis (Weber et al., 1972; Tomino et al., 1974). A similar effect of increased ODC activity on OAT was observed under short-term induction of ODC in rat colon, which was accompanied by an increase in cellular Orn and Arg besides increased Put. Reduced OAT activity has been shown to result in higher accumulation of Orn in both OAT-deficient humans and adult mice with inhibited OAT activity. However, increased arginase activity in skin tumour cells showed an elevated Orn level as well as ODC activity, thus increasing Put production. Han et al. (2001) suggested that such changes in Orn or Arg concentrations in the cells trigger the conversion of a normal cell to a tumour phenotype. While the accumulation of Orn will allow more of it to be available for PA biosynthesis, the depletion of Orn might create metabolic flows in pathways, especially the PA and the amino acid biosynthetic pathways that are connected through Orn.

Demand of Arg, a semi-essential amino acid for mammals, can be met either through endogenous biosynthesis or dietary intake. Besides being a major substrate for PA biosynthesis (via Orn), Arg plays diverse roles in other metabolic pathways, e.g. urea, creatine and NO biosynthesis. Male Swiss albino mice when fed with Arg-rich dietary supplements for short periods did not show dramatic changes in the overall free PA pool (Teixeira et al., 2002). In long-term studies, the major changes observed were in Put content at 15 d and Spd at 30 d; the effects differed among different organs. It was concluded that endogenous Arg biosynthesis was the major contributing factor in maintaining PA pools in the tissues. In another study to elucidate the role of exogenous Orn on Arg metabolism, piglets fed with ^{14}C-Orn showed no significant changes in the plasma Arg concentration or the flux of Arg or Arg synthesis from Orn; Orn had a greater effect on Pro biosynthesis (Urschel et al., 2006). The addition of Orn to an Arg-deficient diet had a significant effect on Orn and Arg metabolism, more so than supplementation with dietary α-ketoglutarate. Analyses of Arg in various organs in mice revealed that decreased Arg level in plasma resulted in a reduced Gln level in the muscle, accompanied by increased renal Arg and gut Cit (Vissers et al., 2007a,b). Reduced plasma Arg concentration was also observed in tumour-bearing mice and human cancer patients (Vissers et al., 2005). The feeding of Orn to young rats through the diet caused a significant increase in tissue protein synthesis by positively affecting growth hormone levels (Tujioka *et al.*, 2012).

In animals, Orn biosynthesis is also critical for the disposal of excess ammonia in the form of urea, thereby detoxifying the cells of excess N. The N in Orn, however, remains intact; as for urea production, the N comes from ammonia and Asp (Krebs, 1973). Thus, endogenous levels of Orn apparently limit the production of urea in animals, and the exogenous supply of Orn is the most effective way to promote urea formation (Sikorska *et al.*, 2010).

Ornithine is a commonly used nutraceutical supplement to increase muscle strength and cardiovascular activity (http://www.tysonnutraceuticals.com/L-Ornithine.htm). In order to investigate the role of L-Orn in alleviating fatigue, a study was conducted with healthy individuals in which Orn was administered orally before exercise (Sugino *et al.*, 2008). Individuals with L-Orn supplements were able to overcome fatigue by increasing their efficiency of energy consumption with efficient excretion of cellular ammonia. Other positive effects of Orn are related to its ability to produce NO. In an earlier study by Alhonen *et al.* (2002), the increased production of PAs from dietary Orn was shown to stimulate hepatic regeneration in rats. Dietary supplementation with Orn also led to enhanced wound breaking strength and collagen deposition (Sikorska *et al.*, 2010).

There are numerous genetic disorders known in humans that are related to the Arg-Orn-Glu or the urea cycle pathway (Leonard and Morris, 2002; Ficicioglu and Yudkoff, 2013) and many of them are manageable by dietary supplementation of amino acids. In disorders such as citrullinaemia, argininosuccinic aciduria and argininaemia, Orn insufficiency and an excess of ammonia also occur (Zieve, 1986). It is recognized that the presence of insufficient Orn for normal urea cycle responsiveness leads to vulnerability to ammonia toxicity under conditions of protein deficiency and malnutrition (Zieve, 1986). A recent review (Bai *et al.*, 2013) has revealed that supplementation with L-Orn-L-Asp (LOLA, a stable salt of two nonessential L-amino acids) could help hepatic encephalopathy patients in the improvement of disease by reducing the serum ammonia concentration compared with the placebo/no-intervention controls. Ammonia excess also occurs in the normal liver when amino acid mixtures lacking Orn, Arg or Cit are introduced/infused or when the gastrointestinal tract contains an excess of protein, urea or NH_4^+; e.g. after a gastrointestinal haemorrhage. It may also occur with the injection of specific amino acids such as Gly, or ammonium salts, urea and urease (Zieve, 1986). Thus, a better understanding of the regulation of Orn biosynthetic pathway should be useful for genetic manipulation of Orn and Arg in plants for enhancement of their nutraceutical properties.

9.8 Exogenous Supply of D- and L-Ornithine

Ornithine is available in two racemic forms – D-Orn and L-Orn, which are interconvertible by the enzyme Orn racemase (EC 5.1.1.12), and are found in several organisms, including humans (Chen *et al.*, 2000). Alternatively, D-Orn can be converted into D-Arg and then into L-Arg by Arg racemase (EC 5.1.1.9; see http://www.genome.jp/kegg-bin/show_pathway?mlo00472) for normal metabolism. Some microbes directly metabolize D-Orn into (2*R*,4*S*)-2,4-diaminopentanoate (by D-Orn 4,5-aminomutase; EC 5.4.3.5), a pathway that allows them to utilize D-Orn as a source of N (Yonaha *et al.*, 1975). Therefore, D-Orn is sold on the market as a nutraceutical for the same usage as L-Orn or Arg; however, in some studies, differences in the physiological effects of the two enantiomers have been reported.

Besides rapid utilization of intracellular Orn and its impact on PAs, amino acids, the TCA cycle metabolites etc., the exogenous supply of Orn (L/D forms) also affects related metabolites in living cells. In intact leaves or leaf discs of cashew plants, exogenous application of L-Orn (10 mM) showed a significant increase in the cellular content of Pro (da Rocha *et al.*, 2012). In 12-d-old *Arabidopsis* seedlings, short-term treatment (24 h) with exogenous L-Orn (0.1 mM) resulted in significant increases in Glu, Gln, Orn and Ser (Majumdar *et al.*, 2013). Another study on the effect of D- and L-Orn on cell viability and other associated metabolites in tobacco (*N. tabacum* L. cv. Burley 21) cell suspension cultures showed differential effects of the two forms of Orn on amino acid biosynthesis and the TCA cycle (Gholami *et al.*, 2014). Cell viability decreased significantly with increasing concentration of L-Orn (beyond 1 mM), whereas D-Orn did not have any such effects. The amino acids that showed a linear increase with increase in L-Orn concentrations (1, 5, 10 mM) were L-Orn, Cit, Glu, Gly, Pro, Leu, Lys, Tyr, Phe, Val, Ile and Trp, with the highest increase observed in L-Orn (40-fold) and Cit (tenfold) with 10 mM L-Orn treatment; there was no change in the same amino acids when treated with the corresponding concentrations of D-Orn. The TCA cycle metabolites that were affected (positively) most by L-Orn were galactonate and malate, in addition to the alkaloid nicotine. In contrast, D-Orn caused a significant increase in Arg content (80-fold higher than the control) and also in urea, lactic acid and glucose.

Different physiological responses of tobacco cells treated with different Orn enantiomers were seen under salt stress (250 mM NaCl). Growth of these cells was adversely affected by L-Orn under normal conditions, with further deterioration under salt stress, whereas the same concentration (1 mM) of D-Orn improved cell growth and ameliorated salt stress. This was apparently achieved through the activation of antioxidant enzymes (namely superoxide dismutase, catalase and peroxidase), increase in Pro content and a lowering of H_2O_2 production in the cells. Interestingly, the same concentration (1 mM) of D-Orn had a positive effect on Put content in these cells when grown under normal conditions. These results together suggest a potential role via unknown mechanisms of D-Orn in plant cells activating the antioxidant system under stress situations (Ghahremani *et al.*, 2013).

9.9 Modelling of Ornithine Metabolism and Associated Flux: Ornithine as a Regulatory Molecule

There are two key steps involved in the regulation of steady state metabolic flux in a cell: a sensory mechanism to regulate flux rate; and the actual control of flux rate. The sensory aspects often require molecular mechanisms that can assess the need to respond to the changing environment and then send messages to the regulatory mechanisms. For decades following the early models of transcriptional control, it was believed that most metabolic pathways are regulated through changes in the concentration of the enzyme(s) that are involved in the specific metabolic step(s), which produce a certain metabolite. Kochanowski *et al.* (2013) have recently summarized the various scenarios under which metabolic flux regulation could occur in microbes at the level of transcription, translation, post-translational protein modifications and metabolites (substrates and products). The discovery of feedback regulation of enzyme activities, the complexities of certain reactions involving multiple substrates, cofactors and coenzymes, which were either limiting or non-limiting, and the stimulation of catabolism by certain metabolites themselves acting as the substrate, made the study of flux rates rather complex.

There are three things that become apparent from the published literature on the regulation of Orn metabolism in plants: (i) rate of its biosynthesis is dependent upon the activity of NAGS, the first enzyme in the pathway; (ii) this enzyme must somehow respond to the need for Orn utilization in multiple pathways (Pro, Arg and PA biosynthesis) because Orn is not the end product of the pathway; and (iii) the other enzymes of the so-called Orn pathway (i.e. NAGK, NAGPR, NAOGAcT, NAOAT and NAOD) are often constitutively available. If one assumes that the biosynthesis of Orn is not regulated by the need for its accumulation nor by a feedback regulation by its products (Arg, Pro and PAs), one can then argue that Orn biosynthesis is largely regulated by the rate of its utilization. The experimental evidence for this line of reasoning comes from the studies involving: (i) genetic manipulations (both upregulation and downregulation via transgene expression, the RNAi approach and mutations) of the cellular contents of the three major products of Orn; and (ii) considerable (many fold) variations in the cellular contents of the three products under conditions of abiotic stress without significant changes in cellular Orn content.

The two main driving forces to increase or decrease metabolic flux through a pathway are: (i) an upsurge in demand for the product(s); and (ii) a need to ameliorate the accumulation of an intermediate. In multi-nodal pathways, such as those involving Orn utilization into the production of Arg, Pro, PAs and alkaloids, at a given time the increase in demand may come from any one or more products. Based upon (i) the results of our own studies with transgenic manipulation of the PA pathway, where the Orn → Put step was targeted, and (ii) several other studies where increased PA biosynthesis in response to abiotic stress is often accompanied by an increase in Pro and Arg as well, it appears that increases in Arg, Pro and PAs often occur concurrently under several conditions. There is also a concomitant increase in GABA accumulation under the same conditions, which can be produced either directly from Glu by Glu decarboxylase (GAD; EC 4.1.1.15) or through the catabolism of Put by diamine oxidase (DAO; EC 1.4.3.6) (Shelp *et al.*, 2012). That makes Glu the primary substrate being used in all of these reactions, so that its demand must be under strict coordination, and its steady state production/availability must be ensured. One obvious common metabolite for all but the GAD node of this pathway is the intermediate Orn. Thus, it can be hypothesized that as long as Glu is not limiting, the Pro, Arg, PA and GABA (and perhaps alkaloid production, where it occurs) pathways can be upregulated concurrently if Orn is presumed to play a regulatory role; this argument has recently been made by Majumdar *et al.* (2013). It is further argued that upregulation of any of the terminal products of these nodal pathways would increase the demand for Orn (by acting as a sink to draw Orn consumption), which would force the flux of Glu to Orn to increase. As Glu is critical for a multitude of metabolic pathways, this situation could also induce mechanisms to increase the production of Glu in plants, perhaps directly from N assimilation. Under normal growth conditions in plants, the C needed for increased Glu production will of course come from the products of photosynthesis.

Therefore, a question that can be raised is: 'Could Orn act as a controlling metabolite to enhance total N uptake through which total C sequestration could also be positively affected'? This would enhance the total biomass of the vegetative organs in the plant, providing a potential way to increase bioenergy production. Increased Glu → Orn flux would also enhance the production of the critical amino acids involved in protein biosynthesis, in plant stress response and, most importantly, in affecting the oxidative state of the cells, suggesting pleiotropic effects of such single-step manipulation on overall cellular metabolism. While increased biomass production (i.e. faster growing plants) through genetic alteration of the major nitrogenous compounds (Pro, Arg, Glu and PAs) could play a positive role in plants for the production of renewable energy, increased use in phytoremediation, reduction in atmospheric CO_2 and perhaps also in reducing N pollution; the cautionary adverse aspect of this approach could be potential changes in the balance of N-rich nutrients for food and feed crops. Our recent analyses of metabolomic changes in WT and the high Put poplar cells have indicated that the genetic manipulation of a single step (i.e. Orn consumption via overexpression of ODC) causes major changes not only in the nitrogenous metabolites but also in numerous other metabolites, including those associated with the TCA cycle (Page *et al.*, unpublished). This is presumably due to the metabolic connection of Put to the various C metabolites via the succinate-GABA shunt (reviewed by Shelp *et al.*, 2012). In the next generation of genetically engineered plants aimed at modifying their nutrient values, this would require a careful analysis of the entire metabolome of the final product, i.e. the transgenic plant. The key questions as to how Orn would control its own biosynthesis without accumulation (flux rate), and what mechanism is involved in Orn sensing and the signal transduction pathway still pose a future challenge for the scientific community.

9.10 Conclusions

Despite the fact that Orn is not involved in protein synthesis, is often present in relatively small quantities and has no direct and specific role in any known physiological function in plants, it plays a critical role in overall N metabolism that directly affects several critical functions in plant growth, development and stress responses. Due to its presence at the crossroads of three important metabolic branches, namely the biosynthesis of Arg, Pro and the PAS (and also alkaloids in some plants), the flux of Orn biosynthesis (from Glu) must be stringently regulated to meet its demands without it accumulating in large quantities. The published literature so far indicates that under the conditions of changing demands for Orn utilization into the various metabolic sub-pathways, most of this regulation occurs via post-translational events for regulation of enzyme activity rather than the level of transcription of the various genes that code for these enzymes. Furthermore, increased demand of Orn that leads to increased utilization of Glu appears to result in replenishing Glu availability to the other competing pathways via its increased biosynthesis. As increased Glu biosynthesis involves additional N as well as C, one can postulate that genetic manipulation of an increased flux of Glu into Pro, Arg and/or PAs could increase the assimilation of both N and C from the external environment. Overall, the increased biosynthesis of Pro, Arg, and PAs (and perhaps of GABA, which is also produced directly from Glu via GAD and from the catabolism of Put as well) would not only increase the stress tolerance of plants (which is a well-known phenotype associated with high Pro, Arg, PAs and GABA) but also enhance the assimilation of N and C in the plants. Unfortunately, the mechanism by which the flux of Glu into Orn is regulated is not currently understood.

Acknowledgements

Partial funding was provided by the New Hampshire Agricultural Experiment Station. This is Scientific Contribution Number 2542. This work was also supported by the US Department of Agriculture Forest Service, Northern Research Station.

References

Adio, A.M., Casteel, C.L., De Vos, M., Kim, J.H., Joshi, V., Li, B., Juéry, C., Daron, J., Kliebenstein, D.J. and Jander, G. (2011) Biosynthesis and defensive function of N^{δ}-acetylornithine, a jasmonate-induced *Arabidopsis* metabolite. *The Plant Cell* 23, 3303–3318.

Alhonen, L., Rasanen, T.L., Sinervirta, R., Parkkinen, J.J., Korhonen, V.P., Pietila, M. and Janne, J. (2002) Polyamines are required for the initiation of rat liver regeneration. *Biochemical Journal* 362, 149–153.

Bai, M., Yang, Z., Qi, X., Fan, D. and Han, G. (2013) L-ornithine-L-aspartate for hepatic encephalopathy in patients with cirrhosis: a meta-analysis of randomized controlled trials. *Journal of Gastroenterology and Hepatology* 28, 783–792.

Bauer, G.A., Bazzaz, F.A., Minocha, R., Long, S., Magill, A., Aber, J. and Berntson, G.M. (2004) Effects of chronic N additions on tissue chemistry, photosynthetic capacity and carbon sequestration potential of a red pine (*Pinus resinosa* Ait.) stand in the NE United States. *Forest Ecology and Management* 196, 173–186.

Beez, S., Fokina, O., Herrmann, C. and Forchhammer, K. (2009) *N*-acetyl-L-glutamate kinase (NAGK) from oxygenic phototrophs: P(II) signal transduction across domains of life reveals novel insights in NAGK control. *Journal of Molecular Biology* 389, 748–758.

Bell, E.A. (2003) Nonprotein amino acids of plants: significance in medicine, nutrition and agriculture. *Journal of Agricultural and Food Chemistry* 51, 2854–2865.

Bhatnagar, P., Glasheen, B.M., Bains, S.K., Long, S.L., Minocha, R., Walter, C. and Minocha, S.C. (2001) Transgenic manipulation of the metabolism of polyamines in poplar cells. *Plant Physiology* 125, 2139–2153.

Bhatnagar, P., Minocha, R. and Minocha, S.C. (2002) Genetic manipulation of the metabolism of polyamines in poplar cells: the regulation of putrescine catabolism. *Plant Physiology* 128, 1455–1469.

Brauc, S., De Vooght, E., Claeys, M., Geuns, J.M., Hofte, M. and Angenon, G. (2012) Overexpression of arginase in *Arabidopsis thaliana* influences defence responses against *Botrytis cinerea*. *Plant Biology* 1, 39–45.

Burillo, S., Luque, I., Fuentes, I. and Contreras, A. (2004) Interactions between the nitrogen signal transduction protein P_{II} and *N*-acetyl glutamate kinase in organisms that perform oxygenic photosynthesis. *Journal of Bacteriology* 186, 3346–3354.

Chellamuthu, V.R., Alva, V. and Forchhammer, K. (2013) From cyanobacteria to plants: conservation of P_{II} functions during plastid evolution. *Planta* 237, 451–462.

Chen, H.P., Lin, C.F., Lee, Y.J., Tsay, S.S. and Wu, S.H. (2000) Purification and properties of ornithine racemase from *Clostridium sticklandii*. *Journal of Bacteriology* 182, 2052–2054.

Crawford, N.M. (2006) Mechanisms for nitric oxide synthesis in plants. *Journal of Experimental Botany* 57, 471–478.

Cunin, R., Glansdorff, N., Piérard, A. and Stalon, V. (1986) Biosynthesis and metabolism of arginine in bacteria. *Microbiological Reviews* 50, 314–352.

Cyr, D.R. and Bewley, J.D. (1989) Carbon and nitrogen reserves of leafy spurge (*Euphorbia esula*) roots as related to overwintering strategy. *Physiologia Plantarum* 77, 67–72.

da Rocha, I.M., Vitorello, V.A., Silva, J.S., Ferreira-Silva, S.L., Viegas, R.A., Silva, E.N. and Silveira, J.A. (2012) Exogenous ornithine is an effective precursor and the δ-ornithine amino transferase pathway contributes to proline accumulation under high N recycling in salt-stressed cashew leaves. *Journal of Plant Physiology* 169, 41–49.

Davis, R.H. (1986) Compartmental and regulatory mechanisms in the arginine pathways of *Neurospora crassa* and *Saccharomyces cerevisiae*. *Microbiological Reviews* 50, 280–313.

DeBoer, K.D., Dalton, H.L., Edward, F.J. and Hamill, J.D. (2011) RNAi-mediated down-regulation of ornithine decarboxylase (ODC) leads to reduced nicotine and increased anatabine levels in transgenic *Nicotiana tabacum* L. *Phytochemistry* 72, 344–355.

DeBoer, K.D., Dalton, H.L., Edward, F.J., Ryan, S.M. and Hamill, J.D. (2013) RNAi-mediated down-regulation of ornithine decarboxylase (ODC) impedes wound-stress stimulation of anabasine synthesis in *Nicotiana glauca*. *Phytochemistry* 86, 21–28.

Docimo, T., Reichelt, M., Schneider, B., Kai, M., Kunert, G., Gershenzon, J. and D'Auria, J.C. (2012) The first step in the biosynthesis of cocaine in *Erythroxylum coca*: the characterization of arginine and ornithine decarboxylases. *Plant Molecular Biology* 78, 599–615.

Durzan, D.J. (2010) Arginine and the shade tolerance of white spruce saplings entering winter dormancy. *Journal of Forest Science* 56, 77–83.

Ferrario-Méry, S., Bouvet, M., Leleu, O., Savino, G., Hodges, M. and Meyer, C. (2005) Physiological characterisation of *Arabidopsis* mutants affected in the expression of the putative regulatory protein P_{II}. *Planta* 223, 28–39.

Ferrario-Méry, S., Besin, E., Pichon, O., Meyer, C. and Hodges, M. (2006) The regulatory P_{II} protein controls arginine biosynthesis in *Arabidopsis*. *FEBS Letters* 580, 2015–2020.

Ficicioglu, C. and Yudkoff, M. (2013) Urea Cycle: Disease Aspects. In: Lennarz, W.J. and Lane, M.D. (eds) *Encyclopedia of Biological Chemistry*. Academic Press, Waltham, Massachusetts, pp. 494–501.

Flores, T., Todd, C.D., Tovar-Mendez, A., Dhanoa, P.K., Correa-Aragunde, N., Hoyos, M.E., Brownfield, D.M., Mullen, R.T., Lamattina, L. and Polacco, J.C. (2008) Arginase-negative mutants of *Arabidopsis* exhibit increased nitric oxide signaling in root development. *Plant Physiology* 147, 1936–1946.

Frémont, N., Riefler, M., Stolz, A. and Schmülling, T. (2013) The *Arabidopsis TUMOR PRONE5* gene encodes an acetylornithine aminotransferase required for arginine biosynthesis and root meristem maintenance in blue light. *Plant Physiology* 161, 1127–1140.

Ghahremani, M., Ghanati, F., Bernard, F., Gholami, M. and Azad, T. (2013) Effects of exogenous ornithine enantiomers on tobacco cells under salinity conditions. *Progress in Biological Sciences* 3, 100–107.

Gholami, M., Boughton, B.A., Fakhari, A.R., Ghanati, F., Mirzaei, H.H., Borojeni, L.Y., Zhang, Y., Breitbach, Z.S., Armstrong, D.W. and Roessner, U. (2014) Metabolomic study reveals a selective accumulation of L-arginine in the D-ornithine treated tobacco cell suspension culture. *Process Biochemistry* 49, 140–147.

Grün, S., Lindermayr, C. and Durner, J. (2006) Nitric oxide and gene regulation in plants. *Journal of Experimental Botany* 57, 507–516.

Han, X., Kazarinoff, M.N., Seiler, N. and Stanley, B.A. (2001) Rat colon ornithine and arginine metabolism: coordinated effects after proliferative stimuli. *American Journal of Physiology – Gastrointestinal and Liver Physiology* 280, G389–G399.

Hanfrey, C., Sommer, S., Mayer, M.J., Burtin, D. and Michael, A.J. (2001) *Arabidopsis* polyamine biosynthesis: absence of ornithine decarboxylase and the mechanism of arginine decarboxylase activity. *The Plant Journal* 27, 551–560.

Hanzawa, Y., Takahashi, T., Michael, A.J., Burtin, D., Long, D., Pineiro, M., Coupland, G. and Komeda, Y. (2000) *ACAULIS*5, an *Arabidopsis* gene required for stem elongation, encodes a spermine synthase. *EMBO Journal* 19, 4248–4256.

Hickok, N.J., Seppanen, P.J., Kontula, K.K., Janne, P.A., Bardin, C.W. and Janne, O.A. (1986) Two ornithine decarboxylase mRNA species in mouse kidney arise from size heterogeneity at their 3′ termini. *Proceedings of the National Academy of Sciences of the United States of America* 83, 594–598.

Hu, W.W., Gong, H. and Pua, E.C. (2005) The pivotal roles of the plant *S*-adenosylmethionine decarboxylase 5′ untranslated leader sequence in regulation of gene expression at the transcriptional and posttranscriptional levels. *Plant Physiology* 138, 276–286.

Huergo, L.F., Chandra, G. and Merrick, M. (2013) P_{II} signal transduction proteins: nitrogen regulation and beyond. *FEMS Microbiology Reviews* 37, 251–283.

Imai, A., Akiyama, T., Kato, T., Sato, S., Tabata, S., Yamamoto, K.T. and Takahashi, T. (2004a) Spermine is not essential for survival of *Arabidopsis*. *FEBS Letters* 556, 148–152.

Imai, A., Matsuyama, T., Hanzawa, Y., Akiyama, T., Tamaoki, M., Saji, H., Shirano, Y., Kato, T., Hayashi, H., Shibata, D., Tabata, S., Komeda, Y. and Takahashi, T. (2004b) Spermidine synthase genes are essential for survival of *Arabidopsis*. *Plant Physiology* 135, 1565–1573.

Jogaiah, S., Oulkur, D.P., Banerjee, K., Sharma, J., Patil, A.G., Maske, S.R. and Somkuwar, R.G. (2013) Biochemically induced variations during some phenological stages in Thompson seedless grapevines grafted on different rootstocks. *South African Journal of Enology and Viticulture* 34, 36–45.

Kakehi, J., Kuwashiro, Y., Niitsu, M. and Takahashi, T. (2008) Thermospermine is required for stem elongation in *Arabidopsis thaliana*. *Plant and Cell Physiology* 49, 1342–1349.

Kalamaki, M.S., Alexandrou, D., Lazari, D., Merkouropoulos, G., Fotopoulos, V., Pateraki, I., Aggelis, A., Carrillo-López, A., Rubio-Cabetas, M.J. and Kanellis, A.K. (2009a) Over-expression of a tomato *N*-acetyl-L-glutamate synthase gene (*SlNAGS1*) in *Arabidopsis thaliana* results in high ornithine levels and increased tolerance in salt and drought stresses. *Journal of Experimental Botany* 60, 1859–1871.

Kalamaki, M.S., Merkouropoulos, G. and Kanellis, A.K. (2009b) Can ornithine accumulation modulate abiotic stress tolerance in *Arabidopsis*? *Plant Signaling and Behavior* 4, 1099–1101.

Kochanowski, K., Volkmer, B., Gerosa, L., Haverkorn van Rijsewijk, B.R., Schmidt, A. and Heinemann, M. (2013) Functioning of a metabolic flux sensor in *Escherichia coli*. *Proceedings of the National Academy of Sciences of the United States of America* 110, 1130–1135.

Krebs, H.A. (1973) The discovery of the ornithine cycle of urea synthesis. *Biochemical Education* 1, 19–23.

Leonard, J.V. and Morris, A.A. (2002) Urea cycle disorders. *Seminars in Neonatology* 7, 27–35.

Lombardo, M.C., Graziano, M., Polacco, J.C. and Lamattina, L. (2006) Nitric oxide functions as a positive regulator of root hair development. *Plant Signaling Behaviour* 1, 28–33.

Magill, A.H., Aber, J.D., Currie, W.S., Nadelhoffer, K.J., Martin, M.E., Mcdowell, W.H., Melillo, J.M. and Steudler, P. (2004) Ecosystem response to 15 years of chronic nitrogen additions at the Harvard Forest LTER, Massachusetts, USA. *Forest Ecology and Management* 196, 7–28.

Majumdar, R. (2011) Polyamine metabolism in *Arabidopsis*: transgenic manipulation and gene expression. PhD thesis, University of New Hampshire, Durham, New Hampshire.

Majumdar, R., Shao, L., Minocha, R., Long, S. and Minocha, S.C. (2013) Ornithine: the overlooked molecule in regulation of polyamine metabolism. *Plant and Cell Physiology* 54, 990–1004.

Mattioli, R., Costantino, P. and Trovato, M. (2009) Proline accumulation in plants: not only stress. *Plant Signaling Behavior* 4, 1016–1018.

Minocha, R., Long, S., Magill, A., Aber, J. and Mcdowell, W. (2000) Foliar free polyamine and inorganic ion content in relation to soil and soil solution chemistry in two fertilized forest stands at the Harvard Forest, Massachusetts. *Plant and Soil* 222, 119–137.

Minocha, R., Long, S., Thangavel, P., Minocha, S.C., Eagar, C. and Driscoll, C.T. (2010) Elevation dependent sensitivity of northern hardwoods to Ca addition at Hubbard Brook Experimental Forest, NH USA. *Forest Ecology and Management* 260, 2115–2125.

Minocha, R., Turlapati, S.A., Long, S. and North, M. (2013) Fuel treatment effects on soil chemistry and foliar physiology of three coniferous species at the Teakettle Experimental Forest, California, USA. *Trees – Structure and Function* 27, 1101–1113.

Mohapatra, S., Cherry, S., Minocha, R., Majumdar, R., Thangavel, P., Long, S. and Minocha, S.C. (2010a) The response of high and low polyamine producing cell lines to aluminum and calcium stress. *Plant Physiology and Biochemistry* 48, 612–620.

Mohapatra, S., Minocha, R., Long, S. and Minocha, S.C. (2010b) Transgenic manipulation of a single polyamine in poplar cells affects the accumulation of all amino acids. *Amino Acids* 38, 1117–1129.

Mollá-Morales, A., Sarmiento-Mañús, R., Robles, P., Quesada, V., Pérez-Pérez, J.M., González-Bayón, R., Hannah, M.A., Willmitzer, L., Ponce, M.R. and Micol, J.L. (2011) Analysis of *ven3* and *ven6* reticulate mutants reveals the importance of arginine biosynthesis in *Arabidopsis* leaf development. *The Plant Journal* 65, 335–345.

Morris, S.M.J. (2006) Arginine: beyond protein. *American Journal of Clinical Nutrition* 83, 508S–512S.

Morris, S.M.J. (2007) Arginine metabolism: boundaries of our knowledge. *Journal of Nutrition* 137, 1602S–1609S.

Mur, L.A., Mandon, J., Persijn, S., Cristescu, S.M., Moshkov, I.E., Novikova, G.V., Hall, M.A., Harren, F.J., Hebelstrup, K.H. and Gupta, K.J. (2013) Nitric oxide in plants: an assessment of the current state of knowledge. *AoB Plants* 5:pls052.

Page, A.F., Mohapatra, S., Minocha, R. and Minocha, S.C. (2007) The effects of genetic manipulation of putrescine biosynthesis on transcription and activities of the other polyamine biosynthetic enzymes. *Physiologia Plantarum* 129, 707–724.

Page, A., Minocha, R. and Minocha, S.C. (2012) Living with high putrescine: expression of ornithine and arginine biosynthetic pathway genes in high and low putrescine producing poplar cells. *Amino Acids* 42, 295–308.

Pegg, A.E. and McCann, P.P. (1982) Polyamine metabolism and function. *American Journal of Physiology – Cell Physiology* 243, C212–C221.

Polacco, J.C. and Holland, M.A. (1993) Roles of urease in plant cells. In: Jeon, K.W. and Jarvik, J. (eds) *International Review of Cytology*. Academic Press, San Diego, California, pp. 65–103.

Quesada, V., Ponce, M.R. and Micol, J.L. (1999) *OTC* and *AUL1*, two convergent and overlapping genes in the nuclear genome of *Arabidopsis thaliana*. *FEBS Letters* 461, 101–106.

Roosens, N., Bitar, F., Loenders, K., Angenon, G. and Jacobs, M. (2002) Overexpression of ornithine-δ-aminotransferase increases proline biosynthesis and confers osmotolerance in transgenic plants. *Molecular Breeding* 9, 73–80.

Shao, L., Bhatnagar, P., Majumdar, R., Minocha, R. and Minocha, S.C. (2014) Putrescine overproduction does not affect the catabolism of spermidine and spermine in poplar and *Arabidopsis*. *Amino Acids* 46, 743–757.

Sharma, S., Shinde, S. and Verslues, P.E. (2013) Functional characterization of an ornithine cyclodeaminase-like protein of *Arabidopsis thaliana*. *BMC Plant Biology* 13:182.

Shelp, B.J., Bozzo, G.G., Trobacher, C.P., Zarei, A., Deyman, K.L. and Brikis, C.J. (2012) Hypothesis/review: contribution of putrescine to γ-aminobutyrate (GABA) production in response to abiotic stress. *Plant Science* 194, 130–135.

Shi, H., Ye, T., Chen, F., Cheng, Z., Wang, Y., Yang, P., Zhang, Y. and Chan, Z. (2013) Manipulation of arginase expression modulates abiotic stress tolerance in *Arabidopsis*: effect on arginine metabolism and ROS accumulation. *Journal of Experimental Botany* 64, 1367–1379.

Sikorska, H., Cianciara, J. and Wiercinska-Drapalo, A. (2010) Physiological functions of L-ornithine and L-aspartate in the body and the efficacy of administration of L-ornithine- L-aspartate in conditions of relative deficiency. *Polski Merkuriusz Lekarski* 28, 490–495.

Slocum, R.D. (2005) Genes, enzymes and regulation of arginine biosynthesis in plants. *Plant Physiology and Biochemistry* 43, 729–745.

Sugino, T., Shirai, T., Kajimoto, Y. and Kajimoto, O. (2008) L-ornithine supplementation attenuates physical fatigue in healthy volunteers by modulating lipid and amino acid metabolism. *Nutrition Research* 28, 738–743.

Sugiyama, K., Hayakawa, T., Kudo, T., Ito, T. and Yamaya, T. (2004) Interaction of *N*-acetylglutamate kinase with a P_{II}-like protein in rice. *Plant and Cell Physiology* 45, 1768–1778.

Teixeira, D., Santaolaria, M.L., Meneu, V. and Alonso, E. (2002) Dietary arginine slightly and variably affects tissue polyamine levels in male Swiss albino mice. *Journal of Nutrition* 132, 3715–3720.

Tomino, I., Katunuma, N., Morris, H.P. and Weber, G. (1974) Imbalance in ornithine metabolism in hepatomas of different growth rates as expressed in behavior of L-ornithine:2-oxoacid aminotransferase (ornithine transaminase, EC 2.6113). *Cancer Research* 34, 627–636.

Trovato, M., Maras, B., Linhares, F. and Costantino, P. (2001) The plant oncogene *rolD* encodes a functional ornithine cyclodeaminase. *Proceedings of the National Academy of Sciences of the United States of America* 98, 13449–13453.

Tujioka, K., Yamada, T., Aoki, M., Morishita, K., Hayase, K. and Yokogoshi, H. (2012) Dietary ornithine affects the tissue protein synthesis rate in young rats. *Journal of Nutritional Science and Vitaminology* 58, 297–302.

Urano, K., Yoshiba, Y., Nanjo, T., Ito, T., Yamaguchi-Shinozaki, K. and Shinozaki, K. (2004) *Arabidopsis* stress-inducible gene for arginine decarboxylase *AtADC2* is required for accumulation of putrescine in salt tolerance. *Biochemical and Biophysical Research Communications* 313, 369–375.

Urano, K., Hobo, T. and Shinozaki, K. (2005) *Arabidopsis ADC* genes involved in polyamine biosynthesis are essential for seed development. *FEBS Letters* 579, 1557–1564.

Urschel, K.L., Shoveller, A.K., Uwiera, R.R., Pencharz, P.B. and Ball, R.O. (2006) Citrulline is an effective arginine precursor in enterally fed neonatal piglets. *Journal of Nutrition* 136, 1806–1813.

Verslues, P.E. and Sharma, S. (2010) Proline metabolism and its implications for plant–environment interaction. *The Arabidopsis Book* 8:e0140.

Vissers, Y.L., Dejong, C.H., Luiking, Y.C., Fearon, K.C., von Meyenfeldt, M.F. and Deutz, N.E. (2005) Plasma arginine concentrations are reduced in cancer patients: evidence for arginine deficiency? *American Journal of Clinical Nutrition* 81, 1142–1146.

Vissers, Y.L., von Meyenfeldt, M.F., Argiles, J.M., Luiking, Y.C., Dejong, C.H. and Deutz, N.E. (2007a) Protein breakdown on whole-body and organ level in non-cachectic tumour-bearing mice undergoing surgery. *Clinical Nutrition* 26, 483–490.

Vissers, Y.L., von Meyenfeldt, M.F., Luiking, Y.C., Dejong, C.H., Buurman, W.A. and Deutz, N.E. (2007b) Presence of tumour inhibits the normal post-operative response in arginine and NO production in non-cachectic mice. *Clinical Science* 112, 527–532.

Wagner, I. and Musso, H. (1983) New naturally occurring amino acids. *Angewandte Chemie (International Edition in English)* 22, 816–828.

Wargo, P.M., Minocha, R., Wong, B.L., Long, R.P., Horsley, S.B. and Hall, T.J. (2002) Measuring changes in stress and vitality indicators in limed sugar maple on the Allegheny Plateau in north-central Pennsylvania. *Canadian Journal of Forest Research* 32, 629–641.

Watson, M.B., Emory, K.K., Piatak, R.M. and Malmberg, R.L. (1998) Arginine decarboxylase (polyamine synthesis) mutants of *Arabidopsis thaliana* exhibit altered root growth. *The Plant Journal* 13, 231–239.

Weber, G., Queener, S.F. and Morris, H.P. (1972) Imbalance in ornithine metabolism in hepatomas of different growth rates as expressed in behavior of L-ornithine carbamyl transferase activity. *Cancer Research* 32, 1933–1940.

Yan, Y., Stolz, S., Chetelat, A., Reymond, P., Pagni, M., Dubugnon, L. and Farmer, E.E. (2007) A downstream mediator in the growth repression limb of the jasmonate pathway. *The Plant Cell* 19, 2470–2483.

Yang, C.-W., Lin, C.-C. and Kao, C.H. (1999) Endogenous ornithine and arginine contents and dark-induced proline accumulation in detached rice leaves. *Journal of Plant Physiology* 155, 665–668.

Yonaha, K., Misono, H., Yamamoto, T. and Soda, K. (1975) D-amino acid aminotransferase of *Bacillus sphaericus*. Enzymologic and spectrometric properties. *The Journal of Biological Chemistry* 250, 6983–6989.

You, J., Hu, H. and Xiong, L. (2012) An ornithine δ-aminotransferase gene *OsOAT* confers drought and oxidative stress tolerance in rice. *Plant Science* 197, 59–69.

Zieve, L. (1986) Conditional deficiencies of ornithine or arginine. *Journal of the American College of Nutrition* 5, 167–176.

Zimmermann, P., Hirsch-Hoffmann, M., Hennig, L. and Gruissem, W. (2004) GENEVESTIGATOR. *Arabidopsis* microarray database and analysis toolbox. *Plant Physiology* 136, 2621–2632.

10 Polyamines in Plants: Biosynthesis From Arginine, and Metabolic, Physiological and Stress-Response Roles

A.K. Mattoo,* T. Fatima, R.K. Upadhyay and A.K. Handa
Beltsville Agricultural Research Center, USDA Agricultural Research Service, Beltsville, Maryland, USA

10.1 Abstract

Biogenic amines affect a myriad of growth and developmental processes in all organisms, including plants. Therefore, there is continued interest in understanding the biosynthesis and functional roles of polyamines in regulating plant metabolism, physiology and development. The role of polyamines in plant responses to environmental stresses has also elicited considerable interest in recent years. Four major biogenic amines and their derivatives commonly found in plants are putrescine (Put), spermidine (Spd), spermine (Spm) and thermospermine (T-Spm). The biosynthesis of these polyamines derives from arginine (Arg) and/or ornithine (Orn). This chapter's focus is on the Arg pathway. Genes encoding enzymes in the various steps of polyamine biosynthesis from Arg have been identified and characterized from a number of plants. The enzymes catalysing different steps in Arg pathway are: Arg decarboxylase, agmatine iminohydrolase/agmatine deiminase, *N*-carbamoylputrescine amidase/*N*-carbamoyl Put amido hydrolase, *S*-adenosylmethionine decarboxylase, Spd synthase, Spm synthase and T-Spm synthase. Many of these have been purified and studied using heterologous expression systems. Advanced genetics tools and mutants have been used, including production of loss-of-function and gain-of-function plants. Thus far, the studies have provided unequivocal evidence of polyamine role in diverse aspects of plant life. Characterizations of isogenic mutants varying in their endogenous levels of different polyamines have revealed that polyamines are central to important signalling and metabolic pathways, and affect grain filling, senescence, nodulation, root development and fruit ripening, as well as having a role in the abiotic stress responses of plants.

10.2 Introduction

Polyamines are a class of biogenic amines that are ubiquitous and are now recognized as having important roles in a myriad of biological processes, including cellular growth and cell division, embryogenesis, organ development, proliferation, memory, apoptosis/senescence and response to abiotic and biotic stressors (Slocum and Flores, 1991; Cohen, 1998; Cassol and Mattoo, 2003; Kaur-Sawhney *et al.*, 2003). In plants, the main polyamines are the diamine putrescine (Put), the triamine spermidine (Spd) and the tetraamine spermine/thermospermine (Spm/T-Spm). Flowering plants also synthesize cadaverine, 1,3-diaminopropane, and other modified forms (Cohen, 1998), while the Spm isomer, T-Spm, is widespread in bacteria and higher plants (Knott *et al.*, 2007). Spm was the

*autar.mattoo@ars.usda.gov

first polyamine discovered; it was found in 1678 by Anton Leeuwenhoek upon microscopic examination of human semen. Two centuries later, Put was discovered; and finally Spd was discovered in 1927 (see Cohen, 1998, for timeline of discovery). The polycationic nature of these biogenic amines catalyses their binding to macromolecules such as nucleic acids (DNA, RNA), proteins and other cellular molecules, such as phospholipids, causing conformational changes that, in turn, may affect one or more biological function (Tabor and Tabor, 1984; Cohen, 1998). Observations of varying concentrations of the different polyamines during plant growth, development and senescence, pharmacological evidence for their effects on the physiology and *in vitro* biochemistry, and correlative evidence for their involvement in cellular response to different stressors, have classified polyamines as plant growth regulators (Galston and Sawhney, 1990; Cohen, 1998). Compelling evidence gathered through studies with loss-of-function and gain-of-function transgenic plants, and with polyamine mutants, including *Arabidopsis*, tomato, pear, wheat, maize and rice, demonstrates that polyamines are important players in plant metabolism and responses to environmental stresses, including drought, salinity and extreme temperatures (Capell *et al.*, 2004; Pang *et al.*, 2007; Mattoo and Handa, 2008; Handa and Mattoo, 2010; Takahashi and Kakehi, 2010; Shukla and Mattoo, 2013).

Levels of Spd and Spm, but not of Put, are associated with higher kernel set and seed development in maize (Sen *et al.*, 1981; Lin *et al.*, 1984; Feng *et al.*, 2011) and grain filling in rice (Yang *et al.*, 2008; Wang *et al.*, 2012). These studies indicated antagonistic cellular interactions between polyamines (Spd and Spm) and ethylene in regulating grain filling. Interestingly, studies with transgenic tomato fruits and poplar cell cultures have shown that the accumulation of several cellular metabolites positively correlates with Spd and Spm levels, while having a negative correlation with endogenous levels of Put (Mattoo *et al.*, 2010). Similarly, transcriptome analyses favoured the conclusion that Spd and Spm act as anabolic regulators, whereas Put acts as a catabolic regulator in tomato fruit (Handa and Mattoo, 2010). Further advancement was made by the observation that a mutation in a gene (*ACL5*) that encodes for T-Spm synthase impaired stem elongation in *Arabidopsis*, which was rescued by the exogenous application of T-Spm (Kakehi *et al.*, 2008). A role for T-Spm in plant development was revealed in its prevention of premature death in xylem differentiation (Clay and Nelson, 2005; Muñiz *et al.*, 2008), which involves the *SAC51* gene that encodes a bHLH-type (basic-helix-loop-helix) transcription factor (Takahashi and Kakehi, 2010). A greater interest in understanding polyamine action has been intensified by the recent developments.

10.3 Substrates and Enzymes Catalysing Polyamine Biosynthesis

10.3.1 The route to the diamine putrescine

As in animals/mammals, amino acid arginine (Arg) is the precursor of the diamine Put, as is ornithine (Orn). In mammals, it was presumed that Orn pathway involving Orn decarboxylase (ODC) was the only biosynthetic route for the synthesis of Put until it was demonstrated that Put could be synthesized from Arg via Arg decarboxylase (ADC; EC 4.1.1.19) (Mistry *et al.*, 2002; Zhu *et al.*, 2004). In plants, both the Orn and Arg pathways are prevalent, except in the model plant *Arabidopsis*, in whose genome the ODC gene was found to be missing. The Orn pathway and its implications in plants are discussed elsewhere in this book (Chapter 9). The current chapter focuses on the Arg pathway, which is diagrammatically presented in Fig. 10.1.

ADC, the first enzyme involved in this pathway is a pyridoxal-phosphate dependent protein, which catalyses conversion of Arg to agmatine. Agmatine is next hydrolysed by agmatine deiminase (also called agmatine iminohydrolase, EC 3.5.3.12; AIH) to *N*-carbamoylputrescine and ammonia. Put formation occurs via removal of the ureido group from *N*-carbamoylputrescine by the enzyme *N*-carbamoylputrescine amidohydrolase (NLP1; EC 3.5.1.53; also called CPA) (Fig. 10.1).

Oat ADC encodes a 66 kDa pre-protein that is enzymatically inactive and is subsequently processed to a mature 24-kDa, enzymatically active ADC (Bell and Malmberg, 1990; Malmberg and Cellino, 1994). The expression of oat ADC in tobacco plants led to an enzymatically active

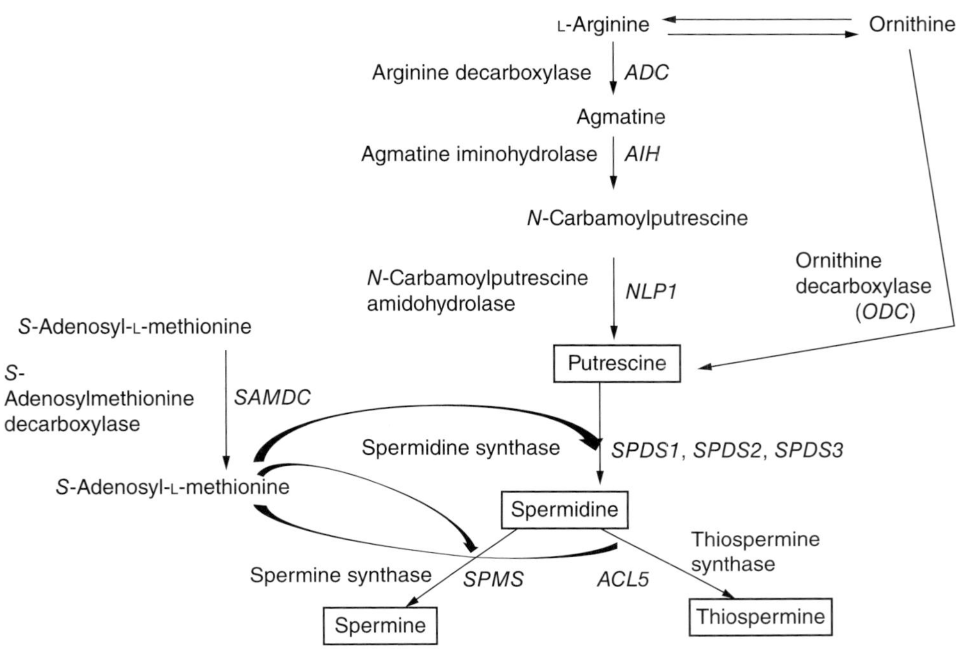

Fig. 10.1. The biosynthesis of polyamines from arginine, indicating the intermediates, enzymes and genes that are involved in the pathway. The genes are presented in italic upper case and follow the (roman) acronyms for the enzymes.

enzyme and consequently to several-fold accumulation of the ADC product agmatine (Burtin and Michael, 1997). The requirement for proteolytic cleavage to produce an active enzyme is a also characteristic of other amino acid decarboxylases, so ADC shares this characteristic with these enzymes; however, oat ADC processing differs from them in not being auto-catalytic and requiring a distinct processing enzyme (Malmberg and Cellino, 1994). In tomato, ADC is encoded by a single gene that encodes a polypeptide of 502 amino acids with a predicted molecular mass of ~55 kDa and 47% sequence similarity with oat ADC (Rastogi *et al.*, 1993).

Plant genomes differ in the number of ADC genes that they harbour; in some, ADC is a single gene member (oat) and in *Arabidopsis* there are two (ADC1, ADC2), but in *Brassica juncea*, in addition to ADC1 and ADC2, a third ADC3 member has been found (Galloway *et al.*, 1998; Mo and Pua, 2002). The duplication of an ancestral ADC gene has led to two ADC paralogues in the *Brassicaceae* family, which are 80% homologous. In *Arabidopsis*, the two gene products probably localize to different subcellular locations (Hanfrey *et al.*, 2001), and the *Arabidopsis* ADC1 and ADC2 gene transcripts differentially accumulate during development and in response to stress in the subantarctic brassicaceous plant *Pringlea antiscorbutica* (Hummel *et al.*, 2004).

Conversion of agmatine to *N*-carbamoylputrescine is catalysed by agmatine iminohydrolase (AIH). AIH activity was reported from groundnut seeds and was found to increase in cotyledons (Sindhu and Desai, 1980). Soybean AIH is a monomer of 70 kDa and prone to inhibition by sulfhydryl (-SH) agents such as *p*-hydroxymercuribenzoate (Park and Cho, 1991). In cereal crops, native AIH is a dimer of 75 kDa made up of two monomers of 43.5 kDa and 44 kDa (Yanagisawa, 2001), while the cucumber enzyme has a mass of 67 kDa and is made of two monomers of 47 kDa and 36 kDa (Sakakibara and Yanagisawa, 2003). AIH purified to homogeneity from maize and cucumber was found to be devoid of Put synthase or Orn and Put transcarbamylase activities, suggesting that AIH is a distinct enzyme. The cloning and characterization of plant AIH showed the recombinant *Arabidopsis* AIH to be a dimer, as had been the case with purified maize enzyme, and converted agmatine to *N*- carbamoylputrescine and ammonia (Janowitz *et al.*, 2003). These studies confirmed the presence of active cysteine residues.

The immediate precursor of Put biosynthesis is *N*-carbamoylputrescine. The conversion of the latter to Put is catalysed by NLP1 (Fig. 10.1), which belongs to a family of C-N hydrolases. This activity was identified from maize shoots (Yanagisawa and Suzuki, 1982) before the identification of plant AIH. The native maize NLPi was found to have a molecular mass of 125 kDa, was inhibited by -SH reagents and had a high affinity for its substrate, *N*-carbamoylputrescine (Yanagisawa and Suzuki, 1982). The recombinant *Arabidopsis* NLP1 protein produced in *Escherichia coli* was found to be active *in vitro*, and confirmed to be highly specific for the substrate *N*-carbamoylputrescine (Piotrowski *et al.*, 2003), like its maize counterpart. NLIP1 was found to be an enzyme of 279 kDa made up of eight monomers, homologous to β-alanine synthase and, like the latter enzyme, shows positive cooperative binding with the substrate (Piotrowski *et al.*, 2003). NLP1 is an important enzyme in urea metabolism and the metabolism of amino groups.

10.3.2 The route to higher polyamines, spermidine and spermine/thermospermine

The biosynthesis of higher polyamines, Spd and Spm/T-Spm, depends on the activity of the pyruvoyl-cofactor containing enzyme *S*-adenosylmethionine (SAM) decarboxylase (SAMDC; EC 4.1.1.50), which is the rate limiting step in polyamine pathway (Yamanoha and Cohen, 1985; Mehta *et al.*, 2002; Capell *et al.*, 2004) and catalyses the synthesis of decarboxylated SAM (dcSAM) and S-methyl-5′-thioadenosine (MTA) from SAM. MTA is recycled back to L-methionine via the methionine cycle. Aminopropyltransferases catalyse the synthesis of Spd and Spm. Spd biosynthesis involves an addition of a propylamine moiety from dcSAM to Put catalysed by aminopropyltransferase Spd synthase (SPDS; EC 2.5.1.16) (Fig. 10.1). Finally, another aminopropyl group, from dcSAM, is incorporated by SPMS into Spd to form Spm, while T-Spm synthase (ACL5) catalyses the synthesis of T-Spm from Spd (Fig. 10.1).

10.3.3 *S*-Adenosylmethionine decarboxylase

SAMDC activity was purified and characterized from Chinese cabbage and found to be different from animal and yeast SAMDC in that the two latter sources require Put (Yamanoha and Cohen, 1985). In terms of the molecular mass, the cabbage SAMDC, with a MW of 35 kDa, very clearly differed from those from *E. coli* (108 kDa), yeast (88 kDa) and rat liver (68 kDa). Later, soybean SAMDC was purified to homogeneity, found to have a MW of 66 kDa and, like the cabbage enzyme, did not require Put (Yang and Cho, 1991); these authors showed that agmatine inhibited soybean SAMDC as well as ADC activities, which led to inhibition of dcSAM and Put levels, respectively. Further insight into plant SAMDCs became apparent after the *Catharanthus* SAMDC gene was cloned and expressed in *E. coli* (Schroder and Schroder, 1995). The largest open reading frame (ORF) of the cloned SAMDC encoded a protein made of 357 amino acids, with a size of approx. 39.7 kDa, and was 74% identical with the potato homologue but less so with the mammalian, yeast and *E. coli* enzymes. The recombinant SAMDC gene produced a 32 kDa protein in *E. coli* following processing from a larger 41 kDa pro-enzyme. Schroder and Schroder (1995) indicated that the processing of plant SAMDC occurred at the same Ser residue as the animal and yeast enzymes. Interestingly, these studies also reported the presence of a conserved upstream ORF (uORF) in the *Cantharanthus* SAMDC in the 5′ RNA leader, and authors hypothesized that this might represent a plant-specific development for the modulation of SAMDC transcript translation.

The uORF in the plant SAMDC gene (Schroder and Schroder, 1995; Franceschetti *et al.*, 2001) was studied using transgenic tobacco plants (Hanfrey *et al.*, 2002). Elimination of the small uORF from SAMDC led to increased relative translational efficiency of the SAM pro-enzyme in transgenic plants. The increased SAMDC activity caused a 400-fold increase in dcSAM levels, accompanied by a disruption of Put and Spm levels (Hanfrey *et al.*, 2002). However, SAM levels were not much affected. Hanfrey *et al.* (2002) suggested that uORF-mediated control of SAMDC translation favours polyamine homeostasis and is important for plant growth and development.

Later, work with mustard SAMDC revealed a gene family encoding this enzyme in which the genes were similarly expressed in response to external stimuli (Hu *et al.*, 2006). This study reported that 5′ leader sequence is responsible for transcriptional regulation of SAMDC expression. Further, the introns present in the 5′ leader sequence upregulated SAMDC activity in situations when endogenous Spd levels drops low.

10.3.4 Spermidine synthase

SPDS purified from Chinese cabbage was found to be 81 kDa in molecular mass (Yamanoha and Cohen, 1985). Soybean SPDS was purified to homogeneity and found to have a molecular mass of 74 kDa (Yoon *et al.*, 2000), close to that of the cabbage enzyme. In the latter study, Put was found to be a preferred substrate and the enzyme activity was significantly stimulated by metals such as cobalt, copper and calcium. Yoon *et al.* (2000) suggested that these metal ions could be the cellular regulators in polyamine biosynthesis. More interesting information on this enzyme was generated after the SPDS gene (*SPDS*) was cloned from three plant species, *Nicotiana sylvestris*, *Hyoscyamus niger* and *Arabidopsis*, and investigations revealed SPDS to be distinct from SPMS (Hashimoto *et al.*, 1998). Two different genes were identified in *H. niger*, *SPSDS1* and *SPDS2*. This study found the molecular masses of the expressed genes to be close to one another: in *Nicotiana*, 34.5 kDa; in *Hyoscyamus*, 34.6 kDa (SPDS1) and 33.9 kDa (SPDS2) (Hashimoto *et al.*, 1998). Interestingly, the amino acid sequence of SPDS was found to be similar to that of tobacco Put methyltransferase (PMT), and the authors proposed that PMT evolved from plant SPDS.

An *Arabidopsis* gene family with three members of SPDS, *SPDS1*, *SPDS2* and *SPDS3*, was characterized (Hanzawa *et al.*, 2002), out of which *SPDS1* and *SPDS2* but not *SPDS3* had SPDS activity and complemented SPDS-deficient yeast mutants. The two closely related *Arabidopsis* SPDS proteins, SPDS1 and SPDS2, were found to exist in a complex (metabolon) with SPMS (Panicot *et al.*, 2002). Using the yeast two-hybrid system, as well as *in vitro* studies, SPDS2 was shown to interact with SPDS1 and SPMS. SPMS complemented the Spm deficiency in yeast, indicating that SPMS is a novel spermine synthase. *In vitro*, the proteins did not homodimerize but were able to form heterodimers. The epitope-labelled SPDS and SPMS protein complexes co-purified with a size range of 650 to 750 kDa (Panicot *et al.*, 2002). In maize also, two SPDS genes were identified, but the *in silico* analysis suggested that the second gene, *ZmSPDS2*, may, in fact, encode for SPMS (Rodríguez-Kessler and Jiménez-Bremont, 2008). The *Arabidopsis SPDS3* gene expressed in heterologous expression systems potentially had both SPDS and SPMS activities (Hanzawa *et al.*, 2002); these authors considered the possibility that the *in vivo* activity may be different because an invariant phenylalanine (Phe) residue in SPDS may contribute to forming a Put-binding cavity (Korolev *et al.*, 2002), which is substituted by a valine residue (Val236) in SPDS3 (Hanzawa *et al.*, 2002).

SPDS was also cloned from the leguminous plant *Lotus japonicus* and its validity shown by complementing the yeast Spd mutant (Efrose *et al.*, 2008). Physiologically, *LjSPDS* transcript levels and polyamine content were found to change systemically in the roots, stems and leaves of nodulated plants, but not in control, uninoculated plants. These studies implicated polyamines in nodule development and symbiotic nitrogen fixation (SNF), roles that need to be further investigated.

10.3.5 Spermine/thermospermine synthases

Initial thoughts that SPMS and SPDS activities are synonymous in plants had to be revised once the two enzyme activities were found to be distinct, with each encoded by a different gene. A similar discrepancy was apparent in distinguishing SPMS, which catalyses the synthesis of Spm from Spd, from the other distinct enzyme ACL5, which catalyses the synthesis of T-Spm (Hanzawa *et al.*, 2000; Imai *et al.*, 2004b; Clay and Nelson, 2005; Yamaguchi *et al.*, 2007). T-Spm is an abundant polyamine in the bacterium *Thermus thermophilus*, while Spm is generally found at very low levels in plants compared with Spd. SPMS is encoded by *SPMS*, and *Arabidopsis* plants with a mutation in this gene were found to grow normally, so Imai *et al.* (2004a) concluded that Spm is not essential for the growth of plants.

However, the very same plants were unable to survive under drought conditions, implicating Spm to have a positive role in plant survival under some abiotic stresses (Yamaguchi *et al.*, 2007).

The *SPMS* cloned from *L. japonicus* was characterized and found to complement an Spm mutation in yeast. *LjSPMS* transcript levels, like those of *LjSPDS*, increased systemically in nodulated plants (in roots, stems and leaves), implicating Spm in nodule development and SNF (Efrose *et al.*, 2008).

The *ACL5* gene, which was initially thought to encode SPMS was later found to actually encode for ACL5, T-Spm synthase (Kakehi *et al.*, 2008; Pegg and Michael, 2009). ACL5 targets the N^1 atom of the aminopropyl moiety of Spd to catalyse the synthesis of T-Spm. T-Spm is required for stem elongation in *Arabidopsis* (Kakehi *et al.*, 2008; Pegg and Michael, 2009).

Table 10.1 lists the enzymes and corresponding *Arabidopsis* genes involved in polyamine biosynthesis.

10.4 Substrate Flux into the Polyamine Versus Ethylene Pathway

Flux distributions allow systematic analysis of the behaviour and regulation of plant metabolic networks. Delineation of the metabolic pathways and study of the substrate fluxes, particularly of metabolites that feed into more than one end product, help to establish cellular metabolic networks. The amino acids Arg and methionine (Met) provide the substrate skeleton for higher polyamines, Arg produces Put while Met provides the essential C-1 donor SAM that is converted by the SAMDC enzyme to dcSAM to catalyse Spd and Spm biosynthesis (Fig. 10.1). SAM carbons are also diverted in plants to the plant hormone ethylene, as well as into the methionine cycle (Fluhr and Mattoo, 1996). In the ethylene pathway, 1-aminocyclopropane-1-carboxylate (ACC) synthase catalyses the synthesis of ACC from SAM, and ACC is, in turn, converted to ethylene

Table 10.1. List of enzymes and corresponding *Arabidopsis* genes involved in polyamine biosynthesis.

Gene symbol	Enzyme description	Gene ID	Sequence length (no. amino acids)	MW polypeptide/ multimer	Reference
ACL5	Thermospermine synthase	AT5G19530	339		Hanzawa *et al.*, 2000
ADC1/SPE1	Arg decarboxylase 1	AT2G16500	702		Urano *et al.*, 2004
ADC2/SPE2	Arg decarboxylase 2	AT4G34710	711		Soyka and Heyer, 1999
AIH	Agmatine iminohydrolase/ Agmatine deiminase	AT5G08170	383	2 Subunits 43 kDa/81 kDa	Janowitz *et al.*, 2003
NLP1/CPA	*N*-Carbamoylputrescine amidase/ *N*-carbamoyl put amido hydrolase	AT2G27450	326/299	8 Subunits/ 279 kDa multimer	Piotrowski *et al.*, 2003
ODC1	Ornithine decarboxylase	XP_003523230 Soybean	434/431		Alabadi and Carbonell, 1998
ODC2		NP_001234616 Tomato			
SAMDC	*S*-Adenosylmethionine decarboxylase	AT3G02470	366	Hetrotetramer 2 Subunits 80 kDa polypeptide	Park and Cho, 1999
SPDS	Spermidine synthase	AT1G23820	386		Korolev *et al.*, 2002
SPMS	Spermine synthase	AT5G53120	359		Hanzawa *et al.*, 2002

by ACC oxidase (Yang and Hoffman, 1984; Fluhr and Mattoo, 1996; Harpaz-Saad *et al.*, 2012).

Ethylene is a ripening and senescence promoter, while PAs are promoters of growth and act as anti-senescence regulators. Because of their generally antagonistic functions in plants, the flux of SAM is expected to be under homeostatic control (Harpaz-Saad *et al.*, 2012). Interestingly, interconnection between ethylene and polyamines in plant processes also arose in studies on rice submergence and potato suberization. Acireductone dioxygenase, which catalyses the reaction of acireductone with dioxygen, to produce 2-keto-4-methylthiobutyrate, the immediate precursor of Met, seems to be an important regulatory step (Sauter *et al.*, 2005; Bürstenbinder *et al.*, 2007, 2010; Rzewuski *et al.*, 2007; Kim *et al.*, 2008).

The availability of 'gain-of-function' transgenic tomato fruit lines expressing yeast *SAMDC*, which accumulate high steady state levels of Spd and Spm concomitant to a decrease in Put (Mehta *et al.*, 2002), has been tested as a system to measure the flux of several substrates into the ethylene and polyamine pathways during fruit ripening (Lasanajak *et al.*, 2013). The question on how plants manage and coordinate SAM flux into ethylene or polyamine pathway in plants is important. Using radiolabelled ^{14}C-Arg, ^{14}C-Orn, l-[U-^{14}C]Met, ^{14}C-SAM and ^{14}C-Put to quantify flux through the two pathways, Lassanajak *et al.* (2013) validated the synthesis of polyamines in tomato fruits from both Arg and Orn, although the Orn pathway declined as fruit ripening progressed. Further, the expression of heterologous SAMDC was found to reverse the ripening-associated decrease in Spd and Spm levels observed in normal control fruit, causing twofold to threefold higher accumulation of labelled Spd in transgenic fruit; also, the decline in ^{14}C-Met incorporation into Spd/Spm during ripening of the control fruit was reversed in the SAMDC fruits. Surprisingly, ^{14}C-Met incorporation into ethylene or ACC during ripening was not altered in the SAMDC fruit. This study showed for the first time that the cellular flux of SAM in plants is homeostatically regulated based on the demand for it in competing pathways. In mammals, studies of glycine *N*-methyltransferase (GNMT)-deficient humans and GNMT knocked-out mice have reported that GNMT drives homeostatic regulation of cellular SAM levels (Luka *et al.*, 2009). It is not known whether plant cells invoke a similar mechanism for regulating SAM concentration.

10.5 Back Conversion of Polyamines and Reactive Oxygen Species Signalling

Polyamines are oxidatively deaminated by the action of polyamine oxidases. Diamine oxidases (DAOs) which are copper-containing amine oxidases (CuAOs), oxidize the diamines in Put and cadaverine at the primary amino groups, thus producing 4-aminobutanal with the release of ammonia and hydrogen peroxide. Polyamine oxidases (PAOs) are flavin-containing oxidases involved in the terminal catabolism of Spd and Spm, producing 4-aminobutanal or *N*-(3-aminopropyl)- 4-aminobutanal,1,3-diaminopropane and H_2O_2 (Cona *et al.*, 2006; Moschou *et al.*, 2008a,b). Another group of PAOs is involved in the back conversion of Spm to Spd, with 3-aminopropanal and H_2O_2 as by products (Moschou *et al.*, 2008c).

The *Arabidopsis* genome contains five known PAO genes, out of which the enzyme AtPAO1 showed activities very similar to those of animal Spm oxidase and is involved in the back conversion of Spm to Spd (Alcázar *et al.*, 2006, 2010b). *In silico* analysis of five AtPAOs identified AtPAO3 as a non-typical PAO compared with other oxidases (Moschou *et al.*, 2008c). AtPAO3 was overexpressed in *E. coli* and found to catalyse a sequential conversion of Spm to Spd and of Spd to Put, which showed it to be similar to the animal counterpart. PAO3 is localized in peroxisomes and the C-terminus of the protein contains the sorting/localization information (Moschou *et al.*, 2008c).

Seven PAOs have been reported from *Oryza sativa* (Ono *et al.*, 2012). Detailed work on OsPAO3, OsPAO4 and OSPAO5 revealed that only OsPAO3 showed full back polyamine conversion activity, with its preference for Spd, followed by T-Spm and Spm, while OsPAO4 and OsPAO5 preferred Spm and T-Spm instead of Spd. OsPAO3, OsPAO4 and OsPAO5 are localized in the peroxisomes, like AtPAO3 (Ono *et al.*, 2012).

H_2O_2 is an integral component of reactive oxygen species (ROS) signalling in plants, which system also includes other ROS. Polyamine oxidation generates H_2O_2 in the apoplast and this contributes towards the defence response against pathogens in plants (Cona *et al.*, 2006). It is also known that some PAOs participate in the hypersensitive response (HR) of tobacco plants against the infection of *Tobacco mosaic virus* (TMV). Polyamine-mediated H_2O_2 formation may also be associated with non-host induced HR responses in tobacco, indicating a role or PAs in both host and non-host induced HR responses in plants (Yoda *et al.*, 2003, 2006, 2009).

10.6 Polyamines have an Impact on Metabolism

Physiological studies have revealed that higher levels of Spd and Spm, but not of Put, are associated with better kernel set and seed development in maize (Sen *et al.*, 1981; Lin *et al.*, 1984; Feng *et al.*, 2011) and grain filling in rice (Yang *et al.*, 2008). In a study in which six rice cultivars differing in grain filling rate were field grown and analysed for polyamine and ethylene levels, a higher ratio of free polyamines (Spd and Spm) to ethylene in spikelets was associated with superior grain filling (Wang *et al.*, 2012). Earlier, it had been shown that a higher ratio of ethylene to polyamines correlated with abortion in maize (Cheng and Lur, 1996) and reduced grain weight in wheat (Beltrano *et al.*, 1994) and rice (Yang *et al.*, 2008).

Metabolomic analyses using NMR profiling to study changes in metabolism in plants overexpressing the higher polyamines Spd and Spm revealed higher accumulation of glutamate (Glu), asparagine (Asn), citrate, fumarate, malate and an unidentified compound A, and decreased accumulation of Val, aspartate (Asp), sucrose and glucose in tomato fruits with higher cellular levels of Spd/Spm (Mattoo *et al.*, 2006). Interestingly, the pathways involved in the accumulation of several of these metabolites indicated a positive correlation with Spd and Spm levels and a negative correlation with Put levels when studies on transgenic tomato fruits and poplar cells were compared and analysed (Mattoo *et al.*, 2010). Along with these changes, the N sensing/signalling and C metabolism indicators were preferentially activated in transgenic fruits with elevated content of Spd/Spm. Mattoo *et al.* (2006) suggested that Spd and Spm as nitrogenous metabolites are perceived as such and that the cells then stimulate C sequestration in fruit cells. Similar results were obtained in a recent study by Neily *et al.* (2011). A graphic representation of some of the effects positively regulated by higher polyamines is presented in Fig. 10.2.

When compared with diamine Put (as a negative regulator) the anabolic role of higher polyamines in plant metabolism was in congruence with the results of transcriptome analyses (Handa and Mattoo, 2010). Other studies showing a role for the polyamine pathway in plant cell processes relates to the target of rapamycin (TOR) signalling effect on growth and lifespan enhancement of *Arabidopsis* plants expressing yeast FKBP12 protein, which is a target of rapamycin

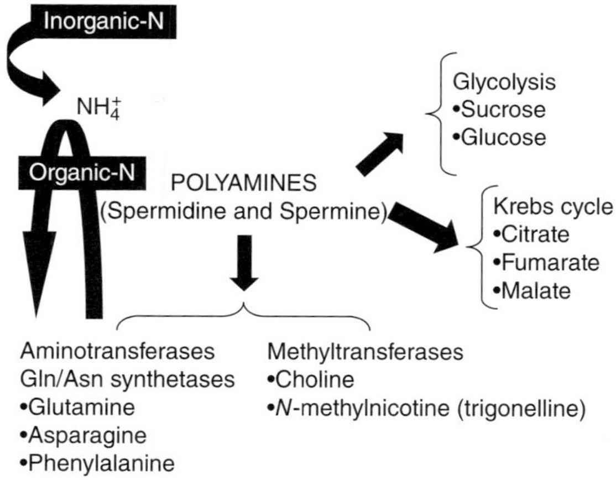

Fig. 10.2. Polyamines and metabolic regulation. Asn, asparagine; Gln, glutamine.

(Ren *et al.*, 2012). Global metabolite and transcript analyses of these plants included features that highlighted the polyamine biosynthetic pathway as well as the SAMDC transcripts (Ren *et al.*, 2012). Interestingly, a redirection of C and N metabolism was also apparent in TOR-inhibited plants. It is reasonable to postulate that a connection exists between TOR and polyamines because polyamines have been found to be associated with anabolic processes and C:N signalling in tomato (Mattoo *et al.*, 2006; Handa and Mattoo, 2010).

10.7 Polyamines and Plant Growth Processes

Polyamines use at least two mechanisms to regulate cellular processes. The first mode of action involves the property of the cationic amine and imine groups present in polyamines that can bind electrostatically to micro- and macro-molecules in the cell. Generally, such binding occurs with the anionic phosphate groups present in DNA, RNA and phosphoproteins, or with the carboxyl groups present in acidic proteins (Lindemose *et al.*, 2005). Such electrostatic interactions can cause conformational changes, altering the activity or function of the bound molecule or membrane. For instance, such a binding was found to stabilize the double helix structure in DNA by creating a polarized hydrogen bond (Lindemose *et al.*, 2005). Likewise, covalent bonds between polyamines and proteins can effect post-translational modification of proteins (Serafini-Fracassini and Del-Duca, 2008; Serafini-Fracasini *et al.*, 2009). The second mode of action attributed to polyamines could invoke plant signal transduction pathways such as the protein phosphorylation cascade by which downstream metabolic processes are regulated (Pandey *et al.*, 2000). It is still too premature to rule out either of these possible modes of action.

Polyamine levels are high during plant growth processes such as cell division, but decline during cell expansion and elongation (reviewed by Kaur-Sawhney, 2003, and references therein). This is also the case in apical shoots and meristems before flowering in many plants (Cabanne *et al.*, 1981; Martin-Tanguy, 1985, 2001). Floral buds in callus cultures from tobacco inflorescence pedicels contain more Spd than vegetative buds (Kaur-Sawhney *et al.*, 1988), as is the case in *Arabidopsis* flowers (Applewhite *et al.*, 2000). In the latter study, exogenously applied Spd accelerated flowering in a delayed-flowering mutant of *Arabidopsis*. Pea root growth and rice coleoptile elongation correlated with high Put concentrations (Lee *et al.*, 1997). Interestingly, the expression of *rol* (hairy root) genes affects polyamine metabolism in root cultures, promoting growth and ramification of roots, which was correlated with the depletion of Put and Put conjugates (Ben-Hayyim *et al.*, 1994, 1996; Couée *et al.*, 2004). In Virginia pine, polyamines promoted root elongation and growth by increasing root cell division (Tang and Newton, 2005). Root development at low temperatures in *Pringlea antiscorbutica* is also correlated with polyamine levels (Hummel *et al.*, 2002), while polyamines suppressed nodule formation in soybean (Terakado *et al.*, 2006).

Fruit maturation is also correlated with higher polyamine content but, during ripening, their levels, especially of Spd and Spm, precipitously decrease in some fruit species, including tomato, avocado and the grape (Mehta *et al.*, 2002; Cassol and Mattoo, 2003; Nambeesan *et al.*, 2010; Agudelo-Romero *et al.*, 2013). Transgenic tomato plants expressing a heterologous SPDS gene in a constitutive manner had higher levels of Spd but not Spm in the leaf and ripening fruits (Nambeesan *et al.*, 2010). The consequences of this event include higher vegetative growth, less shrivelling and less postharvest decay in the transgenic fruits compared with the control fruits. In the grapevine (*Vitis* spp.), ADC is upregulated during ripening and Put and Spd levels were higher at earlier stages of fruit development, but Spm level remained unchanged (Agudelo-Romero *et al.*, 2013).

The anti-senescence role of polyamines is well documented in the literature (Mattoo and Handa, 2004). Cut carnation flower senescence is inhibited by Spm (Lee *et al.*, 1997), and the same phenomenon was shown for rose species (Sood and Nagar, 2003); Spm also prevented leaf senescence in *Lactuca sativa* (Serafini-Fracasini *et al.*, 2010) and tobacco corolla senescence (Serafini-Fracasini *et al.*, 2002). The Spm-mediated prevention of *Lactuca* leaf senescence was related to the preservation of chloroplast photosystems. Spm and Spd also have an impact on differing chlorophyll levels in chloroplasts (reviewed in Sobieszczuk-Nowicka and Legocka 2013, and references therein).

10.8 Polyamines in Plant Responses to Abiotic Stress

Polyamines have been found to be associated with plant responses to abiotic stress for over six decades, first when elevated Put was seen in barley plants upon potassium starvation (Richards and Coleman, 1952), which was later attributed to the decarboxylation of Arg, thus implicating ADC (Smith and Richards, 1964). Since then, many reports have appeared on the association of the ADC pathway with abiotic stress in plants (Flores and Galston, 1982; Galston, 1983). There is now good documentation that polyamines accumulate in response to several unfavourable environmental conditions, such as drought, salinity, chilling, heat, hypoxia, ozone, UV radiation, heavy metals, mechanical wounding and herbicide treatments (Bouchereau *et al.*, 1999; Alcázar *et al.*, 2006; Groppa and Benavides, 2008; Alcázar *et al.*, 2010b; Gill and Tuteja 2010; Bitrián *et al.*, 2012). Transgenic approaches employed for studies with *Arabidopsis*, rice, tobacco and tomato have led to the thinking that polyamines play a protective role in plant stress (Capell *et al.*, 2004; Alcázar *et al.*, 2006; Kusano *et al.*, 2008; Gill and Tuteja, 2010; Gupta *et al.*, 2013); these are summarized in Table 10.2.

Much progress has been made in defining stress response characteristics by developing polyamine-deficient mutants (Watson *et al.*, 1998; Kaur-Sawhney *et al.*, 2003; Urano *et al.*, 2004). *Arabidopsis spe-1* and *spe-2* mutants, which are deficient in polyamine accumulation, exhibit reduced salt tolerance (Watson *et al.*, 1998; Kasinathan and Wingler, 2004), while an *ADC2*-knockout mutant (*adc2-1*) was more sensitive to salt stress, which was reversed by exogenous Put (Urano *et al.*, 2004). An *Arabidopsis spms* mutant defective in Spm production exhibited a sensitive phenotype under drought and salinity stresses but recovered with the exogenous application of Spm (Yamaguchi *et al.*, 2006, 2007). These results demonstrate the use of mutants in defining specific role(s) of polyamines in stress tolerance.

The *Arabidopsis ADC1* and *ADC2* genes are differentially expressed during various stress conditions: *ADC1* is upregulated by cold, while *ADC2* is upregulated in response to various abiotic stresses (Perez-Amador *et al.*, 2002; Urano *et al.*, 2003; Armengaud *et al.*, 2004; Hummel *et al.*, 2004; Alcázar *et al.*, 2006). *SPMS*, *SPDS1* and *SPDS2* transcripts are upregulated during dehydration and under high salinity (Urano *et al.*, 2003; Alcázar *et al.*, 2006). *SAMDC1* is induced by cold while *SAMDC2* is additionally induced by salt (Vergnolle *et al.*, 2005). In transgenic plants, overexpression of *ADC* and *ODC* was found to impart stress tolerance (Capell *et al.*, 1998, 2004; Roy and Wu, 2001; Kumría and Rajam, 2002; Altabella *et al.*, 2009; Alcázar *et al.*, 2010a), as was the case when the *SAMDC* gene from *Tritordeum*, carnation, yeast or *Arabidopsis*, as well as the human gene, was overexpressed in tobacco, rice or tomato (Roy and Wu, 2002; Waie and Rajam, 2003; Wi *et al.*, 2006; Alcázar *et al.*, 2006; Cheng *et al.*, 2009). Similarly, overexpression of the SPDS gene from various sources imparts abiotic stress tolerance (Kasukabe *et al.*, 2004, 2006; He *et al.*, 2008; Wen *et al.*, 2008), as well as metal tolerance (Wen *et al.*, 2009, 2010) in transgenic plants (Table 10.2).

The above description of the association of polyamines with plant tolerance to stress conditions is now supported by studies showing a crosstalk between the polyamine biosynthesis pathway and the stress hormone abscisic acid (ABA) (Alcázar *et al.*, 2010b; Gupta *et al.*, 2012a,b). It was shown that Spd and ABA mediate phosphorylation of a 42 kDa protein in rice, while *in vitro* phosphorylation of OsBZ8, a bZIP type of transcription factor, was mediated by Spd (Gupta *et al.*, 2012a, b). Furthermore, *Arabidopsis* salt-hypersensitive (salt overlay sensitive, or SOS) and polyamine biosynthesis mutants accumulate free Spd and Spm. Similarly, T-Spm was found to provide salt tolerance to *Arabidopsis*. These results indicate that polyamine accumulation is an SOS independent response to salinity (Alet *et al.*, 2012).

10.9 Conclusions

Various facets of the biosynthetic pathway of polyamines from Arg, including the substrate flux studies, indicate a dominance of this pathway

Table 10.2. Catalogue of abiotic stress tolerance in transgenic plants expressing polyamine biosynthesis pathway genes.

Gene name[a]	Gene source	Transfer to	Inducible/constitutive	Overproduction[b]	Tolerance to	Reference
ADC	Oat	Rice	Inducible	Put	Salt	Roy and Wu, 2001
ADC	*Datura stramonium*	Rice	Inducible	Spd and Spm	Drought	Capell *et al*., 2004
ADC1	*Arabidopsis*	*Arabidopsis*	Constitutive	Put	Freezing	Altabella *et al*., 2009
ADC2	*Arabidopsis*	*Arabidopsis*	Constitutive	Put	Drought	Alcázar *et al*., 2010a
ADC	*Arabidopsis*	Aubergine	Constitutive	Put, Spd and Spm	Multiple	Prabhavathi and Rajam, 2007
ODC	Mouse	Tobacco	Constitutive	Put	Salt	Kumría and Rajam, 2002
SAMDC	*Tritordeum*	Rice	Inducible	Spd and Spm	Salt	Roy and Wu, 2002
SAMDC	Human	Tobacco	Constitutive	Put and Spd	Salt, osmotic pressure	Waie and Rajam, 2003
SAMDC	Carnation	Tobacco	Constitutive	Put, Spd and Spm	Multiple	Wi *et al*., 2006
SAMDC1	*Arabidopsis*	*Arabidopsis*	Constitutive	Spm	Multiple	Alcázar *et al*., 2006
SAMDC	Yeast	Tomato	Constitutive	Spd and Spm	Heat	Cheng *et al*., 2009
SPDS	*Cucurbita ficifolia*	*Arabidopsis*	Constitutive	Spd	Multiple	Kasukabe *et al*., 2004
SPDS	*Cucurbita ficifolia*	Sweet Potato	Constitutive	Spd	Drought and salt	Kasukabe *et al*., 2006
SPDS	Apple	Pear	Constitutive	Spd	Multiple	Wen *et al*., 2008
					Oxidative	He *et al*., 2008

[a]Abbreviations: *ADC*, arginine decarboxylase; *ODC*, ornithine decarboxylase; *SAMDC*, *S*-adenosylmethionine decarboxylase; *SPDS*, spermidine synthase.
[b]Put, putresine; Spd, spermidine; Spm, spermine.

in plants when rate-limiting enzymes are active. The studies that have been reviewed here show unequivocally that different polyamines have a wider influence on plant growth and development. It is apparent that many interesting questions arise about how polyamines achieve their effects, and what mechanisms are involved, both in general and, particularly, in plants. It is clear that endogenous levels of polyamines also change during organ development and in response to environmental extremes. For example, during tomato fruit development higher levels of polyamines accumulate during early development, specifically during the cell division phase, before declining at the fruit maturation phase encompassing cell expansion until the fruit achieves its maximal size (Cassol and Mattoo, 2003; Nambeesan *et al.*, 2010). The levels of Put, Spd and Spm vary considerably depending upon the fruit, but it seems that higher polyamines are quite minimal during the ripening stage. Further, each polyamine has been proposed to have a unique influence on the pattern of transcripts (Handa and Mattoo, 2010) and accumulation of metabolites (Mattoo *et al.*, 2010). The questions that arise include: what mechanism(s) is(are) involved that causes an increase or decrease in polyamine biosynthesis, and which signalling molecules are utilized? The involvement of polyamines in anabolic processes is supported by transcriptome (Handa and Mattoo, 2010) and metabolomic (Mattoo *et al.*, 2006) analyses. The elucidation of the fundamental mechanism(s) through which polyamines function is paramount to understanding what cues plants use to upregulate or downregulate the machinery that results in the accumulation of different polyamines in plants. Similarly, significant progress needs to be made in understanding the long-distance transport and cellular uptake of polyamines to further delineate their effects on plant processes. The emerging genomic technologies, especially transcriptome and proteome analyses, are expected to provide an insight into the molecular function of various polyamines.

References

Agudelo-Romero, P., Bortolloti, C., Pais, M.S., Tiburcio, F.A. and Fortes, M.A. (2013) Study of polyamines during grape ripening indicate an important role of polyamine catabolism. *Plant Physiology and Biochemistry* 67, 105–119.

Alabadi, D. and Carbonell, J. (1998) Expression of ornithine decarboxylase is transiently increased by pollination, 2,4-dichlorophenoxyacetic acid, and gibberellic acid in tomato ovaries. *Plant Physiology* 118, 323–328.

Alcázar, R., Marco, F., Cuevas, J.C., Patrón, M., Ferrando, A., Carrasco, P., Tiburcio, A.F. and Altabella, T. (2006) Involvement of polyamines in plant response to abiotic stress. *Biotechnology Letters* 28, 1867–1876.

Alcázar, R., Planas, J., Saxena, T., Zarza, X., Bortolotti, C., Cuevas, J., Bitrián, M., Tiburcio, A.F. and Altabella, T. (2010a) Putrescine accumulation confers drought tolerance in transgenic *Arabidopsis* plants overexpressing the homologous arginine decarboxylase 2 gene. *Plant Physiology and Biochemistry* 48, 547–552.

Alcázar, R., Altabella, T., Marco, F., Bortolotti, C., Reymond, M., Koncz, C., Carrasco, P. and Tiburcio, A. (2010b) Polyamines: molecules with regulatory functions in plant abiotic stress tolerance. *Planta* 231, 1237–1249.

Alet, A.I., Sánchez, D.H., Cuevas, J.C., Marina, M., Carrasco, P., Altabella, T., Tiburcio, A.F. and Ruiz, O.A. (2012) New insights into the role of spermine in *Arabidopsis thaliana* under long-term salt stress. *Plant Science* 182, 94–100.

Altabella, T., Tiburcio, A.F. and Ferrando, A. (2009) Plant with resistance to low temperature and method of production thereof. Spanish patent application WO2010/004070; US Patent Application No:2011/0126,322.

Applewhite, P.B., Kaur-Sawhney, R. and Galston, A.W. (2000) A role of spermidine in the bolting and flowering of *Arabidopsis*. *Physiologia Plantarum* 108, 314–320.

Armengaud, P., Breitling, R. and Amtmann, A. (2004) The potassium dependent transcriptome of *Arabidopsis* reveals a prominent role of jasmonic acid in nutrient signaling. *Plant Physiology* 136, 2556–2576.

Bell, E. and Malmberg, R.L. (1990) Analysis of a cDNA encoding arginine decarboxylase from oat reveals similarity to the *Escherichia coli* arginine decarboxylase and evidence of protein processing. *Molecular and General Genetics* 224, 431–436.

Beltrano, J., Carbone, A., Montaldim, E.R. and Guiamet, J.J. (1994) Ethylene as promoter of wheat grain maturation and ear senescence. *Plant Growth Regulation* 15, 107–112.

Ben-Hayyim, G., Damon, J.P., Martin-Tanguy, J. and Tepfer, D. (1994) Changing root system architecture through inhibition of putrescine and feruloyl putrescine accumulation. *FEBS Letters* 342, 145–148.

Ben-Hayyim, G., Martin-Tanguy, J. and Tepfer, D. (1996) Changing root and shoot architecture with the *rolA* gene from *Agrobacterium rhizogenes*: interactions with gibberellic acid and polyamine metabolism. *Physiologia Plantarum* 96, 237–243.

Bitrián, M., Zarza, X., Altabella, T., Tiburcio, A.F. and Alcázar, R. (2012) Polyamines under abiotic stress: metabolic crossroads and hormonal crosstalks in plants. *Metabolites* 2, 516–528.

Bouchereau, A., Aziz, A., Larher, F. and Martin-Tanguy, J. (1999) Polyamines and environmental challenges: recent development. *Plant Science* 140, 103–125.

Bürstenbinder, K., Rzewuski, G. and Wirtz, M. (2007) The role of methionine recycling for ethylene synthesis in *Arabidopsis*. *The Plant Journal* 49, 238–249.

Bürstenbinder, K., Waduwara, I., Schoor, S., Moffatt, B.A., Wirtz, M., Minocha, S.C., Oppermann, Y., Büouchereau, A., Hell, R. and Sauter, M. (2010) Inhibition of 5′-methylthioadenosine metabolism in the Yang cycle alters polyamine levels, and impairs seedling growth and reproduction in *Arabidopsis*. *The Plant Journal* 62, 977–988.

Burtin, D. and Michael, T. (1997) Over-expression of arginine decarboxylase in transgenic plants. *Biochemical Journal* 325, 331–337.

Cabanne, F., Dalebroux, M.A., Martin-Tanguy, J. and Martin, C. (1981) Hydroxycinnamic acid amides and ripening to flower of *Nicotiana tabacum* L. var. *xanthi* n.c. *Physiologia Plantarum* 53, 399–404.

Capell, T., Escobar, C., Liu, H., Burtin, D., Lepri, O. and Christou, P. (1998) Over-expression of oat arginine decarboxylase cDNA in transgenic rice (*Oryza sativa* L.) affects normal development pattern *in vitro* and results in putrescine accumulation in transgenic plants. *Theoretical and Applied Genetics* 97, 246–254.

Capell, T., Bassie, L. and Christou, P. (2004) Modulation of the polyamine biosynthetic pathway in transgenic rice confers tolerance to drought stress. *Proceedings of the National Academy of Sciences of the United States of America* 101, 990–991.

Cassol, T. and Mattoo, A.K. (2003) Do polyamines and ethylene interact to regulate plant growth, development and senescence? In: Nath, P., Mattoo, A.K., Ranade, S.A. and Weil, J.H. (eds) *Molecular Insight in Plant Biology* Science Publishers, Enfield, New Hampshire, pp. 121–132.

Cheng, C.-Y. and Lur, H.-S. (1996) Ethylene may be involved in abortion of the maize caryopsis. *Physiologia Plantarum* 98, 245–252.

Cheng, L., Zou, Y.J., Ding, S.L., Zhang, J.J., Yu, X.L., Cao, J.S. and Lu, G. (2009) Polyamine accumulation in transgenic tomato enhances the tolerance to high temperature stress. *Journal of Integrative Plant Biology* 51, 489–499.

Clay, N.K. and Nelson, T. (2005) *Arabidopsis* thick vein mutation affects vein thickness and organ vascularization, and resides in a provascular cell-specific spermine synthase involved in vein definition and in polar auxin transport. *Plant Physiology* 138, 767–777.

Cohen, S.S. (1998) *A Guide to the Polyamines*. Oxford University Press, New York, NY.

Cona, A., Rea, G., Angelini, R., Federico, R. and Tavladoraki, P (2006) Functions of amine oxidases in plant development and defence. *Trends Plant Science* 11, 80–88.

Couée, I., Hummel, I., Sulmon, C., Gouesbet, G. and Amrani, A.E. (2004) Involvement of polyamines in root development. *Plant Cell, Tissue and Organ Culture* 76, 1–10.

Efrose, R. C., Flemetakis, E., Sfichi, L., Stedel, C., Kouri, E.D., Udvardi, M.K., Kotzabasis, K. and Katinakis, P. (2008) Characterization of spermidine and spermine synthases in *Lotus japonicus*: induction and spatial organization of polyamine biosynthesis in nitrogen fixing nodules. *Planta* 228, 37–49.

Feng, H. Y., Wang, Z. M., Kong, F. N., Zhang, M. J. and Zhou, S. L. (2011) Roles of carbohydrate supply and ethylene, polyamines in maize kernel set. *Journal of Integrative Plant Biology* 53, 388–398.

Flores, H.E. and Galston, A.W. (1982) Polyamines and plant stress – activation of putrescine biosynthesis by osmotic shock. *Science* 217, 1259–1261.

Fluhr, R. and Mattoo, A.K. (1996) Ethylene – biosynthesis and perception. *Crit. Rev. Plant Sci.* 15, 479–523.

Franceschetti, M., Hanfrey, C., Scaramagli, S., Torrigiani, P., Bagni, N., Burtin, D. and Michael, A.J. (2001) Characterization of monocot and dicot plant *S*-adenosyl-L-methionine decarboxylase gene families including identification in the mRNA of a highly conserved pair of upstream overlapping open reading frames. *Biochemical Journal* 353, 403–409.

Galloway, G.L., Malmberg, R.L. and Price, R.A. (1998) Phylogenetic utility of the nuclear gene arginine decarboxylase: an example from *Brassicaceae*. *Molecular Biology and Evolution* 15, 1312–1320.

Galston, A.W. (1983) Polyamines as modulators of plant development. *BioScience* 33, 382–388.

Galston, A.W. and Sawhney, R.K. (1990) Polyamines in plant physiology. *Plant Physiology* 94, 406–410.

Gill, S.S. and Tuteja, N. (2010) Polyamines and abiotic stress tolerance in plants. *Plant Signaling and Behavior* 5, 26–33.

Groppa, M.D. and Benavides, M.P. (2008) Polyamines and abiotic stress: recent advances. *Amino Acids* 34, 35–45.

Gupta, B., Gupta, K. and Sengupta, D.N. (2012a) Spermidine-mediated *in vitro* phosphorylation of transcriptional regulator OSBZ8 by SNF1-type serine/threonine protein kinase SAPK4 homolog in indica rice. *Acta Physiologiae Plantarum* 34, 1321–1336.

Gupta, K., Gupta, B., Ghosh, B. and Sengupta, D.N. (2012b) Spermidine and abscisic acid-mediated phosphorylation of a cytoplasmic protein from rice root in response to salinity stress. *Acta Physiologiae Plantarum* 34, 29–40.

Gupta, K., Dey, A. and Gupta, B. (2013) Plant polyamines in abiotic stress responses. *Acta Physiologiae Plantarum* 35, 2015–2036.

Handa A.K. and Mattoo A.K. (2010) Differential and functional interactions emphasize the multiple roles of polyamines in plants. *Plant Physiology and Biochemistry* 48, 540–546.

Hanfrey, C., Sommer, S., Mayer, M.J., Burtin, D. and Michael, A.J. (2001) *Arabidopsis* polyamine biosynthesis: absence of ornithine decarboxylase and the mechanism of arginine decarboxylase activity. *The Plant Journal* 27, 551–560.

Hanfrey, C., Franceschetti, M., Mayer, M.J., Illingworth, C. and Michael, A.J. (2002) Abrogation of upstream open reading frame-mediated translational control of a plant *S*-adenosylmethionine decarboxylase results in polyamine disruption and growth perturbations. *The Journal of Biological Chemistry* 277, 44131–44139.

Hanzawa, Y., Takahashi, T., Michael, A.J., Burtin, D., Long, D., Pineiro, M., Coupland, G. and Komeda, Y. (2000) *ACAULIS5*, an *Arabidopsis* gene required for stem elongation, encodes a spermine synthase. *EMBO Journal* 19, 4248–4256.

Hanzawa, Y., Imai, A., Michael, A.J., Komeda, Y. and Takahashi, T. (2002) Characterization of the spermidine synthase-related gene family in *Arabidopsis thaliana*. *FEBS Letters* 527, 176–180.

Harpaz-Saad, S., Yoon, G.M., Mattoo, A.K. and Kieber, J.J. (2012) The formation of ACC and competition between polyamines and ethylene for SAM. *Annual Plant Reviews* 44, 53–81.

Hashimoto, T., Tamaki, K., Suzuki, K-I. and Yamada, Y. (1998) Molecular cloning of plant spermidine synthases. *Plant and Cell Physiology* 39, 73–79.

He, L., Ban, Y., Inoue, H., Matsuda, N., Liu, J. and Moriguchi, T. (2008) Enhancement of spermidine content and antioxidant capacity in transgenic pear shoots over expressing apple spermidine synthase in response to salinity and hyperosmosis. *Phytochemistry* 69, 2133–2141.

Hu, W.W., Gong H. and Pua E.C. (2006) Modulation of SAMDC expression in *Arabidopsis thaliana* alters *in vitro* shoot organogenesis. *Physiologia Plantarum* 128, 740–750.

Hummel, I., Couée, I., Amrani, E.A., Martin-Tanguy, J. and Hennion, F. (2002) involvement of polyamines in root development at low temperature in the cruciferous species *Pringlea antiscorbutica*. *Journal of Experimental Botany* 53, 1463–1473.

Hummel, I., Gouesbet, G., Amrani, A., Ainouche, A. and Couée, I. (2004) Characterization of the two arginine decarboxylase (polyamine biosynthesis) paralogues of the endemic subantarctic cruciferous species *Pringlea antiscorbutica* and analysis of their differential expression during development and response to environmental stress. *Gene* 342, 199–209.

Imai, A., Akiyama, T., Kato, T., Sato, S., Tabata, S., Yamamoto, K. T. and Takahashi, T. (2004a) Spermine is not essential for survival of *Arabidopsis*. *FEBS Letters* 556, 148–152.

Imai, A., Hanzawa, Y., Akiyama, T., Tamaoki, M., Saji, H., Shirano, Y., Kato, T., Hayashi, H., Shibata, D., Tabata, S., Komeda, Y. and Takahashi, T. (2004b) Spermidine synthase genes are essential for survival of *Arabidopsis*. *Plant Physiology* 135, 1565–1573.

Janowitz, T., Kneifel, H. and Piotrowski, M. (2003) Identification and characterization of plant agmatine iminohydrolase, the last missing link in polyamine biosynthesis of plants. *FEBS Letters* 544, 258–261.

Kakehi, J.I., Kuwashiro, Y., Niitsu, M. and Takahashi, T. (2008) Thermospermine is required for stem elongation in *Arabidopsis thaliana*. *Plant and Cell Physiology* 49, 1342–1349.

Kasinathan, V. and Wingler, A. (2004) Effect of reduced arginine decarboxylase activity on salt tolerance and on polyamine formation during salt stress in *Arabidopsis thaliana*. *Plant Physiology* 121, 101–107.

Kasukabe, Y., He, L., Nada, K., Misawa, S., Ihara, I. and Tachibana, S. (2004) Over expression of spermidine synthase enhances tolerance to multiple environmental stresses and up-regulates the expression of various stress-regulated genes in transgenic *Arabidopsis thaliana*. *Plant and Cell Physiology* 45, 712–722.

Kasukabe, Y., He, L., Watakabe, Y., Otani, M., Shimada, T. and Tachibana S. (2006) Improvement of environmental stress tolerance of sweet potato by introduction of genes for spermidine synthase. *Plant Biotechnology* 23, 75–83.

Kaur-Sawhney, R., Tiburcio, A.F. and Galston, A.W. (1988) Spermidine and flower bud differentiation in thin-layer explants of tobacco. *Planta* 173, 282–284.

Kaur-Sawhney, R., Tiburcio, F.A., Altabella T. and Galston, A.W. (2003) Polyamines in plants: an overview. *Journal of Cell and Molecular Biology* 2, 1–12.

Kim, J.H., Kim, H.S. and Lee, Y.H. (2008) Polyamine biosynthesis regulated by *StARD* expression plays an important role in potato wound periderm formation. *Plant and Cell Physiology* 49, 1627–1632.

Knott, J.M., Romer, P. and Sumper, M. (2007) Putative spermine synthases from *Thalassiosira pseudonana* and *Arabidopsis thaliana* synthesize thermospermine rather than spermine, *FEBS Letters* 581, 3081–3086.

Korolev, S., Ikeguchi, Y., Skarina, T., Beasley, S., Arrowsmith, C., Edwards, A., Joachimiak, A., Pegg, A.E. and Savchenko, A. (2002) The crystal structure of spermidine synthase with a multisubstrate adduct inhibitor. *Nature Structural Biology* 9, 27–31.

Kumría, R. and Rajam, M.V. (2002) Ornithine decarboxylase transgene in tobacco affects polyamines, *in vitro* morphogenesis and response to salt stress. *Journal of Plant Physiology* 159, 983–990.

Kusano, T., Berberich, T., Tateda, C. and Takahashi, Y. (2008) Polyamines: essentials factors for growth and survival. *Planta* 228, 367–381.

Lasanajak, Y., Minocha, R., Minocha, S.C., Goyal, R., Fatima, T., Handa, A.K. and Mattoo, A.K. (2013) Enhanced flux of substrates into polyamine biosynthesis but not ethylene in tomato fruit engineered with yeast *S*-adenosylmethionine decarboxylase gene. *Amino Acids* 46, 729–742.

Lee, M., Lee, S.H. and Park, K.Y. (1997) Effects of spermine on ethylene biosynthesis in cut carnation (*Dianthus caryophyllus* L.) flowers during senescence. *Journal of Plant Physiology* 151, 68–73.

Lin, P.P.C., Egli, D.B., Li, G.M. and Meckel, L. (1984) Polyamine titer in the embryonic axis and cotyledons of *Glycine max* (L.) during seed growth and maturation. *Plant Physiology* 103, 273–280.

Lindemose, S., Nielsen, P.E. and Mollegaard, N.E. (2005) Polyamines preferentially interact with bent adenine tracts in double-stranded DNA. *Nucleic Acids Research* 33, 1790–1803.

Luka, Z., Mudd, S.H. and Wagner, C. (2009) Glycine *N*-methyltransferase and regulation of *S*-adenosylmethionine levels. *The Journal of Biological Chemistry* 284, 22507–22511.

Malmberg, R.L. and Cellino, M.L. (1994) Arginine decarboxylase of oats is activated by enzymatic cleavage into two polypeptides. *The Journal of Biological Chemistry* 269, 2703–2706.

Martin-Tanguy, J. (1985) The occurrence and possible function of hydroxy-cinnamoyl acid amides in plants. *Plant Growth Regulation* 3, 383–399.

Martin-Tanguy, J. (2001) Metabolism and function of polyamines in plants: Recent development (new approaches). *Plant Growth Regulation* 34, 135–148.

Mattoo, A.K. and Handa, A.K. (2004) Ethylene signaling in plant cell death. In: Nooden, L. (ed.) *Plant Cell Death Processes*. Academic Press, London, pp. 125–142.

Mattoo, A.K. and Handa, A.K. (2008) Higher polyamines restore and enhance metabolic memory in ripening fruit. *Plant Science* 174, 386–393.

Mattoo, A.K., Sobolev, A.P., Neelam, A., Goyal, R.K., Handa, A.K. and Segre, A. L. (2006) Nuclear magnetic resonance spectroscopy-based metabolite profiling of transgenic tomato fruit engineered to accumulate spermidine and spermine reveals enhanced anabolic and nitrogen–carbon interactions. *Plant Physiology* 142, 1759–1770.

Mattoo, A.K., Minocha, S.C., Minocha, R., Handa, A.K. (2010) Polyamines and cellular metabolism in plants: transgenic approaches reveal different responses to diamine putrescine versus higher polyamines spermidine and spermine. *Amino Acids* 38, 405–413.

Mehta, R.A., Cassol, T., Li, N., Ali, N., Handa, A.K. and Mattoo, A.K. (2002) Engineered polyamine accumulation in tomato enhances phytonutrient content, juice quality, and vine life. *Nature Biotechnology* 20, 613–618.

Mistry, S.K., Burwell, T.J., Chambers, R.M., Rudolph-Owen. L., Spaltmann, F., Cook, W.J. and Morris, S.M. (2002) Cloning of human agmatinase. An alternate path for polyamine synthesis induced in liver by hepatitis B virus. *American Journal of Physiology: Gastrointestinal and Liver Physiology* 282, G375–G381.

Mo, H. and Pua, E.C. (2002) Up-regulation of arginine decarboxylase gene expression and accumulation of polyamines in mustard (*Brassica juncea*) in response to stress. *Physiologia Plantarum* 114, 439–449.

Moschou, P.N., Paschalidis, K.A., Delis, I.D., Andriopoulou, A.H., Lagiotis, G.D., Yakoumakis, D.I. and Roubelakis-Angelakis, K.A. (2008a) Spermidine exodus and oxidation in the apoplast induced by abiotic stress is responsible for H_2O_2 signatures that direct tolerance responses in tobacco. *The Plant Cell* 20, 1708–1724.

Moschou, P.N., Paschalidis, K.A. and Roubelakis-Angelakis, K.A. (2008b) Plant polyamine catabolism: the state of the art. *Plant Signaling and Behavior* 3, 1061–1066.

Moschou, P.N., Sanmartin, M., Andriopoulou, A.H., Rojo, E., Sanchez-Serrano, J.J. and Roubelakis-Angelakis, K.A. (2008c) Bridging the gap between plant and mammalian polyamine catabolism: a novel peroxisomal polyamine oxidase responsible for a full backconversion pathway in *Arabidopsis*. *Plant Physiology* 147, 1845–1857.

Muñiz, L., Minguet, E.G., Singh, S.K., Pesquet, E., Vera-Sirera F., Moreau-Courtois C.L., Carbonell, J., Blázquez, M. A. and Tuominen, H. (2008) *ACAULIS5* controls *Arabidopsis* xylem specification through the prevention of premature cell death. *Development* 135, 2573–2582.

Nambeesan, S., Datsenka, T., Ferruzzi, M.G., Malladi, A., Mattoo, A.K. and Handa A.K. (2010) Overexpression of yeast spermidine synthase impacts ripening, senescence and decay symptoms in tomato. *The Plant Journal* 63, 836–847.

Neily, M.H., Matsukura, C., Maucourt, M., Bernillon, S., Deborde, C., Moing, A., Yin, Y.G., Saito, T., Mori, K., Asamizu, E., Rolin, D., Moriguchi, T. and Ezura, H. (2011) Enhanced polyamine accumulation alters carotenoid metabolism at the transcriptional level in tomato fruit over-expressing spermidine synthase. *Journal of Plant Physiology* 168, 242–252.

Ono, Y., Kim, D.W., Watanabe, K., Sasaki, A., Niitsu, M., Berberich, T., Kusano, T. and Takahashi, Y. (2012) Constitutively and highly expressed *Oryza sativa* polyamine oxidases localize in peroxisomes and catalyze polyamine back conversion. *Amino Acids* 42, 867–876.

Pandey, S., Ranade, S.A., Nagar, P.K. and Kumar, N. (2000) Role of polyamine and ethylene as modulators of plant senescence. *Journal of Bioscience and Bioengineering* 25, 291–299.

Pang, X., Zhang, Z., Wen, X., Ban, Y. and Moriguchi, T. (2007) Polyamines, all–purpose players in response to environmental stresses in plants. *Plant Stress* 1, 173–188.

Panicot, M., Minguet, E.G., Ferrando, A., Alcázar, R., Blázquez, M.A., Carbonell, J., Altabella, T., Koncz, C. and Tiburcio, A.F. (2002) A polyamine metabolon involving aminopropyl transferase complexes in *Arabidopsis*. *The Plant Cell* 14, 2539–2551.

Park, S.J. and Cho, Y.D. (1991) Purification of monomeric agmatine iminohydrolase from soybean. *Biochemical and Biophysical Research Communications* 174, 32–36.

Park, S.J. and Cho, Y.D. (1999) Identification of functionally important residues of *Arabidopsis thaliana* *S*-adenosylmethionine decarboxylase. *The Journal of Biochemistry* 126, 996–1003.

Pegg, A.E. and Michael, A.J. (2009) Spermine synthase. *Cellular and Molecular Life Sciences* 67, 113–121.

Perez-Amador, M.A., Leon, J., Green, P.J. and Carbonell, J. (2002) Induction of the arginine decarboxylase *ADC2* gene provides evidence for the involvement of polyamines in the wound response in *Arabidopsis*. *Plant Physiology* 130, 1454–1463.

Piotrowski, M., Janowitz, T. and Kneifel, H. (2003) Plant C-N hydrolases and the identification of a plant *N*-carbamoylputrescine amidohydrolase involved in polyamine biosynthesis. *The Journal of Biological Chemistry* 278, 1708–1712.

Prabhavathi, V.R. and Rajam, M.V. (2007) Polyamine accumulation in transgenic eggplant enhances tolerance to multiple abiotic stresses and fungal resistance. *Plant Biotechnology* 24, 273–282.

Rastogi, R., Dulson, J. and Rothstein, S.J. (1993) Cloning of tomato (*Lycopersicon esculentum* Mill.) arginine decarboxylase gene and its expression during fruit ripening. *Plant Physiology* 103, 829–834.

Ren, M., Venglat, P., Qiu, S., Feng, L., Cao, Y., Wang, E., Xiang, D., Wang, J., Alexander, D., Chalivendra, S., Logan, D., Mattoo, A.K., Selvaraj, G. and Datla, R. (2012) Target of rapamycin signaling regulates metabolism, growth, and life span in *Arabidopsis*. *The Plant Cell* 24, 4850–4874.

Richards, F.J. and Coleman, R.G. (1952) Occurrence of putrescine in potassium-deficient barley. *Nature* 170, 460.

Rodríguez-Kessler, M. and Jiménez-Bremont, J.F. (2008) *Zmspds2* maize gene: coding a spermine synthase? *Plant Signaling and Behavior* 3, 551–553.

Roy, M. and Wu, R. (2001) *Arginine decarboxylase* transgene expression and analysis of environmental stress tolerance in transgenic rice. *Plant Science* 160, 869–875.

Roy, M. and Wu, R. (2002) Overexpression of *S-adenosylmethionine decarboxylase* gene in rice increases polyamine level and enhances sodium chloride-stress tolerance. *Plant Science* 163, 987–992.
Rzewuski, G., Cornell, K.A. and Rooney, L. (2007) *OsMTN* encodes a 5-methylthioadenosine nucleosidase that is up-regulated during submergence induced ethylene synthesis in rice (*Oryza sativa* L). *Journal of Experimental Botany* 58, 1505–1514.
Sakakibara, Y. and Yanagisawa, H. (2003) Agmatine deiminase from cucumber seedlings is a mono-specific enzyme: purification and characteristics. *Protein Expression and Purification* 30, 88–93.
Sauter, M., Lorbiecke, R. and Ouyang, B. (2005) The immediate-early ethylene response gene *OsARD1* encodes an acireductone dioxygenase involved in recycling of the ethylene precursor *S*-adenosylmethionine. *The Plant Journal* 44, 718–729.
Schroder, G. and Schroder, J. (1995) cDNA for *S*-adenosyl-L-methionine decarboxylase from *Catharanthus roseus*, heterologous expression, identification of the proenzyme-processing site, evidence for the presence of both subunits in the active enzyme, and a conserved region in the 5′ mRNA leader. *European Journal of Biochemistry* 228, 14–18.
Sen, K., Choudhuri, M.M. and Ghosh, B. (1981) Changes in polyamine contents during development and germination of rice seeds. *Phytochemistry* 20, 631–633.
Serafini-Fracassini, D. and Del-Duca, S. (2008) Transglutaminases: widespread cross–linking enzymes in plants. *Annals of Botany* 102, 145–152.
Serafini-Fracassini, D., Del-Duca, S., Monti, F., Poli, F., Sacchetti, G., Bregoli, A.M., Biondi, S. and Della-Mea, M. (2002) Transglutaminase activity during senescence and programmed cell death in the corolla of tobacco (*Nicotiana tabacum*) flowers. *Cell Death and Differentiation* 9, 309–321.
Serafini-Fracasini, D., Della-Mea, M., Tasco, G., Casadio, R. and Del-Duca, S. (2009) Plant and animal transglutaminases: do similar functions imply similar structures? *Amino Acids* 36, 643–657.
Serafini-Fracassini, D., Di-Sandro, A. and Del-Duca, S. (2010) Spermine delays leaf senescence in *Lactuca sativa* and prevents the decay of chloroplast photosystems. *Plant Physiology and Biochemistry* 48, 602–611.
Shukla, V. and Mattoo, A.K. (2013) Developing robust crop plants for sustaining growth and yield under adverse climatic changes. In: Tuteja, N. and Gill, S.S. (eds) *Climate Change and Plant Abiotic Stress Tolerance*. Wiley-VCH Verlag, Weinheim, Germany, pp. 27–56.
Sindhu, R.K. and Desai, H.V. (1980) Agmatine iminohydrolase activity during development and germination of groundnut seeds. *Experientia* 36, 530–531.
Slocum, R.D. and Flores H.E. (1991) *Biochemistry and Physiology of Polyamines in Plants*. CRC Press, Boca Raton, Florida.
Smith, T.A. and Richards, F.J. (1964) The biosynthesis of putrescine in higher plants and its relation to potassium nutrition. *Biochemical Journal* 84, 292–294.
Sobieszczuk-Nowicka, E. and Legocka, J. (2013) Plastid-associated polyamines: their role in differentiation, structure, functioning, stress response and senescence. *Plant Biology* 16, 297–305.
Sood, S. and Nagar, P.K. (2003) The effect of polyamines on leaf senescence in two diverse rose species. *Plant Growth Regulation* 39, 155–160.
Soyka, S. and Heyer, A.G. (1999) *Arabidopsis* knockout mutation of *ADC2* gene reveals inducibility by osmotic stress. *FEBS Letters* 458, 219–223.
Tabor, C.W. and Tabor, H. (1984) Polyamines. *Annual Review of Biochemistry* 53, 749–790.
Takahashi, T. and Kakehi, J. (2010) Polyamines: ubiquitous polycations with unique roles in growth and stress responses. *Annals of Botany* 105, 1–6.
Tang, W. and Newton, J.R. (2005) Polyamines promote root elongation and growth by increasing root cell division in regenerated Virginia pine. *Plant Cell Reports* 24, 581–589.
Terakado, J., Yoneyama, T. and Fujihara, S. (2006) Shoot-applied polyamines suppress nodule formation in soybean (*Glycine max*) *Journal of Plant Physiology* 163, 497–505.
Urano, K., Yoshiba, Y., Nanjo, T., Igarashi, Y., Seki, M., Sekiguchi, F., Yamaguchi-Shinozaki, K. and Shinozaki, K. (2003) Characterization of *Arabidopsis* genes involved in biosynthesis of polyamines in abiotic stress responses and developmental stages. *Plant, Cell and Environment* 26, 1917–1926.
Urano, K., Yoshiba, Y., Nanjo, T., Ito, Y., Seki, M., Yamaguchi-Shinozaki, K. and Shinozaki, K. (2004) *Arabidopsis* stress-inducible gene for arginine decarboxylase *AtADC2* is required for accumulation of putrescine in salt tolerance. *Biochemical and Biophysical Research Communications* 313, 369–375.
Vergnolle, C., Vaultier, M.N., Taconnat, L., Renou, J.P., Kader, J.C., Zachowski, A. and Ruelland, E. (2005) The cold-induced early activation of phospholipase C and D pathways determines the response of two distinct clusters of genes in *Arabidopsis* cell suspensions. *Plant Physiology* 139, 1217–1233.

Waie, B. and Rajam, M.V. (2003) Effect of increased polyamine biosynthesis on stress responses in transgenic tobacco by introduction of human *S-adenosylmethionine* gene. *Plant Science* 164, 727–734.

Wang, Z., Xu, Y., Wang, J., Yang, J. and Zhang, J. (2012) Polyamine and ethylene interactions in grain filling of superior and inferior spikelets of rice. *Plant Growth Regulation* 66, 215–228.

Watson, M.B., Emory, K.K., Piatak, R.M. and Malmberg, R.L. (1998) Arginine decarboxylase (polyamine synthesis) mutants of *Arabidopsis thaliana* exhibit altered root growth. *The Plant Journal* 13, 231–239.

Wen, X.P., Pang, X.M., Matsuda, N., Kita, M., Inoue, M., Hao, Y.J., Honda C. and Moriguchi, T. (2008) Overexpression of the apple spermidine synthase gene in pear confers multiple abiotic stress tolerance by altering polyamine titers. *Transgenic Research* 17, 251–263.

Wen, X.P., Ban, Y., Inoue, H., Matsuda, N. and Moriguchi, T. (2009) Aluminium tolerance in a spermidine synthase-overexpressing transgenic European pear is correlated with the enhanced level of spermidine via alleviating oxidative status. *Environmental and Experimental Botany* 66, 471–478.

Wen, X.P., Ban Y., Inoue, H., Matsuda, N. and Moriguchi, T. (2010) Spermidine levels are implicated in heavy metal tolerance in a spermidine synthase overexpressing transgenic European pear by exerting antioxidant activities. *Transgenic Research* 19, 91–103.

Wi, S.J., Kim, W.T. and Park, K.Y. (2006) Over expression of carnation *S-adenosylmethionine decarboxylase* gene generates a broad-spectrum tolerance to abiotic stresses in transgenic tobacco plants. *Plant Cell Reports* 25, 1111–1121.

Yamaguchi, K., Takahashi, Y., Berberich, T., Imai, A., Miyazaki, A., Takahashi, T., Michael, A. and Kusano, T. (2006) The polyamine spermine protects against high salt stress in *Arabidopsis thaliana*. *FEBS Letters* 580, 6783–6788.

Yamaguchi, K., Takahashi, Y., Berberich, T., Imai, A., Takahashi, T., Michael, A.J. and Kusano, T. (2007) A protective role for the polyamine spermine against drought stress in *Arabidopsis*. *Biochemical and Biophysical Research Communications* 352, 486–490.

Yamanoha, B. and Cohen, S.S. (1985) *S*-adenosylmethionine decarboxylase and spermidine synthase from Chinese cabbage. *Plant Physiology* 78, 784–790.

Yanagisawa, H. (2001) Agmatine deiminase from maize shoots: purification and properties. *Phytochemistry* 56, 643–647.

Yanagisawa, H. and Suzuki, Y. (1982) Preparation and properties of *N*-carbamyl putrescine amidohydrolase from maize shoot. *Phytochemistry* 21, 2201–2203.

Yang, J., Cao, Y., Zhang, H., Liu, L. and Zhang, J. (2008) Involvement of polyamines in the post-anthesis development of inferior and superior spikelets in rice. *Planta* 228, 137–149.

Yang, S.F. and Hoffman, N.E. (1984) Ethylene biosynthesis and its regulation in higher plants. *Annual Review of Plant Physiology* 35, 155–189.

Yang, Y.G. and Cho, Y.D. (1991) Purification of *S*-adenosylmethionine decarboxylase from soybean. *Biochemical and Biophysical Research Communications* 181, 1181–1186.

Yoda, H., Yamaguchi, Y. and Sano, H. (2003) Induction of hypersensitive cell death by hydrogen peroxide produced through polyamine degradation in tobacco plants. *Plant Physiology* 132, 1973–1981.

Yoda, H., Hiroi, Y. and Sano, H. (2006) Polyamine oxidase is one of the key elements for oxidative burst to induce programmed cell death in tobacco cultured cells. *Plant Physiology* 142, 193–206.

Yoda, H., Fujimura, K., Takahashi, H., Munemura, I., Uchimiya, H. and Sano, H. (2009) Polyamines as a common source of hydrogen peroxide in host- and nonhost hypersensitive response during pathogen infection. *Plant Molecular Biology* 70, 103–112.

Yoon, S.O., Lee, Y.S., Lee, S.H. and Cho, Y.D. (2000) Polyamine synthesis in plants: isolation and characterization of spermidine synthase from soybean (*Glycine max*) axes. *Biochimica et Biophysica Acta* 1475, 17–26.

Zhu, M.-Y., Iyo, A. and Regunathan, S. (2004) Expression of human arginine decarboxylase, the biosynthetic enzyme for agmatine. *Biochimica et Biophysica Acta* 1670, 156–164.

11 Serine Acetyltransferase

M. Watanabe,[1]* H.-M. Hubberten,[1] K. Saito[2,3] and R. Hoefgen[1]
[1]*Max Planck Institute of Molecular Plant Physiology, Potsdam-Golm, Germany;* [2]*Graduate School of Pharmaceutical Sciences, Chiba University, Chiba, Japan;* [3]*RIKEN Center for Sustainable Resource Science, Yokohama, Japan*

11.1 Abstract

Sulfur is an essential macronutrient for plants. Sulfate, the major form of inorganic sulfur, is reduced to sulfide by the assimilatory sulfate reduction pathway. Sulfide is integrated into organic compounds via cysteine biosynthesis. Serine acetyltransferase (SERAT, also called SAT; EC 2.3.1.30), which catalyses the formation of *O*-acetylserine (OAS) from serine and acetyl coenzyme A, is responsible for the entry step from glycine-serine metabolism to cysteine biosynthesis. OAS, as activated serine, is the direct precursor of cysteine formation catalysed by *O*-acetylserine (thiol)lyase (OASTL; EC 4.2.99.8). OASTL incorporates sulfide into OAS essentially by exchanging the alcohol group of serine (-CH-OH) against a thiol group (-CH-SH), thus producing cysteine. Both the SERAT and OASTL enzymes are encoded by members of large gene families and are localized in the cytosol, plastids and mitochondria. SERAT plays an important role in the regulation of sulfur assimilation and cysteine synthesis. The product of SERAT, OAS, is not only a limiting factor for cysteine synthesis but has also been suggested to be a signal molecule for gene expression in the sulfur assimilation pathway and to cause induction of a specific gene set, the OAS cluster genes. SERAT activity is modulated by two regulation mechanisms: feedback inhibition by cysteine and the formation of a protein complex with OASTL, which allows the regulation of OAS and cysteine biosynthesis in plant cells. Analysis of SERATs using various approaches in recent years has provided substantially new information concerning the properties and functions of these enzymes in sulfur metabolism. Here we review the available data on SERAT function in *Arabidopsis*.

11.2 Introduction

Sulfur is an essential macronutrient affecting plant growth, plant performance and yield. Inorganic sulfate is taken up through plant roots, reduced and incorporated into organic compounds via cysteine biosynthesis (Fig. 11.1). Cysteine, as the first common organic compound containing reduced sulfur, serves as a precursor for the synthesis of methionine (Met), glutathione (GSH), cofactors, vitamins, sulfur esters and other sulfur derivatives (Beinert, 2000; Leustek *et al.*, 2000; Marquet, 2001; Gerber and Lill, 2002; Mendel and Haensch, 2002; Noctor *et al.*, 2002; Saito, 2004; Hoefgen and Hesse, 2008). Serine acetyltransferase (SERAT, in the past often also termed SAT; EC 2.3.1.30), which catalyses the formation of *O*-acetylserine (OAS) from serine and acetyl coenzyme A (acetyl-CoA), links serine metabolism to cysteine biosynthesis. OAS reacts with sulfide to yield cysteine. This reaction is catalysed by cysteine synthase (*O*-acetylserine

*E-mail address: watanabe@mpimp-golm.mpg.de

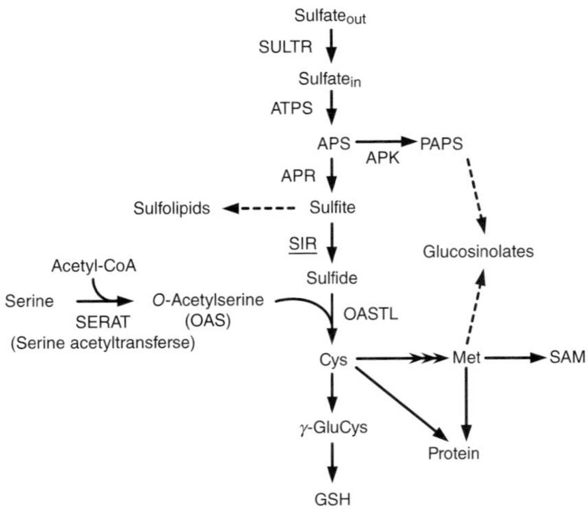

Fig. 11.1. Sulfur metabolism in *Arabidopsis thaliana*. Sulfate is taken up by sulfate transporters (SULTR) and activated by ATP sulfurylase (ATPS) to adenosine 5′-phosphosulfate (APS). APS is further converted to sulfite by APS reductase (APR) and 3′-phosphoadenosine 5′-phosphosulfate (PAPS) by APS kinase (APK). PAPS serves as a substrate for sulfotransferases, producing numerous sulfated metabolites including, glucosinolates. Sulfite is then reduced to sulfide by sulfite reductase (SIR). In a side reaction, sulfite is also utilized to synthesize sulfolipids. Serine acetyltransferase (SERAT) forms *O*-acetylserine (OAS) using serine and acetyl coenzyme A (acetyl-CoA). Cysteine is produced by *O*-acetylserine(thiol)lyase (OASTL), using OAS and sulfide. Cysteine is used for numerous downstream products such as methionine (Met), gamma-glutamylcysteine (γ-GluCys), glutathione (GSH) and *S*-adenosylmethionine (SAM).

(thiol)-lyase, OASTL; EC 4.2.99.8). In contrast to the fact that sulfate reduction to sulfide exclusively occurs in plastids, the biosynthesis of OAS and cysteine is located in three subcellular compartments of the plant cell, namely the cytosol, plastids and mitochondria. SERAT exhibits a regulatory role on sulfur assimilation through the production of OAS, which is a marker metabolite of sulfate starvation and a positive regulator of the sulfur assimilation pathway (Koprivova *et al.*, 2000; Kopriva *et al.*, 2002; Hesse *et al.*, 2003; Hirai *et al.*, 2003; Maruyama-Nakashita *et al.*, 2003; Nikiforova *et al.*, 2003; Maruyama-Nakashita *et al.*, 2004). Further, a recent study by Hubberten *et al.* (2012) revealed an additional role of OAS as an independent regulator of sulfur status during plant growth. In terms of the regulation of SERAT activity, two major mechanisms have been reported: feedback inhibition by cysteine (Noji *et al.*, 1998; Wirtz *et al.*, 2001; Kawashima *et al.*, 2005); and the formation of a protein complex with OASTL (Kredich *et al.*, 1969; Droux *et al.*, 1998).

In several plant species, the SERAT enzymes are encoded by a multi-gene family (Table 11.1). Most studies have been performed on the model species *Arabidopsis thaliana*. In the *Arabidopsis* genome, there are five genes that are subdivided into three classes, termed groups I to III based on amino acid sequence similarities; these genes are named *SERAT1;1*, *SERAT2;1*, *SERAT2;2*, *SERAT3;1* and *SERAT3;2*. Although a high degree of conservation of amino acid sequences is observed within the three groups, they have characteristic exon–intron gene structures. Groups I and II contain one or no introns, while group III isoforms contain nine introns (Fig. 11.2) (Kawashima *et al.*, 2005; Watanabe *et al.*, 2008a). Recent research has provided cumulative insights into the function of SERATs by biochemical characterization *in vitro*, sub-cellular localization studies, *in vivo* functionality studies of the SERAT isoforms and OAS signalling in *Arabidopsis*. Here, we review our current knowledge of the function of SERATs, mainly concentrating on *Arabidopsis*.

Table 11.1. Serine acetyltransferase gene (*SERAT*) family content in plant species. Orthologous *SERAT* genes of were estimated by BLAST search employing the PLAZA web site (http://bioinformatics.psb.ugent.be/plaza/). The number of orthologous genes in the HOM000768 gene family was counted. Genes in two subfamilies (ORTHO006173 and ORTHO027349) were excluded because these genes belong to the GRIP (golgin-97, RanBP2alpha, Imh1p and p230/golgin-245) gene family.

Plant species	Gene no.
Arabidopsis lyrata	5
Arabidopsis thaliana	5
Brachypodium distachyon	4
Carica papaya	3
Chlamydomonas reinhardtii	2
Fragaria vesca	2
Glycine max	11
Lotus japonicus	5
Malus domestica	13
Manihot esculenta	4
Micromonas sp. *RCC299*	3
Medicago truncatula	3
Ostreococcus lucimarinus	3
Oryza sativa subsp. *japonica*	6
Oryza sativa subsp. *indica*	8
Ostreococcus tauri	3
Physcomitrella patens	5
Populus trichocarpa	4
Ricinus communis	3
Sorghum bicolor	4
Selaginella moellendorffii	3
Theobroma cacao	4
Volvox carteri	3
Vitis vinifera	4
Zea mays	4

11.3 Biochemical Properties and Sub-cellular Localization of Serine Acetyltransferases

The five genuine SERAT isoforms in *Arabidopsis* have different biochemical properties in terms of K_m values for the two substrates (serine and acetyl-CoA) used for OAS formation and sensitivities to cysteine feedback inhibition (Table 11.2) (Noji *et al.*, 1998; Wirtz *et al.*, 2001; Kawashima *et al.*, 2005). Recombinant protein studies with *Escherichia coli* revealed that the K_m values for serine and acetyl-CoA were in the ranges of 1.6–2.7 mM and 0.02–0.28 mM, respectively, for SERAT1;1, SERAT2;1 and SERAT2;2, while they were in the ranges of 39.5–121.4 mM and 24.5–45.1 mM, respectively, for SERAT3;1 and SERAT3;2. Because of their low affinities to serine and acetyl-CoA, it has been suggested that SERAT3;1 and SERAT3;2 might either have different or at least additional metabolic functions (Kawashima *et al.*, 2005). However, SERAT3;2 and SERAT1;1 are sensitive to feedback inhibition by cysteine (Kawashima *et al.*, 2005), suggesting that SERAT3;2 has a function in OAS and cysteine metabolism. The activity of these two isoforms was inhibited by cysteine in a non-competitive manner to serine, but in a competitive manner to acetyl-CoA, with K_i values of 10.8 μM and 7.4 μM for SERAT1;1, respectively, and 17.3 μM and 2.5 μM for SERAT3;2, respectively (Kawashima *et al.*, 2005). SERAT2;2 was only inhibited at higher cysteine concentrations (IC_{50} = 53 μM) (Wirtz *et al.*, 2001).

Sub-cellular localization studies with promoter-GFP (green fluorescent protein) constructs demonstrated that SERAT1;1, SERAT3;1 and SERAT3;2 were localized in the cytosol, SERAT2;1 is localized in the plastids, and SERAT2;2 is localized in the mitochondria (Noji *et al.*, 1998; Kawashima *et al.*, 2005). In *Arabidopsis*, the cellular concentrations for serine are in the range of 1–10 mmol g^{-1} fresh weight (Goyer *et al.*, 2005; Watanabe *et al.*, 2008a; Timm *et al.*, 2013) and 5 nmol g^{-1} fresh weight for acetyl-CoA (Roughan, 1997; Tumaney *et al.*, 2004), respectively. Thus, acetyl-CoA might be limiting for SERAT activity. However, there is no report of the compartment-specific concentrations of either metabolite, and investigations of these are required to estimate the endogenous SERAT activity in the plant. The cellular concentration of cysteine is in the range of 7–22 nmol g^{-1} fresh weight (Hirai *et al.*, 2003; Heeg *et al.*, 2008; Watanabe *et al.*, 2008a,b; Krueger *et al.*, 2009; Khan *et al.*, 2010). In the case of cysteine, the compartment-specific concentrations in the cytosol, plastid and vacuole were estimated from metabolite profiles using non-aqueous fractionation, although the mitochondria were contained in the vacuolar and cytosolic fractions (Krueger *et al.*, 2009). In *Arabidopsis*, the concentrations of cysteine are 332 μM in the cytosol, 9 μM in the plastids and 3 μM in the vacuoles, suggesting that feedback sensitivity to cysteine seems to modulate cytosolic SERAT1;1 and SERAT3;2 in this plant.

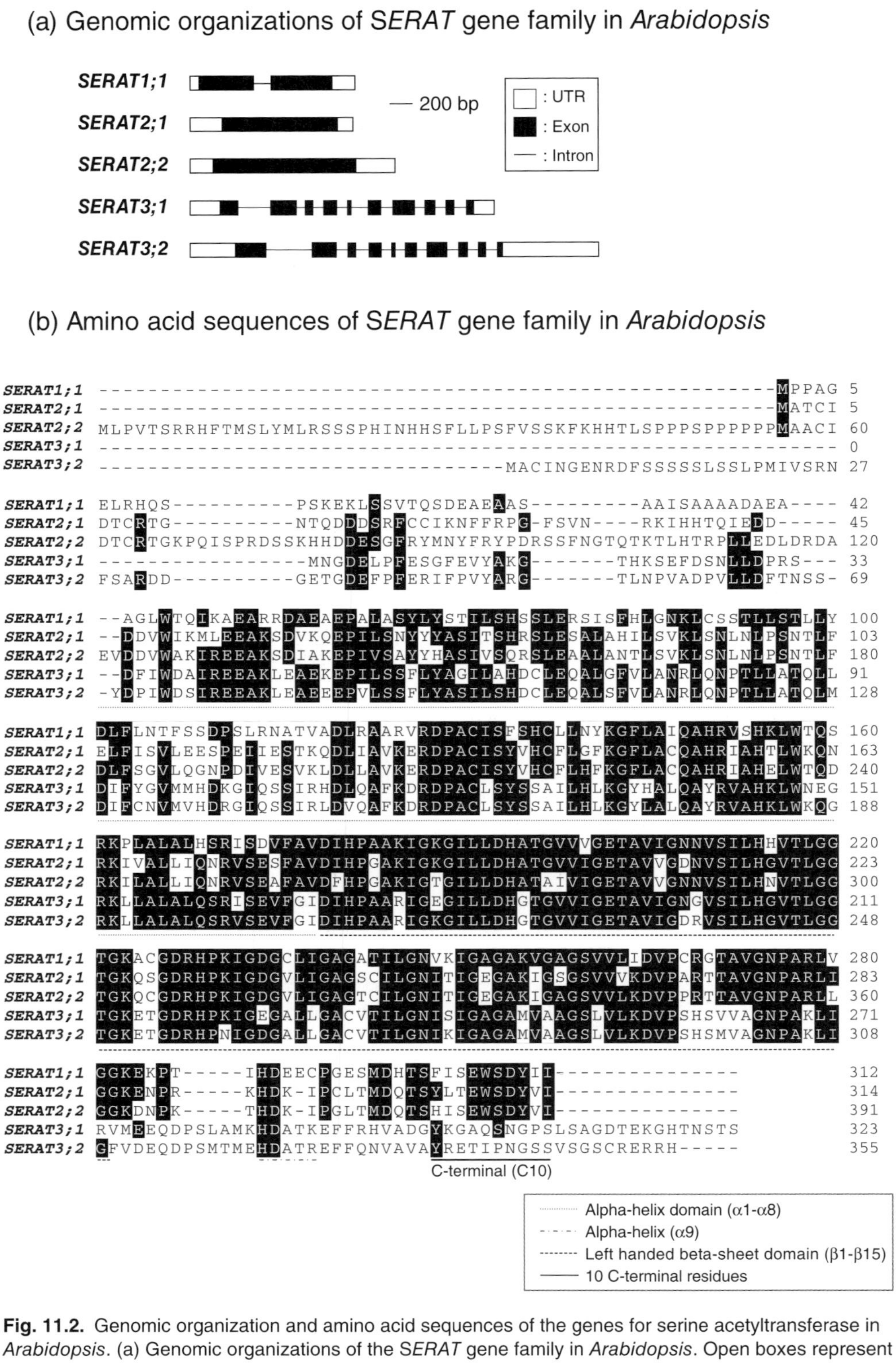

Fig. 11.2. Genomic organization and amino acid sequences of the genes for serine acetyltransferase in *Arabidopsis*. (a) Genomic organizations of the *SERAT* gene family in *Arabidopsis*. Open boxes represent

Table 11.2. Biochemical properties and sub-cellular localization of serine acetyltransferases (SERATs) in *Arabidopsis*.

Class	Isoform	AGI code (unique locus identifier)	Localization	L-Serine K_m (mM)	Acetyl-CoA K_m (mM)	Feedback inhibition (K_i) by cysteine (μM) for L-serine and acetyl-CoA
Class I	SERAT1;1	AT5G56760	Cytosol[a]	2.7[a]	0.28[a]	Sensitive (10.8 and 7.4)[a]
Class II	SERAT2;1	AT1G55920	Plastids[a]	1.6[a]	0.16[a]	Insensitive[a]
	SERAT2;2	AT3G13110	Mitochondria[a]	1.7[a]	0.02[a]	Insensitive[a]
				(2.4)[c]	(0.15)[c]	(Sensitive, IC_{50} = 53 μM)[c]
Class III	SERAT3;1	AT2G17640	Cytosol[b]	121.4[b]	24.50[b]	Insensitive[b]
	SERAT3;2	AT4G35640	Cytosol[b]	39.5[b]	45.10[b]	Sensitive (17.3 and 2.5)[b]

[a]Noji *et al.*, 1998; [b]Kawashima *et al.*, 2005; [c]Wirtz *et al.*, 2001.

In summary, in *Arabidopsis*, the biochemically most active SERAT isoforms, which we therefore term major isoforms (SERAT1;1, SERAT2;1, SERAT2;2), are distributed over all three compartments, while the feedback-sensitive SERAT isoforms are exclusively localized in the cytosol (Fig. 11.3). In other plant species, the distribution of the dominant SERAT activity and the presence of cysteine feedback regulation of SERAT isoforms differs both between species and between sub-cellular compartments. SERAT activity was detected exclusively in the cytosol of *Citrullus vulgaris* (Saito *et al.*, 1995), in the plastids of *Spinacea oleracea* (Brunold and Suter, 1982) and in the mitochondria of *Phaseolus vulgaris* (Smith, 1972). In *Pisum sativum*, SERAT activities were demonstrated to be present in three sub-cellular compartments; 80–90% of total SERAT activity was present in the mitochondria, with 5–10% assigned to the cytosol and plastids, respectively (Ruffet *et al.*, 1995; Droux, 2003). Feedback inhibition by cysteine was reported for *C. vulgaris* (Saito *et al.*, 1995), *Allium tuberosum* (Urano *et al.*, 2000) and *Glycine max* (Chronis and Krishnan, 2004) for cytosolic isoforms, whereas in *S. oleracea* (Noji *et al.*, 2001) and *P. sativum* (Droux, 2003), the plastidial isoforms were feedback sensitive. Thus, SERAT isoforms have been described substantially in terms of biochemical properties and sub-cellular localization, but it has to be borne in mind that the results obtained with *Arabidopsis* have to be scrutinized for other plant species and cannot be generalized in all aspects.

11.4 The Serine Acetyltransferase-*O*-Acetylserine(Thiol)Lyase Complex

SERAT and OASTL form a hetero-oligomeric cysteine synthase complex (CSC). The function of this CSC formation seems not to be metabolic channelling (Fig. 11.4) (Kredich *et al.*, 1969; Droux *et al.*, 1998; Mino *et al.*, 2001). SERATs are activated or stabilized in the CSC, while OASTLs are inactive in the CSC but rather fully active as a dimer in the free form when not associated with the CSC. Hence, OAS diffuses out of the complex for cysteine synthesis via free OASTL. It has been suggested in several publications that SERAT is active only in the complex with OASTL (Sirko *et al.*, 2004; Riemenschneider *et al.*, 2005; Wirtz and Hell, 2006). However, these statements produced the misleading opinion that SERAT is inactive in the free form. Actually,

Fig. 11.2. continued
untranslated sequences (UTRs); solid boxes represent coding regions; and solid lines represent introns. (b) The amino acid sequences of *SERAT* genes from *Arabidopsis* were aligned using the multiple sequence alignment program CLUSTAL X (Thompson *et al.*, 1997). Dots indicate gaps in the sequences for the best alignment. Black shading indicates >60% similarity. The indicated secondary structure domains were predicted from the crystal structure of SERAT in *Escherichia coli* (Wirtz *et al.*, 2010). Asterisks indicate amino acid residues that are predicted to be important for the formation of *O*-acetylserine (OAS). (Wirtz *et al.*, 2010.)

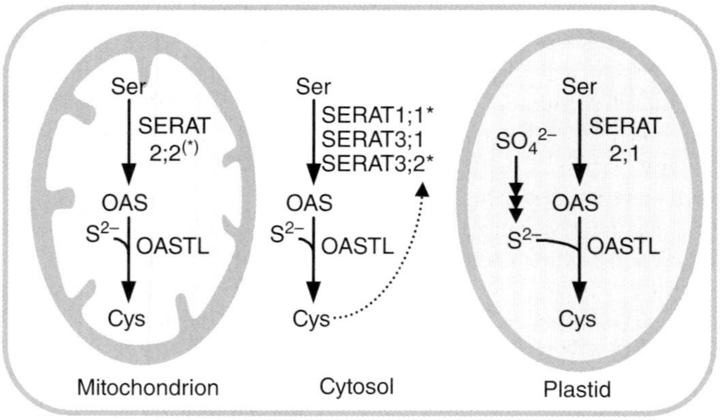

Fig. 11.3. Schematic representation of five serine acetyltransferase (SERAT) isoforms in *Arabidopsis*. *O*-acetylserine (OAS) and cysteine synthesis take place in the cytosol, plastids and mitochondria. Asterisks indicate SERATs sensitive to feedback inhibition by cysteine. Abbreviations: Cys, cysteine; OASTL, *O*-acetylserine (thiol)lyase; Ser, serine. The figure is modified from that of Watanabe *et al.*, 2010 (http://mplant.oxfordjournals. org/content/3/2/438.full.pdf; Copyright Oxford University Press).

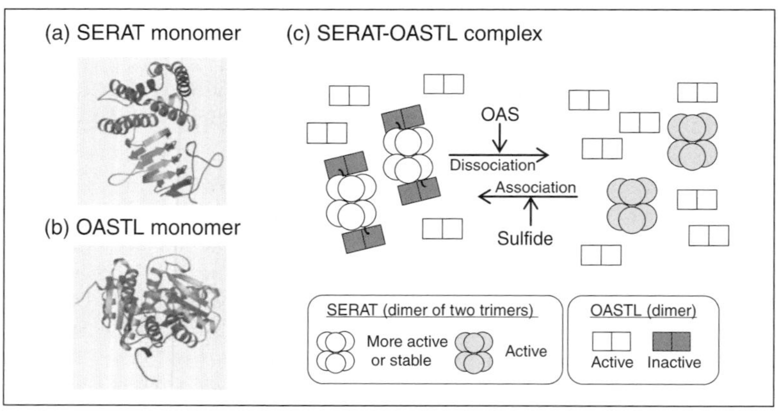

Fig. 11.4. The serine acetyltransferase and *O*-acetylserine(thiol)lyase (SERAT-OASTL) complex. Structural modelling of (a) cytosolic SERAT1;1 and (b) cytosolic OASTL (At4g14880) from *Arabidopsis* using SWISS-MODEL (http://swissmodel.expasy.org/repository/). The SERAT monomer consists of an amino-terminal α-helical domain and a carboxyl-terminal left-handed β-helix. The OASTL monomer consists of two α/β structural domains. The crystal structure of *Arabidopsis* cytosolic OASTL was reported by Bonner *et al.* (2005). (c) Regulation of the SERAT-OASTL complex. The complex contains two trimers of SERAT and two dimers of OASTL. The equilibrium of complex association and dissociation is regulated by *O*-acetylserine (OAS) and sulfide. Sulfide promotes the association of the complex whereas OAS promotes its dissociation. SERATs are activated or stabilized in the complex, whereas OASTLs are inactive in the complex and are active only as free forms. White circles, more active or stable SERAT; grey circles, active SERAT; white squares, active OASTL; grey squares, inactive OASTL.

biochemical properties such as the K_m values of recombinant plant SERATs were investigated using the free form. The SERAT protein is functional in the absence of OASTL *in vitro*, and probably also *in vivo*, judging from the biochemical properties, which are in the range of physiological concentrations in the plant in the case of the major SERAT isoforms in *Arabidopsis*. This was supported by the reports that the free homomer of SERAT2;2 and the complex-bound SERAT2;2 have similar catalytic SERAT activities (Berkowitz *et al.*, 2002) and that C-terminally truncated SERAT proteins that lack OASTL interaction have higher SERAT activity compared with the full-length SERAT protein *in vitro* (Bogdanova and Hell, 1997; Wirtz *et al.*, 2010). Complex formation with OASTL rather modifies SERAT activity through altering the sensitivities of SERAT to feedback inhibition by cysteine or the stability of the SERAT protein (Denk and Bock, 1987; Noji *et al.*, 1998; Mino *et al.*, 2000; Droux, 2003; Francois *et al.*, 2006; Wirtz *et al.*, 2010).

The crystal structures of bacterial SERATs (Gorman and Shapiro, 2004; Olsen *et al.*, 2004; Pye *et al.*, 2004), bacterial OASTLs (Burkhard *et al.*, 1998; Burkhard *et al.*, 1999) and *Arabidopsis* cytosolic OASTL (Bonner *et al.*, 2005) have been reported, but the crystal structure of the CSC is not available yet in any organism. However, biochemical studies, such as application of size exclusion chromatography, analytical ultracentrifugation and isothermal titration calorimetry have helped to elucidate the mechanism of the interaction between SERAT and OASTL (Hell and Wirtz, 2011). These studies suggested that SERAT forms a central dimer of trimers associated with two OASTL homodimers in *Arabidopsis* (Wirtz *et al.*, 2010). This composition might be conserved in plant species because CSCs from *S. oleracea*, *Nicotiana tabacum* and *G. max* possess similar molecular masses (Wirtz and Hell, 2006; Wirtz *et al.*, 2010). The C-terminus of SERAT was identified as an interaction site between SERAT and OASTL (Bogdanova and Hell, 1997; Francois *et al.*, 2006; Kumaran *et al.*, 2009). The ten C-terminal residues of *Arabidopsis* SERAT (C10 peptide) were sufficient to bind to the catalytic centre of the OASTL enzyme, which results in the inactivation of OASTL (Francois *et al.*, 2006). In *Arabidopsis*, complex formation have been studied in detail for the major isoforms (SERAT1;1, SERAT2;1 and SERAT2;2). The minor SERAT isoforms (SERAT3;1 and SERAT3;2), however, have not yet been investigated. The amino acid sequence analysis, though, suggests that SERAT3;1 and SERAT3;2 could not bind to OASTL because of their additional C-terminal extensions (Fig. 11.2b) (Kawashima *et al.*, 2005). Deletion analysis of the C-terminus of SERAT not only revealed that the C-terminus represses SERAT activity but also that it modulates feedback inhibition by cysteine (Denk and Bock, 1987; Noji *et al.*, 1998; Mino *et al.*, 2000; Francois *et al.*, 2006; Wirtz *et al.*, 2010). The activities of the C-terminally truncated proteins SERAT2;2ΔC15 and SERAT2;2ΔC10 were significantly increased compared with that of full-length SERAT2;2 (Wirtz *et al.*, 2010). The importance of the C-terminus of SERAT in the regulation of SERAT activity was further supported by a finding that the feedback sensitivity of SERAT was decreased by CSC formation where the C-terminus of SERAT is bound to OASTL. Furthermore, different feedback sensitivities of the phosphorylated and non-phosphorylated C-terminus of SERAT were reported for the plastidic SERAT of *G. max*, which might indicate additional regulatory mechanisms (Liu *et al.*, 2006).

In order to investigate the function of the CSC and to modify cysteine synthesis, Wawrzynska *et al.* (2013) screened for SERAT-OASTL interaction sites in the amino acid sequences of OASTL. On the basis of computational modelling, five specific linear polypeptides were synthesized that were assumed to bind competitively to SERAT and would therefore result in increased amounts of free OASTLs. In a yeast two-hybrid assay, the polypeptide most strongly interacting with SERAT was transformed into *Arabidopsis*. In the resulting transgenic lines, a moderate increase in total SERAT and OASTL activities was observed, suggesting that it might be possible to regulate *in planta* cysteine synthesis using these synthetic polypeptides.

The binding of SERAT and OASTL is stabilized by the presence of sulfide and dissociated by the addition of OAS (Fig. 11.4c). However, in the presence of sulphide, the dissociation of the CSC requires OAS concentrations in the mM range *in vitro*, which might not be reached *in vivo* as OAS concentrations are in the μM range *in planta* (Droux *et al.*, 1998; Wirtz *et al.*, 2004; Wirtz and Hell, 2006; Krueger *et al.*, 2009). This suggests that under sulfate-sufficient conditions, the CSC is stabilized as sufficient sulfide should be available, whereas under sulfide-depleted conditions caused by, e.g. sulfate deficiency, OAS accumulates and could dissociate the CSC. CSC formation has been suggested to regulate OAS and cysteine synthesis through altering SERAT and OASTL activities, but the universal validity of the CSC model might need amendments owing to the fact that isolated

SERAT isoforms display OAS synthesizing capacity, at least *in vitro*. Furthermore, regulation of the CSC might be different in each cellular compartment because of different metabolite concentrations. Sulfide has been shown to accumulate in the plastids, while OAS accumulated in the mitochondria (Krueger *et al.*, 2009). Moreover, OASTL and SERAT proteins accumulate differentially in the respective compartments, which might affect regulation as well. In *Arabidopsis*, for example, the OASTL/SERAT activity ratio is 200:1 in the cytosol, 300:1 in the plastids and about 4:1 in the mitochondria (Haas *et al.*, 2008; Heeg *et al.*, 2008; Birke *et al.*, 2012). That changes in the OASTL/SERAT ratio affect the flux of cysteine has been shown through antisense repression of OASTL in *Solanum tuberosum* (which resulted in cysteine increases rather than decreases) (Riemenschneider *et al.*, 2005), in *Arabidopsis* T-DNA mutants of OASTL (Haas *et al.*, 2008; Heeg *et al.*, 2008; Watanabe *et al.*, 2008b; Birke *et al.*, 2012) and in miRNA lines of SERAT (Haas *et al.*, 2008) (see Section 11.6). A molar excess of OASTL to SERAT protein is required for efficient cysteine synthesis (Kredich *et al.*, 1969; Ruffet *et al.*, 1994; Droux *et al.*, 1998; Droux, 2003; Wirtz and Hell, 2006; Hell and Wirtz, 2008). Therefore, in order to eventually understand the regulation of the SERAT-OASTL complex, careful studies that consider the differences in sub-cellular localizations and ratios are required.

The formation of the SERAT-OASTL complex to form the cysteine synthase complex has been convincingly proven. Recent studies suggest a possible binding of further proteins to SERAT or to OASTL that might affect the regulation of the complex or the activities of its constituents. Cyclophilin CYP20-3 was found to be a protein interacting partner for SERAT2;1 in yeast two-hybrid screenings (see next section) (Dominguez-Solis *et al.*, 2008). Another report suggests a possible protein interaction between a mutated OASTL and the 'recognition of *Peronospora parasitica* 1' (RPP1)-like disease resistance R gene(s) in specific *Arabidopsis* accessions that may be involved in regulating plant senescence (Tahir *et al.*, 2013). The regulation of the SERAT-OASTL complex is still not resolved and might be more complex than currently depicted, as additional proteins might modulate its function. Moreover, the function of SERAT and OASTL might not be limited to sulfur metabolism.

11.5 Expression Patterns of Serine Acetyltransferase Genes

The five *SERAT* genes in *Arabidopsis* show distinct expression patterns during development and under diurnal/circadian regulation, as well as under stress conditions. The major isoforms, *SERAT1;1*, *SERAT2;1* and *SERAT2;2*, have higher mRNA abundance (approximately tenfold to 100-fold higher) than the minor isoforms, *SERAT3;1* and *SERAT3;2* (Kawashima *et al.*, 2005). During plant development, *SERAT1;1* and *SERAT2;1* showed high expression at the vegetative stage but *SERAT3;1* and *SERAT3;2* showed high expression at the reproductive stage in quantitative RT (real-time)-PCR analyses. *SERAT2;2* showed a biphasic expression pattern with high expression at the young vegetative and late reproductive stages (Kawashima *et al.*, 2005). These data were corroborated by analysing the expression data obtained from microarray data using the *Arabidopsis* eFP browser (Winter *et al.*, 2007) and Genevestigator (https://www.genevestigator.com/gv/plant.jsp). The expression data indicate tissue-specific expression patterns of the *SERAT* genes (Fig. 11.5a). High gene expression levels were observed in mature pollen for *SERAT1;1*, in petals, sepals and senescent leaves for *SERAT2;1*, and in siliques and seeds for *SERAT3;1*, suggesting specific functions in these tissues.

Sulfur assimilation genes, including *SERATs*, have been reported to be circadian/diurnal regulated (Harmer, 2009). Among the *SERATs*, *SERAT2;1* especially showed strong diurnal oscillations in expression (Espinoza *et al.*, 2010) (Fig. 11.5b). The expression phase was affected by environmental conditions such as cold acclimation (Espinoza *et al.*, 2010). Under sulfur starvation, *SERAT3;2*, one of the usually minor isoforms, was highly induced (a 40–50-fold increase) in leaf and root tissues, and *SERAT3;1* was slightly induced in roots (a 2–3-fold increase), but the expression levels of the other three *SERATs* were not altered (Kawashima *et al.*, 2005) (Fig. 11.5c). Moreover, *SERAT3;1* was induced in roots by treatment with selenate, which is an analogue of sulfate (Van Hoewyk *et al.*, 2008) (Fig. 11.5c). In contrast to the specific induction of the group III *SERATs* under sulfur starvation, under other nutrient starvations, for example, under phosphate starvation, *SERAT2;1*

(a) Gene expression patterns of *SERAT*genes in plant development

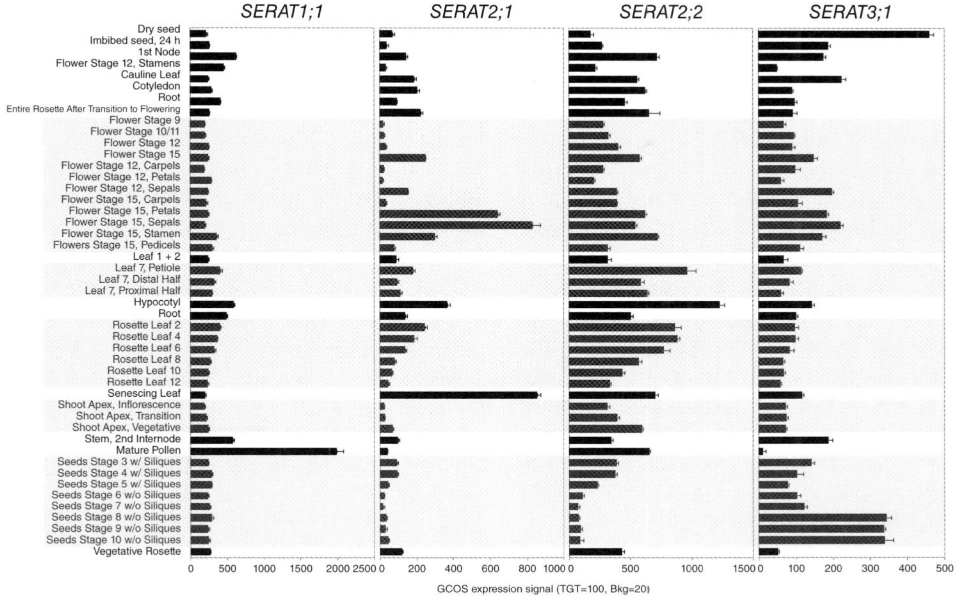

(b) Diurnal regulation of *SERAT* genes

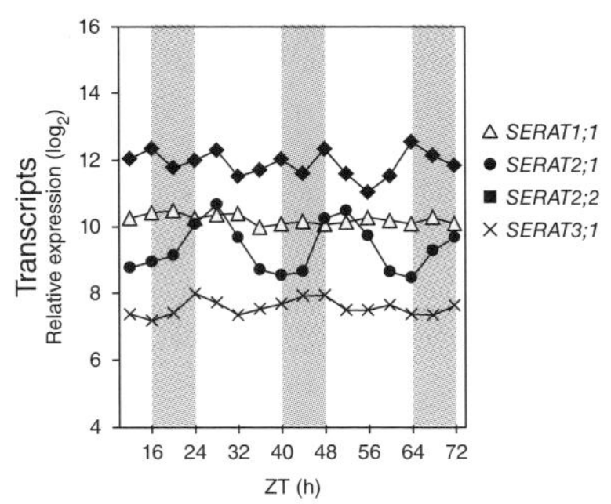

(c) Induction of *SERAT* genes under stress condition

Experiment	Tissue	*SERAT 1;1*	*SERAT 2;1*	*SERAT 2;2*	*SERAT 3;1*	*SERAT 3;2*	Reference
–Sulfate	shoot	0.9	0.8	0.9	1.0	40.0	*Kawashima et al., 2005*
–Sulfate	root	1.2	0.8	0.9	2.0	45.0	*Kawashima et al., 2005*
–Sulfate	root	1.2	1.3	1.1	2.6	nd	*Maruyama-Nakashita et al., 2006*
+Selenate	shoot	1.1	1.0	1.1	1.2	nd	*Van Hoewyk et al., 2008*
+Selenate	root	0.9	1.2	1.0	2.1	nd	*Van Hoewyk et al., 2008*
–Nitrate	seedling	1.1	1.7	1.3	0.7	nd	*Scheible et al., 2004*
–Phosphate	seedling	1.6	2.2	4.5	0.6	nd	*Morcuende et al., 2007*
–Potassium	seedling	1.2	1.4	1.3	0.6	nd	*Armengaud et al., 2004*
+Menadione	seedling	2.1	11.4	9.8	1.0	nd	*Lehmann et al., 2009*
$+H_2O_2$	seedling	0.8	14.7	2.2	0.9	nd	*Davletova et al., 2005*

Fig. 11.5. Gene expression patterns of the serine acetyltransferase *(SERAT)* gene family in *Arabidopsis*. (a) Gene expression patterns in *Arabidopsis* tissues (Arabidopsis eFP Browser; Winter *et al*., 2007). The expression data was normalized with the GCOS (Gene Chip Operating Software) method and a Target intensity (TGT) value of 100. Error bars represent standard deviation. Most tissues were sampled in triplicate. (b) *SERAT* gene expression in diurnal cycles (Espinoza *et al*., 2010). Plants were grown under long-day conditions (16 h light/8 h dark) at 20 °C. Relative expression (log_2) is indicated. White and grey bars indicate the day and night periods, respectively. ZT, zeitgeber time. (c) Induction of *SERAT* genes under several stress conditions. The fold change in each stress condition relative to the control condition is shown. Changes of >2-fold are coloured by grey; nd, not determined. The specific probe for the *SERAT3;2* gene is not found in ATH1 Affymetrix microarray, so this was not included in the analysis.

and *SERAT2;2* were induced (Armengaud *et al*., 2004; Scheible *et al*., 2004; Morcuende *et al*., 2007) (Fig. 11.5c). The transcript levels of *SERAT2;1* and *SERAT2;2* were also increased under oxidative stress conditions caused by menadione treatment (Lehmann *et al*., 2009) (Fig. 11.5c). Similarly, high levels of *SERAT2;1* transcripts were observed in plants treated with H_2O_2 (Davletova *et al.*, 2005) (Fig. 11.5c) and in senescing leaves (Fig. 11.5a), both conditions related to oxidative stress conditions that cause the accumulation of reactive oxygen species (ROS). A pivotal role of *SERAT2;1* under oxidative stress was supported by the report that SERAT2;1 was a protein interacting partner for cyclophilin CYP20-3 (Dominguez-Solis *et al.*, 2008). CYP20-3 has peptidyl prolyl isomerase activity which assists protein folding and assembly, and its activity depends on reduction by thioredoxins. The *Arabidopsis cyp20-3* mutant was hypersensitive to oxidative stress conditions caused by high light, rose bengal, high salt and osmotic stress. In the *cyp20-3* mutant, SERAT activity and thiols were less induced than in wild type plants under light and oxidative stress, suggesting that CYP20-3 links redox regulation to the activity of SERAT2;1 and might affect its folding properties or protein stability, both of which would affect SERAT activity.

11.6 *In Vivo* Functions of Serine Acetyltransferases

The localization of the individual SERAT isoforms in *Arabidopsis* in the three cellular compartments, as well as their different biochemical properties and gene expression patterns, suggest specific functions of each SERAT isoform *in vivo*. To define the *in vivo* function of individual SERAT isoforms, *Arabidopsis* T-DNA mutants of *SERAT* genes, including single and multiple knockout mutants, have been analysed (Watanabe *et al.*, 2008a).

The quintuple mutant lacking all five *SERAT* genes was embryo lethal. This suggests that the *SERAT* genes are essential and that no alternative pathway for cysteine production from OAS exists, at least at the embryo maturation stage. All five quadruple mutants where a single *SERAT* gene was retained were viable (Fig. 11.6) indicating that OAS production in a single compartment is sufficient and that OAS and/or Cys are freely interchangeable among cellular compartments. Additional to the fact that the five single mutants (*serat1;1*, *serat2;1*, *serat2;2*, *serat3;1* and *serat3;2*) did not reveal any morphological phenotype under normal growth conditions, they did not exhibit any visible phenotype under stress conditions such as sulfur deficiency and heavy metal treatment either. Thus, the individual SERAT isoforms are functionally redundant and OAS and cysteine are efficiently transported between the cytosol and organelles, even under stress conditions (Watanabe *et al.*, 2008a).

Among the quadruple mutants, those expressing only *SERAT2;2* (Q2;2) or *SERAT1;1* (Q1;1) showed normal growth compared with wild type plants, whereas those expressing only *SERAT2;1* (Q2;1), *SERAT3;1* (Q3;1) or *SERAT3;2* (Q3;2) showed dwarfism on agar plates (Watanabe *et al.*, 2008a) and on soil under long-day conditions (Fig. 11.6). However, the three dwarf mutants

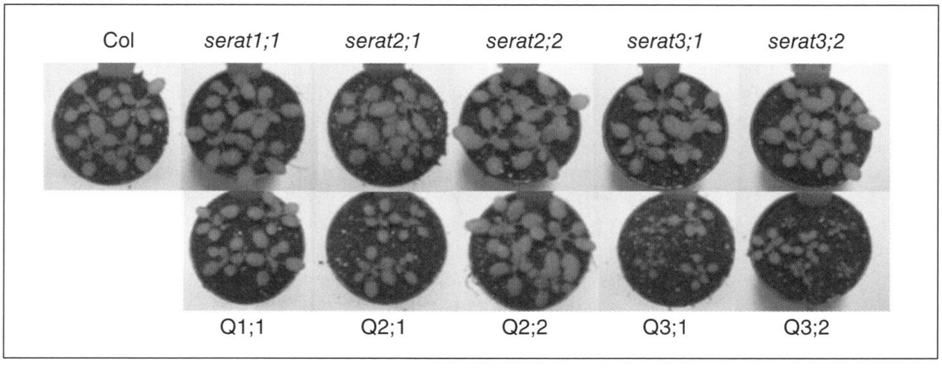

Fig. 11.6. Growth phenotypes of serine acetyltransferase quadruple knockout (*serat*) mutants of *Arabidopsis* showing the phenotypic changes in *serat* mutants. Plants were grown for 4 weeks on soil under long-day conditions. The five quadruple mutants were named as Q1;1 (*serat2;1serat2;2serat3;1serat3;2*), Q2;1 (*serat1;1serat2;2serat3;1serat3;2*), Q2;2 (*serat1;1serat2;1serat3;1serat3;2*), Q3;1 (*serat1;1serat2;1serat2;2serat3;2*) and Q3;2 (*serat1;1serat2;1serat2;2serat3;1*) according to the SERAT gene retained in each mutant. Col indicates the wild type (non-mutant) growth phenotype.

could complete their life cycles, including flowering and seed setting, indicating that even the minor isoforms SERAT3;1 and SERAT3;2 were sufficient for the survival of the mutants. The efficient distribution of OAS and cysteine between sub-cellular compartments was corroborated by non-aqueous fractionation analysis of cytosolic *serat1;1* and plastidic *serat2;1* mutants under normal and sulfur-deficient growth conditions (Krueger *et al.*, 2009, 2010). The growth phenotypes of quadruple mutants were consistent with the total SERAT activity in the single *serat* mutants (Fig. 11.7) (Watanabe *et al.*, 2008a). A significant reduction in total SERAT activity was observed in leaves, stems and roots in the mitochondrial *serat2;2* mutant, and in roots of the cytosolic *serat1;1* mutant. Loss of mitochondrial SERAT2;2 especially, caused the loss of 70–80% of total SERAT activity and a significant reduction of OAS contents in leaf and stem tissues, indicating that the mitochondrial SERAT2;2 plays a predominant role in cellular OAS formation in the photosynthetic leaf and stem tissues. In contrast, in root and silique tissues, the predominance of mitochondrial SERAT2;2 disappeared and the contribution of the cytosolic SERAT1;1 for cellular SERAT activity and OAS formation increased. The importance of the cytosolic isoforms was more obvious in silique tissues, where the contributions of the minor cytosolic isoforms, SERAT3;1 and SERAT3;2, were also increased and hence substantially contributed to the cellular SERAT activity. The increased contribution of SERAT3;1 and SERAT3;2 in silique tissues was consistent with their elevated gene expression levels at the reproductive stage (Kawashima *et al.*, 2005) (Fig. 11.5A). The three cytosolic isoforms, SERAT1;1, SERAT3;1 and SERAT3;2, may thus play

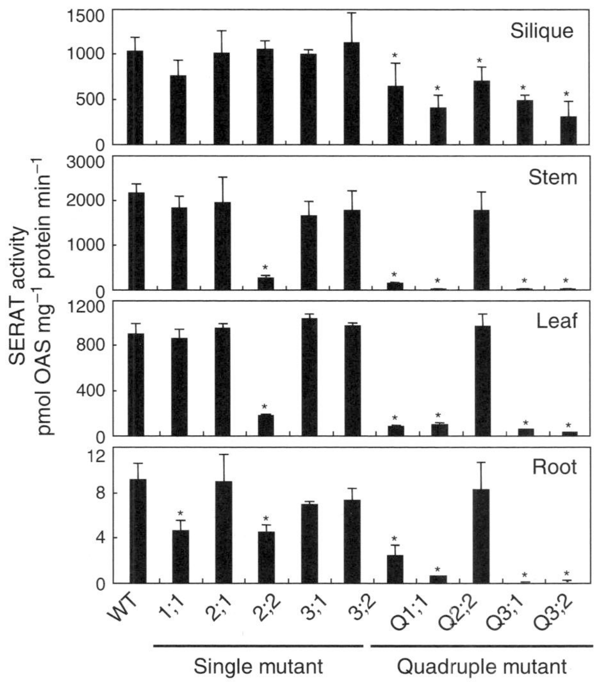

Fig. 11.7. Serine acetyltransferase (SERAT) activity in single and quadruple knockout *serat* mutants of *Arabidopsis*. The SERAT activity of *serat* mutants in leaves and roots of 2-week-old plants and in stems and siliques (5 to 10 d after flowering) of 8-week-old plants was analysed (Watanabe *et al.*, 2008a). Data represent the mean of three experiments, with 15 plants in each experiment. Where indicated, differences between the wild type (WT) and *serat* mutants, as analysed using the Student's *t*-test were statistically significant (*, $P < 0.05$). The naming convention used for the mutants (Q1:1, Q2;1, Q2;2, Q3;1 and Q3;2) is given in the caption to Fig. 11.6. The figure is modified from that of Watanabe *et al.*, 2008 (www.plantcell.org; Copyright American Society of Plant Biologists).

a major role during seed development instead of the mitochondrial SERAT2;2, whose gene expression is significantly decreased during seed maturation (Fig. 11.5a).

The analysis of these *serat* knockout mutants revealed that the plastidic SERAT2;1 contributes less to OAS formation than the major cytosolic SERAT1;1 and the mitochondrial SERAT2;2. This is corroborated by the fact that knockout *oastl* mutant lines revealed a lesser contribution of the plastidic OASTL to cysteine synthesis than the cellular or mitochondrial OASTLs (Heeg *et al.*, 2008; Watanabe *et al.*, 2008b).

Plastids have initially been estimated to be the main compartment for cysteine synthesis because sulfate reduction exclusively takes place in the plastids, but that is not the case, at least in *Arabidopsis*. The reduced contribution of plastidic SERAT and OASTL for the synthesis of OAS and cysteine might enable the efficient distribution of sulfide from plastids to other sub-cellular compartments through the cytosol, as it would allow higher sulfide concentrations to build up in the plastids, thereby facilitating diffusion to the cytosol, as no sulfide transport system is yet known. High activity of SERAT in the mitochondria of photosynthetic tissues allows the efficient use of the serine that is synthesized in these tissues and accumulated in the mitochondria (Ho *et al.*, 1999; Ho and Saito, 2001; Krueger *et al.*, 2009). The predominance of cysteine synthesis (Heeg *et al.*, 2008; Watanabe *et al.*, 2008b) and cysteine accumulation in the cytosol (Krueger *et al.*, 2009) accords with the presence of feedback-sensitive SERAT isoforms in the cytosol, as these would allow keeping tight control of the homeostasis of cysteine synthesis (Kawashima *et al.*, 2005). Further, the predominance of cysteine synthesis in the cytosol might allow the efficient utilization of OAS from the mitochondria and sulfide from the plastids in the cytosol. However, the pertinent findings for *Arabidopsis* should not to be overgeneralized as different distributions of SERAT activities and of feedback-sensitive SERAT isoforms in the organelles are found in different plant species.

As mentioned above, the induction of gene expression of the plastidic *SERAT2;1* in senescing leaves and also by oxidative stresses, together with the possible regulation of the SERAT2;1 protein by CYP20-3 under oxidative stress conditions, suggests a specific function of SERAT in such oxidative stress conditions (Dominguez-Solis *et al.*, 2008). It appears possible that increased *SERAT2;1* expression and, thus, activity, results in higher flux into cysteine and eventually GSH, the latter being involved in the ascorbate-GSH cycle and its role in detoxifying ROS (Blaszczyk *et al.*, 1999; Harms *et al.*, 2000).

In contrast to the finding that no growth retardation was observed in a *serat2;2* knockout mutant (Watanabe *et al.*, 2008a), the downregulation of *SERAT2;2* by a gene-specific artificial miRNA resulted in stunted growth (Haas *et al.*, 2008). In the miRNA lines, a strong correlation between the mRNA levels of *SERAT2;2* and plant biomass was observed. The discrepancy between these studies might be caused by differences in growth conditions as Watanabe *et al.* (2008a) used agar plates containing sucrose under long-day conditions, while Haas *et al.* (2008) grew plants photoautotrophically on soil under short-day conditions (Haas *et al.*, 2008). Constituents of the sulfur pathway, especially APR (adenosine-5′-phosphosulfate reductase), are regulated by light and sugar (Kopriva *et al.*, 1999). Indeed, the gene expression patterns of *SERATs*, especially *SERAT2;1* and *SERAT2;2*, are affected by light (Genvestigator). Moreover, differences between the experiments might also be accounted for by the differences in protein accumulation of SERAT2;2 protein. In the knockout *serat2;2* mutant, the SERAT2;2 protein is completely absent, whereas it might still exist in low abundance in the miRNA lines. Hell and Wirtz (2011) suggested that the CSC has a regulatory function beyond the modulation of SERAT enzyme activity. An imbalance of the SERAT/OASTL ratio in the miRNA lines, rather than the complete lack of complex regulation in knockout mutants, might cause the deregulation of sulfur metabolism in addition to the deregulation of cysteine synthesis. This is supported by antisense repression of OASTL in *S. tuberosum*, in which changing the OASTL/SERAT ratio resulted in deregulation of the system and an effect on cysteine homeostasis (Riemenschneider *et al.*, 2005).

11.7 Serine Acetyltransferase Overexpressors

SERAT overexpressors are useful not only for studying the regulation of cysteine biosynthesis in plants, but also for improving the nutritional quality of crops. Increasing the cysteine and methionine contents of seeds in crops has long since been a goal for breeding and agricultural biotechnology. SERAT overexpressors have been produced in several plant species: *A. thaliana* (Noji and Saito, 2002); *N. tabacum* (Blaszczyk *et al.*, 1999, 2002; Wirtz and Hell, 2003, 2007); *S. tuberosum* (Harms *et al.*, 2000); *Oryza sativa* (Huu Cuong *et al.*, 2012) and *Lupinus angustifolius.* (Tabe *et al.*, 2010). The overexpression of SERAT in general could contribute to increased cysteine and GSH accumulation in the transgenic lines. The induction levels achieved depended on the feedback sensitivity and the sub-cellular localization of the overexpressed SERAT and differed between plant species.

The overexpression of SERATs was mainly achieved using constitutive promoters and was assayed in vegetative plant parts (Sirko *et al.*, 2004). However, from an agricultural point of view, the assay of sink tissues such as seeds, tubers and fruits is more important. In the case of *S. tuberosum*, the transgenic lines overexpressing SERAT with a constitutive 35S promoter displayed increased levels of cysteine and GSH in leaves, but not in tubers (Harms *et al.*, 2000). Transgenic rice, *O. sativa*, overexpressing SERAT under control of a constitutive ubiquitin promoter, succeeded in producing higher levels of cysteine and methionine in the seeds (Huu Cuong *et al.*, 2012), while the transgenic lines of the grain legume narrow leaf lupin (*L. angustifolius*) overexpressing SERAT with a seed-specific promoter had increased levels of cysteine but not methionine in the developing seeds (Tabe *et al.*, 2010). This indicates differences in regulation of sulfur metabolism in different plant species. The modification of cysteine metabolism is of further agricultural interest because of its importance for the production of GSH, which is involved in redox regulation, stress tolerance and heavy metal detoxification (Noctor *et al.*, 2012). The production of OAS and cysteine limits GSH biosynthesis and the maintenance of the GSH pool (Barroso *et al.*, 1995; Meyer and Fricker, 2002). Consistent with these reports, the *SERAT* overexpressors in *N. tabacum* with elevated SERAT activity and subsequently elevated GSH levels showed increased resistance to oxidative stress generated by hydrogen peroxide (Blaszczyk *et al.*, 2002). SERAT overexpressors in several plant species were not assayed for heavy metal detoxification, in contrast to OASTL overexpressors in *Arabidopsis* and *N. tabacum* which displayed increased tolerance to heavy metals such as cadmium (reviewed in Sirko *et al.*, 2004). However, the importance of SERAT for tolerance to heavy metals was indicated by the report that in the Ni hyperaccumulator *Thlaspi goesingense*, the constitutively higher activity of SERAT was responsible for the elevated GSH levels and hence improved Ni tolerance (Freeman and Salt, 2007). Moreover, OAS levels have been shown to correlate with Ni/Zn hyperaccumulation ability across numerous genera in the *Brassicaceae* family (Peer *et al.*, 2006).

11.8 *O*-Acetylserine Signalling

11.8.1 Identification of *O*-acetylserine cluster genes

OAS levels are highly increased under sulfur starvation. Thus, the OAS molecule is a marker metabolite of sulfate starvation and has been assumed to be a signalling molecule. The idea of a signalling function of OAS in plants originally came from *N*-acetylserine (NAS) and OAS as positive regulators for the sulfur pathway in bacteria (Ostrowski and Kredich, 1990, 1991; Lynch *et al.*, 1994). NAS, rather than OAS, strongly induced the cysteine regulon in bacteria (Lynch *et al.*, 1994). In plants though, OAS and not NAS induced the expression of enzymes of the sulfate assimilation pathway (Neuenschwander *et al.*, 1991; Hawkesford *et al.*, 1995). Hopkins *et al.* (2005) reported OAS and NAS contents in *S. tuberosum* leaves and roots, but they mentioned difficulties in measuring these because the conversion from OAS to NAS easily happens during the extraction process that is required for the measurement. NAS is produced by a non-enzymatic conversion reaction from OAS at a rate of about 1% per minute at neutral pH (Flavin and Slaughter, 1965). Therefore, there are no strict data of OAS and NAS contents in plant cell at the moment.

Several OAS adding experiments demonstrated that OAS induced gene expression of *APRs* (Koprivova *et al.*, 2000; Kopriva *et al.*, 2002; Hesse *et al.*, 2003) and of the root-localized sulfate transporters *SULTR1;1* and *SULTR1;2* (Maruyama-Nakashita *et al.*, 2004). Further, comparison of the transcriptome data from sulfate starvation and OAS addition experiments revealed that OAS was probably a global regulator of a large number of genes, which in many cases were the same genes responding to sulfur starvation (Hirai *et al.*, 2003). Moreover, it was mentioned in sulfur starvation studies that a positive correlation might occur between OAS accumulation and the induction of several sulfur pathway genes, such as sulfate transporters and *APRs* (Maruyama-Nakashita *et al.*, 2003; Nikiforova *et al.*, 2003). However, the role of OAS as a signalling molecule is still under debate, mainly based on the argument that the concentrations of other sulfur-related metabolites, including possible signal molecules, for example, sulfide, cysteine and GSH (Kopriva, 2006), were also changed in plants when OAS was supplied or under sulfate starvation conditions. OAS addition might in fact mimic sulfur starvation as it creates an increased demand for reduced sulfur. In support of this, induction of the expression of *SULTR1;1*, which is one of genes responsive to sulfur deficiency, did not correlate with OAS accumulation, but rather with a decrease in sulfate contents (Rouached *et al.*, 2008) suggesting that sulfate is also a possible signal molecule, like nitrate, which regulates nitrogen and carbon metabolism and coordinates plant development (Forde, 2002; Sakakibara *et al.*, 2006; Vidal and Gutierrez, 2008). Moreover, substantial increases of OAS in tissues were observed only after an extended period of sulfur starvation (Buchner *et al.*, 2004; Hopkins *et al.*, 2005).

In order to dissect whether a certain sulfur-related metabolite might act as a signal molecule for the sulfur pathway, it is necessary to devise specific conditions under which only a single metabolite of the potential signal metabolites is changed. Hubberten *et al.* (2012) succeeded in identifying OAS-correlated genes that are specifically co-expressed with OAS. To this end, metabolite-gene correlation analyses of time-series transcriptome and metabolome data were performed; these were published for studies in *Arabidopsis* of metabolite–gene correlations in natural diurnal oscillations (Espinoza *et al.*, 2010) and in response to light and temperature (Caldana *et al.*, 2011) (Fig. 11.8). In both studies, specific and transient OAS accumulations were observed without any changes in other metabolites, including sulfur-related metabolites, such as sulfide, sulfite or thiols. OAS levels peaked in the middle of the night in the diurnal oscillations (Espinoza *et al.*, 2010) and OAS increased transiently within 5–10 min after transferring plants from light to darkness (Caldana *et al.*, 2011). Using these two data sets, an OAS to gene correlation

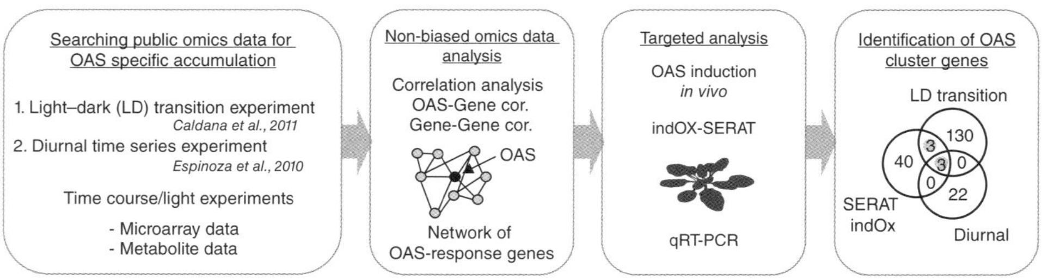

Fig. 11.8. The identification of *O*-acetylserine (OAS)-correlated genes. The workflow illustrates the strategy used to identify OAS-correlated genes (Hubberten *et al.*, 2012) by analysing 'omics data sets from two time course light experiments (Espinoza *et al*, 2010; Caldana *et al.*, 2011). OAS-gene correlations were calculated to identify genes showing similar behaviour to OAS. The identified genes were filtered by partial correlation analysis to exclude those that putatively respond to other metabolites. The inferred genes were then used as guide genes in a co-expression analysis by calculating gene-gene correlations. Derived candidate genes were again filtered by partial correlation. The candidate genes were compared with a gene set derived by analysing SERAT overexpressing lines that accumulate OAS *in vivo*. Results of the analysis were visualized as a Venn diagram. Three genes (*APR3*, *SDI1* and *SDI2*) were overlapped among the two omics experiments and the SERAT experiment. An additional three genes (*LSU1*, *SHM7* and *ChaC*) were overlapped between the light–dark transition experiment and the SERAT experiment. (see Table 11.3.)

analysis allowing time shifts was performed in order to identify OAS-correlated genes. Additionally, correlation analyses for other metabolites overlapping with these responses were conducted to filter for specifically OAS-correlated genes. To evaluate the results of this computational approach, transgenic *Arabidopsis* plants were created that expressed *SERAT* under an inducible promoter in order to be able to control SERAT accumulation, and hence OAS accumulation, in a time-dependent manner. The OAS to gene correlation of the candidate genes was proven by quantitative RT-PCR. The induction of the *SERAT* genes resulted in a specific increase of OAS, but with no changes in other sulfur-related metabolites, for up to 3 h after induction (Hubberten *et al.*, 2012). Comparison of the results from the computational approach and the analysis of the SERAT overexpressing plants resulted in the identification of six genes as core OAS-correlated genes: *APR3*, adenosine-5′-phosphosulfate reductase 3; *SDI1*, sulfur-deficiency-induced 1; *SDI2*, sulfur-deficiency-induced 2; *LSU1*, low-sulfur- induced 1; *SHM7*, serine hydroxymethyltransferase 7; and *ChaC*, ChaC-like protein (Table 11.3). This study suggested a signalling function of OAS for a specific gene set irrespective of the sulfur status of the plant. The set of genes induced by OAS was termed the OAS cluster.

11.8.2 Regulation of *O*-acetylserine cluster genes

In *Arabidopsis*, the core six OAS cluster genes displayed a high degree of co-regulation under various experimental conditions (Obayashi *et al.*, 2007, 2009; Mutwil *et al.*, 2010; Hubberten *et al.*, 2012). When using the six core genes as query genes for co-expression network analysis of 1388 available microarray data sets using the ATTED II software (http://atted.jp), a stable co-expression network that included the core six genes with the sulfur pathway-related genes (*APR1* and *APR2*), the tonoplast-localized sulfate transporters (*SULTR4;1* and *SULTR4;2*) and the *LSU* family gene (*LSU2*, a gene annotated as beta glucosidase 28, *BGLU28*, and an unknown gene, At1g112030) was displayed (Fig. 11.9). Further, the very same co-expression network was also observed in other plant species, namely *O. sativa* and *Populus trichocarpa* (Mutwil *et al.*, 2008, 2011; Hubberten *et al.*, 2012), suggesting that the co-regulation of the OAS cluster genes is conserved in different plant species, and even between monocotyledons and dicotyledons. The tight co-regulation of the OAS cluster genes was corroborated by the report that the promoter regions contained a conserved cis-element, UPE box (UP9 binding element) or UPE-like motif (Wawrzynska *et al.*, 2010). The *UP9* gene was found as a response gene to sulfur deficiency in *N. tabacum*, and the respective homologous genes in *Arabidopsis* are members of the *LSU* family (Nikiforova *et al.*, 2003, 2005; Wawrzynska *et al.*, 2005). The UPE box consists of two parallel TEB boxes (tobacco ethylene-insensitive 3-like binding) and was identified as required for the binding of transcription factors belonging to the EIL (ethylene-insensitive 3-like) family, including SLIM1 (sulfur limitation 1) (Wawrzynska *et al.*, 2010). SLIM1 is a central transcriptional regulator of sulfur response in *Arabidopsis*

Table 11.3. The gene list of the six core *O-acetylserine* (OAS)-correlated genes from various studies.

Gene name	AGI code (unique locus identifier)	Annotation	Screening method[a]
APR3	At4g21990	Adenosine-5′-phosphosulfate reductase 3	LD/Diurnal/SERATindOx
ChaC	At5g26220	ChaC-like protein	LD/SERATindOx
LSU1	At3g49580	Low-sulfur-induced 1	LD/SERATindOx
SDI1	At5g48850	Sulfur-deficiency-induced 1	LD/Diurnal/SERATindOx
SDI2	At1g04770	Sulfur-deficiency-induced 2	LD/Diurnal/SERATindOx
SHM7	At1g36370	Serine hydroxymethyltransferase 7	LD/SERATindOx

LD, light–dark transition experiment (Caldana *et al.*, 2011); Diurnal, diurnal time series experiment (Espinoza *et al.*, 2010); SERATindOx, SERAT overexpressing lines experiment (Hubberten *et al.*, 2012). See also Fig. 11.8.

(Maruyama-Nakashita *et al.*, 2006). The presence of the SLIM1 binding site in the promoter region of OAS cluster genes suggests a regulation by SLIM1. However, transcriptome analysis of the knockout *slim1* mutant suggests the existence of additional regulators or redundancy of the EIL family genes for the regulation of OAS cluster genes because in the *slim1* mutant the induction of only two among the six core genes, *ChaC* and *SDI1*, was repressed under sulfate deficiency (Maruyama-Nakashita *et al.*, 2006). Another possible explanation of this result is that a tissue-specific regulation of SLIM1 exists and only root tissues were investigated. As *SLIM1* itself is not regulated at the gene expression level under sulfur deficiency, the existence of additional factors regulating SLIM1 activity are likely.

11.8.3 Functions of *O*-acetylserine cluster genes

The strong correlation of OAS cluster genes to OAS together with some sulfur pathway genes (Fig. 11.9) suggests that the OAS cluster genes should have a function in sulfur metabolism. However, the molecular functions and physiological consequences of the inductions of the OAS cluster genes are still unknown. Among the six core genes, the function of only *APR3* has been defined. APR is a key limiting enzyme in the assimilatory reduction of sulfate, and the induction of *APR3* directly increases the flux through the sulfur assimilation pathway (Vauclare *et al.*, 2002).

The *SDI* gene was originally identified as an indicator of sulfur status by cDNA-amplified fragment length polymorphism (AFLP) analysis utilizing field-grown, nutrient-deficient wheat (*Triticum aestivum* var. Hereward) (Howarth *et al.*, 2005). The *Arabidopsis SDI1* gene is the one most highly homologous to the wheat *SDI* gene. The OAS cluster genes *SDI1* and *SDI2* belong to the *MS5* (Male-sterile 5) gene family, which consists of five genes in *Arabidopsis*. One of the *MS5* family genes, *MS5* (At4g20900), has been identified to be involved in male fertility, anther development, meiosis and tapetal development (Ross *et al.*, 1997; Bulankova *et al.*, 2013). However, the gene expression of *MS5* is not altered under sulfur starvation. The *Arabidopsis* T-DNA knockout mutant of *SDI1* showed higher sulfate concentrations than wild type plants under sulfate starvation conditions, suggesting a role in the utilization of stored sulfate, but the molecular function of *SDI* still remains unknown (Howarth *et al.*, 2009).

The *LSU* gene was originally identified as a strongly upregulated gene under sulfur starvation in *N. tabacum*, named *UP9* (Wawrzynska *et al.*, 2005). By yeast two-hybrid experiments, a range of plant proteins were identified that potentially interacted with the UP9C protein, which is one of six UP9 proteins in *N. tabacum* and the isoform most highly expressed under sulfur starvation (Lewandowska *et al.*, 2005). UP9C has been suggested to be involved in the synthesis and signalling of phytohormones such as jasmonic acid and ethylene, and also during autophagy through its multiple protein–protein interactions (Zientara-Rytter *et al.*, 2011). Both functions have been predicted as part of the sulfur starvation response (Nikiforova *et al.*, 2003, 2005). However, more studies, including the analysis of *Arabidopsis LSU* genes, are required to fully confirm the suggested UP9/LSU function. In *Arabidopsis*, the *LSU* gene family consists of four genes, of which two (*LSU1* and *LSU2*) belong to the OAS-co-expressed network (Fig. 11.9) and were reported as being strongly induced by sulfur starvation (Nikiforova *et al.*, 2003, 2005). One of the *LSU* family genes, *LSU4*, which is not a sulfur response gene, has been reported to be involved in flower development by analysis of a T-DNA knockout mutant, *lsu4* (Myakushina *et al.*, 2009).

The molecular function of SHM7 also remains unknown. However, there are some data highlighted by ionomics studies with inductively coupled plasma mass spectrometry (ICP-MS) that the *Arabidopsis* T-DNA knockout mutant of *SHM7* accumulated sulfur in the shoot (see http://www.ionomicshub.org/arabidopsis/piims/showIndex.action), suggesting a function in sulfur metabolism. *SHM7* belongs to the serine hydroxyl methyltransferase (*SHM*) gene family, which consists of seven genes in *Arabidopsis*. The SHMs are important enzymes of cellular one-carbon metabolism (McClung *et al.*, 2000; Hanson and Roje, 2001; Bauwe and Kolukisaoglu, 2003). Mitochondrial SHM1 and SHM2 were reported to be essential for the photorespiratory glycine-into-serine conversion in leaf mesophyll

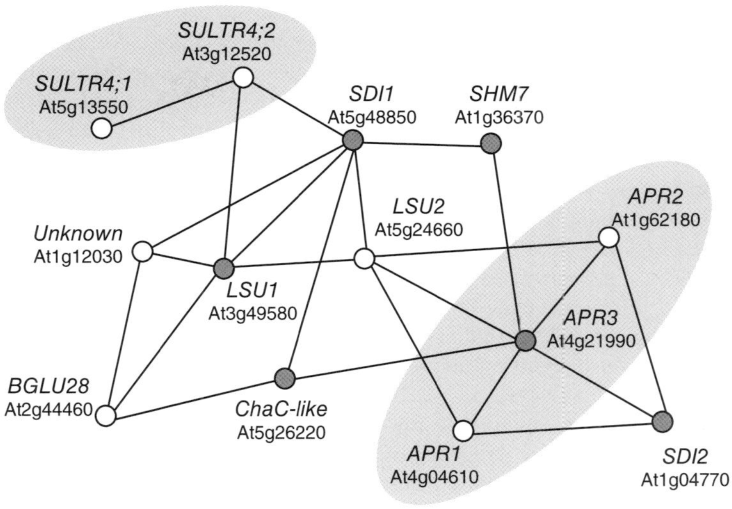

Fig. 11.9. Co-expressed gene network around the *O*-acetylserine (OAS)-correlated genes. The six OAS correlated genes were used as query genes for searching the co-expressed gene network using the software ATTED II (http://atted.jp/). The OAS-correlated genes are in dark grey circles. Sulfur pathway genes (*SULTRs* and *APRs*) are in grey underlay. Abbreviations: *APR,* adenosine-5′-phosphosulfate reductase; *BGLU*, beta-glucosidase; *ChaC*, ChaC-like protein; *LSU*, low-sulfur-induced; *SDI*, sulfur-deficiency-induced; *SHM*, serine hydroxymethyltransferase; *SULTR*, sulfate transporter; Unknown, unknown function.

mitochondria (Voll *et al.*, 2006; Engel *et al.*, 2007). Plastidic SHM3 was identified as catalysing a reversible hydroxymethyl group transfer from serine to tetrahydrofolate, yielding glycine and 5,10-methylenetetrahydrofolate (Zhang *et al.*, 2010). However, other isoforms as SHM4 and SHM5, which are assumed to be localized in the cytosol, and SHM6 and SHM7, which are likely to be localized in nuclei, have not been studied well yet.

ChaC was identified in *E. coli* as a gene of the *Cha* operon (Ca^{2+}/H^+ antiporter), which consists of three genes, *ChaA*, *ChaB* and *ChaC*, and is assumed to be a possible regulator of this operon by adjusting the function of *ChaA* (Ivey *et al.*, 1993; Osborne *et al.*, 2004; Uddin *et al.*, 2008). The mammalian ChaC1 protein was identified as an proapoptotic protein induced during endoplasmic reticulum stress (Mungrue *et al.*, 2009) and also as a prognostic marker for certain types of cancer (Goebel *et al.*, 2012). Despite the possible function of the ChaC protein in cation transport, ChaC has γ-glutamyl cyclotransferase-like domain and γ-glutamyl cyclotransferase activity and degrades GSH to yield 5-oxoproline and cysteinyl-glycine (Kumar *et al.*, 2012). This suggests that the ChaC protein is probably involved in the recycling of GSH during sulfur starvation. There is an interesting report that in profiles of suppression subtractive hybridizations in rice plants, the *OsCTP* (cation transport protein) gene, which displays homologies to the *ChaC* of *E. coli*, and GSH synthetase were identified as submergence-induced genes from the submergence-tolerant rice cultivar (Qi *et al.*, 2005), suggesting that there is a link between ChaC function and GSH metabolism in rice as well.

11.9 Conclusions

SERAT is a key enzyme of the sulfur assimilation pathway. The functions of SERATs are not only restricted to OAS production for cysteine synthesis, but also cover a function in keeping cysteine homeostasis by feedback inhibition of cysteine synthesis and formation of the SERAT-OASTL complex (CSC). Further, the accumulation of OAS has a signalling function for sulfate metabolism but also beyond this. A combination of biochemical studies (Table 11.2), gene expression analysis (Fig. 11.5) and T-DNA knockout mutant analysis (Figs 11.6 and 11.7) revealed the contribution of each SERAT isoform to the total SERAT activity and production of OAS in *Arabidopsis* tissues. The contribution of mitochondrial SERAT2;2 is significantly predominant in photosynthetic tissues such as leaves and stems. The contribution of the cytosolic SERAT1;1 is higher in roots and siliques. The contributions of the two further cytosolic isoforms, SERAT3;1 and SERAT3;2, seem to be higher at the reproductive stage. The survival of quadruple mutants expressing either *SERAT3;1* or *SERAT3;2* only proved their activity as a functional SERAT in plant cells (Fig. 11.6). However, the dwarf phenotypes of the two quadruple mutants (Fig. 11.6), the biochemical properties of higher K_m values for the substrates (Table 11.2) and specific gene expression patterns such as their higher expression at the reproductive stage (Fig. 11.5a) and significant induction under stress conditions (Fig. 11.5c) suggests that there are still additional functions in specific cell types or under specific conditions where higher concentrations of substrates exist in the plant cell. The contribution to the total SERAT activity of the plastidic SERAT2;1 is less than that of the other major isoforms, mitochondrial SERAT2;2 and cytosolic SERAT1;1, but the plastidic SERAT2;1 seems to have a specific function under oxidative stress conditions, as indicated by the regulation of an interacting protein, CYP20-3 (Dominguez-Solis *et al.*, 2008). The identification of a new interaction partner to SERAT2;1 speculatively suggests the existence of further protein partners interacting with SERAT or the SERAT-OASTL complex, and might indicate additional functions of SERAT besides its function in sulfur metabolism.

The survival of single and quadruple *serat* mutants (Fig. 11.6) and metabolite profiles obtained by non-aqueous fractionation of *serat* mutants (Krueger *et al.*, 2009) indicate that sulfide, OAS and cysteine are transported between sub-cellular compartments. However, the transport system of these metabolites between the compartments is currently unclear. The sub-cellular distribution of OAS is an important issue because of its function as a signal molecule. It is not yet clear in which compartment OAS accumulates under sulfur starvation or light–dark shifts, whether the distribution of OAS between compartments is limited or not, and how OAS induces the OAS cluster genes. Identification of the transport system and characterization of the OAS cluster genes will help to elucidate the mechanisms of OAS function as a signal molecule. This might further help in understanding why at all OAS and cysteine synthesis takes place in three compartments in a plant, given that effective transport systems do seem to exist.

References

Armengaud, P., Breitling, R. and Amtmann, A. (2004) The potassium-dependent transcriptome of *Arabidopsis* reveals a prominent role of jasmonic acid in nutrient signaling. *Plant Physiology* 136, 2556–2576.

Barroso, C., Vega, J.M. and Gotor, C. (1995) A new member of the cytosolic *O*-acetylserine(thiol)lyase gene family in *Arabidopsis thaliana*. *FEBS Letters* 363, 1–5.

Bauwe, H. and Kolukisaoglu, U. (2003) Genetic manipulation of glycine decarboxylation. *Journal of Experimental Botany* 54, 1523–1535.

Beinert, H. (2000) A tribute to sulfur. *European Journal of Biochemistry* 267, 5657–5664.

Berkowitz, O., Wirtz, M., Wolf, A., Kuhlmann, J. and Hell, R. (2002) Use of biomolecular interaction analysis to elucidate the regulatory mechanism of the cysteine synthase complex from *Arabidopsis thaliana*. *The Journal of Biological Chemistry* 277, 30629–30634.

Birke, H., Mueller, S.J., Rother, M., Zimmer, A.D., Hoernstein, S.N.W., Wesenberg, D., Wirtz, M., Krauss, G.-J., Reski, R. and Hell, R. (2012) The relevance of compartmentation for cysteine synthesis in phototrophic organisms. *Protoplasma* 249, 147–155.

Blaszczyk, A., Brodzik, R. and Sirko, A. (1999) Increased resistance to oxidative stress in transgenic tobacco plants overexpressing bacterial serine acetyltransferase. *The Plant Journal* 20, 237–243.

Blaszczyk, A., Sirko, L., Hawkesford, M.J. and Sirko, A. (2002) Biochemical analysis of transgenic tobacco lines producing bacterial serine acetyltransferase. *Plant Science* 162, 589–597.

Bogdanova, N. and Hell, R. (1997) Cysteine synthesis in plants: protein–protein interactions of serine acetyltransferase from *Arabidopsis thaliana*. *The Plant Journal* 11, 251–262.

Bonner, E.R., Cahoon, R.E., Knapke, S.M. and Jez, J.M. (2005) Molecular basis of cysteine biosynthesis in plants – structural and functional analysis of *O*-acetylserine sulfhydrylase from *Arabidopsis thaliana*. *The Journal of Biological Chemistry* 280, 38803–38813.

Brunold, C. and Suter, M. (1982) Intracellular-localization of serine acetyltransferase in spinach leaves. *Planta* 155, 321–327.

Buchner, P., Stuiver, C.E.E., Westerman, S., Wirtz, M., Hell, R., Hawkesford, M.J. and De Kok, L.J. (2004) Regulation of sulfate uptake and expression of sulfate transporter genes in *Brassica oleracea* as affected by atmospheric H_2S and pedospheric sulfate nutrition. *Plant Physiology* 136, 3396–3408.

Bulankova, P., Akimcheva, S., Fellner, N. and Riha, K. (2013) Identification of *Arabidopsis* meiotic cyclins reveals functional diversification among plant cyclin genes. *PloS Genetics* 9.

Burkhard, P., Rao, G.S.J., Hohenester, E., Schnackerz, K.D., Cook, P.F. and Jansonius, J.N. (1998) Three-dimensional structure of *O*-acetylserine sulfhydrylase from *Salmonella typhimurium*. *Journal of Molecular Biology* 283, 121–133.

Burkhard, P., Tai, C.H., Ristroph, C.M., Cook, P.F. and Jansonius, J.N. (1999) Ligand binding induces a large conformational change in *O*-acetylserine sulfhydrylase from *Salmonella typhimurium*. *Journal of Molecular Biology* 291, 941–953.

Caldana, C., Degenkolbe, T., Cuadros-Inostroza, A., Klie, S., Sulpice, R., Leisse, A., Steinhauser, D., Fernie, A.R., Willmitzer, L. and Hannah, M.A. (2011) High-density kinetic analysis of the metabolomic and transcriptomic response of *Arabidopsis* to eight environmental conditions. *The Plant Journal* 67, 869–884.

Chronis, D. and Krishnan, H.B. (2004) Sulfur assimilation in soybean (*Glycine max* [L.] Merr.): molecular cloning and characterization of a cytosolic isoform of serine acetyltransferase. *Planta* 218, 417–426.

Davletova, S., Schlauch, K., Coutu, J. and Mittler, R. (2005) The zinc-finger protein Zat12 plays a central role in reactive oxygen and abiotic stress signaling in *Arabidopsis*. *Plant Physiology* 139, 847–856.

Denk, D. and Bock, A. (1987) L-Cysteine biosynthesis in *Escherichia coli* – nucleotide sequence and expression of the serine acetyltransferase (*cysE*) gene from the wild-type and a cysteine-excreting mutant. *Journal of General Microbiology* 133, 515–525.

Dominguez-Solis, J.R., He, Z., Lima, A., Ting, J., Buchanan, B.B. and Luan, S. (2008) A cyclophilin links redox and light signals to cysteine biosynthesis and stress responses in chloroplasts. *Proceedings of the National Academy of Sciences of the United States of America* 105, 16386–16391. [Correction at 106, 1292–1292.]

Droux, M. (2003) Plant serine acetyltransferase: new insights for regulation of sulphur metabolism in plant cells. *Plant Physiology and Biochemistry* 41, 619–627.

Droux, M., Ruffet, M.L., Douce, R. and Job, D. (1998) Interactions between serine acetyltransferase and *O*-acetylserine(thiol)lyase in higher plants – structural and kinetic properties of the free and bound enzymes. *European Journal of Biochemistry* 255, 235–245.

Engel, N., van den Daele, K., Kolukisaoglu, U., Morgenthal, K., Weckwerth, W., Pärnik, T., Keerberg, O. and Bauwe, H. (2007) Deletion of glycine decarboxylase in *Arabidopsis* is lethal under nonphotorespiratory conditions. *Plant Physiology* 144, 1328–1335.

Espinoza, C., Degenkolbe, T., Caldana, C., Zuther, E., Leisse, A., Willmitzer, L., Hincha, D.K. and Hannah, M.A. (2010) Interaction with diurnal and circadian regulation results in dynamic metabolic and transcriptional changes during cold acclimation in *Arabidopsis*. *PloS ONE* 5(11): e14101.

Flavin, M. and Slaughter, C. (1965) Synthesis of succinic ester of homoserine a new intermediate in bacterial biosynthesis of methionine. *Biochemistry* 4, 1370–1375.

Forde, B.G. (2002) Local and long-range signaling pathways regulating plant responses to nitrate. *Annual Review of Plant Biology* 53, 203–224.

Francois, J.A., Kumaran, S. and Jez, J.M. (2006) Structural basis for interaction of *O*-acetylserine sulfhydrylase and serine acetyltransferase in the *Arabidopsis* cysteine synthase complex. *The Plant Cell* 18, 3647–3655.

Freeman, J.L. and Salt, D.E. (2007) The metal tolerance profile of Thlaspi goesingense is mimicked in *Arabidopsis thaliana* heterologously expressing serine acetyl-transferase. *BMC Plant Biology* 7:63.

Gerber, J. and Lill, R. (2002) Biogenesis of iron-sulfur proteins in eukaryotes: components, mechanism and pathology. *Mitochondrion* 2, 71–86.

Goebel, G., Berger, R., Strasak, A.M., Egle, D., Mueller-Holzner, E., Schmidt, S., Rainer, J., Presul, E., Parson, W., Lang, S., Jones, A., Widschwendter, M. and Fiegl, H. (2012) Elevated mRNA expression of CHAC1 splicing variants is associated with poor outcome for breast and ovarian cancer patients. *British Journal of Cancer* 106, 189–198.

Gorman, J. and Shapiro, L. (2004) Structure of serine acetyltransferase from *Haemophilus influenzae* Rd. *Acta Crystallographica Section D – Biological Crystallography* 60, 1600–1605.

Goyer, A., Collakova, E., de la Garza, R.D., Quinlivan, E.P., Williamson, J., Gregory, J.F., Shachar-Hill, Y. and Hanson, A.D. (2005) 5-Formyltetrahydrofolate is an inhibitory but well tolerated metabolite in *Arabidopsis* leaves. *The Journal of Biological Chemistry* 280, 26137–26142.

Haas, F.H., Heeg, C., Queiroz, R., Bauer, A., Wirtz, M. and Hell, R. (2008) Mitochondrial serine acetyltransferase functions as a pacemaker of cysteine synthesis in plant cells. *Plant Physiology* 148, 1055–1067.

Hanson, A.D. and Roje, S. (2001) One-carbon metabolism in higher plants. *Annual Review of Plant Physiology and Plant Molecular Biology* 52, 119–137.

Harmer, S.L. (2009) The circadian system in higher plants. *In Annual Review of Plant Biology,* pp. 357–377.

Harms, K., von Ballmoos, P., Brunold, C., Hofgen, R. and Hesse, H. (2000) Expression of a bacterial serine acetyltransferase in transgenic potato plants leads to increased levels of cysteine and glutathione. *The Plant Journal* 22, 335–343.

Hawkesford, M.J., Schneider, A., Belcher, A.R. and Clarkson, D.T. (1995) Regulation of enzymes involved in the sulfur-assimilatory pathway. *Zeitschrift für Pflanzenernahrung und Bodenkunde* 158, 55–57.

Heeg, C., Kruse, C., Jost, R., Gutensohn, M., Ruppert, T., Wirtz, M. and Hell, R. (2008) Analysis of the *Arabidopsis* *O*-acetylserine(thiol)lyase gene family demonstrates compartment-specific differences in the regulation of cysteine synthesis. *The Plant Cell* 20, 168–185.

Hell, R. and Wirtz, M. (2008) Metabolism of cysteine in plants and phototrophic bacteria. In: Hell, R., Dahl, C., Knaff, D. and Leustek, T. (eds) *Sulfur Metabolism in Phototrophic Organisms*. Advances in Photosynthesis and Respiration Volume 27, Springer, Dordrecht, The Netherlands, pp. 59–91.

Hell, R. and Wirtz, M. (2011) Molecular biology, biochemistry and cellular physiology of cysteine metabolism in *Arabidopsis thaliana*. *The Arabidopsis Book* 9, e0154–e0154.

Hesse, H., Trachsel, N., Suter, M., Kopriva, S., von Ballmoos, P., Rennenberg, H. and Brunold, C. (2003) Effect of glucose on assimilatory sulphate reduction in *Arabidopsis thaliana* roots. *Journal of Experimental Botany* 54, 1701–1709.

Hirai, M.Y., Fujiwara, T., Awazuhara, M., Kimura, T., Noji, M. and Saito, K. (2003) Global expression profiling of sulfur-starved *Arabidopsis* by DNA macroarray reveals the role of *O*-acetyl-L-serine as a general regulator of gene expression in response to sulfur nutrition. *The Plant Journal* 33, 651–663.

Ho, C.L. and Saito, K. (2001) Molecular biology of the plastidic phosphorylated serine biosynthetic pathway in *Arabidopsis thaliana*. *Amino Acids* 20, 243–259.

Ho, C.L., Noji, M., Saito, M. and Saito, K. (1999) Regulation of serine biosynthesis in *Arabidopsis* – crucial role of plastidic 3-phosphoglycerate dehydrogenase in non-photosynthetic tissues. *The Journal of Biological Chemistry* 274, 397–402.

Hoefgen, R. and Hesse, H. (2008) Sulfur and cysteine metabolism. In: Jez, J. (ed.) *Sulfur: A Missing Link between Soils, Crops and Nutrition*. Agronomy Monograph 50, American Society of Agronomy, Madison, Wisconsin, pp. 83–104.

Hopkins, L., Parmar, S., Baszczyk, A., Hesse, H., Hoefgen, R. and Hawkesford, M.J. (2005) *O*-acetylserine and the regulation of expression of genes encoding components for sulfate uptake and assimilation in potato. *Plant Physiology* 138, 433–440.

Howarth, J.R., Barraclough, P.B. and Hawkesford, M.J. (2005) A highly specific sulfate-deficiency induced gene (*sdi1*) from wheat. In: Saitō, K. (ed.) *Sulfur Transport and Assimilation in Plants in the Post Genomic Era*. Backhuys, Leiden, The Netherlands, pp. 161–164.

Howarth, J.R., Parmar, S., Barraclough, P.B. and Hawkesford, M.J. (2009) A sulphur deficiency-induced gene, *sdi1*, involved in the utilization of stored sulphate pools under sulphur-limiting conditions has potential as a diagnostic indicator of sulphur nutritional status. *Plant Biotechnology Journal* 7, 200–209.

Hubberten, H.-M., Klie, S., Caldana, C., Degenkolbe, T., Willmitzer, L. and Hoefgen, R. (2012) Additional role of *O*-acetylserine as a sulfur status-independent regulator during plant growth. *The Plant Journal* 70, 666–677.

Huu Cuong, N., Hoefgen, R. and Hesse, H. (2012) Improving the nutritive value of rice seeds: elevation of cysteine and methionine contents in rice plants by ectopic expression of a bacterial serine acetyltransferase. *Journal of Experimental Botany* 63, 5991–6001.

Ivey, D.M., Guffanti, A.A., Zemsky, J., Pinner, E., Karpel, R., Padan, E., Schuldiner, S. and Krulwich, T.A. (1993) Cloning and characterization of a putative Ca^{2+}/H^{+} antiporter gene from *Escherichia coli* upon functional complementation of Na^{+}/H^{+} antiporter-deficient strains by the overexpressed gene. *The Journal of Biological Chemistry* 268, 11296–11303.

Kawashima, C.G., Berkowitz, O., Hell, R., Noji, M. and Saito, K. (2005) Characterization and expression analysis of a serine acetyltransferase gene family involved in a key step of the sulfur assimilation pathway in *Arabidopsis*. *Plant Physiology* 137, 220–230.

Khan, M.S., Haas, F.H., Samami, A.A., Gholami, A.M., Bauer, A., Fellenberg, K., Reichelt, M., Haensch, R., Mendel, R.R., Meyer, A.J., Wirtz, M. and Hell, R. (2010) Sulfite reductase defines a newly discovered bottleneck for assimilatory sulfate reduction and is essential for growth and development in *Arabidopsis thaliana*. *The Plant Cell* 22, 1216–1231.

Kopriva, S. (2006) Regulation of sulfate assimilation in *Arabidopsis* and beyond. *Annals of Botany* 97, 479–495.

Kopriva, S., Muheim, R., Koprivova, A., Trachsel, N., Catalano, C., Suter, M. and Brunold, C. (1999) Light regulation of assimilatory sulphate reduction in *Arabidopsis thaliana*. *The Plant Journal* 20, 37–44.

Kopriva, S., Suter, M., von Ballmoos, P., Hesse, H., Krahenbuhl, U., Rennenberg, H. and Brunold, C. (2002) Interaction of sulfate assimilation with carbon and nitrogen metabolism in *Lemna minor*. *Plant Physiology* 130, 1406–1413.

Koprivova, A., Suter, M., Op den Camp, R., Brunold, C. and Kopriva, S. (2000) Regulation of sulfate assimilation by nitrogen in *Arabidopsis*. *Plant Physiology* 122, 737–746.

Kredich, N.M., Becker, M.A. and Tomkins, G.M. (1969) Purification and characterization of cysteine synthetase a bifunctional protein complex from *Salmonella typhimurium*. *The Journal of Biological Chemistry* 244, 2428–2439.

Krueger, S., Niehl, A., Lopez Martin, M.C., Steinhauser, D., Donath, A., Hildebrandt, T., Romero, L.C., Hoefgen, R., Gotor, C. and Hesse, H. (2009) Analysis of cytosolic and plastidic serine acetyltransferase mutants and subcellular metabolite distributions suggests interplay of the cellular compartments for cysteine biosynthesis in *Arabidopsis*. *Plant Cell and Environment* 32, 349–367.

Krueger, S., Donath, A., Carmen Lopez-Martin, M., Hoefgen, R., Gotor, C. and Hesse, H. (2010) Impact of sulfur starvation on cysteine biosynthesis in T-DNA mutants deficient for compartment-specific serine-acetyltransferase. *Amino Acids* 39, 1029–1042.

Kumar, A., Tikoo, S., Maity, S., Sengupta, S., Sengupta, S., Kaur, A. and Bachhawat, A.K. (2012) Mammalian proapoptotic factor ChaC1 and its homologues function as γ-glutamyl cyclotransferases acting specifically on glutathione. *EMBO Reports* 13, 1095–1101.

Kumaran, S., Yi, H., Krishnan, H.B. and Jez, J.M. (2009) Assembly of the cysteine synthase complex and the regulatory role of protein–protein interactions. *The Journal of Biological Chemistry* 284, 10268–10275.

Lehmann, M., Schwarzlaender, M., Obata, T., Sirikantaramas, S., Burow, M., Olsen, C.E., Tohge, T., Fricker, M.D., Moller, B.L., Fernie, A.R., Sweetlove, L.J. and Laxa, M. (2009) The metabolic response of *Arabidopsis* roots to oxidative stress is distinct from that of heterotrophic cells in culture and highlights a complex relationship between the levels of transcripts, metabolites and flux. *Molecular Plant* 2, 390–406.

Leustek, T., Martin, M.N., Bick, J.A. and Davies, J.P. (2000) Pathways and regulation of sulfur metabolism revealed through molecular and genetic studies. *Annual Review of Plant Physiology and Plant Molecular Biology* 51, 141–165.

Lewandowska, M., Wawrzynska, A., Kaminska, J., Liszewska, F. and Sirko, A. (2005) Identification of novel proteins of *Nicotiana tabacum* regulated by short term sulfur starvation. In: Saitō, K. (ed.) *Sulfur Transport and Assimilation in Plants in the Post Genomic Era*. Backhuys, Leiden, The Netherlands,, 153–156.

Liu, F., Yoo, B.-C., Lee, J.-Y., Pan, W. and Harmon, A.C. (2006) Calcium-regulated phosphorylation of soybean serine acetyltransferase in response to oxidative stress. *The Journal of Biological Chemistry* 281, 27405–27415.

Lynch, A.S., Tyrrell, R., Smerdon, S.J., Briggs, G.S. and Wilkinson, A.J. (1994) Characterization of the Cysb Protein of *Klebsiella aerogenes* – direct evidence that *N*-acetylserine rather than *O*-acetylserine serves as the inducer of the cysteine regulon. *Biochemical Journal* 299, 129–136.

Marquet, A. (2001) Enzymology of carbon–sulfur bond formation. *Current Opinion in Chemical Biology* 5, 541–549.

Maruyama-Nakashita, A., Inoue, E., Watanabe-Takahashi, A., Yarnaya, T. and Takahashi, H. (2003) Transcriptome profiling of sulfur-responsive genes in *Arabidopsis* reveals global effects of sulfur nutrition on multiple metabolic pathways. *Plant Physiology* 132, 597–605.

Maruyama-Nakashita, A., Nakamura, Y., Yamaya, T. and Takahashi, H. (2004) Regulation of high-affinity sulphate transporters in plants: towards systematic analysis of sulphur signalling and regulation. *Journal of Experimental Botany* 55, 1843–1849.

Maruyama-Nakashita, A., Nakamura, Y., Tohge, T., Saito, K. and Takahashi, H. (2006) *Arabidopsis* SLIM1 is a central transcriptional regulator of plant sulfur response and metabolism. *The Plant Cell* 18, 3235–3251.

McClung, C.R., Hsu, M., Painter, J.E., Gagne, J.M., Karlsberg, S.D. and Salome, P.A. (2000) Integrated temporal regulation of the photorespiratory pathway. Circadian regulation of two *Arabidopsis* genes encoding serine hydroxymethyltransferase. *Plant Physiology* 123, 381–391.

Mendel, R.R. and Haensch, R. (2002) Molybdoenzymes and molybdenum cofactor in plants. *Journal of Experimental Botany* 53, 1689–1698.

Meyer, A.J. and Fricker, M.D. (2002) Control of demand-driven biosynthesis of glutathione in green *Arabidopsis* suspension culture cells. *Plant Physiology* 130, 1927–1937.

Mino, K., Hiraoka, K., Imamura, K., Sakiyama, T., Eisaki, N., Matsuyama, A. and Nakanishi, K. (2000) Characteristics of serine acetyltransferase from *Escherichia coli* deleting different lengths of amino acid residues from the C-terminus. *Bioscience Biotechnology and Biochemistry* 64, 1874–1880.

Mino, K., Imamura, K., Sakiyama, T., Eisaki, N., Matsuyama, A. and Nakanishi, K. (2001) Increase in the stability of serine acetyltransferase from *Escherichia coli* against cold inactivation and proteolysis by forming a bienzyme complex. *Bioscience Biotechnology and Biochemistry* 65, 865–874.

Morcuende, R., Bari, R., Gibon, Y., Zheng, W., Pant, B.D., Blasing, O., Usadel, B., Czechowski, T., Udvardi, M.K., Stitt, M. and Scheible, W.-R. (2007) Genome-wide reprogramming of metabolism and regulatory networks of *Arabidopsis* in response to phosphorus. *Plant Cell and Environment* 30, 85–112.

Mungrue, I.N., Pagnon, J., Kohannim, O., Gargalovic, P.S. and Lusis, A.J. (2009) CHAC1/MGC4504 is a novel proapoptotic component of the unfolded protein response, downstream of the ATF4-ATF3-CHOP cascade. *Journal of Immunology* 182, 466–476.

Mutwil, M., Obro, J., Willats, W.G.T. and Persson, S. (2008) GeneCAT – novel webtools that combine BLAST and co-expression analyses. *Nucleic Acids Research* 36, W320–W326.

Mutwil, M., Usadel, B., Schuette, M., Loraine, A., Ebenhoeh, O. and Persson, S. (2010) Assembly of an interactive correlation network for the *Arabidopsis* genome using a novel heuristic clustering algorithm. *Plant Physiology* 152, 29–43.

Mutwil, M., Klie, S., Tohge, T., Giorgi, F.M., Wilkins, O., Campbell, M.M., Fernie, A.R., Usadel, B., Nikoloski, Z. and Persson, S. (2011) PlaNet: combined sequence and expression comparisons across plant networks derived from seven species. *The Plant Cell* 23, 895–910.

Myakushina, Y.A., Milyaeva, E.L., Romanov, G.A. and Nikiforova, V.Y. (2009) Mutation in *LSU4* gene affects flower development in *Arabidopsis thaliana*. *Doklady Biochemistry and Biophysics* 428, 257–260.

Neuenschwander, U., Suter, M. and Brunold, C. (1991) Regulation of sulfate assimilation by light and *O*-acetyl-L-serine in *Lemna minor* L. *Plant Physiology* 97, 253–258.

Nikiforova, V., Freitag, J., Kempa, S., Adamik, M., Hesse, H. and Hoefgen, R. (2003) Transcriptome analysis of sulfur depletion in *Arabidopsis thaliana*: interlacing of biosynthetic pathways provides response specificity. *The Plant Journal* 33, 633–650.

Nikiforova, V.J., Daub, C.O., Hesse, H., Willmitzer, L. and Hoefgen, R. (2005) Integrative gene–metabolite network with implemented causality deciphers informational fluxes of sulphur stress response. *Journal of Experimental Botany* 56, 1887–1896.

Noctor, G., Gomez, L., Vanacker, H. and Foyer, C.H. (2002) Interactions between biosynthesis, compartmentation and transport in the control of glutathione homeostasis and signalling. *Journal of Experimental Botany* 53, 1283–1304.

Noctor, G., Mhamdi, A., Chaouch, S., Han, Y., Neukermans, J., Marquez-Garcia, B., Queval, G. and Foyer, C.H. (2012) Glutathione in plants: an integrated overview. *Plant Cell and Environment* 35, 454–484.

Noji, M. and Saito, K. (2002) Molecular and biochemical analysis of serine acetyltransferase and cysteine synthase towards sulfur metabolic engineering in plants. *Amino Acids* 22, 231–243.

Noji, M., Inoue, K., Kimura, N., Gouda, A. and Saito, K. (1998) Isoform-dependent differences in feedback regulation and subcellular localization of serine acetyltransferase involved in cysteine biosynthesis from *Arabidopsis thaliana*. *The Journal of Biological Chemistry* 273, 32739–32745.

Noji, M., Takagi, Y., Kimura, N., Inoue, K., Saito, M., Horikoshi, M., Saito, F., Takahashi, H. and Saito, K. (2001) Serine acetyltransferase involved in cysteine biosynthesis from spinach: molecular cloning, characterization and expression analysis of cDNA encoding a plastidic isoform. *Plant and Cell Physiology* 42, 627–634.

Obayashi, T., Kinoshita, K., Nakai, K., Shibaoka, M., Hayashi, S., Saeki, M., Shibata, D., Saito, K. and Ohta, H. (2007) ATTED-II: a database of co-expressed genes and cis elements for identifying co-regulated gene groups in *Arabidopsis*. *Nucleic Acids Research* 35, D863–D869.

Obayashi, T., Hayashi, S., Saeki, M., Ohta, H. and Kinoshita, K. (2009) ATTED-II provides coexpressed gene networks for *Arabidopsis*. *Nucleic Acids Research* 37, D987–D991.

Olsen, L.R., Huang, B., Vetting, M.W. and Roderick, S.L. (2004) Structure of serine acetyltransferase in complexes with CoA and its cysteine feedback inhibitor. *Biochemistry* 43, 6013–6019.

Osborne, M.J., Siddiqui, N., Iannuzzi, P. and Gehring, K. (2004) The solution structure of ChaB, a putative membrane ion antiporter regulator from *Escherichia coli*. *BMC Structural Biology* 4:9.

Ostrowski, J. and Kredich, N.M. (1990) *In vitro* interactions of cysb protein with the cysjih promoter of *Salmonella typhimurium* – inhibitory effects of sulfide. *Journal of Bacteriology* 172, 779–785.

Ostrowski, J. and Kredich, N.M. (1991) Negative autoregulation of cysb in *Salmonella-typhimurium – in vitro* interactions of Cysb protein with the cysb promoter. *Journal of Bacteriology* 173, 2212–2218.

Peer, W.A., Mahmoudian, M., Freeman, J.L., Lahner, B., Richards, E.L., Reeves, R.D., Murphy, A.S. and Salt, D.E. (2006) Assessment of plants from the *Brassicaceae* family as genetic models for the study of nickel and zinc hyperaccumulation. *New Phytologist* 172, 248–260.

Pye, V.E., Tingey, A.P., Robson, R.L. and Moody, P.C.E. (2004) The structure and mechanism of serine acetyltransferase from *Escherichia coli. The Journal of Biological Chemistry* 279, 40729–40736.

Qi, Y.H., Yamauchi, Y., Ling, J.Q., Kawano, N., Li, D.B. and Tanaka, K. (2005) The submergence-induced gene *OsCTP* in rice (*Oryza sativa* L.) is similar to *Escherichia coli* cation transport protein ChaC. *Plant Science* 168, 15–22.

Riemenschneider, A., Riedel, K., Hoefgen, R., Papenbrock, J. and Hesse, H. (2005) Impact of reduced *O*-acetylserine(thiol)lyase isoform contents on potato plant metabolism. *Plant Physiology* 137, 892–900.

Ross, K.J., Fransz, P., Armstrong, S.J., Vizir, I., Mulligan, B., Franklin, F.C.H. and Jones, G.H. (1997) Cytological characterization of four meiotic mutants of *Arabidopsis* isolated from T-DNA-transformed lines. *Chromosome Research* 5, 551–559.

Rouached, H., Wirtz, M., Alary, R., Hell, R., Arpat, A.B., Davidian, J.-C., Fourcroy, P. and Berthomieu, P. (2008) Differential regulation of the expression of two high-affinity sulfate transporters, SULTR1.1 and SULTR1.2, in *Arabidopsis*. *Plant Physiology* 147, 897–911.

Roughan, P.G. (1997) Stromal concentrations of coenzyme A and its esters are insufficient to account for rates of chloroplast fatty acid synthesis: evidence for substrate channelling within the chloroplast fatty acid synthase. *Biochemical Journal* 327, 267–273.

Ruffet, M.L., Droux, M. and Douce, R. (1994) Purification and kinetic properties of serine acetyltransferase free of *O*-acetylserine(thiol)lyase from spinach chloroplasts. *Plant Physiology* 104, 597–604.

Ruffet, M.L., Lebrun, M., Droux, M. and Douce, R. (1995) Subcellular-distribution of serine acetyltransferase from *Pisum sativum* and characterization of an *Arabidopsis thaliana* putative cytosolic isoform. *European Journal of Biochemistry* 227, 500–509.

Saito, K. (2004) Sulfur assimilatory metabolism. The long and smelling road. *Plant Physiology* 136, 2443–2450.

Saito, K., Yokoyama, H., Noji, M. and Murakoshi, I. (1995) Molecular cloning and characterization of a plant serine acetyltransferase playing a regulatory role in cysteine biosynthesis from watermelon. *The Journal of Biological Chemistry* 270, 16321–16326.

Sakakibara, H., Takei, K. and Hirose, N. (2006) Interactions between nitrogen and cytokinin in the regulation of metabolism and development. *Trends in Plant Science* 11, 440–448.

Scheible, W.R., Morcuende, R., Czechowski, T., Fritz, C., Osuna, D., Palacios-Rojas, N., Schindelasch, D., Thimm, O., Udvardi, M.K. and Stitt, M. (2004) Genome-wide reprogramming of primary and secondary metabolism, protein synthesis, cellular growth processes and the regulatory infrastructure of *Arabidopsis* in response to nitrogen. *Plant Physiology* 136, 2483–2499.

Sirko, A., Blaszczyk, A. and Liszewska, F. (2004) Overproduction of SAT and/or OASTL in transgenic plants: a survey of effects. *Journal of Experimental Botany* 55, 1881–1888.

Smith, I.K. (1972) Studies of L-cysteine biosynthetic enzymes in *Phaseolus vulgaris* L. *Plant Physiology* 50, 477–479.

Tabe, L., Wirtz, M., Molvig, L., Droux, M. and Hell, R. (2010) Overexpression of serine acetlytransferase produced large increases in *O*-acetylserine and free cysteine in developing seeds of a grain legume. *Journal of Experimental Botany* 61, 721–733.

Tahir, J., Watanabe, M., Jing, H.-C., Hunter, D.A., Tohge, T., Nunes-Nesi, A., Brotman, Y., Fernie, A.R., Hoefgen, R. and Dijkwel, P.P. (2013) Activation of R-mediated innate immunity and disease susceptibility is affected by mutations in a cytosolic *O*-acetylserine(thiol)lyase in *Arabidopsis*. *The Plant Journal* 73, 118–130.

Thompson, J.D., Gibson, T.J., Plewniak, F., Jeanmougin, F. and Higgins, D.G. (1997) The CLUSTAL_X Windows interface: flexible strategies for multiple sequence alignment aided by quality analysis tools. *Nucleic Acids Research* 25, 4876–4882.

Timm, S., Florian, A., Wittmiss, M., Jahnke, K., Hagemann, M., Fernie, A.R. and Bauwe, H. (2013) Serine acts as a metabolic signal for the transcriptional control of photorespiration-related genes in *Arabidopsis*. *Plant Physiology* 162, 379–389.

Tumaney, A.W., Ohlrogge, J.B. and Pollard, M. (2004) Acetyl coenzyme A concentrations in plant tissues. *Journal of Plant Physiology* 161, 485–488.

Uddin, M.I., Qi, Y., Yamada, S., Shibuya, I., Deng, X.-P., Kwak, S.-S., Kaminaka, H. and Tanaka, K. (2008) Overexpression of a new rice vacuolar antiporter regulating protein OsARP improves salt tolerance in tobacco. *Plant and Cell Physiology* 49, 880–890.

Urano, Y., Manabe, T., Noji, M. and Saito, K. (2000) Molecular cloning and functional characterization of cDNAs encoding cysteine synthase and serine acetyltransferase that may be responsible for high cellular cysteine content in *Allium tuberosum*. *Gene* 257, 269–277.

Van Hoewyk, D., Takahashi, H., Inoue, E., Hess, A., Tamaoki, M. and Pilon-Smits, E.A.H. (2008) Transcriptome analyses give insights into selenium-stress responses and selenium tolerance mechanisms in *Arabidopsis*. *Physiologia Plantarum* 132, 236–253.

Vauclare, P., Kopriva, S., Fell, D., Suter, M., Sticher, L., von Ballmoos, P., Krahenbuhl, U., den Camp, R.O. and Brunold, C. (2002) Flux control of sulphate assimilation in *Arabidopsis thaliana*: adenosine 5′-phosphosulphate reductase is more susceptible than ATP sulphurylase to negative control by thiols. *The Plant Journal* 31, 729–740.

Vidal, E.A. and Gutierrez, R.A. (2008) A systems view of nitrogen nutrient and metabolite responses in *Arabidopsis*. *Current Opinion in Plant Biology* 11, 521–529.

Voll, L.M., Jamai, A., Renne, P., Voll, H., McClung, C.R. and Weber, A.P.M. (2006) The photorespiratory *Arabidopsis shm1* mutant is deficient in SHM1. *Plant Physiology* 140, 59–66.

Watanabe, M., Mochida, K., Kato, T., Tabata, S., Yoshimoto, N., Noji, M. and Saito, K. (2008a) Comparative genomics and reverse genetics analysis reveal indispensable functions of the serine acetyltransferase gene family in *Arabidopsis*. *The Plant Cell* 20, 2484–2496.

Watanabe, M., Kusano, M., Oikawa, A., Fukushima, A., Noji, M. and Saito, K. (2008b) Physiological roles of the β-substituted alanine synthase gene family in *Arabidopsis*. *Plant Physiology* 146, 310–320.

Watanabe, M., Hubberten, H-M., Saito, K. and Hoefgen, R. (2010) General regulatory patterns of plant mineral nutrient depletion as revealed by *serat* quadruple mutants disturbed in cysteine synthesis. *Molecular Plant* 3, 438–466.

Wawrzynska, A., Lewandowska, M., Hawkesford, M.J. and Sirko, A. (2005) Using a suppression subtractive library-based approach to identify tobacco genes regulated in response to short-term sulphur deficit. *Journal of Experimental Botany* 56, 1575–1590.

Wawrzynska, A., Lewandowska, M. and Sirko, A. (2010) *Nicotiana tabacum* EIL2 directly regulates expression of at least one tobacco gene induced by sulphur starvation. *Journal of Experimental Botany* 61, 889–900.

Wawrzynska, A., Kurzyk, A., Mierzwinska, M., Plochocka, D., Wieczorek, G. and Sirko, A. (2013) Direct targeting of *Arabidopsis* cysteine synthase complexes with synthetic polypeptides to selectively deregulate cysteine synthesis. *Plant Science* 207, 148–157.

Winter, D., Vinegar, B., Nahal, H., Ammar, R., Wilson, G.V. and Provart, N.J. (2007) An “Electronic Fluorescent Pictograph” browser for exploring and analyzing large-scale biological data sets. *PloS ONE* (8): e718.

Wirtz, M. and Hell, R. (2003) Production of cysteine for bacterial and plant biotechnology: application of cysteine feedback-insensitive isoforms of serine acetyltransferase. *Amino Acids* 24, 195–203.

Wirtz, M. and Hell, R. (2006) Functional analysis of the cysteine synthase protein complex from plants: structural, biochemical and regulatory properties. *Journal of Plant Physiology* 163, 273–286.

Wirtz, M. and Hell, R. (2007) Dominant-negative modification reveals the regulatory function of the multimeric cysteine synthase protein complex in transgenic tobacco. *The Plant Cell* 19, 625–639.

Wirtz, M., Berkowitz, O., Droux, M. and Hell, R. (2001) The cysteine synthase complex from plants – mitochondrial serine acetyltransferase from *Arabidopsis thaliana* carries a bifunctional domain for catalysis and protein–protein interaction. *European Journal of Biochemistry* 268, 686–693.

Wirtz, M., Droux, M. and Hell, R. (2004) *O*-acetylserine(thiol)lyase: an enigmatic enzyme of plant cysteine biosynthesis revisited in *Arabidopsis thaliana*. *Journal of Experimental Botany* 55, 1785–1798.

Wirtz, M., Birke, H., Heeg, C., Mueller, C., Hosp, F., Throm, C., Koenig, S., Feldman-Salit, A., Rippe, K., Petersen, G., Wade, R.C., Rybin, V., Scheffzek, K. and Hell, R. (2010) Structure and function of the hetero-oligomeric cysteine synthase complex in plants. *The Journal of Biological Chemistry* 285, 32810–32817.

Zhang, Y., Sun, K., Sandoval, F.J., Santiago, K. and Roje, S. (2010) One-carbon metabolism in plants: characterization of a plastid serine hydroxymethyltransferase. *Biochemical Journal* 430, 97–105.

Zientara-Rytter, K., Lukomska, J., Moniuszko, G., Gwozdecki, R., Surowiecki, P., Lewandowska, M., Liszewska, F., Wawrzynska, A. and Sirko, A. (2011) Identification and functional analysis of Joka2, a tobacco member of the family of selective autophagy cargo receptors. *Autophagy* 7, 1145–1158.

12 Cysteine Homeostasis

I. García,* L.C. Romero and C. Gotor
Instituto de Bioquímica Vegetal y Fotosíntesis, Consejo Superior de Investigaciones Científicas and Universidad de Sevilla, Sevilla, Spain

12.1 Abstract

In plant metabolism, cysteine is the first reduced sulfur donor molecule, and it constitutes the metabolic precursor of essential biomolecules, such as vitamins, cofactors, antioxidants and many defence compounds, that are formed in response to adverse environmental conditions in which the sulfur moiety is the functional group. Sulfate is transported inside the plant, reduced to sulfide and incorporated into *O*-acetylserine (OAS) to form cysteine. The enzymes used in the last step of cysteine biosynthesis, serine acetyltransferase (SAT, also called SERAT) and *O*-acetylserine(thiol)lyase (OASTL), form a complex that senses the sulfur nutrient status of the plant and modulates cysteine biosynthesis.

Arabidopsis thaliana contains different OASTL and SAT enzymes in the different cellular compartments, resulting in different sub-cellular pools of cysteine. In recent years, significant progress has been made in determining the specific roles of the different sub-cellular pools of cysteine, as well as its associated metabolites. In chloroplasts, OAS can serve either as the substrate of the OASTL isoform OAS-B in the stroma to produce cysteine, or as the substrate of the OASTL isoform CS26 in the thylakoid lumen, which uses thiosulfate instead of sulfide to produce *S*-sulfocysteine. *S*-sulfocysteine is a signalling molecule that is involved in the regulation and protection of the photosystems of the plant. In the mitochondria, cysteine is essential for detoxifying cyanide by a loop that is established by the activities of the OASTL isoforms CYS-C1 and OAS-C. Mitochondrial cyanide regulates the root hair development and modulates the plant immune response. The main OASTL isoform, OAS-A1, and DES1, an OASTL isoform that acts as a cytosolic desulfhydrase and degrades cysteine contribute to the homeostasis of the cytosolic cysteine and sulfur. Cytosolic sulfide acts as a signalling molecule, regulating the process of autophagy, while cytosolic cysteine is essential for inducing the hypersensitive response to pathogens. Thus, cytosolic cysteine homeostasis contributes to orchestrating the plant response to pathogens.

12.2 Introduction

Sulfur is an essential macronutrient that is present in nature in organic and inorganic forms. In its different oxidation states, it represents one of the most versatile elements, and the most common oxidized form is sulfate (SO_4^{2-}), which contains an S atom with a redox state of +6. Plants, microorganisms and fungi, differently from animals, are able to reduce sulfate to sulfide and to incorporate sulfur into organic metabolites through an energy-dependent process, thereby originating a large variety of organic compounds that are essential for plant

*E-mail address: irene.garcia@ibvf.csic.es

growth and development. Plants are the most important source of sulfur-containing amino acids (mainly methionine and cysteine) for humans and animals (Wirtz and Droux, 2005). Sulfur is present in vitamins such as biotin and thiamine, cofactors such as *S*-adenosylmethionine (SAM), coenzyme A, molybdenum cofactor (CoMo), and lipoic acid and iron–sulfur clusters that participate in electron transport. Glucosinolates, phytoalexins, thionines and defensins are sulfur-containing compounds that are involved in the plant response to pathogens (Rausch and Wachter, 2005). Sulfur is also present in sulfolipids, which are essential for the formation of thylakoid membranes in the chloroplasts. Some plant hormones are inactivated by sulfatation, including brassinosteroids and jasmonic acid; in contrast, sulfatation is essential for the function of peptide hormones that stimulate cellular growth (Hirai and Saito, 2008). Sulfur compounds are also of great importance for food quality and the production of phyto-pharmaceuticals (Moriarty *et al.*, 2007). Sulfur deficiency will result in the loss of plant production, fitness and resistance to environmental stress and pests.

Cysteine is the first reduced sulfur organic compound that is synthesized from the plant photosynthetic assimilation of sulfate. Cysteine is essential not only as a proteinogenic amino acid, but also for stabilizing the tertiary and quaternary structures of proteins, because it participates in the formation of disulfide bonds and regulates the active sites of many enzymes. Cysteine plays a central role in the primary and secondary metabolism of the plant owing to its biochemical functions. It is the precursor metabolite of the antioxidant glutathione, the amino acid methionine and most of the essential biomolecules and above-mentioned defence compounds; the catalytic mechanism of many of these compounds is based on the reactivity of the thiol group (Droux, 2004). The conversion between free thiol groups and disulfide bond formation is a dynamic system that is the basis for protein redox modifications. These changes are involved in regulating the activity of certain enzymes that, in turn, regulate cellular metabolism in response to environmental cues (Wirtz and Droux, 2005).

12.3 Photosynthetic Assimilation of Sulfate in Plants

The use of bacteria and yeast mutants, as well as the complete genome sequence of *Arabidopsis thaliana*, has led to a breakthrough in the understanding of the sulfate assimilation process in plants (Barroso *et al.*, 1997; Hell, 1997; Leustek *et al.*, 2000; Noji *et al.*, 2006; Saito 2004; Takahashi *et al.*, 2011). The biosynthesis of cysteine is the last step of the photosynthetic sulfate assimilation. The three basic steps of the process are as follows: sulfate transport into the cell, the reduction of sulfate to sulfide and the incorporation of the reduced sulfur into the carbon skeletons (Fig. 12.1).

12.3.1 Sulfate transport

Plants take sulfate up from the soil via their roots and it is then distributed to the entire plant. At the root level, sulfate is transported through the plasma membrane by a proton/sulfate co-transport mechanism. It is subsequently loaded into the xylem vessels and transported to the shoot by the transpiration stream using the energy that is driven by a proton ATPase-generated proton gradient. The uptake of sulfate by the roots is strictly controlled and appears to be the most regulated step in the plant sulfur assimilation pathway (Vauclare *et al.*, 2002).

The sulfate transporters, which are proteins, all possess 12 transmembrane domains in the central zone of the polypeptide. The differences in the amino acid sequences of the amino and carboxy terminal ends of the proteins determine the differences between the proteins in their affinity for sulfate and their specific functions (Vidmar *et al.*, 2000). The sulfate transporters also contain in their carboxy terminus a domain called STAS (sulfate transporters and anti-sigma factor antagonists), which may play a regulatory role in controlling the activity and localization of membrane transporters (Shibagaki and Grossman, 2004; Rouached *et al.*, 2005). In *Arabidopsis thaliana*, 14 genes have been identified that form a

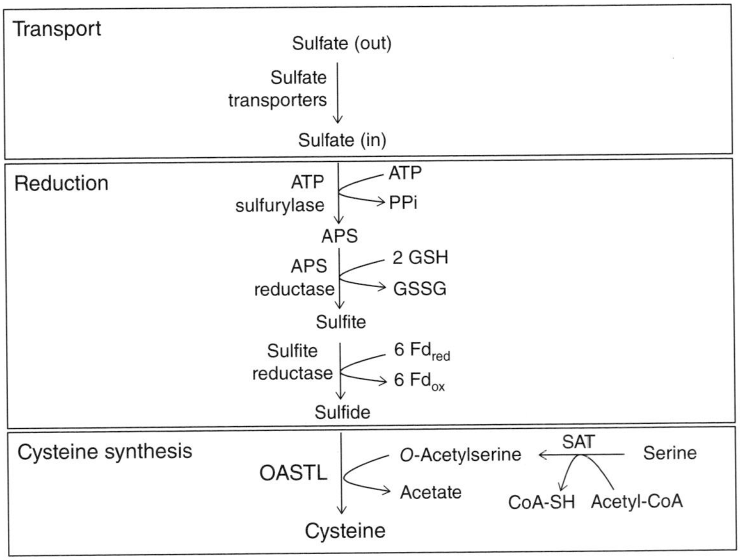

Fig. 12.1. Sulfate assimilation in plants. Abbreviations: APS, adenosine 5′-phosphosulfate; Fd, ferredoxin; GSH, reduced glutathione; GSSG, oxidized glutathione; OASTL, *O*-acetylserine(thiol)lyase; PPi, pyrophosphate; SAT, serine acetyltransferase.

family of sulfate transporters (Howarth *et al.*, 2003b). According to their sequences, these genes can be divided into five groups (SULTR1–SULTR5), depending on both the substrate affinity of the protein that is produced and the function that it exerts (Hawkesford and De Kok, 2006).

12.3.2 Sulfate reduction

In the roots or shoots, sulfate can be stored in the vacuoles, or it can be transported to the shoot chloroplasts, where it is reduced. The intracellular distribution of sulfate is regulated by the available concentrations of cysteine or glutathione (Bolchi *et al.*, 1999).

Sulfate must be activated to adenosine 5′-phosphosulfate (APS) before it is reduced. ATP sulfurylases (EC 2.7.7.4) catalyse the addition of sulfate to the α-phosphate of ATP. In *Arabidopsis*, four genes code for the ATP sulfurylases, three of which are plastidial and one cytosolic (Hatzfeld *et al.*, 2000; Mugford *et al.*, 2009). All of these genes are expressed in the shoots and the roots (Lee and Leustek 1999), even though sulfate activation and reduction take place predominantly in the leaf chloroplast.

APS is reduced to sulfite by APS reductase (APR; EC 1.8.99.2) using glutathione as a reductant. *Arabidopsis* has three genes coding for APR, which encode polypeptides of 50 kDa presenting an amino-terminal catalytic domain that is homologous to the bacterial APS and 3′-phosphoadenosine-5′-phosphosulfate (PAPS) reductases and a carboxy-terminal domain that interacts with the reductant homologue of thioredoxins and glutaredoxins (Gutierrez-Marcos *et al.*, 1996; Setya *et al.*, 1996). The reduction of APS is the key regulation point in sulfate assimilation, and APR activity is highly regulated according to the demand for reduced sulfur (Lee *et al.*, 2011).

Sulfite is reduced to sulfide by sulfite reductases (EC 1.8.7.1) which have a high affinity for the ferredoxin that they use as a reductant (Yonekura-Sakakibara *et al.*, 2000). Sulfite reductases are homo-oligomeric haemoproteins with two to four identical subunits. *Arabidopsis* has a single gene coding for this activity.

12.3.3 Cysteine biosynthesis

The synthesis of cysteine in plants represents the final step of assimilatory sulfate reduction and the almost exclusive entry mechanism for reduced sulfur into the metabolism of not only plants but also into the human food chain in general. In plants, cysteine is synthesized in the cytosol, plastids and mitochondria by the sequential action of the enzyme serine acetyltransferase (SAT, also known as SERAT; EC 2.3.1.30), which synthesizes the intermediary product, *O*-acetylserine (OAS), and *O*-acetylserine(thiol) lyase (OASTL; EC 2.5.1.47), which combines a sulfide molecule with an OAS molecule to produce L-cysteine. The formation of cysteine is the direct coupling step between sulfur and nitrogen assimilation in plants.

In *A. thaliana*, nine OASTL- and five SAT-encoding genes have been identified by sequence homology (Table 12.1), and the functionality of some of these genes has been demonstrated by several studies.

A phylogenetic analysis of the SATs indicates that the isoforms SAT106 and SAT-N, which are located in the cytosol, are very closely related, as are the plastidial SAT53 and the mitochondrial SAT1 SAT isoforms. The cytosolic SAT52 isoform is the most divergent SAT, but it is more closely related to the cytosolic than to the organellar SATs (Fig. 12.2a).

The phylogenetic tree of the OASTLs shows that chloroplastic isoform OAS-B and the mitochondrial OAS-C are very closely related, that the cytosolic isoforms DES1, CYS-D1 and CYS-D2 diverged recently and that the mitochondrial CYS-C1 isoform is the most divergent protein of the family (Fig. 12.2b).

12.4 The Cysteine Synthase Complex: Regulation of Cysteine Biosynthesis

It is well established that SAT and OASTL physically interact to form the hetero-oligomeric cysteine synthase complex (CSC), which is composed of a dimer of SAT trimers that are associated with two homodimers of OASTL ($\alpha_6\beta_4$). The SAT enzymes possess two conserved domains: an N-terminal domain that is rich in α-helix structures that are involved in SAT/SAT interaction and a β-sheet-containing the C-terminal domain that is responsible for the SAT/OASTL interactions, as well as for catalysis (Wirtz and Hell, 2006). OASTL forms stable homodimers with a molecular weight of 68–75 kDa that carry two molecules of pyridoxal-5′-phosphate (PLP) per dimer as a prosthetic group (León *et al.*, 1987). The CSC is a part of the regulatory circuit controlling sulfur homeostasis (Fig. 12.3). While OASTL is only active outside the CSC,

Table 12.1. Serine acetyltransferase (SAT) and *O*-acetylserine(thiol)lyase (OASTL) families in *Arabidopsis thaliana*.

Protein	Gene ID	Subcellular location	References
SATs			
SAT1	At3g13110	Mitochondria	Krueger *et al.*, 2009
SAT106	At2g17640	Cytosol	Howarth *et al.*, 2003a)
SAT52	At5g56760	Cytosol	Krueger *et al.*, 2009
SAT53	At1g55920	Plastids	Krueger *et al.*, 2009
SAT-N	At4g35640	Cytosol	Howarth *et al.*, 2003b
OASTLs			
Atcys-C1	At3g61440	Mitochondria	García *et al.*, 2010
Atcys-D1	At3g04940	Cytosol	Jost *et al.*, 2000
Atcys-D2	At5g28020	Cytosol	Jost *et al.*, 2000
CS26	At3g03630	Plastids	Bermúdez *et al.*, 2010
DES1	At5g28030	Cytosol	Álvarez *et al.*, 2010
OAS-A1	At4g14880	Cytosol	Barroso *et al.*, 1995
OAS-A2	At3g22460	None	Jost *et al.*, 2000
OAS-B	At2g43750	Plastids	Heeg *et al.*, 2008
OAS-C	At3g59760	Mitochondria	Álvarez *et al.*, 2012c

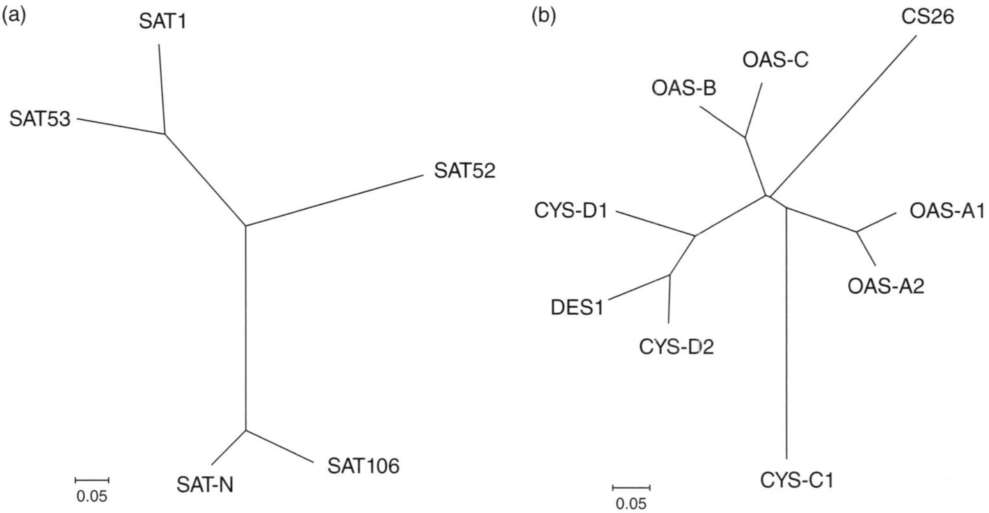

Fig. 12.2. The phylogeny of (a) serine acetyltransferase (SATs) and (b) *O*-acetylserine(thiol)lyases (OASTLs) in *Arabidopsis thaliana*. The phylogenetic analyses were conducted using the Molecular Evolutionary Genetics Analysis software version 5 (MEGA5) (Tamura *et al.*, 2011). The complete protein sequences were aligned using MUSCLE software and the unrooted phylogenetic tree was inferred by the neighbour-joining method; all of the methods were integrated in the MEGA tool.

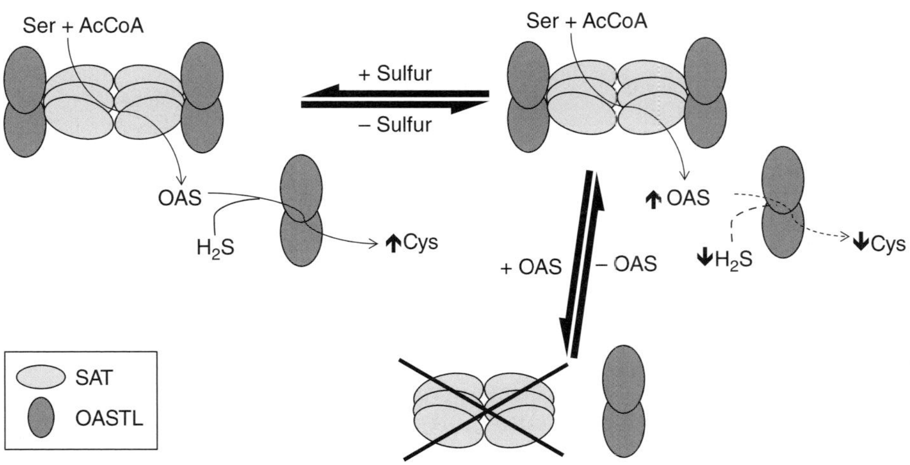

Fig. 12.3. Regulation of the cysteine synthase complex (CSC). Abbreviations: AcCoA, acetyl-CoA; Cys, cysteine; OAS, *O*-acetylserine; OASTL, *O*-acetylserine(thiol)lyase; SAT, serine acetyltransferase; Ser, serine.

SAT activity requires an association with OASTL (Droux *et al.*, 1998). The CSC is stabilized by free sulfide and destabilized by OAS. Furthermore, OASTL activity exceeds SAT activity by 345 times in plastids, 200 times in the cytosol and ten times in the mitochondria, indicating that the total SAT is found in the CSC together with a small part of OASTL (Heeg *et al.*, 2008). So when sulfur is not limiting, the sulfide that is produced in the chloroplasts by the sulfate assimilation pathway stabilizes the CSC, which produces OAS that combines with sulfide via the free OASTL. When OAS is in excess, the CSC dissociates, and the SAT activity decreases drastically. The CSC, then, effectively senses both the OAS and the sulfur availability and self-regulates further OAS production according to the supply and demand.

In addition to its transcriptional regulation and regulation by the CSC, the cytosolic SAT isoform is also retro-inhibited by cysteine. This feature is not observed in the chloroplastic and mitochondrial isoforms, which are insensitive to cysteine levels (Noji *et al.*, 1998; Kumaran *et al.*, 2009).

12.5 Cysteine Synthesis in Cellular Compartments

The fact that the assimilatory reduction of sulfate takes place in the chloroplast suggests that the major plastidial OASTLs and SATs participate in the primary assimilation pathway. The other isoforms (those in a minority or located in organelles other than the chloroplasts) would then exhibit other metabolic and/or signalling functions. Our research group is interested in unravelling the function of each subcellular pool of cysteine. The detailed study of *Arabidopsis* T-DNA insertional mutants in each of these isoforms, the application of '-omics' technologies and the identification of the distribution of the cellular enzymatic activities and metabolites has allowed the determination of the specific functions of the different OASTL and SAT isoforms and, therefore, has also allowed the contribution of the different subcellular pools of cysteine to plant physiology to be determined.

Initially, it was supposed that cysteine does not permeate membranes owing to its thiol group reactivity and that every compartment that is able to synthesize proteins would need its own cysteine production (Lunn *et al.*, 1990). The lack of plastidial cysteine biosynthesis in the mutant of the major chloroplastic isoform OAS-B is completely compensated for by the cysteine that is synthesized in the other compartments; as a result, a membrane transport of sulfide from the chloroplast to the other cellular compartments is required (Heeg *et al.*, 2008). It has been speculated that sulfide reaches the cytosol by diffusion through the chloroplast membranes. Hydrogen sulfide (H_2S) is a lipophilic molecule that freely permeates plasma membranes, although its ionized forms, HS^- and S^{2-}, cannot permeate membranes (Kabil and Banerjee, 2010). H_2S is weakly acidic and dissociates in aqueous solutions. Under physiological pH conditions (pH 7.4), one third of the H_2S that is present is not dissociated, and the remaining two thirds dissociate into H^+ and HS^-. Under high pH conditions, HS^- dissociates into H^+ and S^{2-}, which cannot permeate membranes. So at the pH of 8.5 that is maintained by the chloroplast stroma under illumination, sulfide is mainly present in its charged form and, consequently, is unable to undergo transport across the chloroplast envelope. Therefore, sulfide needs to be transported actively across the chloroplast membrane (Heldt *et al.*, 1973; Wu and Berkowitz 1992; Romero *et al.*, 2013). Additionally, the analysis of SAT-deficient mutants has demonstrated that the mitochondria are the main source of OAS in the cell, as they contain 80% of the total SAT activity and show the lowest OASTL activity of all of the organelles (Haas *et al.*, 2008; Krueger *et al.*, 2009). Finally, as demonstrated by mutants that are deficient in the major OASTLs, most of the cysteine is synthesized in the cytosol by the main cytosolic isoform, OAS-A1. OAS-A1 constitutes 23% of the total OASTL proteins and 44 and 80% of the total OASTL activity in the leaves and roots, respectively (Heeg *et al.*, 2008; López-Martín *et al.*, 2008a; Watanabe *et al.*, 2008). The concentration of cysteine in the cytosol has been estimated to be 300 μM, while in other cellular compartments it is below 10 μM (Krueger *et al.*, 2009).

In summary, the analysis of knockout mutants that are deficient in the different major isoforms of OASTL or SAT in the plastids, mitochondria and/or cytosol has revealed that chloroplasts constitute the main source of sulfide, mitochondria are rich is OAS and cysteine is mainly synthesized in the cytosol (see Fig. 12.4). Moreover, the compartments that are not affected by the mutations are able to provide cysteine to the compartment that is affected. As the viability of the plant is not or is only very slightly affected by the mutation of any of the main OASTLs, efficient transport mechanisms of sulfide, OAS and cysteine between the plastids, mitochondria and cytosol should occur.

12.6 Other Members of the *O*-Acetylserine(Thiol)Lyase Gene Family

In *A. thaliana*, the most transcribed *OASTL* genes encode the cytosolic OAS-A1 (At4g14880), the plastidial OAS-B (At2g43750) and the mitochondrial OAS-C (At3g59760) isoforms.

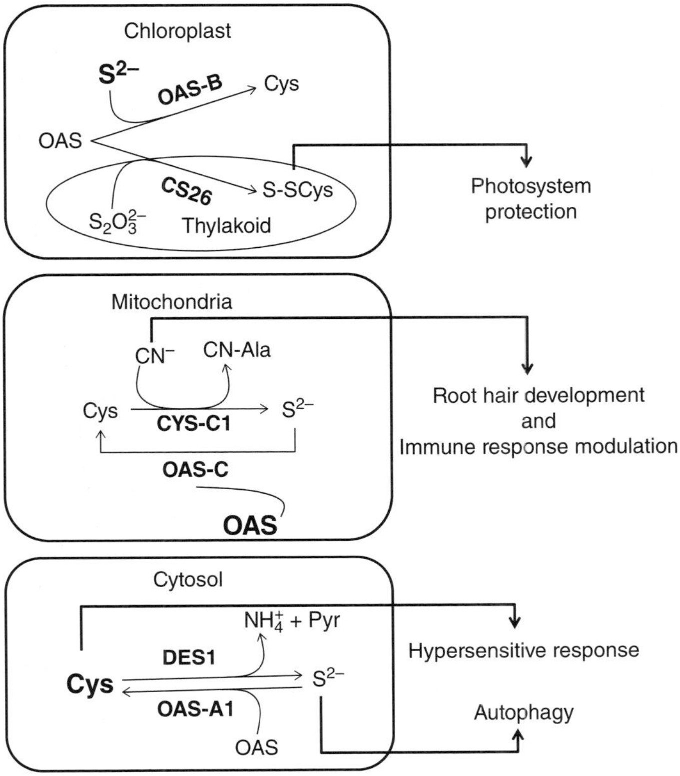

Fig. 12.4. The importance of the compartment-specific pool of cysteine in *Arabidopsis* cells. The function of every pool of cysteine or its associated metabolites is indicated on the right. The most abundant sulfur-associated metabolite in every cellular compartment is indicated with large bold letters. *O*-acetylserine (thiol)lyase (OASTL) isoforms: CS26, CYS-C1, OAS-A1, OAS-B, OAS-C, DES1. Other abbreviations: Cys, cysteine; OAS, *O*-acetylserine; S^{2-}, sulfide; $S_2O_3^{2-}$, thiosulfate.

These proteins are considered to be authentic OASTLs because they catalyse the synthesis of cysteine from OAS and sulfide, and they interact with SAT (Bonner *et al.*, 2005; Heeg *et al.*, 2007). One of the other OASTL genes, *OAS-A2*, does not produce a functional protein owing to the presence of an in-frame stop codon and an unspliced intron. In addition to these major OASTLs, the OASTL enzyme family also includes another isoform, CYS-C1, which is located in the mitochondria; this isoform is the most highly expressed isoform, but actually functions as a *β*-cyanoalanine synthase (Yamaguchi *et al.*, 2000). The remaining OASTL proteins, the cytosolic cys-D1, cys-D2 and DES1, and the plastidic CS26, are expressed at much lower levels, were identified by sequence homology upon the sequencing of the *A. thaliana* genome and were thought to have auxiliary functions with respect to the major isoforms. Our investigations have shed light on to the functions of CS26, CYS-C1 and DES1 as a way to unravel the role of cysteine metabolism in every cellular compartment (see Fig. 12.4).

12.6.1 CS26

In *Arabidopsis* chloroplasts, two OASTL isoforms are present, a true OASTL, OAS-B and CS26, which has *S*-sulfocysteine synthase activity and catalyses the incorporation of thiosulfate into OAS to form *S*-sulfocysteine (S-Cys) (Nakamura *et al.*, 1984; Bermúdez *et al.*, 2010). In photosynthetic organisms, the reductive steps that are catalysed by APS reductase and sulfite reductase are restricted to the plastids; thus, sulfate reduction occurs only in this compartment (Takahashi *et al.*, 2011). Logically, cysteine production via OASTL should be also restricted to the chloroplast; however, OAS-B does not appear to play an essential role in the plant, as plants lacking OAS-B show no apparent phenotype when grown under long- or short-day photoperiods (Bermúdez *et al.*, 2012). In contrast, it has been shown that the minor isoform CS26 plays an essential role in chloroplast redox control (Bermúdez *et al.*, 2012). CS26 loss-of-function mutants display severe growth inhibition and leaf paleness when grown under long-day photoperiod conditions; this phenotype depends on the photoperiod, as under short-day conditions the *cs26* mutant is indistinguishable from the wild type plants. The long-day specific phenotype is accompanied by an accumulation of reactive oxygen species (ROS), such as superoxide radicals and hydrogen peroxide, in the *cs26* leaves, which is not observed under short-day conditions (Bermúdez *et al.*, 2010). The *cs26* mutant plants also show a significant reduction in photosynthetic performance when grown under long-day conditions, manifested as a reduction in the net CO_2 assimilation rate, mesophyll conductance and mitochondrial respiration in darkness, which phenomena are not observed under short-day conditions (Bermúdez *et al.*, 2012). The dependence of the *cs26* mutant phenotype on the photoperiod suggests that plants lacking this protein are not able to deal with the light stress that is produced during periods of prolonged light exposure, which damages the photosynthetic machinery and causes the accumulation of ROS and, ultimately, growth inhibition. Therefore, the chloroplastic *S*-sulfocysteine synthase activity is essential for maintaining the balance between ROS production and detoxification in the chloroplast, and also for preventing photochemical damage to the photosystems and to the electron transport chain. In fact, when grown under low light intensity, the *cs26* mutant shows similar photosynthetic parameters to those of plants that are grown under high-light conditions (Bermúdez *et al.*, 2012).

OAS-B and CS26 are located in different subcellular compartments and have different enzymatic activities, although they share OAS as a substrate. OAS-B is found in the stroma, has OASTL activity and uses sulfur to incorporate itself into the OAS, whereas CS26 is found in the lumen, where it incorporates thiosulfate into the OAS to produce S-Cys (Bermúdez *et al.*, 2010). S-Cys can act as an oxidant, reacting with thiol groups following the reaction:

$$RSSO_3^- + RSH \leftrightarrows RSSR + HSO_3^- \qquad (12.1)$$

This allows the oxidation of protein thiol groups in the thylakoid lumen. CS26 would then act as a photosynthesis sensor. Under normal photosynthetic conditions, sulfite is directly converted into sulfide in the chloroplasts. When the photosynthetic machinery is saturated, the ROS accumulate and sulfite is converted into thiosulfate in the chloroplasts (Khan *et al.*, 2010). Thiosulfate is detected by the CS26 protein and is converted to S-Cys, which, acting as a mild oxidant in the lumen, triggers protection mechanisms in the photosynthetic apparatus by regulating the activity of enzymes that are involved in photosystem repair or reorganization, such as STN7 (a thylakoid-associated kinase), KBP13 (a thylakoid peptidyl-prolyl isomerase)and Deg1 (a thylakoid protease) (Dietz and Pfannschmidt, 2011; Gotor and Romero, 2013).

12.6.2 CYS-C1

In non-cyanogenic plants, cyanide is a co-product of ethylene biosynthesis. It is also produced during the biosynthesis of camalexin, a phytoalexin formed in *Arabidopsis* plants when they are infected by a large variety of microorganisms, including bacteria, fungi and oomycetes (Glawischnig, 2007). The lethality of the anion cyanide lies in its high reactivity with keto compounds and Schiff base intermediates to give cyanohydrins and stable nitrile derivatives, respectively, and because it chelates divalent and trivalent metal ions in the prosthetic groups of several metalloenzymes. In mitochondria, cyanide binds to the haem iron of cytochrome *c* oxidase, thereby blocking the utilization of oxygen in cellular functions (Donato *et al.*, 2007). Hence, cyanide must be rapidly detoxified and metabolized by the plant to keep the concentration below toxic levels. *Arabidopsis* plants carry the mitochondrial β-cyanoalanine synthase (CAS) CYS-C1 (Watanabe *et al.*, 2008), which belongs, together with the OASTLs, to the family of β-substituted alanine synthase enzymes. CAS is a PLP-dependent enzyme that uses cysteine to detoxify cyanide by converting cyanide and cysteine into hydrogen sulfide (H_2S) and β-cyanoalanine, the latter being converted later to asparagine (Asn), aspartate (Asp) and ammonia by NIT4 class nitrilases, thus recycling the nitrogen for the plant (Piotrowski, 2008). H_2S also blocks the mitochondrial respiratory pathway and, for this reason, needs to be detoxified by the true OASTL, OAS-C, which produces cysteine from OAS and H_2S that can be used by CYS-C1 to detoxify cyanide, so completing the cyanide detoxification cycle in mitochondria (Álvarez *et al.*, 2012c).

The *CYS-C1* or *OAS-C* loss-of-function mutations are not toxic for the plant and lead to an increased level of cyanide in *cys-c1* and *oas-c* null mutants as well as a root hairless phenotype, suggesting that cyanide plays a signalling role in root development (García *et al.*, 2010; Álvarez *et al.*, 2012c). In both cases, the genetic complementation of the *cys-c1* and *oas-c* null mutants with the corresponding *CYS-C1* and *OAS-C* gene rescues the root hair elongation impairment, restoring the wild type phenotype and thereby confirming that the observed phenotypes were indeed due to the mutations (García *et al.*, 2010; Álvarez *et al.*, 2012c). Furthermore, the root hair defect is phenocopied in wild type plants by the exogenous addition of cyanide to the growth medium and is reversed by the addition of hydroxocobalamin, which is the most commonly used antidote for severe acute cyanide poisoning in humans (García *et al.*, 2010).

In addition to its role in plant root development, cyanide has been proposed to act as a regulator of other biological processes in plants, such as seed dormancy and germination (Cohn and Hughes, 1986; Bethke *et al.*, 2006; Siegien and Bogatek, 2006), and resistance to viral and fungal pathogens (Chivasa and Carr, 1998; Wong *et al.*, 2002; Seo *et al.*, 2011) using exogenously applied cyanide. The *cys-c1* mutant allows for the investigation of the role of endogenously produced cyanide.

We have definitively observed that *cys-c1* plants present an increased susceptibility to the necrotrophic fungus *Botrytis cinerea* and an increased tolerance to the biotrophic *Pseudomonas syringae* pv. *tomato DC3000* bacterium and the *Beet curly top virus*. The *cys-c1* mutation reduces the respiration rate in the leaves, causes the accumulation of ROS and induces the alternative oxidase *AOX1a* and the pathogenesis-related expression of *PR1*, a protein induced by the salicylic acid-dependent pathway. Interestingly, no patches of dead cells are observed when this mutant is grown under either long- or short-photoperiod conditions. We hypothesize that cyanide, which is transiently accumulated during avirulent bacterial infection and constitutively accumulated in the *cys-c1* mutant, uncouples the respiratory electron chain, which is dependent on cytochrome c oxidase, and that this uncoupling induces the alternative oxidase activity and the accumulation of ROS, which stimulate the salicylic acid-dependent signalling pathway of the plant immune system (García *et al.*, 2013).

Considering these results, we hypothesize that the cysteine mitochondrial pool is essential for maintaining the non-toxic concentrations of mitochondrial cyanide and sulfur to allow their signalling role directly or through the controlled accumulation of ROS in this organelle.

12.6.3 DES1

Cysteine is highly reactive and so it can be a very toxic molecule. Thiols are easily oxidized to form species with sulfur in higher oxidation states that subsequently inhibit enzyme activity; furthermore, cysteine reduces the ferric iron at an exceptional rate and promotes oxidative damage through the Fenton reaction (Jacob *et al.*, 2003; Park and Imlay, 2003). Consequently, cysteine homeostasis must be precisely maintained in the cytosol. Cysteine is mainly produced *de novo* in the cytosol by the action of the OAS-A1 protein (Heeg *et al.*, 2008; Watanabe *et al.*, 2008) and can be degraded in the cytosol by the action of L-cysteine desulfhydrase (DES; EC 4.4.1.1), which catalyses the formation of sulfide, ammonia and pyruvate from cysteine in a stoichiometric ratio of 1:1:1. In *A. thaliana*, the only cytosolic desulfhydrase that has been described to date is encoded by *DES1* (At5g28030). Knockout *des1* plants show increased cysteine levels and decreased accumulation of ROS, leading to enhanced antioxidant defences and tolerance to conditions that promote oxidative stress (Álvarez *et al.*, 2010). In contrast, *oas-a1* knockout mutants show decreased cysteine levels and an imbalance between the generation and removal of ROS, leading to an accumulation of ROS and the induction of programmed cell death (PCD) in patches of cells (López-Martín *et al.*, 2008a). This imbalance results in an increased sensitivity of *oas-a1* mutants to oxidative stress and the conditions promoting it (López-Martín *et al.*, 2008a). The contrasting phenotypes of *des1* and *oas-a1* mutants suggest that cytosolic cysteine is a determinant of the antioxidative capacity of this cellular compartment; therefore, the coordination of the contrasting enzymatic activities of DES1 and OAS-A1 maintains cytosolic cysteine homeostasis and so should influence the processes that are regulated by the cysteine/redox homeostasis (López-Martín *et al.*, 2008b; Gotor *et al.*, 2010). The tolerance to heavy metals such as Cd is affected by cytosolic cysteine homeostasis, as *des1* mutants are more tolerant to increased concentrations of Cd (up to 250 μM) than are wild type plants (Álvarez *et al.*, 2010), and *oas-a1* mutants are more sensitive to low concentrations of this metal (as low as 75 μM) than are wild type plants. Thus, the cysteine levels are directly correlated with plant resistance to Cd-induced stress. In fact, the overexpression of *oas-a1* in *Arabidopsis* and yeast leads to an enhanced tolerance to heavy metals and salinity, respectively (Dominguez-Solis *et al.*, 2001, 2004; Romero *et al.*, 2001).

During the plant immune response, redox signalling plays an essential role. One of the earliest responses to pathogen infection is the production of ROS (reviewed in Lamb and Dixon, 1997), which are essential in the establishment of resistance. In response to some pathogens, plants can induce the hypersensitive response (HR), which causes the rapid death of the infected cells, so limiting the spread of biotrophic pathogens (Heath, 2000). The establishment of the HR is dependent on the generation of ROS and contributes to the activation of defence in adjacent cells and to systemic acquired resistance (SAR), which is a broad-spectrum form of disease resistance that is mediated by the action of the hormone salicylic acid (Vlot *et al.*, 2008). The study of the response of *des1* and *oas-a1* mutant plants to biotrophic and necrotrophic pathogens has shown that the cytosolic cysteine content is directly correlated with the resistance; therefore, an increase of cytosolic cysteine increases the plant resistance to pathogens, while a decrease of cytosolic cysteine decreases the plant resistance to pathogens (Álvarez *et al.*, 2012a). The *des1* mutants behave as constitutive SAR mutants, in which this systemic response is constitutively activated and the mutants show a high level of resistance to biotrophic and necrotrophic pathogens, the accumulation of salicylic acid and the induction of *WRKY54* and *PR1*. Additionally, *oas-a1* mutations enhance the pathogen sensitivity (Álvarez *et al.*, 2012a) and impair the HR that is associated with the infection by the bacteria *Pseudomonas syringae* pv. *tomato DC3000 avrRpm1* (Álvarez *et al.*, 2012a). These data suggest that cytosolic cysteine homeostasis, which is driven by the enzymatic activities of OAS-A1 and DES1, plays an essential role in regulating the plant response to pathogens.

The *des1* mutants also exhibit premature leaf senescence, along with an increased expression of senescence-associated genes, such as *SAG12*, *NAP* and *PR1*, and the presence of senescence- associated vacuoles (SAVs) (Álvarez *et al.*, 2010, 2012b). In contrast, the *oas-a1.1* mutant shows a decreased expression of senescence-associated genes, such as *SAG12* and *SEN1*, and lacks SAVs in its mesophyll cells (Álvarez *et al.*, 2010, 2012b). During senescence, autophagy is essential for maintaining viability via the effective recycling of nutrients (Bassham *et al.*, 2006; Bassham, 2007; Xiong *et al.*, 2007). Autophagy is a universal mechanism present in eukaryotic cells and consists of the degradation of cellular content to recycle nutrients, as well as to degrade damaged or toxic components. Autophagy is a constitutive process occurring at a basal level in growing cells, allowing for protein and organelle recycling. Plants induce autophagy under abiotic stress conditions, such as nutrient limitation or drug presence, and during normal development, the immune response and senescence. At the cellular level, autophagy is characterized by the *de novo* synthesis of double membrane-bound structures called autophagosomes, which engulf and deliver materials to the vacuole for breakdown (Nakatogawa *et al.*, 2009). Autophagy-related (ATG) proteins are essential for the autophagy process and have been used to monitor autophagic activity in plants. The most commonly used protein is ATG8, which is covalently conjugated to phosphatidylethanolamine (PE) and tethered to the autophagosomes (Li and Vierstra, 2012; Pérez-Pérez *et al.*, 2012).

The *des1* mutants accumulate ATG8, especially in its lipidated form, confirming that autophagy is induced in this mutant (Álvarez *et al.*, 2010, 2012b). Mutations in the *DES1* gene impede the degradation of cytosolic cysteine and, concomitantly, the generation of H_2S in the cytosol. H_2S is a known signalling molecule in animal systems. Recently, emerging experimental evidence has shown that H_2S is also a signalling molecule that is as important as NO and H_2O_2 in plant systems (Zhang *et al.*, 2008; García-Mata and Lamattina, 2010; Lisjak *et al.*, 2010; Zhang *et al.*, 2010; Chen *et al.*, 2011). The exogenous application of H_2S to *des1* mutants restores the differences that these mutants have with wild type plants at the phenotypic (the induced senescence is reverted), cellular (the SAVs and lipidated ATG8 are not over-accumulated) and transcriptomic (the transcriptome of *des1-1* plants that are treated with sulfide is very similar to that of the wild type plants) levels, demonstrating that sulfide negatively regulates autophagy in this model plant (Gotor *et al.*, 2013). In summary, DES1 and OAS-A1 modulate cysteine/sulfide homeostasis, which is critical for plant metabolism, development and stress responses in the cytosol (Romero *et al.*, 2013).

12.7 Conclusions

Our recent investigations of the OASTL family in *Arabidopsis* have uncovered functions for cysteine and its related molecules other than those of a mere metabolic role (see Fig. 12.4). Indeed, chloroplastic OAS can be used to form cysteine or S-Cys, which is a metabolite that is involved in the regulation and protection of the photosystems. In the mitochondria, cysteine maintains appropriate levels of cyanide, which regulates root hair development and modulates the plant immune response. Finally, cytosolic cysteine homeostasis orchestrates the plant response to pathogens, and cytosolic sulfide regulates autophagy. The mechanisms underlying these new signalling roles of cysteine and its related molecules are still unknown and constitute a novel and exciting field of research.

Acknowledgements

This work was funded in part by the European Regional Development Fund through the Ministerio de Economía y Competitividad (Grant no. BIO2013-44648) and the Junta de Andalucía (Grant no. CVI-7190).

References

Álvarez, C., Calo, L., Romero, L.C., García, I. and Gotor, C. (2010) An *O*-acetylserine(thiol)lyase homolog with L-cysteine desulfhydrase activity regulates cysteine homeostasis in *Arabidopsis*. *Plant Physiology* 152, 656–669.

Álvarez, C., Bermúdez, M.A., Romero, L.C., Gotor, C. and García, I. (2012a) Cysteine homeostasis plays an essential role in plant immunity. *New Phytologist* 193, 165–177.

Álvarez, C., García, I., Moreno, I., Pérez-Pérez, M.E., Crespo, J.L., Romero, L.C. and Gotor, C. (2012b) Cysteine-generated sulfide in the cytosol negatively regulates autophagy and modulates the transcriptional profile in *Arabidopsis*. *The Plant Cell* 24, 4621–4634.

Álvarez, C., García, I., Romero, L.C. and Gotor, C. (2012c) Mitochondrial sulfide detoxification requires a functional isoform *O*-acetylserine(thiol)lyase C in *Arabidopsis thaliana*. *Molecular Plant* 5, 1217–1226.

Barroso, C., Vega, J.M. and Gotor, C. (1995) A new member of the cytosolic *O*-acetylserine(thiol)lyase gene family in *Arabidopsis thaliana*. *FEBS Letters* 363, 1–5.

Barroso, C., Romero, L.C., Vega, J.M. and Gotor, C. (1997) Molecular characterization of the sulfur metabolism in plants. *Current Topics in Phytochemistry* 1, 19–29.

Bassham, D.C. (2007) Plant autophagy – more than a starvation response. *Current Opinion in Plant Biology* 10, 587–593.

Bassham, D.C., Laporte, M., Marty, F., Moriyasu, Y., Ohsumi, Y., Olsen, L.J. and Yoshimoto, K. (2006) Autophagy in development and stress responses of plants. *Autophagy* 2, 2–11.

Bermúdez, M.A., Paez-Ochoa, M.A., Gotor, C. and Romero, L.C. (2010) *Arabidopsis S*-sulfocysteine synthase activity is essential for chloroplast function and long-day light-dependent redox control. *The Plant Cell* 22, 403–416.

Bermúdez, M.A., Galmés, J., Moreno, I., Mullineaux, P.M., Gotor, C. and Romero, L.C. (2012) Photosynthetic adaptation to length of day is dependent on *S*-sulfocysteine synthase activity in the thylakoid. *Plant Physiology* 160, 274–288.

Bethke, P.C., Libourel, L.G.L., Reinohl, V. and Jones, R.L. (2006) Sodium nitroprusside, cyanide, nitrite, and nitrate break *Arabidopsis* seed dormancy in a nitric oxide-dependent manner. *Planta* 223, 805–812.

Bolchi, A., Petrucco, S., Tenca, P.L., Foroni, C. and Ottonello, S. (1999) Coordinate modulation of maize sulfate permease and ATP sulfurylase mRNAs in response to variations in sulfur nutritional status: stereospecific down-regulation by L-cysteine. *Plant Molecular Biology* 39, 527–537.

Bonner, E.R., Cahoon, R.E., Knapke, S.M. and Jez, J.M. (2005) Molecular basis of cysteine biosynthesis in plants – structural and functional analysis of *O*-acetylserine sulfhydrylase from *Arabidopsis thaliana*. *The Journal of Biological Chemistry* 280, 38803–38813.

Chen, J., Wu, F.H., Wang, W.H., Zheng, C.J., Lin, G.H., Dong, X.J., He, J.X., Pei, Z.M. and Zheng, H.L. (2011) Hydrogen sulphide enhances photosynthesis through promoting chloroplast biogenesis, photosynthetic enzyme expression, and thiol redox modification in *Spinacia oleracea* seedlings. *Journal of Experimental Botany* 62, 4481–4493.

Chivasa, S. and Carr, J.P. (1998) Cyanide restores *N* gene-mediated resistance to tobacco mosaic virus in transgenic tobacco expressing salicylic acid hydroxylase. *The Plant Cell* 10, 1489–1498.

Cohn, M.A. and Hughes, J.A. (1986) Seed dormancy in red rice. 5. Response to azide, hydroxylamine and cyanide. *Plant Physiology* 80, 531–533.

Dietz, K.J. and Pfannschmidt, T. (2011) Novel regulators in photosynthetic redox control of plant metabolism and gene expression. *Plant Physiology* 155, 1477–1485.

Dominguez-Solis, J.R., Gutierrez-Alcala, G., Vega, J.M., Romero, L.C. and Gotor, C. (2001) The cytosolic *O*-acetylserine(thiol)lyase gene is regulated by heavy metals and can function in cadmium tolerance. *The Journal of Biological Chemistry* 276, 9297–9302.

Dominguez-Solis, J.R., López-Martín, M.C., Ager, F.J., Ynsa, M.D., Romero, L.C. and Gotor, C. (2004) Increased cysteine availability is essential for cadmium tolerance and accumulation in *Arabidopsis thaliana*. *Plant Biotechnology Journal* 2, 469–476.

Donato, D.B., Nichols, O., Possingham, H., Moore, M., Ricci, P.F. and Noller, B.N. (2007) A critical review of the effects of gold cyanide-bearing tailings solutions on wildlife. *Environment International* 33, 974–984.

Droux, M. (2004) Sulfur assimilation and the role of sulfur in plant metabolism: a survey. *Photosynthesis Research* 79, 331–348.

Droux, M., Ruffet, M.L., Douce, R. and Job, D. (1998) Interactions between serine acetyltransferase and *O*-acetylserine (thiol) lyase in higher plants – structural and kinetic properties of the free and bound enzymes. *European Journal of Biochemistry* 255, 235–245.

García, I., Castellano, J.M., Vioque, B., Solano, R., Gotor, C. and Romero L.C. (2010) Mitochondrial β-cyanoalanine synthase is essential for root hair formation in *Arabidopsis thaliana*. *The Plant Cell* 22, 3268–3279.

García, I., Rosas, T., Bejarano, E.R., Gotor, C. and Romero, L.C. (2013) Transient transcriptional regulation of the *CYS-C1* gene and cyanide accumulation upon pathogen infection in the plant immune response. *Plant Physiology* 162, 2015–2027.

García-Mata, C. and Lamattina, L. (2010) Hydrogen sulphide, a novel gasotransmitter involved in guard cell signalling. *New Phytologist* 188, 977–984.

Glawischnig, E. (2007) Camalexin. *Phytochemistry* 68, 401–406.

Gotor, C. and Romero, L.C. (2013) *S*-sulfocysteine synthase function in sensing chloroplast redox status. *Plant Signaling and Behavior* 8(3): e23313.

Gotor, C., Álvarez, C., Bermúdez, M.A., Moreno, I., García, I. and Romero, L.C. (2010) Low abundance does not mean less importance in cysteine metabolism. *Plant Signaling and Behavior* 5, 1028–1030.

Gotor, C., García, I., Crespo, J.L. and Romero, L.C. (2013) Sulfide as a signaling molecule in autophagy. *Autophagy* 9, 609–611.

Gutierrez-Marcos, J.F., Roberts, M.A., Campbell, E.I. and Wray, J.L. (1996) Three members of a novel small gene-family from *Arabidopsis thaliana* able to complement functionally an *Escherichia coli* mutant defective in PAPS reductase activity encode proteins with a thioredoxin-like domain and “APS reductase” activity. *Proceedings National Academy Sciences of the United States of America* 93, 13377–13382.

Haas, F.H., Heeg, C., Queiroz, R., Bauer, A., Wirtz, M. and Hell, R. (2008) Mitochondrial serine acetyltransferase functions as a pacemaker of cysteine synthesis in plant cells. *Plant Physiology* 148, 1055–1067.

Hatzfeld, Y., Lee, S., Lee, M., Leustek, T. and Saito, K. (2000) Functional characterization of a gene encoding a fourth ATP sulfurylase isoform from *Arabidopsis thaliana*. *Gene* 248, 51–58.

Hawkesford, M.J. and De Kok, L.J. (2006) Managing sulphur metabolism in plants. *Plant, Cell and Environment* 29, 382–395.

Heath, M.C. (2000) Hypersensitive response-related death. *Plant Molecular Biology* 44, 321–334.

Heeg, C., Muller, C., Throm, C., Rippe, K., Rybin, V., Wirtz, M. and Hell, R. (2007) Protein–protein interactions define regulatory properties of the cysteine synthase complex. *FEBS Journal* 274, 241–241.

Heeg, C., Kruse, C., Jost, R., Gutensohn, M., Ruppert, T., Wirtz, M. and Hell, R. (2008) Analysis of the *Arabidopsis O*-acetylserine(thiol)lyase gene family demonstrates compartment-specific differences in the regulation of cysteine synthesis. *The Plant Cell* 20, 168–185.

Heldt, W.H., Werdan, K., Milovancev, M. and Geller, G. (1973) Alkalization of the chloroplast stroma caused by light-dependent proton flux into the thylakoid space. *Biochimica and Biophysica Acta* 314, 224–241.

Hell, R. (1997) Molecular physiology of plant sulfur metabolism. *Planta* 202, 138–148.

Hirai, M.Y. and Saito, K. (2008) Analysis of systemic sulfur metabolism in plants using integrated '-omics' strategies. *Molecular BioSystems* 4, 967–973.

Howarth, J.R., Dominguez-Solis, J.R., Gutierrez-Alcala, G., Wray, J.L., Romero, L.C. and Gotor, C. (2003a) The serine acetyltransferase gene family in *Arabidopsis thaliana* and the regulation of its expression by cadmium. *Plant Molecular Biology* 51, 589–598.

Howarth, J.R., Fourcroy, P., Davidian, J.C., Smith, F.W. and Hawkesford, M.J. (2003b) Cloning of two contrasting high-affinity sulfate transporters from tomato induced by low sulfate and infection by the vascular pathogen *Verticillium dahliae*. *Planta* 218, 58–64.

Jacob, C., Giles, G.I., Giles, N.M. and Sies, H. (2003) Sulfur and selenium: the role of oxidation state in protein structure and function. *Angewandte Chemie* 42, 4742–4758.

Jost, R., Berkowitz, O., Wirtz, M., Hopkins, L., Hawkesford, M.J. and Hell, R. (2000) Genomic and functional characterization of the *oas* gene family encoding *O*-acetylserine (thiol) lyases, enzymes catalyzing the final step in cysteine biosynthesis in *Arabidopsis thaliana*. *Gene* 253, 237–247.

Kabil, O. and Banerjee, R. (2010) Redox biochemistry of hydrogen sulfide. *The Journal of Biological Chemistry* 285, 21903–21907.

Khan, M.S., Haas, F.H., Samami, A.A., Gholami, A.M., Bauer, A., Fellenberg, K., Reichelt, M., Hansch, R., Mendel, R.R., Meyer, A.J., Wirtz, M. and Hell, R. (2010) Sulfite reductase defines a newly discovered bottleneck for assimilatory sulfate reduction and is essential for growth and development in *Arabidopsis thaliana*. *The Plant Cell* 22, 1216–1231.

Krueger, S., Niehl, A., Lopez Martin, M.C., Steinhauser, D., Donath, A., Hildebrandt, T., Romero, L.C., Hoefgen, R., Gotor, C. and Hesse, H. (2009) Analysis of cytosolic and plastidic serine acetyltransferase mutants and subcellular metabolite distributions suggests interplay of the cellular compartments for cysteine biosynthesis in *Arabidopsis*. *Plant Cell and Environment* 32, 349–367.

Kumaran, S., Yi, H., Krishnan, H.B. and Jez, J.M. (2009) Assembly of the cysteine synthase complex and the regulatory role of protein–protein interactions. *The Journal of Biological Chemistry* 284, 10268–10275.

Lamb, C. and Dixon, R.A. (1997) The oxidative burst in plant disease resistance. *Annual Review of Plant Physiology and Plant Molecular Biology* 48, 251–275.

Lee, B.-R., Koprivova, A. and Kopriva, S. (2011) The key enzyme of sulfate assimilation, adenosine 5′-phosphosulfate reductase, is regulated by HY5 in *Arabidopsis*. *The Plant Journal* 67, 1042–1054.

Lee, M. and Leustek, T. (1999) Identification of the gene encoding homoserine kinase from *Arabidopsis thaliana* and characterization of the recombinant enzyme derived from the gene. *Archives of Biochemistry and Biophysics* 372, 135–142.

León, J., Romero, L.C., Galván, F. and Vega, J.M. (1987) Purification and physicochemical characterization of *O*-acetyl-l-serine sulfhydrilase from *Chlamydomonas reinhardtii*. *Plant Science* 53, 93–99.

Leustek, T., Martin, M.N., Bick, J.A. and Davies, J.P. (2000) Pathways and regulation of sulfur metabolism revealed through molecular and genetic studies. *Annual Review of Plant Physiology and Plant Molecular Biology* 51, 141–165.

Li, F. and Vierstra, R.D. (2012) Autophagy: a multifaceted intracellular system for bulk and selective recycling. *Trends in Plant Sciences* 17, 526–537.

Lisjak, M., Srivastava, N., Teklic, T., Civale, L., Lewandowski, K., Wilson, I., Wood, M.E., Whiteman, M. and Hancock, J.T. (2010) A novel hydrogen sulfide donor causes stomatal opening and reduces nitric oxide accumulation. *Plant Physiology and Biochemistry* 48, 931–935.

López-Martín, M.C., Becana, M., Romero, L.C. and Gotor, C. (2008a) Knocking out cytosolic cysteine synthesis compromises the antioxidant capacity of the cytosol to maintain discrete concentrations of hydrogen peroxide in *Arabidopsis*. *Plant Physiology* 147, 562–572.

López-Martín, M.C., Romero, L.C. and Gotor, C. (2008b) Cytosolic cysteine in redox signaling. *Plant Signaling and Behavior* 3, 880–881.

Lunn, J.E., Droux, M., Martin, J. and Douce, R. (1990) Localization of ATP sulfurylase and *O*-acetylserine (thiol)lyase in spinach leaves. *Plant Physiology* 94, 1345–1352.

Moriarty, R.M., Naithani, R. and Surve, B. (2007) Organosulfur compounds in cancer chemoprevention. *Mini Reviews in Medicinal Chemistry* 7, 827–838.

Mugford, S.G., Yoshimoto, N., Reichelt, M., Wirtz, M., Hill, L., Mugford, S.T., Nakazato, Y., Noji, M., Takahashi, H., Kramell, R., Gigolashvili, T., Flugge, U.I. *et al.* (2009) Disruption of adenosine-5′-phosphosulfate kinase in *Arabidopsis* reduces levels of sulfated secondary metabolites. *The Plant Cell* 21, 910–927.

Nakamura, T., Iwahashi, H. and Eguchi, Y. (1984) Enzymatic proof for the identity of the *S*-sulfocysteine synthase and cysteine synthase B of *Salmonella typhimurium*. *Journal of Bacteriology* 158, 1122–1127.

Nakatogawa, H., Suzuki, K., Kamada, Y. and Ohsumi, Y. (2009) Dynamics and diversity in autophagy mechanisms: lessons from yeast. *Nature Reviews Molecular Cell Biology* 10, 458–467.

Noji, M., Inoue, K., Kimura, N., Gouda, A. and Saito, K. (1998) Isoform-dependent differences in feedback regulation and subcellular localization of serine acetyltransferase involved in cysteine biosynthesis from *Arabidopsis thaliana*. *The Journal of Biological Chemistry* 273, 32739–32745.

Noji, M., Goulart Kawashima, C., Obayashi, T. and Saito, K. (2006) *In silico* assessment of gene function involved in cysteine biosynthesis in *Arabidopsis*: expression analysis of multiple isoforms of serine acetyltransferase. *Amino Acids* 30, 163–171.

Park, S. and Imlay, J.A. (2003) High levels of intracellular cysteine promote oxidative DNA damage by driving the Fenton reaction. *Journal of Bacteriology* 185, 1942–1950.

Pérez-Pérez, M.E., Lemaire, S.D. and Crespo, J.L. (2012) Reactive oxygen species and autophagy in plants and algae. *Plant Physiology* 160, 156–164.

Piotrowski, M. (2008) Primary or secondary? Versatile nitrilases in plant metabolism. *Phytochemistry* 69, 2655–2667.

Rausch, T. and Wachter, A. (2005) Sulfur metabolism: a versatile platform for launching defence operations. *Trends in Plant Science* 10, 503–509.

Romero, L.C., Dominguez-Solis, J.R., Gutierrez-Alcala, G. and Gotor, C. (2001) Salt regulation of *O*-acetylserine(thiol)lyase in *Arabidopsis thaliana* and increased tolerance in yeast. *Plant Physiology and Biochemistry* 39, 643–647.

Romero, L.C., García, I. and Gotor, C. (2013) L-Cysteine desulfhydrase 1 modulates the generation of the signaling molecule sulfide in plant cytosol. *Plant Signaling and Behavior* 8(5): e24007.

Rouached, H., Berthomieu, P., El Kassis, E., Cathala, N., Catherinot, V., Labesse, G., Davidian, J.C. and Fourcroy, P. (2005) Structural and functional analysis of the C-terminal STAS (sulfate transporter and anti-sigma antagonist) domain of the *Arabidopsis thaliana* sulfate transporter SULTR1.2. *The Journal of Biological Chemistry* 280, 15976–15983.

Saito, K. (2004) Sulfur assimilatory metabolism. The long and smelling road. *Plant Physiology* 136, 2443–2450.

Seo, S., Mitsuhara, I., Feng, J., Iwai, T., Hasegawa, M. and Ohashi, Y. (2011) Cyanide, a coproduct of plant hormone ethylene biosynthesis, contributes to the resistance of rice to blast fungus. *Plant Physiology* 155, 502–514.

Setya, A., Murillo, M. and Leustek, T. (1996) Sulfate reduction in higher plants: molecular evidence for a novel 5′-adenylylsulfate reductase. *Proceedings of the National Academy of Sciences of the United States of America* 93, 13383–13388.

Shibagaki, N. and Grossman, A.R. (2004) Probing the function of STAS domains of the *Arabidopsis* sulfate transporters. *The Journal of Biological Chemistry* 279, 30791–30799.

Siegien, I. and Bogatek, R. (2006) Cyanide action in plants – from toxic to regulatory. *Acta Physiologiae Plantarum* 28, 483–497.

Takahashi, H., Kopriva, S., Giordano, M., Saito, H. and Hell, R. (2011) Sulfur assimilation in photosynthetic organisms: molecular functions and regulations of transporters and assimilatory enzymes. *Annual Review of Plant Biology* 62, 157–184.

Tamura, K., Peterson, D., Peterson, N., Stecher, G., Nei, M. and Kumar, S. (2011) mega5: molecular evolutionary genetics analysis using maximum likelihood, evolutionary distance, and maximum parsimony methods. *Molecular Biology and Evolution* 28, 2731–2739.

Vauclare, P., Kopriva, S., Fell, D., Suter, M., Sticher, L., von Ballmoos, P., Krahenbuhl, U., den Camp, R.O. and Brunold, C. (2002) Flux control of sulphate assimilation in *Arabidopsis thaliana*: adenosine 5′-phosphosulphate reductase is more susceptible than ATP sulphurylase to negative control by thiols. *Plant Journal* 31, 729–740.

Vidmar, J.J., Tagmount, A., Cathala, N., Touraine, B. and Davidian, J.E. (2000) Cloning and characterization of a root specific high-affinity sulfate transporter from *Arabidopsis thaliana*. *FEBS Letters* 475, 65–69.

Vlot, A.C., Klessig, D.F. and Park, S.W. (2008) Systemic acquired resistance: the elusive signal(s). *Current Opinion in Plant Biology* 11, 436–442.

Watanabe, M., Kusano, M., Oikawa, A., Fukushima, A., Noji, M. and Saito, K. (2008) Physiological roles of the β-substituted alanine synthase gene family in *Arabidopsis*. *Plant Physiology* 146, 310–320.

Wirtz, M. and Droux, M. (2005) Synthesis of the sulfur amino acids: cysteine and methionine. *Photosynthesis Research* 86, 345–362.

Wirtz, M. and Hell, R. (2006) Functional analysis of the cysteine synthase protein complex from plants: structural, biochemical and regulatory properties. *Journal of Plant Physiology* 163, 273–286.

Wong, C.E., Carson, R.A. and Carr, J.P. (2002) Chemically induced virus resistance in *Arabidopsis thaliana* is independent of pathogenesis-related protein expression and the *NPR1* gene. *Molecular Plant Microbe Interactions* 15, 75–81.

Wu, W. and Berkowitz, G.A. (1992) Stromal pH and photosynthesis are affected by electroneutral K and H exchange through chloroplast envelope ion channels. *Plant Physiology* 98, 666–672.

Xiong, Y., Contento, A.L., Nguyen, P.Q. and Bassham, D.C. (2007) Degradation of oxidized proteins by autophagy during oxidative stress in *Arabidopsis*. *Plant Physiology* 143, 291–299.

Yamaguchi, Y., Nakamura, T., Kusano, T. and Sano, H. (2000) Three *Arabidopsis* genes encoding proteins with differential activities for cysteine synthase and β-cyanoalanine synthase. *Plant and Cell Physiology* 41, 465–476.

Yonekura-Sakakibara, K., Onda, Y., Ashikari, T., Tanaka, Y., Kusumi, T. and Hase, T. (2000) Analysis of reductant supply systems for ferredoxin-dependent sulfite reductase in photosynthetic and nonphotosynthetic organs of maize. *Plant Physiology* 122, 887–894.

Zhang, H., Hu, L.Y., Hu, K.D., He, Y.D., Wang, S.H. and Luo, J.P. (2008) Hydrogen sulfide promotes wheat seed germination and alleviates oxidative damage against copper stress. *Journal of Integrative Plant Biology* 50, 1518–1529.

Zhang, H., Tan, Z.Q., Hu, L.Y., Wang, S.H., Luo, J.P. and Jones, R.L. (2010) Hydrogen sulfide alleviates aluminum toxicity in germinating wheat seedlings. *Journal of Integrative Plant Biology* 52, 556–567.

13 Lysine Metabolism

L.O. Medici,[1] A.C. Nazareno,[2] S.A. Gaziola,[2] D. Schmidt[2] and R.A. Azevedo[2]*

[1]Universidade Federal Rural do Rio de Janeiro, Seropédica, Brazil; [2]Universidade de São Paulo, Piracicaba, Brazil

13.1 Abstract

Lysine is an essential amino acid for growth, development and reproduction in animals and its status as the first-limiting amino acid in cereals is unequivocal. Consequently, there is continuing interest in its metabolism in higher plants. The pathway from which lysine is synthesized is now well established, although advances continue to appear in publications. The major site of regulation appears to be at the enzyme dihydrodipicolinate synthase. There is also strong evidence for the important role of the enzymes involved in lysine catabolism (specifically, in the seed) for the accumulation of this amino acid, and the major advances in lysine metabolism made in more recent years have been made in lysine catabolism. This chapter will cover the study and analysis of mutants and transgenic plants that over-accumulate lysine. A lot of what has been learned about the regulation of lysine metabolism has been obtained from the study of mutants and applied to the production of transgenic plants. The study of such plants also allows an understanding of the emerging concepts on the biosynthesis and catabolism of this key amino acid.

13.2 Introduction

Cereal crops are the basis of food intake for the poorest populations in the world. Wheat, rice, maize, sorghum and barley are excellent sources of energy, and also provide proteins, but their nutritional quality is low, mainly due to lysine deficiency. For instance, in maize, the grain lysine content is around 2%, while 5% is the minimum required value, according to the World Health Organization (WHO, 2007). Deficiency in essential amino acids such as lysine is a very important issue because the WHO estimates that 30% of the people living in the developing world suffer from protein-derived malnutrition.

Lysine is produced by the aspartate metabolic pathway, in which three other essential amino acids, threonine, methionine and isoleucine, are also produced (Fig. 13.1a) (Azevedo *et al.*, 1997). This pathway is characterized by one main feature: feedback inhibition by the end products. However, the presence of bifunctional proteins directly involved in lysine metabolism has also been clearly characterized.

An overview of this pathway reveals two main branches, one leading to lysine formation, and the other with two sub-branches, one leading to threonine synthesis and the other to methionine synthesis. The threonine sub-branch goes further to form isoleucine (Fig. 13.1a). As

*E-mail address: raa@usp.br

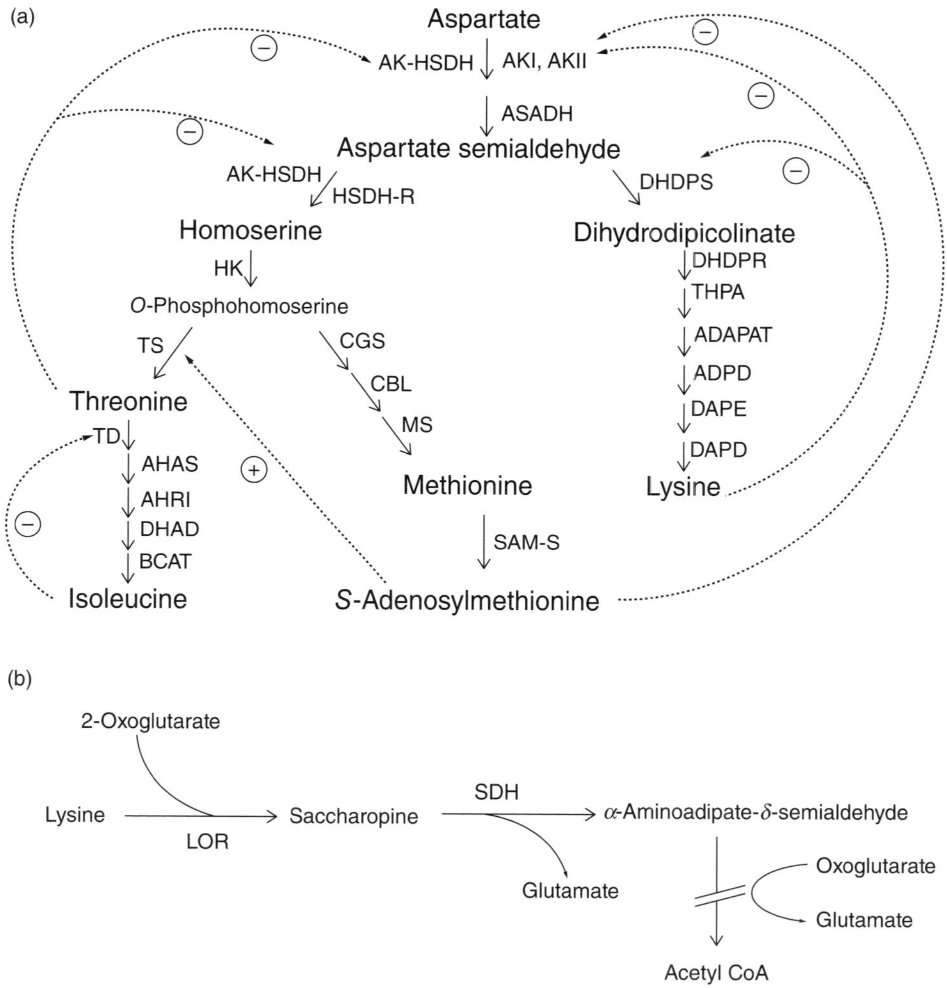

Fig. 13.1. (a) The aspartate metabolic pathway. Abbreviations: ADAPAT, *N*-acyl-L,L-diaminopimelate aminotransferase (EC 2.6.1.17); ADPD, *N*-α-acyl-L,L-diaminopimelate deacylase (EC 3.5.1.18); AHAS, acetohydroxyacid synthase (EC 4.1.3.18); AHRI, acetohydroxy acid isomeroreductase (EC 1.1.1.86); AK-HSDH, bifunctional aspartate kinase-homoserine dehydrogenase; AKI and AKII, two isoforms of monofunctional aspartate kinase (EC 2.7.2.4); ASADH, aspartate semialdehyde dehydrogenase (EC 1.2.1.11); BCAT, branched-chain amino acid aminotransferase (EC 2.6.1.42); CBL, cystathionine-β-lyase (EC 4.4.1.8); CGS, cystathionine-γ-synthase (EC 4.9.99.9); DAPD, diaminopimelate decarboxylase (EC 4.1.1.20); DAPE, diaminopimelate epimerase (EC 5.1.1.7); DHAD, dihydroxy acid dehydratase (EC 4.2.1.9); DHDPR, dihydrodipicolinate reductase (EC 1.3.1.26); DHDPS, dihydrodipicolinate synthase (EC 4.2.1.52); HK, homoserine kinase (EC 2.7.1.39); HSDH-R, monofunctional homoserine dehydrogenase (EC 1.1.1.3); MS, methionine synthase (EC 2.1.1.13); SAM-S, *S*-adenosylmethionine synthetase (EC 2.5.1.6); TD, threonine deaminase (EC 4.2.1.16); THPA, tetrahydrodipicolinate acylase (EC 2.3.1.117); TS, threonine synthase (EC 4.2.99.2); – negative feedback. + positive feedback. (b) Lysine degradation. Abbreviations: LOR, 2-oxoglutarate reductase (EC 1.5.1.8); SDH, saccharopine dehydrogenase (EC 1.5.1.9).

this chapter focuses on lysine metabolism, the second main branch of the pathway will not be covered, but some specific points will be commented upon indirectly. Detailed information about the threonine and methionine sub-branches and, indeed, the entire pathway can be obtained from previous review papers by Azevedo *et al.* (1997, 2006).

The regulation of lysine metabolism has received a great deal of attention and understanding the regulatory mechanisms that are involved has helped the genetic manipulation of plants and allowed the production of cereal crops with higher amounts of lysine in the seeds. The study of the pathway as a whole has also allowed an understanding of the regulatory mechanisms involved, not only in cereal crops, but also in legume crop species, in which methionine deficiency rather than lysine deficiency is a problem (Galili and Amir, 2013). There is another very interesting aspect of this pathway as well, which is the fact that it does not occur in mammalian species, so the enzymes that are involved in it are potential important targets for the development of herbicides and antimicrobial agents (Bareich *et al.*, 2004).

In this chapter, the main enzymes of lysine metabolism and their regulation are covered, with the aim of indicating their complexity and importance, as well as new challenges to be faced.

13.3 Aspartate Kinase and Homoserine Dehydrogenase

Aspartate kinase (AK; EC 2.7.2.4) is the starting point of the aspartate pathway and catalyses the conversion of aspartate to β-aspartyl phosphate. AK was first isolated and characterized in bacteria and was isolated from plant species only after a decade of studies on bacterial AK (Bryan *et al.*, 1970). There are at least two AK isoenzymes in higher plants, a predominant monofunctional form, and a bifunctional protein with AK and homoserine dehydrogenase (HSDH; EC 1.1.1.3) activity domains (AK-HSDH). Nevertheless, the number of isoenzymes varies among plant species. For instance, in *Arabidopsis thaliana* there are three monofunctional AKs and two bifunctional AK-HSDHs (Dumas *et al.*, 2012), whereas in maize two monofunctional AKs and one bifunctional polypeptide AK-HSDH have been identified (Azevedo *et al.*, 1992).

Although bifunctional proteins containing AK and HSDH had already been observed in bacteria (Patte *et al.*, 1966), the bifunctional protein was only initially proposed in plants in the early 1970s by Aarnes and Rognes (1974) as possibly acting in a similar manner to the bacterial enzyme. The confirmation of the occurrence of the bifunctional enzyme in plants was later confirmed by Wilson *et al.* (1991) and Azevedo *et al.* (1992). The monofunctional AK isoenzymes normally comprise at least one that is lysine sensitive and one that is synergistically inhibited by lysine and *S*-adenosylmethionine (SAM) (Rognes *et al.*, 1980; Curien *et al.*, 2007); the other AK isoenzyme, which is part of the bifunctional protein, is sensitive to threonine (Azevedo *et al.*, 1997).

The first enzyme leading to the biosynthesis of methionine, threonine and isoleucine is HSDH. Besides the bifunctional AK-HSDH described above, there is also a monofunctional HSDH, which is insensitive to threonine feedback inhibition. The natural presence of these two isoenzymes occurs possibly to allow flux through the aspartate metabolic pathway in such a way that carbons can be directed to the threonine/methionine branch or the lysine branch. Although the physiological role of HSDH was not clear for a period, its branch point may allow genetic manipulation of the pathway, for example to increase the synthesis of threonine and methionine (Schroeder *et al.*, 2010).

Investigation of the protein structure of AK-HSDH from *A. thaliana* revealed AK and HSDH domains in the N- and C-terminal regions, respectively, and a regulatory domain in the middle (Paris *et al.*, 2002b, 2003). The regulatory domain was shown to have two binding regions for threonine, one of them that can inhibit AK activity and also allows an easier binding of a second threonine which, in turn, inhibits HSDH activity (Paris *et al.*, 2003).

The expression of the two AK-HSDH genes of *A. thaliana*, *akthr1* and *akthr2*, was compared by Rognes *et al.* (2003), who reported that both genes are expressed in meristematic cells, leaves and stamens, while *akthr1* is less expressed in roots and *akthr2* is less expressed in the stem and gynoecium, and during seed formation. Analysis of the expression of one of the three monofunctional AK genes of *A. thaliana*, *aklys3*, was also conducted; this was expressed in a tissue specific manner as well, because it was highly expressed in the xylem of leaves and hypocotyls and in the stele of roots. Significant expression of this gene was also observed in trichomes after bolting, but only slight expression was detected in the vascular bundles and mesophyll cells of cauline leaves, inflorescence stems, sepals, petals and stigmas (Yoshioka *et al.*, 2001).

All of the information obtained over the years on plant AKs has been used in the challenge to produce nutritionally better balanced seeds. As already noted, methionine deficiency is a major problem in legume seeds whereas lysine is the main limiting amino acid in cereal seeds. Two principle approaches were initially employed: the production of transgenic plants, mainly using plant model species such as *A. thaliana* and tobacco; and the use of mutants, mainly for crop plant species (Azevedo, 2002). For instance, the feedback-insensitive AK from *Escherichia coli* was used to produce a transgenic lucerne (*Medicago sativa*). The overexpression of this AK isoenzyme in this forage legume species resulted in enhanced content of both soluble and protein-bound threonine (Galili *et al.*, 2000). This result indicated that AK manipulation can increase the metabolic flux to the threonine branch of aspartate pathway, but this alteration alone is not enough to also and simultaneously increase methionine content. However, an increase in methionine leaf content was possible in tobacco plants expressing both feedback-insensitive AK and the first committed enzyme of the methionine biosynthesis branch pathway, cystathionine-γ-synthase (CGS; EC 4.9.99.9) (Hacham *et al.*, 2008). In this work, the CGS enzyme used was a mutated form that was insensitive to feedback inhibition by SAM.

13.4 Aspartate Semialdehyde Dehydrogenase

The product of AK activity, β-aspartyl phosphate, is converted to aspartate semialdehyde (ASA) by the action of the enzyme aspartate semialdehyde dehydrogenase (ASADH; EC 1.2.1.11). ASA is a branch point substrate, which can be used either for the branch leading to lysine synthesis, or for the other branch leading to the synthesis of threonine, methionine and isoleucine (Fig. 13.1a). ASADH has been studied mainly because of its role in bacterial virulence, so the research has mainly looked for inhibitors that could serve as antibacterial – and also as fungicidal or herbicidal – agents (Hadfield *et al.*, 2001; Luniwal *et al.*, 2012), rather than studied the role or potential regulatory properties of this enzyme for the synthesis of the aspartate-derived amino acids. The activity of ASADH is one of the steps for the production of metabolites involved in the process of quorum sensing that triggers bacterial virulence and also in the formation of the bacterial cell wall (Kumar and Garg, 2013). ASADH has been mainly studied in infectious organisms (Kumar and Garg, 2013) and perhaps the only plant species from which this enzyme has been well characterized is *A. thaliana* (Paris *et al.*, 2002a). The *Arabidopsis* ASADH has a plastid transit sequence and is non-allosteric (Curien *et al.*, 2009).

One interesting aspect of ASADH is its putative role as a bridge in substrate channelling between AK and HSDH activities. The bifunctionality of the protein could be related to metabolic channelling, when both enzymes catalyse consecutive reactions. In that situation, the product of one of them is the substrate for the other; this direct transference is a process that can bring some advantages because the metabolites and enzymes are not mixed with the bulk phase (Zhang, 2011). None the less, the threonine-sensitive AK-HSDH is an unusual bifunctional protein in that these enzyme activities are not involved in consecutive reactions. The explanation for this is still under debate; one hypothesis is about the role of ASADH as a bridge in taking the product from AK, β-aspartyl phosphate, and converting it to the substrate for HSDH, ASA. This hypothesis was confirmed in *E. coli* just over 10 years ago (James and Viola, 2002) but has been rejected in *A. thaliana* (Paris *et al.*, 2002a; Curien *et al.*, 2009) because in *Arabidopsis* the addition of ASADH to the *in vitro* assay did not result in any enhancement of either AK or HSDH activity.

13.5 Homoserine Kinase

The metabolite ASA that is produced by ASADH is converted by HSDH to homoserine, which is then phosphorylated to *O*-phosphohomoserine (OPH) by the action of the enzyme homoserine kinase (HK; EC 2.7.1.39), a member of GHMP kinase superfamily (galactokinase, homoserine kinase, mevalonate kinase and phosphomevalonate kinase) (Zhou *et al.*, 2000). Like ASA, OPH is also at a branch point in the plant aspartate pathway, because this metabolite can be used to form methionine on one branch, or to form threonine or isoleucine on the other branch (Fig. 13.1a). In a similar manner to ASADH, HK was characterized first in bacteria, revealing a high specificity for homoserine (Krishna *et al.*, 2001). In plants, like ASADH, HK has a plastid transit peptide sequence and is non-allosteric.

Attempts to manipulate the regulation of the aspartate pathway at the HK catalytic step have been carried out. For instance, the overexpression of HK in *A. thaliana* and potato did not result in increased contents of the end products, methionine or threonine, but when homoserine was supplied externally an enhanced effect on threonine and methionine accumulation was observed in both plant species, indicating that homoserine, rather than OPH, is the limiting factor (Lee *et al.*, 2005; Rinder *et al.*, 2008). However, in potato, the overexpression of an artificial cytosolic form of HK led to a downregulation of the endogenous HK gene, as well of genes encoding for other enzymes of the pathway, including threonine synthase (TS; EC 4.2.99.2), cystathionine-β-lyase (CbL; EC 4.4.1.8), and methionine synthase (MS; EC 2.1.1.13) (Rinder *et al.*, 2008).

13.6 Dihydrodipicolinate Synthase

The enzyme dihydrodipicolinate synthase (DHDPS; EC 4.2.1.52) is the first enzyme directly involved in lysine biosynthesis, catalysing the conversion of ASA and pyruvate to form dihydrodipicolinate (Azevedo *et al.*, 2006). DHDPS has been very well studied and characterized in a large number of bacterial and plant species. The first characterization efforts were based on biochemical approaches such as isolation, purification and kinetic analysis. In plants, the enzyme was isolated and purified from spinach (Wallsgrove and Mazelis, 1981), wheat (Kumpaisal *et al.*, 1987), tobacco (Ghislain *et al.*, 1990), maize (Frisch *et al.*, 1991a) and pea (Dereppe *et al.*, 1992), just to cite a few. The enzyme is formed by four identical subunits, although the molecular mass has been shown to vary from 115 to 167 kDa among plant species (Azevedo *et al.*, 1997).

Kinetic studies of DHDPS revealed that the enzyme follows the Michaelis–Menten kinetics for both of its substrates, pyruvate and ASA. Blickling *et al.* (1997b) proposed a 'ping-pong' reaction mechanism comprising three steps: first, a Schiff base formation of the active-site lysine residue (K183 in *Nicotiana sylvestris*) with pyruvate; secondly, the addition of ASA to the enamine of the Schiff base; and lastly, dehydration, cyclization and release of the product, dihydrodipicolinate. Although the active-site lysine residue is very important for DHDPS activity, it can be dispensable, at least in *E. coli* mutants lacking K161A and K161R at the active site (da Costa *et al.*, 2010).

The regulation of DHDPS activity occurs through feedback inhibition by the end product of the enzyme. In this sense, the sensitivity of DHDPS to lysine can be classified into three groups: first, the strongly inhibited form of DHDPS from plants (K_i 5–50 μM) (Galili, 2002); secondly, the moderately inhibited form of DHDPS from Gram negative bacteria, e.g. *E. coli* (K_i 0.4 mM) (Karsten, 1997); and finally, those forms from Gram positive bacteria (e.g. *Bacillus subtilis* and *Corynebacterium glutamicum*) that are completely insensitive to lysine inhibition (Cremer *et al.*, 1991).

The quaternary structure of DHDPS has also been determined and, for the large majority of species studied, is of a tetrameric structure. In wild type tobacco (*N. sylvestris*), DHDPS forms a homotetramer or dimer of dimers (Blickling *et al.*, 1997a). Unlike the architecture observed for bacterial tetramers though, the plant enzyme assumes a back-to-back dimer of dimers, which is a 100 times more sensitive to lysine inhibition than the *E. coli* form (Blickling *et al.*, 1997a). This specific information was essential for studies involving the regulatory manipulation of DHDPS in plant species in such a way that any excess of lysine produced would inhibit its own synthesis by inhibiting DHDPS activity; in contrast, if a bacterial enzyme was introduced into the plant, a much higher lysine concentration would be required for the same level of DHDPS activity inhibition, which would potentially lead to lysine overproduction (Shaul and Galili, 1992a). Recently, Atkinson *et al.* (2012) reported a consistent structure of the back-to-back homotetramer of DHDPS from grapevine (*Vitis vinifera*) that was established using crystal, solution and *in silico* studies. Some alternative structures of DHDPS, e.g. a trimeric quaternary structure from pea (Dereppe *et al.*, 1992) and a dimeric structure from *Staphylococcus aureus* (Burgess *et al.*, 2008) and *Pseudomonas aeruginosa* (Kaur *et al.*, 2011), have also been reported.

Molecular approaches have also been employed to characterize and better understand the transcriptional regulation of DHDPS. Up to now, the DHDPS gene has been isolated and cloned from several plant species, including soybean (Silk *et al.*, 1994), poplar (Vauterin and Jacobs, 1994), coix (Dante *et al.*, 1999), wild type rice (*Zizania latifolia*) (Kong *et al.*, 2009), rice (Sikdar and Kim, 2010), grapevine (Atkinson *et al.*, 2011) and *Medicago truncalata* (Erzeel *et al.*, 2013). All of the DHDPS genes that have been cloned comprise a sequence encoding for a transit peptide to plastids. Interestingly, the *Z. latifolia* DHDPS gene was lacking this sequence, and the enzyme is localized in the cytoplasm. However, its transcriptional regulation was similar to that observed for other plant species in being of cell type specific, with high levels of expression in growing tissues (Kong *et al.*, 2009). Kaneko *et al.* (1990) isolated for the first time two different cDNA clones of DHDPS from suspension cultures of wheat by screening a wheat cDNA library with oligonucleotide probes based on the DHDPS amino acid sequence. Later, Frisch *et al.* (1991b) isolated the complete cDNA sequence from maize and observed 86–88% of similarity to the wheat cDNA. Vauterin *et al.* (1999) also cloned and studied the transcriptional regulation of the DHDPS gene of *A. thaliana* (*dhdps1*) and reported that the expression was cell type-specific and mainly in young tissues that exhibit a high rate of growth and protein synthesis, and could also direct cell type-specific expression in a heterologous plant, *N. tabacum*. Craciun *et al.* (2000) isolated and cloned a second gene encoding DHDPS (*dhdps2*) from *A. thaliana*, which exhibited 84% similarity with the previous DHDPS gene. Although both genes exhibited similar transcriptional properties, the DHDPS enzyme encoded by the *dhdps2* gene was shown to be less sensitive to lysine inhibition (Craciun *et al.*, 2000). Moreover, these authors also showed that a T-DNA insertion mutant of *dhdps2* displayed lower lysine synthesis and, as a consequence, a strongly increased synthesis of threonine. Although little is known about the presence of multiple *dhdps* alleles, Stuttmann *et al.* (2011) reported an increased resistance of a *dhdps2* knockout mutant of *A. thaliana* to the biotrophic oomycete *Hyaloperonospora arabidopsidi*. Recently, Jones-Held *et al.* (2012) reported that two mutants lacking *dhdps1* and *dhdps2* are viable, but that while the *dhdps2* mutant accumulated threonine, *dhdps1* plants do not, suggesting that the phenotype was a result of reduced DHDPS activity in *dhdps2*. These authors also showed that DHDPS 1 contributes only 30% of the total DHDPS activity in the leaves of young plants, so that DHDPS 2 contributes 70%.

Owing to the protein structure of DHDPS and its biochemical features, this enzyme plays an important role in regulating metabolite flux into lysine and is accepted as one of the most important targets when manipulating plants for lysine overproduction (Azevedo and Arruda, 2010). In order to produce a DHDPS that is less sensitive to feedback inhibition by lysine, the first approach attempted by researchers was mutagenesis, followed by direct selection in a medium containing the lysine analogue *S*-aminoethylcysteine (AEC). Most of the attempts were frustrating, as the mutants produced exhibited only a slight or no alteration in lysine content. Furthermore, these studies very often did not include a more detailed enzymological study of DHDPS, although in many cases the sensitivity of DHDPS to lysine was not altered. Despite the rather disappointing results, some useful information was obtained from a number of studies. Ghislain *et al.* (1995) generated mutant plants of *N. sylvestris* overproducing soluble lysine in the leaves and seeds due to a single amino acid substitution (asparagine to isoleucine) of the DHDPS mature protein. In the following year, Shaver *et al.* (1996), by mutating the enzyme expressed in *E. coli*, reported the production of a maize mutant with only a single amino acid change in the DHDPS mature protein, but this was enough to eliminate the feedback inhibition by lysine.

Although in plants DHDPS is highly sensitive to lysine inhibition, the *dapA* gene from *E. coli* encodes a less sensitive form of DHDPS that has been used to transform plants. Shaul and Galili (1992a) reported that the constitutive expression of the bacterial gene containing a chloroplast transit peptide produced a significant increase of soluble lysine content in tobacco leaves, though no increase was observed in protein-bound lysine. A second strategy combining the expression of both the *dapA* and *lysC* genes of *E. coli* encoding insensitive DHDPS and AK, respectively, was also attempted. The result demonstrated a large increase of soluble lysine when compared with plants expressing only the *E. coli dapA* gene (Shaul and Galili, 1993).

Kwon *et al.* (1995) also reported studies of the expression of *E. coli* DHDPS in tobacco leaves. The results showed a correlation between soluble lysine content and DHDPS activity, although the absolute value for either varied between plants that were heterozygous and homozygous for the insensitive *DHDPS* allele, being more effective the homozygous plants than in wild tobacco plants. In another study, Karchi *et al.* (1994) expressed the *E. coli dapA* gene under the control of a seed-specific promoter in order to increase the lysine content in mature seeds of tobacco. Although the authors reported that during the early stages of seed development the soluble lysine content was twofold higher than that of the control plant, the experiment was not successful as the once increased lysine content was not observed again later when the mature seeds were analysed.

Subsequent efforts were directed towards increasing the lysine content of plants of agronomic importance. Falco *et al.* (1995) reported the expression of the *C. glutamicum dapA* and *E. coli lysC* genes in soybean and canola under the control of a seed-specific promoter. Besides an increase in soluble lysine, the authors reported a twofold increase in total lysine in the seeds of canola expressing the *C. glutamicum dapA* gene. In soybean, when both genes were expressed simultaneously, there was a fivefold increase in total lysine content. In cereal crops, the expression of deregulated DHDPS was also reported in an attempt to overcome the main shortcoming of cereals, i.e. their lack of sufficient lysine. Huang *et al.* (2005) generated transgenic maize expressing the *C. glutamicum dapA* gene in the embryo; separately, these authors also produced transgenic maize plants with a reduced content of lysine-poor storage proteins (zeins). They then combined these two strategies through genetic crosses and achieved more than a twofold increase in total lysine content in the F_1 progeny compared with each of the single approaches used. Frizzi *et al.* (2008) also used a double strategy, but in a different way. They expressed a single transgene cassette containing the *C. glutamicum dapA* gene and an inverted repeat of the gene encoding the bifunctional enzyme lysine 2-oxoglutarate reductase-saccharopine dehydrogenase (LOR-SDH) in maize seeds. This strategy allowed a 40-fold increase of soluble lysine in transgenic plants compared with the controls. This result is an example of how the study of lysine degradation (discussed in the next section) had been neglected over a period of time in contrast to the rather major importance that had been given to the study of lysine synthesis. Even though the double approaches cited above have contributed important results, the single approach using bacterial feedback-insensitive DHDPS in an embryo-specific manner in maize was also quite successful (Lucas *et al.*, 2007). The genetically modified maize line (LY038) that was used represents a unique case of a commercial transgenic high-lysine cereal; unfortunately, it was subsequently withdrawn due to concerns about its safety for human consumption (Galili and Amir, 2013). It is interesting that the maize line LY038 also had a high seed oil content, suggesting a possible physiological link between oil and lysine in grains, as had been reported and suggested by Medici *et al.* (2009) in a diallelic analysis of lysine and oil contents in maize grains.

13.7 Lysine Catabolism

In addition to the understanding that has been obtained of the synthesis of lysine, great emphasis has been given to the study of its catabolism as well, but only after a significant amount of information had already been obtained on lysine synthesis. Two main enzymes were shown to be part of the lysine degradation pathway in animals, microorganisms and plants. Lysine 2-oxoglutarate reductase (LOR; EC 1.5.1.8) (also known as lysine α-ketoglutarate reductase, LKR) is the first enzyme of the pathway, which condenses lysine and 2-oxoglutarate to form saccharopine; the second enzyme is saccharopine dehydrogenase (SDH; EC 1.5.1.9), which hydrolyses saccharopine to glutamate and α-aminoadipate-δ-semialdehyde (Fig. 13.1b). In a very similar manner to that described earlier in this chapter for the earlier studies on lysine synthesis, investigations on lysine degradation were also initiated much earlier in other species than in plants. For instance, detailed information was already available for microorganisms (Jones and Broquist, 1965, 1966; Saunders and Broquist, 1966; Ramos *et al.*, 1988) and mammals (Hutzler and Dancis, 1968, 1970; Fjellstedt and Robinson, 1975a; Noda and Ichihara, 1978; Markovitz *et al.*, 1984), in which the degradation of lysine has been extensively studied.

The lack of lysine degradation in humans results in a severe genetic disease named familial hyperlysinaemia (Markovitz *et al.*, 1984), which is classified into two types: hyperlysinaemia type I, which is associated with a combined deficiency of LOR and SDH activities; and hyperlysinaemia type II, which is characterized by a deficiency of SDH activity. The purification of the rat liver enzymes showed that LOR and SDH are distinct monofunctional polypeptides (Noda and Ichihara, 1978). However, evidence that the activities of LOR and SDH were part of a single bifunctional polypeptide in humans and animals had been previously provided by studies with the human placenta (Fjellstedt and Robinson, 1975b) and with baboon and bovine livers (Markovitz *et al.*, 1984). The native form of LOR-SDH in mammals was determined to have a molecular mass of 468 kDa, suggesting a tetrameric structure with identical subunits of 115 kDa, which was observed by denaturing PAGE analysis (Markovitz *et al.*, 1984). In fungi and yeast, LOR and SDH are present as monomers of 49 kDa and 73 kDa that are encoded by two genes, *lys9* and *lys1*, respectively (Borell *et al.*, 1984; Ramos *et al.*, 1988).

Initial information about lysine catabolism in higher plants was derived from the studies of amino acid metabolism using compounds labelled with ^{14}C. In developing seeds of maize and barley, ^{14}C-labelled lysine was incorporated into glutamic acid, proline and saccharopine (Sodek and Wilson, 1970; Brandt, 1975; Arruda *et al.*, 1982), revealing that lysine catabolism in plants also occurred via the saccharopine pathway.

In plants, LOR received much more attention than SDH. It was first characterized in maize (Arruda *et al.*, 1982; Arruda and da Silva, 1983; Brochetto-Braga *et al.*, 1992) and, much later, in other plant species such as rice (Gaziola *et al.*, 1997), soybean (Miron *et al.*, 2000), coix (Lugli *et al.*, 2002), beans (Cunha Lima *et al.*, 2003) and sorghum (Fornazier *et al.*, 2005). These studies showed that the expression of LOR was tissue specific, with high rates of activity during endosperm development. Another major finding, also initially reported in maize and soon after in rice, was that for the majority of plant species LOR and SDH are part of a single bifunctional polypeptide (LOR-SDH) (Goncalves-Butruille *et al.*, 1996; Epelbaum *et al.*, 1997; Gaziola *et al.*, 1997; Miron *et al.*, 1997; Tang *et al.*, 1997). The evidence supporting this hypothesis was obtained from the co-purification of both enzymes using distinct chromatographic techniques and non- denaturating PAGE. Nevertheless, monofunctional forms of both enzymes were also found (Zhu *et al.*, 2000; Tang *et al.*, 2002). In *A. thaliana*, Tang *et al.* (1997) provided evidence of the presence of two isoenzymes of SDH, which were demonstrated by two peaks of activity on anion exchange chromatography fractionation, one comprising LOR-SDH activity and the other SDH activity. Furthermore, hybridization analysis of genomic DNA and mRNA suggested that these two isoenzymes were translated from two different mRNAs and produced by a single gene through an unknown mechanism (Tang *et al.*, 1997). In maize, immature seeds exhibited LOR and SDH activities located on two functionally independent domains, and the native form existed as a dimer of 260 kDa (Goncalves-Butruille *et al.*, 1996). In SDS-PAGE, the LOR and SDH proteins co-migrated in a single band of 125 kDa (Goncalves-Butruille *et al.*, 1996). LOR and SDH were isolated and partially purified from immature rice seeds and the molecular masses estimated as 203 kDa for both enzymes by gel-filtration chromatography (Sephacryl S200) and 202 kDa for both enzymes by non-denaturing PAGE using Ferguson plots (Gaziola *et al.*, 1997).

Where molecular studies are concerned, Epelbaum *et al.* (1997) reported the first cloning of the gene encoding LOR and SDH from *A. thaliana*. The gene occupies 6.2 kb of the genome and contains a regulatory element (opaque-2), which appears to be involved in LOR regulation in maize endosperm (Brochetto-Braga *et al.*, 1992). In immature maize endosperm, sequence analysis of the cDNA that encodes the bifunctional LOR-SDH enzyme revealed a protein of 117 kDa and an open reading frame (ORF) of 1056 amino acids (Kemper *et al.*, 1999).

Kinetic analyses have been carried out for the enzymes in several plant species. For instance, in *A. thaliana* the K_m of the monofunctional enzyme for lysine was approximately tenfold higher than that of the bifunctional LOR-SDH (Azevedo *et al.*, 2006). In rice, the K_m of the LOR enzyme was determined as 4.48 mM for lysine, 0.79 mM for 2-oxoglutarate and 0.032 mM for NADPH, whereas the K_m of the SDH enzyme was 0.13 mM for saccharopine and 0.023 mM for NAD (Gaziola *et al.*, 1997).

The regulation of lysine catabolism occurs not only in the mature proteins but also by their expression patterns. Sequence analysis of the *sdhlor* gene of maize and *A. thaliana* showed the presence of the sequence CCAAT and a TATA box within the promoter region and the inner region of the gene, which possibly controls the expression of the LOR-SDH bifunctional form and the SDH monofunctional enzyme. In addition, the type GCN4 sequence was found upstream of the promoter region and the internal region of the gene. However, binding sites for the transcription factor opaque-2 are present in the region upstream of the promoter and in the internal region of the gene of *A. thaliana*, but only in the region upstream of the promoter in maize (Arruda *et al.*, 2000).

In contrast to the biosynthesis of lysine, which seems to operate in plastids, the enzymes involved in the first two steps of lysine catabolism in plants appear to be located in the cytoplasm (Kemper *et al.*, 1999). The cytosolic localization of plant LOR and SDH enzymes may be different from that of the mammalian enzymes which have been suggested to be localized in the mitochondria (Blemings *et al.*, 1994). The maximum activity of LOR occurs at a neutral pH, while the maximum activity of SDH occurs at pH 9.0 or higher. As the bifunctional polypeptide is located in the cytoplasm (Zhu *et al.*, 2000), where the pH is neutral, the SDH domain activity can be limiting, so an additional monofunctional SDH enzyme may be required to maintain effective levels of enzyme for lysine degradation (Stepansky *et al.*, 2006). This hypothesis is supported by the work of Stepansky *et al.* (2005), which reported that in *A. thaliana* the mRNA levels of the SDH monofunctional enzyme and also of the mature protein were always found in excess of those of the bifunctional polypeptide. At the post-transcriptional level, the regulation of LOR-SDH occurs by interaction with the functional domain of SDH, which is mediated by the intermediate region connecting the two domains (Zhu *et al.*, 2000). Post-translational regulation studies revealed that the enzymes are differently modulated by a cascade of intracellular signalling pathways involving mainly Ca^{2+}, phosphorylation–dephosphorylation and ionic strength (Karchi *et al.*, 1995; Kemper *et al.*, 1998; Gaziola *et al.*, 2000).

Besides the biochemical and molecular studies on LOR and SDH, several efforts have been carried out to reach an understanding of the role of lysine catabolism in plants, and how to manipulate this pathway in order to increase the lysine content of plants. The role of lysine degradation and its importance in lysine accumulation was initially proposed by Arruda and da Silva (1979). Their initial work was confirmed when analysis of the opaque-2 maize mutant, which contains higher concentrations of lysine, was shown to have very low activity of LOR in the endosperm (Brochetto-Braga *et al.*, 1992). Additional strong evidence was obtained from an analysis of LOR and SDH activity in rice (Gaziola *et al.*, 1997), a cereal crop species with a higher lysine concentration than that of other cereal crops, and in canola (Falco *et al.*, 1995), which has a naturally higher lysine content; these were both shown to contain lower LOR and SDH activity in the seeds. The importance of lysine degradation was further indicated when Karchi *et al.* (1994) expressed DHDPS and AK insensitive to lysine feedback inhibition in transgenic *N. tabacum*. Although the transgenic plant exhibited higher soluble lysine content during seed development, no lysine increase was observed in mature seeds, suggesting the degradation of the excess lysine that had been produced. The authors explained that due to the higher LOR activity in the transgenic plants, no lysine accumulation could be achieved in mature seeds, suggesting that the enzymes involved in lysine synthesis and catabolism are coordinately regulated during seed development.

In addition to the previous results, the analysis of transgenic soybean and canola expressing the genes *dap*A and *lys*C encoding, respectively, the enzymes AK and DHDPS showed large increases in the accumulation of soluble lysine in the seeds (Falco *et al.*, 1995), unlike the results obtained in *N. tabacum*, in which the high lysine was shown only during seed development (Karchi *et al.*, 1994). Although LOR or SDH activities had not been measured, Falco *et al.* (1995) observed the accumulation of intermediates of the lysine catabolism, such as α-aminoadipate-δ-semialdehyde and saccharopine in canola and soybean, respectively. Therefore, studies using high-lysine mutants and transgenic plants expressing altered regulation of lysine biosynthetic enzymes (Azevedo *et al.*, 1990; Shaul and Galili, 1992a,b; Heremans and Jacobs, 1997; Lee *et al.*, 2001) corroborated that the rate of lysine breakdown regulates the accumulation of lysine in the seeds (Azevedo, 2002).

One special case was the discovery of the maize mutant opaque-2 in 1963 (Mertz *et al.*, 1964), which contains higher contents of the essential amino acids lysine and tryptophan. Extensive biochemical and molecular studies were carried out with this mutant in order to understand the high lysine trait and also to solve some agronomic disadvantages of this mutant, such as low yield and higher susceptibility to pathogens (Azevedo and Arruda, 2010). In maize, the gene *opaque-2* encodes a transcription factor that regulates the gene expression of LOR-SDH and zeins, the most abundant (and lysine-poor) storage proteins of maize. An increased rate of zein synthesis promoted by the opaque-2 factor is accompanied by an increase in the expression of LOR-SDH, resulting in higher levels of lysine degradation and consequently low levels of lysine in the seeds. The high-lysine maize mutant contains the *opaque*-2 homozygous recessive allele. LOR activity is greatly reduced in the opaque-2 mutant, which also has decreased levels of zeins (Arruda *et al.*, 1982; Brochetto-Braga *et al.*, 1992). Biochemical and molecular studies of the opaque-2 mutant showed that there was a decrease in mRNA levels (>90%) of the mature protein and also in LOR-SDH activity, which confirmed the role of the transcription factor opaque-2 in lysine catabolism (Brochetto-Braga *et al.*, 1992; Gaziola *et al.*, 1999; Kemper *et al.*, 1999). Moreover, in studies comparing normal maize varieties, the opaque-2 mutant and QPM (Quality Protein Maize) lines produced by the introduction of endosperm modifier genes (to restore the vitreous features of the endosperm) maintained the characteristics of opaque-2 mutant, suggesting that the introduction of these genes does not affect the *opaque-2* gene or LOR and SDH activities. The reduction of the activity of LOR-SDH in QPM lines has also led to an increase in the availability of soluble lysine in these varieties (Gaziola *et al.*, 1999).

Overall, it is clear that it is possible to increase lysine synthesis by the manipulation of some key enzymes involved in lysine synthesis, such as AK and DHDPS. However, for the accumulation of the excess lysine produced, especially in the seeds, the catabolism of lysine has to be decreased, and this can be achieved by manipulating gene expression and/or enzyme activity of LOR and SDH. Other strategies that involve redistribution of the amino acid between the soluble and protein form will still depend upon how much lysine is produced and made available for soluble accumulation and protein synthesis.

13.8 What Next?

Back in 1997 and again in 2006, Azevedo *et al.* (1997, 2006), when reviewing the aspartate pathway, indicated how important it would be to be able to use the information that is available on the regulation of the pathway to produce economically viable plants containing elevated contents of all the aspartate-derived amino acids – lysine, threonine, isoleucine and methionine – in seed production for farmers. Based on what has been reported in this chapter, very few, if any, commercial lines are currently available and being used in seed production and, consequently, for animal and human food production. It is also somewhat disappointing to see the reduced number of groups that are still strongly involved in the study of this pathway.

The use of mutants and transgenic plants associated with plant breeding techniques appears to be the way to go if high-lysine cereal crops are to be definitively produced with economic viability. In addition to the transgenic approaches already discussed in this chapter, several other transgenic plants have been produced over the years that have explored the key regulatory points of the aspartate pathway. Overall, there have been over 20 years of studies consistently trying distinct combinations of pathway manipulation (Fig. 13.2).

Other approaches have been employed that have also used transgenic plants. For instance, the production of plants containing adequate concentrations of lysine and methionine in the seeds was also possible from transgenic plants overexpressing natural or recombinant methionine- and/or lysine-rich storage proteins, and/or transgenic plants expressing less methionine- and/or lysine-poor proteins (Galili and Amir, 2013). One typical example is the work carried out by Lange *et al.* (2007), who generated transgenic barley plants expressing an antisense construct of the C-hordein genes in order to suppress the expression of the major group of lysine-poor proteins in the barley grain; this resulted in increases of 16, 13 and 11% in lysine, threonine and methionine content, respectively. A combined approach that involved manipulation

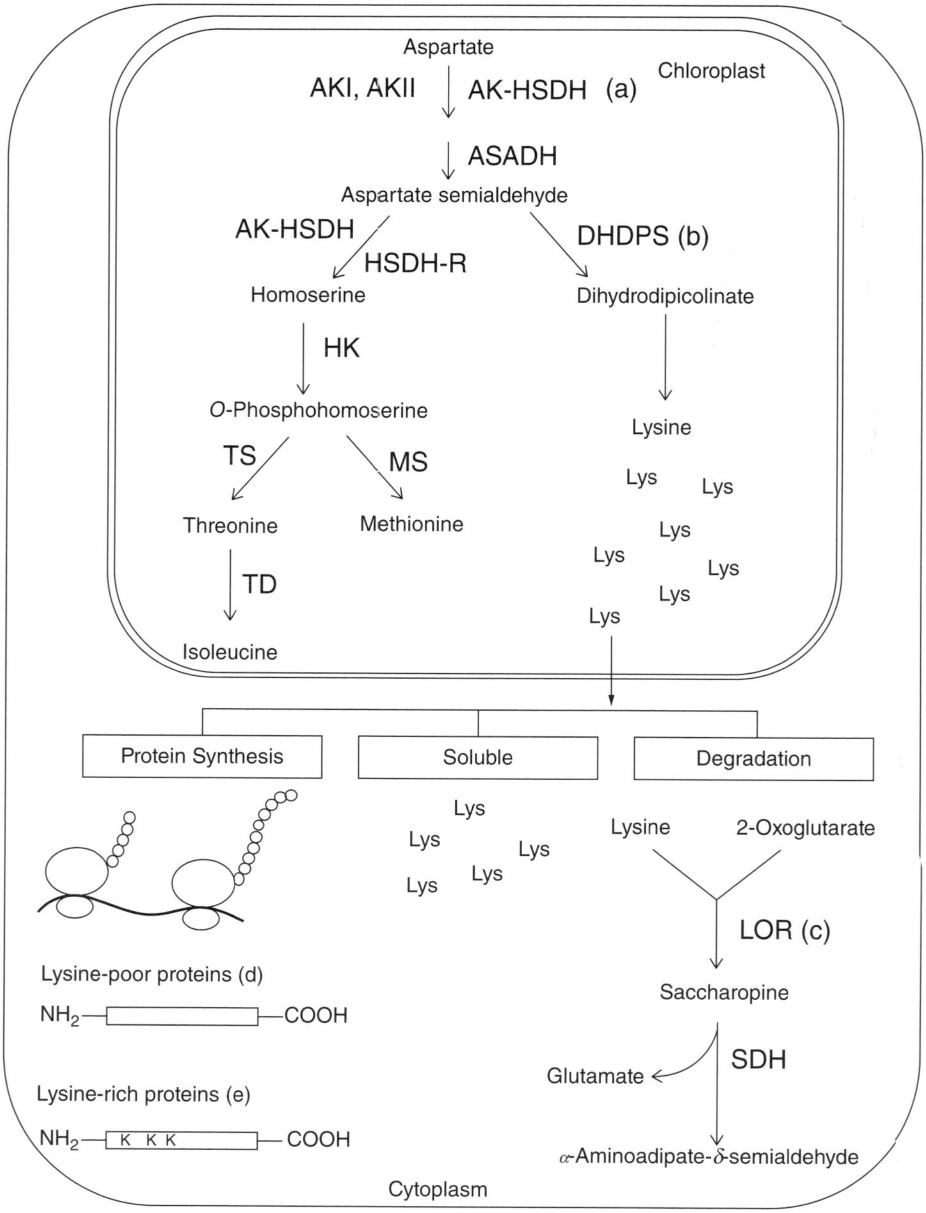

Fig. 13.2. Lysine metabolism. An overview of the main approaches made to increasing the lysine content of seeds: (a) the expression of aspartate kinase (AK); (b) the expression of dihydrodipicolinate synthase (DHDPS) insensitive to feedback inhibition; (c) the repression of genes encoding lysine 2-oxoglutarate reductase-saccharopine dehydrogenase (LOR-SDH) or reduced LOR-SDH activities; (d) the suppression of lysine-poor storage protein; and (e) the expression of natural or new lysine-rich storage proteins or lysine-rich proteins.

of the storage protein pool and altered AK regulation has also been attempted, as in the work reported by Demidov *et al.* (2003), who used a feedback-insensitive AK and the Brazil nut 2S albumin in a transgenic bean (*Vicia narbonensis*), unfortunately with limited success. Indeed, with the advent of genetic engineering techniques it is now possible to use biotechnology to develop further studies to unravel the complex mechanisms that control the flow of amino acids in grains.

Another strategy that has been attempted is based on the expression of a synthetic protein(s) with higher amounts of lysine, which is then introduced and expressed as a new seed protein. An example of such an approach is the production of maize transgenic plants expressing a chimeric protein, the 'zeolin', under the control of an endosperm-specific promoter (Ambrozevicius *et al.*, 2009; Ambrozevicius, 2010). The research carried out with maize was based on the original work on zeolin proposed and developed by the group of A. Vitale (Mainieri *et al.*, 2004). The zeolin is a combination of 421 amino acids of bean phaseolin with 89 amino acids of the maize γ-zein inserted into the maize genome under the control of an endosperm-specific promoter isolated from the protein γ-kafirin from sorghum. The transformed plants expressing zeolin, which acts as a sink for lysine, can be used to study in future works the metabolic pathways involved in the metabolism of lysine and the accumulation of storage proteins in maize seeds. The maize transgenic plants produced revealed variations among the distinct transformation events, but some exhibited higher lysine concentration and changes in the storage protein distribution.

13.9 Conclusions

Lysine is an important and key amino acid, particularly for monogastric animals, which do not have the aspartic metabolic pathway operating, so cannot synthesize lysine. Study of the route leading to lysine synthesis and degradation in plants started 44 years ago after a great deal of information was already available for microorganisms. The advances have been tremendous and it is nice to see how the research has developed with the development of a number of new technologies. Initially, biochemical studies were carried out to isolate and study the enzymes. A second stage involved the production and study of mutants with altered regulation for some key enzymes; this predominated in the 1970s and 1980s. The 1990s was dominated by the use of plant transformation and transgenic plants were produced and studied. Such techniques allowed a combination of strategies from the overexpression of enzymes to the manipulation of storage proteins, reaching a stage in which major seed companies have now been able to produce better balanced seeds. All of these studies, strategies and efforts were also possible by the study of a wide range of plant species, not only those that were deficient in lysine. Another major and key finding was the understanding of the regulation of lysine catabolism, which was neglected for a period. Finally, all of the new plant lines produced, transgenic or not, mutant or not, have to be evaluated agronomically, and any future research must be assisted by plant breeders. Today, the challenge of producing better nutritional crops is not over, but the advances witnessed in the past 40 years have been astonishing.

References

Aarnes, H. and Rognes, S.E. (1974) Threonine-sensitive aspartate kinase and homoserine dehydrogenase from *Pisum sativum*. *Phytochemistry* 13, 2717–2724.

Ambrozevicius, L.P. (2010) Maize genetic transformation with a new protein: the zeolin. PhD thesis, University of São Paulo, São Paulo, Brazil. [In Portuguese.]

Ambrozevicius, L.P., Martins, P.F., Carvalho, G., Schmidt, D., Rolão, M.B., Peters, L.P., Gaziola, S.A. and Azevedo, R.A. (2009) High-lysine maize lines expressing a new protein in the endosperm. *Amino Acids* 37(Suppl 1), S45.

Arruda, P. and da Silva, W.J. (1979) Amino acid composition of vascular sap of maize ear peduncle. *Phytochemistry* 18, 409–410.

Arruda, P. and da Silva, W.J. (1983) Lysine-ketoglutarate reductase activity in maize: its possible role in lysine metabolism of developing endosperm. *Phytochemistry* 22, 2687–2689.

Arruda, P., Sodek, L. and da Silva, W.J. (1982) Lysine-ketoglutarate reductase activity in developing maize endosperm. *Plant Physiology* 69, 988–989.

Arruda, P., Kemper, E.L., Papes, F. and Leite, A. (2000) Regulation of lysine catabolism in higher plants. *Trends in Plant Science* 5, 324–330.

Atkinson, S.C., Dogovski, C., Newman, J., Dobson, R.C.J. and Perugini, M.A. (2011) Cloning, expression, purification and crystallization of dihydrodipicolinate synthase from the grapevine *Vitis vinifera*. *Acta Crystallographica Section F – Structural Biology and Crystallization Communications* 67, 1537–1541.

Atkinson, S.C., Dogovski, C., Downton, M.T., Pearce, F.G., Reboul, C.F., Buckle, A.M., Gerrard, J.A., Dobson, R.C.J., Wagner, J. and Perugini, M.A. (2012) Crystal, solution and *in silico* structural studies of dihydrodipicolinate synthase from the common grapevine. *PLoS ONE* 7(6): e38318.

Azevedo, R.A. (2002) Analysis of the aspartic acid metabolic pathway using mutant genes. *Amino Acids* 22, 217–230.

Azevedo, R. and Arruda, P. (2010) High-lysine maize: the key discoveries that have made it possible. *Amino Acids* 39, 979–989.

Azevedo, R.A., Arana, J.L. and Arruda, P. (1990) Biochemical genetics of the interaction of the lysine plus threonine resistant mutant Ltr*1 with opaque-2 maize mutant. *Plant Science* 70, 81–90.

Azevedo, R.A., Smith, R.J. and Lea, P.J. (1992) Aspartate kinase regulation in maize: evidence for co-purification of threonine-sensitive aspartate kinase and homoserine dehydrogenase. *Phytochemistry* 31, 3731–3734.

Azevedo, R.A., Arruda, P., Turner, W.L. and Lea, P.J. (1997) The biosynthesis and metabolism of the aspartate derived amino acids in higher plants. *Phytochemistry* 46, 395–419.

Azevedo, R.A., Lancien, M. and Lea, P.J. (2006) The aspartic acid metabolic pathway, an exciting and essential pathway in plants. *Amino Acids* 30, 143–162.

Bareich, D., Koteva, K., Nazi, I. and Wright, G.D. (2004) Small molecule functional discrimination of the kinases required for the microbial synthesis of threonine and isoleucine. *Bioorganic and Medicinal Chemistry* 12, 807–815.

Blemings, K.P., Crenshaw, T.D., Swick, R.W. and Benevenga, N.J. (1994) Lysine-α-ketoglutarate reductase and saccharopine dehydrogenase are located only in the mitochondrial matrix in rat liver. *Journal of Nutrition* 124, 1215–1221.

Blickling, S., Beisel, H.-G., Bozic, D., Knäblein, J., Laber, B. and Huber, R. (1997a) Structure of dihydrodipicolinate synthase of *Nicotiana sylvestris* reveals novel quaternary structure. *Journal of Molecular Biology* 274, 608–621.

Blickling, S., Renner, C., Laber, B., Pohlenz, H.-D., Holak, T.A. and Huber, R. (1997b) Reaction mechanism of *Escherichia coli* dihydrodipicolinate synthase investigated by X-ray crystallography and NMR spectroscopy. *Biochemistry* 36, 24–33.

Borell, C.W., Urrestarazu, L.A. and Bhattacharjee, J.K. (1984) Two unlinked lysine genes (*LYS9* and *LYS14*) are required for the synthesis of saccharopine reductase in *Saccharomyces cerevisiae*. *Journal of Bacteriology* 159, 429–432.

Brandt, A.B. (1975) *In vivo* incorporation of lysine-C^{14} into the endosperm of wild type and high lysine barley. *FEBS Letters* 52, 288–291.

Brochetto-Braga, M.R., Leite, A. and Arruda, P. (1992) Partial purification and characterization of lysine-ketoglutarate reductase in normal and opaque-2 maize endosperms. *Plant Physiology* 98, 1139–1147.

Bryan, P.A., Cawley, R.D., Brunner, C.E. and Bryan, J.K. (1970) Isolation and characterization of a lysine-sensitive aspartokinase from a multicellular plant. *Biochemical and Biophysical Research Communications* 41,1211–1217.

Burgess, B.R., Dobson, R.C.J., Bailey, M.F., Atkinson, S.C., Griffin, M.D.W., Jameson, G.B., Parker, M.W., Gerrard, J.A. and Perugini, M.A. (2008) Structure and evolution of a novel dimeric enzyme from a clinically important bacterial pathogen. *The Journal of Biological Chemistry* 283, 27598–27603.

Craciun, A., Jacobs, M. and Vauterin, M. (2000) *Arabidopsis* loss-of-function mutant in the lysine pathway points out complex regulation mechanisms. *FEBS Letters* 487, 234–238.

Cremer, J., Eggeling, L. and Sahm, H. (1991) Control of the lysine biosynthesis sequence in *Corynebacterium glutamicum* as analyzed by overexpression of the individual corresponding genes. *Applied and Environmental Microbiology* 57, 1746–1752.

Cunha Lima, S.T., Azevedo, R.A., Santoro, L.G., Gaziola, S.A. and Lea, P.J. (2003) Isolation of the bifunctional enzyme lysine 2-oxoglutarate reductase-saccharopine dehydrogenase from *Phaseolus vulgaris* L. *Amino Acids* 24, 179–186.

Curien, G., Laurencin, M., Robert-Genthon, M. and Dumas, R. (2007) Allosteric monofunctional aspartate kinases from *Arabidopsis*. *FEBS Journal* 274, 164–176.

Curien, G., Bastien, O., Robert-Genthon, M., Cornish-Bowden, A., Cardenas, M.L. and Dumas, R. (2009) Understanding the regulation of aspartate metabolism using a model based on measured kinetic parameters. *Molecular Systems Biology* 5, 1–14.

da Costa, T.P.S., Muscroft-Taylor, A.C., Dobson, R.C.J., Devenish, S.R.A., Jameson, G.B. and Gerrard, J.A. (2010) How essential is the 'essential' active-site lysine in dihydrodipicolinate synthase? *Biochimie* 92, 837–845.

Dante, R.A., Neto, G.C., Leite, A., Yunes, J.A. and Arruda, P. (1999) The *DapA* gene encoding the lysine biosynthetic enzyme dihydrodipicolinate synthase from *Coix lacryma-jobi*: cloning, characterization, and expression analysis. *Plant Molecular Biology* 41, 551–561.

Demidov, D., Horstmann, C., Meixner, M., Pickardt, T., Saalbach, I., Galili, G. and Müntz, K. (2003) Additive effects of the feed-back insensitive bacterial aspartate kinase and the Brazil nut 2S albumin on the methionine content of transgenic narbon bean (*Vicia narbonensis* L.). *Molecular Breeding* 11, 187–201.

Dereppe, C., Bold, G., Ghisalba, O., Ebert, E. and Schar, H.P. (1992) Purification and characterization of dihydrodipicolinate synthase from pea. *Plant Physiology* 98, 813–821.

Dumas, R., Cobessi, D., Robin, A.Y., Ferrer, J.-L. and Curien, G. (2012) The many faces of aspartate kinases. *Archives of Biochemistry and Biophysics* 519, 186–193.

Epelbaum, S., McDevitt, R. and Falco, S.C. (1997) Lysine-ketoglutarate reductase and saccharopine dehydrogenase from *Arabidopsis thaliana*: nucleotide sequence and characterization. *Plant Molecular Biology* 35, 735–748.

Erzeel, E., Bochaute, P., Thu, T. and Angenon, G. (2013) *Medicago truncatula* dihydrodipicolinate synthase (DHDPS) enzymes display novel regulatory properties. *Plant Molecular Biology* 81, 401–415.

Falco, S.C., Guida, T., Locke, M., Mauvais, J., Sanders, C., Ward, R.T. and Webber, P. (1995) Transgenic canola and soybean seeds with increased lysine. *Bio/Technology* 13, 577–582.

Fjellstedt, T.A. and Robinson, J.C. (1975a) Purification and properties of L-lysine-α-ketoglutarate reductase from human placenta. *Archives of Biochemistry and Biophysics* 168, 536–548.

Fjellstedt, T.A. and Robinson, J.C. (1975b) Properties of partially purified saccharopine dehydrogenase from human placenta. *Archives of Biochemistry and Biophysics* 171, 191–196.

Fornazier, R.F., Gaziola, S.A., Helm, C.V., Lea, P.J. and Azevedo, R.A. (2005) Isolation and characterization of enzymes involved in lysine catabolism from sorghum seeds. *Journal of Agricultural and Food Chemistry* 53, 1791–1798.

Frisch, D.A., Gengenbach, B.G., Tommey, A.M., Sellner, J.M., Somers, D.A. and Myers, D.E. (1991a) Isolation and characterization of dihydrodipicolinate synthase from maize. *Plant Physiology* 96, 444–452.

Frisch, D.A., Tommey, A.M., Gengenbach, B.G. and Somers, D.A. (1991b) Direct genetic selection of a maize cDNA for dihydrodipicolinate synthase in an *Escherichia coli* dapA$^-$ auxotroph. *Molecular and General Genetics* 228, 287–293.

Frizzi, A., Huang, S., Gilbertson, L.A., Armstrong, T.A., Luethy, M.H. and Malvar, T.M. (2008) Modifying lysine biosynthesis and catabolism in corn with a single bifunctional expression/silencing transgene cassette. *Plant Biotechnology Journal* 6, 13–21.

Galili, G. (2002) New insights into the regulation and functional significance of lysine metabolism in plants. *Annual Review of Plant Biology* 53, 27–43.

Galili, G. and Amir, R. (2013) Fortifying plants with the essential amino acids lysine and methionine to improve nutritional quality. *Plant Biotechnology Journal* 11, 211–222.

Galili, S., Guenoune, D., Wininger, S., Hana, B., Schupper, A., Ben-Dor, B. and Kapulnik, Y. (2000) Enhanced levels of free and protein-bound threonine in transgenic alfalfa *(Medicago sativa* L.) expressing a bacterial feedback-insensitive aspartate kinase gene. *Transgenic Research* 9, 137–144.

Gaziola, S.A., Teixeira, C.M.G., Lugli, J., Sodek, L. and Azevedo, R.A. (1997) The enzymology of lysine catabolism in rice seeds : isolation, characterization, and regulatory properties of a lysine 2-oxoglutarate reductase saccharopine dehydrogenase bifunctional polypeptide. *European Journal of Biochemistry* 247, 364–371.

Gaziola, S.A., Alessi, E.S., Guimaraes, P.E.O., Damerval, C. and Azevedo, R.A. (1999) Quality protein maize: a biochemical study of enzymes involved in lysine metabolism. *Journal of Agricultural and Food Chemistry* 47, 1268–1275.

Gaziola, S.A., Sodek, L., Arruda, P., Lea, P.J. and Azevedo, R.A. (2000) Degradation of lysine in rice seeds: effect of calcium, ionic strength, *S*-adenosylmethionine and *S*-2-aminoethyl-L-cysteine on the lysine 2-oxoglutarate reductase-saccharopine dehydrogenase bifunctional enzyme. *Physiologia Plantarum* 110, 164–171.

Ghislain, M., Frankard, V. and Jacobs, M. (1990) Dihydrodipicolinate synthase of *Nicotiana sylvestris* a chloroplast-localized enzyme of the lysine pathway. *Planta* 180, 480–486.

Ghislain, M., Frankard, V. and Jacobs, M. (1995) A dinucleotide mutation in dihydrodipicolinate synthase of *Nicotiana sylvestris* leads to lysine overproduction. *The Plant Journal* 8, 733–743.

Goncalves-Butruille, M., Szajner, P., Torigoi, E., Leite, A. and Arruda, P. (1996) Purification and characterization of the bifunctional enzyme lysine-ketoglutarate reductase-saccharopine dehydrogenase from maize. *Plant Physiology* 110, 765–771.

Hacham, Y., Matityahu, I., Schuster, G. and Amir, R. (2008) Overexpression of mutated forms of aspartate kinase and cystathionine γ-synthase in tobacco leaves resulted in the high accumulation of methionine and threonine. *The Plant Journal* 54, 260–271.

Hadfield, A., Shammas, C., Kryger, G., Ringe, D., Petsko, G.A., Ouyang, J. and Viola, R.E. (2001) Active site analysis of the potential antimicrobial target aspartate semialdehyde dehydrogenase. *Biochemistry* 40, 14475–14483.

Heremans, B. and Jacobs, M. (1997) A mutant of *Arabidopsis thaliana* (L.) Heynh. with modified control of aspartate kinase by threonine. *Biochemical Genetics* 35, 139–153.

Huang, S.S., Kruger, D.E., Frizzi, A., D'Ordine, R.L., Florida, C.A., Adams, W.R., Brown, W.E. and Luethy, M.H. (2005) High-lysine corn produced by the combination of enhanced lysine biosynthesis and reduced zein accumulation. *Plant Biotechnology Journal* 3, 555–569.

Hutzler, J. and Dancis, J. (1968) Conversion of lysine to saccharopine by human tissues. *Biochimica et Biophysica Acta – General Subjects* 158, 62–69.

Hutzler, J. and Dancis, J. (1970) Saccharopine cleavage by a dehydrogenase of human liver. *Biochimica et Biophysica Acta – Enzymology* 206, 205–214.

James, C.L. and Viola, R.E. (2002) Production and characterization of bifunctional enzymes. Substrate channeling in the aspartate pathway. *Biochemistry* 41, 3726–3731.

Jones, E.E. and Broquist, H.P. (1965) Saccharopine, an intermediate of the aminoadipic acid pathway of lysine biosynthesis: II. Studies in *Saccharomyces cereviseae. The Journal of Biological Chemistry* 240, 2531–2536.

Jones, E.E. and Broquist, H.P. (1966) Saccharopine, an intermediate of the aminoadipic acid pathway of lysine biosynthesis: III. Aminoadipic semialdehyde-glutamate reductase. *The Journal of Biological Chemistry* 241, 3430–3434.

Jones-Held, S., Ambrozevicius, L.P., Campbell, M., Drumheller, B., Harrington, E. and Leustek, T. (2012) Two *Arabidopsis thaliana* dihydrodipicolinate synthases, DHDPS1 and DHDPS2, are unequally redundant. *Functional Plant Biology* 39, 1058–1067.

Kaneko, T., Hashimoto, T., Kumpaisal, R. and Yamada, Y. (1990) Molecular cloning of wheat dihydrodipicolinate synthase. *The Journal of Biological Chemistry* 265, 17451–17455.

Karchi, H., Shaul, O. and Galili, G. (1994) Lysine synthesis and catabolism are coordinately regulated during tobacco seed development. *Proceedings of the National Academy of Sciences of the United States of America* 91, 2577–2581.

Karchi, H., Miron, D., Ben-Yaacov, S. and Galili, G. (1995) The lysine-dependent stimulation of lysine catabolism in tobacco seed requires calcium and protein phosphorylation. *The Plant Cell* 7, 1963–1970.

Karsten, W.E. (1997) Dihydrodipicolinate synthase from *Escherichia coli*: pH dependent changes in the kinetic mechanism and kinetic mechanism of allosteric inhibition by L-lysine. *Biochemistry* 36, 1730–1739.

Kaur, N., Gautam, A., Kumar, S., Singh, A., Singh, N., Sharma, S., Sharma, R., Tewari, R. and Singh, T.P. (2011) Biochemical studies and crystal structure determination of dihydrodipicolinate synthase from *Pseudomonas aeruginosa*. *International Journal of Biological Macromolecules* 48, 779–787.

Kemper, E.L., Cord-Neto, G., Capella, A.N., Goncalves-Butruile, M., Azevedo, R.A. and Arruda, P. (1998) Structure and regulation of the bifunctional enzyme-lysine-oxoglutarate reductase-saccharopine dehydrogenase in maize. *European Journal of Biochemistry* 253, 720–729.

Kemper, E.L., Neto, G.C., Papes, F., Moraes, K.C.M., Leite, A. and Arruda, P. (1999) The role of opaque-2 in the control of lysine-degrading activities in developing maize endosperm. *The Plant Cell* 11, 1981–1994.

Kong, F., Jiang, S., Meng, X., Song, C., Shi, J., Jin, D., Jiang, S. and Wang, B. (2009) Cloning and characterization of the *DHDPS* gene encoding the lysine biosynthetic enzyme dihydrodipocolinate synthase from *Zizania latifolia* (Griseb). *Plant Molecular Biology Reporter* 27, 199–208.

Krishna, S.S., Zhou, T., Daugherty, M., Osterman, A. and Zhang, H. (2001) Structural basis for the catalysis and substrate specificity of homoserine kinase. *Biochemistry* 40, 10810–10818.

Kumar, R. and Garg, P. (2013) Molecular modeling and active site binding mode characterization of aspartate β-semialdehyde dehydrogenase family. *Molecular Informatics* 32, 377–383.

Kumpaisal, R., Hashimoto, T. and Yamada, Y. (1987) Purification and characterization of dihydrodipicolinate synthase from wheat suspension-cultures. *Plant Physiology* 85, 145–151.

Kwon, T., Sasahara, T. and Abe, T. (1995) Lysine accumulation in transgenic tobacco expressing dihydrodipicolinate synthase of *Escherichia coli*. *Journal of Plant Physiology* 146, 615–621.

Lange, M., Vincze, E., Wieser, H., Schjoerring, J.K. and Holm, P.B. (2007) Suppression of C-hordein synthesis in barley by antisense constructs results in a more balanced amino acid composition. *Journal of Agricultural and Food Chemistry* 55, 6074–6081.

Lee, M., Martin, M.N., Hudson, A.O., Lee, J., Muhitch, M.J. and Leustek, T. (2005) Methionine and threonine synthesis are limited by homoserine availability and not the activity of homoserine kinase in *Arabidopsis thaliana*. *The Plant Journal* 41, 685–696.

Lee, S.I., Kim, H.U., Lee, Y.-H., Suh, S.-C., Lim, Y.P., Lee, H.-Y. and Kim, H.-I. (2001) Constitutive and seed-specific expression of a maize lysine-feedback-insensitive dihydrodipicolinate synthase gene leads to increased free lysine levels in rice seeds. *Molecular Breeding* 8, 75–84.

Lucas, D.M., Taylor, M.L., Hartnell, G.F., Nemeth, M.A., Glenn, K.C. and Davis, S.W. (2007) Broiler performance and carcass characteristics when fed diets containing Lysine maize (LY038 or LY038 × MON 810), control or conventional reference maize. *Poultry Science* 86, 71–72.

Lugli, J., Campbell, A., Gaziola, S.A., Smith, R.J., Lea, P.J. and Azevedo, R.A. (2002) Enzymes of lysine metabolism from *Coix lacryma-jobi* seeds. *Plant Physiology and Biochemistry* 40, 25–32.

Luniwal, A., Wang, L., Pavlovsky, A., Erhardt, P.W. and Viola, R.E. (2012) Molecular docking and enzymatic evaluation to identify selective inhibitors of aspartate semialdehyde dehydrogenase. *Bioorganic and Medicinal Chemistry* 20, 2950–2956.

Mainieri, D., Rossi, M., Archinti, M., Bellucci, M., De Marchis, F., Vavassori, S., Pompa, A., Arcioni, S. and Vitale, A. (2004) Zeolin. A new recombinant storage protein constructed using maize gamma-zein and bean phaseolin. *Plant Physiology* 136, 3447–3456.

Markovitz, P.J., Chuang, D.T. and Cox, R.P. (1984) Familial hyperlysinemias. Purification and characterization of the bifunctional aminoadipic semialdehyde synthase with lysine-ketoglutarate reductase and saccharopine dehydrogenase activities. *The Journal of Biological Chemistry* 259, 11643–11646.

Medici, L.O., Gaziola, S.A., Varisi, V.A., Paula J.A.C., de, Ferreira, R.R. and Azevedo, R.A. (2009) Diallelic analysis for lysine and oil contents in maize grains. *Scientia Agricola* 66, 204–209.

Mertz, E.T., Bates, L.S. and Nelson, O.E. (1964) Mutant gene that changes protein composition and increases lysine content of maize endosperm. *Science* 145, 279–280.

Miron, D., Ben-Yaacov, S., Karchi, H. and Galili, G. (1997) *In vitro* dephosphorylation inhibits the activity of soybean lysine-ketoglutarate reductase in a lysine-regulated manner. *The Plant Journal* 12, 1453–1458.

Miron, D., Ben-Yaacov, S., Reches, D., Schupper, A. and Galili, G. (2000) Purification and characterization of bifunctional lysine-ketoglutarate reductase/saccharopine dehydrogenase from developing soybean seeds. *Plant Physiology* 123, 655–664.

Noda, C. and Ichihara, A. (1978) Purification and properties of L-lysine-α-ketoglutarate reductase from rat liver mitochondria. *Biochimica et Biophysica Acta – Enzymology* 525, 307–313.

Paris, S., Wessel, P.M. and Dumas, R. (2002a) Overproduction, purification, and characterization of recombinant aspartate semialdehyde dehydrogenase from *Arabidopsis thaliana*. *Protein Expression and Purification* 24, 99–104.

Paris, S., Wessel, P.M. and Dumas, R. (2002b) Overproduction, purification, and characterization of recombinant bifunctional threonine-sensitive aspartate kinase-homoserine dehydrogenase from *Arabidopsis thaliana*. *Protein Expression and Purification* 24, 105–110.

Paris, S., Viemon, C., Curien, G. and Dumas, R. (2003) Mechanism of control of *Arabidopsis thaliana* aspartate kinase-homoserine dehydrogenase by threonine. *The Journal of Biological Chemistry* 278, 5361–5366.

Patte, J.-C., Truffa-Bachi, P. and Cohen, G.N. (1966) The threonine-sensitive homoserine dehydrogenase and aspartokinase activities of *Escherichia coli*: I. Evidence that the two activities are carried by a single protein. *Biochimica et Biophysica Acta – Enzymology and Biological Oxidation* 128, 426–439.

Ramos, F., Dubois, E. and Piérard, A. (1988) Control of enzyme synthesis in the lysine biosynthetic pathway of *Saccharomyces cerevisiae*. *European Journal of Biochemistry* 171, 171–176.

Rinder, J., Casazza, A.P., Hoefgen, R. and Hesse, H. (2008) Regulation of aspartate-derived amino acid homeostasis in potato plants (*Solanum tuberosum* L.) by expression of *E. coli* homoserine kinase. *Amino Acids* 34, 213–222.

Rognes, S.E., Lea, P.J. and Miflin, B.J. (1980) *S*-adenosylmethionine – a novel regulator of aspartate kinase. *Nature* 287, 357–359.

Rognes, S.E., Dewaele, E., Aas, S.F., Jacobs, M. and Frankard, V. (2003) Transcriptional and biochemical regulation of a novel *Arabidopsis thaliana* bifunctional aspartate kinase-homoserine dehydrogenase gene isolated by functional complementation of a yeast *hom6* mutant. *Plant Molecular Biology* 51, 281–294.

Saunders, P.P. and Broquist, H.P. (1966) Saccharopine, an intermediate of the aminoadipic acid pathway of lysine biosynthesis: IV. Saccharopine dehydrogenase. *The Journal of Biological Chemistry* 241, 3435–3440.

Schroeder, A.C., Zhu, C., Yanamadala, S.R., Cahoon, R.E., Arkus, K.A.J., Wachsstock, L., Bleeke, J., Krishnan, H.B. and Jez, J.M. (2010) Threonine-insensitive homoserine dehydrogenase from soybean: genomic organization, kinetic mechanism, and *in vivo* activity. *The Journal of Biological Chemistry* 285, 827–834.

Shaul, O. and Galili, G. (1992a) Increased lysine synthesis in tobacco plants that express high levels of bacterial dihydrodipicolinate synthase in their chloroplasts. *The Plant Journal* 2, 203–209.

Shaul, O. and Galili, G. (1992b) Threonine overproduction in transgenic tobacco plants expressing a mutant desensitized aspartate kinase of *Escherichia coli*. *Plant Physiology* 100, 1157–1163.

Shaul, O. and Galili, G. (1993) Concerted regulation of lysine and threonine synthesis in tobacco plants expressing bacterial feedback-insensitive aspartate kinase and dihydrodipicolinate synthase. *Plant Molecular Biology* 23, 759–768.

Shaver, J.M., Bittel, D.C., Sellner, J.M., Frisch, D.A., Somers, D.A. and Gengenbach, B.G. (1996) Single amino acid substitutions eliminate lysine inhibition of maize dihydrodipicolinate synthase. *Proceedings of the National Academy of Sciences of the United States of America* 93, 1962–1966.

Sikdar, M.S.I. and Kim, J.S. (2010) Characterization of a gene encoding for dihydrodipicolinate synthase from rice. *Australian Journal of Crop Science* 4, 461–466.

Silk, G.W., Matthews, B.F., Somers, D.a. [sic] and Gengenbach, B.G. (1994) Cloning and expression of the soybean *DapA* gene encoding dihydrodipicolinate synthase. *Plant Molecular Biology* 26, 989–993.

Sodek, L. and Wilson, C.M. (1970) Incorporation of ^{14}C-leucine and ^{14}C-lysine into protein in the developing endosperm of normal and opaque-2 corn. *Archives of Biochemistry and Biophysics* 140, 29–38.

Stepansky, A., Yao, Y., Tang, G. and Galili, G. (2005) Regulation of lysine catabolism in *Arabidopsis* through concertedly regulated synthesis of the two distinct gene products of the composite *AtLKR/SDH* locus. *Journal of Experimental Botany* 56, 525–536.

Stepansky, A., Less, H., Angelovici, R., Aharon, R., Zhu, X. and Galili, G. (2006) Lysine catabolism, an effective versatile regulator of lysine level in plants. *Amino Acids* 30, 121–125.

Stuttmann, J., Hubberten, H.-M., Rietz, S., Kaur, J., Muskett, P., Guerois, R., Bednarek, P., Hoefgen, R. and Parker, J.E. (2011) Perturbation of *Arabidopsis* amino acid metabolism causes incompatibility with the adapted biotrophic pathogen *Hyaloperonospora arabidopsidis*. *The Plant Cell* 23, 2788–2803.

Tang, G., Miron, D., Zhu-Shimoni, J.X. and Galili, G. (1997) Regulation of lysine catabolism through lysine-ketoglutarate reductase and saccharopine dehydrogenase in *Arabidopsis*. *The Plant Cell* 9, 1305–1316.

Tang, G., Zhu, X., Gakiere, B., Levanony, H., Kahana, A. and Galili, G. (2002) The bifunctional *LKR/SDH* locus of plants also encodes a highly active monofunctional lysine-ketoglutarate reductase using a polyadenylation signal located within an intron. *Plant Physiology* 130, 147–154.

Vauterin, M. and Jacobs, M. (1994) Isolation of a poplar and an *Arabidopsis thaliana* dihydrodipicolinate synthase cDNA clone. *Plant Molecular Biology* 25, 545–550.

Vauterin, M., Frankard, V. and Jacobs, M. (1999) The *Arabidopsis thaliana dhdps* gene encoding dihydrodipicolinate synthase, key enzyme of lysine biosynthesis, is expressed in a cell-specific manner. *Plant Molecular Biology* 39, 695–708.

Wallsgrove, R.M. and Mazelis, M. (1981) Spinach leaf dihydrodipicolinate synthase: partial purification and characterization. *Phytochemistry* 20, 2651–2655.

WHO (2007) *Protein and Amino Acid Requirements in Human Nutrition. Report of a Joint WHO/FAO/UNU Expert Consultation.* WHO Technical Report Series 935, World Health Organization, Geneva, Switzerland.

Wilson, B.J., Gray, A.C. and Matthews, B.F. (1991) Bifunctional protein in carrot contains both aspartokinase and homoserine dehydrogenase activities. *Plant Physiology* 97, 1323–1328.

Yoshioka, Y., Kurei, S. and Machida, Y. (2001) Identification of a monofunctional aspartate kinase gene of *Arabidopsis thaliana* with spatially and temporally regulated expression. *Genes and Genetic Systems* 76, 189–198.

Zhang, Y.H.P. (2011) Substrate channeling and enzyme complexes for biotechnological applications. *Biotechnology Advances* 29, 715–725.

Zhou, T., Daugherty, M., Grishin, N.V., Osterman, A.L. and Zhang, H. (2000) Structure and mechanism of homoserine kinase: prototype for the GHMP kinase superfamily. *Structure* 8, 1247–1257.

Zhu, X., Tang, G. and Galili, G. (2000) Characterization of the two saccharopine dehydrogenase isozymes of lysine catabolism encoded by the single composite *AtLKR/SDH* locus of *Arabidopsis*. *Plant Physiology* 124, 1363–1372.

14 Histidine

R.A. Ingle*
University of Cape Town, South Africa

14.1 Abstract

Histidine (His) is a proteinogenic amino acid, and is thus critical for plant growth and development. The chemical properties of the imidazole side group of His allow this amino acid to participate in acid–base catalysis and in the coordination of metal ions, including zinc (Zn) and nickel (Ni). Despite the biological importance of His, this was the last amino acid biosynthetic pathway to be characterized in plants, and a number of important questions remain concerning it. While the eight enzymes involved in the biosynthesis of His have now been identified, the metabolic route by which His catabolism occurs in plants remains unknown. ATP-phosphoribosyl transferase, which catalyses the first reaction in the pathway, has been shown to play an important role in the regulation of His biosynthesis, yet we still know little about the transcriptional regulation of the His biosynthetic genes, nor do we understand how the synthesis of this amino acid is integrated with other metabolic demands in plants.

14.2 Introduction

Histidine (His) is one of the standard amino acids found in proteins, and is essential for plant growth and development (Muralla *et al.*, 2007; Bikard *et al.*, 2009). It is found in the active sites of many enzymes owing to its ability to participate in general acid–base catalysis (Fersht, 1999). This is a result of the imidazole side group of His, which has a pK_a ~6, allowing it to alternate between the protonated and unprotonated states under physiologically relevant conditions (Fig. 14.1). The unprotonated imidazole group is nucleophilic, and can therefore act as a general base, for example in catalytic triads in which the basic N abstracts protons from threonine (Thr), cysteine (Cys) or serine (Ser) residues, while the protonated form is a general acid. In addition, the unprotonated imidazole group of His plays important roles as a nucleophile in phosphoryl transfer and in the coordination of metal ions, for example in zinc finger proteins (Fraústo da Silva and Williams, 2001; Harding, 2004).

His is the fourth most metabolically expensive amino acid to synthesize after the aromatic amino acids phenylalanine (Phe), tryptophan (Trp) and tyrosine (Tyr). Estimates of the number of ATP molecules required for the synthesis range from 31 to 41 per molecule of His synthesized (Alifano *et al.*, 1996; Akashi and Gojobori, 2002; Swire, 2007). It has been suggested that

*E-mail address: Robert.Ingle@uct.ac.za

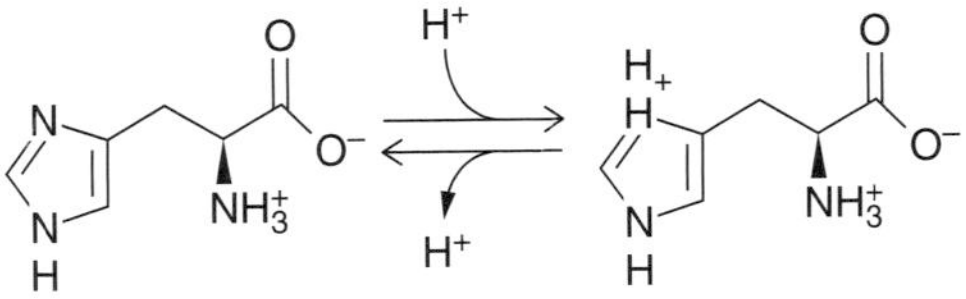

Fig. 14.1. L-Histidine (His). The imidazole side group of His has a pK_a of approximately 6, allowing it to switch between the protonated and unprotonated states under physiological conditions, as shown here.

the high energetic cost of His biosynthesis may account for the generally low abundance of this amino acid in proteins (in which, on average, it is the third or fourth least abundant amino acid) in line with the predictions of the cost minimization hypothesis (Seligmann, 2003).

The genetics of the His biosynthetic pathway have been regarded as a paradigm for metabolic gene regulation in microorganisms (Winkler, 1987; Alifano *et al.*, 1996; Meinke *et al.*, 2008). For example, research on His biosynthesis in bacteria was instrumental in the discovery of operon structure and the regulation of amino acid biosynthesis by attenuation (Ames *et al.*, 1960; Roth and Ames, 1966). In contrast, His biosynthesis in plants was largely neglected and it was the last amino acid biosynthetic pathway to be characterized. Despite early reports of the detection of the activities of three His biosynthetic enzymes in plant protein extracts (Wiater *et al.*, 1971), it has only recently been established that His biosynthesis in plants (see Fig. 14.2) proceeds via the same metabolic route as in microorganisms (Ohta *et al.*, 2000; Stepansky and Leustek, 2006). Although His is an essential amino acid in humans, this pathway was not the focus of attempts to increase His concentrations for nutritional purposes, as the His contents of crop plants are not generally regarded as limiting. Instead, a number of the His biosynthetic enzymes in plants were first identified through efforts to develop selective herbicides to inhibit this metabolic pathway (Mori *et al.*, 1995; Ward and Ohta, 1999).

Genes encoding all eight of the His biosynthetic enzymes (HISNs) have now been identified from *Arabidopsis*, with histidinol-phosphate phosphatase (*HISN7*) identified as recently as 2010 (Petersen *et al.*, 2010). While genetic redundancy is common in the majority of amino acid biosynthetic pathways in plants, only the *HISN1*, *HISN5* and *HISN6* genes are duplicated in the *Arabidopsis* genome. The remaining five HISN enzymes are encoded by single-copy genes (Stepansky and Leustek, 2006; Petersen *et al.*, 2010); individuals homozygous for null mutations in any of these five genes (generated by the selfing of heterozygous plants) abort at the pre-globular stage of embryonic development (Muralla *et al.*, 2007; Petersen *et al.*, 2010). These data indicate that the capacity for His biosynthesis is essential for plant reproduction, and that endogenous His concentrations in the surrounding heterozygous maternal tissues are not sufficient to support the embryonic development of null mutants (Muralla *et al.*, 2007). In contrast, individuals homozygous for weak *hisn* alleles are viable, but may display abnormal seedling phenotypes, including stunted root development or reduced chlorophyll content (Noutoshi *et al.*, 2005; Mo *et al.*, 2006).

This chapter outlines the His biosynthetic pathway in plants, focusing on our current knowledge of the enzymes involved, their evolutionary origin and the regulation of this important metabolic pathway. The important roles played by His in plants that hyperaccumulate the metal ion nickel (Ni) in their shoot tissues is also discussed.

14.3 Histidine Biosynthesis in Plants

The His biosynthetic pathway in plants consists of ten reactions catalysed by eight enzymes (Fig. 14.2), beginning with the condensation of 5-phosphoribosyl-1-pyrophosphate (PRPP) and ATP to form *N′*-5′-phosphoribosyl-ATP. (Ohta *et al.*, 2000). This reaction is catalysed by the enzyme ATP-phosphoribosyl transferase (HISN1). While HISN1 activity was first reported in 1971 in total protein extracts from barley, oat and pea plants (Wiater *et al.*, 1971), only 21 years later was the first HISN1 enzyme purified, from wheat (Münzer *et al.*, 1992). The native wheat HISN1 enzyme was found to be a hexameric protein requiring Mg^{2+} as a cofactor for activity, as do the *Escherichia coli* and *Salmonella typhimurium* enzymes (Voll *et al.*, 1967; Münzer *et al.*, 1992). While the HISN1 protein was first isolated from wheat, the first *HISN1* cDNA to be cloned was

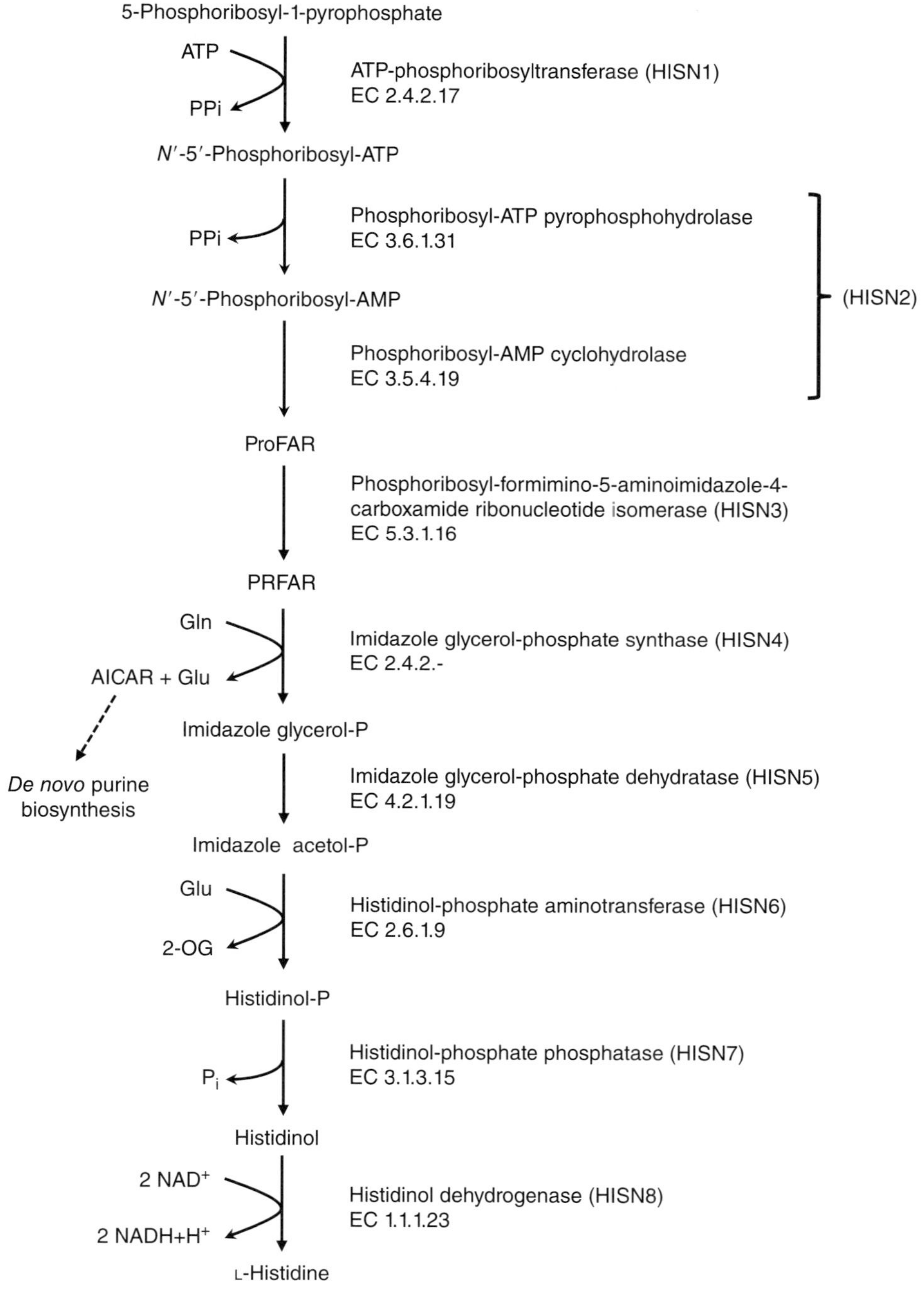

Fig. 14.2. The histidine biosynthetic pathway in plants. Abbreviations used for enzyme names in the text are indicated in parentheses and the corresponding enzyme commission numbers are provided. Abbreviations used for metabolic intermediates are: AICAR, 5′-phosphoribosyl-4-carboximide-5-aminoimidazole; Gln, glutamine; Glu, glutamate; 2-OG, 2-oxoglutarate; PPi, pyrophosphate; PRFAR *N*′-[(5′-phosphoribulosyl)formimino]-5-aminoimidazole-4-carboxamide ribonucleotide; ProFAR, *N*′-[(5′-phosphoribosyl)formimino]-5-aminoimidazole-4-carboxamide ribonucleotide.

from the Ni-hyperaccumulating species *Thlaspi goesingense* (Persans *et al.*, 1999). Southern blot analysis indicated the presence of two genes encoding HISN1 in the *T. goesingense* genome (Persans *et al.*, 1999) and, shortly afterwards, two HISN1 encoding genes were identified in *Arabidopsis*, *HISN1A* (At1g58080) and *HISN1B* (At1g09795) (Ohta *et al.*, 2000). The presence of two HISN1 paralogues has subsequently been reported from several other members of the *Brassicaceae*, including *Alyssum lesbiacum* and *A. montanum* (Ingle *et al.*, 2005).

In vitro analysis of recombinant *Arabidopsis* HISN1A protein determined that this enzyme had similar kinetic constants to those of the HISN1 enzymes from wheat and *S. typhimurium*, with K_m values of 0.6 mM and 0.13 mM for ATP and PRPP, respectively (Ohta *et al.*, 2000). The recombinant HISN1B protein was, however, found to be unstable; while enzyme activity could be detected in crude protein extracts from *E. coli* expressing HISN1B, the recombinant protein could not be affinity purified (Ohta *et al.*, 2000). Interestingly, while null *hisn1a Arabidopsis* mutants display a severely stunted phenotype and do not survive to maturity, *hisn1b* mutants are viable and display no obvious phenotype (Muralla *et al.*, 2007). This observation suggests that *HISN1A* and *HISN1B* are not fully redundant in *Arabidopsis* (Muralla *et al.*, 2007).

The next two reactions in the His biosynthetic pathway in plants are catalysed by a single enzyme, HISN2 (Fig. 14.2). The C-terminal domain of the HISN2 protein catalyses the hydrolysis of N′-5′-phosphoribosyl-ATP to N′-5′-phosphoribosyl-AMP. Subsequently, the N-terminal cyclohydrolase domain of HISN2 opens the purine ring of N′-5′-phosphoribosyl-AMP to generate N′-[(5′-phosphoribosyl)formimino]-5-aminoimidazole-4-carboxamide ribonucleotide (ProFAR) (Fujimori and Ohta, 1998a). The domain organization of the HISN2 protein in plants is identical to that observed in a number of bacteria, including *E. coli* and *S. typhimurium*, which also possess a single bifunctional enzyme. In contrast, in the *Actinobacteria*, these two reactions are catalysed by separate enzymes (Chiariotti *et al.*, 1986). A third type of organization is observed in *Saccharomyces cerevisiae* where a trifunctional protein, which also possesses histidinol dehydrogenase activity, is present (Donahue *et al.*, 1982). A single HISN2 encoding gene, At1g31860, is present in the *Arabidopsis* genome (Fujimori and Ohta, 1998a). No data on the kinetic properties or structure of any HISN2 protein from plants have been reported up to now.

The fourth reaction in the pathway is the isomerization of the aminoaldose ProFAR to the aminoketose N′-[(5′-phosphoribulosyl)formimino]-5-aminoimidazole-4-carboxamide ribonucleotide (PRFAR) catalysed by the enzyme ProFAR isomerase (HISN3). In *Arabidopsis*, this monofunctional enzyme is encoded by a single copy gene on chromosome II, At2g36230, and was identified through functional complementation of an *E. coli hisA*$^-$ mutant (Fujimori *et al.*, 1998). As with HISN2, the HISN3 protein has not yet been characterized in plants.

Imidazole glycerol-phosphate synthase (HISN4) catalyses the branch point reaction of the His biosynthetic pathway, which involves the transfer of an amide group from glutamine (Gln) to PRFAR, followed by a cyclization reaction that generates the imidazole ring. This results in the formation of imidazole glycerol phosphate (IGP) and a by-product, 5′-phosphoribosyl-4-carboximide-5-aminoimidazole (AICAR), which enters the *de novo* purine biosynthetic pathway (Ward and Ohta, 1999). In bacteria, the glutamine amidotransferase and cyclase activities are found on separate polypeptides, hisH and hisF, which form a heterodimer, while in *S. cerevisiae* a single bifunctional protein is present (Kuenzler *et al.*, 1993). The isolation of a single cDNA from *Arabidopsis* that could complement the His auxotrophy of an *S. cerevisiae his7* mutant demonstrated that the HISN4 protein is also bifunctional in plants (Fujimori and Ohta, 1998b). As is the case with the *S. cerevisiae* HIS7 enzyme, the N-terminal domain of HISN4 (encoded by At4g26900) catalyses the amidotransferase reaction, while the C-terminal domain catalyses the cyclase reaction (Fujimori and Ohta, 1998b).

The next reaction in the His biosynthesis pathway is the dehydration of IGP to imidazole acetol phosphate (IAP) by IGP dehydratase (HISN5). HISN5 activity was first detected in plants by Wiater *et al.* (1971) and an HISN5 protein was subsequently purified from wheat by Mano *et al.* (1993). The wheat enzyme did not display histidinol-phosphate phosphatase activity, indicating that it was a monofunctional enzyme, as is the HIS4 protein in *S. cerevisiae*, in contrast to the bifunctional hisB protein in *E. coli*, which has both enzymatic activities (Alifano *et al.*, 1996). *Arabidopsis* has two genes encoding HISN5 enzymes, *HISN5A* (At3g22425) and *HIS5NB* (At4g14910) which display 82% identity at the amino acid level (Tada *et al.*, 1994). While null *hisn5a* mutants are viable, no *hisn5b* mutants have been characterized to date, so it is unclear whether HISN5A and HISN5B are fully functionally redundant (Muralla *et al.*, 2007). As is the case in microorganisms, the *Arabidopsis* HISN5 enzymes require Mn^{2+} for the formation of catalytically active multimers (Mano *et al.*, 1993; Tada *et al.*, 1995). The removal of Mn^{2+} causes the 24 subunit holoenzyme to dissociate into inactive trimers, while its re-addition leads to reassembly of the active holoenzyme (Mano *et al.*, 1993; Tada *et al.*, 1995; Wilkinson *et al.*, 1995; Glynn *et al.*, 2005). Crystallization of the HISN5A protein has demonstrated that Mn^{2+} is also required for formation of the active site, and that the reaction mechanism probably proceeds through the binding of the imidazole ring of IGP to a di-Mn^{2+} cluster, followed by conversion through diazafulvene and Δ2-enol intermediates to IAP (Glynn *et al.*, 2005). Both the wheat germ and *Arabidopsis* HISN5 enzymes also require a high concentration of the reducing agent β-mercaptoethanol for maximal activity (Mano *et al.*, 1993; Tada *et al.*, 1995). The biochemical basis for this requirement is unclear as the mature HISN5 proteins do not contain Cys residues (Tada *et al.*, 1994).

The transamination of IAP to histidinol phosphate is then catalysed by histidinol-phosphate aminotransferase (HISN6) using glutamate (Glu) as an amino donor. The first plant cDNA encoding HISN6 was identified from *Nicotiana tabacum* through functional complementation of an *E. coli hisC*⁻ mutant (El Malki *et al.*, 1998). Two HISN6 encoding genes are present in the *N. tabacum* genome, and also in *Arabidopsis*: *HISN6A* (At5g10330) and *HISN6B* (At1g71920). The coding sequences of the *Arabidopsis* paralogues differ by only two synonymous single nucleotide polymorphisms (SNPs), but the two genes have highly divergent 5′ upstream regions (Mo *et al.*, 2006; Bikard *et al.*, 2009). Intraspecific sequence comparisons with *Arabidopsis lyrata* have shown that *HISN6B* is the ancestral locus, with *HISN6A* resulting from a recent 3.3 kb duplication event (Moore and Purugganan, 2003). Whether both *HISN6* genes in *Arabidopsis* are functional is unclear, as while *HISN6A* was found to be expressed in all tissues, *HISN6B* transcripts could not be detected (Moore and Purugganan, 2003; Bikard *et al.*, 2009). However, while null *hisn6a* mutants display embryonic lethality, null *hisn6a* seeds (obtained by supplying *HISN6A/hisn6a* heterozygotes with His) can germinate and form a small rosette without supplemental His, suggesting that *HISN6B* may be functional in *Arabidopsis* (Muralla *et al.*, 2007).

The dephosphorylation of histidinol phosphate to histidinol is catalysed by histidinol-phosphate phosphatase (HPP; HISN7). This was the last His biosynthetic enzyme to be identified in plants (Petersen *et al.*, 2010), although enzyme activity in plant extracts was first reported almost 40 years earlier (Wiater *et al.*, 1971). The *Arabidopsis* HISN7 enzyme does not belong to either the DDDD (containing four conserved aspartate (Asp) residues) or PHP (polymerase and histidinol phosphatase) superfamilies. Instead, like several HPP enzymes recently identified from *Actinobacteria*, HISN7 has high sequence identity to myoinositol monophosphatase enzymes (IMPs) (Mormann *et al.*, 2006; Marineo *et al.*, 2008; Petersen *et al.*, 2010). The *Arabidopsis* genome contains three putative IMPs, VTC4, myoinositol monophosphatase-like 1 (IMPL1) and IMPL2 (Torabinejad *et al.*, 2009). Heterologous expression of IMPL2 but not IMPL1 was sufficient to rescue the His auxotrophy of an *Streptomyces coelicolor hisN*⁻ mutant, and homozygous *impl2* null mutants displayed embryonic lethality (Petersen *et al.*, 2010), suggesting that IMPL2/HISN7 is most likely to be the only protein with HPP activity in *Arabidopsis*. IMPL2 has previously been shown to catalyse the dephosphorylation of D-inositol-1(or 3)-P and L-galactose-1-P *in vitro* (Torabinejad *et al.*, 2009), but whether this enzyme is involved in myoinositol or ascorbate metabolism *in planta* is unknown.

The final reactions of the His biosynthetic pathway comprise the two-step oxidation of histidinol to histidine (via the aldehyde intermediate histidinal) by histidinol dehydrogenase (HISN8). HISN8 activity was detected in extracts from several plant species by Wiater *et al.* (1971) and a HISN8 protein from wheat was partially purified by Wong and Mazelis (1981). The HISN8 purified from cabbage has been well characterized. This enzyme functions as a homodimer (Nagai and Scheidegger, 1991) and produces 1 mol His/2 mol NAD^+ reduced. Despite being localized in the plastid, the enzyme displays a 870-fold preference for NAD^+ over $NADP^+$ (Nagai *et al.*, 1993). The reaction proceeds via a bi-uni-uni-bi-ping-pong mechanism whereby histidinol binds first and NAD^+ second, with His the last product to be released (Kheirolomoom *et al.*, 1994). The oxidation of histidinol to histidinal is reversible, while that of histidinal to His is not, and a low K_d for the histidinal-HISN8 complex explains the failure to detect free histidinal (Kheirolomoom *et al.*, 1994). Site-directed mutagenesis has established that His261 is absolutely required as a Zn^{2+} binding ligand in the active site (Nagai *et al.*, 1993; Nagai and Ohta, 1994; Kanaori *et al.*, 1996). HISN8 is encoded by a single-copy gene in *Arabidopsis* (At5g63890), but the protein product has not been characterized to date.

14.4 Links Between Histidine Biosynthesis and Other Metabolic Pathways in Plants

His biosynthesis in plants is tightly linked to nucleotide metabolism through PRPP. PRPP is derived from ribose-5-phosphate generated by the pentose phosphate pathway and, in addition to serving as the first intermediate in His biosynthesis, is also required for the *de novo* biosynthesis and salvaging of purines, pyrimidines and the pyridine nucleotide cofactors NAD and NADP (Alifano *et al.*, 1996; Koslowsky *et al.*, 2008). In addition, the by-product AICAR generated at the branch point of the His biosynthetic pathway by HISN4 (see Fig. 14.2) enters the *de novo* purine biosynthetic pathway (Alifano *et al.*, 1996; Ward and Ohta, 1999).

In contrast, His biosynthesis is relatively isolated from other amino acid biosynthetic pathways. Besides His, the only other amino acid that derives part of its C skeleton from PRPP is Trp, as PRPP is required for the synthesis of *N*′-5′-phosphoribosyl anthranilate from anthranilate (Koslowsky *et al.*, 2008). However, the His biosynthetic pathway is also linked to Glu and Gln, as both are required as amino donors during the production of His (Fig. 14.2).

14.5 Sub-Cellular Localization and Evolution of Plant Histidine Biosynthetic Enzymes

While the HISN proteins are encoded by nuclear genes in plants, they all contain putative N-terminal plastid transit peptides, and the HISN5, HISN7 and HISN8 proteins have been demonstrated to localize to the plastid (Nagai *et al.*, 1993; Tada *et al.*, 1995; Petersen *et al.*, 2010), suggesting that His biosynthesis is confined to this organelle. Given that plastids evolved from cyanobacterial endosymbionts (Sagan, 1967), it might be expected that the His biosynthetic genes in plants are derived from cyanobacterial genes that were transferred to the host nuclear genome. However, a recent phylogenetic study has suggested that the His biosynthetic pathway is instead a mosaic comprising enzymes of different evolutionary origins. Maximum likelihood (ML) analyses of HISN protein sequences from bacteria, *Archaea* and eukaryotes revealed an unambiguous cyanobacterial origin only for HISN5. While a cyanobacterial provenance for HISN1 and HISN7 could not be excluded owing to poor resolution of the ML trees for these proteins, a non-cyanobacterial origin for the remaining HISN enzymes was well supported (Reyes-Prieto and Moustafa, 2012). For example, HISN6 appears to be derived from the *Chloroflexi* bacteria, while HISN8 is a host-derived enzyme (Reyes-Prieto and Moustafa, 2012), perhaps explaining its strong preference for NAD^+ over $NADP^+$ as a cofactor. Similar results were obtained for the enzymes involved in the synthesis of aromatic and Asp-derived amino acids (among others), suggesting that this phenomenon is not unique to His biosynthesis.

14.6 Regulation of Histidine Biosynthesis in Plants

While comparatively little is known yet about the regulation of His biosynthesis in plants, several lines of evidence have indicated that HISN1 plays a major role in this process. The majority of enzymes catalysing the first committed step of an amino acid biosynthetic pathway are subject to allosteric inhibition by the end product of the pathway (Less and Galili, 2008) and, accordingly, the activities of recombinant *Arabidopsis* HISN1A and HISN1B proteins were found to be inhibited by His (Ohta *et al.*, 2000). However, an eightfold difference in the IC_{50} value was observed, with HISN1A displaying greater sensitivity to feedback inhibition by His with an IC_{50} value of 40 μM His versus an IC_{50} value of 320 μM His for HISN1B (Ohta *et al.*, 2000). Constitutive overexpression of either *HISN1A* or *HISN1B* in transgenic *Arabidopsis* plants was sufficient to increase the levels of free His >40-fold in shoot tissues, while overexpression of any other *HISN* cDNA had no effect on free His concentrations (Rees *et al.*, 2009). Lastly, transgenic *Arabidopsis* plants constitutively expressing a PRPP synthase from *Ashbya gossypii* had increased levels of Trp but not of His (Koslowsky *et al.*, 2008), suggesting that PRPP was not limiting for His biosynthesis.

These data suggest that HISN1 controls the size of the free His pool, at least in *Arabidopsis*. The concentrations of free amino acids in *Arabidopsis* plastids have not been reported, so it is currently unclear as to whether free His levels in the plastids would be sufficient to exert feedback inhibition on HISN1 activity *in vivo*. Nevertheless, it appears that a sufficiently high degree of HISN1 overexpression is sufficient to overcome any such feedback regulation, resulting in a substantial increase in the levels of free His (Ingle *et al.*, 2005; Rees *et al.*, 2009). This is in marked contrast with the Trp and lysine (Lys) biosynthetic pathway, where the use of feedback insensitive forms of enzymes was required in order to increase free levels of these amino acids (Ufaz and Galili, 2008).

As is the case with the majority of amino acid biosynthetic pathways, little is known about the transcriptional regulation of His biosynthesis in plants. In *Arabidopsis*, the *HISN* genes are constitutively expressed in all tissues and at all developmental stages, suggesting that His is produced throughout the plant and at all stage of the life cycle (Muralla *et al.*, 2007). The transcriptional regulation of catabolic genes is thought to play an important role in the modulation of amino acid metabolism in response to environmental stimuli (Less and Galili, 2008), although the metabolic route of His catabolism in plants is currently unknown. In animals and microorganisms, His is decarboxylated to histamine by the enzyme His decarboxylase, or deaminated to urocanate by His ammonia lyase and subsequently metabolized to glutamate via 4-imidazolone-5-propionate. A putative His decarboxylase from *Arabidopsis* was later shown to have activity against Ser but not His (Rontein *et al.*, 2001), and no His ammonia lyase enzyme has yet been reported from plants.

While the His biosynthetic pathway is intimately linked to nucleotide metabolism, it is unclear to what extent His biosynthesis is coordinated with nucleotide biosynthesis in plants. In fungi, the regulation of His and purine biosynthesis is linked via the BAS1p-BAS2p transcriptional complex, which mediates the de-repression of several His biosynthetic genes in response to adenine limitation (Springer *et al.*, 1996). The biosynthesis of His in fungi is also coordinated with that of other amino acids via the general nitrogen control mechanism. The bZIP transcription factor GCN4 mediates the coordinated de-repression of genes encoding amino acid biosynthetic enzymes (from all pathways except Cys) in response to limitation for a single amino acid (Niederberger *et al.*, 1981). Six of the *HISN* genes in *Arabidopsis* contain GCN4-like binding elements in their upstream regions (Stepansky and Leustek, 2006), which may be indicative of coordinate regulation of expression, but the regulatory protein(s) that bind to this motif have not been identified, and no GCN4 homologue has yet been identified in plants.

The existence of a general nitrogen control mechanism in plants has been proposed, based on the observations that treatment of *Arabidopsis* with an inhibitor of HISN5 activity led to increased expression of *HISN8* and several genes involved in the biosynthesis of Lys, aromatic amino acids and purines (Guyer *et al.*, 1995), while the *apg10* mutant (a weak *HISN3* allele) had increased concentrations of most amino acids (Noutoshi *et al.*, 2005). Similarly, a correlation between the concentrations of free His and eight other minor amino acids has been observed in potato shoot tissue, though not in wheat (Noctor *et al.*, 2002). However, transgenic *Arabidopsis* plants with greatly elevated free His content (up to 20% of the total free amino acid pool) displayed little or no differences in the concentrations of any other amino acid (Rees *et al.*, 2009).

14.7 Role of Histidine in Nickel Hyperaccumulation in Plants

The threshold concentration for Ni hyperaccumulation in plants is 0.1% on a dry biomass basis (Reeves and Baker, 2000). Over 300 species of plant have been demonstrated to hyperaccumulate Ni in their above-ground tissues, with the genus *Alyssum* constituting the largest group, with 48 species capable of accumulating Ni to over 3% of dry shoot biomass identified to date (Reeves and Baker, 2000). The ecological function of metal hyperaccumulation remains controversial, but has been suggested to protect plants against herbivory and pathogen attack (Fones *et al.*, 2010). A role for His in Ni hyperaccumulation was first suggested by Krämer *et al.* (1996), who demonstrated a linear and proportional increase of free His in the xylem of *A. lesbiacum* and *A. murale* across a range of Ni concentrations. *In vivo* coordination of Ni by His was demonstrated in the xylem, root and shoot tissues of these hyperaccumulators, while the application of exogenous His to the non-accumulator *A. montanum* reduced the toxic effects of Ni on this species, and led to increased flux of Ni into the xylem (Krämer *et al.*, 1996). Measurement of steady state His levels in the root tissues of congeneric hyperaccumlator and non-accumulator species has demonstrated that the Ni hyperaccumulators have constitutively elevated His contents (Persans *et al.*, 1999; Ingle *et al.*, 2005, Richau *et al.*, 2009). In *A. lesbiacum*, mRNA levels of *HIS1NA* and *HIS1NB* were found to be constitutively elevated in the root tissue compared with those in *A. montanum*, while the other *HISN* genes analysed showed only modest differences in expression (Ingle *et al.*, 2005). Constitutive expression of the *A. lesbiacum HISN1B* cDNA in *Arabidopsis* was sufficient to increase free His levels by up to 15-fold, supporting the hypothesis that high constitutive expression of *HISN1* is responsible for the larger pool of free His in the root tissue of *A. lesbiacum*. His therefore appears to function as a ligand for Ni and promotes the xylem loading of this metal ion in Ni hyperaccumulators. The observation that His inhibits vacuolar uptake of Ni in root-derived tonoplast vesicles of *Thlaspi caerulescens* (Richau *et al.*, 2009), suggests one possible mechanism by which this may occur.

14.8 Conclusions

The enzymes catalysing the reactions of His biosynthesis have now been identified in plants, and several lines of evidence have implicated HISN1 as an important regulator of this metabolic pathway. However, a number of questions remain unanswered; for example, how is the biosynthesis of His balanced with other metabolic demands in plants, and what is the metabolic route of His catabolism in plants? Answering these questions is a priority for future research on this amino acid.

References

Akashi, H. and Gojobori, T. (2002) Metabolic efficiency and amino acid composition in the proteomes of *Escherichia coli* and *Bacillus subtilis*. *Proceedings of the National Academy of Sciences of the United States of America* 99, 3695–3700.

Alifano, P., Fani, R., Liò, P., Lazcano, A., Bazzicalupo, M., Carlomagno, M.S. and Bruni, C.B. (1996) Histidine biosynthetic pathway and genes: structure, regulation, and evolution. *Microbiological Reviews* 60, 44–69.

Ames, B.N., Garry, B. and Herzenberg, L.A. (1960) The genetic control of the enzymes of histidine biosynthesis in *Salmonella typhimurium*. *Journal of Molecular Biology* 33, 533–546.

Bikard, D., Patel, D., Le Mette, C., Giorgi, V., Camilleri, C., Bennett, M.J. and Loudet, O. (2009) Divergent evolution of duplicate genes leads to genetic incompatibilities within *A. thaliana*. *Science* 323, 623–626.

Chiariotti, L., Alifano, P., Carlomagno, M.S. and Bruni, C.B. (1986) Nucleotide sequence of the *Escherichia coli hisD* gene and of the *Escherichia coli* and *Salmonella typhimurium hisIE* region. *Molecular and General Genetics* 203, 382–388.

Donahue, T.F., Farabaugh, P.J. and Fink, G.R. (1982) The nucleotide sequence of the *HIS4* region of yeast. *Gene* 18, 47–59.

El Malki, F., Frankard, V. and Jacobs, M. (1998) Molecular cloning and expression of a cDNA sequence encoding histidinol phosphate aminotransferase from *Nicotiana tabacum*. *Plant Molecular Biology* 37, 1013–1022.

Fersht, A. (1999) *Structure and Mechanism in Protein Science*. W.H. Freeman, New York.

Fones, H., Davis, C.A.R., Rico, A., Fang, F., Smith, J.A.C. and Preston, G.M. (2010) Metal hyperaccumulation armors plants against disease. *PLoS Pathogens* 6: e1001093.

Fraústo da Silva, J.J.R. and Williams, R.J.P. (2001) *The Biological Chemistry of the Elements: The Inorganic Chemistry of Life*. Oxford University Press, Oxford, UK.

Fujimori, K. and Ohta, D. (1998a) Isolation and characterization of a histidine biosynthetic gene in *Arabidopsis* encoding a polypeptide with two separate domains for phosphoribosyl-ATP pyrophosphohydrolase and phosphoribosyl-AMP cyclohydrolase. *Plant Physiology* 118, 275–283.

Fujimori, K. and Ohta, D. (1998b) An *Arabidopsis* cDNA encoding a bifunctional glutamine amidotransferase/cyclase suppresses the histidine auxotrophy of a *Saccharomyces cerevisiae his7* mutant. *FEBS Letters* 428, 229–234.

Fujimori, K., Tada, S., Kanai, S. and Ohta, D. (1998) Molecular cloning and characterization of the gene encoding N′-[(5′-phosphoribosyl)-formimino]-5-aminoimidazole-4-carboxamide ribonucleotide (BBM II) isomerase from *Arabidopsis thaliana. Molecular and General Genetics* 259, 216–223.

Glynn, S.E., Baker, P.J., Sedelnikova, S.E., Davies, C.L., Eadsforth, T.C., Levy, C.W., Rodgers, H.F., Blackburn, G.M., Hawkes, T.R., Viner, R. and Rice, D.W. (2005) Structure and mechanism of imidazoleglycerol-phosphate dehydratase. *Structure* 13, 1809–1817.

Guyer, D., Patton, D. and Ward, E. (1995) Evidence for cross-pathway regulation of metabolic gene expression in plants. *Proceedings of the National Academy of Sciences of the United States of America* 92, 4997–5000.

Harding, M.M. (2004) The architecture of metal coordination groups in proteins. *Acta Crystallographica Section D* 60. 849–859.

Ingle, R.A., Mugford, S.T., Rees, J.D., Campbell, M.M. and Smith, J.A.C. (2005) Constitutively high expression of the histidine biosynthetic pathway contributes to nickel tolerance in hyperaccumulator plants. *The Plant Cell* 17, 2089–2106.

Kanaori, K., Uodome, N., Nagai, A., Ohta, D., Ogawa, A., Iwasaki, G. and Nosaka, A.Y. (1996) Cd-113 nuclear magnetic resonance studies of cabbage histidinol dehydrogenase. *Biochemistry* 35, 5949–5954.

Kheirolomoom, A., Mano, J., Nagai, A., Ogawa, A., Iwasaki, G. and Ohta, D. (1994) Steady-state kinetics of cabbage histidinol dehydrogenase. *Archives of Biochemistry and Biophysics* 312, 493–500.

Koslowsky, S., Riegler, H., Bergmüller, E. and Zrenner, R. (2008) Higher biomass accumulation by increasing phosphoribosylpyrophosphate synthetase activity in *Arabidopsis thaliana* and *Nicotiana tabacum*. *Plant Biotechnology Journal* 6, 281–294.

Krämer, U., Cotter-Howells, J.D., Charnock, J.M., Baker, A.J.M. and Smith, J.A.C. (1996) Free histidine as a metal chelator in plants that accumulate nickel. *Nature* 379, 635–638.

Kuenzler, M., Balmelli, C., Egli, C.M., Paravinci, G. and Braus, G.H. (1993) Cloning, primary structure, and regulation of the *HIS7* gene encoding a bifunctional glutamine amidotransferase:cyclase from *Saccharomyces cerevisiae*. *Journal of Bacteriology* 175, 5548–5558.

Less, H. and Galili, G. (2008) Principal transcriptional programs regulating plant amino acid metabolism in response to abiotic stresses. *Plant Physiology* 147, 316–330.

Mano, J., Hatano, M., Koizumi, S., Tada, S., Hashimoto, M. and Scheidegger, A. (1993) Purification and properties of a monofunctional imidazoleglycerol-phosphate dehydratase from wheat. *Plant Physiology* 103, 733–739.

Marineo, S., Cusimano, M.G., Limauro, D., Coticchio, G. and Puglia, A.M. (2008) The histidinol phosphate phosphatase involved in histidine biosynthetic pathway is encoded by *SCO5208* (*hisN*) in *Streptomyces coelicolor* A3(2). *Current Microbiology* 56, 6–13.

Meinke, D., Muralla, R., Sweeney, C. and Dickerman, A. (2008) Identifying essential genes in *Arabidopsis thaliana*. *Trends in Plant Science* 13, 483–491.

Mo, X., Zhu, Q., Li, X., Li, J., Zeng, Q., Rong, H., Zhang, H. and Wu, P. (2006) The *hpa1* mutant of *Arabidopsis* reveals a crucial role of histidine homeostasis in root meristem maintenance. *Plant Physiology* 141, 1425–1435.

Moore, R.C. and Purugganan, M.D. (2003) The early stages of genome duplication. *Proceedings of the National Academy of Sciences of the United States of America* 100, 15682–15687.

Mori, I., Fonnepfister, R., Matsunaga, S., Tada, S., Kimura, Y., Iwasaki, G., Mano, J., Hatano, M., Nakano, T., Koizumi, S., Scheidegger, A., Hayakawa, K. and Ohta, D. (1995) A novel class of herbicides – specific inhibitors of imidazoleglycerol phosphate dehydratase. *Plant Physiology* 107, 719–723.

Mormann, S., Lömker, A., Rückert, C., Gaigalat, L., Tauch, A., Pühler, A. and Kalinowski, J. (2006) Random mutagenesis in *Corynebacterium glutamicum* ATCC 13032 using an IS*6100*-based transposon vector identified the last unknown gene in the histidine biosynthesis pathway. *BMC Genomics* 7, 205.

Münzer, S., Hashimoto-Kumpaisal, R., Scheidegger, A. and Ohta, D. (1992) Purification and properties of ATP-phosphoribosyl transferase from wheat germ. In: Murata, N. (ed.) *Research in Photosynthesis*. Kluwer, Dordrecht, The Netherlands, pp. 91–94.

Muralla, R., Sweeney, C., Stepansky, A., Leustek, T. and Meinke, D. (2007) Genetic dissection of histidine biosynthesis in *Arabidopsis*. *Plant Physiology* 144, 890–903.

Nagai, A. and Ohta, D. (1994) Histidinol dehydrogenase loses its catalytic function through the mutation of His261→Asn due to its inability to ligate the essential Zn. *The Journal of Biochemistry* 115, 22–25.

Nagai, A. and Scheidegger, A. (1991) Purification and characterization of histidinol dehydrogenase from cabbage. *Archives of Biochemistry and Biophysics* 284, 127–132.

Nagai, A., Suzuki, K., Ward, E., Moyer, M., Mano, J., Beck, J., Tada, S., Hashimoto, M., Chang, J., Ryals, J., Scheidegger, A. and Ohta, D. (1993) Histidinol dehydrogenase in higher plants. Purification, cloning and expression. In: Murata, N. (ed.) *Research in Photosynthesis*. Kluwer, Dordrecht, The Netherlands, pp. 95–98.

Niederberger, P., Miozzari, G. and Hutter, R. (1981) Biological role of the general control of amino acid biosynthesis in *Saccharomyces cerevisiae*. *Molecular and Cellular Biology* 1, 584–593.

Noctor, G., Novitskaya, L., Lea, P.J. and Foyer, C.H. (2002) Co-ordination of leaf minor amino acid contents in crop species: significance and interpretation. *Journal of Experimental Botany* 53, 939–945.

Noutoshi, Y., Ito, T. and Shinozaki, K. (2005) *ALBINO AND PALE GREEN 10* encodes BBMII isomerase involved in histidine biosynthesis in *Arabidopsis thaliana*. *Plant and Cell Physiology* 46, 1165–1172.

Ohta, D., Fujimori, K., Mizutani, M., Nakayama, Y., Kunpaisal-Hashimoto, R., Munzer, S. and Kozaki, A. (2000) Molecular cloning and characterization of ATP-phosphoribosyl transferase from *Arabidopsis*, a key enzyme in the histidine biosynthetic pathway. *Plant Physiology* 122, 907–914.

Persans, M.W., Yan, X., Patnoe, J.-M.M.L., Krämer, U. and Salt, D.E. (1999) Molecular dissection of the role of histidine in nickel hyperaccumulation in *Thlaspi goesingense* (Halacsy). *Plant Physiology* 121, 1117–1126.

Petersen, L.N., Marineo, S., Mandala, S., Davids, F., Sewell, B.T. and Ingle, R.A. (2010) The missing link in plant histidine biosynthesis: *Arabidopsis myoinositol monophosphatase-like2* encodes a functional histidinol-phosphate phosphatase. *Plant Physiology* 152, 1186–1196.

Rees, J.D., Ingle, R.A. and Smith, J.A.C. (2009) Relative contributions of nine genes in the pathway of histidine biosynthesis to control of free histidine concentrations in *Arabidopsis thaliana*. *Plant Biotechnology Journal* 7, 499–511.

Reeves, R.D. and Baker, A.J.M. (2000) Metal-accumulating plants. In: Raskin, I. and Ensley, B.D. (eds) *Phytoremediation of Toxic Metals: Using Plants to Clean Up the Environment*. Wiley, New York, pp. 193–229.

Reyes-Prieto, A. and Moustafa, A. (2012) Plastid-localized amino acid biosynthetic pathways of *Plantae* are predominantly composed of non-cyanobacterial enzymes. *Scientific Reports* 2, 955.

Richau, K.H., Kozhevnikova, A.D., Seregin, I.V., Vooijs, R., Koevoets, P.L.M., Smith, J.A.C., Ivanov, V.B. and Schat, H. (2009) Chelation by histidine inhibits the vacuolar sequestration of nickel in roots of the hyperaccumulator *Thlaspi caerulescens*. *New Phytologist* 183, 106–116.

Rontein, D., Nishida, I., Tashiro, G., Yoshioka, K., Wu, W.-I., Voelker, D.R., Basset, G. and Hanson, A.D. (2001) Plants synthesize ethanolamine by direct decarboxylation of serine using a pyridoxal phosphate enzyme. *The Journal of Biological Chemistry* 276, 35523–35529.

Roth, J.R. and Ames, B.N. (1966) Histidine regulatory mutants having altered histidyl-tRNA synthetase. *Journal of Molecular Biology* 22, 325–334.

Sagan, L. (1967) On the origin of mitosing cells. *Journal of Theoretical Biology* 14, 225–274.

Seligmann, H. (2003) Cost-minimization of amino acid usage. *Journal of Molecular Evolution* 56, 151–161.

Springer, C., Künzler, M., Balmelli, T. and Braus, G.H. (1996) Amino acid and adenine cross-pathway regulation act through the same 5′-TGACTC-3′ motif in the yeast *HIS7* promoter. *The Journal of Biological Chemistry* 271, 29637–29643.

Stepansky, A. and Leustek, T. (2006) Histidine biosynthesis in plants. *Amino Acids* 30, 127–142.

Swire, J. (2007) Selection on synthesis cost affects interprotein amino acid usage in all three domains of life. *Journal of Molecular Evolution* 64, 558–571.

Tada, S., Hatano, M., Nakayama, Y., Volrath, S., Guyer, D., Ward, E. and Ohta, D. (1995) Insect cell expression of recombinant imidazoleglycerolphosphate dehydratase of *Arabidopsis* and wheat and inhibition by triazole herbicides. *Plant Physiology* 109, 153–159.

Tada, S., Volrath, S., Guyer, D., Scheidegger, A., Ryals, J., Ohta, D. and Ward, E. (1994) Isolation and characterization of cDNAs encoding imidazoleglycerolphosphate dehydratase from *Arabidopsis thaliana*. *Plant Physiology* 105, 579–583.
Torabinejad, J., Donahue, J.L., Gunesekera, B.N., Allen-Daniels, M.J. and Gillaspy, G.E. (2009) VTC4 is a bifunctional enzyme that affects myoinositol and ascorbate biosynthesis in plants. *Plant Physiology* 150, 951–961.
Ufaz, S. and Galili, G. (2008) Improving the content of essential amino acids in crop plants: goals and opportunities. *Plant Physiology* 147, 954–961.
Voll, M.J., Appella, E. and Martin, R. (1967) Purification and composition studies of phosphoribosyladenosine triphosphate: pyrophosphate phosphoribosyltransferase, the first enzyme of histidine biosynthesis. *The Journal of Biological Chemistry* 242, 1760–1767.
Ward, E. and Ohta, D. (1999) Histidine biosynthesis. In: Singh, B.K. (ed.) *Plant Amino Acids: Biochemistry and Biotechnology*. Marcel Dekker, New York, pp. 293–303.
Wiater, A., Krajewska-Grynkiewicz, K. and Kloptowski, T. (1971) Histidine biosynthesis and its regulation in plants. *Acta Biochimica Polonica* 18, 299–307.
Wilkinson, K.W., Baker, P.J., Rice, D.W., Rodgers, H.F., Stillman, T.J., Hawkes, T., Thomas, P. and Edwards, L. (1995) Crystallization and analysis of the subunit assembly and quaternary structure of imidazoleglycerol phosphate dehydratase from *Saccharomyces cerevisiae*. *Acta Crystallographica Section D* 51, 845–847.
Winkler, M. (1987) Biosynthesis of histidine. In: Neidhart, F., Ingraham, J., Low, K., Magasanik, B., Schaechter, M. and Umbarger, H. (eds) Escherichia coli *and* Salmonella typhimurium: *Cellular and Molecular Biology*. American Society for Microbiology, Washington, DC, pp. 395–411.
Wong, Y.S. and Mazelis, M. (1981) Detection and properties of L-histidinol dehydrogenase in wheat germ. *Phytochemistry* 20, 1831–1834.

15 Amino Acid Synthesis Under Abiotic Stress

E. Planchet* and A.M. Limami
University of Angers, Angers, France

15.1 Abstract

Almost all plants have to deal with climatic factors and to develop strategies to adapt their metabolism, which may have been adversely affected, in order to acclimate and survive under these unfavourable growth conditions. Mechanisms of plant stress tolerance involve some changes in gene expression, protein modification and metabolic pathways. More particularly, plant amino acid (AA) composition is modified by environmental conditions and characterized by an elevated accumulation of specific AAs involved in plant stress tolerance. These free AAs are synthesized by various distinct metabolic networks and accumulate differentially in plant species exposed to various stresses. In response to these environmental constraints, AA metabolism plays an important regulatory role, not only because AAs are constituents of proteins but also because free AAs are potential regulatory and signalling molecules, and precursors for energy-associated metabolites, as well numerous secondary metabolites, that have several functions in plant growth and adaptive responses to various stresses. The AA synthesis pathways are complex and undergo tight regulation, particularly in respect of the metabolism of three specific AA family pathways when plants are exposed to abiotic stresses. First, the glutamate family pathway is strongly activated under stress, leading to the accumulation of proline and γ-aminobutyric acid (GABA), two stress-related molecules involved in metabolic responses to stress. Secondly, stimulation of the metabolism of the pyruvate family pathway mainly allows the production and accumulation of alanine, which is the most abundant amino acid under oxygen deficiency; it also leads to the accumulation of branched-chain amino acids, such as leucine and valine, which could play the role of compatible solutes and provide an alternative source of respiratory substrates. Thirdly, regulation of biosynthetic and catabolic fluxes through the aspartate family pathway appears to be essential for generating energy, mainly via lysine catabolism, as stress causes energy deprivation.

15.2 Introduction

In the natural environment, plants often cope with adverse environmental constraints, such as drought, salinity, cold, heat, flooding, nutrient deficiency, radiation, chemicals, pollutants and oxidative stress. These abiotic stresses can delay plant growth and development, reduce agronomic yields and, sometimes in extreme ways, cause the plant death. To grow under unfavourable conditions, plants have to sense and respond to stress by adapting their physiological processes, in particular their carbohydrate and nitrogen metabolism; this requires extensive changes, including alterations in the metabolic networks associated with amino acids (AAs). Among these AAs,

*E-mail address: elisabeth.planchet@univ-angers.fr

glutamate, glutamine, asparagine and aspartate are the major organic nitrogen compounds in higher plants in that they provide storage and transport forms for nitrogen, and also act as the precursors of several metabolic pathways. An overall accumulation of AAs is also the most common response to abiotic stresses, and in plant tissues exposed to such stresses, concentrations of AAs may rise up to the millimolar range (Singh *et al.*, 1972; Planchet *et al.*, 2011). This increase in AAs could be due to AA synthesis or interconversion, stress-reduced protein synthesis and/or stress-induced protein degradation. Indeed, abiotic stresses induce irreversible damage to plant proteins, which are then degraded by proteases (Vierstra, 1996), although a gene expression study by Less and Galili (2008) in *Arabidopsis* concluded that increased protease production is probably not responsible for the increase in free AA accumulation under stress conditions.

In many plant species, a characteristic linear increase in AAs is observed under abiotic stresses, followed by a reduction of AA concentration upon recovery from stress. Although this AA accumulation is ubiquitous in higher plants, the concentration of AAs is dependent on the plant species, tissue, growth and developmental status, and the nature and longevity of the specific abiotic stress. Free AAs have been shown to have functional roles in plant stress tolerance by: (i) acting as compatible osmolytes; (ii) being involved in pH regulation and detoxification of reactive oxygen species (ROS); (iii) being a nitrogen and carbon reserve – mainly for the synthesis of specific enzymes and precursors of various secondary metabolites (such as flavonoids and lignins); and (iv) being an available stock of AAs that would be useful during recovery from stress. Free AAs may also act as regulatory and signalling molecules. These protecting roles of AAs are also supported by the fact that salt-tolerant plants accumulate higher free AA contents than salt-sensitive plants and show a correlation between stress tolerance and AA levels. However, if the increase in the level of some AAs has been reported to lead to beneficial effects during stress acclimation, in some cases AAs accumulate as a sink of excess nitrogen as a result of a stress-induced reduction of growth rate. In the latter case, the increase of AA concentrations indicates damage to cell functioning rather than an adaptive response to environmental stresses (Munns, 2002). The AAs that are accumulated include not only the 20 involved in protein synthesis, but also free AAs from AA metabolism, such as polyamines (putrescine, spermine and spermidine are the most common). It has been suggested that increased polyamine levels following various environmental stresses are involved in the acquisition of stress tolerance (Gill and Tuteja, 2010). The synthesis and roles of polyamines are discussed in the Chapters 9 and 10.

Following the disruption and alteration of plant metabolic homeostasis due to environmental stresses, the expression of numerous stress-inducible genes and their associated protein activities are modified (upregulated and/or downregulated), leading to changes in plant metabolism. Genes encoding AA catabolic enzymes have been reported to be stimulated and to respond faster to abiotic stresses than genes encoding biosynthetic enzymes (Less and Galili, 2008). These catabolic enzymes would then play major regulatory roles in AA metabolism and contribute to changes in fluxes through distinct AA metabolic pathways to allow metabolic adjustments. It is important to note that following different stress conditions, several metabolic pathways respond differentially, according to their different physiological functions. However, plant stress tolerance is intimately related to energy availability, as most of the stresses affect the overall energy status of the plants. To avoid a large reduction in the cellular energy pool, AA metabolism acts in concert with the glycolysis in the cytoplasm and the Krebs tricarboxylic acid (TCA) cycle in the mitochondria, two pathways that provide the carbon skeletons for AA biosynthesis. AA synthetic pathways are particularly complex because they are subjected to strict controls and are very often 'connected', i.e. several AAs can be derived from the same precursor. This chapter presents an overview of the synthesis and the involvement of AAs in plant response to abiotic stresses, by focusing on the metabolism of three major AA family pathways in plants: the glutamate, pyruvate and aspartate family pathways (Fig. 15.1). The involvement of AAs from the serine family (glycine, serine and cysteine) and the aromatic amino acid family (called also the shikimate family, which includes tyrosine, phenylalanine and tryptophan) pathways in response to abiotic stresses will not be discussed.

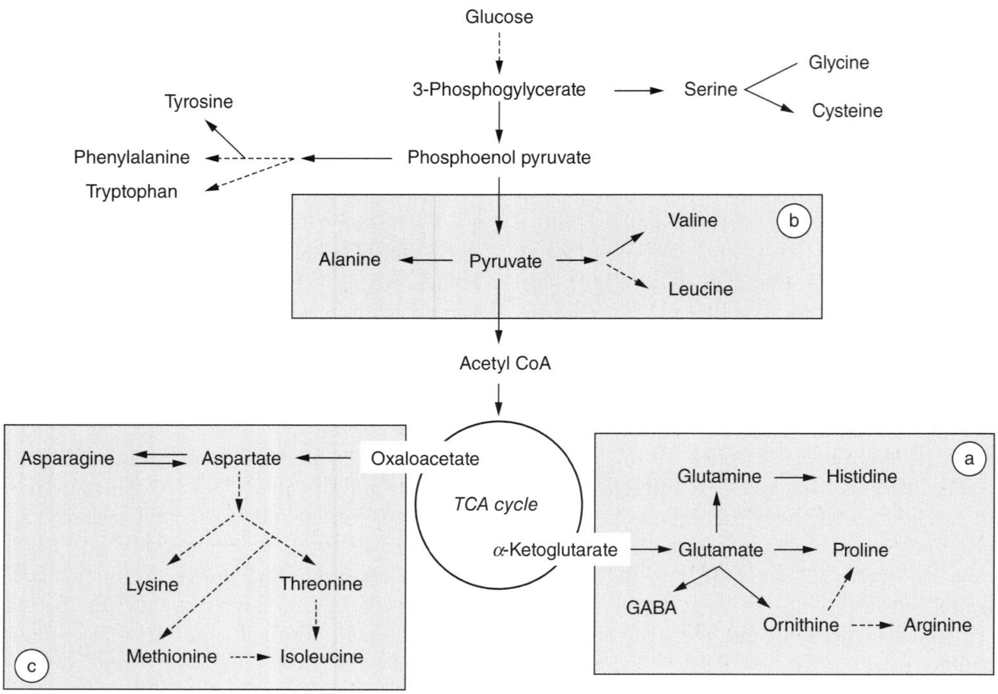

Fig. 15.1. Pathways involved in the biosynthesis of amino acids in higher plants. Grey boxes represent: (a) the glutamate family pathway; (b) the pyruvate family pathway; and (c) the aspartate family pathway. GABA, γ-aminobutyric acid; TCA, tricarboxylic acid cycle. (Adapted from Mohapatra *et al*., 2010.)

15.3 The Glutamate Family Pathway

Amino acids belonging to the glutamate family are glutamate itself and AAs that are synthetized from the carbon skeleton and α-amino group of glutamate, i.e. glutamine, proline, arginine and γ-aminobutyric acid (GABA). Glutamate is a key amino acid in cellular nitrogen metabolism in higher plants due to its central position in nitrogen assimilation and its role as a substrate for the synthesis of various important metabolites. Pathways of glutamate biosynthesis and metabolism in plants have been reviewed in detail (Forde and Lea, 2007). In addition to its key role in nitrogen metabolism, glutamate has also been reported to function as a protective osmolyte and to act as a signalling and stress-related molecule in plants. Glutamate homeostasis, by playing a vital role in carbon and nitrogen balance, appears to be essential for the primary metabolism and the functioning of plant cells. The concentrations of glutamate in plant tissues have been reported to change as a function of the stage of development, the diurnal cycle, nitrogen nutrition and/or metabolite feeding, but few data are available concerning the pool of free glutamate in cells after response to abiotic stress. Although glutamate metabolism is centrally involved in the process of assimilation and remobilization of nitrogen, it also has an important function in the development of adaptive metabolites, such as proline and GABA, developing in response to environmental stress.

15.3.1 Proline accumulation and adaptive responses to stress

Proline, the most water soluble of the plant AAs, is the only one of the proteinogenic amino acids in which the α-amino group is present as a secondary amine. The metabolism of proline in plants has been mainly studied in response to environmental stresses. Proline accumulation was first reported to occur after wilting in perennial rye grass (*Lolium perenne*) (Kemble and MacPherson, 1954). Since then, many studies have shown an increase in proline content in most of the higher plants under various environmental constraints, including wilting and other stresses such as drought, salinity, high and low temperature, nutrient deficiency, anaerobiosis, heavy metals, UV irradiation and oxidative stress (Munns, 2002; Sharma and Dietz, 2006; Verbruggen and Hermans, 2008; Planchet *et al.*, 2011). According to the plant species, proline content can accumulate to high levels, more than 100-fold above the control level, in seedlings exposed to low water potential (Verslues and Sharma, 2010) and also in larger amounts than other plant AAs. This capacity to accumulate proline has been correlated with improved plant performance under abiotic stresses (Singh *et al.*, 1972; Nanjo *et al.*, 1999).

Although there is a debate about whether proline accumulation is a stress symptom and/or a result of reduced growth and metabolism rather than a stress-adaptive response (Verbruggen and Hermans, 2008), several likely roles of proline have been reported in plant stress tolerance. These functions are mainly dependent on the amount of proline accumulated under various environmental constraints. Owing to its properties as a zwitterion and its solubility, proline can be assimilated as a compatible osmolyte, a putative role supported by the fact that it has been reported to accumulate in the cytosol in response to hyperosmotic stress (Yoshiba *et al.*, 1997). Indeed, proline accumulation induces a decrease in cellular osmotic potential that enables the maintenance of an equilibrium in plant water potential. There is evidence that in its role as an osmoprotectant, proline is mainly accumulated in the cytosol and the chloroplasts, while other solutes (organic acids, sugars, K^+) are accumulated in the vacuole (Verslues and Sharp, 1999). Proline has also been shown to act as a regulator of cellular redox potential by contributing to redox homeostasis, owing to the involvement of proline synthesis in mechanisms for alleviating cytosolic acidosis and in maintaining $NADP^+$/NADPH ratios (Hare and Cress, 1997). Proline plays a role as a molecular chaperone as well in that it protects the cellular structure of proteins and membranes during dehydration (Verbruggen and Hermans, 2008; Szabados and Savouré, 2010). As a contribution to heavy metal tolerance, proline accumulation has been reported to protect the activities of some enzyme (for example, nitrate reductase and glucose-6-phosphate dehydrogenase) against the effects of specific toxic heavy metals by forming proline-metal complexes (Sharma and Dietz, 2006). In addition, proline has been demonstrated to scavenge free radicals (ROS) and protect cells from oxidative damage (Alia *et al.*, 2001). Recently, Signorelli *et al.* (2013) have questioned this capacity for proline to act as a non-enzymatic antioxidant under stress, as these authors demonstrated that proline accumulation does not play any role in the scavenging of singlet oxygen in plants.

Apart from this functional role in stress tolerance, proline could, in addition, act as a player in signalling systems. Certainly, it has been recognized to be a signalling molecule in plant development and also as a component of signalling networks influencing adaptive responses and controlling mitochondrial functions (Szabados and Savouré, 2010). In agreement with this, proline has been shown to induce the expression of some salt-stress responsive genes and to improve the salt tolerance of *Pancratium maritimum* by upregulating stress-protective proteins and protecting the protein turnover machinery against stress damage (Khedr *et al.*, 2003).

Interestingly, although under most abiotic stresses the increase in proline level leads to stress tolerance, proline accumulation induced by heat stress results in plant sensitivity in tobacco and *Arabidopsis* seedlings (Rizhsky *et al.*, 2004; Lv *et al.*, 2011). These data demonstrate a contradictory relation between proline accumulation and thermotolerance. Remarkably, the proline accumulation that is observed under drought stress did not occur during a combination of drought and heat stress. This was due to an increased toxicity to plant cells of the proline that was induced by heat stress. Under these conditions, it has been suggested that sucrose replaced proline as the major osmoprotectant (Rizhsky *et al.*, 2004). A similar observation was made for barley in which proline content was not correlated with salt tolerance (Chen *et al.*, 2007), suggesting that the relationship between stress tolerance and proline accumulation is more complex than it may at first seem.

15.3.2 The regulation of proline metabolism during stress

Intracellular proline content depends on a tight regulation between its biosynthesis and catabolism (Hare *et al.*, 1999; Szabados and Savouré, 2010). In plants, proline is synthesized from two precursors, glutamate and arginine/ornithine. Based on *in vivo* labelling experiments with ^{14}C-labelled precursors (Morris *et al.*, 1969), $^{15}NH_4^+/^{15}NO_3^-$ (Rhodes *et al.*, 1986) or [^{13}C]-glutamate (Heyser *et al.*, 1989), the glutamate pathway has been described as the main pathway of proline accumulation in plants during osmotic stress. The glutamate pathway allows the conversion of glutamate to proline by the following successive reactions: (i) the reduction of glutamate to glutamate semialdehyde (GSA) catalysed by the bifunctional enzyme Δ^1-pyrroline-5-carboxylate synthetase (P5CS; EC 2.7.2.11/1.2.1.41); (ii) the spontaneous conversion of GSA to pyrroline-5-carboxylate (P5C); and (iii) the further reduction of P5C to proline catalysed by P5C reductase (P5CR; EC 1.5.1.2). Two *P5CS* genes, *P5CS1* and *P5CS2*, have been reported to be present in *Arabidopsis thaliana* and regulated differentially according to plant organ and osmotic stress, suggesting that these P5CS isoforms have different metabolic functions (Strizhov *et al.*, 1997). The proline synthesis reactions take place in the cytosol or chloroplasts, while proline catabolism occurs in the mitochondria. Proline is degraded following a reversed process of proline synthesis in a reaction catalysed by proline dehydrogenase (ProDH; EC 1.5.99.8; also referred to as proline oxidase) and P5C dehydrogenase (P5CDH; EC 1.5.1.12).

Under abiotic stress, proline accumulation appears to be the concomitant result of an up-regulation of proline biosynthesis and a down-regulation of proline catabolism (Savouré *et al.*, 1995). Indeed, under stress conditions, proline content is correlated with the expression of the *P5CS* gene, but not the *P5CR* gene, and with an increased stress tolerance (Yoshiba *et al.*, 1995; Székely *et al.*, 2008). It is well known that P5CS is the rate-limiting enzyme in proline synthesis in higher plants, and that its activity is inhibited by proline through feedback inhibition (Hong *et al.*, 2000). Using transgenic plants overexpressing the *P5CS* gene, several studies have shown that these *P5CS* overexpressors exhibit an increase in intracellular proline and a better tolerance to various stresses, such as high salinity, freezing and osmotic stress (Kishor *et al.*, 1995; Nanjo *et al.*, 1999; Hong *et al.*, 2000). In agreement with this finding, *Arabidopsis P5CS1* knockout plants displayed a lower proline content and were more sensitive to salinity (Székely *et al.*, 2008). A net proline accumulation during abiotic stresses could be also attributed to a decrease in proline utilization in protein synthesis and a decrease in proline oxidation to glutamate. Accordingly, using an alternative strategy for increasing intracellular proline content, Nanjo *et al.* (1999) have demonstrated that antisense suppression of proline catabolism (ProDH antisense) in *Arabidopsis* plants resulted in the constitutive accumulation of proline and improved tolerance to freezing and salinity. It should be noted that proline degradation is enhanced during recovery from stress. This proline catabolism provides reducing equivalents to maintain mitochondrial oxidative phosphorylation and, therefore, to generate ATP for recovery from stress and to repair damage induced by environmental cues (Hare and Cress, 1997).

Apart from the glutamate pathway, the ornithine/arginine pathway offers an alternative for proline production. In mitochondria, ornithine, a common intermediate in proline biosynthesis and degradation, can be converted into P5C via the action of ornithine δ-aminotransferase (OAT; EC 2.6.1.13). Usually, the conversion of ornithine to glutamate is an essential way to recover nitrogen stored or transported as arginine. However, the involvement of OAT in stress-induced proline accumulation appears to be contradictory and questionable (Xue *et al.*, 2009). On the one hand, transgenic *Arabidopsis* overexpressing the *δ-OAT* gene in tobacco and rice accumulated a higher level of proline and showed a better stress tolerance (Roosens *et al.*, 1998). On the other hand, *Arabidopsis* mutants deficient in the *δ-oat* gene accumulated proline at a similar level and exhibited a salt stress tolerance (Funck *et al.*, 2008), indicating that the biosynthesis of proline occurs predominantly via the glutamate pathway and appears to be independent of glutamate produced by the ornithine/arginine pathway.

The regulation of proline accumulation has been also described to be dependent on: (i) signalling molecules, such the hormone abscisic acid (ABA), that are known to play important roles in stress-induced metabolic adjustments (Strizhov *et al.*, 1997; Planchet *et al.*, 2011); and (ii) regulatory proteins, such protein kinases, under drought, cold and/or salt stress. Some calcium-responsive protein kinases (CDPKs), SNF (sucrose non-fermenting)-related protein kinases (SnRKs) and/or mitogen-activated protein kinases (MAPK) have been reported to be involved in signalling towards proline accumulation and to regulate, in a positive way, plant stress tolerance. Indeed, some studies have shown that SnRK2s induced the increase of proline content through an ABA-dependent pathway (Boudsocq *et al.*, 2004) and that SnRK3s interacted with calcium-binding protein interacting kinases (CIPK) to modulate proline contents (Xiang *et al.*, 2007) in response to stress. These data indicate that transcriptional control and signal transduction are crucial in stress-induced metabolic changes.

15.3.3 Accumulation of γ-aminobutyric acid (GABA) in response to plant stresses

The non-protein amino acid GABA (Chapter 7) is considered to be a stress metabolite and a signalling molecule in many plants, and is known to rapidly accumulate to high levels in response to environmental stress conditions such as oxygen deficiency (waterlogging and anoxia), acidification, drought, salinity, heat shock, low temperature and osmotic and mechanical stimulation (Shelp *et al.*, 1999; Kinnersley and Turano, 2000; Bouché and Fromm, 2004). GABA was first discovered in plants (potato tubers) (Steward *et al.*, 1949) and was later reported in mammalian systems to play a major role in neurotransmission and, more particularly, in signalling (Nathan *et al.*, 1994). GABA may be derived from polyamines and more particularly from putrescine, but it is mainly synthesized from glutamate in the cytoplasm via the action of glutamate decarboxylase (GAD; EC 4.1.1.15) and is then transported to the mitochondria (Fig. 15.2). Subsequently, GABA is converted into succinate via, respectively, GABA transaminase (GABA-T) and succinic semialdehyde dehydrogenase (SSADH), thus allowing feeding (as succinate) into the TCA cycle (Shelp *et al.*, 1999; Fait *et al.*, 2008).

The accumulation of GABA in response to various environmental stresses has been speculated to be dependent on plant developmental age and organ, e.g. shoots, roots, seedlings, flowers and fruits (Snedden and Fromm, 1999; Kinnersley and Turano, 2000), as two distinct GAD isoforms have been characterized and are expressed in separate tissues (Lähdesmäki, 1968). The accumulation of GABA occurs under abiotic stress concomitantly with an increase in calcium levels and a decrease in cytosolic pH (a common event under stress conditions), and also is dependent on enzyme activation involved in GABA metabolism. Indeed, the GAD reaction is proton consuming and stimulated by cytosolic acidification demonstrating that GABA synthesis is an adaptive response leading to pH cytoplasmic regulation.

The impairment of GABA-T in an *Arabidopsis* loss-of-function mutant has been reported to be responsible for a decrease in salt tolerance and, consequently, to enhanced levels of AAs and decreased carbohydrate levels under salt stress (Renault *et al.*, 2010), showing that GABA catabolism is important in stress responses associated with carbon and nitrogen metabolism (Bouché and Fromm, 2004). A functional GABA shunt appears to be important for stress tolerance, but the functions of GABA are still a matter of debate (Bouché and Fromm, 2004; Fait *et al.*, 2008). Various GABA functions have been reported, such a role in the control of glutamate homeostasis (Masclaux-Daubresse *et al.*, 2002), the regulation of plant growth and development (Snedden and Fromm, 1999), nitrate uptake (Beuve *et al.*, 2004), osmoregulation (Shelp *et al.*, 1999), pH regulation (Crawford *et al.*, 1994), antioxidant scavenging (Liu *et al.*, 2011), TCA cycle bypassing (Studart-Guimarães *et al.*, 2007) and cell signalling (Kinnersley and Turano, 2000).

15.4 The Pyruvate Family Pathway

Changes in AA levels under abiotic stress are intimately linked to primary carbon metabolism (glycolysis and the TCA cycle). When the oxygen is deficient (hypoxia) or absent (anoxia) in plant tissues, the pyruvate produced via glycolysis is metabolized to either lactate, via lactate dehydrogenase (LDH), or to ethanol, via pyruvate decarboxylase (PDC) and alcohol dehydrogenase (ADH), to allow the regeneration of NAD^+ to maintain glycolysis during fermentation. These metabolic changes are also accompanied by cytosolic acidification as a result of lactate accumulation, a loss of carbon due to the fermentative production of ethanol, a downregulation of respiratory activity and a decrease of the adenylate energy charge due to the inhibition of oxidative phosphorylation (van Dongen *et al.*, 2009). One consequence of pyruvate accumulation under hypoxic/anoxic conditions is an increase in alanine content, which represents one of the most important products (with ethanol and lactate) of anaerobic metabolism; there are also increases in the contents of leucine and valine, two branched-chain amino acids which can be *de novo* synthesized, mainly from pyruvate.

15.4.1 Alanine accumulation: a universal phenomenon under stress

Under oxygen deficiency, alanine accumulates in relatively large amounts, mainly as a result of increases in the levels and interconversions of free AAs, in which transaminases such as alanine aminotransferase (Chapter 2) play an important role (Reggiani *et al.*, 1988; Good and Muench 1993). In detached roots of rice under anoxia, the origin of alanine accumulation has been estimated to be 77.5% from the interconversion of free AAs and 22.5% from proteolysis (Reggiani *et al.*, 1998). This increase in alanine content, which is reported to occur rapidly after oxygen deprivation (a short-term adaptation), takes place at the expense of acidic AAs such as glutamate and aspartate, and concomitantly with the accumulation of GABA (Good and Muench, 1993).

Alanine, even accumulated at high levels, does not lead to detrimental effects or cell toxicity, an essential requirement for tolerance to hypoxia/anoxia. The increase of alanine accumulation in plant tissues exposed to low oxygen concentration can be explained by the stimulation of three different biosynthetic pathways (Fig. 15.2): (i) the conversion, in a reversible reaction, of glutamate and pyruvate to alanine and 2-oxoglutarate, catalysed by alanine aminotransferase (AlaAT; EC 2.6.1.2); (ii) the production from GABA and pyruvate of alanine and succinic semi-aldehyde, a precursor of succinate, catalysed by GABA-T; and (iii) an indirect pathway via the conversion of 2-ketoglutarate and aspartate into oxaloacetate and glutamate (afterwards used directly by AlaAT), catalysed by aspartate aminotransferase (AspAT; EC 2.6.1.1).

The key enzyme, AlaAT, has been reported to be responsible for the main source of alanine production under low oxygen, because: (i) waterlogging and flooding (but not anoxia) induce the expression of the *AlaAT* gene; and (ii) the increase in AlaAT activity is correlated with alanine accumulation under hypoxic stress conditions in various plant species (Muench and Good, 1994; Miyashita *et al.*, 2007; Limami *et al.*, 2008; Rocha *et al.*, 2010). AlaAT catalyses the reversible conversion of alanine and 2-oxoglutarate into pyruvate and glutamate, so this pyridoxal phosphate-dependent enzyme plays an important role in interconnecting primary carbon metabolism and AA synthesis. AlaAT is widely distributed in animals, plants, yeasts and bacteria, and its activity has been found in various organs and species. Four genes encoding AlaAT have been identified in the model plants *Arabidopsis thaliana* and *Medicago truncatula*. Among these four, two of them, *AlaAT1* and *AlaAT2*, encode AlaAT, while the two other genes have glutamate:glyoxalate aminotransferase (GGAT; EC 2.6.1.4) activities, which are mainly located in the peroxisomes. *In silico* predictions suggest that AlaAT1 is located in the cytosol and AlaAT2 in the mitochondria. The presence of several AlaAT activities in different compartments would indicate different metabolic roles. The key enzyme, AlaAT, appears to play an important role during a hypoxic/anoxic phase (Ricoult *et al.*, 2006), but also during the recovery phase after oxygen deprivation. In roots of soybean exposed to low oxygen concentrations, the activity of AlaAT was still increasing during hypoxia, whereas alanine accumulation had reached its maximum (de Sousa and Sodek, 2003). It was observed that hypoxia induced AlaAT activity earlier in roots (after 24 h of hypoxia treatment) than in shoots (after 48 h of hypoxia treatment). The return to normoxia induced a drastic decrease of alanine levels to a pre-hypoxic level within 24 h, suggesting that AlaAT, owing to its reversible function, is involved in alanine degradation during the recovery period. These results have been confirmed by Miyashita *et al.* (2007), who demonstrated that the *Arabidopsis AlaAT1* knockout mutant *alaat1-1* accumulated more alanine during hypoxia in a similar way to wild type plants, whereas the subsequent decrease in alanine was delayed

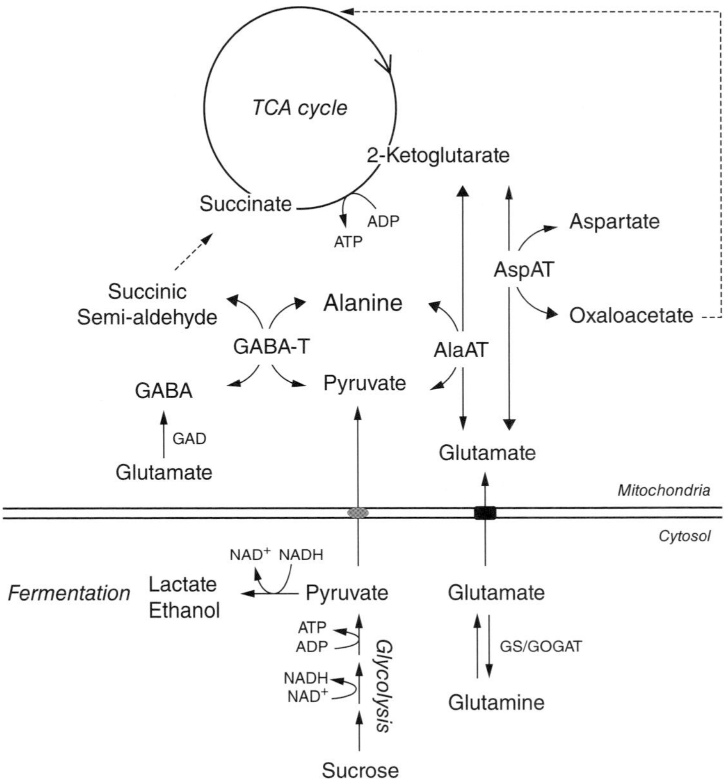

Fig. 15.2. Model of plant primary metabolism under hypoxia. During hypoxia, the induction of mitochondrial AlaAT leads, via the conversion of pyruvate and glutamate, to the accumulation of alanine and of 2-ketoglutarate. The induction of GABA-T (due to GABA accumulation) allows the feeding of succinate into the TCA cycle. In addition, fermentative pathways, through the induction of alcohol dehydrogenase and lactate dehydrogenase, allow the regeneration of NAD^+ to maintain glycolysis. During recovery from low-oxygen stress, AlaAT catalyses the rapid conversion of alanine to pyruvate. Abbreviations : AlaAT, alanine aminotransferase; AspAT, aspartate aminotransferase; GABA, γ-aminobutyric acid; GABA-T, GABA transaminase; GAD, glutamate decarboxylase; GOGAT, glutamate synthase; GS, glutamine synthetase; TCA, tricarboxylic acid.

during the re-oxygenation period as a result of the large reduction in AlaAT activity. From these findings, expression of the *AlaAT* gene appears to be upregulated during hypoxia, thus demonstrating a limited role in alanine synthesis under hypoxia; during the recovery phase, however, the enzyme AlaAT is more involved in the breakdown of alanine (by rapid conversion of the accumulated alanine into glutamate and pyruvate) than its synthesis.

Although this increase of levels of alanine under low oxygen conditions is proven in many plant species and organs, its beneficial effects have not been yet completely elucidated. Alanine accumulation in plants during waterlogging would offer several advantages: (i) saving carbon skeletons by competing with the ethanol fermentation pathway for the common precursor pyruvate; (ii) indirectly contributing to pH homeostasis by competing with the lactic acid fermentation pathway for the common precursor pyruvate (Good and Muench, 1993; Ricoult *et al.*, 2005; Rocha *et al.*, 2010); (iii) through the alanine shunt, the synthesis of alanine generates 2-ketoglutarate that can be further metabolized into succinate via the TCA cycle enzyme succinate CoA ligase, thus providing additional ATP per molecule of sucrose utilized; and (iv) allowing the detoxification of ammonium to form a pool of nitrogen that could be used in AA synthesis during the recovery phase in normoxia (Vanlerberghe *et al.*, 1991).

15.4.2 Leucine and valine: the importance of branched-chain amino acid accumulation in response to stress

Besides alanine, two other amino acids, leucine and valine, from the pyruvate family pathway have been reported to be overproduced in response to abiotic stresses. Leucine and valine, like isoleucine which is derived from the aspartate family pathway, form the small group of branched-chain amino acids (BCAAs, classified by their small branched hydrocarbon residues) that serve as precursors of plant secondary metabolites. The biosynthesis of BCAAs in higher plants occurs only in plastids, whereas the catabolism of BCAAs might occur in the mitochondria and/or peroxisomes (for details, see Binder *et al.*, 2007). The production of these three AAs needs the same four biosynthetic enzymes, even though the substrates are different. The final transamination step in BCAA biosynthesis is catalysed by BCAA transferases (BCAT; EC 2.6.1.42), which are also involved in the first step in the catabolic pathway.

Several reports have demonstrated that the contents of leucine and valine increased significantly in leaf tissues from wheat and in seedlings from *Arabidopsis* in response to drought stress (Bowne *et al.*, 2012; Sharma *et al.*, 2013). In particular, Nambara *et al.* (1998) showed that leucine was increased by around 150-fold under drought stress in *Arabidopsis* foliage compared with normal conditions, while the relative increase in proline was only 80-fold. These data suggest that leucine and valine could play also a role as compatible solutes under stress conditions. The relative increases of leucine and valine concentration could be also due to the corresponding increases in substrate level, as observed for pyruvate, which accumulated and doubled after several days of drought stress in *Brassica napus* (Good and Zaplachinski, 1994). Under stress conditions, BCAA accumulation has been reported to be regulated at the transcriptional level and BCAAs to act as signalling molecules in the regulation of gene expression (Urano *et al.*, 2009). Indeed, an increase of free BCAA levels has been described to be correlated with the upregulation of *BCAT2* gene (located in plastids) expression after treatment with ABA and under various stresses. These increased free BCAAs could serve as substrate for the synthesis of proteins that are induced in response to stress, as some genes encoding proteins rich in BCAAs are upregulated under osmotic stress and low temperature. This could be explained by the fact that BCAAs are hydrophobic amino acids consisting of unsubstituted aliphatic side chains with branched alkyl groups, and mainly situated in the protein core; these properties offer the possibility that BCAAs could play an important role in the structure of globular proteins and in facilitating the interaction of domains of transmembrane proteins with phospholipid bilayers. Interestingly, Malatrasi *et al.* (2006) have reported that the level of BCAAs increased in barley exposed to drought stress, but that this accumulation also led to the activation of their catabolism through the upregulation of BCAT. To avoid excessive accumulation and cytotoxicity of BCAAs, it has been suggested that their increased degradation would provide an alternative source of respiratory substrates for the TCA cycle or serve as a detoxification mechanism for maintaining the free BCAA pool at levels that are compatible with cellular homeostasis during severe plant stress (Taylor *et al.*, 2004).

15.5 The Aspartate Family Pathway

The amino acid aspartate is the precursor of three main metabolic pathways, which are modified by various abiotic stresses. The first pathway leads to the synthesis of asparagine, an AA involved in transport and storage of nitrogen. The second synthesizes *de novo* NAD^+, a pyridine nucleotide that plays a major role in redox recycling. The third pathway, called also the aspartate-derived amino acid pathway in higher plants, leads to the biosynthesis of four essential AAs, lysine, methionine, threonine and isoleucine, via three different branches (Fig. 15.1). The first branch leads to the biosynthesis of lysine, which is further catabolized into glutamate and acetyl-CoA, showing that this aspartate family pathway is intricately linked to the TCA cycle. In terms of nutrition, lysine is considered to be the most important essential AA in human and animal food/feed due to its high nutritional value, but in cereal grains, the amount present is limited (see also Chapter 13). The second branch of the aspartate-derived AA pathway results in the biosynthesis of methionine (the substrate for glucosinolates, *S*-adenosyl methionine and, therefore, polyamines) and isoleucine. The last branch leads to the biosynthesis of threonine, which can be metabolized to glycine (involved in photorespiration) and isoleucine, which is necessary to energy production. Under non-stress growth conditions, the major energy source for plants is provided by photosynthesis during the daytime, whereas during the dark period, AAs are the main precursors for energy generation via glycolysis, AA metabolism and the TCA cycle. It is known that stress-induced energy deprivation, retarded growth and cell death are the result of a reduction in photosynthesis and/or respiration, showing that the energy availability and stress tolerance are strongly linked (Baena-González and Sheen, 2008).

15.5.1 Stress-induced asparagine accumulation

Asparagine, the first AA to be isolated from plants, is known to be a major nitrogen-transport compound in the xylem and phloem and to act as a store of reduced nitrogen. For nitrogen fixation in leguminous plants, asparagine is the major end product and accumulates in high levels in the nodules and in root xylem exudates. It has been also reported that asparagine, which is accumulated in germinating seedlings, is quantitatively the most important protein AA present in seedlings (Planchet *et al.*, 2011); it is also a major product during the metabolism in senescing leaves (Lea *et al.*, 2007). Plants exposed to environmental changes and constraints have been shown to accumulate asparagine to a considerable extent during drought and salt stress (Fougère *et al.*, 1991; Planchet *et al.*, 2011), concomitantly with the increase in proline. A medium deficient in mineral ions (such as phosphorus, sulfur, potassium and magnesium) also induces an increase in asparagine content as, for example, in nodules under phosphorus deficiency (Almeida *et al.*, 2000) or under sulfur-deficient conditions, where a concomitant increase in soluble asparagine to 50% of the total pool of AAs and an alteration in the proportion of various proteins formed in maize seeds have been reported (Baudet *et al.*, 1986). Among the effects of abiotic stresses that have been reported, toxic metals and pollutants, which are deleterious to plant growth and development, induced an accumulation of asparagine, for example, in both the roots and the shoots of lupin in the presence of cadmium in *in vitro* culture (Lea *et al.*, 2007). This AA has this ability to bind to metals such as zinc, cadmium and copper by forming intracellular asparagine-metal complexes, which are useful for reducing the metal toxicity.

The synthesis of asparagine is via the ATP-dependent transfer of the nitrogen from the amide group of the glutamine to aspartate to generate glutamate and asparagine, and is catalysed by asparagine synthetase (AS; EC 6.3.5.4; Chapter 6). Three distinct genes encoding AS (*ASN1*, *ASN2* and *ASN3*) have been identified in *A. thaliana* and *Helianthus annuus* (sunflower), and these appear to be regulated in different ways by light and dark conditions (Lam *et al.*, 1998; Herrera-Rodriguez *et al.*, 2004). In *Arabidopsis*, the level of expression of *ASN1* was correlated with the increase in free asparagine in dark-adapted plants, showing that asparagine is an important nitrogen carrier, especially under carbon-limiting conditions (Lam *et al.*, 1998). The expression of *ASN2* is induced by light and is correlated with high ammonium stress and the reassimilation of that ammonium (Wong *et al.*, 2004). Similarly, an *Arabidopsis* mutant with a disrupted *ASN2* gene has been reported to be less tolerant to salt stress than the wild type plant; the *asn2-1* mutant was impaired in nitrogen assimilation and translocation under salt treatment (Maaroufi-Dguimi *et al.*, 2011). Asparagine content can also be modulated by the relative instability of AS and the catabolism of asparagine via the rapid turnover of asparagine by L-asparaginase (ASP; EC 3.5.1.1), which hydrolyses the amide group to form aspartate and ammonium. A low temperature-inducible cDNA sequence encoding ASP (named GmASP1) has been isolated from leaves of *Glycine max* (soybean) (Cho *et al.*, 2007); the expression of the *GmASP1* gene (measured as *GmASP1*mRNA) was also transiently induced by cold stress in soybean leaves, indicating a potential role of ASP in the early response to low temperature stress by modulating the asparagine and aspartate content.

15.5.2 Aspartate-derived amino acids in response to stress

Most of the essential AAs in human food and livestock feeds are from plants and, among these, AAs of nutritional interest are isoleucine, methionine, threonine and lysine, the aspartate-derived amino acids. These AAs are involved in protein synthesis and contribute via their catabolism to cellular energy metabolism via the TCA cycle (Galili *et al.*, 2001).

Under some physiological and growth conditions, the availability of isoleucine is influenced by the synthesis and the catabolism of threonine and methionine, two AAs that serve as substrates for isoleucine biosynthesis and are derived from aspartate via enzymes localized in the plastids. Isoleucine, a branched-chain amino acid, is synthesized from the precursor 2-ketobutyrate, which is produced predominantly by the conversion of threonine via threonine deaminase/dehydratase (EC 4.3.1.16) and, to a lesser extent, from the conversion of methionine via methionine γ-lyase (EC 4.4.1.11). In *Arabidopsis*, isoleucine accumulation was observed in reproductive tissue in response to osmotic stress; this was correlated with the induced expression of the gene coding for threonine deaminase, but also methionine γ-lyase, which offers an alternative pathway for increasing the isoleucine in response to demand (Nambara *et al.*, 1998; Less and Galili, 2008; Joshi and Jander, 2009). Further, in the presence of high salt concentrations, an increase of threonine deaminase activity in *Arabidopsis* roots also contributed to an accumulation of isoleucine (Joshi and Jander, 2009). This significant amount of accumulated isoleucine, which is toxic to the cell, like all accumulated BCAAs, represents an alternative substrate for respiration.

As a general response to environmental stresses, cellular energy status is profoundly affected, leading to an energy limitation. Consequently, the expression of genes encoding biosynthetic enzymes of the aspartate family pathway is partially repressed in order to preserve energy (Galili, 2011). In accord with this suggestion, lysine catabolism has been shown to be stimulated to create additional energy. This catabolic pathway is tightly regulated and may play important roles in physiological and developmental processes. By combining it with α-ketoglutarate, lysine is catabolized into saccharopine via lysine ketoglutarate reductase (LKR; EC 1.5.1.8). Saccharopine dehydrogenase (SDH; EC 1.5.1.9) then converts saccharopine into glutamate and α-amino-adipic semialhehyde, a precursor of an osmolyte, pipecolic acid, and also a substrate of acetyl-CoA. These catabolic genes and activities (LKR/SDH), which are dependent of the intracellular lysine content, are upregulated mainly by salt and osmotic stress, as well as in response to the ABA (Stepansky and Galili, 2003; Moulin *et al.*, 2006). This increase of lysine catabolism can be explained by the fact that the upregulation of *SDH* in response to stress conditions leads to the formation of glutamate, a precursor of important stress-related metabolites such as proline and GABA.

15.6 Conclusions

Understanding the molecular, cellular and metabolic mechanisms of plants subjected to various environmental constraints (abiotic stress and biotic interactions) is of much importance in improving the stress tolerance of economically important crop species. As some specific AAs, accumulated in large amounts, are involved in the metabolic processes of regulation for stress response strategies, AA pathways represent important targets for the enhancement of abiotic stress tolerance via plant metabolic engineering. The impact of RNA interference (RNAi) on metabolic engineering through transgenic approaches by the use of knockout mutants of individual isoenzymes involved in AA metabolism will provide additional insights on the regulation and interconnection of AA pathways. At present, metabolome analysis has become an invaluable tool for improving our knowledge of changes in plant metabolic networks after exposure to abiotic stress conditions, in particular through the accumulation of known and novel metabolites. In concert with this, fluxomics, a recently developed and more informative approach, has been designed to follow metabolite fluxes through pathways (Wiechert *et al.*, 2007) and will provide further elucidation. In addition, further investigations via bioinformatic and transcriptomic approaches to the transcriptional regulation of genes in response to environmental stresses are also required to provide new openings for understanding specific metabolic perturbations (including increases in or depletion of AAs) and for unravelling changes in metabolic networks relating to plant responses and tolerance to stress.

References

Alia, Mohanty, P. and Matysik, J. (2001) Effect of proline on the production of singlet oxygen. *Amino Acids* 21, 195–200.

Almeida, J.P., Hartwig, U.A., Frehner, M., Nösberger, J. and Lüscher, A. (2000) Evidence that P deficiency induces N feedback regulation of symbiotic N_2 fixation in white clover (*Trifolium repens* L.). *Journal of Experimental Botany* 51, 1289–1297.

Baena-González, E. and Sheen, J. (2008) Convergent energy and stress signaling. *Trends in Plant Science* 13, 474–482.

Baudet, J., Huet, J.C., Jolivet, E., Lesaint, C., Mossé, J. and Pernollet, J.C. (1986) Changes in accumulation of seed nitrogen compounds in maize under conditions of sulphur deficiency. *Physiologia Plantarum* 68, 608–614.

Beuve, N., Rispail, N., Laine, P., Cliquet, J-B., Ourry, A. and Le Deunff, E. (2004) Putative role of γ-aminobutyric acid (GABA) as a long-distance signal in up-regulation of nitrate uptake in *Brassica napus* L. *Plant, Cell and Environment* 27, 1035–1046.

Binder, S., Knill, T. and Schuster, J. (2007) Branched-chain amino acid metabolism in higher plants. *Physiologia Plantarum* 129, 68–78.

Bouché, N. and Fromm, H. (2004) GABA in plants: just a metabolite? *Trends in Plant Science* 9, 110–115.

Boudsocq, M., Barbier-Brygoo, H. and Laurière, C. (2004) Identification of nine sucrose nonfermenting 1-related protein kinases 2 activated by hyperosmotic and saline stresses in *Arabidopsis thaliana*. *The Journal of Biological Chemistry* 279, 41758–41766.

Bowne, J.B., Erwin, T.A., Juttner, J., Schnurbusch, T., Langridge, P., Bacic, A. and Roessner, U. (2012) Drought responses of leaf tissues from wheat cultivars of differing drought tolerance at the metabolite level. *Molecular Plant* 5, 418–429.

Chen, Z., Cuin, T.A., Zhou, M., Twomey, A., Naidu, B.P. and Shabala, S. (2007) Compatible solute accumulation and stress-mitigating effects in barley genotypes contrasting in their salt tolerance. *Journal of Experimental Botany* 58, 4245–4255.

Cho, C.W., Lee, H.J., Chung, E., Kim, K.M., Heo, J.E., Kim, J.I., Chung, J., Ma, Y., Fukui, K., Lee, D.W., Kim, D.H., Chung, Y.S. and Lee, J.H. (2007) Molecular characterization of the soybean L-asparaginase gene induced by low temperature stress. *Molecules and Cells* 23, 280–286.

Crawford, L.A., Bown, A.W., Breitkreuz, K.E. and Guinel, F.C. (1994) The synthesis of γ-aminobutyric acid in response to treatments reducing cytosolic pH. *Plant Physiology* 104, 865–871.

de Sousa, C.A.F. and Sodek, L. (2003) Alanine metabolism and alanine aminotransferase activity in soybean (*Glycine max*) during hypoxia of the root system and subsequent return to normoxia. *Environmental and Experimental Botany* 50, 1–8.

Fait, A., Fromm, H., Walter, D., Galili, G. and Fernie, A.R. (2008) Highway or byway: the metabolic role of the GABA shunt in plants. *Trends in Plant Science* 13, 14–19.

Forde, B.G. and Lea, P.J. (2007) Glutamate in plants: metabolism, regulation, and signalling. *Journal of Experimental Botany* 58, 2339–2358.

Fougère, F., Le Rudulier, D. and Streeter, J.G. (1991) Effects of salt stress on amino acid, organic acid, and carbohydrate composition of roots, bacteroids, and cytosol of alfalfa (*Medicago sativa* L.). *Plant Physiology* 96, 1228–1236.

Funck, D., Stadelhofer, B. and Koch, W. (2008) Ornithine-δ-aminotransferase is essential for arginine catabolism but not for proline biosynthesis. *BMC Plant Biology* 8:40.

Galili, G. (2011) The aspartate-family pathway of plants: linking production of essential amino acids with energy and stress regulation. *Plant Signaling and Behavior* 6, 192–195.

Galili, G., Tang, G., Zhu, X. and Gakiere, B. (2001) Lysine catabolism: a stress and development super-regulated metabolic pathway. *Current Opinion in Plant Biology* 4, 261–266.

Gill, S.S. and Tuteja, N. (2010) Polyamines and abiotic stress tolerance in plants. *Plant Signaling and Behavior* 5, 26–33.

Good, A.G. and Muench, D.G. (1993) Long-term anaerobic metabolism in root tissue (metabolic products of pyruvate metabolism). *Plant Physiology* 101, 1163–1168.

Good, A.G. and Zaplachinski, S.T. (1994) The effects of drought stress on free amino-acid accumulation and protein-synthesis in *Brassica napus*. *Physiologia Plantarum* 90, 9–14.

Hare, P.D. and Cress, W.A. (1997) Metabolic implications of stress-induced proline accumulation in plants. *Plant Growth and Regulation* 21, 79–102.

Hare, P.D., Cress, W.A. and van Staden, J. (1999) Proline synthesis and degradation: a model system for elucidating stress-related signal transduction. *Journal of Experimental Botany* 50, 413–434.

Herrera-Rodriguez, M.B., Maldonado, J.M. and Pérez-Vicente, R. (2004) Light and metabolic regulation of *HAS1*, *HAS1.1* and *HAS2*, three asparagine synthetase genes in *Helianthus annuus*. *Plant Physiology and Biochemistry* 42, 511–518.

Heyser, J.W., Chacon, M.J. and Warren, R.C. (1989) Characterization of L-[5-^{13}C]-proline biosynthesis in halophytic and nonhalophytic suspension cultures by ^{13}C NMR. *Journal of Plant Physiology* 135, 459–466.

Hong, Z., Lakkineni, K., Zhang, Z. and Verma, D.P. (2000) Removal of feedback inhibition of Δ^1-pyrroline-5-carboxylate synthetase results in increased proline accumulation and protection of plants from osmotic stress. *Plant Physiology* 122, 1129–1136.

Joshi, V. and Jander, G. (2009) *Arabidopsis* methionine γ-lyase is regulated according to isoleucine biosynthesis needs but plays a subordinate role to threonine deaminase. *Plant Physiology* 151, 367–378.

Kemble, A.R. and MacPherson, H.T. (1954) Liberation of amino acids in perennial rye grass during wilting. *Biochemical Journal* 58, 46–50.

Khedr, A.H., Abbas, M.A., Wahid, A.A., Quick, W.P. and Abogadallah, G.M. (2003) Proline induces the expression of salt-stress-responsive proteins and may improve the adaptation of *Pancratium maritimum* L. to salt-stress. *Journal of Experimental Botany* 54, 2553–2562.

Kinnersley, A.M. and Turano F.J. (2000) Gamma aminobutyric acid (GABA) and plant responses to stress. *Critical Reviews in Plant Science* 19, 479–509.

Kishor, K.P.B., Hong, Z., Miao, G.H., Hu, C.A.A. and Verma, D.P.S. (1995) Overexpression of Δ^1-pyrroline-5-carboxylate synthetase increases proline production and confers osmotolerance in transgenic plants. *Plant Physiology* 108, 1387–1394.

Lähdesmäki, P. (1968) The amount of γ-amino butyric acid and the activity of glutamic decarboxylase in aging leaves. *Physiologia Plantarum* 21, 1322–1327.

Lam, H.M., Hsieh, M.H. and Coruzzi, G. (1998) Reciprocal regulation of distinct asparagine synthetase genes by light and metabolites in *Arabidopsis thaliana*. *The Plant Journal* 16, 345–353.

Lea, P.J., Sodek, L., Parry, M.A.J., Shewry, P.R. and Halford, N.G. (2007) Asparagine in plants. *Annals of Applied Biology* 150, 1–26.

Less, H. and Galili, G. (2008) Principal transcriptional programs regulating plant amino acid metabolism in response to abiotic stresses. *Plant Physiology* 147, 316–330.

Limami, A.M., Glévarec, G., Ricoult, C., Cliquet, J.B. and Planchet, E. (2008) Concerted modulation of alanine and glutamate metabolism in young *Medicago truncatula* seedlings under hypoxic stress. *Journal of Experimental Botany* 59, 2325–2335.

Liu, C., Zhao, L. and Yu, G. (2011) The dominant glutamic acid metabolic flux to produce γ-amino butyric acid over proline in *Nicotiana tabacum* leaves under water stress relates to its significant role in antioxidant activity. *Journal of Integrative Plant Biology* 53, 608–618.

Lv, W.T., Lin, B., Zhang, M. and Hua, X.J. (2011) Proline accumulation is inhibitory to *Arabidopsis* seedlings during heat stress. *Plant Physiology* 156, 1921–1933.

Maaroufi-Dguimi, H., Debouba, M., Gaufichon, L., Clément, G., Gouia, H., Hajjaji, A. and Suzuki, A. (2011) An *Arabidopsis* mutant disrupted in *ASN2* encoding asparagine synthetase 2 exhibits low salt stress tolerance. *Plant Physiology and Biochemistry* 6, 623–628.

Malatrasi, M., Corradi, M., Svensson, J.T., Close, T.J., Gulli, M. and Marmiroli, N. (2006) A branched-chain amino acid aminotransferase gene isolated from *Hordeum vulgare* is differentially regulated by drought stress. *Theoretical and Applied Genetics* 113, 965–976.

Masclaux-Daubresse, C., Valadier, M.H., Carrayol, E., Reisdorf-Cren, M. and Hirel, B. (2002) Diurnal changes in the expression of glutamate dehydrogenase and nitrate reductase are involved in the C/N balance of tobacco source leaves. *Plant, Cell and Environment* 25, 1451–1462.

Miyashita, Y., Dolferus, R., Ismond, K.P. and Good, A.G. (2007) Alanine aminotransferase catalyses the breakdown of alanine after hypoxia in *Arabidopsis thaliana*. *The Plant Journal* 49, 1108–1121.

Mohapatra, S., Minocah, R., Long, S. and Minocha, S.C. (2010) Transgenic manipulation of a single polyamine in poplar cells affects the accumulation of all amino acids. *Amino Acids* 38, 1117–1129.

Morris, C.J., Thompson, J.F. and Johnson, C.M. (1969) Metabolism of glutamic acid and *N*-acetylglutamic acid in leaf discs and cell-free extracts of higher plants. *Plant Physiology* 44, 1023–1026.

Moulin, M., Deleu, C., Larher, F. and Bouchereau, A. (2006) The lysine-ketoglutarate reductase-saccharopine dehydrogenase is involved in the osmo-induced synthesis of pipecolic acid in rapeseed leaf tissues. *Plant Physiology and Biochemistry* 44, 474–482.

Muench, D.G. and Good, A.G. (1994) Hypoxically inducible barley alanine aminotransferase: cDNA cloning and expression analysis. *Plant and Molecular Biology* 24, 417–427.

Munns, R. (2002) Comparative physiology of salt and water stress. *Plant, Cell and Environment* 25, 239–250.

Nambara, E., Kawaide, H., Kamiya, Y. and Naito S. (1998) Characterization of an *Arabidopsis thaliana* mutant that has a defect in ABA accumulation: ABA-dependent and ABA-independent accumulation of free amino acids during dehydration. *Plant and Cell Physiology* 39, 853–858.

Nanjo, T., Kobayashi, M., Yoshiba, Y., Sanada, Y., Wada, K., Tsukaya, H, Kakubari, Y., Yamaguchi-Shinozaki, K. and Shinozaki, K. (1999) Biological functions of proline in morphogenesis and osmotolerance revealed in antisense transgenic *Arabidopsis thaliana*. *The Plant Journal* 18, 185–193.

Nathan, B., Bao, J., Hsu, J.C., Aguilar, P., Wu, R., Yarom, M., Kuo, C.Y. and Wu, J.Y. (1994) A membrane form of brain L-glutamate decarboxylase: identification, isolation, and its relation to insulin-dependent diabetes mellitus. *Proceedings of the National Academy of Sciences of the United States of America* 91, 242–246.

Planchet, E., Rannou, O., Ricoult, C., Boutet-Mercey, S., Maia-Grondard, A. and Limami, A.M. (2011) Nitrogen metabolism responses to water deficit act through both abscisic acid (ABA)-dependent and independent pathways in *Medicago truncatula* during post-germination. *Journal of Experimental Botany* 62, 605–615.

Reggiani, R., Cantú, C.A., Brambilla, I. and Bertani, A. (1988) Accumulation and interconversion of amino acids in rice roots under anoxia. *Plant and Cell Physiology* 29, 982–987.

Renault, H., Roussel, V., El Amrani, A., Arzel, M., Renault, D., Bouchereau, A. and Deleu, C. (2010) The *Arabidopsis pop2-1* mutant reveals the involvement of GABA transaminase in salt stress tolerance. *BMC Plant Biology* 10:20.

Rhodes, D., Handa, S. and Bressan, R.A. (1986) Metabolic changes associated with adaptation of plant cells to water stress. *Plant Physiology* 82, 890–903.

Ricoult, C., Cliquet, J.B. and Limami, A.M. (2005) Stimulation of alanine amino transferase (AlaAT) gene expression and alanine accumulation in embryo axis of the model legume *Medicago truncatula* contribute to anoxia stress tolerance. *Physiologia Plantarum* 123, 30–39.

Ricoult, C., Echeverria, L.O., Cliquet, J.B. and Limami, A.M. (2006) Characterization of alanine aminotransferase (AlaAT) multigene family and hypoxic response in young seedlings of the model legume *Medicago truncatula*. *Journal of Experimental Botany* 57, 3079–3089.

Rizhsky, L., Liang, H., Shuman, J., Shulaev, V., Davletova, S. and Mittler, R. (2004) When defense pathways collide. The response of *Arabidopsis* to a combination of drought and heat stress. *Plant Physiology* 134, 1683–1696.

Rocha, M., Licausi, F., Araújo, W.L., Nunes-Nesi, A., Sodek, L., Fernie, A.R. and van Dongen, J.T. (2010) Glycolysis and the tricarboxylic acid cycle are linked by alanine aminotransferase during hypoxia induced by waterlogging of *Lotus japonicus*. *Plant Physiology* 152, 1501–1513.

Roosens, N.H., Thu, T.T., Iskandar, H.M. and Jacobs, M. (1998) Isolation of the ornithine-δ-aminotransferase cDNA and effect of salt stress on its expression in *Arabidopsis thaliana*. *Plant Physiology* 117, 263–271

Savouré, A., Jaoua, S., Hua, X.J., Ardiles, W., Van Montagu, M. and Verbruggen, N, (1995) Isolation, characterization, and chromosomal location of a gene encoding the Δ^1-pyrroline-5-carboxylate synthetase in *Arabidopsis thaliana*. *FEBS Letters* 372, 13–19.

Sharma, S.S. and Dietz, K.J. (2006) The significance of amino acids and amino acid-derived molecules in plant responses and adaptation to heavy metal stress. *Journal of Experimental Botany* 57, 711–726.

Sharma, S., Lin, W., Villamor, J.G. and Verslues, P.E. (2013) Divergent low water potential response in *Arabidopsis thaliana* accessions Landsberg *erecta* and Shahdara. *Plant, Cell and Environment* 36, 994–1008.

Shelp, B.J., Bown, A.W. and McLean, M.D (1999) Metabolism and functions of γ-aminobutyric acid. *Trends in Plant Science* 4, 446–452.

Signorelli, S., Arellano, J.B., Melo, T.B., Borsani, O. and Monza, J. (2013) Proline does not quench singlet oxygen: evidence to reconsider its protective role in plants. *Plant Physiology and Biochemistry* 64, 80–83.

Singh, T.N., Aspinall, D. and Paleg, L.G. (1972) Proline accumulation and varietal adaptability to drought in barley; a potential metabolic measure of drought resistance. *Nature* 286, 188–190.
Snedden, W.A. and H. Fromm, H. (1999) Regulation of the γ-aminobutyrate-synthesizing enzyme, glutamate decarboxylase, by calcium-calmodulin: a mechanism for rapid activation response to stress. In: Lerner, H.R. (ed.) *Plant Responses to Environmental Stresses: From Phytohormones to Genome Reorganization*. Marcel Dekker, New York, pp. 549–574.
Stepansky, A. and Galili, G. (2003) Synthesis of the *Arabidopsis* bifunctional lysine-ketoglutarate reductase/saccharopine dehydrogenase enzyme of lysine catabolism is concertedly regulated by metabolic and stress-associated signals. *Plant Physiology* 133, 1407–1415.
Steward, F.C., Thompson, J.F. and Dent, D.E. (1949) γ-Aminobutyric acid, a constituent of potato tuber. *Science* 110, 439–440.
Strizhov, N., Abraham, E., Okresz, L., Blickling, S., Zilberstein, A., Schell, J., Koncz, C. and Szabados, L. (1997) Differential expression of two *P5CS* genes controlling proline accumulation during salt-stress requires ABA and is regulated by *ABA1*, *ABI1* and *AXR2* in *Arabidopsis*. *The Plant Journal* 12, 557–569.
Studart-Guimarães, C., Fait, A., Nunes-Nesi, A., Carrari, F., Usadel, B. and Fernie, A.R. (2007) Reduced expression of succinyl-coenzyme A ligase can be compensated for by up-regulation of the γ-aminobutyrate shunt in illuminated tomato leaves. *Plant Physiology* 145, 626–639.
Szabados, L. and Savouré, A. (2010) Proline: a multifunctional amino acid. *Trends in Plant Science* 15, 89–97.
Székely, G., Abrahám, E., Cséplo, A., Rigó, G., Zsigmond, L., Csiszár, J., Ayaydin, F., Strizhov, N., Jásik, J., Schmelzer, E., Koncz, C. and Szabados, L. (2008) Duplicated *P5CS* genes of *Arabidopsis* play distinct roles in stress regulation and developmental control of proline biosynthesis. *The Plant Journal* 53, 11–28.
Taylor, N.L., Heazlewood, J.L., Day, D.A. and Millar, A.H. (2004) Lipoic acid dependent oxidative catabolism of α-keto acids in mitochondria provides evidence for branched-chain amino acid catabolism in *Arabidopsis*. *Plant Physiology* 134, 838–848.
Urano, K., Maruyama, K., Ogata, Y., Morishita, Y., Takeda, M., Sakurai, N., Suzuki, H., Saito, K., Shibata, D., Kobayashi, M., Yamaguchi-Shinozaki, K. and Shinozaki, K. (2009) Characterization of the ABA-regulated global responses to dehydration in *Arabidopsis* by metabolomics. *The Plant Journal* 57, 1065–1078.
van Dongen, J.T., Fröhlich, A., Ramírez-Aguilar, S.J., Schauer, N., Fernie, A.R., Erban, A., Kopka, J., Clark, J., Langer, A. and Geigenberger, P. (2009) Transcript and metabolite profiling of the adaptive response to mild decreases in oxygen concentration in the roots of *Arabidopsis* plants. *Annals of Botany* 103, 269–280.
Vanlerberghe, G.C., Joy, K.W. and Turpin, D.H. (1991) Anaerobic metabolism in the N-limited green alga *Selenastrum minutum*. III. Alanine is the product of anaerobic ammonium assimilation. *Plant Physiology* 95, 655–658.
Verbruggen, N. and Hermans, C. (2008) Proline accumulation in plants: a review. *Amino Acids* 35, 753–759.
Verslues, P.E. and Sharma, S. (2010) Proline metabolism and its implications for plant–environment interaction. *The Arabidopsis Book* 8: e0140.
Verslues, P.E and Sharp, R.E. (1999) Proline accumulation in maize (*Zea mays* L.) primary roots at low water potentials. II. Metabolic source of increased proline deposition in the elongation zone. *Plant Physiology* 119, 1349–1360.
Vierstra, R.D. (1996) Proteolysis in plants: mechanisms and functions. *Plant Molecular Biology* 32, 275–302.
Wiechert, W., Schweissgut, O., Takanaga, H. and Frommer, W.B. (2007) Fluxomics: mass spectrometry versus quantitative imaging. *Current Opinion in Plant Biology* 10, 323–330.
Wong, H.K., Chan, H.K, Coruzzi, M. and Lam, H.M. (2004) Correlation of *ASN2* gene expression with ammonium metabolism in *Arabidopsis*. *Plant Physiology* 134, 332–338.
Xiang, Y., Huang, Y. and Xiong, L. (2007) Characterization of stress-responsive *CIPK* genes in rice for stress tolerance improvement. *Plant Physiology* 144, 1416–1428.
Xue, X., Liu, A. and Hua, X. (2009) Proline accumulation and transcriptional regulation of proline biosynthesis and degradation in *Brassica napus*. *Biochemistry and Molecular Biology Reports* 42, 28–34.
Yoshiba, Y., Kiyosue, T., Katagiri, T., Ueda, H., Mizoguchi, T., Yamaguchi-Shinozaki, K., Wada, K., Harada, Y. and Shinozaki, K. (1995) Correlation between the induction of a gene for Δ^1-pyrroline-5-carboxylate synthetase and the accumulation of proline in *Arabidopsis thaliana* under osmotic stress. *The Plant Journal* 7, 751–760.
Yoshiba, Y., Kiyosue, T., Nakashima, K., Yamaguchi-Shinozaki, K. and Shinozaki, K. (1997) Regulation of levels of proline as an osmolyte in plants under water stress. *Plant and Cell Physiology* 38, 1095–1102.

16 The Central Role of Glutamate and Aspartate in the Post-translational Control of Respiration and Nitrogen Assimilation in Plant Cells

B. O'Leary[1] and W.C. Plaxton[2]*

[1]University of Oxford, Oxford, UK; [2]Queen's University, Kingston, Ontario, Canada

16.1 Abstract

The assimilation of inorganic N, in the form of NH_4^+, into glutamine and glutamate via glutamine synthetase (GS) and glutamine oxoglutarate aminotransferase (GOGAT) is a pivotal metabolic process in plants. This is because glutamate and glutamine are the primary amino donors for the biosynthesis of all nitrogenous compounds needed by plants. N assimilation must closely interact with glycolytic and respiratory C metabolism, because GS requires ATP and GOGAT requires C skeletons and reductant in the form of 2-oxoglutarate and reduced ferredoxin or NADH, respectively. Over 50% of net plant C may be committed to N assimilation in some tissues in order to generate sufficient 2-oxoglutarate, ATP and reductant for the GS/GOGAT system. Control of the cytosolic phosphoenolpyruvate (PEP) branch point by the tightly regulated enzymes cytosolic pyruvate kinase (PK_c) and phosphoenolpyruvate carboxylase (PEPC) exerts a major influence on the overall rate of plant respiration and N assimilation. Glutamate and aspartate are important allosteric effectors of PEP carboxylase and PK_c isoforms in plant tissues that are active in NH_4^+ assimilation. The coordinate feedback control of PEPC and PK_c by glutamate and aspartate provides a direct link between N assimilation via GS/GOGAT and the control of respiratory C metabolism.

16.2 Introduction: The Metabolic Organization of N Assimilation

The assimilation of inorganic nitrogen (N), as ammonium (NH_4^+), into organic molecules is a crucial metabolic process in plants and can represent a major constraint to vegetative growth. Nitrate (NO_3^-) and NH_4^+ represent the two predominant forms of N available for uptake from the environment. To be assimilated, NO_3^- must first be reduced to NH_4^+ by nitrate reductase within the chloroplast. In legume root nodules, *Rhizobium* bacteria reduce N_2 to NH_4^+, which is supplied to the plant in exchange for a carbon (C) source (Vance and Heichel, 1991). In all cases, NH_4^+ is primarily assimilated into the biocompatible amino acids glutamine (Gln) and glutamate (Glu) via glutamine synthetase (GS) and glutamine 2-oxoglutarate aminotransferase (GOGAT, aka glutamate synthetase; see Fig. 16.1). Together, GS and GOGAT constitute a metabolic cycle with their respective reactions driven by ATP and reductant in the form of reduced ferredoxin or NADH. The GS/GOGAT cycle also

*E-mail address: plaxton@queensu.ca

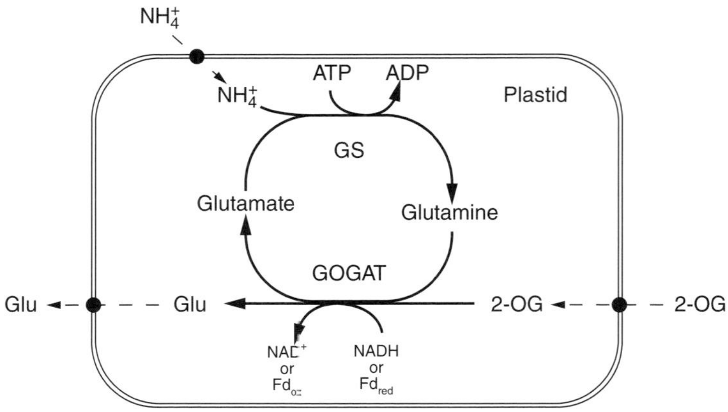

Fig. 16.1. The plastid glutamine synthetase/glutamine 2-oxoglutarate aminotransferase (aka glutamate synthetase) or GS/GOGAT cycle functions in the assimilation of NH_4^+ into glutamate and requires ATP, a reductant (either NADH or reduced ferredoxin) and 2-oxoglutarate (2-OG). Other abbreviations: Fd_{ox}, oxidized ferredoxin; Fd_{red}, reduced ferredoxin; Gln, glutamine; Glu, glutamate.

requires 2-oxoglutarate (2-OG) as a C backbone for each NH_4^+ assimilated. The Glu and Gln produced are consequently the primary NH_4^+ donors for the biosynthesis of all nitrogenous compounds needed by growing plants, including amino acids, nucleic acids, chlorophylls, polyamines and important secondary metabolites such as alkaloids. Synthesis of the other major amino acids from Glu requires aminotransferase activity and a corresponding α-keto acid carbon backbone produced by primary C metabolism (Fig. 16.2). The carboxylation of phosphoenolpyruvate (PEP) to oxaloacetate (OAA) is needed as the substrate for aspartate (Asp) synthesis by aspartate aminotransferase (AAT), while the glycolytic intermediates pyruvate and 3-phosphogylcerate, respectively, can be transaminated to form alanine (Ala) and serine (Ser) (Fig. 16.2).

In photosynthesizing leaves, previously assimilated NH_4^+ can be released at a high rate by the photorespiratory activity of the mitochondrial glycine decarboxylase complex (Leegood *et al.*, 1995). The release of NH_4^+ produced by photorespiration and its reassimilation into Glu can be viewed as a cycle not requiring a net addition of C skeletons. None the less, photorespiration is energetically costly, has very important implications for how amino acids are metabolized in the light, and is a key interface of C and N metabolism in plants (Leegood *et al.*, 1995; Nunes-Nesi *et al.*, 2010).

Owing to its metabolic organization, N assimilation places a large demand on the cell for ATP, carbohydrates and reductant, and must therefore closely interact with photosynthesis and respiratory metabolism. Furthermore, as NH_4^+ is toxic, it must be rapidly assimilated once it has been transported into the cell via NH_4^+ transporters, or produced by chloroplastic nitrate reductase or the mitochondrial glycine decarboxylase complex (Britto *et al.*, 2001). Consequently, the control of photosynthesis, respiration and N assimilation are tightly interwoven to ensure rapid and timely assimilation of available NH_4^+ and avoid the accumulation of toxic NH_4^+. In green plant tissues, photosynthesis can supply the ATP and reductant needed for N assimilation, whereas in darkened leaves or non-photosynthetic tissues these energy requirements are fulfilled by C respiration. The C skeletons needed for N assimilation and amino acid synthesis are derived ultimately from photosynthesis. This C requirement can be substantial: for example, in N-limited green algae resupplied with NH_4^+, over 50% of the net C gain is required to generate C skeletons and energy for the GS/GOGAT system (Huppe and Turpin, 1994). In multicellular plants, the 3-phosphoglycerate produced by photosynthetic CO_2 fixation can be stored as starch, used for amino acid synthesis within the cell, or converted to sucrose and transported to other tissues. Either way, all cells active in primary N assimilation require an enhanced respiratory C

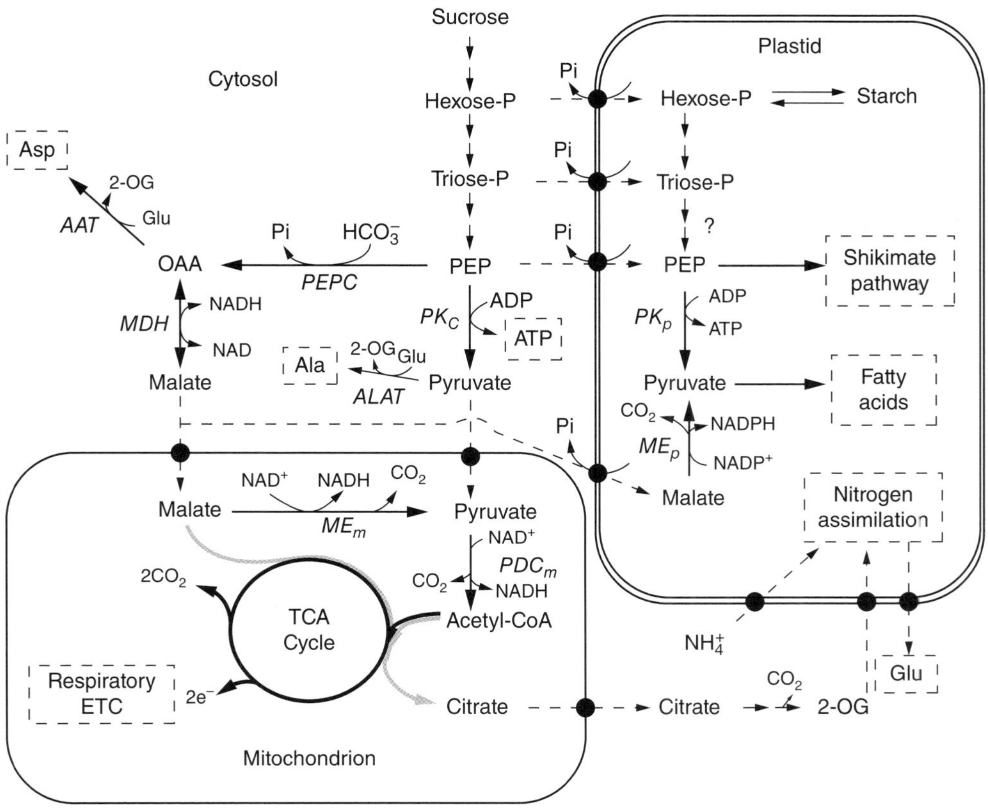

Fig. 16.2. In plants, the interactions between carbon (C) and nitrogen (N) metabolism involve cytosolic, mitochondrial and plastidial compartments. This scheme represents the organization of primary C metabolism surrounding the key glycolytic phosphoenolpyruvate (PEP) branch point where carbohydrates from glycolysis are partitioned to support the pathways shown in dashed boxes. Some plastids do not contain the glycolytic isoenzymes necessary for conversion of triose phosphate to PEP as denoted by the '?'. The two terminal enzymes of cytosolic glycolysis, pyruvate kinase (PK_c) and PEP carboxylase (*PEPC*), have a pivotal role in controlling the provision of respiratory substrates (pyruvate and malate) to the mitochondria, as well in the generation of the 2-oxoglutarate (2-OG) required for NH_4^+-assimilation via GS/GOGAT (glutamine synthetase/glutamine 2-oxoglutarate aminotransferase) in the plastids and the oxaloacetate (OAA) required for NH_4^+-assimilation via aspartate (Asp) aminotransferase (*AAT*) in the cytosol. The pathway of carbon flux through the tricarboxylic acid (TCA) cycle in support of NH_4^+ assimilation is shown in grey, and the (cyclic) TCA cycle flux is shown in black. Alternative metabolic routes exist for the conversion of PEP into fatty acids. Plastidial pyruvate kinase (PK_p) generates pyruvate and ATP for plastidial fatty acid synthesis. The import of malate into the plastid can also supply pyruvate and reducing power (NADPH) via malic enzyme for fatty acid synthesis. Other abbreviations: Ala, alanine; *ALAT,* alanine aminotransferase; ETC, electron transport chain; Glu, glutamate; *MDH*, malate dehydrogenase; ME_m/ME_p, mitochondrial and plastidial isoenzymes of malic enzyme; Pi, phosphate; PDC_m, mitochondrial pyruvate dehydrogenase complex.

flux to supply C skeletons needed for NH_4^+ assimilation into Glu and subsequent transamination to produce other amino acids.

Importantly, net formation of the C_4 and C_5 backbones (OAA and 2-OG) needed for the production of Asp, Glu, Gln and their derived amino acids, requires the anaplerotic carboxylation of PEP to OAA by phosphoenolpyruvate carboxylase (PEPC; Fig. 16.2). Subsequently, the generation of 2-OG requires the condensation of OAA and acetyl-CoA (derived from pyruvate) into citrate, which is catalysed by citrate synthase.

Citrate can be metabolized to 2-OG via the mitochondrial or cytosolic isoenzymes of aconitase and isocitrate dehydrogenase. The salient features of 2-OG synthesis are that it requires two separate C fluxes (pyruvate and OAA) originating from the glycolytic PEP branch point, and the operation of some tricarboxylic acid (TCA) cycle enzymes in a non-cyclic mode of C flux (Fig. 16.2). The enzymes and pathways involved in the formation of 2-OG and OAA are not dedicated to the task of amino acid synthesis and must simultaneously support other aspects of primary metabolism, such as energy generation via the (cyclic) TCA cycle flux (Fig. 16.2). This provides an interesting challenge to metabolic biochemistry in terms of elucidating how flux through partially overlapping pathways is controlled to meet changing metabolic demands. The efficient control of respiration during N assimilation occurs at transcriptional and post-translational levels and our knowledge in both areas continues to grow rapidly (Nunes-Nesi *et al.*, 2010). This review focuses on how the acute post-translational control of carbohydrate flux through glycolysis and respiratory pathways is achieved to ensure that the necessary supply of C skeletons and energy needed for amino acid synthesis matches the net rate of N assimilation.

16.2.1 The pivotal role of phosphoenolpyruvate metabolism in the control of plant glycolysis and respiration

A prominent feature of plant metabolism is the presence of the complete or nearly complete glycolytic pathways in both the cytosolic and plastid compartments which can interact through the action of specific metabolite transporters of the inner plastid envelope (see Fig. 16.2) (Plaxton and Podestá, 2006). These transporters include the PEP translocator, which mediates the counter-exchange of the glycolytic intermediate PEP and inorganic phosphate (Pi) between the plastid and cytosol (Weber and Linka, 2011). The metabolism of PEP is a crucial branch point in the partitioning of respiratory C flux to support different metabolic processes (Fig. 16.2) (Plaxton and Podestá, 2006; Plaxton and O'Leary, 2012). PEP metabolism by cytosolic pyruvate kinase (PK_c) generates cytosolic ATP and pyruvate. Pyruvate transported into the mitochondria serves to generate acetyl-CoA, citrate and ATP via the pyruvate dehydrogenase complex, the TCA cycle and oxidative phosphorylation, respectively. Cytosolic pyruvate can also be transaminated to Ala via alanine aminotransferase (ALAT). Plants, unlike animals and fungi, have additional flexibility at the PEP node because they express PEPC and a plastidial pyruvate kinase (PK_p). PK_p has a key function within plastids to produce pyruvate and ATP in support of plastidial anabolic pathways, particularly fatty acid synthesis (Andre *et al.*, 2007). PEPC is a tightly controlled cytosolic enzyme that catalyses the irreversible *β*-carboxylation of PEP to OAA. C_4 and Crassulacean acid metabolism (CAM) photosynthetic PEPCs have been widely studied owing to their central role in atmospheric CO_2 fixation (Chollet *et al.*, 1996). However, the fundamental purpose of PEPC in all plant cells is the anaplerotic replenishment of the organic acid pool that is consumed by biosynthetic reactions such as NH_4^+ assimilation via GS/GOGAT (O'Leary *et al.*, 2011b). Furthermore, when coupled with malate dehydrogenase and the mitochondrial NAD malic enzyme or the plastidial NADP malic enzyme, PEPC forms an alternative flux mode that can function in place of either PK_c or PK_p (Schwender *et al.*, 2006; Alonso *et al.*, 2007). *In vivo* metabolic flux studies of developing *Brassica napus* embryos demonstrated that approximately 40% of mitochondrial pyruvate was derived via NAD-malic enzyme, with the rest being imported from the cytosol (Schwender *et al.*, 2006). The plastidial shikimate pathway also draws directly from the plastidial pool of PEP and is the source of aromatic amino acids and secondary metabolites.

In non-plant systems, such as mammalian liver, the glycolytic flux to pyruvate is controlled primarily at the level of the conversion of fructose-6-phosphate (Fru-6-P) to fructose-1,6-bisphosphate (Fru-1,6-P_2) by ATP-dependent phosphofructokinase (ATP-PFK), with secondary control at the level of PEP catabolism by PK (Plaxton, 1996; Plaxton and Podestá, 2006). This 'top-down' control occurs because Fru-1,6-P_2, which increases upon ATP-PFK activation, is a potent feed-forward allosteric activator of the majority of non-plant PKs examined to date. In contrast, plant glycolysis is controlled from the 'bottom up' with primary and secondary regulation exerted at the levels of PEP and Fru-6-P utilization, respectively (Fig. 16.3). This has been consistently demonstrated in a variety of green algal and vascular plant systems by quantifying changes in the levels of glycolytic intermediates that immediately follow the stimulation of respiration (Plaxton, 1996; Plaxton and Podestá, 2006). In concurrence with these metabolite studies, *in vitro* analyses of plant ATP-PFKs and pyrophosphate-dependent PFKs (PPi-PFK) demonstrated the potent allosteric feedback inhibition of these enzymes by PEP, whereas plant PKs are insensitive to Fru-1,6-P_2 (Plaxton and Podestá, 2006). Therefore, an enhancement in the activity of PK_c and/or PEPC lowers PEP levels, thereby relieving the feedback inhibition of the ATP- and PPi-PFKs by PEP, leading to an overall stimulation of the glycolytic flux from hexose phosphates (hexose-Ps) (Fig. 16.3). Furthermore, a reduction in cytosolic PEP, triose phosphate (triose-P) and 3-phosphoglycerate (3-PGA) levels all stimulate the production of the potent signal metabolite fructose-2,6-bisphosphate (Fru-2,6-P_2) by 6-phosphofructo-2-kinase (Nielsen *et al.*, 2004). Increased Fru-2,6-P_2 levels serve to activate PPi-PFK and inhibit cytosolic fructose bisphosphatase, thereby activating glycolysis and inhibiting gluconeogenesis (Fig. 16.3). In these ways, the activities of PEPC and PK_c play a central role in determining the rate of mobilization of starch or sucrose for respiration, while simultaneously controlling the provision of pyruvate for mitochondrial ATP synthesis and organic acids for N assimilation.

The integration of cellular metabolism necessitates controlled interactions between pathways sequestered in different sub-cellular compartments. This is certainly true of plant N assimilation, which requires reactions specific to the cytosolic, mitochondrial and plastidic

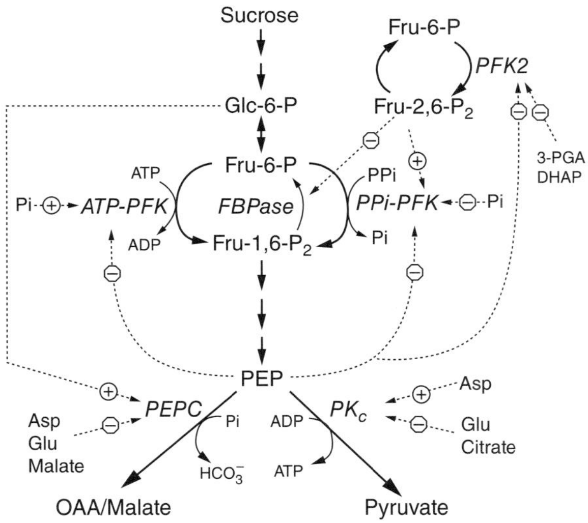

Fig. 16.3. The control of plant glycolysis includes potent allosteric effectors of key cytosolic enzymes. The central role of phosphoenolpyruvate (PEP) in providing a 'bottom up' type of control is an important feature of plant glycolysis. The metabolism of PEP by cytosolic pyruvate kinase (PK_c) and PEP carboxylase (PEPC) is tightly controlled by allosteric effectors, mainly from downstream metabolism, including reciprocal control by aspartate (Asp). The potent glycolytic regulator fructose-2,6-biphosphate (Fru-2,6-P_2) activates pyrophosphate-dependent phosphofructokinase (PPi-PFK) and inhibits fructose bisphosphatase (FBPase) but does not affect ATP-dependent phosphofructokinase (ATP-PFK). The levels of Fru-2,6-P_2 are controlled by the bifunctional 6-phosphofructo-2-kinase/fructose-2,6-bisphosphatase (PFK2) which, in turn, is controlled by allosteric effectors, of which key inhibitors are PEP, dihydroxyacetone phosphate (DHAP) and 3- phosphoglycerate (3-PGA). Dotted arrows with a circled plus and minus sign indicate enzyme activation and inhibition, respectively, by allosteric effectors. Other abbreviations: Asp, aspartate; Fru-6-P, fructose-6-phosphate; Glc-6-P, glucose-6-phosphate; Glu, glutamate; OAA, oxaloacetate; Pi, phosphate; PPi, pyrophosphate.

compartments. It has been hypothesized that Glu and Asp are important *in vivo* allosteric effectors of PEPC and PK_c activity that signal the demand for the 2-OG and OAA C skeletons needed to assimilate and transfer currently available NH_4^+. The following sections provide evidence to support this model of glycolytic control, based upon metabolomic data, the *in vitro* enzymology of PEPC, PK_c and PK_p, and the characterization of genetically modified plants that display alterations in the control of PEP and overall respiratory metabolism by Asp and Glu.

16.3 Metabolic Effects of N Resupply in Unicellular Green Algae and Vascular Plants

16.3.1 The response of primary C metabolism to N resupply in N-starved green microalgae

An elegant series of papers published in the early 1990s by David Turpin and colleagues utilized algal chemostats to analyse in detail the metabolic response of N-starved algae to N resupply (Weger and Turpin, 1989; Turpin *et al.*, 1990; Vanlerberghe *et al.*, 1990, 1992; Amory *et al.*, 1991; Huppe and Turpin, 1994). Chemostat cultures of unicellular green algae are a very useful model system for studies of C/N interactions in plant cells as the presence of multiple cell types in vascular plant tissues can confound metabolic measurements because, ultimately, they represent averages across cell types that may differ metabolically. A homogeneous population of unicellular green algae in culture helps to avoid this major issue. Furthermore, the nutrient media that sustain algal cultures can be very accurately and rapidly manipulated within chemostat incubators. In this environment, the algae display a uniform and rapid response to N resupply that can be exploited to reveal key aspects of the control and integration of plant C/N interactions.

Fundamentally, it was known that N-limited algae can assimilate NH_4^+ in the dark in a process that required CO_2 and carbohydrate (starch) (Amory *et al.*, 1991; Huppe and Turpin, 1994). N-sufficient algal cells, in contrast, would only assimilate NH_4^+ during active photosynthesis as they do not accumulate sufficient amounts of starch (Huppe and Turpin, 1994). Therefore, the use of dark, N-limited algal cells enabled the study of C respiration during N assimilation in isolation from photosynthetic C metabolism. This was advantageous because it showed that the assimilation of N by NH_4^+-resupplied algae was accompanied by a rapid decrease in starch stores and a concomitant upregulation of dark CO_2 fixation by PEPC at a rate of 0.3 CO_2 fixed per N assimilated (Vanlerberghe *et al.*, 1990). These two events were attributed to the activation of glycolysis and anaplerosis, respectively, to generate C skeletons needed for N assimilation via GS/GOGAT. However, the mechanisms that allowed C respiration to be functionally integrated with N assimilation remained unknown.

Comparison of time course metabolite extractions from the green algae *Selenastrum minutum* during the first few seconds after NH_4^+ resupply identified potential mechanisms of control which were hypothesized to rapidly activate glycolytic C flux in response to N resupply. A distinct pattern of changes in metabolite levels was observed that began with a drop in Glu, 2-OG and ATP levels, and a sharp rise in Gln, AMP and ADP levels. This reflected the rapid assimilation of NH_4^+ by GS/GOGAT (Fig. 16.1; Turpin *et al.*, 1990; Vanlerberghe *et al.*, 1992; Huppe and Turpin, 1994). These rapid changes in metabolite concentrations were proposed to act as a signal to enhance upstream C metabolism at the level of PEP metabolism. Within seconds of NH_4^+ resupply, there was a dramatic drop in PEP levels and an increase in pyruvate, indicating a rapid activation of PK_c (Schuller *et al.*, 1990a; Turpin *et al.*, 1990; Vanlerberghe *et al.*, 1992). The rapid and massive induction of dark CO_2 fixation and decrease in the PEP/malate ratio also demonstrated the concurrent activation of PEPC (Schuller *et al.*, 1990a; Vanlerberghe *et al.*, 1992). Simultaneously, the build-up of pyruvate via PK_c, combined with the reduction of *in vivo* adenylate energy charge, would serve to activate the mitochondrial pyruvate dehydrogenase complex, which would explain the increase in citrate levels (Turpin *et al.*, 1990; Vanlerberghe *et al.*, 1992). The induction of the TCA cycle (or partial cycle) was then suggested to explain a rapid increase in the $NADH/NAD^+$ ratio

(Vanlerberghe *et al.*, 1992). As mentioned above, the reduction in PEP levels will act to relieve feedback inhibition of ATP-PFK (*S. minutum* contains plastidic PFK, but not cytosolic ATP-PFK or PPi-PFK) and thus enhance flux from hexose-Ps derived from stored starch through glycolysis. Increased activity of ATP-PFK was reflected by increased Fru-1,6-P_2 levels and a decrease in hexose-Ps (i.e., a so-called 'positive cross-over') (Turpin *et al.*, 1990; Vanlerberghe *et al.*, 1992). The large increase observed for triose-P levels, combined with decreased levels of 3-phosphoglycerate, were consistent with an increased flux through the lower half of glycolysis to PEP. Lastly, the export of glycolytic intermediates from the plastids to the cytosol occurs in exchange for Pi, and increased plastidial levels of Pi favour the breakdown of starch in the dark (Heldt *et al.*, 1977). Therefore, the drop in PEP levels caused by activation of PK_c and PEPC ultimately serves to activate C flux from starch through glycolysis, anaplerosis and all or part of the TCA cycle. It is important to note that all these changes in metabolite levels, which are suggestive of changes in respiratory pathway flux, are occurring in the order of seconds or less and are, consequently, exclusively due to the post-translational control of pre-existing enzymes rather than to any *de novo* change in enzyme abundances. Thus, this model relies heavily on the sensitive allosteric control of both PEPC and PK_c to molecules that are indicative of the N status of the cell.

Similar experiments performed with NO_3^- resupply revealed that the same general pattern of metabolite changes occurs as with NH_4^+, but with a time delay attributed to the need to reduce NO_3^- to NH_3^+ via NO_3^- reductase (Vanlerberghe *et al.*, 1992). The consequence of these prerequisite NO_3^- reductase reactions is that the supply of NH_4^+ for assimilation is less volatile. It follows that the amplitude of the changes in metabolite levels was also dampened with NO_3^- compared with NH_4^+, except for a rapid increase in 6-phosphogluconate levels that indicated increased flux through the oxidative pentose phosphate pathway to generate reductant (NADPH) needed to support plastidial NO_3^- reduction in the dark (Vanlerberghe *et al.*, 1992; Huppe and Turpin, 1994).

16.3.2 The response of primary C metabolism to N resupply in vascular plants

It could be expected that similar regulatory signals operate in vascular plants as in green algae to control efficient NH_4^+ assimilation. However, the situation in plants is complicated because of a diversity in the location and timing of NH_4^+ assimilation. Also, several key differences exist between the physiology/metabolic organization of algal and plant cells: (i) the dynamics of the algal liquid environment create a feast and famine relationship with N, emphasizing rapid NO_3^- or NH_4^+ assimilation; (ii) green algae such as *S. minutum* do not appear to contain cytosolic ATP- or PPi-PFKs; and (iii) sucrose synthesis and export is a major function of photosynthetic plant cells, but not of algal cells (Geigenberger and Stitt, 1991). Still, in N-starved, photosynthetic poppy (*Papaver somniferum*) and spinach (*Spinacia oleracea*) leaf cells (Paul *et al.*, 1978; Larsen *et al.*, 1981; Lawyer *et al.*, 1981) and in germinating castor cotyledons (Geigenberger and Stitt, 1991), rapid changes in metabolite levels occur upon NH_4^+ resupply that mirror the changes in green algal cultures. In particular, the addition of NH_4^+ leads to a rapid decrease in PEP/pyruvate and PEP/malate ratios, indicating a sudden activation of PEPC and PK. In the isolated poppy and spinach cells, the Glu levels diminish directly upon addition of NH_4^+, combined with an increase in Gln, Asp, citrate and Ala. In the germinating castor cotyledons, the observed increases in Fru-1,6-P_2, Fru-2,6-P_2, triose-Ps and CO_2 production clearly demonstrated a rapid increase in glycolytic and respiratory C flux upon NH_4^+ resupply (Geigenberger and Stitt, 1991). Interestingly, the flow of C from Glu into γ-aminobutyric acid (GABA) ceased immediately after NH_4^+ resupply to spinach cells (Larsen *et al.*, 1981). This observation further suggests that the metabolically available pool of Glu has been sharply reduced (Larsen *et al.*, 1981).

Similar metabolite perturbations do not occur in vascular plant cells resupplied with NO_3^- instead of NH_4^+. There are several reasons why this may be so. High levels of NO_3^- can be stored in the vacuole, which normally allows NO_3^- reduction to NH_4^+ to occur at a scheduled pace throughout the diurnal cycle. Upon NO_3^- resupply to N-starved plants, a change in metabolite levels does occur, but only after several hours, and it is accompanied by changes in the transcripts and proteins of the corresponding enzymes (Scheible *et al.*, 2004). Furthermore, evidence is mounting that the organic acids used to assimilate NH_4^+ in the leaves during the day originate at least partially from a pool of citrate that was formed the previous night and is separate from the organic acids being created by PEPC during the day (Tcherkez *et al.*, 2009; Gauthier *et al.*, 2010; Nunes-Nesi *et al.*, 2010). Therefore, it does not appear that the same coordination of respiration that occurs in green algae in response to rapid, transient NO_3^- metabolism is occurring in the leaves of vascular plants.

16.4 The Post-translational Control of Plant Phosphoenolpyruvate Carboxylase and Cytosolic Pyruvate Kinase is Often Geared to NH_4^+ Assimilation

The organization of primary plant metabolism, as depicted in Fig. 16.2, places much importance on the coordinate post-translational control of PEP metabolism via PEPC and PK (Plaxton and Podestá, 2006). Allosteric effectors, post-translational modifications, adenylate energy charge, pH and protein–protein interactions have all been described as mechanisms of fine metabolic control that determine the C flux through these enzymatic reactions (Huppe and Turpin, 1994; Plaxton and Podestá, 2006; O'Leary *et al.*, 2011b). During NH_4^+ assimilation, both PEPC and PK_c must respond in a concerted and balanced fashion to provide 2-OG and OAA needed for GS/GOGAT and AAT, respectively. In the absence of photosynthesis especially, respiration from PEP through the TCA cycle also has a critical function to supply ATP and reductant for other cellular processes (Weger *et al.*, 1988; Weger and Turpin, 1989). A biochemical mechanism to balance these fluxes is the differential regulation of PEPC and PK_c by allosteric effectors, in particular Glu and Asp. Here we discuss the outcome of several decades of research into the biochemical properties of PEPC and PK_c across multiple plant and algal species and how these studies have presented a working model for the integration of plant C respiration and N assimilation.

16.4.1 The functional diversity of plant phosphoenolpyruvate carboxylase isoenzymes reflects their complex mechanisms of post-translational control

Organic acids play a number of important physiological roles *in vivo* in addition to being obligate substrates for biosynthetic reactions such as N assimilation: they act as counter-ions, osmoticums, acidifying agents, respiratory substrates, transporters of redox potential, transporters/stores of CO_2 in CAM and C_4 photosynthesis, and chelators of metal ions (Martinoia and Rentsch, 1994; Finkemeier and Sweetlove, 2009). The actions of organic acids are inextricably linked to PEPC activity because this carboxylation reaction is the predominant anaplerotic pathway in plant cells, i.e. the means of *de novo* production of organic acids from sugars and starch. Therefore, PEPC must operate in a number of different metabolic contexts, all involving the creation of organic acids (O'Leary *et al.*, 2011b). To fulfil these functions, PEPC activity is regulated at several levels. First, PEPCs belong to a small multi-gene family encoding several plant-type PEPCs (PTPCs), along with at least one distantly related bacterial-type PEPC (BTPC; Sánchez and Cejudo 2003; Mamedov *et al.*, 2005; O'Leary *et al.*, 2011b). These genes have distinct expression patterns and encode catalytically active PEPC subunits with differing kinetic parameters. Secondly, the PEPC subunits associate into two different functional PEPC complexes, known as Class-1 PEPC and Class-2 PEPC, which have very different physical and regulatory properties (Rivoal *et al.*, 2001; Blonde and Plaxton, 2003; Mamedov *et al.*, 2005; Gennidakis *et al.*, 2007). Thirdly, the functional PEPC complexes are each subjected to several different regulatory post-translational modifications (PTMs), as summarized in Fig. 16.4 (O'Leary *et al.*, 2011a).

These multiple tiers of regulation have been the subject of several decades of molecular and biochemical analyses of the PEPCs of vascular plants and green algae. To summarize, Class-1 PEPCs typically exist as homotetramers composed

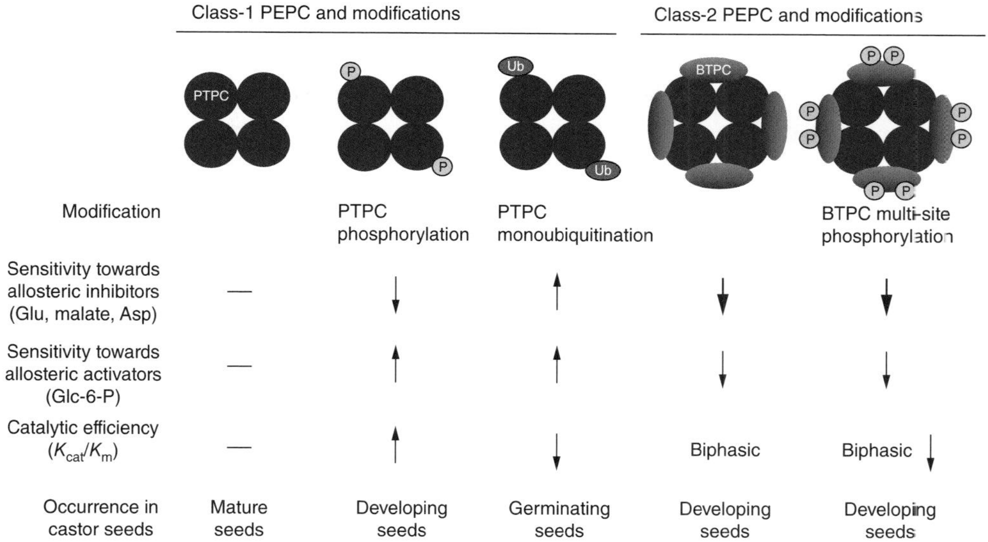

Fig. 16.4. Model illustrating the biochemical complexity of vascular plant phosphoenolpyruvate carboxylase (PEPC) as revealed through detailed analyses of novel PEPC isoforms from developing versus germinating castor beans. The qualitative effects are compared to known reversible regulatory post-translational modifications (PTMs) to Class-1 PEPC homotetramers of PTPC (plant type PEPC) subunits and to the Class-2 PEPC hetero-octameric complex of a PTPC tetramer with four BTPC (bacterial type PEPC) subunits. In developing castor seeds, the Class-1 PEPC is activated *in vivo* by sucrose-dependent phosphorylation at Ser-11 (Tripodi *et al.*, 2005). Also in developing seeds, a tight association between the same Class-1 PEPC and distantly related BTPC subunits forms the allosterically desensitized Class-2 PEPC hetero-octameric complex that dynamically associates with the mitochondrial outer envelope *in vivo* (Blonde and Plaxton, 2003; Gennidakis *et al.*, 2007; Park *et al.*, 2012). Within Class-2 PEPCs, both subunit types are active but differ in catalytic properties, resulting in biphasic phosphoenolpyruvate (PEP) saturation kinetics (O'Leary *et al.*, 2009). The BTPC subunits are subject to *in vivo* multi-site phosphorylation at Ser-425 and Ser-451, which are both inhibitory (O'Leary *et al.*, 2011a; Dalziel *et al.*, 2012). Castor bean maturation is accompanied by disappearance of the Class-2 PEPC complex and BTPC polypeptides and transcripts and a marked reduction in the amount of Class-1 PEPC coupled with dephosphorylation of its PTPC subunits (Tripodi *et al.*, 2005; Gennidakis *et al.*, 2007). Seed imbibition and germination is accompanied by increased PTPC gene expression and PEPC activity. In germinating seeds, 50% of the PTPC subunits become monoubiquitinated at Lys-628, which has an inhibitory effect on Class-1 PEPC activity (Uhrig *et al.*, 2008b). Based on the occurrence and effects of each isoform and modification, phosphorylation of Class-1 PEPC appears to be geared towards supporting the production of C skeletons needed for storage protein and lipid biosynthesis in castor seeds (Tripodi *et al.*, 2005).

of identical 105–110 kDa PTPC subunits. Plant genomes encode several different PTPC isoenzymes which all contain a distinctive N-terminal seryl-phosphorylation domain and a C-terminal QNTG motif (Chollet *et al.*, 1996; Izui *et al.*, 2004; O'Leary *et al.*, 2011b). PTPCs can be further classified as being either photosynthetic (i.e. C_4 and CAM PEPCs) or non-photosynthetic (C_3) isoenzymes, based on the presence of distinct amino acid substitutions (Svensson *et al.*, 2003; Izui *et al.*, 2004). All plant and algal genomes also encode at least one *BTPC* gene, which together constitute a monophyletic group, separate from either the PTPCs or bacterial and archaeal PEPCs, and appear to have evolved in green algae (Sánchez and Cejudo, 2003; Mamedov *et al.*, 2005; O'Leary *et al.*, 2011b). The BTPCs are larger ~120–130 kDa polypeptides that lack the N-terminal phosphorylation motif characteristic of PTPCs and have a prokaryotic-like (R/K)NTG C-terminal motif. The BTPC subunits have only been isolated as subunits of a larger hetero-octameric complex known as a Class-2 PEPC (Fig. 16.4). The ~910 kDa Class-2 PEPC complex consists of a PTPC homotetramer tightly associated with four additional BTPC subunits (Gennidakis *et al.*, 2007; O'Leary *et al.*, 2009; Park *et al.*, 2012).

It has been known since the early 1980s that Class-1 PEPCs show allosteric inhibition by malate and activation by glucose-6-phosphate (Glc-6-P) (Chollet *et al.*, 1996; O'Leary *et al.*, 2011b). Additional effectors, especially the inhibitors Glu and Asp, have been identified for many Class-1 PEPCs, as well as several regulatory PTMs. Reversible phosphorylation of PTPC subunits at their conserved N-terminal Ser residue has been well studied in many systems and serves to activates Class-1 PEPCs by relieving inhibition by malate, Asp and Glu, while increasing activation by Glc-6-P and slightly lowering the K_m(PEP) (Chollet *et al.*, 1996; Vidal and Chollet, 1997; Nimmo, 2000; Izui *et al.*, 2004; Tripodi *et al.*, 2005). Other regulatory post-translational PEPC modifications have been discovered and characterized more recently. Monoubiquitination of PTPC subunits at a conserved lysine residue is prevalent in many plant tissues and inhibits Class-1 PEPCs by increasing their K_m(PEP) and sensitivity to allosteric inhibitors, including malate and Asp (Uhrig *et al.*, 2008a; Igawa *et al.*, 2010; O'Leary *et al.*, 2011a; Shane *et al.*, 2013; Ruiz-Ballesta *et al.*, 2014). Similarly, the *in vivo* phosphorylation of developing castor seed BTPC subunits at two distinct Ser residues inhibits the Class-2 PEPC complex by increasing the K_m(PEP) of the BTPC and its sensitivity to allosteric inhibition by malate and Asp (Uhrig *et al.*, 2008b; O'Leary *et al.*, 2011c; Dalziel *et al.*, 2012). Recent studies have also revealed the dynamic localization of the Class-2 PEPC, but not the Class-1 PEPC, with the mitochondrial outer membrane in developing castor seeds (Park *et al.*, 2012) and the import of a divergent Class-1 PEPC isoenzyme into the chloroplast stroma in rice leaves (Masumoto *et al.*, 2010). The physiological function of PEPC isoenzymes could be tightly linked to their subcellular localization, and future research in this area is needed. Altogether, the specific substrate kinetics, allosteric properties, protein–protein interactions and PTMs of PEPCs are quite complex (Fig. 16.4), and, as discussed below, appear to be well adapted to the physiological role(s) of specific PEPC isoenzymes and the nature of their cellular environment.

The best studied example of the specialization of regulatory properties for a PEPC is that of the C_4 Class-1 PEPCs which display a decreased affinity for PEP and a decreased sensitivity to malate inhibition relative to typical C_3 Class-1 PEPCs (Svensson *et al.*, 2003). The post-translational control of photosynthetic C_4 PEPCs is largely achieved by allosteric activation and inhibition by Glc-6-P and malate, respectively, combined with activation by phosphorylation in the light (Chollet *et al.*, 1996). Therefore, the fine tuning of effector and substrate sensitivities probably allow C_4 PEPCs to be an efficient and controllable step of the C_4 cycle, which operates in mesophyll cells where cytosolic concentrations of PEP and malate are thought to be higher than in C_3 leaves (Svensson *et al.*, 2003).

In contrast to photosynthetic PEPCs, the control of non-photosynthetic PEPCs from many tissues appears to be geared towards a role in N assimilation. A key series of papers by Jean Rivoal and co-workers reported the purification and characterization of native PEPCs from the green algae *S. minutum* and *Chlamydomonas reinhardtii* (Schuller *et al.*, 1990a; Rivoal *et al.*, 1996, 1998, 2001, 2002). These studies are particularly informative given the aforementioned wealth of data on the metabolomic responses of these algal systems to N resupply. Also, the remarkable Class-2 PEPC complexes were first discovered and characterized from these algal systems alongside the typical Class-1 PEPCs. In both green algae and vascular plants, the Class-1 PEPCs exhibit a much greater sensitivity to allosteric effectors, including the inhibitors Glu and Asp, relative to the corresponding Class-2 PEPCs (Fig. 16.4) (Schuller *et al.*, 1990a; Rivoal *et al.*, 1998, 1996, 2001). The allosteric activation of algal Class-1 PEPCs by dihydroxyacetone-P (DHAP) and Gln, and inhibition by Glu, Asp, 2-OG and malate, matched well with the known changes in metabolites that occur after N resupply to dark grown, N-starved algae (see above) (Vanlerberghe *et al.*, 1992; Huppe and Turpin, 1994). In particular, the potent activation of Class-1 PEPCs by Gln and DHAP appears to be rather specific to green algae. The allosteric activation of Class-1 PEPCs resulting from these metabolite changes offers an explanation for the rapid decrease in PEP levels and the orders of magnitude increase in the rate of dark CO_2 fixation that immediately follow N resupply

(Schuller *et al.*, 1990a; Turpin *et al.*, 1990; Vanlerberghe *et al.*, 1992). Later, the identification and characterization of the *C. reinhardtii PEPC* genes by Raymond Chollet's laboratory revealed that green algal Class-1 PEPC transcripts and protein are also upregulated during acclimation to low N levels (Mamedov *et al.*, 2005; Moellering *et al.*, 2007). It was hypothesized that N-starved algal cells are primed for high N-assimilation capacity once N sources become available. From the combined biochemical and molecular characterization of algal Class-1 and Class-2 PEPCs, it was concluded that the Class-1 PEPCs in algae are specialized for anaplerotic fixation of CO_2 to support N assimilation (Rivoal *et al.*, 2001; Moellering *et al.*, 2007).

A second important model for PEPC research has been castor (*Ricinus communis*) oil seeds, which accumulate large amounts of both triglycerides and storage proteins to sustain later embryo germination before the commencement of photoautotrophic growth. PEPC activity is commonly linked to storage protein synthesis because the seeds import N mainly as Asn and Gln. These amino acids contain two amino groups per C skeleton and their transamination into the other amino acids sustains the need for OAA and 2-OG C skeletons. Also, amino acids such as Ala and Glu can be deamidated and respired through the TCA cycle, yielding NH_4^+ that must be rapidly reassimilated. Reminiscent of the studies in green algae, a Class-2 PEPC complex was purified and characterized alongside a Class-1 PEPC from developing castor endosperm (Blonde and Plaxton, 2003; Tripodi *et al.*, 2005; Gennidakis *et al.*, 2007). In this system it was again shown that the potent inhibition of Class-1 PEPC by Asp and Glu was largely relieved in the Class-2 PEPC (Blonde and Plaxton, 2003; O'Leary *et al.*, 2009). Therefore, the castor seed Class-1 PEPC appears to be well suited to support anaplerotic flux towards N assimilation (Blonde and Plaxton, 2003). The physiological role of mitochondrial associated Class-2 PEPC remains obscure but has been hypothesized to partition C towards plastid fatty acid synthesis via plastidial NADP-dependent malic enzyme, while simultaneously facilitating refixation of respired CO_2 (Fig. 16.2; Blonde and Plaxton, 2003; O'Leary *et al.*, 2009, 2011b; Park *et al.*, 2012).

Non-photosynthetic Class-1 PEPCs have been characterized from a number of other vascular plant tissues where an anaplerotic flux is expected to support N assimilation, such as developing bean (*Vicia faba*) seeds (Golombek *et al.*, 1999), germinating castor cotyledons (Podestá and Plaxton, 1994), rice leaves (Masumoto *et al.*, 2010), ripening banana fruit (Law and Plaxton, 1995), soybean nodules (Schuller *et al.*, 1990b), and *Arabidopsis* and *Brassica napus* suspension cell cultures (Gregory *et al.*, 2009, Moraes *et al.*, 2000). In these tissues the Class-1 PEPC is typically subject to tight allosteric control by the feedforward activator Glc-6-P and the feedback inhibitors Asp, Glu and malate (rice leaf PEPC is an interesting exception, see below). These findings support the existence of a mechanism, similar to that in green algae, for the integration of C/N metabolism by the allosteric regulation of PEPC in vascular plant tissues. Regulatory phosphorylation of Class-1 PEPCs by PEPC kinase is a second important layer of control in plant C/N interactions. Legume species express a root-nodule specific PTPC that is activated by phosphorylation via PEPC kinase whose expression is dependent upon the supply of photosynthate to the root (Xu *et al.*, 2003). In C_3 leaves, castor oil seeds and non-legume roots, PEPC kinase transcripts, activity and Class-1 PEPC phosphorylation–activation increase during transient N assimilation (Le Van Quy *et al.*, 1991; Champigny and Foyer, 1992; Duff and Chollet, 1995; Scheible *et al.*, 1997; Pasqualini *et al.*, 2001; Stitt *et al.*, 2002; Tripodi *et al.*, 2005; Murmu and Plaxton, 2007).

In the case of PEPC, the exceptions can also prove the rule. Non-photosynthetic Class-1 PEPCs from developing citrus and tomato fruits (Guillet *et al.*, 2002; Perotti *et al.*, 2010) and rice leaves (Masumoto *et al.*, 2010) all display properties that differ from the abovementioned 'typical' non-photosynthetic PEPCs in that they have a lower affinity for PEP and a reduced sensitivity to inhibition by typical allosteric inhibitors. For the tomato and orange fruits, these differences are consistent with their likely role in the maintenance of very high concentrations of organic acids rather than N assimilation. A surprising recent discovery was the presence of a chloroplast-targeted Class-1 PEPC, OsPPC4, in the mesophyll of rice leaves (Masumoto *et al.*, 2010). Heterologously expressed OsPPC4 displayed an increased K_m(PEP) and a reduced inhibition by malate, Asp and Glu compared with other C_3

Class-1 PEPCs. These differences were linked to the dissimilar metabolite concentrations found in the chloroplast stroma and the cytosol. This chloroplastic PEPC appears to provide a novel pathway towards organic acid production for the purpose of N assimilation because knock-outs of *OsPPC4* led to stunted growth as a result of impaired NH_4^+ assimilation and lowered levels of organic acids (Masumoto *et al.*, 2010). It is unknown whether chloroplastic PEPC exists in other genera, and further research in this area is anticipated.

16.4.2 The allosteric features of plant cytosolic pyruvate kinase isoenzymes help to synchronize C/N interactions in different tissues

PK catalyses the irreversible conversion of PEP and ADP into pyruvate and ATP (Fig. 16.2), and is thus a major source of cytosolic pyruvate and ATP. Along with OAA formed via the action of PEPC, equal amounts of pyruvate are needed for *de novo* synthesis of 2-OG from carbohydrate to support NH_4^+ assimilation by GS/GOGAT. Citrate is a key intermediate in this process and is formed in the mitochondria via the combined actions of the pyruvate dehydrogenase complex and citrate synthase. As with the organic acids OAA and 2-OG, pyruvate supports important metabolic fluxes besides N assimilation (Fig. 16.2). The oxidation of pyruvate to CO_2 via the pyruvate dehydrogenase complex and TCA cycle fuels the mitochondrial electron transport chain and the consequent oxidative phosphorylation of ADP to ATP. Pyruvate is also a major carbon precursor for plastidial fatty acid synthesis (Andre *et al.*, 2007; Baud *et al.*, 2007), as well as for the synthesis of alanine, valine, leucine and isoleucine. In addition, pyruvate can be fermented to ethanol or lactate during hypoxia or anoxia. Understanding the production of pyruvate is complicated by the fact that it is synthesized in multiple cellular compartments and by multiple enzymes. Plants and green algae contain both PK_c and PK_p isoenzymes which show dramatic differences in their physical, immunological and kinetic/regulatory properties. *Arabidopsis*, for example, has eight annotated genes for PK_c and four for PK_p (Andre *et al.*, 2007). Genetic studies have clearly shown that PK_p in developing *Arabidopsis* seeds functions in support of fatty acid synthesis (Baud *et al.*, 2007), whereas PK_c has been suggested to exert control over the glycolytic flux to pyruvate in the plant cytosol (Knowles *et al.*, 1998; Plaxton and Podestá, 2006; Oliver *et al.*, 2008). Additionally, malic enzyme isoenzymes have been shown to produce pyruvate *in vivo* by the oxidative decarboxylation of malate in both the mitochondria and plastids (Schwender *et al.*, 2006; Alonso *et al.*, 2007). As shown in Fig. 16.2, the combination of PEPC, malate dehydrogenase and mitochondrial malic enzyme can form a functional bypass to PK_c.

The biochemical properties of PK_c and PK_p have been characterized from a number of different tissues. In particular, the concurrent characterization of PK_p and PK_c isoenzymes from *S. minutum* and *B. napus* cells and various castor plant tissues provides a set of informative comparisons (Tables 16.1 and 16.2) (Lin *et al.*, 1989a,b; Smith *et al.*, 2000; Plaxton *et al.*, 2002). The plant and algal PK_ps and PK_cs characterized to date show a variety of allosteric effectors; however, it has been repeatedly shown that Glu is a potent allosteric inhibitor specific to most PK_c isoenzymes and that Asp functions as an activator by effectively alleviating the inhibition of the enzyme by Glu (Fig. 16.5). The inhibition of PK_c by Glu and its reversal by Asp was found to be specific for these particular amino acids, which differ by a single C atom in their functional group. As a result, the cytosolic [Asp]:[Glu] ratio can be expected to act as a major determinant of *in vivo* PK_c activity in many tissues (Fig. 16.5). In contrast, the PK_c isoenzyme isolated from the endosperm of germinating castor beans was insensitive to Glu and Asp (Table 16.2) (Podestá and Plaxton, 1991). The subunit composition of PK_cs from developing and germinating castor seeds are also different and may reflect the expression of tissue-specific PK_c isoenzymes to suit the distinctive metabolic requirements of developing versus germinating oil seeds. Allosteric regulation of PK_c by Asp and Glu may be beneficial during N-assimilation or storage protein accumulation, but is not required during the mobilization of storage proteins by germinating seeds, when the plant PK_c appears to be controlled by levels of adenylates and TCA cycle intermediates (Turner *et al.*, 2005; Plaxton and Podestá, 2006).

Table 16.1. Comparison of physical, kinetic and regulatory properties of purified plastidial pyruvate kinase (PK_p) and cytosolic pyruvate kinase (PK_c) isoenzymes from the green alga *Selenastrum minutum* and the vascular plant *Brassica napus*.

	Selenastrum minutum		*Brassica napus* cell culture	
Property	PK_c[a]	PK_p[a]	PK_c[b]	PK_p[c]
Subunit size(s) (kDa)	57	56	56	58 and 64
Subunit structure	Homodecamer	Homotetramer	Homotetramer	Heterohexamer
K_m for PEP (mM)[d]	0.09	0.18	0.12	0.052
K_m for ADP (mM)[d]	0.05	0.20	0.075	0.14
Most significant allosteric inhibitors	Glu, citrate, Pi	Citrate, ATP, 2,3-diphosphoglycerate	Glu, oxalate, quercetin	2-OG, oxalate
Most significant allosteric activators	Asp[e], DHAP, Glc-1-P, Glc-6-P	DHAP	Asp[e]	6-Phospho-gluconate

Abbreviations: Asp, aspartate; DHAP, dihydroxyacetone-P; Glc-1-P, glucose-1-phosphate; Glc-6-P, glucose-6-phosphate; Glu, glutamate; 2-OG, 2-oxoglurate; PEP, phosphoenolpyruvate; Pi, inorganic phosphate.
[a]Lin *et al.*, 1989a,b; [b]Smith *et al.*, 2000; [c]Plaxton *et al.*, 2002; [d]Determined at optimum pH; [e]Activation also by relieving inhibition by Glu and/or Pi.

Table 16.2. Comparison of physical, kinetic and regulatory properties of purified cytosolic pyruvate kinase (PK_c) isoenzymes from various tissues of *Ricinus communis* (castor oil plant).

	Tissue			
Property	Expanding leaves	Developing endosperm	Germinating endosperm	Germinating cotyledon
Subunit size(s) (kDa)	56 and 57	56	56 and 57	56 and 57
Subunit structure	Heterotetramer	Homotetramer	Heterotetramer	Heterotetramer
K_m for PEP (mM)[a]	0.089	0.082	0.152	0.034
K_m for ADP (mM)[a]	0.087	0.090	0.078	0.051
Most significant allosteric inhibitors	Glu, citrate	Glu	ATP, malate, isocitrate	Glu, citrate, ATP
Most significant allosteric activators	Asp[b]	Asp[b]	None identified	None identified

Abbreviations: Asp, aspartate; Glu, glutamate.
[a]Determined at optimum pH; [b]Activation mainly by relieving Glu inhibition.

16.4.3 Glutamate and aspartate play a central role in the coordinate allosteric control of phosphoenolpyruvate carboxylase and cytosolic pyruvate kinase during NH_4^+ assimilation

The stimulation of respiration that occurs in vascular plants and green algae during periods of active N uptake has been ascribed to an initial activation of PK_c and PEPC leading to increased production of the C skeletons needed for NH_4^+ assimilation and transamination reactions. Figure 16.3 presents a model summarizing the allosteric features of PEPC and PK_c that are believed to be important in coordinating C/N metabolism in plant cells active in NH_4^+ assimilation. The regulatory features of both enzymes appear to be well suited to their central role in providing the mitochondria with respiratory substrates as well as generating C skeletons for NH_4^+ assimilation via GS/GOGAT and AAT (Fig. 16.2).

Glc-6-P activation of PEPC serves to balance sucrose availability with the flux of PEP carboxylation to dicarboxylic acids via PEPC. Malate inhibition of PEPC provides tight feedback regulation that closely links PEPC activity with the overall

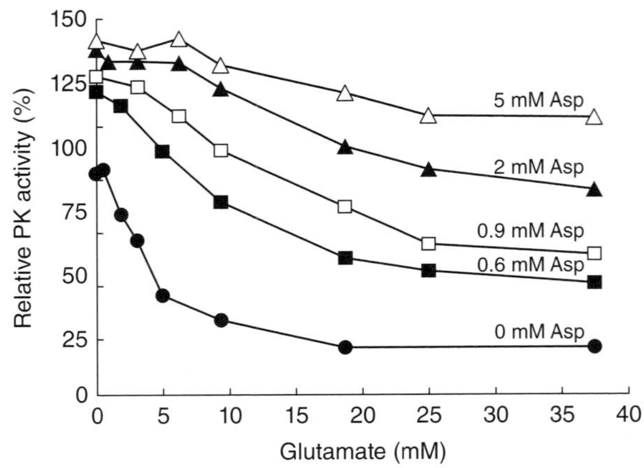

Fig. 16.5. The influence of levels of aspartate (Asp) and glutamate (Glu) on the *in vitro* activity of *Brassica napus* cytosolic pyruvate kinase PK_c. Asp activates PK_c mainly by relieving its inhibition by Glu. Assays were conducted at subsaturating substrate (phosphoenolpyruvate, PEP; and ADP) concentrations. Glu and Asp levels needed to modulate *B. napus* PK_c activity *in vitro* are well within the physiological range, as the cytosolic Glu and Asp concentrations of N-sufficient spinach leaves have both been estimated to be at least 20 mM (Winter *et al.*, 1994). (Figure adapted from Smith *et al.*, 2000.)

rate of malate metabolism. In particular, Asp and Glu appear to play a central role in the coordinate allosteric control of the PEP branch point in plant tissues active in NH_4^+ assimilation. The addition of NH_4^+ to spinach leaves has been estimated to increase metabolically active Asp levels by as much as 300%, while decreasing Glu levels by as much as 70% (Baysdorfer and Bassham, 1984). Feedback inhibition of PK_c and PEPC by Glu thus provides a rationale for the known activation of the two enzymes that occurs *in vivo* during periods of enhanced NH_4^+ assimilation. In contrast to PEPC, Asp functions as an allosteric activator of PK_c by effectively relieving inhibition of the enzyme by Glu (Fig. 16.5) (Baysdorfer and Bassham, 1984; Hu and Plaxton, 1996; Smith *et al.*, 2000; Turner and Plaxton, 2000; Turner *et al.*, 2005). Reciprocal control of PK_c and PEPC by Asp provides an intriguing mechanism for decreasing flux from PEP to Asp (via PEPC and AAT), while promoting PK_c activity when cytosolic Asp levels become elevated. This may occur because the demands of the cell for N are satisfied, and the overall rate of protein synthesis becomes more dependent upon ATP availability, rather than the supply of amino acids. Indeed a gradual elevation of Asp concentrations occurs when the level of amino acid synthesis exceeds amino acid depletion by protein synthesis (Fritz *et al.*, 2006). In the absence of photosynthesis, this situation would necessitate increased flux via PK_c to support the large ATP requirement of protein synthesis rather than the production of 2-OG. The production of transgenic plants expressing allosteric mutants of PEPC and PK_c may help to further define the precise roles of Asp and Glu in coordinating glycolytic flux and PEP partitioning with NH_4^+ assimilation.

16.5 Transgenic Plants with Altered Phosphoenolpyruvate or Glutamate Metabolism Display an Altered C/N Balance

The generation of specific knock-down and overexpression lines has shed light on the importance of the proper regulation of PEPC and PK_c activities and Glu levels in the context of balancing C and N metabolism.

16.5.1 Mutants with phosphoenolpyruvate metabolism perturbed by cytosolic pyruvate kinase or phosphoenolpyruvate carboxylase

Increasing the *in vivo* activity of PEPC has been a long-term goal of metabolic engineering, even though it had been clearly shown that straightforward overexpression of a PTPC gene had minimal impact on *in vivo* metabolism, despite appreciable increases in total extractable PEPC activity (Miyao and Fukayama, 2003). These results can be attributed to the tight allosteric and post-translational regulation of PTPC *in vivo*, which is capable of inhibiting ectopically expressed PTPC as well as endogenous PTPC. The importance of post-translational PEPC control was ultimately confirmed by a series of metabolic engineering efforts overexpressing various forms of PEPC with diminished feedback inhibition. The thermostable PEPC from *Synechococcus vulcanus* displays regulatory properties that differ from plant PTPCs because it is greatly desensitized to inhibition by malate and Glu and potently activated by Fru-1,6-P_2 (Chen *et al.*, 2002). Similar to PTPCs, *S. vulcanis* PEPC is inhibited by Asp; however, unusually, the inhibition by Asp decreases at lower pH values (with K_i values of 2 mM and 0.025 mM at pH 7.5 and 9.0, respectively; Chen *et al.*, 2002). *Arabidopsis* plants overexpressing *svPEPC* displayed stunted growth, a major accumulation of Gln, Asn and Arg, and a depletion of aromatic acids that was attributable to a low cellular concentration of PEP (the precursor for the shikimate pathway; Chen *et al.*, 2004). In a second example, the PEPC from the cyanobacterium *Corynebacterium glutamicum* is not inhibited by the levels of malate or Glu that are expected in the plant cytosol (Delaunay *et al.*, 2004; Rolletschek *et al.*, 2004). Seed-specific overexpression of this enzyme in *Vicia narbonensis* resulted in increased N and C uptake by the seeds, increased anaplerotic flux into protein and increased amino acid levels throughout seed development (Rolletschek *et al.*, 2004; Radchuk *et al.*, 2007). In a third example, a PTPC from potato was metabolically engineered to have a lowered inhibition by malate (other effectors were not reported) and a higher affinity for PEP (Rademacher *et al.*, 2002). Expression of this PEPC in potato correlated with stunted growth, reduced tuber yield and increased C flux into organic acids and amino acids at the expense of starch and soluble sugars (Rademacher *et al.*, 2002). In all three cases, the overexpression of PEPC with altered allosteric properties increased flux into organic/amino acids and was highly deleterious to the plants; during seed-specific expression of *C. glutamicum PEPC* in *V. narbonensis*, Seventy percent of transformed plants did not produce viable seeds (Rolletschek *et al.*, 2004). Although the exact mechanism(s) of incompatibility are unknown, this pattern emphasizes the need for proper allosteric control to balance glycolytic/anaplerotic C metabolism in plant cells. Here, it will be interesting to study the effects of overexpressing the BTPC subunits in plants because, although they too are greatly desensitized to allosteric effectors, they are somehow integrated into normal metabolism in several plant cell types (Uhrig *et al.*, 2008b; Igawa *et al.*, 2010; O'Leary *et al.*, 2011a; Shane *et al.*, 2013; Ruiz-Ballesta *et al.*, 2014).

A large decrease in PK_c activity has been achieved in several instances. Knock-out of tobacco leaf PK_c resulted in increased PEP levels but otherwise apparently normal plant growth under standard conditions (Gottlob-McHugh *et al.*, 1992). When grown under low light, however, these leaves displayed a decreased night-time export of sucrose to the root, resulting in stunted root growth, with a simultaneous increase in respiratory CO_2 release (Knowles *et al.*, 1998; Grodzinski *et al.*, 1999). These results indicate that: (i) plants are capable of using a metabolic bypass to circumvent the cytosolic production of pyruvate by PK_c; and (ii) this metabolic bypass does not control glycolytic flux towards dark respiration to the same extent as PK_c (Knowles *et al.*, 1998; Grodzinski *et al.*, 1999). In potato, RNA interference gene silencing of tuber PK_c resulted in an increased PEP/pyruvate ratio along with a reduction in Asp, glycerate and several TCA cycle intermediates, and a conspicuous increase in Lys (Oliver *et al.*, 2008). Calculations of *in situ* fluxes from ^{14}C-glucose in tuber discs indicated that reduced PK_c activity diverted C away from glycolysis and into starch, consistent with a reduced labelling of protein. Further insight into the functioning of PK_c in the context of N metabolism could be gained from the creation and analysis of mutant lines overexpressing PK_c isoforms with altered allosteric sensitivity to Asp and Glu.

16.5.2 Effect of mutations that perturb glutamate levels

After performing a meta-analysis of cellular amino acid measurements in plants across a range of metabolic perturbations and the diurnal cycle, Forde and Lea (2007) pointed out that Glu levels appear to be maintained at relatively static levels; a property that sets Glu apart from the other amino acids. This tight control of Glu levels is obviously needed if it is to function as a key allosteric effector of glycolytic flux at the PEP branch point. The homeostasis of Glu levels is remarkable considering that the metabolism of Glu is possible via several different major routes: (i) GOGAT produces Glu from 2-OG and Gln; (ii) GS consumes Glu in the production of Gln; (iii) transaminase reactions function in reversible interconversions of Glu and 2-OG with other amino acid/α-keto acid pairings; (iv) Glu dehydrogenase functions in the deamidation of Glu to 2-OG (Purnell and Botella, 2007); and (v) Glu can be irreversibly decarboxylated to GABA by Glu decarboxylase. Perturbations to cellular Glu levels caused by the genetic manipulation of GS and GOGAT enzymes tend to have a large effect of N metabolism and the levels of other amino acids (Forde and Lea, 2007). As GS and GOGAT directly participate in NH_4^+ assimilation, these findings do not inform us as to the role of Glu as a PK_c and PEPC allosteric inhibitor during NH_4^+ assimilation, because NH_4^+ assimilation has itself has been distorted. However, the targeted overexpression of a non-regulated Glu decarboxylase in *Arabidopsis* seeds proved more revealing because Glu decarboxylase does not directly participate in N assimilation but still consumes Glu (Fait *et al.*, 2011). Although it can serve to bypass several steps of the TCA cycle, the physiological purpose of GABA formation by Glu decarboxylase and the GABA shunt remains obscure (Fait *et al.*, 2008). Nevertheless, the enhanced Glu to GABA conversion caused by overexpression of the non-regulated Glu decarboxylase had a complex effect on seed C and N metabolism that ultimately resulted in increased in protein and amino acid content at the expense of sugars, fatty acids and organic acids (Fait *et al.*, 2011). Although these effects cannot be definitely attributed to reduced Glu levels, they are in line with the expected outcome of a reduced allosteric inhibition of PK_c and PEPC causing increased respiratory flux towards amino and organic acid synthesis.

16.6 Conclusions and Future Directions

While the central importance of PEP metabolism within plant cells has been appreciated for many years (Plaxton and O'Leary, 2012), recent research has provided numerous insights into the complex biochemical and molecular mechanisms that underpin the organization and control of PEP metabolism in plants. Analyses at the genomic, transcriptomic, proteomic, enzymological/biochemical and cellular levels have revealed the biochemical complexity of the key PEP metabolizing enzymes and their close integration with NH_4^+ assimilation. Theoretical and experimental studies with mutant plant lines have solidified the view that the majority of metabolic control within plant respiration, including that of PK and PEPC activities, occurs at the post-transcriptional level (Sweetlove and Fernie, 2005; Plaxton and Podestá, 2006). Therefore, information about the sub-cellular location of PEP metabolizing isoenzymes, their protein–protein interactions, allosteric effectors and PTMs will remain essential for determining linkages between the mRNA transcript expression data of individual PEPC and PK isoenzymes and their specific physiological/metabolic functions. A comprehensive understanding of the synchronous regulation of PEPC with PK_c, PK_p and other enzymes of the PEP branch point (e.g. PEP carboxykinase, pyruvate Pi-dikinase, the shikimate pathway) represents a major challenge. Functional genomic studies of PEP metabolizing enzymes should continue to be combined with knowledge of their allosteric effectors and PTMs. In particular, overexpression studies of allosterically modified PK_c or the BTPC subunit of Class-2 PEPCs are obvious strategies to modulate PEP metabolism and respiration *in vivo*. Subtle modifications to the regulatory properties of PEP metabolizing enzymes may prove more effective than simple overexpression in achieving a desirable redistribution of respiratory C flux to fatty acids, proteins or aromatic compounds in plant sink organs such as developing oilseeds.

Acknowledgements

W.C.P. is greatly indebted to past and present members of his laboratory who have made interesting discoveries concerning the molecular and biochemical characteristics of plant and algal PEP metabolizing enzymes. Research in W.C.P.'s laboratory has been generously supported by grants from the Natural Sciences and Engineering Research Council of Canada (NSERC) and the Queen's Research Chair programme.

References

Alonso, A.P., Goffman, F.D., Ohlrogge, J.B. and Shachar-Hill, Y. (2007) Carbon conversion efficiency and central metabolic fluxes in developing sunflower (*Helianthus annuus* L.) embryos. *The Plant Journal* 52, 296–308.

Amory, A.M., Vanlerberghe, G.C. and Turpin, D.H. (1991) Demonstration of both a photosynthetic and a nonphotosynthetic CO_2 requirement for NH_4^+ assimilation in the green alga *Selenastrum minutum*. *Plant Physiology* 95, 192–196.

Andre, C., Froehlich, J.E., Moll, M.R. and Benning, C. (2007) A heteromeric plastidic pyruvate kinase complex involved in seed oil biosynthesis in *Arabidopsis*. *The Plant Cell* 19, 2006–2022.

Baud, S., Wuillème, S., Dubreucq, B., de Almeida, A., Vuagnat, C., Lepiniec, L., Miquel, M. and Rochat, C. (2007) Function of plastidial pyruvate kinases in seeds of *Arabidopsis thaliana*. *The Plant Journal* 52, 405–419.

Baysdorfer, C. and Bassham, J.A. (1984) Spinach pyruvate kinase isoforms: partial purification and regulatory properties. *Plant Physiology* 74, 374–379.

Blonde, J.D. and Plaxton, W.C. (2003) Structural and kinetic properties of high and low molecular mass phosphoenolpyruvate carboxylase isoforms from the endosperm of developing castor oilseeds. *The Journal of Biological Chemistry* 278, 11867–11873.

Britto, D.T., Siddiqi, M.Y., Glass, A.D. and Kronzucker, H.J. (2001) Futile transmembrane NH_4^+ cycling: a cellular hypothesis to explain ammonium toxicity in plants. *Proceedings of the National Academy of Sciences of the United States of America* 98, 4255–4258.

Champigny, M.L. and Foyer, C. (1992) Nitrate activation of cytosolic protein kinases diverts photosynthetic carbon from sucrose to amino acid biosynthesis: basis for a new concept. *Plant Physiology* 100, 7–12.

Chen, L.-M., Omiya, T., Hata, S. and Izui, K. (2002) Molecular characterization of a phosphoenolpyruvate carboxylase from a thermophilic cyanobacterium, *Synechococcus vulcanus* with unusual allosteric properties. *Plant and Cell Physiology* 43, 159–169.

Chen, L.-M., Li, K.-Z., Miwa, T. and Izui, K. (2004) Overexpression of a cyanobacterial phosphoenolpyruvate carboxylase with diminished sensitivity to feedback inhibition in *Arabidopsis* changes amino acid metabolism. *Planta* 219, 440–449.

Chollet, R., Vidal, J. and O'Leary, M.H. (1996) Phosphoenolpyruvate carboxylase : a ubiquitous, highly regulated enzyme in plants. *Annual Review of Plant Physiology and Plant Molecular Biology* 47, 273–298.

Dalziel, K.J., O'Leary, B., Brikis, C., Rao, S.K., She, Y.-M., Cyr, T. and Plaxton, W.C. (2012) The bacterial-type phosphoenolpyruvate carboxylase isozyme from developing castor oil seeds is subject to *in vivo* regulatory phosphorylation at serine-451. *FEBS Letters* 586, 1049–1054.

Delaunay, S., Daran-Lapujade, P., Engasser, J.-M. and Goergen, J.-L. (2004) Glutamate as an inhibitor of phosphoenolpyruvate carboxylase activity in *Corynebacterium glutamicum*. *Journal of Industrial Microbiology and Biotechnology* 31, 183–188.

Duff, S. and Chollet, R. (1995) *In vivo* regulation of wheat-leaf phosphoenolpyruvate carboxylase by reversible phosphorylation. *Plant Physiology* 107, 775–782.

Fait, A., Fromm, H., Walter, D., Galili, G. and Fernie, A.R. (2008) Highway or byway: the metabolic role of the GABA shunt in plants. *Trends in Plant Science* 13, 14–19.

Fait, A., Nesi, A.N., Angelovici, R., Lehmann, M., Pham, P.A., Song, L., Haslam, R.P., Napier, J.A., Galili, G. and Fernie, A.R. (2011) Targeted enhancement of glutamate-to-γ-aminobutyrate conversion in *Arabidopsis* seeds affects carbon-nitrogen balance and storage reserves in a development-dependent manner. *Plant Physiology* 157, 1026–1042.

Finkemeier, I. and Sweetlove, L.J. (2009) The role of malate in plant homeostasis. *F1000 Biology Reports* 1: 47.

Forde, B.G. and Lea, P.J. (2007) Glutamate in plants: metabolism, regulation, and signalling. *Journal of Experimental Botany* 58, 2339–2358.

Fritz, C., Mueller, C., Matt, P., Feil, R. and Stitt, M. (2006) Impact of the C-N status on the amino acid profile in tobacco source leaves. *Plant, Cell and Environment* 29, 2055–2076.

Gauthier, P.P.G., Bligny, R., Gout, E., Mahé, A., Nogués, S., Hodges, M. and Tcherkez, G.G.B. (2010) *In folio* isotopic tracing demonstrates that nitrogen assimilation into glutamate is mostly independent from current CO_2 assimilation in illuminated leaves of *Brassica napus*. *New Phytologist* 185, 988–999.

Geigenberger, P. and Stitt, M. (1991) Regulation of carbon partitioning between sucrose and nitrogen assimilation in cotyledons of germinating *Ricinus communis* L. seedlings. *Planta* 185, 563–568.

Gennidakis, S., Rao, S., Greenham, K., Uhrig, R.G., O'Leary, B., Snedden, W.A., Lu, C. and Plaxton, W.C. (2007) Bacterial- and plant-type phosphoenolpyruvate carboxylase polypeptides interact in the hetero-oligomeric Class-2 PEPC complex of developing castor oil seeds. *The Plant Journal* 52, 839–849.

Golombek, S., Heim, U., Horstmann, C., Wobus, U. and Weber, H. (1999) Phosphoenolpyruvate carboxylase in developing seeds of *Vicia faba* L.: gene expression and metabolic regulation. *Planta* 208, 66–72.

Gottlob-McHugh, S.G., Sangwan, R.S., Blakeley, S.D., Vanlerberghe, G.C., Ko, K., Turpin, D.H., Plaxton, W.C., Miki, B.L. and Dennis, D.T. (1992) Normal growth of transgenic tobacco plants in the absence of cytosolic pyruvate kinase. *Plant Physiology* 100, 820–825.

Gregory A.L., Hurley B.A., Tran H.T., Valentine A.J., She Y-M, Knowles V.L. and Plaxton W.C. (2009) *In vivo* regulatory phosphorylation of the phosphoenolpyruvate carboxylase AtPPC1 in phosphate-starved *Arabidopsis thaliana*. *Biochemical Journal* 420, 57–65

Grodzinski, B., Jiao, J., Knowles, V. and Plaxton, W. (1999) Photosynthesis and carbon partitioning in transgenic tobacco plants deficient in leaf cytosolic pyruvate kinase. *Plant Physiology* 120, 887–896.

Guillet, C., Just, D., Bénard, N., Destrac-Irvine, A., Baldet, P., Hernould, M., Causse, M., Raymond, P. and Rothan, C. (2002) A fruit-specific phosphoenolpyruvate carboxylase is related to rapid growth of tomato fruit. *Planta* 214, 717–726.

Heldt, H.W., Chon, C.J. and Maronde, D. (1977) Role of orthophosphate and other factors in the regulation of starch formation in leaves and isolated chloroplasts. *Plant Physiology* 59, 1146–1155.

Hu, Z. and Plaxton, W.C. (1996) Purification and characterization of cytosolic pyruvate kinase from leaves of the castor oil plant. *Archives of Biochemistry and Biophysics* 333, 298–307.

Huppe, H.C. and Turpin, D.H. (1994) Integration of carbon and nitrogen metabolism in plant and algal cells. *Annual Review of Plant Physiology and Plant Molecular Biology* 45, 577–607.

Igawa, T., Fujiwara, M., Tanaka, I., Fukao, Y. and Yanagawa, Y. (2010) Characterization of bacterial-type phosphoenolpyruvate carboxylase expressed in male gametophyte of higher plants. *BMC Plant Biology* 10:200.

Izui, K., Matsumura, H., Furumoto, T. and Kai, Y. (2004) Phosphoenolpyruvate carboxylase: a new era of structural biology. *Annual Review of Plant Biology* 55, 69–84.

Knowles, V.L., McHugh, S.G., Hu, Z., Dennis, D.T., Miki, B.L. and Plaxton, W.C. (1998) Altered growth of transgenic tobacco lacking leaf cytosolic pyruvate kinase. *Plant Physiology* 116, 45–51.

Larsen, P.O., Cornwell, K.L., Gee, S.L. and Bassham, J.A. (1981) Amino acid synthesis in photosynthesizing spinach cells. *Plant Physiology* 68, 292–299.

Law, R.D. and Plaxton, W.C. (1995) Purification and characterization of a novel phosphoenolpyruvate carboxylase from banana fruit. *Biochemical Journal* 307, 807–816.

Lawyer, A.L., Cornwell, K.L., Larsen, P.O. and Bassham, J.A. (1981) Effects of carbon dioxide and oxygen on the regulation of photosynthetic carbon metabolism by ammonia in spinach mesophyll cells. *Plant Physiology* 68, 1231–1236.

Leegood, R.C., Lea, P.J., Adcock, M.D. and Hausler, R.E. (1995) The regulation and control of photorespiration. *Journal of Experimental Botany* 46, 1397–1414.

Le Van Quy, Foyer, C. and Champigny, M.L. (1991) Effect of light and NO_3^- on wheat leaf phosphoenolpyruvate carboxylase activity: evidence for covalent modulation of the C_3 enzyme. *Plant Physiology* 97, 1476–1482.

Lin, M., Turpin, D.H. and Plaxton, W.C. (1989a) Pyruvate kinase isozymes from the green alga, *Selenastrum minutum*. I. Purification and physical and immunological characterization. *Archives of Biochemistry and Biophysics* 269, 219–227.

Lin, M., Turpin, D.H. and Plaxton, W.C. (1989b) Pyruvate kinase isozymes from the green alga, *Selenastrum minutum*. II. Kinetic and regulatory properties. *Archives of Biochemistry and Biophysics* 269, 228–238.

Mamedov, T.G., Moellering, E.R. and Chollet, R. (2005) Identification and expression analysis of two inorganic C- and N-responsive genes encoding novel and distinct molecular forms of eukaryotic phosphoenolpyruvate carboxylase in the green microalga *Chlamydomonas reinhardtii*. *The Plant Journal* 42, 832–843.

Martinoia, E. and Rentsch, D. (1994) Malate compartmentation-responses to a complex metabolism. *Annual Review of Plant Physiology and Plant Molecular Biology* 45, 447–467.

Masumoto, C., Miyazawa, S.-I., Ohkawa, H., Fukuda, T., Taniguchi, Y., Murayama, S., Kusano, M., Saito, K., Fukayama, H. and Miyao, M. (2010) Phosphoenolpyruvate carboxylase intrinsically located in the chloroplast of rice plays a crucial role in ammonium assimilation. *Proceedings of the National Academy of Sciences of the United States of America* 107, 5226–5231.

Miyao, M. and Fukayama, H. (2003) Metabolic consequences of overproduction of phosphoenolpyruvate carboxylase in C_3 plants. *Archives of Biochemistry and Biophysics* 414, 197–203.

Moellering, E.R., Ouyang, Y., Mamedov, T.G. and Chollet, R. (2007) The two divergent PEP-carboxylase catalytic subunits in the green microalga *Chlamydomonas reinhardtii* respond reversibly to inorganic-N supply and co-exist in the high-molecular-mass, hetero-oligomeric Class-2 PEPC complex. *FEBS Letters* 581, 4871–4876.

Moraes T. and Plaxton W.C. (2000) Purification and characterization of phosphoenolpyruvate carboxylase from *Brassica napus* (rapeseed) suspension cell cultures. Implications for phosphoenolpyruvate carboxylase regulation during phosphate starvation, and the integration of glycolysis with nitrogen assimilation. *European Journal of Biochemistry* 267, 4465–4476.

Murmu, J. and Plaxton, W.C. (2007) Phosphoenolpyruvate carboxylase protein kinase from developing castor oil seeds: partial purification, characterization, and reversible control by photosynthate supply. *Planta* 226, 1299–1310.

Nielsen, T.H., Rung, J.H., and Villadsen, D. (2004) Fructose-2,6-bisphosphate: a traffic signal in plant metabolism. *Trends in Plant Science* 9, 556–563.

Nimmo, H.G. (2000) The regulation of phosphoenolpyruvate carboxylase in CAM plants. *Trends in Plant Science* 5, 75–80.

Nunes-Nesi, A., Fernie, A.R. and Stitt, M. (2010) Metabolic and signaling aspects underpinning the regulation of plant carbon nitrogen interactions. *Molecular Plant* 3, 973–996.

O'Leary, B., Rao, S.K., Kim, J. and Plaxton, W.C. (2009) Bacterial-type phosphoenolpyruvate carboxylase (PEPC) functions as a catalytic and regulatory subunit of the novel class-2 PEPC complex of vascular plants. *The Journal of Biological Chemistry* 284, 24797–24805.

O'Leary, B., Rao, S.K. and Plaxton, W.C. (2011a) Phosphorylation of bacterial-type phosphoenolpyruvate carboxylase at Ser^{425} provides a further tier of enzyme control in developing castor oil seeds. *Biochemical Journal* 433, 65–74.

O'Leary, B., Park, J. and Plaxton, W.C. (2011b) The remarkable diversity of plant PEPC (phosphoenolpyruvate carboxylase): recent insights into the physiological functions and post-translational controls of non-photosynthetic PEPCs. *Biochemical Journal* 436, 15–34.

O'Leary, B., Fedosejevs, E.T., Hill, A.T., Bettridge, J., Park, J., Rao, S.K., Leach, C.A and Plaxton, W.C. (2011c) Tissue-specific expression and post-translational modifications of plant- and bacterial-type phosphoenolpyruvate carboxylase isozymes of the castor oil plant, *Ricinus communis* L. *Journal of Experimental Botany* 62, 5485–5495.

Oliver, S.N., Lunn, J.E., Urbanczyk-Wochniak, E., Lytovchenko, A., van Dongen, J.T., Faix, B., Schmälzlin, E., Fernie, A.R. and Geigenberger, P. (2008) Decreased expression of cytosolic pyruvate kinase in potato tubers leads to a decline in pyruvate resulting in an *in vivo* repression of the alternative oxidase. *Plant Physiology* 148, 1640–1654.

Park, J., Khuu, N., Howard, A.S.M., Mullen, R.T. and Plaxton, W.C. (2012) Bacterial- and plant-type phosphoenolpyruvate carboxylase isozymes from developing castor oil seeds interact *in vivo* and associate with the surface of mitochondria. *The Plant Journal* 71, 251–262.

Pasqualini, S., Ederli, L., Piccioni, C., Batini, P., Bellucci, M., Arcioni, S. and Antonielli, M. (2001) Metabolic regulation and gene expression of root phosphoenolpyruvate carboxylase by different nitrogen sources. *Plant, Cell and Environment* 24, 439–447.

Paul, J.S., Cornwell, K.L. and Bassham, J.A. (1978) Effects of ammonia on carbon metabolism in photosynthesizing isolated mesophyll cells from *Papaver somniferum* L. *Planta* 142, 49–54.

Perotti, V.E., Figueroa, C.M., Andreo, C.S., Iglesias, A.A. and Podestá, F.E. (2010) Cloning, expression, purification and physical and kinetic characterization of the phosphoenolpyruvate carboxylase from orange (*Citrus sinensis* osbeck var. Valencia) fruit juice sacs. *Plant Science* 179, 527–535.

Plaxton, W.C. (1996) The organization and regulation of plant glycolysis. *Annual Review of Plant Physiology and Plant Molecular Biology* 47, 185–214.

Plaxton, W.C. and O'Leary, B. (2012) The central role of phosphoenolpyruvate in developing oilseeds. In: Agrawal, G.K. and Rakwal, R. (eds)

Plaxton, W.C. and Podestá, F.E. (2006) The functional organization and control of plant respiration. *Critical Reviews in Plant Sciences* 25, 159–198.

Plaxton, W.C., Smith, C.R. and Knowles, V.L. (2002) Molecular and regulatory properties of leucoplast pyruvate kinase from *Brassica napus* (rapeseed) suspension cells. *Archives of Biochemistry and Biophysics* 400, 54–62.

Podestá, F.E. and Plaxton, W.C. (1991) Kinetic and regulatory properties of cytosolic pyruvate kinase from germinating castor oil seeds. *Biochemical Journal* 279(Pt 2), 495–501.

Podestá, F.E. and Plaxton, W.C. (1994) Regulation of cytosolic carbon metabolism in germinating *Ricinus communis* cotyledons. *Planta* 194, 381–387.

Purnell, M.P. and Botella, J.R. (2007) Tobacco isoenzyme 1 of NAD(H)-dependent glutamate dehydrogenase catabolizes glutamate *in vivo*. *Plant Physiology* 143, 530–539.

Radchuk, R., Radchuk, V., Götz, K.-P., Weichert, H., Richter, A., Emery, R.J.N., Weschke, W. and Weber, H. (2007) Ectopic expression of phosphoenolpyruvate carboxylase in *Vicia narbonensis* seeds: effects of improved nutrient status on seed maturation and transcriptional regulatory networks. *The Plant Journal* 51, 819–839.

Rademacher, T., Häusler, R.E., Hirsch, H.-J., Zhang, L., Lipka, V., Weier, D., Kreuzaler, F. and Peterhänsel, C. (2002) An engineered phosphoenolpyruvate carboxylase redirects carbon and nitrogen flow in transgenic potato plants. *The Plant Journal* 32, 25–39.

Rivoal, J., Dunford, R., Plaxton, W.C. and Turpin, D.H. (1996) Purification and properties of four phosphoenolpyruvate carboxylase isoforms from the green alga *Selenastrum minutum*: evidence that association of the 102-kDa catalytic subunit with unrelated polypeptides may modify the physical and kinetic properties of the enzyme. *Archives of Biochemistry and Biophysics* 332, 47–57.

Rivoal, J., Plaxton, W.C. and Turpin, D.H. (1998) Purification and characterization of high- and low-molecular-mass isoforms of phosphoenolpyruvate carboxylase from *Chlamydomonas reinhardtii*. *Biochemical Journal* 331, 201–209.

Rivoal, J., Trzos, S., Gage, D.A, Plaxton, W.C. and Turpin, D.H. (2001) Two unrelated phosphoenolpyruvate carboxylase polypeptides physically interact in the high molecular mass isoforms of this enzyme in the unicellular green alga *Selenastrum minutum*. *The Journal of Biological Chemistry* 276, 12588–12597.

Rivoal, J., Turpin, D.H. and Plaxton, W.C. (2002) *In vitro* phosphorylation of phosphoenolpyruvate carboxylase from the green alga *Selenastrum minutum*. *Plant and Cell Physiology* 43, 785–792.

Rolletschek, H., Borisjuk, L., Radchuk, R., Miranda, M., Heim, U., Wobus, U. and Weber, H. (2004) Seed-specific expression of a bacterial phosphoenolpyruvate carboxylase in *Vicia narbonensis* increases protein content and improves carbon economy. *Plant Biotechnology Journal* 2, 211–219.

Ruiz-Ballesta, I., Feria, A.-B., Hong, N., Plaxton, W.C. and Echevarria, C. (2014) *In Vivo* monoubiquitination of anaplerotic phosphoenolpyruvate carboxylase occurs at lysine-624 in germinating sorghum seeds. *Journal of Experimental Botany* 65, 443–451.

Sánchez, R. and Cejudo, F.J. (2003) Identification and expression analysis of a gene encoding a bacterial-type phosphoenolpyruvate carboxylase from *Arabidopsis* and rice. *Plant Physiology* 132, 949–957.

Scheible, W.R., Gonzalez-Fontes, A., Lauerer, M., Muller-Rober, B., Caboche, M. and Stitt, M. (1997) Nitrate acts as a signal to induce organic acid metabolism and repress starch metabolism in tobacco. *The Plant Cell* 9, 783–798.

Scheible, W.-R., Morcuende, R., Czechowski, T., Fritz, C., Osuna, D., Palacios-Rojas, N., Schindelasch, D., Thimm, O., Udvardi, M.K. and Stitt, M. (2004) Genome-wide reprogramming of primary and secondary metabolism, protein synthesis, cellular growth processes and the regulatory infrastructure of *Arabidopsis* in response to nitrogen. *Plant Physiology* 136, 2483–2499.

Schuller, K.A., Plaxton, W.C. and Turpin, D.H. (1990a) Regulation of phosphoenolpyruvate carboxylase from the green alga *Selenastrum minutum*: properties associated with replenishment of tricarboxylic acid cycle intermediates during ammonium assimilation. *Plant Physiology* 93, 1303–1311.

Schuller, K.A, Turpin, D.H. and Plaxton, W.C. (1990b) Metabolite regulation of partially purified soybean nodule phosphoenolpyruvate carboxylase. *Plant Physiology* 94, 1429–1435.

Schwender, J., Shachar-Hill, Y. and Ohlrogge, J.B. (2006) Mitochondrial metabolism in developing embryos of *Brassica napus*. *The Journal of Biological Chemistry* 281, 34040–34047.

Shane, M.W., Fedosejevs, E.T. and Plaxton, W.C. (2013) Reciprocal control of anaplerotic phosphoenolpyruvate carboxylase by in vivo monoubiquitination and phosphorylation in developing proteoid roots of phosphate-deficient harsh hakea. *Plant Physiology* 161, 1634–1644.

Smith, C.R., Knowles, V.L. and Plaxton, W.C. (2000) Purification and characterization of cytosolic pyruvate kinase from *Brassica napus* (rapeseed) suspension cell cultures: implications for the integration of glycolysis with nitrogen assimilation. *European Journal of Biochemistry* 267, 4477–4485.

Stitt, M., Müller, C., Matt, P., Gibon, Y., Carillo, P., Morcuende, R., Scheible, W.-R. and Krapp, A. (2002) Steps towards an integrated view of nitrogen metabolism. *Journal of Experimental Botany* 53, 959–970.

Svensson, P., Bläsing, O.E. and Westhoff, P. (2003) Evolution of C_4 phosphoenolpyruvate carboxylase. *Archives of Biochemistry and Biophysics* 414, 180–188.

Sweetlove, L.J. and Fernie, A.R. (2005) Regulation of metabolic networks: understanding metabolic complexity in the systems biology era. *The New Phytologist* 168, 9–24.

Tcherkez, G., Mahé, A., Gauthier, P., Mauve, C., Gout, E., Bligny, R., Cornic, G. and Hodges, M. (2009) *In folio* respiratory fluxomics revealed by ^{13}C isotopic labeling and H/D isotope effects highlight the noncyclic nature of the tricarboxylic acid "cycle" in illuminated leaves. *Plant Physiology* 151, 620–630.

Tripodi, K.E., Turner, W.L., Gennidakis, S. and Plaxton, W.C. (2005) *In vivo* regulatory phosphorylation of novel phosphoenolpyruvate carboxylase isoforms in endosperm of developing castor oil seeds. *Plant Physiology* 139, 969–978.

Turner, W.L. and Plaxton, W.C. (2000) Purification and characterization of cytosolic pyruvate kinase from banana fruit. *The Biochemical Journal* 352(Pt 3), 875–882.

Turner, W.L., Knowles, V.L. and Plaxton, W.C. (2005) Cytosolic pyruvate kinase: subunit composition, activity, and amount in developing castor and soybean seeds, and biochemical characterization of the purified castor seed enzyme. *Planta* 222, 1051–1062.

Turpin, D.H., Botha, F.C., Smith, R.G., Feil, R., Horsey, A.K. and Vanlerberghe, G.C. (1990) Regulation of carbon partitioning to respiration during dark ammonium assimilation by the green alga *Selenastrum minutum*. *Plant Physiology* 93, 166–175.

Uhrig, R.G., She, Y.-M., Leach, C.A. and Plaxton, W.C. (2008a) Regulatory monoubiquitination of phosphoenolpyruvate carboxylase in germinating castor oil seeds. *The Journal of Biological Chemistry* 283, 29650–29657.

Uhrig, R.G., O'Leary, B., Spang, H.E., MacDonald, J.A., She, Y.-M. and Plaxton, W.C. (2008b) Coimmunopurification of phosphorylated bacterial- and plant-type phosphoenolpyruvate carboxylases with the plastidial pyruvate dehydrogenase complex from developing castor oil seeds. *Plant Physiology* 146, 1346–1357.

Vance, C.P. and Heichel, G.H. (1991) Carbon in N_2 Fixation: limitation or exquisite adaptation. *Annual Review of Plant Physiology and Plant Molecular Biology* 42, 373–390.

Vanlerberghe, G.C., Schuller, K.A., Smith, R.G., Feil, R., Plaxton, W.C. and Turpin, D.H. (1990) Relationship between NH_4^+ assimilation rate and *in vivo* phosphoenolpyruvate carboxylase activity : regulation of anaplerotic carbon flow in the green alga *Selenastrum minutum*. *Plant Physiology* 94, 284–290.

Vanlerberghe, G.C., Huppe, H.C., Vlossak, K.D. and Turpin, D.H. (1992) Activation of respiration to support dark NO_3^- and NH_4^+ assimilation in the green alga *Selenastrum minutum*. *Plant Physiology* 99, 495–500.

Vidal, J. and Chollet, R. (1997) Regulatory phosphorylation of C_4 PEP carboxylase. *Trends in Plant Science* 2, 230–237.

Weber, A.P.M. and Linka, N. (2011) Connecting the plastid: transporters of the plastid envelope and their role in linking plastidial with cytosolic metabolism. *Annual Review of Plant Biology* 62, 53–77.

Weger, H.G. and Turpin, D.H. (1989) Mitochondrial respiration can support NO_3^- and NO_2^- reduction during photosynthesis : interactions between photosynthesis, respiration, and N assimilation in the N-limited green alga *Selenastrum minutum*. *Plant Physiology* 89, 409–415.

Weger, H.G., Birch, D.G., Elrifi, I.R. and Turpin, D.H. (1988) Ammonium assimilation requires mitochondrial respiration in the light : A study with the green alga *Selenastrum minutum*. *Plant Physiology* 86, 688–692.

Winter, H., Robinson, D.G. and Heldt, H.W. (1994) Subcellular volumes and metabolite concentrations in spinach leaves. *Planta* 193, 530–535.

Xu, W., Zhou, Y. and Chollet, R. (2003) Identification and expression of a soybean nodule-enhanced PEP-carboxylase kinase gene (*NE-PpcK*) that shows striking up-/down-regulation *in vivo*. *The Plant Journal* 34, 441–452.

17 Amino Acid Export in Plants

M.B. Price and S. Okumoto*
Virginia Tech, Blacksburg, Virginia, USA

17.1 Abstract

Amino acids are the main form of nitrogen transported in the plant body, and the transport of amino acids across cellular and sub-cellular membranes is considered essential for nitrogen economy in plants. In particular, the long-distance transport of amino acids through the phloem and xylem, and the transfer of amino acids between the phloem and xylem during nitrogen cycling, requires both cellular importers and exporters. Even though some of the amino acid carrier proteins were identified more than 15 years ago, the physiological roles of many amino acid transporters are still unclear. There is even less known about the proteins that mediate the cellular efflux of amino acids. Recently, however, there have been reports of amino acid carrier proteins that can mediate both export and import, and might play roles in nitrogen cycling and other physiological processes. In this chapter, we will summarize what is known about plant amino acid exporters and their physiological roles.

17.2 Introduction

As sessile organisms, land plants acquire mineral nutrients, including nitrogen (N) from the rhizosphere. As clearly indicated by the classical experiment started in 1843 at the then Rothamsted Experimental Station (now Rothamsted Research) in the UK, plants have a capacity to absorb and utilize inorganic N. Depending on the molecular species of the N, the inorganic ions are absorbed and incorporated into organic molecules either in the root or leaf tissues (Andrews, 1986; Marschner, 1995; Kruse *et al.*, 2002). A large part of the organic N synthesized in the source tissue is further transported through the plant's long-distance transport system (i.e. the xylem and phloem) to the sink tissues that are dependent on the imported organic N (Pate, 1973; Pate *et al.*, 1981).

Amino acids are the main forms of N found in the xylem and phloem, and hence are considered to be the major N carrier in plants (Herschbach *et al.*, 2012). Long-distance transport of amino acids is, therefore, critical for the optimal distribution of N (Tegeder and Rentsch, 2010). The circulation of amino acids through the xylem and phloem, as well as their exchange between intracellular compartments, calls for a number of steps in which amino acids cross the separating lipid bilayer. For example, phloem loading at the source tissue would require active import into the phloem tissue if the phloem is symplasmically isolated from the cells that synthesize amino acids. Amino acids are also known to cycle between the xylem and phloem (Schobert and Komor, 1987; Jeschke and Hartung, 2000), and would require a transporter to enter or exit

*E-mail address: sokumoto@vt.edu

the phloem tissue. To feed the growing embryonic tissues, which are symplasmically isolated, amino acids need to leave the maternal cells and enter the symplasm of the growing embryo. Amino acid exchange between the plant and the rhizosphere would require an amino acid carrier on the membrane of the cells at the root surface. The seasonal remobilization of amino acids in perennial species would require amino acid export from the cells in the source tissues.

Moreover, a number of studies have suggested the possibility that amino acids function as signalling molecules as well. Amino acid feeding induces a massive change in the transcript levels of a large number of genes, many of which are involved in the assimilation and uptake of inorganic and organic N (Vincentz *et al.*, 1993; Rawat *et al.*, 1999; Oliveira *et al.*, 2001; Nazoa *et al.*, 2003; Fritz *et al.*, 2006; Hirner *et al.*, 2006; Gutierez *et al.*, 2008). In addition, amino acids are capable of triggering rapid ion fluxes at the plasma membrane, some of which are mediated by amino acid-gated channels (Price *et al.*, 2012; Vincill *et al.*, 2012; Tapken *et al.*, 2013). Taken together, the regulation of amino acid transport in plants is critical, both as the source of N and also for intracellular communication.

Over the past two decades, a large number of proteins that are involved in the import of amino acids across the plasma membrane have been identified (reviewed in: Wipf *et al.*, 2002; Rentsch *et al.*, 2007; Tegeder and Rentsch, 2010; Tegeder, 2012). The discoveries were largely facilitated by the use of heterologous expression systems (mainly the yeast *Saccharomyces cerevisiae*) that are auxotrophic for amino acids as the sole nitrogen source; many of the plant membrane importers, when expressed in such heterologous systems, retained their import capacity and, as a result, complemented the auxotrophy of the yeast. As predicted from physiological studies, these importer proteins were found to be critical in processes such as phloem loading (Hirner *et al.*, 2006), the loading of amino acids into growing embryos (Schmidt *et al.*, 2007; Sanders *et al.*, 2009), xylem–phloem transfer (Zhang *et al.*, 2010) and uptake from the soil (Svennerstam *et al.*, 2008, 2011).

In contrast, we have very little information on the molecular mechanisms of the amino acid export process in plants. As discussed above, critical processes in the circulation of amino acids in the plant body (e.g. xylem–phloem transfer, seed loading, efflux of amino acids into the rhizosphere) require amino acid exporters at the cellular membrane. Yet the exporters that are responsible for such activities are largely unknown. This is primarily due to the fact that so far there has been no efficient screening process that would identify membrane carrier proteins with exclusively exporter activities. Nevertheless, some clues are starting to emerge; one family of transporters capable of bidirectional amino acid transport has been identified (see Section 17.4.1); proteins that play a role in the regulation of amino acid export have also been discovered (Section 17.5).

In this chapter, we summarize what is known about the amino acid export process – from physiological and genetic experiments to what can be predicted from the bioinformatics approach, which uses information from amino acid porter carriers in other organisms. The molecular mechanisms for amino acid import have been extensively reviewed elsewhere, and therefore will not be covered in this article (Wipf *et al.*, 2002; Rentsch *et al.*, 2007; Tegeder and Rentsch, 2010; Tegeder, 2012). Likewise, the efflux carriers of plant hormone auxin, which is synthesized from the amino acid tryptophan, have been covered extensively by recent excellent references (Křeček *et al.*, 2009; Cho and Cho, 2012; Forestan and Varotto, 2012) and so will not be covered here.

17.3 Physiology of Amino Acid Export

A number of processes in plant development require amino acid export from cells, although detecting the amino acid export process is, in general, technically more challenging than detecting the import process. It is expected that the same population of plant cells will have strong amino acid import activities on top of any amino acid export activities, and that this reuptake of amino acids into the cells makes it harder to detect exporter activity. The issues with reuptake can be partially addressed by labelling the plant tissue with one N isotope (e.g. ^{15}N) and adding an N source labelled with another isotope (e.g. ^{14}N) to the extracellular medium. However, such a method is only useful in detecting the most abundant amino acids owing to the low amounts that are secreted in the media and the low sensitivity of detection. Preloading with radiolabelled amino acid can increase the sensitivity, but the results have to be interpreted with caution because this technique typically does not discriminate between different labelled compounds. In addition, determining the amino acid concentration in the extracellular milieu is technically challenging. While techniques for obtaining apoplasmic wash have been useful in determining the amino acid composition in the leaves (Pilot *et al.*, 2004; Hirner *et al.*, 2006), they cannot be used to determine absolute concentrations. Also, the use of this technique is not suitable for tissues that cannot be infiltrated with medium (i.e. roots).

Due to such technical difficulties, amino acid export activities have been tested in only a limited number of tissues. This section summarizes the physiological characteristics of the amino acid export activities that have been found in those tissues that have been tested.

17.3.1 Amino acid export from the seed coat

The developing embryo is symplasmically isolated from the maternal tissue, so the metabolites that it requires need first to be released from the seed coat cells to the apoplasm before they can be taken up into the developing seeds. Amino acid secretion from seed coats was reported as early as in 1985 (Wolswinkel and Deruiter, 1985) and the efflux rates of amino acids were predicted to be four to five orders of magnitude higher than those at the lipid bilayer, suggesting the involvement of a carrier protein that facilitates the translocation (de Jong *et al.*, 1997). The 'Empty seed coat technique' (Thorne and Rainbird, 1983), in which developing embryos are removed from the seed coat and the cavity filled with a standardized buffer, proved tremendously useful in investigating the biochemical properties of amino acid export from the seed coat.

The efflux rates of sugars and amino acids measured from the seed coat still attached to the mother plant and in the detached seed coat were near identical for the initial (up to 2 h) period of measurement, indicating that the sugars and amino acids released are not directly supplied from the phloem connected to the seed coat, but are released from the seed coat itself (De Jong and Wolswinkel, 1995). This view is also supported by studies that examined the amino acids in seed coat exudates, and found that their chemical composition was different from that of the amino acids found in the phloem sap (Wolswinkel and Deruiter, 1985; Lanfermeijer *et al.*, 1992). Further experiments using symplasmic dyes suggested that solutes are transported symplasmically from the phloem to seed coat (Grusak and Minchin, 1988; Offler *et al.*, 1989), where the efflux presumably happens, although some variations were observed among specific cultivars (Liu *et al.*, 2012).

Studies using various inhibitors of transport processes provided insights into the molecular mechanisms mediating amino acid efflux from the seed coat. The export was sensitive to *p*-chloromercuribenzenesulfonic acid (PCMBS), a thiol reagent that modifies the extracellular sulfhydryl groups of carrier proteins, thus further suggesting the involvement of carrier proteins in the process (De Jong and Wolswinkel, 1995).

Amino acid efflux from the seed coat was also found to be insensitive to the protonophore cyanide-*m*-chlorophenylhydrazone (CCCP) or to change in the pH gradient, suggesting that the process is not proton-gradient driven (De Jong and Wolswinkel, 1995). The influx of amino acids into the seed coat was found to be proton-gradient insensitive as well (de Jong *et al.*, 1997; van Dongen *et al.*, 2001). Taken together, the physiological data obtained so far suggest that poorly selective pores that are capable of bidirectional transport are serving in unloading of a range of solutes in these tissues.

17.3.2 Amino acid export into the xylem

Amino acids represent a major fraction of the N content in the xylem in many plant species. Their chemical composition varies depending on the inorganic N source available, generally increasing with the availability of ammonium in the rhizosphere owing to the fact that that the ammonium needs to be assimilated locally in roots to avoid toxicity, whereas nitrate is often translocated to the leaf tissue. In addition to newly synthesized N compounds, it is known that amino acids constantly circulate between the xylem and phloem tissues. In *Ricinus* roots, for example, about half of the amino acids supplied by the descending phloem are transferred to the ascending xylem (Jeschke and Hartung, 2000). Amino acids (arginine, Arg; alanine, Ala; asparagine, Asn; and glutamine, Gln) provided to the root by the phloem have been shown to inhibit nitrate uptake, suggesting that root–shoot cycling of amino acids enables the adjustment of inorganic nitrogen uptake to shoot demand (Touraine *et al.*, 1992; Imsande and Touraine, 1994; Gansel *et al.*, 2001).

Not surprisingly, the amino acid content in the xylem sap responds to myriad of environmental stimuli, such as N availability (Peoples *et al.*, 1986; Tonin *et al.*, 1990; Bi *et al.*, 2009; Balint and Rengel, 2011), the light/dark cycle (Lam *et al.*, 1995), mineral nutrition stress (Rellan-Alvarez *et al.*, 2011) and other abiotic stresses (Fernandez-Garcia *et al.*, 2011; Liu *et al.*, 2013). The mechanism through which the export of amino acids into the xylem happens is almost completely unknown. Unlike the seed coat, it is impractical to apply inhibitors locally to investigate the mode of amino acid transport in the xylem tissue, so the driving mechanisms for this process are not known. Several mutants with altered xylem amino acid concentrations have been reported so far. The overexpression of *Glutamine Dumper1* (*GDU1*) and its homologues causes a massive increase in amino acids, especially of Gln, in the xylem sap and apoplasm (Pilot *et al.*, 2004; Pratelli *et al.*, 2010) (see Section 17.5). Also, overexpression of the noduline-93 gene (*OsENOD93-1*), which contains two transmembrane domains and localizes to mitochondria, increases xylem amino acid contents (Bi *et al.*, 2009). In either case, the exact mechanism that leads to the increase in xylem amino acids is not completely understood.

17.3.3 Amino acid exchange with the rhizosphere

Plant roots secrete a large number of primary and secondary metabolites. The composition of the exudates is controlled by plants according to their needs, as exemplified by the legume plant root secreting flavonoids under conditions of N deficiency to attract rhizobia (Wasson *et al.*, 2006; Zhang *et al.*, 2009) and by iron deficiency in grasses inducing the secretion of a phytosiderophore that chelates Fe(III) ions (Mori, 1999; Schmidt, 2003). The composition of root exudates regulates the diversity and biomass of the soil microbial community (Broeckling *et al.*, 2008).

There is a constant flux of amino acids between the plant root and the rhizosphere in both directions. Plants are capable of taking up amino acids into the roots, and in some situations such as in boreal forest, it is expected that amino acids constitute a substantial part of the N source for the plants (Nasholm *et al.*, 2009; Inselsbacher and Nasholm, 2012). Traditionally it was thought that amino acids underwent a passive efflux from the plant root as a result of the large concentration gradient, and that plants and soil microorganisms compete for amino acids as a nutrient source. However, studies from the past decade collectively indicate that the amino acid flux through the root plasma membrane is not always a passive process.

A recent study using aseptically grown *Arabidopsis* demonstrated that amino acid secretion from the root is developmentally controlled, increasing in the later growth phases (Chaparro *et al.*, 2013). Likewise, a study using a bacterial biosensor demonstrated that tryptophan is more abundant in the older part of the root (12 to 16 cm from the tip) than in the root tip zone (Jaeger *et al.*, 1999). Curiously, in both cases, the distribution patterns of sugars were very different from those of amino acids, suggesting a different mode of regulation for these two classes of primary metabolites. In addition to developmental control, the nutritional status of plants has profound effects on the organic molecules that they secrete, including amino acids. It has been shown that deficiency of N, P, K and Fe in axenically grown maize alters the amino acid composition of root exudates; the efflux of glutamate (Glu) and γ-aminobutyric acid (GABA) increases with Fe and P deficiency, respectively, while N deficiency decreases amino acid exudation (Carvalhais *et al.*, 2011).

The molecular mechanism through which amino acid export into the rhizosphere happens is not known. A recent study (Lesuffleur and Cliquet, 2010) investigated the effect of various inhibitors on the influx and efflux of amino acids in white clover roots. It found that compared with the composition of amino acids in root tissues, glycine (Gly) is largely over-represented among the effluxed amino acids, and also that glycine efflux was not strongly affected by the application of various inhibitors: the protonophore CCCP, vanadate (an inhibitor of plasma membrane phosphotyrosyl phosphatases) or *N*-ethylmaleimide (NEM, an inhibitor of sulfhydryl proteins). Collectively, the results indicated that while a transporter that is selective for Gly is involved in the efflux process, the transporter is not likely to be mediated by proton antiporters or ABC-type transporters (Lesuffleur and Cliquet, 2010).

17.3.4 Vascular amino acid transport

Many families of transporters include members that are targeted both to the plasma membrane and the vacuolar membrane, e.g. ABC transporters (Kretzschmar *et al.*, 2011), sucrose transporters (Reinders *et al.*, 2008), aquaporin family proteins (Maurel *et al.*, 2008) and ion channels (Apse and Blumwald, 2007; Dunkel *et al.*, 2008), to name a few. Topologically speaking, the transporter that would reside at the plasma membrane and mediate solute export would mediate transport into the vacuole. In fact, transporters that are involved in the detoxification of metals and xenobiotics operate in both modes: excretion from the plasma membrane and accumulation into the vacuole, which is termed 'internal excretion' (Martinoia *et al.*, 1993). Additionally, vacuoles can be physically isolated from the plant tissue, so that solute import through the tonoplast (i.e. solute export from the cytosol) can be studied. For this reason, amino acid import into the vacuole is of particular interest in the context of amino acid export.

It has long been known that plant vacuoles accumulate amino acids (Homeyer and Schultz, 1988; Dietz *et al.*, 1990). The amino acid composition of the vacuoles in higher plants, determined either from isolated vacuoles (Homeyer and Schultz, 1988; Dietz *et al.*, 1990) or by non-aqueous fractionation techniques (Leidreiter *et al.*, 1995; Lohaus *et al.*, 1999), was significantly different from that of the cytosol, suggesting selective transport mechanisms for the tonoplast. Not surprisingly, pharmacological studies using various inhibitors suggested the contribution of multiple components to vacuolar amino acid transport.

ATP or its non-hydrolysable analogue stimulated both import (Dietz *et al.*, 1990; Goerlach and Willmshoff, 1992) and export (Dietz *et al.*, 1989; Thume and Dietz, 1991) of neutral amino acids (Ala; leucine, Leu; Gln; and glycine, Gly), indicating that the porter or the channel of these amino acids is stimulated by ATP, but not driven by the hydrolysis of ATP. The transport of these amino acids was insensitive to protonophores and to a pH gradient, indicating that the proton gradient is not the driving force. In contrast, import of phenylalanine (Phe) into the vacuole was dependent on the proton gradient and the ionophore, indicating that the transport of aromatic amino acids is mediated by a system that is driven by an electrochemical proton gradient (Homeyer and Schultz, 1988; Homeyer *et al.*, 1989). The transport systems for the basic amino acids show yet another property; they are stimulated by ATP in a similar fashion to the system for neutral amino acids, yet have a distinct substrate specificity, as demonstrated by competition assays (Martinoia *et al.*, 1991).

17.4 Amino Acid Export Proteins in Plants and Other Systems

Relatively little information is available about the molecular mechanisms of amino acid export in plants, although there are a number of transporter families from kingdoms other than plants that include carriers that can transport amino acids outwards, and some of these families have plant members. Even though most of these transporters are uncharacterized, it is quite likely that these plant proteins do function as amino acid exporters. In this section, we will summarize the characteristics of transporter families that (i) include either characterized plant amino acid exporters; or (ii) include amino acid exporters from other organisms *and* have plant members.

17.4.1 The drug/metabolite transporter (DMT) superfamily

Currently, 32 families of transporters are recognized within the drug/metabolite transporter (DMT) superfamily (see the Transporter

Classification Database at http://www.tcdb.org). The founding members of this family of small multi-drug resistance proteins were identified as efflux carriers for multiple lipophilic cationic drugs in prokaryotes (Paulsen *et al.*, 1996). The substrates transported by this superfamily of proteins are diverse and include sugars, nucleotide-sugars and amino acids apart from various drug molecules (Jack *et al.*, 2001).

Nine families within the DMT superfamily include plant members: plant drug/metabolite exporters (P-DMEs), triose phosphate transporters (TPTs), CMP-sialate:CMP antiporters (CSAs), UDP-glucose transporters (UGAs), GDP-mannose:GMP antiporters (GMAs), plant organocation permeases (POPs), nucleobase uptake transporters (NBUTs), thiamine pyrophosphate transporters (TPPTs) and NIPA Mg^{2+} uptake permeases (NIPAs). Among these nine families, P-DMEs are of particular interest as an amino acid export system in plants. The P-DME family consists exclusively of plant proteins, most of which are uncharacterized. Ladwig *et al.* (2012) recently demonstrated that the Siliques Are Red1 (SiAR1)/UMAMI18 protein, which belongs to the P-DME family, is capable of transporting amino acids bidirectionally across the plasma membrane. T-DNA insertion lines in the *SiAR1/UMAMI18* gene showed reduced amino acid content in developing siliques, suggesting that amino acid export mediated by the SiAR1/UMAMI18 protein is important in providing organic N to the developing siliques. A T-DNA insertion in the gene of another member of this family, *Walls-Are-Thin1 (WAT1)/UMAMI5*, increased sensitivity to a toxic analogue of tryptophan, 5-Methyltryptophan (5-Me-Trp) (Ranocha *et al.*, 2010). Although the exact mechanism of increased sensitivity is unknown, it is possible that this vacuolar transporter is involved in the sequestration of 5-Me-Trp into the vacuole. Later, the same group demonstrated that tryptophan metabolism and auxin signalling in this mutant are disrupted, and demonstrated that the WAT1 protein functions as a vascular auxin carrier (Denancé *et al.*, 2013; Ranocha *et al.*, 2013). The P-DME genes belong to a large gene family (44 genes are found in *Arabidopsis* alone), so it is quite possible that some other members of this family also function in the efflux of amino acids. Curiously, one of the first P-DME members to be discovered was Nodulin 21 from *Medicago truncatula*, which was identified as a nodulation-specific protein (Delauney *et al.*, 1990). It is tempting to hypothesize that P-DMEs play a role in exporting fixed organic N from nodules in legumes.

17.4.2 The amino acid-polyamine-organocation (APC) superfamily

Eleven families of transporters have been reported as members of the APG superfamily (see Table 17.1 and Wong *et al.*, 2012). Among these families, four include plant members, two of which include amino acid transporters.

Plant members of the amino acid-polyamine-organocation (APC) family are of particular interest in terms of amino acid export processes. Cationic amino acid transporters (CATs), amino acid/choline transporters (ACTs), and polyamine H^+-symporters (PHSs) from the plant kingdom belong to this family. One of the founding members of this family, the cationic amino acid transporter, was initially discovered because of its ability to mediate amino acid uptake in a heterologous expression system (Frommer *et al.*, 1995). Since then, several amino acid transporters belonging to this family in *Arabidopsis thaliana* (CAT1, 5, 6, 8) have been biochemically characterized. Interestingly, while AtCAT1, 5, and 6 were found to be proton symporters (Frommer *et al.*, 1995; Su *et al.*, 2004; Hammes *et al.*, 2006; Rentsch *et al.*, 2007), another member, AtCAT8, was suggested to be a proton-gradient-independent facilitator (Yang *et al.*, 2010). The proteomics of vacuolar membrane and fluorescent protein-tag experiments showed that AtCAT8 localizes to both the plasma membrane and the tonoplast membrane (Yang *et al.*, 2010). Whether the CAT8-mediated transport of amino acids across the plasma and vacuolar membranes are indeed bidirectional remains to be seen.

The bidirectional amino acid transport1 (BAT1)/GABA permease (GABAP) from *Arabidopsis* is the only plant member of the amino acid/choline transporters (ACTs) family characterized to date. When expressed in yeast cells, BAT1/GABAP increased the accumulation of Arg and Ala while decreasing the accumulation of lysine (Lys) and Glu. From these results, it was suggested that BAT1/GABAP is capable of mediating bidirectional transport. Later studies

Table 17.1. Transporters belonging to the amino acid-polyamine-organocation (APC) superfamily.

Families	Acronym	Found in	Plant members
Amino acid-polyamine-organocation	APC	Archaea, bacteria, fungi (including yeast), slime moulds, unicellular eukaryotic protists, plants and animals	Cationic amino acid transporters (CATs); amino acid/choline transporters (ACTs); polyamine H^+-symporters (PHSs); amino acid/choline transporters (ACTs)
Alanine or glycine: cation symporter	AGCS	Archaea, bacteria	
Amino acid/auxin permease	AAAP	Fungi (including yeast), plants and animals	Amino acid permeases (AAPs), lysine/histidine transporters (LHTs), proline transporters (ProTs), γ-aminobutyric acid transporters (GATs), auxin transporters (AUX/LAXs)
Betaine/carnitine/choline transporter	BCCT	Archaea, bacteria	
Cation-chloride cotransporter	CCC	Bacteria, yeast, plants and animals	Cation–Cl^- cotransporters (CCCs)
Hydroxy/aromatic amino acid permease	HAAAP	Bacteria	
Neurotransmitter: sodium symporter	NSS	Archaea, bacteria, animals	
Nucleobase:cation symporter-1	NCS1	Archaea, bacteria, fungi (including yeast), plants	Nucleobase cation symporters (NCSs)
Nucleobase:cation symporter-2	NCS2	Archaea, bacteria, fungi, plants and animals	
Solute:sodium symporter	SSS	Archaea, bacteria, animals	
Sulfate permease	SulP	Archaea, bacteria, fungi, plants and animals	

indicated that this protein is a mitochondrial GABA carrier that provides GABA to mitochondrial enzymes involved in the GABA-shunt pathway (Michaeli *et al.*, 2011).

The amino acid/auxin permease (AAAP) family includes a number of plant transporter families, including well-characterized amino acid transporters (see Table 17.1). All of the previously characterized amino acid carriers belonging to this family – amino acid permeases (AAPs), lysine/histidine transporters (LHTs), proline transporters (ProTs), γ-aminobutyric acid transporters (GATs) and auxin transporters (AUX/LAXs) – are proton symporters (reviewed in: Wipf *et al.*, 2002; Rentsch *et al.*, 2007; Tegeder and Rentsch, 2010; Tegeder, 2012). While symporters play a wide variety of roles in the long-distance transport of amino acids, they are unlikely to mediate cellular export owing to the steep proton gradient between the apoplast and the cytosol. Interestingly, the AAAP family also includes amino acid vacuolar transporters (AVTs) from yeast and vesicular GABA and glycine transporters (VGATs) from animals. There are 19 genes that belong to this family from *Arabidopsis* (Sekito *et al.*, 2008). One of these is the aromatic and neutral transporter (ANT1), which is capable of mediating amino acid transport when expressed in yeast cells, although the subcellular localization of this protein is not yet known (Chen *et al.*, 2001). Several *Arabidopsis* AVT-like proteins have been identified in proteomic study of vacuoles, indicating that they might indeed be vacuolar amino acid transporters (Carter *et al.*, 2004; Jaquinod *et al.*, 2007; Tanz *et al.*, 2013).

17.4.3 The ATP-binding cassette (ABC) transporter superfamily

ABC transporters are characterized by a conserved monophyletic ATP-hydrolysing domain that energizes the transport process. ABC transporters are one of the largest superfamilies of transporters found in all domains of life; they are divided into nine families (ABCA–ABCI). The plant genome in particular encodes a large number of ABC transporters (130 in *Arabidopsis*, including members of ABCA, B, C, D, E/F, G, I).

The ABC transporters can mediate both efflux and import, and the substrates are extremely diverse (sugars, amino acids, organic acids, peptides, inositol phosphates, metals/metalloids, phytochelatins, plant hormones, nucleotides, lipids, flavonoids, alkaloids, xenobiotic drugs, vitamins, glutathione conjugates and glucans) (Higgins, 1992; Davidson and Chen, 2004; Kretzschmar *et al.*, 2011). Many of the bacterial ABC transporters are involved in solute-binding protein-dependent (SBP) amino acid transport. There is no established plant ABC transporter that is capable of transporting amino acids. A T-DNA insertion in *AtMRP2*, a gene encoding a vacuolar transporter of glutathione conjugates and possibly chlorophyll catabolites, causes an increase in amino acids secreted from roots (Badri *et al.*, 2008). The mechanism of this increase in amino acid secretion is so far unknown.

17.4.4 The major facilitator superfamily (MFS)

The MFS superfamily is the largest family of transporters and includes members from all domains of life. Examples are amino acid transporters from animals, such as the aromatic amino acid carrier TAT1 (Ramadan *et al.*, 2006), vesicular glutamate transporters (Juge *et al.*, 2006) and system L amino acid transporters (LAT) 3 and 4 (Babu *et al.*, 2003; Bodoy *et al.*, 2005), as well as vacuolar basic amino acid transporters (VBAs) from yeast (Shimazu *et al.*, 2005), the vacuolar amino acid carrier Atg22 from fission yeast (Yang *et al.*, 2006) and phagosomal transporters PHTs from proteobacteria (Sauer *et al.*, 2005). However, none of these MFS transporters have apparent homologues in plants, and whether any of the uncharacterized plant transporters belonging to this family are involved in amino acid efflux remains to be seen.

17.5 Regulation of Amino Acid Export

As discussed in Section 17.3.2, the amino acid concentration and composition in the xylem are influenced by various environmental and developmental factors. While it is expected that some of the changes that occur are caused by altered enzymatic activities (Lam *et al.*, 1995; Hsieh *et al.*, 1996; Scheible *et al.*, 2004; Krouk *et al.*, 2010; Nunes-Nesi *et al.*, 2010; Funayama *et al.*, 2013), an amino acid export process into the xylem sap could also be important in such regulation.

In this context, the phenotype of the *GDU1*-overexpressing mutant is of particular interest. Originally isolated in an activation tagging screen, the overexpression of GDU1 causes a large increase of Gln and other amino acids in the xylem sap, to the extent to which Gln starts to crystallize at the leaf hydathodes (Pilot *et al.*, 2004). Studies using radiolabelled amino acids demonstrated that the efflux of amino acids is increased in plants overexpressing GDU1 or its paralogues, and that the efflux is passive and specific for amino acids showing preference for Gln over other amino acids (Pratelli *et al.*, 2010). Interestingly, these characteristics match the ones that were observed for amino acid export systems found in seed coat studies (discussed in Section 17.3.1), suggesting that GDU1 is involved in the regulation of amino acid export activities.

The exact mechanism through which the overexpression of *GDU1* enhances amino acid export is unknown. As GDU1 is a small, single transmembrane (type I) protein, it is unlikely that the protein functions as an exporter on its own. Rather, it is likely that protein–protein interactions mediated by GDU1 control the export of amino acids. Recently, a RING-type E3 ubiquitin ligase, LOSS OF GDU2 (LOG2) has been identified as a physical and genetic interactor of GDU1 in *Arabidopsis* (Pratelli *et al.*, 2012). Interestingly, the RING domain of LOG2 shares homology with that of Mammalian Mahogunin Ring Finger 1 (MGRN1) protein, which interacts with and controls the activity of a membrane G protein-coupled receptor. The expression of MGRN1 in mammals functionally replaces that of LOG2 in plants, suggesting that the function of RING domain is evolutionarily conserved (Guerra *et al.*, 2013). Analogous to the activity of MGRN1 in the trafficking of the receptors,

LOG2 and GDU1 might be involved in the trafficking of amino acid exporters in plant cells.

17.6 Amino Acids in Inter-Organism Interactions

17.6.1 Amino acid secretion into the rhizosphere

Not surprisingly, ample evidence suggests that the secretion of amino acids from the plant functions as a signal for soil microorganisms. Many soil bacteria, including *Pseudomonas fluorescens* and *Bacillus subtilis*, show chemotactic responses to all 20 proteinogenic amino acids (Ordal and Gibson, 1977; de Weert *et al.*, 2002; Oku *et al.*, 2012). Plant root exudates are capable of inducing transcriptional changes in soil bacteria, and the genes involved in amino acid metabolism are induced upon the addition of root extracts to the bacteria (Mark *et al.*, 2005; Ramachandran *et al.*, 2011; Carvalhais *et al.*, 2013). Furthermore, exudates from nutrient-deficient plants induced global transcriptome changes in plant growth-promoting bacteria, and concentrations of the amino acids aspartate (Asp), valine (Val) and Glu in root exudates were found to correlate significantly with the transcriptional changes (Carvalhais *et al.*, 2013).

Moreover, soil microorganisms are capable of promoting the secretion of organic molecules from plant roots. Specifically, compounds such as 2,4-diacetylphloroglucinol (DAPG) and phenazine from *Pseudomonas* bacteria, and zearalenone from *Fusarium* fungi, were found to promote the efflux of amino acids in many plants (Phillips *et al.*, 2004). The same study showed that myristic and palmitic acids, known for their protonophoric effects, did not have any effect on amino acid secretion, suggesting that the effects of DAPG, phenazine and zearalenone are specific. In another study, rice roots inoculated with *Corynebacterium* sp. (Sb26) and *Rhizobium* sp. (Sb16) secreted more sugar and amino acids into the growth medium (Naher *et al.*, 2009). Taken together, these results indicate that the secretion of amino acids is likely to be a part of complex chemical language that is exchanged between the plants and soil organisms, which could function in recruiting beneficial microorganisms or deterring pathogens (Matilla *et al.*, 2010).

The molecular mechanisms through which amino acids are secreted to the rhizosphere are relatively unknown. Two mutants of ABC transporters from *Arabidopsis*, *Atpdr6* and *Atmrp2*, were found to secrete low levels of organic acids and high levels of amino acids compared with the wild type (Badri *et al.*, 2008). The reported substrates for these transporters, however, are either unknown (AtPDR6) or are not amino acids (glutathione conjugates, AtMRP2). Also, the relatively broad change in the secreted organic molecules (i.e. changes in both sugars and amino acids, as well as secondary metabolites) suggest that the changes in secreted amino acids in these mutants were indirectly caused by altered cellular biochemistry resulting from the mutations in the respective ABC transporters.

17.6.2 Amino acid transport during nodulation

Symbiotic nitrogen fixation (SNF) between legumes and rhizobia happens in a structure called the symbiosome, which consists of the plant-derived symbiosome membrane and the enclosed bacteria (Udvardi and Poole, 2013). Originally, it was thought that the exchange between the plant and the bacteroids was simply 'carbon-in, nitrogen-out'; plants provide carbon sources as dicarboxylic acid, whereas the ammonium made by bacteroids is exported from the symbiosome into the cytoplasm of infected cells. Later, it was proven that the provision of branched chain amino acids from plants is necessary for the growth of N-fixing bacteria, and that in bacteroids the bacteria shut down ammonium assimilation and essentially function as a sub-cellular organelle (Lodwig et al., 2003; Prell *et al.*, 2009). This so-called 'symbiotic auxotrophy' for amino acids has been observed in both definite and indefinite nodules, although in some combinations of rhizobia and plants, symbiotic auxotrophy was not observed (Prell *et al.*, 2010). In addition, it has been proposed that phloem-borne amino acids function as a signal to control N-fixation activities in nodules (Sulieman *et al.*, 2010; Sulieman and Schulze, 2010).

The mechanism through which the plant-borne amino acids are delivered to the bacteroids is not known. Earlier studies evaluating uptake of radiolabelled amino acids through the symbiosome membrane indicated that the influx of Gln, Asp and Ala into the symbiosome was very slow, if it occurred at all (Udvardi *et al.*, 1988; Herrada *et al.*, 1989; Whitehead *et al.*, 1998). Proteomics studies have so far not identified known or putative amino acid carriers on the symbiosome membrane (Panter *et al.*, 2000; Wienkoop and Saalbach, 2003). Hence, whether amino acids cross the symbiosome membrane through a specific carrier protein remains to be seen.

17.6.3 Amino acids in plant–pathogen interactions

Amino acids have been implicated in signalling between plant hosts and infectious pathogens. In response to *Agrobacterium tumefaciens* infection, plants activate the synthesis of GABA (Chevrot *et al.*, 2006; Haudecoeur *et al.*, 2009). High concentrations of GABA in the plant tissue were found to promote the degradation of the quorum-sensing signal OC8HSL (*N*-(3-oxooctanoyl) homoserine lactone). This, in turn, reduces the conjugation and amplification of the Ti (tumour-inducing) plasmid and alleviates tumoral symptoms during infection (Chevrot *et al.*, 2006; Haudecoeur *et al.*, 2009). Plant-derived GABA is taken up by the bacteria via the branched chain amino acid (Bra) transporter system and activates the *Agrobacterium tumefaciens* lactonase AttM, leading to the degradation of the quorum sensing signal by an unknown mechanism; this suggests that GABA may be used by the plant as a way to control bacterial growth (Hosie *et al.*, 2002; Chevrot *et al.*, 2006). Moreover, another amino acid, proline (Pro), antagonizes the GABA-induced degradation of OC8HSL, thus demonstrating a system in which two amino acids serve antagonistically as signalling molecules during plant–pathogen interactions (Haudecoeur *et al.*, 2009). During infection with the fungal pathogen *Cladosporium fulvum*, the concentrations of most amino acids and the total nitrogen content increase (Solomon and Oliver, 2001).

Whether the regulation of amino acid levels in tissues attacked by pathogens involves control at membrane transporter levels is not known. Recently, it was shown that a reduction in the expression of OsSWEET11, a sugar efflux transporter in rice (*Oryza sativa*), results in resistance to *Xanthomonas oryzae* pv. *oryzae* (Xoo) strain PXO99^A, suggesting that OsSWEET11 is exploited by the pathogen to provide the sugars necessary for its growth (Chen *et al.*, 2010). It is tempting to hypothesize that a similar mechanism is in place for amino acid transporters. Interestingly, AtLHT1, a general amino acid transporter in *Arabidopsis*, is induced by pathogen infection, and the plants carrying a T-DNA insertion in the *AtLHT1* gene show broad resistance to biotrophic and hemibiotrophic pathogens (Liu *et al.*, 2010). It has been suggested that depletion of cytosolic Gln, caused by the lack of functional AtLHT1, functions as a signal to activate the defence response.

17.7 Conclusions

Studies of amino acid efflux from plant cells are still in their infancy, and most of the molecular components are unknown. Technological breakthroughs that enable the analysis of export processes on a larger scale would assist in the discovery of additional components of the amino acid export system (Okumoto *et al.*, 2008). Also, additional discoveries of the genetic and biochemical interactions that are involved will shed light on the amino acid export process per se. A number of studies have demonstrated that the quantity and quality of N, supplied as amino acids, has fundamental effects on the growing of organs such as seeds (Riens *et al.*, 1991; Lohaus *et al.*, 1995; Lohaus and Moellers, 2000). Manipulation of the amino acid export process would offer an alternative strategy to altering the composition of seeds.

References

Andrews, M. (1986) The partitioning of nitrate assimilation between the root and shoot of higher plants. *Plant, Cell and Environment* 9, 511–519.

Apse, M.P. and Blumwald, E. (2007) Na+ transport in plants. *FEBS Letters* 581, 2247–2254.

Babu, E., Kanai, Y., Chairoungdua, A., Kim, D.K., Iribe, Y., Tangtrongsup, S., Jutabha, P., Li, Y., Ahmed, N., Sakamoto, S., Anzai, N., Nagamori, S. and Endou, H. (2003) Identification of a novel system L amino acid transporter structurally distinct from heterodimeric amino acid transporters. *The Journal of Biological Chemistry* 278, 43838–43845.

Badri, D.V., Loyola-Vargas, V.M., Broeckling, C.D., De-la-Peña, C., Jasinski, M., Santelia, D., Martinoia, E., Sumner, L.W., Banta, L.M., Stermitz, F. and Vivanco, J.M. (2008) Altered profile of secondary metabolites in the root exudates of *Arabidopsis* ATP-binding cassette transporter mutants. *Plant Physiology* 146, 762–771.

Balint, T. and Rengel, Z. (2011) Amino acid composition of xylem and phloem sap varies in canola genotypes differing in nitrogen- and sulfur-use efficiency. *Crop and Pasture Science* 62, 198–207.

Bi, Y.M., Kant, S., Clark, J., Gidda, S., Ming, F., Xu, J.Y., Rochon, A., Shelp, B.J., Hao, L.X., Zhao, R., Mullen, R.T., Zhu, T. and Rothstein, S.J. (2009) Increased nitrogen-use efficiency in transgenic rice plants over-expressing a nitrogen-responsive early nodulin gene identified from rice expression profiling. *Plant, Cell and Environment* 32, 1749–1760.

Bodoy, S., Martín, L., Zorzano, A., Palacín, M., Estévez, R. and Bertran, J. (2005) Identification of LAT4, a novel amino acid transporter with system L activity. *The Journal of Biological Chemistry* 280, 12002–12011.

Broeckling, C.D., Broz, A.K., Bergelson, J., Manter, D.K. and Vivanco, J.M. (2008) Root exudates regulate soil fungal community composition and diversity. *Applied and Environmental Microbiology* 74, 738–744.

Carter, C., Pan, S.Q., Jan, Z.H., Avila, E.L., Girke, T. and Raikhel, N.V. (2004) The vegetative vacuole proteorne of Arabidopsis thaliana reveals predicted and unexpected proteins. *The Plant Cell* 16, 3285–3303.

Carvalhais, L.C., Dennis, P.G., Fedoseyenko, D., Hajirezaei, M.R., Borriss, R. and von Wirén, N. (2011) Root exudation of sugars, amino acids, and organic acids by maize as affected by nitrogen, phosphorus, potassium, and iron deficiency. *Journal of Plant Nutrition and Soil Science* 174, 3–11.

Carvalhais, L.C., Dennis, P.G., Fan, B., Fedoseyenko, D., Kierul, K., Becker, A., von Wirén, N. and Borriss, R. (2013) Linking plant nutritional status to plant–microbe interactions. *PLoS ONE* 8(7): e68555.

Chaparro, J.M., Badri, D.V., Bakker, M.G., Sugiyama, A., Manter, D.K. and Vivanco, J.M. (2013) Root exudation of phytochemicals in *Arabidopsis* follows specific patterns that are developmentally programmed and correlate with soil microbial functions. *PLoS ONE* 8(2): e55731.

Chen, L., Ortiz-Lopez, A., Jung, A. and Bush, D.R. (2001) ANT1, an aromatic and neutral amino acid transporter in *Arabidopsis*. *Plant Physiology* 125, 1813–1820.

Chen, L.Q., Hou, B.H., Lalonde, S., Takanaga, H., Hartung, M.L., Qu, X.Q., Guo, W.J., Kim, J.G., Underwood, W., Chaudhuri, B., Chermak, D., Antony, G. *et al.* (2010) Sugar transporters for intercellular exchange and nutrition of pathogens. *Nature* 468, 527–532.

Chevrot, R., Rosen, R., Haudecoeur, E., Cirou, A., Shelp, B.J., Ron, E. and Faure, D. (2006) GABA controls the level of quorum-sensing signal in *Agrobacterium tumefaciens*. *Proceedings of the National Academy of Sciences of the United States of America* 103, 7460–7464.

Cho, M. and Cho, H.-T. (2012) The function of ABCB transporters in auxin transport. *Plant Signaling and Behavior* 8(2): e22990.

Davidson, A.L. and Chen, J. (2004) ATP-binding cassette transporters in bacteria. *Annual Review of Biochemistry* 73, 241–268.

De Jong, A. and Wolswinkel, P. (1995) Differences in release of endogenous sugars and amino acids from attached and detached seed coats of developing pea seeds. *Physiologia Plantarum* 94, 78–86.

de Jong, A., Koerselman-Kooij, J., Schuurmans, J. and Borstlap, A. (1997) The mechanism of amino acid efflux from seed coats of developing pea seeds as revealed by uptake experiments. *Plant Physiology* 114, 731–736.

de Weert, S., Vermeiren, H., Mulders, I.H.M., Kuiper, I., Hendrickx, N., Bloemberg, G.V., Vanderleyden, J., De Mot, R. and Lugtenberg, B.J.J. (2002) Flagella-driven chemotaxis towards exudate components is an important trait for tomato root colonization by *Pseudomonas fluorescens*. *Molecular Plant–Microbe Interactions* 15, 1173–1180.

Delauney, A.J., Cheon, C.I., Snyder, P.J. and Verma, D.P.S. (1990) A nodule-specific sequence encoding a methionine-rich polypeptide, nodulin-21. *Plant Molecular Biology* 14, 449–451.

Denancé, N., Ranocha, P., Oria, N., Barlet, X., Rivière, M.P., Yadeta, K.A., Hoffmann, L., Perreau, F., Clement, G., Maia-Grondard, A., van den Berg, G.C.M., Savelli, B. *et al.* (2013) *Arabidopsis wat1* (*walls are thin1*)-mediated resistance to the bacterial vascular pathogen, *Ralstonia solanacearum*, is accompanied by cross-regulation of salicylic acid and tryptophan metabolism. *The Plant Journal* 73, 225–239.

Dietz, K.-J., Martinoia, E. and Heber, U. (1989) Mobilization of vacuolar amino acids in leaf cells as affected by ATP and the level of cytosolic amino acids: ATP regulates but appears not to energize vacuolar amino acid release. *Biochimica et Biophysica Acta* 984, 57–62.

Dietz, K.-J., Jäger, R., Kaiser, G. and Martinoia, E. (1990) Amino acid transport across the tonoplast of vacuoles isolated from barley mesophyll protoplasts: uptake of alanine, leucine, and glutamine. *Plant Physiology* 92, 123–129.

Dunkel, M., Latz, A., Schumacher, K., Müller, T., Becker, D. and Hedrich, R. (2008) Targeting of vacuolar membrane localized members of the TPK channel family. *Molecular Plant* 1, 938–949.

Fernandez-Garcia, N., Hernandez, M., Casado-Vela, J., Bru, R., Elortza, F., Hedden, P. and Olmos, E. (2011) Changes to the proteome and targeted metabolites of xylem sap in *Brassica oleracea* in response to salt stress. *Plant, Cell and Environment* 34, 821–836.

Forestan, C. and Varotto, S. (2012) The role of PIN auxin efflux carriers in polar auxin transport and accumulation and their effect on shaping maize development. *Molecular Plant* 5, 787–798.

Fritz, C., Mueller, C., Matt, P., Feil, R. and Stitt, M. (2006) Impact of the C-N status on the amino acid profile in tobacco source leaves. *Plant, Cell and Environment* 29, 2055–2076.

Frommer, W.B., Hummel, S., Unseld, M. and Ninnemann, O. (1995) Seed and vascular expression of a high-affinity transporter for cationic amino acids in *Arabidopsis*. *Proceedings of the National Academy of Sciences of the United States of America* 92, 12036–12040.

Funayama, K., Kojima, S., Tabuchi-Kobayashi, M., Sawa, Y., Nakayama, Y., Hayakawa, T. and Yamaya, T. (2013) Cytosolic glutamine synthetase1;2 is responsible for the primary assimilation of ammonium in rice roots. *Plant and Cell Physiology* 54, 934–943.

Gansel, X., Munos, S., Tillard, P. and Gojon, A. (2001) Differential regulation of the NO_3^- and NH_4^+ transporter genes *AtNrt2.1* and *AtAmt1.1* in *Arabidopsis*: relation with long-distance and local controls by N status of the plant. *The Plant Journal* 26, 143–155.

Goerlach, J. and Willmshoff, I. (1992) Glycine uptake into barley mesophyll vacuoles is regulated but not energized by ATP. *Plant Physiology* 99, 134–139.

Grusak, M.A. and Minchin, P.E.H. (1988) Seed coat unloading in *Pisum sativum* – osmotic effects in attached versus excised empty ovules. *Journal of Experimental Botany* 39, 543–559.

Guerra, D.D., Pratelli, R., Kraft, E., Callis, J. and Pilot, G. (2013) Functional conservation between mammalian MGRN1 and plant LOG2 ubiquitin ligases. *FEBS Letters* 587, 3400–3405.

Gutierez, R.A., Stokes, T.L., Thum, K., Xu, X., Obertello, M., Katari, M.S., Tanurdzic, M., Dean, A., Nero, D.C., McClung, C.R. and Coruzzi, G.M. (2008) Systems approach identifies an organic nitrogen-responsive gene network that is regulated by the master clock control gene *CCA1*. *Proceedings of the National Academy of Sciences of the United States of America* 105, 4939–4944.

Hammes, U.Z., Nielsen, E., Honaas, L.A., Taylor, C.G. and Schachtman, D.P. (2006) AtCAT6, a sink-tissue-localized transporter for essential amino acids in *Arabidopsis*. *The Plant Journal* 48, 414–426.

Haudecoeur, E., Planamente, S., Cirou, A., Tannieres, M., Shelp, B.J., Morera, S. and Faure, D. (2009) Proline antagonizes GABA-induced quenching of quorum-sensing in *Agrobacterium tumefaciens*. *Proceedings of the National Academy of Sciences of the United States of America* 106, 14587–14592.

Herrada, G., Puppo, A. and Rigaud, J. (1989) Uptake of metabolites by bacteroid-containing vesicles and by free bacteroids from French bean nodules. *Journal of General Microbiology* 135, 3165–3171.

Herschbach, C., Gessler, A. and Rennenberg, H. (2012) Long-distance transport and plant internal cycling of N- and S-compounds. *Progress in Botany* 73, 161–188.

Higgins, C.F. (1992) ABC transporters: from microorganisms to man. *Annual Review of Cell Biology* 8, 67–113.

Hirner, A., Ladwig, F., Stransky, H., Okumoto, S., Keinath, M., Harms, A., Frommer, W.B. and Koch, W. (2006) *Arabidopsis* LHT1 is a high-affinity transporter for cellular amino acid uptake in both root epidermis and leaf mesophyll. *The Plant Cell* 18, 1931–1946.

Homeyer, U. and Schultz, G. (1988) Transport of phenylalanine into vacuoles isolated from barley mesophyll protoplasts *Planta* 176, 378–382.

Homeyer, U., Litek, K., Huchzermeyer, B. and Schultz, G. (1989) Uptake of phenylalanine into isolated barley vacuoles is driven by both tonoplast adenosine-triphosphatase and pyrophosphatase – evidence for a hydrophobic L-amino-acid carrier system. *Plant Physiology* 89, 1388–1393.

Hosie, A.H., Allaway, D., Galloway, C.S., Dunsby, H.A. and Poole, P.S. (2002) *Rhizobium leguminosarum* has a second general amino acid permease with unusually broad substrate specificity and high similarity to branched-chain amino acid transporters (Bra/LIV) of the ABC family. *Journal of Bacteriology* 184, 4071–4080.

Hsieh, M.H., Lam, H.M. and Coruzzi, G. (1996) Metabolic regulation of nitrogen assimilatory genes in *Arabidopsis*. *Plant Physiology* 111, 615–615.

Imsande, J. and Touraine, B. (1994) N demand and the regulation of nitrate uptake. *Plant Physiology* 105, 3–7.

Inselsbacher, E. and Nasholm, T. (2012) The below-ground perspective of forest plants: soil provides mainly organic nitrogen for plants and mycorrhizal fungi. *New Phytologist* 195, 329–334.

Jack, D.L., Yang, N.M. and Saier, M.H. (2001) The drug/metabolite transporter superfamily. *European Journal of Biochemistry* 268, 3620–3639.

Jaeger, C.H., Lindow, S.E., Miller, S., Clark, E. and Firestone, M.K. (1999) Mapping of sugar and amino acid availability in soil around roots with bacterial sensors of sucrose and tryptophan. *Applied and Environmental Microbiology* 65, 2685–2690.

Jaquinod, M., Villiers, F., Kieffer-Jaquinod, S., Hugouvieu, V., Bruley, C., Garin, J. and Bourguignon, J. (2007) A proteomics dissection of *Arabidopsis thaliana* vacuoles isolated from cell culture. *Molecular and Cellular Proteomics* 6, 394–412.

Jeschke, W.D. and Hartung, W. (2000) Root–shoot interactions in mineral nutrition. *Plant and Soil* 226, 57–69.

Juge, N., Yoshida, Y., Yatsushiro, S., Omote, H and Moriyama, Y. (2006) Vesicular glutamate transporter contains two independent transport machineries. *The Journal of Biological Chemistry* 281, 39499–39506.

Křeček, P., Skůpa, P., Libus, J., Naramoto, S., Tejos, R., Friml, J. and Zažímalova, E. (2009) The PIN-FORMED (PIN) protein family of auxin transporters. *Genome Biology* 10, 249.

Kretzschmar, T., Burla, B., Lee, Y., Martinoia, E. and Nagy, R. (2011) Functions of ABC transporters in plants. *Essays in Biochemistry* 50, 145–160.

Krouk, G., Mirowski, P., LeCun, Y., Shasha, D.E. and Coruzzi, G.M. (2010) Predictive network modeling of the high-resolution dynamic plant transcriptome in response to nitrate. *Genome Biology* 11, R123.

Kruse, J., Hetzger, I., Hansch, R., Mendel, R., Walch-Liu, P., Engels, C. and Rennenberg, H. (2002) Elevated pCO(2) favours nitrate reduction in the roots of wild-type tobacco (*Nicotiana tabacum* cv. Gat.) and significantly alters N-metabolism in transformants lacking functional nitrate reductase in the roots. *Journal of Experimental Botany* 53, 2351–2367.

Ladwig, F., Stahl, M., Ludewig, U., Hirner, A.A., Hammes, U.Z., Stadler, R., Harter, K. and Koch, W. (2012) *Siliques Are Red1* from *Arabidopsis* acts as a bidirectional amino acid transporter that is crucial for the amino acid homeostasis of siliques. *Plant Physiology* 158, 1643–1655.

Lam, H.M., Coschigano, K., Schultz, C., Melo-Oliveira, R., Tjaden, G., Oliveira, I., Ngai, N., Hsieh, M.H. and Coruzzi, G. (1995) Use of *Arabidopsis* mutants and genes to study amide amino acid biosynthesis. *The Plant Cell* 7, 887–898.

Lanfermeijer, F.C., Vanoene, M.A. and Borstlap, A.C. (1992) Compartmental analysis of amino-acid release from attached and detached pea seed coats. *Planta* 187, 75–82.

Leidreiter, K., Kruse, A., Riens, B., Winter, H., Lohaus, G., Robinson, D., Heineke, D. and Heldt, H.W. (1995) Subcellular compartmentation of metabolites in plant cells. In: Mathis, P. (ed.) *Photosynthesis: From Light to Biosphere. Volume 5*. Kluwer, Dordrecht, The Netherlands, pp. 483–486.

Lesuffleur, F. and Cliquet, J.B. (2010) Characterisation of root amino acid exudation in white clover (*Trifolium repens* L.). *Plant and Soil* 333, 191–201.

Liu, G., Ji, Y., Bhuiyan, N.H., Pilot, G., Selvaraj, G., Zou, J. and Wei, Y. (2010) Amino acid homeostasis modulates salicylic acid-associated redox status and defense responses in *Arabidopsis*. *The Plant Cell* 22, 3845–3863.

Liu, X.-P., Gong, C.M., Fan, Y.-Y., Eiblmeier, M., Zhao, Z., Han, G. and Rennenberg, H. (2013) Response pattern of amino compounds in phloem and xylem of trees to soil drought depends on drought intensity and root symbiosis. *Plant Biology* 15(Supplement s1), 101–108.

Liu, Y.-H., Cao, J.-S., Li, G.-J., Wu, X.-H., Wang, B.-G., Xu, P., Hu, T.-T., Lu, Z.-F., Patrick, J.W. and Ruan, Y.-L. (2012) Genotypic differences in pod wall and seed growth relate to invertase activities and assimilate transport pathways in asparagus bean. *Annals of Botany* 109, 1277–1284.

Lodwig, E.M., Hosie, A.H., Bourdes, A., Findlay, K., Allaway, D., Karunakaran, R., Downie, J.A. and Poole, P.S. (2003) Amino-acid cycling drives nitrogen fixation in the legume–*Rhizobium* symbiosis. *Nature* 422, 722–726.

Lohaus, G. and Moellers, C. (2000) Phloem transport of amino acids in two *Brassica napus* L. genotypes and one *B. carinata* genotype in relation to their seed protein content. *Planta* 211, 833–840.

Lohaus, G., Winter, H., Riens, B. and Heldt, H.W. (1995) Further studies of the phloem loading process in leaves of barley and spinach: the comparison of metabolite concentrations in the apoplastic compartment with those in the cytosolic compartment and in the sieve tubes. *Botanica Acta* 108, 270–275.

Lohaus, G., Heineke, D., Kruse, A., Leidreiter, K., Riens, B., Robinson, D.G., Winter, H., Winzer, T. and Heldt, H.W. (1999) Compartmentation of metabolites between the subcellular compartments of leaves, the apoplast, the phloem and the storage tissue of different crop plants. In: Kruger, N.J., Hill, S.A. and Ratcliffe, R.G. (eds) *Regulation of Primary Metabolic Pathways in Plants. Proceedings of the Phytochemical Society of Europe Volume 42*. Kluwer, Dordrecht, The Netherlands, pp. 159–172.

Mark, G.L., Dow, J.M., Kiely, P.D., Higgins, H., Haynes, J., Baysse, C., Abbas, A., Foley, T., Franks, A., Morrissey, J. and O'Gara, F. (2005) Transcriptome profiling of bacterial responses to root exudates identifies genes involved in microbe–plant interactions. *Proceedings of the National Academy of Sciences of the United States of America* 102, 17454–17459.

Marschner, H. (1995) *Mineral Nutrition of Higher Plants*, 2nd edn. Academic Press (imprint of Elsevier), London and San Diego, California.

Martinoia, E., Thume, M., Vogt, E., Rentsch, D. and Dietz, K.J. (1991) Transport of arginine and aspartic-acid into isolated barley mesophyll vacuoles. *Plant Physiology* 97, 644–650.

Martinoia, E., Grill, E., Tommasini, R., Kreuz, K. and Amrhein, N. (1993) ATP-dependent glutathione *S*-conjugate 'export' pump in the vacuolar membrane of plants. *Nature* 364, 247–249.

Matilla, M.A., Ramos, J.L., Bakker, P.A.H.M., Doornbos, R., Badri, D.V., Vivanco, J.M. and Ramos-Gonzalez, M.I. (2010) *Pseudomonas putida* KT2440 causes induced systemic resistance and changes in *Arabidopsis* root exudation. *Environmental Microbiology Reports* 2, 381–388.

Maurel, C., Verdoucq, L., Luu, D.T. and Santoni, V. (2008) Plant aquaporins: membrane channels with multiple integrated functions. *Annual Review of Plant Biology* 59, 595–624.

Michaeli, S., Fait, A., Lagor, K., Nunes-Nesi, A., Grillich, N., Yellin, A., Bar, D., Khan, M., Fernie, A.R., Turano, F.J. and Fromm, H. (2011) A mitochondrial GABA permease connects the GABA shunt and the TCA cycle, and is essential for normal carbon metabolism. *The Plant Journal* 67, 485–498.

Mori, S. (1999) Iron acquisition by plants. *Current Opinion in Plant Biology* 2, 250–253.

Naher, U.A., Othman, R., Mohd Saud, H., Shamsuddin, Z. and Ismail, M.R. (2009) Influence of root exudate carbon compounds of three rice genotypes on rhizosphere and endophytic diazotrophs. *Pertanika* 32, 209–223.

Nasholm, T., Kielland, K. and Ganeteg, U. (2009) Uptake of organic nitrogen by plants. *New Phytologist* 182, 31–48.

Nazoa, P., Vidmar, J.J., Tranbarger, T.J., Mouline, K., Damiani, I., Tillard, P., Zhuo, D.G., Glass, A.D.M. and Touraine, B. (2003) Regulation of the nitrate transporter gene *AtNRT2.1* in *Arabidopsis thaliana*: responses to nitrate, amino acids and developmental stage. *Plant Molecular Biology* 52, 689–703.

Nunes-Nesi, A., Fernie, A.R. and Stitt, M. (2010) Metabolic and signaling aspects underpinning the regulation of plant carbon nitrogen interactions. *Molecular Plant* 3, 973–996.

Offler, C.E., Nerlich, S.M. and Patrick, J.W. (1989) Pathway of photosynthate transfer in the developing seed of *Vicia faba* L. Transfer in relation to seed anatomy. *Journal of Experimental Botany* 40, 769–780.

Oku, S., Komatsu, A., Tajima, T., Nakashimada, Y. and Kato, J. (2012) Identification of chemotaxis sensory proteins for amino acids in *Pseudomonas fluorescens* Pf0-1 and their involvement in chemotaxis to tomato root exudate and root colonization. *Microbes and Environments* 27, 462–469.

Okumoto, S., Takanaga, H. and Frommer, W.B. (2008) Quantitative imaging for discovery and assembly of the metabo-regulome. *New Phytologist* 180, 271–295.

Oliveira, I.C., Brenner, E., Chiu, J., Hsieh, M.H., Kouranov, A., Lam, H.M., Shin, M.J. and Coruzzi, G. (2001) Metabolite and light regulation of metabolism in plants: lessons from the study of a single biochemical pathway. *Brazilian Journal of Medical and Biological Research* 34, 567–575.

Ordal, G.W. and Gibson, K.J. (1977) Chemotaxis toward amino acids by *Bacillus subtilis*. *Journal of Bacteriology* 129, 151–155.

Panter, S., Thomson, R., de Bruxelles, G., Laver, D., Trevaskis, B. and Udvardi, M. (2000) Identification with proteomics of novel proteins associated with the peribacteroid membrane of soybean root nodules. *Molecular Plant–Microbe Interactions* 13, 325–333.

Pate, J.S. (1973) Uptake, assimilation and transport of nitrogen compounds by plants. *Soil Biology and Biochemistry* 5, 109–119.

Pate, J.S., Atkins, C.A., Herridge, D.F. and Layzell, D.B. (1981) Synthesis, storage, and utilization of amino compounds in white lupin (*Lupinus albus* L.). *Plant Physiology* 67, 37–42.

Paulsen, I.T., Brown, M.H. and Skurray, R.A. (1996) Proton-dependent multidrug efflux systems. *Microbiology Reviews* 60, 575–608.

Peoples, M.B., Pate, J.S., Atkins, C.A. and Bergersen, F.J. (1986) Nitrogen nutrition and xylem sap composition of peanut (*Arachis hypogaea* L. cv Virginia Bunch). *Plant Physiology* 82, 946–951.

Phillips, D.A., Fox, T.C., King, M.D., Bhuvaneswari, T.V. and Teuber, L.R. (2004) Microbial products trigger amino acid exudation from plant roots. *Plant Physiology* 136, 2887–2894.

Pilot, G., Stransky, H., Bushey, D.F., Pratelli, R., Ludewig, U., Wingate, V.P. and Frommer, W.B. (2004) Overexpression of *GLUTAMINE DUMPER1* leads to hypersecretion of glutamine from hydathodes of *Arabidopsis* leaves. *The Plant Cell* 16, 1827–1840.

Pratelli, R., Voll, L.M., Horst, R.J., Frommer, W.B. and Pilot, G. (2010) Stimulation of nonselective amino acid export by glutamine dumper proteins. *Plant Physiology* 152, 762–773.

Pratelli, R., Guerra, D.D., Yu, S., Wogulis, M., Kraft, E., Frommer, W.B., Callis, J. and Pilot, G. (2012) The ubiquitin E3 ligase LOSS OF GDU2 is required for GLUTAMINE DUMPER1-induced amino acid secretion in *Arabidopsis*. *Plant Physiology* 158, 1628–1642.

Prell, J., White, J.P., Bourdes, A., Bunnewell, S., Bongaerts, R.J. and Poole, P.S. (2009) Legumes regulate *Rhizobium* bacteroid development and persistence by the supply of branched-chain amino acids. *Proceedings of the National Academy of Sciences of the United States of America* 106, 12477–12482.

Prell, J., Bourdès, A., Kumar, S., Lodwig, E., Hosie, A., Kinghorn, S., White, J. and Poole, P. (2010) Role of symbiotic auxotrophy in the *Rhizobium*–legume symbioses. *PLoS ONE* 5, e13933.

Price, M.B., Jelesko, J. and Okumoto, S. (2012) Glutamate receptor homologs in plants: functions and evolutionary origins. *Frontiers in Plant Science* 3:235.

Ramachandran, V.K., East, A.K., Karunakaran, R., Downie, J.A. and Poole, P.S. (2011) Adaptation of *Rhizobium leguminosarum* to pea, alfalfa and sugar beet rhizospheres investigated by comparative transcriptomics. *Genome Biology* 12:R106.

Ramadan, T., Camargo, S.M., Summa, V., Hunziker, P., Chesnov, S., Pos, K.M. and Verrey, F. (2006) Basolateral aromatic amino acid transporter TAT1 (Slc16a10) functions as an efflux pathway. *Journal of Cell Physiology* 206, 771–779.

Ranocha, P., Denancé, N., Vanholme, R., Freydier, A., Martinez, Y., Hoffmann, L., Köhler, L., Pouzet, C., Renou, J.P., Sundberg, B., Boerjan, W. and Goffner, D. (2010) *Walls are thin 1* (*WAT1*), an *Arabidopsis* homolog of *Medicago truncatula NODULIN21*, is a tonoplast-localized protein required for secondary wall formation in fibers. *The Plant Journal* 63, 469–483.

Ranocha, P., Dima, O., Nagy, R., Felten, J., Corratgé-Faillie, C., Novák, O., Morreel, K., Lacombe, B., Martinez, Y., Pfrunder, S., Jin, X., Renou, J.P. *et al.* (2013) *Arabidopsis* WAT1 is a vacuolar auxin transport facilitator required for auxin homoeostasis. *Nature Communications* 4: 2625.

Rawat, S.R., Silim, S.N., Kronzucker, H.J., Siddiqi, M.Y. and Glass, A.D.M. (1999) *AtAMT1* gene expression and $NH4^+$ uptake in roots of *Arabidopsis thaliana*: evidence for regulation by root glutamine levels. *The Plant Journal* 19, 143–152.

Reinders, A., Sivitz, A.B., Starker, C.G., Gantt, J.S. and Ward, J.M. (2008) Functional analysis of LjSUT4, a vacuolar sucrose transporter from *Lotus japonicus*. *Plant Molecular Biology* 68, 289–299.

Rellan-Alvarez, R., El-Jendoubi, H., Wohlgemuth, G., Abadia, A., Fiehn, O., Abadia, J. and Alvarez-Fernandez, A. (2011) Metabolite profile changes in xylem sap and leaf extracts of strategy I plants in response to iron deficiency and resupply. *Frontiers in Plant Science* 2:66.

Rentsch, D., Schmidt, S. and Tegeder, M. (2007) Transporters for uptake and allocation of organic nitrogen compounds in plants. *FEBS Letters* 581, 2281–2289.

Riens, B., Lohaus, G., Heineke, D. and Heldt, H.W. (1991) Amino acid and sucrose content determined in the cytosolic, chloroplastic, and vacuolar compartments and in the phloem sap of spinach leaves. *Plant Physiology* 97, 227–233.

Sanders, A., Collier, R., Trethewy, A., Gould, G., Sieker, R. and Tegeder, M. (2009) AAP1 regulates import of amino acids into developing *Arabidopsis* embryos. *The Plant Journal* 59, 540–552.

Sauer, J.D., Bachman, M.A. and Swanson, M.S. (2005) The phagosomal transporter A couples threonine acquisition to differentiation and replication of *Legionella pneumophila* in macrophages. *Proceedings of the National Academy of Sciences of the United States of America* 102, 9924–9929.

Scheible, W.R., Morcuende, R., Czechowski, T., Fritz, C., Osuna, D., Palacios-Rojas, N., Schindelasch, D., Thimm, O., Udvardi, M.K. and Stitt, M. (2004) Genome-wide reprogramming of primary and secondary metabolism, protein synthesis, cellular growth processes, and the regulatory infrastructure of *Arabidopsis* in response to nitrogen. *Plant Physiology* 136, 2483–2499.

Schmidt, R., Stransky, H. and Koch, W. (2007) The amino acid permease AAP8 is important for early seed development in *Arabidopsis thaliana*. *Planta* 226, 805–813.

Schmidt, W. (2003) Iron solutions: acquisition strategies and signaling pathways in plants. *Trends in Plant Science* 8, 188–193.

Schobert, C. and Komor, E. (1987) Amino acid uptake by *Ricinus communis* roots: characterization and physiological significance. *Plant, Cell and Environment* 10, 493–500.

Sekito, T., Fujiki, Y., Ohsumi, Y. and Kakinuma, Y. (2008) Novel families of vacuolar amino acid transporters. *IUBMB Life* 60, 519–525.

Shimazu, M., Sekito, T., Akiyama, K., Ohsumi, Y. and Kakinuma, Y. (2005) A family of basic amino acid transporters of the vacuolar membrane from *Saccharomyces cerevisiae*. *The Journal of Biological Chemistry* 280, 4851–4857.

Solomon, P.S. and Oliver, R.P. (2001) The nitrogen content of the tomato leaf apoplast increases during infection by *Cladosporium fulvum*. *Planta* 213, 241–249.

Su, Y.H., Frommer, W.B. and Ludewig, U. (2004) Molecular and functional characterization of a family of amino acid transporters from *Arabidopsis*. *Plant Physiology* 136, 3104–3113.

Sulieman, S. and Schulze, J. (2010) Phloem-derived γ-aminobutyric acid (GABA) is involved in upregulating nodule N_2 fixation efficiency in the model legume *Medicago truncatula*. *Plant, Cell and Environment* 33, 2162–2172.

Sulieman, S., Fischinger, S.A., Gresshoff, P.M. and Schulze, J. (2010) Asparagine as a major factor in the N-feedback regulation of N_2 fixation in *Medicago truncatula*. *Physiologia Plantarum* 140, 21–31.

Svennerstam, H., Ganeteg, U. and Nasholm, T. (2008) Root uptake of cationic amino acids by *Arabidopsis* depends on functional expression of amino acid permease 5. *New Phytologist* 180, 620–630.

Svennerstam, H., Jamtgard, S., Ahmad, I., Huss-Danell, K., Nasholm, T. and Ganeteg, U. (2011) Transporters in *Arabidopsis* roots mediating uptake of amino acids at naturally occurring concentrations. *New Phytologist* 191, 459–467.

Tanz, S.K., Castleden, I., Hooper, C.M., Vacher, M., Small, I. and Millar, H.A. (2013) SUBA3: a database for integrating experimentation and prediction to define the SUBcellular location of proteins in *Arabidopsis*. *Nucleic Acids Research* 41, D1185–D1191.

Tapken, D., Anschutz, U., Liu, L.H., Huelsken, T., Seebohm, G., Becker, D. and Hollmann, M. (2013) A plant homolog of animal glutamate receptors is an ion channel gated by multiple hydrophobic amino acids. *Science Signaling* 6, ra47.

Tegeder, M. (2012) Transporters for amino acids in plant cells: some functions and many unknowns. *Current Opinion in Plant Biology* 15, 315–321.

Tegeder, M. and Rentsch, D. (2010) Uptake and partitioning of amino acids and peptides. *Molecular Plant* 3, 997–1011.

Thorne, J.H. and Rainbird, R.M. (1983) An *in vivo* technique for the study of phloem unloading in seed coats of developing soybean seeds. *Plant Physiology* 72, 268–271.

Thume, M. and Dietz, K.J. (1991) Reconstitution of the tonoplast amino-acid carrier into liposomes – evidence for an ATP-regulated carrier in different species. *Planta* 185, 569–575.

Tonin, G.S., Wheeler, C.T. and Crozier, A. (1990) Effect of nitrogen nutrition on amino acid composition of xylem sap and stem wood in *Alnus glutinosa*. *Physiologia Plantarum* 79, 506–511.

Touraine, B., Muller, B. and Grignon, C. (1992) Effect of phloem-translocated malate on NO(3) uptake by roots of intact soybean plants. *Plant Physiology* 99, 1118–1123.

Udvardi, M. and Poole, P.S. (2013) Transport and metabolism in legume–rhizobia symbioses. *Annual Review of Plant Biology* 64, 781–805.

Udvardi, M.K., Salmon, C.J. and Day, D.A. (1988) Transport of L-glutamine across the bacteroid membrane but not the peribacteroid membrane from soybean root nodules. *Molecular Plant–Microbe Interactions* 1, 250–254.

van Dongen, J.T., Laan, R.G.W., Wouterlood, M. and Borstlap, A.C. (2001) Electrodiffusional uptake of organic cations by pea seed coats. Further evidence for poorly selective pores in the plasma membrane of seed coat parenchyma cells. *Plant Physiology* 126, 1688–1697.

Vincentz, M., Moureaux, T., Leydecker, M.T., Vaucheret, H. and Caboche, M. (1993) Regulation of nitrate and nitrite reductase expression in *Nicotiana plumbaginifolia* leaves by nitrogen and carbon metabolites. *The Plant Journal* 3, 315–324.

Vincill, E.D., Bieck, A.M. and Spalding, E.P. (2012) Ca(2+) conduction by an amino acid-gated ion channel related to glutamate receptors. *Plant Physiology* 159, 40–46.

Wasson, A.P., Pellerone, F.I. and Mathesius, U. (2006) Silencing the flavonoid pathway in *Medicago truncatula* inhibits root nodule formation and prevents auxin transport regulation by rhizobia. *The Plant Cell* 18, 1617–1629.

Whitehead, L.F., Young, S. and Day, D.A. (1998) Aspartate and alanine movement across symbiotic membranes of soybean nodules. *Soil Biology and Biochemistry* 30, 1583–1589.

Wienkoop, S. and Saalbach, G. (2003) Proteome analysis. Novel proteins identified at the peribacteroid membrane from *Lotus japonicus* root nodules. *Plant Physiology* 131, 1080–1090.

Wipf, D., Ludewig, U., Tegeder, M., Rentsch, D., Koch, W. and Frommer, W.B. (2002) Conservation of amino acid transporters in fungi, plants and animals. *Trends in Biochemical Sciences* 27, 139–147.

Wolswinkel, P. and Deruiter, H. (1985) Amino acid release from the seed coat of developing seeds of *Vicia faba* and *Pisum sativum*. *Annals of Botany* 55, 283–287.

Wong, F.H., Chen, J.S., Reddy, V., Day, J.L., Shlykov, M.A., Wakabayashi, S.T. and Saier, M.H. Jr (2012) The amino acid-polyamine-organocation superfamily. *Journal of Molecular Microbiology and Biotechnology* 22, 105–113.

Yang, H., Bogner, M., Stierhof, Y.-D. and Ludewig, U. (2010) H^+-independent glutamine transport in plant root tips. *PLoS ONE* 5, e8917.

Yang, Z., Huang, J., Geng, J., Nair, U. and Klionsky, D.J. (2006) Atg22 recycles amino acids to link the degradative and recycling functions of autophagy. *Molecular Biology of the Cell* 17, 5094–5104.

Zhang, J., Subramanian, S., Stacey, G. and Yu, O. (2009) Flavones and flavonols play distinct critical roles during nodulation of *Medicago truncatula* by *Sinorhizobium meliloti*. *The Plant Journal* 57, 171–183.

Zhang, L., Tan, Q., Lee, R., Trethewy, A., Lee, Y.-H. and Tegeder, M. (2010) Altered xylem–phloem transfer of amino acids affects metabolism and leads to increased seed yield and oil content in *Arabidopsis*. *The Plant Cell* 22, 3603–3620.

18 Uptake, Transport and Redistribution of Amino Nitrogen in Woody Plants

S. Pfautsch,[1]* T.L. Bell[2] and A. Gessler[3]
[1]*Hawkesbury Institute of the Environment, University of Western Sydney, Richmond, New South Wales, Australia;* [2]*University of Sydney, Eveleigh, New South Wales, Australia;* [3]*Swiss Federal Institute for Forest, Snow and Landscape Research WSL, Birmensdorf, Switzerland*

18.1 Abstract

Nitrogen (N) is the main limiting macronutrient for plant growth and N deficits lead to severe negative effects in plant metabolism and development. Hence, knowledge about the complex dynamics of plant N nutrition is essential to comprehend plant functioning in natural and managed ecosystems. The availability of N varies widely among these systems, and this has a significant effect on how successful plants compete with soil microorganisms for available sources of N. It is widely accepted that plants incorporate, transport and recycle inorganic and organic forms of N, including amino-N. Uptake of amino-N can be direct via shuttle mechanisms that transport intact amino acids, but amino-N can also be gained through plant-fungal (i.e. mycorrhizal) or plant-bacterial (i.e. rhizobial) symbioses, yet it can also be lost through parasitism. Regardless of origin, once amino-N has entered the plant root, synthesized amino-N compounds play a key role in long-distance transport of N, metabolism, N storage and the regulation of N uptake. This chapter provides an overview how plants, in particular woody plants, acquire, transport and recycle amino-N. Further, it summarizes our current understanding of why organic N is of varying importance for plant nutrition across biomes and also outlines future directions for research that can help to further complete our understanding of the role of amino-N in ecosystem functioning.

18.2 Introduction

Some 5 billion metric tonnes of nitrogen (N) are contained in the natural world, including oceans, the atmosphere, terrestrial and marine biota, soil organic matter and sedimentary rocks (Mackenzie, 1998). Yet only 2% of this vast pool is accessible to organisms. In natural ecosystems, this situation leads to competition for N and results in the widely accepted notion that N is a major limiting element for plant growth in terrestrial biomes, given that the availability of sunlight and water is sufficient. The evolutionary consequence of this limitation is the relatively closed N cycle in natural systems, the development of highly efficient uptake strategies by plants and the internal cycling of N compounds within plants (Rennenberg *et al.*, 2010; Rennenberg and Schmidt, 2010). Anthropogenic alteration of the natural global N cycle has been recognized as the second most important driver – next to the global carbon (C) cycle – for climate change (Galloway *et al.*, 2008). Organic N in the form of amino acids has been identified as a critical component of the global N cycle (Schimel and Bennett, 2004; Chapin *et al.*, 2011).

*E-mail address: s.pfautsch@uws.edu.au

Environmental conditions across biomes are predicted to change dramatically (IPCC, 2007) and understanding how these changes may alter the movement of N as amino acids through terrestrial ecosystems requires a sound understanding of the function of amino acids in plant nutrition.

For many decades, researchers have studied the dynamics of N in terrestrial ecosystems, focusing on processes that involve the most important forms of inorganic N for agriculture – nitrate (NO_3^-) and ammonium (NH_4^+). The processes most often studied included mineralization, turnover, plant and microbial uptake, consumption and inputs and losses of these two compounds. However, in the late 1980s and throughout the 1990s, evidence gradually made it apparent that our understanding of N dynamics in plant nutrition was incomplete. Studies that have investigated rates of N mineralization in a range of boreal, alpine and arctic ecosystems suggest that annual rates of net mineralization of N were insufficient to account for the amounts of N being taken up by plants (e.g. Shaver *et al.*, 1991; Fisk and Schmidt, 1995; Ruess *et al.*, 1996; Kaye and Hart, 1997). What was being missed?

In retrospect, the research that was described in several influential papers has helped to solve the mystery. First, Chapin *et al.* (1993), using laboratory-based experiments, documented that roots of arctic sedge exhibited a preference for taking up amino acids (from here on collectively referred to as 'amino-N') over inorganic N. Kielland (1994) followed with another experiment done under controlled conditions. This work shed light on the wider ecological implications of soil N dynamics in arctic ecosystems by showing that uptake of amino-N existed across several plant forms.

Up to this point, all studies had used excised roots to test the capacity of plants to take up amino-N. The next logical step was for researchers to go into the field to understand whether experiments *ex situ* had neglected the natural dynamics of competition between soil microorganisms and plant roots for amino-N. One of the first studies that documented how well plant roots were able to compete for amino-N *in situ* was that of Schimel and Chapin (1996). However, it was the field-based study of Näsholm *et al.* (1998) that finally 'inaugurated' a new field of plant physiological research: *in situ* uptake of amino-N by plants. It can be argued that the greater impact of *in situ* over *ex situ* studies is reflected in how often these studies have been cited (according to the Web of Science in November 2014, Näsholm *et al.* (1998) has been cited 485 times, Chapin *et al.* (1993) 391 times, Kielland (1994) 348 times and Schimel and Chapin (1996) 171 times). Systematic experimental progress made during the 1990s triggered a rapid increase in the amount and level of sophistication of the empirical and theoretical studies investigating the role of amino-N in plant nutrition. Of course, we should pause to remember that Melin and Nilsson (1953) and, even earlier, Virtanen and Linkola (1946), published evidence that plant roots were theoretically capable of taking up organic forms of N.

Initially, it was thought that the uptake of amino-N is restricted to ecosystems where rates of N mineralization were low (i.e. at high altitude and latitude). This can be explained in part by the origin of the research teams that predominantly worked in this field. The most active teams had research foci on boreal and other cold climate systems. Under cold-climate conditions, mineralization of N is slow and plants have little chance to access small pools of inorganic N as they are outcompeted by soil microorganisms. Thus, plants need to scavenge N sources as they become available, for example, during the proteolysis of organic matter. In doing so, plants reduce their competition with soil microorganisms. This line of argument has often been presented as 'by-passing-the-bottleneck' theory (Chapin *et al.*, 1993; Schimel and Bennett, 2004; Chapman *et al.*, 2006; Warren, 2006).

Within a few years, it became clear that plants growing in tropical climates were also able to take up amino-N (Schmidt and Stewart, 1998). The novelty of work done using non-cold-climate vegetation was not only proof-of-concept for a juxtaposed climatic zone, the authors also concentrated exclusively on woody plants. Although the levels of soil N in the systems studied by Schmidt and Stewart (1998) were still relatively low, it became obvious that not only sedges or heath plants but also woody shrubs and trees have the capacity to take up amino-N from soil. Since then, the capacity of tree roots to take up amino-N has been shown for *Eucalyptus* spp. from various climates in Australia (Warren 2006, 2009; Warren and Adams, 2007; Pfautsch *et al.*, 2009a), boreal conifers (e.g. Nordin *et al.*, 2001; Öhlund and Näsholm, 2004), temperate conifers and broad-leaved species (e.g. Hofmockel *et al.*, 2007; Gallet-Budynek *et al.*, 2009; McFarland *et al.*, 2010; Rennenberg *et al.*, 2010) and tropical trees (e.g. Arndt *et al.*, 2002). Today, it is clear that most woody plants are capable of accessing amino-N through root-bound transporters.

The previous chapter provides detailed information about the export of amino-N into the rhizosphere. This chapter will follow the pathway of amino-N from the soil into plant roots, onwards into the transpiration stream, and upwards to the leaves where it is metabolized and incorporated into organic matter or is redistributed to other sink organs through transport in the phloem. Depending on the species and environmental conditions, amino-N compounds undergo various metabolic transformations in roots before being loaded into the transpiration stream. These transformations will be explained. At the root level, we will provide insights into the benefits and costs related to the uptake of amino-N via symbiotic associations with root-colonizing mycorrhizal fungi or soil bacteria. We will also shed light on to the fate of amino-N being scavenged by root hemiparasites. After describing the loading of amino-N into the transpiration stream, we explain why some amino-N compounds are preferred for long-distance transport over others. Towards the end of the chapter, we describe the functions of amino-N in phloem. To close the loop, we will also describe the current understanding how 'recycled' amino-N in the phloem has an impact on the uptake of N by woody plants. Finally, we conclude the chapter with an outlook into emerging fields within the wide area of amino-N nutrition in (woody) plants.

18.3 Uptake of Amino-N by Plant Roots

The uptake of inorganic and organic sources of N, including amino-N, by plant roots is primarily mediated by root hairs located within a few centimetres behind the root tip (Waisel *et al.*, 2002) and follows the process of mass flow and diffusion. Due to root growth and the short-lived nature of root hairs, direct uptake of N sources from newly accessed volumes of soil only lasts a few days (Thaler and Pagès, 1998). The main benefit for plant roots in taking up and metabolizing amino-N from soil is believed to be twofold. First, amino-N is already in the required state of reduction, which decreases the energetic investment in anabolic metabolism compared with that which would be required if N assimilation was via inorganic N. Secondly, 'amino-C' as 'by-product' does not need to be supplied from photosynthesis, leading to lower construction costs for the whole plant (Zerihun *et al.*, 1998). However, in contradiction of this belief, when the uptake efficiency of N by the roots of trees was measured using a calorimetric method, it was found that the energetic costs associated with the uptake of amino-N (supplied as glutamine) were mostly greater than those associated with the uptake of inorganic N (supplied as NH_4^+ or NH_4NO_3) (Pfautsch *et al.*, 2009a). At this point it is, therefore, necessary to reflect on the general principles of the mineralization, uptake and assimilation of N by plant roots in order to better understand the possible advantages or disadvantages that the uptake of amino-N represents.

18.3.1 Principles of N uptake

In the first step of the process, N is released into the soil through the decay of organic material. In the second step, N-containing molecules undergo several transformational processes before becoming available for uptake by plant roots. Research suggests that parts of this N-acquisition process can be actively influenced by plants through the polymeric composition and polymer:N ratio of litter, indicating a feedback mechanism of plants on their nutritional status (Schweitzer *et al.*, 2004; Chapman *et al.*, 2006). In addition, N flows in the rhizosphere are actively influenced by the excretion of organic compounds (termed rhizodeposits or root exudates) from plant roots that enhance the turnover of microbial biomass, resulting in increased availability of soil N (Kuzyakov and Xu, 2013; see also the previous chapter). Nevertheless, the starting point of N mineralization is the breakdown of organic N into amino acids (termed proteolysis) and nucleic acids, and it is completed with the release of ammonium (NH_4^+). This final step is termed ammonification and is primarily mediated by soil microorganisms. Depending on the quantity and demand, NH_4^+ can be taken up by plants, or oxidized by soil microorganisms to nitrate (NO_3^-). The latter process is referred to as nitrification. Nitrite (NO_2^-) is produced in an intermediate step and the final product, NO_3^-, can also be taken up and assimilated by plant roots.

Plants and heterotrophic soil microorganisms are constantly competing for soil NH_4^+, NO_3^- and amino-N. The degree of this competition is not yet clear (Schimel and Bennett, 2004; Kuzyakov and Xu, 2013). However, it is believed that in the short term, the relationship between soil microorganisms and plant roots is mutualistic, while over longer timescales, plant roots have an advantage in scavenging N due to their longevity (Kuzyakov and Xu, 2013). In addition to NH_4^+, NO_3^- and amino-N, other N-containing compounds are present in the soil solution and are thought to be involved in plant nutrition. Low-molecular weight organic substances (LMWOS) include amino-sugars, organic acids, peptides and quaternary ammonium compounds (e.g. Kuzyakov and Jones, 2006; Warren, 2006, 2013a,b; Hill *et al.*, 2011; Soper *et al.*, 2011). Their role in plant nutrition is yet to be disentangled, and we discuss these and other interesting research directions for the future in Section 18.6.

Once mineralization has produced N compounds that are available for plant uptake, translocation of these compounds is mediated by specialized transport proteins, which are located in the plasma membrane of root cells. Regulation of these transporters depends not only on the abundance of organic or inorganic N compounds, but also on abiotic factors (e.g. soil pH, temperature, light) and the nutritional status of the plant. Here, the presence (or absence) of specific C- and N-containing molecules plays an important role in regulating the rates of uptake and expression of transporters for specific sources of N compounds. Low-affinity transport systems (LATS) predominate when N compounds are available in high abundance, while high-affinity transport systems (HATS) operate when there are low concentrations of N compounds in the rhizosphere. The gene families of AMT1 and AMT2 (NH_4^+ transporters) encode a putative NH_4^+ transporter protein (Loqué and von Wirén, 2004) and their activity is mostly dependent on the endogenous concentration of N compounds near the root tip. In contrast, any uptake of NO_3^- is regulated by signals related to the N status of the whole plant (Gessler *et al.*, 2004; Jackson *et al.*, 2008; see also Section 18.5). Once taken up, NO_3^- is reduced to NO_2^- by nitrate reductase (NR) before further reduction to NH_4^+ by nitrite reductase (NiR). Interestingly, the uptake of NH_4^+ and NO_3^- can be suppressed by glutamine in the cytosol of root cells (Jackson *et al.*, 2008). We will come back to this regulatory mechanism (Section 18.5).

When plant roots compete successfully for either NH_4^+ or NO_3^- that is reduced to NH_4^+, the resulting increase in cytosolic NH_4^+ may become toxic to root cells and can dissipate transmembrane proton gradients if it accumulates in high concentrations (Britto and Kronzucker, 2002). Consequently, NH_4^+ is quickly transformed into amino-N before being incorporated into other metabolic processes (Lea and Azevedo, 2006). The initial transformation from inorganic to organic forms of N is controlled by several enzymes, which are found in either the roots or shoots (Lam *et al.*, 1996). The enzymes glutamine synthetase (GS) and glutamate synthase (GOGAT) act to form glutamine and glutamate, while asparagine and aspartate are synthesized with the help of aspartate aminotransferase (AspAT) and asparagine synthetase (AS) (Lam *et al.*, 1996). The role of the enzyme glutamate dehydrogenase (GDH) in primary NH_4^+ assimilation remains controversial (Lea and Ireland, 1999; Yamaya and Oaks, 2004), but has been included here for completeness. Table 18.1 outlines the reaction pathways involved in substrate-to-product relationships mediated by the relevant enzymes. Reactions 1 and 2 are commonly described as the GS/GOGAT cycle, indicating the interdependency of these reactions. It is important to note that the GS/GOGAT cycle does not operate in root cells exclusively, but is also known to take place in leaves if NO_3^- has been transported in the vascular system. However, due to its electrical charge, NH_4^+ is not loaded into the transpiration stream and it follows that all NH_4^+ is metabolized to amino-N in the roots before it is translocated. Returning to the energetic costs related to N uptake, experiments with a crop species have revealed that up to 14% of total root C catabolism is devoted to the

Table 18.1. Enzymes involved in synthesis of glutamine and asparagine, the two most important amino-N compounds for long distance transport of organic N in woody plants. All five enzymes can be found in the roots and/or leaves of plants. Reactions outlined under 1 and 2 are commonly described as the GS/GOGAT cycle owing to their cyclic and interdependent nature.

Enzyme	Acronym	Reaction catalysed[a]
1. Glutamine synthetase	GS	L-Glu + NH_4^+ + ATP → L-Gln + H_2O + ADP + Pi
2. Glutamate synthase	GOGAT	L-Gln + α-KG + 2 Fd_{red} (or 2 NADH + H^+) → L-Glu + 2 Fd_{ox} (or 2 NAD^+)
3. Glutamate dehydrogenase	GDH	L-Glu + H_2O + NAD^+ ↔ α-KG + NH_4^+ + NADH + H^+
4. Aspartate aminotransferase	AspAT	L-Asp + α-KG ↔ Oxaloacetate + L-Glu
5. Asparagine synthetase	AS	L-Gln + L-Asp + ATP ↔ L-Asn + L-Glu + AMP + PPi

[a]Abbreviations: Asn, asparagine; Asp, aspartate; Fd, ferredoxin; Glu, glutamate; Gln, glutamine; α-KG, α-ketogluratate; Pi, phosphate; PPi, pyrophosphate.

assimilation of NH_4^+ and 23% to the assimilation of NO_3^- (Bloom *et al.*, 1992). The lower energetic efficiency of uptake of NO_3^- over that of NH_4^+ is also apparent in tree roots (Pfautsch *et al.*, 2009a).

When focusing on the uptake of amino-N by roots, it is important to note that the presence of plant-available amino-N has been confirmed for soils from natural ecosystems around the globe. However, the omnipresence of amino-N compounds does not necessarily mean that amino-N is (i) bioavailable and (ii) taken up as an intact molecule, or even taken up at all. On bioavailability, Hobbie and Hobbie (2012) concluded that a large part of soil amino-N is chemically protected, hidden in pores or even artificially released in extracts from bacteria and roots during sample preparation. On uptake, we have to consider that although plant roots are capable of taking up whole amino-N molecules via plasma membrane-bound transport systems similar to those that mediate the uptake of NH_4^+ or NO_3^-, intense competition with microorganisms (e.g. Owen and Jones, 2001), as well as plant-fungal or plant-bacterial symbioses can have profound effects on how amino-N is incorporated into root tissues. The ability to acquire amino-N, especially that aided by mycorrhizal fungal associations, is thought to represent a large advantage in the competition of plant roots with soil microorganisms over the same resource pool (Harrison *et al.*, 2007, 2008; Van Der Heijden *et al.*, 2007). In the following section, the processes involved will be described separately, although in the natural world they can potentially occur simultaneously.

18.3.2 Capacity and importance of uptake of amino-N

Numerous studies have provided evidence that the roots of woody plants are capable of taking up a range of amino-N species, yet our understanding of the ecological significance of this capacity remains uncertain. A comprehensive review of this topic has recently been provided by Näsholm *et al.* (2009). The authors conclude that 'the capacities [of plant roots to take up amino-N] are fully comparable to, and share many features with, inorganic N uptake capacities', but 'direct evidence that organic N contributes significantly to plant N nutrition is still lacking'. The vast majority of bioavailable amino-N in soils is in the form of L-enantiomers and it is believed that D-enantiomers play a negligible role in plant nutrition because of their low abundance and possible toxicity to plants. While a recent review by Vranova *et al.* (2012) concentrated exclusively on the importance of D-enantiomers to plant nutrition and growth performance, the ecological importance of D-enantiomers is not well understood. We therefore concentrate on uptake of amino-N as L-enantiomers.

Amino-N is usually taken up via proton symport shuttles that organize simultaneous transport of a single molecule of amino-N and one or two protons (Bush, 1999). The first functional description of plasma membrane-bound amino-N shuttles originates from experiments with yeast mutants (see Ludewig and Koch, 2008, and references therein). Improvements in techniques used in molecular biology allowed Rentsch *et al.* (2007) to identify 67 putative amino-N transporters and genome sequencing has now classified 134 amino-N transporters in a species of poplar (Tuskan *et al.*, 2006). These findings document the complexity of amino-N uptake and transport in higher plants.

The main research on specific amino-N transporters in plants has been done using *Arabidopsis thaliana* and thus the most detailed information on root-specific expression is available for this (non-woody) species. In the root hairs and tips of *A. thaliana*, AtAAP1, which belongs to the amino acid permease (AAP) family is expressed. This transporter is known to belong to a moderate-affinity uptake system transporting acidic and neutral amino acids at concentrations higher than 50 μM (Fischer *et al.*, 2002; Lee *et al.*, 2007; Svennerstam *et al.*, 2011). At lower concentrations, AtAAP5 and AtLHT1 (transporters belonging to the lysine/histidine-type transporter family) are involved in the uptake of acidic, basic and neutral amino acids into root tissues (Svennerstam *et al.*, 2008, 2011). Expression of the two latter transporters seems to depend on the early developmental stage of the plant (Tegeder, 2012). Another root-expressed transporter in *A. thaliana* is AtProT2 (a proline/compatible solute transporter), which is specific for proline uptake (Lehmann *et al.*, 2011).

How do we know whether amino-N is actually taken up by plant roots as intact molecules and how can we quantify that uptake? A widely used approach to answer these questions is by feeding ^{13}C and ^{15}N labelled amino-N compounds to plant roots. After the provision of a labelled amino-N source, plant roots are analysed for their C and N isotopic composition using isotope-mass-spectrometry (e.g. Näsholm *et al.*, 1998). Using a stoichiometric approach based on the known ratio of ^{13}C:^{15}N of the source provided and the ratio of the ^{13}C:^{15}N recovered from root tissues, the quantity of amino-N being taken up as intact molecules can be calculated (e.g. Warren and Adams, 2007; Pfautsch *et al.*, 2009a). There are, however, some issues that need to be taken into account when interpreting the uptake data (Rasmussen and Kuzyakov, 2009; Warren, 2012). Lower ^{13}C:^{15}N ratios in the root organic matter than in the amino-N compound(s) supplied might not be an exclusive indicator of partial deamination before uptake and thus may be due to mixed uptake of the amino-N compound (labelled with C and N) and the NH_4^+ (labelled with N only). It may also be a result of fast post-uptake decarboxylation of the amino-N compound with concomitant release of $^{13}CO_2$. Simultaneous uptake of labelled inorganic ^{13}C in the form of $H^{13}CO_3^-$ dissolved in the soil solution or from root-respired $^{13}CO_2$ should also be considered when assessing ^{13}C:^{15}N ratios. Moreover, we need to be aware that stable isotope-labelling studies mainly assess net uptake rates of amino-N (which is the difference between gross influx and efflux) and that amino-N influx could be significantly higher. For inorganic N it is known that the efflux of NO_3^- and NH_4^+ can be up to 50% (ter Steege *et al.*, 1999) and 85% (Kronzucker *et al.*, 2003), respectively, of the influx. Even though there is evidence that amino-N efflux from roots can occur (Phillips *et al.*, 2006; Lesuffleur *et al.*, 2007), it is not well quantified, so the difference between amino-N net uptake and gross influx is an area for future investigation.

Two recent and very comprehensive reviews dealing with the uptake of organic N by roots (Lipson and Näsholm, 2001; Näsholm *et al.*, 2009) conclude that the kinetic properties of the uptake systems of plants are in a range that would allow, at least in principal, competition with soil microorganisms. There is some evidence that the relative importance of organic N for tree nutrition increases with an increase in amino-N pools in the soil (Finzi and Berthrong, 2005) which, in turn, are dependent on the environmental conditions driving N mineralization (e.g. Gallet-Budynek *et al.*, 2009; Averill and Finzi, 2011). There are also indications that the importance of the uptake of organic N might depend on the presence of a particular amino-N compound and whether it is preferentially taken up by plants or microorganisms. For example, glycine is taken up faster by plants than other amino-N sources of higher molecular weight, whereas the opposite is true for soil microorganisms (e.g. Schmidt and Stewart, 1997; Lipson *et al.*, 1999). Sauheitl *et al.* (2009) also found that from a variety of amino-N compounds provided to plant roots (glycine, lysine, tyrosine, valine), glycine was preferred. Contrary to the studies mentioned previously, these authors found that the preference and overall capacity of roots to incorporate intact amino-N decreased with increasing concentration of amino-N. This downregulation seems to be controlled by the concentration of inorganic N in the soil solution, which under high amino-N should be increasing as a result of the elevated rates of turnover in flourishing microbial populations. We note that this mechanism may not apply to woody plants growing in ecosystems at high latitude and altitude, for the reasons explained earlier.

However, what is lacking at present is a detailed quantification of the contribution of organic N to the N nutrition of trees (and plants in general) and of the real partitioning of organic N between plants and microorganisms in the soil. The lack of such quantitative data under real-world conditions is mainly due to fast turnover of microorganisms together with a strong interaction between roots and soil/rhizosphere microorganisms. It is possible that competition between plants and microorganisms for organic-N resources over the short term results in cooperation in and facilitation of the N supply over the long term. The most obvious example is from mycorrhizal symbioses, where the fungal hyphae and not the plant root take up organic N more efficiently but, in the end, the woody plant (also) benefits from this fungal uptake (see Section 18.3.3). The same might be true for rhizosphere bacteria, especially when we take into account a new hypothesis which proposes that plant roots can incorporate rhizosphere bacteria by endocytosis (Schmidt *et al.*, 2013).

18.3.3 Uptake involving mycorrhizal associations

At least 80% of vascular plants form symbiotic relationships with mycorrhizal fungi (Brundrett, 2009). In exchange for carbon and energy, the mycorrhizal fungi provide their partner plants with the advantage of a massively increased absorptive area via extramatrical hyphae (or mycelium) compared with their own root systems. Fungal hyphae have a capacity similar to the roots of woody plants to absorb NH_4^+ and NO_3^- from the soil (Smith and Read, 2008), giving a mycorrhizal plant a distinct competitive advantage over a non-mycorrhizal plant, particularly when N is scarce. In addition, a range of extracellular enzymes may be exuded by extramatrical hyphae to increase the release of inorganic and organic N from otherwise chemically inaccessible sources (Read *et al.*, 2004).

Benefits from mycorrhizal associations come at a cost and when the sink strength of C by mycorrhizal fungi becomes disproportionally large, the host plant will be negatively affected. In the literature, this balance is usually coupled with N availability, but the outcomes for ectomycorrhizal fungi are conflicting and have been attributed to a variety of causes (Corrêa *et al.*, 2011). For example, the host plant may invest more or less C in ectomycorrhizal fungi under conditions of low N availability, but this may or may not be dependent on actual plant N gain. This suggests that the plant-fungal interaction is under control of the plant. For ericoid mycorrhizal associations, is it thought that the transfer of N to the host plant is controlled by the fungus rather than the plant, but it may still be influenced by the availability of C from the plant (Grelet *et al.*, 2009). The relationship between fungal demand for C and the N requirements of the host plant is at present poorly understood, at least in part owing to a lack of standard methodologies.

Inorganic N is readily taken up by fungal hyphae. Excess NH_4^+ is as toxic to fungal hyphae as it is to plant roots so is generally converted into glutamine in ericoid mycorrhizal (Bell and Pate, 1996) and ectomycorrhizal fungi, and to arginine in arbuscular mycorrhizal fungi (Jin *et al.*, 2005). Amino-N is an important source of N for ericoid and ectomycorrhizal fungi but not for arbuscular mycorrhizae (Smith and Read, 2008). While arbuscular mycorrhizae can access N originating from organic sources (Hodge *et al.*, 2001; Jin *et al.*, 2005, 2012; Hodge and Fitter, 2010), it is not yet clear whether they have the ability to stimulate the decomposition of N from organic material in the same way as ericoid and ectomycorrhizal fungi. Soil was traditionally thought to contain 20–50% total N as amino-N, but estimates of a staggering 80%

of soil N as amino-N have recently been made (Inselsbacher and Näsholm, 2012), at least in some ecosystems (but see Jones and Kielland, 2012). Given these revised amounts, a limited capacity for arbuscular mycorrhizae to augment the release of amino-N from organic pools may not be unexpected (McFarland *et al.*, 2010).

One of the earliest indications that ericoid and ectomycorrhizal fungi were involved in the utilization of amino-N came from the work of Kender and Childers (1959). Since then, uptake of amino-N has been demonstrated innumerable times using fungal isolates grown in pure culture in the laboratory and supplied with a variety of amino-N compounds. Using this system, it has been shown that mycorrhizal fungi can readily absorb amino-N, particularly glutamine, glutamate, glycine and alanine (e.g. Abuzinadah and Read, 1988; Keller, 1996). The membrane transporter systems involved (see review by Chalot and Brun, 1998) and the kinetics of amino-N uptake (Wallenda and Read, 1999) are well studied, at least for pure cultures of mycorrhizal fungi.

The nuances of the uptake of amino-N by ericoid and ectomycorrhizal fungi have also been investigated extensively under field conditions (see review by Lipson and Näsholm, 2001; also Persson *et al.*, 2003). Similar to the techniques used in studies investigating the capacity of roots to take up amino-N, isotopic tracers are the preferred weapons of choice when assessing the aptitude of mycorrhizae to take up amino-N (^{15}N:^{14}N; Hobbie and Högberg, 2012). Studies have included the direct uptake of ^{15}N-labelled amino-N compounds by mycorrhizal roots of woody plants, both excised and *in situ* (e.g. Lipson and Monson, 1998; Näsholm *et al.*, 1998). Another popular method traces the movement of ^{15}N released during the mineralization of labelled organic residues (e.g. Zeller *et al.*, 2000; Perelo *et al.*, 2006). Measurement of the transformation of ^{15}N:^{14}N ratios at its natural abundance has also been used in field studies to indicate sources of amino-N for hyphal uptake. For this technique, ^{15}N:^{14}N ratios in leaves from different mycorrhizal and non-mycorrhizal woody plant species are purported to reflect differences in access to different sources of N, including amino-N (e.g. Michelsen *et al.*, 1996; Hobbie *et al.*, 1999; Hobbie and Högberg, 2012), and in the movement of N in hyphal networks (e.g. He *et al.*, 2009). The genes involved in N utilization are becoming an increasingly active area of research (Avolio *et al.*, 2012), and a new technique that uses fluorescing nanoparticles (termed quantum dots) has been developed to estimate the uptake of amino-N by arbuscular mycorrhizal fungi (Whiteside *et al.*, 2012).

Field studies investigating amino-N uptake but not involving ^{15}N tracers are reasonably rare. Dannenmann *et al.* (2009) found a positive correlation between amino-N uptake by mature beech trees and the abundance of the dominant mycorrhizal fungi. Similarly, Avolio *et al.* (2009) found increased ectomycorrhizal root tip abundances on some host plants but not on others when amino-N was applied. These studies suggest that the role of mycorrhizal fungi in amino-N acquisition is highly variable. Despite the obvious benefits of mycorrhizal symbioses for woody plants, evidence is mounting that mycorrhizal fungi may contribute to the immobilization of N and limit rather than improve plant N uptake (e.g. Alberton *et al.*, 2007; Corrêa *et al.*, 2008; Näsholm *et al.*, 2012). In ecosystems with abundant and well-formed mycorrhizal associations, it is hypothesized that there would be considerable allocation of C below ground with less C release directly to the soil (i.e. C is mostly directed to the hyphae) and, considering the superior spatial exploitation of fungal hyphae, a higher depletion of soil N. This is likely to alter the interactions among microorganisms living in the rhizosphere and plant roots, and would, consequently, alter the cycling of N and C (Kuzyakov and Xu, 2013). Such modifications could have a significant impact on ecosystem productivity and so additional research is needed to clarify the long-term effects of plant-fungal associations on soil pools of C and N.

Study of the uptake and utilization of amino-N by mycorrhizal fungi has developed steadily over the past 50 years and has shown a natural progression from investigations involving the uptake of inorganic N by woody plants. For the mycorrhizal fungus itself, there is still much to be understood about the regulation of N-transport systems and metabolic pathways. Both laboratory- and field-based studies have revealed the very complex nature of the interactions of fungal hyphae with the soil matrix and with plant roots. New techniques and new molecular perspectives are being developed to selectively study competition among members of the greater microbial community (in which mycorrhizal fungi participate), and to extend these studies to investigating the benefits that could be gained from genetically different individuals of the same fungal species.

18.3.4 'Uptake' involving an N_2-fixing association

Another important source of amino-N for woody plants originates from the symbiotic relationship between root cells and soil bacteria that have the capacity to fix atmospheric N_2. We have adopted the term 'woody legumes' for woody plants that form such a relationship. Two bacterial strains are known to team up with woody legumes. While *Frankia* spp. develop a symbiosis with actinorhizal plants (e.g. *Alnus* spp., *Myrica* spp. and others), bacteria of the *Rhizobiaceae* (collectively termed rhizobia) form this relationship with many members of the plant family *Fabaceae*. Here, we will concentrate on the key process known as biological nitrogen fixation (BNF) that is associated with rhizobia and leads to the synthesis and translocation of amino-N.

As low availability of N can limit plant growth, an increased capacity to meet N requirements will clearly increase the competitiveness of woody legumes over woody plants that do not have the capacity for BNF. However, the energetic costs of supporting BNF are much greater than those supporting the uptake of inorganic N sources (Burris and Roberts, 1993). Up to 50% of C assimilated during photosynthesis may be supplied to rhizobia (Lynch and Whipps, 1990). This strong sink effect represents a considerable disadvantage to the host, as less C can be used to accumulate biomass, thereby reducing plant competitiveness. Hence, a tight regulatory feedback mechanism within root cells will suppress BNF as soon as sufficient N is available from the soil (Chalk, 2000). This important cost–benefit relationship seemed to be a good candidate to help explain why woody (tree) legumes are abundant in wet and arid tropical regions but almost completely absent in temperate climates (Crews, 1999), as it neatly mirrored the decreasing availability of soil N from temperate to tropical ecosystems. However, as reviewed by Adams *et al.* (2010), tree legumes also play an important role in temperate and tropical regions in Australia, southern Africa and South America. Those 'southern candidates' seemingly dominate landscapes where old, nutrient-poor soils and low rainfall dominate, and woody legumes have developed sophisticated mechanisms to cope with frequent droughts and other natural disturbances (e.g. fire). Under such stressful environmental conditions, it seems less advantageous to invest in the accumulation of biomass rather than maximizing the capacity to accumulate N per se.

The cascade of reactions involved in the initial infection of fine roots, nodule formation, nodule functioning and, ultimately, the exchange of amino-N between the nodule and the plant, is highly complex. Here, we can only provide basic information and refer the interested reader to excellent in-depth reviews by Udvardy and Day (1997), Day *et al.* (2001) and Sprent (2001, 2009). In the following paragraphs, we concentrate on describing the main biochemical processes that lead to the production and export of asparagine and possibly alanine from the nodule into the plant root.

Plant roots initiate the symbiotic relationship between root cells and soil-borne rhizobia by releasing inductors such as luteolin and chrysoeriol into the surrounding soil. These inductors trigger the release of 'Nod-factors' by rhizobia which, in turn, react with receptors in the epidermal cells of roots. The docking of Nod-factors is a prerequisite for the successful colonization of the epidermal surface by rhizobia (Sprent, 2001). Following cell deformation and controlled hydrolysis of the epidermal cell wall, the rhizobia enter the protoplast of the cell and begin extensive multiplication. At the same time, new root cells develop to encapsulate the infected cell, leading to the formation of the nodule that contains the rhizobia. Once encapsulated, the rhizobia morph from bacteria to bacteroids. One important aspect of this process is the production of a peribacteroid membrane (PBM) that separates the bacteroid from the cytosol of the plant cell. At this developmental stage, bacteroids have become endosymbiotic, N_2-fixing organelles located inside the root cell (Mylona *et al.*, 1995).

Within the bacteroids, the enzyme nitrogenase reduces N_2 to two NH_3 molecules (Yandulov and Schrock, 2003). The complex sequence of reactions involved requires six protons, six electrons and energy in the form of ATP. Protons, electrons and ATP are generated during the citric acid cycle within the bacteroid; the 'fuel' required to drive this cycle is provided by the host plant in the form of malate and succinate (C_4 carboxylic acids). These two compounds originate from photosynthetic processes in the leaves that generate sucrose (a C_{12} sugar). This product is loaded into the phloem, transported to the root, where it is fragmented, and subsequently imported into the nodules. Transporters located in the PBM shuttle malate and succinate from the cytosol of the nodule into the bacteroid, where they are metabolized (Day *et al.*, 2001; Lodwig *et al.*, 2003). The provision of malate and succinate represents the investment of the plant into the symbiotic relationship. In return, the bacteroid exports NH_3 into the PBM, where high concentrations of protons lead to the formation of NH_4^+ that, in turn, is released into the cytosol of the nodule. Here, and also in the plastid of the nodule, NH_4^+ is used to synthesize glutamine from glutamate during the GS/GOGAT reaction, followed by transformation to asparagine (Crawford *et al.*, 2001; also described in Section 18.3.1). Finally, asparagine is exported from the nodules into the root tissues where it is loaded into the transpiration stream and distributed to N sinks (Section 18.4). Amino-N may be exported in the form of ureides, particularly in woody legumes from subtropical climates.

18.3.5 'Double-dipping' or how root hemiparasites access amino-N

About 1% of angiosperms (approximately 4100 species) are considered to be parasitic, with the majority of species being hemiparasitic, meaning they are only partially dependent on their hosts (Nickrent and Musselman, 2004). Hemiparasites retain some capacity for photosynthesis and only acquire mineral nutrients and water from their hosts, although some may have a functioning root system and grow to maturity without a host (facultative parasitism). Holoparasitic plants lack chlorophyll and must obtain all of their requirements for growth from their hosts. Parasitic plants that attach to the roots of host plants are most common (60% of species), with those that attach to stems (aerial parasites or mistletoes) making up the balance (Musselman and Press, 1995). While aerial hemiparasites must gain their N from a single host, facultative root hemiparasites have the potential to satisfy their N requirements autotrophically with functional roots (hence our pun on 'double-dipping'), at least for part of their life cycle. Even if a root hemiparasite is strictly obligate (e.g. *Striga*), sources of N can potentially be 'mixed' with the parasite forming haustoria on a range of host species (Musselman and Press, 1995). Because of these two features, we consider the 'uptake' of amino-N by root hemiparasites here. N gain by woody hemiparasites, both aerial and root parasites, has been regularly reviewed by Pate (2001), Shen *et al.* (2006) and Bell and Adams (2011).

The nature of the exchange of N-containing compounds between root hemiparasites and their hosts has been well studied, mainly due to the economic losses caused by associations between non-woody hosts and parasites. *Striga* and *Rhinanthus* are examples of genera of root hemiparasites that have had a considerable impact on agriculture (Hautier *et al.*, 2010). As hemiparasites are generally 'xylem feeders', comparisons of inorganic N and amino-N in the xylem sap from root hemiparasites and their associated hosts have been used to determine the relative extent of N uptake and, in some cases, the level of dependence of the hemiparasite on the host (Pate and Bell, 2000). Despite this ostensibly useful application, the results have been equivocal. Some studies have indicated considerably higher concentrations, but similar compositions, of certain amino-N compounds in parasitic tissue compared with host plants (e.g. Pate *et al.*, 1994; Tennakoon *et al.*, 1997), some have shown similar concentrations and compositions (e.g. Radomiljac *et al.*, 1998; Pate and Bell, 2000) and others have demonstrated mutually different concentrations and compositions (e.g. Tennakoon *et al.*, 1997). Studies using isotopically labelled ^{15}N- and ^{13}C-containing compounds or particular amino-N compounds supplied to the host plant (e.g. *S*-ethenyl cysteine, Thumfort *et al.*, 1993; djenkolic acid, Calladine *et al.*, 2000) have also been used to determine the extent of uptake of amino-N from the host xylem and its subsequent transformations. It seems that, depending on the species of host and root hemiparasite involved, amino-N can either be absorbed indiscriminately or selectively taken up from host tissues by the haustoria of hemiparasites

(Pate, 1995, 2001; Hibberd and Jeschke, 2001; Irving and Cameron, 2009).

The form in which N is available to the parasite will be determined by the types of N-containing compounds acquired and transported by the host. Commonly, parasitic angiosperms are more productive when grown *in vitro* on amino-N than on inorganic N (Graves, 1995), indicating some level of reliance of hemiparasites on amino-N. If the host plant assimilates inorganic N in its roots, then the parasite will have access to amino-N transported in the transpiration stream. If the host plant assimilates inorganic N in its shoots, then N is transported in the xylem as NO_3^- and haustorial tissues should have the capacity to reduce this NO_3^- (Pate, 2001; Irving and Cameron, 2009). In support of this generalization, it was possible to induce NR in haustorial tissue of the woody root hemiparasites *Olax phyllanthi* (Pate *et al.*, 1994) and *Santalum acuminatum* (Tennakoon *et al.*, 1997). For annual root hemiparasites in the family *Scrophulariaceae*, it has been found that the induction of NR in parasitic tissue is inversely proportional to the reliance of the hemiparasite on the host for N (Graves, 1995). This is likely to be the case for woody root hemiparasites as well. The presence of GS and amino transferases in the haustorial tissue of *O. phyllanthi* implies the additional capacity for the assimilation of NH_4^+ and amino-N (Pate *et al.*, 1994). The ability to metabolize a wide range of inorganic and organic forms of N would benefit the root hemiparasite, particularly if it is profligate in its host partners.

There are many examples of the enhanced productivity of root hemiparasites on N_2-fixing hosts for both herbaceous annual (e.g. Press *et al.*, 1993; Seel *et al.*, 1993) and woody (e.g. Tennakoon *et al.*, 1997) hemiparasites. Along with increased growth, there is an inevitable increase in N as amino-N in parasitic tissues. For example, proline was found to be prevalent in tissues of *S. accumulatum* parasitizing the N_2-fixing host *Acacia rostellifera* (Tennakoon *et al.*, 1997), while asparagine was prevalent in *Parentucellia viscosa* growing on the herbaceous N_2-fixing host *Lotus angustissimus* (Pate and Bell, 2000). It has been suggested that an abundance of certain amino-N compounds such as proline, and also of mannitol, in parasitic tissues acts as an osmotica (along with other organic and inorganic solutes), which is important for lowering the water potential of the parasite to facilitate water uptake from host tissues (Press and Graves, 1995; Tennakoon *et al.*, 1997).

It can been hypothesized that C is taken up from host plants in preference to N simply on the basis of the considerably higher C:N ratio of the xylem sap of the parasite compared with its host (Tennakoon *et al.*, 1997). This is despite hemiparasites retaining their capacity for photosynthesis. Amino-N is easy to assimilate and can have particularly high C:N ratios, so perhaps the C gain is a windfall rather than the primary reason for solute uptake. Given the financial impact of certain root hemiparasites, future research is likely to focus on the C economies of host plants and the concomitant heterotrophic gain of C by root hemiparasites. Fortunately, as C uptake goes hand in hand with N uptake, particularly when N is delivered in the form of amino-N, there will also be continued interest in the N economies of both host and parasitic plants.

18.4 Transporting Amino-N in the Xylem

Being the bio-factories of woody plants, non-mature leaves are the primary sinks for amino-N, whereas the roots are recognized as the primary source. Hence, amino-N is loaded into the transpiration stream in roots, transported upwards in the xylem of the stem and branches until it reaches the leaves, where it is unloaded from the xylem. To sustain the source–sink relationship between roots and shoots, amino-N is constantly loaded into the transpiration stream. This loading process is also responsible for maintaining the concentration gradient of amino-N between the root tissues and the xylem, which allows continuous uptake and assimilation of available N sources from the soil. To understand the mechanism of long-distance amino-N transport it is necessary to briefly reflect on how water transport is organized in woody plants.

18.4.1 Transpiration – the upward 'conveyor belt' for amino-N

To a large degree, the transpiration stream in plants is dependent on the water potential gradient between soil moisture and atmospheric water vapour, and from the perspective of the plant is considered to be a passive process. Under most environmental conditions, the water potential of the atmosphere exceeds that of the soil manifold. Plants use the gradient along which water moves from low (close to zero) to high (increasingly negative) water potential to passively facilitate the transport of water (and nutrients) from the soil to the leaves. Organization of the transpiration stream involves water passing through a sequence of plant tissues and moving it in and out of the xylem. A water potential gradient between the soil water and the water inside the root draws water into the root hairs and through the epidermal and cortical cell layers via apoplastic, transmembrane and symplastic pathways. At the Casparian strip, water molecules must pass through the symplast of the endodermis in the root before entering the xylem parenchyma. Once in the xylem, water is transported upwards in the vascular system along the root–stem–branch–twig–petiole–leaf vascular continuum. The water potential gradient between the water in the vascular system of the leaves and the atmosphere will draw water out of the vascular bundle onto hydrophilic wall components of the mesophyll cells, where it feeds into a water film. The water film on the mesophyll cells stretches through the air-filled space inside the leaf where it is in direct contact with the atmosphere. A change from the liquid to the gas phase represents the final process before water molecules leave the leaf via the sub-stomatal cavity and through the stomatal pore into the atmosphere. One can depict the transpiration stream as a continuously upward moving conveyor belt that delivers 'building materials' (e.g. amino-N and other compounds) to 'construction sites'.

18.4.2 Loading amino-N into the xylem

Once assimilated into the cytosol of the root cells, amino-N compounds are loaded onto the conveyor belt. However, it should be mentioned here that inorganic N is also conveyed in the transpiration stream, particularly when NO_3^- is the major N source. This is the case for many agricultural crops and grassland species. In herbaceous species, the reduction of NO_3^- occurs mainly in the leaves and, consequently, high concentrations of NO_3^- have been measured in the xylem sap of annual and perennial herbaceous species, e.g. *Ricinus communis* (Peuke *et al.*, 2001), tobacco (Kruse *et al.*, 2002), tomato (Bialczyk *et al.*, 2004), sunflower (Azedo-Silva *et al.*, 2004) and grasses (Macduff and Bakken, 2003), but only in traces in woody plants (e.g. Martin and Amraoui, 1989; Dambrine *et al.*, 1995; Gessler *et al.*, 1998a).

The loading of amino-N into the transpiration stream firstly requires symplastic/transmembrane transport. Several possible mechanisms have been reported in the literature, for example transport of amino-N across plasma membranes organized by co-transport (Li and Bush, 1990a,b). These transporters can be more or less specialized (Fischer *et al.*, 1998). Recently membrane-bound proteins have been identified that facilitate amino-N transport and operate as proton symporters with hydrophobic characteristics. Their molecular mass ranges from 30 to 40 kDa and they have 9–14 membrane spans (Ludewig and Koch, 2008). Some evidence suggests that the loading of amino-N and other nutrients into the xylem is independent of the rate of transpiration, but rather is regulated by loading processes in the xylem parenchyma cells (Siebrecht *et al.*, 2003). Clearly, the loading of amino-N into the xylem remains a topic with many unknowns and researchers currently support both active and passive mechanisms. Regardless of the discussion of active or passive transport, it is generally accepted that the transport of any solute, including amino-N, requires some form of charge gradient across the membrane separating the xylem parenchyma and the xylem member cells. This can be mediated by either chemical or electrical gradients, or both (de Boer and Volkov, 2003). Several other requirements are well established. The loading and long-distance transport of amino-N: (i) involves a wide range of membrane-bound transport proteins; (ii) always uses water as the transport medium (Ludewig and Koch, 2008); and (iii) is expected to be as or even more complex than the long-distance transport of carbohydrates (Lalonde *et al.*, 2004).

18.4.3 Amino-N composition of xylem sap

Compared with the very high diversity of amino-N compounds found in the leaves and phloem (e.g. Grassi *et al.*, 2002; Pfautsch *et al.*, 2009b), the diversity of amino-N in the xylem is considerably limited. Across biomes and the tree species therein, the amides glutamine and asparagine, and the amino acids glutamate and aspartate are the most abundant molecules for the long-distance transport of N in the xylem sap (Fischer *et al.*, 1998). Arginine, γ-aminobutyric acid (GABA), proline, threonine and valine have also been detected at lower concentrations in the xylem sap. Different forms of ureides can also be found in the xylem sap of subtropical woody legumes, including allantoic acid and allantoin (Schmidt and Stewart, 1998), and citrulline (Cramer *et al.*, 2002). A broad selection of studies that characterize the amino-N profiles of the xylem sap of woody plants is presented in Table 18.2 and underlines the omnipresence of the aforementioned compounds.

Table 18.2. The most abundant amino-N compounds detected in the xylem sap of trees. Amino acids[a] are listed in order of importance to the total amino acid concentration in the xylem sap. Modified from Pfautsch *et al.* (2009b).

Tree species	Dominant N compounds in xylem sap[a]	Reference
Abies alba	Gln, Asn, Arg	Escher *et al.*, 2004b
Acacia dealbata	Asn, Asp, Arg	Pfautsch *et al.*, 2009a
Acacia melanoxylon	Asn, Asp	Pfautsch *et al.*, 2009a
Actinidia deliciosa	Gln	Clark and Smith, 1991
Betula pendula	Gln, Cit	Millard *et al.*, 1998
Casuarina glauca	Asn, Pro, Arg, Cit, Gln	Cramer *et al.*, 2002
Citrus unshiu	Asn, Gln, Arg	Kato *et al.*, 1981
Coffea arabica	Asn, Gln, Ser	Mazzafera and Gončalves, 1999
Eucalyptus spp.	Gln	Adams *et al.*, 1995
Eucalyptus camaldulensis	Gln, Glu, Asn, Arg	Radomiljac *et al.*, 1998
Eucalyptus regnans	Gln, Glu	Adams and Atkinson, 1991 Pfautsch *et al.* (2009a)
Eucalyptus tereticornis	Gln, Glu, Pro, Asp, Arg	Marsh and Adams, 1995
Fagus sylvatica	Gln, Asn, Arg	Nahm *et al.*, 2006a,b, 2007; Gessler *et al.*, 1998a, 2004
Juglans regia	GABA, Cit, Gln, Asp	Frak *et al.*, 2002
Macadamia integrifolia	Asn, Gln, Arg	Fletcher *et al.* (2010)
Malus domestica	Arg, Asn, Asp, Gln	Tromp and Ovaa, 1985; Malaguti *et al.*, 2001; Guak *et al.*, 2003
Picea abies	Gln, Arg, Asp	Ericson *et al.*, 1993; Dambrine *et al.*, 1995; Schneider *et al.*, 1996; Weber *et al.*, 1998; Gessler *et al.*, 1998a, 2003
Pinus spp.	Gln	Barnes, 1963; Plassard *et al.*, 2000
Pinus sylvestris	Arg, Gln	Nordin *et al.*, 2001
Populus spp.	Gln, Asn, Arg	Sauter and van Cleve, 1994; Schneider *et al.*, 1994; Siebrecht and Tischner, 1999; Escher *et al.*, 2004a,b
Prunus spp.	Gln	Andersen *et al.* (1995)
Prunus avium	Asn, Asp, Glu, Arg	Grassi *et al.*, 2002; Millard *et al.*, 2006
Prunus dulcis	Asn	Youseffi *et al.*, 2000
Quercus ilex	Asn, Asp, Gln	Nabais *et al.*, 2005
Salix spp.	Gln, Asn, Arg	Sauter, 1981
Santalum album	Asn, Arg, Gln, Glu	Radomiljac *et al.*, 1998
Sorbus aucuparia	Asn	Prima Putra and Botton, 1998
Various Australian tree species	Gln, Arg, Asn, Asp, ureides	Schmidt and Stewart, 1997
Various Brazilian tree species	Asn	Aidar *et al.* (2003)

[a]Abbreviations of amino acids: Asn, asparagine; Asp, aspartate; Arg, arginine; Cit, citrulline; GABA, γ-amino-butyric acid; Gln, glutamine; Glu, glutamate; Pro, proline; Thr, threonine; Val, valine.

The concentration and composition of amino-N in the xylem sap can vary throughout the diel and seasonal cycle. Diurnal variation is mostly the result of enzymatic activity and the light-dependent availability of precursor metabolites and concentration of inorganic forms of N (Ludewig and Koch, 2008). For example, Lam *et al.* (1996) showed that the availability of primary products from photosynthesis (i.e. simple carbohydrates such as sucrose) had a significant impact on the synthesis of glutamine and asparagine, resulting in downregulation of amino-N synthesis during the night. The seasonal variability of amino-N concentration in the xylem sap of trees is driven by remobilization from storage in the spring (e.g. Grassi *et al.*, 2002; Guak *et al.*, 2003; Millard *et al.*, 2006). The xylem pH – the electrochemical gradient perhaps necessary for xylem loading – has been reported to also vary with season, being slightly acidic in the spring and close to neutrality during the winter (e.g. Fromard *et al.*, 1995), which has an impact on the composition of amino-N in the xylem. In addition, species-dependent genetic control over the composition of amino-N profiles in the xylem sap has been suggested (Pfautsch *et al.*, 2009b).

The dominant form of amino-N for long-distance transport has been shown to reflect the ecological aspects of woody plants, such as the social status of a tree within the forest matrix (Pfautsch *et al.*, 2009b). In this context, the composition of xylem amino-N in light-limited understorey trees was dominated by asparagine, whereas non-light-limited overstorey plants used mostly glutamine for long-distance transport. As light limits the manufacture of carbon skeletons (i.e. sugars) produced during photosynthesis, understorey trees use available carbon more efficiently. This is achieved by synthesizing amino-N compounds of the aspartate group, which have a preferable C:N ratio (e.g. asparagine 4:2) compared with amino-N compounds of the glutamate group (e.g. glutamine 5:2).

Not all of the amino-N present in the xylem is used immediately at sink locations and it has been suggested that specific amino-N compounds represent a 'mobile N pool'. The most prominent and logical candidate for a mobile N pool is arginine, as this has the highest C:N ratio (6:4) of all the amino-N compounds detected in woody plants. Large concentrations of arginine have been detected in the xylem of woody plants from the northern (Gessler *et al.*, 1998a; Nahm *et al.*, 2006a,b; Fotelli *et al.*, 2009) and southern (Schmidt and Stuart, 1998; Pfautsch *et al.*, 2009b) hemispheres. All of the studies listed here conclude that this specific amino-N compound is cycled between the xylem and phloem, where it remains readily available for metabolic processes once availability of soil N becomes limited.

18.5 Exchange of Amino Acids Between Xylem and Phloem and Integration of N Transport and Plant N Metabolism

Xylem transport alone is not adequate to supply the whole plant with N, especially in plants where NO_3^- assimilation occurs to a large extent in the leaves (e.g. Jeschke and Hartung, 2000; Peuke, 2010). Moreover, developing leaves, which are not yet coupled to the transpiration stream, cannot be supplied via the xylem (Gessler *et al.*, 1998a). In woody plants, amino-N as well as proteins are stored in the trunk and remobilized in the spring, often before significant uptake of soil N occurs (Millard, 1996). At the beginning of the growing season, the xylem cannot support transport from these storage tissues in a basipetal direction to the roots.

In all cases, the phloem transport of amino-N plays a crucial role, either by the direct loading of amino-N from the tissues in which it was synthesized/remobilized or by the occurrence of loading from the xylem to the phloem. It becomes increasingly obvious that bidirectional exchange of amino-N between the xylem and phloem is important for a coordinated N supply in woody plants. In young Norway spruce trees, different amino-N compounds, such as glutamine and aspartate, are subject to a bidirectional exchange between the xylem and phloem of twigs (Gessler *et al.*, 2003). In *Prunus*, *Malus* and *Fagus* species, glutamine was shown to be intensively exchanged between the xylem and phloem (Gessler *et al.*, 1998a; Malaguti *et al.*, 2001; Grassi *et al.*, 2003), and for herbaceous species, a similar exchange has been observed for asparagine between the two transport systems (Atkins *et al.*, 1980).

Empirical evidence for the involvement of AAP transporters in the long-distance transport of amino-N was documented by Koch *et al.* (2003) in an experiment in which the production of the mRNA of an AAP transporter in a source leaf was suppressed. As a consequence, the expression of AAP transporters was downregulated which, in turn, led to a lower concentration of amino-N in the developing root. Although it is very likely that similar results would be observed in woody plants, the empirical tests required to evaluate this have not yet been conducted. Activity of the high-affinity transporter AAP6 has been identified in xylem parenchyma cells, indicating the involvement of this particular transporter in the translocation of amino-N from the xylem to the phloem. Transporters located in vacuoles and vesicles are expected to participate in the storage and remobilization of amino-N. During seed germination, when metabolism relies entirely on the remobilization of stored compounds, large quantities of amino-N are exported from the vacuole into the cytoplasm as vacuoles are an important storage compartment for proteins. The concentration of amino-N in the cytosol can be 10–100 times higher than that in the vacuole (Winter *et al.*, 1994); this indicates that large amounts of amino-N produced during vacuolar degradation of storage proteins must be exported through transporters in the tonoplast of vacuoles before long-distance transport to sink tissues can be accomplished. Likewise, a substantial proportion of amino-N can be mobilized from proteolysis in mature leaves, reloaded into the phloem and returned to the roots or other heterotrophic tissues (e.g. Bush, 1999). Amino-N can also be directly loaded from root tissue into the phloem. The amino-N transporter AAP3 has been marked as a possible candidate for the translocation of amino-N taken up by roots and loaded into the phloem for transport to actively growing root tips. According to Okumoto *et al.* (2004), the expression of APP3 is restricted to the companion cells of root phloem and endodermal cells.

In addition, to optimize N supply, the bidirectional exchange of amino-N between the phloem and xylem allows amino-N compounds to cycle through the plant and to regulate root N uptake to the actual demand of the whole plant (Gessler *et al.*, 2004). In annual or perennial herbaceous plants, signalling of N status and adapting the regulation of N uptake is important during particular developmental stages, such as seed production or flowering (Imsande and Touraine, 1994). In long-lived woody plants, the ability to adjust N uptake to changing environmental conditions, the supply of other nutrients and the seasonal dynamics of N storage and remobilization, is crucial during their whole lifespan and needs to be fine-tuned at different developmental stages (Gessler *et al.*, 2004).

It has been hypothesized that for the pool of amino-N compounds (or particular amino-N compounds within this pool), cycling between the shoot and roots serves as a signal to the roots indicating the internal N status of the whole plant (Cooper and Clarkson, 1989; Muller and Touraine, 1992; Kreuzwieser *et al.*, 1997; Gessler *et al.*, 1998a,b). The pool of amino-N is elevated at sites of uptake and remobilization and depleted in actively growing and storage tissues. When the N supply of plants exceeds N demand, the pool of amino-N compounds that is reallocated from the shoot to the roots via the phloem increases, thereby providing a signal that leads to the suppression of N uptake by fine roots (Gessler *et al.*, 2004).

18.6 Future Research Directions

There is a myriad of opportunities in research on plant amino-N and we will outline only a selection of those discussed in the literature. An excellent summary of past, present and future research on the role of organic N, including amino-N, in plant nutrition has been provided by Paungfoo-Lonhienne *et al.* (2012). The focus of their summary is mostly on agroecosystems. In light of questions about the capacity to feed the growing human population, such a summary is very timely and necessary, and the authors conclude that increasing the use of organic N as fertilizer will improve crop yield. Here, however, we focus on possible future research avenues that relate to natural ecosystems and the processes therein. Obviously, some of these avenues cross over into both types of systems and the plant physiological functions within them, but others exclusively relate to natural systems. Future research can range from the nanoscale, where unknown amino-N transporters that link fungal hyphae with plant roots should be identified (e.g. Rennenberg *et al.*, 2010), to understanding how widely observed shifts in dominant plant species have an impact on amino-N availability at the ecosystem scale (e.g. Cramer *et al.*, 2013).

Our understanding of the functional and regulatory mechanisms involved in the transport of amino-N across membranes, and within cells and tissues, remains limited (Tegeder, 2012). Our current knowledge about these processes was derived mostly from work with *Arabidopsis* and it remains to be evaluated whether higher plants such as woody perennials operate in a similar or a different fashion. As stated by Tegeder (2012), further work is also required to assess 'physiological functions such as the role of amino acid loading into the collection phloem or reproductive floral tissues for growth and reproduction'.

Research on the capacity of plants to take up organic N has concentrated on only a few species of amino-N. New analytical methods (e.g. capillary electrophoresis–mass spectrometry (CE-MS) with MS^2 and MS^3, i.e. multiple-stage mass spectrometry, for the identification of unknown peaks) are being developed that are capable of profiling organic-N compounds at great accuracy (Warren, 2013a). Results indicate that besides amino-N, a vast pool of other organic-N compounds is present in the soil solution (e.g. non-protein amino acids and quaternary ammonium compounds) and is utilized by plants (Warren, 2013b). It is now necessary to move forward from studies on the uptake of single amino acids and include a wider range of organic-N compounds to better quantify the contribution of these compounds to ecosystem N cycling.

For plants in most ecosystems, the importance of amino-N uptake remains unclear, as N fluxes have generally been studied indirectly. A novel technique, based on the principle of microdialysis, now allows *in situ* measurement of N fluxes in the rhizosphere, including those of amino-N (Inselbacher *et al.*, 2011; Inselbacher and Näsholm, 2012). Using this technique in a range of ecosystems will add important information to the long-standing debate of how important amino-N really is in plant nutrition (e.g. Kahmen *et al.*, 2009). Moreover, such an approach can also help to solve questions about how the availability of and plant-specific preferences for particular sources of N change during plant ontogenesis. The existence of such changes in amino-N availability has been documented for a chronosequence of a conifer forest (LeDuc and Rothstein, 2010), but is missing for other ecosystems and species therein.

In a wider context, the effects of global change on rates of soil N mineralization and, consequently, on the availability of amino-N for plants, have not been studied sufficiently. Although artificially increased concentrations of atmospheric CO_2 did not result in additional uptake of amino-N in a pine forest (Hofmockel *et al.*, 2007), elevated temperatures and reduced soil moisture had a negative effect on the activity of soil microorganisms (Castro *et al.*, 2010) possibly resulting in reduced quantities of plant available amino-N. How the effects of global change will alter competitiveness among microorganisms, plant roots and fungal hyphae for soil amino-N is unknown.

18.7 Conclusions

To date, it is clear that amino-N is utilized by many woody plant species, seemingly more so in ecosystems in which the rates of N mineralization are low. Over the short term (hours to days), soil microorganisms appear to be generally superior when competing for amino-N with plant roots. Yet over the long term (days to years), plants have developed sophisticated mechanisms that guarantee that their N requirements are met, including the niche of root parasitism. These can be *active* mechanisms, in which root exudates are fed to soil microorganisms to intensify the turnover of the entire microbial population, leading to increased release of amino-N. The provision of C and energy to fungi in return for amino-N that is otherwise unavailable to plants represents another *active* mechanism. The most important *passive* mechanism is the unidirectional (net) flow of nutrients, including amino-N, from soil into roots, in combination with the longevity of the whole plant. Once successfully competed for, amino-N compounds play multiple roles in plants as outlined in this chapter, ranging from the long-distance transport of N required for anabolic reactions to the suppression of additional N uptake by roots. Although studied intensively, many aspects of the role of amino-N in plant nutrition, its ecological significance, and the possible effects of global change on the abundance and availability of soil amino-N, are unknown. We literally need to 'start digging deeper' if we want to comprehensively understand the importance of amino-N as a plant nutrient because the competition between soil microorganisms and plant roots for amino-N (and all other sources of N!) remains to be *excavated*.

References

Abuzinadah, R.A. and Read, D.J. (1988) Amino acids as nitrogen sources for ectomycorrhizal fungi: utilisation of individual amino acids. *Transactions of the British Mycological Society* 91, 473–479.

Adams, M.A. and Atkinson, P.I. (1991) Nitrogen supply and insect herbivory in eucalypts: the role of nitrogen assimilation and transport processes. In: Ryan, P.J. (ed.) *Productivity in Perspective. Proceedings of the Third Australian Forest Soils and Nutrition Conference, Melbourne, 7–11 October 1991*. Forestry Commission of New South Wales, Sydney, Australia, pp. 239–241.

Adams, M.A., Attiwill, P.M. and Wang, L.M. (1995) Effects of phosphorus supply on growth and nitrogen fractions in xylem sap and foliage of *Eucalyptus regnans* (F. Muell.), *E. nitens* (Maiden) and *E. globulus* (Labill.) seedlings: implications for herbivory. *Trees* 9, 324–331.

Adams, M.A., Simon, J. and Pfautsch, S. (2010) Woody legumes: a (re-)view from the south. *Tree Physiology* 30, 1072–1082.

Aidar, M.P.M., Schmidt, S., Moss, G., Stewart, G.R. and Joly, C.A. (2003) Nitrogen use strategies of neotropical rainforest trees in threatened Atlantic forest. *Plant, Cell and Environment* 26, 389–399.

Alberton, O., Kuyper, T.W. and Gorissen, A. (2007) Competition for nitrogen between *Pinus sylvestris* and ectomycorrhizal fungi generates potential for negative feedback under elevated CO_2. *Plant and Soil* 296, 159–172.

Andersen, P.C., Brodbeck, B.V. and Mizel, R.F. (1995) Diurnal variations in tension, osmolarity and the composition of nitrogen and carbon assimilated in xylem fluid of *Prunus persica*, *Vitis* hybrid and *Pyrus communis*. *Journal of the American Society of Horticultural Science* 120, 600–606.

Arndt, S.K., Wanek, W., Hoch, G., Richter, A. and Popp, M. (2002) Flexibility of nitrogen metabolism in the tropical C_3-crassulacean acid metabolism tree species, *Clusia minor*. *Functional Plant Biology* 29, 741–747.

Atkins, C.A., Pate, J.S. and McNeil D.L. (1980) Phloem loading and metabolism of xylem-borne amino compounds in fruiting shoots of a legume. *Journal of Experimental Botany* 31, 1509–1520.

Averill, C. and Finzi, A. (2011) Increasing plant use of organic nitrogen with elevation is reflected in nitrogen uptake rates and ecosystem $\delta^{15}N$. *Ecology* 92, 883–891.

Avolio, M.L., Tuininga, A.R., Lewis, J.D. and Marchese, M. (2009) Ectomycorrhizal responses to organic and inorganic nitrogen sources when associating with two host species. *Mycological Research* 113, 897–907.

Avolio, M., Müller, T., Mpangara, A., Fitz, M., Becker, B., Pauck, A., Kirsch, A. and Wipf, D. (2012) Regulation of genes involved in nitrogen utilization on different C/N ratios and nitrogen sources in the model ectomycorrhizal fungus *Hebeloma cylindrosporum*. *Mycorrhiza* 22, 515–524.

Azedo-Silva, J., Osorio, J., Fonseca, F. and Correia, M.J. (2004) Effects of soil drying and subsequent re-watering on the activity of nitrate reductase in roots and leaves of *Helianthus annuus*. *Functional Plant Biology* 31, 611–621.

Barnes, R.L. (1963) Organic nitrogen compounds in tree xylem sap. *Forest Science* 9, 98–102.

Bell, T.L. and Adams, M.A. (2011) Attack on all fronts: functional relationships between aerial and root parasitic plants and their woody hosts and consequences for ecosystems. *Tree Physiology* 31, 3–15.

Bell, T.L. and Pate, J.S. (1996) Nitrogen and phosphorus nutrition in mycorrhizal *Epacridaceae* of south-west Australia. *Annals of Botany* 77, 389–397.

Bialczyk, J., Lechowski, Z. and Dziga, D. (2004) Composition of the xylem sap of tomato seedlings cultivated on media with HCO_3^- and nitrogen source as NO_3^- or NH_4^+. *Plant and Soil* 263, 265–272.

Bloom, A.J., Sukrapanna, S.S. and Warner, R.L. (1992) Root respiration associated with ammonium and nitrate absorption and assimilation by barley. *Plant Physiology* 99, 1294–1301.

Britto, D.T. and Kronzucker, H.J. (2002) NH_4^+ toxicity in higher plants: a critical review. *Journal of Plant Physiology* 159, 567–584.

Brundrett, M.C. (2009) Mycorrhizal associations and other means of nutrition of vascular plants: understanding the global diversity of host plants by resolving conflicting information and developing reliable means of diagnosis. *Plant and Soil* 320, 37–77.

Burris, R.H. and Roberts, G.P. (1993) Biological nitrogen fixation. *Annual Review of Nutrition* 13, 317–335.

Bush, D.R. (1999) Amino acid transport. In: Singh, B.J. (ed.) *Plant Amino Acids – Biochemistry and Biotechnology*. Marcel Dekker, New York, pp. 305–318.

Calladine, A., Pate, J.S. and Dixon, K.W. (2000) Haustorial development and growth benefit to seedlings of the root hemiparasitic tree *Nuytsia floribunda* (Labill.) R.Br. in association with various hosts. *Annals of Botany* 85, 733–740.

Castro, H.F., Classen, A.T., Austin, E.E., Norby, R.J. and Schadt, C.W. (2010) Soil microbial community responses to multiple experimental climate change drivers. *Applied and Environmental Microbiology* 76, 999–1007.

Chalk, P.M. (2000) Integrated effects of mineral nutrition on legume performance. *Soil Biology and Biochemistry* 32, 577–579.

Chalot, M. and Brun, A. (1998) Physiology of organic nitrogen acquisition by ectomycorrhizal fungi and ectomycorrhizas. *FEMS Microbiology Reviews* 22, 21–44.

Chapin, F.S., Moilanen, L. and Kielland, K. (1993) Preferential use of organic nitrogen for growth by a non-mycorrhizal arctic sedge. *Nature* 361, 150–153.

Chapin, F.S., Matson, P.A. and Vitousek, P.M. (eds) (2011) *Principles of Terrestrial Ecosystem Ecology*. Springer, New York.

Chapman, S.K., Langley, J.A., Hart, S.C. and Koch, G.W. (2006) Plants actively control nitrogen cycling: uncorking the microbial bottleneck. *New Phytologist* 169, 27–34.

Clark, C.J. and Smith, G.S. (1991) Seasonal variation of nitrogenous compounds in components of the kiwifruit vine. *Annals of Botany*, 68: 441–450.

Cooper, H.D. and Clarkson, D.T. (1989). Cycling of amino nitrogen and other nutrients between shoot and roots in cereals: a possible mechanism integrating shoot and root in the regulation of nutrient uptake. *Journal of Experimental Botany* 40, 753–762.

Corrêa, A., Strasser, R.J. and Martins-Loução, M.A. (2008) Response of plants to ectomycorrhizae in N-limited conditions: which factors determine its variation? *Mycorrhiza* 18, 413–427.

Corrêa, A., Hampp, R., Magel, E. and Martins-Loução, M.-A. (2011) Carbon allocation in ectomycorrhizal plants at limited and optimal N supply: an attempt at unravelling conflicting theories. *Mycorrhiza* 21, 35–51.

Cramer, V.A., Schmidt, S., Stewart, G.R. and Thorburn, P.J. (2002) Can nitrogenous composition of xylem sap be used to assess salinity stress in *Casuarina glauca*? *Tree Physiology* 22, 1019–1026.

Cramer, C.A., Filley, T.R., Olk, D.C., Scott, D.E., Dooling, V. and Button, T.W. (2013) Changes to soil organic N dynamics with leguminous woody plant encroachment into grasslands. *Biogeochemistry* 113, 307–321.

Crawford, N.M., Kahn, M.L., Leustek, T. and Long, S.R. (2001) Nitrogen and sulfur. In: Buchanan, B.B., Gruissem, W. and Jones, R.L. (eds) *Biochemistry and Molecular Biology of Plants*. American Society of Plant Physiologists, Rockville, Maryland, pp. 786–846.

Crews, T.E. (1999) The presence of nitrogen fixing legumes on terrestrial communities: evolutionary vs ecological considerations. *Biogeochemistry* 46, 233–246.

Dambrine, E., Martin, F., Cariey, N., Granier, A., Hällgren, J. and Bishop, K. (1995) Xylem sap composition: a tool for investigating mineral uptake and cycling in adult spruce. *Plant and Soil* 168–169, 233–241.

Dannenmann, M., Simon, J., Gasche, R., Holst, J., Naumann, P.S., Kögel-Knabner, I., Knicker, H., Mayer, H., Schloter, M., Pena, R., Polle, A., Rennenberg, H. and Papen, H. (2009) Tree girdling provides insight in the role of labile carbon in nitrogen partitioning between soil microorganisms and adult European beech. *Soil Biology and Biochemistry* 41, 1622–1631.

Day, D.A., Poole, P.S., Tyerman, S.D. and Rosendahl, L. (2001) Ammonia and amino acid transport across symbiotic membranes in nitrogen-fixing legume nodules. *Cellular and Molecular Life Sciences* 58, 61–71.

de Boer, A.H. and Volkov, V. (2003) Logistics of salt and water transport through the plant: structure and functioning of the xylem. *Plant, Cell and Environment* 26, 87–101.

Ericson, A., Nordén, L.G., Näsholm, T. and Walheim, M. (1993) Mineral nutrient imbalances and arginine concentrations in needles of *Picea abies* L., Karst from two areas with different levels of airborne deposition. *Trees – Structure and Function* 8, 67–74.

Escher, P., Eiblmeier, M. and Rennenberg, H. (2004a) Differences in the influx of glutamine and nitrate into *Viscum album* from the xylem sap of its host. *Plant Physiology and Biochemistry* 42, 739–744.

Escher, P., Eiblmeier, M., Hetzger, I. and Rennenberg, H. (2004b) Spatial and seasonal variation in amino compounds in the xylem sap of a mistletoe (*Viscum album*) and its hosts (*Populus* spp. and *Abies alba*). *Tree Physiology* 24, 639–650.

Finzi, A.C. and Berthrong, S.T. (2005) The uptake of amino acids by microbes and trees in three cold-temperate forests. *Ecology* 86, 3345–3353.

Fischer, W.-N., André, B., Rentsch, D., Krolkiewicz, S., Tegeder, M., Breitkreuz, K. and Frommer, W.B. (1998) Amino acid transport in plants. *Trends in Plant Science* 3, 188–195.

Fischer W.-N., Loo, D.D.F., Koch, W., Ludewig, U., Boorer, K.J., Tegeder, M., Rentsch, D., Wright, E.M. and Frommer, W.B. (2002) Low and high affinity amino acid H^+-cotransporters for cellular import of neutral and charged amino acids. *The Plant Journal* 29, 717–731.

Fisk, M.C. and Schmidt, S.K. (1995) Nitrogen mineralization and microbial biomass nitrogen dynamics in three alpine tundra communities. *Soil Science Society of America Journal* 59, 1036–1043.

Fletcher, A., Rennenberg, H. and Schmidt, S. (2010) Nitrogen partitioning in orchard-grown *Macadamia integrifolia*. *Tree Physiology* 30, 244–256.

Fotelli, M.N., Radoglou, K., Nahm, M. and Rennenberg, H. (2009) Climate effects on the N balance of beech (*Fagus sylvatica*) at its south-eastern distribution limit in Europe. *Plant Biosystems* 143, 34–45.

Frak, E., Millard, P., Le Roux, X., Guillaumie, S. and Wendler, R. (2002) Coupling sap flow velocity and amino acid concentrations as an alternative method to ^{15}N labelling for quantifying nitrogen remobilization by walnut trees. *Plant Physiology* 130, 1043–1053.

Fromard, L., Babin, V., Fleuratlessard, P., Fromont, J.C., Serrano, R. and Bonnemain, J.L. (1995) Control of vascular sap pH by the vessel-associated cells in woody species physiological and immunological studies. *Plant Physiology* 108, 913–918.

Gallet-Budynek, A., Brzostek, E., Rogers, V.L., Talbot, J.M., Hyzy, S. and Finzi, A.C. (2009) Intact amino acid uptake by northern hardwood and conifer trees. *Oecologia* 160, 129–138.

Galloway, J.N., Townsend, A.R., Erisman, J.W., Bekunda, M. Cai, Z., Freney, J.R., Martinelli, L.A., Seitzinger, S.P. and Sutton, M.A. (2008) Transformation of the nitrogen cycle: recent trends, questions and potential solutions. *Science* 320, 889–892.

Gessler, A., Schneider, S., Weber, P., Hanemann, U. and Rennenberg, H. (1998a) Soluble N compounds in trees exposed to high loads of N: a comparison between the roots of Norway spruce (*Picea abies*) and beech (*Fagus sylvatica*) trees grown under field conditions. *New Phytologist* 138, 385–399.

Gessler, A., Schultze, M., Schrempp, S. and Rennenberg, H. (1998b) Interaction of phloem-translocated amino compounds with nitrate net uptake by the roots of beech (*Fagus sylvatica*) seedlings. *Journal of Experimental Botany* 49, 1529–1537.

Gessler, A., Weber, P., Schneider, S. and Rennenberg, H. (2003) Bidirectional exchange of amino compounds between phloem and xylem during long-distance transport in Norway spruce trees (*Picea abies* [L.] Karst). *Journal of Experimental Botany* 54, 1389–1397.

Gessler, A., Kopriva, S. and Rennenberg, H. (2004) Regulation of nitrate uptake at the whole-tree level: interaction between nitrogen compounds, cytokinins and carbon metabolism. *Tree Physiology* 24, 1313–1321.

Grassi, G., Millard, P., Wendler, R., Minota, G. and Tagliavini, M. (2002) Measurement of xylem sap amino acid concentrations in conjunction with whole tree transpiration estimates spring N remobilization by cherry (*Prunus avium* L.) trees. *Plant, Cell and Environment* 25, 1689–1699.

Grassi, G., Millard, P., Gioacchini, P. and Tagliavini, M. (2003) Recycling of nitrogen in the xylem of *Prunus avium* trees starts when spring remobilization of internal reserves declines. *Tree Physiology* 23, 1061–1068.

Graves, J.D. (1995) Host-plant responses to parasitism. In: Press, M.C. and Graves, J.D. (eds) *Parasitic Plants*. Chapman and Hall, London, pp. 206–225.

Grelet, G.-A., Meharg, A.A., Duff, E.I., Anderson, I.C. and Alexander, I.J. (2009) Small genetic differences between ericoid mycorrhizal fungi affect nitrogen uptake by *Vaccinium*. *New Phytologist* 181, 708–718.

Guak, S., Neilsen, D., Millard, P., Wendler, R. and Neilsen, G.H. (2003) Determining the role of N remobilization for growth of apple (*Malus domestica* Borkh.) trees by measuring xylem-sap N flux. *Journal of Experimental Botany* 54, 2121–2131.

Harrison, K.A., Bol, R. and Bardgett, R.D. (2007) Preferences for different nitrogen forms by coexisting plant species and soil microbes. *Ecology* 88, 989–999.

Harrison, K.A., Bol, R. and Bardgett, R.D. (2008) Do plant species with different growth strategies vary in their ability to compete with soil microbes for chemical forms of nitrogen? *Soil Biology and Biochemistry* 40, 228–237.

Hautier, Y., Hector, A., Vojtech, E., Purves, D. and Turnbull, L.A. (2010) Modelling the growth of parasitic plants. *Journal of Ecology* 98, 857–866.
He, X., Xu, M., Qui, G.Y. and Zhou, J. (2009) Use of ^{15}N stable isotope to quantify nitrogen transfer between mycorrhizal plants. *Journal of Plant Ecology* 2, 107–118.
Hibberd, J.M. and Jeschke, W.D. (2001) Solute flux into parasitic plants. *Journal of Experimental Botany* 52, 2043–2049.
Hill, P.W., Quilliam, R.S., DeLuca, T.H., Farrar, J., Farrell, M., Roberts, P., Newsham, K.K., Hopkins, D.W., Bradgett, R.D. and Jones, D.L. (2011) Acquisition and assimilation of nitrogen as peptide-bound and D-enantiomers of amino acids by wheat. *PLoS ONE* 6(4): e19220.
Hobbie, J.E. and Hobbie, E.A. (2012) Amino acid cycling in plankton and soil microbes studied with radioisotopes: measured amino acids in soil do not reflect bioavailability. *Biogeochemistry* 107, 339–360.
Hobbie, E.A. and Högberg, P. (2012) Nitrogen isotopes link mycorrhizal fungi and plants to nitrogen dynamics. *New Phytologist* 196, 367–382.
Hobbie, E.A., Macko, S.A. and Shugart, H.H. (1999) Insights into nitrogen and carbon dynamics of ectomycorrhizal and saprotrophic fungi from isotopic evidence. *Oecologia* 118, 353–360.
Hodge, A. and Fitter, A.H. (2010) Substantial nitrogen acquisition by arbuscular mycorrhizal fungi from organic material has implications for N cycling. *Proceedings of the National Academy of Sciences of the United States of America* 107, 13754–13759.
Hodge, A., Campbell, C.D. and Fitter, A.H. (2001) An arbuscular mycorrhizal fungus accelerates decomposition and acquires nitrogen directly from organic material. *Nature* 413, 297–299.
Hofmockel, K.S., Schlesinger, W.H. and Jackson, R.B. (2007) Effects of elevated atmospheric carbon dioxide on amino acid and NH_4^+-N cycling in a temperate pine ecosystem. *Global Change Biology* 13, 1950–1959.
Imsande, J. and Touraine, B. (1994) N-demand and regulation of nitrate uptake. *Plant Physiology* 105, 3–7.
Inselbacher, E. and Näsholm, T. (2012) The below-ground perspective of forest plants: soil provides mainly organic nitrogen for plants and mycorrhizal fungi. *New Phytologist* 195, 329–334.
Inselsbacher, E., Öhlund, J., Jämtgård, S., Huss-Danell, K. and Näsholm, T. (2011) The potential of microdialysis to monitor organic and inorganic nitrogen compounds in soil. *Soil Biology and Biochemistry* 43, 1321–1332.
IPCC (2007) *Climate Change 2007: Synthesis Report. Contribution of Working Groups I, II and III to the Fourth Assessment Report of the Intergovernmental Panel on Climate Change*. IPCC, Geneva, Switzerland.
Irving, L.J. and Cameron, D.D. (2009) You are what you eat: interactions between root parasitic plants and their hosts. *Advances in Botanical Research* 50, 87–138.
Jackson, L.E., Burger, M. and Cavagnaro, T.R. (2008) Roots, nitrogen transformations, and ecosystem services. *Annual Review of Plant Biology* 59, 341–363.
Jeschke, W.D. and Hartung W. (2000) Root–shoot interactions in mineral nutrition. *Plant and Soil* 226, 57–69.
Jin, H., Pfeffer, P.E., Douds, D.D., Piotrowski, E., Lammers, P.J. and Shachar-Hill, Y. (2005) The uptake, metabolism, transport and transfer of nitrogen in an arbuscular mycorrhizal symbiosis. *New Phytologist* 168, 687–696.
Jin, H.R., Liu, J., Liu, J. and Huang, X.W. (2012) Forms of nitrogen uptake, translocation, and transfer via arbuscular mycorrhizal fungi: a review. *Science China Life Sciences* 55, 474–482.
Jones, D.L. and Kielland, K. (2012) Amino acid, peptide and protein mineralization dynamics in a taiga forest soil. *Soil Biology and Biochemistry* 55, 60–69.
Kahmen, A., Livesley, S.J. and Arndt, S.K. (2009) High potential, but low actual, glycine uptake of dominant plant species in three Australian land-use types with intermediate N availability. *Plant and Soil* 325, 109–121.
Kato, T. (1981) Major nitrogen compounds transported in vessels from roots to top in citrus trees. *Physiologia Plantarum* 52, 275–279.
Kaye, J.P. and Hart, S.C. (1997) Competition for nitrogen between plants and soil microorganisms. *Trends in Ecology and Evolution* 12, 139–143.
Keller, G. (1996) Utilization of inorganic and organic nitrogen sources by high-subalpine ectomycorrhizal fungi of *Pinus cembra* in pure culture. *Mycological Research* 100, 989–998.
Kender, W.J. and Childers, N.F. (1959) Growth of cranberry plants (*Vaccinium macrocarpon*) with various sources of nitrogen. *Proceedings of the American Society for Horticultural Science* 73, 407–413.
Kielland K. (1994) Amino acid absorption by arctic plants: implications for plant nutrition and nitrogen cycling. *Ecology* 75, 2373–2383.

Koch, W., Kwart, M., Laubner, M., Heineke, D., Stransky, H., Frommer, W.B. and Tegeder, M. (2003) Reduced amino acid content in transgenic potato tubers due to antisense inhibition of the H^+/amino acid symporter StAAP1. *Plant Journal* 33, 211–220.

Kreuzwieser, J., Herschbach, C., Stulen, I., Wiersema, P., Vaalburg, W. and Rennenberg, H. (1997) Interactions of NH_4^+ and L-glutamate with NO_3^- transport processes of non-mycorrhizal *Fagus sylvatica* roots. *Journal of Experimental Botany* 48, 1431–1438.

Kronzucker, H.J., Siddiqi, M.Y., Glass, A.D.M. and Britto, D.T. (2003) Root ammonium transport efficiency as the determinant in forest colonization patterns: an hypothesis. *Physiologia Plantarum* 117, 164–170.

Kruse, J., Hetzger, I., Hänsch, R., Mendel, R.R., Walch-Liu, P., Engels, C. and Rennenberg, H. (2002) Elevated pCO_2 favors nitrate reduction in the roots of wild-type tobacco (*Nicotiana tabacum* cv. Gat.) and significantly alters N-metabolism in transformants lacking functional nitrate reductase in the roots. *Journal of Experimental Botany* 53, 2351–2367.

Kuzyakov, Y. and Jones, D.L. (2006) Glucose uptake by maize roots and its transformation in the rhizosphere. *Soil Biology and Biochemistry* 38, 851–860.

Kuzyakov, Y. and Xu, X. (2013) Competition between roots and microorganisms for nitrogen: mechanisms and ecological relevance. *New Phytologist* 198, 656–669.

Lalonde, S., Wipf, D. and Frommer, W.B. (2004) Transport mechanisms for organic forms of carbon and nitrogen between source and sink. *Annual Review of Plant Biology* 55, 341–372.

Lam, H.-M., Coschigano, K.T., Oliveira, I.C., Melo-Oliveira, R. and Coruzzi, G.M. (1996) The molecular-genetics of nitrogen assimilation into amino acids in higher plants. *Annual Review of Plant Physiology and Plant Molecular Biology* 47, 569–593.

Lea, P.J. and Azevedo, R.A. (2006) Nitrogen use efficiency. 1. Uptake of nitrogen from the soil. *Annals of Applied Biology* 149, 243–247.

Lea, P.J. and Ireland, R.J. (1999) Nitrogen metabolism in higher plants. In: Singh, B.K. (ed.) *Plant Amino Acids – Biochemistry and Biotechnology*. Marcel Dekker, New York, pp. 1–47.

LeDuc, S.D. and Rothstein, D.E. (2010) Plant available organic and mineral nitrogen shift in dominance with forest stand age. *Ecology* 91, 708–720.

Lee, Y.H., Foster, J., Chen, J., Voll, L.M., Weber, A.P.M. and Tegeder, M. (2007) AAP1 transports uncharged amino acids into roots of *Arabidopsis*. *The Plant Journal* 50, 305–319.

Lehmann, S., Gumy, C., Blatter, E., Boeffel, S., Fricke, W. and Rentsch, D. (2011) *In planta* function of compatible solute transporters of the AtProT family. *Journal of Experimental Botany* 62, 787–796.

Lesuffleur, F., Paynel, F., Bataillé, M.-P., Le Deunff, E. and Cliquet, J.-B. (2007) Root amino acid exudation: measurement of high efflux rates of glycine and serine from six different plant species. *Plant and Soil* 294, 235–246.

Li, Z.C. and Bush, D.R. (1990a) pH-dependent amino acid transport into plasma membrane vesicles isolated from sugar beet leaves. 1. Evidence for carrier-mediated, electrogenic flux through multiple transport systems. *Plant Physiology* 94, 268–277.

Li, Z.C. and Bush, D.R. (1990b) pH-dependent amino acid transport into plasma membrane vesicles isolated from sugar beet leaves. 2. Evidence for multiple aliphatic, neutral amino acid symports. *Plant Physiology* 96, 1338–1344.

Lipson, D.A. and Monson, R.K. (1998) Plant–microbe competition for soil amino acids in the alpine tundra: effects of freeze–thaw and dry–rewet events. *Oecologia* 113, 406–414.

Lipson, D. and Näsholm, T. (2001) The unexpected versatility of plants: organic nitrogen use and availability in terrestrial ecosystems. *Oecologia* 128, 305–316.

Lipson, D.A., Raab, T.K., Schmidt, S.K. and Monson R.K. (1999) Variation in competitive abilities of plants and microbes for specific amino acids. *Biology and Fertility of Soils* 29, 257–261.

Lodwig, E.M., Hosie, A.H.F., Bourdès, A., Findlay, K., Allaway, D., Karunakaran, R., Downie, J.A. and Poole, P.S. (2003) Amino acid cycling drives nitrogen fixation in legume–*Rhizobium* symbiosis. *Nature* 422, 722–726.

Loqué, D. and von Wirén, N. (2004) Regulatory levels for the transport of ammonium in plant roots. *Journal of Experimental Botany* 55, 1293–1305.

Ludewig, U. and Koch, W. (2008) Amino acid transporters in plants. In: Jaiwal, P.K., Singh, R.P. and Dhankher, O.P. (eds) *Plant Membrane and Vacuolar Transporters*. CAB International, Wallingford, UK, pp. 267–282.

Lynch, J.M. and Whipps, J.M. (1990) Substrate flow in the rhizosphere. *Plant and Soil* 129, 1–10.

Macduff, J.H. and Bakken, A.K. (2003) Diurnal variation in uptake and xylem contents of inorganic and assimilated N under continuous and interrupted N supply to *Phleum pratense* and *Festuca pratensis*. *Journal of Experimental Botany* 54, 431–444.

Mackenzie FT (1998) *Our Changing Planet*, 2nd edn. Prentice Hall, Upper Saddle River, New Jersey.
Malaguti, D., Millard, P., Wendler, R., Hepburn, A. and Tagliavini, M. (2001) Translocation of amino acids in the xylem of apple (*Malus domestica* Borkh.) trees in spring as a consequence of both N remobilization and root uptake. *Journal of Experimental Botany* 52, 1665–1671.
Marsh, N.R. and Adams, M.A. (1995) Decline of *Eucalyptus tereticornis* near Bairndale, Victoria: insect herbivory and nitrogen fractions in sap and foliage. *Australian Journal of Botany* 43, 39–50.
Martin, F. and Amraoui, M.B. (1989) Partitioning of assimilated nitrogen in beech (*Fagus sylvatica*). *Annales des Sciences Forestières* 46, S660–S662.
Mazzafera, P. and Gončalves, K.V. (1999) Nitrogen compounds in xylem sap of coffee. *Phytochemistry* 50, 383–386.
McFarland, J.W., Ruess, R.W., Kielland, K., Pregitzer, K., Hendrick, R. and Allan, M. (2010) Cross-ecosystem comparisons of *in situ* plant uptake of amino acid-N and NH_4^+. *Ecosystems* 13, 177–193.
Melin, E. and Nilsson, H. (1953) Transfer of labelled nitrogen from glutamic acid to pine seedlings through the mycelium of *Boletus variegatus* (Sw.) Fr. *Nature* 171, 134.
Michelsen, A., Schmidt, I.K., Jonasson, S., Quarmby, C. and Sleep, D. (1996) Leaf ^{15}N abundance of subarctic plants provides field evidence that ericoid, ectomycorrhizal and non- and arbuscular mycorrhizal species access different sources of soil nitrogen. *Oecologia* 105, 53–63.
Millard, P. (1996). Ecophysiology of internal cycling of nitrogen for tree growth. *Zeitschrift für Pflanzenernährung und Bodenkunde* 159, 1–10.
Millard, P., Wendler, R., Hepburn, A. and Smith, A. (1998) Variations in the amino acid composition of xylem sap of *Betula pendula* Roth. trees due to remobilization of stored N in the spring. *Plant, Cell and Environment* 21, 715–722.
Millard, P., Wendler, R., Grassi, G., Grelet, G.-A. and Tagliavini, M. (2006) Translocation of nitrogen in the phloem of field-grown cherry and poplar trees during remobilization. *Tree Physiology* 26, 527–536.
Muller, B. and Touraine, B. (1992) Inhibition of nitrate uptake by various phloem-translocated amino acids in soybean seedlings. *Journal of Experimental Botany* 43, 617–623.
Musselman, L.J. and Press, M.C. (1995) Introduction to parasitic plants. In: Press, M.C. and Graves, J.D. (eds) *Parasitic Plants*. Chapman and Hall, London, pp. 103–118.
Mylona, P., Pawlowski, K. and Bisseling, T. (1995) Symbiotic nitrogen fixation. *The Plant Cell* 7, 869–885.
Nabais, C., Hagemeyer, J. and Freitas, H. (2005) Nitrogen transport in the xylem sap of *Quercus ilex*: the role of ornithine. *Journal of Plant Physiology* 162, 603–606.
Nahm, M., Holst, T., Matzarakis, A., Mayer, H., Rennenberg, H. and Gessler, A. (2006a) Soluble N compound profiles and concentrations in European beech (*Fagus sylvatica* L.) are influenced by local climate and thinning. *European Journal of Forest Research* 125, 1–14.
Nahm, M., Radoglou, K., Halyvopulos, G., Gessler, A., Rennenberg, H. and Fotelli, M.N. (2006b) Physiological performance of beech (*Fagus sylvatica* L.) at its southeastern distribution limit in Europe: seasonal changes in nitrogen, carbon and water balance. *Plant Biology* 8, 52–63.
Nahm, M., Matzarakis, A., Rennenberg, H. and Gessler, A. (2007) Seasonal courses of key parameters of nitrogen, carbon and water balance in European beech (*Fagus sylvatica* L.) grown on four different study sites along a European north–south climate gradient during the 2003 drought. *Trees* 21, 79–92.
Näsholm, T., Ekblad, A., Nordin, A., Giesler, R., Högberg, M. and Högberg, P. (1998) Boreal forest plants take up organic nitrogen. *Nature* 392, 914–916.
Näsholm, T., Kielland, K. and Ganetek, U. (2009) Uptake of organic nitrogen by plants. *New Phytologist* 182, 31–48.
Näsholm, T., Högberg, P., Franklin, O., Metcalfe, D., Keel, S.G., Campbell, C., Hurry, V., Linder, S. and Högberg, M.N. (2012) Are ectomycorrhizal fungi alleviating or aggravating nitrogen limitation of tree growth in boreal forests? *New Phytologist* 198, 214–221.
Nickrent, D.L. and Musselman, L.J. (2004) *Introduction to Parasitic Flowering Plants*. The Plant Health Instructor. The American Phytopathological Society, St Paul, Minnesota. Available at: http://www.apsnet.org/edcenter/intropp/pathogengroups/pages/parasiticplants.aspx (accessed 28 August 2014).
Nordin, A., Uggla, C. and Näsholm, T. (2001) Nitrogen forms in bark, wood and foliage of nitrogen-fertilized *Pinus sylvestris*. *Tree Physiology* 21, 59–64.
Öhlund, J. and Näsholm, T. (2004) Regulation of organic and inorganic nitrogen uptake in Scots pine (*Pinus sylvestris*) seedlings. *Tree Physiology* 24, 1397–1402.
Okumoto, S., Koch, W., Tegeder, M., Fischer, W.N., Biehl, A., Leister, D., Stierhof, Y.D. and Frommer, W.B. (2004) Root phloem-specific expression of the plasma membrane amino acid proton co-transporter AAP3. *Journal of Experimental Botany* 55, 2155–2168.

Owen, A.G. and Jones, D.L. (2001) Competition for amino acids between wheat roots and rhizosphere microorganisms and the role of amino acids in plant N acquisition. *Soil Biology and Biochemistry* 33, 651–657.

Pate, J.S. (1995) Mineral relationships of parasitic plants and their hosts. In: Press, M.C. and Graves, J.D. (eds) *Parasitic Plants*. Chapman and Hall, London, pp. 80–102.

Pate, J.S. (2001) Haustoria in action: case studies of nitrogen acquisition by woody xylem-tapping hemiparasites from their hosts. *Protoplasma* 215, 204–217.

Pate, J.S. and Bell, T.L. (2000) Host associations of the introduced annual root hemiparasite *Parentucellia viscosa* in agricultural and bushland settings in Western Australia. *Annals of Botany* 85, 203–213.

Pate, J.S., Woodall, G., Jeschke, W.D. and Stewart, G.R. (1994) Root xylem transport of amino acids in the root hemiparasitic shrub *Olax phyllanthi* (Labill.) R Br (*Olacaceae*) and its multiple hosts. *Plant, Cell and Environment* 17, 1263–1273.

Paungfoo-Lonhienne, C., Visser, J., Lonhienne, T.G.A. and Schmidt, S. (2012) Past, present and future of organic nutrients. *Plant and Soil* 359, 1–18.

Perelo, L.W., Jimenez, M. and Munch, J.C. (2006) Microbial immobilisation and turnover of ^{15}N labelled substrates in two arable soils under field and laboratory conditions. *Soil Biology and Biochemistry* 38, 912–922.

Persson, J., Högberg, P., Ekblad, A., Högberg, M.N., Nordgren, A. and Näsholm, T. (2003) Nitrogen acquisition from inorganic and organic sources by boreal forest plants in the field. *Oecologia* 137, 252–257.

Peuke, A.D. (2010) Correlations in concentrations, xylem and phloem flows, and partitioning of elements and ions in intact plants. *A summary and statistical re-evaluation of modeling experiments in Ricinus communis. Journal of Experimental Botany* 61, 635–655.

Peuke, A.D., Rokitta, M., Zimmermann, U., Schreiber, L. and Haase, A. (2001) Simultaneous measurement of water flow velocity and solute transport in xylem and phloem of adult plants of *Ricinus communis* over a daily time course by nuclear magnetic resonance spectrometry. *Plant, Cell and Environment* 24,491–503.

Pfautsch, S., Rennenberg, H., Bell, T.L. and Adams M.A. (2009a) Nitrogen uptake by *Eucalyptus regnans* and *Acacia* spp. preferences, resource overlap and energetic costs. *Tree Physiology* 29, 389–399.

Pfautsch, S., Gessler, A., Adams, M.A. and Rennenberg, H. (2009b) Using amino-nitrogen pools and fluxes to identify contributions of understory *Acacia* spp. to overstorey *Eucalyptus regnans* and stand nitrogen uptake in temperate Australia. *New Phytologist* 183, 1097–1113.

Phillips, D.A., Fox, T.C. and Six, J. (2006) Root exudation (net efflux of amino acids) may increase rhizodeposition under elevated CO_2. *Global Change Biology* 12, 561–567.

Plassard, C., Bonafos, B. and Touraine, B. (2000) Differential effects of mineral and organic N sources, and of ectomycorrhizal infection by *Hebeloma cylindrosporum*, on growth and N utilization in *Pinus pinaster*. *Plant, Cell and Environment* 23, 1195–1205.

Press, M.C. and Graves, J.D. (1995) *Parasitic Plants*. Chapman and Hall, London.

Press, M.C., Parsons, A.N., Mackay, A.W., Vincent, C.A., Cochrane, V. and Seel, W.E. (1993) Gas exchange characteristics and nitrogen relations of two Mediterranean root hemiparasites: *Bartsia trixago* and *Parentuceltia viscosa*. *Oecologia* 95, 145–151.

Prima Putra, D. and Botton, B. (1998) Organic and inorganic compounds of xylem exudates from five woody plants at the stage of bud breaking. *Journal of Plant Physiology* 153, 670–676.

Radomiljac, A.M., McComb, J.A., Pate, J.S. and Tennakoon, K.U. (1998) Xylem transfer of organic solutes in *Santalum album* L. (Indian sandalwood) in association with legume and non-legume hosts. *Annals of Botany* 82, 675–682.

Rasmussen, J. and Kuzyakov, Y. (2009) Carbon isotopes as proof for plant uptake of organic nitrogen: relevance of inorganic uptake. *Soil Biology and Biochemistry* 41, 1586–1587.

Read, D.J., Leake, J.R. and Perez-Moreno, J. (2004) Mycorrhizal fungi as drivers of ecosystem processes in heathland and boreal forest biomes. *Canadian Journal of Botany* 82, 1243–1263.

Rennenberg, H. and Schmidt, S. (2010) Perennial lifestyle – an adaptation to nutrient limitation? *Tree Physiology* 30, 1047–1049.

Rennenberg, H., Wildhagen, H. and Ehlting, B. (2010) Nitrogen nutrition of poplar trees. *Plant Biology* 12, 275–291.

Rentsch, D., Schmidt S. and Tegeder, M. (2007) Transporters for uptake and allocation of organic nitrogen compounds in plants. *FEBS Letters* 581, 2281–2289.

Ruess, R.W., Van Cleve, K., Yarie, J. and Vierck, L.A. (1996) Contributions of fine root production and turnover to the carbon and nitrogen cycling in the taiga forests of the Alaskan interior. *Canadian Journal of Forest Research* 26, 1326–1336.

Sauheitl, L., Glaser, B. and Weigelt, A. (2009) Uptake of intact amino acids depends on soil amino acid concentrations. *Environmental and Experimental Botany* 66, 145–152.
Sauter, J.J. (1981) Seasonal variation of amino acids and amides in the xylem sap of *Salix*. *Zeitschrift für Pflanzenphysiologie* 101, 399–411.
Sauter, J.J. and van Cleve, B. (1994) Storage, mobilization and interrelations of starch, sugars, protein and fat in the ray storage tissue of poplar trees. *Trees* 8, 297–304.
Schimel, J.P. and Bennett, J. (2004) Nitrogen mineralization: challenges of a changing paradigm. *Ecology* 85, 591–602.
Schimel, J.P. and Chapin, F.S. III (1996) Tundra plant uptake of amino acid and NH_4^+ nitrogen *in situ*: plants compete well for amino acid N. *Ecology* 77, 2142–2147.
Schmidt, S. and Stewart, G.R. (1997) Waterlogging and fire impact on nitrogen availability and utilization in a subtropical wet heathland (wallum). *Plant, Cell and Environment* 20, 1231–1241.
Schmidt, S. and Stewart, G.R. (1998) Transport, storage and mobilization of nitrogen by trees and shrubs in the dry/wet tropics of northern Australia. *Tree Physiology* 18, 403–410.
Schmidt, S., Raven, J.A. and Paungfoo-Lonhienne, C. (2013) The mixotrophic nature of photosynthetic plants. *Functional Plant Biology* 40, 425–438.
Schneider, A., Kreuzwieser, J., Schrupp, R., Sauter, J.J. and Rennenberg, H. (1994) Thiol and amino-acid-composition of the xylem sap of poplar trees (*Populus canadensis* × *robusta*). *Canadian Journal of Botany* 72, 347–351.
Schneider, S., Gessler, A., Weber, P., von Sengbusch, D., Haneman, U. and Rennenberg, H. (1996) Soluble N compounds in trees exposed to high loads of N: a comparison of spruce (*Picea abies*) and beech (*Fagus sylvatica*) grown under field conditions. *New Phytologist* 134, 103–114.
Schweitzer, J.A., Baily, J.K., Rehill, B.J., Martinsen, G.D., Hart, S.C., Lindroth, R.L., Keim, P. and Witham, T.G. (2004) Genetically based trait in a dominant tree affects ecosystem processes. *Ecology Letters* 7, 127–134.
Seel, W.E., Cooper, R.E. and Press, M.C. (1993) Growth, gas exchange and water use efficiency in the facultative hemiparasite *Rhinanthus minor* associated with hosts differing in foliar nitrogen concentration. *Physiologia Plantarum* 89, 64–70.
Shaver, G.R., Nadelhoffer, K.J. and Giblin, A.E. (1991) Biogeochemical diversity and element transport in a heterogeneous landscape, the North Slope of Alaska. In: Turner, M.G. and Gardner, R.H. (eds) *Quantitative Methods in Landscape Ecology*. Springer, New York, pp 105–126.
Shen, H., Ye, W., Hong, L., Huang, H., Wang, Z., Deng, X., Yang, Q. and Xu, Z. (2006) Progress in parasitic plant biology: host selection and nutrient transfer. *Plant Biology* 8, 175–185.
Siebrecht, S. and Tischner, R. (1999) Changes in the xylem exudate composition of poplar (*Populus tremula* × *P. alba*) dependent on the nitrogen and potassium supply. *Journal of Experimental Botany* 50, 1797–1806.
Siebrecht, S., Herdel, K., Schurr, U. and Tischner, R. (2003) Nutrient translocation in the xylem of poplar: diurnal variations and spatial distribution along the shoot axis. *Planta* 217, 783–793.
Smith, S.E. and Read, D.J. (2008) *Mycorrhizal Symbiosis*, 3rd edn. Academic Press, London.
Soper, F.M., Paungfoo-Lonhienne, C., Brackin, R., Rentsch, D., Schmidt, S. and Robinson, N. (2011) *Arabidopsis* and *Lobelia anceps* access small peptides as a nitrogen source for growth. *Functional Plant Biology* 38, 788–796.
Sprent, J.I. (2001) *Nodulation in Legumes*. Royal Botanic Gardens, Kew, UK.
Sprent, J.I. (2009) *Legume Nodulation: a Global Perspective*. Wiley-Blackwell, Oxford, UK.
ter Steege, M., Stulen, I., Wiersema, P., Posthumus, F. and Vaalburg, W. (1999) Efficiency of nitrate uptake in spinach: impact of external nitrate concentration and relative growth rate on nitrate influx and efflux. *Plant and Soil* 208, 125–134.
Svennerstam, H., Ganeteg, U. and Näsholm, T. (2008) Root uptake of cationic amino acids by *Arabidopsis* depends on functional expression of amino acid permease. *New Phytologist* 180, 620–630.
Svennerstam, H., Jämtgård, S., Ahmad, I., Huss-Danell, K., Näsholm, T. and Ganeteg, U. (2011) Transporters in *Arabidopsis* roots mediating uptake of amino acids at naturally occurring concentrations. *New Phytologist* 191, 459–467.
Tegeder, M. (2012) Transporters for amino acids in plant cells: some functions and many unknowns. *Current Opinion in Plant Biology* 15, 315–321.
Tennakoon, K.U., Pate, J.S. and Stewart, G.R. (1997) Haustorium-related uptake and metabolism of host xylem solutes by the root hemiparasitic shrub *Santalum acuminatum* (R.Br) A.DC. (*Santalaceae*). *Annals of Botany* 80, 257–264.

Thaler, P. and Pagès, L. (1998) Modelling the influence of assimilate availability on root growth and architecture. *Plant and Soil* 201, 307–320.

Tromp, J. and Ovaa, J.C. (1985) Response of young apple trees to time of nitrogen-fertilization with respect to the nitrogen, potassium, and calcium levels in xylem sap, new growth, and the tree as a whole. *Journal of Plant Physiology* 119, 301–309.

Thumfort, P.-P., Pate, J. and Ghisalberti, E. (1993) *S*-ethyl cysteine: an amino acid from *Olax phyllanthi*. *Phytochemistry* 34, 657–659.

Tuskan, G.A., Difazio, S., Jansson, S., Bohlmann, J., Grigoriev, I., Hellsten, U., Putnam, N., Ralph, S., Rombauts, S., Salamov, A., Schein, J., Sterck, L. *et al.* (2006) The genome of black cottonwood, *Populus trichocarpa* (Torr & Gray). *Science* 313, 1596–1604.

Udvardy, M.K. and Day, D.A. (1997) Metabolite transfer across symbiotic membranes of legume nodules. *Annual Review of Plant Physiology and Plant Molecular Biology* 48, 493–523.

Van Der Heijden, M.G.A., Bardgett, R.D. and van Straalen, N.M. (2007) The unseen majority: soil microbes as drivers of plant diversity and productivity in terrestrial ecosystems. *Ecology Letters* 11, 296–310.

Virtanen, A.I. and Linkola, H. (1946) Organic nitrogen compounds as nitrogen nutrition for higher plants. *Nature* 158, 515.

Vranova, V., Zahradnickova, H., Janous, D., Skene, K.R., Matharu, A.S., Rejsek, K. and Formanek, P. (2012) The significance of D-amino acids in soil, fate and utilization by microbes and plants: review and identification of knowledge gaps. *Plant and Soil* 354, 21–39.

Waisel, Y., Eshel, A. and Kafkafi, U. (2002) *Plant Roots: The Hidden Half*, 3rd edn. Marcel Dekker, New York.

Wallenda, T. and Read, D.J. (1999) Kinetics of amino acid uptake by ectomycorrhizal roots. *Plant, Cell and Environment* 22, 179–187.

Warren, C.R. (2006) Potential organic and inorganic N uptake by six *Eucalyptus* species. *Functional Plant Biology* 33, 653–660.

Warren, C.R. (2009) Why does temperature affect relative uptake rates of nitrate, ammonium and glycine: a test with *Eucalyptus pauciflora*. *Soil Biology and Biochemistry* 41, 778–784.

Warren, C.R. (2012) Post-uptake metabolism affects quantification of amino acid uptake. *New Phytologist* 193, 522–531.

Warren C.R. (2013a) High diversity of small organic N observed in soil water. *Soil Biology and Biochemistry* 57, 444–450.

Warren, C.R. (2013b) Quaternary ammonium compounds can be abundant in some soils and are taken up as intact molecules by plants. *New Phytologist* 198, 476–485.

Warren, C.R. and Adams, P.R. (2007) Uptake of nitrate, ammonium and glycine by plants of Tasmanian wet eucalypt forests. *Tree Physiology* 27, 413–419.

Weber, P., Stoermer, H., Gessler, A., Schneider, S., von Sengenbusch, D., Hanemann, U. and Rennenberg, H. (1998) Metabolic responses of Norway spruce (*Picea abies* L. Karst.) trees to long-term forest management practices and acute $(NH_4)_2SO_4$ fertilization: transport of soluble non-protein N compounds in xylem and phloem. *New Phytologist* 140, 461–475.

Whiteside, M.D., Digman, M.A., Gratton, E. and Treseder, K.K. (2012) Organic nitrogen uptake by arbuscular mycorrhizal fungi in a boreal forest. *Soil Biology and Biochemistry* 55, 7–13.

Winter, H., Robinson, D.G. and Heldt, H.W. (1994) Subcellular volumes and metabolite concentrations in spinach leaves. *Planta* 193, 530–535.

Yamaya, T. and Oaks, A. (2004) Metabolic regulation of ammonium uptake and assimilation. In: Amâncio, S. and Stulen, I. (eds) *Nitrogen Acquisition in Higher Plants*. Kluwer, Dordrecht, The Netherlands, pp. 35–63.

Yandulov, D.V. and Schrock, R.R. (2003) Catalytic reduction of dinitrogen to ammonia in a single molybdenum centre. *Science* 301, 76–78.

Youseffi, F., Brown, P.H. and Weinbaum, S.A. (2000) Relationship between tree nitrogen status, xylem and phloem sap amino acid concentrations, and apparent soil nitrogen uptake by almond trees (*Prunus dulcis*). *Journal of Horticultural Science and Biotechnology* 75, 62–68.

Zeller, B., Colin-Belgrand, M., Dambrine, E., Martin, F. and Bottner, P. (2000) Decomposition of ^{15}N-labelled beech litter and fate of nitrogen derived from litter in a beech forest. *Oecologia* 123, 550–559.

Zerihun, A., McKenzie, B.A. and Morton, J.D. (1998) Photosynthate costs associated with the utilization of different nitrogen-forms: influence on the carbon balance of plants and shoot–root biomass partitioning. *New Phytologist* 138, 1–11.

19 Auxin Biosynthesis

J.W. Chandler*
Institute of Developmental Biology, Cologne Biocenter, Cologne University, Cologne, Germany

19.1 Abstract

Auxin is arguably one of the most important plant hormones and is implicated in almost every aspect of development, including the specification of the apical–basal body axis, lateral organ formation, organogenesis, tropisms and environmental responses. Although the majority of the auxin in plants is produced from L-tryptophan (Trp), tryptophan-independent biosynthesis might also occur. Auxin biology has enjoyed a renaissance in recent years, with the recognition and elucidation of its role in gene regulation, and an understanding at the cell biology level of how gradients and concentration maxima are created in tissues via polar transport. However, many questions remain open, especially concerning the function of auxin as a morphogen and how auxin self-regulates aspects of its own transport.

The major auxin in plants is indole-3-acetic acid (IAA), though other active forms exist as well, including 4-chloroindole-3-acetic acid and phenylacetic acid; there are also storage forms, which include indole-3-butyric acid. Only some auxin exists as free IAA; the reversible or non-reversible conjugation of IAA to amino acids, sugars or proteins plays an important role in auxin homeostasis and metabolism, and can either act as a storage pool for active auxin when required, or target it for degradation.

In bacteria, auxin synthesis consists of a relatively simple, two-step pathway. In plants, most of the information on auxin synthesis derives from the model species *Arabidopsis thaliana*, in which mutants have partially elucidated some of these pathways, but other species, such as pea, maize and tobacco have also been informative. Four main biosynthetic pathways have been proposed in plants, and the prevailing view for many years has been that these function in parallel, with species- or tissue-specific differences. Auxin synthesis is thus complex, and little understood, as some of the enzymatic steps remain uncharacterized; also, some of the pathways of synthesis converge with those that produce other secondary metabolites, and it is unclear whether or not these are important for the majority of auxin production. Furthermore, there have been problems in correlating *in vitro* studies with *in vivo* substrates for some pathways. Nevertheless, recent work has clarified one complete pathway from Trp to IAA that is highly conserved across the plant kingdom and is, therefore, thought to be the major pathway for auxin production in higher plants. This pathway consists of two steps: the first is catalysed by tryptophan aminotransferase to convert Trp to indole-3-pyruvic acid; the second rate-limiting step uses the YUCCA (YUC) flavin monooxygenases to create IAA (Zhao, 2012).

Several important conclusions about auxin synthesis have been reached recently. First, its local synthesis in specific tissues is tightly regulated in response to the environment, and in tissue-specific and developmentally important time windows (Zhao, 2008; Chandler *et al.*, 2009a). A further level of regulation includes genetic redundancy, as many enzymes for different biosynthetic steps are encoded by multi-gene families in numerous species. Compartmentation is also important for synthesis: tryptophan is synthesized in the chloroplast and

*E-mail address: john.chandler@uni-koeln.de

auxin in the cytoplasm, although one metabolic step is regulated by an alternatively spliced gene that encodes an enzyme that affects the intracellular compartmentation of auxin synthesis. Current understanding of the principal route to auxin production in plants is summarized here, together with the evidence for alternative routes. The regulation of auxin biosynthesis is described at genetic, transcriptional and environmental levels, and at the level of cross-talk by other hormones; there is also discussion of how homeostasis and conjugation determines the size of the active auxin pool. An awareness of auxin synthesis in different plant species allows aspects of conservation to be identified and informs evolutionary developmental biology debates on the origin of auxin synthesis. Understanding the conservation of the major auxin biosynthetic pathway might accelerate knowledge in crop species, for which less is currently known of the genes encoding the catalytic steps than in *Arabidopsis*; such an understanding might also facilitate targeted manipulation to optimize crop productivity.

19.2 Introduction

The aromatic amino acid L-tryptophan (Trp) is synthesized via the shikimate pathway and subsequently by the branched aromatic amino acid pathway, via chorismate (Tzin and Galili, 2010; Maeda and Dudareva, 2012). In addition to being a substrate for protein synthesis, Trp is a precursor for a spectrum of plant secondary metabolites and other molecules, one of the most developmentally important of which is auxin. Thus, the Trp-to-auxin conversion represents an important interface that translates metabolism into signalling molecules, which then effect gene transcription and ultimately coordinate development.

The story of the identification and characterization of auxin as a plant hormone (Abel and Theologis, 2010) represents a huge milestone in plant developmental biology and is one of the oldest areas of plant research; yet only very recently have major breakthroughs been made in understanding its synthesis and recognizing its role in gene regulation. The existence of a diffusible substance, subsequently identified as auxin, involved in phototropism and geotropism (the differential growth of seedlings that causes them to bend towards a light source or in response to gravity) was postulated by Darwin (1880). The major auxin, indole-3-acetic acid (IAA), was first chemically identified in the 1930s (Went and Thimann, 1937) and its structure was shown to be very similar to that of Trp in consisting of a 3-substituted indole, which was the first suggestion that it might be derived from Trp. IAA was first isolated from plant tissue by Haagen-Smit *et al.* (1946). Auxin is one of the most important plant hormones; it affects cell division, elongation and differentiation in almost every aspect of growth and development, including processes as diverse as tissue culture plant regeneration, growth tropisms, leaf vascularization, apical dominance, fruit development and dehiscence, and abscission (Ellis *et al.*, 2005; Leyser, 2005; Chandler, 2009a; Sorefan *et al.*, 2009; Vanneste and Friml, 2009). The existence of these developmental contexts suggest that auxin might convey positional information or act as an instructive signal for hypophysis specification (Friml *et al.*, 2003), lateral root initiation (Dubrovsky *et al.*, 2008) or leaf primordium initiation (Reinhardt *et al.*, 2003).

Auxin can act as a classical hormone, which is defined as a chemical substance that is synthesized in one place and exerts a specific effect after transportation elsewhere, and also as a morphogen, which influences tissue morphogenesis across a concentration gradient. The contemporary emphasis on the function of auxin in development has rested on its differential accumulation by polar transport and its recognition by recently characterized receptors, even though the size of the active pool of auxin at any point in a specific tissue is not only determined by polar transport, which generates gradients and concentration maxima, but also by overall auxin homeostasis, which includes both synthesis and degradation.

No plant or plant mutant exists that produces no auxin, underlining its fundamental importance, but also suggesting that it might be synthesized by diverse pathways. Tools that have allowed the elucidation of auxin biosynthesis include radiolabelled chase experiments and biochemical studies in parallel with genetic approaches, using mutants mainly in the model species *Arabidopsis thaliana*, although several other plants have been used for studies as well, such as maize (Kriechbaumer *et al.*, 2006) and pea (Quittenden *et al.*, 2009). Until now, it has been thought that auxin in plants is produced by several redundant pathways operating in parallel. This proposed complexity of diverse auxin biosynthesis pathways was deduced from the presence of many indole intermediates in different species. However, further elucidation of some of the proposed pathways has depended on *in vitro* biochemical assays that have often been misleading in reflecting *in vivo* substrates, owing to broad substrate specificities, and limited by the failure to identify either enzymes for key steps or the genes encoding them. Furthermore, some of these steps have not been corroborated genetically by mutant or overexpression studies. The understanding of auxin biosynthesis has, therefore, been a dynamic process, though rapid advances have recently been made using genetic and biochemical approaches, mainly in *Arabidopsis*, that have refined our understanding of *de novo* auxin biosynthesis and shown that a single two-step biosynthetic route is predominantly responsible for the majority of auxin synthesis. This route is essential for growth and will be outlined in this chapter, together with other postulated alternative pathways for which there remains strong evidence. Differences exist across the plant kingdom within these pathways, suggesting that plants have optimized their auxin production mechanisms at several key levels, so one major challenge is to understand auxin biosynthetic differences that exist between plant species and families. The importance of understanding auxin metabolism is that its elucidation would create opportunities to manipulate endogenous levels of auxin and thereby to regulate growth in crop plants.

19.3 Sites of Auxin Synthesis in Plants and Cells

The traditional view that auxin is mainly synthesized in shoot tips and young leaf primordia and is then transported basipetally to all plant tissues is now known not to be correct. Experiments to monitor the destination of deuterium-labelled indole-ring intermediates in the shikimate pathway show that *de novo* IAA synthesis occurs in embryos, and in seedling shoots and root tips, with the highest production rate in meristematic tissue containing actively dividing cells (Ljung *et al.*, 2001a). It is now known that local auxin biosynthesis in almost all developing tissues is as important as its long-range redistribution by the polar transport of shoot-derived auxin to generate gradients and concentration maxima of auxin within tissues. For example, gradients of auxin concentration in several *Arabidopsis* tissues reflect local auxin synthesis in the female gametophyte (Pagnussat *et al.*, 2009), the gynoecium (Sohlberg *et al.*, 2006) and the root (Petersson *et al.*, 2009); they are also responsible for generating embryo apical–basal polarity (Cheng *et al.*, 2007) and polarity in root hair differentiation (Ikeda *et al.*, 2009). Within cells, Trp is synthesized in the chloroplast and is then transported across the chloroplast envelope via amino acid transporters within the envelope (Pudelski *et al.*, 2010) into the cytoplasm, where most auxin is synthesized. The amounts of free IAA in plant tissues have been measured in extracts, using isotope dilution analysis with [$^{13}C_6$]IAA as an internal standard and followed by gas chromatography-mass spectroscopy (GC-MS) (Ljung *et al.*, 2001b; Barkawi et al., 2010). Mature leaf tissue of *A. thaliana* plants contains approximately 20 ng free IAA g^{-1} fresh weight, but in developing leaves this can vary from the tip to the base, with up to 140 ng g^{-1} fresh weight at the base (Ljung *et al.*, 2001a). IAA levels can vary dramatically depending on the developmental stage and tissue, and also between individuals of a species and between species (Ljung *et al.*, 2001a; Barkawi *et al.*, 2010); in *Arabidopsis*, the highest *de novo* IAA synthesis rate is in the youngest leaves, but the roots contain the highest IAA pool size (Ljung *et al.*, 2001a).

19.4 Pathways of Auxin Synthesis

There is evidence from several plants, including orange, carrot and maize, that IAA can be synthesized from Trp precursors such as indole or indole-3-glycerol phosphate (IGP) (Normanly *et al.*, 1993; Östin *et al.*, 1999; Ehlert *et al.*, 2008; Rosquete *et al.*, 2011) in a Trp-independent manner. Furthermore, IAA production in *Arabidopsis* Trp biosynthesis mutants was actually higher than in wild type plants, but lower in plants with reduced IGP, suggesting that IGP might represent a branch point for Trp-dependent and Trp-independent auxin biosynthetic pathways (Ouyang *et al.*, 2000) In support of this, an indole synthase (INS) gene (*INS*) was characterized from *Arabidopsis* which was a homologue of the Trp synthase α (TSA1) gene (*TSA1*); INS catalysed the conversion of IGP to indole in the Trp-dependent pathway and was capable of cleaving IGP; the gene was also able to complement the Trp auxotrophy of an *Escherichia coli* strain deficient in Trp synthesis (Zhang *et al*., 2008), suggesting that INS might be involved in Trp-dependent IAA synthesis in *Arabidopsis*.

This review, however, concentrates on the Trp-dependent pathways of IAA synthesis that have been elucidated in plants, or for which strong evidence exists. Auxin biosynthetic pathways have been investigated by feeding radioactively labelled compounds (^{2}H or ^{14}C) to plants or plant extracts and monitoring the fate of the label via GC-MS (Ljung *et al*., 2005; Quittenden *et al*., 2009; Sugawara *et al*., 2009). Together with genetic and biochemical approaches, which have measured intermediates at increasingly better resolution, this research has enabled four distinct pathways for Trp-dependent IAA synthesis to be proposed (see reviews by Woodward and Bartel, 2005; Normanly, 2010; Ljung, 2013). These are named by their intermediates and are outlined below: the indole-3-pyruvate (IPA) pathway (Section 19.4.1); the indole-3-acetaldoxime (IAOx) pathway (Section 19.4.3); the indole-3-acetamide (IAM) pathway (Section 19.4.4); and the tryptamine (TAM) pathway (Section 19.4.5). Recent evidence suggests that despite the apparent complexity of these alternative auxin biosynthetic routes, a relatively simple two-step pathway via IPA is the predominant biosynthetic route in *Arabidopsis* and appears to be highly conserved in many plant species; this aspect is briefly discussed in Section 19.4.2.

19.4.1 The indole-3-pyruvate (IPA) pathway

The two-step pathway that uses IPA as an intermediate to synthesize auxin is currently the only fully elucidated complete pathway from Trp to auxin and has been derived from genetic and biochemical experiments with *Arabidopsis*. The first step is catalysed by tryptophan aminotransferase (TAA), which transfers an amino group from Trp to a recipient such as pyruvate, to generate IPA, an α-keto acid. The second and rate-limiting step is catalysed by the YUCCA (YUC) class of flavin monooxygenases (FMOs) to create IAA (Figs 19.1 and 19.2b). The biochemical pathway of the IPA to IAA conversion, including the intermediates, has now been elucidated in *Arabidopsis* for YUC6 (Dai *et al*., 2013) and this is the first biochemical characterization of a plant FMO. The reaction depends on the flavin (FAD) cofactor within the YUC protein, which is reduced by NADPH and then reacts with oxygen to form the intermediate C4a-hydroperoxyflavin, which is not very stable, with a half-life of 20 s. C4a-hydroperoxyflavin reacts with IPA to perform the oxidative decarboxylation of IPA to IAA (Fig. 19.1; Dai *et al*., 2013).

The TAA class of enzymes was identified genetically from several independent screens for mutants altered in responses to ethylene or to *N*-1-naphthylphthalamic acid (NPA) in *Arabidopsis*. The gene for *TRYPTOPHAN AMINOTRANSFERASE OF ARABIDOPSIS1* (*TAA1*) is also known as *WEAK ETHYLENE INSENSITIVE8* (*WEI8*; Stepanova *et al*., 2008), *SHADE AVOIDANCE3* (*SAV3*; Tao *et al*., 2008) and *TRANSPORT INHIBITOR RESPONSE2* (*TIR2*; Yamada *et al*., 2009). In *Arabidopsis*, the *TAA* gene family also includes *TRYPTOPHAN AMINOTRANSFERASE-RELATED1–4* (*TAR1–4*). Protein domain comparisons show that TAR1 and TAR2 are structurally similar to TAA1 and are likely to share homologous enzymatic functions (Stepanova *et al*., 2008), although the different family members appear to be differentially localized. TAA1 is localized in the cytoplasm; TAR1 is potentially a TAA1 paralogue from its sequence homology, but is expressed at extremely low levels (Stepanova *et al*., 2008); and TAR2, TAR3 and TAR4 contain a predicted signal peptide that might target them to the vacuole (Stepanova *et al*., 2008).

Apart from root-specific ethylene insensitivity and a reduction in gravitropism, *wei8* has no developmental phentoypes and neither do any of the single *tar* mutants (Stepanova *et al*., 2008). However, *wei8 tar2* double mutants have reduced auxin levels and developmental defects in the root meristems, apical hooks and gynoecia (Stepanova *et al*., 2008), showing genetic redundancy that reflects their overlapping expression patterns in these tissues. Seedlings of *wei8 tar1 tar2* triple mutants show auxin-related phenotypes and seedlings that often lack hypocotyls and cotyledons, showing that all genes are necessary for auxin production during

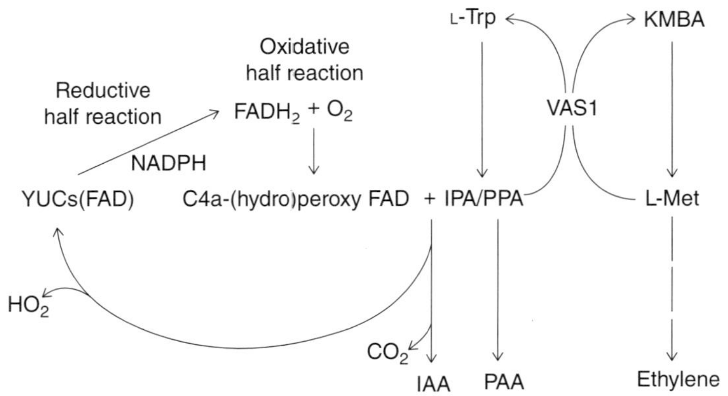

Fig. 19.1. Outline of the indole-3-pyruvic acid (IPA) pathway of production of indole-3-acetic acid (IAA), including the mechanism of the YUCCA (YUC) flavin monooxygenase (FMO) catalytic activity, and the link between auxin synthesis via the IPA pathway and ethylene biosynthesis via the pyridoxal phosphate-dependent aminotransferase (VAS1 or reversal of *sav3* phenotype 1) pathway. The flavin (FAD) cofactor of the YUC proteins is first reduced by NADH in a reductive half reaction and then reacts with molecular oxygen in an oxidative half reaction to form the short-lived compound C4a-(hydro)peroxy FAD. C4a-(hydro)peroxy FAD reacts with IPA or phenylacetic acid (PPA) to form IAA or 3-phenylpropionic acid (PAA), respectively, and regenerates the YUC(FAD) moiety. VAS1 is capable of recycling IPA to L-tryptophan (L-Trp) or using L-methionine (L-Met) as an amino acid donor to produce 2-oxo-4-methylthiobutyric acid (KMBA) in the Yang cycle, which generates ethylene, thus linking the IAA and ethylene biosynthetic pathways.

embryogenesis, perhaps in different cellular compartments, and are genetically redundant.

Constitutive expression of *YUC* genes or dominant gain-of-function mutants leads to auxin overproduction phenotypes such as long hypocotyls, cotyledon epinasty and strong apical dominance, and also to elevated auxin levels (Zhao *et al.*, 2001; Kim *et al.*, 2007a,b).

The *YUC* gene family has 11 members that are genetically redundant, because individual *yuc* mutants have no developmental defects, whereas increasingly higher order mutants have correspondingly more severe defects. For example, mutations in all four *YUC1 YUC4 YUC10* and *YUC11* genes, which have overlapping expression domains in the embryo, lead to the production of a seedling that lacks a hypophysis and hypocotyl and often has single cotyledons and lacks vasculature (Cheng *et al.*, 2007). Thus, local auxin biosynthesis by this *YUC* gene complement is essential for embryonic pattern formation and the establishment of the apical–basal axis. *YUC1*, *YUC2*, *YUC4* and *YUC6* redundantly control venation in leaves and flowers and the initiation or outgrowth of floral organs in each whorl (Cheng *et al.*, 2006); *YUC1* and *YUC4* have the greatest contribution and increasingly higher order mutants for these four genes have increasingly severe developmental defects.

YUC gene function was historically placed within the tryptamine branch of auxin biosynthesis, based on the ability of *YUC1* to convert tryptamine to *N*-hydroxyl tryptamine *in vitro* (Zhao *et al.*, 2001), but as flavin monooxygenases have broad substrate specificity (Ziegler, 1990) and the *N*-hydroxyl tryptamine intermediate has not been identified in plants, the evidence that *YUC* genes function *in planta* in the IPA pathway is now overwhelming. First, genetic evidence for TAA and YUC functioning in the same pathway is that *taa* and *yuc* mutants phenocopy each other in many major developmental processes (Won *et al.*, 2011), though due to redundancy and tissue specificity within each family, the appropriate mutant combinations must be compared. Overexpression studies combined with mutant analyses show that *YUC*-mediated auxin production is dependent on TAA function (Stepanova *et al.*, 2011) and that IPA accumulates in *yuc* mutants and is decreased in *taa* mutants, placing both sets of functions in the same pathway (Mashiguchi *et al.*, 2011; Won *et al.*, 2011). Specifically, biochemical experiments have shown that TAA can convert Trp to IPA and that YUCs can use IPA as a substrate to produce IAA (Mashiguchi *et al.*, 2011).

YUC gene families are present in all sequenced plant genomes (Schlaich, 2007) and have been characterized in many species apart from *Arabidopsis*, including rice (Yamamoto *et al.*, 2007), maize (Gallavotti, 2008) and *Petunia* (Tobeña-Santamaria *et al.*, 2002) (see Table 19.1). However, the difficulty of reverse genetic approaches in many of these species means that mutant phenotypes have only been described for *Arabidopsis* and *Petunia*, in which the *FLOOZY* gene is necessary for floral organ initiation in the outer three floral whorls and also for leaf vascularization (Tobeña-Santamaria *et al.*, 2002). The conservation of *YUC* gene function supports a predominant role for the IPA pathway as the major auxin biosynthetic route in plants, as it is to be expected that a core pathway should share components that are widely conserved across the plant kingdom. Similarly, the existence of a *TAA* co-orthologue, *VANISHING TASSEL2* (*VT2*) in maize, also suggests wide conservation of the first step in the pathway; this has been suggested to function genetically in the same pathway as *SPARSE INFLORESCENCE1*, a maize *YUC* homologue (Phillips *et al.*, 2011). Several TAA homologues have now been characterized from several species (see Table 19.1). Greater redundancy exists among the *YUC* components of the second step than among the *TAA* genes, which also affect other processes than auxin production, though the strong phenotypes of *Arabidopsis* plants mutant for either step of the IPA pathway show that this route is both essential and non-redundant with other postulated pathways.

19.4.2 Alternative biosynthetic routes

Despite the recognition of the IPA pathway as the predominant auxin synthetic route, probably in most plants, there is evidence that alternative

Table 19.1. Summary of published *TAA* or *YUC* genes from different species encoding enzymes in the indole-3-pyruvic acid (IPA) auxin biosynthesis pathway.

Species	Gene name	Reference
TAA genes encoding tryptophan aminotransferases characterized from different species:		
Arabidopsis thaliana	*SHADE AVOIDANCE3* (*SAV3*)	Tao *et al.*, 2008
	WEI8/TAA1	Stepanova *et al.*, 2008
	TIR2	Yamada *et al.*, 2009
	TAR1-4	Stepanova *et al.*, 2008
Oryza sativa	*OsTAR1*	Abu-Zaitoon *et al.*, 2012
Solanum tuberosum	*StTAR1 to 7*	Kim *et al.*, 2013
Zea mays	*VANISHING TASSEL2*	Phillips *et al.*, 2011
	ZmTAR1	Chourey *et al.*, 2010
	ZmTAR3	Bernardi *et al.*, 2012
YUC genes encoding flavin monooxygenases characterized from different species:		
Arabidopsis thaliana	*AtYUC1* to *11*	Cheng *et al.*, 2006, 2007
Fragaria × ananassa	*FaYUC1* to 2	Liu *et al.*, 2012
Hordeum vulgare	*YUC* genes 31933 and 5729	Sakata *et al.*, 2010
Oryza sativa	*OsYUC1* to *7*	Yamamoto *et al.*, 2007
	OsYUC1 to *14*	Abu-Zaitoon *et al.*, 2012
	CONSTITUTIVELY WILTED1	Woo *et al.*, 2007
Petunia hybrida	*FLOOZY*	Tobeña-Santamaria *et al.*, 2002
Pisum sativum	*PsYUC1* to 2	Tivendale *et al.*, 2010
Solanum lycopersicum	*ToFZY1* to *6*	Expósito-Rodríguez *et al.*, 2007, 2011
Solanum tuberosum	*StYUC-like1*	Roumeliotis *et al.*, 2012
	StYUC1-8	Kim *et al.*, 2013
Zea mays	*SPARSE INFLORESCENCE1*	Gallavotti *et al.*, 2008
	ZmYUC	LeClere *et al.*, 2010
	ZmYUC1	Bernardi *et al.*, 2012

routes contribute to auxin synthesis, although there is no estimate of how much auxin is produced by the IPA route compared with via these other pathways. Many open questions remain concerning these pathways, as gaps exist in either demonstrating intermediates or isolating the enzymes that are proposed to catalyse certain steps. These pathways might only function in some species, or only in tissue-specific or particular developmental contexts. Problems in elucidating these pathways have included mild or pleiotropic phenotypes for mutants in which the enzymatic steps are known, and the fact that plants have a large number of indolic intermediates that interconnect other metabolic pathways, so it is not always clear to what extent these are intermediate substrates for auxin biosynthesis. Furthermore, problems exist in extrapolating demonstrated *in vitro* enzyme activities to *in vivo* situations, and in clarifying substrate specificities and convergence points between auxin and non-auxin pathways.

19.4.3 The indole-3-acetaldoxime (IAOx) pathway

IAOx is postulated to be an intermediate in the production of IAA via indole-3-acetonitrile (IAN) (see Fig. 19.2c). Trp is converted to IAOx by the cytochrome P450 enzymes, CYP79B2 and CYP79B3, which are predicted to localize to the chloroplast (Hull *et al.*, 2000). The genes encoding these enzymes are specific to the *Brassicaeae*, and have been found only in *Arabidopsis*, *Brassica* and *Sinapis* (Bak *et al.*, 1998). Similarly, IAOx has only been found in *Arabidopsis* and *Brassica*, so this pathway is probably not widespread across the plant kingdom. One problem in characterizing the pathway is that IAOx is also a key intermediate in the production of secondary metabolites important for plant defence, such as indole glucosinolate (IG), which is produced via the activity of another cytochrome P450 enzyme, CYP83B1, and camalexin, which is produced via the activity cytochrome P450 CYP71A13; thus, it has been thought to represent a branch point for the synthesis of auxin and secondary metabolites. Furthermore, the expression of *CYP79B2* can be induced by pathogens (Hull *et al.*, 2000) and *CYP79B3* is thought to predominantly act in IG biosynthesis (Glawischnig, 2007), suggesting that auxin biosynthesis is only a secondary consequence of this metabolic defence pathway.

The *superoot1* (*sur1*) and *sur2* mutants of *Arabidopsis* that encode *C-S* lyase (Mikkelsen *et al.*, 2004) and CYP83B1 (Barlier *et al.*, 2000), respectively, within the IG synthetic pathway downstream from IAOx were initially isolated as mutants with high auxin levels, a characteristic that has been ascribed to a block in the conversion of IAOx into IGs (Boerjan *et al.*, 1995; Delarue *et al.*, 1998; Seo *et al.*, 1998; Barlier *et al.*, 2000; Morant *et al.*, 2010). However, metabolomic and transcriptional studies have since shown that the phenotype at least of *sur2* acts in a more complex way than simply affecting IAOx and auxin levels, and also affects ethylene and abscisic acid responses (Morant *et al.*, 2010), again showing IAOx to be an intermediate for several interacting pathways.

Although the levels of IAOx, IG, IAN and camalexin are significantly reduced in double knockout mutants of *CYP79B2* and *CYP79B3*, or elevated resulting from the overexpression of these two genes, IAA content was not altered in roots and its production was significantly slower (Ljung *et al.*, 2005). It has been suggested that IAN is converted to IAA by nitrilase (NIT) activity (Bartel and Fink, 1994); tobacco, which does not normally contain IAN, when transgenic for *AtNIT2*, was capable of metabolizing radiolabelled IAN to IAA and gave auxin-overproducing phenotypes when supplied with high IAN levels (Schmidt *et al.*, 1996). More recent studies have cast doubt on the substrate specificity of the *Arabidopsis* nitrilases for IAN, implicating their activity more in glucosinolate turnover or degradation (Vorwerk *et al.*, 2001). Thus, the existence of the catalytic step from IAN to IAA, or whether the IAOx pathway is a functionally relevant method of IAA production, is uncertain.

19.4.4 The indole-3-acetamide (IAM) pathway

In the IAM pathway, Trp is postulated to be converted to IAM by oxidoreduction, and IAM is then hydrolysed to IAA (Fig. 19.2a). This pathway was thought to be confined to bacteria, but IAM is widespread in all plant species studied

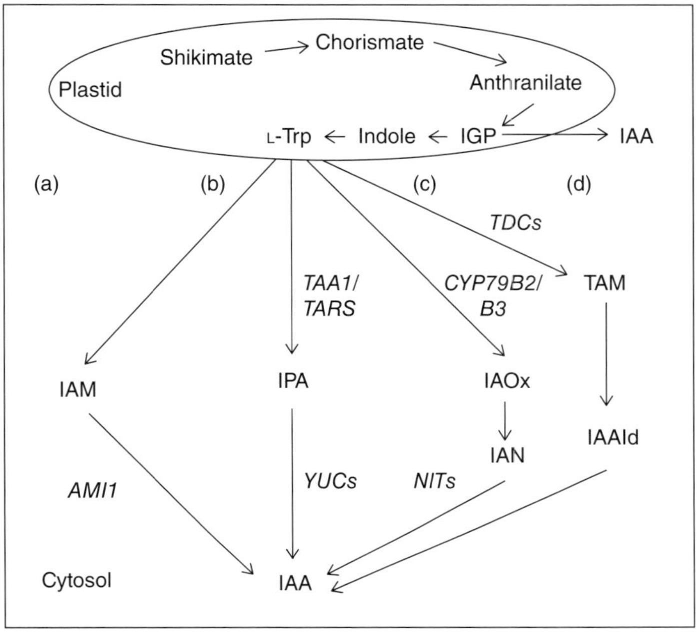

Fig. 19.2. Summary of indole-3-acetic acid (IAA) biosynthesis pathways in plants. L-Tryptophan (L-Trp) synthesis occurs in the chloroplast (top) and via the four main IAA synthetic pathways in the cytosol. The postulated Trp-independent synthesis from indole-3-glycerol phosphate (IGP) is also shown (top). Pathways are: (a), the indole-3-acetamide (IAM) pathway; (b), the indole-3-pyruvate (IPA) pathway: (c), the indole-3-acetaldoxime (IAOx) pathway; and (d), the tryptamine (TAM) pathway. The enzymes that have been characterized for the known steps of each pathway are shown in italics. Abbreviations: *AMI1*, amidase1; *CYP79B2*, cytochrome p450 monooxygenase 79B2; *CYP79B3*, cytochrome p450 monooxygenase 79B3; IAAld, indole-3-acetaldehyde; IAN, indole-3-acetonitrile; *NITs*, nitrilases; *TAA1*, TRYPTOPHAN AMINOTRANSFERASE OF ARABIDOPSIS1; *TARs*, TRYPTOPHAN AMINOTRANSFERASE RELATED1–4; *TDCs*, tryptophan decarboxylases; *YUCs*, YUCCA flavin monooxygenases.

(Mano and Nemoto, 2012). The gene for amidase1 (*AMI1*), which encodes an IAM hydrolase, exists in many species (Mano and Nemoto, 2012) and has been isolated from *Arabidopsis* (Pollmann *et al.*, 2003) and tobacco (Nemoto *et al.*, 2009), and shown to have activity *in vitro*. IAM hydrolase activity in cell-free extracts of many plants, including monocotyledons (Mano and Nemoto, 2012), the widespread presence of *AMI1* gene families throughout the plant kingdom (Mano *et al.*, 2010) and the demonstration that cell-free extracts can synthesize IAA via IAM (Pollmann *et al.*, 2009), provides strong evidence for the existence of this pathway and its high functional conservation in plants. To date though, how IAM is synthesized in plants remains unclear. It might involve a tryptophan-2-monooxygenase, the gene for which is homologous to the *aux1* gene that performs the synthesis of IAM from Trp in bacteria (Mano and Nemoto, 2012), although no such plant enzyme has yet been characterized.

19.4.5 The tryptamine (TAM) pathway

Trp decarboxylases (TDCs) can convert Trp to TAM and both *TDC* genes and TAM have been found in monocotyledons and dicotyledons (Mano and Nemoto, 2012). Despite the finding that YUC1 can metabolize tryptamine to *N*-hydroxytryptamine in *Arabidopsis in vitro* (Zhao *et al.*, 2001), this has not been confirmed *in vivo*, and YUC function has now been firmly placed in the IPA pathway. In pea, however, labelling studies have shown that Trp can be converted to IAA via TAM and indole-3-acetaldehyde (IAAld) (Fig. 19.2d; Quittenden *et al.*, 2009). Even though the level of the intermediates is low, this confirms that the TAM pathway is active in some plants, but might also be used in parallel for the production of serotonin derivatives and indole alkaloids (Quittenden *et al.*, 2009; Mano and Nemoto, 2012).

19.5 Endogenous Auxins

In addition to IAA as the primary endogenous auxin in plants, three other endogenous auxins have been identified. These are the active auxins, 4-chloroindole-3-acetic acid (4-Cl-IAA) and phenylacetic acid (PAA), and the inactive auxin, indole-3-butyric acid (IBA) (Fig. 19.3). 4-Cl-IAA is restricted to certain species, for example it is found in seeds of *Pinus sylvestris* (Ernstsen and Sandberg, 1986) and legumes including *Pisum* and *Vicia* (Pless *et al.*, 1984; Katamaya *et al.*, 1987; Magnus *et al.*, 1997), but not in *Arabidopsis* (Reinecke, 1999). 4-Cl-IAA is both more stable and more active than IAA (Marumo *et al.*, 1974; Reinecke, 1999) and in pea, it has been shown to be synthesized independently of IAA, in a parallel pathway, from 4-Cl-Trp, by the aminotransferases PsTAR1 and 2, via 4-chloroindole-3-pyruvic acid (Tivendale *et al.*, 2012).

PAA is a phenyl derivative that was thought to derive from benzylglucosinolate within the nitrilase pathway (Ludwig-Müller and Cohen, 2002), but the affinity of YUC6 for 3-phenylpropionic acid (PPA) *in vitro* and its ability to convert it to PAA (Dai *et al.*, 2013) suggests that PAA is derived in parallel to the IPA pathway (see Fig. 19.1). PAA has been shown to exist in several species (see Korasick *et al.*, 2013), but not in *Arabidopsis* (Simon and Petrášek, 2010). Apart from the postulated effects of 4-Cl-IAA in seed and fruit development, and a role for PAA in plant–soil interactions, the direct effects of these two auxins are unknown, as well as their specificities and mechanisms of recognition (Simon and Petrášek, 2010). Auxinic activity was also shown in wheat and nasturtium coleoptiles by PPA and 2-phenylproprionitrile (PPN) (Wheeler, 1980), probably because of their subsequent conversion to PAA. Similarly, other compounds shown to have auxinic effects are the IAA precursors IAN and IPA, presumably because of their conversion within plants to IAA.

One major contributor to the active auxin pool derives from the native but inactive auxin,

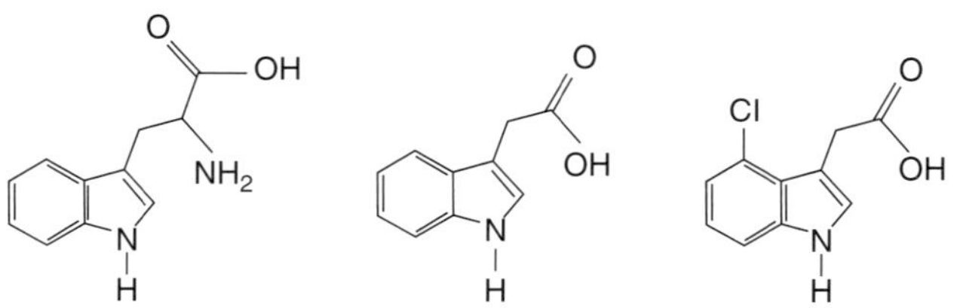

Tryptophan (Trp) | Indole-3-acetic acid (IAA) | 4-Chloro-indole-3-acetic acid (4-Chl-IAA)

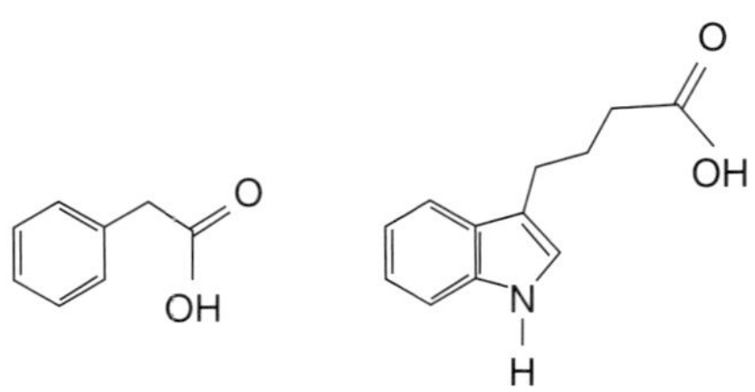

Phenylacetic acid (PAA) | Indole-3-butyric acid (IBA)

Fig. 19.3. Structural formulae of tryptophan, the precursor of auxin synthesis, and the native plant auxins. IAA, 4-Cl-IAA and PAA are active auxins and IBA is an inactive storage form of IAA that contributes IAA to the active pool following IBA synthase activity.

IBA, which comprises up to 30% of the total auxin in *Arabidopsis* seedlings (Ludwig-Müller *et al.*, 1993). It is synthesized directly from IAA, via lengthening of the IAA side chain by IBA synthase (Lüdwig-Müller and Hilgenberg, 1995) and its reconversion to IAA by β-oxidation in the peroxisome is dependent on several enzymes (Zolman *et al.*, 2008; Strader *et al.*, 2011). IBA has no auxinic activity in its own right (Simon and Petrášek, 2010), but is thought to represent a stable storage form of IAA that contributes to the active auxin pool in specific developmental or environmental contexts. For example, IBA synthesis is regulated by drought stress in maize roots (Ludwig-Müller and Hilgenberg, 1995) and by drought and salinity stress in *Arabidopsis* (Tognetti *et al.*, 2010) and during *Arabidopsis* seedling development (Strader *et al.*, 2011).

19.6 Auxin Synthesis via the IPA Pathway is Transcriptionally and Spatio-temporally Regulated

A key feature of auxin biosynthesis is that it is local. The distinct but overlapping expression profiles for each *YUC* gene within the IPA pathway in *Arabidopsis* mean that different combinations of genes contribute to different developmental process. For example, *YUC1*, *YUC2*, *YUC4* and *YUC6* predominantly control auxin biosynthesis in shoots (Cheng *et al.*, 2006), whereas *YUC3*, *YUC5*, *YUC7*, *YUC8* and *YUC9* co-regulate IAA synthesis in roots (Cheng *et al.*, 2006; Won *et al.*, 2011). The rice and tomato *YUC* gene family members are also differentially expressed (Yamamoto *et al.*, 2007; Expósito-Rodríguez *et al.*, 2011) and similarly regulate local auxin biosynthesis in a tissue-specific and temporally controlled way. Despite redundancy among the *TAA* gene family in *Arabidopsis*, these genes are responsible for auxin biosynthesis in both roots and shoots (Stepanova *et al.*, 2008; Tao *et al.*, 2008).

Local auxin depletion in *Arabidopsis yuc* mutants is shown by a decrease in auxin response maxima measured by the *DR5* reporter in the corresponding tissues where the *YUC* genes are expressed, which can be restored by the expression of auxin from the bacterial *iaaM* gene expressed in the same tissue (Cheng *et al.*, 2006). Similarly, auxin produced in other floral organs cannot substitute for the inactivation of *YUC2* and *YUC6* in stamens, showing the local nature of auxin synthesis (Cheng *et al.*, 2006). However, species-specific differences exist in *YUC* function; *Petunia*, for example, only has a single *YUC* gene, *FLOOZY* (Tobeña-Santamaria *et al.*, 2002) that has non cell-autonomous effects and, thus, auxin produced locally can be used by surrounding cells or by surrounding organs.

Evidence suggests that the *YUC* genes are transcriptionally and differentially regulated. Local auxin synthesis in the gynoecium helps to generate an apical–basal gradient that regulates development (Balanzá *et al.*, 2006) and two classes of transcription factors positively regulate *YUC* activity in this process. The STYLISH1 (STY1) transcription factor, and possibly other members of the redundant *STY* family, directly bind to the *YUC4* promoter and positively regulate its transcription (Eklund *et al.*, 2010a); similarly, the members of the NGATHGA protein family regulate *YUC2* and *YUC4* expression, possibly in concert with STY proteins (Trigueros *et al.*, 2009). Apart from in the gynoecium, the *SPOROCYTELESS* (*SPL*) gene negatively regulates *YUC2* and *YUC6* in lateral organs (Li *et al.*, 2008) and in embryogenesis, the *LEAFY COTYLEDON2* (*LEC2*) gene positively regulates *YUC1*, *YUC2*, *YUC4* and *YUC10* transcription in different contexts (Stone *et al.*, 2008; Wójcikowska *et al.*, 2013), in the case of *YUC4*, by directly binding to its promoter (Stone *et al.*, 2008). The *TERMINAL FLOWER2* (*TFL2*) gene of *Arabidopsis* is a homologue of animal *HETEROCHROMATIN PROTEIN1* (*HP1*) genes, although by analogy, it is likely that *TFL2* regulates auxin biosynthesis by affecting euchromatin genes as part of a transcriptional protein complex involving other transcription factors (Rizzardi *et al.*, 2011). *TFL2* directly regulates a subset of *YUC* genes, comprising *YUC5*, *YUC8* and *YUC9* by directly binding their promoters and transcriptionally activating their expression. Correspondingly, the *tfl2* mutant has lower rates of auxin synthesis (Rizzardi *et al.*, 2011). Methyl jasmonate treatment also transcriptionally activates the expression of both *YUC8* and *YUC9* (Hentrich *et al.*, 2013), showing a direct point of contact between auxin synthesis and jasmonic acid.

A further facet of *YUC* gene regulation is at the level of compartmentation and alternative splicing. The *YUC4* transcript exists as two splice variants that encode functional protein isoforms (Kriechbaumer *et al.*, 2012). Alternative splicing is tissue-specific, with a short YUC4 isoform present in all tissues and another isoform that is found in flowers, which possesses a C-terminal transmembrane domain that anchors it to the membrane of the endoplasmic reticulum (ER), where *YUC4* is predominantly active. This isoform is oriented so that the FAD domain can support auxin biosynthesis in the cytoplasm. The implications of this auxin compartmentation at the ER membrane are poorly understood but might relate to conjugation efficiency or to PIN5 (a cytosol-to-ER auxin transporter protein)-directed auxin transport (Kriechbaumer *et al.*, 2012).

19.7 Environmental Regulation of Auxin Synthesis

The modulation of auxin synthesis and response is one important way in which plants respond to changing environmental conditions, for example, light and temperature. The mechanisms by which plants respond to these variables has pinpointed direct regulation at the level of transcription of the *YUC* genes, at least for *Arabidopsis*, though it is unknown whether these mechanisms are present in other species. In *Arabidopsis*, in high red:far-red (R:FR) light ratios, phytochrome B (PHYB) is physically bound to the basic helix-loop-helix (bHLH) transcription factor phytochrome interacting factor7 (PIF7). However, in shade conditions, this interaction dissociates, PIF7 is rapidly dephosphorylated and binds to promoters of *YUC8* and *YUC9*, via multiple G-boxes (Li *et al.*, 2012). The expression of *YUC2* and *YUC5* is also upregulated by shade conditions (Li *et al.*, 2012) and so the rapid accumulation of auxin by *YUC* genes in the shade within 1 h is responsible for the elongated hypocotyl shade response phenotype. This response is also dependent on the upregulation of *TAA1* under shade conditions (Tao *et al.*, 2008).

A similar regulation of auxin synthesis exists in response to temperature in *Arabidopsis*, where at high temperatures (29°C), *PIF4* and *PIF5* bind to the G-box in the *YUC8* promoter to activate its expression (Hornitschek *et al.*, 2012; Sun *et al.*, 2012). PIF4 also binds to the promoters of the auxin biosynthesis genes *TAA1* and *CYP79B2* (Franklin *et al.*, 2011), and *TIR2* expression increases with higher temperature throughout *Arabidopsis* seedlings (Yamada *et al.*, 2009). Despite these studies linking increased auxin synthesis with higher temperatures, high temperature has been shown to lower endogenous auxin levels in developing *Arabidopsis* anthers by downregulating *YUC2*, *YUC6* and *TAA1*, and in barley panicles via the downregulation of two *YUC* homologous genes (Sakata *et al.*, 2010). The mechanism of this regulation is unknown, but temperature can clearly regulate auxin biosynthesis differently in various developmental and tissue-specific contexts.

ZmYUC expression was reduced in kernels of the *miniature* maize mutant, which has elevated sugar levels (LeClere *et al.*, 2010), suggesting that auxin biosynthesis in this tissue might be regulated by sugar metabolism. Further evidence linking auxin biosynthesis with sugars is that the rate of *de novo* auxin synthesis and its metabolism varies on a diurnal cycle in *Arabidopsis*, with the lowest IAA levels at the end of the dark period and highest at the end of the light period, a diurnal pattern that coincides with a minimum and maximum of soluble sugars, respectively. The lack of IAA induction with mannitol shows that the effect of soluble sugars is a metabolic one, not an osmotic one. Auxin biosynthesis was upregulated by low physiological concentrations of exogenously applied glucose (Glc), as well as several IAA precursors downstream of Trp (Sairanen *et al.*, 2012), and *YUC9* was transcriptionally upregulated by Glc in a concentration-dependent manner, suggesting that it is a central regulator of Glc-controlled auxin biosynthesis, although *YUC2* transcript levels in roots are also induced by Glc (Mishra *et al.*, 2009). In addition, Glc affected auxin homeostasis by increasing levels of catabolized auxin and conjugated auxin, probably as a result of higher auxin levels (Sairanen *et al.*, 2012). Evidence suggests that IAA induction is even dependent on the phosphorylation status of sugars and on sugar signalling as well as metabolism (Sairanen *et al.*, 2012).

The induction by Glc of IAA biosynthesis in a multiple *pif* mutant lacking the function of *PIF1*, *PIF3*, *PIF4* and *PIF5* was higher than in the wild type, and was lower in a *PIF5* overexpressing plant, suggesting that the PIF proteins are negative regulators of Glc-induced IAA biosynthesis. *PIF* gene expression is positively regulated by sucrose (Stewart *et al.*, 2011), and PIF proteins are implicated in various aspects of environmental response (Leivar and Quail, 2011). Hence, they represent one way in which, via soluble sugars, auxin biosynthesis and homeostasis are directly linked to light and photosynthesis. This comprises part of the dynamic manner in which plants can modulate their growth to changing environmental conditions by altering auxin metabolism.

19.8 Hormonal Regulation of Auxin Biosynthesis

The function of all plant hormones is dependent on the co-activity of other hormones and auxin biosynthesis is integrated into a complex matrix of hormone cross-talk interactions at a number of key steps (Chandler, 2009b; Vanstraelen and Benková, 2012). These mechanisms involve feedback and feedforward loops, and multiple points for interaction with other hormones exist at the metabolic level of IAA synthesis. As auxin and cytokinin synthesis are often co-localized in the shoot tip and roots, and the two hormones often have antagonist effects on development, a homeostatic feedback loop is one way of critically regulating the local ratio of both hormones in developmental contexts.

Increasing cytokinin levels, either exogenously or endogenously, leads to higher rates of IAA synthesis in both shoots and roots; conversely, lowering endogenous cytokinin levels reduces the rate of IAA biosynthesis (Jones *et al.*, 2010). Transcriptional profiling of cytokinin- treated roots showed that several genes encoding enzymes in the Trp biosynthetic pathway were upregulated, together with genes involved in *de novo* auxin biosynthesis, such as *YUC6* and *TAA1* from the IPA pathway and *CYP79B2*, *CYP79B3*, *NIT2* and *NIT3* from the *IAOx* pathway (Jones *et al.*, 2010). Furthermore, genes encoding IAA conjugating or deconjugating enzymes were affected, suggesting that IAA homeostasis is also fine-tuned by cytokinin. The MYB transcription factor *ATR1*, which regulates genes in several auxin biosynthesis pathways, was also positively regulated by cytokinin (Jones *et al.*, 2010; Jones and Ljung, 2011). Thus, cytokinin regulates IAA biosynthesis and homeostasis at many levels. The role of ethylene in the regulation of auxin synthesis is shown by the genes *WEI2* and *WEI7*, which encode the α and β subunits of anthranilate synthase (ASA and ASB), a rate-limiting enzyme in Trp biosynthesis, and by *TAA1*, which encodes the tryptophan aminotransferase of the IPA pathway. All of these genes, when mutated, show root-specific ethylene insensitivity (Stepanova *et al.*, 2005, 2008) and so represent points of cross-talk between ethylene response and auxin biosynthesis. Consistent with this, ethylene stimulates IAA synthesis in root tips (Růžička *et al.*, 2007; Swarup *et al.*, 2007). A second-site suppressor of the *SHADE AVOIDANCE3* (*SAV3*) gene (also known as *TAA1*), was characterized, called VAS1, whose mutation was found to increase auxin levels and, simultaneously, the level of the ethylene biosynthesis precursor 1-aminocyclopropane-1-carboxylate (Zheng *et al.*, 2013). VAS1 can use L-methionine (L-Met) as an amino acid donor to catalyse the conversion of 3-IPA to L-Trp and deaminate L-Met to 2-oxo-4-methylthiobutyric acid (KMBA) (see Fig. 19.1). As L-Met is an essential precursor of ethylene biosynthesis, its activity intimately links the auxin and ethylene biosynthetic pathways. A further mechanism of cross-talk between auxin synthesis and ethylene was revealed by L-kynurenine (Kyn), a specific inhibitor of the TAA1/TAR proteins. IAA inhibits the degradation of the ETHYLENE INSENSITIVE3 (EIN3) transcription factor by ethylene signalling as well as synthesis in the roots, thereby stabilizing its levels and creating a positive feedback mechanism to increase its own synthesis (He *et al.*, 2011). Due to its similar structure to Trp, Kyn can function as an alternative substrate of TAA/TAR proteins and, therefore, as a Trp antagonist. Kyn is a natural degradation product of Trp in animals (He *et al.*, 2011) although it is unknown whether it exists in plants; if it does, it could represent an *in vivo* mechanism to regulate auxin production via aminotransferases. Additionally, anthranilate synthase α1 (ASA1) and indole-2-glycerol phosphate synthase, two enzymes involved in the synthesis of Trp from chorismate, are responsive to methyl jasmonate (Sun *et al.*, 2009; Soto *et al.*, 2012).

Auxin feeds back on its own synthesis in several ways. The *YUC* genes do not have canonical auxin responsive elements (AuxREs) in their promoters, but the transcription of *YUC5*, *YUC8* and *YUC9* is negatively regulated by auxin, possibly via TFL2, as the binding of TFL2 to their promoters was also dependent on auxin (Rizzardi *et al.*, 2011). Feedback inhibition of auxin biosynthesis in root tips was also shown when roots were treated with the auxin transport inhibitor NPA, which blocked the basipetal transport of newly synthesized auxin away from the root tip, so allowing auxin levels to accumulate (Ljung *et al.*, 2005). It is not known more precisely at which level this mechanism operates. Several *GH3* (auxin early response) genes in many species are under transcriptional control by auxin via auxin response factors in their promoters (Mallory *et al.*, 2005; Paponov *et al.*, 2008), thus demonstrating feedback regulation of auxin homeostasis.

19.9 Conjugation Contributes to Auxin Homeostasis

The size of the pool of active free IAA in a plant is not only dependent on synthesis, but also on degradation and conjugate formation with amino acids, peptides or proteins. It is estimated that in some tissues, free IAA only comprises up to 25% of the total IAA (Ludwig-Müller, 2011), and it is thought that conjugate forms partially represent a storage pool, with reversible conjugation under relevant environment or developmental stimuli able to release free auxin. IAA can be variously conjugated to low molecular weight monosaccharides or polysaccharides, amino acids or methyl groups, or to high molecular weight peptides or proteins (Ludwig-Müller, 2011). It is thought that all endogenous auxins can be conjugated, although not enough information on the forms that are derived from IBA, PAA and 4-Cl-IAA is available to allow functional conclusions to be drawn (Ludwig-Müller, 2011).

Different plant species have different profiles of auxin conjugates, for example, monocotyledons have predominantly ester conjugates and dicotyledons, predominantly amide conjugates (Bajguz and Piotrowska, 2009). Ester conjugates of IAA with sugar groups such as glucose and *myo*-inositol are found in monocotyledons and dicotyledons (Ludwig-Müller, 2011), and in the germination of Scots pine (*Pinus sylvestris*) seedlings, the hydrolysis of IAA esters was shown to be functionally important in releasing free IAA for root elongation (Ljung *et al.*, 2001b); the reversible nature of this conjugation, at least in maize (Jakubowska and Kowalczyk, 2004), readily represents a mechanism for contributing to the free IAA pool when necessary. In *Arabidopsis*, the conversion of IAA to its inactive methyl ester form by IAA carboxyl methyltransferase1 (IAMT1; Qin *et al.*, 2005) and its esterase hydrolysis (Yang *et al.*, 2008) occurs (Fig. 19.4). A further example of an IAA-protein conjugate thought to be a storage form for auxin is the attachment of IAA to the PvIAP1 protein (the major IAA-modified protein in bean seeds) (Walz *et al.*, 2002). The amide conjugation of IAA with a wide variety of amino acids is widespread in plants, and the amino acids involved include aspartate (Asp), glutamate (Glu), alanine (Ala), glycine (Gly), valine (Val) and leucine (Leu) (Bajguz and Piotrowska, 2009) (Fig. 19.4). Some of these conjugates, such as IAA-Ala and IAA-Leu, can be readily hydrolysed by auxin amidohydrolases to free IAA, and so probably represent storage IAA forms that are functionally relevant, for example, during germination (Rampey *et al.*, 2004). In contrast, the formation of IAA-Asp and IAA-Glu irreversibly leads to the oxidation and catabolism of IAA (Fig. 19.4). The IAA-Trp conjugate appears to be an auxin antagonist and to inhibit auxin responses (Staswick, 2009). IAA-amide conjugates are formed by some members of the GH3 family of synthases (Staswick *et al.*, 2005) (Fig. 19.4). Genes encoding both the GH3 auxin amide synthetases and auxin amidohydrolases have varying specificities and are known to be spatially and temporally regulated in plants; they are, therefore, thought to fine-tune auxin homeostasis in response to environmental and developmental stimuli (Ludwig-Müller, 2011). Ongoing studies suggest that different species and tissues have different spectra of protein moieties attached to IAA (Ludwig-Müller, 2011).

19.10 The Evolution of Auxin Synthesis in Plants

Auxin is present in all higher plants in the green plant lineage and has been found in unicellular

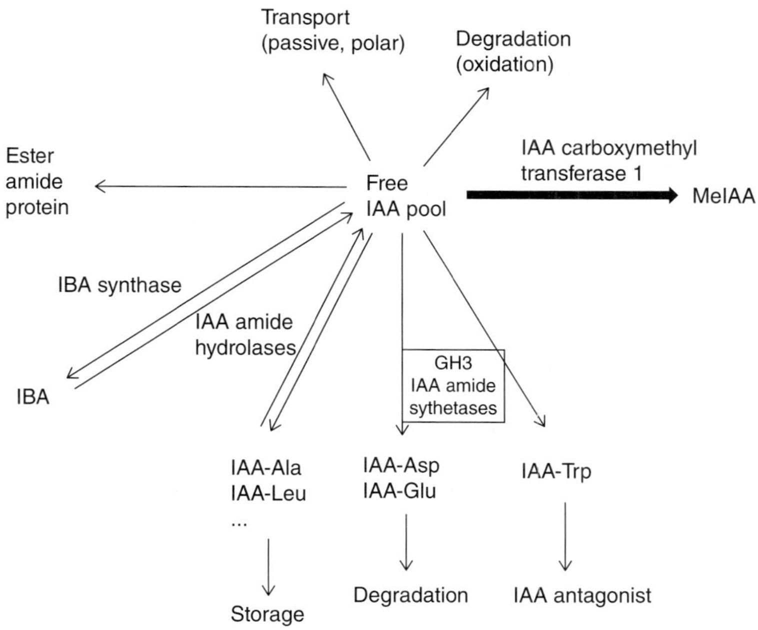

Fig. 19.4. Scheme of factors affecting the free indole-3-acetic acid (IAA) pool in plants, including conjugation, interconversion, degradation and transport. Abbreviations: Ala, alanine; Asp, aspartate; GH3, auxin early response protein; Glu, glutamate; IBA, indole-5-butyric acid; Leu, leucine; MeIAA, methyl indole-3-acetic acid; Trp, tryptophan.

and multicellular algae (Cooke *et al.*, 2002), though there is debate as to whether the low auxin concentrations in algae have any physiological relevance (Lau *et al.*, 2009). It has also been estimated that 80% of rhizosphere bacteria are capable of synthesizing auxin (Spaepen and Vanderleyden, 2011). Bacteria synthesize auxin via at least five different pathways. The main route is a two-step pathway in which tryptophan monooxygenase, encoded by *iaaM*, converts Trp to IAM, which is then hydrolysed by IAM hydrolase, the product of the *iaaH* gene (Spaepen and Vanderleyden, 2011). Also important is a three-step pathway in which Trp is converted to IPA, which is then converted in a second, rate-limiting step to (IAAld), which is finally oxidized to IAA (Spaepen and Vanderleyden, 2011). Thus, there are similarities between plant and bacterial pathways in terms of shared intermediates and multiple potential synthesis pathways.

Many studies have addressed the evolution of auxin synthesis, aided by genome-sequencing projects that facilitate *in silico* analyses of genes encoding biosynthetic enzymes.

For example, the class of AMI1 proteins, which hydrolyse IAM to IAA, are widely conserved across the whole plant kingdom (Lehmann *et al.*, 2010), suggesting that the IAM pathway might be conserved from bacteria across algae and higher plants. Further phylogenetic compilations of auxin biosynthesis genes will aid interpretations of which synthesis pathways have been conserved and which have been more prone to adaptation, though it should be noted that *in silico* searches for homologous genes in plants are often uninformative due to presumed evolutionary divergence of the plant enzymes compared with those of lower organisms.

Given the importance of the IPA pathway in higher plants, the *YUC* genes have not only been found in all higher plants for which genome sequence data are available, but also in the moss *Physcomitrella patens* and the lycophyte *Selaginella moellendorffii* (Finet and Jaillais, 2012). The conservation of this ancient gene family is also reflected in its regulators such as the *SHI* genes, which are also found in *Physcomitrella* (Eklund *et al.*, 2010b). However, there is evidence that auxin in the apices of liverworts and charophytes is produced via a Trp-independent pathway, and that IAA levels are regulated by the balance between synthesis and degradation (Cooke *et al.*, 2002). Although conjugation mechanisms exist in moss, it appears that the evolution of land plants has been accompanied by a finer control of auxin levels by conjugation mechanisms, which are related to specific developmental stages such as embryogenesis or vasculature development; the conjugation and hydrolysis of auxin conjugates has also probably evolved several times independently (Ludwig-Müller, 2011).

19.11 Conclusions

The majority of the auxin in higher plants is synthesized via a highly conserved two-step pathway that metabolizes Trp and uses IPA as an intermediate. The enzymes that regulate both of these steps are each encoded in many different species by multi-gene families that exhibit genetic redundancy; further, the differential spatial and temporal regulation of these genes ensures that auxin is locally synthesized rapidly in appropriate tissues and at appropriate developmental stages or in response to environmental stimuli. The sub-cellular compartmentation of auxin synthesis in *Arabidopsis* is effected by alternative splicing of *YUC4*, and auxin biosynthesis is dependent on other hormones, specifically ethylene and cytokinin. The IPA pathway integrates auxin synthesis to modulate plant development in response to environmental conditions, as aminotransferases act as a point of cross-talk between ethylene response and auxin synthesis. More is increasingly being understood on the transcription regulation of *YUC* genes by the environment, so that auxin biosynthesis is finely regulated on many levels, which include feedback regulation by auxin itself. Further knowledge of the extent of conservation or adaptation of the IPA pathway is currently being obtained by evolutionary developmental approaches in diverse species. The analysis of *YUC* and *TAA* gene expression and regulation in these species, together with more complex genetic interaction studies in *Arabidopsis*, will elucidate the gene transcriptional networks that fine-tune auxin biosynthesis.

In addition to the IPA pathway, alternative routes of auxin synthesis from Trp are postulated to exist, although these additional pathways either function in only some species, such as the IAOx pathway, which is restricted to the *Brassicaceae*, or probably play a minor role in total auxin production by having a more important role in the production of other secondary metabolites. Open questions are the biological relevance of multiple auxin synthesis pathways in different species and how these are integrated; the confirmation of their function is dependent on the increased resolution of and ability to detect biochemical intermediates, as well as the characterization of the enzymes that are hypothesized to catalyse some pathway steps, combined with further mutant and genetic analyses in species such as *Arabidopsis*.

The expansion of auxinology into a comparative evolutionary molecular biology approach is also necessary to assess the diversity of IAA biosynthetic pathways and intermediates in the plant kingdom and to identify the major conserved components and those that have evolved more freely.

Acknowledgement

This work was initiated with funding from the Deutsche Forschungsgemeinschaft via SFB572 at the University of Cologne.

References

Abel, S. and Theologis, A. (2010) Odyssey of auxin. *Cold Spring Harbor Perspectives in Biology* 2(10): a004572.

Abu-Zaitoon, Y.M., Bennett, K., Normanly, J. and Nonhebel, H.M. (2012) A large increase in IAA during development of rice grains correlates with the expression of tryptophan aminotransferase *OsTAR1* and a grain-specific *YUCCA*. *Physiologia Plantarum* 146, 487–499.

Bajguz, A. and Piotrowska, A. (2009) Conjugates of auxin and cytokinin. *Phytochemistry* 70, 957–969.

Bak, S., Nielsen, H.L. and Halkier, B.A. (1998) The presence of CYP79 homologues in glucosinolate-producing plants shows evolutionary conservation of the enzymes in the conversion of amino acid to aldoxime in the biosynthesis of cyanogenic glucosides and glucosides. *Plant Molecular Biology* 38, 725–734.

Balanzá, V., Navarrete, M., Triguerso, M. and Ferrándiz, C. (2006) Patterning the female side of *Arabidopsis*: the importance of hormones. *Journal of Experimental Botany* 57, 3457–3469.

Barkawi, L.S., Tam, Y.-Y., Tillman, J.A., Normanly, J. and Cohen, J.D. (2010) A high-throughput method for the quantitative analysis of auxins. *Nature Protocols* 5, 1609–1618.

Barlier, I., Kowalczyk, M., Marchant, A., Ljung, K., Bhalerao, R., Bennett, M., Sandberg, G. and Bellini, C. (2000) The *SUR2* gene of *Arabidopsis thaliana* encodes the cytochrome P450 CYP83B1, a modulator of auxin homeostasis. *Proceedings of the National Academy of Sciences of the United States of America* 97, 14819–14824.

Bartel, B. and Fink, G.R. (1994) Differential regulation of an auxin-producing nitrilase gene family in *Arabidopsis thaliana*. *Proceedings of the National Academy of Sciences of the United States of America* 91, 6649–6653.

Bernardi, J., Lanubile, A., Li, Q.-B., Kumar, D., Kladnik, A., Cook, S.D., Ross, J.J., Marocco, A. and Chourey, P.S. (2012) Impaired auxin biosynthesis in the *defective endosperm18* mutant is due to mutational loss of expression in the *ZmYuc1* gene encoding endosperm-specific YUCCA1 protein in maize. *Plant Physiology* 160, 1318–1328.

Boerjan, W., Cervera, M.-T., Delarue, M., Beeckman, T., Dewitte, W., Bellini, C., Caboche, M., Van Onckelen, H., Van Montagu, M. and Inzé, D. (1995) *superroot*, a recessive mutation in *Arabidopsis*, confers auxin overproduction. *The Plant Cell* 7, 1405–1419.

Chandler, J.W. (2009a) Local auxin production: a small contribution to a big field. *BioEssays* 31, 60–70.

Chandler, J.W. (2009b) Auxin as compère in plant hormone crosstalk. *Planta* 231, 1–12.

Cheng, Y., Dai, X. and Zhao, Y. (2006) Auxin biosynthesis by the YUCCA flavin monooxygenases controls the formation of floral organs and vascular tissues in *Arabidopsis*. *Genes and Development* 20, 1790–1799.

Cheng, Y., Dai, X. and Zhao, Y. (2007) Auxin synthesized by the YUCCA flavin monooxygenases is essential for embryogenesis and leaf formation in *Arabidopsis*. *The Plant Cell* 19, 2430–2439.

Chourey, P.S., Li, Q.-B. and Kumar, D. (2010) Sugar–hormone cross-talk in seed development: two redundant pathways of IAA biosynthesis are regulated differentially in the invertase-deficient *miniature* (*mn1*) seed mutant in maize. *Molecular Plant* 3, 1026–1036.

Cooke, T.J., Poli, D.B., Sztein, A.E. and Cohen, J.D. (2002) Evolutionary patterns in auxin action. *Plant Molecular Biology* 49, 319–338.

Dai, X., Mashiguchi, K., Chen, Q., Kasahara, H., Kamiya, Y., Ojha, S., DuBois, J., Ballou, D. and Zhao, Z. (2013) The biochemical mechanisms of auxin biosynthesis by an *Arabidopsis* YUCCA flavin-containing monooxygenase. *The Journal of Biological Chemistry* 288, 1448–1457.

Darwin, C. (1880) *The Power of Movement in Plants*. John Murray, London.

Delarue, M., Prinsen, E., Van Onckelen, H., Caboche, M. and Bellini, C. (1998) *Sur2* mutations of *Arabidopsis thaliana* define a new locus involved in the control of auxin homeostasis. *The Plant Journal* 14, 603–611.

Dubrovsky, J.G., Sauer, M., Napsucialy-Mendivil, S., Ivanchenko, M.G., Friml, J., Shixhkova, S., Celenza, J. and Benková, E. (2008) Auxin acts as a local morphogenetic trigger to specify lateral root founder cells. *Proceedings of the National Academy of Sciences of the United States of America* 105, 8790–8794.

Ehlert, B., Schöttler, M.A., Tischendorf, G., Ludwig-Müller, J. and Bock, R. (2008) The paramutated *SULFUREA* locus of tomato is involved in auxin biosynthesis. *Journal of Experimental Botany* 59, 3635–3647.

Eklund, D.M., Ståldal, V., Valsecchi, I., Cierlik, I., Eriksson, C., Hiratsu, K., Ohme-Takagi, M., Sunström, J.F., Thelander, M., Ezcurra, I. and Sundberg, E. (2010a) The *Arabidopsis thaliana* STYLISH protein acts as a transcriptional activator regulating auxin biosynthesis. *The Plant Cell* 22, 349–363.

Eklund, D.M., Thelander, M., Landberg, K., Ståldal, V., Nilsson, A., Johansson, M., Valsecchi, I., Pederson, E.R.A., Kowalczyk, M., Ljung, K., Ronne, H. and Sundberg, E. (2010b) Homologues of the *Arabidopsis thaliana SHI/STY/LRP1* genes control auxin biosynthesis and affect growth and development in the moss *Physcomitrella patens*. *Development* 137, 1275–1284.

Ellis, C.M., Nagpal, P., Young, J.C., Hagen, G., Guilfoyle, T.J. and Reed J. (2005) *AUXIN RESPONSE FACTOR1* and *AUXIN RESPONSE FACTOR2* regulate senescence and floral organ abscission in *Arabidopsis thaliana*. *Development* 132, 4563–4574.

Ernstsen, A. and Sandberg, G. (1986) Identification of 4-chloroindole-3-acetic acid and indole-3-aldehyde in seeds of *Pinus sylvestris*. *Physiologia Plantarum* 68, 511–518.

Expósito-Rodríguez, M., Borges, A.A., Borges-Pérez, M., Hernandez, M and Pérez, J.A. (2007) Cloning and biochemical characterization of *ToFZY*, a tomato gene encoding a flavin monooxygenase involved in a tryptophan-dependent auxin biosynthesis pathway. *Journal of Plant Growth Regulation* 26, 329–340.

Expósito-Rodríguez M., Borges, A.A., Borges-Pérez, A. and Pérez, J.A (2011) Gene structure and spatio-temporal expression profile of tomato genes encoding YUCCA-like flavin monooxygenases: the *ToFZY* gene family. *Plant Physiology and Biochemistry* 49, 782–791.

Finet, C. and Jaillais, Y. (2012) AUXOLOGY: when auxin meets plant evo-devo. *Developmental Biology* 369, 19–31.

Franklin, K.A., Lee, S.H., Patel, D., Kumar, S.V., Spartz, A.K., Gu, C., Ye, S., Yu, P., Breen, G., Cohen, J.D., Wigge, P.A. and Gray, W.M. (2011) PHYTOCHROME-INTERACTING FACTOR 4 (PIF4) regulates auxin biosynthesis at high temperatures. *Proceedings of the National Academy of Sciences of the United States of America* 108, 20231–20235.

Friml, J., Vieten, A., Sauer, M., Weijers, D., Schwarz, H., Hamann, T., Offringa, R. and Jürgens, G. (2003) Efflux-dependent auxin gradients establish the apical–basal axis of *Arabidopsis*. *Nature* 426, 147–153.

Gallavotti, A., Barazesh, S., Malcomber, S., Hall, D., Jackson, D., Schmidt, R. and McSteen, P. (2008) *sparse inflorescence1* encodes a monocot-specific *YUCCA*-like gene required for vegetative and reproductive development in maize. *Proceedings of the National Academy of Sciences of the United States of America* 105, 15196–15201.

Glawischnig, R. (2007) Camalexin. *Phytochemistry* 68, 401–406.

Haagen-Smit, A.J., Dandeliker, W.B., Wittwer, S.H. and Murneek, A.E. (1946) Isolation of indole-3-acetic acid from immature corn kernels. *American Journal of Botany* 33, 118–120.

He, W., Brumos, J., Li, H., Ji, Y., Ke, M., Gong, M., Zeng, Q., Li, W., Zhang, X., An, F., Wen, X., Li, P. *et al*. (2011) A small-molecule screen identifies L-kynurenine as a competitive inhibitor of TAA1/TAR activity in ethylene-directed auxin biosynthesis and root growth in *Arabidopsis*. *The Plant Cell* 23, 3944–3960.

Hentrich, M., Böttcher, C., Düchting, P., Cheng, Y., Zaho, Y., Berkowitz, O., Masle, J., Medina, J. and Pollmann, S. (2013) The jasmonic acid signaling pathway is linked to auxin homeostasis through the modulation of *YUCCA8* and *YUCCA9* gene expression. *The Plant Journal* 74, 626–637.

Hornitschek, P., Kohnen, M.V., Lorrain, S., Rougemont, J., Ljung, K., López-Vidriero, I., Franco-Zorilla, J.M., Solano, R., Trevisan, M., Pradervand, S., Xenarios, I. and Fankhauser, C. (2012) Phytochrome interacting factors 4 and 5 control seedling growth in changing light conditions by directly controlling auxin signaling. *The Plant Journal* 71:699–711.

Hull, A.K., Vij, R. and Celenza, J.L. (2000) *Arabidopsis* cytochrome P450s that catalyze the first step of tryptophan-dependent indole-3-acetic acid biosynthesis. *Proceedings of the National Academy of Science U.S.A.* 97, 2380–2384.

Ikeda, Y., Men, S., Fischer, U., Stepanova, A.N., Alonso, J.M., Ljung, K. and Grebe, M. (2009) Local auxin biosynthesis modulates gradient-directed planar polarity in *Arabidopsis*. *Nature Cell Biology* 11, 731–738.

Jakubowska, A. and Kowalczyk, S. (2004) The auxin conjugate 1-*O*-indole-3-acetic acid β-D-glucose is synthesized in immature legume seeds by IAGlc synthase and may be used for modification of some high molecular weight compounds. *Journal of Experimental Botany* 55, 791–801.

Jones, B. and Ljung, K. (2011) Auxin and cytokinin regulate each other´s levels via a metabolic feedback loop. *Plant Signaling and Behavior* 66, 901–904.

Jones, B., Gunnerås, S.A., Peterson, S.V., Tarkowski, P., Graham, N., May, S., Dolezal, K., Sandberg, G. and Ljung, K. (2010) Cytokinin regulation of auxin synthesis in *Arabidopsis* involves a homeostatic feedback loop regulated via auxin and cytokinin signal transduction. *The Plant Cell* 22, 2956–2969.

Katayama, M., Thiruvikraman, S.V. and Marumo, S. (1987) Identification of 4-chloroindole-3-acetic acid and its methyl ester in immature seeds of *Vicia amurensis* (the tribe *Vicieae*), and their absence from three species of *Phaseoleae*. *Plant and Cell Physiology* 28, 383–386.

Kim, J.I., Murphy, A.S., Baek, D., Lee, S.-W., Yun, D.-J., Bressan, R.A. and Narasimhan, M.L. (2007a) *YUCCA6* over-expression demonstrates auxin function in delaying leaf senescence in *Arabidopsis thaliana*. *Journal of Experimental Botany* 62, 3981–3992.

Kim, J.I., Sharkhuu, A., Jin, J.B., Li, P., Jeong, J.C., Baek, D., Lee, S.Y., Blakeslee, J.J., Murphy, A., Bohnert, H.J., Hasegawa, P.M., Yun, D.-J., and Bressan, R.A. (2007b) *yucca6*, a dominant mutation in *Arabidopsis*, affects auxin accumulation and auxin-related phenotypes. *Plant Physiology* 145, 722–735.

Kim, J.I., Baek, D., Park, H.C., Chun, H.J., Oh, D.-H., Lee, M.K., Cha, J.-Y., Kim, W.-Y., Kim, M.C., Chung, W.S., Bohnert, H.J., Lee, S.Y. *et al.* (2013) Overexpression of *Arabidopsis YUCCA6* in potato results in high-auxin development phenotypes and enhanced resistance to water deficit. *Molecular Plant* 6, 337–349.

Korasick, D.A., Enders, T.A. and Strader, L.C. (2013) Auxin biosynthesis and storage forms. *Journal of Experimental Botany* 9, 2541–2555.

Kriechbaumer, V., Park, W.J., Gierl, A. and Glawischnig, E. (2006) Auxin biosynthesis in maize. *Plant Biology* 8, 334–339.

Kriechbaumer, V., Wang, P., Hawes, C. and Abell, B.M. (2012) Alternative splicing of the auxin biosynthesis gene *YUCCA4* determines its subcellular compartmentation. *The Plant Journal* 70, 292–302.

Lau, S., Shao, N., Bock, R., Jürgens, G. and de Smet, I. (2009) Auxin signaling in algal lineages: fact or myth? *Trends in Plant Science* 14, 182–188.

LeClere, S., Schmelz, E.A. and Chourey, P.S. (2010) Sugar levels regulate tryptophan-dependent auxin biosynthesis in developing maize kernels. *Plant Physiology* 153, 306–318.

Lehmann, T., Hoffmann, M., Hentrich, M. and Pollmann, S. (2010) Indole-3-acetamide-dependent auxin biosynthesis: a widely distributed way of indole-3-acetic acid production? *European Journal of Cell Biology* 89, 895–905.

Leivar, P. and Quail, P.H. (2011) PIFs: pivotal components in a cellular signaling hub. *Trends in Plant Science* 16, 19–28.

Leyser, O. (2005) The fall and rise of apical dominance. *Current Opinion in Genetics and Development* 15, 468–471.

Li, L., Ljung, K., Breton, G., Schmitz, R.J., Pruneda-Paz, J., Cowing-Zitron, C., Cole, B.J., Ivans, L.J., Pedmale, U.V., Jung, H.S., Ecker, J.R., Kay, S.A. and Chory, J. (2012) Linking photoreceptor excitation to changes in plant architecture. *Genes and Development* 26, 785–790.

Li, L.-C., Qin, G.-J., Tsuge, T., Hou, X.-H., Ding, M.-Y., Aoyama, T., Oka, A., Chen, Z., Gu, H., Zhao, Y. and Qu, L.-J. (2008) *SPOROCYTELESS* modulates *YUCCA* expression to regulate the development of lateral organs in *Arabidopsis*. *New Phytologist* 179, 751–764.

Liu, H., Ying, Y.-Y., Zhang, L., Gao, Q.-H., Li, J., Zhang, Z., Fang, J.-G. and Duan, K. (2012) Isolation and characterization of two YUCCA flavin monooxygenase genes from cultivated strawberry (*Fragaria* × *ananassa* Duch.) *Plant Cell Reports* 31, 1425–1435.

Ljung, K. (2013) Auxin metabolism and homeostasis during plant development. *Development* 140, 943–950.

Ljung, K., Bhalerao, R.P. and Sandberg, G. (2001a) Sites and homeostatic control of auxin biosynthesis in *Arabidopsis* during vegetative growth. *The Plant Journal* 28, 465–473.

Ljung, K., Östin, A., Lioussanne, L. and Sandberg, G. (2001b) Developmental regulation of indole-3-acetic acid turnover in Scots pine seedlings. *Plant Physiology* 125, 464–475.

Ljung, K., Hull, A.K., Celenza, J., Yamada, M., Estelle, M., Normanly, J. and Sandberg, G. (2005) Sites and regulation of auxin biosynthesis in *Arabidopsis* roots. *The Plant Cell* 17, 1090–1104.

Ludwig-Müller, J. (2011) Auxin conjugates: their role for plant development and in the evolution of land plants. *Journal of Experimental Botany* 62, 1757–1773.

Ludwig-Müller, J. and Cohen, J.D. (2002) Identification and quantification of three active auxins in different tissues of *Tropaeolum majus*. *Physiologia Plantarum* 115, 320–329.

Ludwig-Müller, J. and Hilgenberg, W. (1995) Characterization and partial purification of indole-3-butyric acid synthetase from maize (*Zea mays*). *Physiologia Plantarum* 94, 651–660.

Ludwig-Müller, J., Sass, S., Sutter, E.G., Wodner, E. and Epstein, E. (1993) Indole-3-butyric acid in *Arabidopsis thaliana*. I. Identification and quantification. *Plant Growth Regulation* 13, 179–187.

Maeda, H. and Dudareva, N. (2006) The shikimate pathway and aromatic amino acid biosynthesis in plants. *Annual Review of Plant Biology* 63, 73–105.

Magnus, V., Ozga, J.A., Reinecke, D.M., Pierson, G.L., Larue, T.A., Cohen, J.D. and Brenner, M.L. (1997) 4-Chloroindole-3-acetic and indole-3-acetic acids in *Pisum sativum*. *Phytochemistry* 46, 675–681.

Mallory, A.C., Bartel, D.P. and Bartel, B. (2005) MicroRNA-directed regulation of *Arabidopsis AUXIN RESPONSE FACTOR17* is essential for proper development and modulates expression of early auxin responsive genes. *The Plant Cell* 17, 1360–1375.

Mano, Y. and Nemoto, K. (2012) The pathway of auxin biosynthesis in plants. *Journal of Experimental Botany* 63, 285–372.

Mano, Y., Nemoto, K., Suzuki, M., Seki, H., Fujii, I. and Muranaka, T. (2010) The *AMI* gene family: indole-3-acetamide hydrolase functions in auxin biosynthesis in plants. *Journal of Experimental Botany* 61, 25–32.

Marumo, S., Hattori, A. and Yamamoto, A. (1974) Biological activity of 4-chloroindolyl-3-acetic acid. In: Sumiki, Y. (ed.) *Plant Growth Substances 1973: Proceedings 8th International Conference on Plant Growth Substances*. Publishing Co., Tokyo, pp. 419–428.

Mashiguchi, K., Tanaka, K., Sakai, T., Sugawara, S., Kawaide, H., Natsume, M., Hanada, A., Yaeno, T., Shirasu, K., Yao, H., McSteen, P., Zhao, Y. *et al.* (2011) The main auxin biosynthesis pathway in *Arabidopsis*. *Proceedings of the National Academy of Sciences of the United States of America* 108, 18512–18517.

Mikkelsen, M.D., Naur, P. and Halkier, B.A. (2004) *Arabidopsis* mutants in the *C-S* lyase glucosinolate biosynthesis establish a critical role for indole-3-acetic acetaldoxime in auxin homeostasis. *The Plant Journal* 37, 770–777.

Mishra, B.S., Singh, M., Aggrasael, P. and Laxmi, A. (2009) Glucose and auxin signaling interaction in controlling *Arabidopsis thaliana* seedlings root growth and development *PLoS ONE* 4(2): e4502.

Morant, M., Ekstrøm, C., Ulvskov, P., Kristensen, C., Rudemo, M., Olsen, C.E., Hansen, J., Jørgensen, K., Jørgensen, B., Møller, B.L. and Bak, S. (2010) Metabolomic, transcriptional, hormonal, and signaling cross-talk in *Superroot2*. *Molecular Plant* 3, 192–211.

Nemoto, K., Hara, M., Suzuki, M., Seki, H., Muranaka, T. and Mano, Y. (2009) The *AMI* gene functions in cell division of tobacco BY-2 cells in the presence of indole-3-acetamide. *FEBS Letters* 583, 487–492.

Normanly, J. (2010) Approaching cellular and molecular resolution of auxin biosynthesis and metabolism. *Cold Spring Harbor Perspectives in Biology* 2(1): a001594.

Normanly J., Cohen J.D. and Fink G.R. (1993) *Arabidopsis thaliana* auxotrophs reveal a tryptophan-independent biosynthetic pathway for indole-3-acetic acid. *Proceedings of the National Academy of Sciences of the United States of America* 90, 10355–10359.

Östin, A., Ilić, N. and Cohen, J.D. (1999) An *in vitro* system from maize seedlings for tryptophan-independent indole-3-acetic acid biosynthesis. *Plant Physiology* 119, 173–178.

Ouyang, J., Shap, X. and Li, J. (2000) Indole-3-glycerol phosphate, a branchpoint of indole-3-acetic acid biosynthesis from the tryptophan biosynthetic pathway in *Arabidopsis thaliana*. *The Plant Journal* 24, 327–333.

Pagnussat, G.C., Alandete-Saez, M., Bowman, J.L. and Sundaresan, V. (2009) Auxin-dependent patterning and gamete specification in the *Arabidopsis* female gametophyte. *Science* 324, 1684–1689.

Paponov, I.A., Paponov, M., Teale, W., Menges, M., Chakrabortee, S., Murray, J.A.H. and Palme, K. (2008) Comprehensive transcriptome analysis of auxin responses in *Arabidopsis*. *Molecular Plant* 1, 321–337.

Petersson, S.V., Johansson, A.I., Kowalczyk, M., Makovychuk, A., Wang, J.Y., Moritz, T., Grebe, M., Benfey, P.N., Sandberg, G. and Ljung, K. (2009) An auxin gradient and maximum in the *Arabidopsis* root apex shown by high-resolution cell-specific analysis of IAA distribution and synthesis. *The Plant Cell* 21, 1659–1668.

Phillips, K.A., Skirpan, A.L., Liu, X., Christensen, A., Slewinski, T.L., Hudson, C., Barazesh, S., Cohen, J.D., Malcomber, S. and McSteen, P. (2011) *vanishing tassel2* encodes a grass-specific tryptophan aminotransferase required for vegetative and reproductive development in maize. *The Plant Cell* 23, 550–566.

Pless, T., Boettger, M., Hedden, P. and Graebe, J. (1984) Occurrence of 4-Cl-indoleacetic acid in broad beans and correlation of its levels with seed development. *Plant Physiology* 74, 320–233.

Pollmann, S., Neu, D. and Weiler, E.W. (2003) Molecular cloning and characterization of an amidase from *Arabidopsis thaliana* capable of converting indole-3-acetamide into the plant growth hormone, indole-3-acetic acid. *Phytochemistry* 62, 293–300.

Pollmann, S., Düchting, P. and Weiler, E.W. (2009) Tryptophan-dependent indole-3-acetic acid biosynthesis by 'IAA-synthase' proceeds via indole-2-acetamide. *Phytochemistry* 70, 523–531.

Pudelski, B., Kraus, S., Soll, J. and Philippar, K. (2010) The plant PRAT proteins – preprotein and amino acid transport in mitochondria and chloroplasts. *Plant Biology* 12, 42–55.

Qin, G., Gu, H., Zhao, A., Ma, Z., Shi, G., Yang, A., Pichersky, E., Chen, H., Liu, M. and Chen, Z. (2005) Regulation of *Arabidopsis* leaf development by an indole-3-acetic acid carboxyl methyltransferase in *Arabidopsis*. *The Plant Cell* 17, 2963–2704.

Quittenden, L.J., Davies, N.W., Smith, J.A., Molesworth, P.P., Tivendale, N.D. and Ross, J.J. (2009) Auxin biosynthesis in pea: characterization of the tryptamine pathway. *Plant Physiology* 151, 1130–1138.

Rampey, R.A., LeClere, S., Kowalczyk, M., Ljung, K., Sandberg, G. and Bartel, B. (2004) A family of auxin-conjugate hydrolases that contributes to free indole-3-acetic acid levels during *Arabidopsis* germination. *Plant Physiology* 135, 978–988.

Reinecke, D.M. (1999) 4-Chloroindole-3-acetic acid and plant growth. *Plant Growth Regulation* 27, 3–13.

Reinhardt, D., Pesce, E.R., Stieger, P., Mandel, T., Baltenspergere, K., Bennett, M., Traas, J., Friml, J. and Kuhlemeier, C. (2003) Regulation of phyllotaxis by polar auxin transport. *Nature* 426, 255–260.

Rizzardi, K., Landberg, K., Nilsson, L., Ljung, K. and Sundås-Larsson, A. (2011) *TFL2/LHP1* is involved in auxin biosynthesis through positive regulation of *YUCCA* genes. *The Plant Journal* 65, 897–906.

Rosquete, M.R., Barbez, E. and Kleine-Vehn, J. (2011) Cellular auxin homeostasis: gatekeeping is housekeeping. *Molecular Plant* 5, 772–786.

Roumeliotis, E., Kloosterman, B., Oortwijn, M., Kohlen, W., Bouwmeester, H.J., Visser, R.G.F. and Bachem, C.W.B. (2012) The effects of auxin and strigolactones on tuber initiation and stolon architecture in potato. *Journal of Experimental Botany* 63, 4539–4547.

Růžička, K., Ljung, K., Vanneste, S., Podhorská, R., Beeckman, T., Friml, J. and Benková, E. (2007) Ethylene regulates root growth through effects on auxin biosynthesis and transport-dependent auxin distribution. *The Plant Cell* 19, 2197–2212.

Sairanen, I., Novák, O., Pencik, A., Ikeda, Y., Jones, B., Sandberg, G. and Ljung, K. (2012) Soluble carbohydrates regulate auxin biosynthesis via PIF proteins in *Arabidopsis*. *The Plant Cell* 24, 4907–4916.

Sakata, T., Oshino, T., Miura, S., Tomabechi, M., Tsunaga, Y., Higashitani, N., Miyazawa, Y., Takahashi, H., Watanabe, M. and Higashitani, A. (2010) Auxins reverse plant male sterility caused by high temperatures. *Proceedings of the National Academy of Sciences of the United States of America* 107, 8569–8574.

Schlaich, N.L. (2007) Flavin-containing monooxygenases in plants: looking beyond detox. *Trends in Plant Science* 12, 412–418.

Schmidt, R.C., Müller, A., Hain, R., Bartling, D. and Weiler, E.W. (1996) Transgenic tobacco plants expressing the *Arabidopsis thaliana* nitrilase II enzyme. *The Plant Journal* 9, 683–691.

Seo, M., Akaba, S., Oritani, T., Delarue, M., Bellini, C., Caboche, M. and Koshiba, T. (1998) Higher activity of an aldehyde oxidase in the auxin-overproducing *superroot1* mutant of *Arabidopsis thaliana*. *Plant Physiology* 116, 687–693.

Simon, S. and Petrášek, J. (2010) Why plants need more than one type of auxin. *Plant Science* 180, 454–460.

Sohlberg, J.J., Myrenås, M., Kuusk, S., Lagercrantz, U., Kowalczyk, M., Sandberg, G. and Sundberg, E. (2006) *STY1* regulates auxin homeostasis and affects apical–basal patterning of the *Arabidopsis* gynoecium. *The Plant Journal* 47, 112–123.

Sorefan, K., Girin, T., Liljegren, S.J., Ljung, K., Robles, P., Galván-Ampudia, C.S., Offringa, R., Friml, J., Yanovsky, M.F. and Østergaard, L. (2009) A regulated auxin minimum is required for seed dispersal in *Arabidopsis*. *Nature* 459, 583–586.

Soto, A., Ruiz, K.B., Ziosi, V., Costa, G. and Toorigiani, P. (2012) Ethylene and auxin biosynthesis and signaling are impaired by methyl jasmonate leading to a transient slowing down of ripening in peach fruit. *Journal of Plant Physiology* 169, 1858–1865.

Spaepen, S. and Vanderleyden, J. (2011) Auxin and plant–microbe interactions. *Cold Spring Harbor Perspectives in Biology* 3(4): a001438.

Staswick, P.E. (2009) The tryptophan conjugates of jasmonic acid and indole-3-acetic acids are endogenous auxin inhibitors. *Plant Physiology* 150, 1310–1321.

Staswick, P.E., Serban, B., Rowe, M., Tiryaki, I., Maldonado, M.T., Maldonado, M.C. and Suza, W. (2005) Characterization of an *Arabidopsis* enzyme family that conjugates amino acids to indole-3-acetic acid. *The Plant Cell* 17, 616–627.

Stepanova, A.N., Hoyt, J.M., Hamilton, A.A. and Alonso, J.M. (2005) A Link between ethylene and auxin uncovered by the characterization of two root-specific ethylene-insensitive mutants in *Arabidopsis*. *The Plant Cell* 17, 2230–2242.

Stepanova, A.N., Robertson-Hoyt, J., Yun, J., Benavente, L.M., Xie, D.Y., Dolezal, K., Schlereth, A., Jürgens, G. and Alonso J.M. (2008) *TAA1*-mediated auxin biosynthesis is essential for hormone crosstalk and plant development. *Cell* 133, 177–191.

Stepanova, A.N., Yun, J., Robies, L.M., Novak, O., He, W., Guo, H., Ljung, K. and Alonso, J. (2011) The *Arabidopsis* YUCCA1 flavin monooxygenase functions in the indole-3-pyruvic acid branch of auxin biosynthesis. *The Plant Cell* 23, 3961–3973.

Stewart, J.L., Maloof, J.N. and Nemhauser, J.L. (2011) *PIF* genes mediate the effect of sucrose on seedling growth dynamics. *PLoS ONE* 6(5): e19894.

Strader, L.C., Wheeler, D.L., Christensen, S.E., Berens, J.C., Cohen, J.D., Rampey, R.A. and Bartel, B. (2011) Multiple facets of *Arabidopsis* seedling development require indole-3-butyric acid-derived auxin. *The Plant Cell* 23, 984–999.

Stone, S.L., Braybrook, S.A., Paula, S.L., Kwong, L.W., Meuser, J., Pelletier, J., Hsieh, T.F., Fischer, R.L., Goldberg, R.B. and Harada, J.J. (2008) *Arabidopsis* LEAFY COTYLEDON2 induces maturation traits and auxin activity: Implications for somatic embryogenesis. *Proceedings of the National Academy of Sciences of the United States of America* 105, 3151–3156.

Sugawara, S., Hishiyama, S., Jikumaru, Y., Hanada, A., Nixhimura, T., Koshiba, T., Zhao, Y., Kamiya, Y. and Kasahara, H. (2009) Biochemical analyses of indole-3-acetaldoxime-dependent auxin biosynthesis in *Arabidopsis*. *Proceedings of the National Academy of Sciences of the United States of America* 106, 5430–5435.

Sun, J., Xu, Y., Ye, S., Jiang, H., Chen, Q., Liu F., Zhou, W., Chen, R., Li, X., Tietz, O., Wu, X., Cohen, J.D. *et al.* (2009) *Arabidopsis ASA1* is important for jasmonate-mediated regulation of auxin biosynthesis and transport during lateral root formation. *The Plant Cell* 21, 1495–1511.

Sun, J., Qi, L., Li, Y., Chu, J. and Li, C. (2012) PIF4-mediated activation of *YUCCA8* expression integrates temperature into the auxin pathway in regulating *Arabidopsis* hypocotyl growth. *PLoS Genetics* 8, e1002594.

Swarup, R., Perry, P., Hagenbeek, D., Van Der Straeten, D., Beemster, G.T.S., Sandberg, G., Bhalerao, R., Ljung, K. and Bennett, M.J. (2007) Ethylene upregulates auxin biosynthesis in *Arabidopsis* seedlings to enhance inhibition of root cell elongation. *The Plant Cell* 19, 2186–2196.

Tao, Y., Ferrer, J.-L., Ljung, K., Pojer, F., Hong, F., Long, J.A., Li, L., Moreno, J.E., Bowman, M.E., Ivans, L.J., Cheng, Y., Lim, J. *et al.* (2008) Rapid synthesis of auxin via a new tryptophan-dependent pathway is required for shade avoidance in plants. *Cell* 133, 164–174.

Tivendale, N.D., Davies, N.W., Molesworth, P.P., Davidson, S.E., Smith, J.A., Lowe, E.K., Reid, J.B. and Ross, J.J. (2010) reassessing the role of *N*-hydroxytryptamine in auxin biosynthesis. *Plant Physiology* 154, 1957–1965.

Tivendale, N.D., Davidson, S.E., Davies, N.W., Smith, J.A., Dalmais, M., Bendahmene, A.I., Quittenden, L.J., Sutton, L., Bala, R.K., Le Signor, C., Thompson, R., Horne, J. *et al.* (2012) Biosynthesis of the halogenated auxin, 4-chloroindole-3-acetic acid. *Plant Physiology* 159, 1055–1063.

Tobeña-Santamaria, R., Bliek, M., Ljung, K., Sandberg, G., Mo, J.M.M., Souer, E. and Koes, R. (2002) FLOOZY of petunia is a flavin mono-oxygenase-like protein required for the specification of leaf and flower architecture. *Genes and Development* 16, 753–763.

Tognetti, V.B., Van Aken, O., Morreel, K., Vandenbroucke, B., van de Cotte, I., De Clerq, S., Chiwocha, R., Fenske, E., Prinsen, W., Boerjan, B., Gentry, K.A., Stubbs, D. *et al.* (2010) Perturbation of indole-3-butyric acid homeostasis by the UDP-glucosyltransferase *UGT74E2* modulates *Arabidopsis* architecture and water stress tolerance. *The Plant Cell* 22, 2660–2679.

Trigueros, M., Navarrete-Gómez, M., Sato, S., Christensen, S.K., Pelaz, S., Weigel, D., Yanaovsky, M.F. and Ferrándiz, C. (2009) The *NGATHA* genes direct style development in the *Arabidopsis* gynoecium. *The Plant Cell* 21, 1394–1409.

Tzin, V. and Galili, G. (2010) New insights into the shikimate and aromatic amino acids biosynthesis pathways in plants. *Molecular Plant* 3, 956–972.

Vanneste, S. and Friml, J. (2009) Auxin: a trigger for change in plant development. *Cell* 136, 1005–1016.

Vanstralen, M. and Benková, E. (2012) Hormonal interactions in the regulation of plant development. *Annual Review of Cell and Developmental Biology* 28, 463–487.

Vorwerk, S., Biernacki, S., Hillebrand, H., Janzik, I., Müller, A., Weiler, E.W. and Piotrowski, M. (2001) Enzymatic characterization of the recombinant *Arabidopsis thaliana* nitrilase subfamily encoded by the *NIT2/NIT1/NIT3* gene cluster. *Planta* 212, 508–516.

Walz, A., Park, S., Slovin, J.P., Ludwig-Müller, J., Momonoki, Y. and Cohen, J.D. (2002) A gene encoding a protein modified by the phytohormone indoleacetic acid. *Proceedings of the National Academy of Sciences of the United States of America* 99, 1718–1723.

Went, F.W. and Thimann, K.V. (1937) *Phytonormones*. Macmillan, New York.

Wheeler, A.W. (1980) Auxin-like growth activity of 3-phenylpropionitrile from water-cress (*Nasturtium officinale* R. Br.). *Annals of Botany* 46, 1–5.

Wójcikowska, B., Jaskóla, K., Gasiorek, P., Meus, M., Nowak, K. and Gai, M.D. (2013) *LEAFY COTYLEDON2* (*LEC2*) promotes embryogenic induction in somatic tissues of *Arabidopsis*, via *YUCCA*-mediated auxin biosynthesis. *Planta* 238, 425–440.

Won, C., Shen, X., Mashiguchi, K., Zheng, Z., Dai, X., Cheng, Y., Kasahara, H., Kamiya, Y., Chory, J. and Zhao, Y. (2011) Conversion of tryptophan to indole-3-acetic acid by TRYPTOPHAN AMINOTRANSFERASES OF *ARABIDOPSIS* and YUCCAs in *Arabidopsis*. *Proceedings of the National Academy of Sciences of the United States of America* 108, 18518–18523.

Woo, Y.-M., Park, H.-J., Suúdi, M., Yang, J.-I., Park, J.-J., Back, K., Park, Y.-M. and An, G. (2007) Constitutively wilted 1, a member of the rice YUCCA gene family, is required for maintaining water homeostasis and an appropriate root to shoot ratio. *Plant Molecular Biology* 65, 125–136.

Woodward, A.W. and Bartel, B. (2005) Auxin: regulation, action, and interaction. *Annals of Botany* 95, 707–735.

Yamada, M., Greenham, K., Prigge, M.J., Jensen, P.J. and Estelle, M. (2009) The *TRANSPORT INHIBITOR RESPONSE2* gene is required for auxin synthesis and diverse aspects of plant development. *Plant Physiology* 151, 168–179.

Yamamoto, Y., Kamiya, N., Morinaka, Y., Matsuoka, M. and Sazuka, S. (2007) Auxin biosynthesis by the *YUCCA* genes in rice. *Plant Physiology* 143, 1362–1371.

Yang, Y., Xu, R., Ma, C.J., Vlot, A.C., Klessig, D.F. and Pichersky, E. (2008) Inactive methyl-3-acetic acid ester can be hydrolyzed and activated by several esterases belonging to the AtMES esterase family of *Arabidopsis*. *Plant Physiology* 147, 1034–1045.

Zhang, R., Wang, B., Ouyang, J., Li, J. and Wang, Y. (2008) *Arabidopsis* indole synthase, a homolog of tryptophan synthase alpha, is an enzyme involved in the Trp-independent indole-containing metabolite biosynthesis. *Journal of Integrative Plant Biology* 50, 1070–1077.

Zheng, Z., Guo, Y., Novák, O., Dai, X., Zaho, Y., Ljung, K., Noel, P. and Chory, J. (2013) Coordination of auxin and ethylene biosynthesis by the aminotransferase VAS1. *Nature Chemical Biology* 9, 244–248.

Zhao, Y. (2008) The role of local biosynthesis of auxin and cytokinin in plant development. *Current Opinion in Plant Biology* 11, 16–22.

Zhao, Y. (2012) Auxin biosynthesis: a simple two-step pathway converts tryptophan to indole-3-acetic acid in plants. *Molecular Plant* 5, 334–338.

Zhao, Y., Christensen, S.K., Fankhauser, C., Cashman, J.R., Cohen, J.D., Weigel, D. and Chory, J. (2001) A role for flavin monooxygenase-like enzymes in auxin biosynthesis. *Science* 291, 306–309.

Ziegler, D.M. (1990) Flavin-containing monooxygenases: enzymes adapted for multi-substrate specificity. *Trends in Pharmacological Sciences* 11, 321–324.

Zolman, B.K., Martinez, M., Millius, A., Adham, A.R. and Bartel, B. (2008) Identification and characterization of *Arabidopsis* indole-3-butyric acid response mutants defective in novel peroxisomal enzymes. *Genetics* 180, 237–251.

20 Involvement of Tryptophan-Pathway-Derived Secondary Metabolism in the Defence Responses of Grasses

A. Ishihara,[1*] T. Matsukawa,[2] T. Nomura,[3] M. Sue,[4] A. Oikawa,[5] Y. Okazaki[6] and S. Tebayashi[7]

[1]*Tottori University, Koyama, Tottori, Japan;* [2]*Kinki University, Kinokawa, Japan;* [3]*Toyama Prefectural University, Imizu, Toyama, Japan;* [4]*Tokyo University of Agriculture, Setagaya, Tokyo, Japan;* [5]*Yamagata University, Tsuruoka, Japan;* [6]*RIKEN Center for Sustainable Resource Science, Yokohama, Japan;* [7]*Kochi University, Nangoku, Japan*

20.1 Abstract

The tryptophan (Trp) pathway diverges from the shikimate pathway at chorismate. The first committed enzyme for Trp synthesis is anthranilate synthase. Anthranilate is then converted to Trp in five steps. The secondary metabolites derived from this pathway have been shown to play important roles in the defence systems of members of the grass family (*Poaceae*). The most intensively investigated group of the secondary metabolites derived from this pathway is the benzoxazinones (Bxs) in maize, wheat and rye. Detailed analysis of Bx biosynthesis has been conducted using molecular genetic and biochemical techniques. Because Bxs are present as glucosides, and their active forms are considered to be aglycones and degraded compounds from these, the enzymes involved in the sequestration and release of aglycones have been investigated in depth. In addition, the multiple functions of Bxs in defending against pathogens and herbivores have been elucidated. In oats, anthranilate is converted to avenanthramide (Av) phytoalexins. Analysis of the metabolic fates of Avs has revealed their conversion to dimers and incorporation into cell walls for the purpose of reinforcement. Thus, Avs have functions in both the chemical and physical defence systems of oats in their interactions with pathogens. In rice, Trp was found to be converted into serotonin in leaves infected by pathogenic fungi. Serotonin is also incorporated into cell walls, and it has been suggested that it may be involved in the reinforcement of cell walls. The accumulation of serotonin was also found in multiple *Poaceae* family species, indicating that the accumulation of serotonin in response to pathogenic attack was acquired early in the evolution of this family.

20.2 General Introduction to Secondary Metabolism Derived From the Tryptophan Pathway

The tryptophan (Trp) pathway diverges from the shikimate pathway at chorismate, which is the common precursor of the other aromatic amino acids, phenylalanine (Phe) and tyrosine (Tyr). Anthranilate synthase (AnS) catalyses the conversion of chorismate to anthranilate, and this is the first committed step in the Trp pathway. Next, anthranilate is converted to Trp by five

*E-mail address: aishihara@muses.tottori-u.ac.jp

additional reactions catalysed by phosphoribosylanthranilate transferase, phosphoribosylanthranilate isomerase, indole-3-glycerol phosphate synthase and tryptophan synthase α (TSA) and β (TSB) subunits.

The Trp biosynthetic pathway provides precursors for secondary metabolites. In mammals, Trp serves as a precursor of serotonin and melatonin, which play essential roles in the nervous system, but the number of Trp-derived metabolites is limited. In contrast, in plants, numerous secondary metabolites are derived from the Trp biosynthetic pathway, as shown in Fig. 20.1. Because secondary metabolites derived from this pathway comprise various biologically important compounds, the regulation of the pathway has been intensively investigated.

AnS is the regulatory point of the metabolic flux to Trp which, in turn, regulates AnS activity through feedback inhibition. In tobacco, application of the false feedback inhibitor of AnS, 5-methyltryptophan, suppressed the accumulation of Trp, while a 5-methyltryptophan-resistant mutation of AnS significantly increased the concentration of Trp, indicating that AnS is the rate-limiting step in Trp biosynthesis and, therefore, plays a crucial role in the regulation of this pathway (Widholm, 1972, 1974). The enzyme is a heterotetramer composed of two large α subunits (AnSα) and two small β subunits (AnSβ) (Niyogi *et al.*, 1993; Poulsen *et al.*, 1993). The function of AnSβ is to liberate ammonia from glutamine, whereas AnSα catalyses the conversion of chorismate to anthranilate and pyruvate. The Trp allosteric site for feedback inhibition is present in AnSα.

The expression of AnS is under the control of environmental factors. There are two types of AnSα in higher plants, and their expression is regulated in different ways: one type of AnSα is constitutively expressed, and the other type is induced in response to environmental stimuli (Niyogi and Fink, 1992; Bohlmann *et al.*, 1995; Tozawa *et al.*, 2001). In the case of *Ruta graveolens*, the elicitor-inducible AnSα is feedback insensitive, suggesting that this isoenzyme is involved in secondary metabolism (Bohlmann *et al.*, 1996). In other species, such as oats, rice and *Arabidopsis*, two isoenzymes have been isolated, one of which is inducible by exogenous stimuli, but both are

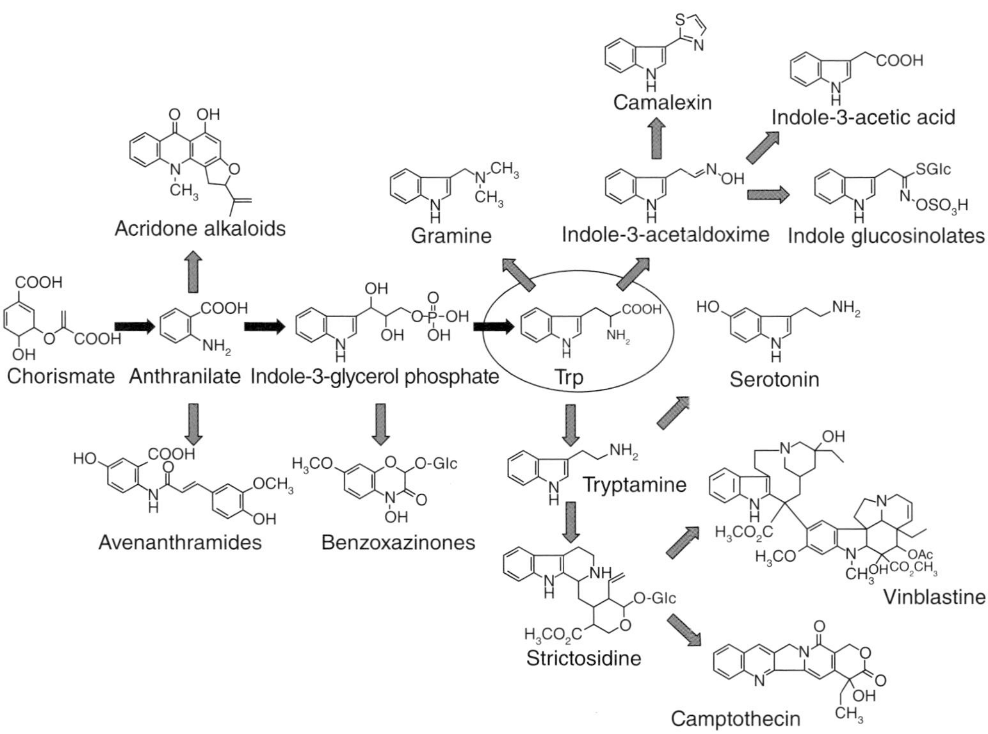

Fig. 20.1. Major secondary metabolites derived from the tryptophan pathway.

feedback sensitive, so feedback-sensitive isoenzymes have been indicated to be involved in secondary metabolism (Niyogi and Fink, 1992; Zhao and Last, 1996; Tozawa *et al.*, 2001; Matsukawa *et al.*, 2002b).

The Trp pathway has been the target of metabolic engineering with a view towards increasing the amount of Trp and/or Trp-derived metabolites. A mutant *trp5* that possesses relaxed feedback sensitivity was acquired by selection based on the resistance of *Arabidopsis* seedlings to α-methyltryptophan and 6-methylanthranilate, which are, respectively, the false feedback inhibitor and the precursor of the false feedback inhibitor (Kreps *et al.*, 1996; Li and Last, 1996). This *Arabidopsis* gene was introduced into hairy root cultures of *Catharanthus roseus*, and the transgenic hairy root cultures that resulted accumulated elevated levels of Trp and tryptamine (Hughes *et al.*, 2004). However, the concentrations of monoterpenoid indole alkaloids derived from the Trp pathway did not significantly increase. A similar mutation was introduced into the gene encoding AnSα in rice, and the mutated enzyme was also shown to have diminished feedback sensitivity (Tozawa *et al.*, 2001; Wakasa *et al.*, 2006). The effects of such diminished feedback inhibition on secondary metabolism were investigated in rice and *Arabidopsis* expressing the mutated AnSα gene of rice (Ishihara *et al.*, 2006; Dubouzet *et al.*, 2007; Matsuda *et al.*, 2009). In transgenic rice seedlings, metabolic profiling with mass spectrometry revealed increases in the contents of several Trp-derived metabolites (10- to 50-fold increases), but considering their small peak intensity and the larger increase in Trp (162-fold), the effects on secondary metabolism are limited. In *Arabidopsis*, the contents of Trp indicated around a 200-fold increase, but the amount of indole glucosinolate increased by no more than 2.1-fold. Thus, the regulation of the metabolic flux from the Trp pathway to secondary metabolism is tightly regulated.

Many species in the grass family (*Poaceae*) have been shown to accumulate secondary metabolites from the Trp pathway, although their biosynthetic pathways branch off at different compounds. Wheat, rye and maize accumulate benzoxazinones (Bxs) as constitutive defence compounds (Niemeyer, 1988). Bxs are derived from indole-3-glycerol phosphate (Frey *et al.*, 1997). The avenanthramides (Avs), which are synthesized by the condensation of hydroxycinnamates and anthranilate, are oat phytoalexins (Ishihara *et al.*, 1999a; Matsukawa *et al.*, 2002b). Furthermore, it has been shown that serotonin levels significantly increased after the inoculation of rice leaves with fungal pathogens, suggesting that serotonin is involved in defence responses in rice (Ishihara *et al.*, 2008a). In this chapter, the biosynthesis and physiological roles of the secondary metabolites derived from the Trp pathway in the grasses are highlighted.

20.3 The Biosynthesis and Functions of Benzoxazinones in Wheat, Rye and Maize

20.3.1 Molecular genetics of the benzoxazinone pathway

Bxs are one of the better studied classes of plant secondary metabolites in terms of information on their distribution, biological activities and biosynthesis collected using biochemical and molecular genetic approaches (Niemeyer, 1988, 2009; Sicker *et al.*, 2000; Frey *et al.*, 2009). These compounds are produced in many species of *Poaceae*, including the major agricultural crops, wheat (*Triticum aestivum*), rye (*Secale cereale*) and maize (*Zea mays*). Other than in the *Poaceae*, Bxs are found in two distant orders of the eudicots, the *Ranunculales*, including the family *Ranunculaceae*, and the *Lamiales*, including the families *Acanthaceae*, *Plantaginaceae* and *Lamiaceae*. Representative Bxs are 2,4-dihydroxy-1,4-benzoxazin-3-one (DIBOA) and its 7-methoxy derivative (DIMBOA), which are stored in the form of 2-*O*-β-D-glucopyranosides, designated DIBOA-Glc and DIMBOA-Glc, respectively (Fig. 20.2). The Bx-Glcs, which have lower toxicity than the aglycones, are stored in the vacuole. The toxic aglycones are released by a specific β-glucosidase in the plastid when cells are disrupted by pathogen infection or herbivore predation. Thus, the primary function of Bxs in plants is considered to be chemical defence.

The Bx biosynthetic pathway has been almost fully elucidated in maize and partly characterized in wheat, rye and wild barley (Frey *et al.*, 1995, 1997, 2003; von Rad *et al.*, 2001; Nomura

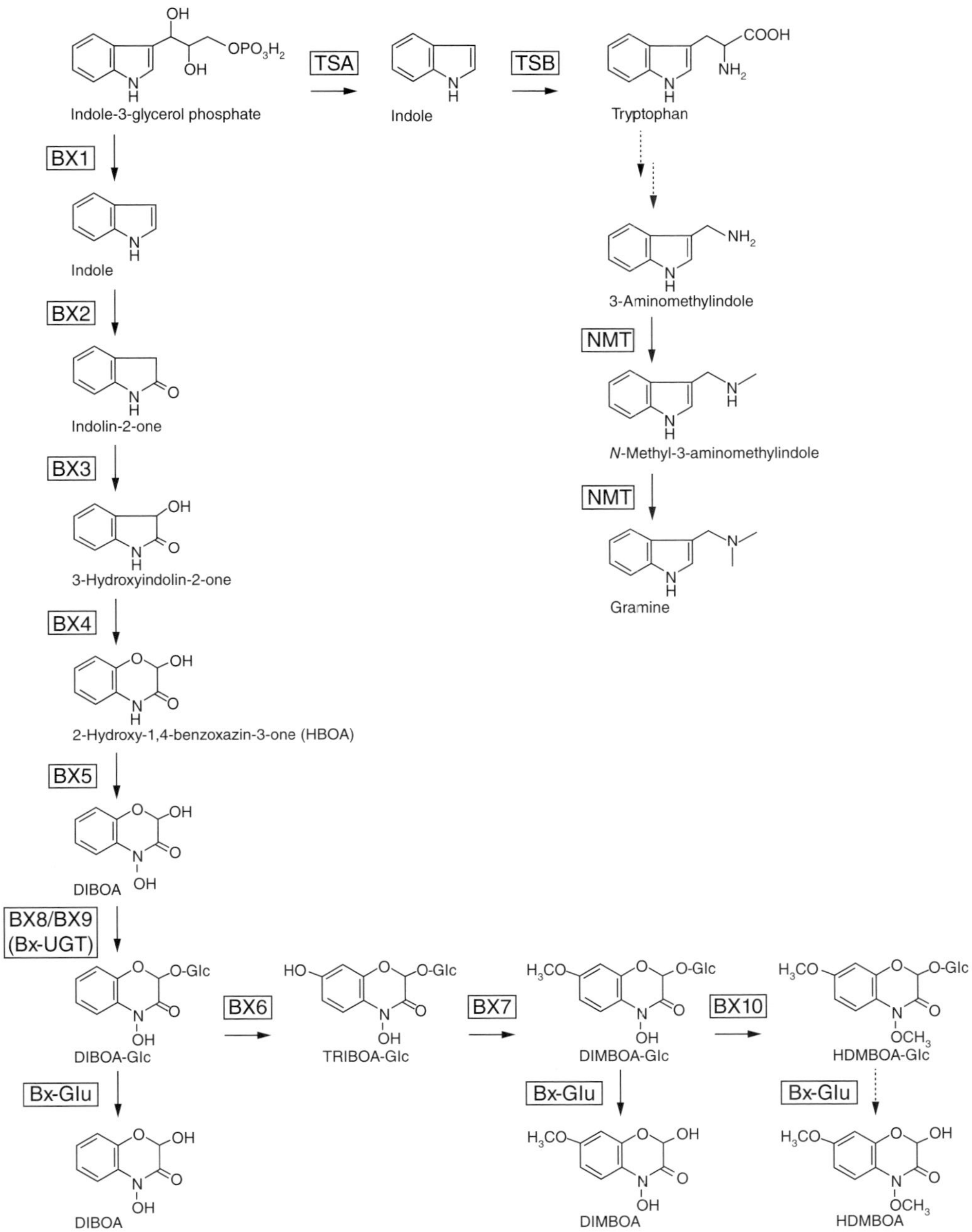

Fig. 20.2. Biosynthetic pathways of benzoxazinones and gramine derived from the tryptophan pathway. Abbreviations: BX1–10, enzymes of the benzoxazinone synthetic pathway; Bx-Glu, benzoxazinone-specific glucosidase enzymes; DIBOA, 2,4-dihydroxy-1,4-benzoxazin-3-one; DIMBOA, 7-methoxy-DIBOA; Glc, glucose; HDMBOA, 2-hydroxy-4,7-dimethoxy-1,4-benzoxazin-3-one; NMT, *N*-methyltransferase; TRIBOA, 2,4,7-trihydroxy-1,4-benzoxazin-3-one; TSA and TSB, tryptophan synthase α and β subunits. See text for further details.

et al., 2002, 2003, 2005, 2007; Grün *et al.*, 2005; Jonczyk *et al.*, 2008; Sue *et al.*, 2011b; Meihls *et al.*, 2013). The Bx biosynthetic pathway branches off from the Trp pathway at indole-3-glycerol phosphate (Frey *et al.*, 1997, 2000; Melanson *et al.*, 1997) (Fig. 20.2). The branch-point reaction was defined by the isolation of the *Bx1* gene, which is characterized by the Bx-less phenotype of mutant maize. During Trp biosynthesis, indole-3-glycerol phosphate is converted to indole by TSA, which reaction is followed by the conversion of indole to Trp catalysed by TSB. Because TSA and TSB form a complex, indole generated by TSA immediately serves as a substrate for TSB without any release of free indole. In contrast, the BX1 enzyme functions as a monomer to form free indole, while the *Bx1* gene is considered to have evolved via gene duplication and modification from the *TSA* gene (Frey *et al.*, 2000). Next, indole is converted to DIBOA by consecutive reactions to introduce four oxygen atoms; these reactions are catalysed by four independent cytochrome P450 monooxygenases, BX2–BX5 (Frey *et al.*, 1995, 1997). These P450s belong to the CYP71C subfamily, and have almost certainly arisen from a common ancestral sequence by tandem gene duplication and the acquisition of new functions. It has long been unclear whether the conversion of DIBOA to DIMBOA occurs on the aglycone or the glucoside, but gene identification and enzymatic characterization of a 2-oxoglutarate-dependent dioxygenase, BX6, and an *O*-methyltransferase, BX7, in maize has revealed that 2-*O*-glucosylation of DIBOA by UDP-Glc:Bx glucosyltransferase (Bx-UGT) precedes the 7-hydroxylation step, which is then followed by 7-*O*-methylation of TRIBOA (2,4,7-trihydroxy-1,4-benzoxazin-3-one)-Glc to form DIMBOA-Glc (Frey *et al.*, 2003; Jonczyk *et al.*, 2008). Hence, the glucosylation reaction catalysed by the Bx-UGTs (BX8 and BX9) has been demonstrated to occur for DIBOA and led to the formation of DIBOA-Glc, although the Bx-UGTs do exhibit enzymatic activity towards both DIBOA and DIMBOA *in vitro* (von Rad *et al.*, 2001). The resulting DIBOA-Glc and DIMBOA-Glc are deglucosylated for reactivation by specific glucosidase enzymes (Bx-Glus), for which the *glu1* and *glu2* genes in maize (Cicek and Esen, 1999) and the *glu1* orthologues in wheat (Sue *et al.*, 2006; 2011b) and rye (Nikus *et al.*, 2003) have been identified. Most recently, a *Bx10* gene encoding an *O*-methyltransferase that catalyses 4-*O*-methylation of DIMBOA-Glc to form HDMBOA (2-hydroxy-4,7-dimethoxy-1, 4-benzoxazin-3-one) has been identified in maize (Meihls *et al.*, 2013) (Fig. 20.2).

The Bx biosynthetic pathway has attracted attention owing to the discovery of a gene cluster for the Bx biosynthetic loci in maize; this is the first reported example of a secondary metabolic gene cluster in plants (Frey *et al.*, 2009). Of the ten *Bx* genes (*Bx1*–*Bx10*) in maize, *Bx1*–*Bx8* are clustered on the short arm of chromosome 4, whereas *Bx9* (a highly identical homologue of *Bx8*) and *Bx10* are situated on chromosome 1 (Frey *et al.*, 1997; von Rad *et al.*, 2001; Jonczyk *et al.*, 2008; Meihls *et al.*, 2013). The *glu1* and *glu2* genes are located outside the cluster on chromosome 10 (see http://www.maizesequence.org/index.html). As grass genomes are supposed to have evolved from a common ancestor, it is natural that the Bx biosynthetic genes in grasses are of monophyletic origin (Frey *et al.*, 2009; Dutartre *et al.*, 2012). Meanwhile, the basic (haploid) chromosome number differs between maize and other Bx-containing species: maize has ten chromosomes, whereas wheat and rye, belonging to the tribe *Triticeae*, have a basic chromosome number of seven. Therefore, the chromosomal locations of the Bx-pathway loci have been examined in detail in wheat and rye to determine whether or not the *Bx* gene cluster found in maize is common to all Bx-producing grass species.

Common wheat (bread wheat, *T. aestivum*) is hexaploid with the genome constitution AABBDD ($2n = 6\times = 42$), originated through successive chromosome doubling of hybrids involving three ancestral diploid progenitors ($2n = 2\times = 14$). The A genome came from *T. urartu* (AA), the B genome from *Aegilops speltoides* (SS) or another species in the genus *Aegilops* (section *Sitopsis*), and the D genome from *A. tauschii* (Huang *et al.*, 2002; Feldman and Levy, 2005; Salse *et al.*, 2008). In most cases, orthologous genes in each diploid progenitor were retained following the allopolyploidization, and such duplicated genes are referred to as homoeologues. Consequently, the hexaploid wheat genome contains triplicated homoeologues for most genes, although elimination and/or amplification of homoeologues has also occurred for some genes

(Ozkan *et al.*, 2001). Interestingly, unlike in maize, the *Bx* gene cluster is split across more than one chromosome in hexaploid wheat. The *Bx1* and *Bx2* genes co-localize on the homoeologous group-4 chromosomes (4A, 4B and 4D), while the *Bx3–Bx5* genes co-localize on homoeologous group-5 chromosomes (5A, 5B and 5D) (Nomura *et al.*, 2002, 2003, 2005). Recently, a *Bx8* gene encoding a Bx-UGT was found to be located on group-7 chromosomes 7A, 7B (two loci) and 7D, and the *glu1* gene encoding a Bx-Glu on group-2 chromosomes 2A, 2B (two loci) and 2D (Sue *et al.*, 2011b). Dispersal of the Bx-pathway loci is also the case in rye ($2n = 2\times = 14$; RR): the *Bx1* and *Bx2* genes are located on chromosome 7R, the *Bx3–Bx5* genes on 5R, the *Bx8* gene on 4R and the *glu1* gene on 2R (Nomura *et al.*, 2003; Sue *et al.*, 2011b).

As the chromosomal locations of orthologous loci parallel the chromosomal synteny among wheat, rye and maize, it is highly likely that the Bx-pathway loci originally arose as a cluster in an ancient grass genome, following which the cluster was split into different chromosomes through chromosomal rearrangement during the evolutionary process. However, it is also possible that a complete cluster evolved only in the maize lineage. To complete the comparative analysis of Bx-pathway loci in wheat, rye and maize, the characterization of the *Bx6* and *Bx7* loci remains to be elucidated in wheat and rye. The *Bx-Glu* and *Bx10* loci are not included in the *Bx* gene cluster even in maize, so it is unclear whether or not they were originally included in the cluster. The analysis of wild grass species may clarify the precise sequence of events (Takos and Rook, 2012). Genes in the Bx pathway have also been analysed to some degree in certain eudicot (eudicotyledon) plants, with respect to *Bx1*, *Bx-UGT* and *Bx-Glu* genes (Schullehner *et al.*, 2008; Dick *et al.*, 2012). The pathway begins from indole-3-glycerol phosphate, as in the grasses, but the sequence comparison of the three enzymes of grasses and eudicots indicates that the pathway evolved independently, i.e. through convergent evolution, and so it is not of monophyletic origin (Schullehner *et al.*, 2008; Dick *et al.*, 2012). It would be interesting to determine whether the Bx-pathway genes in eudicots are clustered together.

In hexaploid wheat, all of the Bx-pathway genes identified thus far are present on all three genomes A, B and D, but the transcript levels differ greatly among the genomes in that it is the B-genome loci that are predominantly transcribed (Nomura *et al.*, 2005; Sue *et al.*, 2011b). This is the case in tetraploid wheat with the genome constitution AABB as well. Furthermore, among the three diploid progenitors of hexaploid wheat, transcript levels of the Bx-pathway genes are the highest in *A. speltoides*, the B-genome donor to hexaploid wheat. These facts suggest that the transcriptional bias of the Bx-pathway genes in hexaploid wheat originated at the diploid level and was retained through polyploidization. Nomura *et al.* (2008) compared promoter activities of the *Bx3* and *Bx4* genes among the three genomes in hexaploid wheat, but detected no significant differences. To date, the mechanism of the preferential transcription of the B-genome loci for all Bx-pathway genes is unknown.

In addition, the mechanism of co-expression of the Bx pathway genes in each genome remains to be elucidated. The selective advantages promoting the organization of biosynthetic gene clusters are hypothesized to be co-expression and co-inheritance. Clustering allows the coordinated transcriptional regulation of pathway genes because it enables localized changes in chromatin structure, facilitates the co-inheritance of beneficial allele combinations and maintains a functional pathway (Osbourn and Field, 2009; Wegel *et al.*, 2009; Osbourn, 2010; Chu *et al.*, 2011; Kliebenstein and Osbourn, 2012; Takos and Rook, 2012). This model explained the existence of the clustered *Bx1–Bx8* genes in maize, but in wheat and rye, all Bx-pathway genes were co-expressed despite dispersal of the pathway genes to different chromosomes (Nomura *et al.*, 2005; Sue *et al.*, 2011b), suggesting that the coordinated transcription of Bx-pathway genes does not necessarily depend on the characteristics of the cluster. It is likely that there are *cis*-elements and transcription factors commonly controlling the expression of Bx-pathway genes (Nomura *et al.*, 2008). Transcription of some or all of the genes in a secondary metabolic pathway can be regulated by a small number of transcription factors, such as the OsTGAP1 transcription factor involved in diterpenoid phytoalexin biosynthesis in rice (*Oryza sativa*) (Okada *et al.*, 2009). Since the discovery of the *Bx* gene cluster, additional gene clusters have been shown in diverse plant species. Detailed analysis of transcriptional regulation mechanisms through the identification of transcription factor(s) and chromatin-related machinery for these gene clusters may provide important clues on the function and significance of gene clusters in plant secondary metabolism.

Wheat, rye and barley (*Hordeum vulgare*), belonging to the tribe *Triticeae*, have evolved from a common ancestor, share an identical basic chromosome number and have highly similar gene sequences. However, in contrast to the occurrence of Bxs in wheat and rye, barley is unable to produce them because of loss of the Bx-pathway loci during its evolution (Gierl and Frey, 2001; Nomura *et al.*, 2003; Grün *et al.*, 2005; Sue *et al.*, 2011b). In contrast, a simple indole alkaloid 3-(dimethylaminomethyl)indole (gramine), which is biosynthesized from Trp (Grower and Leete, 1963), occurs in some barley cultivars (Fig. 20.2). Although the gramine biosynthetic pathway has not yet been fully elucidated, an *N*-methyltransferase (NMT) that catalyses two sequential *N*-methylation reactions has been identified (Larsson *et al.*, 2006).

Grün *et al.* (2005) examined the occurrence of Bxs and gramine in some wild barley species, and showed that one species (*H. spontaneum*) lacks Bxs but contains gramine instead, whereas gramine is not detectable in Bx-containing *Hordeum* spp. (e.g. *H. lechleri*). Both Bxs and gramine are linked to Trp biosynthesis, and function in defence (Frey *et al.*, 2009). On the basis of the fact that the biosynthesis of Bxs and gramine is mutually exclusive in wild *Hordeum* species, it was suggested that during evolution selection occurred against the simultaneous expression of the two pathways so as to avoid competition for substrate flux from the same primary biosynthetic pathway (Grün *et al.*, 2005). Nomura *et al.* (2007) identified Bx-deficient variants of the A-genome diploid wheat *T. boeoticum*, and proposed that the degeneration of coding sequences, gene silencing or loss of one *Bx* locus triggers the loss of other *Bx* loci, which finally leads to the total elimination of the Bx-pathway loci. Eliminations of the Bx-pathway loci and the gramine-pathway loci in Bx-deficient and gramine-deficient *Hordeum* species, respectively, might have occurred in a similar manner to that described for *T. boeoticum*. The Bx-gramine system may represent a good model to study the process of selective evolution of a secondary metabolic pathway and also to address the physiological and ecological significance of secondary metabolites in plants.

20.3.2 Detoxification and reactivation of benzoxazinones

During the biosynthesis of plant secondary metabolites, both the intermediates and the final products often undergo glycosylation, which alters their chemical stability and bioavailability. For example, a 3′-*O*-glucoside isomer of catechins was shown to be more stable at a neutral pH at 37 °C than other isomers glucosylated at different hydroxyl groups (Raab *et al.*, 2010). The attachment of rhamnose to the 6-hydroxyl group of the glucose moiety of naringenin-7-*O*-glucoside introduces a bitter taste in certain citrus species (e.g. grapefruit) while rhamnosylation of its 2-hydroxyl group generates a tasteless flavonoid (Rouseff *et al.*, 1987). The hydrolysis of ginsenosides (saponins of ginseng and one of the most commonly used components of herbal medicines) in the human stomach and intestine was shown to result in varied absorption rates in the intestine (Tawab *et al.*, 2003). The formation of acyl glycosides by some glycosyltransferases functions to activate carboxyl groups, as does conjugation with CoA (Shirley and Chapple, 2003).

The glycosylation of plant secondary metabolites plays important roles in plant growth and maintenance. For example, most of plant hormones, including cytokinins, auxins, gibberellins, abscisic acid, jasmonic acid, brassinosteroids and salicylic acid, partly exist as glycosylated conjugates, and their biological functions are modified by such glycosylation (Kleczkowski *et al.*, 1995; Lim and Bowles, 2004; Gachon *et al.*, 2005; Miersch *et al.*, 2008).

Other significant examples of glycosylated conjugates exist in plant defence systems. The attachment of sugars to aglycones enhances the water solubility of these compounds and facilitates their accumulation in plants, often within specific sub-cellular compartments such as vacuoles. The deglycosylation of avenacins and avenacosides, which are defensive terpenoids in oats, attenuated their antifungal activity, and so the glycoside hydrolases derived from pathogens and/or the plant itself play crucial roles in the resistance promoted by these substances (Osbourn *et al.*, 1991; Gus-Mayer *et al.*, 1994; Morrissey *et al.*, 2000). Plants belonging to the *Brassicales* produce glucosinolates that contain *S*-glucosides in their structures, attached through thioglucosidic bonds. *β*-Glucosidases known as myrosinases hydrolyse these bonds, followed by the release of isothiocyanates, thiocyanates and nitriles that act as defensive factors and attractants for insects (Rask *et al.*, 2000; Fahey *et al.*, 2001). Cyanogenic glucosides are also well-studied defensive secondary metabolites (Poulton, 1990). In cassava, linamarin is hydrolysed by linamarase, a *β*-glucosidase acting on linamarin, in response to disruption in cell integration. This results in the release in glucose and cyanohydrin. Next, an *α*-hydroxynitrile lyase converts the cyanohydrin into acetone and HCN to exert toxic effects on pathogens and herbivores (Hughes *et al.*, 1992). Such cyanogenesis occurs in a variety of plants, for example from amygdalin and prunasin in *Prunus* and from dhurrin in sorghum.

Another well-characterized group of glycosylated secondary metabolites is Bxs in the *Poaceae* (Niemeyer, 1988, 2009; Sicker *et al.*, 2000). As described in the previous section, the main Bx species in these plants are DIBOA and DIMBOA. Both DIBOA and DIMBOA are considered to accumulate in vacuoles as the corresponding *β*-glucosides (DIBOA-Glc and DIMBOA-Glc). Hydrolysis of the glucosidic bond in response to cell damage liberates the toxic aglycones to defend the plant from pathogens and/or insects. In addition, it is of interest to note that *Mythimna separata*, a serious pest of grasses, metabolized DIMBOA mixed in the diet, and excreted it as DIMBOA-Glc and HMBOA-Glc. Thus the insect employs the same detoxification strategy, that is glucosylation of DIMBOA by GT, with grasses accumulating Bxs as a result of convergent evolution (Sasai *et al.*, 2009).

An extremely wide variety of glycosides exists in plants, reflecting divergent evolution to adapt to the environment. Such a variety in structures is, of course, derived from differences in the enzyme structures of the plants concerned. Precise characterization of the enzymes involved in glycosylation and deglycosylation (glycosyltransferases and glycosidases, respectively) would help in understanding the evolutionary processes behind plant secondary metabolism, as well as on the mechanisms controlling their biological activity in plants. Typically, the glycosyltransferases (GTs) that are implicated in plant secondary metabolism utilize UDP-activated sugars as the glycosyl donors and, therefore, are members of UDP-glycosyltransferases (UGTs). GTs are classified into around 100 families based on their primary structures, which are available in the CAZy (Carbohydrate-Active enZYmes) database (see http://www.cazy.org/). They can also be classified into several groups based on their tertiary structures, with two major groups, the GT-A and GT-B folds. Their mechanism of reaction is another method by which GTs can be categorized as either inverting or retaining, a reference to the stereochemistry of the anomeric carbon of glycose through the reaction. The plant UGTs involved in secondary metabolism are the members of family 1 GTs, and the members of this family possess GT-B folds and an inverting reaction mechanism. Despite the occurrence of conserved folds among plant UGTs, they share low sequence similarity with one another. Extensive information on plant GTs has been summarized in several recent reviews (Lairson *et al.*, 2008; Osmani *et al.*, 2009; Wang, 2009).

The genes encoding Bx-UGTs have been isolated from maize (*ZmBx8* and *ZmBx9*), wheat (*TaGTa–TaGTd*; *TaBx8*s) and rye (*ScGT*; *ScBx8*) (von Rad *et al.*, 2001; Sue *et al.*, 2011b). High amino acid sequence identities have been observed among the enzymes (over 90% between the TaBX8s and ScBX8 and about 70% among the TaBX8s and ZmBX8, and between ZmBX8 and ZmBX9). Every Bx-UGT contains a PSPG box in the C-terminal domain, which is a highly conserved region comprising 44 amino acids among plant secondary products glycosyltransferases (hence PSPG) (Mackenzie *et al.*, 1997), and is proposed to participate in selective binding to the corresponding sugar donors (Osmani *et al.*, 2009; Caputi *et al.*, 2012). In the N-terminal domain, the most important and conserved residue is the catalytic residue. Plant UGTs for secondary metabolism contain a histidine residue as the catalytic centre in the N-terminal region. The histidine residue is activated by a neighbouring glutamate (Glu) or aspartate (Asp) and deprotonates the sugar acceptor, after which the activated sugar acceptor nucleophilically attacks the anomeric carbon of a UDP-glycoside (Offen *et al.*, 2006). From the sequence alignment, His19 is surmised to be the catalytic centre in ZmBX8, TaBX8s and ScBX8, while His28 performs the role in ZmBX9. Although ZmBX8 displays comparable activity towards DIBOA and DIMBOA, ZmBX9 has much lower affinity towards DIBOA. The ZmBX8 and TaBX8s display only negligible activity towards HBOA and HMBOA (4-dehydroxy derivatives of DIBOA and DIMBOA, respectively), suggesting that they better recognize the 4-hydroxyl groups than the 7-methoxy groups of Bxs (von Rad *et al.*, 2001; Sue *et al.*, 2000a). In contrast, ZmBX9 exhibits lower but moderate activity towards HMBOA, so it is suggested that ZmBX9 is able to recognize 7-methoxy groups better than 4-hydroxyl groups. Because the N-terminal domain of plant UGTs is considered to participate primarily in binding with sugar acceptors, these domains display lower sequence similarities than the C-terminal domains, reflecting the structural diversity of the sugar acceptors. Despite worldwide extensive research on plant UGTs, only a few crystal structures have been obtained – and from only three plant species, *A. thaliana*, *Medicago truncatula* and *Vitis vinifera* (Shao *et al.*, 2005; Offen *et al.*, 2006; Brazier-Hicks *et al.*, 2007; Li *et al.*, 2007; Modolo *et al.*, 2009). Currently, it remains difficult to determine the structural determinants of Bx-UGTs governing the observed substrate specificity towards the Bx species.

Although *ZmBx8* is clustered with other Bx-related genes on the short arm of chromosome 4 of maize, *ZmBx9* is located outside the cluster, on chromosome 1. Furthermore, all the Bx-UGTs in wheat and rye show comparable substrate specificity with ZmBX8 (TaBX8s, Sue *et al.*, 2011b; ScBX8, unpublished data). Consequently, ZmBX9 can be considered to have acquired a unique function through evolution, even though its genetic origin has not yet been discovered. The Bx-UGTs, including ZmBX9, have substrate preferences for DIMBOA over DIBOA. However, the physiological significance of this DIMBOA-UGT activity is uncertain because it has been shown that DIMBOA is not the immediate precursor of DIMBOA-Glc in the Bx biosynthetic pathway (Jonczyk *et al.*, 2008) (Fig. 20.2). In this study, it is unlikely that the DIMBOA that was liberated after cell disruption was recovered and translocated to other intact cells. Therefore, this enzyme may possess other, as yet unknown, functions.

As glucose conjugation of plant secondary metabolites changes their chemical and biological features, *β*-glucosidases, as well as GTs, play important roles in plant defence systems. The most prominent examples can be found in the activation of phytoanticipins such as cyanogenic glucosides, glucosinolates and Bxs. The glucosylation of phytoanticipins makes them less toxic to plants and enables their accumulation at high concentrations. *β*-Glucosidases act as a detonator for the quick release of toxic compounds in response to the attack by pathogens, insects and/or herbivores (Morant *et al.*, 2008).

β-Glucosidases are present in all organisms, ranging from microbes to animals. The *β*-glucosidases that act on plant secondary metabolites are the members of the GH1 glycoside hydrolase family, one of the largest of over 100 families of glycoside hydrolases; these are also listed in the CAZy database. GH1 has a $(\beta/\alpha)_8$-barrel (TIM barrel) structure and catalyses hydrolysis by a retaining (double-displacement) mechanism (McIntosh *et al.*, 1996). Plant GH1 *β*-glucosidases, except for myrosinase, have two catalytic carboxylate residues, one as an acid/base and the other as a nucleophile, and the primary structures around these residues are highly conserved. Until recently, glucosidases were considered to possess a broad substrate range, but the existence of many glucosidases with great structural diversity, even within a plant species, indicates that each enzyme may have a specific physiological role. The sorghum (*Sorghum bicolor*) dhurrinase 1 (SbDhr1) was shown to hydrolyse only the natural substrate, dhurrin (Hösel *et al.*, 1987). One of the SbDhr1 homologues in maize (ZmGlu1), which shares about 70% identity with SbDhr, exhibits activity towards several artificial substrates as well as towards its natural substrates (Bx-Glcs), but not towards dhurrin (Cicek and Esen, 1999; Cicek *et al.*, 2000). The raucaffricine glucosidase in *Rauvolfia* spp. predominantly hydrolyses raucaffricine, whereas the structurally related strictosidine glucosidase in the same plant acts only on strictosidine (Xia *et al.*, 2012).

The Bx-Glc *β*-glucosidases have been analysed crystallographically in three major Bx-containing plants, maize, wheat and rye. The wheat and rye glucosidases (TaGlu1 and ScGlu, respectively) display more than 90% identity with each other, and about 70% similarity to the maize glucosidase (ZmGlu1). The latter exists *in vivo* as a dimer, as is the case with the other plant *β*-glucosidases with known structures that hydrolyse the glucosidic bonds between glucose and non-sugar secondary metabolites. In contrast, TaGlu1 and ScGlu occur in a hexameric form (Sue *et al.*, 2006), though it has not been elucidated why they exhibit their activity only in that form. TaGlu1 and, particularly, ZmGlu1 act on DIMBOA-Glc with a much higher efficiency than they do on DIBOA-Glc, while ScGlu hydrolyses both DIBOA-Glc and DIMBOA-Glc at similar rates. Such substrate preferences are considered to reflect the major Bx species in these plants: DIMBOA-Glc in maize and wheat, and DIBOA-Glc in the rye shoots (Sue *et al.*, 2000b). Substrate binding is considered to be weaker in the Bx-Glc glucosidases than in SbDhr1, a hypothesis that is strengthened by the clarity of electron density maps and the conformations of the sugar moieties in the substrate-binding pockets. The aglycone moieties of the substrates are primarily stabilized by π–π interactions between the aromatic aglycone and the Trp residues (Trp378 in ZmGlu1 and Trp379 in TaGlu1 and ScGlu). The Trp residue is highly conserved among the plant GH1 enzymes. It has been proposed that the conformation of this residue, which is directed by hydrogen bonds with the neighbouring Ser/Thr/Asn (serine/threonine/asparagine) residue(s) is essential for catalytic activity (Xia *et al.*, 2012). In ZmGlu1, three Phe residues are located opposite the Trp378, and the aglycone moiety is sandwiched between and stabilized by aromatic rings of these residues. Two of the three Phe are substituted by His (histidine) and Ser/Gly (glycine) in TaGlu1 and ScGlu, making the aglycone-binding sites more commodious. Among the three enzymes, the spatial restrictions and increased hydrophobicity in the aglycone-binding pocket of ZmGlu1 may be responsible for its highest specificity towards DIMBOA-Glc. The entrance of its substrate-binding pocket opens to a greater extent than that of TaGlu1 and ScGlu, because the proline (Pro377) that is positioned at the entrance in ZmGlu1 is substituted by the bulky Tyr residues (Tyr378) in the latter two enzymes (Sue *et al.*, 2006, 2011a). The presence of an aromatic residue is important in determining substrate binding for TaGlu1 and ScGlu, and furthermore, the oxygen atom of the 7-methoxy group of DIMBOA-Glc is positioned at a distance where it is able to interact with the hydroxyl group of the Tyr residue. Comparison of substrate-binding pockets between TaGlu1 and ScGlu reveals that only two residues (in TaGlu1, Ser464 and leucine 465 (Leu465); and in ScGlu, Gly464 and Ser465) near the entrance are different. Mutation of these residues affects substrate preferences for DIBOA-Glc or DIMBOA-Glc, indicating that these two residues contribute to the recognition of the 7-methoxy group of Bx.

20.3.3 Inducible defence response associated with benzoxazinones

The concentration of Bxs is known to be affected by environmental factors such as temperature, water stress and light intensity. Considering that the temperature-dependent changes in Bx concentrations have been attributed to differences in the plant growth rates (Gianoli and Niemeyer, 1997a), Bx accumulation in plants can be considered to be regulated primarily by plant development. The accumulation of Bx in wild wheat (*T. uniaristatum*) triggered by an infestation of the aphid *Rhopalosiphum padi* was shown to be derived from the translocation of Bxs from the stem and root to the infested leaf (Gianoli and Niemeyer, 1997b), suggesting that the plant coped with the aphid attack with the least energy cost, thereby avoiding a sacrifice in plant growth. Similar strategies may be employed at the local tissue or cellular level as well as in the whole plant.

In addition to constitutive defence mechanisms mediated by the accumulation of Bx glucosides, inducible defence responses associated with Bxs have also been characterized. Several reports have indicated an increase in the total amount of Bxs in maize and wheat after insect feeding and/or pathogen infections (Gutierrez *et al.*, 1988; Niemeyer *et al.*, 1989; Morse *et al.*, 1991; Weibull and Niemeyer, 1995). Further, DIMBOA-Glc was converted into HDMBOA-Glc by methylation of the 4-hydroxyl group in wheat cultivars with moderate resistance against infection by stem rust (Bücker and Grambow, 1990). Oikawa *et al.* (2001, 2004) reported a decrease in the level of constitutively accumulated DIMBOA-Glc and an increase in the level of HDMBOA-Glc in maize leaves after wounding and treatment with elicitors and jasmonic acid (JA). Feeding by herbivorous insects and infection by plant pathogens also caused an increase in the levels of HDMBOA-Glc. In addition, JA treatment induced specific DIMBOA-Glc 4-*O*-methyltransferase activity linked with HDMBOA-Glc biosynthesis in maize leaves (Oikawa *et al.*, 2002). Recently, the *Bx10* gene encoding this methyltransferase was identified by a quantitative trait locus (QTL) analysis (Meihls *et al.*, 2013) (Fig. 20.2). In addition to HDMBOA-Glc, 2-*O*-D-glucopyranosyloxy-4,7,8-trimethoxy-1,4-benzoxazin-3-one (HDM_2BOA-Glc) was also found to accumulate after fungal infection (Glauser *et al.*, 2011). Marti *et al.* (2013) applied a metabolomic approach to detect herbivory-induced secondary metabolites in maize leaves and roots, and demonstrated the specific accumulation of HDMBOA-Glc and HDM_2BOA-Glc, together with increases in the levels of phospholipids, *N*-hydroxycinnamoyltyramines, azealic acids and Trp.

The bioactive peptide ZmPep1, encoded by the maize gene *ZmPROPEP1*, which is an orthologue of the *Arabidopsis AtPROPEP1* gene, has been reported to induce the accumulation of HDMBOA-Glc together with the expression of genes encoding defence-related proteins such as endochitinase A and PR-4 (Huffaker *et al.*, 2011). This peptide also stimulates the expression of *Bx1* gene, along with the induction of JA and ethylene synthesis. In *Arabidopsis*, AtPep1, together with its receptors, AtPEPR1 and AtPEPR2, activates and amplifies innate immune responses, enhancing resistance to pathogens. ZmPep1 may also regulate maize innate immunity and enhance disease resistance by stimulating the accumulation of HDMBOA-Glc. The accumulated HDMBOA-Glc is generated not only by the methylation of constitutively accumulated DIMBOA-Glc (Oikawa *et al.*, 2001) but also by *de novo* synthesis (Huffaker *et al.*, 2011). These findings indicate that the inducible production of HDMBOA-Glc plays an important role in defence responses in maize and wheat.

The antimicrobial activities of Bxs and benzoxazolinones have been well investigated (Virtanen *et al.*, 1957; Elnaghy and Linko, 1962; Corcuera *et al.*, 1978; Woodward *et al.*, 1978; Nakagawa *et al.*, 1995; Oikawa *et al.*, 2004). For example, 6-methoxybenzoxazolin-2(3H)-one (MBOA) generated from DIMBOA and HDMBOA has been shown to be biocidal (Virtanen *et al.*, 1957; Elnaghy and Linko, 1962; Oikawa *et al.*, 2004). The glucoside of HDMBOA itself has no biocidal activity, like other Bx-Glcs such as DIMBOA-Glc (Oikawa *et al.*, 2004). Because HDMBOA is chemically more unstable than DIMBOA, and thus is quickly degraded into the benzoxazolinone, MBOA, under natural conditions (Grambow *et al.*, 1986; Maresh *et al.*, 2006), the conversion of DIMBOA-Glc to HDMBOA-Glc is suggested to be a mechanism that enables the quick release of the biocide MBOA.

If this is the case, a pertinent question arises: why does maize not store HDMBOA-Glc in advance? The recent QTL analysis by Meihls *et al.* (2013) may provide an answer; it demonstrated that several maize recombinant inbred lines that showed specific susceptibility to the maize leaf aphid (*Rhopalosiphum maidis*) accumulated HDMBOA-Glc instead of DIMBOA-Glc. The candidate gene responsible for this susceptibility encoded a DIMBOA-Glc 4-*O*-methyltransferase (BX10). Although HDMBOA-Glc was more toxic to *R. maidis* than DIMBOA-Glc *in vitro*, the methyltransferase reduced the degree of callose accumulation that was necessary for defence against aphid feedings. This indicated that apoplastic DIMBOA, but not HDMBOA-Glc, induced callose deposition (Ahmad *et al.*, 2011), suggesting a role for DIMBOA in signal transduction for the induction of physical defence responses based on callose deposition. On the basis of these findings, the major Bx species accumulating in maize may have been determined by trade-offs between the benefit from HDMBOA-Glc accumulation acting as a direct defence via its toxicity (but with the associated loss of callose deposition) and the benefit of callose deposition induced by DIMBOA. The complex defence systems associated with Bxs in maize may have evolved through its various interactions with its multiple enemies.

20.4 Significance of the Metabolic Processes of Avenanthramides in the Defence Response of Oats

20.4.1 Biosynthesis of avenanthramide phytoalexins in oats

Avs, a series of substituted hydroxycinnamic acid amides with anthranilates (Fig. 20.3), were originally discovered in groats and hulls

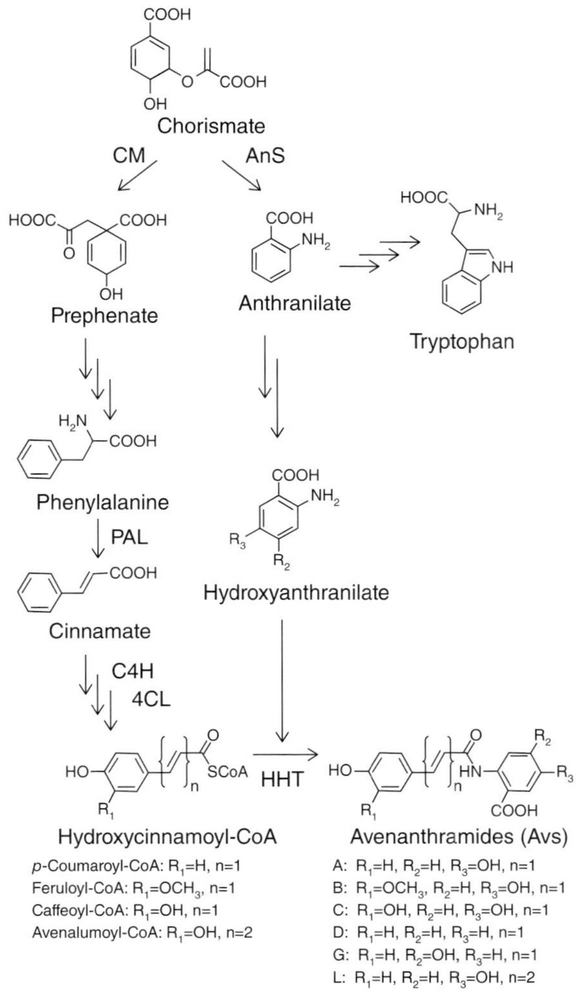

Fig. 20.3. Major biosynthetic pathway of avenanthramides (Avs). The anthranilate moieties of avenanthramides are derived from the tryptophan pathway. Note that the biosynthesis of avenanthramide L (avenalumoyl 5-hydroxyanthranilate) is different from that of the other avenanthramides in that it involves a reaction to elongate the C2 unit by using malonyl-CoA and *p*-coumaroyl-CoA (Ishihara *et al.*, 1999a). Abbreviations: AnS, anthranilate synthase; C4H, cinnamate 4-hydroxylase; 4CL, 4-coumarate:CoA ligase; CM, chorismate mutase; HHT, hydroxycinnamoyl-CoA:hydroxyanthranilate *N*-hydroxycinnamoyltransferase; PAL, phenylalanine ammonia-lyase.

of oats (*Avena sativa*) (Collins, 1989). These compounds have attracted attention because of their strong antioxidant activity in relation to human health (Bryngelsson *et al.*, 2002; Fagerlund *et al.*, 2009). The same compounds were identified as phytoalexins in oat leaves infected with crown rust fungus (*Puccinia coronata* f. sp. *avenae*) (Mayama *et al.*, 1981, 1982). The accumulation of Avs is also evoked by the treatment of oat leaves with various elicitors, including chitin fragments (Bordin *et al.*, 1991), a host-specific toxin victorin C (Mayama *et al.*, 1986), heavy metal ions (Fink *et al.*, 1990), calcium ionophores (Ishihara *et al.*, 1996) and benzothiadiazole (Wise, 2011).

The biosynthesis of Avs has been investigated through feeding experiments with putative precursors and measurements of enzyme activities. The feeding of elicited oat leaves with labelled Phe and anthranilate revealed that Avs are produced *de novo* from these precursors (Ishihara *et al.*, 1999a). Anthranilate and Phe are derived from chorismate synthesized via the shikimate pathway. The enzymes AnS and chorismate mutase (CM) catalyse the first committed reactions leading to anthranilate and Phe, respectively, and have been shown to be upregulated by the treatment with chitin fragments, namely oligo-*N*-acetylchitooligosaccharides (Matsukawa *et al.*, 2002a, 2002b), along with downstream enzymes in the phenylpropanoid pathway including Phe ammonia-lyase (PAL), cinnamate 4-hydroxylase (C4H), and 4-coumarate:CoA ligase (4CL) (Ishihara *et al.*, 1999b). In addition, hydroxycinnamoyl-CoA:hydroxyanthranilate *N*-hydroxycinnamoyltransferase (HHT), which catalyses the final condensation reaction, was inducibly detected in oat leaves treated with oligo-*N*-acetylchitooligosaccharides (Ishihara *et al.*, 1998) and with victorin C (Ishihara *et al.*, 1997). The genes encoding HHT were also identified (Yang *et al.*, 2004).

20.4.2 Metabolism of avenanthramides in elicitor-treated oat leaves

While investigating the fundamental mechanisms of Av biosynthesis, a discrepancy between the substrate specificity of HHT and the composition of the Avs accumulated in the elicited oat leaves was noticed: the most preferable acyl donor for HHT is feruloyl-CoA, but the Av B-bearing feruloyl moiety is not the most major Av species (Ishihara *et al.*, 1999b). In addition, once the levels of Avs produced by elicitor-treated leaf segments reached a maximum, they gradually decreased (Ishihara *et al.*, 1996, 1999a). These findings strongly suggested that Avs are actively metabolized. If this is the case, Av B is biosynthesized at a high rate, but a high metabolic rate may negate this.

The metabolism of Avs in oat leaves treated with a chitin elicitor has been investigated using labelled Avs (Okazaki *et al.*, 2004b). When oat leaf segments with the lower epidermis peeled off were floated on the elicitor solution, most of the Avs accumulated in the solution. The transfer of elicited oat leaves to solutions containing stable-isotope-labelled Avs led to a rapid decrease in the amount of labelled Avs in the solution, suggesting active metabolism of Avs. The rate of decrease was enhanced by elicitor treatment and was dependent on the species of Avs, with Av B decreasing the most rapidly. This result was in accordance with the hypothesis that Avs were actively metabolized by the plant itself, and also suggested that the metabolism of Avs was a part of the elicitor-inducible defence reactions of oats. The rates of biosynthesis and metabolism of Avs A and B were also determined using a model of isotope-labelling dynamics based on the administration of the small amounts of labelled Avs to elicited oat leaves. As a result, Av B was found to be more actively biosynthesized and metabolized than was Av A.

The metabolic fates of Avs were analysed by the administration of radiolabelled Av B to elicitor-treated oat leaves. In elicitor-treated leaf segments, significant radioactivity was incorporated into the cell wall fraction, and the activity was twice that of water-treated control leaves. The radioactivity was released by the treatment of the segments by 3 mol l^{-1} KOH. Gel filtration of the released radioactive fraction obtained by 3 mol l^{-1} KOH treatment indicated that radioactivity was present in the high-molecular-weight compound fraction. Thus, Av B is not simply linked to the cell wall components by ester bonds. In addition, about half of the radioactivity was detected in the elicitor solution, of which radioactivity from Av B accounted for only a part. The data clearly indicated that water-soluble metabolites derived from Av B were produced in elicitor solutions. These metabolites were dehydrodimers of Av B and are referred to as bisavenanthramides (bisAvs) B1–B6 (Okazaki *et al.*, 2004a, 2007). Because these dehydrodimers were considered to be formed by radical coupling reactions, the involvement of peroxidases in the metabolism of Avs was suggested. Consistent with this expectation, the decrease in stable-isotope-labelled Avs was suppressed by inhibitors of peroxidase, such as catalase, salicylhydroxamic acid and ascorbate. Further, peroxidase activity that accepts Av B as a substrate was enhanced in apoplastic fractions by elicitor treatment. The appearance of multiple basic isoperoxidases was also observed by activity staining with 3-amino-9-ethylcarbazole coupled with isoelectric focusing of proteins from elicitor-treated leaves. These findings suggest that accumulated Avs are further metabolized in apoplasts of the oat leaves by inducible isoperoxidases.

The metabolism of phytochemicals is sometimes involved in the enhancement of certain biological functions in plants. For example, in grapevine (*V. vinifera*), viniferins produced by the dimerization of resveratrol in response to a fungal infection show stronger antifungal activity than the monomer (Langcake and Pryce, 1977). As described above, Avs are actively metabolized in the plant itself, suggesting a possibility that metabolites derived from Avs have stronger antifungal activity than Avs. In this context, the antifungal activity of bisAv B-6, the major dehydrodimer of Avs, was investigated. This compound inhibited the germination of spores of both *Helminthosporium victoriae* and *Curvularia lunata* in a dose-dependent manner, and completely inhibited spore germination when it was present at 5 mM (Table 20.1). Spore germination was also inhibited by Av B, but this was less effective than bisAv B. Dimerization of Avs by peroxidase thus contributes to the enhancement of the toxicity of oat phytoalexins as observed in the case of the viniferins in grapevine.

Table 20.1. Antifungal activities of the bisavenanthramide (bisAv B-6) and the avenanthramide (Av B) as demonstrated by the inhibition of conidial germination of *Helminthosporium victoriae* and *Curvularia lunata*.[a]

	Germination rate of conidia (%)	
	H. victoriae	*C. lunata*
Water (1.5% DMSO)	100	100
BisAv B-6 (mM)		
0.1	91	99
1	61	85
3	15	41
5	0	0
Av B (mM)		
0.1	86	96
1	87	95
3	85	70
5	69	67

[a]Spores of *H. victoriae* and *C. lunata* were suspended in a water containing 4% sucrose (2×10^5 spores ml^{-1}). To these suspensions, bisAv B-6 and Av B dissolved in water containing 3% DMSO (dimethylsulfoxide) were added. After incubation at 20°C for 12 h, the germinated spores were manually counted under a light microscope using a haemocytometer.

Avs can be classified as hydroxycinnamic acid amides based on their chemical structure. These amides have frequently been shown to play a role in the reinforcement of cell walls. For instance, hydroxycinnamic acid amides of tyramine in solanaceous plants have been indicated to be a component of the suberin that accumulates in mechanically damaged tissues (Negrel *et al.*, 1993; Keller *et al.*, 1996). Hydroxycinnamoylagmatine has also been indicated to be incorporated into cell walls after pathogen infection in barley (von Röpenack *et al.*, 1998) and in *Arabidopsis* (Muroi *et al.*, 2009, 2012). To analyse the relationship between Av accumulation and cell wall reinforcement, elicitor-treated oat leaves were treated with cellulase and pectinase, and the number of released protoplasts was counted. A 6 h incubation in the presence of cell wall degradation enzymes led to the formation of protoplasts from water-treated and control oat leaves, and almost all of the leaf tissues except the upper epidermis were digested (Fig. 20.4). In contrast, elicitor treatment completely suppressed protoplast formation from oat leaves. So it is very likely that Avs incorporated into cell walls are involved in cell wall reinforcement.

Hydroxycinnamic acid amides are frequently suggested to play a role in relation to the reinforcement of cell walls. Avs may have been recruited as phytoalexins from those compounds that function to reinforce cell walls because of their toxicity to pathogens, and may still have three roles, namely as phytoalexins, in chemical defence, as precursors of more toxic dimers and as substrates for the reinforcement of cell walls in physical defence. The existence of differences in rates of metabolism among Avs suggests that their degree of contribution to these roles is dependent on the Av species.

20.5 Accumulation of Serotonin in Rice in Response to Biological Stimuli

20.5.1 Occurrence of serotonin and its putative ecological roles in plants

Serotonin is a Trp-derived amine widely distributed in the plant kingdom and has been suggested to have defensive roles. It was first

Number of released protoplasts (cells g^{-1} fresh weight)

None 1.63×10^7 1.61×10^7

Fig. 20.4. The induction of cell wall resistance to degradation enzymes after treatment with the elicitor deacetylated chitin. After peeling off the lower epidermis, the segments of first oat leaves were floated in a solution of partially deacetylated chitin (a) or water (b). After a 48 h incubation, the leaves were treated with 30 mM potassium phosphate buffer (pH 5.6) containing 2% cellulase, 0.1% pectolyase Y-23 (cell wall-degrading enzymes) and 0.7 M mannitol. After a further 6 h incubation, the numbers of released protoplasts were manually counted under a light microscope using a haemocytometer. Leaf segments not treated with either the elicitor or water were also incubated with the cell wall-degrading enzyme mixture (c).

identified from cowhage (*Mucuna pruriens*) (Bowden *et al.*, 1954), and was later demonstrated as occurring in several edible plants, including tomato (*Solanum lycopersicum*), paprika (*Capsicum annuum*), aubergine (*S. melongena*), spinach (*Spinacia oleracea*), lettuce (*Lactuca serriola*), garden pea (*Pisum sativum*), pineapple (*Ananas comosus*), papaya (*Carica papaya*), banana (*Musa sapientum*), strawberry (*Fragaria × ananassa*), coffee (*Coffea canephora* and *C. arabica*) and walnuts (*Juglans ailanthifolia*, *J. mandshurica* and *J. nigra*) (Udenfriend *et al.*, 1955, 1959; Waalkes *et al.*, 1958; West 1960; Erspamer, 1961; Applewhite, 1973; Grobe, 1982; Stranc, 1993; Ly *et al.*, 2008). Some vegetables have been shown to contain high concentrations of serotonin (Ly *et al.*, 2008). Tomato and Chinese cabbage (*Brassica rapa*) contain serotonin at concentrations of 221.9 μg g^{-1} and 110.9 μg g^{-1} fresh weight, respectively. Furthermore, it was shown that griffonia (*Griffonia simplicifolia*), sedum (*Sedum pachyphyllum*), mimosa (*Mimosa tenuiflora* and *M. pudica*), wood nettle (*Urtica cubensis*) and stinging bush (*Laportea moroides*) all contain serotonin in their shoots (Collier and Chesher, 1956; Fellows and Bell, 1971; Regula and Devide, 1980; Kimbrough *et al.*, 1987; Nicasio *et al.*, 2005; Ramakrishna *et al.*, 2009). Seeds of griffonia, which are used as herbal supplements, contain a high concentration of serotonin at 18 mg g^{-1} fresh weight, while leaves of the plant contain a negligible amount of serotonin at less than 1.7 ng g^{-1} fresh weight (Fellows and Bell, 1971).

Serotonin is a monoamine neurotransmitter in the central nervous systems of animals, including humans (Berger *et al.*, 2009), where it assists in regulating mood, appetite, sleep and pain (Jouvet, 1999; Sommer, 2004; Jann and Slade, 2007; Ruhé *et al.*, 2007; Yadav *et al.*, 2009). In humans, serotonin is located primarily in enterochromaffin cells and secreted into the blood. The pain experienced by mammals after a sting by a wasp or a honeybee is enhanced because of the existence of serotonin in the venoms (Jaques and Schachter, 1954; Rash and Hodgson, 2001; Weisel-Eichler and Libersat, 2004). Furthermore, insects also use serotonin as a neurotransmitter that regulates circadian rhythms, feeding behaviour, the visual and olfactory systems, and memory (Nässel, 1988; Tomioka *et al.*, 1993; Kaufmann *et al.*, 2004; Sitaraman, *et al.*, 2008; Ellen and Mercer, 2012). Therefore, a high intake of serotonin from food resources could disrupt their endocrine systems. For instance, feeding experiments involving ants (*Camponotus mus*) fed with serotonin revealed that the feeding behaviour of those ants, especially the sucking-pump behaviour that contributes to sugar intake through nectar, was suppressed (Falibene *et al.*, 2012). Another feeding experiment on the cabbage looper (*Trichoplusia ni*) showed that leaf-trenching behaviour was induced by serotonin feeding (Dussourd, 2003). Leaf trenching is a defensive response of phytophagous insects against plant chemical barriers; when the cabbage loopers encounter plant latex, unpalatable sap or other exudates, they stop feeding and initiate leaf trenching. Another type of effect is the interruption of insect reproduction. The female cabbage butterfly (*Pieris rapae*) fed with serotonin exhibited decreased rates of mating in a field experiment (Obara *et al.*, 2011). The occurrence of serotonin in trichomes – which act as a physical barrier against herbivorous insects – was reported in the bull nettle (*U. dioica*), Himalayan nettle (*Girardinia heterophylla*) and stinging bush (*Laportea moroides*) (Lookadoo and Pollard, 1991; Upton, 2013). These findings suggest that serotonin functions in defence against herbivores in plants. In addition, as described below, the involvement of serotonin in the defence response against pathogenic infection has also been demonstrated.

20.5.2 Critical role of serotonin accumulation in the interaction between rice and its pathogens

The involvement of secondary metabolites derived from the Trp pathway in plant defence responses has been observed in rice, as in other grasses. In the case of rice, Trp metabolites are involved in inducible defence responses against pathogen attack, in a manner similar to that of Avs in oats (Ishihara *et al.*, 1999b; Matsukawa *et al.*, 2002b). To examine the possibility of the involvement of the Trp pathway in the defence responses of rice, the first step was to analyse the expression of the genes encoding AnS subunits in leaves inoculated with a causal agent of rice brown spot, *Bipolaris oryzae* (Ishihara *et al.*, 2008a). The transcripts of the genes encoding one isoform of AnSα and two isoforms of AnSβ displayed a marked increase in leaves infected with *B. oryzae*. Increases in the amounts of AnS transcripts accompanied the enhancement in AnS enzyme activity. In addition, the amounts of two intermediates in the Trp biosynthesis pathway, anthranilate and indole, increased in parallel with Trp. These findings indicated the upregulation of the Trp pathway in response to pathogen attack.

To identify the secondary metabolites that are synthesized in tandem with the upregulation of the Trp pathway already identified, the leaves that had been inoculated with *B. oryzae* were extracted and the contents of indole compounds analysed by LC-MS/MS (liquid chromatography-tandem mass spectrometry) in the MRM (multiple reaction monitoring) mode. The accumulated compounds included tryptamine, serotonin, 5-hydroxytryptophan, 5-methoxytryptamine, feruloyltryptamine, *p*-coumaroylserotonin and feruloylserotonin. Serotonin content significantly exceeded that of other compounds, with the maximum serotonin concentration being approximately 250 nmol g^{-1} fresh weight. The accumulation of serotonin was preceded by that of tryptamine, suggesting that serotonin is synthesized from tryptamine.

The metabolic fate of the accumulated serotonin was analysed by adding ^{14}C-labelled serotonin to the culture medium, after which the leaves were inoculated with *B. oryzae* conidia. The leaves were repeatedly extracted by methanol to remove soluble substances, following which radioactivity was detected by autoradiography. Strong radioactivity was found at lesions, indicating the incorporation of serotonin into cell walls at the sites of infection. In addition, peroxidase that accepted serotonin as a substrate was activated in the *B. oryzae*-infected leaves. Thus, peroxidase-mediated polymerization of serotonin is the mechanism of incorporation of serotonin into cell walls. The polymerized serotonin is considered to serve as a phenolic barrier against pathogen infection. This polymer may have different features from other Phe-derived polymers, such as lignin and suberin. The detailed characterization of the phenolic polymers that are derived from serotonin is an important goal to be addressed in the future.

To examine the significance of serotonin accumulation, a rice mutant that lacks serotonin accumulation was utilized; *sl* (Sekiguchi-Asahi) is a lesion mimic mutant, and forms spontaneously propagating lesions on old leaves (Kiyosawa, 1970). Similar lesions are also formed at sites of pathogen infection on the *sl* leaves (Ueno *et al.*, 2003). The concentrations of tryptamine and serotonin were determined in *B. oryzae*-infected leaves (Ishihara *et al.*, 2008a). The leaves accumulated a high concentration of tryptamine (Ueno *et al.*, 2003), but the level of serotonin was low, whereas the wild type (Asahi) leaves accumulated serotonin at a high concentration and the level of tryptamine was low. Therefore, *sl* is a mutant defective in the step that converts tryptamine into serotonin. An additional difference was found in the appearance of the lesions: the colour of the lesions on *sl* leaves were pale yellow, while those on the wild type were dark brown. Microscopic inspection showed that the infection sites on the wild type leaves were surrounded by rings of dark brown materials, whereas these materials were absent on *sl* leaves. The non-accumulation of dark brown materials at the infection sites in *sl* leaves was reversed after feeding the seedlings with serotonin from their roots, suggesting that the brown materials were derived from serotonin. Taking into consideration the inducible activation of peroxidase, the brown materials were considered to be polymers formed by the peroxidase-catalysed oxidation of serotonin. In addition, in the *sl* leaves, the average length of fungal hyphae was longer than that in the wild type leaves, indicating the effectiveness of serotonin accumulation in suppressing hyphal development in the leaf tissues.

The role of serotonin in the defence response of rice was further investigated by using an inhibitor of tryptophan decarboxylase (TDC), *S*-α-fluorotryptophan (*S*-α-FMT) (Ishihara *et al.*, 2011). This compound serves as a suicide substrate of TDC. Treatment of rice seedlings with *S*-α-FMT resulted in suppression of the accumulation of tryptamine, serotonin and dark brown materials in the *B. oryzae*-infected leaves. Leaves treated with this compound also displayed enhanced foliar damage upon *B. oryzae* infection, but this damage, as well as the accumulation of serotonin and brown materials, was reversed after feeding with tryptamine. On the basis of these findings, the accumulation of serotonin and serotonin-derived brown materials is a component of effective antifungal defence mechanisms in rice.

The biosynthetic pathway for serotonin has been elucidated by the analysis of TDC genes. The rice genome contains genes encoding seven aromatic amino acid decarboxylases (Kang *et al.*, 2007). From these, the genes for two (AK069031 and AK103253) were expressed in *Escherichia coli*; the purified gene products specifically catalysed the decarboxylation of Trp. The two TDC genes were also expressed in rice plants under the control of the maize ubiquitin promoter. Transgenic rice lines expressing these genes at high levels accumulated serotonin as well as tryptamine, indicating that serotonin is synthesized from Trp via tryptamine (Kang *et al.*, 2007; Dubouzet *et al.*, 2013). In addition, this finding suggested that tryptamine 5-hydroxylase (T5H), which converts tryptamine to serotonin, is constitutively expressed and is not a rate-limiting step in serotonin synthesis. This biosynthetic route was supported by the enhanced accumulation of the transcript of a TDC gene (AK069031) and tryptamine accumulation preceding serotonin accumulation in *B. oryzae*-infected leaves (Ishihara *et al.*, 2008a). More recently, an additional TDC gene (BAD11583) was characterized (Kanjanaphachoat *et al.*, 2012) and shown to encode a functional TDC; the plants overexpressing this gene also accumulated high concentrations of serotonin. These transgenic plants also exhibited the stunted growth, dark brown pigmentation, low fertility and accumulation of a serotonin dimer. The gene encoding T5H was identified by analysis of the *sl* mutant (Fujiwara *et al.*, 2010). The causal mutation was identified by map-based cloning and found in a gene encoding a cytochrome P450 enzyme, CYP71P1. The recombinant wild type protein catalysed the hydroxylation of tryptamine to serotonin. It has been reported that foliar treatments of barley and rice seedlings with some indole compounds, including tryptamine, enhance resistance against the blast fungus (*Magnaporthe grisea*) (Ueno *et al.*, 2004, 2011). Considering the constitutive expression of the T5H gene, tryptamine may act on plants after being converted into serotonin.

The infection-induced accumulation of serotonin is not limited to rice. The conidia of *B. oryzae* and *B. maydis* were inoculated to the leaves of other grasses, including barley, wheat, oats, maize, sorghum, proso millet (*Panicum miliaceum*), Japanese barnyard millet (*Echinochloa esculenta*) and foxtail millet (*Setaria italica*). Among these barley, wheat, foxtail millet and Japanese barnyard millet accumulated serotonin in response to pathogen infection, together with hydroxycinnamic acid amides of serotonin (Fig. 20.5). In other species, the accumulated amounts of serotonin were quite small although the amount increased in infected leaves. According to Bouchenak-Khelladi *et al.* (2008), the species analysed could be classified into two clades in the *Poaceae*, the BEP clade (rice, barley, wheat and oats) and the PACCMAD clade (maize, sorghum,

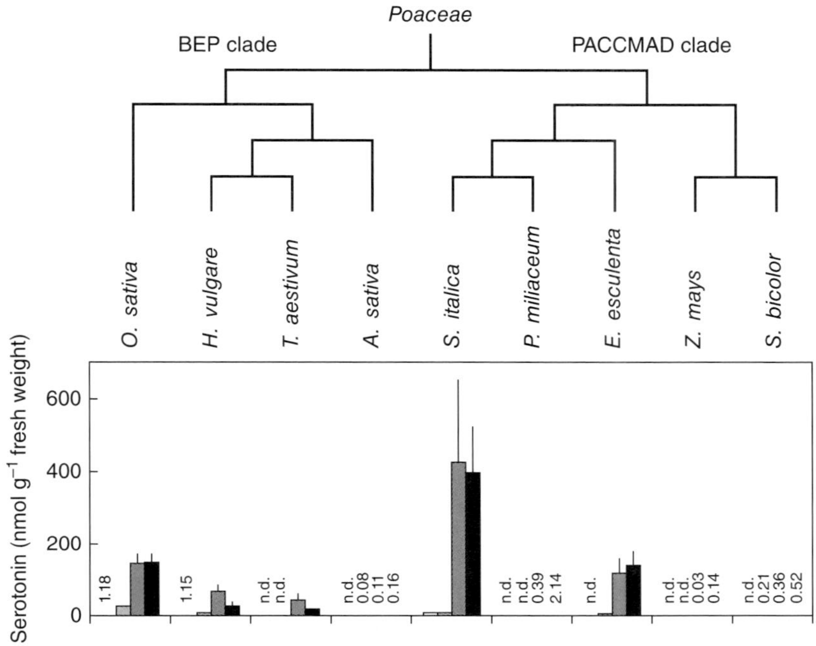

Fig. 20.5. Schematic representation of a phylogenetic tree of some species of *Poaceae* species and their accumulation of serotonin in response to pathogen infection. The phylogenetic tree was generated according to Bouchenak-Khelladi *et al.* (2008). Droplets of water (grey bars), and suspensions of conidia of *Bipolaris oryzae* (dark grey bars) and of *B. maydis* (black bars) were placed on the leaves, and serotonin contents were determined 48 h after inoculation. Serotonin contents in intact leaves (white bars) were also determined. Data represent means with SD of three independent replicates. The species used (in order of display) are: *Oryza sativa*, *Hordeum vulgare*, *Triticum aestivum*, *Avena sativa*, *Setaria italica*, *Panicum miliaceum*, *Echinochloa esculenta*, *Zea mays*, *Sorghum bicolor*.

proso millet, Japanese barnyard millet and foxtail millet). It is therefore likely that the accumulation of serotonin and its amides was a trait acquired by a common ancestor of the BEP and PACCMAD clades in the early evolution of species of *Poaceae*.

In addition to pathogen infection, serotonin production is induced by various stimuli. For example, JA was observed to induce the accumulation of serotonin and feruloylserotonin in rice leaves (Ishihara *et al.*, 2008a). As JA has been shown to be an effective inducer of phytoalexins in rice (Nojiri *et al.*, 1996; Tamogami *et al.*, 1997), serotonin accumulation is considered to be under a similar control as the accumulation of other phytoalexins. Leaf feeding by larvae of the rice striped stem borer (*Chilo suppressalis*) also induced the accumulation of serotonin, together with hydroxycinnamic acid amides of serotonin (Ishihara *et al.*, 2008b). These compounds may function to form a phenolic polymer at the site wounded by these larvae to prevent the loss of water and/or opportunistic infections by microorganisms. The accumulation of serotonin has also been suggested to play a role in senescence (Kang *et al.*, 2009). Serotonin and Trp accumulate to a significant extent in rice leaves undergoing senescence induced by either nutrient deprivation or leaf detachment. Another experiment on genetically manipulated plants accumulating either high or low levels of serotonin indicated later and faster senescence progression, respectively, than in the wild type, indicating the role of serotonin in the attenuation of senescence. An increase in methanol concentrations has been suggested to function as an endogenous signal that induces serotonin and Trp accumulation (Kang *et al.*, 2011). On the basis of a strong correlation between pectin methyltransferase activity and methanol concentrations during senescence, cell wall pectin is considered to be the major source of this methanol.

In the interaction between rice and *B. oryzae*, serotonin serves as a substrate for the reinforcement of cell walls. In addition, it may serve as a signal molecule that activates the expression of genes involved in defending against pathogen infection. Fujiwara *et al.* (2010) showed that serotonin treatment of rice seedlings resulted in the formation of smaller lesions when rice leaves were inoculated with the compatible blast fungus. Further, serotonin treatment of rice cell cultures induced the expression of defensive genes such as *PBZ1*, *CHT1*, *PAL1*, and *Cht3*. The induction of these genes by serotonin treatment was compromised in rice cell cultures expressing RNA interference knockdown constructs for OsRac1 and Sti/Hop, and in the cell culture of a dwarf1 mutant, which lacks a functional Gα subunit of the heterotrimeric G protein (Suharsono *et al.*, 2002). Thus, these proteins are involved in signal transduction for the expression defence genes from serotonin accumulation. In *Arabidopsis*, serotonin has been suggested to act as auxin inhibitor (Pelagio-Flores *et al.*, 2011). Exogenously applied serotonin stimulated lateral root development at concentrations from 10 to 160 μM. At higher concentrations, serotonin repressed lateral root growth and root hair development, and stimulated adventitious root hair development. Inhibition of the expression of DR5:uidA and BA3:uidA genes in roots and abolition of the effects of IAA and naphthalene acetic acid on auxin-regulated genes by serotonin application indicated anti-auxin activity of serotonin. This characteristic of serotonin may affect plant–microorganism interactions and, indeed, the effect of auxin on such interactions has been reported in various studies (Spaepen *et al.*, 2007; Bari and Jones, 2009).

20.6 Concluding Remarks

In this chapter, we have described the biosynthesis, regulation, physiological and ecological significance, and evolution of three groups of compounds, Bxs, Avs and serotonin. The biosynthetic pathways for these compounds diverge from the Trp pathway at different intermediates or at Trp itself. These secondary metabolites all serve for defensive purposes, but in different ways. Bxs are phytoanticipins that accumulate before attacks by herbivores or pathogens (Niemeyer, 1988). After infection or herbivory occurs, the corresponding toxic aglycones are released as a chemical weapon. In addition, analysis of metabolic changes in the pathogen-infected leaves has shown that the generation of HDMBOA-Glc is a unique system for preparing against pathogen infection (Oikawa *et al.*, 2001, 2002). Avs have been extensively characterized as phytoalexins of oats (Mayama *et al.*, 1981), and their significance in a chemical defence system has been proven by analysis of phytoalexin accumulation in various combinations of oat cultivars and races of crown rust fungus (Mayama *et al.*, 1982). In the resistant combinations, Avs accumulated rapidly and in large amounts when compared with susceptible combinations. Analysis of the metabolic fates of Avs showed that Avs are further metabolized by oat leaves and are incorporated into cell walls, resulting in reinforcement of those cell walls (Okazaki *et al.*, 2004b). Serotonin does not exert strong antifungal activity, but it also is incorporated into cell walls to form a phenolic polymer (Ishihara *et al.*, 2008a). Analyses both with the *sl* mutant and with an inhibitor of serotonin biosynthesis clearly indicated the significance of serotonin in the interaction of rice with *B. oryzae* (Ishihara *et al.*, 2008a) and *M. grisea* (Fujiwara *et al.*, 2010). Because the phenolic polymers that have been characterized in plants so far are exclusively derived from Phe, the serotonin polymer represents an entirely new class of phenolic polymer. In other words, the formation of a phenolic polymer is considered to be a discovery of a new function of Trp. As shown in these examples, investigations into the metabolic flow of secondary metabolites in grasses revealed multifunctional roles for such metabolites derived from the Trp pathway.

Bxs are distributed among species that belong to both BEP and PACCMAD clades in the *Poaceae* (Niemeyer, 1988). The biosynthetic genes in wheat and maize are orthologous, indicating that they share a common origin for Bx biosynthesis (Nomura *et al.*, 2002). However, some species lack this trait, for example *T. boeticum* and *H. vulgare* (Nomura *et al.*, 2007). These species are considered to have lost the biosynthetic ability for Bxs during their evolution. A similar situation exists in the case of serotonin. Infection-induced accumulation of serotonin does not necessarily occur in all species of *Poaceae*. Species that have lost these traits may have acquired the ability to biosynthesize other defensive metabolites. Recruiting the Trp pathway for the synthesis of different secondary metabolites may have been an easier route for a species to follow if the pathway continued to exhibit a potentially large capacity for metabolic flux. One such example might be the synthesis of Avs in oats.

Recently, the genes encoding glucosidase and glucosyltransferase that are involved in Bx metabolism have been isolated and characterized in a dicot species, larkspur (*Consolida orientalis*) (Dick *et al.*, 2012). The genes were not orthologous to the corresponding genes in species of *Poaceae*, indicating that the accumulation of Bxs in this species is a result of convergent evolution. Nevertheless, it may be of interest to systematically explore the species in other monocotyledonous families in terms of their accumulation of Bxs. Similarly, the origin of serotonin accumulation may be older than the divergence of the ancestor of the grass family from other monocotyledonous species. Through analysis of these metabolites in species inside and outside the *Poaceae*, it may be possible to elucidate a bigger picture of the evolution of defensive secondary metabolism derived from the Trp pathway.

References

Ahmad, S., Veyrat, N., Gordon-Weeks, R., Zhang, Y., Martin, J., Smart, L., Glauser, G., Erb, M., Flors, V., Frey, M. and Ton, J. (2011) Benzoxazinoid metabolites regulate innate immunity against aphids and fungi in maize. *Plant Physiology* 157, 317–327.

Applewhite, P.B. (1973) Serotonin and norepinephrine in plant tissues. *Phytochemistry* 12, 191–192.

Bari, R. and Jones, J.D.G. (2009) Role of plant hormones in plant defence responses. *Plant Molecular Biology* 69, 473–488.

Berger, M., Gray, J.A. and Roth, B.L. (2009) The expanded biology of serotonin. *Annual Review of Medicine* 60, 355–366.

Bohlmann, J., DeLuca, V., Eilert, U. and Martin, W. (1995) Purification and cDNA cloning of anthranilate synthase from *Ruta graveolens*: modes of expression and properties of native and recombinant enzymes. *The Plant Journal* 7, 491–501.

Bohlmann, J., Lins, T., Martin, W. and Eilert, U. (1996) Anthranilate synthase from *Ruta graveolens*. *Plant Physiology* 111, 507–514.

Bordin, A.P.A., Mayama, S. and Tani, T. (1991) Potential elicitors for avenalumin accumulation in oat leaves. *Annals of the Phytopathological Society of Japan* 57, 688–695.

Bouchenak-Khelladi, Y., Salamin, N., Savolainen, V., Forest, F., van der Bank, M., Chase, M.W. and Hodkinson, T.R. (2008) Large multi-gene phylogenetic trees of the grasses (*Poaceae*): progress towards complete tribal and generic level sampling. *Molecular Phylogenetics and Evolution* 47, 488–505.

Bowden, K., Brown, B.G. and Batty, J.E. (1954) 5-Hydroxytryptamine: its occurrence in cowhage. *Nature* 174, 925–926.

Brazier-Hicks, M., Offen, W.A., Gershater, M.C., Revett, T.J., Lim, E.K., Bowles, D.J., Davies, G.J. and Edwards, R. (2007) Characterization and engineering of the bifunctional *N*- and *O*-glucosyltransferase involved in xenobiotic metabolism in plants. *Proceedings of the National Academy of Sciences of the United States of America* 104, 20238–20243.

Bryngelsson, S., Mannerstedt-Fogelfors, B., Kamal-Eldin, A., Andersson, R. and Dimberg, L. H. (2002) Lipids and antioxidants in groats and hulls of Swedish oats (*Avena sativa* L.). *Journal of the Science of Food and Agriculture* 82, 606–614.

Bücker, C. and Grambow, H.J. (1990) Alterations in 1,4-benzoxazinone levels following inoculation with stem rust in wheat leaves carrying various alleles for resistance and their possible role as phytoalexins in moderately resistant leaves. *Zeitschrifut für Naturforschung C* 45, 1151–1155.

Caputi, L., Malnoy, M., Goremykin, V., Nikiforova, S. and Martens, S. (2012) A genome-wide phylogenetic reconstruction of family 1 UDP-glycosyltransferases revealed the expansion of the family during the adaptation of plants to life on land. *The Plant Journal* 69, 1030–1042.

Chu, H.Y., Wegel, E. and Osbourn, A. (2011) From hormones to secondary metabolism: the emergence of metabolic gene clusters in plants. *The Plant Journal* 66, 66–79.

Cicek, M. and Esen, A. (1999) Expression of soluble and catalytically active plant (monocot) β-glucosidases in *E. coli*. *Biotechnology and Bioengineering* 63, 392–400.

Cicek, M., Blanchard, D., Bevan, D.R. and Esen, A. (2000) The aglycone specificity-determining sites are different in 2,4-dihydroxy-7-methoxy-1,4-benzoxazin-3-one (DIMBOA)-glucosidase (maize β-glucosidase) and dhurrinase (sorghum β-glucosidase). *The Journal of Biological Chemistry* 275, 20002–20011.

Collier, H.O.J. and Chesher, G.B. (1956) Identification of 5-hydroxytryptamine in the sting of the nettle (*Urtica dioica*). *British Journal of Pharmacology and Chemotherapy* 11, 186–189.

Collins, F.W. (1989) Oat phenolics: avenanthramides, novel substituted *N*-cinnamoylanthranilate alkaloids from oat groats and hulls. *Journal of Agricultural and Food Chemistry* 37, 60–66.

Corcuera, L.J., Woodward, M.D., Helgeson, J.P., Kelman, A. and Upper, C.D. (1978) 2,4-Dihydroxy-7-methoxy-2*H*-1,4-benzoxazin-3(4*H*)-one, an inhibitor from *Zea mays* with differential activity against soft rotting *Erwinia* species. *Plant Physiology* 61, 791–795.

Dick, R., Rattei, T., Haslbeck, M., Schwab, W., Gierl, A. and Frey, M. (2012) Comparative analysis of benzoxazinoid biosynthesis in monocots and dicots: independent recruitment of stabilization and activation functions. *The Plant Cell* 24, 915–928.

Dubouzet, J.G., Ishihara, A., Matsuda, F., Miyagawa, H., Iwata, H. and Wakasa K (2007) Integrated metabolomic and transcriptomic analyses of high-tryptophan rice expressing a mutant anthranilate synthase alpha subunit. *Journal of Experimental Botany* 58, 3309–3321.

Dubouzet, J.G., Matsuda, F., Ishihara, A., Miyagawa, H. and Wakasa, K. (2013) Production of indole alkaloids by metabolic engineering of the tryptophan pathway in rice. *Plant Biotechnology Journal* 11, 1103–1111.

Dussourd, D.E. (2003) Chemical stimulants of leaf-trenching by cabbage loopers: natural products, neurotransmitters, insecticides, and drugs. *Journal of Chemical Ecology* 29, 2023–2047.

Dutartre, L., Hilliou, F. and Feyereisen, R. (2012) Phylogenomics of the benzoxazinoid biosynthetic pathway of *Poaceae*: gene duplications and origin of the Bx cluster. *BMC Evolutionary Biology* 12:64.

Ellen, C.W. and Mercer, A.R. (2012) Modulatory actions of dopamine and serotonin on insect antennal lobe neurons: insights from studies *in vitro*. *Journal of Molecular Histology* 43, 401–404.

Elnaghy, M.A. and Linko, P. (1962) The role of 4-*O*-glucosyl-2,4-dihydroxy-7-methoxy-1,4-benzoxazin-3-one in resistance of wheat to stem rust. *Physiologia Plantarum* 15, 764–771.

Erspamer, V. (1961) Recent research in the field of 5-hydroxytryptamine and related indolealkylamines. In: Jucker, E. (ed.) *Progress In Drug Research*. Birkhäuser, Basel, Germany, pp. 151–367.

Fagerlund, A., Sunnerheim, K. and Dimberg, L.H. (2009) Radical-scavenging and antioxidant activity of avenanthramides. *Food Chemistry* 113, 550–556.

Fahey, J.W., Zalcmann, A.T. and Talalay, P. (2001) The chemical diversity and distribution of glucosinolates and isothiocyanates among plants. *Phytochemistry* 56, 5–51.

Falibene, A., Rössler, W. and Josens, R. (2012) Serotonin depresses feeding behaviour in ants. *Journal of Insect Physiology* 58, 7–17.

Feldman, M. and Levy, A.A. (2005) Allopolyploidy: a shaping force in the evolution of wheat genomes. *Cytogenetic and Genome Research* 109, 250–258.

Fellows, L.E. and Bell, E.A. (1971) Indole metabolism in *Piptadenia peregrina*. *Phytochemistry* 10, 2083–2091.

Fink, W., Liefland, M. and Mendgen, K. (1990) Comparison of various stress responses in oat in compatible and nonhost resistant interactions with rust fungi. *Physiological and Molecular Plant Pathology* 37, 309–321.

Frey, M., Kliem, R., Saedler, H. and Gierl, A. (1995) Expression of a cytochrome P450 gene family in maize. *Molecular and General Genetics* 246, 100–109.

Frey, M., Chomet, P., Glawischnig, E., Stettner, C., Grün, S., Winklmair, A., Eisenreich, W., Bacher, A., Meeley, R.B., Briggs, S.P., Simcox, K. and Gierl, A. (1997) Analysis of a chemical defense mechanism in grasses. *Science* 277, 696–699.

Frey, M., Stettner, C., Paré, P.W., Schmelz, E.A., Tumlinson, J.H. and Gierl, A. (2000) An herbivore elicitor activates the gene for indole emission in maize. *Proceedings of the National Academy of Sciences of the United States of America* 97, 14801–14806.

Frey, M., Huber, K., Park, W.J., Sicker, D., Lindberg, P., Meeley, R.B., Simmons, C.R., Yalpani, N. and Gierl, A. (2003) A 2-oxoglutarate-dependent dioxygenase is integrated in DIMBOA-biosynthesis. *Phytochemistry* 62, 371–376.

Frey, M., Schullehner, K., Dick, R., Fiesselmann, A. and Gierl, A. (2009) Benzoxazinoid biosynthesis, a model for evolution of secondary metabolic pathways in plants. *Phytochemistry* 70, 1645–1651.

Fujiwara, T., Maisonneuve, S., Isshiki, M., Mizutani, M., Chen, L., Wong, H.L., Kawasaki, T. and Shimamoto, K. (2010) Sekiguchi lesion gene encodes a cytochrome P450 monooxygenase that catalyzes conversion of tryptamine to serotonin in rice. *The Journal of Biological Chemistry* 285, 11308–11313.

Gachon, C.M., Langlois-Meurinne, M. and Saindrenan, P. (2005) Plant secondary metabolism glycosyltransferases: the emerging functional analysis. *Trends in Plant Science* 10, 542–549.

Gianoli, E. and Niemeyer, H.M. (1997a) Environmental effects on the accumulation of hydroxamic acids in wheat seedlings: the importance of plant growth rate. *Journal of Chemical Ecology* 23, 543–551.

Gianoli, E. and Niemeyer, H.M. (1997b) Lack of costs of herbivory-induced defenses in a wild wheat: integration of physiological and ecological approaches. *Oikos* 80, 269–275.

Gierl, A. and Frey, M. (2001) Evolution of benzoxazinone biosynthesis and indole production in maize. *Planta* 213, 493–498.

Glauser, G., Marti, G., Villard, N., Doyen, G.A., Wolfender, J.-L., Turlings, T.C.J. and Erb, M. (2011) Induction and detoxification of maize 1,4-benzoxazin-3-ones by insect herbivores. *The Plant Journal* 68, 901–911.

Grambow, H.J., Lückge, J., Klausener, A. and Müller, E. (1986) Occurrence of 2-(2-hydroxy-4,7-dimethoxy-2H-1,4-benzoxazin-3-one)-β-D-glucopyranoside in *Triticum aestivum* leaves and its conversion into 6-methoxy-benzoxazolinone. *Zeitschrift für Naturforschung C* 41, 684–690.

Grobe, W. (1982) Function of serotonin in seeds of walnuts. *Phytochemistry* 21, 819–822.

Grower, B.G. and Leete, E. (1963) Biosynthesis of gramine: the immediate precursors of the alkaloid. *Journal of the American Chemical Society* 20, 3683–3685.

Grün, S., Frey, M. and Gierl, A. (2005) Evolution of the indole alkaloid biosynthesis in the genus *Hordeum*: distribution of gramine and DIBOA and isolation of the benzoxazinoid biosynthesis genes from *Hordeum lechleri*. *Phytochemistry* 66, 1264–1272.

Gus-Mayer, S., Brunner, H., Schneider-Poetsch, H.A. and Rüdiger, W. (1994) Avenacosidase from oat: purification, sequence analysis and biochemical characterization of a new member of the BGA family of β-glucosidases. *Plant Molecular Biology* 26, 909–921.

Gutierrez, C., Castañera, P. and Torres, V. (1988) Wound-induced changes in DIMBOA (2,4-dihydroxy-7-methoxy-2*H*-1,4-benzoxazin-3(4*H*)-one) concentration in maize plants caused by *Sesamia nonagrioides* (Lepidoptera: Noctuidae). *Annals of Applied Biology* 113, 447–454.

Hösel, W., Tober, I., Eklund, S.H. and Conn, E.E. (1987) Characterization of β-glucosidases with high specificity for the cyanogenic glucoside dhurrin in *Sorghum bicolor* (L.) Moench seedlings. *Archives of Biochemistry and Biophysics* 252, 152–162.

Huang, S., Sirikhachornkit, A., Su, X., Faris, J., Gill, B., Haselkorn, R. and Gornicki, P. (2002) Genes encoding plastid acetyl-CoA carboxylase and 3-phosphoglycerate kinase of the *Triticum/Aegilops* complex and the evolutionary history of polyploid wheat. *Proceedings of the National Academy of Sciences of the United States of America* 99, 8133–8138.

Huffaker, A., Dafoe, N.J. and Schmelz, E.A. (2011) ZmPep1, an ortholog of *Arabidopsis* elicitor peptide 1, regulates maize innate immunity and enhances disease resistance. *Plant Physiology* 155, 1325–1338.

Hughes, E.H., Hong, S.B., Gibson, S.I., Shanks, J.V. and San, K.Y. (2004) Expression of a feedback-resistant anthranilate synthase in *Catharanthus roseus* hairy roots provides evidence for tight regulation of terpenoid indole alkaloid levels. *Biotechnology and Bioengineering* 86, 718–727.

Hughes, M.A., Brown, K., Pancoro, A., Murray, B.S., Oxtoby, E. and Hughes, J. (1992) A molecular and biochemical analysis of the structure of the cyanogenic β-glucosidase (linamarase) from cassava (*Manihot esculenta* Cranz). *Archives of Biochemistry and Biophysics* 295, 273–279.

Ishihara, A., Miyagawa, H., Kuwahara, Y., Ueno, T. and Mayama, S. (1996) Involvement of Ca^{2+} ion in phytoalexin induction in oats. *Plant Science* 115, 9–16.

Ishihara, A., Matsukawa, T., Miyagawa, H., Ueno, T., Mayama, S. and Iwamura, H. (1997) Induction of hydroxycinnamoyl-CoA: hydroxyanthranilate *N*-hydroxycinnamoyltransferase (HHT) activity in oat leaves by victorin C. *Zeitschrift für Naturforschung C* 52, 756–760.

Ishihara, A., Miyagawa, H., Matsukawa, T., Ueno, T., Mayama, S. and Iwamura, H. (1998) Induction of hydroxyanthranilate hydroxycinnamoyl transferase activity by oligo-*N*-acetylchitooligosaccharides in oats. *Phytochemistry* 47, 969–974.

Ishihara, A., Ohtsu, Y. and Iwamura, H. (1999a) Biosynthesis of oat avenanthramide phytoalexins. *Phytochemistry* 50, 237–242.

Ishihara, A., Ohtsu, Y. and Iwamura, H. (1999b) Induction of biosynthetic enzymes for avenanthramides in elicitor-treated oat leaves. *Planta* 208, 512–518.

Ishihara, A., Asada, Y., Takahashi, Y., Yabe, N., Komeda, Y., Nishioka, T., Miyagawa, H. and Wakasa, K. (2006) Metabolic changes in *Arabidopsis thaliana* expressing the feedback-resistant anthranilate synthase α subunit gene OASA1D. *Phytochemistry* 67, 2349–2362.

Ishihara, A., Hashimoto, Y., Miyagawa, H. and Wakasa, K. (2008a) Induction of serotonin accumulation by feeding of rice striped stem borer in rice leaves. *Plant Signaling and Behavior* 3, 714–716.

Ishihara, A., Hashimoto, Y., Tanaka, C., Dubouzet, J.G., Nakao, T., Matsuda, F., Nishioka, T., Miyagawa, H. and Wakasa, K. (2008b) The tryptophan pathway is involved in the defense responses of rice against pathogenic infection via serotonin production. *The Plant Journal* 54, 481–495.

Ishihara, A., Nakao, T., Mashimo, Y., Murai, M., Ichimaru, N., Tanaka, C., Nakajima, H., Wakasa, K. and Miyagawa, H. (2011) Probing the role of tryptophan-derived secondary metabolism in defense responses against *Bipolaris oryzae* infection in rice leaves by a suicide substrate of tryptophan decarboxylase. *Phytochemistry* 72, 7–13.

Jann, M.W. and Slade, J.H. (2007) Antidepressant agents for the treatment of chronic pain and depression. *Pharmacotherapy* 27, 1571–1587.

Jaques, R. and Schachter, M. (1954) The presence of histamine, 5-hydroxytryptamine and a potent, slow contracting substance in wasp venom. *British Journal of Pharmacology and Chemotherapy* 9, 53–58.

Jonczyk, R., Schmidt, H., Osterrieder, A., Fiesselmann, A., Schullehner, K., Haslbeck, M., Sicker, D., Hofmann, D., Yalpani, N., Simmons, C., Frey, M. and Gierl, A. (2008) Elucidation of the final reactions of DIMBOA-glucoside biosynthesis in maize: characterization of *Bx6* and *Bx7*. *Plant Physiology* 146, 1053–1063.

Jouvet, M. (1999) Sleep and serotonin: an unfinished story. *Neuropsychopharmacology* 21, 24S–27S.

Kang, K., Kim, Y.S., Park, S. and Back, K. (2009) Senescence-induced serotonin biosynthesis and its role in delaying senescence in rice leaves. *Plant Physiology* 150, 1380–1393.

Kang, K., Park, S., Natsagdorj, U., Kim, Y.S. and Back, K. (2011) Methanol is an endogenous elicitor molecule for the synthesis of tryptophan and tryptophan-derived secondary metabolites upon senescence of detached rice leaves. *The Plant Journal* 66, 247–257.

Kang, S., Kang, K., Lee, K. and Back, K., (2007) Characterization of rice tryptophandecarboxylases and their direct involvement in serotonin biosynthesis in transgenic rice. *Planta* 227, 263–272.

Kanjanaphachoat, P., Wei, B.Y., Lo, S.F., Wang, I.W., Wang, C.S., Yu, S.M., Yen, M.L., Chiu, S.H., Lai, C.C. and Chen, L.J. (2012) Serotonin accumulation in transgenic rice by over-expressing tryptophan decarboxylase results in a dark brown phenotype and stunted growth. *Plant Molecular Biology* 78, 525–543.

Kaufmann, L., Schürmann, F., Yiallouros, M., Harrewijn, P. and Kayser, H. (2004) The serotonergic system is involved in feeding inhibition by pymetrozine. Comparative studies on a locust (*Locusta migratoria*) and an aphid (*Myzus persicae*), *Comparative Biochemistry and Physiology – Part C: Toxicology and Pharmacology* 138, 469–483.

Keller, H., Hohlfeld, H., Wray, V., Hahlbrock, K., Scheel, D. and Strack, D. (1996) Changes in the accumulation of soluble and cell wall-bound phenolics in elicitor-treated cell suspension cultures and fungus-infected leaves of *Solanum tuberosum*. *Phytochemistry* 42, 389–396.

Kimbrough, T.D., Reynolds, J.D., Humphreys, K.J. and Weekley, L.B. (1987) Diurnal changes in tissue leaf levels of tryptophan, tyrosine and amine metabolites in *Sedum morganianum* and *Sedum pachyphyllum*. *Biochemie und Physiologie der Pflanzen* 182, 67–72.

Kiyosawa, S. (1970) Inheritance of a particular sensitivity of the rice variety, Sekiguchi Asahi, to pathogens and chemicals, and linkage relationship with blast resistance genes. *Bulletin of the National Institute of Agricultural Sciences (Japan). Series D: Plant Physiology, Genetics and Crops in General* 21, 61–71.

Kleczkowski, K., Schell, J. and Bandur, R. (1995) Phytohormone conjugates: nature and function. *Critical Reviews in Plant Sciences* 14, 283–298.

Kliebenstein, D.J. and Osbourn, A. (2012) Making new molecules – evolution of pathways for novel metabolites in plants. *Current Opinion in Plant Biology* 15, 415–423.

Kreps, J.A, Ponappa, T., Dong, W. and Town, C.D. (1996) Molecular basis of α-methyltryptophan resistance in *amt-1*, a mutant of *Arabidopsis thaliana* with altered tryptophan metabolism. *Plant Physiology* 110, 1159–1165.

Lairson, L.L., Henrissat, B., Davies, G.J. and Withers, S.G. (2008) Glycosyltransferases: structures, functions, and mechanisms. *Annual Review of Biochemistry* 77, 521–555.

Langcake, P. and Pryce, R.J. (1977) A new class of phytoalexins from grapevines. *Experientia* 33, 151–152.

Larsson, K.A.E., Zetterlund, I., Delp, G. and Jonsson, L.M.V. (2006) *N*-Methyltransferase involved in gramine biosynthesis in barley: cloning and characterization. *Phytochemistry* 67, 2002–2008.

Li, J. and Last, R.L. (1996) The *Arabidopsis thaliana trp5* mutant has a feedback-resistant anthranilate synthase and elevated soluble tryptophan. *Plant Physiology* 110, 51–59.

Li, L., Modolo, L.V., Escamilla-Trevino, L.L., Achnine, L., Dixon, R.A. and Wang, X. (2007) Crystal structure of *Medicago truncatula* UGT85H2 – insights into the structural basis of a multifunctional (iso)flavonoid glycosyltransferase. *Journal of Molecular Biology* 370, 951–963.

Lim, E.-K. and Bowles, D.J. (2004) A class of plant glycosyltransferases involved in cellular homeostasis. *The EMBO Journal*, 23, 2915–2922.

Lookadoo, S.E. and Pollard, A.J. (1991) Chemical contents of stinging trichomes of *Cnidoscolus texanus*. *Journal of Chemical Ecology* 17, 1909–1916.

Ly, D., Kang, K., Choi, J.-Y., Ishihara, A., Back, K. and Lee, S.-G. (2008) HPLC analysis of serotonin, tryptamine, tyramine, and the hydroxycinnamic acid amides of serotonin and tyramine in food vegetables. *Journal of Medicinal Food* 11, 385–389.

Mackenzie, P., Owens, I., Burchell, B., Bock, K., Bairoch, A., Bélanger, A., Fournel-Gigleux, S., Green, M., Hum, D., Iyanagi, T., Lancet, D., Louisot, P. *et al.* (1997) The UDP glycosyltransferase gene superfamily: recommended nomenclature update based on evolutionary divergence. *Pharmacogenetics* 7, 255–269.

Maresh, J., Zhang, J. and Lynn, D.G. (2006) The innate immunity of maize and the dynamic chemical strategies regulating two-component signal transduction in *Agrobacterium tumefaciens*. *ACS Chemical Biology* 1, 165–175.

Marti, G., Erb, M., Boccard, J., Glauser, G., Doyen, G.R., Villard, N., Robert, C.A.M., Turlings, T.C.J., Rudaz, S. and Wolfender, J.-L. (2013) Metabolomics reveals herbivore-induced metabolites of resistance and susceptibility in maize leaves and roots. *Plant, Cell and Environment* 36, 621–639.

Matsuda, F., Ishihara, A., Takanashi, K., Morino, K., Miyazawa, H., Wakasa, K. and Miyagawa, H. (2009) Metabolic profiling analysis of genetically modified rice seedlings that overproduce tryptophan reveals the occurrence of its inter-tissue translocation. *Plant Biotechnology* 27, 17–27.

Matsukawa, T., Ishihara, A. and Iwamura, H. (2002a) Differential induction of chorismate mutase isoforms by elicitors in oat leaves. *Plant Physiology and Biochemistry* 40, 795–802.

Matsukawa, T., Ishihara, A. and Iwamura, H. (2002b) Induction of anthranilate synthase activity by elicitors in oats. *Zeitschrift für Naturforschung C* 57, 121–128.

Mayama, S., Tani, T., Matsuura, Y., Ueno, T. and Fukami, H. (1981) The production of phytoalexins by oat in response to crown rust, *Puccinia coronata* f. sp. *avenae*. *Physiological Plant Pathology* 19, 217–226.

Mayama, S., Matsuura, Y., Iida, H. and Tani, T. (1982) The role of avenalumin in the resistance of oat to crown rust, *Puccinia coronata* f. sp. *avenae*. *Physiological Plant Pathology* 20, 189–199.

Mayama, S., Tani, T., Ueno, T., Midland, S.L., Sims, J.J. and Keen, N.T. (1986) The purification of victorin and its phytoalexin elicitor activity in oat leaves. *Physiological and Molecular Plant Pathology* 29, 1–18.

McIntosh, L.P., Hand, G., Johnson, P.E., Joshi, M.D., Körner, M., Plesniak, L.A., Ziser, L., Wakarchuk, W.W. and Withers, S.G. (1996) The p*Ka* of the general acid/base carboxyl group of a glycosidase cycles during catalysis: a ^{13}C-NMR study of *Bacillus circulans* xylanase. *Biochemistry* 35, 9958–9966.

Meihls, L.N., Handrick, V., Glauser, G., Barbier, H., Kaur, H., Haribal, M.M., Lipka, A.E., Gershenzon, J., Buckler, E.S., Erb, M., Köllner, T.G. and Jander, G. (2013) Natural variation in maize aphid resistance is associated with 2,4-dihydroxy-7-methoxy-1,4-benzoxazin-3-one glucoside methyltransferase activity. *The Plant Cell* 25, 2341–2355.

Melanson, D., Chilton, M.-D., Masters-Moore, D. and Chilton, W.S. (1997) A deletion in an indole synthase gene is responsible for the DIMBOA-deficient phenotype of *bxbx* maize. *Proceedings of the National Academy of Sciences of the United States of America* 94, 13345–13350.

Miersch, O., Neumerkel, J., Dippe, M., Stenzel, I. and Wasternack, C. (2008) Hydroxylated jasmonates are commonly occurring metabolites of jasmonic acid and contribute to a partial switch-off in jasmonate signaling. *New Phytologist* 177, 114–127.

Modolo, L.V., Li, L., Pan, H., Blount, J.W., Dixon, R.A. and Wang, X. (2009) Crystal structures of glycosyltransferase UGT78G1 reveal the molecular basis for glycosylation and deglycosylation of (iso)flavonoids. *Journal of Molecular Biology* 392, 1292–1302.

Morant, A.V., Jørgensen, K., Jørgensen, C., Paquette, S.M., Sánchez-Pérez, R., Møller, B.L. and Bak, S. (2008) *β*-Glucosidases as detonators of plant chemical defense. *Phytochemistry* 69, 1795–1813.

Morrissey, J.P., Wubben, J.P. and Osbourn, A.E. (2000) *Stagonospora avenae* secretes multiple enzymes that hydrolyze oat leaf saponins. *Molecular Plant–Microbe Interaction* 13, 1041–1052.

Morse, S., Wratten, S.D., Edwards, P.J. and Niemeyer, H.M. (1991) Changes in the hydroxamic acid content of maize leaves with time and after artificial damage; implications for insect attack. *Annals of Applied Biology* 119, 239–249.

Muroi, A., Ishihara, A., Tanaka, C., Ishizuka, A., Takabayashi, J., Miyoshi, H. and Nishioka, T. (2009) Accumulation of hydroxycinnamic acid amides induced by pathogen infection and identification of agmatine coumaroyltransferase in *Arabidopsis thaliana*. *Planta* 230, 517–527.

Muroi, A., Matsui, K., Shimoda, T., Kihara, H., Ozawa, R., Ishihara, A., Nishihara, M. and Arimura, G.-i. (2012) Acquired immunity of transgenic torenia plants overexpressing agmatine coumaroyltransferase to pathogens and herbivore pests. *Scientific Reports* 2, Article number: 689.

Nakagawa, E., Amano, T., Hirai, N. and Iwamura, H. (1995) Non-induced cyclic hydroxamic acids in wheat during juvenile stage of growth. *Phytochemistry* 38, 1349–1354.

Nässel, D.R. (1988) Serotonin and serotonin-immunoreactive neurons in the nervous system of insects. *Progress in Neurobiology* 30, 1–85.

Negrel, J., Javelle, F. and Paynot, M. (1993) Wound-induced tyramine hydroxycinnamoyl transferase in potato (*Solanum tuberosum*) tuber disks. *Journal of Plant Physiology* 142, 518–524.

Nicasio, M.D.P., Villarreal, M.L., Gillet, F., Bensaddek, L. and Fliniaux, M.-A. (2005) Variation in the accumulation levels of *N,N*-dimethyltryptamine in micropropagated trees and in *in vitro* cultures of *Mimosa tenuiflora*. *Natural Product Research* 19, 61–67.

Niemeyer, H.M. (1988) Hydroxamic acids (4-hydroxy-1,4-benzoxazin-3-ones): defence chemicals in the *Gramineae*. *Phytochemistry* 27, 3349–3358.

Niemeyer, H.M. (2009) Hydroxamic acids derived from 2-hydroxy-2*H*-1,4-benzoxazin-3(4*H*)-one: key defense chemicals of cereals. *Journal of Agricultural and Food Chemistry* 57, 1677–1696.

Niemeyer, H.M., Pesel, E., Copaja, S.V., Bravo, H.R., Franke, S. and Francke, W. (1989) Changes in hydroxamic acid levels of wheat plants induced by aphid feeding. *Phytochemistry* 28, 447–449.

Nikus, J., Esen, A. and Jonsson, L.M.V. (2003) Cloning of a plastidic rye (*Secale cereale*) *β*-glucosidase cDNA and its expression in *Escherichia coli*. *Physiologia Plantarum* 118, 337–345.

Niyogi, K.K. and Fink, G.R. (1992) Two anthranilate synthase genes in *Arabidopsis*; defense-related regulation of the tryptophan pathway. *The Plant Cell* 4, 721–733.

Niyogi, K.K., Last, L.R., Fink, G.R. and Keith, B. (1993) Suppressors of *trp1* fluorescence identify a new *Arabidopsis* gene, *TRP4*, encoding the anthranilate synthase beta subunit. *The Plant Cell* 5, 1011–1027.

Nojiri, H., Sugimori, M., Yamane, H., Nishimura, Y., Yamada, A., Shibuya, N., Kodama, O., Murofushi, N. and Omori, T. (1996) Involvement of jasmonic acid in elicitor-induced phytoalexin production in suspension-cultured rice cells. *Plant Physiology* 110, 387–392.

Nomura, T., Ishihara, A., Imaishi, H., Endo, T.R., Ohkawa, H. and Iwamura, H. (2002) Molecular characterization and chromosomal localization of cytochrome P450 genes involved in the biosynthesis of cyclic hydroxamic acids in hexaploid wheat. *Molecular Genetics and Genomics* 267, 210–217.

Nomura, T., Ishihara, A., Imaishi, H., Ohkawa, H., Endo, T.R. and Iwamura, H. (2003) Rearrangement of the genes for the biosynthesis of benzoxazinones in the evolution of *Triticeae* species. *Planta* 217, 776–782.

Nomura, T., Ishihara, A., Yanagita, R.C., Endo, T.R. and Iwamura, H. (2005) Three genomes differentially contribute to the biosynthesis of benzoxazinones in hexaploid wheat. *Proceedings of the National Academy of Sciences of the United States of America* 102, 16490–16495.

Nomura, T., Ishihara, A., Iwamura, H. and Endo, T.R. (2007) Molecular characterization of benzoxazinone-deficient mutation in diploid wheat. *Phytochemistry* 68, 1008–1016.

Nomura, T., Nasuda, S., Kawaura, K., Ogihara, Y., Kato, N., Sato, F., Kojima, T., Toyoda, A., Iwamura, H. and Endo, T.R. (2008) Structure of the three homoeologous loci of wheat benzoxazinone biosynthetic genes *TaBx3* and *TaBx4* and characterization of their promoter sequences. *Theoretical and Applied Genetics* 116, 373–381.

Obara, Y., Fukano, Y., Watanabe, K., Ozawa, G. and Sasaki, K. (2011) Serotonin-induced mate rejection in the female cabbage butterfly, *Pieris rapae crucivora*. *Naturwissenschaften* 98, 989–993.

Offen, W., Martinez-Fleites, C., Yang, M., Kiat-Lim, E., Davis, B.G., Tarling, C.A., Ford, C.M., Bowles, D.J. and Davies, G.J. (2006) Structure of a flavonoid glucosyltransferase reveals the basis for plant natural product modification. *The EMBO Journal* 25, 1396–1405.

Oikawa, A., Ishihara, A., Hasegawa, M., Kodama, O. and Iwamura, H. (2001) Induced accumulation of 2-hydroxy-4,7-dimethoxy-1,4-benzoxazin-3-one glucoside (HDMBOA-Glc) in maize leaves. *Phytochemistry* 56, 669–675.

Oikawa, A., Ishihara, A. and Iwamura, H. (2002) Induction of HDMBOA-Glc accumulation and DIMBOA-Glc 4-*O*-methyltransferase by jasmonic acid in poaceous plants. *Phytochemistry* 61, 331–337.
Oikawa, A., Ishihara, A., Tanaka, C., Mori, N., Tsuda, M. and Iwamura, H. (2004) Accumulation of HDMBOA-Glc is induced by biotic stresses prior to the release of MBOA in maize leaves. *Phytochemistry* 65, 2995–3001.
Okada, A., Okada, K., Miyamoto, K., Koga, J., Shibuya, N., Nojiri, H. and Yamane, H. (2009) OsTGAP1, a bZIP transcription factor, coordinately regulates the inductive production of diterpenoid phytoalexins in rice. *The Journal of Biological Chemistry* 284, 26510–26518.
Okazaki, Y., Ishihara, A., Nishioka, T. and Iwamura, H. (2004a) Identification of a dehydrodimer of avenanthramide phytoalexin in oats. *Tetrahedron* 60, 4765–4771.
Okazaki, Y., Isobe, T., Iwata, Y., Matsukawa, T., Matsuda, F., Miyagawa, H., Ishihara, A., Nishioka, T. and Iwamura, H. (2004b) Metabolism of avenanthramide phytoalexins in oats. *The Plant Journal* 39, 560–572.
Okazaki, Y., Ishizuka, A., Ishihara, A., Nishioka, T. and Iwamura, H. (2007) New dimeric compounds of avenanthramide phytoalexin in oats. *Journal of Organic Chemistry* 72, 3830–3839.
Osbourn, A. (2010) Gene clusters for secondary metabolic pathways: an emerging theme in plant biology. *Plant Physiology* 154, 531–535.
Osbourn, A.E. and Field, B. (2009) Operons. *Cellular and Molecular Life Sciences* 66, 3755–3775.
Osbourn, A.E., Clarke, B.R., Dow, J.M. and Daniels, M.J. (1991) Partial characterization of avenacinase from *Gaeumannomyces graminis* var. *avenae*. *Physiological and Molecular Plant Pathology* 38, 301–312.
Osmani, S.A., Bak, S. and Møller, B.L. (2009) Substrate specificity of plant UDP-dependent glycosyltransferases predicted from crystal structures and homology modeling. *Phytochemistry* 70, 325–347.
Ozkan, H., Levy, A.A. and Feldman, M. (2001) Allopolyploidy-induced rapid genome evolution in the wheat (*Aegilops–Triticum*) group. *The Plant Cell* 13, 1735–1747.
Pelagio-Flores, R., Ortíz-Castro, R., Méndez-Bravo, A., Macías-Rodríguez, L. and López-Bucio, J. (2011) Serotonin, a tryptophan-derived signal conserved in plants and animals, regulates root system architecture probably acting as a natural auxin inhibitor in *Arabidopsis thaliana*. *Plant and Cell Physiology* 52, 490–508.
Poulsen, C., Bongaerts, R.J.M. and Verpoorte, R. (1993) Purification and characterization of anthranilate synthase from *Catharanthus roseus*. *European Journal of Biochemistry* 212, 431–440.
Poulton, J.E. (1990) Cyanogenesis in plants. *Plant Physiology* 94, 401–405.
Raab, T., Barron, D., Vera, F.A., Crespy, V., Oliveira, M. and Williamson, G. (2010) Catechin glucosides: occurrence, synthesis, and stability. *Journal of Agricultural and Food Chemistry* 58, 2138–2149.
Ramakrishna, A., Giridhar, P. and Ravishankar, G.A. (2009) Indoleamines and calcium channels influence morphogenesis in *in vitro* cultures of *Mimosa pudica* L. *Plant Signaling and Behavior* 4, 1–6.
Rash, L.D. and Hodgson, W.C. (2001) Pharmacology and biochemistry of spider venoms. *Toxicon* 40, 225–254.
Rask, L., Andrésson, E., Ekbom, B., Eriksson, S., Pontoppidan, B. and Meijer, J. (2000) Myrosinase: gene family evolution and herbivore defense in *Brassicaceae*. *Plant Molecular Biology* 42, 93–113.
Regula, I. and Devide, Z. (1980) The presence of serotonin in some species of genus *Urtica*. *Acta Botanica Croatica* 39, 47–50.
Rouseff, R.L., Martin, S.F. and Youtsey, C.O. (1987) Quantitative survey of narirutin, naringin, hesperidin, and neohesperidin in citrus. *Journal of Agricultural and Food Chemistry* 35, 1027–1030.
Ruhé, H.G., Mason, N.S. and Schene, A.H. (2007) Mood is indirectly related to serotonin, norepinephrine and dopamine levels in humans: a meta-analysis of monoamine depletion studies. *Molecular Psychiatry* 12, 331–359.
Salse, J., Chagué, V., Bolot, S., Magdelenat, G., Huneau, C., Pont, C., Belcram, H., Couloux, A., Gardais, S., Evrard, A., Segurens, B., Charles, M. *et al.* (2008) New insights into the origin of the B genome of hexaploid wheat: evolutionary relationships at the *SPA* genomic region with the S genome of the diploid relative *Aegilops speltoides*. *BMC Genomics* 9:555.
Sasai, H., Ishida, M., Tadokoro, N., Ishihara, A. and Mori, N. (2009) Species-specific glucosylation of DIMBOA in larvae of the rice armyworm. *Bioscience, Biotechnology and Biochemistry* 73, 1333–1338.
Schullehner, K., Dick, R., Vitzthum, F., Schwab, W., Brandt, W., Frey, M. and Gierl, A. (2008) Benzoxazinoid biosynthesis in dicot plants. *Phytochemistry* 69, 2668–2677.
Shao, H., He, X., Achnine, L., Blount, J.W., Dixon, R.A., and Wang, X. (2005) Crystal structures of a multifunctional triterpene/flavonoid glycosyltransferase from *Medicago truncatula*. *The Plant Cell* 17, 3141–3154.
Shirley, A.M. and Chapple, C. (2003) Biochemical characterization of sinapoylglucose:choline sinapoyltransferase, a serine carboxypeptidase-like protein that functions as an acyltransferase in plant secondary metabolism. *The Journal of Biological Chemistry* 278, 19870–19877.

Sicker, D., Frey, M., Schulz, M. and Gierl, A. (2000) Role of natural benzoxazinones in the survival strategy of plants. *International Review of Cytology* 198, 319–346.

Sitaraman, D., Zars, M., LaFerriere, H., Chen, Y.-C., Sable-Smith, A., Kitamoto, T., Rottinghaus, G.E. and Zars, T. (2008) Serotonin is necessary for place memory in *Drosophila*. *Proceedings of the National Academy of Sciences of the United States of America* 105, 5579–5584.

Sommer, C. (2004) Serotonin in pain and analgesia: actions in the periphery. *Molecular Neurobiology* 30, 117–125.

Spaepen, S., Vanderleyden, J. and Remans, R. (2007) Indole-3-acetic acid in microbial and microorganism–plant signaling. *FEMS Microbiology Reviews* 31, 425–448.

Stranc, A. (1993) Astra – a natural coffee with a reduced irritant content. *Przemysł Spożywczy* 47, 76–77.

Sue, M., Ishihara, A. and Iwamura, H. (2000a) Occurrence and characterization of a UDP-glucose: hydroxamic acid glucosyltransferase isolated from wheat (*Triticum aestivum*) seedlings. *Zeitschrift für Naturforschung C* 55, 701–707.

Sue, M., Ishihara, A. and Iwamura, H. (2000b) Purification and characterization of a β-glucosidase from rye (*Secale cereale* L.) seedlings. *Plant Science* 155, 67–74.

Sue, M., Yamazaki, K., Yajima, S., Nomura, T., Matsukawa, T., Iwamura, H. and Miyamoto, T. (2006) Molecular and structural characterization of hexameric β-D-glucosidases in wheat and rye. *Plant Physiology* 141, 1237–1247.

Sue, M., Nakamura, C., Miyamoto, T. and Yajima, S. (2011a) Active-site architecture of benzoxazinone-glucoside β-D-glucosidases in *Triticeae*. *Plant Science* 180, 268–275.

Sue, M., Nakamura, C. and Nomura, T. (2011b) Dispersed benzoxazinone gene cluster: molecular characterization and chromosomal localization of glucosyltransferase and glucosidase genes in wheat and rye. *Plant Physiology* 157, 985–997.

Suharsono, U., Fujisawa, Y., Kawasaki, T., Iwasaki, Y., Satoh, H. and Shimamoto, K. (2002) The heterotrimeric G protein α subunit acts upstream of the small GTPase Rac in disease resistance of rice. *Proceedings of the National Academy of Sciences of the United States of America* 99, 13307–13312.

Takos, A.M. and Rook, F. (2012) Why biosynthetic genes for chemical defense compounds cluster. *Trends in Plant Science* 17, 383–388.

Tamogami, S., Rakwal, R. and Kodama, O. (1997) Phytoalexin production elicited by exogenously applied jasmonic acid in rice leaves (*Oryza sativa* L.) is under the control of cytokinins and ascorbic acid. *FEBS Letters* 412, 61–64.

Tawab, M.A., Bahr, U., Karas, M., Wurglics, M. and Schubert-Zsilavecz, M. (2003) Degradation of ginsenosides in humans after oral administration. *Drug Metabolism and Disposition* 31, 1065–1071.

Tomioka, K., Ikeda, M., Nagao, T. and Tamotsu, S. (1993) Involvement of serotonin in the circadian rhythm of an insect visual system. *Naturwissenschaften* 80, 137–139.

Tozawa, Y., Hasegawa, H., Terakawa, T. and Wakasa, K. (2001) Characterization of rice anthranilate synthase α subunit genes *OASA1* and *OASA2*. Tryptophan accumulation in transgenic rice expressing a feedback-insensitive mutant of *OASA1*. *Plant Physiology* 126, 1493–1506.

Udenfriend, S., Weissbach, H. and Clark, C.T. (1955) The estimation of 5-hydroxytryptamine (serotonin) in biological tissues. *The Journal of Biological Chemistry* 215, 337–344.

Udenfriend, S., Lovenberg, W. and Sjoerdsma, A. (1959) Physiologically active amines in common fruits and vegetables. *Archives of Biochemistry and Biophysics* 85, 487–490.

Ueno, M., Shibata, H., Kihara, J., Honda, Y. and Arase, S. (2003) Increased tryptophan decarboxylase and monoamine oxidase activities induce Sekiguchi lesion formation in rice infected with *Magnaporthe grisea*. *The Plant Journal* 36, 215–228.

Ueno, M., Kihara, J., Honda, Y. and Arase, S. (2004) Indole-related compounds induce the resistance to rice blast fungus, *Magnaporthe grisea* in barley. *Journal of Phytopathology* 152, 606–612.

Ueno, M., Kumura, Y., Ueda K., Kihara, J., Honda, Y. and Arase, S. (2011) Indole derivatives enhance resistance against the rice blast fungus *Magnaporthe orizae*. *Journal of General Plant Pathology* 77, 209–213.

Upton, R. (2013) Stinging nettles leaf (*Urtica dioica* L.): extraordinary vegetable medicine. *Journal of Herbal Medicine* 3, 9–38.

Virtanen, A.I., Hietala, P.K. and Wahlroos, Ö. (1957) Antimicrobial substances in cereals and fodder plants. *Archives of Biochemistry and Biophysics* 69, 486–500.

von Rad, U., Hüttl, R., Lottspeich, F., Gierl, A. and Frey, M. (2001) Two glucosyltransferases are involved in detoxification of benzoxazinoids in maize. *The Plant Journal* 28, 633–642.

von Röpenack, E., Parr, A. and Schulze-Lefert, P. (1998) Structural analyses and dynamics of soluble and cell wall-bound phenolics in a broad spectrum resistance to the powdery mildew fungus in barley. *The Journal of Biological Chemistry* 273, 9013–9022.

Waalkes, T.P., Sjoerdsma, A., Creveling, C.R., Weissbach, H. and Udenfriend, S. (1958) Serotonin, norepinephrine, and related compounds in bananas. *Science* 127, 648–650.

Wakasa, K., Hasegawa, H., Nemoto, H., Matsuda, F., Miyazawa, H., Tozawa, Y., Morino, K., Komatsu, A., Yamada, T., Terakawa, T. and Miyagawa, H. (2006) High-level tryptophan accumulation in seeds of transgenic rice and its limited effects on agronomic traits and seed metabolite profile. *Journal of Experimental Botany* 57, 3069–3078.

Wang, X. (2009) Structure, mechanism and engineering of plant natural product glycosyltransferases. *FEBS Letters* 583, 3303–3309.

Wegel, E., Koumproglou, R., Shaw, P. and Osbourn, A. (2009) Cell type-specific chromatin decondensation of a metabolic gene cluster in oats. *The Plant Cell* 21, 3926–3936.

Weibull, J. and Niemeyer, H.M. (1995) Changes in dihydroxymethoxybenzoxazinone glycoside content in wheat plants infected by three plant pathogenic fungi. *Physiological and Molecular Plant Pathology* 47, 201–212.

Weisel-Eichler, A. and Libersat, F. (2004) Venom effects on monoaminergic systems. (2004) *Journal of Comparative Physiology A: Neuroethology, Sensory, Neural, and Behavioral Physiology* 190, 683–690.

West, G.B. (1960) Carcinoid tumours and pineapples. *The Journal of Pharmacy and Pharmacology* 12, 768.

Widholm, J.M. (1972) Tryptophan biosynthesis in *Nicotiana tabacum* and *Daucus carota* cell cultures: site of action of inhibitory tryptophan analogs. *Biochimica et Biophysica Acta* 261, 44–51.

Widholm, J.M. (1974) Control of aromatic amino acid biosynthesis in cultured plant tissues: effect of intermediates and aromatic amino acids on free levels. *Physiologia Plantarum* 30, 13–18.

Wise, M.L. (2011) Effect of chemical systemic acquired resistance elicitors on avenanthramide biosynthesis in oat (*Avena sativa*). *Journal of Agricultural and Food Chemistry* 59, 7028–7038.

Woodward, M.D., Corcuera, L.G., Helgeson, J.P., Kelman, A. and Upper, C.D. (1978) Factors that influence the activity of 2,4-dihydroxy-7-methoxy-2*H*-1,4-benzoxazin-3(4*H*)-one on *Erwinia* species in growth assays. *Plant Physiology* 61, 803–805.

Xia, L., Ruppert, M., Wang, M., Panjikar, S., Lin, H., Rajendran, C., Barleben, L. and Stöckigt, J. (2012) Structures of alkaloid biosynthetic glucosidases decode substrate specificity. *ACS Chemical Biology* 7, 226–234.

Yadav, V.K., Oury, F., Suda, N., Liu, Z.-W., Gao, X.-B., Confavreux, C., Klemenhagen, K.C., Tanaka, K.F., Gingrich, J.A., Guo, X.E., Tecott, L.H., Mann, J.J. *et al.* (2009) A serotonin-dependent mechanism explains the leptin regulation of bone mass, appetite, and energy expenditure. *Cell* 138, 976–989.

Yang, Q., Trinh, H.X., Imai, S., Ishihara, A., Zhang, L., Nakayashiki, H., Tosa, Y. and Mayama, S. (2004) Analysis of the involvement of hydroxyanthranilate hydroxycinnamoyltransferase and caffeoyl-CoA 3-*O*-methyltransferase in phytoalexin biosynthesis in oat. *Molecular Plant–Microbe Interactions* 17, 81–89.

Zhao, J. and Last, R.L. (1996) Coordinate regulation of the tryptophan biosynthetic pathway and indolic phytoalexin accumulation in *Arabidopsis*. *The Plant Cell* 8, 2235–2244.

21 Melatonin: Synthesis From Tryptophan and its Role in Higher Plants

M.B. Arnao* and J. Hernández-Ruiz
University of Murcia, Murcia, Spain

21.1 Abstract

Melatonin, a well-known neurohormone, was discovered in animal tissues in 1958. In plants, its presence was detected in 1995, and its biosynthesis, from tryptophan, seems to follow similar enzymatic steps in plant and animal cells, although with some very particular characteristics. Even though our knowledge of the biosynthetic enzymes of melatonin in plants is still taking its first steps, a greater plasticity has been observed in plant cells than in animal cells, as was to be expected. As they have been in animals, some catabolic products of the kynuric pathway have been determined in plants. Initially, melatonin was investigated to establish whether it had the same physiological function, as a chronoregulator, in plants as in animals. Its antioxidant properties have marked much of the research carried out into its possible physiological roles in both plants and animals. In plants, various studies have suggested more specific physiological actions, such as, for example, its roles as a plant growth regulator, an activator of rhizogenesis and a signal molecule/antioxidant in many plant stress situations. The melatonin levels that have been observed in different plant species vary widely – from picograms to micrograms per gram of tissue. The search for plant materials with high levels of melatonin has opened the door for its possible use as a nutraceutical for both humans and livestock. In addition, there has been great interest in the application of melatonin in agriculture to obtain more resistant crops and better harvests in adverse situations.

21.2 Introduction

Melatonin (*N*-acetyl-5-methoxytryptamine) is the main hormone synthesized in the pineal gland of animals. This indoleamine is a pleiotropic molecule and has numerous cellular and physiological actions throughout the diverse kingdoms. Its biological functions range from direct action as a free radical scavenger to a regulatory role as a hormone in the central nervous system of animals, to lesser known roles in bacteria, fungi, plants and insects. In all of the organisms that have been studied, this biogenic amine is synthesized from the amino acid tryptophan in a pathway that is well known in humans. In contrast, some important differences in its biosynthetic pathway have appeared in plants. In the following sections, we present an overview of the most relevant aspects of this interesting molecule: from its discovery in animals and plants and its biosynthesis and catabolism, to its physiological actions in animals and plants, focusing on the most novel and interesting aspects in higher plants.

*E-mail address: marino@um.es

21.2.1 Discovery of melatonin

The first data about the existence of melatonin was published by Lerner and co-workers in 1958 (Lerner *et al.*, 1958). They isolated from the bovine pineal gland the active factor that lightens skin colour in frogs, tadpoles, toads and certain fish, but not in mammals, as a consequence of the aggregation of the melanin granules of melanocytes. For this reason, the authors proposed that the substance be called *melatonin*. The chromatographed substance showed a positive response with Ehrlich's reagent, presenting ultraviolet and fluorescent properties characteristic of hydroxyindoles. It was identified in 1959 as *N*-acetyl-5-methoxytryptamine, a derivative of *N*-acetylserotonin (Lerner *et al.*, 1959a,b). In 1960, its biosynthetic pathway from serotonin was discovered, as well as its main catabolite, 6-hydroxymelatonin, and the sulfate form of this (Axelrod and Weissbach, 1960; Kopin *et al.*, 1960; Weissbach *et al.*, 1960). These authors highlighted the progress made by Drs Hoffman and Reiter in the 1960s on the pineal gland in mammals and the effects of melatonin as the possible 'hibernation factor' sought by the US Army to prolong the duration of travel into space by astronauts. These investigations showed that the lack of melatonin synthesis in pinealectomic hamsters prevented the reproductive failure associated with darkness, data that were published in *Science* and *Nature* in 1965 (Hoffman and Reiter, 1965a,b). Shortly, afterwards came the studies demonstrating the role of melatonin in seasonal changes in the reproductive capacity of mammals, which indicated the functionality of the pineal gland.

21.2.2 Physiological roles of melatonin

The lightening of skin colour studied by Lerner and co-workers (Lerner *et al.*, 1958) and by previous authors was the first action attributed to melatonin. Today, melatonin is known, broadly, as a biological modulator of circadian rhythms, mood, sleep, body temperature, locomotor activity, food intake, retinal physiology, sexual behaviour, seasonal reproduction and immunological system. Figure 21.1 (lower diagram) shows a scheme that incorporates the most relevant hormonal functions attributed to melatonin in animals. Added to this, melatonin is an excellent antioxidant, at physiological concentrations, and has a wide range of roles at a cellular level (Fig. 21.1, upper diagram).

In mammals, melatonin is secreted by the pineal gland to the cerebrospinal fluid and to the bloodstream, presenting maximal levels during the middle of the night. This circadian pattern of melatonin secretion is regulated by the biological clock that resides in mammals within the hypothalamic suprachiasmatic nucleus (SCN) (Klein and Moore, 1979). The SCN is synchronized to the environmental light–dark cycle by the light perceived by the retina. Also, the SCN regulates pineal secretion through a polysynaptic network in the paraventricular nucleus of the hypothalamus (Buijs *et al.*, 1988). Melatonin liberation occurs when the noradrenaline released by paraventricular nucleus fibres activates the pineal β-adrenergic receptors, stimulating, through via cyclic AMP, serotonin *N*-acetyltransferase (SNAT) expression in the dark. Brief exposure to light provokes a decrease in melatonin production because of the degradation of SNAT in the pineal gland (Klein *et al.*, 1971; Krause and Dubocovich, 1990; Gastel *et al.*, 1998; Maronde and Stehle, 2007).

Accordingly, melatonin is of particular importance as a chronobiological agent (Fig. 21.1, lower diagram), acting as a signal of darkness that provides information to the brain and peripheral organs. It is an endogenous synchronizer for both endocrine (i.e. via neurotransmitter release) and other physiological rhythms, regulating sleep–wake cycles and synchronizing life activity with seasonal periods and reproductive functions (Reiter, 1993b; De la Iglesia *et al.*, 2000; Reppert and Weaver, 2002; Perreau-Lenz *et al.*, 2003; Stehle *et al.*, 2003; Saper *et al.*, 2005; Silver and Schwartz, 2005; Fuller *et al.*, 2006; Jan *et al.*, 2009).

Melatonin plays an important role in the regulation of sleep, body temperature, the state of alertness and degree of concentration or performance, and cortisol rhythms. It is a sleep initiator by opening the circadian sleep gate, thus acting as a sleep regulator. Melatonin adjusts the timing of or reinforces oscillators of the central biological clock. The administration of exogenous melatonin alters the timing of bodily rhythms, including sleep, whereby phase delays are observed with the morning administration of melatonin,

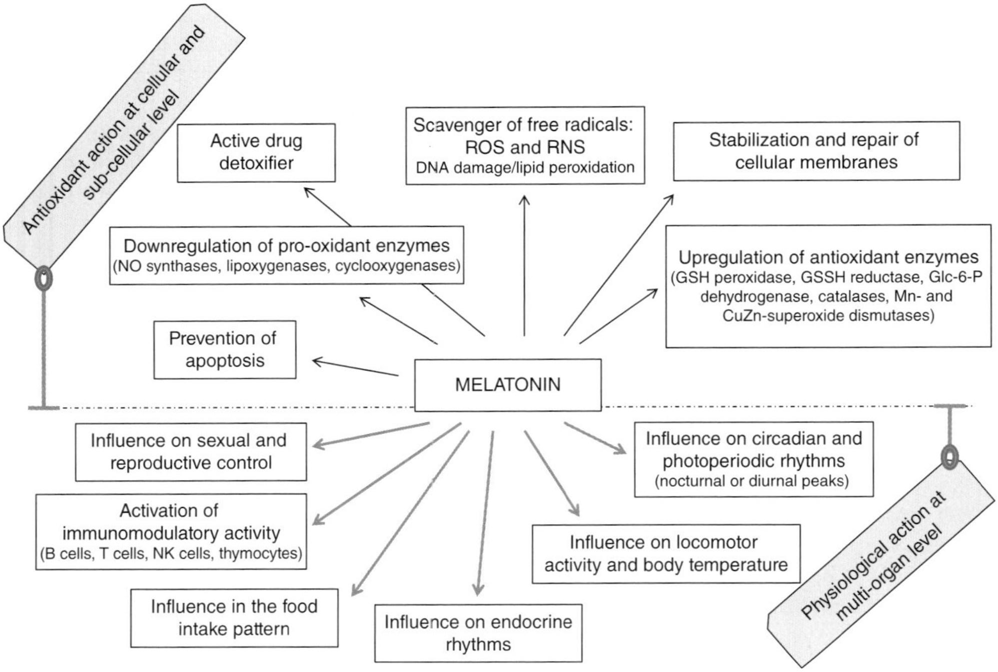

Fig. 21.1. Roles attributed to melatonin in animals. The upper diagram shows the wide range of roles at cellular and sub-cellular levels related mainly to its antioxidative properties. The lower diagram shows the hormonal functions of melatonin, mainly in mammals at a multi-organ level. Abbreviations: GSH, reduced glutathione; GSSH, oxidized glutathione; NK, natural killer; NO, nitric oxide; RNS, reactive nitrogen species; ROS, reactive oxygen species.

while phase advances are found after evening administration (Lewy *et al.*, 1992). Many sleep disorders have been treated with melatonin: delayed sleep phase syndrome, night shift-work sleep disorder, seasonal affective disorder, sleep disorders in the blind and ageing and pathophysiological disorders of children, with notable improvements in 'sleep quality' (Fig. 21.2). The most widespread disorder treated with melatonin is jet lag, a de-phasing in the sleep–wake rhythms caused by transoceanic flights (Waterhouse *et al.*, 1997; Takahashi *et al.*, 2000; Herxheimer, 2005).

Neurological and other disorders have been related to alterations in melatonin oscillations or their receptors (Fig. 21.2), including Alzheimer's and Parkinson's diseases, glaucoma and macular degeneration, chronic fatigue syndrome, multiple sclerosis, depression, schizophrenia, anxiety, menopause, gastrointestinal, coronary and cardiac diseases, food intake patterns and obesity, osteoporosis, ageing and age-related diseases, etc. The provision of information on these particular aspects is beyond the scope of this chapter, but relevant papers on the subjects, and the works cited therein, can be consulted (Chyan *et al.*, 1999; Reiter *et al.*, 2002; Reiter, 2003; Willis and Robertson, 2004; Claustrat *et al.*, 2005; Poeggeler, 2005; Srinivasan *et al.*, 2005; Wu and Swaab, 2005; Witt-Enderby *et al.*, 2006; Hussein *et al.*, 2007; Maldonado *et al.*, 2007; Pandi-Perumal *et al.*, 2008; Agorastos and Huber, 2011; Wilhelmsen *et al.*, 2011; Hardeland, 2012; Hardeland *et al.*, 2012).

Melatonin exerts many of its physiological actions through its interaction with various membrane receptors and a nuclear receptor. In vertebrates, melatonin interacts with three different subtypes of G protein-coupled receptors, so-called MT1 and MT2 (also called Mel1a and Mel1b), which have been cloned and characterized (Reppert *et al.*, 1994, 1995; von Gall *et al.*, 2002; Dubocovich and Markowska, 2005). A subtype, Mel1c, has been identified in fish, amphibians and birds (Li *et al.*, 2013). Melatonin also binds to a

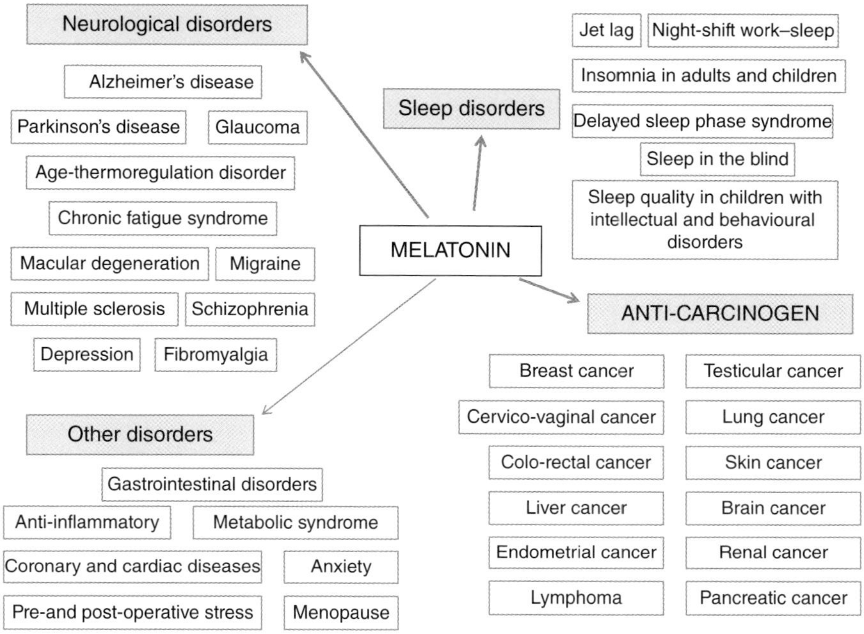

Fig. 21.2. Beneficial effects of melatonin on different diseases and disorders of animals.

nuclear hormone receptor called RZR/ROR, which includes three subtypes (α, β, γ) (Becker-André *et al.*, 1994). Interactions of melatonin with calmodulin, calcireticulin and tubulins are known as well. Receptor mRNA levels vary with a circadian rhythms, expression levels being affected by light and the concentration of melatonin in the blood. Data show that melatonin downregulates some of its receptor population, thereby regulating its own receptors (Masson-Pevet *et al.*, 1996; Gilad *et al.*, 1997; Schuster *et al.*, 2001). Melatonin receptors in humans have been detected in the SCNs, hypothalamus, paraventricular and periventricular nuclei, and many other brain areas; they have also been detected in the retina, cardiovascular system, gastrointestinal tract, parotid gland, exocrine and endocrine pancreas, hepatic cells, kidney, lymphocytes, adipocytes, prostate, breast and ovary/granulosa cells, the myometrium and skin (Li *et al.*, 2013). Alteration in the expression of melatonin receptors, as well as unexpected changes in the endogenous production of melatonin, have been associated with the emergence of many diseases and disorders, some of which are listed above. There are excellent reviews on this topic by Pandi-Perumal *et al.* (2008) and Hardeland (2012).

A decrease in melatonin levels has also been associated with susceptibility to the occurrence of certain types of cancers. Melatonin production by the pineal gland is persistent until the age of 20, after which it declines with age (Humbert and Pévet, 1994), and it has also been related to disorders resulting from alterations in day timing, mainly through changes in the sleep–wake cycles. Since the pioneering work of Hamilton (1969) and Taylor and Pocock (1972), which indicated a possible relationship between the light environment and the appearance of mammary tumours, a report by Cohen *et al.* (1978) that described a relationship between the pineal gland and breast cancer, and also the relationship between low urinary melatonin levels and the occurrence of human breast cancer (Bartsch *et al.*, 1981), many studies have pointed to a high breast cancer incidence in women night-shift workers due to excessive exposure to light at night (Tamarkin *et al.*, 1982; Hansen, 2001). Thus, a continuous reduction in melatonin levels might contribute to the probability of initiation and progression of breast cancer. The possible link of melatonin with cancer risk has been associated with its function as a downregulator of the pituitary and gonadal hormones, which control mammary gland development, among other factors (Cos and Sanchez-Barceló, 2000; Davis *et al.*, 2001; Knutsson, 2003; Kamdar *et al.*, 2013) (Fig. 21.2). These same indications of the effect of alterations in the sleep–wake cycle and cancer risk were studied in reference to the proliferation of colorectal cancer in female nurses (Schernhammer *et al.*, 2003).

Many other cancers have been associated with low melatonin levels in the bloodstream, or have been treated with this indoleamine to stop or decelerate the growth of cancerous cells (Blask *et al.*, 2002; Reiter, 2004). In general, melatonin might act as an anti-carcinogen and antitumoral agent via several mechanisms. These include: regulation of hormone secretion; its ability to reduce DNA damage by reactive oxygen and nitrogen species (ROS/RNS) as a consequence of its excellent free-radical scavenging properties (see below); the upregulation of antioxidant enzymes; alteration of the expression of growth and differentiation-related genes; inhibition, via membrane receptors, of the uptake of fatty acids in hepatoma cells, thereby delaying the conversion of linoleic acid into 13-hydroxyoctadecadienoic acid (a mitogenic signalling molecule); reduction of the metastatic capacity of tumour cells and inhibition of telomerase activity in the same cells; and action on some oncogenesis-related genes that control the cell cycle and cell defence. Complementary to all of these actions, melatonin is an immunomodulator capable of activating lymphocytes and monocytes/macrophages and, by means of immunosurveillance, cancer cells can be detected and eliminated, thus preventing tumour development (Blask *et al.*, 2002; Reiter 2004; Rodríguez *et al.*, 2004, 2013; Tomás-Zapico and Coto-Montes, 2005; Anisimov *et al.*, 2006; Miller *et al.*, 2006; Pandi-Perumal *et al.*, 2008; Jung-Hynes *et al.*, 2010; Carrillo-Vico *et al.*, 2012; Seely *et al.*, 2012; Di Bella *et al.*, 2013).

Melatonin can also be produced by animal organs other than the pineal gland. This so-called 'extra-pineal melatonin' is generated by many different tissues/cells, such as the gastrointestinal tract, retina, several types of leucocytes, bone marrow, cerebral cortex, liver, thymus, spleen, heart, muscle, placenta, testis and skin (Cardinali and Rosner, 1971; Rakhlin *et al.*, 1975; Abe *et al.*, 1999; Tan *et al.*, 1999; Stefulj *et al.*, 2001; Slominski *et al.*, 2007; Venegas *et al.*, 2012). Although in quantitative terms extra-pineal melatonin is a 100 times more plentiful than that generated in the pineal gland, it seems that it is either not released into the bloodstream or is only released for very short time periods. However, it has been noted that large amounts of melatonin from the gastrointestinal tract enter the blood coming during the post-prandial response, and food intake results in an increase in the plasma concentration of melatonin. These pulses of melatonin seem to have little effect on the circadian system owing to their brevity and the fact that they occur during the silent point of the phase-response curve. Gastrointestinal melatonin prevents ulceration of the gastrointestinal mucosa, reduces hydrochloric acid secretion, stimulates the immunological system, increases microcirculation and promotes epithelial regeneration, among other actions (Huether, 1994; Bubenik, 2002; Konturek *et al.*, 2007).

In 1993, investigations into melatonin undertook a major development. This was a through a door that had opened through which a large part of future research on this molecule would pass. The research group of Prof. Russel J. Reiter, at the University of San Antonio in Texas, demonstrated that melatonin acts as an excellent hydroxyl radical scavenger, and published a cascade of reports on the possible relevance of the antioxidant potential of melatonin in multiple physiological and pathological roles (Hardeland *et al.*, 1993; Poeggeler *et al.*, 1993, 1994; Reiter, 1993a, 1994; Reiter *et al.*, 1993, 1994a,b,c, 1995, 2002; Tan *et al.*, 1993a,b). These workers used the DMPO (5,5-dimethyl-1-pyrroline *N*-oxide)-hydroxyl radical trapping method and measured the adducts by electron spin resonance (ESR) (Finkelstein *et al.*, 1980). They demonstrated that the presence of melatonin inhibited the production of •OH when H_2O_2 was exposed to UV light. Melatonin scavenged the •OH through a near-diffusion-controlled rate reaction, with a constant rate (k_a) of $2.7 \cdot 10^{10}$ M·s^{-1} (Poeggeler *et al.*, 1996; Reiter *et al.*, 2001). Melatonin presents higher k_a values than many other well-known •OH-scavengers, in the following sequence: β-carotene > melatonin > *N*-acetylcysteine > ascorbic acid ≈ α-tocopherol > reduced glutathione > desferroxamine > salicylic acid > mannitol. In the reaction between melatonin and •OH, a melatoninyl radical is formed, which evolves to cyclic-3-hydroxymelatonin in a very complex catabolic pathway (see Section 21.4). At this point, we should not overlook parallel work, also published in 1993, by the French group of Dr Laborit. In a study with mouse brain homogenate, melatonin and 6-hydromelatonin showed a remarkable antioxidant capacity in preventing experimental membrane lipoperoxidation induced by thiobarbituric acid (Pierrefiche *et al.*, 1993; Pierrefiche and Laborit, 1995).

Melatonin is also capable of scavenging hydrogen peroxide (H_2O_2), with a high degree of efficiency and in a dose-dependent manner (Zang *et al.*, 1998; Tan *et al.*, 2000a). A k_{Mel} value of $2.52 \cdot \times 10^5$ $M^{-1} \cdot s^{-1}$ was calculated for melatonin scavenging H_2O_2 using the HRP (horseradish peroxidase) compound I formation reaction with H_2O_2. Melatonin reacts directly with H_2O_2, generating N^1-acetyl-N^2-formyl-5-methoxykynuramine (AFMK), an oxidized catabolite of melatonin (see below). This catabolite of melatonin, among others, showed interesting antioxidant characteristics (Tan *et al.*, 2001). Using these data, a large number of ROS/RNS were tested. Superoxide anion radicals, singlet oxygen, hypochlorous acid, nitric oxide, the peroxynitrite anion, peroxynitrous acid and the peroxyl radical, among others, were scavenged in the presence of melatonin. As regards lipid peroxidation, melatonin showed a low scavenging capacity against LOO• (lipid peroxide) radicals, working better as a retardant of lipid peroxidation initiation than as a chain-breaking antioxidant (Reiter *et al.*, 1995, 2001; Antunes *et al.*, 1999; Tan *et al.*, 2000a). Within the family of indoles, antioxidant activity is not exclusive to melatonin. Many other indole compounds have excellent antioxidant activity *in vitro*, although the notable qualities of melatonin *in vivo* should also be noted (Christen *et al.*, 1990; Fuster, 1992; Poeggeler *et al.*, 1996, 1999; Goda *et al.*, 1999; Cano *et al.*, 2003).

An interesting aspect of the complex oxidant/antioxidant property was that melatonin can be easily regenerated by other classical antioxidants such as ascorbate, NADH, Trolox (6-hydroxy-2,5,7,8-tetramethylchromane-2-carboxylic acid), glutathione and urate. Thus, a synergistic action is observed when melatonin is combined with classical antioxidants using the blue/green ABTS (2,2′-azino-bis(3-ethylbenzothiazoline-6-sulfonic acid)) cation radical assay and also in animal tissues (Poeggeler *et al.*, 1995; Morreale and Livrea, 1997; Gitto *et al.*, 2001). One melatonin molecule can scavenge up to four ABTS radicals, an excellent ratio compared with classical antioxidants (ratio 1:2). Furthermore, melatonin scavenging activity occurs gradually in a reaction that has two phases: a rapid initial phase (similar to that which occurs with classical antioxidants), and another gradual (and slow phase) resulting from the continuous generation of reaction products, which themselves also have antioxidant activity (Tan *et al.*, 2001, 2003, 2007c; Rosen *et al.*, 2006; Schaefer and Hardeland, 2008; Galano *et al.*, 2011, 2013). In résumé, melatonin and its related catabolites act as excellent free radical scavengers, its antioxidant potential being greater than those of the most significant antioxidant molecules, such as ascorbic acid, Trolox, tocopherols and glutathione.

In vitro data obtained for the antioxidant properties of melatonin have been corroborated by a multitude of studies with different tissues in both animals and other species. The excellent *in vivo* antioxidant properties of melatonin suggested a plethora of roles at sub-cellular and cellular levels (Fig. 21.1, upper diagram) (Garcia *et al.*, 1997; Benot *et al.*, 1999; Crespo *et al.*, 1999; Cabrera *et al.*, 2000; Karbownik *et al.*, 2000a,b,c; Terron *et al.*, 2001; Teixeira *et al.*, 2003; Rodrígues de Lima *et al.*, 2004, 2007; Mayo *et al.*, 2005, 2013; Schaffazick *et al.*, 2005; Guajardo *et al.*, 2006; Mauriz *et al.*, 2007; López *et al.*, 2008; Fischer *et al.*, 2013). These include roles: as a stabilizer of biological membranes, especially at mitochondrial level; as an antioxidant against free radicals protecting from DNA damage, protein oxidation, and in the prevention of cell apoptosis; as an anti-inflammatory agent as downregulator of pro-oxidant enzymes such as NO synthases and lipoxygenases; and in the upregulation of antioxidant enzymes such as catalases, peroxidases and superoxide dismutases. Excellent reviews on these actions are available (Reiter *et al.*, 1994c, 2000; Rodríguez *et al.*, 2004; Hardeland and Pandi-Perumal, 2005; Tomás-Zapico and Coto- Montes, 2005; Benítez-King, 2006; Hardeland *et al.*, 2006; Catalá, 2007; Radogna *et al.*, 2010; Carrillo-Vico *et al.*, 2013; Tan *et al.*, 2013).

21.2.3 1995: a critical year for plants

Since the discovery of melatonin in 1958, for more than 30 years, researchers thought that melatonin was a hormone of an animal nature, a neurohormone. The first data that altered our way of looking at this methoxyindole appeared in a congress communication in 1989 (Poeggeler *et al.*, 1989), and as a full report in 1991 (Poeggeler *et al.*, 1991). In the work reported here, the addition of exogenous melatonin provoked an oscillating bioluminescent behaviour in the unicellular dinoflagellate *Lingulodinium polyedrum* (syn. *Gonyaulax polyedra*), which had also been observed previously in another dinoflagellate, *Pyrocystis acuta* (Fischer, 1987). Many other data, such as the concentration of melatonin, its metabolism and its effect on the biorhythms of dinoflagellates, euglenoids, phaeophyceans, rhodophyceans and chlorophyceans were presented by Hardeland's group in the following years (Balzer *et al.*, 1990; Balzer and Hardeland, 1991; Hardeland, 1999). Table 21.1 presents a summary of the pioneering works in the identification of melatonin in different organisms.

Possibly the first identification of endogenous melatonin in a higher plant was made in 1993 by van Tassel and co-workers (van Tassel and O'Neill, 1993). These authors detected melatonin by radioimmunoassay (RIA) in the Japanese morning glory plant (*Pharbitis nil*), although these results were not published extensively until 1995 (van Tassel *et al.*, 1995).

It was also in 1995 when two papers, published simultaneously, demonstrated the presence of melatonin and established its content in higher plants. In Germany, Dubbels and co-workers used RIA and GC-MS (gas chromatography-mass spectrometry) to measure melatonin levels in extracts of *Nicotiana tabacum* and in five edible plants (see Table 21.1) (Dubbels *et al.*, 1995). A publication by the Japanese group of Hattori *et al.* (1995) appeared 2 months later that quantified the presence of melatonin in extracts of a large number of edible plants by RIA and HPLC (high-performance liquid chromatography) with fluorescence detection (Table 21.1). Curiously, Prof. Reiter appears in both of these studies. Also in 1995, another communication appeared in which a Czech research group identified the presence of melatonin in *Chenopodium rubrum* by LC-MS/MS (liquid chromatography tandem mass spectrometry) (Kolar *et al.*, 1995).

Since this time, successive studies have quantified the presence of melatonin in plants (Table 21.1), and it is now fully accepted that melatonin is present in all kingdoms, from prokaryotes to eukaryotes, from animals to plants. However, there are gaps where melatonin has not been studied, such as in lichens, or where scant information is available, as in the fungi, bacteria and non-vascular plants.

21.3 Biosynthesis of Melatonin

The biosynthetic pathway of melatonin has been clearly determined in mammals, including humans and other groups (amphibians, reptiles and birds) (Reiter, 1999) (Fig. 21.3, right), in which melatonin biosynthesis is under the sequential control of four enzymes. Tryptophan is first transformed into 5-hydroxytryptophan by the tryptophan 5-hydroxylase (T5H, EC 1.14.16.4). Then, tryptophan decarboxylase (TDC, EC 4.1.1.28) generates serotonin (5-hydroxy-tryptamine), which can act as a neurotransmitter, besides having other functions. Serotonin is then acetylated into *N*-acetylserotonin by serotonin N-acetyltransferase (SNAT, EC 2.3.1.87); finally, hydroxindole-*O*-methyltransferase (HIOMT, EC 2.1.1.4) methylates *N*-acetylserotonin into melatonin (*N*-acetyl-5-methoxytryptamine) (Axelrod, 1974; Namboodiri *et al.*, 1987; Sudgen *et al.*, 1987; Falcón *et al.*, 2009; Nisembaum *et al.*, 2013). Some characteristics of these four enzymes involved in the melatonin biosynthetic pathway are summarized in Table 21.2.

21.3.1 Melatonin-related enzymes and their regulation

21.3.1.1 Tryptophan 5-hydroxylase (T5H)

T5H (or TPH) catalyses the 5-hydroxylation of tryptophan (Fig. 21.3 and Table 21.2), the first step in the biosynthesis of indoleamines (serotonin and melatonin). T5H belongs to the family of the aromatic amino acid hydroxylases, which include phenylalanine hydroxylase and tyrosine hydroxylase (Martínez *et al.*, 2001). All of these enzymes share considerable structural similarity and require the same cofactor (Fe^{2+}) and co-substrates (tetrahydrobiopterin and molecular oxygen) to function (Fitzpatrick, 1999; Zhang *et al.*, 2006). The enzymes are monooxygenases, incorporating one atom of oxygen from molecular

Table 21.1. Some data on the presence and discovery of melatonin in different organisms.

Kingdom	Class/Family/Species/Name	Tissue/cell	Discovery year	References
Animal (Vertebrates)				
	Bovine	Pineal gland	1958	Lerner *et al.*, 1958
	Human	Pineal gland	1959	Lerner *et al.*, 1959a,b
	Rat	Pineal gland	1960	Weissbach *et al.*, 1960
	Hamster	Pineal gland	1965	Hoffman and Reiter, 1965a,b
	Chick	Pineal body	1968	Lauber *et al.*, 1968
	Amphibian	Eye, brain	1965	Baker *et al.*, 1965
	Fish	Pineal body	1970	Fenwick, 1970
Animal (Invertebrates)				
	Insecta:			
	Locusta migratoria	Eye	1984	Vivien-Roels *et al.*, 1984
	Drosophila melanogaster	Eye	1991	Callebert *et al.*, 1991
	Platyhelminths:			
	Dugesia japonica	Body	1987	Morita *et al.*, 1987
	Opisthobranchia:			
	Aplysia californica	Eye, brain	1994	Abran *et al.*, 1994
	Crustacea:			
	Procambarus clarkii	Eye	1997	Balzer *et al.*, 1997
Protists				
	Dinoflagellates:			
	Lingulodinium polyedrum	Cell	1989	Poeggeler *et al.*, 1989, 1991
	Phaeophyceae:			
	Pterygophora californica	Stipe	1996	Fuhrberg *et al.*, 1996
	Trypanosomatida:			
	Trypanosoma cruzi	Cell	1999	Macias *et al.*, 1999
Fungi				
	Saccharomyces cerevisiae	Cell culture	1999	Sprenger *et al.*, 1999
Prokaryotes				
	Rhodospirillum rubrum	Cell culture	1995	Manchester *et al.*, 1995
	Erythrobacter longus		1997	Tilden *et al.*, 1997
Plants				
	Pharbitis nil	Shoot	1993,1995	van Tassel, 1997
	Solanum lycopersicum	Fruit	1997	van Tassel and O'Neill, 1993; van Tassel *et al.*, 1995
	Solanum sp.	Fruit	1995	Dubbels *et al.*, 1995
	Musa sp.			
	Cucumis sativus			
	Beta vulgaris	Root		
	Solanum tuberosum			
	Nicotiana tabacum	Leaf		
	Basella alba	Leaf	1995	Hattori *et al.*, 1995
	Actinidia chinensis	Fruit		
	Brassica sp.	Leaf		
	Raphanus sativus	Root		
	Daucus carota			

Continued

Table 21.1. Continued.

Kingdom	Class/Family/Species/Name	Tissue/cell	Discovery year	References
	Fragaria magna	Fruit		
	Malus domestica			
	Oryza sativa	Seed		
	Hordeum vulgare			
	Avena sativa			
	Allium cepa	Root		
	Chenopodium rubrum	Shoot	1995	Kolar *et al.*, 1995, 1997
	Tanacetum parthenium	Leaf	1997	Murch *et al.*, 1997
	15 edible plants	Seed	2000	Manchester *et al.*, 2000
	Hypericum perforatum	Shoot	2000	Murch *et al.*, 2000
	Prunus cerasus	Fruit	2001	Burkhardt *et al.*, 2001
	108 Chinese medicinal herbs	Whole plant	2003	Chen *et al.*, 2003

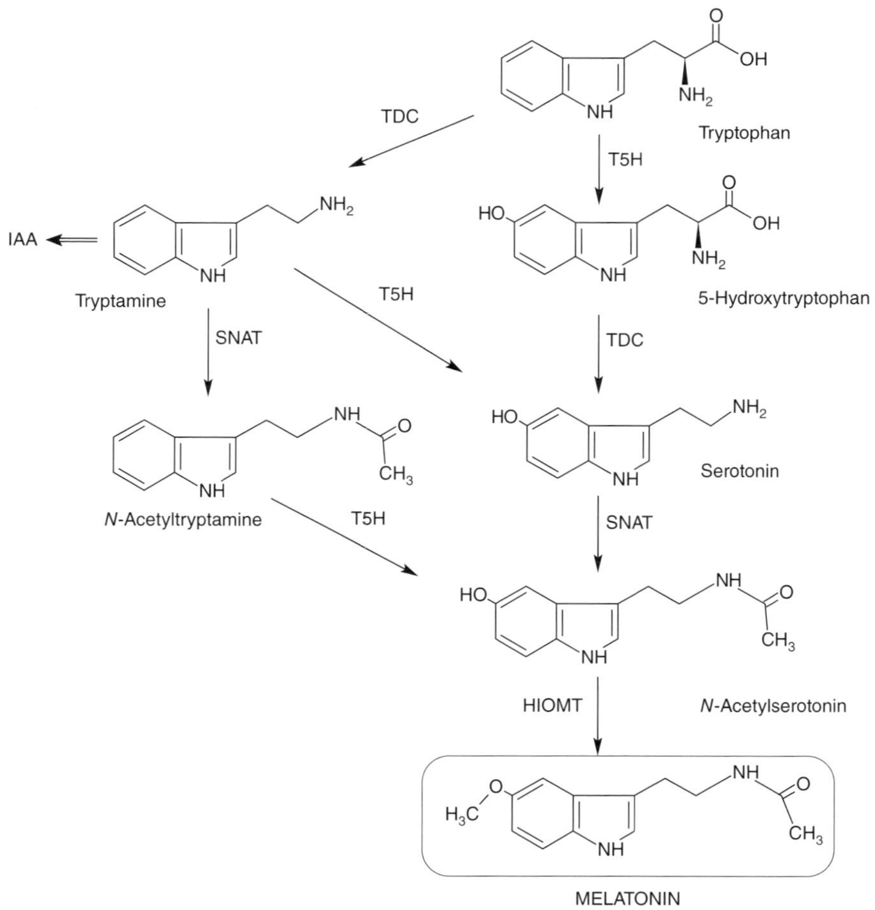

Fig. 21.3. Biosynthetic pathway of melatonin from tryptophan. Abbreviations, including the enzymes of the various steps are: HIOMT, hydroxyindole-*O*-methyltransferase; IAA, indole-3-acetic acid; SNAT, serotonin *N*-acetyltransferase; TDC, tryptophan decarboxylase; T5H, tryptophan 5-hydroxylase.

Table 21.2. Characteristics of the four enzymes involved in the biosynthetic pathway of melatonin from tryptophan.

Name	Other names	Class	Reaction, cofactor and comments	First or relevant references	Crystallographic structure	
					Organism source	References
		EC 1: Oxidoreductases				
	Tryptophan 5-monooxygenase	EC 1.14: Acting on paired donors, with incorporation or reduction of O_2	*Reaction*: L-Tryptophan + Tetrahydrobiopterin + O_2 → 5-Hydroxy-L- tryptophan + Dihydropteridine + H_2O	Jequier *et al*., 1969		Wang *et al*., 2002
Tryptophan 5-hydroxylase (T5H)	L-Tryptophan hydroxylase	EC 1.14.16: With reduced pteridine as one donor, and incorporation of one atom oxygen	*Cofactor:* Iron	Ichiyama *et al*., 1970	*Homo sapiens*	Jin *et al*., 2009
	Indoleacetic acid-5-hydroxylase		*Comments*: Activated by phosphorylation, catalysed by a Ca^{2+}-activated protein kinase	Friedman *et al*., 1972		Cianchetta *et al*., 2010
	Tryptophanhydroxylase (TPH)	EC 1.14.16.4: Tryptophan 5-monooxygenase		Hamon *et al*., 1979	*Gallus gallus*	Windahl *et al*., 2008
	Aromatic-L-amino-acid decarboxylase DOPA decarboxylase	EC 4: Lyases	*Reaction*: L-Tryptophan → Tryptamine + CO_2	McGilvery and Cohen, 1948	*Homo sapiens*	Giardina *et al*., 2011
	Hydroxytryptophan decarboxylase	EC 4.1: Carbon–carbon lyases	*Cofactor:* Pyridoxal phosphate	Weissbach *et al*., 1957		
Tryptophan decarboxylase (TDC)	5-Hydroxytryptophan decarboxylase			Lovenberg *et al*., 1962	*Sus scrofa*	Burkhad *et al*., 2001
	Aromatic amino acid decarboxylase (AADC)	EC 4.1.1: Carboxy-lyases	*Comments*: Also acts on L-tryptophan, 5-hydroxy-L-tryptophan and dihydroxy-L-phenylalanine (DOPA)	Christenson *et al*., 1970		
	Aromatic-L-amino-acid carboxylyase (tryptamine forming)	EC 4.1.1.28: Aromatic-L-amino-acid decarboxylase		De Luca *et al*., 1989	*Drosophila melanogaster*	Han *et al*., 2010

Continued

Table 21.2. Continued.

Name	Other names	Class	Reaction, cofactor and comments	First or relevant references	Crystallographic structure: Organism source	Crystallographic structure: References
Serotonin *N*-acetyltransferase (SNAT)	Aralkylamine *N*-acetyltransferase (AANAT) Serotonin acetyltransferase Serotonin acetylase	EC 2: Transferases EC 2.3: Acyltransferases EC 2.3.1: Transferring groups other than aminoacyl groups EC 2.3.1.87: Aralkylamine *N*-acetyltransferase	*Reaction*: Acetyl-CoA + Aralkylamine → CoA + *N*-acetylaralkylamine *Cofactor:* None *Comments*: Narrow specificity towards aralkylamines, including serotonin	Weissbach *et al*., 1960 Voisin *et al*., 1984 Khalil and Cole, 1998 Ferry *et al*., 2000	*Ovis aries* *Drosophila melanogaster* *Homo sapiens*	Hickman *et al*., 1999a,b Scheibner *et al*., 2002 Wolf *et al*., 2002 Cheng *et al*., 2012 Obsil *et al*., 2001
Hydroxyindole *O*-methyltransferase (HIOMT)	Acetylserotonin *O*-methyltransferase Hydroxyindole methyltransferase *N*-acetylserotonin *O*-methyltransferase Acetylserotonin methyltransferase (ASMT)	EC 2: Transferases EC 2.1: Transferring one-carbon groups EC 2.1.1: Methyltransferases EC 2.1.1.4: Acetylserotonin *O*-methyltransferase	*Reaction:* *S*-Adenosyl-L-methionine + *N*-Acetylserotonin → *S*-Adenosyl-L-homocysteine + *N*-Acetyl-5-methoxytryptamine *Cofactor:* None *Comments:* Some other hydroxyindoles also act as acceptor, but more slowly	Axelrod and Weissbach, 1960 Axelrod and Weissbach, 1961 Cardinali and Rosner, 1971	*Homo sapiens*	Botros *et al*., 2013

oxygen into the substrate and reducing the other atom to water. The two electrons required to reduce the second atom to water are supplied by the tetrahydrobiopterin. Catecholamines are reported to act as inhibitors of T5H (Johansen *et al.*, 1991). The values of the Michaelis constant (K_m) of T5H for tryptophan can oscillate from 22.8 μM in the human (McKinney *et al.*, 2005) to 119 μM in the rat (*Rattus norvegicus*) (Nakata and Fujisawa, 1982).

While phenylalanine hydroxylase and tyrosine hydroxylase are each encoded by a single gene, there is evidence suggesting that both of these genes also encode T5H (Haycock *et al.*, 2002; Kim *et al.*, 1991). Other authors have confirmed the identification of a second gene, called *Tph2* (Walther *et al.*, 2003). The two isoforms of tryptophan hydroxylase that are found are then derived from different genes, which have been called *Tph1* and *Tph2*. Both are found in vertebrates, and *Tph2* has been reported in the genome of the human, mouse, rat, chicken, zebra fish and torafugu (or tiger blowfish, *Takifugu rubripes*), and also in the fruit fly (*Drosophila melanogaster*) (Walther and Bader, 2003; Coleman and Neckameyer, 2005; Zhang *et al.*, 2006; Teraoka *et al.*, 2013). The product of *Tph2*, called TPH2, is predominantly expressed in the brain and enteric neurons (McKinney *et al.*, 2005; Walther and Bader, 2003; Walther *et al.*, 2003; Zhang *et al.*, 2004; Zill *et al.*, 2004). In contrast, the product of *Tph1*, i.e. TPH1, which is classically called TPH (i.e. T5H), and had already been identified 4 decades ago (Grahame-Smith, 1964), is expressed in the gut, pineal gland, spleen and thymus (Walther and Bader, 2003).

The three-dimensional structures of T5H have been determined for the human (Wang *et al.*, 2002) and chicken (*Gallus gallus*) (Windahl *et al.*, 2008) (Table 21.2). It has a tetrameric structure with four identical subunits and displays a high molecular mass, which can vary from 30 kDa in bovines (*Bos taurus*) (Nukiwa *et al.*, 1974) up to 300 kDa in *R. norvegicus* (Nakata and Fujisawa, 1982). In humans, TPH1 and TPH2 have been calculated to weigh 234 kDa and 274 kDa, respectively, values that correspond to tetramers composed of subunits of about 58.5 kDa and 68.5 kDa (McKinney *et al.*, 2005).

In the case of humans, the TPH1 and TPH2 proteins have been reported to contain 444 and 490 amino acids, respectively. They share an overall 71% amino acid identity and a similar three-dimensional structure, but differ in some properties, including solubility, binding affinity to the substrate tryptophan and efficiency of 5-hydroxytryptophan formation (Walther and Bader, 2003; McKinney *et al.*, 2005). However, the most striking difference between the two isoforms lies in their tissue distribution: TPH1 is expressed in the periphery and in the pineal body, whereas TPH2 appears to be responsible for the synthesis of serotonin in the rest of the brain (Sugden, 2003).

It has been established that TPH2 exhibits a flexible, context-dependent gene expression under the control of numerous internal and external factors, including the biological clock, stressors, hormones, growth factors and exogenous chemicals (Chen and Miller, 2013).

21.3.1.2 Tryptophan decarboxylase (TDC)

TDC (also called aromatic L-amino acid decarboxylase, AADC), is an essential enzyme for the synthesis of the monoamine neurotransmitters serotonin and dopamine. The enzyme converts 5-hydroxytryptophan (5-HTP) into serotonin (5-hydroxytryptamine) and L-DOPA (3,4-dihydroxyphenylalanine) into dopamine by a decarboxylation reaction (Brodie *et al.*, 1962; Bowsher and Henry, 1986; Barth *et al.*, 2012). A role for TDC in the synthesis of serotonin was also postulated previously by different authors (Udenfriend *et al.*, 1953; Clark *et al.*, 1954). In 1962, TDC was found to have affinity for a broad range of substrates, catalysing the decarboxylation of a number of aromatic L-amino acids and their methylated derivatives, hence the proposal of the name AADC (Lovenberg *et al.*, 1962) (Table 21.2). TDC is also considered to be involved in the synthesis of trace amines, such as tyramine from tyrosine, 2-phenylethylamine from phenylalanine and tryptamine from tryptophan (Saavedra, 1974; Juorio and Boulton, 1982; Dyck *et al.*, 1983).

The first purification of TDC was from hog kidney (Christenson *et al.*, 1970); subsequently, it was also purified from rat liver (Ando-Yamamoto *et al.*, 1987), rat kidney (Shirota and Fujisawa, 1988), bovine brain (Nishigaki *et al.*, 1988), rat pheochromocytoma (Coge *et al.*, 1990) and human pheochromocytoma (Shirota and Fujisawa, 1988; Maneckjee and Baylin, 2013). It has been reported that TDC has two subunits (Lancaster and Sourkes, 1972; Borri Voltatorni *et al.*, 1979;

Ichinose *et al.*, 1985; Ando-Yamamoto *et al.*, 1987). The enzyme has a high molecular mass, and similar values per subunit have been obtained by SDS PAGE from different species, tissues and cells. Reported subunit weights are 54 kDa in human leucocytes (Kokkinou *et al.*, 2009), 50 kDa in human kidney (Bossinakou and Fragoulis, 1996), bovine brain (Nishigaki *et al.*, 1988) and 48 kDa in rat kidney (Shirota and Fujisawa, 1988). However, *Penicillium raistrickii* has been described as a monomer of 125 kDa (Niedens *et al.*, 1999). TDC from rat kidney was reported to be a homodimeric 52 kDa protein that binds one molecule of pyridoxal phosphate (pyridoxal-5-phosphate) per molecule of dimer (Rudd and Thanassi, 1981; Borri Voltatorni *et al.*, 1987). It is now accepted that TDC is a homodimer binding one pyridoxal phosphate molecule (Kokkinou *et al.*, 2009). Pyridoxal phosphate acts as a co-factor and is tightly bound to the ε-amino group of a lysine residue of TDC – Lys_{303} in animal TDC and Lys_{314} in *Catharanthus roseus* (Bossa *et al.*, 1977; Ishii *et al.*, 1996). Site-directed mutagenesis studies have shown that the pyridoxal phosphate-binding Lys is essential for product release, but not for the decarboxylation step (Nishino *et al.*, 1997). The Michaelis constants (K_m) of TDC for 5-HTP can vary from 18 μM in the rat to 830 μM in bovines, or 47 μM in the human (Allen *et al.*, 2010).

The first sequencing of the TDC gene was reported in 1986 for *Drosophila* (Eveleth *et al.*, 1986). Later, cDNA encoding for the rat (Tanaka *et al.*, 1989), bovine (Kang and Joh, 1990), human (Ichinose *et al.*, 1989) and plants such as *C. roseus* (De Luca *et al.*, 1989) were described. More information can be obtained in an excellent review by Zhu and Juorio (1995). The crystal structure has been obtained for the human (Giardina *et al.*, 2011), pig (*Sus scrofa*) (Burkhad *et al.*, 2001) and *D. melanogaster* (Han *et al.*, 2010) (Table 21.2).

21.3.1.3 Serotonin N-acetyltransferase (SNAT)

SNAT, also called aralkylamine (or arylalkylamine) *N*-acetyltransferase (AANAT), is the penultimate enzyme in melatonin synthesis (Fig. 21.3 and Table 21.2). The enzyme catalyses the conversion of serotonin to *N*-acetylserotonin (Weissbach *et al.*, 1960; Voisin *et al.*, 1984), which is believed to be the rate-limiting step in melatonin synthesis (Boutin *et al.*, 2005). SNAT recognizes a large number of aralkylamines as substrate, including synthetic aralkylamines (Shen *et al.*, 1996; Ferry *et al.*, 2000) and has a narrow specificity. It uses acetyl coenzyme A to preferentially acetylate arylethylamines, including the naturally occurring amines, serotonin, tyramine, dopamine, tryptamine and phenylethylamine, and the synthetic amines, naphthylethylamine, benzofurylethylamine and benzothienylethylamine. Diamines and monoamines are relatively poor substrates of SNAT (Coon *et al.*, 1995; Khalil and Cole, 1998; Ferry *et al.*, 2000, 2004; Mesangeau *et al.*, 2003).

Family members of SNAT have been found in Gram-positive bacteria, fungi, algae, cephalochordates and vertebrates, but are not found in higher plants, insects, nematodes or urochordates (Iyer *et al.*, 2004). In many vertebrates, SNAT is expressed mainly in the pineal gland, but also in the retina (Klein, 2007). The possible expression and functional roles of this enzyme are being investigated in liver and gut in the rat (Sánchez-Hidalgo *et al.*, 2009) and goldfish (Nisembaum *et al.*, 2013). SNAT has been cloned from various species, including humans (Coon *et al.*, 1996). Several research groups have ascertained the crystal structure of the enzyme: in the sheep (*Ovis aries*) (Hickman *et al.*, 1999a,b; Scheibner *et al.*, 2002; Wolf *et al.*, 2002), human (Obsil *et al.*, 2001) and *D. melanogaster* (Cheng *et al.*, 2012) (Table 21.2).

SNAT has a molecular mass that can vary from 26 kDa in *R. norvegicus* (Morrissey *et al.*, 1977) to 100 kDa in *G. gallus* (Pozdeyev *et al.*, 2006). A single SNAT gene (*Aant1*) has been found in mammals and birds, although the whole genome duplication of ancestral teleosts generated at least two SNAT genes encoding for two different enzymes. The *Aant1* gene is mostly expressed in the retina, where it acetylates several amines and is involved in multiple local functions. The *Aant2* gene, mainly expressed in the pineal gland, appears to have acquired a higher selectivity for serotonin to enhance the efficiency of melatonin production (Coon *et al.*, 1999; Zilberman-Peled *et al.*, 2011).

SNAT is broadly considered as the melatonin rhythm-generating enzyme and exhibits a daily rhythm with high nocturnal levels in the pineal gland and retina of many vertebrates

(Klein, 2007). It has been observed that nocturnal increases in the level and activity of this enzyme drive nocturnal increases in melatonin production, which underlines the influence of the internal circadian clock and light. The magnitude of the nocturnal increase in enzyme activity ranges from 7- to 150-fold on a species-to-species basis among vertebrates. In all cases, the nocturnal levels of SNAT decrease very rapidly following exposure to light (Borjigin *et al.*, 1995; Klein *et al.*, 1996; Roseboom *et al.*, 1996; Gastel *et al.*, 1998; Garidou *et al.*, 2001). Although the next step in melatonin synthesis, the conversion of *N*-acetylserotonin to melatonin, which is catalysed by HIOMT (Fig. 21.3), is a step that is generally constitutive and not rate limiting for melatonin production (Khalil *et al.*, 1999), current studies have shown that HIOMT may also be involved in limiting melatonin production in certain circumstances (Tan *et al.*, 2007c). Some observations suggest that part of the *N*-acetylserotonin synthesized by SNAT is not used by HIOMT to produce melatonin, and that in conditions of marked pineal stimulation, HIOMT, rather than SNAT, limits melatonin synthesis (Simonneaux and Ribelayga, 2003).

As regards the modulation of SNAT production, the increase in the level of SNAT activity in the rat pineal gland at night is mainly due to the neo-transcription of its gene and a neo-synthesis of its protein (Roseboom *et al.*, 1996; Gastel *et al.*, 1998). The signal transduction pathway that regulates this process has been studied (Klein *et al.*, 1997; Ganguly *et al.*, 2002; Maronde and Stehle, 2007). The process is primarily regulated by cAMP-mediated noradrenergic stimulation of *Aanat* gene expression and the translation of its messenger into active molecules (Klein *et al.*, 1996), whereby noradrenaline from the sympathetic nerve stimulates both α- and β-adrenergic receptors, resulting in a 100-fold increase in the level of cAMP (Klein, 1985), which controls the transcription of *Aanat*. The mechanism involves the protein kinase A-dependent phosphorylation of CREB (cyclic AMP regulatory element-binding protein). Phosphorylated CREB then binds to the CRE (cAMP response element) site in the promoter region of the *Aanat* gene, where it appears to be critical for the full stimulation of *Aanat* gene expression and for initiating transcription (Roseboom *et al.*, 1996; Baler *et al.*, 1997, 1999; Burke *et al.*, 1999; Schomerus *et al.*, 2000; Coon *et al.*, 2001). This results in a >100 fold increase in *Aanat* mRNA level within 2 h of the onset of darkness, followed shortly after by a similar magnitude of increase in the SNAT protein level and enzyme activity. The amplitude of this increase is modulated by inducible cAMP early repressor, which competes with CREB for CRE sites (Foulkes *et al.*, 1997; Korf *et al.*, 2003; Stehle *et al.*, 2003). This environmental lighting regulation of *Aanat* transcription ensures that melatonin production is triggered only after the onset of darkness and reflects changes in the duration of the environmental lighting conditions.

21.3.1.4 Hydroxyindole O-methyltransferase (HIOMT)

HIOMT (also called acetylserotonin *O*-methyltransferase, ASMT) catalyses the *O*-methylation of *N*-acetylserotonin to form melatonin (Axelrod and Weissbach, 1960) (Fig. 21.3 and Table 21.2). This enzyme not only catalyses the final step in the synthesis of melatonin, but also that of the other 5-methoxyindoles (Axelrod and Weissbach, 1961). HIOMT transfers a methyl group from *S*-adenosyl-L-methionine to its indolic substrate (Baldessarini and Kopin, 1966).

In the rat pineal gland, the best substrate for HIOMT is *N*-acetylserotonin. In relative values, the affinity of the enzyme for *N*-acetylserotonin is between 50 and 80%, for 5-hydroxytryptophol between 15 and 30%, for 5-hydroxytryptamine around 10% and for other substrates (5-HTP, 5-hydroxyindole acetic acid) less than 5% (Axelrod and Weissbach, 1961; Morton, 1987; Simonneaux and Ribelayga, 2003). In the different species that have been studied, the enzyme has a molecular mass of 76 to 78 kDa, and is a dimer consisting of two similar subunits of molecular weight of about 39 kDa each. Immunochemical experiments point to a large heterogeneity in the structure of the protein and its enzymatic properties among species (Nakane *et al.*, 1983; Sudgen *et al.*, 1987).

The cDNA coding for HIOMT was first isolated from the cow (Ishida *et al.*, 1987), then from the chicken (Voisin *et al.*, 1992), human (Donohue *et al.*, 1993), rat (Gauer and Craft, 1996) and monkey (Coon *et al.*, 2002), among others. The human *Hiomt* gene is the best studied (Donohue *et al.*, 1993; Rodríguez *et al.*, 1994;

Bernard *et al.*, 1995). The protein contains 367 amino acids, while the crystal structure has only been described for human HIOMT (Botros *et al.*, 2013) (Table 21.2).

In relation to the regulation of HIOMT expression, it is known that nocturnal stimulation of the pineal gland leads to a twofold increase in HIOMT mRNA levels (Gauer and Craft, 1996) and an increase in HIOMT activity of 30–50% (Axelrod *et al.*, 1965; Nagle *et al.*, 1972; Balemans *et al.*, 1981; Ribelayga *et al.*, 1997), while SNAT mRNA expression undergoes a significant increase of up to 150-fold (Borjigin *et al.*, 1995; Klein *et al.*, 1996; Roseboom *et al.*, 1996) leading to a 70- to 100-fold increase in SNAT activity (Klein and Weller, 1970). It has been suggested that the gentle rhythm of HIOMT activity is not involved in the daily rhythm of melatonin synthesis (Klein, 1985).

It is still not clear which enzyme (HIOMT or SNAT) is responsible for controlling the synthesis of melatonin. Classically, it is considered that SNAT is the enzyme that limits the rate of melatonin production, but other studies indicate that HIOMT could be the rate-limiting enzyme in melatonin production in the pineal gland. This is the case with the Siberian hamster pineal gland, where the level of melatonin is more in accordance with the activity of HIOMT that with the activity of SNAT (Ribelayga *et al.*, 2000). During the day, SNAT activity is lower than HIOMT activity and would be the limiting factor for the synthesis of melatonin. The increase in SNAT activity at the beginning of the night thus induces an increase in melatonin synthesis. During the night, however, HIOMT activity is lower than SNAT activity and would, therefore, become the limiting enzyme for melatonin production (Simonneaux and Ribelayga, 2003). In conclusion, which enzyme actually determines melatonin synthesis may depend on the physiological situation.

All of the enzymes mentioned have been isolated and characterized in animal tissues, but far less information exists for plant material.

21.3.2 Characteristic features of melatonin-related enzymes in plants

In plants, the melatonin biosynthesis pathway is not clearly understood, although it is known to start with tryptophan, while serotonin acts as intermediate and *N*-acetylserotonin as direct precursor of melatonin (Fig. 21.3), in a similar way to that seen in animals. However, it seems that the biosynthetic pathways of melatonin in plants are more complex than that in animals (Arnao and Hernández-Ruiz, 2006; Tan *et al.*, 2012). An additional pathway appears in plants, in which tryptamine is formed from tryptophan by the action of TDC (Fig. 21.3, left). Tryptamine is then transformed into serotonin by T5H acting with tryptamine hydroxylase activity (Schröder *et al.*, 1999). In other words, the first enzymatic product is not 5-HTP, but tryptamine. This additional pathway implies that the order in which the enzymes act to generate serotonin in plants differs from that observed in animals. Another pathway not observed in animals concerns tryptamine, which also acts as substrate of SNAT to form *N*-acetyltryptamine, which can be hydroxylated to form *N*-acetylserotonin via the action of T5H. So there appear to be many more alternatives for the biosynthesis of melatonin in plants (Schröder *et al.*, 1999; Arnao and Hernández-Ruiz, 2006; Posmyk and Janas, 2009; Fujiwara *et al.*, 2010; Park, 2011; Kang *et al.*, 2013) than in animals. These melatonin biosynthetic pathways and the plant enzymes involved in them are described and discussed below.

21.3.2.1 Tryptophan 5-hydroxylase (T5H)

The enzymes T5H and TDC which participate in the pathway of melatonin biosynthesis in animals have also been identified in plants, as mentioned above (Facchini *et al.*, 2000; Fujiwara *et al.*, 2010). T5H has been widely studied and characterized in rice by Kang and co-workers (Kang *et al.*, 2007b, 2009a,b). T5H was found as a soluble enzyme in several organs of rice plants, where it showed high activity in leaves, stems and roots. The maximal activity of T5H in roots was related to high biosynthesis of serotonin. Although T5H is constitutively expressed in rice, the levels vary among the tissues. The enzyme presents a low K_m (20 μM) for tryptamine as substrate.

It has been suggested that T5H belongs to the cytochrome P450 family (Schröder *et al.*, 1999). However, rice T5H activity decreased by 70% in the absence of the co-substrate tetrahydrobiopterin (Kang *et al.*, 2007b), which is required for the activity of amino acid hydroxylases such as phenylalanine hydroxylase, tyrosine hydroxylase, etc. This suggests that the T5H enzyme belongs to the family of amino acid hydroxylases (Fitzpatrick, 1999).

Based on Fig. 21.3, there are two possible pathways for serotonin biosynthesis from tryptophan in plants, one started by T5H, as occurs in animals, and another started by TDC. It remains to be determined whether the two pathways coexist or whether one is dominant in plant melatonin synthesis. Studies to date suggest that the more important route is that started by TDC to convert tryptophan to tryptamine (Grobe, 1982; Facchini *et al.*, 2000). However, the results of other authors from rice plants indicate that melatonin can also be synthesized via 5-HTP, but at low levels. For example, Park and co-workers studied the biosynthesis of melatonin in *Sekiguchi* mutant rice, which lacks functional tryptamine 5-hydroxylase (T5H) activity (Park *et al.*, 2012). On the one hand, they found that tryptamine did not accumulate in wild type rice plants because it is rapidly converted into serotonin by the T5H enzyme, which has a very low K_m (7.3 μM) towards tryptamine (Fujiwara *et al.*, 2010). On the other hand, and as expected, in the *Sekiguchi* rice mutant, which does not express T5H, a very high level of tryptamine was detected but a very low level of serotonin. These results suggest that rice has an alternative serotonin biosynthetic pathway similar to that of animals, but operating at a lower rate.

21.3.2.2 Tryptophan decarboxylase (TDC)

Plant TDCs belong to the family of aromatic L-amino acid decarboxylases and catalyses the conversion of tryptophan into tryptamine in the presence of the coenzyme pyridoxal-5-phosphate (Kang *et al.*, 2007a); the tryptamine that has been formed then takes part in the formation of the serotonin (5-hydroxytryptamine) via the action of T5H, as one of the pathways for serotonin synthesis (Facchini *et al.*, 2000) (Fig. 21.3). Although this action of TDC does not represent the main biosynthetic route, the enzyme has been implicated in the biosynthesis of the plant hormone indole-3-acetic acid (IAA) (Bartel, 1977) and of simple alkaloids such as the *β*-carbolines, which are found in several plant families (Berlin *et al.*, 1994).

The first-known TDC gene was isolated from *C. roseus* (De Luca *et al.*, 1989), in which the cDNA clone encoded a protein of 500 amino acids with a calculated molecular mass of 56,142 Da. In rice plants, two TDC genes are involved in serotonin synthesis, of which TDC1 plays a key role in the induced synthesis of serotonin upon senescence (Kang *et al.*, 2009b). The enzyme has been characterized in cells of *C. roseus* as a soluble cytosolic protein (Stevens *et al.*, 1993), with an isoelectric point of 5.9 and a native MW of 115 kDa; it consists of two identical subunits. The enzyme displays high substrate specificity for L-tryptophan and is able to decarboxylate different tryptophan analogues, including 5-HTP and others. Tryptamine was found to be a competitive inhibitor (K_i = 0.31 mM) (Facchini *et al.*, 2000).

TDC from *C. roseus* has a low K_m for tryptophan (0.072 mM) (Noé *et al.*, 1984), but a tenfold higher K_m for TDC isolated from rice *Oryza sativa* (K_m = 0.69 mM) (Kang *et al.*, 2007a) and from *Ophiorrhiza pumila* (*Rubiaceae*) (K_m = 0.72 mM) (Yamazaki *et al.*, 2003); in tomato, the K_m was found to be as much as 42-fold higher (3.0 mM) (Gibson *et al.*, 1972). The high K_m values of TDC enzymes compared with those of the enzymes that follow it in the melatonin biosynthesis pathway (such as T5H in rice, with a K_m of 20 μM for tryptamine) suggest that TDC works actively in cells where there are high tryptophan accumulations. Thus, TDC plays a rate-limiting role because of its high K_m for the substrate tryptophan, while T5H seems to be constitutively expressed.

Considering that the tryptophan pathway in plants is tightly feedback regulated by micromolar concentrations of tryptophan (RadWanski and Last, 1995) and that the K_m values of TDC (for example in rice and tomato) are high towards tryptophan, TDC overexpression could be insufficient to increase serotonin levels because tryptophan levels are very low (<15 μM) compared with those of other amino acids (Kang *et al.*, 2009b). This is corroborated by the interesting work of Kang and co-workers (Kang *et al.*, 2009a), who obtained two types of transgenic rice plants, one overexpressing TDC and the other in which the expression of TDC was suppressed through an RNA interference (RNAi) system. These authors suggested that TDC plays a rate-limiting role for serotonin accumulation, although the synthesis of serotonin depends on an absolute amount of tryptophan accumulation as a result of the coordinated induction of the tryptophan biosynthetic genes.

The enzymes T5H and TDC are possibly regulated by light, as in animals. Thus, in assays with detached rice leaves, Byeon and co-workers (Byeon *et al.*, 2012) showed that during the senescence process, rice leaves maintained under constant light increase their melatonin level 150-fold with respect to leaves under constant darkness; the levels of tryptamine, serotonin and *N*-acetylserotonin also strongly increased. The transcripts and proteins of T5H and TDC were high in constant light, but low in darkness, indicating that the expression of these enzymes was regulated by light.

21.3.2.3 Serotonin N-acetyltransferase (SNAT)

SNAT catalyses the penultimate step in melatonin synthesis, the acetylation of serotonin to *N*-acetylserotonin (Fig. 21.3). Various studies have been undertaken to better understand the function and activity of SNAT in plants. For example, several authors have obtained transgenic plants that overexpressed the SNAT gene. The first SNAT gene in plants was reported by Okazaki and co-workers in 2009, in the unicellular green algae *Chlamydomonas reinhardtii*, and it was designated *CrAANT*. Furthermore, they transferred this gene to a model tomato cultivar (*Solanum lycopersicum*, var. Micro-Tom). The resulting transgenic plant had a higher melatonin content than the wild type (from 2- to 9-fold more in the different transgenic lines obtained), demonstrating the involvement of the *CrAANAT* gene in melatonin biosynthesis and confirming that the melatonin biosynthetic pathway in the Micro-Tom tomato plant is similar to that of vertebrates (Okazaki *et al.*, 2009).

To study the function of SNAT in plants, Kang and co-workers obtained transgenic rice plants that overexpressed the human SNAT gene and then measured *N*-acetylserotonin and melatonin levels in etiolated rice seedlings, thereby demonstrating that melatonin levels depend on serotonin levels. The levels of *N*-acetylserotonin were 2- to 9-fold higher in the transgenic lines (similar to the values obtained with transgenic tomato expressing CrAANT), which led to the enhanced synthesis of melatonin (nearly 7 times more than in the etiolated seedlings of wild type rice), but the level of melatonin synthesis, even in the presence of excess serotonin and the ectopic expression of SNAT, was relatively low, implying that the terminal enzyme HIOMT is rate-limiting in rice plants (Kang *et al.*, 2010).

Recently, Kang's group cloned a SNAT gene in rice for the first time (Kang *et al.*, 2013). As no SNAT gene homologous to the animal SNAT gene exists in published plant genomes, they consider it possible that plants possess a SNAT enzyme different from animal SNAT homologues. Based on this premise, these workers screened for a SNAT activity belonging to one of the superfamily of *N*-acetyltransferase (GNAT) genes that is responsible for catalysing the transfer of the acetyl group from acetyl coenzyme A to a series of molecules such as histones, aminoglycosides and aralkylamines (Dyda *et al.*, 2000). So far, they have found several rice GNATs in *Escherichia coli* and obtained one protein with SNAT activity. The kinetic properties of the purified recombinant SNAT obtained differed from those observed in animal SNATs. For example, the K_m values of rice SNAT for serotonin, tryptamine and 5-methoxytryptamine were 385, 836 and 375 μM, respectively, values that are about 4.5-, 10- and 13-fold, respectively, higher than those of the SNAT enzyme in sheep (Voisin *et al.*, 1984).

Another possible pathway in melatonin biosynthesis in which SNAT could participate is in the reaction that uses tryptamine as a substrate to produce *N*-acetylserotonin (Fig. 21.3). Park and co-workers (Park *et al.*, 2012) reported the use of tryptamine as substrate by SNAT in leaves of the *Sekiguchi* mutant rice lacking functional T5H. They produced high levels of *N*-acetyltryptamine, whereas the wild type rice had low levels of *N*-acetyltryptamine. This indicates that the putative rice SNAT enzyme has high catalytic activity toward tryptamine as well as serotonin. This is similar to the situation in animals, where a broad substrate specificity toward tryptamine and serotonin has also been found (Klein, 2007). Park *et al.* (2012) considered than the bottleneck appears to be SNAT because the conversion rate of serotonin to *N*-acetylserotonin in plant cells is the lowest among the melatonin biosynthetic enzyme reactions.

As regards the possible regulation of SNAT expression by light in plants (as occurs in animals), a study in which rice seedlings were subjected to various light conditions showed that light does not alter the expression of SNAT mRNA levels, suggesting that SNAT mRNA is expressed constitutively in plants (Kang *et al.*, 2013).

21.3.2.4 Hydroxyindole O-methyltransferase (HIOMT)

HIOMT catalyses the last step in melatonin synthesis, the conversion of *N*-acetylserotonin into melatonin (Fig. 21.3). When the gene of the HIOMT from rice was cloned and expressed in *E. coli*, the protein showed enzyme activity *in vitro* and was inhibited by compounds such as *N*-acetyltryptamine (Kang *et al.*, 2011). Rice HIOMT (called *N*-acetylserotonin *O*-methyltransferase) has also been purified and its activity studied by Park and co-workers (Park *et al.*, 2013a). They obtained a K_m value for *N*-acetylserotonin of 864 μM, which is significantly higher than that measured in chickens (5 μM) (Bernard *et al.*, 1995), humans (7 μM) (Ben-Abdallah *et al.*, 2011) and bovine pineal glands (54 μM) (Voisin *et al.*, 1992), but only 3.6-fold higher than that for human ovarian HIOMT (240 μM) (Itoh *et al.*, 1999). Also, when purified rice HIOMT was incubated in the absence of *S*-adenosyl-L-methionine, no melatonin was formed. In comparison to the situation in animals, rice HIOMT was not inhibited by the substrate. In terms of substrate specificity, it was observed that tryptamine increased rice HIOMT when both tryptamine and *N*-acetylserotonin were simultaneously incubated in the presence of *S*-adenosyl-L-methionine. Moreover, rice HIOMT exhibits low catalytic activity compared with the animal enzyme.

The effect has been measured of light intensity on HIOMT mRNA and melatonin levels in rice seedlings grown under various light conditions, including continuous light and darkness. The level of melatonin was several-fold higher in shoots grown in continuous darkness than in those grown in continuous light, and this was accompanied by an increase in HIOMT mRNA. Rice HIOMT seems to play a major role in dark-induced melatonin biosynthesis. The data obtained suggested to the authors that the low level of melatonin in rice plants may result from the low catalytic activity of HIOMT, as well as the light inhibition of melatonin biosynthesis. Melatonin levels in plants are probably not only controlled by light and/or darkness, but also by other environmental stimuli, although the physiological state of the plant may also participate (see Section 21.5) (Arnao and Hernández-Ruiz, 2009b, 2013a,b; Kang *et al.*, 2010; Vitalini *et al.*, 2011a; Pelagio-Flores *et al.*, 2012).

21.4 Catabolism of Melatonin: Enzymatic and Non-enzymatic Pathways

Melatonin is mainly involved in two catabolic pathways: the indolic and the kynuric pathways (Fig. 21.4). In the indolic pathway (Fig. 21.4, upper diagram), which has been described in animals, melatonin is converted to 6-hydroxymelatonin by cytochrome P450 (CYP1A1, CYP1A2 and CYP2C19) in the liver (Boutin *et al.*, 2005; Ma *et al.*, 2005; Slominski *et al.*, 2007; Ritzenthaler *et al.*, 2013). 6-Hydroxymelatonin is then conjugated with sulfate to form 6-sulfatoxymelatonin, in a reaction catalysed by sulfotransferase (Aust *et al.*, 2005), and is later excreted in the urine. Melatonin can also be oxidized by an extrahepatic cytochrome P450 such as CYP1B1, when high levels of 6-sulfatoxymelatonin are reached in the cerebral cortex, kidney and heart (Lahiri *et al.*, 2004). Melatonin glucuronide is another solubilized form of melatonin that is in a form that is ready to be excreted. In this same indolic pathway, melatonin acts as a substrate of a melatonin deacetylase, resulting in the formation of 5-methoxytryptamine, which is oxidized by an amino oxidase to produce 5-methoxyindolil-3-acetaldehyde. This may be oxidized by an alcohol dehydrogenase to form 5-methoxytryptophol or by an aldehyde dehydrogenase to generate 5-methoxyindolil-3-acetic acid (Slominski *et al.*, 1996, 2002, 2007; Zawilska *et al.*, 2002, 2003; Ceinos *et al.*, 2005; Fischer *et al.*, 2006).

Melatonin can also be degraded through the kynuric pathway, in which melatonin is transformed into several oxygenated compounds such as 2-hydroxymelatonin, cyclic-3-hydroxymelatonin and melatonin dioxetane (Fig. 21.4, lower diagram). These compounds can be formed by reaction with different ROS and, in some cases, by the action of UV radiation (Slominski *et al.*, 2007; Tan *et al.*, 2007c). AFMK, the main catabolic product of melatonin in the kynuric pathway, can be formed in three ways: (i) through the formation of melatonin dioxetane, which is generated from 2-hydroxymelatonin and its related indolinones (Tan *et al.*, 2000a; Fischer *et al.*, 2006); (ii) through the formation of cyclic-3-hydroxymelatonin, which is generated from a melatoninyl radical by the action of hydroxyl radicals (•OH) (Tan *et al.*, 1998, 2001; Horstman *et al.*, 2002; De Almeida *et al.*, 2003); and

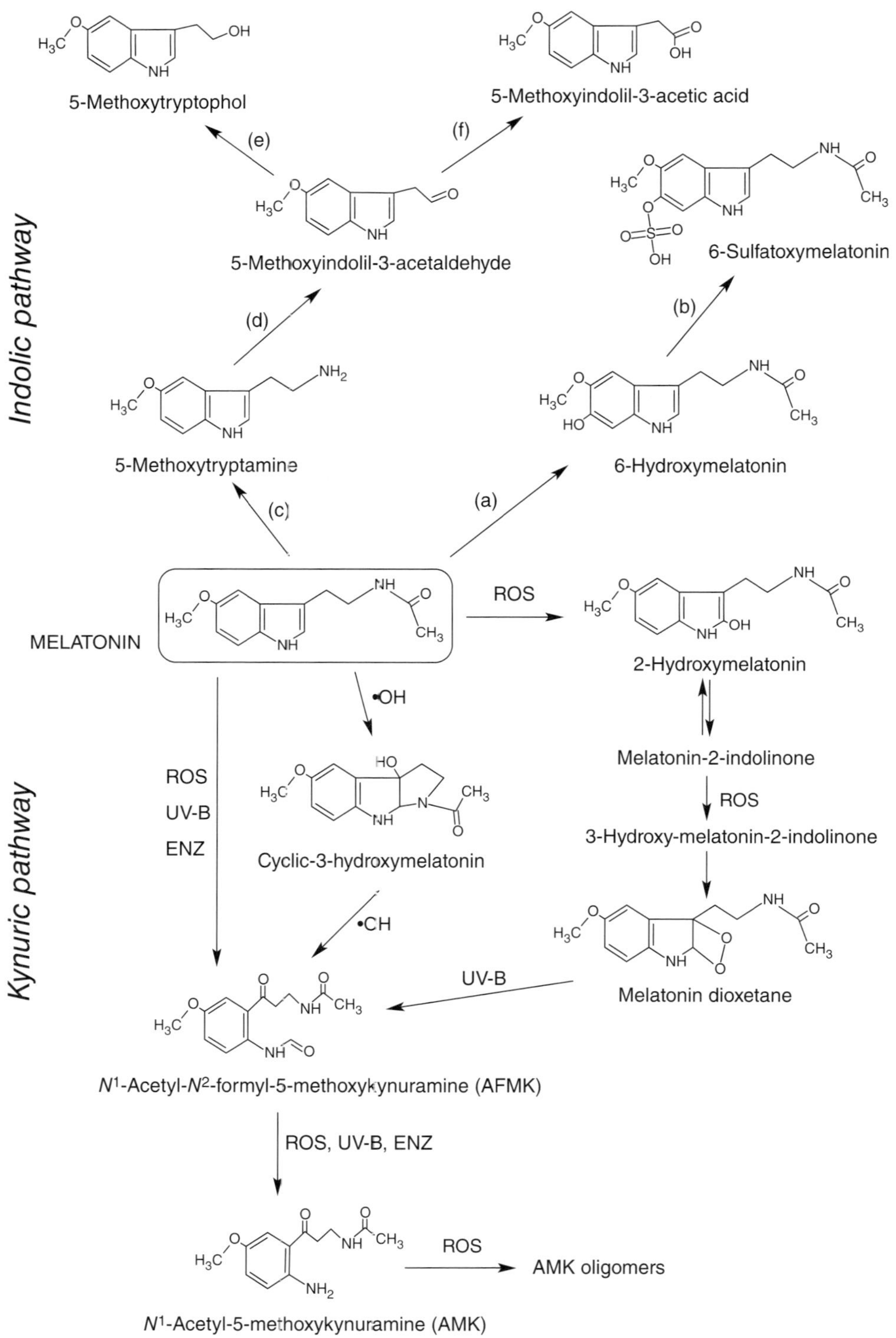

Fig. 21.4. Catabolic pathways for melatonin: the kynuric and indolic pathways. The enzymes involved in the respective reactions are: (a) cytochrome P450; (b) sulfotransferase; (c) melatonin deacetylase; (d) amino oxidase; (e) alcohol dehydrogenase; (f) aldehyde dehydrogenase. Other agents involved are: ENZ, multiple enzymes (see text); •OH, hydroxyl radicals; ROS, reactive oxygen species; UV-B, ultraviolet B (short wave) radiation.

(iii) directly from melatonin through the action of multiple enzymes such as aryl-acylamidase, myeloperoxidase, HRP, haemoperoxidase, indolamine dioxygenase, cytochrome c, oxoferryl haemoglobin or cytochrome P450 (de Oliveira Silva *et al.*, 2000; Tan *et al.*, 2001; Ferry *et al.*, 2005; Semak *et al.*, 2005; Ximenes *et al.*, 2007). The main kynuric product accumulated, AFMK, has been detected in many animal tissues, in dinoflagellates, chlorophyceans, ciliates and rotifers (Poeggeler and Hardeland, 2001). The presence of AFMK has also been determined in the water hyacinth (*Eichhornia crassipes*) (Tan *et al.*, 2007a). AFMK is converted to N^1-acetyl-5-methoxykynuramine (AMK) through the action of a variety of enzymes (catalase, arylamine formamidase, peroxidases), stimulated by UV radiation and/or ROS (Tan *et al.*, 2000a,b; Rozov, 2008; Schaefer and Hardeland 2008). Ultimately, AMK may form AMK oligomers when it reacts with free radicals such as the ABTS cation radical (Tan *et al.*, 2003). It also reacts with RNS such as peroxynitrite, to form other compounds such as 3-acetamido-methyl-6-methoxycinnoline and N^1-acetyl-5-methoxy-3-nitrokynuramine. Other minor catabolites are 1-nitromelatonin and 1-nitrosomelatonin; these are generated *in vitro* when melatonin reacts with peroxynitrite, the nitrogen dioxide radical and nitric oxide (Zhang *et al.*, 1999; Tan *et al.*, 2000b).

Many of the catabolic melatonin compounds have antioxidant activity that is similar to or even higher than that of melatonin itself. This effect means that the whole 'melatonin metabolite pool' works cooperatively as an excellent endogenous and synergistic antioxidant complex (Tan *et al.*, 2000b, 2002; Hardeland and Pandi-Perumal, 2005; Schaefer and Hardeland, 2008; Galano *et al.*, 2011).

21.5 Physiological Actions of Melatonin in Plants

A few years after 1995, once it had been accepted that a neurohormone was present in plant tissues, objectives were set to identify their possible role in the physiology of plants. In higher plants, the existence of a complex secondary metabolism that generates hundreds of biological compounds without a seemingly cellular or specific physiological function was already known (Vogt, 2010; Selmar and Kleinwächter, 2013). Was melatonin one of these secondary metabolites? A few research groups decided to investigate the properties of melatonin, which in animals, had known and specific functions, but about which nothing was known in plants. The first challenge was the correct detection and quantification of melatonin in plant extracts. Today, we know that immunodetection techniques (ELISA, RIA) are not to be recommended for the quantification of melatonin in plants because of the many interferences that occur when using plant extracts. Analysis by liquid chromatography (LC) and detection by mass spectrometry (LC-MS) are the most used and recommended techniques for the detection and quantification of melatonin in plants. However, less specific but very sensitive techniques such as LC with electrochemical or fluorimetric detection are also excellent if accompanied by identification by LC-MS (Kolar and Machackova, 2005; Arnao and Hernández-Ruiz, 2006, 2009a; Cao *et al.*, 2006; Hernández-Ruiz and Arnao, 2008a; Garcia-Parrilla *et al.*, 2009; Paredes *et al.*, 2009; Huang and Mazza, 2011b; Tan *et al.*, 2012; Gómez *et al.*, 2013).

Melatonin has been detected and quantified in roots, shoots, leaves, flowers, fruits and seeds of a considerable variety of plant species, especially in edible plants. These include: fruits, such as the apple, banana, date, fig, grape (from different grapevine varieties), olive (oil), orange, pear, peach, pineapple, plum, pomegranate, kiwi, strawberry, sweet and tart cherry (Dubbels *et al.*, 1995; Hattori *et al.*, 1995; Burkhardt *et al.*, 2001; Iriti *et al.*, 2006; Reiter *et al.*, 2007; González-Gómez *et al.*, 2009; Garrido *et al.*, 2010; Huang and Mazza, 2011a; Sturtz *et al.*, 2011; Vitalini *et al.*, 2011b; Zohar *et al.*, 2011; Mena *et al.*, 2012; Sae-Teaw *et al.*, 2012); vegetables, such as asparagus, bell pepper, beetroot, cabbage, carrot, cucumber, garlic, ginger, onion, pea, spinach and tomato (Pape and Lüning, 2006; Posmyk *et al.*, 2009; Arnao and Hernández-Ruiz, 2013b); seeds and nuts, such as alfalfa, almond, anise, barley, canary grass, cardamom, celery, maize, fennel, fenugreek, lupin, oat, pistachio, rice, sunflower, tall fescue, walnut, wheat, white and black mustard (Manchester *et al.*, 2000; Hernández-Ruiz *et al.*, 2004, 2005; Reiter *et al.*, 2005; Hernández-Ruiz and Arnao, 2008a; Paredes *et al.*, 2009); in many medicinal herbs, such as feverfew, laurel, liquorice, many Chinese herbs, many Mediterranean herbs, purslane, sagebrush and St John's wort (Murch *et al.*, 1997; Conti *et al.*, 2002; Murch and Saxena, 2002a; Chen *et al.*, 2003; Simopoulos *et al.*, 2005; Afreen *et al.*, 2006; Marioni *et al.*, 2008); beverages, such as beer, coffee, different teas, mulberry and different wines (Iriti *et al.*, 2010; Stege *et al.*, 2010; Pothinuch and Tongchitpakdee, 2011; Rodríguez-Naranjo *et al.*, 2011; Garcia-Moreno *et al.*, 2012; Ramakrishna *et al.*, 2012); other plants, such as *C. rubrum*, devil's trumpet (*Datura matel*), *Chara australis*, hyacinth, Japanese morning glory (*Ipomoea nil*), myrtle, tobacco (Kolar *et al.*, 1995; van Tassel, 1997; Tan *et al.*, 2007a; Murch *et al.*, 2009; Lazár *et al.*, 2013); and edible mushrooms such as *Agaricus*, *Chantarellus*, *Lactarius* and *Leccinum* (Muszynska *et al.*, 2011).

Generally, seeds and leaves present the highest level of melatonin and fruits the lowest, but the levels of melatonin in plant organs vary considerably, from picograms to micrograms per gram of plant material. Variables such as the variety, the cultivation conditions and the extraction protocols used are important aspects for an accurate quantification of the actual melatonin levels in tissues and plant organs (Arnao and Hernández-Ruiz, 2009a, 2013a,b; Paredes *et al.*, 2009).

The possible roles that melatonin was first thought to have in plants tended to reflect the roles that had been established in animals. Thus, the first approach was that melatonin in plants would behave as a regulator of endogenous or exogenous biological rhythms. The Czech group working with *C. rubrum* were pioneers in this area (Kolar *et al.*, 1995). The coincidence of the discovery of melatonin in plants (1995) with publications on the antioxidant properties of melatonin (1993–1994) led to the search for the role(s) of melatonin in the physiology of plants. Figure 21.5 presents a diagram that schematically shows the possible roles that melatonin has or may have in roots, stems, leaves, fruits, flowers and seeds. Research on the physiological functions of melatonin in plants can be classified from two different perspectives: (i) those investigations that have sought some physiological role similar to that known in animals; and (ii) work that have sought specific physiological roles for melatonin in plants, although, in some cases, it is difficult to establish this classification. Both perspectives will be developed below.

21.5.1 Searching for roles of melatonin in plants similar to those observed in animals

The role of melatonin as a regulator of light–dark cycles has been clearly established in mammals (see Section 21.2.2). This role as chronoregulator was the starting point for the research into melatonin in plants and, more specifically, its possible involvement as a regulatory molecule in circadian rhythms and in aspects connected with photoperiodicity, such as flowering. In 1997, the Czech group of Kolar and co-workers demonstrated the presence of melatonin in cultivated 15-day-old plants of *C. rubrum* and observed an oscillating behaviour of melatonin levels in light/dark cycles of 12 h (Kolar *et al.*, 1997). Measurements were made of the low or undetectable melatonin during the light period, and a considerable increase to reach maximum levels of around 500 pg g^{-1} fresh weight was detected in the darkness. This increase in melatonin during the dark period was similar to that observed in mammals. In the next work of this same group, plants were exposed to different photoperiodic cycles, but no relationship between the photoperiod applied and the duration of melatonin increase was observed (Wolf *et al.*, 2001). The authors concluded that although the maximum of melatonin always occurred after lights off, melatonin biosynthesis was not directly light regulated, but showed a circadian rhythm, as in mammals. At the same time, van Tassel and colleagues, investigating *Pharbitis nil* seedlings, concluded that no differences in melatonin content appeared with respect to the light/dark cycle applied (van Tassel *et al.*, 2001).

A more concise and decisive study was presented by the Tan group, working with water hyacinth (*Eichhornia crassipes*) (Tan *et al.*, 2007a). Plants grown under natural conditions presented maximum levels of melatonin, with a peak occurring late in the light phase of the light–dark cycle. Also, the catabolite AFMK (determined for the first time in plants) showed a similar rhythm. Interestingly, when the plants were cultivated in artificial light, melatonin levels were considerably

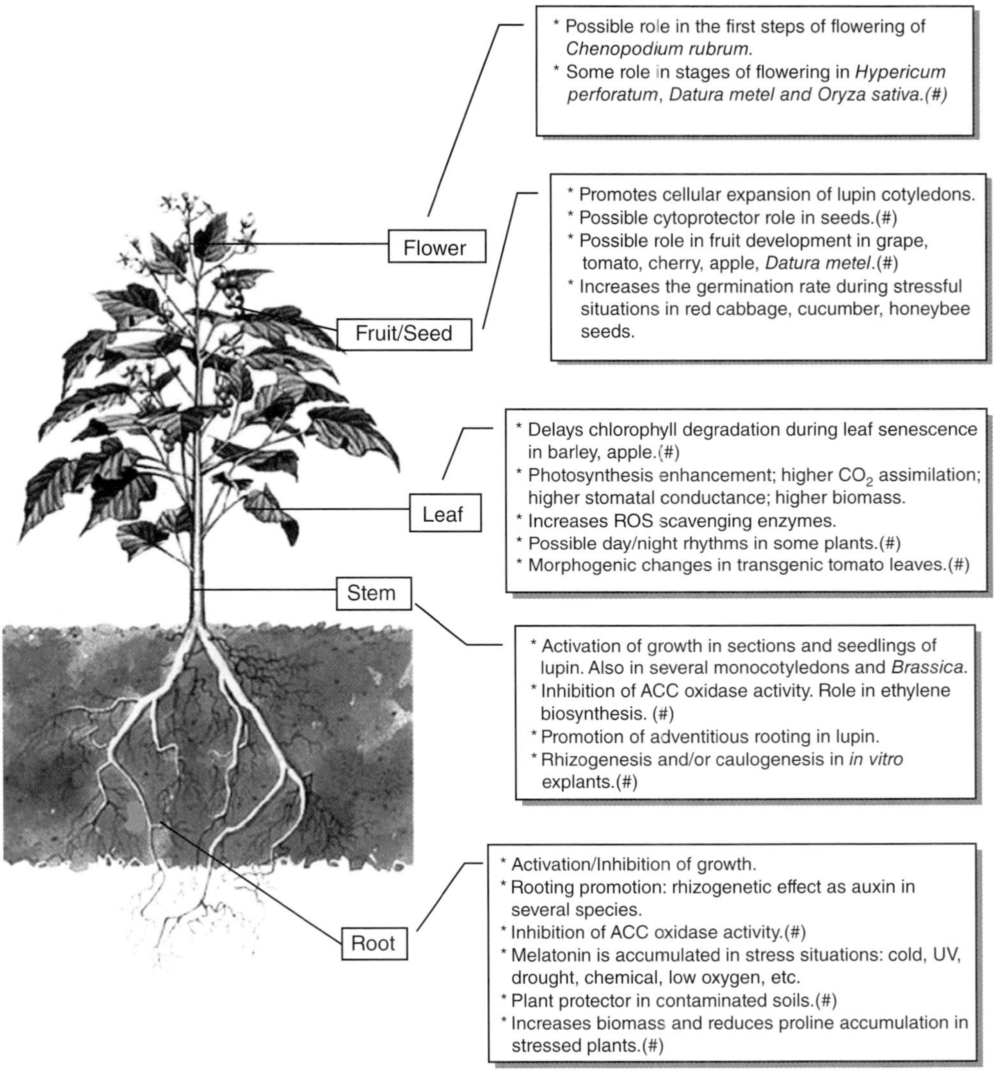

Fig. 21.5. Proposed functions for melatonin in different organs of higher plants. Functions supported by relevant experimental data are presented; the symbol (#) indicates that the attributed role requires more experimental data and/or application to more plant species. Abbreviations: ACC oxidase, 1-aminocyclopropane-1-carboxylic acid oxidase; ROS, reactive oxygen species.

lower, indicating the importance of the quality of light for the melatonin biosynthetic capacity of plants. Similarly, the macroalga *Ulva* sp. showed a melatonin rhythm, with a maximum at night in long-photoperiod day (16 h light:8 h dark) (Tal *et al.*, 2011).

An oscillating behaviour of melatonin levels was observed in the berry skin of *Vitis vinifera* cv. Malbec during the night/day cycle and under field conditions (Boccalandro *et al.*, 2011). The grapes showed a melatonin peak of 158 ng g^{-1} dry weight (DW) at sunrise (local time 09.30 h), decreasing in the next 4 h to undetectable values both at noon and during the afternoon. During the night, melatonin reached low levels of around 10 ng g^{-1} DW. The author's interpretation was that melatonin levels were probably controlled by the circadian clock, and that the decrease in melatonin during the light period would be due to melatonin consumption in grapes as an antioxidant response against solar radiation. More recently, two peaks of melatonin production in two sweet cherry cultivars, at 05.00 h and 14.00 h during a 24 h period, have been described. This double peak matched the expression of TDC (see Fig. 21.3), suggesting that it acts as a rate-limiting enzyme in melatonin synthesis in plants (Zhao *et al.*, 2012). The results of the four studies cited above (in *Chenopodium*, *Eichhornia*, *Vitis* and *Prunus*) indicate that the time of day at which the sample is taken seems to be relevant for melatonin levels in plant tissues, because depending on the time of day or night, there may be differences of several orders of magnitude. Thus, in *Chenopodium* and *Eichhornia*, maximum melatonin levels were reached at night or in the late afternoon; in contrast, this maximum occurred at sunrise in *Vitis*, with zero melatonin during the day. In this last case, the suggested protective role of 'melatonin consumption' in the grape skin as a protective measure against light should be contrasted with studies on melatonin transport in tissues. At present, there are no data available on the possible place (organ/s, tissue/s, cell/s) where melatonin is biosynthesized in plants. Neither is much known about its possible transport by vascular bundles, xylem and/or phloem elements (Arnao and Hernández- Ruiz, 2013b).

The excellent properties of melatonin as a natural antioxidative molecule have been widely studied in plants. Its role as cytoprotector and abiotic stress protector due to its relevant properties as a scavenger of free radicals, especially against ROS and RNS, has been the aim of many investigations in plants. Thus, exogenous melatonin application to the soil enhanced the tolerance and survival of pea plants (*Pisum sativum*) against copper contamination (Tan *et al.*, 2007b). Seed pretreatment with melatonin, at micromolar concentrations, reduces copper toxicity in red cabbage seedlings (*Brassica oleracea rubrum*) (Posmyk *et al.*, 2008). The relationship between chemical-induced stress and endogenous melatonin was clearly described in barley (*Hordeum* vulgare) and lupin (*Lupinus* albus) by Arnao and co-workers (Arnao and Hernández-Ruiz, 2009b, 2013a). In both plants, treatment with different chemical stressors such as zinc salt, hydrogen peroxide and sodium chloride provoked an increases in endogenous melatonin levels. Such induction of melatonin biosynthesis was time and concentration dependent, zinc salt being the best inducer in roots, where it multiplied melatonin levels by six times in barley and 12 times in lupin. Exogenous melatonin treatments improve the vegetative development of plants, probably owing to an improved redox state caused by the reduction of free radicals and a stabilization of biological membranes in the presence of chemical stressors. Furthermore, these same stressors were able to induce melatonin biosynthesis, possibly through the activation of some of the enzymes of its biosynthetic pathway. This interesting aspect was also studied in melatonin-rich transgenic rice (Park *et al.*, 2013b). A transgenic rice line expressing sheep SNAT (the penultimate biosynthesis enzyme; see Fig. 21.3), which produces more melatonin than wild type plants, exhibited a higher resistance against oxidative stress induced by the presence of the herbicide butafenacil (singlet-oxygen-generating herbicide) than did wild type plants. These results demonstrate that endogenous melatonin, even at very low concentrations (ng g^{-1} fresh weight) was critical in ROS scavenging and serves as a first line of defence against stressors.

Another abiotic stressor that influences the development of plants is ambient temperature. Lupin plants grown in a cold environment (6°C) showed a 2.5-fold increase in their melatonin levels, compared with control plants grown at 24 °C (Arnao and Hernández-Ruiz, 2013a). Also, cold stress led to an increase in melatonin as a response. Other chilling stress situations have been related to a protective role for melatonin. For example, in carrot (*Daucus carota*) culture cells, the presence of exogenous melatonin attenuated cold-induced apoptosis (Lei *et al.*, 2004), and cucumber seeds pretreated with melatonin improved their germination rate during chilling stress compared with untreated seeds (Posmyk *et al.*, 2009). Melatonin also improved the survival of cryopreserved callus of the perennial crassulacean *Rhodiola crenulata*, a traditional Tibetan herb that grows in extreme conditions of cold, low oxygen, high altitude and intense UV radiation (Zhao *et al.*, 2011). Similarly, the better cold resistance of transgenic rice seedlings that overproduce melatonin by ectopic overexpression of human serotonin *N*-acetyltransferase than that of wild type plants has been described by Kang *et al.* (2010).

A further abiotic stressor, drought/water deficit, has been studied in plants in relation to melatonin. Water restrictions during the growth of lupin seedlings provoked an increase in endogenous melatonin (Arnao and Hernández-Ruiz, 2013a). In another case, water-stressed cucumber plants treated with melatonin showed a clear improvement in the seed germination rate and root growth, which indicates that the application of melatonin minimized induced water stress (Zhang *et al.*, 2013). With regard to light, UV and solar radiation have received special attention. For example, *Glycyrrhyza uralensis* roots increased their melatonin content to about 80 μg g^{-1} fresh weight after UV-B (short wave UV) treatment (Afreen *et al.*, 2006). In 1995, Dubbels and coworkers had already proposed that the different susceptibility of two cultivars of *N. tabacum* to ozone could be a reflection in their different content of melatonin (Dubbels *et al.*, 1995). The influence of solar radiation on the melatonin levels of plant organs was clearly demonstrated in tomato and lupin plants by Arnao and Hernández-Ruiz (2013a,b); in both cases, plants grown in natural conditions contained more than three times the melatonin content of those grown in culture chambers. In these and in another case mentioned above (Tan *et al.*, 2007a), UV radiation appears to trigger increases in melatonin levels in plants. All of this data seem to indicate that melatonin may have a physiological role associated with stress situations.

21.5.2 Searching for specific roles of melatonin in plants

One of the most studied actions of melatonin in plants is its possible role as a plant regulator and, more specifically, as an auxinic agent. In plants, auxin (Chapter 19) is a hormone involved in many physiological actions, but mainly it has the capacity to stimulate plant growth, and is identified with the compound IAA. Structurally, melatonin and IAA present with their only similarity as the existence of the indole ring; IAA has neither the 5-methoxy nor the *N*-acetyl group. However, both compounds are biosynthesized from L-tryptophan, with tryptamine as a common intermediate (see Fig. 21.3). The experimental demonstration that melatonin acts as a plant growth promoter was made by Arnao and co-workers in 2004 (Hernández-Ruiz *et al.*, 2004). The incubation of etiolated hypocotyl sections of lupin in the presence of a range of concentrations of melatonin provoked an increase in the size of sections, similarly to the same bioassay carried out with IAA. In lupin plants with an excised aerial meristem, growth was inhibited, but reversed when IAA or melatonin were applied exogenously. Also, IAA and melatonin were distributed in a similar concentration gradient in tissues, and produced active growth at micromolar concentrations but had an inhibitory effect at high concentrations (Hernández-Ruiz *et al.*, 2004). This same group of workers reported similar effects of melatonin in several monocotyledons (oat, wheat, barley, canary grass), which demonstrated the auxin-like action of melatonin in plants; the growth-promoting activity of melatonin ranged from 10% in oat to 55% in barley coleoptiles, or 63% in lupin, all with respect to 100% IAA (Hernández- Ruiz *et al.*, 2005). Similar effects have been described in *Brassica juncea*, in which melatonin showed both actions, a stimulatory effect on growth at low concentrations and an inhibitory effect on growth at high levels (Chen *et al.*, 2009), and in *B. oleracea rubrum* (Posmyk *et al.*, 2008). Melatonin has also been associated with cellular expansion in plant cells. Thus, etiolated cotyledons of lupin can be stimulated to expand in the presence of melatonin, acting in a similar role to IAA, but with characteristic aspects (Hernández-Ruiz and Arnao, 2008b).

Also mirroring the effect of auxin, melatonin was capable of modifying the organogenic pattern of plants in a similar way to auxin. Melatonin induced root primordia from pericycle cells in lupin, resulting in the generation of adventitious or lateral roots (Arnao and Hernández-Ruiz, 2007). In this significant work, the number and length of lateral and adventitious roots, their distribution pattern, the time course, the optimal dose of melatonin, and comparisons of these with IAA, were studied. Later, this rhizogenic effect was confirmed in species such as cucumber (Zhang *et al.*, 2013), two cherry rootstocks (Sarropoulou *et al.*, 2012a,b) and rice (Park and Back, 2012). Similarly, in elegant work with *Arabidopsis thaliana*, melatonin and IAA were seen to be involved in the architecture of roots, with some similarities but also with important particularities in their action (Pelagio-Flores *et al.*, 2012). In *in vitro* culture explants of St John's wort (*Hypericum perforatum*), the amount of melatonin in the culture media modulates plant morphogenesis, changing the rates of rhizogenesis and caulogenesis, which points to a possible role for melatonin as plant regulator (Murch *et al.*, 2001; Murch and Saxena, 2002b). In another interesting piece of work, transgenic tomato plants were modified by overexpression of the rice indoleamine 2,3-dioxygenase (IDO) gene (Okazaki *et al.*, 2010). IDO is a well- documented enzyme that cleaves the indole ring of several indolic compounds, such as indoleamines (tryptophan and related metabolites, including melatonin and serotonin) and also IAA. Tomato IDO-transgenic plants showed lower levels of melatonin in their leaves than did the wild type, owing to the action of the overexpressed IDO protein. In these transgenic tomato plants, significant changes in the lateral leaflet pattern (in the number of leaflets, and the more flattened and less serrated pattern than the wild type) appeared in plant lines that overexpressed IDO. Nevertheless, the indiscriminate action of IDO on indolic compounds in plant tissues makes it difficult to draw conclusions as to whether the lowest level of melatonin is the cause of these morphogenetic changes in tomato leaves.

Another important contribution to this field appeared in 2009, when melatonin was associated with the capacity to delay senescence in leaves (Arnao and Hernández-Ruiz, 2009c). In barley leaves treated with melatonin at different concentrations, the induced-senescence process was slowed down, delaying the loss of total chlorophyll with respect to control leaves incubated in water. A possible action on chlorophyllases, red chlorophyll catabolite reductase and/or pheophorbide a oxygenase was suggested, but there was also the possibility that this effect of melatonin could be due to its action as an antioxidant. Based on these findings, many studies have focused on the possible action of melatonin in photosynthesis. Thus, exogenous melatonin delayed dark-induced senescence in apple leaves (*Malus domestica*) through the enhancement of some ROS scavenging enzymes activities, which contributed to the elimination of the excess H_2O_2 generated in stressed leaves, while maintaining the ascorbic acid and glutathione content higher than in control leaves (Wang *et al.*, 2012). Also, in the long-term application of melatonin to 1-year-old apple trees under drought conditions, leaf senescence was delayed with a significant reduction in chlorophyll degradation. The suppression of the upregulation of the senescence marker, *senescence-associated gene 12* (*SAG12*), and of the monooxygenase senescence-related pheophorbide a oxygenase clearly indicated the role of melatonin as a regulating factor in induced-foliar senescence (Wang *et al.*, 2013).

In addition to the protective role of melatonin against leaf senescence, melatonin may also increase photosynthetic efficiency in plants. In apple trees, melatonin contributed to a better efficiency of Photosystem II under dark and light conditions, alleviating the inhibition of photosynthesis caused by drought stress and also allowing the leaves to maintain a higher capacity for CO_2 assimilation and stomatal conductance (Wang *et al.*, 2013). Similar data were obtained from water-stressed cucumber seedlings. Melatonin treatment reduced chlorophyll degradation, increased the photosynthetic rate and the activities of ROS scavenging enzymes, so reversing the adverse effect of water stress (Zhang *et al.*, 2013). Also, in shoot tip explants of cherry rootstock PHL-C (*Prunus avium* × *Prunus cerasus*), the application of exogenous melatonin at low concentrations, in addition to its rhizogenic effects, slightly enhanced the content of photosynthetic pigments, total biomass and total carbohydrates, while reducing the proline content of roots, indicating a role for melatonin in plant stress metabolism (Sarropoulou *et al.*, 2012a). Some conservational effect of melatonin on chlorophyll content has also been described in the macroalga *Ulva* sp. (Tal *et al.*, 2011) and in the fresh water characid *Chara australis*, where a clear increase in the efficiency of the reaction centres of Photosystem II was observed (Lazár *et al.*, 2013).

Seed germination is a another relevant step in the life cycle of plants, and the addition of exogenous melatonin seems to have a positive effect on this process. For example, presowing seed treated with melatonin protects red cabbage seedlings (*Brassica oleracea rubrum*) against copper ion toxicity, possibly reducing cell injury by membrane peroxidation and improving the DNA endoreplication pattern (Posmyk *et al.*, 2008). Pretreating cucumber seeds with melatonin improved their germination rate during chilling stress (Posmyk *et al.*, 2009). Also in cucumber, polyethylenglycol water-stressed seeds recovered their initial rate of germination when primed with melatonin (Zhang *et al.*, 2013). In annual honeybee plants (*Phacelia tanacetifolia*), which produce negatively photoblastic and thermosensitive seeds, the presence of melatonin in primed seeds clearly reversed the effects of light and high temperature (Tiryaki and Keles, 2012). In spite of this interesting observation, the authors did not suggest any hypothesis on the possible role of melatonin in the process studied, but rather a connection between photoreceptors (phytochromes and/or blue-light receptors), hormones (auxin, gibberellins and abscisic acid); so here, the role of melatonin may be assumed.

From the early studies on the short-day plant *C. rubrum*, some data have been obtained on the possible role of melatonin in flowering (Machackova and Krekule, 2002; Kolar *et al.*, 2003). In these studies, melatonin at high concentrations inhibited flower induction by an average of 40–50% when applied between 2 h before and 2h after the beginning of the inductive dark period. An agonist of melatonin used in animal studies, CGP-52608 (1-[3-allyl-4-oxothiazolidine-2-ylidene]-4-methyl-thio semicarbazone) was effective in inhibiting flower induction. Melatonin treatment had no effect on the period or phase of the flowering, so it should control some process before floral transition, according to the authors (Kolar and Machackova, 2005). In a similar study in the long-day plant *A. thaliana*, flowering was slightly delayed in plants treated with 100 or 500 μM melatonin, which acted on leaf number, rosette number and anthesis. These data are part of a PhD thesis and, unfortunately, seem to have not been published in any journal (Kolar, 2003). During the floral development of St John's wort, the appearance of a melatonin peak in the intermediate stages of flower development has been described. This melatonin peak matches the increased regenerative potential of isolated anthers, so melatonin accumulation in specific tissues and stages was interpreted as a signal in gametophytic development, which resulted in viable microspores (Murch and Saxena, 2002a). Also in flowers of devil's trumpet, a narcotic plant used in natural medicine around the world, melatonin was found at the highest concentrations as the flower buds matured, suggesting for melatonin a protective role in the reproductive tissues (Murch *et al.*, 2009). More recently, melatonin synthesis was induced in the panicle (flower) of rice plants during the reproductive stages. This induction was parallel to the induction of melatonin biosynthetic enzymes (TDC, T5H and HIOMT, see Fig. 21.3), suggesting that melatonin may participate in flower development (Park *et al.*, 2013c).

Fruit development and ripening has also been studied in relation to the presence of melatonin and its possible role in this physiological process. In devil's trumpet fruits, melatonin was present at an elevated level for the first 10 days after anthesis, with the highest concentrations in the ovule and the least in the fleshy fruit (Murch *et al.*, 2009). During the development of cherry fruits, the highest levels of melatonin were reached in stage II, embryo development and endocarp lignification, which has also been related to ROS accumulation in 'watery' seeds, indicating that the oxidative stress induced a rapid melatonin synthesis (Zhao *et al.*, 2012). In Micro-Tom tomato fruits, melatonin was also accumulated during fruit development, with a maximum in the last stages, when the fruits turn to a pink-red colour (Okazaki and Ezura, 2009). It is significant that tomato seeds showed the highest melatonin concentration, which suggests a possible protective role for melatonin in seeds, as was earlier proposed by Manchester *et al.* (2000). A similar conclusion appeared in a study of fruit development (véraison) in Merlot wine grapes, in which the maximum melatonin was reached in green fruits (Murch *et al.*, 2010). In either case, at present it is unknown whether melatonin peaks act as a signal of/for the fruit development process or as a response to ensure a proper redox status of the fruit and seed during their development.

21.6 Future Perspectives and Concluding Remarks

Various aspects of melatonin in plants should be noted for potential future studies. First, the presence of melatonin in plants quickly aroused interest in the use of this natural compound as a nutraceutical. The high or considerable concentrations of melatonin in plant extracts (μg or ng melatonin g^{-1} fresh weight plant tissue) compared with the levels in human blood plasma (pg ml^{-1} plasma), aroused interest in how the intake of plant products (plant foodstuffs) might interfere with the natural levels or with circadian oscillations of melatonin in plasma. Melatonin from plant foods is absorbed from the gastro-intestinal tract and incorporated into the blood-stream; it also crosses the blood–brain barrier and the placenta. Thus, the possibility of modulating blood melatonin levels in mammals through the ingestion of plant-derived foods has led to some studies in which melatonin is considered to be a health-related phytochemical. However, the peculiar circadian rhythm of melatonin in blood and the different customs and habits of each country and/or region mean that more detailed studies on the influence of the intake of plant foods rich in melatonin are required. In quite another way, melatonin is used in livestock as an inducer of oestrus for greater effectiveness in the reproductive cycles of sheep, goats, etc. The consumption, at certain times of the year, of melatonin-rich herbs could be used as a substitute for treatment by injection or dermal patches of exogenous melatonin.

The second aspect to note is the possible use of melatonin in agriculture and horticulture. There are many experimental data about how melatonin improves the adaptation of plants to various abiotic stress situations such as drought, chemical pollutants, cold, radiations, etc. It can be confidently predicted that the coming years will see a dramatic increase in field trials on the beneficial role of melatonin in the development of plants. Also, the cultivation of plants with high endogenous levels of melatonin to improve environmental adaptation is an interesting focus. As an example, the use of melatonin-rich plants to recover contaminated soils could become a reality in the near future, complementing other phytoremediation techniques.

In conclusion, several relevant aspects of melatonin in plants remain to be studied in depth. Among these, there should be a study of the enzymes involved in their biosynthesis: their isolation, characterization and regulation. The similarities and differences between the catabolism of melatonin in animals and plants, including the possibility of a major common conjugation catabolism in plants, should also be studied. The detection and characterization of potential melatonin receptor(s) in plants is also a primary objective. From the physiological aspect, it is necessary to extend studies to more species on the role of melatonin as a plant growth regulator, with an emphasis on its auxin-like action and its anti-senescence properties. In addition, further biochemical and physiological studies on the antioxidant role of melatonin in plant stress situations are essential before work can be carried out in field trials.

References

Abe, M., Itoh, M., Miyata, M., Ishikawa, S. and Sumi, Y. (1999) Detection of melatonin, its precursor and related enzyme activities in rabbit lens. *Experimental Eye Research* 68, 255–262.

Abran, D., Anctil, M. and Ali, M. (1994) Melatonin activity rhythms in eyes and cerebral ganglia of *Aplysia californica*. *General and Comparative Endocrinology* 96, 215–222.

Afreen, F., Zobayed, S.M.A. and Kozai, T. (2006) Melatonin in *Glycyrrhiza uralensis*: response of plant roots to spectral quality of light and UV-B radiation. *Journal of Pineal Research* 41, 108–115.

Agorastos, A. and Huber, C. (2011) The role of melatonin in glaucoma: implications concerning pathophysiological relevance and therapeutic potential. *Journal of Pineal Research* 50, 1–7.

Allen, G., Neergheen, V., Oppenheim, M., Fitzgeral, J., Footitt, E., Hyland, K., Clayton, P., Land, J. and Heales, S. (2010) Pyridoxal 5′-phosphate deficiency causes a loss of aromatic L-amino acid decarboxylase in patients and human neuroblastoma cells, implications for aromatic L-amino acid decarboxylase and vitamin B_6 deficiency states. *Journal of Neurochemistry* 114, 87–96.

Ando-Yamamoto, M., Hayashi, H., Sugiyama, T., Fukui, H., Watanabe, T. and Wada, H. (1987) Purification of L-DOPA decarboxylase from rat liver and production of polyclonal and monoclonal antibodies against it. *The Journal of Biochemistry* 101, 405–414.

Anisimov, V., Popovich, I., Zabezhinski, M., Anisimov, S., Vesnushkin, G. and Vinogradova, I. (2006) Melatonin as antioxidant, geroprotector and anticarcinogen. *Biochimica et Biophysica Acta* 1757, 573–589.

Antunes, F., Barclay, L., Ingold, K., King, M., Norris, J., Scaiano, J. and Xi, F. (1999) On the antioxidant activity of melatonin. *Free Radical Biology and Medicine* 26, 117–128.

Arnao, M.B. and Hernández-Ruiz, J. (2006) The physiological function of melatonin in plants. *Plant Signaling and Behavior* 1, 89–95.

Arnao, M.B. and Hernández-Ruiz, J. (2007) Melatonin promotes adventitious- and lateral root regeneration in etiolated hypocotyls of *Lupinus albus* L. *Journal of Pineal Research* 42, 147–152.

Arnao, M.B. and Hernández-Ruiz, J. (2009a) Assessment of different sample processing procedures applied to the determination of melatonin in plants. *Phytochemical Analysis* 20, 14–18.

Arnao, M.B. and Hernández-Ruiz, J. (2009b) Chemical stress by different agents affects the melatonin content of barley roots. *Journal of Pineal Research* 46, 295–299.

Arnao, M.B. and Hernández-Ruiz, J. (2009c) Protective effect of melatonin against chlorophyll degradation during the senescence of barley leaves. *Journal of Pineal Research* 46, 58–63.

Arnao, M.B. and Hernández-Ruiz, J. (2013a) Growth conditions determine different melatonin levels in *Lupinus albus* L. *Journal of Pineal Research* 55, 149–155.

Arnao, M. and Hernández-Ruiz, J. (2013b) Growth conditions influence the melatonin content of tomato plants. *Food Chemistry* 138, 1212–1214.

Aust, S., Jaeger, W., Klimpfinger, M., Mayer, K., Baravalle, G., Ekmekcioglu, C. and Thalhammer, T. (2005) Biotransformation of melatonin in human breast cancer cell lines: role of sulfotransferase 1A1. *Journal of Pineal Research* 39, 276–282.

Axelrod, J. (1974) The pineal gland: a neurochemical transducer. *Science* 184, 1341–1348.

Axelrod, J. and Weissbach, H. (1960) Enzymatic *O*-methylation of *N*-acetylserotonin to melatonin. *Science* 131, 1312.

Axelrod, J. and Weissbach, H. (1961) Purification and properties of hydroxy-indole-*O*-methyl transferase. *The Journal of Biological Chemistry* 236, 211.

Axelrod, J., Wartman, R. and Snyder, S. (1965) Control of hydroxyindole-*O*-methyltransferase activity in the rat pineal gland by environmental lighting. *The Journal of Biological Chemistry* 240, 949–995.

Baker, P., Quay, W. and Axelrod, J. (1965) Development of HIOMT activity in the eye and brain of the amphibian *Xenopus laevis*. *Life Science* 4, 1981–1987.

Baldessarini, R. and Kopin, I. (1966) Tritiated norepinephrine: release from brain slices by electrical stimulation. *Science* 152, 1630–1631.

Balemans, M., Bary, F., van Benthem, J. and Legerstee, W. (1981) Seasonal variations in HIOMT activity during the night in the pineal gland of Wistar rats of several ages. In: Birau, N and Schloot, W (eds) *Advances in the Biosciences: Melatonin – Current Status and Perspectives*. Pergamon Press, Oxford, UK, pp. 207–211.

Baler, R., Covington, S. and Klein, D. (1997) The rat arylalkylamine *N*-acetyltransferase gene promoter: cAMP activation via a cAMP-responsive element-CCAT complex. *The Journal of Biological Chemistry* 272, 6979–6985.

Baler, R., Covington, S. and Klein, D. (1999) Rat arylalkylamine *N*-acetyltransferase gene: upstream and intronic components of a bipartite promoter. *Biology of the Cell* 91, 699–705.

Balzer, I. and Hardeland, R. (1991) Photoperiodism and effects of indoleamines in a unicellular alga, *Gonyaulax polyedra*. *Science* 253, 795–797.

Balzer, I., Poeggeler, B. and Hardeland, R. (1990) 5-Methoxylated indoleamines in the dinoflagellate *Gonyaulax polyedra*: presence, stimulation of bioluminescence, and circadian variations in responsiveness. *European Journal of Cell Biology* 51(Suppl. 30), 52.

Balzer, I., Espinola, I. and Fuentes-Pardo, B. (1997) Daily variations in immunoreactive melatonin in the visual system of crayfish. *Biological Cell* 89, 539–543.

Bartel, B. (1977) Auxin biosynthesis. *Annual Review of Plant Physiology and Plant Molecular Biology* 48, 51–66.

Barth, M., Serre, V., Hubert, L., Chaabouni, Y., Bahi-Buisson, N., Cadoudal, M., Rabier, D., Tich, S., Ribeiro, M., Ricquier, D., Munnich, A., Bonneau, D. *et al.* (2012) Kinetic analyses guide the therapeutic decision in a novel form of moderate aromatic acid decarboxylase deficiency. *Journal of Inherited Metabolic Disease Reports* 3, 25–32.

Bartsch, C., Bartsch, H., Jain, A., Laumas, K. and Wetterberg, I. (1981) Urinary melatonin levels in human breast cancer patients. *Journal of Neural Transmission* 52, 281–294.

Becker-André, M., Wiesenberg, I., Schaeren-Wiemers, N., Andre, E., Missbach, M., Saurat, J. and Carlberg, C. (1994) Pineal gland hormone melatonin binds and activates an orphan of the nuclear receptor superfamily. *The Journal of Biological Chemistry* 269, 28531–28534.

Ben-Abdallah, M., Bondet, V., Fauchereau, F., Beguin, P., Goubran-Botros, H., Pagan, C., Bourgeron, T. and Bellalou, J. (2011) Production of soluble, active acetyl serotonin methyl transferase in *Leishmania tarentolae*. *Protein Expression and Purification* 75, 114–118.

Benítez-King, G. (2006) Melatonin as a cytoskeletal modulator: implications for cell physiology and disease. *Journal of Pineal Research* 40, 1–9.

Benot, S., Goberna, R., Reiter, R., Garcia-Mauriño, S., Osuna, C. and Guerrero, J. (1999) Physiological levels of melatonin contribute to the antioxidant capacity of human serum. *Journal of Pineal Research* 27, 59–64.

Berlin, J., Ruegenhagen, C., Kuzovkina, I., Fecker, L. and Sasse, F. (1994) Are tissue cultures of *Peganum harmala* a useful model system for studying how to manipulate the formation of secondary metabolites? *Plant Cell, Tissue and Organ Culture* 38, 289–297.

Bernard, M., Donohue, S. and Klein, D. (1995) Human hydroxyindole-*O*-methyltransferase in pineal gland, retina and Y79 retinoblastoma cells. *Brain Research* 696, 37–48.

Blask, D., Sauer, L. and Dauchy, R. (2002) Melatonin as a chronobiotic/anticancer agent: cellular, biochemical and molecular mechanisms of action and their implications for circadian-based cancer therapy. *Current Topics in Medical Chemistry* 2, 113–132.

Boccalandro, H., Gonzalez, C., Wunderlin, D. and Silva, M. (2011) Melatonin levels, determined by LC-ESI-MS/MS, fluctuate during the day/night cycle in *Vitis vinifera* cv Malbec: evidence of its antioxidant role in fruits. *Journal of Pineal Research* 51, 226–232.

Borjigin, J., Wang, M. and Snyder, S. (1995) Diurnal variation in mRNA encoding serotonin *N*-acetyltransferase in pineal gland. *Nature* 378, 783–785.

Borri Voltatorni, C., Minella, A., Vecchini, P., Fiori, A. and Turano, C. (1979) Purification and characterization of 3,4-dihydroxyphenylalanine decarboxylase from pig kidney. *The Journal of Biochemistry* 93, 181–188.

Borri Voltatorni, C., Giartosio, A. and Turano, C. (1987) Aromatic L-amino acid decarboxylase from pig kidney. In: Kaufman, S (ed.) *Methods in Enzymology*. Academic Press, New York, 179–187.

Bossa, F., Martini, F., Barra, D., Borri Voltatorni, C., Minelli, A. and Turano, C. (1977) The chymotryptic phosphopyridoxal peptide of DOPA decarboxylase from pig kidney. *Biochemical and Biophysical Research Communications* 78, 177–184.

Bossinakou, K. and Fragoulis, E. (1996) Purification and characterisation of L-DOPA decarboxylase from pharate pupae of *Ceratitis capitata*. A comparison with the enzyme purified from the white prepupae. *Comparative Biochemistry and Physiology Part B: Biochemistry and Molecular Biology* 113, 213–220.

Botros, H., Legrand, P., Pagan, C., Bondet, V., Weber, P., Ben-Abdallah, M., Lemière, N., Huget, G., Bellalou, J., Maronde, E., Beguin, P., Haouz, A. *et al.* (2013) Crystal structure and functional mapping of human ASMT, the last enzyme of the melatonin synthesis pathway. *Journal of Pineal Research* 54, 46–57.

Boutin, J.A., Audinot, V., Ferry, G. and Delagrange, P. (2005) Molecular tools to study melatonin pathways and actions. *Trends in Pharmacological Sciences* 26, 412–419.

Bowsher, R. and Henry, D. (1986) Aromatic L-amino acid decarboxylase: biochemistry and functional significance. In: Boulton, A.A.A., Baker, G.B. and Yu, P.H. (ed.) *Neuromethods*. Humana Press, Clifton, New Jersey, 33–37.

Brodie, B., Kuntzman, R., Hitch, C. and Costa, E. (1962) Effects of decarboxylase inhibition on the biosynthesis of brain monoamines. *Life Sciences* 1, 81–84.

Bubenik, G. (2002) Gastrointestinal melatonin: localization, function, and clinical relevance. *Digestive Diseases and Sciences* 47, 2336–2348.

Buijs, R., Hermes, M. and Kalsbeek, A. (1988) The suprachiasmatic nucleus–paraventricular nucleus interactions: a bridge to the neuroendocrine and autonomic nervous system. *Progress in Brain Research* 119, 365–382.

Burke, Z., Wells, T., Carter, D., Klein, D. and Baler, R. (1999) Genetic targeting: the serotonin *N*-acetyltransferase promoter imparts circadian expression selectively in the pineal gland and retina of transgenic rats. *Journal of Neurochemistry* 73, 1343–1349.

Burkhad, P., Dominici, P., Borri-Voltattorni, C., Jansonius, J. and Malashkevich, V. (2001) Structural insight into Parkinson's disease treatment from drug-inhibited DOPA decarboxylase. *Nature Structural Biology* 8, 963–967.

Burkhardt, S., Tan, D., Manchester, L., Hardeland, R. and Reiter, R. (2001) Detection and quantification of the antioxidant melatonin in Montmorency and Balaton tart cherries (*Prunus cerasus*). *Journal of Agricultural and Food Chemistry* 49, 4898–4902.

Byeon, Y., Park, S., Kim, Y., Park, D., Lee, S. and Back, K. (2012) Light-regulated melatonin biosynthesis in rice during the senescence process in detached leaves. *Journal of Pineal Research* 53, 107–111.

Cabrera, J., Reiter, R.J., Tan, D.-X., Qi, W., Sainz, R.M., Mayo, J.C., Garcia, J.J., Kim, S.J. and El-Sokkary, G. (2000) Melatonin reduces oxidative neurotoxicity due to quinolinic acid: *in vitro* and *in vivo* findings. *Neuropharmacology* 39, 507–514.

Callebert, J., Jaunay, J. and Jallon, J. (1991) Control of *Drosophilla* biorhythms. *Advances in Pineal Research* 5, 81–84.

Cano, A., Alcaraz, O. and Arnao, M. (2003) Free radical-scavenging activity of indolic compounds in aqueous and ethanolic media. *Analytical and Bioanalytical Chemistry* 376, 33–37.

Cao, J., Murch, S., O'Brien, R. and Saxena, P. (2006) Rapid method to accurate analysis of melatonin, serotonin and auxin in plant samples using liquid chromatography-tandem mass spectrometry. *Journal of Chromatography A* 1134, 333–337.

Cardinali, D. and Rosner, J. (1971) Retinal localization of hydroxyindole-*O*-methyl transferase (HIOMT) in the rat. *Endocrinology* 89, 301–303.

Carrillo-Vico, A., Guerrero, J., Lardone, P. and Reiter, R. (2012) A review of the multiple actions of melatonin on the immune system. *Endocrine* 27, 189–200.

Carrillo-Vico, A., Lardone, P., Alvarez-Sanchez, N., Rodríguez-Rodriguez, A. and Guerrero, J. (2013) Melatonin: buffering the immune system. *International Journal of Molecular Sciences* 14, 8638–8683.

Catalá, A. (2007) The ability of melatonin to counteract lipid peroxidation in biological membranes. *Current Molecular Medicine* 7, 638–649.

Ceinos, R.M., Rábade, S., Soengas, J.L. and Míguez, J.M. (2005) Indoleamines and 5-methoxyindoles in trout pineal organ *in vivo*: daily changes and influence of photoperiod. *General and Comparative Endocrinology* 144, 67–77.

Chen, G. and Miller, G. (2013) Tryptophan hydroxylase-2: an emerging therapeutic target for stress disorders. *Biochemical Pharmacology* 85, 1227–1233.

Chen, G., Huo, Y., Tan, D., Liang, Z., Zhang, W. and Zhang, Y. (2003) Melatonin in Chinese medicinal herbs. *Life Science* 73, 19–26.

Chen, Q., Qi, W., Reiter, R., Wei, W. and Wang, B. (2009) Exogenously applied melatonin stimulates root growth and raises endogenous IAA in roots of etiolated seedling of *Brassica juncea*. *Journal of Plant Physiology* 166, 324–328.

Cheng, K., Liao, J. and Lyu, P. (2012) Crystal structure of the dopamine *N*-acetyltransferase-acetyl-CoA complex provides insights into the catalytic mechanism. *Biochemical Journal* 446, 395–404.

Christen, S., Peterhans, E. and Stocker, R. (1990) Antioxidant activities of some tryptophan metabolites: possible implication for inflammatory diseases. *Proceedings of National Academy of Sciences of the United States of America* 87, 2506–2510.

Christenson, J., Daiman, W. and Udenfriend, S. (1970) Preparation and properties of a homogeneous aromatic L-amino acid decarboxylase from hog kidney. *Archives of Biochemistry and Biophysics* 141, 356–367.

Chyan, Y., Poeggeler, B., Omar, R., Chain, D., Frangione, B., Ghiso, J. and Pappolla, M. (1999) Potent neuroprotective properties against the Alzheimer β-amyloid by an endogenous melatonin-related indole structure, indole-3-propionic acid. *The Journal of Biological Chemistry* 274, 21937–21942.

Cianchetta, G., Stouch, T., Yu, W., Shi, Z., Tari, L., Swanson, R., Hunter, M., Hoffman, I. and Liu, Q. (2010) Mechanism of inhibition of novel tryptophan hydroxylase inhibitors revealed by co-crystal structures and kinetic analysis. *Current Chemical Genomics* 4, 19–26.

Clark, C., Weissbach, H. and Udenfriend, S. (1954) 5-Hydroxytryptophan decarboxylase: preparation and properties. *The Journal of Biological Chemistry* 210, 139–148.

Claustrat, B., Brun, J. and Chazot, G. (2005) The basic physiology and pathophysiology of melatonin. *Sleep Medical Review* 9, 11–24.

Coge, F., Krieger-Poullet, M, Gross, F. and Thibault, J. (1990) Comparative and quantitative study of L-DOPA decarboxylase mRNA in rat neuronal and non-neuronal tissues. *Biochemical and Biophysical Research Communications* 170, 1006–1012.

Cohen, M., Lippman, M. and Chabner, B. (1978) Role of the pineal gland in aetiology and treatment of breast cancer. *The Lancet* 2, 814–816.

Coleman, C. and Neckameyer, W. (2005) Serotonin synthesis by two distinct enzymes in *Drosophila melanogaster*. *Archives of Insect Biochemistry and Physiology* 59, 12–31.

Conti, A., Tettamanti, C., Singaravel, M., Haldar, C., Pandi-Perumal, S. and Maestroni, G. (2002) Melatonin: an ubiquitous and evolutionary hormone. In: Haldar, C., Singaravel, M. and Maitra, S.K. (eds) *Treatise on Pineal Gland and Melatonin*. Science Publishers, Enfield, New Hampshire, pp. 105–143.

Coon, S., Roseboom, P., Baler, R., Weller, J., Namboodiri, M., Koonin, E. and Klein, D. (1995) Pineal serotonin *N*-acetyltransferase: expression cloning and molecular analysis. *Science* 270, 1681–1683.

Coon, S., Mazuruk, K., Bernard, M., Roseboom, P., Klein, D. and Rodriguez, I. (1996) The human serotonin *N*-acetyltransferase (EC 2.3.1.87) gene (AANAT): structure, chromosomal localization, and tissues expression. *Genomics* 34, 76–84.

Coon, S., Bégay, V., Deurloo, D., Falcón, J. and Klein, D. (1999) Two arylalkylamine *N*-acetyltransferase genes mediate melatonin synthesis in fish. *The Journal of Biological Chemistry* 274, 9076–9082.
Coon, S., Weller, J., Korf, H., Namboodiri, M., Rollag, M. and Klein, D. (2001) cAMP regulation of arylalkylamine *N*-acetyltransferase (AANAT, EC 2.3.1.87): a new cell line (1E7) provides evidence of intracellular AANAT activation. *The Journal of Biological Chemistry* 276, 24097–24107.
Coon, S., Del Olmo, E., Young, W. and Klein, D. (2002) Melatonin synthesis enzymes in *Macaca mulatta*: focus on arylalkylamine *N*-acetyltransferase (EC 2.3.1.87). *The Journal of Clinical Endocrinology and Metabolism* 87, 4699–4706.
Cos, S. and Sanchez-Barceló, E.J. (2000) Melatonin and mammary pathological growth. *Frontiers in Neuroendocrinology* 21, 133–170.
Crespo, E., Macias, M., Pozo, D., Escames, G., Martin, M., Vives, F., Guerrero, J. and Acuña-Castroviejo, D. (1999) Melatonin inhibits expression of the inducible NO synthase II in liver and lung and prevents endotoxemia in lipopolysaccharide-induced multiple organ dysfunction syndrome in rats. *The FASEB Journal* 13, 1537–1546.
Davis, S., Mirick, D. and Stevens, R. (2001) Night shift work, light at night, and risk of breast cancer. *Journal of National Cancer Institute* 93, 1557–1562.
De Almeida, E., Martinez, G., Klitzke, C., De Medeiros, M. and Di Mascio, P. (2003) Oxidation of melatonin by singlet molecular oxygen produces AFMK. *Journal of Pineal Research* 35, 131–137.
De la Iglesia, H., Meyer, J., Carpino, A.J. and Schwartz, W. (2000) Antiphase oscillation of the left and right suprachiasmatic nuclei. *Science* 290, 799–801.
De Luca, V., Marineau, C. and Brisson, N. (1989) Molecular cloning and analysis of cDNA encoding a plant tryptophan decarboxylase: comparison with animal DOPA decarboxylase. *Proceedings of the National Academy of Sciences of the United States of America* 86, 2582–2586.
de Oliveira Silva, S., Ximenes, V.F., Catalani, L.H. and Campa, A. (2000) Myeloperoxidase-catalyzed oxidation of melatonin by activated neutrophils. *Biochemical and Biophysical Research Communications* 279, 657–662.
Di Bella, G., Mascia, F., Gualano, L. and Di Bella, L. (2013) Melatonin anticancer effects: review. *International Journal of Molecular Sciences* 14, 2410–2430.
Donohue, S., Roseboom, P., Illenerova, H., Weller, J. and Klein, D. (1993) Human hydroxyindole-*O*-methyltransferase: presence of LINE-1 fragment in a cDNA clone and pineal mRNA. *DNA Cell Biology* 12, 715–727.
Dubbels, R., Reiter, R., Kienke, E., Goebel, A., Schnakenberg, E., Ehlers, C., Schiwara, H. and Schloot, W. (1995) Melatonin in edible plants identified by radioimmunoassay and by HPLC-MS. *Journal of Pineal Research* 18, 28–31.
Dubocovich, M. and Markowska, M. (2005) Functional MT1 and MT2 melatonin receptors in mammals. *Endocrine* 27, 101–110.
Dyck, L., Yang, C. and Boulton, A. (1983) The biosynthesis of tyramine, tyramine, and phenylethylamine by rat striatal slices. *Journal of Neuroscience Research* 10, 211–220.
Dyda, F., Klein, D. and Hickman, A. (2000) GCN5-related *N*-acetyltransferases: a structural overview. *Annual Review of Biophysics and Biomolecular Structure* 29, 81–103.
Eveleth, D., Gietz, R., Spencer, C., Nargang, E., Hodgetts, R. and Marsh, J. (1986) Sequence and structure of the DOPA decarboxylase gene of *Drosophila*: evidence for novel RNA splicing variants. *The EMBO Journal* 5, 2663–2672.
Facchini, P., Huber-Allanach, K. and Tari, L. (2000) Plant aromatic L-amino acid decarboxylases: evolution, biochemistry, regulation, and metabolic engineering applications. *Phytochemistry* 54, 121–138.
Falcón, J., Besseau, M., Fuentès, S., Sauzet, E., Magnanou, E. and Boeuf, G. (2009) Structural and functional evolution of the pineal melatonin system in vertebrates. *Annals of the New York Academy of Sciences* 1163, 101–111.
Fenwick, J. (1970) Demonstration and effect of melatonin in fish. *General and Comparative Endocrinology* 14, 86–97.
Ferry, G., Loynel, A., Kucharczyk, N., Bertin, S., Rodriguez, M., Delagrange, P., Galizzi, J.P., Jacoby, E., Volland, J.P., Lesieur, D., Renard, P., Canet, E. *et al.* (2000) Substrate specificity and inhibition studies of human serotonin-*N*-acetyltransferase. *The Journal of Biological Chemistry* 275, 8794–8805.
Ferry, G., Ubeaud, C., Dauly, C., Mozo, J., Guillard, S., Berger, S., Jiménez, S., Scoul, C., Leclerc, G., Yous, S., Delagrange, P. and Boutin, J. (2004) Purification of the recombinant human serotonin *N*-acetyltransferase (EC 2.3.1.87): further characterization of and comparison with AANAT from other species. *Protein Expression and Purification* 38, 84–98.

Ferry, G., Ubeaud, C., Lambert, P., Bertin, S., Coge, F., Chomarat, P., Delagrange, P., Serkiz, B., Bouchet, J., Truscottj, R. and Boutin, J.A. (2005) Molecular evidence that melatonin is enzymatically oxidized in a different manner than tryptophan: investigations with both indoleamine-2,3-dioxygenase and myeloperoxidase. *Biochemical Journal* 388, 205–215.

Finkelstein, E., Rosen, G. and Rauckman, E. (1980) Spin trapping. Kinetics of the reaction of superoxide and hydroxyl radicals by nitrones. *Journal of the American Chemical Society* 102, 4994–4999.

Fischer, J. (1987) Effekte von Metaboliten Aromatischer Aminosäuren auf *Pyrocystis* (Dinoflagellata). Diploma thesis, University of Göttingen. Germany.

Fischer, T., Sweatman, T., Semak, I., Sayre, R., Worstman, J. and Slominski, A. (2006) Constitutive and UV-induced metabolism of melatonin in keratinocytes and cell-free systems. *The FASEB Journal* 20, 1564–1575.

Fischer, T., Kleszcynski, K., Hardkop, L., Kruse, N. and Zillikens, D. (2013) Melatonin enhances antioxidative enzyme gene expression (CAT, GPx, SOD), prevents their UVR-induced depletion, and protects against the formation of DNA damage in *ex vivo* human skin. *Journal of Pineal Research* 54, 303–312.

Fitzpatrick, P. (1999) Tetrahydropterin-dependent amino acid hydroxylases. *Annual Review of Biochemistry* 68, 355–381.

Foulkes, N., Borjigin, J., Snyder, S. and Sassone-Corsi, P. (1997) Rhythmic transcription: the molecular basis of circadian melatonin synthesis. *Trends in Neurosciences* 20, 487–492.

Friedman, P., Kappelman, A. and Kaufman, S. (1972) Partial purification and characterization of tryptophan hydroxylase from rabbit hindbrain. *The Journal of Biological Chemistry* 247, 4165–4173.

Fuhrberg, B., Balzer, I., Hardeland, R., Werner, A. and Lüning, K. (1996) The vertebrate pineal hormone melatonin is produced by the brown alga *Pterygophora californica* and mimics dark effects on growth rate in the light. *Planta* 200, 125–131.

Fujiwara, T., Maisonneuve, S., Isshiki, M., Mizutani, M., Chen, L., Wong, H., Kawasaki, T. and Shimamoto, K. (2010) *Sekiguchi* lesion gene encodes a cytochrome P450 monooxygenase that catalyzes conversion of tryptamine to serotonin in rice. *The Journal of Biological Chemistry* 285, 11308–11313.

Fuller, P., Gooley, J. and Saper, C. (2006) Neurobiology of the sleep–wake cycle: sleep architecture, circadian regulation, and regulatory feedback. *Journal of Biological Rhythms* 21, 482–493.

Fuster, M. (1992) Study of the decarboxylative route of indole-3-acetic acid: stability and reactivity of indole-3-carbinol. Graduation thesis, Faculty of Biology, University of Murcia, Spain.

Galano, A., Tan, D.-X. and Reiter, R. (2011) Melatonin as a natural ally against oxidative stress: a physicochemical examination. *Journal of Pineal Research* 51, 1–16.

Galano, A., Tan, D.-X. and Reiter, R.J. (2013) On the free radical scavenging activities of melatonin's metabolites, AFMK and AMK. *Journal of Pineal Research* 54, 245–257.

Ganguly, S., Coon, S. and Klein, D. (2002) Control of melatonin synthesis in the mammalian pineal gland: the critical role of serotonin acetylation. *Cell and Tissue Research* 309, 127–137.

Garcia, J., Reiter, R., Guerrero, J., Escames, G., Yu, B., Oh, C. and Muñoz-Hoyos, A. (1997) Melatonin prevents changes in microsomal membrane fluidity during induced lipid peroxidation. *FEBS Letters* 408, 297–300.

Garcia-Moreno, H., Calvo, J.R. and Maldonado, M.D. (2012) High levels of melatonin generated during the brewing process. *Journal of Pineal Research* 55, 26–30.

Garcia-Parrilla, M., Cantos, E. and Troncoso, A. (2009) Analysis of melatonin in foods. *Journal of Food Composition and Analysis* 22, 177–183.

Garidou, M., Bartol, I., Calgari, C., Pévet, P. and Simonneaux, V. (2001) *In vivo* observation of a non-noradrenergic regulation of arylalkylamine *N*-acetyltransferase gene expression in the rat pineal complex. *Neuroscience* 105, 721–729.

Garrido, M., Paredes, S.D., Cubero, J., Lozano, M., Toribio-Delgado, A.F., Muñoz, J.L., Reiter, R.J., Barriga, C. and Rodríguez, A.B. (2010) Jerte valley cherry-enriched diets improve nocturnal rest and increase 6-sulfatoxymelatonin and total antioxidant capacity in the urine of middle-aged and elderly humans. *The Journals of Gerontology A: Biological Sciences and Medical Sciences* 65, 909–914.

Gastel, J., Roseboom, P., Rinaldi, P., Weller, J. and Klein, D. (1998) Melatonin production: proteasomal proteolysis in serotonin *N*-acetyltransferase regulation. *Science* 279, 1358–1360.

Gauer, F. and Craft, C. (1996) Circadian regulation of hydroxyindole-*O*-methyltransferase mRNA levels in rat pineal and retina. *Brain Research* 737, 99–109.

Giardina, G., Montioli, R., Gianni, S., Cellini, B., Paiardini, A., Voltattorni, C. and Cutruzzola, F. (2011) Open conformation of human DOPA decarboxylase reveals the mechanism of PLP addition to Group II decarboxylases. *Proceedings of the National Academy of Sciences of the United States of America* 108, 20514–20519.

Gibson, R., Barret, G. and Wightman, F. (1972) Biosynthesis and metabolism of indol-3yl-acetic acid: III. Partial purification and properties of a tryptamine-forming L-tryptophan decarboxylase from tomato shoots. *Journal of Experimental Botany* 23, 775–786.

Gilad, E., Matzkin, H. and Zisapel, N. (1997) Inactivation of melatonin receptors by protein kinase C in human prostate epithelial cells. *Endocrinology* 138, 4255–4261.

Gitto, E., Tan, D.-X., Reiter, R., Karbownik, M., Manchester, L.C., Cuzzocrea, S., Fulia, F. and Barberi, I. (2001) Individual and synergistic antioxidative actions of melatonin: studies with vitamin E, vitamin C, glutathione and desferrioxamine (desferoxamine) in rat liver homogenates. *Journal of Pharmacy and Pharmacology* 53, 1393–1401.

Goda, K., Hamane, Y., Kishimoto, R. and Ogishi, Y. (1999) Radical scavenging properties of tryptophan metabolites. Estimation of their radical reactivity. *Advances in Experimental Medicine and Biology* 467, 397–402.

Gómez, F., Hernández, I.G., Martinez, L.D., Silva, M.F. and Cerutti, S. (2013) Analytical tools for elucidating the biological role of melatonin in plants by liquid chromatography-tandem mass spectrometry. *Electrophoresis* 34, 1749–1756.

González-Gómez, D., Lozano, M., Fernández-León, M., Ayuso, M., Bernalte, M. and Rodríguez, A. (2009) Detection and quantification of melatonin and serotonin in eight sweet cherry cultivars (*Prunus avium* L.). *European Food Research and Technology* 229, 223–229.

Grahame-Smith, D. (1964) Tryptophan hydroxylation in brain. *Biochemical and Biophysical Research Communications* 16, 586–592.

Grobe, W. (1982) Function of serotonin in seeds of walnuts. *Phytochemistry* 21, 819–822.

Guajardo, M., Terrasa, A. and Catalá, A. (2006) Lipid–protein modifications during ascorbate-Fe(II) peroxidation of photoreceptor membranes: protective effect of melatonin. *Journal of Pineal Research* 41, 201–210.

Hamilton, T. (1969) Influence of environmental light and melatonin upon mammary tumor induction. *British Journal of Surgery* 56, 764–766.

Hamon, M., Bourgoin, S., Artaud, F. and Glowinski, J. (1979) The role of intraneuronal 5-HT and of tryptophan hydroxylase activation in the control of 5-HT synthesis in rat brain slices incubated in K+-enriched medium. *Journal of Neurochemistry* 33, 1031–1042.

Han, Q., Ding, H., Robinson, H., Christensen, B. and Li, J. (2010) Crystal structure and substrate specificity of *Drosophila* 3,4-dihydroxyphenylalanine decarboxylase. *PLoS ONE* 5 (1): e8826.

Hansen, J. (2001) Increased breast cancer risk among women who work predominantly at night. *Epidemiology* 12, 74–77.

Hardeland, R. (1999) Melatonin and 5-methoxytrypyamine in non-metazoans. *Reproduction, Nutrition, Development* 39, 399–408.

Hardeland, R. (2012) Melatonin in aging and disease. Multiple consequences of reduced secretion, options and limits of treatment. *Aging and Disease* 3, 194–225.

Hardeland, R. and Pandi-Perumal, S. (2005) Melatonin, a potent agent in antioxidative defense: actions as a natural food constituent, gastrointestinal factor, drug and prodrug. *Nutrition and Metabolism* 2, 1–15.

Hardeland, R., Reiter, R.J., Poeggeler, B. and Tan, D.-X. (1993) The significance of the metabolism of the neurohormone melatonin: antioxidative protection and formation of bioactive substances. *Neuroscience and Biobehavioral Reviews* 17, 347–357.

Hardeland, R., Pandi-Perumal, S. and Cardinali, D. (2006) Melatonin. *International Journal of Biochemistry and Cell Biology* 38, 313–316.

Hardeland, R., Madrid, J., Tan, D. and Reiter, R. (2012) Melatonin, the circadian multioscillator system and health: the need for detailed analysis of peripheral melatonin signal. *Journal of Pineal Research* 52, 139–166.

Hattori, A., Migitaka, H., Iigo, M., Yamamoto, K., Ohtani-Kaneko, R., Hara, M., Suzuki, T. and Reiter, R.J. (1995) Identification of melatonin in plants and its effects on plasma melatonin levels and binding to melatonin receptors in vertebrates. *Biochemistry and Molecular Biology International* 35, 627–634.

Haycock, J., Kumer, S., Lewis, D., Vrana, K. and Stockmeier, C. (2002) A monoclonal antibody to tryptophan hydroxylase: applications and identification of the epitope. *Journal of Neuroscience Methods* 114, 205–212.

Hernández-Ruiz, J. and Arnao, M.B. (2008a) Distribution of melatonin in different zones of lupin and barley plants at different ages in the presence and absence of light. *Journal of Agricultural and Food Chemistry* 56, 10567–10573.

Hernández-Ruiz, J. and Arnao, M.B. (2008b) Melatonin stimulates the expansion of etiolated lupin cotyledons. *Plant Growth Regulation* 55, 29–34.
Hernández-Ruiz, J., Arnao, M.B. and Cano, A. (2004) Melatonin: growth-stimulating compound present in lupin tissues. *Planta* 220, 140–144.
Hernández-Ruiz, J., Cano, A. and Arnao, M.B. (2005) Melatonin acts as a growth-stimulating compound in some monocot species. *Journal of Pineal Research* 39, 137–142.
Herxheimer, A. (2005) Jet lag. *Clinical Evidence* 13, 2178–2183.
Hickman, A., Klein, D. and Dyda, F. (1999a) Melatonin biosynthesis: the structure of serotonin *N*-acetyltransferase at 2.5 Å resolution suggests a catalytic mechanism. *Molecular Cell* 3, 23–32.
Hickman, A., Namboodiri, M., Klein, D. and Dyda, F. (1999b) The structural basis of ordered substrate binding by serotonin *N*-acetyltransferase: enzyme complex at 1.8 Å resolution with a bisubstrate analog. *Cell* 97, 361–369.
Hoffman, R. and Reiter, R. (1965a) Influence of compensatory mechanisms and the pineal upon light-induced gonadal atrophy in male hamsters. *Nature* 207, 658–659.
Hoffman, R. and Reiter, R. (1965b) Pineal gland: influence on gonads of male hamsters. *Science* 148, 1609–1611.
Horstman, J., Wrona, W. and Dryhurst, G. (2002) Further insights into the reaction of melatonin with hydroxyl radical. *Bioinorganic Chemistry* 30, 371–382.
Huang, X. and Mazza, G. (2011a) Simultaneous analysis of serotonin, melatonin, piceid and resveratrol in fruits using liquid chromatography tandem mass spectrometry. *Journal of Chromatography A* 1218, 3890–3899.
Huang, X. and Mazza, G. (2011b) Application of LC and LC-MS to the analysis of melatonin and serotonin in edible plants. *Critical Reviews in Food Science and Nutrition* 51, 269–284.
Huether, G. (1994) Melatonin synthesis in the gastrointestinal tract and the impact of nutritional factors on circulating melatonin. *Annals of the New York Academy of Sciences* 719, 146–158.
Humbert, W. and Pévet, P. (1994) The decrease of pineal melatonin production with age. *Annals of the New York Academy of Sciences* 719, 43–61.
Hussein, M., Ahmed, O., Hassan, A. and Ahmed, M. (2007) Intake of melatonin is associated with amelioration of physiological changes, both metabolic and morphological pathologies with obesity: an animal model. *International Journal of Experimental Pathology* 88, 19–29.
Ichinose, H., Kurosawa, Y., Togari, A., Kato, Y., Parvez, S. and Nagatsu, Y. (1985) Simple purification of AADC from human pheochromocytoma using high-performance liquid chromatography. *Analytical Biochemistry* 50, 408–414.
Ichinose, H., Kurosawa, Y., Titani, K., Fujita, K. and Nagatsu, T. (1989) Isolation and characterization of a cDNA clone encoding human aromatic L-amino acid decarboxylase. *Biochemical and Biophysical Research Communications* 164, 1024–1030.
Ichiyama, A., Nakamura, S., Nishizuka, Y. and Hayaishi, O. (1970) Enzymic studies on the biosynthesis of serotonin in mammalian brain. *The Journal of Biological Chemistry* 245, 1699–1709.
Iriti, M., Rossoni, M. and Faoro, F. (2006) Melatonin content in grape: myth or panacea? *Journal of Science of the Science of Food and Agriculture* 86, 1432–1438.
Iriti, M., Varoni, E. and Vitalini, S. (2010) Melatonin in traditional Mediterranean diets. *Journal of Pineal Research* 49, 101–105.
Ishida, I., Obinata, M. and Deguchi, T. (1987) Molecular cloning and nucleotide sequence of cDNA encoding hydroxyindole-*O*-methyltransferase of bovine pineal glands. *The Journal of Biological Chemistry* 262, 2895–2899.
Ishii, S., Mizuguchi, H., Nishino, J., Hayashi, H. and Kagamiyama, H. (1996) Functionally important residues of aromatic L-amino acid decarboxylase probed by sequence alignment and site-directed mutagenesis. *The Journal of Biochemistry* 120, 369–376.
Itoh, M., Ishizuka, B., Kuribayashi, Y., Amemiya, A. and Sumi, Y. (1999) Melatonin, its precursors, and synthesizing enzyme activities in the human ovary. *Molecular Human Reproduction* 5, 402–408.
Iyer, L.M., Aravind, L., Coon, S.L., Klein, D.C. and Koonin, E.V. (2004) Evolution of cell–cell signaling in animals: did late horizontal gene transfer from bacteria have a role? *Trends in Genetics* 20, 292–299.
Jan, J., Reiter, R., Wasdell, M. and Bax, M. (2009) The role of the thalamus in sleep, pineal melatonin production, and circadian rhythm sleep disorders. *Journal of Pineal Research* 46, 1–7.
Jequier, E., Robinson, B., Lovengerg, W. and Sjoerdsma, A. (1969) Further studies on tryptophan hydroxylase in rat brainstem and beef pineal [glands]. *Biochemical Pharmacology* 18, 1071–1081.

Jin, H., Cianchetta, G., Devasagayaraj, A., Gu, K., Marinelli, B., Samala, L., Scott, S., Stouch, T., Tunoori, A., Wang, Y., Zang, Y., Zhang, C. *et al.* (2009) Substituted 3-(4-(1,3,5-triazin-2-yl)-phenyl)-2-aminopropanoic acids as novel tryptophan hydroxylase inhibitors. *Bioorganic and Medicinal Chemistry Letters* 19, 5229–5232.

Johansen, P., Wolf, W. and Kuhn, D. (1991) Inhibition of tryptophan hydroxylase by benserazide and other catechols. *Biochemical Pharmacology* 41, 625–628.

Jung-Hynes, B., Reiter, R.J. and Ahmad, N. (2010) Sirtuins, melatonin and circadian rhythms: building a bridge between aging and cancer. *Journal of Pineal Research* 48, 9–19.

Juorio, A. and Boulton, A. (1982) The effect of some amino acid precursor enzyme inhibitors on the mouse striatal concentration of tyramines and homovanillic acid. *Journal of Neurochemistry* 39, 859–863.

Kamdar, B., Tergas, A., Mateen, F., Bhayani, N. and Oh, J. (2013) Night-shift work and risk of breast cancer: a systematic review and meta-analysis. *Breast Cancer Research and Treatment* 138, 291–301.

Kang, K., Kim, Y., Park, S. and Back, K. (2009a) Senescence-induced serotonin biosynthesis and its role in delaying senescence in rice leaves. *Plant Physiology* 150, 1380–1393.

Kang, K., Park, S., Kim, Y., Lee, S. and Back, K. (2009b) Biosynthesis and biotechnological production of serotonin derivatives. *Applied Microbiology and Biotechnology* 83, 27–34.

Kang, K., Lee, K., Park, S., Kim, Y.S. and Back, K. (2010) Enhanced production of melatonin by ectopic overexpression of human serotonin *N*-acetyltransferase plays a role in cold resistance in transgenic rice seedlings. *Journal of Pineal Research* 49, 176–182.

Kang, K., Kong, K., Park, S., Natsagdori, U., Kim, Y. and Back, K. (2011) Molecular cloning of a plant *N*-acetylserotonin methyltransferase and its expression characteristics in rice. *Journal of Pineal Research* 50, 304–309.

Kang, K., Lee, K., Park, S., Byeon, Y. and Back, K. (2013) Molecular cloning of rice serotonin *N*-acetyltransferase, the penultimate gene in plant melatonin biosynthesis. *Journal of Pineal Research* 55, 7–13.

Kang, S., Kang, K., Lee, K. and Back, K. (2007a) Characterization of rice tryptophan decarboxylases and their direct involvement in serotonin biosynthesis in transgenic rice. *Planta* 227, 263–272.

Kang, S., Kang, K., Lee, K. and Back, K. (2007b) Characterization of tryptamine 5-hydroxylase and serotonin synthesis in rice plants. *Plant Cell Reports* 26, 2009–2015.

Kang, U. and Joh, T. (1990) Deduced amino acid sequence of bovine aromatic L-amino acid decarboxylase: homology to other decarboxylases. *Molecular Brain Research* 8, 83–87.

Karbownik, M., Reiter, R.J., Qi, W., Garcia, J.J., Tan, D.-X., Manchester, L.C. and Vijayalaxmi (2000a) Protective effects of melatonin against oxidation of guanine bases in DNA and decreased microsomal membrane fluidity in rat liver induced by whole body ionizing radiation. *Molecular and Cellular Biochemistry* 211, 137–144.

Karbownik, M., Reiter, R.J., Garcia, J.J. and Tan, D.-X. (2000b) Melatonin reduces phenylhydrazine-induced oxidative damage to cellular membranes: evidence for the involvement of iron. *The International Journal of Biochemistry and Cell Biology* 32, 1045–1054.

Karbownik, M., Tan, D.-X. and Reiter, R.J. (2000c) Melatonin reduces the oxidation of nuclear DNA and membrane lipids induced by the carcinogen α-aminolevulinic acid. *International Journal of Cancer* 88, 7–11.

Khalil, E. and Cole, P. (1998) A potent inhibitor of the melatonin rhythm enzyme. *Journal of the American Chemical Society* 120, 6195–6196.

Khalil, E., De Angelis, J., Ishii, M. and Cole, P. (1999) Mechanism-based inhibition of the melatonin rhythm enzyme: pharmacologic exploitation of active site functional plasticity. *Proceedings of the National Academy of Sciences of the United States of America* 96, 12418–12423.

Kim, K., Wessel, T., Stone, D., Carver, C., Joh, T. and Park, D. (1991) Molecular cloning and characterization of cDNA encoding tryptophan hydroxylase from rat central serotonergic neurons. *Molecular Brain Research* 9, 277–283.

Klein, D. (1985) Photoneural regulation of the mammalian pineal gland. In: Evered, D. and Clark, S. (eds) *Photoperiodism, Melatonin and the Pineal Gland*. Ciba Foundation Symposium, Pitman Press, London, pp. 38–56.

Klein, D. (2007) Arylalkylamine *N*-acetyltransferase: "the Timezyme". *The Journal of Biological Chemistry* 282, 4233–4237.

Klein, D. and Moore, R. (1979) Pineal *N*-acetyltransferase and hydroxyindole-*O*-methyltransferase: control by the retinohypothalamic tract and the suprachiasmatic nucleus. *Brain Research* 174, 245–262.

Klein, D. and Weller, J. (1970) Indole metabolism in the pineal gland: a circadian rhythm in *N*-acetyltransferase. *Science* 169, 1093–1095.

Klein, D., Weller, J. and Moore, R. (1971) Melatonin metabolism: neural regulation of pineal serotonin acetyl coenzyme A *N*-acetyltransferase activity. *Proceedings of National Academy of Sciences of the United States of America* 68, 3107–3110.

Klein, D., Coon, S. and Roseboom, P. (1996) New light is shining on the melatonin rhythm enzyme. The first postcloning view. *Trends in Endocrinology and Metabolism* 7, 106–112.

Klein, D., Coon, S., Roseboom, P., Weller, J., Bernard, M., Gastel, J., Zatz, M., Iuvone, P., Rodriguez, I., Bégay, V., Falcón, J., Cahill, G. *et al.* (1997) The melatonin rhythm-generating enzyme: molecular regulation of serotonin *N*-acetyltransferase in the pineal gland. *Recent Progress in Hormone Research* 52, 307–358.

Knutsson, A. (2003) Health disorders in shift workers. *Occupational Medicine* 53, 103–108.

Kokkinou, I., Nikolouzou, E., Hatzimanolis, A., Fragoulis, E. and Vassilacopoulou, D. (2009) Expression of enzymatically active L-DOPA decarboxylase in human peripheral leukocytes. *Blood Cells, Molecules and Diseases* 42, 92–98.

Kolar, J. (2003) *Effects of melatonin on circadian rhythms and photoperiodism in higher plants*. PhD thesis, Faculty of Natural Sciences, Charles University, Prague.

Kolar, J. and Machackova, I. (2005) Melatonin in higher plants: occurrence and possible functions. *Journal of Pineal Research* 39, 333–341.

Kolar, J., Machackova, I., Illnerova, H., Prinsen, E., van Dongen, W. and van Onckelen, H. (1995) Melatonin in higher plant determined by radioimmunoassay and liquid chromatography-mass spectrometry. *Biological Rhythm Research* 26, 406–409.

Kolar, J., Machackova, I., Eder, J., Prinsen, E., van Dongen, W., van Onckelen, H. and Illnerova, H. (1997) Melatonin: occurrence and daily rhythm in *Chenopodium rubrum*. *Phytochemistry* 44, 1407–1413.

Kolar, J., Johnson, C. and Machackova, I. (2003) Exogenously applied melatonin affects flowering of the short-day plant *Chenopodium rubrum*. *Physiologia Plantarum* 118, 605–612.

Konturek, S., Konturec, P., Brzozowski, T. and Bubenik, G. (2007) Role of melatonin in upper gastrointestinal tract. *Journal of Physiology and Pharmacology* 58, 23–52.

Kopin, I., Pare, C. and Axelrod, J. (1960) 6-Hydroxylation, the major metabolic pathway for melatonin. *Biochimica et Biophysica Acta* 40, 377–378.

Korf, H., von Gall, C. and Stehle, J. (2003) The circadian system and melatonin: lessons from rats and mice. *Chronobiology International* 20, 697–710.

Krause, D. and Dubocovich, M. (1990) Regulatory sites in the melatonin system of mammals. *Trends in Neuroscience* 13, 464–470.

Lahiri, D., Ge, Y., Sharman, E. and Bondy, S. (2004) Age-related changes in serum melatonin in mice: higher levels of combined melatonin and 6-hydroxymelatonin sulfate in the cerebral cortex than serum, heart, liver and kidney tissues. *Journal of Pineal Research* 36, 217–223.

Lancaster, G. and Sourkes, T. (1972) Purification and properties of hog kidney 3,4-dihydroxyphenylalanine decarboxylase in the human brain. *Journal of Neurochemistry* 19, 1549–1559.

Lauber, J., Boyd, J. and Axelrod, J. (1968) Enzymatic synthesis of melatonin in avian pineal body: extraretinal response to light. *Science* 161, 489–490.

Lazár, D., Murch, S.J., Beilby, M.J. and Al Khazaaly, S. (2013) Exogenous melatonin affects photosynthesis in characeae *Chara australis*. *Plant Signaling and Behavior* 8(3): e23279.

Lei, X., Zhu, R., Zhang, G. and Dai, Y. (2004) Attenuation of cold-induced apoptosis by exogenous melatonin in carrot suspension cells: the possible involvement of polyamines. *Journal of Pineal Research* 36, 126–131.

Lerner, A., Case, J., Takahashi, Y., Lee, T. and Mori, W. (1958) Isolation of melatonin, a pineal factor that lightens melanocytes. *Journal of the Chemical American Society* 80, 2587.

Lerner, A., Case, J. and Heinzelmann, R. (1959a) Structure of melatonin. *Journal of the American Chemical Society* 81, 6084–6085.

Lerner, A., Case, J., Mori, W. and Wright, M. (1959b) Melatonin in peripheral nerve[s]. *Nature* 183, 1821.

Lewy, A., Ahmed, S., Jackson, J. and Sack, R. (1992) Melatonin shifts human circadian rhythms according to a phase-response curve. *Chronobiology International* 9, 380–392.

Li, D., Smith, D., Hardeland, R., Yang, M., Xu, H., Zhang, L., Yin, H. and Zhu, Q. (2013) Melatonin receptor genes in vertebrates. *International Journal of Molecular Sciences* 14, 11208–11223.

López, A., García, J.A., Escámes, G., Venegas, C., Ortíz, F., López, L.C. and Acuña-Castroviejo, D. (2008) Melatonin protects the mitochondria from oxidative damage reducing oxygen consumption, membrane potential, and superoxide anion production. *Journal of Pineal Research* 46, 188–198.

Lovenberg, W., Weissbach, H. and Udenfriend, S. (1962) Aromatic L-amino acid decarboxylase. *The Journal of Biological Chemistry* 237, 89–93.

Ma, X., Idle, J., Krausz, K. and González, F. (2005) Metabolism of melatonin by human cytochromes P450. *Drug Metabolism and Disposition* 33, 489–494.

Machackova, I. and Krekule, J. (2002) Sixty-five years of searching for the signals that trigger flowering. *Russian Journal of Plant Physiology* 49, 451–459.

Macias, M., Rodriguez-Cabezas, M., Reiter, R., Osuna, A. and Acuña-Castroviejo, D. (1999) Presence and effects of melatonin in T*rypanosoma cruzi*. *Journal of Pineal Research* 27, 86–94.

Maldonado, M.D., Murillo-Cabezas, F., Terrón, M.P., Flores, L.J., Tan, D.X., Manchester, L.C. and Reiter, R.J. (2007) The potential of melatonin in reducing morbidity–mortality after craniocerebral trauma. *Journal of Pineal Research* 42, 1–11.

Manchester, L.C., Poeggeler, B., Alvares, F.L., Ogden, G.B. and Reiter, R.J. (1995) Melatonin immunoreactivity in the photosynthetic prokaryote *Rhodospirillum rubrum*: implications for an ancient antioxidant system. *Cellular and Molecular Biology Research* 41, 391–395.

Manchester, L.C., Tan, D.-X., Reiter, R.J., Park, W., Monis, K. and Qi, W. (2000) High levels of melatonin in the seeds of edible plants. Possible function in germ tissue protection. *Life Sciences* 67, 3023–3029.

Maneckjee, R. and Baylin, S. (2013) Use of radiolabeled monofluromethyl DOPA to define subunit structure of human l-DOPA decarboxylase. *Biochemistry* 12, 6018–6025.

Marioni, F., Bertoli, A. and Pistelli, L. (2008) A straightforward procedure to biosynthesise melatonin using freshly chopped *Achillea millefolium* L. as reagent. *Phytochemistry Letters* 1, 107–110.

Maronde, E. and Stehle, J. (2007) The mammalian pineal gland: known facts, unknown facets. *Trends in Endocrinology and Metabolism* 18, 142–149.

Martínez, A., Knappskog, P. and Haavik, J. (2001) A structural approach into human tryptophan hydroxylase and its implications for the regulation of serotonin biosynthesis. *Current Medicinal Chemistry* 8, 1077–1091.

Masson-Pevet, M., Bianchi, L. and Pévet, P. (1996) Circadian photic regulation of melatonin receptor density in rat suprachiasmatic nuclei: comparison with light induction in fos-related protein. *Journal of Neuroscience Research* 43, 632–637.

Mauriz, J., Molpeceres, V., Garcia-Mediavilla, M., González, P., Barrio, J. and Gonzalez-Gallego, J. (2007) Melatonin prevents oxidative stress and changes in antioxidant enzyme expression and activity in the liver of aging rats. *Journal of Pineal Research* 42, 222–230.

Mayo, J.C., Sainz, R.M., Tan, D.-X., Hardeland, R., Leon, J., Rodríguez, C. and Reiter, R.J. (2005) Anti-inflammatory actions of melatonin and its metabolites, *N*1-acetyl-*N*2-formyl-5-methoxykynuramine (AFMK) and *N*1-acetyl-5-methoxykynuramine (AMK), in macrophages. *Journal of Neuroimmunology* 165, 139–149.

Mayo, J., Tan, D.X., Sainz, R.M., Natarajan, M., López-Burillo, S. and Reiter, R.J. (2013) Protection against oxidative protein damage induced by metal-catalyzed reaction or alkylperoxyl radicals: comparative effects of melatonin and other antioxidants. *Biochimica et Biophysica Acta* 1620, 139–150.

McGilvery, R. and Cohen, P. (1948) The decarboxylation of L-phenylalanine by *Streptococcus faecalis*. *The Journal of Biological Chemistry* 174, 813–816.

McKinney, J., Knappskog, P.M. and Haavik, J. (2005) Different properties of the central and peripheral forms of human tryptophan hydroxylase. *Journal of Neurochemistry* 92, 311–320.

Mena, P., Gil-Izquierdo, A., Moreno, D.A., Martí, N. and García-Viguera, C. (2012) Assessment of the melatonin production in pomegranate wines. *LWT – Food Science and Technology* 47, 13–18.

Mesangeau, C., Yous, S., Chavatte, P., Ferry, G., Audinot, V., Boutin, J., Delagrange, P., Bennejean, C., Renard, P. and Lesieur, D. (2003) Design, synthesis and *in vitro* evaluation of novel benzo-thiophene derivatives as serotonin *N*-acetyltransferase (AANAT) inhibit. *Journal of Enzyme Inhibition and Medicinal Chemistry* 18, 119–125.

Miller, S., Pandi-Perumal, S., Esquifino, A., Cardinali, D. and Maestroni, G. (2006) The role of melatonin in inmmunoenhancement: potential application in cancer. *International Journal of Experimental Pathology* 87, 81–87.

Morita, M., Hall, F., Best, J. and Gern, W. (1987) Photoperiodic modulation of cephalic melatonin in planarians. *Journal of Experimental Zoology* 241, 383–388.

Morreale, M. and Livrea, M. (1997) Synergistic effect of glycolic acid on the antioxidant activity of α-tocopherol and melatonin in lipid bilayers and in human skin homogenates. *Biochemistry and Molecular Biology International* 42, 1093–1102.

Morrissey, J., Edwards, S. and Lovenberg, W. (1977) Comparison of rat pineal gland and rat liver serotonin-*N*-acetyltransferase. *Biochemical and Biophysical Research Communications* 77, 118–123.

Morton, D. (1987) Hydroxyindole-*O*-methyltransferase catalyses the production of methoxyindoles in rat pineal gland dependent on the concentration of hydroxyl precursors and their affinity for the enzyme. *Journal of Endocrinology* 115, 455–458.

Murch, S. and Saxena, P. (2002a) Mammalian neurohormones: potential significance in reproductive physiology of St John's wort (*Hypericum perforatum* L.). *Naturwissenschaften* 89, 555–560.

Murch, S. and Saxena, P. (2002b) Melatonin: a potential regulator of plant growth and development? *In Vitro Cellular and Developmental Biology – Plant* 38, 531–536.

Murch, S., Simmons, C. and Saxena, P. (1997) Melatonin in feverfew and other medicinal plants. *The Lancet* 350, 1598–1599.

Murch, S., Krishna Raj, S. and Saxena, P. (2000) Tryptophan is a precursor for melatonin and serotonin biosynthesis in *in vitro* regenerated St John's wort (*Hypericum perforatum* L.) plants. *Plant Cell Reports* 19, 698–704.

Murch, S., Campbell, S. and Saxena, P. (2001) The role of serotonin and melatonin in plant morphogenesis. Regulation of auxin-induced root organogenesis in *in vitro*-cultured explants of *Hypericum perforatum* L. *In Vitro Cellular and Developmental Biology – Plant* 37, 786–793.

Murch, S.J., Alan, A.R., Cao, J. and Saxena, P.K. (2009) Melatonin and serotonin in flowers and fruits of *Datura metel* L. *Journal of Pineal Research* 47, 277–283.

Murch, S., Hall, B., Le, C. and Saxena, P. (2010) Changes in the levels of indoleamine phytochemicals during veraison and ripening of wine grapes. *Journal of Pineal Research* 49, 95–100.

Muszynska, B., Sulkowska-Ziaja, K. and Ekiert, H. (2011) Indole compounds in fruiting bodies of some edible Basidiomycota species. *Food Chemistry* 125, 1306–1308.

Nagle, C., Cardinali, D. and Rosner, J. (1972) Light regulation of rat retinal hydroxyindole-*O*-methyltransferase (HIOMT) activity. *Endocrinology* 91, 423–426.

Nakane, M., Yokoyama, E. and Deguchi, T. (1983) Species heterogeneity of pineal hydroxyindole-*O*-methyltransferase. *Journal of Neurochemistry* 40, 790–796.

Nakata, H. and Fujisawa, H. (1982) Purification and properties of tryptophan 5-monooxygenase from rat brain-stem. *European Journal of Biochemistry* 122, 41–47.

Namboodiri, M., Dubbels, R. and Klein, D. (1987) Arylalkylamine *N*-acetyltransferase from mammalian pineal gland. *Methods of Enzymology* 142, 583–590.

Niedens, B., Parker, S., Stierle, D. and Stierle, A. (1999) First fungal aromatic L-amino acid decarboxylase from a paclitaxel-producing *Penicillium raistrickii*. *Mycologia* 91, 619–626.

Nisembaum, L., Tinoco, A., Moure, A., Alonso-Gómez, A., Delgado, M. and Valenciano, A. (2013) The arylalkylamine-*N*-acetyltransferase (AANAT) acetylates dopamine in the digestive tract of goldfish: a role in intestinal motility. *Neurochemistry International* 62, 873–880.

Nishigaki, H., Ichinose, H., Tamai, T. and Nagatsu, T. (1988) Purification of aromatic L-amino acid decarboxylase from bovine brain with a monoclonal antibody. *Biochemical Journal* 252, 331–335.

Nishino, J., Hayashi, H., Ishii, S. and Kagamiyama, H. (1997) An anomalous side reaction of the Lys303 mutant aromatic L-amino acid decarboxylase unravels the role of the residue in catalysis. *The Journal of Biochemistry* 121, 604–611.

Noé, W., Mollenschott, C. and Berlin, J. (1984) Tryptophan decarboxylase from *Catharanthus roseus* cell suspension cultures: purification, molecular and kinetic data of the homogenous protein. *Plant Molecular Biology* 3, 281–288.

Nukiwa, T., Toyama, C., Okita, C., Kataoka, T. and Ichiyama, A. (1974) Purification and some properties of bovine pineal tryptophan 5-monooxygenase. *Biochemical and Biophysical Research Communications* 60, 1029–1035.

Obsil, T., Ghirlando, R., Klein, D., Ganguly, S. and Dyda, F. (2001) Crystal structure of the 14-3-3:serotonin *N*-acetyltransferase complex. A role for scaffolding in enzyme regulation. *Cell* 105, 257–267.

Okazaki, M. and Ezura, H. (2009) Profiling of melatonin in the model tomato (*Solanum lycopersicum* L.) cultivar Micro-Tom. *Journal of Pineal Research* 46, 338–343.

Okazaki, M., Higuchi, K., Hanawa, Y., Shiraiwa, Y. and Ezura, H. (2009) Cloning and characterization of a *Chlamydomonas reinhardtii* cDNA arylalkylamine *N*-acetyltransferase and its use in the genetic engineering of melatonin content in the Micro-Tom tomato. *Journal of Pineal Research* 46, 373–382.

Okazaki, M., Higuchi, K., Aouini, A. and Ezura, H. (2010) Lowering intercellular melatonin levels by transgenic analysis of indoleamine 2,3-dioxygenase from rice in tomato plants. *Journal of Pineal Research* 49, 239–247.

Pandi-Perumal, S., Trakht, I., Srinivasan, V., Spence, D., Maestroni, G., Zisapel, N. and Cardinali, D. (2008) Physiological effects of melatonin: role of melatonin receptors and signal transduction pathways. *Progress in Neurobiology* 85, 335–353.

Pape, C. and Lüning, K. (2006) Quantification of melatonin in phototrophic organisms. *Journal of Pineal Research* 41, 157–165.

Paredes, S.D., Korkmaz, A., Manchester, L.C., Tan, D.-X. and Reiter, R.J. (2009) Phytomelatonin: a review. *Journal of Experimental Botany* 60, 57–69.

Park, S. and Back, K. (2012) Melatonin promotes seminal root elongation and root growth in transgenic rice after germination. *Journal of Pineal Research* 53, 385–389.

Park, S., Lee, K., Kim, Y. and Back, K. (2012) Tryptamine 5-hydroxylase-deficient *Sekiguchi* rice induces synthesis of 5-hydroxytryptophan and *N*-acetyltryptamine but decreases melatonin biosynthesis during senescence process of detached leaves. *Journal of Pineal Research* 52, 211–216.

Park, S., Byeon, Y., Kim, Y.S. and Back, K. (2013a) Kinetic analysis of purified recombinant rice *N*-acetylserotonin methyltransferase and peak melatonin production in etiolated rice shoots. *Journal of Pineal Research* 54, 139–144.

Park, S., Lee, D., Jang, H., Byeon, Y., Kim, Y. and Back, K. (2013b) Melatonin-rich transgenic rice plants exhibit resistance to herbicide-induced oxidative stress. *Journal of Pineal Research* 54, 258–263.

Park, S., Le, T.N.N., Byeon, Y., Kim, Y.S. and Back, K. (2013c) Transient induction of melatonin biosynthesis in rice (*Oryza sativa* L.) during the reproductive stage. *Journal of Pineal Research* 55, 40–45.

Park, W. (2011) Melatonin as an endogenous plant regulatory signal: debates and perspectives. *Journal of Plant Biology* 54, 143–149.

Pelagio-Flores, R., Muñoz-Parra, E., Ortiz-Castro, R. and Lopez-Bucio, J. (2012) Melatonin regulates *Arabidopsis* root system architecture likely acting independently of auxin signaling. *Journal of Pineal Research* 53, 279–288.

Perreau-Lenz, S., Kalsbeek, A., Garidou, M., Wortel, J., van der Vliet, J., van Heijningen, C., Simonneaux, V., Pévet, P. and Buijs, R. (2003) Suprachiasmatic control of melatonin synthesis in rats: inhibitory and stimulatory mechanisms. *European Journal of Neuroscience* 17, 221–228.

Pierrefiche, G. and Laborit, H. (1995) Oxygen free radicals, melatonin, and aging. *Experimental Gerontology* 30, 213–227.

Pierrefiche, G., Topall, G., Courboin, I., Henriet, I. and Laborit, H. (1993) Antioxidant activity of melatonin in mice. *Research Communications in Chemical Pathology and Pharmacology* 80, 211–223.

Poeggeler, B. (2005) Melatonin, aging, and age-related diseases: perspectives for prevention, intervention, and therapy. *Endocrine* 27, 201–212.

Poeggeler, B. and Hardeland, R. (2001) Observations on melatonin oxidation and metabolite release by unicellular organisms and small aquatic metazoans. In: Hardeland, R (ed.) *Actions and Redox Properties of Melatonin and other Aromatic Amino Acid Metabolites*. Cuviller Verlag, Göttingen, Germany, pp. 66–69.

Poeggeler, B., Balzer, I., Fischer, J., Behrmann, G. and Hardeland, R. (1989) A role of melatonin in dinoflagellates? *Acta Endocrinologica (Copenhagen)* 120(S1), 97.

Poeggeler, B., Balzer, I., Hardeland R and Lerchl A. (1991) Pineal hormone melatonin oscillates also in dinoflagellates? *Naturwissenschaften* 78, 268–269.

Poeggeler, B., Reiter, R.J., Tan, D.-X., Chen, L. and Manchester, L. (1993) Melatonin, hydroxyl radical-mediated oxidative damage, and aging: a hypothesis. *Journal of Pineal Research* 14, 151–168.

Poeggeler, B., Saarela, S., Reiter, R.J., Tan, D.-X., Chen, L.D., Manchester, L.C. and Barlow-Walden, L.R. (1994) Melatonin: a highly potent endogenous radical scavenger and electron donor: new aspects of the oxidation chemistry of this indole accessed *in vitro*. *Annals of the New York Academy of Sciences* 738, 419–420.

Poeggeler, B., Reiter, R.J., Hardeland, R., Sewerynek, E., Melchiorri, D. and Barlow-Walden, L.R. (1995) Melatonin, a mediator of electron transfer and repair reactions, acts synergistically with the chain-breaking antioxidants ascorbate, trolox and glutathione. *Neuroendocrinology Letters* 17, 87–92.

Poeggeler, B., Reiter, R.J., Hardeland, R., Tan, D.X. and Barlow-Walden, L.R. (1996) Melatonin and structurally-related, endogenous indoles act as potent electron donors and radical scavengers *in vitro*. *Redox Report* 2, 179–184.

Poeggeler, B., Pappolla, M., Hardeland, R., Rassoulpour, A., Hodgkins, P., Guidatti, P. and Schwarcz, R. (1999) Indole-3-propionate: a potent hydroxyl radical scavenger in rat brain. *Brain Research* 815, 382–388.

Posmyk, M. and Janas, K. (2009) Melatonin in plants. *Acta Physiologica Plantarum* 31, 1–11.

Posmyk, M.M., Kuran, H., Marciniak, K. and Janas, K.M. (2008) Presowing seed treatment with melatonin protects red cabbage seedlings against toxic copper ion concentrations. *Journal of Pineal Research* 45, 24–31.

Posmyk, M.M., Balabusta, M., Wieczorek, M., Sliwinska, E. and Janas, K.M. (2009) Melatonin applied to cucumber (*Cucumis sativus* L.) seeds improves germination during chilling stress. *Journal of Pineal Research* 46, 214–223.

Pothinuch, P. and Tongchitpakdee, S. (2011) Melatonin contents in mulberry (*Morus* spp.) leaves: effects of sample preparation, cultivar, leaf age and tea processing. *Food Chemistry* 128, 415–419.

Pozdeyev, N., Taylor, C., Haque, R., Chaurasia, S., Visser, A., Thazyeen, A., Du, Y., Fu, H., Weller, J., Klein, D. and Iuvone, P. (2006) Photic regulation of arylalkylamine *N*-acetyltransferase binding to 14-3-3 proteins in retinal photoreceptor cells. *Journal of Neuroscience* 26, 9153–9161.

Radogna, F., Diederich, M. and Ghibelli, L. (2010) Melatonin: a pleiotropic molecule regulating inflammation. *Biochemical Pharmacology* 80, 1844–1852.

RadWanski, E. and Last, R. (1995) Tryptophan biosynthesis and metabolism: biochemical and molecular genetics. *The Plant Cell* 7, 921–934.

Rakhlin, N., Kvetnoy, I. and Tolkachev, V. (1975) Melatonin may be synthesized in enterochromafinne cells. *Nature* 255, 344–345.

Ramakrishna, A., Giridhar, P., Sankar, K. and Ravishankar, G. (2012) Melatonin and serotonin profiles in beans of *Coffea* species. *Journal of Pineal Research* 52, 470–476.

Reiter, R.J. (1993a) Interactions of the pineal hormone melatonin with oxygen-centered free-radicals. A brief review. *Brazilian Journal of Medical and Biological Research* 26, 1141–1155.

Reiter, R.J. (1993b) The melatonin rhythm: both a clock and a calendar. *Experientia* 49, 654–664.

Reiter, R.J. (1994) Free-radicals, melatonin, and cellular antioxidative defense-mechanisms. In: Gupta, D., Wollman, H.A. and Fedor-Freybergh, P.G (eds) *Pathophysiology of Inmuno-neuroendocrine Communication Network*. Volkswagenstift, Tubingen, Germany, pp. 135–160.

Reiter, R.J. (1999) Pineal melatonin: cell biology of its synthesis and of its physiological interactions. *Endocrine Reviews* 12, 151–180.

Reiter, R.J. (2003) Melatonin: clinical relevance. *Best Practice and Research Clinical Endocrinology and Metabolism* 17, 273–285.

Reiter, R.J. (2004) Mechanisms of cancer inhibition by melatonin. *Journal of Pineal Research* 37, 213–214.

Reiter, R.J., Poeggeler, B., Tan, D.-X., Chen, L., Manchester, L. and Guerrero, J. (1993) Antioxidant capacity of melatonin. A novel action not requiring a receptor. *Neuroendocrinology Letters* 15, 103–116.

Reiter, R.J., Tan, D.-X., Poeggeler, B., Chen, L. and Menendez-Pelaez, A. (1994a) Melatonin, free radicals and cancer initiation. In: Maestroni, G.J.M., Conti, A., and Reiter, R.J. (eds) *Advances in Pineal Research, Volume 7*. John Libbey, London, pp. , 211–228.

Reiter, R.J., Menendez-Pelaez, A., Poeggeler, B., Tan, D.-X., Pablos, M. and Acuña-Castroviejo, D. (1994b) The role of melatonin in the pathophysiology of oxygen radical damage. In: Møller, M. and Pévet, P. (eds) *Advances in Pineal Research, Volume 8. Proceedings of the VIth Colloquium of the European Pineal Society Held in Copenhagen, Denmark, 23–27 July, 1993*. John Libbey, London, pp. 403–412.

Reiter, R.J., Tan, D., Poeggeler, B., Menendez-Pelaez, A., Chen, L. and Saarela, S. (1994c) Melatonin as a free-radical scavenger. Implications for aging and age-related diseases. *Annals of the New York Academy of Sciences* 719, 1–12.

Reiter, R.J., Melchiorri, D., Sewerynek, E., Poeggeler, B., Barlow-Walden, L., Chuang, J., Ortiz, G.G. and Acuña Castroviejo, D. (1995) A review of the evidence supporting melatonin's role as an antioxidant. *Journal of Pineal Research* 18, 1–11.

Reiter, R.J., Tan, D.-X., Qi, W., Manchester, L.C., Karbownik, M. and Calvo, J.R. (2000) Pharmacology and physiology of melatonin in the reduction of oxidative stress *in vivo*. *Biological Signals and Receptors* 9, 160–171.

Reiter, R.J., Tan, D.-X., Manchester, L.C. and Qi, W. (2001) Biochemical reactivity of melatonin with reactive oxygen and nitrogen species. A review of the evidence. *Cell Biochemistry and Biophysics* 34, 237–256.

Reiter, R.J., Burkhardt, S., Cabrera, J. and Garcia, J.J. (2002) Beneficial neurobiological effects of melatonin under conditions of increased oxidative stress. *Current Medical Chemistry – Central Nervous System Agents* 2, 45–58.

Reiter, R.J., Manchester, L.C. and Tan, D.-X. (2005) Melatonin in walnuts: influence on levels of melatonin and total antioxidant capacity of blood. *Nutrition* 21, 920–924.

Reiter, R.J., Tan, D.-X., Manchester, L.C., Simopoulos, A.P., Maldonado, M.D., Flores, L.J. and Terron, M.P. (2007) Melatonin in edible plants (phytomelatonin): identification, concentrations, bioavailability and proposed functions. *World Review of Nutrition and Dietetics* 97, 211–230.

Reppert, S. and Weaver, D. (2002) Coordination of circadian timing in mammals. *Nature* 418, 935–941.

Reppert, S., Weaver, D. and Ebisawa, T. (1994) Cloning and characterization of a mammalian melatonin receptor that mediates reproductive and circadian responses. *Neuron* 13, 1177–1185.

Reppert, S., Godson, C., Mahle, C., Weaver, D., Slaugenhaupt, S. and Gusella, J. (1995) Molecular characterization of a second melatonin receptor expressed in human retina and brain: the Mel1b receptor. *Proceedings of National Academy of Sciences of the United States of America* 92, 8734–8738.

Ribelayga, C., Pevet, P. and Simonneaux, V. (1997) Adrenergic and peptidergic regulations of hydroxyindole-*O*-methyltransferase in rat pineal gland. *Brain Research* 777, 247–250.

Ribelayga, C., Pevet, P. and Simonneaux, V. (2000) HIOMT drives the photoperiodic change in the amplitude of the melatonin peak in the Siberian hamster. *American Journal of Physiology* 278, R1339–R1345.

Ritzenthaler, T., Lhommeau, I., Douillard, S., Cho, T.H., Brun, J., Patrice, T., Nighoghossian, N. and Claustrat, B. (2013) Dynamics of oxidative stress and urinary excretion of melatonin and its metabolites during acute ischemic stroke. *Neuroscience Letters* 544, 1–4.

Rodríguez, C., Mayo, JC. Sainz, R.M., Antolín, I., Herrera, F., Martin, V. and Reiter, R.J. (2004) Regulation of antioxidant enzymes: a significant role for melatonin. *Journal of Pineal Research* 36, 1–9.

Rodríguez, C., Martin, V., Herrera, F., García-Santos, G., Rodríguez-Blanco, J., Casado-Zapico, S., Sanchez-Sanchez, A., Suárez, S., Puente-Moncada, N., Anítua, M. and Antolín, I. (2013) Mechanisms involved in the pro-apoptotic effect of melatonin in cancer cells. *International Journal of Molecular Sciences* 14, 6597–6613.

Rodríguez, I., Mazuruk, K., Schoen, T. and Chader, G. (1994) Structural analysis of the human hydroxyindole-*O*-methyltransferase gene. *The Journal of Biological Chemistry* 269, 31969–31977.

Rodrígues de Lima, V., Morfim, M., Teixeira, A. and Creczynski-Pasa, T. (2004) Relationship between the action of reactive oxygen and nitrogen species on bilayer membranes and antioxidants. *Chemistry and Physics of Lipids* 132, 197–208.

Rodrígues de Lima, V., Balparda-Caro, M., Bruno-Tavares, M. and Creczynski-Pasa, T. (2007) Melatonin location in egg phosphatidylcholine liposomes: possible relation to its antioxidant mechanisms. *Journal of Pineal Research* 43, 276–282.

Rodríguez-Naranjo, M., Gil-Izquierdo, A., Troncoso, A., Cantos, E. and Garcia-Parrilla, M. (2011) Melatonin: a new bioactive compound in wine. *Journal of Food Composition and Analysis* 24, 603–608.

Roseboom, P., Coon, S., Baler, R., McCune, S., Weller, J. and Klein, D. (1996) Melatonin synthesis: analysis of the more than 150-fold nocturnal increase in serotonin *N*-acetyltransferase messenger ribonucleic acid in the rat pineal gland. *Endocrinology* 137, 3033–3045.

Rosen, J., Than, N., Koch, D., Poeggeler, B., Laatsch, H. and Hardeland, R. (2006) Interactions of melatonin and its metabolites with the ABTS cation radical: extension of the radical scavenger cascade and formation of a novel class of oxidation products, C-2substituted 3-indolinones. *Journal of Pineal Research* 41, 374–378.

Rozov, S. (2008) Features of melatonin catabolism in chicks. *Neurochemical Journal* 2, 188–192.

Rudd, E. and Thanassi, J. (1981) Inhibition of AADC by coenzyme amino acid adducts. *Biochemistry* 20, 7469–7475.

Saavedra, J. (1974) Enzymatic isotopic assay for the presence of β-phenylethylamine in brain. *Journal of Neurochemistry* 22, 211–216.

Sae-Teaw, M., Johns, J., Johns, N.P. and Subongkot, S. (2012) Serum melatonin levels and antioxidant capacities after consumption of pineapple, orange, or banana by healthy male volunteers. *Journal of Pineal Research* 55, 58–64.

Sánchez-Hidalgo, M., de la Lastra, C., Carrascosa-Salmoral, M., Naranjo, M., Gómez-Corvera, A., Caballero, B. and Guerrero, J. (2009) Age-related changes in melatonin synthesis in rat extrapineal tissues. *Experimental Gerontology* 44, 328–334.

Saper, C., Scammell, T. and Lu, J. (2005) Hypothalamic regulation of sleep and circadian rhythms. *Nature* 437, 1257–1263.

Sarropoulou, V., Dimassi-Theriou, K., Therios, I. and Koukourikou-Petridou, M. (2012a) Melatonin enhances root regeneration, photosynthetic pigments, biomass, total carbohydrates and proline content in the cherry rootstock PHL-C (*Prunus avium* × *Prunus cerasus*). *Plant Physiology and Biochemistry* 61, 162–168.

Sarropoulou, V., Therios, I. and Dimassi-Theriou, K. (2012b) Melatonin promotes adventitious root regeneration in *in vitro* shoot tip explants of the commercial sweet cherry rootstocks CAB-6P (*Prunus cerasus* L.), Gisela 6 (*P. cerasus* × *P. canescens*), and M×M 60 (*P. avium* × *P.mahaleb*). *Journal of Pineal Research* 52, 38–46.

Schaefer, M. and Hardeland, R. (2008) The melatonin metabolite *N*-acetyl-5-methoxykynuramine is a potent singlet oxygen scavenger. *Journal of Pineal Research* 46, 49–52.

Schaffazick, S., Pohlmann, A., de Cordova, C., Creczynski-Pasa, T. and Guterres, S. (2005) Protective properties of melatonin-loaded nanoparticles against lipid peroxidation. *International Journal of Pharmaceutics* 289, 209–213.

Scheibner, K., De Angelis, J., Burley, S. and Cole, P. (2002) Investigation of the roles of catalytic residues in serotonin *N*-acetyltransferase. *The Journal of Biological Chemistry* 277, 18118–18126.

Schernhammer, E., Laden, F., Speizer, F., Willet, W., Hunter, D., Kawachi, I., Fuchs, C. and Colditz, G. (2003) Night-shift work and risk of colorectal cancer in the nurses' health study. *Journal of National Cancer Institute* 95, 825–828.

Schomerus, C., Korf, H.-W., Laedtke, E., Weller, J.L. and Klein, D.C. (2000) Selective adrenergic/cyclic AMP dependent switch-off of proteasomal proteolysis alone switches on neural signal transduction: an example from the pineal gland. *Journal of Neurochemistry* 75, 2123–2132.

Schröder, P., Abele, C., Gohr, P., Stuhlfauth-Roisch, U. and Grosse, W. (1999) Latest on the enzymology of serotonin biosynthesis in walnut seeds. *Advances in Experimental Medicine and Biology* 467, 637–644.

Schuster, C., Gauer, F., Malan, A., Recio, J., Pévet, P. and Masson-Pevet, M. (2001) The circadian clock, light/dark cycle and melatonin differentially involved in the expression of daily and photoperiodic variations in MT1 melatonin receptors in the Siberian and Syrian hamsters. *Neuroendocrinology* 74, 55–68.

Seely, D., Wu, P., Fritz, H., Kennedy, D.A., Tsui, T., Seely, A.J.E. and Mills, E. (2012) Melatonin as adjuvant cancer care with and without chemotherapy: a systematic review and meta-analysis of randomized trials. *Integrative Cancer Therapies* 11, 293–303.

Selmar, D. and Kleinwächter, M. (2013) Stress enhances the synthesis of secondary plant products: the impact of stress-related over-reduction on the accumulation of natural products. *Plant and Cell Physiology* 54, 817–826.

Semak, I., Naumova, M., Korik, E., Terekhovich, V., Worstman, J. and Slominski, A. (2005) A novel metabolic pathway of melatonin: oxidation by cytochrome c. *Biochemistry* 44, 9300–9307.

Shen, S., Brémont, B., Serraz, I., Andrieux, J., Poncet, A., Mathé-Allainmat, M., Chanut, E., Trouvin, J. and Langlois, M. (1996) Structure–activity relationships for substrates and inhibitors of pineal 5-hydroxytryptamine-*N*-acetyltransferase: preliminary studies. *European Journal of Pharmacology* 307, 133–140.

Shirota, K. and Fujisawa, H. (1988) Purification and characterization of aromatic L-amino acid decarboxylase from rat kidney and monoclonal antibody to the enzyme. *Journal of Neurochemistry* 51, 428–434.

Silver, R. and Schwartz, W. (2005) The suprachiasmatic nucleus is a functionally heterogenous timekeeping organ. *Methods of Enzymology* 393, 451–465.

Simonneaux, V. and Ribelayga, C. (2003) Generation of the melatonin endocrine message in mammals: a review of the complex regulation of melatonin synthesis by norepinephrine, peptides, and other pineal transmitters. *Pharmacological Reviews* 55, 325–395.

Simopoulos, A.P., Tan, D.-X., Manchester, L.C. and Reiter, R.J. (2005) Purslane: a plant source of omega-3 fatty acids and melatonin. *Journal of Pineal Research* 39, 331–332.

Slominski, A., Baker, J., Rosano, T., Guisti, L., Ermak, G., Grande, M. and Gaudet, S. (1996) Metabolism of serotonin to *N*-acetylserotonin, melatonin, and 5-methoxytryptamine in hamster skin culture. *The Journal of Biological Chemistry* 271, 12281–12286.

Slominski, A., Pisarchik, A., Semak, I., Sweatman, T., Worstman, J., Szczesniewski, A., Slugocki, G., McNulty, J., Kauser, S., Tobin, D., Jing, C. and Johansson, O. (2002) Serotoninergic and melatoninergic systems are fully expressed in human skin. *The FASEB Journal* 16, 896–898.

Slominski, A., Tobin, D., Zmijewski, M., Worstman, J. and Paus, R. (2007) Melatonin in the skin: synthesis, metabolism and functions. *Trends in Endocrinology and Metabolism* 19, 17–24.

Sprenger, J., Hardeland, R., Fuhrberg, B. and Han, S. (1999) Melatonin and other 5-methoxylated indoles in yeast: presence in high concentrations and dependence on tryptophan availability. *Cytologia* 64, 209–213.

Srinivasan, V., Pandi-Perumal, S., Maestroni, G., Esquifino, A., Hardeland, R. and Cardinali, D. (2005) Role of melatonin in neurodegenerative diseases. *Neurotoxicity Research* 7, 293–318.

Stefulj, J., Hortner, M., Ghosh, M., Schauenstein, K., Rinner, I., Wolfler, A., Semmler, J. and Liebmann, P. (2001) Gene expression of the key enzymes of melatonin synthesis in extrapineal tissues of the rat. *Journal of Pineal Research* 30, 243–247.

Stege, P.W., Sombra, L.L., Messina, G., Martinez, L.D. and Silva, M.a.F. (2010) Determination of melatonin in wine and plant extracts by capillary electrochromatography with immobilized carboxylic multi-walled carbon nanotubes as stationary phase. *Electrophoresis* 31, 2242–2248.

Stehle, J., von Gall, C. and Korf, H. (2003) Melatonin: a clock-output, a clock-input. *Journal of Neuroendocrinology* 15, 383–389.

Stevens, L., Blom, T. and Verpoorte, R. (1993) Subcellular localization of tryptophan decarboxylase, strictosidine synthase and strictosidine glucosidase in suspension cultured cells of *Catharanthus roseus* and *Tabernaemontana divaricata*. *Plant Cell Reports* 12, 573–576.

Sturtz, M., Cerezo, A., Cantos-Villar, E. and Garcia-Parrilla, M. (2011) Determination of the melatonin content of different varieties of tomatoes (*Lycopersicon esculentum*) and strawberries (*Fragaria ananassa*). *Food Chemistry* 127, 1329–1334.

Sudgen, D., Cena, V. and Klein, D. (1987) Hydroxyindole-*O*-methyltransferase. *Methods in Enzymology* 142, 592–596.

Sugden, D. (2003) Comparison of circadian expression of tryptophan hydroxylase isoform mRNAs in the rat pineal gland using real-time PCR. *Journal of Neurochemistry* 86, 1308–1311.

Takahashi, T., Sasaki, M., Itoh, H., Ozone, M., Yamadera, W., Hayshida, K.I., Ushijima, S., Matsunaga, N., Obuchi, K. and Sano, H. (2000) Effect of 3 mg melatonin on jet lag syndrome in an 8-h eastward flight. *Psychiatry and Clinical Neurosciences* 54, 377–378.

Tal, O., Haim, A., Harel, O. and Gerchman, Y. (2011) Melatonin as an antioxidant and its semi-lunar rhythm in green macroalga *Ulva* sp. *Journal of Experimental Botany* 62, 1903–1910.

Tamarkin, I., Danforth, D., Lichter, A., DeMoss, E., Cohen, M., Chabner, B. and Lippman, M. (1982) Decreased nocturnal plasma melatonin peak in patients with estrogen receptor positive breast cancer. *Science* 216, 1005.

Tan, D.-X., Poeggeler, B., Reiter, R.J., Chen, L.D., Chen, S., Lucien, M.C. and Barlow-Walden, L.R. (1993a) The pineal hormone melatonin inhibits DNA-adduct formation induced by the chemical carcinogen safrole *in vivo*. *Cancer Letters* 70, 65–71.

Tan, D.-X., Chen, L.D., Poeggeler, B., Manchester, L.C. and Reiter, R.J. (1993b) Melatonin: a potent, endogenous hydroxyl radical scavenger. *Endocrine Journal* 1, 57–60.

Tan, D.-X., Manchester, L.C., Reiter, R.J., Plummer, B.F., Hardies, L.J., Weintraub, S.T., Vijayalaxmi and Shepherd, A.M.M. (1998) A novel melatonin metabolite, cyclic 3-hydroxymelatonin: a biomarker of *in vivo* hydroxyl radical generation. *Biochemical and Biophysical Research Communications* 253, 614–620.

Tan, D.-X., Manchester, L.C., Reiter, R.J., Qi, W., Zhang, M., Weintraub, S.T., Cabrera, J., Saínz, R.M. and Mayo, J.C. (1999) Identification of highly elevated levels of melatonin in bone marrow: its origin and significance. *Biochimica et Biophysica Acta* 1472, 206–214.

Tan, D.-X., Manchester, L.C., Reiter, R.J., Plummer, B.F., Limson, F., Weintraub, S.T. and Qi, W. (2000a) Melatonin directly scavenges hydrogen peroxide: a potentially new metabolic pathway of melatonin biotransformation. *Free Radical Biology and Medicine* 29, 1177–1185.

Tan, D.-X., Manchester, L.C., Reiter, R.J., Qi, W., Karbownik, M. and Calvo, J.R. (2000b) Significance of melatonin in antioxidative defense system: reactions and products. *Biological Signals and Receptors* 9, 137–159.

Tan, D.-X., Manchester, L.C., Burkhardt, S., Sainz, R.M., Mayo, J.C., Kohen, R., Shohami, E., Huo, Y.-S., Hardeland, R. and Reiter, R.J. (2001) AFMK, a biogenic amine and melatonin metabolite, functions as a potent antioxidant. *The FASEB Journal* 15, 2294–2296.

Tan, D.-X., Reiter, R.J., Manchester, L.C., Yan, M.T., El-Sawi, M., Sainz, R.M., Mayo, J.C., Kohen, R., Allegra, M.C. and Hardeland, R. (2002) Chemical and physical properties and potential mechanisms: melatonin as a broad spectrum antioxidant and free radical scavenger. *Current Topics in Medicinal Chemistry* 2, 181–197.

Tan, D.-X., Hardeland, R., Manchester, L.C., Poeggeler, B., López-Burillo, S., Mayo, J.C., Sainz, R.M. and Reiter, R.J. (2003) Mechanistic and comparative studies of melatonin and classic antioxidants in terms of their interactions with the ABTS cation radical. *Journal of Pineal Research* 34, 249–259.

Tan, D.-X., Manchester, L.C., Di Mascio, P., Martinez, G.R., Prado, F.M. and Reiter, R.J. (2007a) Novel rhythms of *N*[1]-acetyl-*N*[2]-formyl-5-methoxykynuramine and its precursor melatonin in water hyacinth: importance for phytoremediation. *The FASEB Journal* 21, 1724–1729.

Tan, D.-X., Manchester, L.C., Helton, P. and Reiter, R.J. (2007b) Phytoremediative capacity of plants enriched with melatonin. *Plant Signaling and Behavior* 2, 514–516.

Tan, D.-X., Manchester, L.C., Terron, M.P., Flores, L.J. and Reiter, R.J. (2007c) One molecule, many derivates: a never-ending interaction of melatonin with reactive oxygen and nitrogen species? *Journal of Pineal Research* 42, 28–42.

Tan, D.-X., Hardeland, R., Manchester, L.C., Korkmaz, A., Ma, S., Rosales-Corral, S. and Reiter, R.J. (2012) Functional roles of melatonin in plants, and perspectives in nutritional and agricultural science. *Journal of Experimental Botany* 63, 577–597.

Tan, D.-X., Manchester, L.C., Liu, X., Rosales-Corral, S.A., Acuña-Castroviejo, D. and Reiter, R.J. (2013) Mitochondria and chloroplasts as the original sites of melatonin synthesis: a hypothesis related to melatonin's primary function and evolution in eukaryotes. *Journal of Pineal Research* 54, 127–138.

Tanaka, T., Horio, Y., Taketoshi, M., Imamura, I., Ando-Yamamoto, M., Kangawa, K., Matsuo, H., Kuroda, M. and Wasa, H. (1989) Molecular cloning and sequencing of a cDNA of DOPA decarboxylase: partial amino acid homologies with other enzymes synthesising catecholamines. *Proceedings of the National Academy of Sciences of the United States of America* 86, 8142–8146.

Taylor, P. and Pocock, S. (1972) Mortality of shift and day workers 1956–68. *British Journal of Industrial Medicine* 29, 201–207.

Teixeira, A., Morfim, M., de Cordova, C., Charao, C., Rodrigues de Lima, V. and Creczynski-Pasa, T. (2003) Melatonin protects against pro-oxidant enzymes and reduces lipid peroxidation in distinct membranes induced by the hydroxyl and ascorbyl radicals and by peroxynitrite. *Journal of Pineal Research* 35, 262–268.

Teraoka, H., Russell, C., Regan, J., Chandrasekhar, A., Concha, M. and Yokoyama, R. (2013) Hedgehog and Fgf signaling pathways regulate the development of tphR-expressing serotonergic raphe neurons in zebrafish embryos. *Journal of Neurobiology* 60, 275–288.

Terron, M., Marchena, J., Shadi, F., Harvey, S., Lea, R. and Rodriguez, A. (2001) Melatonin: an antioxidant at physiological concentrations. *Journal of Pineal Research* 31, 95–96.

Tilden, A.R., Becker, M.A., Amma, L.L., Arciniega, J. and McGaw, A.K. (1997) Melatonin production in an aerobic photosynthetic bacterium: an evolutionarily early association with darkness. *Journal of Pineal Research* 22, 102–106.

Tiryaki, I. and Keles, H. (2012) Reversal of the inhibitory effect of light and high temperature on germination of *Phacelia tanacetifolia* seeds by melatonin. *Journal of Pineal Research* 52, 332–339.

Tomás-Zapico, C. and Coto-Montes, A. (2005) A proposed mechanism to explain the stimulatory effect of melatonin on antioxidative enzymes. *Journal of Pineal Research* 39, 99–104.

Udenfriend, S., Clark, C. and Titus, E. (1953) 5-Hydroxytryptophan decarboxylase: a new route of metabolism of tryptophan. *Journal of the American Chemical Society* 75, 501–502.

van Tassel, D. (1997) Identification and quantification of melatonin in higher plants. PhD thesis, University of California, Davis, California.

van Tassel, D., and O'Neill, S. (1993) Melatonin: identification of a potential dark signal in plants. *Plant Physiology* 102(supplement 1), article 659.

van Tassel, D., Roberts, N. and O'Neill, S. (1995) Melatonin from higher plants: isolation and identification of *N*-acetyl-5-methoxytryptamine. *Plant Physiology* 108(supplement 2), article 101.

van Tassel, D., Roberts, N., Lewy, A. and O'Neill, S. (2001) Melatonin in plant organs. *Journal of Pineal Research* 31, 8–15.

Venegas, C., García, J.A., Escames, G., Ortiz, F., López, A., Doerrier, C., García-Corzo, L., López, L.C., Reiter, R.J. and Acuña-Castroviejo, D. (2012) Extrapineal melatonin: analysis of its subcellular distribution and daily fluctuations. *Journal of Pineal Research* 52, 217–227.

Vitalini, S., Gardana, C., Zanzotto, A., Fico, G., Faoro, F., Simonetti, P. and Iriti, M. (2011a) From vineyard to glass: agrochemicals enhance the melatonin and total polyphenol contents and antiradical activity of red wines. *Journal of Pineal Research* 51, 278–285.

Vitalini, S., Gardana, C., Zanzotto, A., Simonetti, P., Faoro, F. and Iriti, M. (2011b) The presence of melatonin in grapevine (*Vitis vinifera* L.) berry tissues. *Journal of Pineal Research* 51, 331–337.

Vivien-Roels, B., Pévet, P., Beck, O. and Fevre-Montange, M. (1984) Identification of melatonin in the compound eyes of an insect, the locust (*Locusta migratoria*) by radioimmunoassay and gas chromatography-mass spectrometry. *Neuroscience Letter* 49, 153–157.

Vogt, T. (2010) Phenylpropanoid biosynthesis. *Molecular Plant* 3, 2–20.

Voisin, P., Namboodiri, M. and Klein, D. (1984) Arylamine *N*-acetyltransferase and arylalkylamine N-acetyltransferase in the mammalian pineal gland. *The Journal of Biological Chemistry* 259, 10913–10918.

Voisin, P., Guerlotte, J., Bernard, M., Collin, J. and Cogne, M. (1992) Molecular cloning and nucleotide sequence of a cDNA encoding hydroxyindole-*O*-methyltransferase from chicken pineal gland. *Biochemical Journal* 282, 571–576.

von Gall, C., Stehle, J. and Weaver, D. (2002) Mammalian melatonin receptors: molecular biology and signal transduction. *Cell and Tissue Research* 309, 151–162.

Walther, D. and Bader, M. (2003) A unique central tryptophan hydroxylase isoform. *Biochemical Pharmacology* 66, 1673–1680.

Walther, D., Peter, J., Bashammakh, S., Hortnagl, H., Voits, M. and Fink, H. (2003) Synthesis of serotonin by a second tryptophan hydroxylase isoform. *Science* 299, 76.

Wang, L., Erlandsen, H., Haavik, J., Knappskog, P. and Stevens, R. (2002) Three-dimensional structure of human tryptophan hydroxylase and its implications for the biosynthesis of the neurotransmitters serotonin and melatonin. *Biochemistry* 41, 12569–12574.

Wang, P., Yin, L., Liang, D., Li, C., Ma, F. and Yue, Z. (2012) Delayed senescence of apple leaves by exogenous melatonin treatment: toward regulating the ascorbate-glutathione cycle. *Journal of Pineal Research* 53, 11–20.

Wang, P., Sun, X., Li, C., Wei, Z., Liang, D. and Ma, F. (2013) Long-term exogenous application of melatonin delays drought-induced leaf senescence in apple. *Journal of Pineal Research* 54, 292–302.

Waterhouse, J., Atkinson, G. and Reilly, T. (1997) Jet lag. *The Lancet* 350, 1611–1616.

Weissbach, H., Bogdanski, D., Redfield, B. and Udenfriend, S. (1957) Studies on the effect of vitamin B_6 on 5-hydroxytryptamine (serotonin) formation. *The Journal of Biological Chemistry* 227, 617–624.

Weissbach, H., Redfield, B. and Axelrod, J. (1960) Biosynthesis of melatonin: enzyme conversion of serotonin to *N*-acetylserotonin. *Biochimica et Biophysica Acta* 43, 352–353.

Wilhelmsen, M., Amiriam, I., Reiter, R., Rosenberg, J. and Gögenur, I. (2011) Analgesic effects of melatonin: a review of current evidence from experimental and clinical studies. *Journal of Pineal Research* 51, 270–277.

Willis, G. and Robertson, A. (2004) Recovery of experimental Parkinson's disease with melatonin analogues ML-23 and S-20928 in a chronic, bilateral 6-OHDA model: a new mechanism involving antagonism of the melatonin receptor. *Pharmacology, Biochemistry and Behavior* 79, 413–429.

Windahl, M., Petersen, C., Christensen, H. and Harris, P. (2008) Crystal structure of tryptophan hydroxylase with bound amino acid substrate. *Biochemistry* 47, 12087–12094.

Witt-Enderby, P., Radio, N., Doctor, J. and Davies, V. (2006) Therapeutics treatments potentially mediated by melatonin receptors: potential clinics uses in the prevention of osteoporosis, cancer and as an adjuvant therapy. *Journal of Pineal Research* 41, 297–305.

Wolf, E., De Angelis, J., Khalil, E., Cole, P. and Burley, S. (2002) X-ray crystallographic studies of serotonin *N*-acetyltransferase catalysis and inhibition. *Journal of Molecular Biology* 317, 215–224.

Wolf, K., Kolar, J., Witters, E., van Dongen, W., van Onckelen, H. and Machackova, I. (2001) Daily profile of melatonin levels in *Chenopodium rubrum* L. depends on photoperiod. *Journal of Plant Physiology* 158, 1491–1493.

Wu, Y. and Swaab, D. (2005) The human pineal gland and melatonin in aging and Alzheimer's disease. *Journal of Pineal Research* 38, 145–152.

Ximenes, V., Catalani, N. and Campa, A. (2007) The effect of pH on HRP-catalyzed oxidation of melatonin: production of AFMK versus radical-mediated degradation. *Journal of Pineal Research* 42, 291–296.

Yamazaki, Y., Sudo, H., Yamazaki, M., Aimi, N. and Saito, K. (2003) Camptothecin biosynthesis genes in hairy roots of *Ophiorrhiza pumila*: cloning, characterization and differential expression in tissues and by stress compounds. *Plant and Cell Physiology* 44, 395–403.

Zang, L., Cosma, G., Gardner, H. and Vallyathan, V. (1998) Scavenging of reactive oxygen species by melatonin. *Biochimica et Biophysica Acta* 1425, 469–477.

Zawilska, J.B., Rosiak, J., Vivien-Roels, B., Skene, D.J., Pévet, P. and Nowak, J.Z. (2002) Daily variation in the concentration of 5-methoxytryptophol and melatonin in the duck pineal gland and plasma. *Journal of Pineal Research* 32, 214–218.

Zawilska, J.B., Berezinska, M., Rosiak, J., Vivien-Roels, B., Skene, D.J., Pévet, P. and Nowak, J.Z. (2003) Daily variation in the concentration of melatonin and 5-methoxytryptophol in the goose pineal gland, retina, and plasma. *General and Comparative Endocrinology* 134, 296–302.

Zhang, H., Squadrito, G., Uppu, R. and Pryor, W. (1999) Reaction of peroxynitrite with melatonin: a mechanistic study. *Chemical Research in Toxicology* 12, 526–534.

Zhang, N., Zhao, B., Zhang, H.J., Weeda, S., Yang, C., Yang, Z.C., Ren, S. and Guo, Y.D. (2013) Melatonin promotes water-stress tolerance, lateral root formation, and seed germination in cucumber (*Cucumis sativus* L.). *Journal of Pineal Research* 54, 15–23.

Zhang, X., Beaulieu, J., Sotnikova, R., Gainetdinov, R. and Caron, M. (2004) Tryptophan hydroxylase-2 controls brain serotonin synthesis. *Science* 305, 217.

Zhang, X., Beaulieu, J., Gainetdinov, R. and Caron, M. (2006) Functional polymorphisms of the brain serotonin synthesizing enzyme tryptophan hydroxylase-2. *Cellular and Molecular Life Sciences* 63, 6–11.

Zhao, Y., Qi, L., Wang, W., Saxena, P. and Liu, C. (2011) Melatonin improves the survival of cryopreserved callus of *Rhodiola crenulata*. *Journal of Pineal Research* 50, 83–88.

Zhao, Y., Tan, D.-X., Lei, Q., Chen, H., Wang, L., Li, Q.t., Gao, Y. and Kong, J. (2012) Melatonin and its potential biological functions in the fruits of sweet cherry. *Journal of Pineal Research* 55, 79–88.

Zhu, M. and Juorio, A. (1995) Aromatic L-amino acid decarboxylase: biological characterization role and functional role. *General Pharmacology* 26, 681–696.

Zilberman-Peled, B., Bransburg-Zabary, S., Klein, D.C. and Gothilf, Y. (2011) Molecular evolution of multiple arylalkylamine *N*-acetyltransferase (AANAT) in fish. *Marine Drugs* 9, 906–921.

Zill, P., Buttner, A., Eisenmenger, W., Bondy, B. and Ackenheil, M. (2004) Regional mRNA expression of a second tryptophan hydroxylase isoform in postmortem tissue samples of two human brains. *European Neuropsychopharmacology* 14, 282–284.

Zohar, R., Izhaki, I., Koplovich, A. and Ben-Shlomo, R. (2011) Phytomelatonin in the leaves and fruits of wild perennial plants. *Phytochemistry Letters* 4, 222–226.

22 Glucosinolate Biosynthesis From Amino Acids

H.U. Stotz[1]* P.D. Brown[2] and J. Tokuhisa[3]

[1]University of Hertfordshire, Hatfield, UK; [2]Trinity Western University, Langley, British Columbia, Canada; [3]Virginia Tech, Blacksburg, Virginia, USA

22.1 Abstract

Glucosinolates are natural plant products that are derived from aliphatic and aromatic amino acids by certain plants. Glucosinolates may be metabolized to a variety of potentially toxic products via myrosinase-catalysed hydrolysis. The enzyme and substrates are sequestered in different cell compartments and come into contact only upon tissue disruption. Glucosinolate-derived breakdown products are biologically active and influence interactions with insects and microbes as well as human health. In addition, indole glucosinolates contribute to the innate immune system of plants via the regulation of callose deposition in response to pathogen infection. Transcriptional regulation of glucosinolate biosynthesis is an important area of research, as is the spatial distribution of glucosinolates using imaging techniques.

22.2 Introduction: Evolution of Glucosinolate Biosynthesis

Glucosinolates are synthesized from two groups of primary amino acids; the aliphatic amino acids methionine (Met) and alanine (Ala) together with the branched-chain amino acids (BCAAs) leucine (Leu), isoleucine (Ile) and valine (Val) as well as the aromatic amino acids phenylalanine (Phe), tyrosine (Tyr) and tryptophan (Trp) can act as precursors for glucosinolate biosynthesis (Halkier and Gershenzon, 2006). Glucosinolate biosynthesis is a derived trait, occurring only in the order *Brassicales* and a single distantly related genus, *Drypetes*, in the family *Euphorbiaceae* of the order *Malpighiales* (Fig. 22.1a). Glucosinolate biosynthesis is probably derived from cyanogenic glycoside biosynthesis. The two pathways share similar amino acid precursors. The occurrence of cyanogenic glycosides derived from BCAAs and Phe is widespread in the plant kingdom. The basal members of the *Brassicales* consistently produce glucosinolates from BCAAs and Phe, whereas the more derived members belonging to the core *Brassicales* also produce chain-elongated glucosinolates from Met (Mithen *et al.*, 2010). Amino acid precursors are converted to aldoximes by cytochrome P450 CYP79 enzymes in both the glucosinolate and cyanogenic glycoside biosynthetic pathways (Fig. 22.1b). The metabolism of this aldoxime via the action of a CYP71 enzyme results in the formation of hydroxynitrile, another intermediate in the biosynthesis of cyanogenic glycosides. The CYP83 enzymes, which metabolize aldoxime in the glucosinolate biosynthetic pathway,

*E-mail address: h.stotz@herts.ac.uk

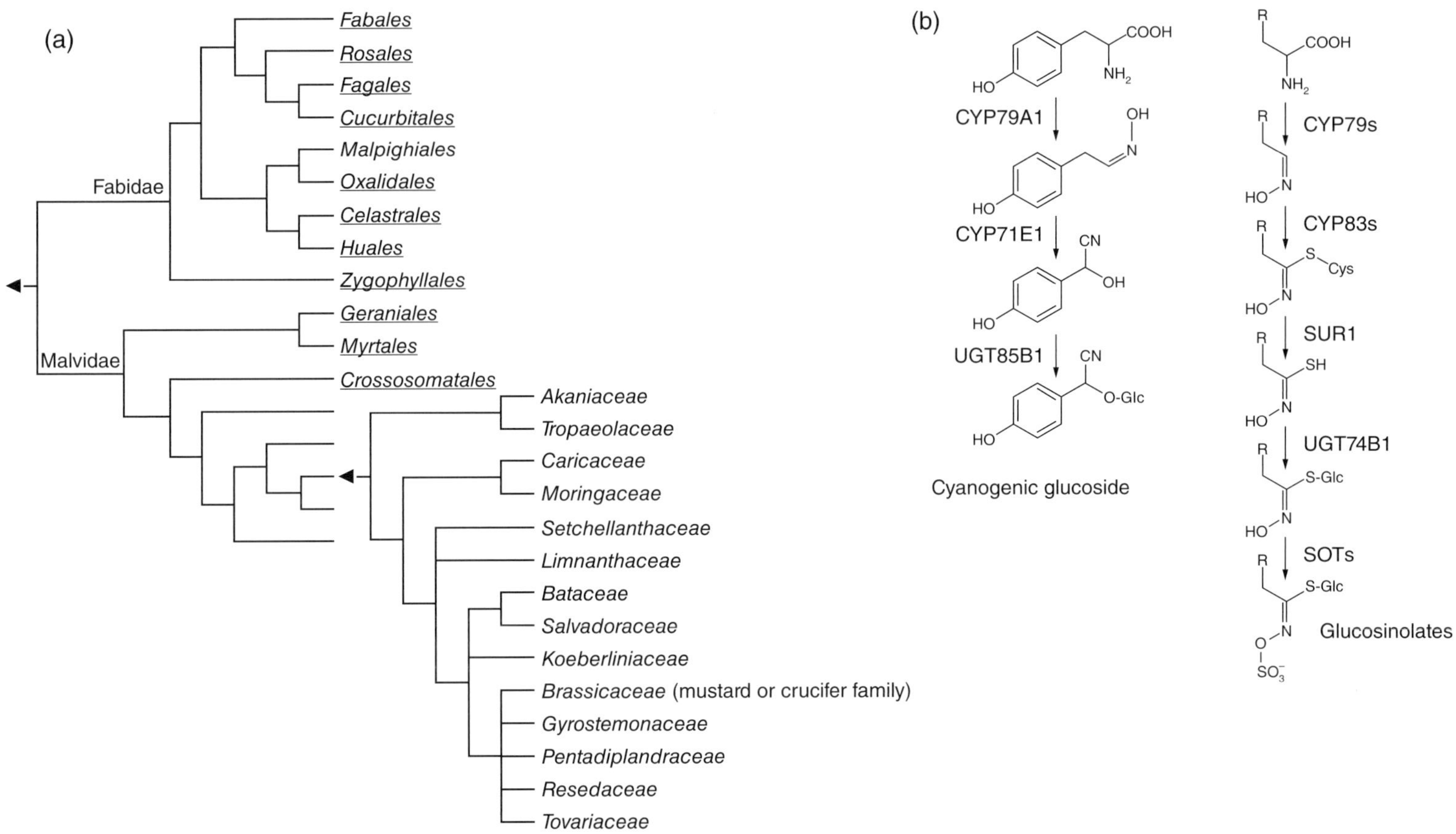

Fig. 22.1. The evolution of glucosinolate biosynthesis. (a) Glucosinolate (GSL) biosynthesis is a monophyletic trait as only plants in the order *Brassicales* produce these natural products. The only other plant producing GSL is *Drypetes* in the family *Euphorbiaceae* and order *Malpighiales*. Glucosinolate-producing orders and families are in plain italic and other orders are in italic and underlined. The phylogeny is based on the tree of life (http://tolweb.org/tree). (b) Comparison of cyanogenic glucoside and glucosinolate biosynthetic pathways. Enzyme names are shown next to arrows. In both pathways, an aldoxime is produced by the action of CYP79 (cytochrome P450) biosynthetic enzymes, but the metabolism of this intermediate differs between the two pathways. Abbreviations: CYP71E1, a cytochrome P450 enzyme; CYP83s, cytochrome P450 enzymes that have recently been merged into the CYP71 group of enzymes; SOTs, sulfotransferases; SUR1, SUPERROOT1 C-S lyase; UGT74B1, glucosyltransferase; UGT85B1, cyanohydrin glucosyltransferase. See text for further information.

have now been merged into the ever-expanding CYP71 family (Nelson and Werck-Reichhart, 2011). It is presumed that a mutation in the gene of a CYP71 enzyme resulted in the production of a toxic *aci*-nitro compound instead of a hydroxynitrile. Subsequent recruitment of detoxification enzymes such as glutathione-*S*-transferases, glucosyltransferases and sulfotransferases then resulted in a controlled disposal of the final glucosinolate products (Halkier and Gershenzon, 2006).

22.3 Cellular and Tissue Distribution of Glucosinolate Metabolism

Glucosinolate biosynthesis occurs in different cellular compartments (Fig. 22.2a). The first step in aliphatic glucosinolate biosynthesis is a transamination reaction that occurs in the cytoplasm and is catalysed by branched-chain amino acid transaminases (BCATs), for example BCAT4 in the case of methionine (Schuster *et al.*, 2006). The resulting product, an α-keto acid, can enter the chain elongation pathway, which consists of a condensation with acetyl-CoA catalysed by methylthioalkylmalate (MAM) synthases such as MAM1 and MAM3 (Textor *et al.*, 2007). The gene for methylthioalkylmalate synthase 1 (*MAM1*) was identified by Kroymann *et al.* (2001) and shown to encode a chloroplast enzyme (Textor *et al.*, 2007). Isopropylmalate isomerases (IPMIs) are heterodimeric enzymes that are located in the plastids (Knill *et al.*, 2009). The action of isopropylmalate dehydrogenase (IPMDH) completes the chain elongation pathway, resulting in a net gain of one carbon (Sawada *et al.*, 2009a). The transaminase BCAT3 converts the chain-elongated keto acid into an amino acid derivative that is exported from the chloroplast into the cytoplasm (Knill *et al.*, 2008). The bile acid:sodium symporter family protein 5 is responsible for the transport of precursors to and/or products out of plastids to accomplish side chain elongation of glucosinolate precursors (Sawada *et al.*, 2009b). The subsequent steps of biosynthesis are shared among aliphatic and aromatic glucosinolates and take place at the endoplasmic reticulum.

The *CYP79* gene family encodes cytochrome P450 enzymes that catalyse two consecutive *N*-hydroxylations of amino acids, followed by dehydration and decarboxylation reactions. Several members of this family are involved in glucosinolate biosynthesis. CYP79B2 and CYP79B3 convert tryptophan to indole-3-acetaldoxime, which is a central metabolic branch point (Glawischnig *et al.*, 2004). CYP79A2 is a benzyl glucosinolate biosynthetic enzyme converting phenylalanine (Wittstock and Halkier, 2000). CYP79F1 and CYP79F2 are involved in aliphatic glucosinolate biosynthesis (Chen *et al.*, 2003). However, CYP79D2 from the cyanogenic plant cassava metabolizes Val and Ile to the corresponding oximes which, in glucosinolate-producing species, can promote aliphatic glucosinolate biosynthesis (Mikkelsen and Halkier, 2003). CYP83 family members catalyse the next step in glucosinolate biosynthesis. CYP83B1 catalyses the initial conversion of indole-3-acetaldoxime to the corresponding *S*-alkyl-thiohydroximate in indole glucosinolate biosynthesis (Bak *et al.*, 2001). In the absence of thiol compounds, the product of the enzymatic reaction inactivates the enzyme. CYP83A1 has the corresponding activity in aliphatic glucosinolate biosynthesis (Naur *et al.*, 2003). The final steps in glucosinolate biosynthesis take place in the cytoplasm.

Glutathione-*S*-transferases are thought to conjugate the *aci*-nitro compound to glutathione, but definitive enzymes for this step have not yet been identified. Based on transcriptomic investigations, the genes for *GSTU20* and *GSTF11* were suggested to be involved in glucosinolate biosynthesis by Hirai *et al.* (2005, 2007). *SUR1* encodes a C-S lyase, which converts *S*-alkylthiohydroximate to thiohydroximate (Mikkelsen *et al.*, 2004). Excessive root production occurs in the *sur1* mutant because of an increase in auxin biosynthesis (Sugawara *et al.*, 2009). The glucosyltransferase UGT74B1 generates desulfoglucosinolate by forming a thioglucoside bond (Grubb *et al.*, 2004). The final step in glucosinolate biosynthesis is then carried out by sulfotransferases (SOTs), with substrate specificity varying among SOT isoforms (Klein and Papenbrock, 2009).

The profiles of glucosinolates at the suborgan and cellular levels were identified in crucifers using cryosectioning-scanning electron microscopy-energy dispersive X-ray analysis (Koroleva *et al.*, 2010); specific glucosinolates were detected using the recently applied technique of

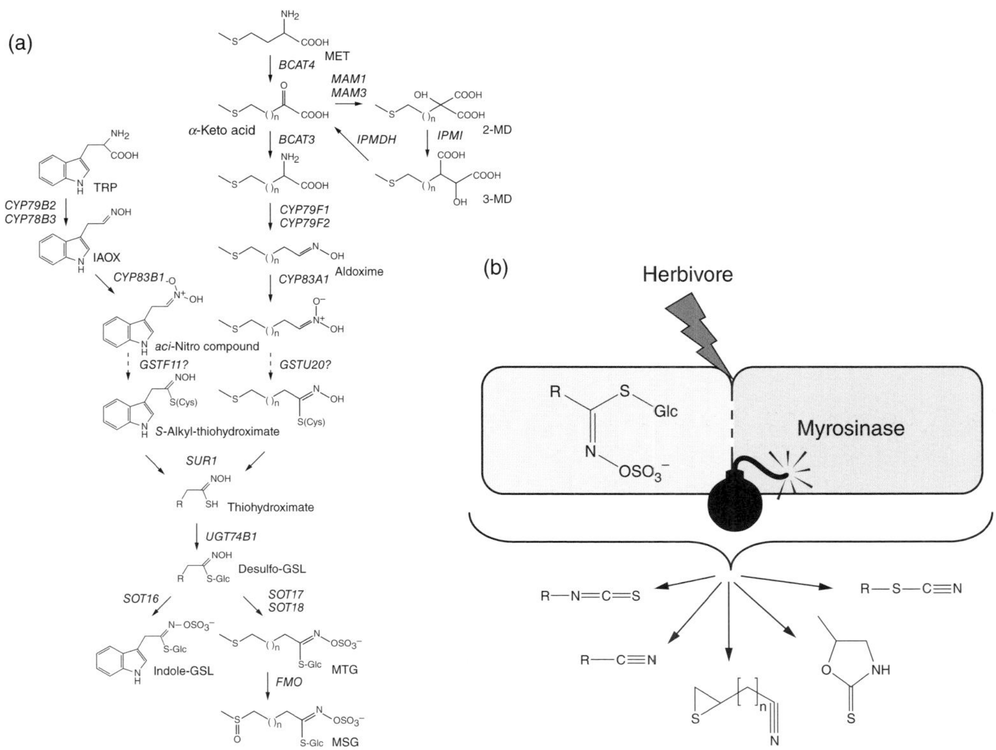

Fig. 22.2. Glucosinolate (GSL) metabolism. (a) Biosynthesis of aromatic and aliphatic glucosinolates (GSLs). Enzyme names are shown in italics next to arrows. Abbreviations: BCAT3, branched-chain amino acid transaminase 3; BCAT4, branched-chain amino acid transaminase 4; CYP enzymes, cytochrome P450 biosynthetic enzymes; Cys, cysteine; FMO, flavin-monooxygenase; GSTF11 and GSTU20, glutathione-*S*-transferases; IAOX, indole-3-acetaldoxime; IPMDH, isopropylmalate dehydrogenase; IPMI, isopropylmalate isomerase; 2-MD, 2-malate derivative; 3-MD, 3-malate derivative; MET, methionine; MAM1, methylthioalkylmalate synthase 1; MAM3, methylthioalkylmalate synthase 3; MSG, methylsulfinylalkyl-GSL; MTG, methylthioalkyl-GSL; SOT16, 17 and 18, sulfotransferases 16, 17 and 18; SUR1, SUPERROOT1 C-S lyase; TRP, tryptophan; UGT74B1, UDP-glucosyltransferase. See text for further information. (b) The mustard oil bomb. Myrosinase, the enzyme that hydrolyses glucosinolates, and its glucosinolate substrates are compartmentalized in separate but neighbouring cells. Chewing herbivores disrupt this compartmentalization and permit mixing of the enzyme with its substrates, resulting in the release of toxic products including isothiocyanates, nitriles, epithionitriles, oxazolidine-2-thiones and thiocyanates.

matrix-free laser desorption/ionisation mass spectrometric imaging (Shroff *et al.*, 2008). High concentrations of glucosinolates were demonstrated to accumulate in specialized S-cells (Koroleva *et al.*, 2010), which are associated peripherally with the phloem in the aerial portions of *Arabidopsis* (Koroleva *et al.*, 2010; Sarsby *et al.*, 2012). The *Arabidopsis* leaf blade has an uneven distribution of glucosinolates with high concentrations in the mid-vein and the outer lamina relative to the inner lamina (Shroff *et al.*, 2008). The elevated concentrations of glucosinolates in the S-cells and the outer lamina are due to aliphatic and indolic glucosinolates, respectively.

Several historic lines of evidence have indicated significant contributions of long-range glucosinolate transport to the profiles observed during the life cycle of *Arabidopsis* (Nour-Eldin and Halkier, 2009). Reciprocal crosses in *Brassica* or *Arabidopsis* involving parental lines with contrasting glucosinolate profiles consistently produced F_1 seed with maternal rather than intermediate profiles. The levels of transcripts of genes encoding enzymes and transcription factors associated with aliphatic glucosinolate biosynthesis in developing seeds are barely detectable compared with those in siliques. In contrast, the transcripts of genes encoding enzymes for modifying the side-chain derived groups of aliphatic glucosinolates have a much higher abundance.

A functional screen was developed for transport proteins synthesized from *Arabidopsis* cDNA co-expression constructs that facilitate glucosinolate uptake in *Xenopus laevis* oocytes (Nour-Eldin *et al.*, 2006). A member of the nitrate/peptide transporter family, designated glucosinolate transporter 1 (GTR1, At3g47960), actively facilitates the uptake of 4-methylthiobutyl glucosinolate (4MTB); a homologue, GTR2, shares 80% amino acid similarity and has 75% of the uptake activity (Nour-Eldin *et al.*, 2012). The two transporters are high affinity H^+-glucosinolate symporters. With increasing 4MTB concentrations, they exhibit saturation curves with a best fit by a Michaelis–Menten double-reciprocal plot generating apparent affinity constants of 20 ± 1.2 μM for GTR1 and 18.7 ± 4.2 μM for GTR2. The recombinant transporters in *Xenopus* respond similarly to additions of short-chained aliphatic methylsulfinyl- and 8-octylmethylthio-glucosinolates and the aromatic *p*-hydroxybenzyl-glucosinolate. The *Arabidopsis* double mutant *gtr1gtr2*, which is deficient in these transporter activities shows barely detectable levels of all glucosinolates in the seed and conversely higher levels of aliphatic but not indolic glucosinolates in rosettes and siliques.

The role of glucosinolate transport was assessed in the source–sink relationship of glucosinolates in the above- and below-ground organs in vegetative *Arabidopsis* by Andersen *et al.* (2013). These authors quantified glucosinolates and their distribution between the roots and rosette in various rootstock–scion graft combinations of the quadruple knockout mutant, which is deficient in aliphatic and indolic glucosinolate biosynthesis (*myb28 myb29 cyp79b2 cyp79b3*) (Muller *et al.*, 2010), in the *gtr1gtr2* gluosinolate transport double mutant and in wild type *Arabidopsis*. Their results indicate that GTR1 and GTR2 have a role in the translocation of glucosinolates from the rosette to the roots and in the retention of root-synthesized aliphatic glucosinolates in the root. They proposed a model of proton-dependent GTR1 and GTR2-facilitated glucosinolate transport from the apoplast to the phloem in both roots and rosettes, and an additional upward GTR-independent transport. The distribution and the abundance of long-chained aliphatic glucosinolates were most affected by the double *gtr* mutations.

The compartmentalization of glucosinolate biosynthesis has still not yet been clarified, although several studies have provided information on various aspects of glucosinolate metabolism. Hoglund *et al.* (1991) investigated the distribution of myrosinase, the enzyme that breaks down glucosinolates into toxic metabolites, in rapeseed tissues. Myrosinase is found in myrosin cells, which are idioblasts, while its glucosinolate substrates are located in separate but neighbouring S-cells (Fig. 22.2b). Chewing herbivores cause tissue damage that results in mixing of the contents of the myrosin cells and S-cells, resulting in the mustard oil bomb that triggers the release of toxic chemicals. The composition of the resulting toxic products is dependent on proteins that interact with myrosinase. For instance, in the presence of epithiospecifier protein, myrosinase produces epithionitriles and nitriles instead of isothiocyanates (Lambrix *et al.*, 2001). Similarly, nitrile specifier proteins are responsible for the accumulation of nitriles instead of isothiocyanates, which are the default products of glucosinolate degradation (Burow *et al.*, 2009; Kissen and Bones, 2009). Lastly, thiocyanate-forming proteins have the capacity to promote the formation of thiocyanates and nitriles instead of isothiocyanates (Burow and Wittstock, 2009; Kuchernig *et al.*, 2011). The composition of the resulting toxic products influences chemical ecology (Burow and Wittstock, 2009), and it is therefore relevant that the expression of specifier proteins is regulated (Kissen and Bones, 2009). All of these findings demonstrate that glucosinolate metabolism is a highly regulated process.

22.4 Connections of Glucosinolate Metabolism to Amino Acid Biosynthesis

Glucosinolate secondary metabolism is intertwined with many metabolic and signalling pathways (Chen *et al.*, 2012). Rather than being regarded as a linear biosynthetic pathway, glucosinolate biosynthesis and metabolism seems better described as a 'more complex integrated network' (Sønderby *et al.*, 2010b). For example, overlaps have been described between the leucine and glucosinolate biosynthetic pathways in *Arabidopsis* leaves (Chen *et al.*, 2012; He *et al.*, 2013) and in roots of radish (*Raphanus sativus*) (Zou *et al.*, 2013). Glucosinolate biosynthesis of aliphatic glucosinolates in transgenic *Arabidopsis* lines was suppressed by using inducible RNA interference (RNAi) to repress the expression of *CYP79F1* and *CYP79F2*, which also resulted in higher amounts of at least 13 amino acids compared with controls (Chen *et al.*, 2012). An inverse relationship was suggested to occur between Met-derived glucosinolates in *Arabidopsis* roots and Ile biosynthesis in root plastids of Ile-deficient mutants (Yu *et al.*, 2013). Interesting work is being done but, clearly, much exploration remains to be done to elucidate the nodes and consequences of interaction in the network.

22.5 Regulation of Glucosinolate Biosynthesis

Glucosinolate biosynthesis is regulated at the transcriptional level. Different types of transcription factors regulate indole (aromatic) and aliphatic glucosinolate biosynthesis.

ATR1, also known as MYB34, is a key homeostatic regulator of Trp metabolism (Celenza *et al.*, 2005); it induces the expression of *CYP79B2*, *CYP79B3* and *CYP83B1* (Fig. 22.2a) and regulates indole glucosinolate homeostasis. Similarly, the DOF transcription factor OBP2, also known as AtDof1, confers jasmonate, wound and herbivore-inducible indole glucosinolate accumulation by activating the expression of *CYP79B2*, *CYP79B3* and *CYP83B1* (Skirycz *et al.*, 2006). MYB51 is a third transcription factor that positively regulates indole glucosinolate biosynthesis (Gigolashvili *et al.*, 2007). Interestingly, *MYB51* expression is responsive not only to wounding but also to touch. In contrast to ATR1 and AtDof1, however, MYB51 does not upregulate auxin levels and the expression of the Trp biosynthetic gene, *ASA1*. The expression of MYB51 was shown to reduce leaf consumption by the generalist lepidopteran herbivore *Spodoptera exigua* (Gigolashvili *et al.*, 2007).

A similar network of transcription factors regulates the biosynthesis of aliphatic glucosinolates. Specifically, MYB28, MYB29 and MYB76 control aliphatic glucosinolate production (Hirai *et al.*, 2007; Sønderby *et al.*, 2007). Of these, MYB28 is the dominant regulator responsible for basal level production of aliphatic glucosinolates (Hirai *et al.*, 2007; Sønderby *et al.*, 2010a). MYB29 acts as a rheostat responsible for the jasmonate-mediated induction of aliphatic glucosinolate biosynthetic genes. MYB76, in contrast, plays a role in determining the spatial distribution of aliphatic glucosinolates within the leaf, thus suggesting a potential role of MYB76 in transport regulation (Sønderby *et al.*, 2010a).

The induction of glucosinolate biosynthetic genes in response to the necrotrophic pathogen *Sclerotinia sclerotiorum* is not strictly correlated with changes in levels of glucosinolate metabolite (Stotz *et al.*, 2011). This implies that post-transcriptional processes regulate glucosinolate production or that glucosinolate catabolism occurs in pathogen-challenged leaf tissues. Aliphatic glucosinolates were systemically induced in pathogen- and mock-inoculated plants, suggesting a possible response to touch, which confirms the exquisite sensitivity of the glucosinolate biosynthetic system to minor perturbations (Stotz *et al.*, 2011). MYB28 and, to a lesser extent, MYB29 mediate the accumulation of aliphatic glucosinolates in response to glucose (Miao *et al.*, 2013). Conversely, brassinosteroids have a negative effect on glucosinolate biosynthesis (Guo *et al.*, 2013). These studies have demonstrated that glucosinolate biosynthesis is sensitively and tightly regulated during development (Sønderby *et al.*, 2010a) and in response to phytohormones and exogenous perturbations.

22.6 Biological Activities of Glucosinolate Metabolites

Glucosinolates undergo enzymatic hydrolysis to form bioactive compounds such as isothiocyanates (ITCs), oxaloidinethiones, nitriles, epthionitriles, organic thiocyanates and ionic thiocyanate (SCN^-) (Fig. 22.3), although the glucosinolates themselves

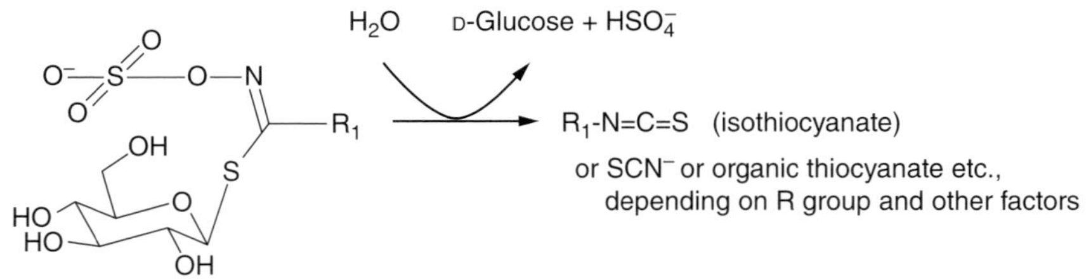

Fig. 22.3. Glucosinolate hydrolysis to produce bioactive metabolites. The β-thioglucoside linkage of the glucosinolate is cleaved to release D-glucose, and the resulting aglycone undergoes rearrangement to release a sulfate and toxic products, including isothiocyanates and thiocyanates (SCN^-).

are not known for their bioactivity. Like many other plant secondary metabolites, a major hypothesis to help explain their existence is that they function in plant defence; this is based on many well-known observations that the products of glucosinolate hydrolysis have toxic or other biological activities towards many organisms (Brown and Morra, 1997). This kind of biological function may be thought of as occurring within two broad categories:

1. *Defensive* actions *internal* to the plant as a *constitutive* (always on) process or as an *induced* response to attack from an external source (e.g. herbivory or pathogenic infection).
2. *Offensive, pre-emptive*, or *proactive* actions *external* to the plant, with natural products being secreted, released through mechanical disruption of cells, emitted as volatiles, or otherwise entering the environment to chemically influence plant–plant or plant–other organism interactions. Such a role could fit an extended concept of *allelopathy*. Such actions could also be constitutive or induced.

Within the first category – defensive actions – the biological activities of glucosinolate breakdown products fall into two subcategories, penetration resistance against biotrophic pathogens, and action that has been described as the 'mustard oil bomb'. Resistance against biotrophic pathogens has recently been shown to involve living plant cells and to be mediated via vesicle flow (Bednarek *et al.*, 2009). This glucosinolate metabolic pathway converts indol-3-ylmethylglucosinolate into isothiocyanate–glutathione adducts with the help of CYP81F2 and the atypical myrosinase PEN2 (Bednarek *et al.*, 2009). The pathway is also required for regulated callose deposition in response to the recognition of pathogen-associated molecular patterns (PAMPs), for example, the flagellin peptide flg22 (Clay *et al.*, 2009; Millet *et al.*, 2010).

The defensive pathway is fundamentally different from the one that involves the mustard oil bomb (Fig. 22.2b). Chewing herbivores and necrotrophic pathogens are likely to be controlled by this latter mechanism, which involves the release of isothiocyanates and other glucosinolate breakdown products (Kliebenstein, 2004; Wittstock *et al.*, 2004; Burow *et al.*, 2006; de Vos *et al.*, 2008; Stotz *et al.*, 2011). Isothiocyanates are generally more toxic than nitriles and, in fact, the herbivore specialist cabbage white butterfly (*Pieris rapae*) expresses a nitrile specifier protein in the gut to ingest less toxic glucosinolate metabolites (Wittstock *et al.*, 2004). Isothiocyanates are also more toxic to microbial pathogens (Tierens *et al.*, 2001). Their toxicity against the necrotrophic fungal pathogen *Sclerotinia sclerotiorum* is a function of side chain length, with 8-methylsulfinyloctyl isothiocyanate being the most toxic compound (Stotz *et al.*, 2011), although indole-3-acetonitrile can also display antifungal activity (Pedras *et al.*, 2002). The importance of the side chain is also demonstrated by the fact that allyl glucosinolate was the only glucosinolate that was negatively associated with susceptibility to *S. sclerotiorum*, when 96 *A. thaliana* ecotypes were compared (Guo and Stotz, unpublished). Epithionitriles are derived from alkenyl glucosinolates, and this group of compounds was reported to be more deterrent to the generalist herbivore *Trichoplusia ni* than regular nitriles (Kliebenstein, 2004). Conversely, nitrile production promotes direct and indirect defence mechanisms against the cabbage butterfly *P. rapae* through reduction of its oviposition and attraction of the parasitoid *Cotesia rubecula* (Hymenoptera, Braconidae), respectively (Mumm *et al.*, 2008). These findings also demonstrate that the structure and chemistry of the side chain is important for biological activity.

An additional feature of several hydrolysis products of glucosinolates, including sulforaphane, 7-methylsulfinylheptyl isothiocyanate and 8-methylsulfinyloctyl isothiocyanate, is that they are potent inducers of phase II enzymes, which are important contributors to xenobiotic resistance and anticancer activity in mammalian cells (Rose *et al.*, 2000; Clarke *et al.*, 2008). Isothiocyanates act as electrophiles that activate redox signalling via the modification of Keap1 (Kelch-like ECH-associated protein 1), which releases Nrf2 (transcription factor NF-E2-related factor-2), permitting this transcription factor to accumulate in the nucleus and to activate antioxidant response elements (Hu *et al.*, 2012).

Finally, oxazolidine-2-thiones have goitrogenic properties, resulting in morphological and physiological changes to the thyroid and leading to goitre formation (Borgen *et al.*, 2010). Similarly, thiocyanates interfere with thyroid function by interfering with iodine availability (Han and Kwon, 2009; Borgen *et al.*, 2010). Breakdown of glucosinolate therefore results in a toxic mix that can affect herbivores and microbes in multiple ways.

Support for the second category of biological functions of glucosinolates – the offensive strategy of allelopathy – has been less consistent. For example, several studies have examined root or rhizosphere compounds from *Arabidopsis* and while many have been found (Walker *et al.*, 2003), including proteins (De-la-Peña *et al.*, 2010), volatile terpenes (Steeghs *et al.*, 2004), indolylic compounds (Bednarek *et al.*, 2005) and acids (Rudrappa *et al.*, 2008), glucosinolates or their degradation products are often conspicuously absent. The results of some of our own work also agree with these findings (Brown and Tokuhisa, unpublished). Yet at times, and in various species, glucosinolates or their products do seem to show up or, alternatively, tantalizing evidence of an indirect nature suggests their presence in the rhizosphere. For example, glucosinolates were found in a hydroponic system using a much larger plant than *Arabidopsis* – the turnip (*Brassica rapa*) (Schreiner *et al.*, 2011). Isothiocyanate degradation products have also been detected in the rhizosphere of *Carica papaya* (Tang and Takenaka, 1983) and *Rorippa indica* (Yamane *et al.*, 1992). Attempts to measure phenylethyl isothiocyanate in the rhizosphere of *B. napus* by removing soil from the roots of field-dug plants caused some root damage, which would release isothiocyanates into the rhizosphere that were not secreted. Nevertheless, a correlation between different soil microbial population markers and the isothiocyanate concentration measured was observed (Rumberger and Marschner, 2004). Rhizophere production of phenylethyl isothiocyanate from *B. napus* was also implicated in stimulation of the germination of parasitic plants (Auger *et al.*, 2012). Similarly, while no glucosinolates or hydrolysis products were detected in the rhizosphere soil, a comparison of the microbial communities (using ^{13}C isotope probing approach) in the rhizosphere of a wild type *Arabidopsis* and a mutant with an altered glucosinolate root profile implicated glucosinolate-associated root secretions as responsible for the differences that were found in microbial communities of the two types of plants (Bressan *et al.*, 2009).

Based on the concentrations of glucosinolates in the periphery of the root structure, it has been suggested that some glucosinolates could be released as roots expand circumferentially, and that mechanical damage to roots or ephemeral growth and decay of fine roots might release bioactive products (McCully *et al.*, 2008; Van Dam *et al.*, 2009). This may help to explain some of the variability among the various observations that have been made, but also seems to reinforce an idea that the role of glucosinolates in roots may be more protective (defensive) than alleopathic (offensive).

22.7 Conclusions

Glucosinolates are natural plant products that are derived from aliphatic or aromatic amino acids; they are metabolized to a variety of potentially toxic products via myrosinase-catalysed hydrolysis. The biological activities of glucosinolate breakdown products in defence against herbivores and microbial pathogens and in human health have been determined. Glucosinolate transporters (GTR1 and GTR2) are involved in translocation of glucosinolates from the siliques to the seeds and from rosettes to roots. Much has been learned about the transcriptional regulation of glucosinolate biosynthesis, but regulatory network connections to amino acid metabolism remain to be explored.

References

Andersen, T.G., Nour-Eldin, H.H., Fuller, V.L., Olsen, C.E., Burow, M. and Halkier, B.A. (2013) Integration of biosynthesis and long-distance transport establish organ-specific glucosinolate profiles in vegetative *Arabidopsis*. *The Plant Cell* 25, 3133–3145.

Auger, B., Pouvreau, J.B., Pouponneau, K., Yoneyama, K., Montiel, G., Le Bizec, B., Yoneyama, K., Delavault, P., Delourme, R. and Simier, P. (2012) Germination stimulants of *Phelipanche ramosa* in the rhizosphere of *Brassica napus* are derived from the glucosinolate pathway. *Molecular Plant–Microbe Interactions* 25, 993–1004.

Bak, S., Tax, F.E., Feldmann, K. A., Galbraith, D.W. and Feyereisen, R. (2001) CYP83B1, a cytochrome P450 at the metabolic branch point in auxin and indole glucosinolate biosynthesis in *Arabidopsis*. *The Plant Cell* 13, 101–111.

Bednarek, P., Schneider, B., Svatoš, A., Oldham, N.J. and Hahlbrock, K. (2005) Structural complexity, differential response to infection, and tissue specificity of indolic and phenylpropanoid secondary metabolism in *Arabidopsis* roots. *Plant Physiology* 138, 1058–1070.

Bednarek, P., Piślewska-Bednarek, M., Svatoš, A., Schneider, B., Doubský, J., Mansurova, M., Humphry, M., Consonni, C., Panstruga, R., Sanchez-Vallet, A., Molina, A. and Schulze-Lefert, P. (2009) A glucosinolate metabolism pathway in living plant cells mediates broad-spectrum antifungal defense. *Science* 323, 101–106.

Borgen, B.H., Thangstad, O.P., Ahuja, I., Rossiter, J.T. and Bones, A.M. (2010) Removing the mustard oil bomb from seeds: transgenic ablation of myrosin cells in oilseed rape (*Brassica napus*) produces *MINELESS* seeds. *Journal of Experimental Botany* 61, 1683–1697.

Bressan, M., Roncato, M.A., Bellvert, F., Comte, G., Haichar, F.Z., Achouak, W. and Berge, O. (2009) Exogenous glucosinolate produced by *Arabidopsis thaliana* has an impact on microbes in the rhizosphere and plant roots. *The ISME Journal* 3, 1243–1257.

Brown, P.D. and Morra, M.J. (1997) Control of soil-borne plant pests using glucosinolate-containing plants. In: Sparks, D.L. (ed.) *Advances in Agronomy*. Academic Press, New York, pp. 167–231.

Burow, M. and Wittstock, U. (2009) Regulation and function of specifier proteins in plants. *Phytochemistry Reviews* 8, 87–99.

Burow, M., Müller, R., Gershenzon, J. and Wittstock, U. (2006) Altered glucosinolate hydrolysis in genetically engineered *Arabidopsis thaliana* and its influence on the larval development of *Spodoptera littoralis*. *Journal of Chemical Ecology* 32, 2333–2349.

Burow, M., Losansky, A., Müller, R., Plock, A., Kliebenstein, D. J. and Wittstock, U. (2009) The genetic basis of constitutive and herbivore-induced ESP-independent nitrile formation in *Arabidopsis*. *Plant Physiology* 149, 561–574.

Celenza, J.L., Quiel, J.A., Smolen, G.A., Merrikh, H., Silvestro, A.R., Normanly, J. and Bender, J. (2005) The *Arabidopsis* ATR1 Myb transcription factor controls indolic glucosinolate homeostasis. *Plant Physiology* 137, 253–262.

Chen, S.X., Glawischnig, E., Jorgensen, K., Naur, P., Jorgensen, B., Olsen, C.E., Hansen, C.H., Rasmussen, H., Pickett, J.A. and Halkier, B.A. (2003) CYP79F1 and CYP79F2 have distinct functions in the biosynthesis of aliphatic glucosinolates in *Arabidopsis*. *The Plant Journal* 33, 923–937.

Chen, Y.z., Pang, Q.-Y., He, Y., Zhu, N., Branstrom, I., Yan, X.-F. and Chen, S. (2012) Proteomics and metabolomics of *Arabidopsis* responses to perturbation of glucosinolate biosynthesis. *Molecular Plant* 5, 1138–1150.

Clarke, J.D., Dashwood, R.H. and Ho, E. (2008) Multi-targeted prevention of cancer by sulforaphane. *Cancer Letters* 269, 291–304.

Clay, N.K., Adio, A.M., Denoux, C., Jander, G. and Ausubel, F.M. (2009) Glucosinolate metabolites required for an *Arabidopsis* innate immune response. *Science* 323, 95–101.

De-la-Peña, C., Badri, D.V., Lei, Z., Watson, B.S., Brandão, M.M., Silva-Filho, M.C., Sumner, L.W. and Vivanco, J.M. (2010) Root secretion of defense-related proteins is development-dependent and correlated with flowering time. *The Journal of Biological Chemistry* 285, 30654–30665.

de Vos, M., Kriksunov, K.L. and Jander, G. (2008) Indole-3-acetonitrile production from indole glucosinolates deters oviposition by *Pieris rapae*. *Plant Physiology* 146, 916–926.

Gigolashvili, T., Yatusevich, R., Berger, B., Müller, C. and Flugge, U.I. (2007) The R2R3-MYB transcription factor HAG1/MYB28 is a regulator of methionine-derived glucosinolate biosynthesis in *Arabidopsis thaliana*. *The Plant Journal* 51, 247–261.

Glawischnig, E., Hansen, B.G., Olsen, C.E. and Halkier, B.A. (2004) Camalexin is synthesized from indole-3-acetaldoxime, a key branching point between primary and secondary metabolism in *Arabidopsis*. *Proceedings of the National Academy of Sciences of the United States of America* 101, 8245–8250.

Grubb, C.D., Zipp, B.J., Lüdwig-Müller, J., Masuno, M.N., Molinski, T.F. and Abel, S. (2004) *Arabidopsis* glucosyltransferase UGT74B1 functions in glucosinolate biosynthesis and auxin homeostasis. *The Plant Journal* 40, 893–908.

Guo, R., Qian, H., Shen, W., Liu, L., Zhang, M., Cai, C., Zhao, Y., Qiao, J. and Wang, Q. (2013) BZR1 and BES1 participate in regulation of glucosinolate biosynthesis by brassinosteroids in *Arabidopsis*. *Journal of Experimental Biology* 64, 2401–2412.

Halkier, B.A. and Gershenzon, J. (2006) Biology and biochemistry of glucosinolates. *Annual Review of Plant Biology* 57, 303–333.

Han, H. and Kwon, H. (2009) Estimated dietary intake of thiocyanate from *Brassicaceae* family in Korean diet. *Journal of Toxicology and Environmental Health A* 72, 1380–1387.

He, Y., Dai, S.J., Dufresne, C.P., Zhu, N., Pang, Q.Y. and Chen, S.X. (2013) Integrated proteomics and metabolomics of *Arabidopsis* acclimation to gene-dosage dependent perturbation of isopropylmalate dehydrogenases. *PLoS ONE* 8 (3): e57118.

Hirai, M.Y., Klein, M., Fujikawa, Y., Yano, M., Goodenowe, D.B., Yamazaki, Y., Kanaya, S., Nakamura, Y., Kitayama, M., Suzuki, H., Sakurai, N., Shibata, D. *et al.* (2005) Elucidation of gene-to-gene and metabolite-to-gene networks in *Arabidopsis* by integration of metabolomics and transcriptomics. *The Journal of Biological Chemistry* 280, 25590–25595.

Hirai, M.Y., Sugiyama, K., Sawada, Y., Tohge, T., Obayashi, T., Suzuki, A., Araki, R., Sakurai, N., Suzuki, H., Aoki, K., Goda, H., Nishizawa, O.I. *et al.* (2007) Omics-based identification of *Arabidopsis* Myb transcription factors regulating aliphatic glucosinolate biosynthesis. *Proceedings of the National Academy of Sciences of the United States of America* 104, 6478–6483.

Hoglund, A.S., Lenman, M., Falk, A. and Rask, L. (1991) Distribution of myrosinase in rapeseed tissues. *Plant Physiology* 95, 213–221.

Hu, C., Nikolic, D., Eggler, A.L., Mesecar, A.D. and van Breemen, R.B. (2012) Screening for natural chemoprevention agents that modify human Keap1. *Analytical Biochemistry* 421, 108–114.

Kissen, R. and Bones, A.M. (2009) Nitrile-specifier proteins involved in glucosinolate hydrolysis in *Arabidopsis thaliana*. *The Journal of Biological Chemistry* 284, 12057–12070.

Klein, M. and Papenbrock, J. (2009) Kinetics and substrate specificities of desulfo-glucosinolate sulfotransferases in *Arabidopsis thaliana*. *Physiologia Plantarum* 135, 140–149.

Kliebenstein, D.J. (2004) Secondary metabolites and plant/environment interactions: a view through *Arabidopsis thaliana* tinged glasses. *Plant, Cell and Environment* 27, 675–684.

Knill, T., Schuster, J., Reichelt, M., Gershenzon, J. and Binder, S. (2008) *Arabidopsis* branched-chain aminotransferase 3 functions in both amino acid and glucosinolate biosynthesis. *Plant Physiology* 146, 1028–1039.

Knill, T., Reichelt, M., Paetz, C., Gershenzon, J. and Binder, S. (2009) *Arabidopsis thaliana* encodes a bacterial-type heterodimeric isopropylmalate isomerase involved in both Leu biosynthesis and the Met chain elongation pathway of glucosinolate formation. *Plant Molecular Biology* 71, 227–239.

Koroleva, O.A., Gibson, T.M., Cramer, R. and Stain, C. (2010) Glucosinolate-accumulating S-cells in *Arabidopsis* leaves and flower stalks undergo programmed cell death at early stages of differentiation. *The Plant Journal* 64, 456–469.

Kroymann, J., Textor, S., Tokuhisa, J.G., Falk, K.L., Bartram, S., Gershenzon, J. and Mitchell-Olds, T. (2001) A gene controlling variation in *Arabidopsis* glucosinolate composition is part of the methionine chain elongation pathway. *Plant Physiology* 127, 1077–1088.

Kuchernig, J.C., Backenköhler, A., Lübbecke, M., Burow, M. and Wittstock, U. (2011) A thiocyanate-forming protein generates multiple products upon allylglucosinolate breakdown in *Thlaspi arvense*. *Phytochemistry* 72, 1699–1709.

Lambrix, V., Reichelt, M., Mitchell-Olds, T., Kliebenstein, D.J. and Gershenzon, J. (2001) The *Arabidopsis* epithiospecifier protein promotes the hydrolysis of glucosinolates to nitriles and influences *Trichoplusia ni* herbivory. *The Plant Cell* 13, 2793–2807.

McCully, M.E., Miller, C., Sprague, S.J., Huang, C.X. and Kirkegaard, J.A. (2008) Distribution of glucosinolate and sulphur-rich cells in roots of field-grown canola (*Brassica napus*). *New Phytologist* 180, 193–205.

Miao, H., Wei, J., Zhao, Y., Yan, H., Sun, B., Huang, J. and Wang, Q. (2013) Glucose signalling positively regulates aliphatic glucosinolate biosynthesis. *Journal of Experimental Botany* 64, 1097–1109.

Mikkelsen, M.D. and Halkier, B.A. (2003) Metabolic engineering of valine- and isoleucine-derived glucosinolates in *Arabidopsis* expressing CYP79D2 from cassava. *Plant Physiology* 131, 773–779.

Mikkelsen, M.D., Naur, P. and Halkier, B.A. (2004) *Arabidopsis* mutants in the C-S lyase of glucosinolate biosynthesis establish a critical role for indole-3-acetaldoxime in auxin homeostasis. *The Plant Journal* 37, 770–777.

Millet, Y.A., Danna, C.H., Clay, N.K., Songnuan, W., Simon, M.D., Werck-Reichhart, D. and Ausubel, F.M. (2010) Innate immune responses activated in *Arabidopsis* roots by microbe-associated molecular patterns. *The Plant Cell* 22, 973–990.

Mithen, R., Bennett, R. and Marquez, J. (2010) Glucosinolate biochemical diversity and innovation in the *Brassicales*. *Phytochemistry* 71, 2074–2086.

Müller, R., de Vos, M., Sun, J.Y., Sønderby, I.E., Halkier, B.A., Wittstock, U. and Jander, G. (2010) Differential effects of indole and aliphatic glucosinolates on lepidopteran herbivores. *Journal of Chemical Ecology* 36, 905–913.

Mumm, R., Burow, M., Bukovinszkine'kiss, G., Kazantzidou, E., Wittstock, U., Dicke, M. and Gershenzon, J. (2008) Formation of simple nitriles upon glucosinolate hydrolysis affects direct and indirect defense against the specialist herbivore, *Pieris rapae*. *Journal of Chemical Ecology* 34, 1311–1321.

Naur, P., Petersen, B.L., Mikkelsen, M.D., Bak, S., Rasmussen, H., Olsen, C.E. and Halkier, B.A. (2003) CYP83A1 and CYP83B1, two nonredundant cytochrome P450 enzymes metabolizing oximes in the biosynthesis of glucosinolates in *Arabidopsis*. *Plant Physiology* 133, 63–72.

Nelson, D. and Werck-Reichhart, D. (2011) A P450-centric view of plant evolution. *The Plant Journal* 66, 194–211.

Nour-Eldin, H.H. and Halkier, B.A. (2009) Piecing together the transport pathway of aliphatic glucosinolates. *Phytochemistry Reviews* 8, 53–67.

Nour-Eldin, H.H., Nørholm, M.H. and Halkier, B.A. (2006) Screening for plant transporter function by expressing a normalized *Arabidopsis* full-length cDNA library in *Xenopus* oocytes. *Plant Methods* 2:17.

Nour-Eldin, H.H., Andersen, T.G., Burow, M., Madsen, S.R., Jørgensen, M.E., Olsen, C.E., Dreyer, I., Hedrich, R., Geiger, D. and Halkier, B.A. (2012) NRT/PTR transporters are essential for translocation of glucosinolate defence compounds to seeds. *Nature* 488, 531–534.

Pedras, M.S., Nycholat, C.M., Montaut, S., Xu, Y. and Khan, A.Q. (2002) Chemical defenses of crucifers: elicitation and metabolism of phytoalexins and indole-3-acetonitrile in brown mustard and turnip. *Phytochemistry* 59, 611–625.

Rose, P., Faulkner, K., Williamson, G. and Mithen, R. (2000) 7-Methylsulfinylheptyl and 8-methylsulfinyloctyl isothiocyanates from watercress are potent inducers of phase II enzymes. *Carcinogenesis* 21, 1983–1988.

Rudrappa, T., Czymmek, K.J., Paré, P.W. and Bais, H.P. (2008) Root-secreted malic acid recruits beneficial soil bacteria. *Plant Physiology* 148, 1547–1556.

Rumberger, A. and Marschner, P. (2004) 2-Phenylethylisothiocyanate concentration and bacterial community composition in the rhizosphere of field-grown canola. *Functional Plant Biology* 31, 623–631.

Sarsby, J., Towers, M.W., Stain, C., Cramer, R. and Koroleva, O.A. (2012) Mass spectrometry imaging of glucosinolates in *Arabidopsis* flowers and siliques. *Phytochemistry* 77, 110–118.

Sawada, Y., Kuwahara, A., Nagano, M., Narisawa, T., Sakata, A., Saito, K. and Hirai, M.Y. (2009a) Omics-based approaches to methionine side chain elongation in *Arabidopsis*: characterization of the genes encoding methylthioalkylmalate isomerase and methylthioalkylmalate dehydrogenase. *Plant and Cell Physiology* 50, 1181–1190.

Sawada, Y., Toyooka, K., Kuwahara, A., Sakata, A., Nagano, M., Saito, K. and Hirai, M.Y. (2009b) *Arabidopsis* bile acid:sodium symporter family protein 5 is involved in methionine-derived glucosinolate biosynthesis. *Plant and Cell Physiology* 50, 1579–1586.

Schreiner, M., Krumbein, A., Knorr, D. and Smetanska, I. (2011) Enhanced glucosinolates in root exudates of *Brassica rapa* ssp. *rapa* mediated by salicylic acid and methyl jasmonate. *Journal of Agricultural and Food Chemistry* 59, 1400–1405.

Schuster, J., Knill, T., Reichelt, M., Gershenzon, J. and Binder, S. (2006) Branched-chain aminotransferase4 is part of the chain elongation pathway in the biosynthesis of methionine-derived glucosinolates in *Arabidopsis*. *The Plant Cell* 18, 2664–2679.

Shroff, R., Vergara, F., Muck, A., Svatoš, A. and Gershenzon, J. (2008) Nonuniform distribution of glucosinolates in *Arabidopsis thaliana* leaves has important consequences for plant defense. *Proceedings of the National Academy of Sciences of the United States of America* 105, 6196–6201.

Skirycz, A., Reichelt, M., Burow, M., Birkemeyer, C., Rolcik, J., Kopka, J., Zanor, M.I., Gershenzon, J., Strnad, M., Szopa, J., Mueller-Roeber, B. and Witt, I. (2006) DOF transcription factor AtDof1.1 (OBP2) is part of a regulatory network controlling glucosinolate biosynthesis in *Arabidopsis*. *The Plant Journal* 47, 10–24.

Sønderby, I.E., Hansen, B.G., Bjarnholt, N., Ticconi, C., Halkier, B.A. and Kliebenstein, D.J. (2007) A systems biology approach identifies a R2R3 MYB gene subfamily with distinct and overlapping functions in regulation of aliphatic glucosinolates. *PLoS ONE* 2(12): e1322.

Sønderby, I.E., Burow, M., Rowe, H.C., Kliebenstein, D.J. and Halkier, B.A. (2010a) A complex interplay of three R2R3 MYB transcription factors determines the profile of aliphatic glucosinolates in *Arabidopsis*. *Plant Physiology* 153, 348–363.

Sønderby, I.E., Geu-Flores, F. and Halkier, B.A. (2010b) Biosynthesis of glucosinolates – gene discovery and beyond. *Trends in Plant Science* 15, 283–290.

Steeghs, M., Bais, H.P., de Gouw, J., Goldan, P., Kuster, W., Northway, M., Fall, R. and Vivanco, J.M. (2004) Proton-transfer-reaction mass spectrometry as a new tool for real time analysis of root-secreted volatile organic compounds in *Arabidopsis*. *Plant Physiology* 135, 47–58.

Stotz, H.U., Sawada, Y., Shimada, Y., Hirai, M.Y., Sasaki, E., Krischke, M., Brown, P.D., Saito, K. and Kamiya, Y. (2011) Role of camalexin, indole glucosinolates, and side chain modification of glucosinolate-derived isothiocyanates in defense of *Arabidopsis* against *Sclerotinia sclerotiorum*. *The Plant Journal* 67, 81–93.

Sugawara, S., Hishiyama, S., Jikumaru, Y., Hanada, A., Nishimura, T., Koshiba, T., Zhao, Y., Kamiya, Y. and Kasahara, H. (2009) Biochemical analyses of indole-3-acetaldoxime-dependent auxin biosynthesis in *Arabidopsis*. *Proceedings of the National Academy of Sciences of the United States of America* 106, 5430–5435.

Tang, C.-S. and Takenaka, T. (1983) Quantitation of a bioactive metabolite in undisturbed rhizosphere – benzyl isothiocyanate from *Carica papaya* L. *Journal of Chemical Ecology* 9, 1247–1253.

Textor, S., de Kraker, J.W., Hause, B., Gershenzon, J. and Tokuhisa, J.G. (2007) MAM3 catalyzes the formation of all aliphatic glucosinolate chain lengths in *Arabidopsis*. *Plant Physiology* 144, 60–71.

Tierens, K.-J., Thomma, B., Brower, M., Schmidt, J., Kistner, K., Porzel, A., Mauch-Mani, B., Cammue, B. and Broekaert, W. (2001) Study of the role of antimicrobial glucosinolate-derived isothiocyanates in resistance of *Arabidopsis* to microbial pathogens. *Plant Physiology* 125, 1688–1699.

Van Dam, N.M., Tytgat, T.O.G. and Kirkegaard, J.A. (2009) Root and shoot glucosinolates: a comparison of their diversity, function and interactions in natural and managed ecosystems. *Phytochemistry Reviews* 8, 171–186.

Walker, T., Bais, H.P., Halligan, K., Stermitz, F.R. and Vivanco, J.M. (2003) Metabolic profiling of root exudates of *Arabidopsis thaliana*. *Journal of Agricultural and Food Chemistry* 51, 2548–2554.

Wittstock, U. and Halkier, B.A. (2000) Cytochrome P450 CYP79A2 from *Arabidopsis thaliana* L. catalyzes the conversion of L-phenylalanine to phenylacetaldoxime in the biosynthesis of benzylglucosinolate. *The Journal of Biological Chemistry* 275, 14659–14666.

Wittstock, U., Agerbirk, N., Stauber, E.J., Olsen, C.E., Hippler, M., Mitchell-Olds, T., Gershenzon, J. and Vogel, H. (2004) Successful herbivore attack due to metabolic diversion of a plant chemical defense. *Proceedings of the National Academy of Sciences of the United States of America* 101, 4859–4864.

Yamane, A., Fujikura, J., Ogawa, H. and Mizutani, J. (1992) Isothiocyanates as alleopathic compounds from *Rorippa indica* Hiern. (*Cruciferae*) roots. *Journal of Chemical Ecology* 18, 1941–1954.

Yu, H., Zhang, F., Wang, G., Liu, Y. and Liu, D. (2013) Partial deficiency of isoleucine impairs root development and alters transcript levels of the genes involved in branched-chain amino acid and glucosinolate metabolism in *Arabidopsis*. *Journal of Experimental Botany* 64, 599–612.

Zou, Z., Ishida, M., Li, F., Kakizaki, T., Suzuki, S., Kitashiba, H. and Nishio, T. (2013) QTL analysis using SNP markers developed by next-generation sequencing for identification of candidate genes controlling 4-methylthio-3-butenyl glucosinolate contents in roots of radish, *Raphanus sativus* L. *PLoS ONE* 8 (1): e53541.

23 Natural Toxins that Affect Plant Amino Acid Metabolism

S.O. Duke* and F.E. Dayan
USDA Agricultural Research Service, Natural Products Utilization Research Unit, Oxford, Mississippi, USA

23.1 Abstract

A diverse range of natural compounds interfere with the synthesis and other aspects of amino acid metabolism. Some are amino acid analogues, but most are not. This review covers a number of specific natural phytotoxic compounds by molecular target site. The inhibition of glutamine synthetase is of particular interest because a major herbicide is based on the microbial inhibitor of this enzyme, i.e. phosphinothricin. Natural inhibitors of plant asparagine, glutamate, ornithine, methionine and lysine synthesis are also discussed. Glutamate is one of the most abundant and significant amino acids in living organisms. Ornithine is a key intermediate in the urea cycle and is a precursor of polyamines. Methionine is an essential amino acid for animals and humans, but its metabolism to cysteine and use as a sulfur and methyl donor is of universal importance. Lysine is also an essential amino acid for animals and humans. The discussion includes the secondary effects of inhibitors on various aspects of plant physiology. Finally, natural compounds that may cause toxicity at a L-phenylalanine binding site, that appear to interfere with amino acid transporters and that can cause specific photodegradation of L-phenylalanine are mentioned.

23.2 Introduction

Many natural compounds are toxic to plants, and many of these are toxic by means of their interference with amino acid metabolism or the functions of amino acids (e.g. Copping and Duke, 2007; Dayan and Duke, 2009; Duke and Dayan, 2011a). Three enzymes that are concerned with amino acid metabolism are target sites for a major part of the synthetic herbicide market (Duke and Dayan, 2011b), and natural phytotoxins that kill plants via other amino acid-related target sites can provide valuable leads for much needed new herbicides with new modes of action (Duke, 2012; Dayan *et al.*, 2012). Clearly, amino acid analogues are good candidates for inhibitors of enzymes involved in amino acid metabolism, so many of these compounds are non-protein amino acids, although there are many others that are neither amino acids nor even compounds with an amine substituent. These phytotoxins are largely produced by plants and microbes. Plant pathogens are an especially rich source of such compounds, as these organisms often weaken or kill host plant tissues before using them as a food source.

The molecular target sites of most natural phytotoxins are unknown, so this review probably represents only a small fraction of the natural phytotoxins that act through direct effects

*E-mail address: stephen.duke@ars.usda.gov

on plant amino acid metabolism. A lethal phytotoxin will eventually affect all metabolic processes, and many authors take advantage of this fact to publish papers on the secondary and tertiary effects of phytotoxins. The claims by such papers of the effects of phytotoxins on amino acid metabolism are often specious, in that phytotoxins with target sites unrelated to amino acid metabolism might also generate similar effects. Thus, this literature will not be discussed unless there is a good rationale to connect the compound under investigation with a primary effect on amino acid metabolism, such as the compound being a protein amino acid analogue.

This chapter is prefaced with an experimental approach section, which is followed by sections on enzymatic targets or suspected target sites of phytotoxins.

23.3 Approaches to the Discovery of Phytotoxin Mode of Action

Determination of the mode of action of a phytotoxin can be difficult, especially if the molecular target site is not one already known to be inhibited by other phytotoxins. Dayan *et al.* (2000) provided a strategy for obtaining first approximations of the modes of action of natural phytotoxins. 'Omics' approaches (e.g. transcriptomics, proteomics, metabolomics and physionomics) can be useful in finding hints to the mode of action of a phytotoxic compound (Duke *et al.*, 2013), although the discovery of only one new mode of action has been documented from such a strategy (Grossmann *et al.*, 2012). Perhaps, the most direct approach to finding whether a compound inhibits the synthesis of one or more amino acids is to supplement growth media or nutrient media with amino acids to determine if the inhibitory effect can be ameliorated (e.g. Amagasa *et al.*, 1994). This can be done with plant cell or tissue cultures (e.g. Kilmer *et al.*, 1981) or with whole plants, such as *Lemna* spp., grown in nutrient solutions (e.g. Michel *et al.*, 2004; Duke *et al.*, 2011). One assumes that if the phytotoxin causes a deficiency in certain amino acid(s), the effect will be at least partially reversed by supplementation with that amino acid(s). Such indirect approaches only point to possible enzyme targets, which must then be verified by *in vitro* enzyme studies. This can be complicated if the phytotoxin is a protoxin that must be converted to the enzyme-active compounds *in vivo*, often by enzymatic activation. Several phytotoxins, both natural and synthetic, are protoxins that have no activity at the enzyme target in the protoxin form. An example of this that is discussed later in the chapter (Section 23.7) is the natural protoxin bialaphos, which is converted to the active form, phosphinothrin, *in vivo* (Wild and Ziegler, 1989).

23.4 Inhibitors of Aminotransferases

Secondary compounds from several microorganisms are or appear to be aminotransferase inhibitors. Perhaps the best known of these is gabaculine (Fig. 23.1), a product of *Streptomyces toyacaenis* (Kobayashi *et al.*, 1976; Rando, 1977). This inhibits several aminotransferases

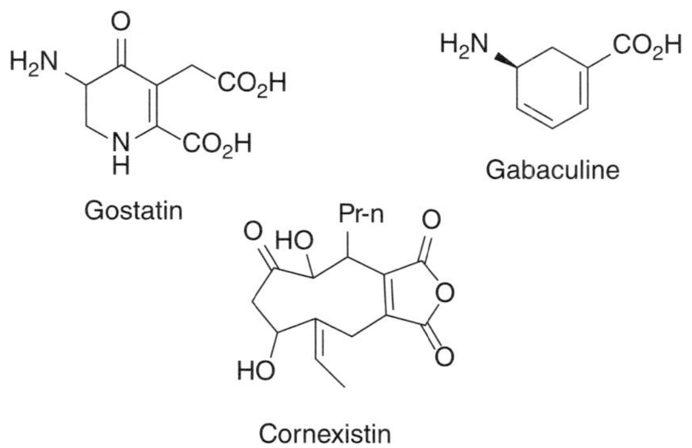

Fig. 23.1. Structures of three microbial compounds known or thought to inhibit aminotransferases.

of plants, microbes and animals (e.g. Rando, 1977; Nair *et al.*, 1990; Rajaram *et al.*, 2008). In plants, it strongly inhibits glutamate 1-semialdehyde aminotransferase, an enzyme required for 5-aminolevulinate synthesis and thus porphyrin and chlorophyll synthesis (Rosellini *et al.*, 2007; Giancaspro *et al.*, 2012). Gabaculine has been a useful tool for studying the regulation of chlorophyll biosynthesis (e.g. Becerril and Duke, 1989; Demko *et al.*, 2010).

Cornexistin (Fig. 23.1), a fungal metabolite from *Paecilomyces variotii*, is highly phytotoxic (Amagasa *et al.*, 1994) and has been patented as a herbicide. The synthetic aminotransferase inhibitor aminooxyacetate causes identical herbicidal symptoms in duckweed. Cornexistin inhibits aspartate aminotransferase activity at high concentrations only after incubation in a plant cellular extract, suggesting that cornexistin is a proherbicide that must be metabolized to an aminotransferase inhibitor. Gostatin (Fig. 23.1), a product of *Streptomyces sumanensis* (Nishino and Murao, 1983; Nishino *et al.*, 1984), is an inhibitor of both aspartate and alanine aminotransferase from mammals and aspartate aminotransferase from plants.

23.5 An Inhibitor of β-Cystathionase (Cystathionine β-lyase)

Some *Bradyrhizobium* strains produce the phytotoxin rhizobitoxine (Fig. 23.2) (Owens *et al.*, 1968). This inhibits β-cystathionase, which is required for methionine synthesis (Owens *et al.*, 1968; Giovanelli *et al.*, 1973). At one time, rhizobitoxine was considered to be sufficiently phytotoxic to be proposed as a commercial herbicide (Owens, 1973). Synthesis of the essential plant hormone ethylene is dependent on methionine, so one could assume that ethylene synthesis would be greatly inhibited in plants treated with this compound. This is true, but rhizobitoxine also directly inhibits the production of ethylene

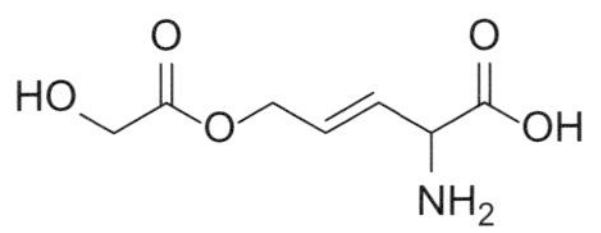

Fig. 23.2. Structure of rhizobitoxine.

from methionine (Owens *et al.*, 1971) by inhibiting 1-aminocyclopropane-1-carboxylate synthase (Yasuta *et al.*, 1999). The inhibition of ethylene synthesis by rhizobitoxine can enhance nodulation in host legumes that are hosts to *Bradyrhizobium* spp. that produce rhizobitoxine (Duodu *et al.*, 1999; Ma *et al.*, 2002; Sugawara *et al.*, 2006). Plants treated with rhizobitoxine become chlorotic, and this symptom can be alleviated by treatment with methionine (Okazaki *et al.*, 2007). Rhizobitoxine and its ethoxy analogue (L-2-amino-4-[2′-aminoethoxy]-trans-3-butenoic acid) have been routinely used as ethylene synthesis inhibitors to study the physiological roles of ethylene in plants (e.g. Aharoni, 1978; Zimmerman *et al.*, 1977).

23.6 Inhibitors of Glutamate Synthase and Asparagine Synthetase

Acivicin (Fig. 23.3) is a product of *Streptomyces sviceu* (Gould and Ju, 1989) that has been patented as a herbicide (Brunner *et al.*, 2007). It has not been well studied in plants, but has been well researched as a pharmaceutical. Acivicin is an analogue of glutamine and inhibits a number of glutamine-dependent enzymes, including glutamate synthase (Conti *et al.*, 2003). It also inhibits amidophosphoribosyltransferase, phosphoribosylformyl-glycinamidine synthase, GMP synthase, and γ-glutamyltranspeptidase (Natsumeda *et al.*, 1989; Grzam *et al.*, 2007; Rodriguez-Suerez *et al.*, 2007; Williams *et al.*, 2009). Acivicin inhibits both purine and pyrimidine biosynthesis in plants (Roberts *et al.*, 2003).

Albizziin (Fig. 23.3), a glutamine analogue from *Albizia* (Gmelin *et al.*, 1958; Brown and Turan, 1996) and *Acacia* (Gmelin, 1959; McSweeney *et al.*, 2005) species, is phytotoxic to plants (Gulati *et al.*, 1981; Lam *et al.*, 1994) through its inhibition of asparagine synthetase (AS) and glutamate synthase (Lea and Fowden, 1975; Oaks *et al.*, 1979; Oaks and Ross, 1984; Singh and Srivastava, 1987; Turnbull *et al.*, 1996; Lea *et al.*, 2006), as well as perhaps other enzymes. Not all plant AS is inhibited by albizziin, as the root AS of maize was found to be insensitive to it (Oaks and Ross, 1984). Feeding asparagine, glutamine and glutamate to *Arabidopsis thaliana* plants

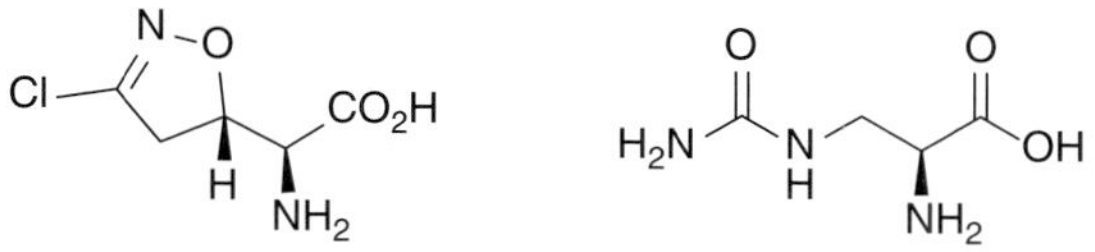

Fig. 23.3. Structures of acividin (left) and albizziin (right).

exhibiting albizziin-inhibited root growth will reverse the effects of the phytotoxin (Lam *et al.*, 1994). The toxin also inhibits glutaminyl-RNA synthetase of higher plants, including the enyzme from a species of *Albizia* that produces albizziin (Lea and Fowden, 1973). Some microbial glutamate synthases are also inhibited by albizziin (Turnbull *et al.*, 1996). The AS of *Escherichia coli* is apparently insensitive to the inhibitor, making the gene for AS a useful selective marker gene for transformants of organisms previously with albizziin-sensitive AS (Cartier *et al.*, 1987). Mammalian AS is also inhibited by the toxin (Andrulis *et al.*, 1983). Albizziin protects the enzymes glucosamine-6-phosphate synthetase from *Salmonella typhimurium* (Chmara *et al.*, 1986) and glutaminase from *E. coli* (Trotta *et al.*, 1974) from inhibitors. At very high concentrations, albizziin inhibits plant 4-methyleneglutamine amidohydrolase, an enzyme thought to be involved in supplying N to the growing shoots of plants (Winter and Dekker, 1991).

23.7 Inhibitors of Glutamine Synthetase

Plants have two isoenzymes of glutamine synthetase (GS), a cytoplasmic form and a chloroplastic form. The latter form is the major GS isoenzyme in leaves. GS activity is critical for the synthesis of glutamine and glutamic acid, and the chloroplastic GS plays a particularly important role in recycling ammonia during photorespiration. Consequently, the inhibition of chloroplastic GS interrupts photorespiration and causes the accumulation of ammonia to toxic levels, leading to the photooxidation of critical chloroplast components (Wild and Wendler, 1993). While GS is the target site of only one commercial herbicide (glufosinate or D,L-phosphinothricin), it is also inhibited by a wide range of natural products (Fig. 23.4a). Consequently, a large amount of information is available on the discovery of GS inhibitors, their mode of action and their biosynthesis.

Bialaphos, ((L-2-amino-4-[(hydroxy) (methyl) phosphinoyl]-butyryl-L-alanyl-L-alanine) (Fig. 23.4a), was discovered in the early 1970s by two separate groups working on related actinomycete species, *Streptomyces viridiochromogenes* and *S. hygroscopicus* (Bayer *et al.*, 1972; Kondo *et al.*, 1972). Its structure consists of two molecules of L-alanine and one molecule of the non-protein amino acid phosphinothricin. Bialaphos is produced commercially by fermentation by Meiji Seika and is sold as the herbicide Herbiace®. This natural herbicide is also called bilanafos, phosphinothricyl-alanyl-alanine and phosphinothricin-alanyl-alanine. Other natural GS inhibitors containing phosphinothricin have been discovered (e.g. trialaphos and phosalacine) (Omura *et al.*, 1984a). Glufosinate (D,L-phosphinothricin), the synthetic form of the active component of bialaphos, is commercialized by Bayer CropScience as a non-selective broad-spectrum herbicide under the trade names of Basta® in most of the world and of Ignite® and Liberty® (initially Hoechst Hoe 039866) in North America. Both glufosinate and bialaphos are low environmental impact herbicides because they rapidly degrade and have a relatively low toxicity to most organisms other than plants (Duke and Cerdiera, 2005: Ebert *et al.*, 1990).

As mentioned above, a large number of natural products that inhibit GS have been discovered. Tabtoxin (Fig. 23.4a), also known as wildfire toxin, is produced by various *Pseudomonas syringae* var. *tabaci* pathovars. As with bialaphos, tabtoxin is a proherbicide that is hydrolysed *in planta* to yield its biologically active form, tabtoxinine-*β*-lactam (2-amino-4-(3-hydroxy-2-oxo-azacyclobutan-3-yl)-butanoic acid) (Fig. 23.4a) (Sinden and Durbin, 1968). Oxetin ((2*R*,3*S*)-3-amino-2-oxetane carboxylic acid) (Fig. 23.4a) was isolated from a fermentation broth of a soil *Streptomyces* sp. OM-2317 (Omura *et al.*, 1984b).

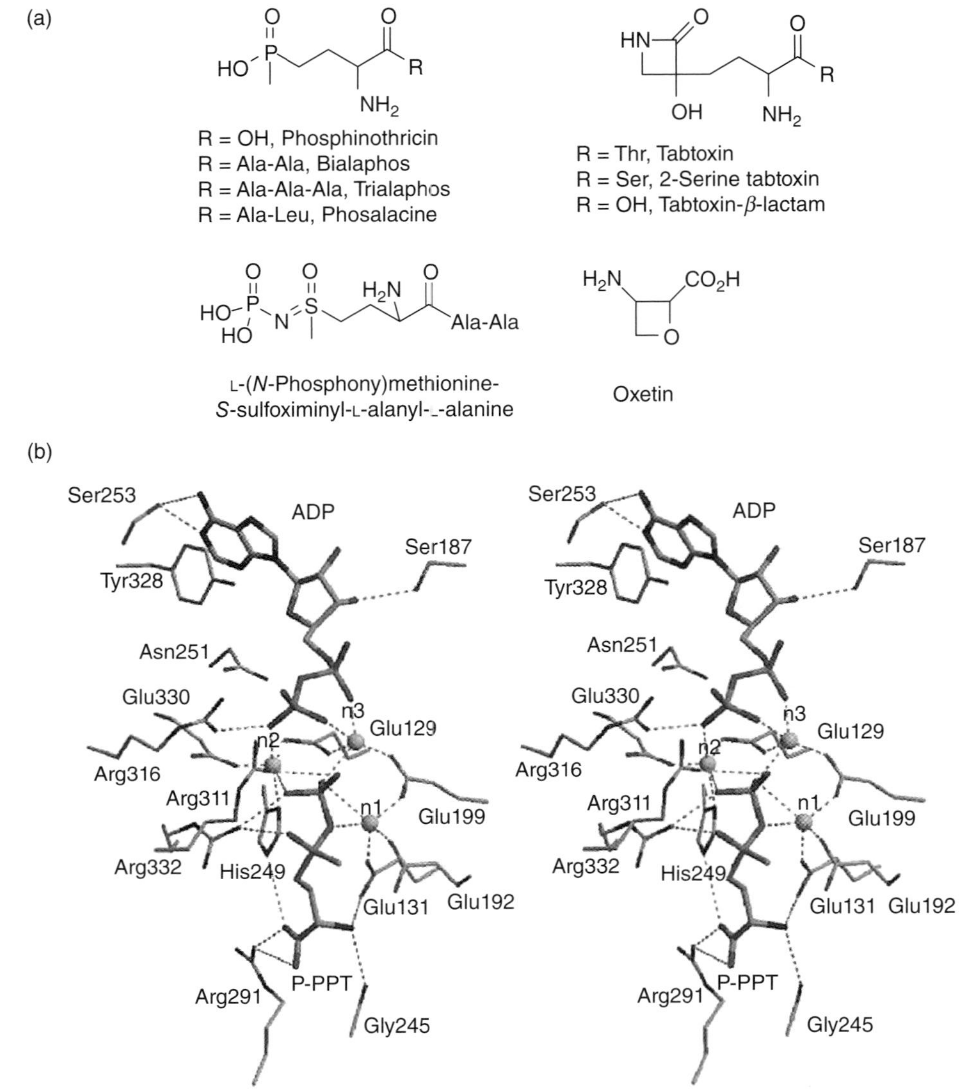

Fig. 23.4. (a) Inhibitors of glutamine synthase (GS) from natural sources. (b) Representations (stick models) of the interactions between GS, ADP and phosphinothricin (shown as P-PPT at the bottom of the figure (Unno *et al.*, 2006). Carbon, oxygen, nitrogen, phosphorus and sulfur atoms are coloured different shades of grey. Three Mn^{2+} ions are indicated in the grey spheres labelled n1, n2 and n3. The dotted lines designate hydrogen bonds and coordination bonds to the Mn^{2+} ions. Residues without dotted lines have hydrophobic interactions with the substrate. Abbreviations: Arg, arginine; Asn, asparagine; Glu, glutamate; Gly, glycine; His, histidine; Ser, serine; Tyr, tyrosine.

Methionine sulfoximine (Fig. 23.4a) was first discovered as a synthetic GS inhibitor, but was later identified in the plant tissue and bark of *Cnestis glabra*, *Cnestis polyphylla* and *Rourea orientalis*. Methionine sulfoximine is also a proherbicide, but it is activated by being phosphorylated by GS rather than as a result of hydrolysis (Ronzio and Meister, 1968; Ronzio *et al.*, 1969).

Most of the GS inhibitors mentioned above act as competitive inhibitors and form irreversible complexes with GS. The inhibition of GS results in a rapid decrease in the free pools of glutamine, glutamate, aspartate, alanine, serine and glycine (Lydon and Duke, 1999) and an accumulation of ammonia. This latter phenomenon can serve as a biological marker of GS inhibition (Dayan *et al.*, in preparation).

A maize GS has been crystallized and its three-dimensional structure has recently been determined (Unno *et al.*, 2006). The active enzyme consists of a decameric structure with little resemblance to the structure of the bacterial GS. GS was also crystallized with phosphinothricin complexed to the active domain (Fig. 23.4b). The crystallized pentamer contains ten phosphinothricin molecules, one bound to each of the ten substrate-binding sites. Analysis of this structure has been an important step in understanding the reaction mechanism of the enzyme. The structure is also a useful template for guiding the development of potential new herbicides with greater activity against GS (Occhipinti *et al.*, 2010).

Glufosinate has a wide spectrum of weed control. Its lack of selectivity initially hindered its commercialization in row crops. However, transgenic plants expressing either a bialaphos resistance (*bar*) gene or a phosphinothricin *N*-acetyltransferase (*pat*) gene from different *Streptomyces* spp. are extremely resistant to glufosinate (Strauch *et al.*, 1988; Thomas *et al.*, 2004; Áy *et al.*, 2012). A number of glufosinate-resistant transgenic varieties of several crops have now been developed, including cotton, canola, corn, soybean, sugarbeet and rice. However, only the varieties of canola, cotton, soybean and maize have been commercialized. The *bar* gene is used in glufosinate-resistant cotton whereas the *pat* gene is used in glufosinate-resistant maize and soybean. Glufosinate-resistant crops are commercialized as LibertyLink® crops by Bayer CropScience.

23.8 Inhibitors of Ornithine Transcarbamoylase

Ornithine transcarbamoylase (OCTase) is involved in arginine production in that its product, citrulline, is a precursor of arginine. Phaseolotoxin (Fig. 23.5) is a tripeptide produced by *Pseudomonas syringae* pv. *phaseolicola*. It is a protoxin which, in addition to acting as a strong phytotoxin, has potent antimicrobial activity through its

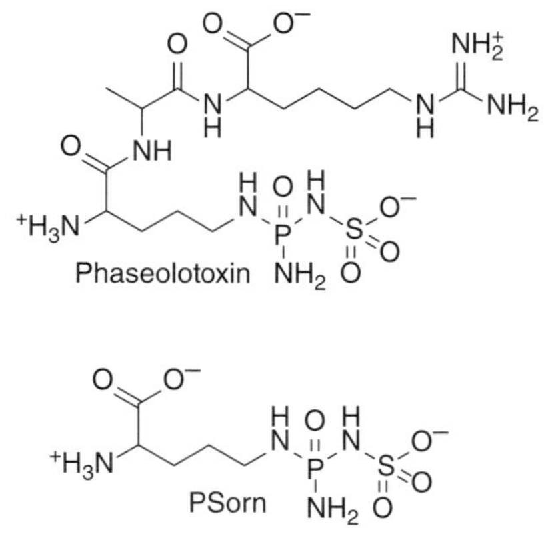

Fig. 23.5. Structures of phaseolotoxin and PSorn (N^{δ}-(N^{1}-sulfodiaminophospinhyl)-L-ornithine).

inhibition of microbial OCTase (Bender *et al.*, 1999). Plant peptidases convert it to N^{δ}-(N^{1}-sulfodiaminophospinhyl)-L-ornithine (PSorn) (Fig. 23.5), which is a much more potent inhibitor of OCTase than the original phaseolotoxin (Templeton *et al.*, 1985, 2005). Also, while phaseolotoxin is a reversible OCTase inhibitor, PSorn inhibits the enzyme irreversibly (Templeton *et al.*, 1985).

Phaseolotoxin causes the symptoms of halo blight disease in beans (*Phaseolus vulgaris*) (Volksch *et al.*, 1984) and has been proposed as a model for future herbicides (Fritche *et al.*, 1990). The chlorosis associated with halo blight has been attributed to reduced chlorophyll synthesis caused by an arginine deficit (Turner and Mitchell, 1985). Phaseolotoxin-inhibited carrot cell growth can be reversed by citrulline, but not by arginine (Jacques and Sung, 1981).

Mangotoxin, a hydrophilic oligopeptide of unknown structure produced by *Pseudomonas syringae* pv. *syringae*, is also an inhibitor of OCTase (Arrebola *et al.*, 2003). The genetics of the biosynthesis of this phytotoxin is being probed (e.g. Carrión *et al.*, 2013), even though its structure is still unknown.

23.9 Inhibitor of Dihydrodipicolinate Synthase

Dihydrodipicolinate synthase (DHDPS) catalyses the first committed step in lysine biosynthesis in plants and microorganisms. Specifically, it catalyses the aldol condensation between aspartate *β*-semialdehyde and pyruvate (Dobson *et al.*, 2004). Lysine is an essential amino acid and the inhibition of lysine synthesis is lethal, so inhibitors of DHDPS have been intensively studied as herbicides and antimicrobials.

DHDPS is a homotetramer that possesses a lysine allosteric site that strongly downregulates the activity of the enzyme (feedback inhibition) by causing large changes in the quaternary structure (Blickling *et al.*, 1997). Mutations in DHDPS that render the enzyme insensitive to the L-lysine feedback mechanism result in the accumulation of L-lysine in plants (Ghislain *et al.*, 1995). In contrast, *S*-aminoethyl-L-cysteine (Fig. 23.6), a lysine analogue, which is a natural product produced by the edible gypsy mushroom (*Cortinarius caperatus*), strongly inhibits DHDPS (Matsumoto, 1984). This natural herbicide is superior to hygromycin as a selectable marker for the selection of transgenic plants (Rao *et al.*, 2009).

23.10 Potential Inhibitors of Amino Acid Metabolism

Certain subspecies of fescue (*Festuca* spp.) are highly allelopathic due to root exudation of *m*-tyrosine (Fig. 23.7) into their environment (Bertin *et al.*, 2007, 2009). *m*-Tyrosine inhibits plant growth at submillimolar concentrations. *Festuca* produces *m*-tyrosine by hydroxylation of L-phenylalanine near the root tip, while the spurge (*Euphorbia myrsintes*) transaminates *m*-hyroxyphenylpyruvate to produce the compound (Huang *et al.*, 2012). Although the mode of action of this phytotoxin is unknown, mutants of *A. thaliana* with greatly enhanced levels of phenylalanine are resistant to *m*-tyrosine (Huang *et al.*, 2010), indicating that phenylalanine may compete for a target binding site of *m*-tyrosine.

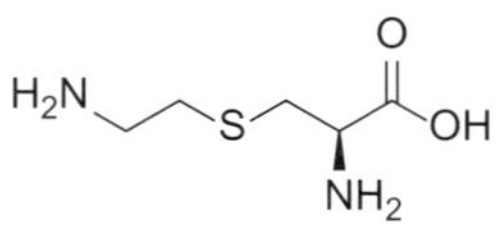

Fig. 23.6. Structure of *S*-aminoethyl-L-cysteine.

23.11 Ascaulitoxin Aglycone – A Potential Aminotransferase Inhibitor

Ascochyta caulina, a fungus being studied as a potential mycoherbicide (Vurro *et al.*, 2001), produces ascaulitoxin aglycone (Fig. 23.8a). It is a potent phytotoxin at submicromolar concentrations that has profound effects on amino acid metabolism as determined by metabolomics studies with *Lemna paucicostata* (Duke *et al.*, 2011) (Fig. 23.8b); which showed reduced levels of alanine, serine, asparagine, aspartate, glutamine, glutamate and arginine, but increased levels of most other amino acids. Feeding treated plants with all amino acids except for glutamine, valine and leucine reversed the inhibition of growth caused by the toxin (Fig. 23.8c). However, *in vitro* assays found that the toxin did not inhibit alanine aminotransferase nor alanine: glyoxylate aminotransferase, two enzymes that were predicted to be affected from analysis of the metabolomics results. Even supplementation with the non-protein amino acids ornithine, citrulline, D-lysine, and D-arginine reduced the activity of the phytotoxin. This led the authors to speculate that the phytotoxin might inhibit another aminotransferase or one or more amino acid transporters.

23.12 Enhanced Photodegradation of L-Phenylalanine

Aspergillomarasmine A (Fig. 23.9), produced by the plant pathogens *Pyrenophora teres* (Weiergang *et al.*, 2002), *Colletotrichum gloeosporoides* (Bousquet *et al.*, 1971), *Aspergillus flavus oryzae* (Robert *et al.*, 1962) and *Fusarium oxysporum* (Trouvelot *et al.*, 1971), chelates Fe^{3+}, and in the chelated form, it induces the photodecomposition of L-phenylalanine (alone or in peptides) to benzaldehyde (Barbier, 1987).

23.13 Final Thoughts

The compounds discussed above are probably only a small fraction of the natural phytotoxins that interfere with amino acid metabolism. This statement is based on the facts that the modes of

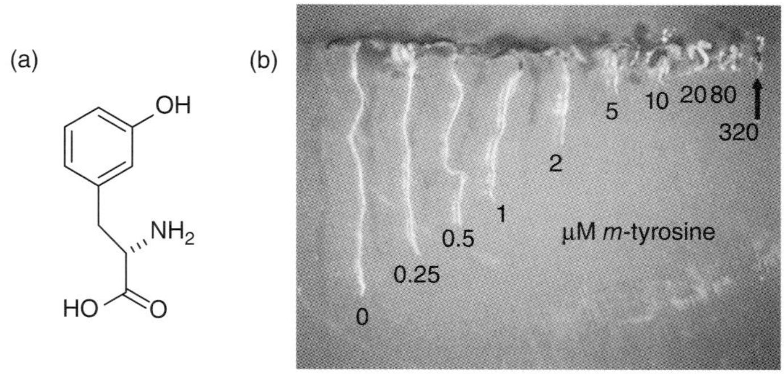

Fig. 23.7. (a) Structure of *m*-tyrosine; and (b) and its effect on root growth of *Arabidopsis thaliana* at steadily increasing concentrations (from Bertin *et al*., 2007).

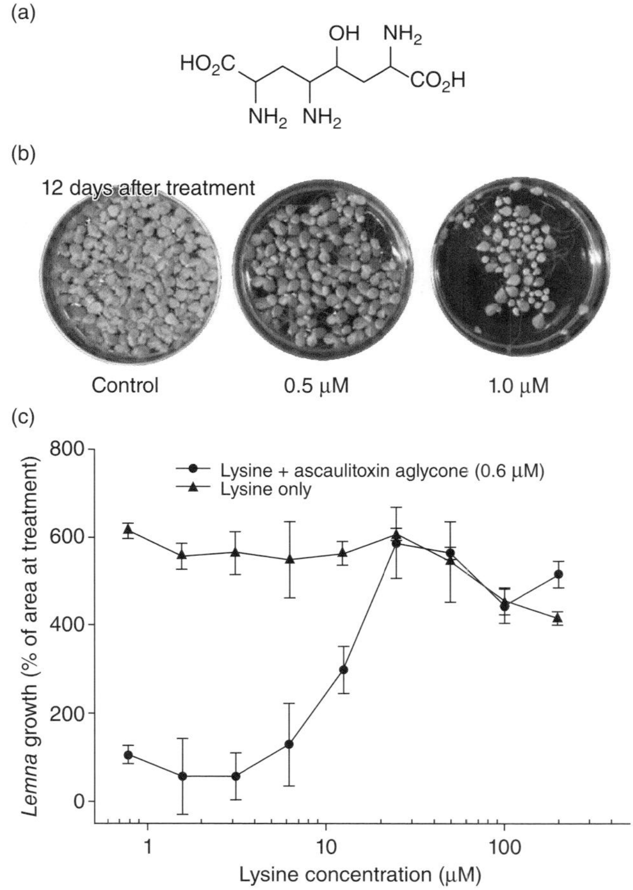

Fig. 23.8. (a) Structure of ascaulitoxin aglycone. (b) Effects of ascaulitoxin aglycone on growth of *Lemna paucicostata* after 12 days of exposure. (c) Effects of different concentrations of L-lysine on ascaulitoxin aglycone (AscA)-inhibited growth of *L. paucicostata* after 7 days exposure to both chemicals. Error bars are 1 SE (from Duke *et al*., 2011).

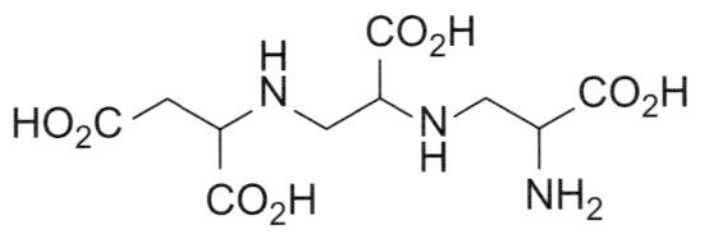

Fig. 23.9. Structure of aspergillomarasmine A.

action of most natural phytotoxins are unknown and that interference with amino acid metabolism is involved in a significant fraction of the known modes of action of natural phytotoxins (e.g. Duke and Dayan, 2011a,b). Furthermore, the phytotoxins associated with most plant pathogens (especially those that infect non-crop species) and soil microbes have not yet been discovered. Herbicides that inhibit amino acid synthesis account for the majority of worldwide herbicide usage, even though these herbicides affect only three target sites, GS, 5-enolpyruvylshikimate-3-phosphate synthase (EPSPS) and acetolactate synthase (ALS) (Duke and Dayan, 2011b; Duke, 2012). Efforts to discover new compounds with new modes of action are under way, partly because such herbicides are badly needed because of the resistance that has evolved to almost all of the 20 or so modes of action of current commercial herbicides, including their effects on GS, EPSPS and ALS (Duke, 2012). We hope that these efforts will lead to further discovery of natural phytotoxins that interfere with amino acid metabolism.

References

Aharoni, N. (1978) Relationship between leaf water status and endogenous ethylene in detached leaves. *Plant Physiology* 61, 658–662.

Amagasa, T., Paul, R.N., Heitholt, J.J. and Duke, S.O. (1994) Physiological effects of cornexistin on *Lemna pauscicostata*. *Pesticide Biochemistry and Physiology* 49, 37–52.

Andrulis, I.L., Duff, C., Evans-Blackler, S., Worton, R. and Siminovitch, L. (1983) Chromosomal alterations associated with overproduction of asparagine synthetase in albizziin-resistant Chinese hamster ovary cells. *Molecular and Cellular Biology* 3, 391–398.

Arrebola, E., Cazorla, F.M., Durán, V.E., Rivera, E., Olea, F., Codina, J.C., Pérez-Garciá, A. and de Vicente, A. (2003) Mangotoxin: a novel antimetabolite toxin produced by *Pseudomonas syringae* inhibiting ornithine/arginine biosynthesis. *Physiological and Molecular Plant Pathology* 63, 117–127.

Áy, Z., Mihály, R., Cserháti, M., Kótai, E. and Pauk, J. (2012) The effect of high concentrations of glufosinate ammonium on the yield components of transgenic spring wheat (*Triticum aestivum* L.) constitutively expressed with the *bar* gene. *The Scientific World Journal* 2012, Article ID 657945.

Barbier, M. (1987) Remarks on the biological activity of aspergillomarasmine A Fe3+ chelate and other iron transporting phytotoxins with reference to their role in the photodegradation of aromatic amino acids in infected plant leaves. *Journal of Phytopathology* 120, 365–368.

Bayer, E., Gugel, K.H., Hägele, K., Hagenmaier, H., Jassipow, S., König, W.A. and Zähner, H. (1972) Stoffwechselprodukte von mikroorganismen. Phosphinothricin und phosphinothricyl-alanyl-alanine. *Helvetica Chimica Acta* 55, 224–239.

Becerril, J.M. and Duke, S.O. (1989) Acifluorfen effects on intermediates of chlorophyll synthesis in green cucumber cotyledon tissues. *Pesticide Biochemistry and Physiology* 35, 119–126.

Bender, C.L., Alarcon-Chaidez, F. and Gross, D.C. (1999) *Pseudomonas syringae* phytotoxins: mode of action, regulation, and biosynthesis by peptide and polyketide synthetases. *Microbiology and Molecular Biology Reviews* 63, 266–292.

Bertin, C., Weston, L.A., Huang, T., Jander, G. Owens, T., Meinwald, J and Schroeder, F.C. (2007) Grass roots chemistry: *meta*-tyrosine, an herbicidal nonprotein amino acid. *Proceedings of the National Academy of Sciences of the United States of America* 104, 16964–16964.

Bertin, C., Harmon, R., Akaogi, M., Weidenhamer, J.D. and Weston, L.A. (2009) Assessment of the phytotoxic potential of *m*-tyrosine in laboratory soil bioassays. *Journal of Chemical Ecology* 35, 1288–1294.

Blickling, S., Beisel, H.-G., Bozic, D., Knäblein, J., Laber, B. and Huber, R. (1997) Structure of dihydrodipicolinate synthase of *Nicotiana sylvestris* reveals novel quaternary structure. *Journal of Molecular Biology* 274, 608–621.

Bousquet, J.F., Vegh, I., Pouteau-Thouvenot, M. and Barbier, M. (1971) Isolation of aspergillomarasimine A from cultures of *Colletotrichum gloeosporiodes*, willow pathogens. *Annales de Phytopathologie* 3, 407–408.

Brown, E.G. and Turan, Y. (1996) Formation of albizziine and 2,3-diaminopropanoic acid from uracil in *Albizia* seedlings. *Phytochemistry* 41, 1491–1495.

Brunner, H.G., Chemla, P., Dobler, M.R., O'Sullivan, A.C., Pachlatko, P., Pillonel, C. and Stierli, D. (2007) Fungicidal properties of acivicin and its derivatives. *American Chemical Society Symposium Series* 948, 121–135.

Carrión, V.J., Arrebola, E., Cazorla, F.M., Murillo, J. and de Vicente, A. (2013) The *mbo* operon is specific and essential for biosynthesis of mangotoxin in *Pseudomonas syringae*. *PLoS ONE* 7(5): e36709.

Cartier, M., Chang, M.W.M. and Stanners, C.P. (1987) Use of the *Escherichia coli* gene for asparagine synthetase as a selective marker in shuttle vector capable of dominant transfection and amplification in animal cells. *Molecular and Cellular Biology* 7, 1623–1628.

Chmara, H., Andruszkiewicz, R. and Borowski, E. (1986) Inactivation of glucosamine-6-phosphate synthetase from *Salmonella typhimurium* LT2 by fumaroyl diaminopropanoic acid derivatives, a novel group of glutamine analogs. *Biochimica et Biophysica Acta* 870, 357–366.

Conti, P., Roda, G., Stabile, H., Vanoni, M.A., Curti, B. and De Amici, M. (2003) Synthesis and biological evaluation of new amino acids structurally related to the antitumor agent acivicin. *Farmaco* 58, 683–690.

Copping, L.G. and Duke. S.O. (2007) Natural products that have been used commercially as crop protection agents – a review. *Pest Management Science* 63, 524–554.

Dayan, F.E. and Duke, S.O. (2009) Biological activity of allelochemicals. In: Osbourn, A. and Lanzotti, V. (eds) *Plant-Derived Natural Products – Synthesis, Function and Application*. Springer, Dordrecht, The Netherlands, pp. 361–384.

Dayan, F.E., Romagni, J.G. and Duke, S.O. (2000) Investigating the mode of action of natural phytotoxins. *Journal of Chemical Ecology* 26, 2079–2094.

Dayan, F.E., Owens, D.K. and Duke, S.O. (2012) Rationale for a natural products approach to herbicide discovery. *Pest Management Science* 68, 519–528.

Demko, V., Pavlovic, A. and Hudak, J. (2010) Gabaculine alters plastid development and differentially affects abundance of plastid-encoded DPOR and nuclear-encoded GluTR and FLU-like proteins in spruce cotyledons. *Journal of Plant Physiology* 167, 693–700.

Dobson, R.C.J., Gerrard, J.A. and Grant, P. (2004) Dihydrodipicolinate synthase is not inhibited by its substrate, (*S*)-aspartate β-semialdehyde. *Biochemical Journal* 377, 757–762.

Duke, S.O. (2012) Why have no new herbicide modes of action appeared in recent years? *Pest Management Science* 68, 505–512.

Duke, S.O. and Cerdeira, A.L. (2005) Potential environmental impacts of herbicide-resistant crops. In: *Collection of Biosafety Reviews, Vol.* 2. International Centre for Genetic Engineering and Biotechnology, Trieste, Italy, pp. 66–143.

Duke, S.O. and Dayan, F.E. (2011a) Modes of action of microbially-produced phytotoxins. *Toxins* 3, 1038–1064.

Duke, S.O. and Dayan, F.E. (2011b) Bioactivity of herbicides. In: Moo-Young, M. (ed) *Comprehensive Biotechnology Vol. 4*, 2nd edn. Elsevier, Amsterdam, pp. 23–35.

Duke, S.O., Evidente, A., Fiore, M., Rimando, A.M., Dayan, F.E., Vurro, M., Christiansen, N., Looser, R., Hutzler, J. and Grossmann, K. (2011) Effects of the aglycone of ascaulitoxin on amino acid metabolism in *Lemna paucicostata*. *Pesticide Biochemistry and Physiology* 100, 41–50.

Duke, S.O., Bajsa, J. and Pan, Z. (2013) Omics methods for probing the mode of action of natural and synthetic phytotoxins. *Journal of Chemical Ecology* 39, 333–347.

Duodu, S., Bhuvaneswari, T.V., Stokkermans, T.J. and Kent, P.N. (1999) A positive role for rhizobitoxine in *Rhizobium*-legume symbiosis. *Plant–Microbe Interactions* 12, 1082–1089.

Ebert, E., Leist, K.H. and Mayer, D. (1990) Summary of safety evaluation toxicity studies of glufosinate ammonium. *Food Chemistry and Toxicology* 28, 339–349.

Fritche, W., Voelksch, B., Nueske, J., Schmauder, H.P., Guenther, T. and Weiss, I. (1990) [Microbial ecology as the basis for the development of biotechnology. Microbial phytoeffectors.] *Wissenschaftliche Zeitschrift der Friedrich-Schiller-Universität Jena/Thüringen: Mathematisch-Naturwissenschaftliche* 39, 327–336.

Ghislain, M., Frankard, V. and Jacobs, M. (1995) A dinucleotide mutation in dihydrodipicolinate synthase of *Nicotiana sylvestris* leads to lysine overproduction. *The Plant Journal* 8, 733–743.

Giancaspro, A., Rosellini, D., Blanco, A. and Gadaleta, A. (2012) Gabaculine selection using bacterial and plant marker genes (GSA-AT) in durum wheat transformation. *Plant Cell, Tissue and Organ Culture* 109, 447–455.

Giovanelli, J., Owens, L.D. and Mudd, S.H. (1973) β-cystathionase. *In vivo* inactivation by rhizobitoxine and role of the enzyme in methionine biosynthesis in corn seedlings. *Plant Physiology* 51, 492–503.

Gmelin, R. (1959) Free amino acids in seeds of *Acacia willardiana*. Isolation of willardine, a new plant amino acid, probably L-uracil-1-(α-aminoproprionic acid). *Hoppe-Seyler's Zeitschrift für Physiologie Chemie* 316, 164–169.

Gmelin, R., Strauss, G. and Hasenmaier, G. (1958) The isolation of two new plant amino acids, *S*-(2-carboxyethyl)-L-cysteine and albizzin, from seeds of *Albizzia julibrissin. Zeithschrift für Naturforschung* 13b, 252–256.

Gould, S.J. and Ju, S. (1989) The biosynthesis of acividin and 4-hydroxyacivicin from N$^{\delta}$-hydroxyornithine. *Journal of the American Chemical Society* 111, 2329–2331.

Grossmann, K., Hutzler, J., Tresch, S., Christiansen, N., Looser, R. and Ehrhardt, T. (2012) On the mode of action of the herbicides cinmethylin and 5-benzyloxymethyl-1,2-isoxazolinones: putative inhibitors of plant tyrosine aminotransferase. *Pest Management Science* 68, 482–492.

Grzam, A., Martin, M.N., Hell, R. and Meyer, A.J. (2007) γ-Glutamyl transpeptidase GGT4 initiates vacuolar degradation of glutathione *S*-conjugates in *Arabidopsis*. *FEBS Letters* 58, 3131–3138.

Gulati, D.K., Chambers, C.L., Rosenthal, G.A. and Sabharwal, P.S. (1981) Comparative toxicity of some naturally occurring and synthetic nonprotein amino acids. *Environmental and Experimental Botany* 21, 225–230.

Huang, T., Tohge, T, Lytovchenko, A., Fernie, A.R. and Jander, G. (2010) Pleiotrophic physiological consequences of feedback-insensitive phenylalanine biosynthsis in *Arabidopsis thaliana*. *The Plant Journal* 63, 823–835.

Huang, T., Rehak, L. and Jander, G. (2012) *meta*-Tyrosine in *Festuca rubra* ssp. *commutata* (Chewings fescue) is synthesized by hydroxylation of phenylalanine. *Phytochemistry* 75, 60–66.

Jacques, S. and Sung, Z.R. (1981) Regulation of pyrimidine and arginine biosynthesis investigated by the use of phaseolotoxin and 5-fluorouracil. *Plant Physiology* 67, 287–291.

Kilmer, J.L., Widholm, J.M. and Slife, F.W. (1981) Reversal of glyphosate inhibition of carrot cell culture growth by glycolytic intermediates and organic and amino acids. *Plant Physiology* 68, 1299–1302.

Kobayashi, K., Miyazawa, S., Terahara, A., Mishima, H. and Kurihara, H. (1976) Gabaculine: γ-aminobutyrate aminotransferase inhibitor of microbial origin. *Tetrahedron Letters* 7, 537–540.

Kondo, K., Shomura, Ogawa, Y., Tsuruoka, T., Watanabe, H., Totsukawa, K., Suzuki, T., Moriyama, C., Yoshida, J., Inouye, S. and Niida, T. (1972) Studies on a new antibiotic SF-1293. I. Isolation and physico-chemical and biological characterization of SF-1293 substances. *Scientific Reports of Meiji Seika Kaisha* 13, 34–41.

Lam, H.-M., Peng, S. S.-Y. and Coruzzi, G.M. (1994) Metabolic regulation of the gene encoding glutamine-dependent asparagine synthetase in *Arabidopsis thaliana*. *Plant Physiology* 106, 1347–1357.

Lea, P.J. and Fowden, L. (1973) Amino acid substrate specificity of asparaginyl-, aspartyl-, and glutaminyl-tRNA synthetase isolated from higher plants. *Phytochemistry* 12, 1903–1916.

Lea, P.J. and Fowden, L. (1975) The purification and properties of glutamine-dependent asparagine synthetase isolated from *Lupinus albus*. *Proceedings of the Royal Society of London. Series B, Biological Science* 192, 13–26.

Lea, P.J., Sodek, L., Parry, M.A.J., Shewry, P.R. and Halford, N.G. (2006) Asparagine in plants. *Annals of Applied Biology* 150, 1–26.

Lydon, J. and Duke, S.O. (1999) Inhibitors of glutamine biosynthesis. In: Singh, B.K. (ed.) *Plant Amino Acids: Biochemistry and Biotechnology*. Marcel Dekker, New York, pp. 445–464.

Ma, W., Penroe, D. and Glick, B.R. (2002) Strategies use by rhizobia to lower plant ethylene levels and increase nodulation. *Canadian Journal of Microbiology* 48, 947–954.

Matsumoto, N. (1984) Isolation and identification of *S*-2-aminoethyl-L-cysteine from *Rozites caperta* and 2-amino-3-butenoic acid from *Rhodophyllus crassipes* and their antibacterial activity. *Toho Igakkai Zasshi* 31, 249–264.

McSweeney, C.S., Gough, J., Conlan, L.L., Hegarty, M.P., Palmer, B. and Krause, D.O. (2005) Nutritive value assessment of the tropical shrub legume *Acacia angustissima*: anti-nutritional compounds an *in vitro* digestibility. *Animal Feed Science and Technology* 121, 175–190.

Michel, A., Johnson, R.D., Duke, S.O. and Scheffler, B.E. (2004) Dose–response relationships between herbicides with different modes of action and growth of *Lemna paucicostata* – an improved ecotoxicological method. *Environmental and Toxicological Chemistry* 23, 1074–1079.

Nair, S.P., Kannangara, C., Gamini, C., Harwood, J.L. and John, R.A. (1990) Inhibition studies on 5-aminolevulinate biosynthesis in *Pisum sativum* L. (pea). *Biochemical Society Transactions* 18, 656–657.

Natsumeda, Y., Ikegami, T., Olah, E. and Weber G. (1989) Significance of purine salvage in circumventing the action of antimetabolites in rat hepatoma cells. *Cancer Research* 49, 88–92.

Nishino, T. and Murao, S. (1983) Isolation and some properties of an aspartate aminotransferase inhibitor. *Agricultural and Biological Chemistry* 47, 1961–1966.
Nishino, T., Murao, S. and Wada, H. (1984) Mechanism of inactivation of pyridoxal phosphate-linked aspartate transaminase by gostatin. *The Journal of Biochemistry* 95, 1283–1288.
Oaks, A. and Ross, D.W. (1984) Asparagine synthetase in *Zea mays*. *Canadian Journal of Botany* 62, 68–73.
Oaks, A., Jones, K. and Misra, S. (1979) A comparison of glutamate synthase obtained from maize endosperms and root. *Plant Physiology* 63, 793–795.
Occhipinti, A., Berlicki, Ł., Giberti, S., Dzięedzioła, G., Kafarski, P. and Forlani, G. (2010) Effectiveness and mode of action of phosphonate inhibitors of plant glutamine synthetase. *Pest Management Science* 66, 51–58.
Okazaki, S., Sugawara, M., Yuhashi, K.-I. and Minamisawa, K. (2007) Rhizobitoxine-induced chlorosis occurs in coincidence with methionine deficiency in soybeans. *Annals of Botany* 100, 55–59.
Omura, S., Murata, M., Hanaki, H., Hinitozawa, K., Oiwa, R. and Tanaka, H. (1984a) Phosalacine, a new herbicidal antibiotic containing phosphinothricin. Fermentation, isolation, biological activity and mechanism of action. *Journal of Antibiotics* 37, 829–835.
Omura, S., Murata, M., Imamura, N., Iwai, Y. and Tanaka, H. (1984b). Oxetin, a new antimetabolite from an actinomycete. Fermentation, isolation, structure and biological activity. *Journal of Antibiotics* 37, 1324–1332.
Owens, L.D. (1973) Herbicidal potential of rhizobitoxine. *Weed Science* 21, 63–66.
Owens, L.D., Guggenheim, S. and Hilton, J.L. (1968) *Rhizobium*-synthesized phytotoxin: an inhibitor of *β*-cystathionase in *Salmonella typhimurium*. *Biochimica et Biophysica Acta* 158, 219–225.
Owens, L.D., Lieberman, M. and Kunishi, A. (1971) Inhibition of ethylene production by rhizobitoxine. *Plant Physiology* 48, 1–4.
Rajaram, V., Prasuna, P.R., Savithri, H.S. and Murthy, M.R.N. (2008) Structure of biosynthetic *N*-acetylornithine aminotransferase from *Salmonella typhimurium*: studies on substrate specificity and inhibitor binding. *Proteins* 70, 429–441.
Rando, R.R. (1977) Mechanism of the irreversible inhibition of gamma-aminobutyric acid-alpha ketoglutaric acid transaminase by the neurotoxin gabaculin. *Biochemistry* 16, 4604–4610.
Rao, S.S., Mamadou, L., McConnell, M., Polisetty, R., Kwanyuen, P. and Hildebrand, D. (2009) Non-antibiotic selection systems for soybean somatic embryos: the lysine analog aminoethyl-cysteine as a selection agent. *BMC Biotechnology* 9:94.
Robert, M., Barbier, M., Lederer, E., Roux, L., Biemann, K. and Vetter, W. (1962) Two new natural phytoxins: aspergillosmarasmine A and B and their relation to lycomarasmine and its derivatives. *Bulletin de la Société Chimique de France* 1962, 187–188.
Roberts, R.L., Metz, M., Monks, D.E., Mullaney, M.L., Hall, T. and Nester, E.W. (2003) Purine synthesis and increased *Agrobacterium tumefaciens* transformation of yeast and plant. *Proceedings of the National Academy of Sciences of the United States of America* 100, 6634–6639.
Rodriguez-Suerez, R., Xu, D., Veillette, K., Davison, J., Sillaots, S., Kauffman, S., Hu, W., Bowman, H., Martel, N., Trosok, S., Wang, H., Zhang, L. *et al.* (2007) Mechanism-of-action determination of GMP synthase inhibitors and target validation in *Candida albicans* and *Aspergillus fumigatus*. *Chemistry and Biology* 14, 1163–1175.
Ronzio, R.A. and Meister, A. (1968) Phosphorylation of methionine sulfoximine by glutamine synthetase. *Proceedings of the National Academy of Sciences of the United States of America* 59, 164–170.
Ronzio, R.A., Rowe, W.B. and Meister, A. (1969) Mechanism of inhibition of glutamine synthetase by methionine sulfoximine. *Biochemistry* 8, 1066–1075.
Rosellini, D., Capamaccio, S., Ferradini, N., Savo Sardaro, M.L., Nicolia, A. and Veronesi, F. (2007) Non-antibiotic, efficient selection for alfalfa genetic engineering. *Plant Cell Reports* 26, 1035–1044.
Sinden, S.L. and Durbin, R.D. (1968) Glutamine synthetase inhibition: possible mode of action of wildfire toxin from *Pseudomonas tabaci*. *Nature* 219, 379–380.
Singh, R.P. and Srivastava, H.S. (1987) *In vivo* effect of some metabolic inhibitors on glutamate dehydrogenase and glutamate synthase activities in excised maize tissues. *Current Science* 56, 93–94.
Strauch, E., Wohlleben, W. and Pühler, A. (1988). Cloning of a phosphinothricin *N*-acetyltransferase gene from *Streptomyces viridochromogenes* Tü494 and its expression in *Streptomyces lividans* and *Escherichia coli*. *Gene* 63, 65–74.
Sugawara, M., Okazaki, S., Nukui, N., Ezura, H., Mitsui, H. and Minamisawa, K. (2006) Rhizobitoxine modulates plant–microbe interactions by ethylene inhibition. *Biotechnology Advances* 24, 382–388.
Templeton, M.D., Mitchell, R.E., Sullivan, P.A. and Shepherd, M.G. (1985) The inactivation of ornithine transcarbamoylase by *N′*-(*N′*-sulpho-diaminophosphinyl)-L-ornithine. *Biochemical Journal* 228, 347–352.

Templeton, M.D., Reinhardt, L.A., Collyer, C.A., Mitchell, R.E. and Cleland, W.W. (2005) Kinetic analysis of the L-ornithine transcarbamolylase from *Pseudomonas savastanoi* pv. *phaseolicola* that is resistant to the transition state analogue (*R*)-N^{δ}-(*N*′- sulfodiaminophosphinyl)-L-ornithine. *Biochemistry* 44, 4408–4415.
Thomas, W.E., Pline, W.A., Wilcut, J.W., Edmisten, K.L., Wells, R., Viator, R.P. and Paulsgrove, M.D. (2004) Glufosinate does not affect floral morphology and pollen viability in glufosinate-resistant cotton. *Weed Technology* 18, 258–262.
Trotta, P.P., Pinkus, L.M. and Meister, A. (1974) Inhibition by dithiothreitol of the utilization of glutamine by carbamyl phosphate synthetase. Evidence for formation of hydrogen peroxide. *The Journal of Biological Chemistry* 249, 1915–1921.
Trouvelot, A., Camporata, P., Barbier, M. and Pouteau-Thouvenot, M. (1971) [Isolation of aspergillomarasimine A from culture media of *Fusarium oxysporum*]. *Comptes Rendus Hebdomadaires des Séances de'Académie des Sciences. Série D: Sciences Naturelles* 272, 754–756.
Turnbull, M.H., Goodall, R. and Steward, G.R. (1996) Evaluating the contribution of glutamate deydrogenase and glutamate synthase cycle to ammonia assimilation by four ectomycorrhizal fungal isolates. *Australian Journal of Plant Physiology* 23, 151–159.
Turner, J.G. and Mitchell, R.E. (1985) Association between symptom development and inhibition of ornithine carbamoyltransferase in bean leaves treated with phaseolotoxin. *Plant Physiology* 79, 468–473.
Unno, H., Uchida, T., Sugawara, H., Kurisu, G., Sugiyama, T., Yamaya, T., Sakakibara, H., Hase, T. and Kusunoki, M. (2006) Atomic structure of plant glutamine synthetase – a key enzyme for plant productivity. *The Journal of Biological Chemistry* 281, 29287–29296.
Volksch, B., Laplace, F. and Fritsche, W. (1984) Studies on the variability of the phaseolotoxin production by *Pseudomonas syringae* pv. *phaseolicola*. *Zentralblatt für Mikrobiologie* 139, 109–118.
Vurro, M., Zonno, M.C., Evidente, A., Andolfi, A. and Montemurro, P. (2001) Enhancement of efficacy of *Ascochyta caulina* to control *Chenopodium album* by use of phytotoxins and reduced rates of herbicides. *Biological Control* 21, 182–190.
Weiergang, I., Lyngs Jorgensen, H.J., Moller, I.M., Friis, P. and Smedegaard-Petersen, V. (2002) Optimization of *in vitro* growth condition of *Pyrenophera teres* for production of the phytotoxin aspergillomarasmine A. *Physiological and Molecular Plant Pathology* 60, 131–140.
Wild, A. and Wendler, C. (1993) Inhibitory action of glufosinate on photosynthesis. *Zeitschrift für Naturforschung* 48c, 369–373.
Wild, A. and Ziegler, C. (1989) The effect of bialaphos and ammonium on ammonium-assimilation and photosynthesis. I. Effect on the enzymes of ammonium-assimilation. *Zeitschrift für Naturforschung* 44c, 97–102.
Williams, K., Cullati, S., Sand, A., Biterova, E.I. and Barycki, J.J. (2009) Crystal structure of acivicin-inhibited γ-glutamyltranspeptidase reveals critical roles for its C-terminus in autoprocessing and catalysis. *Biochemistry* 48, 2459–2467.
Winter, H.C. and Dekker, E.E. (1991) 4-Methyleneglutamine amidohydrolase from peanut leaves. Preparation, catalytic properties, and immunological responses of a highly purified form of the enzyme. *Plant Physiology* 95, 206–212.
Yasuta, T., Satoh, S. and Minamisawa, K. (1999) New assay for rhizobitoxine based on inhibition of 1-aminocyclopropane-1-carboxylate synthase. *Applied Environmental Microbiology* 65, 849–852.
Zimmerman, R.H., Lieberman, M. and Broome, O.C. (1977) Inhibitory effect of rhizobitoxine analog on bud growth after release from dormancy. *Plant Physiology* 59, 158–160.

24 Glyphosate: The Fate and Toxicology of a Herbicidal Amino Acid Derivative

D.A. Saltmiras,* D.R. Farmer, A. Mehrsheikh and M.S. Bleeke
Monsanto, St Louis, Missouri, USA

24.1 Abstract

Glyphosate was first registered in the USA in 1974 for non-crop and industrial use as a herbicide. With the advent of glyphosate-tolerant crops and the adoption of minimum and no-till practices, it has become one the world's foremost herbicides. Glyphosate inhibits a key step in the shikimate acid pathway and thereby the synthesis of the aromatic amino acids tyrosine, phenylalanine and tryptophan in plants. This is achieved by glyphosate binding to the enzyme 5-enolpyruvylshikimate-3-phosphate synthase (EPSPS). As the shikimate pathway is not present in the animal kingdom, animals cannot synthesize these three aromatic amino acids, which are therefore considered essential for humans and must be acquired via the diet.

Following foliar application, glyphosate penetrates the cuticle and is rapidly absorbed and translocated throughout the plant. Aminomethylphosphonic acid (AMPA) is the predominant metabolite within plants. Environmental degradation of glyphosate occurs primarily via microbial degradation, which also produces AMPA on the pathway to full mineralization. *In vitro* metabolic studies in mammals have demonstrated that nearly all glyphosate is excreted unchanged, and what little AMPA is formed is most likely attributable to the metabolism of gastrointestinal microflora.

Glyphosate has undergone extensive toxicology testing over the last 40 years, with at least six separate toxicology data sets generated by different registering companies across the globe. These data are remarkably consistent in demonstrating an absence of concern for developmental and reproductive toxicity, and no carcinogenicity, no genotoxicity, no endocrine disruption potential, no neurotoxicity and no immunotoxicity. AMPA has been shown to be of equal or lesser toxicity to the parent compound. In contrast, published literature and the Internet are bestrewn with allegations of adverse toxic effects attributable to glyphosate, but these are largely data collected from high-dosed *in vitro* test systems, from studies using inappropriate routes of exposure or by unvalidated test methods.

24.2 Introduction

Glyphosate, the active ingredient in the Roundup products produced by Monsanto and other formulations produced by dozens of different companies, has arguably become the most widely applied herbicide in the history of the world. Significant societal benefits accompanying the widespread adoption of glyphosate as an agricultural tool to control weeds in glyphosate-tolerant crops include: (i) greater food security through higher crop yields; (ii) increased farm incomes resulting from higher productivity; and (iii) increased adoption of

*E-mail address: david.a.saltmiras@monsanto.com

sustainable agricultural practices such as no/low-till farming and thus reduced soil erosion and reduced greenhouse gases (Brookes and Barfoot, 2008). For 40 years since the first registration of a glyphosate-based herbicide, select non-governmental organizations (NGOs) and activist groups have staunchly opposed the use of glyphosate. This chapter, however, focuses on the environmental fate and toxicology of glyphosate, and demonstrates that these sciences provide a solid basis for the policies and regulations driving the adoption or rejection of such technologies in our modern society.

24.3 History of Glyphosate

Glyphosate is a phosphonomethyl derivative of the amino acid glycine. The compound was synthesized and tested in the greenhouse for herbicidal activity for the first time by Monsanto in 1970. The herbicidal active ingredient initially advanced through greenhouse screening and field testing, and the first Roundup brand herbicide, was introduced by Monsanto in 1974 for the control of perennial and annual weeds in non-crop and industrial areas. The uses of Roundup-branded herbicides began to expand in 1976, with the first crop use label allowing perennial weed control in perennial crops, and use before planting or following the harvest of annual crops. In the years to follow, spot spraying allowed the control of perennial weeds in annual crops such as cotton and soybeans. The use of selective application methods, such as recirculating sprayers or rope wick applicators, allowed the control of annual and perennial weeds in annual crops. Spraying glyphosate to control annual weeds before planting annual crops allowed growers to move towards reduced tillage or no-tillage systems in around 1986. The introduction of Roundup Ready technology permitted the direct application of Roundup-branded herbicides for weed control in glyphosate-tolerant crops beginning in 1996. With the advent of such crops and the adoption of minimum and no-tillage practices, glyphosate has become one the world's foremost herbicides. Over the last decade or more, dozens of companies throughout the world have taken advantage of the expiry of the glyphosate patent and developed numerous formulations of this highly successful non-selective herbicide. Over 750 separate glyphosate-based products are registered for the US market alone (NPIC, 2010).

24.4 Herbicidal Mode of Action of Glyphosate

Glyphosate affects plant growth by inhibiting the enzyme 5-enolpyruvylshikimate-3-phosphate synthase (EPSPS) in the shikimate acid pathway. This enzyme controls a key step in the synthesis of the aromatic amino acids tyrosine, phenylalanine and tryptophan in plants. When blocked, the result is starvation of these amino acids, the proteins that should include them and downstream plant products, manifesting as chlorosis, necrosis and plant death. The shikimate acid pathway is present in plants, fungi and some bacteria, but not in animals, thus providing a specific selective toxicity of glyphosate towards plant species. In humans, tyrosine, phenylalanine and tryptophan are considered to be essential and are acquired via the diet.

24.5 Physico-Chemical Properties of Glyphosate

Glyphosate (*N*-phosphonomethylglycine) is an herbicidal phosphonomethyl derivative of the amino acid glycine. It is a white and odourless crystalline solid that has one basic amino function and three ionizable acidic sites. Structural and spectroscopic studies show that glyphosate, both in crystalline form and in aqueous solution, has a zwitterionic character with a positive charge at the protonated nitrogen centre and a negative charge on the phosphonate moiety, as shown in Fig. 24.1. The pK_a values for the dissociation of the three acidic protons of glyphosate in aqueous solutions are reported as 10.14 for the

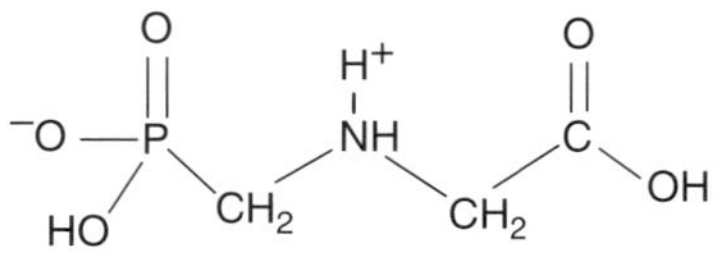

Fig. 24.1. Structural formula of glyphosate.

ammonium proton, 5.46 for the second phosphonate proton and 2.23 for the carboxylic acid functional proton (Motekaitis *et al.*, 1985).

Information on specific physico-chemical properties of chemicals has often proven useful in predicting their behaviour or fate in the environment. The physical/chemical properties of glyphosate indicate a potential for favourable commercial and environmental profiles. Owing to its amphoteric nature, glyphosate is readily dissolved in dilute aqueous bases and strong aqueous acids to produce anionic and cationic salts, respectively. The free acid of glyphosate is modestly water soluble (1.16 g l^{-1} at 25 °C) at room temperature, but when converted into monobasic salts, its solubility increases substantially. Glyphosate-based herbicidal products are generally formulated as concentrated water solutions of approximately 30–50% glyphosate in the form of the more soluble monobasic salt (isopropylamine, sodium, potassium or ammonium) because of the limited aqueous solubility of the free acid. Neither glyphosate acid nor the commercial salts are significantly soluble in common organic solvents. It is not surprising, then, that the log octanol/water partition coefficient (log K_{ow} or log P_{ow}) of glyphosate is –4.1, which predicts a low potential for its bioaccumulation in higher organisms. The lack of solubility of glyphosate in non-aqueous solvents has been attributed to the strong intermolecular hydrogen bonding in the molecule (Knuuttila and Knuuttila, 1985). Furthermore, because of its high polarity and intermolecular hydrogen bonding, glyphosate and its salts also have low volatility (2.59×10^{-5} Pa at 25 °C for free glyphosate acid). Glyphosate's low volatility and high density (1.75 g cm^{-3}) suggest that it is unlikely to evaporate from treated surfaces and therefore it will neither move through the air to injure non-target sources nor remain suspended in the air for a long time after application; it is, therefore, unlikely to produce significant exposure through inhalation. Upon heating in its neat form, it softens and undergoes a cyclo-dehydration transformation to form a diketo-piperazine dimeric anhydride at 200–230 °C, which is stable up to 316 °C. However, this cyclic anhydride is readily hydrolysed back to glyphosate under mild basic aqueous conditions or in strong acidic media. Glyphosate is non-flammable and has no critical explosive or oxidizing properties.

24.6 Glyphosate in the Environment

Glyphosate is one of the most comprehensively studied herbicides in the world, presumably because of its wide range of use in agricultural, ornamental, aquatic and wildlife habitats, and in residential environments. With the advent of glyphosate-tolerant crops and the widespread use of glyphosate products in many different crops (Duke and Powles, 2008), interest in studying the fate and behaviour of glyphosate in the environment grew significantly. In addition to the required regulatory studies, additional investigations have been undertaken by many independent research groups and institutes to look for potential adverse effects of glyphosate in the environment. The fate and behaviour of glyphosate have been reviewed by scientists from industry (Franz *et al.*, 1997), government regulatory agencies in several countries (US EPA, 1993; European Commission, 2002; and see a summary at http://www.glyphosate.eu/regulatory-documents), scientific institutions (Giesy *et al.*, 2000) and international organizations (WHO, 1994; FAO, 2005). The overall weight of evidence indicates that, when used according to recommendations, glyphosate has a low potential for direct impact on the environment. The uptake and metabolism of glyphosate by plants and its environmental fate, as summarized in these reviews, are discussed below.

24.6.1 Uptake and metabolism in plants

Investigations into the metabolic behaviour of glyphosate in intact plants are challenging because of its herbicidal activity. The slow degradation of glyphosate to aminomethylphosphonic acid (AMPA) has been reported in a number of studies conducted with a wide variety of plants or detached plant tissues (Stock *et al.*, 1991; Gout *et al.*, 1992; Komoba *et al.*, 1992). The metabolism or lack of significant metabolism of glyphosate in various conventional, i.e. non-genetically modified, plant species has been reviewed (Coupland, 1985; Duke, 1988; Franz *et al.*, 1997). The development of crops tolerant to glyphosate has made it possible to investigate glyphosate metabolism in whole plants in the absence of its potent phytotoxicity and allowed for the determination of the nature and magnitude of glyphosate residues in raw agricultural commodities (RACs) while simulating normal agricultural uses of the herbicide (Duke, 2011). Monsanto has conducted separate metabolism studies in conventional and glyphosate-tolerant varieties of soybean, cotton, sugarbeet and wheat, as well as other plants, as required for regulatory purposes (FAO, 2005). In all of these plants, whether conventional or glyphosate tolerant, the chief residue arising from glyphosate herbicide use is glyphosate itself. AMPA is often detected but, typically for most crops, levels of AMPA are about 10% of those of glyphosate, although for certain specific commodities (i.e. soybean grain) the proportion of AMPA can be higher.

Conducting metabolic studies in glyphosate-susceptible conventional crops is actually very challenging because glyphosate tightly binds to agricultural soils, which effectively prevents significant root uptake. In order to conduct a whole-plant metabolism study, it is necessary to bring an actively growing plant into contact with a low-level (non-toxic) dose of glyphosate, which makes the detection and quantification of glyphosate and its metabolites very difficult. To balance the potent phytotoxicity of glyphosate with the requirement for sufficient uptake to permit the characterization of residues, the metabolism of glyphosate in conventional crops should be studied using ^{14}C-labelled glyphosate. In earlier work conducted at Monsanto, the metabolism of glyphosate was investigated following root uptake from soil and hydroponic solutions, and also application to stems and trunks, and sub-herbicidal foliar applications to conventional crops. The metabolism and distribution of ^{14}C-glyphosate, labelled on each of the three carbon atoms in the molecule, in more than 20 varieties of glyphosate-susceptible conventional crops has been investigated and the results summarized (FAO, 2005). The results from these studies show that little degradation of glyphosate occurs in most plants, but that when it occurs, AMPA is the primary degradation product.

For investigations of the metabolism and distribution of glyphosate following root uptake from the soil in soybean, cotton, wheat, maize, coffee, lucerne, clover and pasture grasses, the soil surface of pots planted with the requisite crop was treated before crop emergence with [^{14}C]glyphosate. In all cases, uptake was less than 0.3% of the total applied radioactivity. Because of the very low root uptake of radioactivity from the soil, plants grown in hydroponic solutions containing [^{14}C]glyphosate were utilized to determine the distribution and metabolism of glyphosate following root uptake. For these studies, plants of conventional crops such as soybean, cotton, wheat, maize and coffee were grown in hydroponic solutions and continuously exposed via the roots to sublethal dose levels of [^{14}C]glyphosate.

Uptake of radioactivity from hydroponic solutions was two to three orders of magnitude greater than that observed from soil, and ranged from 1 to 33% of the applied ^{14}C-radioactivity. In general, glyphosate was found to be rapidly and extensively translocated throughout the plant tissues. However, the amount that was translocated into the aerial portions was generally less than that found in roots and varied depending upon the plant species (FAO, 2005). Aqueous extraction released the majority of plant-contained radioactivity in almost all plant tissues. Chromatographic analyses of extracts of plant tissues show that glyphosate is partially metabolized to AMPA. Glyphosate was the major ^{14}C-containing compound in the aqueous extracts in all samples except that from maize forage, where comparable amounts of glyphosate and AMPA were found. In all other plants, AMPA accounted for less than 10% of the radioactive residues. With the exception of AMPA, no other metabolites of glyphosate were present at greater than 1% of the total radioactive residues. The results from investigations with crops exposed to [^{14}C]glyphosate via hydroponic solution also indicate that [^{14}C]glyphosate, presumably via the intermediacy of AMPA, is degraded to carbon dioxide and other one-carbon fragments that are either respired or incorporated into natural plant metabolic processes. It should be noted, however, that the detection of AMPA in plant tissues may have occurred via the microbial degradation of glyphosate to AMPA due to microbial contamination of the hydroponic solutions. Indeed, the authors of two of the studies noted that glyphosate in the hydroponic nutrient media slowly degraded to AMPA (Rueppel *et al.*, 1975; Franz *et al.*, 1997).

To simulate potential inadvertent exposure to and uptake of glyphosate following the application of post-emergence, directed sprays applied according to good agricultural practices (GAP), a solution of [^{14}C]glyphosate was applied to the stems of coffee plants and the treated area was shielded to prevent the test substance from being washed off during subsequent watering. Less than 3% of the applied radioactivity was incorporated into the plants after 35 days. To investigate the metabolism of glyphosate following inadvertent foliar applications, sublethal dose levels of a formulated solution of [^{14}C]glyphosate were applied by micro-syringe to the surfaces of an appropriate number of leaves of coffee plants. The treated leaves were analysed separately from the remainder of the plant. Glyphosate was found to be rapidly and extensively translocated throughout the coffee plant via leaf uptake. In the case of a mature bean-producing coffee tree, approximately 48% of the applied ^{14}C-radioactivity was absorbed and translocated from the treated leaves. However, less than 2.4% of the applied ^{14}C-radioactivity was translocated to the developing coffee beans. At 23 weeks after treatment, the percentage of applied ^{14}C-radioactivity found in the green coffee beans and pods, ripe beans and ripe pods was 0.94, 0.62 and 0.68%, respectively (FAO, 2005).

The development of crops tolerant to glyphosate has provided a new opportunity to investigate glyphosate metabolism in plants at normal herbicide application rates (Duke, 2011). Whole-plant metabolism studies have also been completed in glyphosate-tolerant crops that have been developed by inducing the expression of a glyphosate-tolerant EPSPS gene, the CP4-EPSPS gene. The crops include soybean, sugarbeet, wheat and cotton (FAO, 2005). These studies used higher treatment rates of glyphosate than are possible with non-tolerant crops, involved direct applications of glyphosate to plants at various stages of development and have allowed a thorough understanding of glyphosate metabolism in plants. The timings and application rates used simulated GAP for in-crop treatments to tolerant plants. In all cases, glyphosate was applied as a formulated solution that simulated commercial formulations. In order to distinguish between radioactive residues resulting from the metabolism of [^{14}C]glyphosate within the plant and those resulting from uptake of $^{14}CO_2$ formed by the degradation of [^{14}C]glyphosate in the soil, the plants were divided into two treatment groups. In one group, the soil was covered during the application of [^{14}C]glyphosate to the plant in order to minimize contact with the soil, and in the other group the soil was left uncovered. Additionally, control plants, untreated or treated with non-labelled glyphosate using the same rates and timings, were grown in close proximity to the ^{14}C-treated plants to monitor the amount of radioactivity due to the incorporation of $^{14}CO_2$.

In tolerant soybean containing the CP4-EPSPS gene, glyphosate was metabolized substantially to AMPA. Glyphosate and AMPA were the major radiolabelled compounds present in soybean RACs, accounting for 96, 66 and 74% of the total radioactive residues (TRR) in forage, hay and grain, respectively. The portion of the residues present as intact glyphosate declines over time, from approximately 90% of the TRR in forage to about 25% of the TRR in grain. Concomitantly, AMPA accounts for only 6.8% of the TRR in forage, but is the major radiolabelled compound in grain, accounting for up to 49% of the TRR. In addition, trace-level metabolites present in grain were identified as *N*-glyceryl-AMPA, *N*-acetyl-AMPA, *N*-malonyl-AMPA and N-methyl-AMPA. No single trace-level metabolite accounted for greater than 2% of the total radioactive residues in any RAC. Radioactive residues in hexane-extracted soybean oil were shown to be associated with naturally occurring fatty acids, and no glyphosate or glyphosate-related metabolites were found in soybean oil. Acid-released radioactivity in soybean meal was shown to be closely associated with natural products such as amino acids and organic acids formed as a result of the incorporation of $^{14}CO_2$ and the breakdown of [^{14}C]glyphosate to one-carbon fragments (FAO, 2005).

The metabolism of [^{14}C]glyphosate in tolerant cotton, sugarbeet and wheat, all containing the CP4-EPSPS gene, was very similar to that found in tolerant soybean except that the metabolism of glyphosate to AMPA in these crops occurred at slower rates than in the tolerant soybean. As in tolerant soybean, glyphosate was partially metabolized to AMPA in these crops and AMPA was then further converted to a limited degree to give low levels of simple conjugates. In cotton forage, glyphosate accounted for >91% of TRRs, with AMPA accounting for <1.6% of the TRR. In cotton seeds, glyphosate was the major significant residue (12–24% of

TRR) but only trace levels of AMPA (<2% of TRR) were present. A significant fraction of the TRR in the cotton seeds was attributed to the breakdown of glyphosate/AMPA to one-carbon fragments and their subsequent incorporation into natural products; 10–12% of the TRR was characterized as saponifiable fatty acids in oil and 54–75% of the TRR was characterized as unextracted natural products (FAO, 2005). Glyphosate and AMPA together accounted for at least 99 and 81% of the total radioactive residues in sugarbeet roots and tops, respectively. The most significant residue in all sugarbeet commodities was glyphosate (80–95% of TRR), whereas the levels of AMPA were at 1.8 and 3.8% of TRR in the tops and roots, respectively. Sugarbeet tops and roots also contained trace levels of glyphosate/AMPA-acetylated conjugates and small amounts of ^{14}C-labelled natural products (<2% of TRR) (FAO, 2005). The most significant radioactive component of the aqueous extracts in all the wheat commodities was glyphosate, which constituted 90, 84, 69 and 72% of TRR in wheat forage, hay, straw and grain, respectively. In contrast, AMPA accounted for 0.8–10.8% of TRR in the wheat RACs. In addition to being subject to conjugation, glyphosate and AMPA are further degraded to one-carbon fragments that become broadly incorporated into a wide variety of natural products and plant constituents in wheat (Monsanto, unpublished results).

The ability of certain plant species to metabolize glyphosate to AMPA has been discussed in a review article on glyphosate degradation in glyphosate-resistant and -susceptible crops and weeds (Duke, 2011). It has been suggested that a plant-specific glyphosate oxidoreductase may be responsible for the degradation of glyphosate in plants (Duke *et al.*, 2003; Duke, 2011). In microorganisms, the nature of the enzyme responsible for the cleavage of glyphosate to AMPA has been established by the isolation of a glyphosate oxidoreductase enzyme (GOX) from a naturally occurring bacterium, *Achromobacter* (Barry *et al.*, 1992). A similar mechanism for the formation of AMPA in plants is conceivable, although no plant-specific GOX or gene encoding it have been isolated or identified from plants (Duke, 2011). It has also been suggested that the glyphosate to AMPA transformation in plants may be mediated by monoamine oxidase (Gout *et al.*, 1992).

In summary, the results of the metabolic studies that have been carried out using ^{14}C-labelled glyphosate demonstrate that in most crops, whether conventional or glyphosate-tolerant varieties, the major residue arising from glyphosate herbicide use is intact glyphosate. AMPA is often present as a result of N-C bond cleavage within the glycine side of the glyphosate molecule. Typically, for most crops, AMPA levels are about 10% of the glyphosate levels, although for certain specific commodities the proportion of AMPA can be higher. The metabolic pathway of glyphosate in plants can be summarized in Fig. 24.2, and is the same for both non-glyphosate tolerant and tolerant crops. The nature of the metabolites is also the same, only the relative distribution varies depending on the extent to which glyphosate is converted to AMPA. AMPA can be further modified via conjugation with naturally occurring small molecular weight organic compounds to give trace-level components that make up less than 7% of the total radioactive residues in all cases. The results of various pulsed-dose experiments, transpiration experiments and experiments with glyphosate-tolerant crops show that glyphosate, via the intermediates of AMPA, is ultimately degraded to carbon dioxide and other one-carbon fragments that are either respired or incorporated in natural plant metabolic processes.

24.6.2 Environmental fate

Chemical decomposition does not contribute to the degradation of glyphosate in the environment because the compound is stable to hydrolytic degradation in sterile water in most environmentally relevant pH ranges. Furthermore, unlike many other amino acid derivatives, glyphosate has a remarkable stability in strong acid or base solutions with extreme pH ranges, even at elevated temperatures.

Glyphosate does not absorb light significantly at wavelengths longer than 230 nm. Therefore, in highly purified sterile water, in which direct photolysis is the only mechanism for phototransformation, glyphosate is expected to be photo-stable. Indeed, aqueous photolysis studies have shown that glyphosate is relatively stable to photodegradation in distilled water, confirming

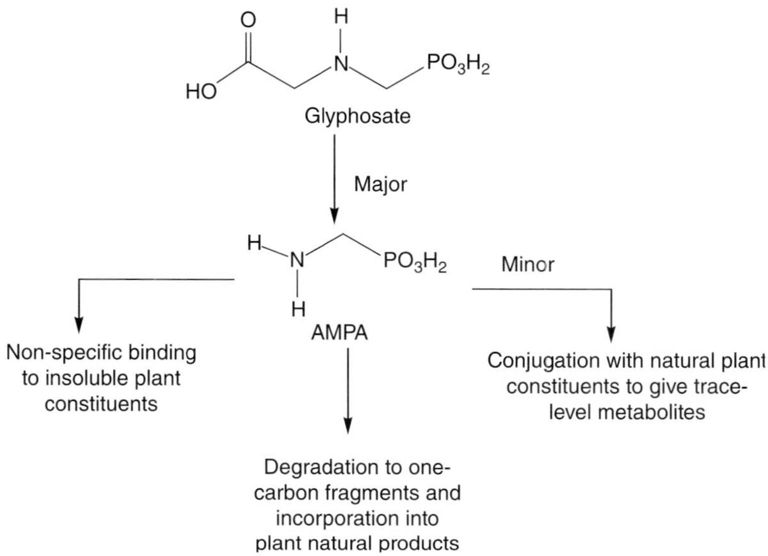

Fig. 24.2. Pathway for the metabolism of glyphosate in plants. AMPA, aminomethylphosphonic acid.

that it does not absorb incident radiation directly, as would be expected from its UV spectrum (Franz *et al.*, 1997). However, photo-induced degradation of glyphosate can occur in water under certain conditions. Studies using artificial light and solutions containing calcium or manganese ions show that glyphosate is susceptible to a slow indirect photodegradation (Bronstad and Friestad, 1985). Similarly, under intense artificial light, glyphosate in natural water is susceptible to degradation via oxidative transformation induced by the photochemical excitation of humic acids, as reported for other pesticides (Aguer and Richard, 1996).

Naturally occurring organic and inorganic solutes such as humic acid, tryptophan, tyrosine, organic peroxides and various metal ions are known to absorb strongly in the ultraviolet and visible region to form reactive species such as singlet molecular oxygen, hydrogen peroxide, hydroxyl radicals, organic peroxy radicals and other free radicals. Photo-oxidants like these could then react with oxidizable compounds such as glyphosate in the aqueous environment. Sunlight-absorbing substances like those mentioned above have been implicated as naturally occurring photosensitizers in the transformation of xenobiotics in natural waters (Draper and Crosby, 1983). Indeed, in recent years, heterogeneous photocatalysed advanced oxidation processes, such as using a UV/TiO_2 treatment system, have received significant attention (Muneer and Boxall, 2008; Assalin *et al.*, 2010) as an effective method for the removal of glyphosate from surface water. Although indirect photodegradation of glyphosate in water can occur, under normal environmental conditions photolysis is expected to be a slow process and, compared with microbial degradation, is, at most, a very minor pathway for the degradation of glyphosate in the environment.

In contrast, glyphosate is readily degraded by microorganisms in soil, non-sterile water and water/sediment systems. In soil, indigenous microflora degrade glyphosate under both aerobic and anaerobic conditions. The main metabolic pathway involves the formation of carbon dioxide and, similar to plant metabolism, the formation of AMPA as the only major metabolite. AMPA typically represents 15–50% of the applied glyphosate as it is further degraded by the soil microflora, although at a slower rate than is glyphosate. Glyphosate degradation in soil is a microbiological process and typically shows a high degree of mineralization, depending on the soil type and its biological activity. Studies have demonstrated that in the soil up to 79–86% of the glyphosate present is biodegraded to carbon dioxide during a 6 month period (Franz *et al.*, 1997). Several authors have studied the mineralization of glyphosate in topsoils, and in most cases the level of mineralization was quite extensive (Doublet *et al.*, 2009; Zablotowicz *et al.*, 2009). Less than 10% mineralization was only observed in a few sand or sandy loam soils after 92–112 days of laboratory incubation at 15 °C (Gimsing *et al.*, 2004). The differences between soils in glyphosate mineralization rate have been attributed to differences in the biological activity of the soils. Gimsing *et al.* (2004) observed a good correlation between the mineralization of glyphosate and the abundance of *Pseudomonas* bacterial species in the soil, although a poor correlation was observed with the overall soil respiration activity. Several authors have also reported an inverse correlation with the strength of soil adsorption, hypothesizing that if glyphosate is strongly adsorbed to soil, it may not be available for biodegradation.

Half-lives (DT_{50}) of glyphosate degradation in laboratory soils under aerobic conditions reported in the literature range from <1 to 151 days, but are typically <25 days (Bergström *et al.*, 2011; Ghafoor *et al.*, 2011). Mamy *et al.* (2005) also evaluated the DT_{50} of AMPA dissipation under lab conditions and reported DT_{50} values ranging from 25 to 75 days. Numerous field studies have been performed to assess the dissipation of glyphosate in soil, covering various soil and climatic conditions, and these have resulted in a wide range of DT_{50} values. The results of over 93 field trials conducted in Europe, Canada and the USA show that glyphosate dissipates with field half-lives in all cases of less than 1 year, and typically less than 38 days (Giesy *et al.*, 2000). The highest DT_{50} values were usually observed in cold regions, where the soil remained frozen for a long period of time. Laboratory and field studies have also demonstrated that dissipation times are not affected by the rate of application and that glyphosate and AMPA do not accumulate following multiple applications either during the same year or over tens of years (Giesy *et al.*, 2000).

Glyphosate shows little herbicidal activity when applied directly to soil, because it strongly adsorbs to the soil particles (Giesy *et al.*, 2000) and is not then available for uptake by plant roots. For this reason, glyphosate is only effective when used as a post-emergence herbicide. Studies have been conducted to investigate the uptake of radiolabelled glyphosate into crops following soil application. The maximum uptake into plants grown in soil treated with glyphosate was in all cases less than 1% of the total applied (Franz *et al.*, 1997). There have been many publications on the strong adsorption of glyphosate to agricultural soils. In laboratory batch equilibrium studies, the values of the sorption partition coefficient K_F or K_{Foc}, a measure of soil-binding properties for glyphosate, ranged from 9.4 to 897 ml g^{-1} (mean: 259 ml g^{-1}) and from 1600 to 60000 ml g^{-1} (mean: 16810 ml g^{-1}), respectively. The higher the K_F or K_{Foc} values, the stronger the tendency is for a chemical to bind with soil. Similarly, the K_F and K_{Foc} values for AMPA based on multiple concentration tests ranged from 10 to 509 ml g^{-1} (mean: 112 ml g^{-1}) and from 1119 to 45900 ml g^{-1} (mean: 9749 ml g^{-1}), respectively. Published K_F values for AMPA have been reported to range from 32.0 to 496 ml g^{-1}, confirming that AMPA also strongly adsorbs to soil (Huang, 2004; Gjettermann, 2009; Mamy, 2010).

In general, the sorption of glyphosate and AMPA by soil is not influenced by a single or dominating soil physico-chemical property (Autio *et al.*, 2004). Most pesticides adsorb to soil organic matter and the degree of adsorption is related to the organic matter content (Kah and Brown, 2006). Sorption studies with purified soil humic substances have shown a strong affinity of glyphosate for adsorption to organic matter (Albers *et al.*, 2009). Adsorption was explained by hydrogen bonding among the various acidic and oxygen-containing groups in glyphosate and humic materials. However, other published literature suggests that the organic matter content of soils alone could not predict glyphosate adsorption to the same extent as that of other herbicides. Research has shown that in addition to the organic carbon content of the soils, other soil properties (e.g. clay mineral type and content) are also important in regulating the sorption of glyphosate. It has further been suggested that the adsorption of glyphosate is influenced by the soil clay content, as the phosphonic acid moiety of glyphosate forms strong complexes with metal cations that are present at the surface of clay minerals. Several mechanisms influenced by different soil properties have been postulated to explain some of the observed variations. In general, it has been demonstrated that high levels of metallic cations in the soil, especially iron and aluminium ions (e.g. Fe^{3+} and Al^{3+}) increase the amount of glyphosate that is adsorbed (Gimsing *et al.*, 2004). Jacobsen (2003) also reported high adsorption of glyphosate to subsoils where there was a positive correlation between glyphosate adsorption and the iron content of soil. It has also been suggested that, in general, glyphosate sorption is inversely correlated with the pH of the soil, as decreased soil sorption of glyphosate at increasing pH has been reported by several investigators (Gimsing *et al.*, 2004; Wang *et al.*, 2005; Al-Rajab *et al.*, 2008; Zhao *et al.*, 2009). The authors described these observations in term of more net negative charge and the occurrence of electrostatically unfavourable conditions between negatively charged soil oxides and the negatively charged ionized glyphosate molecule at the higher pH values. However, this is not always the case, because increasing pH by liming seems to have the opposite effect, i.e. increased glyphosate sorption at increasing pH, although this increase was attributed to the ease of formation of glyphosate aluminium and iron oxide complexes at higher pH values (De Jonge *et al.*, 2001). In summary, glyphosate is unique in that its high affinity for adsorption to soils is influenced by several mechanisms that are controlled by different soil physico-chemical properties, and no single soil property could predict its adsorption/desorption behaviour.

Glyphosate is not likely to leach into groundwater precisely because of its high affinity to bind very tightly to most soils and sediments in the environment. For the same reason, only limited amounts of glyphosate are found in surface water as a result of runoff. Studies examining the fate of glyphosate over several years found that, typically, less than 1% of glyphosate applied was lost as runoff from agricultural fields (Edwards, *et.al.*, 1980), and this has been confirmed by a large number of European monitoring studies. Glyphosate and AMPA are rarely found in groundwater and when they are, this is usually only under exceptional circumstances, for example in very porous soil prone to preferential flow conditions for the rapid and uneven movement of water (Gjettermann *et al.*, 2009), or in groundwater in direct contact with surface water (Sanchís *et al.*, 2012). Although it is generally assumed that pesticides with high adsorption coefficients are virtually immobile in soil, several investigators have suggested potential vertical transport of highly sorbable compounds such as glyphosate and AMPA via bonding to soil colloids or suspended on soil particles (colloid-facilitated or particle-bonded transport) (Gjettermann *et al.*, 2011a,b; Styczen *et al.*, 2011).

The diverse and intensive use and the wide application window of glyphosate herbicides, however, imply that glyphosate has the potential to reach surface water throughout the year from direct application to aquatic vegetation or from indirect routes of entry such as spray drift, run-off and drainage, as well as point-source contamination. Although the detection of glyphosate is more widespread in surface water than in groundwater, levels in surface water are typically seasonal and dissipate over time, and they rarely exceed the threshold of environmental concern (Giesy *et al.*, 2000). The occasional presence of glyphosate in surface water does not pose a threat to human health, as the potential human exposure to surface water is through finished drinking water abstracted from surface water. Numerous studies have shown that glyphosate is readily removed by chemical/oxidative disinfections, which are standard surface water treatment processes used for the production of drinking water (Brosillon *et al.*, 2006; Mehrsheikh *et al.*, 2006; Assalin *et al.*, 2010). In addition, according to the World Health Organization (WHO), the observed concentrations of glyphosate in environmental waters are several orders of magnitude lower than the permitted safety threshold. The WHO concludes that the establishment of a numerical guideline value for glyphosate is not deemed necessary. The occasional incidence of glyphosate and/or AMPA in water would not represent a hazard to human health.

As in soil, glyphosate in natural water and sediments is degraded over time by microorganisms. Water/sediment studies have demonstrated the degradation of glyphosate to AMPA and carbon dioxide, and the subsequent degradation of AMPA to carbon dioxide. A proposed pathway for the degradation of glyphosate in the environment (soil, sediment and natural water) is illustrated in Fig. 24.3.

24.7 Glyphosate in Mammals

Glyphosate was first discovered and subsequently registered as an herbicidal active ingredient several years before the US Environmental

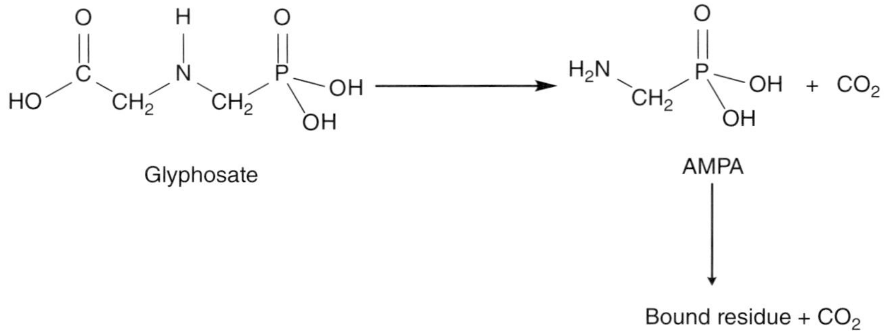

Fig. 24.3. Proposed degradation pathway of glyphosate in the environment.

Protection Agency (EPA) or Organisation for Economic Co-operation and Development (OECD) had developed detailed *in vivo* kinetic and toxicology test guidelines. Despite the lack of guidelines, Monsanto scientists designed studies, developed protocols and undertook numerous investigations to identify potential toxicological end points. Subsequent to initial registration in the USA, regulatory test guidelines were developed and many studies were therefore repeated by Monsanto to align with these guidelines.

These new studies added to the body of scientific evidence that glyphosate did not present any serious toxicological concerns in mammals or unreasonable risks to humans. As mentioned in Section 24.3, as the expiry of the patent on the molecule drew near, multiple companies undertook the necessary toxicology tests to support their product registrations in multiple world regions. These repeat data have demonstrated overwhelmingly consistent results and provided regulators with a high level of confidence in glyphosate as a herbicide of low toxicological concern.

24.7.1 Mammalian absorption, distribution, metabolism and excretion (ADME) studies

Dozens of ADME studies on glyphosate have been generated for regulatory purposes, while relatively few have been published from academic or government scientific institutions (Chan and Mahler, 1992; Anadón *et al.*, 2009). Although conducted over several decades in different world regions, under different laboratory conditions and across a wide range of doses, the results are very consistent. Gastrointestinal absorption is approximately 20–40%. Metabolism is very limited, with the only identified metabolite being small quantities of AMPA in faeces, lending to the hypothesis that glyphosate metabolism is attributable to gastrointestinal microflora. The elimination of systemic glyphosate is fast as the unmetabolized polar compound, which is rapidly filtered by the kidneys and excreted in the urine; the plasma elimination half-life is approximately 1 h. Bioaccumulation is not observed. After 168 h, residual glyphosate was highest in bone, more so in male than female rats. The elimination half-life from bone marrow was 4.2 h and 7.6 h in females and males, respectively (WHO/FAO, 2005, 2006).

The chemical structure of glyphosate (Fig. 24.1) lends explanation to why systemically absorbed glyphosate is rapidly excreted in the urine. This relatively simple molecule comprises an amino acid glycine and a methylphosphonate moiety. Glycine is utilized in mammals for phase II metabolism, wherein the amino acid conjugates to a chemical. Conjugation typically reduces toxicity, and with the increased polarity of the conjugate, also enhances urinary excretion (Parkinson *et al.*, 2013). In the case of glyphosate, the structure presents itself ready for excretion in the form of an amino acid conjugate.

24.7.2 Toxicology

Much of the industry-sponsored toxicology data over the last 40 years have been considered proprietary information. Transparency in disclosure of glyphosate toxicology data to the scientific community and general public was initiated in the late 1990s, when Monsanto granted third-party experts in toxicology and risk assessment full access to its proprietary toxicology data, culminating in the seminal paper in *Regulatory Toxicology and Pharmacology* (Safety evaluation and risk assessment of the herbicide roundup and its active ingredient, glyphosate, for humans; Williams *et al.*, 2000). This review identified acceptable no observable adverse effects levels (NOAELs) in sub-chronic, chronic, developmental and reproductive studies for glyphosate, AMPA and a widely used surfactant, polyethoxylated tallow amine (POEA). With the active ingredient no longer under patent protection, more recent collaborative efforts by about two dozen companies in consortia as glyphosate registrants in Europe (the Glyphosate Task Force, GTF; see http://www.glyphosatetaskforce.org) and in North America (the Joint Glyphosate Task Force, JGTF [no web site]) have further increased transparency of the broader glyphosate toxicology data set. In addition to industry-generated data, a large volume of academic and NGO-sponsored studies and opinions have been published in both the scientific and the non-scientific literature, including the media.

Several industry-sponsored review manuscripts have been published to evaluate regulatory studies designed for application to human health risk assessment in conjunction with the vast number of published scientific research papers. These key toxicology publications are discussed in the relevant sections below.

24.7.2.1 Acute toxicity

Extensive acute toxicity testing has clearly demonstrated that glyphosate is of very low acute toxicological concern. Dozens of acute oral LD_{50} values in rodents (rats and mice) have been greater than the highest doses tested, typically 2000 mg kg^{-1}, but also at 5000 mg kg^{-1} and higher. Similar outcomes are reported for acute dermal rat and rabbit toxicity values, where the LD_{50} values exceed the typical limit dose of 2000 mg kg^{-1}, as well as 5000 mg kg^{-1} and even higher. Likewise, acute rat inhalation LC_{50} values exceed the limit dose of 5 mg l^{-1}. Given the lack of acute oral, dermal and inhalation toxicity, classification and labelling for these end points are not required for glyphosate technical acid and its salts.

Glyphosate acid and its salts were either non-irritating or slightly irritating to the skin of rabbits, and did not require classification as a skin irritant. The results of eye irritation testing, however, differ between technical glyphosate acid, which is the active ingredient manufacturing output, and the more neutral pH glyphosate salts, which are found in commercially formulated herbicide products. Glyphosate acid is classified under the GHS (UN Globally Harmonized System of Classification and Labeling of Chemicals) as a category 1 eye irritant, showing 'irreversible effects on the eye/serious damage to eyes'. Glyphosate salts, in contrast, have been shown to not require classification for eye irritation.

Formulated products containing inert components such as surfactants, defoamers and/or stabilizers (termed inert because they are not pesticidally active towards the target pest), are evaluated using the acute toxicity battery on a case-by-case basis. If classification for acute end points is warranted, these are typically attributable to a surfactant ingredient, not to glyphosate.

The last 40 years of glyphosate research and the subsequent lack of concern about the acute toxicity of this herbicide is reflected in multiple regulatory and authoritative reviews, (US EPA, 1993; European Commission, 2002; WHO/FAO, 2005, 2006), wherein acute reference doses (aRfDs) for comparison with glyphosate acute exposures in acute human health risk assessments were considered unnecessary.

24.7. 2.2 Repeat dose toxicity

Repeat dose toxicological investigations with glyphosate include standard subacute (~28 days), sub-chronic (~90 days) and 1-year chronic studies in several species, including mice, rats and dogs. At least 45 separate repeat-dose glyphosate studies have been submitted across different world regions to support glyphosate registrations (not including more specialized studies such as reproductive and developmental toxicity studies, discussed in later sections). The results of these studies are remarkably consistent, demonstrating low toxicological concern for glyphosate with high NOAELs in the ranges of 200–500 mg kg^{-1} body weight (bw) day^{-1} and higher, across the different species and study durations tested. Typical effects reported at the high doses above the NOAELs include gastrointestinal irritation findings such as diarrhoea, mild liver effects and reduced body weight gain.

24.7. 2.3 Genotoxicity

In 2012, the European GTF agreed to improve data transparency between companies and within the scientific community by publishing the results of the proprietary genotoxicity study results that support product registrations. Sixty five *in vitro* and *in vivo* regulatory genotoxicity study reports on glyphosate and glyphosate-based herbicide formulations were provided to two independent and internationally recognized experts in the field of genetic toxicology. These experts conducted a comprehensive review of both the industry-sponsored regulatory studies and the growing volume of peer-reviewed scientific literature, culminating in a lengthy expert review of glyphosate genotoxicity published in *Critical Reviews in Toxicology* (Kier and Kirkland, 2013). Study summaries are available via open access through the journal as online data supplements to the publication.

The weight of evidence for genotoxic end points supports that glyphosate and its salts do not induce gene mutations in mammalian cells, do not typically induce chromosomal effects *in vitro*, are not clastogenic and are not genotoxic to mammals *in vivo*. Published reports of DNA damage in select Comet Assays with formulated products may be considered secondary to surfactant toxicity (the Comet Assay is a single-cell gel electrophoresis method for studying DNA damage induced by physical and chemical agents). The conclusion of the comprehensive expert review (Kier and Kirkland, 2013) supports the last 40 years of regulatory reviews of glyphosate: that glyphosate does not pose a human health risk for genotoxicity end points.

24.7. 2.4 Carcinogenicity

Carcinogenicity studies, which typically involve daily dosing of large groups of mice for at least 18 months, or rats for 2 years, are an integral component of the standard pesticide toxicology data set for regulatory review. Internationally accepted OECD test guidelines for the rodent carcinogenicity assay have been in place since 1981. A remarkable number of glyphosate carcinogenicity assays have been undertaken by different glyphosate registrants, at least five in mice and eight in rats. Throughout the commercial life of glyphosate-based herbicides, authorities have consistently determined from these studies that glyphosate is not a carcinogen.

Many years of popular science in the media have alleged pesticide use is responsible for a variety cancer types. Significant funding into this area of public concern has fuelled well over a decade of prospective epidemiological research with nearly 90,000 individuals under the umbrella of the Agricultural Health Study (AHS), sponsored by the US National Institutes of Health (NIH), the US EPA and the US National Institute of Occupational Safety and Health (NIOSH). Conclusions of this research note no statistically significant positive association between glyphosate use and any cancer. De Roos *et al.* (2005) from the AHS mention a 'suggested association between glyphosate and multiple myeloma', which was further evaluated by an epidemiologist from Boston University School of Public Health, who considered the analysis leading to the 'suggested association' likely to be biased (Lash, 2007). Raw data subsequently obtained by Monsanto from the AHS under the Freedom of Information Act were evaluated by Tom Sorahan, Professor of Occupational Epidemiology at the University of Birmingham, UK, the lead author of over 60 peer-reviewed publications on the identification and quantification of occupational cancer risks. Sorahan noted that the 'suggested association' was based on a selected subset of data, and that when this was analysed appropriately, the data set unequivocally reflected no indication of any association between glyphosate use and multiple myeloma (Sorahan, 2012). It is of interest to note that no follow-up manuscript to the 2005 AHS assertions has been published detailing further information associating glyphosate with multiple myeloma.

In order to better appreciate the body of literature on glyphosate epidemiology studies and cancer, four highly regarded experts in the field of epidemiological research undertook a comprehensive review of the relevant scientific publications (Mink *et al.*, 2012). The methodologies and results of seven cohort studies and 14 case-control studies were critically evaluated and discussed. The experts 'found no consistent pattern of positive associations indicating a causal relationship between total cancer (in adults or children) or any site-specific cancer and exposure to glyphosate'. This is consistent with the last 40 or more years of regulatory evaluations around the world, which typically evaluated both toxicology data and published epidemiology studies, uniformly concluding that glyphosate is not classifiable as a human carcinogen.

24.7. 2.5 Developmental and reproductive toxicity

Over the product life of glyphosate-based herbicides, numerous developmental and reproductive toxicology (DART) studies have been undertaken; at least 11 multigenerational studies (one to three generations of offspring per study) in rats and 15 developmental toxicity studies in rats and rabbits. These studies consistently demonstrate that glyphosate is not a reproductive toxicant and not a selective developmental toxicant. In contrast, several *in vitro* publications claim the contrary. Williams *et al.* (2012) provided expert evaluation of the Monsanto DART toxicology data set and included technical critiques of the individual publications within this body of scientific literature, and concluded that potential human exposures to glyphosate were considered to be safe and that developmental or reproductive effects were not anticipated based on the body of known science.

Epidemiological publications on glyphosate and non-cancer end points were collectively evaluated and discussed in a detailed review by Mink *et al.* (2011). This paper included assessment of all reported human reproductive and developmental outcomes. The evaluation of the scientific peer-reviewed literature concluded that there was no evidence of a consistent pattern of positive associations indicating a causal relationship between any disease, including DART end points, and exposure to glyphosate.

However, one research publication, which gained significant traction in the public media, popular social networks and anti-pesticide groups, captured the immediate attention of Germany, the Rapporteur Member State responsible for glyphosate Annex I Renewal procedures in Europe. Paganelli *et al.* (2010) claimed to identify a molecular mechanism for teratogenicity resulting from direct embryonic exposures to a glyphosate-formulated herbicide. Within weeks, the European Commission sought expert input on the findings of this research, which emanated from a laboratory at the University of Buenos Aires in Argentina. The German authorities concluded that: (i) the studies were 'performed under highly artificial conditions, extremely different from what can be expected in agricultural circumstances, and that it is hardly possible to predict adverse effect on mammals on this basis'; (ii) the research contradicted 'a comprehensive and reliable toxicological database for glyphosate and the effects observed have not been revealed in mammalian studies, nor evidenced epidemiologically in humans'; and (iii) 'the findings of the studies do not put into question the current EU risk assessment for the substance and the products used in the EU that contain this substance with regard to human health' (European Commission, 2011).

Other scientists challenged the study results too (Mulet, 2011; Palma, 2011; Saltmiras *et al.*, 2011), noting three fundamental issues in the paper that challenge the relevance of reported findings: (i) the effects were not consistent with multiple reproductive and developmental studies in multiple mammalian and ecological species; (ii) the basis for the research described by the authors was completely unsubstantiated; and (iii) the doses administered were both inappropriately high and of irrelevant routes of exposure to human health risk assessment.

In 2010, the NGO Earth Open Source, cited selective scientific literature, and focused on rabbit developmental toxicity data from regulatory reviews, to claim that glyphosate was teratogenic and caused heart malformations. To thoroughly assess these claims, the European GTF engaged a team of third-party experts in the field of DART, and with specific expertise in heart developmental morphology. All the developmental toxicity study reports owned by GTF member companies were provided for thorough review, which culminated in a detailed publication containing evaluations of all of the individual studies and an overall evaluation of the entire data set (Kimmel *et al.*, 2013). The reviewing experts concluded that 'the entire body of the developmental toxicity data reviewed fails to support a potential risk for increased cardiovascular defects as a result of glyphosate exposure during pregnancy'.

Expert opinions on the reproductive and developmental toxicity of glyphosate rendered by DART scientists, epidemiologists and regulatory agencies continue to be consistent in saying that glyphosate is not a reproductive toxicant and is not a selective developmental toxicant. In spite of several scientific papers claiming new-found adverse effects attributable to glyphosate or glyphosate-based formulations, subsequent expert critiques of these reports have repeatedly found that the findings were neither relevant to, nor had an impact on, human health risk assessment.

24.7. 2.6 *Endocrine disruption*

In 2009, the US EPA included glyphosate on their first list of 67 compounds to undergo a Tier 1 Screening under the Endocrine Disruptor Screening Program (EDSP). It is important that the Agency noted 'this list should not be construed as a list of known or likely endocrine disruptors' (US EPA, 2009). Criteria for inclusion on this list were not based on suspected endocrine disruption activity, but rather on potential human exposures via four different pathways: food consumption (residues), drinking water (unlikely, given glyphosate's high sorption to sediment and rapid microbial degradation), residential use and occupational exposure.

Upon receipt of the EDSP Tier 1 Test Orders from the US EPA, the North American JGTF consortium undertook efforts to address these regulatory requirements. Subsequent to the issuance of these Test Orders, validation of the Steroidogenesis Assay (which is to detect chemicals with the potential to inhibit or induce the production of the sex steroid hormones testosterone and 17β-[o]estradiol), co-sponsored by the US EPA and the OECD, was published (Hecker *et al.*, 2011). This publication details that glyphosate was ring tested in the multi-laboratory validation and was clearly demonstrated to have no effects on the steroidogenesis pathway. Therefore, the Agency waived the Steroidogenesis Assay EDSP Test Order.

All remaining EDSP Tier 1 Test Orders for glyphosate were undertaken by the JGTF. These assays included *in vitro* tests (Androgen Receptor Binding, Estrogen receptor Binding, Estrogen Receptor Transcriptional Activation and Aromatase), *in vivo* mammalian tests (Hershberger, Uterotrophic, Male Pubertal and Female Pubertal) and *in vivo* ecological species (Fish Short-Term Reproduction and Amphibian Metamorphosis). In all cases, glyphosate demonstrated a compelling lack of potential to interact with the endocrine pathways under evaluation (Levine *et al.*, 2012; Saltmiras and Tobia, 2012; Bailey *et al.*, 2013; Webb *et al.*, 2013). The US EPA weight of evidence reviews of all List 1 Tier 1 EDSP assays, including those for glyphosate, are scheduled for completion in 2014–2015 (US EPA, 2014).

24.7. 2.7 *Neurotoxicity*

Multiple long-term animal studies with glyphosate did not demonstrate any evidence of neurotoxic effects. In addition, specific acute and subchronic neurotoxicity assays with glyphosate support the observations reported in general toxicity studies: that glyphosate is not neurotoxic.

While some publications have suggested statistical associations between Parkinson's disease and general pesticide exposures, there are no data suggesting an association between glyphosate and this disease. The large epidemiological AHS study of farmers reports no increased risk of Parkinson's disease in association with reported glyphosate use (Kamel *et al.*, 2007). Furthermore, the human non-cancer epidemiological outcomes review manuscript previously cited (Mink *et al.*, 2011) reports there is no convincing evidence for an increased incidence of Parkinson's disease or other neurological disorders in individuals reporting glyphosate exposure.

24.7.3 Human dietary exposures to glyphosate

A fundamental principle of toxicology is that dose determines whether or not something is toxicologically relevant. For example, small doses of vitamin A are not toxic and are in fact essential for healthy vision and immune systems. Literally millions of children in poor regions of the world have grown blind and/or died from vitamin A deficiency. Ironically, vitamin A is classified as both a mutagen and teratogen, and large doses may cause sickness, birth defects and death. Yet Vitamin A is essential for human health and well-being. While glyphosate is not a vitamin essential to maintaining good health, it has a much more favourable toxicological profile than many substances considered to be dietary requirements. Clearly, small amounts of glyphosate residue are of no health concern, which contradicts much of what is expressed in today's popular press and social media. How much glyphosate is acceptable in the diet? This question is answered through risk assessment, in which the toxicity profile is characterized, then an acceptable level at which no adverse outcome is expected is determined. This framework is addressed in Section 24.7.4, below.

A common misunderstanding surrounding human dietary exposures relates to how maximum residue levels (MRLs) or tolerances for an agricultural chemical are calculated. MRLs are determined by residue analysis from field trials following GAP. They are not calculated on the basis of toxicology or human health end points. Before a new use pattern for a pesticide is approved, field trials are conducted and the pesticide residues measured. If the amount of residue measured in the crop potentially contributes to unacceptable dietary exposures to the pesticide, then the new use is not permitted. However, if the new residue level does not increase dietary exposures to an unacceptable level of pesticide, then the new use may be approved. Over the course of time, as glyphosate use expanded and glyphosate-tolerant crops were approved, field trials were conducted for new use patterns. These new uses did not result in unacceptable increases in dietary exposures to glyphosate and, therefore, new glyphosate MRL values were justifiably established to support those new glyphosate use patterns.

Pesticide residues in animal products are also considered in human dietary exposure assessment. As such, MRLs are also established for animal feed, meat, dairy products and eggs. Examples of glyphosate MRLs established by the Codex Alimentarius, measured in milligrams of glyphosate per kilogram of food/feed (mg kg^{-1} or ppm) include: lucerne fodder, 500; maize fodder (dry), 150; cereal grains, 30; sunflower seeds, 7; beans (dry), 2; eggs, milk and meat, 0.05 (FAO/WHO, 2013). Again, these numbers are based on measured residues from trials following GAP. Exceeded MRLs are not triggers measuring unacceptable human health risk but, rather, are a reflection that GAPs have not been followed. MRLs are not health or toxicology derived safety limits.

24.7.4 Human health risk assessments

Risk assessment is the process of evaluating hazardous/non-hazardous doses of a substance through toxicological evaluations, then determining acceptable human dose levels or exposures at which adverse effects are not encountered. Substances with a very low hazard are unlikely to present unreasonable risks unless under very high exposure scenarios. Similarly, substances with very low exposure potential are unlikely to present unreasonable risks, unless they are highly toxic. In the case of glyphosate, both the hazard and human exposure potential are low. Therefore, human health risk in relation to glyphosate is very low.

The regulatory agencies responsible for pesticide approvals in their regions evaluate toxicology data and determine an acceptable daily intake (ADI) or chronic reference dose (cRfD) for each pesticide. This is the amount of active ingredient that a human may ingest every day of their life without an adverse outcome. The WHO in conjunction with the Food and Agricultural Organization (FAO) routinely review active pesticide ingredients with panels of independent experts in toxicology and risk assessment in order to establish ADI values. Figure 24. 4 is a histogram of WHO/FAO herbicide ADI values established through 2013; the higher the ADI, the greater the amount of pesticide in the diet that may be consumed daily without adverse effect.

The WHO also determines Drinking Water Quality Guidelines for a variety of compounds. Their review of glyphosate, which also included its metabolite and environmental degradation product, AMPA, explicitly states:

> Because of their low toxicity, the health-based value derived for AMPA alone or in combination with glyphosate is orders of magnitude higher than concentrations of glyphosate or AMPA normally found in drinking-water. Under usual conditions, therefore, the presence of glyphosate and AMPA in drinking-water does not represent a hazard to human health. For this reason, the establishment of a numerical guideline value for glyphosate and AMPA is not deemed necessary.
>
> WHO, 2005

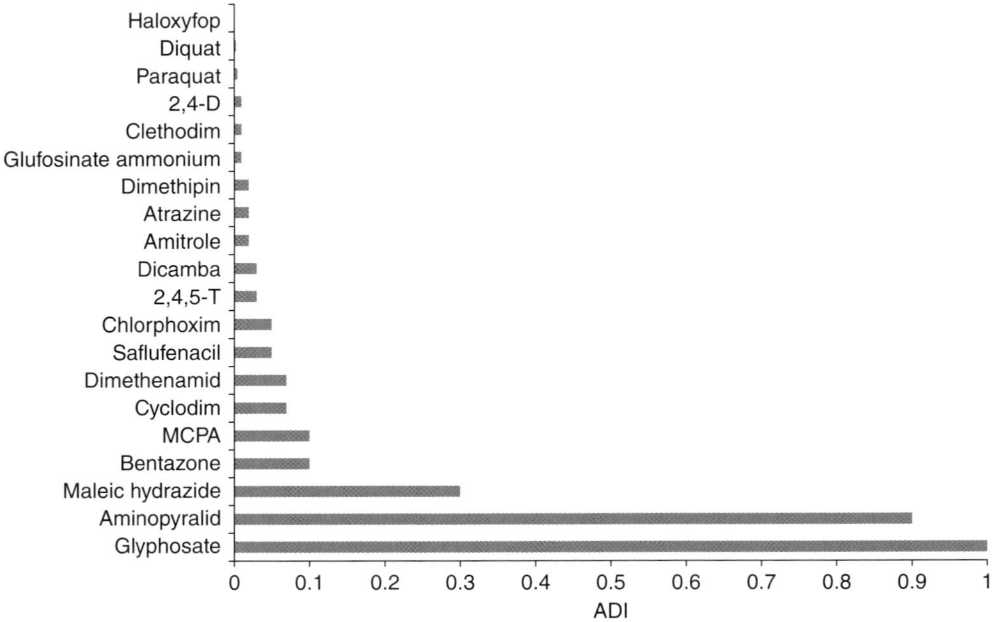

Fig. 24.4. World Health Organization (WHO) herbicide acceptable daily intake (ADI) values (as mg kg^{-1} body weight (bw) day^{-1}) established up to 2013.

24.8 Conclusions

Glyphosate has been the subject of extensive toxicological testing for over 40 years. These investigations have consistently demonstrated that this herbicide is of very low toxicological concern; it has low acute toxicity, is not a reproductive toxicant or a selective developmental toxicant, is non-genotoxic, non-carcinogenic, non-neurotoxic and non-immunotoxic, and is not an endocrine disruptor. Although glyphosate is widely used, its physico-chemical properties, environmental fate and pharmacokinetics all contribute to very low human exposures. With its inherently low hazard and low exposure potential, glyphosate presents negligible human health risk to farmers, pesticide applicators and consumers.

References

Aguer, J. and Richard, C. (1996) Transformation of fenuron induced by photochemical excitation of humic acids. *Pesticide Science* 46, 151–155

Albers, C.N., Banta, G.T., Hansen, P.E. and Jacobsen, O.S. (2009) The influence of organic matter on sorption and fate of glyphosate in soil – comparing different soils and humic substances. *Environmental Pollution* 157, 2865–2870.

Al-Rajab, A.J., Amellal, S. and Schiavon, M. (2008) Sorption and leaching of C-14-glyphosate in agricultural soils. *Agronomy for Sustainable Development* 28, 419–428.

Anadón, A., Martínez-Larrañaga, M.R., Martínez, M.A., Castellano, V.J., Martínez, M., Martin, M.T., Nozal, M.J. and Bernal, J.L. (2009) Toxicokinetics of glyphosate and its metabolite aminomethyl phosphonic acid in rats. *Toxicology Letters* 190, 91–95.

Assalin, M.R., De Mores, S.G., Queiroz, S.C.N., Ferracini, V.L. and Duran, N. (2010) Studies on degradation of glyphosate by several oxidative chemical processes: ozonation, photolysis and heterogeneous photocatalysis. *Journal of Environmental Science and Health, Part B* 45, 89–94.

Autio, S., Siimes, K., Laitinen, P., Ramo, S., Oinonen, S. and Erone, L. (2004) Adsorption of sugar beet herbicides to Finnish soils. *Chemosphere* 55, 215–226.

Bailey, J., Hauswirth, J. and Stump, D. (2013) No evidence of endocrine disruption by glyphosate in Male and Female Pubertal assays (conference abstract). In: *SOT (Society of Toxicology) 52nd Annual Meeting and ToxExpo™, March 10–14, 2013, San Antonio, Texas. The Toxicologist (Supplement to Toxicological Sciences)* 132(1), 412, abstract PS 1937.

Barry, G., Kishore, G., Padgette, S., Taylor, M., Kolacz, K., Weldon, M., Re, D., Eichholtz, D., Fincher, K. and Hallas, L. (1992) Inhibitors of amino acid biosynthesis: strategies for imparting glyphosate tolerance to crop plants. In: Singh, B.K., Flores, H.E. and Shannon, J.C. (eds) *Biosynthesis and Molecular Regulation of Amino Acids in Plants*. American Society of Plant Physiologists, Rockville, Maryland, pp. 139–145.

Bergström, L., Börjesson, E. and Stenström, J. (2011) Laboratory and lysimeter studies of glyphosate and aminomethylphosphonic acid in a sand and a clay soil. *Journal of Environmental Quality* 40, 98–108.

Bronstad, J.O. and Friestad, H.O. (1985) Chapter 13: Behaviour of glyphosate in the aquatic environment. In: Grossbard, E. and Atkinson, D. (eds) *The Herbicide Glyphosate*. Butterworth, London, pp. 201–205.

Brookes, G. and Barfoot, P. (2008) *GM Crops: Global Socio-economic and Environmental Impacts 1996–2006*. PG Economics Ltd, Dorchester, Dorset, UK.

Brosillon, S., Wolbert, D., Lemasle, M., Roche, P. and Mehrsheikh, A. (2006) Chlorination kinetics of glyphosate and its by-products: modeling approach. *Water Research* 40, 2113–2124.

Chan, P.C. and Mahler, J.F. (1992) *NTP Technical Report on Toxicology Studies of Glyphosate (CAS No. 1071-83-6) Administered Dose Feed in F344/N Rats and B6C3F1 Mice*. National Toxicology Program Toxicity Reports Series Number 16. NIH Publication 92-3135, July 1992, US National Institutes of Health, Bethesda, Maryland. Available at: http://ntp.niehs.nih.gov/ntp/htdocs/st_rpts/tox016.pdf (accessed 10 September 2014).

Coupland, D. (1985) Metabolism of glyphosate in plants. In: Grossbard, E. and Atkinson, D. (eds) *The Herbicide Glyphosate*. Butterworth, London, pp. 25–34.

De Jonge, H., De Jonge, L.W., Jacobsen, O.H., Yamaguchi, T. and Moldrup, P. (2001) Glyphosate sorption in soils of different pH and phosphorus content. *Soil Science* 166, 230–238.

De Roos, A.J., Blair, A., Rusiecki, J.A., Hoppin, J.A., Svec, M., Dosemeci, M., Sandler, D.P. and Alavanja, M.C. (2005) Cancer incidence among glyphosate-exposed pesticide applicators in the Agricultural Health Study. *Environmental Health Perspectives* 113, 49–54.

Doublet, J., Mamy, L. and Barriuso, E. (2009) Delayed degradation in soil of foliar herbicides glyphosate and sulcotrione previously absorbed by plants: consequences on herbicide fate and risk assessment. *Chemosphere* 77, 582–589.

Draper, W.M. and Crosby, D.G. (1983) The photochemical generation of hydrogen peroxide in natural waters. *Archives of Environmental Contamination and Toxicology* 12, 121–126.

Duke, S.O. (1988) Glyphosate. In: Kearney, P.C. and Kaufman, D.D. (eds) *Herbicides – Chemistry, Degradation, and Mode of Action, Vol. III*. Marcel Dekker, New York, pp. 1–70.

Duke, S.O. (2011) Glyphosate degradation in glyphosate-resistant and -susceptible crops and weeds. *Journal of Agricultural and Food Chemistry* 59, 5835–5841.

Duke, S.O. and Powles, S. (2008) Glyphosate: a once-in-a-century herbicide. *Pest Management Science* 64, 319–325.

Duke, S.O., Rimando, A.M., Pace, P.F., Reddy, K.N. and Smeda, R.J. (2003) Isoflavone, glyphosate, and aminomethylphosphonic acid levels in seeds of glyphosate-treated, glyphosate-resistant soybean. *Journal of Agricultural and Food Chemistry* 51, 340–344.

Edwards, W.M., Triplett, G.B. Jr. and Kramer, R.M. (1980) A watershed study of glyphosate transport in runoff. *Journal of Environmental Quality* 9, 661–665.

European Commission (2002) *Review Report for the Active Substance Glyphosate. Finalised in the Standing Committee on Plant Health at its Meeting on 29 June 2001 in View of the Inclusion of Glyphosate in Annex I of Directive 91/414/EEC. Glyphosate 6511/VI/99-final, 21 January 2002.* European Commission Health and Consumer Protection Directorate-General, Brussels. Available at: http://ec.europa.eu/food/fs/sfp/ph_ps/pro/eva/existing/list1_glyphosate_en.pdf (accessed 10 September 2014).

European Commission (2011) *Standing Committee on the Food Chain and Animal Health, Section Phytopharmaceuticals – Plant Protection products – Legislation – 22–23 November 2010, Summary Report.* European Commission Health and Consumer Protection Directorate-General, SANCO D, Brussels. Available atl: http://ec.europa.eu/food/plant/standing_committees/sc_phytopharmaceuticals/docs/sum_2223112010_en.pdf (accessed 10 September 2014).

FAO (2005) Glyphosate (158). Draft document prepared by MacLachlan, D. for the Food and Agriculture Organization of the United Nations, Rome. Available at: http://www.fao.org/fileadmin/templates/agphome/documents/Pests_Pesticides/JMPR/Evaluation05/2005_Glyphosate1.pdf (accessed 10 September 2014).

FAO/WHO (2013) Codex Alimentarius [online]: Pesticide Residues in Food and Feed, Pesticide Details, 158 Glyphosate, Maximum Residue Limits for Glyphosate. Food and Agriculture Organization of the United Nations, Rome and World Health Organization, Geneva, Switzerland. Available at: http://www.codexalimentarius.net/pestres/data/pesticides/details.html?d-16497-o=2&d-16497-s=3&id=158&print=true (accessed 10 September 2014).

Franz, J.E., Mao, M.K. and Sikorski, J.A. (eds) (1997) *Glyphosate: A Unique Global Herbicide*. ACS Monograph No. 189, American Chemical Society, Washington, DC.

Ghafoor, A., Jarvis, N.J., Thierfelder, T. and Stenstrom, J. (2011) Measurements and modeling of pesticide persistence in soil at the catchment scale. *Science of the Total Environment* 409, 1900–1908.

Giesy, J.P., Dobson, S. and Solomon, K.R.. (2000) Ecotoxicological risk assessment for Roundup® herbicide. *Reviews of Environmental Contamination and Toxicology* 167, 35–120.

Gimsing, A.L., Borggaard, O.K., Jacobsen, O.S., Aamand, J. and Sorensen, J. (2004) Chemical and microbiological soil characteristics controlling glyphosate mineralisation in Danish surface soils. *Applied Soil Ecology* 27, 233–242.

Gjettermann, B., Petersen, C.T., Koch, C.B., Spliid, N.H., Grøn, C., Baun, D.L. and Styczen, M. (2009) Particle-facilitated pesticide leaching from differently structured soil monoliths. *Journal of Environmental Quality* 38, 2382–2393.

Gjettermann, B., Petersen, C.T., Hansen, S., Koch, C.B. and Styczen, M. (2011a) Kinetics of glyphosate desorption from mobilized soil particles. *Soil Science Society of America Journal* 75, 434–443.

Gjettermann, B., Styczen, M., Koch, C.B., Hansen, S. and Petersen, C.T. (2011b) Evaluation of sampling strategies for pesticides in a macroporous sandy loam soil. *Soil and Sediment Contamination* 20, 986–994.

Gout, E., Bligny, R., Genix, P., Tissut, M. and Douce, R. (1992) Effect of glyphosate on plant cell metabolism. 13P and 14C NMR studies. *Biochimie* 74, 875–882.

Hecker, M., Hollert, H., Cooper, R., Vinggaard, A.M., Akahori, Y., Murphy, M., Nellemann, C., Higley, E., Newsted, J., Laskey, J., Buckalew, A., Grund, S. *et al.* (2011) The OECD validation program of the H295R steroidogenesis assay: Phase 3. Final inter-laboratory validation study. *Environmental Science and Pollution Research.* 18, 503–515.

Huang, X., Pedersen, T., Fischer, M., White, R. and Young, T.M. (2004) Herbicide runoff along highways. 2. Sorption control. *Environmental Science and Technology* 38, 3272–3278.

Jacobsen, O.S. (2003) Degradation, sorption and persistence of glyphosate and the metabolite AMPA in a fractured clay soil profile. In: Del Re, A.A.M., Capri, E., Padovani, L. and Trevisan, M. (eds) *Pesticide[s] in Air, Plant, Soil and Water System[s]. Proceedings of the XII Symposium [on] Pesticide Chemistry, Piacenza, Italy, 4–6 June 2003*. La Goliardica Pavese, Pavia, Italy, pp.137–144.

Kah, M. and Brown, C.D. (2006) Adsorption of ionisable pesticides in soils. *Reviews of Environmental Contamination and Toxicology* 188, 149–217.

Kamel, F., Tanner, C.M., Umbach, D.M., Hoppin, J.A., Alavanja, M.C.R., Blair, A., Comyns, K., Goldman, S.M., Korell, M., Langston, J.W., Ross, G.W. and Sandler, D.P. (2007) Pesticide exposure and self-reported Parkinson's disease in the Agricultural Health Study. *American Journal of Epidemiology* 165, 364–374.

Kier, L.D. and Kirkland, D.J. (2013) Review of genotoxicity studies of glyphosate and glyphosate-based formulations. *Critical Reviews in Toxicology* 43, 283–315.

Kimmel, G.L., Kimmel, C.A., Williams, A.L. and DeSesso, J.M. (2013) Evaluation of developmental toxicity studies of glyphosate with attention to cardiovascular development. *Critical Reviews in Toxicology* 43, 79–95.

Knuuttila, P. and Knuuttila, H. (1985) Molecular and crystalline structure of glyphosate. In: Grossbard, E. and Atkinson, D. (eds) *The Herbicide Glyphosate*. Butterworth, London, pp. 18–22.

Komoba, D., Gennity, I. and Sandermann, H. Jr (1992) Plant metabolism of herbicides with P-C bonds: glyphosate. *Pesticide Biochemistry and Physiology* 43, 85–94

Lash, T.L. (2007) Bias analysis applied to Agricultural Health Study publications to estimate non-random sources of uncertainty. *Journal of Occupational Medicine and Toxicology* 2:15.

Levine, S., Saltmiras, D., Webb, E., Holmes, C., Mortensen, S., Honegger, J. and Bailey, J. (2012) Tier 1 EDSP Assays and Regulatory Safety Studies Provide a Weight of Evidence that Glyphosate is Not an

Endocrine Disruptor (conference abstract). *Abstract Book, SETAC North America 33rd Annual Meeting, Long Beach, California. 11–15 November 2012*. Society of Environmental Toxicology and Chemistry (SETAC), Pensacola, Florida, p. 129, abstract 529. Available at: http://longbeach.setac.org/sites/default/files/SETAC-abstract-book-2012.pdf (accessed 10 September 2014).

Mamy, L., Barriuso, E. and Gabrielle, B. (2005) Environmental fate of herbicides trifluralin, metazachlor, metamitron and sulcotrione compared with that of glyphosate, a substitute broad spectrum herbicide for different glyphosate-resistant crops. *Pest Management Science* 61, 905–916.

Mamy, L., Gabrielle, B. and Berriuso, E. (2010) Comparative environmental impacts of glyphosate and conventional herbicides when used with glyphosate-tolerant and non-tolerant crops. *Environmental Pollution* 158, 3172–3178.

Mehrsheikh, A., Bleeke, M., Brosillon, S., Laplanche, A. and Roche P. (2006) Investigation of the mechanism of chlorination of glyphosate and glycine in water. *Water Research* 40, 3003–3014.

Mink, P.J., Mandel, J.S., Lundin, J.I. and Sceuman, B.K. (2011) Epidemiologic studies of glyphosate and non-cancer health outcomes: a review. *Regulatory Toxicology and Pharmacology* 61, 172–184.

Mink, P.J., Mandel, J.S., Sceuman, B.K. and Lundin, J.I. (2012) Epidemiologic studies of glyphosate and cancer: a review. *Regulatory Toxicology and Pharmacology* 63, 440–452.

Motekaitis, R.J. and Martell, A.E. (1985) Metal chelate formation by *N*-phosphonomethylglycine and related ligands. *Journal of Coordination Chemistry* 14, 139–149.

Mulet, J.M. (2011) Letter to the editor regarding the article by Paganelli *et al*. *Chemical Research in Toxicology* 24, 609.

Muneer, M. and Boxall C. (2008) Photocatalyzed degradation of a pesticide derivative glyphosate in aqueous suspension of titanium dioxide. *International Journal of Photoenergy* 2008, Article ID 197346.

NPIC (2010) *Glyphosate General Fact Sheet*. National Pesticide Information Center, Corvallis, Oregon. Available at: http://npic.orst.edu/factsheets/glyphogen.html (accessed 6 November 2013).

Paganelli, A., Gnazzo, V., Acosta, H., Lopez, S.L. and Carrasco, A.E. (2010) Glyphosate-based herbicides produce teratogenic effects on vertebrates by impairing retinoic acid signaling. *Chemical Research in Toxicology* 23, 1586–1595.

Palma, G. (2011) Letter to the editor regarding the article by Paganelli *et al*. *Chemical Research in Toxicology* 24, 775–776.

Parkinson, A., Ogilvie, B.W., Buckley, D.B., Kamzi, F., Czerwinski, M. and Parkinson, O. (2013) Chapter 6: biotransformation of xenobiotics. In: Klaassen, C.D. (ed.) *Casarett and Doull's Toxicology, The Basic Science of Poisons*, 8th edn. McGraw-Hill, New York, pp. 185–366.

Rueppel, M.L., Marvel, J.T. and Suba, L.A. (1975) The metabolism of *N*-phosphonomethylglycine in corn, soybeans and wheat. In: *Abstracts of Papers to be Presented at the 170th National Meeting, American Chemical Society: PEST 26*. American Chemical Society, Washington, DC.

Saltmiras, D. and Tobia, A. (2012) No evidence of endocrine disruption by glyphosate in Hershberger and Uterotrophic assays (conference abstract). In: *SOT (Society of Toxicology) 51st Annual Meeting and ToxExpo™, March 11–15, 2012, San Francisco, California. The Toxicologist (Supplement to Toxicological Sciences)* 126(1), 474, abstract PS 2198.

Saltmiras, D., Bus, J.S., Spanogle, T., Hauswirth, J., Tobia, A. and Hill, S. (2011) Letter to the editor regarding the article by Paganelli *et al. Chemical Research in Toxicology* 24, 607–608.

Sanchís, J., Kantiani, L., Llorca, M., Rubio, F., Ginebreda, A., Fraile, J., Garrido, T. and Farré M. (2012) Determination of glyphosate in groundwater samples using an ultrasensitive immunoassay and confirmation by on-line solid-phase extraction followed by liquid chromatography coupled to tandem mass spectrometry. *Analytical and Bioanalytical Chemistry* 402, 2335–2345.

Sorahan, T. (2012) Multiple myeloma and glyphosate use: a re-analysis of US Agricultural Health Study data. *Toxicology Letters* 211S, S127, abstract P27-02.

Stock, M., Hieke, M., Weidhase, R.A. and Schutte, H.R. (1991) Degradation of glyphosate in excised leaves of tobacco and sugar beet. *Journal of Plant Physiology* 139, 171–174.

Styczen, M., Petersen, C.T., Koch C.B. and Gjettermann, B. (2011) Macroscopic evidence of sources of particles for facilitated transport during intensive rain. *Vadose Zone Journal* 10, 1151–1161.

US EPA (1993) *Re-registration Eligibility Decision: Glyphosate. Office of Prevention, Pesticides and Toxic Substances*, US Environmental Protection Agency, Washington, DC. Available at http://www.epa.gov/oppsrrd1/REDs/old_reds/glyphosate.pdf (accessed 24 September 2013).

US EPA (2009) Final list of initial pesticide active ingredients and pesticide inert ingredients to be screened under the Federal Food, Drug, and Cosmetic Act. *Federal Register* 74(71), Wednesday,

April 15, 2009, Notices, p. 17579. Available at: http://www.epa.gov/endo/pubs/final_list_frn_041509.pdf (accessed 10 September 2014).

US EPA (2014) *U.S. Environmental Protection Agency Endocrine Disruptor Screening Program Comprehensive Management Plan, June 2012*. Washington, DC. Available at: http://www.epa.gov/endo/pubs/EDSP-comprehensive-management-plan.pdf (accessed 1 March 1 2014).

Wang, Y.J., Zhou, D.M. and Sun, R.J. (2005) Effects of phosphate on the adsorption of glyphosate on three different types of Chinese soils. *Journal of Environmental Sciences-China* 17, 711–715.

Webb, E., Saltmiras, D. and Levine, S. (2013) Endocrine Disruptor Screening Program (EDSP) Tier 1 *in vitro* assays indicate glyphosate does not interact with estrogen and androgen receptor nor inhibit steroidogenesis *International Journal of Toxicology* 32, p. 77. [Poster abstract P500 *from American College of Toxicology 33rd Annual Meeting, November 4–7, 2012*.] Available at: http://ijt.sagepub.com/content/32/1/58.full.pdf (accessed 10 September 2014).

Williams, A.L., Watson, R.E. and DeSesso, J.M. (2012) Developmental and reproductive outcomes in humans and animals after glyphosate exposure: a critical analysis. *Journal of Toxicology and Environmental Health, Part B* 15, 39–96.

Williams, G.M., Kroes, R. and Munco, I.C. (2000) Safety evaluation and risk assessment of the herbicide roundup and its active ingredient, glyphosate, for humans. *Regulatory Toxicology and Pharmacology* 31, 117–165.

WHO (1994) *International Program on Chemical Safety. Environmental Health Criteria 159: Glyphosate*. First draft prepared by Mensink, H. and Janssen, P. Published under the joint sponsorship of the United Nations Environment Programme, the International Labour Organisation and the World Health Organization. World Health Organization, Geneva, Switzerland. Available at http://www.inchem.org/documents/ehc/ehc/ehc159.htm (accessed 24 September 2013).

WHO (2005) *Glyphosate and AMPA in Drinking-Water. Background Document for Development of WHO Guidelines for Drinking Water Quality*. WHO/SDE/WSH/03.04/97 (updated June 2005), World Health Organization, Geneva, Switzerland. Available at: http://www.who.int/water_sanitation_health/dwq/chemicals/glyphosateampa290605.pdf (accessed 10 September 2014).

WHO/FAO (2005) *Pesticide Residues in Food – 2004. Report of the Joint Meeting of the FAO Panel of Experts on Pesticide Residues in Food and the Environment and the WHO Core Assessment Group on Pesticide Residues, Rome, Italy, 20–29 September 2004.* FAO Plant Production and Protection Paper 178, World Health Organization and Food and Agriculture Organization of the United Nations. Rome. Available at: http://www.fao.org/fileadmin/templates/agphome/documents/Pests_Pesticides/JMPR/Reports_1991-2006/report2004jmpr.pdf (accessed 10 September 2014).

WHO/FAO (2006) *Pesticide Residues in Food – 2004. Report of the Joint FAO/WHO Meeting on Pesticide Residues. Evaluations 2004: Part II Toxicological. Sponsored jointly by FAO and WHO with the Support of the International Programme on Chemical Safety (IPCS). Joint Meeting of the FAO Panel of Experts on Pesticide Residues in Food and the Environment and the WHO Core Assessment Group, Rome, Italy, 20–29 September 2004*. World Health Organization, Geneva, Switzerland. Available at: http://whqlibdoc.who.int/publications/2006/9241665203_eng.pdf (accessed 10 September 2014).

Zablotowicz, R.M., Accinelli, C., Krutz, L.J. and Reddy, K.N. (2009) Soil depth and tillage effects on glyphosate degradation. *Journal of Agricultural and Food Chemistry* 57, 4867–4871.

Zhao, B.Z., Zhang, J.B., Gong, J.D., Zhang, H. and Zhang, C.Z. (2009) Glyphosate mobility in soils by phosphate application: laboratory column experiments. *Geoderma* 149, 290–297.

25 Amino Acid Analysis of Plant Products

S.M. Rutherfurd*
Riddet Institute, Massey University, Palmerston North, New Zealand

25.1 Abstract

Amino acid analysis has been used as a tool in protein chemistry laboratories, where obtaining information about the amino acid composition of a protein is an important step in the characterization of protein structure. Within scientific disciplines such as food and nutrition, amino acid analysis has gained greater importance over the traditional nitrogen determination methods as a means of assessing dietary protein quality. The need to make this assessment is largely due to an increasing recognition that it is the amino acid balance of a food, in relation to the amino acid requirement of the consumer, rather than the amount of protein per se, that is important from a nutritional standpoint. Indeed, amino acid analysis forms the cornerstone of dietary protein quality assessment methods such as the digestible indispensable amino acid score (DIAAS) method which was recently recommended by the Food and Agriculture Organization (FAO) as the preferred method for assessing dietary protein quality (FAO, 2013). The aim of this chapter is to discuss amino acid analysis both in general terms and within the context of plant products. While there are many examples of different types of plant materials for which amino acid analysis is undertaken, the main focus of this review will be on foods and feedstuffs derived from plants.

25.2 Introduction

Amino acids are a chemically diverse group of zwitterions containing an N-terminal amino group and a C-terminal carboxyl group; they can therefore be positively charged, negatively charged or neutral depending on the pH of their environment. Amino acids also contain a side chain, the structure of which is unique for each amino acid; these include large aromatic groups, small aliphatic groups, amides, carboxylic acids, amines, guanidino groups and sulfydryl groups. Some amino acid side chains are chemically inert, though others possess considerable reactivity. It is this diversity in the chemical structure of the amino acid side chains that, on the one hand, permits their individual determination but, on the other hand, makes their accurate determination challenging. A procedure to quantify amino acids was first developed by the workers Moore and Stein in the late 1940s and early 1950s (Moore and Stein, 1951). This procedure consisted of an acid hydrolysis step to hydrolyse the peptide bonds within the proteins and liberate the amino acids, followed by an ion-exchange chromatography step to separate the liberated amino acids, and derivatization with ninhydrin to detect and quantify the separated amino acids. Since the

*E-mail address: S.M.Rutherford@massey.ac.nz

original Moore and Stein method was first published, advances have been made in the separation and detection of amino acids in which new technologies such as capillary electrophoresis, mass spectrometry (MS) and nuclear magnetic resonance (NMR) are being used more often and in addition to the traditional high performance liquid chromatography (HPLC)-based separation methods. In contrast, however, the original hydrolysis procedure has largely remained unchanged despite considerable work in this area having been undertaken over the last 60 years.

25.3 Sample Preparation

Amino acid analysis is a technically demanding procedure and considerable effort has been made to reduce the variability in amino acid composition data so obtained (Finley, 1985). Foods and feedstuffs derived from plants are often counted among the more difficult materials to analyse, largely because of the presence of the fibrous material that is common in many plant-based foods and feedstuffs. To prepare samples for amino acid analysis, samples may need to be dried and finely ground to ensure reproducibility when taking aliquots for analysis. Yet the presence of fibrous material makes sample grinding difficult. For example, when analysing breakfast cereals, which may contain significant amounts of fibrous material, grinding methods that cut rather than pulverize the samples can be beneficial. The situation is further complicated for plant materials such as grains, which may comprise the starchy kernel, which is easily ground, and an outer fibrous layer, which is not easily ground. During the grinding process, the kernel and the fibrous layer may be unintentionally separated. This is undesirable as each fraction will contain different amounts and types of proteins. For plant materials such as nuts that contain significant amounts of fat (>10%) grinding should be carried out such that sample temperatures are maintained at a point below the melting point of the fat because melted fat can clog the grinder, making further grinding difficult. Alternatively, samples can be defatted before grinding. Defatting samples that contain significant amounts of fat also improves the recoveries of some of the more labile amino acids (Finley, 1985). The importance of sample homogeneity cannot be overemphasized as high sample heterogeneity can be a leading cause of the variability of amino acid composition data (Rutherfurd and Gilani, 2009).

25.4 Amino Acid Analysis

The amino acids within a protein are held together with peptide bonds. To determine the amino acid composition of protein, including plant proteins, the peptide bonds must first be hydrolysed to release the constituent amino acids. The liberated amino acids are then separated, often using chromatography, and the separated amino acids are then quantified. Each step of the amino acid analysis procedure is discussed below.

25.4.1 The hydrolysis step

Peptide bonds are very stable and require severe chemical treatment to cleave them. Typically, hydrolysis in strong acid (6 M HCl) at high temperature (110 °C) for 24 h in an oxygen-free environment is required to hydrolyse proteins for amino acid analysis (Ozols, 1990; Darragh and Moughan, 2005). However, even under these rigorous conditions, not all peptide bonds are broken. For example, peptide bonds involving valine and isoleucine may require hydrolysis for up to 72 h or longer for complete hydrolysis (Rayner, 1985).

The chemical structures of the amino acid side chains are highly diverse and consequently have very different stabilities during HCl hydrolysis. In particular, the recoveries of tryptophan and cysteine, and to some degree of methionine, are sufficiently poor to warrant separate approaches for their accurate analysis. Typically, cysteine and methionine are analysed using quantitative performic acid oxidation before HCl hydrolysis, while tryptophan is analysed using alkali hydrolysis. The remaining amino acids, sometimes referred to as the acid-stable amino acids, can also be susceptible to hydrolytic losses during HCl hydrolysis; those that are most susceptible are discussed below.

Asparagine and glutamine are the amide derivatives of aspartic acid and glutamic acid, and during acid hydrolysis they are deaminated to aspartic acid and glutamic acid, respectively. Consequently, amino acid analysis data for aspartic acid and glutamic acid represents the sum of both the acid and the amide derivatives. Several methods for analysing asparagine and glutamine have been reported (Tower, 1967; Wilcox, 1967; Carraway and Koshland, 1972; Soby and Johnson, 1981), and are based on a range of chemical strategies, but few are used routinely. From a nutritional point of view, the collective analysis of the acid and amide derivatives is not inappropriate given that transamination reactions occur within the body that can convert the acid to the amide and vice versa.

Tyrosine can undergo halogenation during HCl hydrolysis, but this can be largely overcome by the inclusion of phenol in the 6 M HCl that is used (Nissen, 1992). Defatting samples that contain significant amounts of fat (>5%) can also improve tyrosine yields (Finley, 1985). The use of constant boiling HCl is also important to remove impurities such as iron or copper ions, the presence of which can lead to low tyrosine recoveries (Finley, 1985).

The recovery of serine and threonine after HCl hydrolysis is also often less than complete, although thorough removal of oxygen during hydrolysis will minimize these losses. Typical hydrolytic losses of threonine and serine in our laboratory for plant-based foods or feedstuffs are 3 and 4.5%, respectively, but losses as high as 10–15% in non-plant protein sources have been reported (Darragh *et al.*, 1996). To overcome hydrolytic losses for serine and threonine, hydrolysis for less than 24 h has also been used (Rowan *et al.*, 1992), as has the use of multiple hydrolysis times and the extrapolation of threonine and serine yields back to time zero. Correction factors have been developed to adjust for the hydrolytic losses of these amino acids, but these may not be universally applicable to all protein sources or across laboratories.

HCl hydrolysis has also been conducted using microwave ovens as a means of decreasing hydrolysis times and thereby improving amino acid recoveries. Kabaha *et al.* (2011) determined the amino acid content of wheat flour using microwave hydrolysis, but the maximum recovery of amino acids in relation to total protein was only 85%. In contrast, Marconi *et al.* (1995) reported higher recoveries (87–109%) for amino acids from wheat semolina using much shorter hydrolysis times than those used by Kabaha *et al.* (2011). While there is clearly potential for the use of microwave-based hydrolysis for amino acid analysis, there appears, overall, to be very little adoption of the practice in analytical laboratories. It has been suggested that the quality of microwave ovens, the non-uniformity of the irradiation within the oven and the need to have identical sample volumes are important considerations when using microwave hydrolysis.

Cysteine can be destroyed during HCl hydrolysis and methionine can be oxidized to methionine sulfoxide and further to methionine sulfone. Methionine sulfone is relatively stable during HCl hydrolysis, but methionine sulfoxide can also revert back to methionine during hydrolysis (Rutherfurd and Moughan, 2008). Even though good recoveries of methionine (but not cysteine) can be obtained using HCl hydrolysis with rigorous oxygen removal (Finley, 1985), cysteine and methionine are typically analysed by HCl hydrolysis only after they have been quantitatively oxidized to cysteic acid and methionine sulfone, respectively, with performic acid. Rutherfurd *et al.* (2007) reported that for several complex feedstuffs, including those of plant origin, cysteic acid and methionine sulfone were stable during HCl hydrolysis. In contrast, other workers have reported that in samples of human milk and pure lysozyme, hydrolytic losses of cysteic acid were greater than the losses for any other amino acid (Darragh *et al.*, 1996). Other strategies for cysteine determination have also been reported, for example iodoacetylation or vinylpyridation, but these strategies are not commonly used.

When present in pure proteins, tryptophan has been reported to be relatively stable during HCl hydrolysis (Molnár-Perl and Khalifa, 1992), though in the presence of carbohydrates, as is the case for many plant-based foods and feedstuffs, tryptophan recoveries after HCl acid hydrolysis tend to be poor (Finley, 1985). Consequently, tryptophan is usually determined using alkali hydrolysis, typically for 20 h at 110 °C in sodium hydroxide, barium hydroxide or lithium hydroxide and in an oxygen-free environment. Once hydrolysed, tryptophan can be easily separated using reversed-phase HPLC and detected by either absorbance or fluorescence because tryptophan is the most hydrophobic amino acid and has an intense absorbance maximum at 280 nm. It can also fluoresce with an emission wavelength of 350 nm, although this can vary depending on the environment. The recovery of tryptophan can be poor, even when using base hydrolysis, so an internal standard (usually 5-methyltryptophan or α-methyltryptophan) can be added to the hydrolysis mixture and its recovery used to correct for the hydrolytic losses of tryptophan itself (Fontaine, 2003). This 'internal standard correction' assumes that tryptophan and the internal standard are degraded at the same rate during hydrolysis. Using least-squares non-linear regression (see Section 25.4.2) to model losses of 5-methyltryptophan and tryptophan in different matrices (Table 25.1), the loss rate of 5-methyltryptophan was found to be similar to that of tryptophan when no other material was present. However, this was not the case when different matrices were compared. For example, tryptophan was degraded much more rapidly than 5-methyltryptophan in wheatmeal but was degraded much more slowly than 5-methyltryptophan in soybean meal. Clearly, correcting for the hydrolytic losses of tryptophan using an internal standard, such as 5-methyltryptophan, must be done with a great deal of caution as their relative stabilities are matrix dependent. An alternative approach for the accurate analysis of tryptophan, or of any other labile amino acid, is the least squares non-linear regression method, which can be used to model the release and degradation of amino acids during hydrolysis. The latter method is described below.

Table 25.1. Loss rate coefficient[a] for 5-methyltryptophan and tryptophan during alkali hydrolysis[b] in different sample matrices modelled using least-squares non-linear regression.

Sample matrix	Loss rate coefficient[c]	
	5-Methyl-tryptophan	Tryptophan
None[d]	0.0078	0.0083
Soybean meal	0.0116	0.0087
Wheatmeal	0.0135	0.0157

[a]Based on the model derived from that used by Rutherfurd (2009): $B(t) = B_0 e^{(-lt)}$, where $B(t)$ is the concentration of 5-methyltryptophan or tryptophan after hydrolysis for t h, l is the loss rate of 5-methyltryptophan or tryptophan and B_0 is the initial 5-methyltryptophan or tryptophan concentration.
[b]Hydrolysis carried out in freshly prepared 4.5 M NaOH containing 5% (w/v) maltodextrin in nitrogen-purged sealed Teflon vials at 110 °C for 20 h.
[c]Proportion of 5-methyltryptophan or tryptophan destroyed hourly.
[d]Synthetic 5-methyltryptophan or tryptophan added directly to the NaOH solution.

25.4.2 Least-squares non-linear regression

When accurate amino acid composition data are required, least-squares non-linear regression can be used to estimate the amino acid content of a protein source. This method, originally developed by Robel and Crane (1972) and further modified by Darragh *et al.* (1996), corrects for losses of amino acids during hydrolysis based on the amino acid yield determined using a range of hydrolysis times. It uses the model below (Eqn 25.1):

$$B(t) = \frac{A_0 h\,(e^{-lt} - e^{-ht})}{h - l} + B_0(e^{-lt}) \qquad (25.1)$$

where e is the exponential function, $B(t)$ is the amino acid concentration at time t, B_0 is the free amino acid content prior to hydrolysis, h is the hydrolysis rate of the amino acids (the rate at which the amino acids are released from the protein during hydrolysis), l is the loss rate of the amino acids (the rate at which the amino acids are destroyed during hydrolysis) and A_0 is the actual protein-bound amino acid content. This compartmental model takes into account the simultaneous hydrolysis and loss of amino acids during the hydrolytic process which other predictive methods, such as using multiple hydrolysis times and extrapolating amino acid yields back to time zero (0), do not take into account.

The application of the least-squares non-linear regression method is demonstrated for tryptophan in Fig. 25.1 using soybean meal and wheatmeal as examples. The marked disappearance of tryptophan during alkali hydrolysis can be clearly observed. As a result, the determined tryptophan content of soybean meal and wheatmeal after a 20 h alkali hydrolysis was respectively 20 and 37% lower than the actual tryptophan content (A_0) derived using the least-squares non-linear regression model (Table 25.2). It is of note that even after correction for hydrolytic

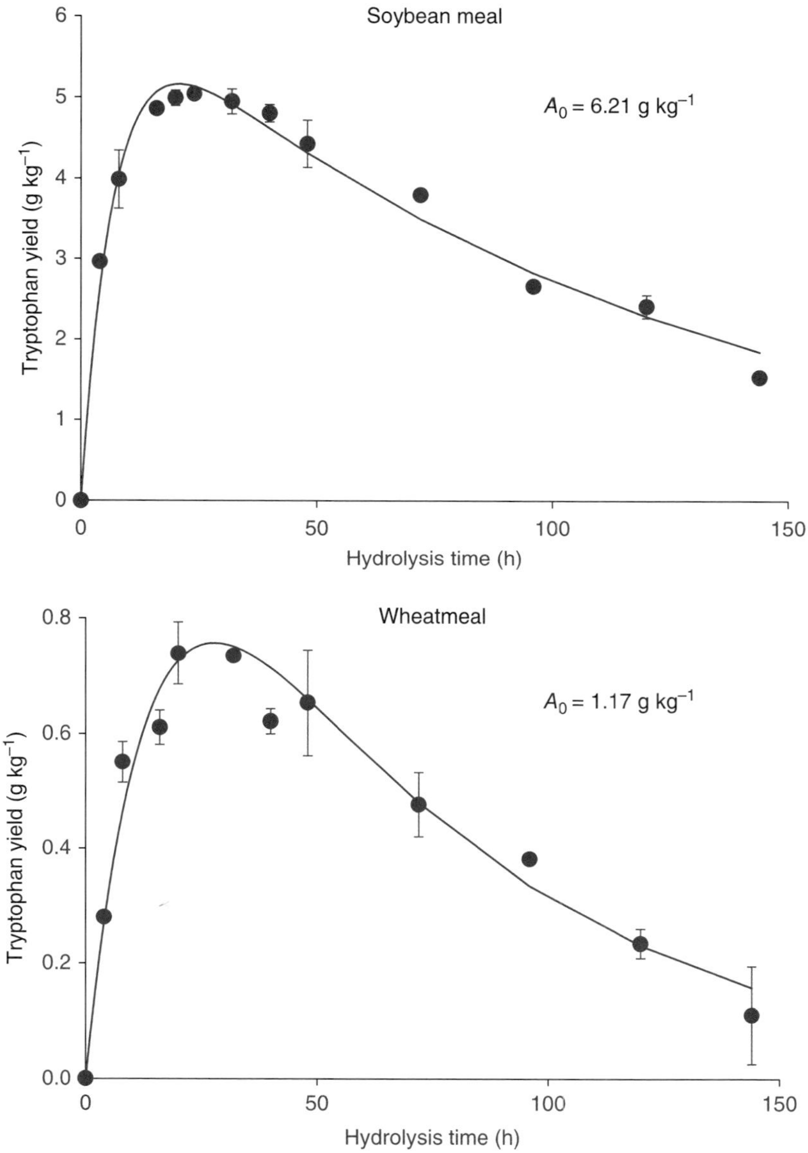

Fig. 25.1. The mean (± SD) yield of tryptophan (●) in soybean meal and wheatmeal over multiple alkali hydrolysis times. The least-squares non-linear regression line (—) is plotted along with the line of best fit derived using Eqn 25.1 (see text) based on A_0, *h* and *l* predicted for the data set. Internal standard correction of the tryptophan data was not used. The analysis was conducted in triplicate.

losses based on the recovery of 5-methyltryptophan, the tryptophan concentration determined in the soybean meal and wheatmeal samples still underestimated the tryptophan concentration determined using least-squares non-linear regression by 4 and 19% respectively. While an underestimation of 4% may be deemed acceptable, an underestimation of 19% is clearly unacceptable.

The least-squares non-linear regression method has been applied to several plant-based feedstuffs including canola meal, maize and soybean meal when using HCl hydrolysis (Rutherfurd *et al.*, 2007; Rutherfurd, 2009).

Table 25.2. Tryptophan concentration (g kg^{-1}) of soybean meal and wheatmeal determined using a single 24 h alkali hydrolysis and using least-squares non-linear regression.

	24 h Hydrolysis[a]		
	Uncorrected[b]	Corrected[c]	A_0[d]
Soybean meal	4.99 (19.6)	5.94 (4.2)	6.21
Wheatmeal	0.739 (37.0)	0.950 (19.0)	1.17

[a]Hydrolysis carried out in freshly prepared 4.5 M NaOH containing 5% (w/v) of maltodextrin in nitrogen-purged sealed Teflon vials at 110 °C for 20 h. Values in parentheses are % underestimation compared with the A_0 value.
[b]Not corrected for the recovery of an internal standard.
[c]Corrected for the recovery of an internal standard (5-methyltryptophan) included prior to hydrolysis.
[d]A_0, actual protein-bound amino acid content.

In general, the estimates of amino acid concentration based on a single 24 h hydrolysis value agreed well (less than a 3% difference) with estimates based on the least-squares non-linear regression method, even for amino acids that are purported to be susceptible to degradation during hydrolysis, such as threonine and serine. The main exceptions were serine in maize, and valine and isoleucine in soybean meal, where the concentrations were 5.5–8.9% lower when determined using a single 24 h hydrolysis compared with least-squares non-linear regression estimates.

Given that the least-squares non-linear regression method requires hydrolysis over multiple hydrolysis time periods, it is unlikely to be adopted routinely. However, this approach may be useful when highly accurate amino acid composition data are required.

25.4.3 The chromatography step

Once the samples have been hydrolysed, the resulting mixture of free amino acids must be separated and quantified. Historically, this has been most commonly carried out by HPLC using either ion-exchange chromatography followed by post-column derivatization with either ninhydrin or *o*-phthalaldehyde (OPA), or pre-column derivatization followed by reversed-phase chromatography. Derivatization is usually required because amino acids generally do not absorb light at any useful wavelength, although tryptophan and to some degree tyrosine and phenylalanine absorb significantly at 280 nm. Some of the more common pre-column derivatizing reagents include OPA, phenylisothiocyanate (PITC; Bidlingmeyer *et al.*, 1984), 6-aminoquinolyl- *N*-hydroxysuccinimidyl carbamate (AQC; Cohen and Michaud, 1993) and 9-fluorenylmethyl chloroformate (FMOC, Einarsson *et al.*, 1983) or a combination of OPA and FMOC (Bütikofer *et al.*, 1991). A thorough discussion of these derivatizing reagents is given by Rutherfurd and Gilani (2009). More recently, several new derivatizing reagents have been reported, including 2,7-dimethy-3,8-dinitrodi pyrazolo[1,5-a:1′,5′-d]pyrazine-4, 9-dione (Gioia *et al.*, 2006) and 2,5-dimethyl- 1*H*-pyrrole-3,4-dicarbaldehyde (Gatti *et al.*, 2010). These and other new candidates have been described by Otter (2012) and will not be discussed further here. Essentially, any derivatizing reagent for amino acid analysis should react rapidly with primary amines (and preferably secondary amines because proline is a secondary amine), be stable, produce stable derivatives and absorb (or, preferentially, fluoresce) intensely.

25.4.4 Mass spectrometry and nuclear magnetic resonance

One main drawback of optical detection is the relative lack of selectivity, particularly for absorbance-based systems. An alternative approach for amino acid detection is the use of non-optical systems such as MS or NMR. Recent advances in the sensitivity and reliability of MS, along with reducing instrument costs, have seen the increased adoption of MS for amino acid analysis as it permits the detection of amino acids without the need to derivatize them.

Generally, amino acids are not well retained by reversed-phase chromatography columns, often used at the front end of mass spectrometers, owing to the low affinity between the hydrophobic columns and the generally hydrophilic amino acids. However, ion-pairing reversed-phase HPLC does permit the reversed-phase separation of underivatized amino acids. Ion-pairing reagents have both an ionic and a non-polar moiety in which the non-polar end of the ion-pairing reagent is strongly associated with the hydrophobic stationary phase, and the ionic end interacts with the charged amino acids in the mobile phase, thereby permitting the retention of amino acids on the column that would otherwise not be retained. By using volatile ion-pairing reagents, liquid chromatography (LC) can be coupled to mass spectrometers using electrospray ionization. Once inside the mass spectrometer, the amino acids can be fragmented within a collision zone within the mass spectrometer to yield fragments specific to each amino acid, and the amino acids being determined can be identified and quantified based on the presence of these amino acid-specific fragment ions. There are a number of different collision techniques, including collision-induced dissociation (CID), in which neutral atoms or molecules are used to fragment parent ions, and electron capture dissociation (ECD), in which electrons are used to provide the energy required for fragmentation. Using multiple reaction monitoring (MRM), whereby multiple parent ions (i.e. each of the amino acids) along with selected fragmentation ions specific for each amino acid are identified and quantified, facilitates the unique identification of amino acids even if complete separation of the amino acids on the front-end LC is not achieved. This provides a potentially very powerful platform for amino acid analysis. Ion-pairing-LC-tandem or multiple step MS (MS/MS) has been used in LCMS (liquid chromatography-mass spectrometry) systems to determine the free amino acid content of biological fluids (Piraud *et al.*, 2005), fertilizers (Alarcón-Flores *et al.*, 2010) and some plant extracts (Thiele *et al.*, 2012). However, to the author's knowledge, it has not been used in conjunction with HCl hydrolysis to determine the total (free plus protein bound) amino acids in plant materials.

NMR has also been used for amino acid analysis and this method has the advantage of being able to determine all of the amino acids within a mixture without requiring separation. The main disadvantage of NMR is its lack of sensitivity, although the use of cyro-probes has greatly increased this. Moreover, multidimensional NMR focusing on both ^{1}H and ^{13}C gives greatly improved resolution compared with single-dimensional approaches. Yang *et al.* (2010) recently compared the amino acid composition of mushrooms and spirulina determined using either NMR or a traditional amino acid analyser (ion-exchange chromatography/ninhydrin post-column detection) after 6 M HCl hydrolysis. They reported very good agreement between the two methods with mean differences across all determined amino acids of 1.9% for mushrooms and 1.3% for spirulina. To the author's knowledge, amino acid analysis using NMR has not yet been reported for products derived from higher plants.

The main drawback of both LCMS and NMR technologies is that the instruments tend to be expensive and require highly skilled operators to ensure that they are performing optimally and that data generated from them is reproducible and reliable. As a result, relatively few studies describing LCMS- and NMR-based amino acid analysis methods have been published. Despite this, both LCMS and NMR appear to hold considerable promise for amino acid analysis, and more work in this area is desperately needed.

25.4.5 Determination of free amino acids

The determination of free amino acids, often referred to as physiological amino acids, is considerably more straightforward than that of their protein-bound counterparts. As no hydrolysis is required, the losses in amino acids associated with the hydrolysis step are not of concern when determining free amino acids. The main difficulty is that free amino acid extracts contain a greater diversity of amino acids than protein acid hydrolysates and are therefore more difficult to separate using traditional HPLC approaches. The main preparative steps involve extraction and deproteinization of the samples; several such deproteinization methods are described by Prosser and Rutherfurd (2012). The free amino acid content of plant materials has usually been undertaken in the context of describing the content of only selected amino acids rather than the content of all of the free amino acids present. One such example is asparagine, which is prevalent in many plants and acts as a nitrogen storage compound. The determination of free asparagine in plants has gained interest in recent times since it was discovered that heat processing of carbohydrate-rich plant-based foods can lead to the formation of acrylamide, a known neurotoxin and possible carcinogen, via the reaction of asparagine and carbonyl-containing compounds such as reducing sugars (Zyzak *et al.*, 2003).

25.4.6 Presentation of amino acid composition data

When presenting amino acid compositional data it is common practice to give the values in units of g 100 g^{-1} or g kg^{-1} of analysed material. As amino acids are usually determined on a molar basis (depending on the amino acid standard used to quantify them), the molecular weight of each amino acid is required to convert the moles of each into the weight of each amino acid. Amino acid weights can be calculated using either the free amino acid molecular weights or the bound (in-chain) molecular weights. Free amino acid molecular weights are 18 Da heavier than the equivalent bound amino acid molecular weights, as the weights reflect the water molecule (water of hydrolysis) added across the peptide bond when the peptide bond is hydrolysed. Free amino acid molecular weights are relevant when describing the amount of amino acids that would be present after digestion of the protein, and so these are commonly used for the analysis of foods. Bound amino acid molecular weights reflect the molecular weights of the amino acids when present within the intact protein and are relevant when using the sum of all the amino acids to estimate the amount of total protein within a sample; these molecular weights are more commonly used in protein chemistry laboratories (Rutherfurd and Dunn, 2011). Both types of molecular weights are valid, depending on how the amino acid composition data is to be used, but the consequences of the choice of molecular weight (free or bound) used and the appropriateness of the two different molecular weights are often not appreciated (Rutherfurd and Moughan, 1993). By using molecular weights inappropriately, the inaccuracy of amino acid composition data can be significant, ranging from 9.7% for tryptophan to 31.5% glycine. It is typically approximately 16% for proteins, although the exact overestimation will vary depending on the amino acid composition of the protein source.

25.4.7 Internal and external standards

Amino acid analysis is a complex and technically demanding methodology and maintaining acceptable repeatability and reproducibility limits is a constant challenge. For plant materials, sample heterogeneity is arguably the most important factor to consider in order to minimize analytical variability, but the use of internal and external standards is also important for achieving this goal. Internal standards can be any compound that can be detected by the detection method used, is stable to the hydrolysis procedure and is not naturally present in the sample. For HCl hydrolysis, norleucine or norvaline are commonly used. The internal standard can be added either before or after acid hydrolysis, but is most often added after (Rutherfurd and Gilani, 2009).

25.5 Determination of the Amino Acid Composition of Processed Plant Products

The processing of plant materials is commonplace, particularly in the food and feedstuff industries. Food processing is used to improve functionality, shelf life and acceptability. As applied to plant products, processing can include cooking, extrusion, grinding, alkali treatment, oxidation and pelleting. During processing and also during long-term storage, proteins can undergo physical and chemical modifications. The ε-amino group of lysine is susceptible to Maillard reactions with other compounds that may be present in plant materials, such as reducing sugars, polyphenols and oxidized lipids (Hurrell and Carpenter, 1981). While the early Maillard reaction (the formation of deoxyketosyl-lysine) during relatively mild heating is well understood, more severe heating can lead to a plethora of end products of which only relatively few have been characterized (Friedman, 1996). Methionine, cysteine and tryptophan can be oxidized, although for plant materials, the presence of plant phenolic compounds, has been shown to inhibit the oxidation of tryptophan, at least in model systems (Salminen and Heinonen, 2008). Threonine and serine can be phosphorylated (Shih, 1996), and asparagine and glutamine can be deamidated (Shih, 1996). Moreover, free asparagine, which can be relatively abundant in plants, can form acrylamide (Zyzak *et al.*, 2003). In addition, many amino acids can undergo racemization during processing, particularly under alkaline conditions (Liardon and Hurrell, 1983).

Many of the modifications that can occur to amino acids in foods and feedstuffs during processing can have a negative impact on their nutritional quality because almost all of the modified amino acid derivatives cannot be used for protein synthesis in the body. Moreover, when present in processed plant materials, some of the modified amino acid derivatives also pose a problem for amino acid analysts. The most well-known examples of such derivatives include the early Maillard product deoxyketosyl-lysine and oxidized sulfur amino acids. These, along with phosphorylated threonine and serine, are discussed below.

25.5.1 Lysine

Lysine is an acid stable amino acid and can be quantitatively recovered after HCl hydrolysis. However, if it has undergone Maillard reactions during processing, then the resulting derivatives can interfere with the determination of lysine during amino acid analysis. The most well-known example is the early Maillard compound deoxyketosyl-lysine. This derivative is not stable during HCl hydrolysis; a portion of the deoxyketosyl-lysine reverts back to lysine during hydrolysis and leads to an overestimate of the unmodified lysine. Generally, the lysine determined using HCl hydrolysis is referred to as total lysine and consists of the unmodified lysine (referred to as reactive lysine) and the reverted lysine (lysine that has reverted from early Maillard products during hydrolysis). The degree of overestimation of the reactive lysine content in total lysine estimates can vary considerably across protein sources and processing conditions. In one study focusing on the reactive and total lysine contents of a range of ready-to-eat cereal- based breakfast foods, the overestimation ranged from 24 to 120% (Torbatinejad *et al.*, 2005). Consequently, it is important that reactive lysine rather than total lysine be determined in plant-based foods that have undergone processing, particularly heat processing.

A number of methods have been published for determining the reactive lysine in processed protein sources. These include the fluorodinitrobenzene (FDNB) method (Hurrell and Carpenter, 1974), the guanidination method (Mauron and Bujard, 1964; Mao *et al.*, 1993; Torbatinejad *et al.*, 2005), the trinitrobenzene sulfonic acid method (Kakade and Leiner, 1969), the determination of OPA-reactive lysine (Vigo *et al.*, 1992; Malec *et al.*, 2005) and of ninhydrin-reactive lysine (Friedman *et al.*, 1984), the FDNB-difference method (Rao *et al.*, 1963), the sodium borohydride method (Hurrell and Carpenter, 1974), the furosine method (Desrosiers *et al.*, 1989) and dye-binding methods (Hendriks *et al.*, 1994). The most commonly reported methods used are the FDNB and guanidination methods. However, all methods (with the exception of the furosine method, which is based on the determination of furosine after HCl hydrolysis, where furosine is one of the breakdown products of deoxyketosyl-lysine during hydrolysis) are based around the reaction with the ε-amino group of lysine, which will be unavailable for reacting if chemical modification has occurred to lysine during processing.

While all the above methods have their inherent advantages and disadvantages, all are suitable (at least from a theoretical perspective) for determining reactive lysine when it is bound within proteins. However, most of the reagents listed above are not specific for the ε-amino group of lysine and will react with the α-N-terminal of lysine as well, thereby rendering most of these methods inaccurate for determining reactive lysine in materials that contain free lysine. This may not be a serious drawback when analysing plant proteins generally, as little free lysine is likely to be present, but it may be of concern for the analysis of formulated feedstuffs for intensive livestock production, because they almost always contain large amounts of plant products (for example, maize, soybean meal, wheat and barley) and are often supplemented with synthetic lysine. If lysine is present at the N-terminal position of proteins, then it may also be doubly derivatized and therefore not identified as reactive lysine when it is determined using most reactive lysine assays. This situation may not be of great concern for large intact proteins but may be of concern for protein hydrolysates, including those of plant origin, in which the likelihood of N- terminal lysine being present is much greater.

Of the derivatization methods described above, only the guanidination method is suitable for the analysis of reactive lysine in plant products that contain free or N-terminal bound lysine. The guanidination method involves the reaction of *O*-methylisourea with lysine to produce homoarginine, an amino acid that is stable during HCl hydrolysis-based amino acid analysis. The molar concentration of reactive lysine is then directly equivalent to the molar concentration of homoarginine. *O*-methylisourea is specific for the ε-amino group of lysine and does not react with the N-terminal amino group (Moughan and Rutherfurd, 1996; Zhang *et al.*, 2006), thereby permitting the determination of reactive lysine in samples containing free lysine or peptides that contain N-terminal lysine. As the guanidination reaction is specific for the ε-amino group of lysine, it can also be used to determine the amino acid content of digesta collected from the terminal ileum (used for determining true ileal amino acid digestibility), which will inevitably contain not only protein but also peptides and amino acids. By coupling the guanidination reaction with the true ileal amino acid digestibility assay, the true ileal digestible reactive lysine (available lysine) can be determined (Moughan and Rutherfurd, 1996). True ileal digestible total lysine can greatly overestimate the true ileal digestible reactive lysine (available lysine), particularly for processed plant products such as breakfast cereals (Rutherfurd *et al.*, 2006), wheat, soya and maize products, for which overestimations of up to 90% have been observed (Rutherfurd and Moughan, 2007). The true ileal amino acid digestibility assay forms the basis of DIAAS, the method for dietary protein quality assessment recently endorsed by the FAO (2013). The FAO also recognizes the shortcomings of the traditional amino acid analysis approach for lysine when applied to processed protein sources (FAO, 2013).

25.5.2 Methionine and cysteine

When plant-based foods or feedstuffs undergo processing, even relatively mild processing, cysteine can be oxidized to cysteic acid, and methionine can be oxidized to methionine sulfoxide and even to methionine sulfone under more extreme processing conditions (Todd *et al.*, 1984; Hayashi and Suzuki, 1985). Significant amounts of oxidized methionine have been reported in several plant-based food products (Table 25.3). Inevitably, the presence of oxidized cysteine and methionine derivatives in plant-based products as a result of processing will interfere with the determination of these amino acids because they are generally determined after quantitative oxidation (with performic acid) to cysteic acid and methionine sulfone, respectively. Given that cysteic acid and methionine sulfone are generally not nutritionally available (i.e. they cannot be used to synthesize body protein), and that methionine sulfoxide is only available under certain circumstances (Anderson *et al.*, 1976; Friedman and Gumbmann, 1988), the inability to discriminate between the unoxidized and oxidized forms of cysteine and methionine in foods and feedstuffs is of concern. Ideally, cysteine, methionine and methionine sulfoxide should be determined separately from cysteic acid and methionine sulfone in processed foods and feedstuffs. Several methods that avoid the use of quantitative oxidation have been used to determine methionine. These include reaction with cyanogen bromide (CNBr) and then determination of methyl thiocyanate (Ellinger and Duncan, 1976) or reaction with CNBr and then determination of homoserine (Elias *et al.*, 2005). For cysteine, derivatization with iodoacetic acid or vinylpyridine has been reported. However, none of these methods are commonly used.

Table 25.3. Reported relative content of methionine and its oxidized derivatives in selected plant-based foods. (Source: Todd *et al.*, 1984; Hayashi and Suzuki, 1985.)

	Relative amounts (%)		
Food type	Methionine	Methionine sulfoxide	Methionine sulfone
Soy flour	75	14	11
Soy infant formula	76	12	12
Soy isolate	35	59	6
Udon noodles	69	31	–

Methionine sulfoxide is unstable during HCl hydrolysis, as it can be reduced to methionine or oxidized to methionine sulfone (Rutherfurd and Moughan, 2008). These two reactions appear to be contradictory, and it has been suggested that the relative flux of the two may be related to the degree of oxygen removal from the hydrolysis tubes (Rutherfurd and Moughan, 2008). Several hydrolysis methods have been reported to be suitable for determining methionine sulfoxide in foods and feedstuffs. These include hydrolysis with methanesulfonic acid (Chiou and Wang, 1988; Puchala *et al.*, 1994; Sochaski *et al.*, 2001), *p*-toluenesulfonic acid (Hayashi and Suzuki, 1985) and alkaline hydrolysis (Gjøen and Njaa, 1977; Sjøgerg and Bostrøm, 1977; Cuq *et al.*, 1978; Nakamura *et al.*, 1983; Todd *et al.*, 1984). The suitability of these methods has been questioned because after hydrolysis with methanesulfonic acid (the most commonly reported hydrolysis method for methionine sulfoxide determination), the recoveries of methionine sulfoxide in fishmeal have been reported to be incomplete when they are compared with the predicted methionine sulfoxide content determined using least-squares non-linear regression (Rutherfurd and Moughan, 2008). To date, the methionine sulfoxide content of plant products determined using methanesulfonic acid hydrolysis has not been compared with least-squares non-linear regression estimates, but such a comparison may be warranted. A more detailed discussion of the analysis of cysteine, methionine and methionine sulfoxide in foods has been presented by Rutherfurd and Moughan (2008), who, overall, concluded that more work is required in this area.

25.5.3 Threonine and serine

When foods undergo heat processing, threonine can react with sugars to produce a range of compounds, including pyrazines, pyridines and pyrroles (Ho and Chen, 1999). From a nutritional standpoint, such amino acid modifications are undesirable, but from an analytical standpoint the impact of the presence of such compounds is less clear. With respect to threonine analysis, given that the structure and chemistry of these threonine derivatives is very different from those of threonine, there is likely to be little interference from processing-induced modified threonine derivatives.

In some dietary proteins, for example casein proteins, threonine and serine residues may be naturally phosphorylated. Moreover, food proteins such as soy proteins, wheat gluten and yeast proteins are commonly phosphorylated as part of the food manufacturing process in order to alter the functional properties of the proteins (Shih, 1996). In addition, other amino acids such as lysine and histidine can also be phosphorylated. The presence of these phosphorylated amino acids will inevitably have an impact on the analysis of their non-phosphorylated counterparts (Rutherfurd and Moughan, 2012).

Phosphothreonine and phosphoserine have been reported to be unstable during both alkali (Kemp, 1980) and HCl hydrolysis (Bylund and Huang, 1976), with a portion of the phosphothreonine and phosphoserine being recovered as threonine and serine, respectively. As phospho-amino acids can be dephosphorylated by alkaline phosphatases present in enterocytes and utilized within the body, determining phosphothreonine and phosphoserine as the sum of their phosphorylated or non-phosphorylated forms may be acceptable from a nutritional standpoint. However, for threonine and serine, collective recovery of the phosphorylated or non-phosphorylated forms from peptides containing either phosphothreonine or phosphoserine have been reported to be only 86 and 73% for phosphothreonine and phosphoserine, respectively, after 6 M HCl hydrolysis for 24 h at 110 °C (Bylund and Huang, 1976). Hydrolysis in more dilute HCl (Bylund and Huang, 1976) or for shorter hydrolysis times (Bylund and Huang, 1976; Kamps and Sefton, 1989), as well as alkali hydrolysis (Kamps and Sefton, 1989) and the use of correction factors to account for hydrolytic losses (Capony and Demaille, 1983), have been suggested as alternative strategies that may provide more accurate phospho-amino acid compositional data. However, these analytical methods and strategies do not appear to have been used for the analysis of plant materials to any significant degree. Overall, further study of the impact of phosphorylation on the accurate determination of threonine and serine in plant-derived foods and feedstuffs, and the development of suitable analytical strategies, may be warranted (Rutherfurd and Moughan, 2012).

25.5.4 Amino acid racemization

While L-amino acids predominate in higher plants, the presence of D-amino acids in some has been reported, although the amounts appear to be low in relation to their L-counterparts (Brückner and Westhauser, 2003). In addition, many foods, including those of plant origin, have been reported to contain D-amino acids (Friedman, 1999; Csapó *et al.*, 2009). The racemization of amino acids, that is the conversion of the L-enantiomer of the amino acid to the D-enantiomer, can also occur during processing, particularly at high pH

(Liardon and Hurrell, 1983), but the rate at which amino acids are racemized varies depending on the amino acid, alkali concentration and protein source (Masters and Friedman, 1979). In addition, racemization occurs about ten times more rapidly for amino acids bound within protein compared with their free counterparts (Liardon and Ledermann, 1986). There are a number of methods reported in the literature for determining D-amino acids. Most generally involve some form of chromatographic separation using either GC or HPLC (capillary electrophoresis – CE – has also been used) in which enantiomers are separated after derivatization with a chiral derivatization reagent, such as 1,5-difluoro-2,4- dinitrobenzene (Marfeys reagent; Marfey, 1984) or OPA/2,3,4,6-tetra-*O*-acetyl-1-thio-*β*-glucopyranoside (Einarsson *et al.*, 1987) for reversed-phase HPLC chromatography, or ethylchloroformate and chloropropanol (Bertrand *et al.*, 2008) for GC chromatography. Alternatively, enantiomers can be separated directly without prior derivatization by using chiral columns.

Recently, Eto *et al.* (2011) described an HPLC-based method whereby amino acids are derivatized with OPA, a common derivatizing agent for amino acid analysis. The derivatized individual amino acids are then separated using reversed-phase HPLC in which the D- and L-enantiomers of each amino acid are co-eluted. Circular dichroism is then used to distinguish the D- and L-enantiomers present in each peak. This approach offers an elegant solution to the determination of D-amino acids within samples that primarily contain L-amino acids, such as those derived from higher plants.

Other non-chromatographic methods have also been reported for the analysis of D-amino acids. For example, Sacchi *et al.* (2012) reported a method based on the enzyme D-amino acid oxidase, a stereospecific enzyme derived from the yeast *Rhodotorula gracilis*, for detecting D-amino acids. Moreover, these workers developed a biosensor by immobilizing the enzyme within a flow cell and used the biosensor to determine free D-alanine in spiked milk samples. This method shows potential as a rapid approach for the quantitation of total D-amino acids in foods, including those of plant origin, though it does not give information about the specific D-amino acids present.

25.6 Conclusions

Amino acid analysis is not straightforward and the analysis of plant-based products can be made more difficult due to the heterogeneous nature of plant materials as well as their high content of carbohydrates. The acid hydrolysis procedure that is used has changed little over the last 50 years and more work is needed to develop hydrolysis conditions that lead to the quantitative recovery of amino acids. LCMS/MS and NMR technologies offer considerable promise as alternatives to traditional optical detection systems, but much more work is needed in this area to justify the considerable expense and time required to set these technologies up within the laboratory. Finally, a number of amino acids are susceptible to chemical modification during the processing of foods and feedstuffs, and for processed plant materials analytical strategies beyond those conventionally used for amino acid analysis are required to obtain accurate amino acid compositional data.

References

Alarcón-Flores, M.I., Romero-González, R., Frenich, A.G., Vidal, J.L.M. and Reyes, R.C. (2010) Rapid determination of underivatized amino acids in fertilizers by ultra high performance liquid chromatography coupled to tandem mass spectrometry. *Analytical Methods* 2, 1745–1751.

Anderson, G.H., Ashley, D.V.M. and Jones, J.D. (1976) Utilization of L-methionine sulphoxide, L-methionine sulfone and cysteic acid by the weanling rat. *Journal of Nutrition* 106, 1108–1114.

Bertrand, M., Chabin, A., Brack, A. and Westall, F. (2008) Separation of amino acid enantiomers via chiral derivatisation and non-chiral gas chromatography. *Journal of Chromatography A* 1180, 131–137.

Bidlingmeyer, B.A., Cohen, S.A. and Tarvin, T.A. (1984) Rapid analysis of amino acids using pre-column derivatisation. *Journal of Chromatography* 336, 93–104.

Brückner, H. and Westhauser, T. (2003) Chromatographic determination of L- and D-amino acids in plants. *Amino Acids* 24, 43–55.

Bütikofer, U., Fuchs, D., Bosset, J.O. and Gmür, W. (1991) Automated HPLC-amino acid determination of protein hydrolysates by precolumn derivatisation with OPA and FMOC and comparison with classical ion exchange chromatography. *Chromatographia* 31, 441–447.

Bylund, D.B. and Huang, T-S. (1976) Decomposition of phosphoserine and phosphothreonine during acid hydrolysis. *Analytical Biochemistry* 73, 477–485.

Capony, J-P. and Demaille, J.G. (1983) A rapid microdetermination of phosphoserine, phosphothreonine, and phosphotyrosine in proteins by automatic cation exchange on a conventional amino acid analyzer. *Analytical Biochemistry* 128, 206–212.

Carraway, K.L. and Koshland, D.E. (1972) Carbodiimide modification of proteins. *Methods in Enzymology* 25, 616–623.

Chiou, S-H. and Wang, K-T. (1988) Peptide and protein hydrolysis by microwave irradiation. *Journal of Chromatography* 448, 404–410.

Cohen, S.A. and Michaud, D.P. (1993) Synthesis of a fluorescent derivatizing reagent, 6-aminoquinolyl-*N*-hydroxysuccinimidyl carbamate, and its application for the analysis of hydrolysate amino acids via high-performance liquid chromatography. *Analytical Biochemistry* 211, 279–287.

Csapó, J., Albert, C. and Csapó-Kiss, Z. (2009) The D-amino acid content of foodstuffs (a review). *Acta Universitatis Sapientiae Alimentaria* 2, 5–30.

Cuq, J-L., Besancon, P., Chartier, L. and Cheftel, C. (1978) Oxidation of methionine residues of food proteins and nutritional availability of protein-bound methionine sulphoxide. *Food Chemistry* 3, 85–102.

Darragh, A.J. and Moughan, P.J. (2005) The effect of hydrolysis time on amino acid analysis. *Journal of the AOAC International* 88, 888–893.

Darragh, A.J., Garrick, D.J., Moughan, P.J. and Hendriks, W.H. (1996) Correction for amino acid loss during acid hydrolysis of a purified protein. *Analytical Biochemistry* 236, 199–207.

Desrosiers, T., Savoie, L., Bergeron, G. and Parent, G. (1989) Estimation of lysine damage in heated whey proteins by furosine determinations in conjunction with the digestion cell technique. *Journal of Agricultural and Food Chemistry* 37, 1385–1391.

Einarsson, S., Josefsson, B. and Lagerkvist S. (1983) Determination of amino acids with 9-fluorenylmethyl chloroformate and reversed-phase high-performance liquid chromatography *Journal of Chromatography A* 282, 609–618.

Einarsson, S., Folestad, S. and Josefsson, B. (1987) Separation of amino acid enantiomers using precolumn derivatization with *o*-phthalaldehyde and 2,3,4,6-tetra-*O*-acetyl-1-thio-β-glucopyranoside. *Journal of Liquid Chromatography* 10, 1589–1601.

Elias, R.J., McClements, J. and Decker, E.A. (2005) Antioxidant activity of cysteine, tryptophan, and methionine residues in continuous phase β-lactoglobulin in oil-in-water emulsions. *Journal of Agricultural and Food Chemistry* 53, 10248–10253.

Ellinger, G.M. and Duncan, A. (1976) The determination of methionine in proteins by gas-liquid chromatography. *Biochemical Journal* 155, 615–621.

Eto, S., Yamaguchi, M., Bounoshita, M., Mizukoshi, T. and Miyano, H. (2011) High-throughput comprehensive analysis of D- and L-amino acids using ultra-high performance liquid chromatography with a circular dichroism (CD) detector and its application to food samples. *Journal of Chromatography B* 879, 3317–3325.

FAO (2013) *Dietary Protein Quality Evaluation in Human Nutrition. Report of an FAO Expert Consultation, 31 March–2 April, 2011, Auckland, New Zealand*. FAO Food and Nutrition Paper 92, Food and Agriculture Organization of the United Nations, Rome.

Finley, J.W. (1985) Reducing variability in amino acid analysis. In: Finley, J.W. and Hopkins, D.T. (eds) *Digestibility and Amino Acid Availability in Cereals and Oilseed*. American Association of Cereal Chemists, St Paul, Minnesota, pp. 15–30.

Fontaine, J. (2003) Amino acid analysis in feeds. In: D'Mello, J.P.F. (ed.) *Amino Acids in Animal Nutrition*, 2nd edn. CAB International, Wallingford, UK, pp. 15–40.

Friedman, M. (1996) Food browning and its prevention: an overview. *Journal of Agricultural and Food Chemistry* 44, 631–653.

Friedman, M. (1999) Chemistry, nutrition, and microbiology of D-amino acids. *Journal of Agricultural and Food Chemistry* 47, 3457–3479.

Friedman, M. and Gumbmann, M.R. (1988) Nutritional value and safety of methionine derivatives, isomeric dipeptides and hydroxy analogs in mice. *Journal of Nutrition* 118, 388–397.

Friedman, M., Pang, J. and Smith, G.A. (1984) Ninhydrin-reactive lysine in food proteins. *Journal of Food Science* 49, 10–13.

Gatti, R., Gioia, M.G., Leoni, A. and Andreani, A. (2010) 2,5-Dimethyl-1*H*-pyrrole-3,4-dicarbaldehyde as a precolumn derivatization reagent for HPLC/UV detection of amino acids. *Journal of Pharmaceutical and Biomedical Analysis* 53, 207–211.

Gioia, M.G., Cacciari, B, Leoni, A, and Gatti, R. (2006) 2,7-Dimethyl-3,8-dinitrodipyrazolo[1,5-a:1′,5′-d]pyrazine-4,9-dione: a new labelling reagent for liquid chromatographic analysis of amino acids. *Analytica Chimica Acta* 579, 152–157.

Gjøen, A.U. and Njaa, L.R. (1977) Methionine sulphoxide as a source of sulphur-containing amino acids for the young rat. *British Journal of Nutrition* 37, 93–105.

Hayashi, R. and Suzuki, F. (1985) Determination of methionine sulphoxide in protein and food by hydrolysis with *p*-toluenesulfonic acid. *Analytical Biochemistry* 149, 521–528.

Hendriks, W.H., Moughan, P.J., Boer, H. and van der Poel, A.F.B. (1994) Effects of extrusion on the dye-binding, fluorodinitrobenzene-reactive and total lysine content of soybean meal and peas. *Animal Feed Science and Technology.* 48, 99–109.

Ho, C-T. and Chen, J. (1999) Generation of volatile compounds from Maillard reaction of serine, threonine, and glutamine with monosaccharides. In: Teranishi, R. (ed.) *Flavor Chemistry: 30 Years of Progress*. Kluwer Academic/Plenum Publishers, New York, pp. 327–334.

Hurrell, R.F. and Carpenter, K.J. (1974) Mechanisms of heat damage in proteins. 4: The reactive lysine content of heat-damaged material as measured in different ways. *British Journal of Nutrition* 32, 589–604.

Hurrell, R.F. and Carpenter, K.J. (1981) The estimation of available lysine in foodstuffs after Maillard reactions. *Progress in Food and Nutrition Science* 5, 159–176.

Kabaha, K., Taralp, A., Cakmak, I. and Ozturk, L. (2011) Accelerated hydrolysis method to estimate the amino acid content of wheat (*Triticum durum* Desf.) flour using microwave irradiation. *Journal of Agricultural and Food Chemistry* 59, 2958–2965.

Kakade, M.L. and Leiner, I.E. (1969) Determination of available lysine in proteins. *Analytical Biochemistry* 27, 273–280.

Kamps, M.P. and Sefton, B.M. (1989) Acid and base hydrolysis of phosphoproteins bound to Immobilon facilitates analysis of phosphoamino acids in gel-fractionated proteins. *Analytical Biochemistry*, 176, 22–27.

Kemp, B.E. (1980) Relative alkali stability of some peptide *o*-phosphoserine and *o*-phosphothreonine esters. *FEBS Letters* 110, 308–312.

Liardon, R. and Hurrell, R.F. (1983) Amino acid racemization in heated and alkali-treated proteins. *Journal of Agricultural and Food Chemistry* 31, 432–437.

Liardon, R. and Ledermann, S. (1986) Racemization kinetics of free and protein-bound amino acids under moderate alkaline treatment. *Journal of Agricultural and Food Chemistry* 34, 557–565.

Malec, L.S., Llosa, R.A., Naranjo, G.B. and Vigo, M.S. (2005) Loss of available lysine during processing of different *dulce de leche* formulations. *International Journal of Dairy Technology* 58, 164–168.

Mao, L-C., Lee, K-H. and Erbersbobler, H.F. (1993) Effects of heat treatment on lysine in soya protein. *Journal of the Science of Food and Agriculture* 62, 307–309.

Marconi, E., Panfili, G., Bruschi, L., Vivanti, V. and Pizzoferrato, L. (1995) Comparative study on microwave and conventional methods for protein hydrolysis in food. *Amino Acids* 8, 201–208.

Marfey, P. (1984) Determination of D-amino acids II. Use of a bifunctional reagent, 1,5-difluoro-2,4-dinitrobenzene. *Carlsberg Research Communications* 49, 591–596.

Masters, P.M. and Friedman, M. (1979) Racemization of amino acids in alkali-treated food proteins. *Journal of Agricultural and Food Chemistry* 27, 507–511.

Mauron, J. and Bujard, E. (1964) Guanidination, an alternative approach to the determination of available lysine in foods. In: Mills, C.F. and Passmore, R. (eds) *Proceedings of the 6th International Nutrition Congress*. Livingstone, Edinburgh, UK, pp. 489–490.

Molnár-Perl, I. and Khalifa, M. (1992) Tryptophan analysis simultaneously with other amino acids in gas phase hydrochloric acid hydrolyzates using the Pico-Tag™ Work Station. *Chromatographia* 36, 43–46.

Moore, S. and Stein, W.H. (1951) Chromatography of amino acids on sulfonated polystyrene resins. *The Journal of Biological Chemistry* 192, 663–681.

Moughan, P.J. and Rutherfurd, S.M. (1996) A new method for determining digestible reactive lysine in foods. *Journal of Agricultural and Food Chemistry* 44, 2202–2209.

Nakamura, A., Higuchi, M., Iwami, K. and Iwai, K. (1983) Digestion and absorption of oxidized casein. *Agricultural and Biological Chemistry* 47, 2395–2396.

Nissen, S. (1992) Amino acid analysis in food and physiological samples. In: Nissen, S. (ed.) *Modern Methods in Protein Nutrition and Metabolism*. Academic Press, San Diego, California, pp. 1–8.

Otter, D. (2012) Standardised methods for amino acid analysis of food. *British Journal of Nutrition* 108, S230–S237.

Ozols, J. (1990) Amino acid analysis. *Methods in Enzymology* 182, 587–601.

Piraud, M., Vianey-Saban, C., Petritis, K., Elfakir, C., Steghens, J.P. and Bouchu, D. (2005) Ion-pairing reversed-phase liquid chromatography/electrospray ionization mass spectrometric analysis of 76 underivatized amino acids of biological interest: a new tool for the diagnosis of inherited disorders of amino acid metabolism. *Rapid Communications in Mass Spectrometry* 19, 1587–1602.

Prosser, C.G. and Rutherfurd, S.M. (2012) Glutamine and taurine. In: Nollet, L.M.L. and Toldrá, F. (eds) *Handbook of Analysis of Active Compounds in Functional Foods*. CRC Press, Boca Raton, Florida, pp. 25–39.

Puchala, R., Piór, H., von Keyserlingk, M., Shelford, J.A. and Barej, W. (1994) Determination of methionine sulphoxide in biological materials using HPLC and its degradability in the rumen of cattle. *Animal Feed Science and Technology* 48, 121–130.

Rao, S.R., Carter, F.L. and Frampton, V.L. (1963) Determination of available lysine in oilseed meal proteins. *Analytical Chemistry* 35, 1927–1930.

Rayner, C.J. (1985) Protein hydrolysis of animal feeds for amino acid content. *Journal of Agricultural and Food Chemistry* 33, 722–725.

Robel, E.J. and Crane, A.B. (1972) An accurate method for correcting unknown amino acid losses from protein hydrolysates. *Analytical Biochemistry* 48, 233–246.

Rowan, A.M., Moughan, P.J. and Wilson, M.N. (1992) Effect of hydrolysis time on the determination of the amino acid composition of diet, ileal digesta, and feces samples and on the determination of dietary amino acid digestibility coefficients. *Journal of Agricultural and Food Chemistry* 40, 981–985.

Rutherfurd, S.M. (2009) Accurate determination of the amino acid content of selected feedstuffs. *International Journal of Food Sciences and Nutrition* 60, 53–62.

Rutherfurd, S.M. and Dunn, B.M. (2011) Quantitative amino acid analysis. *Current Protocols in Protein Science* 63, 3.2.1–3.2.6.

Rutherfurd, S.M. and Gilani, G.S. (2009) Amino acid analysis. *Current Protocols in Protein Science* 58, 11.9.1–11.9.37.

Rutherfurd, S.M. and Moughan, P.J. (1993) Use of free or bound amino acid molecular weights in the determination of amino acid compositions. In: Batterham, E.S. (ed.) *Manipulating Pig Production IV.* Australasian Pig Science Association, Attwood, Victoria, Australia, pp. 229.

Rutherfurd, S.M. and Moughan, P.J. (2007) Development of a novel bioassay for determining the available lysine contents of foods and feedstuffs. *Nutrition Research Reviews* 20, 3–16.

Rutherfurd, S.M. and Moughan, P.J. (2008) The determination of sulfur amino acids in foods as related to bioavailability. *Journal of the AOAC International* 91, 907–913.

Rutherfurd, S.M. and Moughan, P.J. (2012) Available versus digestible dietary amino acids. *British Journal of Nutrition* 108, S298–S305.

Rutherfurd, S.M., Torbatinejad, N.M. and Moughan, P.J. (2006) Available (ileal digestible reactive) lysine in selected cereal-based food products. *Journal of Agricultural and Food Chemistry* 54, 9453–9457.

Rutherfurd, S.M., Schneuwly, A. and Moughan, P.J. (2007) Analyzing sulfur amino acids in selected feedstuffs using least-squares nonlinear regression. *Journal of Agricultural and Food Chemistry* 55, 8019–8024.

Sacchi, S., Rosini, E., Caldinelli, L. and Pollegioni, L. (2012) Biosensors for D-amino acid detection. In: Pollegioni, L. and Servi, S. (eds) *Unnatural Amino Acids: Methods and Protocols. Methods in Molecular Biology 794*. Humana Press (imprint of Springer), New York, pp. 313–324.

Salminen, H. and Heinonen, M. (2008) Plant phenolics affect oxidation of tryptophan. *Journal of Agricultural and Food Chemistry* 56, 7472–7481.

Shih, F.F. (1996) Deamidation and phosphorylation for food protein modification. In Magdassi, S. (ed.) *Surface Activity of Proteins: Chemical and Physiochemical Modifications*. Marcel Dekker, New York, pp. 91–113.

Sjøgerg, L.B. and Bostrøm, S.L. (1977) Studies in rats on the nutritional value of hydrogen peroxide-treated fish protein and the utilization of oxidised sulphur-amino acids. *British Journal of Nutrition* 38, 189–205.

Soby, L.M. and Johnson, P. (1981) Determination of asparagine and glutamine in polypeptides using bis(1,1-trifluoroacetoxy)iodobenzene. *Analytical Biochemistry* 113, 149–153.

Sochaski, M.A., Jenkins, A.J., Lyons, T.J., Thorpe, S.R. and Baynes, J.W. (2001) Isotope dilution gas chromatography/mass spectrometry method for the determination of methionine sulphoxide in protein. *Analytical Chemistry* 73, 4662–4667.

Thiele, B., Stein, N., Oldiges, M. and Hofmann, D. (2012) Direct analysis of underivatized amino acids in plant extracts by LC-MS-MS. *Methods in Molecular Biology* 828, 317–328.

Todd, J.M., Marable, N.L. and Kehrberg, N.L. (1984) Methionine sulphoxide determination after alkaline hydrolysis of amino acid mixtures, model protein systems, soy products and infant formulas. *Journal of Food Science* 49, 1547–1551.

Torbatinejad, N.M., Rutherfurd, S.M. and Moughan, P.J. (2005) Total and reactive lysine contents in selected cereal-based food products. *Journal of Agricultural and Food Chemistry* 53, 4454–4458.

Tower, D.B. (1967) Enzymatic determination of glutamine and asparagine. *Methods in Enzymology* 11, 77–93.

Vigo, M.S., Malec, L.S., Gomez, R.G. and Llosa, R.A. (1992) Spectrophotometric assay using *o*-phthaldialdehyde for determination of reactive lysine in dairy products. *Food Chemistry* 44, 363–365.

Wilcox, P.E. (1967) Determination of amide residues by chemical analysis. *Methods in Enzymology* 11, 63–76.

Yang, Y., Chang, L., Lu, Y. and Yang, L. (2010) Analysis of amino acids in foodstuff by nuclear magnetic resonance. In: Yu, W., Zhang, M., Wang, L. and Song, Y. (eds) *Proceedings of the 3rd International Conference on Biomedical Engineering and Informatics (BMEI).* Institute of Electrical and Electronics Engineers, New York, New York, pp. 750–754.

Zhang, H.L., Li, D.F., Qiao, S.Y., Wang, F.L., Chen, X.J. and Thacker, P.A. (2006) The effect of dietary homoarginine derived from guanidination of synthetic lysine on endogenous amino acid loss and apparent and true ileal amino acid digestibility in the pig. *Animal Science* 82, 23–30.

Zyzak, D.V., Sanders, R.A., Stojanovic, M., Tallmadge, T.H., Eberhart, B.L., Ewald, D.K., Gruber, D.C., Morsch, T.R., Strothers, M.A., Rizzi, R.P. and Villagran, M.D. (2003) Acrylamide formation mechanism in heated foods. *Journal of Agricultural and Food Chemistry* 51, 4782–4787.

26 Metabolic Amino Acid Availability in Foods of Plant Origin: Implications for Human and Livestock Nutrition

C.L. Levesque*
South Dakota State University, Brookings, South Dakota, USA

26.1 Abstract

The bioavailability of amino acids is the proportion of ingested amino acids that are digested and absorbed in a form suitable for protein synthesis. However, it cannot be measured directly, so various methods have been developed to estimate the bioavailability of amino acids and protein in food of plant origin. Digestibility, the proportion of the ingested amino acids that disappear from the digestive tract, is most commonly used to estimate bioavailability although, by definition, it does not necessarily reflect the proportion of digested and absorbed amino acids that can be utilized for metabolic processes, such as amino acids in the diet that have been damaged during the heat processing of foods or feedstuffs. Therefore, new methods need to be developed that provide an estimate of the proportion of amino acids available for protein synthesis. Metabolic availability (MA) is a new method of evaluating amino acid bioavailability that is based on changes in protein retention due to changes in the bioavailability of amino acids and thus reflects the true metabolic availability of amino acids in foodstuffs. The MA method is based on the indicator amino acid oxidation (IAAO) technique, in which the oxidation of the indicator amino acid is inversely proportional to whole body protein synthesis and, hence, accounts for all losses of amino acids during digestion, absorption and metabolic utilization. The new method has been used to determine the MA of amino acids in protein sources fed to growing and adult pigs, as well as human adults. The method has also been demonstrated to be sensitive to changes in lysine availability due to heat processing and, consequently, is a superior method for estimating the bioavailability of amino acids in foods of plant origin.

26.2 Introduction

The quality of dietary proteins is determined by the capacity to satisfy the amino acid requirements of humans and other vertebrates. Protein quality evaluation includes a measure of the amount of individual amino acids and protein in the food source as well as a measure of the bioavailability of the constituent amino acids and protein. The proportion of amino acids in different food ingredients is reasonably well characterized (USDA ARS, 2006; CVB, 2008; National Research Council, 2012), though variation in amino acid content of the same ingredient can occur as a result of variety, environmental, and analytical differences (National Research Council, 2012). Amino acid bioavailability is defined as 'the proportion of the total amino acid that is digested and absorbed in a form suitable for protein synthesis' (Batterham, 1992) and is, therefore, dependent not only on the digestibility of the dietary proteins, but also on the subsequent absorption and

*E-mail address: crystal.levesque@sdstate.edu

potential utilization of the absorbed amino acids for metabolic processes. The bioavailability of amino acids in foods cannot be measured directly, so extensive research has been conducted on establishing methods that can estimate it adequately. Protein efficiency ratio, biological value and protein digestibility-corrected amino acid score are all methods for evaluating protein quality and bioavailability, but the most common method is digestibility. Digestibility is the proportion of ingested amino acids that disappear from the digestive tract (dietary amino acid intake minus total amino acid outflow divided by dietary amino acid intake; Stein *et al.*, 2007) and may be determined as faecal (i.e. total tract) or ileal digestibility. Although digestibility is most commonly used to estimate bioavailability, by definition, it does not necessarily reflect the proportion of digested and absorbed amino acids that can be utilized for metabolic processes. Consequently, new methods need to be developed that also provide estimates of the proportion of amino acids that are available for metabolic processes.

This chapter will review a new method of evaluating amino acid bioavailability, termed metabolic availability (MA), which is based on changes in protein retention due to changes in the bioavailability of amino acids, and thus reflects the true metabolic availability of amino acids in foodstuffs.

26.3 Amino Acid Digestibility and its Limitations

Given that amino acid digestibility is the most commonly used method to estimate bioavailability, a brief discussion of this technique and its limitations is necessary to understand the need to develop alternative estimation methods. Amino acid digestibility is most commonly determined as ileal digestibility so that the effect of hindgut fermentation of amino acids is minimized and because negligible amino acid absorption occurs in the lower gut (Sauer and Ozimek, 1986). Total ileal amino acid outflow includes undigested dietary amino acids and endogenous amino acid losses. Apparent, standardized and true ileal digestibility are used to indicate how ileal endogenous amino acid losses are reflected in digestibility estimates. Endogenous ileal amino acid losses include basal losses (which are diet independent) and specific losses (which are diet dependent in that they are influenced by feed ingredient characteristics such as antinutritional factors and fibre; Stein *et al.*, 2007). Standardized ileal digestibility accounts for basal endogenous amino acid losses and is recommended for routine use in pig diet formulation (Stein *et al.*, 2007).

There are two main limitations to standardized ileal digestibility: (i) it corrects for basal endogenous losses only; and (ii) it is unable to account for amino acids that are digested and absorbed in a form that is unsuitable for use in metabolic processes. The error associated with specific endogenous losses is dependent on the nutrient composition of the foodstuff. For example, specific endogenous losses are minimal for highly digestible or purified foodstuffs such as casein or egg protein (Stein *et al.*, 2007). However, for foodstuffs high in fibre or antinutritional factors, specific endogenous losses may contribute 50% or more of the total ileal endogenous amino acid losses (Souffrant, 1991; Moughan, 2003). Furthermore, the associated error is not consistent across all amino acids and is typically greatest for threonine due to the high proportion of threonine in gastrointestinal mucin and mucous proteins (Lien *et al.*, 2001), which increases as the dietary fibre or antinutritional factor content increases (Nyachoti *et al.*, 1997; Lallès and Jansman, 1998; Myrie, 2004). Heat/alkaline processing is used to improve food flavour and texture or for sterilization/pasteurization, but these processes may cause the formation of compounds that render some amino acids unavailable for protein synthesis, although they are still digested and absorbed (examples include Maillard compounds, oxidized forms of sulfur-containing amino acids and D-amino acids; Gilani *et al.*, 2005). The chemical reaction between lysine and carbohydrates during heat processing (the Maillard reaction) is of particular interest because lysine is often the first limiting amino acid in diets.

In spite of these limitations, standardized ileal digestibility provides an adequate estimate of amino acid bioavailability for common animal feedstuffs and human foods. In addition, there are large, publicly available, databases published in North America and Europe of the digestibility values for feedstuffs used in pig diets (CVB, 2008; National Research Council, 2012). The digestibility of amino acids in human foods, as determined in pigs, is similar to that determined in humans (Rowan *et al.*, 1994), so the use of the pig as a model, as well as existing databases of digestibility values in pigs, could rapidly advance this area of human nutrition. However, in the swine industry, the use of alternative feed ingredients in pig diets (by-products of the biofuel and human food industry) is increasing as pressure for the use of traditional pig feed ingredients (maize, barley) in the biofuels industry rises (Schmit *et al.*, 2008). These by-products are typically high in fibre and anti-nutritional factors. In human nutrition, there is an increased awareness of the importance of daily fibre intake on weight loss/management and the management of diabetes and other metabolic diseases (Chutkan *et al.*, 2012) and, accordingly, the American Diabetes Association (2000) recommends a daily intake of dietary fibre up to 35 g daily. Moreover, the consumption of fast foods and snack foods by adolescent humans is often high, which means an increase in the supply of amino acids damaged by heat (Seiquer *et al.*, 2006). As the levels of alternative ingredients in pig diets and the consumption of fibre and heat-damaged amino acids in humans steadily rise, the need to develop an alternative method of estimating amino acid bioavailability becomes more critical.

26.4 Metabolic Availability of Amino Acids

The metabolic availability (MA) method is based on the indicator amino acid oxidation (IAAO) technique and the slope-ratio assay, and measures changes in protein retention due to changes in the bioavailability of amino acids (Levesque *et al.*, 2012). The IAAO technique is based on the principal that excess amino acids are not stored and, as a result, must either be utilized for protein synthesis or oxidized and excreted. When one indispensable amino acid is deficient for protein synthesis, all other indispensable amino acids are in excess and must be oxidized (Elango *et al.*, 2008). The adequacy of the supply of the deficient amino acid is measured indirectly by the oxidation of another carboxy-labelled (^{13}C or ^{14}C) indispensable amino acid (Kim *et al.*, 1983), which is called the indicator amino acid. As the intake of the deficient amino acid increases, the oxidation of the indicator amino acid decreases linearly until the requirement for the limiting amino acid is reached (Ball and Bayley, 1986; Fig. 26.1). The decrease in oxidation of the indicator amino acid has been demonstrated to be inversely proportional to its incorporation into liver proteins (Ball and Bayley, 1986; Fig. 26.1);

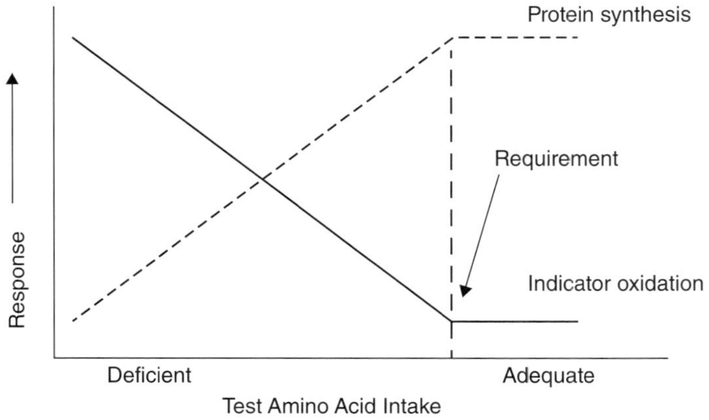

Fig. 26.1. The indicator amino acid oxidation (IAOO) technique. There is an inverse relationship between the oxidation of an indicator amino acid (AA) and protein synthesis with increasing intake of a limiting (test) amino acid. Oxidation of the indicator AA decreases with increasing intake of the test AA. No further oxidation of the indicator AA occurs once the requirement for the test amino acid has been met. Conversely, with increasing intake of the deficient test AA, protein synthesis increases to a point where there is no further change at test AA intakes above the dietary requirement. The inflection point at which no further oxidation of the indicator AA occurs represents the requirement for the test AA. (Adapted from Ball and Bayley, 1986.)

a decrease in IAAO, then, reflects an increase in whole body protein synthesis.

Assays of amino acid bioavailability that measure parameters of growth (slope-ratio assays) are considered to be the standard against which other methods of amino acid availability (i.e. digestibility) are judged (Lewis and Bayley, 1995). The advantages of slope-ratio assays that measure parameters of protein synthesis are that they measure a response that has practical and economic consequences, and accounts for the effect of digestion, absorption and metabolic utilization of the amino acid provided by the protein source (Lewis and Bayley, 1995). Therefore, such assays measure the effect of protein quality on the entire metabolism and so are likely to be equally applicable in animal or human nutrition.

The slope-ratio assay is based on regression equations determined for both a test and a reference nutrient source and is calculated as the ratio of the slopes of the regression equations, hence 'slope-ratio' assay (Fig. 26.2; Littell *et al.*, 1995). In the case of determination of MA, because IAAO is inversely proportional to the rate of protein synthesis, at a given amino acid intake, the relative difference in the IAAO rate between test and reference proteins will be proportional to the whole body MA of the test amino acid for protein metabolism. Consequently, MA accounts for all losses of dietary amino acids during digestion, absorption and cellular metabolism. The total losses taken into account by the IAAO method include: incomplete digestion and absorption, gut endogenous amino acid losses and the unavailability of absorbed amino acids from foods due to the presence of anti-nutritional factors (e.g. Maillard compounds formed due to processing of foods, D-amino acids and cross-linked proteins such as lysinoalanine). Thus, the higher the oxidation of the indicator amino acid, the lower the MA of the test amino acid for protein metabolism and vice versa (Fig. 26.2). The MA method uses incremental increases in crystalline test amino acid to determine the 'reference' slope. The reference slope then represents the maximal unit increase in protein synthesis and is equivalent to 100% MA, because the true digestibility of crystalline amino acids is essentially 100% (Baker, 1992). Similarly, increasing intake of the test protein is used to determine the 'test amino acid' slope. Each subject receives each level of amino acid supplied by the crystalline amino acid and the test protein, and mean reference and test amino acid slopes are determined. Intake of the test amino acid (from either the reference or test protein source) is limited to a maximum of 2 SD below the subject's requirement to ensure that the response to increasing

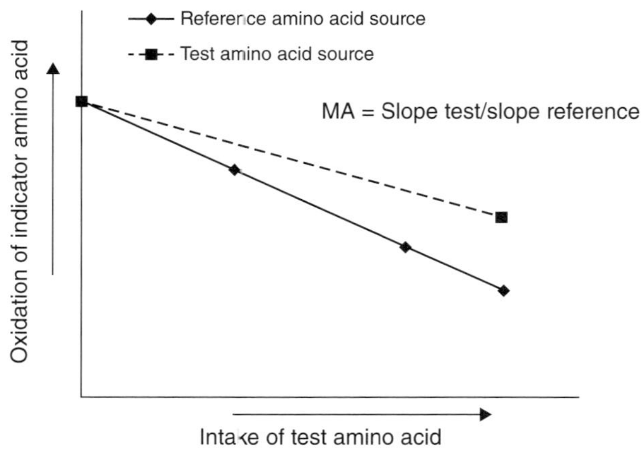

Fig. 26.2. Conceptual application of the metabolic availability (MA) of amino acids from food sources based on indicator amino acid oxidation and the slope-ratio assay. The oxidation of the indicator amino acid is inversely proportional to the rate of whole body protein synthesis. Metabolic availability is calculated as the ratio of the slope of the regression equation for the test amino acid source to the slope of the regression equation for the reference (indicator) amino acid source. Thus, the higher the oxidation of the indicator amino acid, the lower the MA of the test amino acid for protein synthesis and vice versa. (Adapted from Littell *et al.*, 1995 and Moehn *et al.*, 2005.)

amino acid intake is linear (Moehn *et al.*, 2005). The reader is directed to Levesque *et al.* (2012) for further details on the key criteria involved in determining MA.

26.4.1 Metabolic availability in protein sources

The MA of amino acids has been determined in protein sources fed to growing pigs and sows (Moehn *et al.*, 2005; 2007; Levesque, 2010; Levesque *et al.*, 2011) and humans (Humayun *et al.*, 2006; Prolla *et al.*, 2013).

The concept of MA was first tested with lysine in peas fed to growing pigs and the MA for this amino acid was determined to be 88.8% (Moehn *et al.*, 2005), similar to the value from previously published standardized ileal digestibility analysis (88%; National Research Council, 1998). Moehn *et al.* (2005) further tested the ability of MA to detect the effect of heat treatment on the bioavailability of lysine. Peas were heated in a manner similar to that described by Van Barneveld *et al.* (1994), and the lysine in the heated peas was determined to be 54.8% available for protein synthesis, a substantially lower amount than that in unheated peas (88.8%). Using the slope-ratio growth assay, Van Barneveld *et al.* (1994) reported a similar availability of lysine in heated peas (48%). This demonstrated that the MA method accounts for amino acids digested and absorbed in a form unsuitable for metabolic processes. In a later study, also with growing pigs, Moehn *et al.* (2007) measured the MA of lysine in peas heated using a less severe regimen, and found that it was reduced by 10% (68%) compared with that for unheated peas (76%); this contrasted with a 38% reduction in lysine availability due to heat damage as reported in the earlier study (Moehn *et al.*, 2005). When free lysine (an amount calculated to be lost during heating) was added to the heated peas, the MA of lysine was 76.5%, similar to that of the raw peas (Moehn *et al.*, 2007). This demonstrated that MA was also sensitive to small differences in lysine availability due to differences in the degree of heat damage. The MA of lysine in soybean meal (87.5%), cottonseed meal (75.1%) and canola meal (71.4%) fed to growing pigs was also determined (Moehn *et al.*, 2007).

The MA method was further tested in an assessment of the bioavailability of threonine in two common cereal grains (maize and barley). The value of the MA for threonine in maize fed to growing pigs was 82.3% (Levesque, 2010), which was similar to the standardized ileal digestibility values (National Research Council, 1998). In contrast, when the same source of maize was fed to pregnant sows, the MA of threonine was 88%, a 7% increase (Fig. 26.3a; Levesque *et al.*, 2011). Similarly, the MA of threonine in barley fed to pregnant sows was 89.3% (Levesque *et al.*, 2011), 9% greater than the standardized ileal digestibility values determined in growing pigs (National Research Council, 1998). So the quality of a protein source may not be constant if the physiological state of the subject changes, because the whole body amino acid profile (the relative requirements for difference amino acids) changes with physiological state (Reeds and Beckett, 1996). The greater MA of maize and barley fed to pregnant sows suggests that age may alter the availability of food amino acids.

In this same study (Levesque, 2010), the MA of threonine in barley fed to growing pigs was determined to be >100%. It was suggested that the value of >100% MA may reflect a higher demand for threonine for the synthesis of mucin and digestive enzymes owing to a higher proportion of soluble fibre in barley than in maize or casein (the base protein source in the reference diet), which may increase the secretion of endogenous protein and intestinal cell proliferation.

Humayun *et al.* (2006) and Prolla *et al.* (2013) adapted the MA method for use in humans to determine the MA of sulfur-containing amino acids in soy protein isolate and casein, and the MA of lysine in rice. The MAs of sulfur amino acids in soy protein isolate and casein were 72 and 87%, respectively, similar to published net protein utilization values of 80–85% for milk proteins and 71–78% for soy proteins (Humayun *et al.*, 2006). The MA of lysine in cooked white rice was determined to be 97% (Fig. 26.3b), similar to the digestibility value of 100% determined in animals (Eggum *et al.*, 1993). However, as described earlier with the lysine in peas, when the rice was browned before cooking, the MA of lysine was reduced to 70% (Fig. 26.3b; Prolla *et al.*, 2013), thus demonstrating again that the MA method accounts for

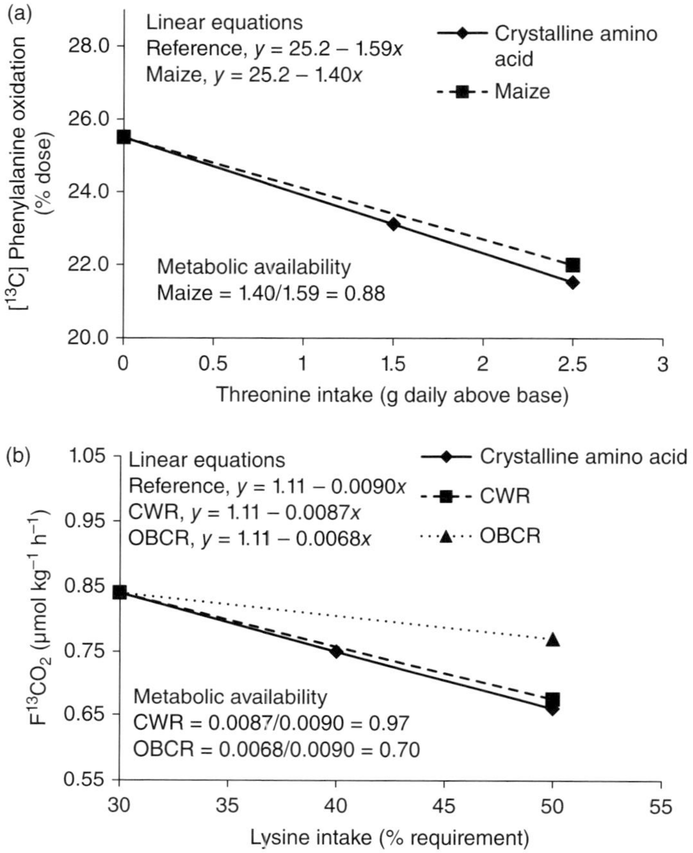

Fig. 26.3. Metabolic availability (MA) of (a) threonine in maize fed to sows and (b) lysine in unheated (CWR) and oven-browned (OBCR) cooked white rice consumed by adult men. MA is calculated as the ratio of the slope of the test protein source (maize or cooked white rice) to the slope of the reference crystalline amino acid source. Phenylalanine (Phe) was used as the indicator amino acid in both studies, with [^{13}C]Phe as the tracer; the *y*-axes show (a) phenylanine oxidation measured as $^{13}CO_2$ released as % tracer administered, and (b) rate of release of $^{13}CO_2$ from oxidation of [^{13}C]Phe ($F^{13}CO_2$) as µmol kg^{-1} h^{-1}. (Adapted from Levesque *et al.*, 2011 and Prolla *et al.*, 2013.)

amino acids digested and absorbed in a form unsuitable for protein synthesis.

26.4.2 Advantages of the metabolic availability method

The measurement of amino acid bioavailability by the slope-ratio assay is regarded as the 'gold standard' (Lewis and Bayley, 1995), although the practical application of amino acid bioavailability growth assays has been said to be limited by time, cost, high standard error and the examination of only one amino acid at a time (Batterham, 1992). The conventional response criteria used in slope-ratio assays (comparative slaughter, body weight or the N-balance technique) require large numbers of animals and prolonged periods of time and are in some cases unethical (cf. any comparative slaughter in humans). Estimates of bioavailability also demonstrate a high standard error due to the effect of individual variation among test subjects on the slope of the response and the accuracy of the total amino acid analysis of the test protein (Batterham, 1992). These problems have been addressed by using the IAAO technique instead of conventional methods. The short adaptation time required for the IAAO technique (under 2 days, Moehn *et al.*, 2004) allows repeated measurement of indicator oxidation at all levels of reference and test amino acid intake within subjects in rapid succession. The within-subject repeated measurements reduce the effect of individual variability and increase the sensitivity of the estimate by accounting for between-subject variation, which is ~10% in pigs (Bertolo *et al.*, 2005) and as high as 30% in humans (Zello *et al.*, 1993). The number of subjects, inter-subject variation and time-frame of a study can be reduced drastically and, typically, the MA of an amino acid in the test feed can be determined within 14 days. What is more, the MA in individual subjects can be determined by the IAAO technique, thus providing a measure of the inter-subject variability of MA. Even though MA still only estimates the bioavailability of one amino acid at a time, the short time needed to complete one trial means that analyses for multiple amino acids can be completed in the same subjects in a relatively short period of time.

The MA method provides a measure of bioavailability that could be applied routinely in humans, whereas the standardized ileal digestibility method used in animals to estimate bioavailability is not practical for routine application in humans due to the techniques required to collect ileal digesta, which are often invasive. Hence, there are very few data on the amino acid bioavailability of foodstuffs for humans. Furthermore, the MA method is particularly well suited to periods of rapid physiological changes, such as pregnancy and early childhood, and its non-invasive nature lends itself particularly to its use in vulnerable subjects, such as the very young and the elderly.

26.5 Future Research and Conclusions

As the utilization of alternative ingredients increases in animal agriculture, the effect of heat damage on losses in amino acid availability needs to be to be explored in greater detail. The effect of heat processing on lysine is well accepted and methods are available for estimating the loss of lysine bioavailability, such as MA (Moehn *et al.*, 2005; Prolla *et al.*, 2013) and the reactive lysine assay (Rutherfurd and Moughan, 1997). However, an understanding of the effect of heat on the bioavailability of other amino acids that are also damaged by heat processing (cysteine or threonine) is necessary. Although the degree of disulfide cross-linking due to heating has been correlated with minor reductions in amino acid digestibility (Opstvedt *et al.*, 1984), the MA method provides a direct method of assessing changes in amino acid bioavailability due to heat in amino acids other than lysine.

It has been clearly demonstrated that the MA method is able to detect differences in bioavailability due to heat processing, but the effect of fibre or anti-nutritional factors on MA needs to be determined too. The influence of fibre on digestive processes has been extensively studied (Zebrowska *et al.*, 1983; Jensen and Jørgensen, 1994; Johansen *et al.*, 1996; Leterme *et al.*, 2000; Myrie *et al.*, 2008; Hooda *et al.*, 2010), but the effect of fibre on whole body amino acid utilization (i.e. IAAO) has not been examined. The over 100% MA of threonine from barley fed to growing pigs indicates that further work is necessary to understand the impact of fibre and other anti-nutritional factors on whole body amino acid utilization and, therefore, on IAAO and MA. Further work is also needed to assess the effect of age on the MA across multiple amino acids, as well as in humans.

The results from studies in pigs and humans indicate that MA can be determined for individual amino acids of plant origin, including amino acid losses due to heat processing. The MA method is superior to digestibility as an estimate of bioavailability because it provides estimates of amino acid availability subsequent to the digestion and absorption of test proteins, though additional work is necessary to validate the method at various physiological stages and in the presence of anti-nutritional factors.

References

American Diabetes Association (2000) Nutrient recommendations and principles for people with diabetes mellitus. *Diabetes Care* 23, S43–S46.

Baker, D.H. (1992) Applications of chemically defined diets to the solution of nutrition problems. *Amino Acids* 2, 1–12.

Ball, R.O. and Bayley, H.S. (1986) Influence of dietary protein concentration on the oxidation of phenylalanine by the young pig. *British Journal of Nutrition* 55, 651–658.

Batterham, E.S. (1992) Availability and utilization of amino acids for growing pigs. *Nutrition Research Reviews* 5, 1–18.

Bertolo, R.F., Moehn, S., Pencharz, P.B. and Ball, R.O. (2005) Estimate of the variability of the lysine requirement of growing pigs using the IAAO technique. *Journal of Animal Science* 83, 2535–2542.

Chutkan, R., Fahey, G., Wright, W.L. and McRorie, J. (2012) Viscous versus nonviscous soluble fiber supplements: mechanisms and evidence for fiber-specific health benefits. *Journal of the American Academy of Nurse Practitioners* 24, 476–487.

CVB (2008) The Dutch Productschap Diervoeder (Product Board Animal Feed) CVB (feed value) feedstuff database. Available at: http://www.pdv.nl/english/Voederwaardering/about_cvb/index.php (accessed 27 September 2012).

Eggum, B.O., Cabrera, M.I. and Juliano, B.O. (1993) Protein and lysine digestibility and protein quality of cooked Filipino rice diets and milled rice in growing rate. *Plant Foods in Human Nutrition* 43, 163–170.

Elango, R., Ball, R.O. and Pencharz, P.B. (2008) Indicator amino acid oxidation: concept and application. *Journal of Nutrition* 138, 243–246.

Gilani, G.S., Cockell, K.A. and Sepehr, E. (2005) Effects of antinutritional factors on protein digestibility and amino acid availability in foods. *Journal of AOAC International* 88, 967–987.

Jensen, B.B. and Jørgensen, H. (1994) Effect of dietary fiber on microbial activity and microbial gas production in various regions of the gastrointestinal tract of pigs. *Applied and Environmental Microbiology* 60, 1897–1904.

Johansen, H.N., Bach Knudsen, K.E., Sandstrom, B. and Skjøth, F. (1996) Effect of varying content of soluble dietary fibre from wheat flour and oat milling fractions on gastric emptying in pigs. *British Journal of Nutrition* 75, 339–351.

Hooda, S., Metzler-Zebeli, B.U., Vasanthan, T. and Zijlstra, R.T. (2010) Effects of viscosity and fermentability of purified non-starch polysaccharides on ileal and total tract nutrient digestibility in ileal-cannulated grower pigs. *Livestock Science* 134, 79–81.

Humayun, M.A., Elango, R., Moehn, S., Ball, R.O. and Pencharz, P.B. (2006) Application of the indicator amino acid oxidation technique for the determination of metabolic availability of sulfur amino acids from casein versus soy protein isolate in adult men. *Journal of Nutrition* 137, 1874–1879.

Kim, K., McMillan, I. and Bayley, H.S. (1983) Determination of amino acid requirements of young pigs using and indicator amino acid. *British Journal of Nutrition* 50, 369–382.

Lallès, J.P. and Jansman, A.J.M. (1998) Recent progress in the understanding of the mode of action and effects of antinutritional factors from legume seeds in nonruminant farm animals. In: Jansman, A.J.M., Hill, G.D., Huisman, J. and van der Poel, A.F.B. (eds) *Recent Advances of Research in Antinutritional Factors in Legume Seeds and Rapeseed. Proceedings of the 3rd International Workshop*. EAAP (European Association for Animal Production) Publication 93. Wageningen Academic Publishers, Wageningen, The Netherlands, pp. 219–232.

Leterme, P., Souffrant, W.B. and Théwis, A. (2000) Effect of barley fibres and barley intake on the ileal endogenous nitrogen losses in piglets. *Journal of Cereal Science* 31, 229–239.

Levesque, C.L. (2010) Evaluation of the threonine requirement and the bioavailability of threonine in feedstuffs in pregnant sows. PhD thesis. University of Alberta, Edmonton, Alberta, Canada.

Levesque, C.L., Moehn, S., Pencharz, P.B. and Ball, R.O. (2011) The metabolic availability of threonine in common feedstuffs fed to adult sows is higher than published ileal digestibility estimates. *Journal of Nutrition* 141, 406–410.

Levesque, C.L., Elango, R. and Ball, R.O. (2012) Metabolic availability of amino acids in food protein: new methodology. In: D'Mello, J.P.F. (ed.) *Amino Acids in Human Nutrition and Health*. CAB International, Wallingford, UK, pp. 256–266.

Lewis, A.J. and Bayley, H.S. (1995) Amino acid bioavailability. In: Ammerman, C.B., Baker, D.H. and Lewis, A.J. (eds) *Bioavailability of Nutrients for Animals – Amino acids, Minerals, Vitamins*. Academic Press, San Diego, California, pp. 35–50.

Lien, K.A., Sauer, W.C. and He, J.M. (2001) Dietary influences on the secretion into and degradation of mucin in the digestive tract of monogastric animals and human. *Journal of Animal Feed Science* 10, 223–245.

Littell, R.C., Lewis, A.J. and Henry, P.R. (1995) Statistical evaluation of bioavailability assays. In: Ammerman, C.B., Baker, D.H. and Lewis, A.J. (eds) *Bioavailability of Nutrient for Animals – Amino Acids, Minerals, Vitamins*. Academic Press, San Diego, California, pp. 5–34.

Moehn, S., Bertolo, R.F.P., Pencharz, P.B. and Ball, R.O. (2004) Indicator amino acid oxidation responds rapidly to changes in lysine or protein intake in growing and adult pigs. *Journal of Nutrition* 134, 836–841.

Moehn, S., Bertolo, R.F.P., Pencharz, P.B. and Ball, R.O. (2005) Development of the indicator amino acid oxidation technique to determine the availability of amino acids from dietary protein in pigs. *Journal of Nutrition* 135, 2866–2870.

Moehn, S., Bertolo, R.F.P., Martinazzo-Dallagnol, E., Pencharz, P.B. and Ball, R.O. (2007) Metabolic availability of lysine in feedstuffs determined using oral isotope delivery. *Livestock Science* 109, 24–26.

Moughan, P.J. (2003) AA digestibility and availability in food and feedstuffs. In: Ball, R.O. (ed.) *Digestive Physiology in Pigs.* Proceedings 9th International Symposium on Digestive Physiology in Pigs, Banff, Canada, May 14–18, 2003. University of Alberta, Edmonton, Alberta, Canada, pp. 199–221.

Myrie, S.B. (2004) The effects of anti-nutritional factors in feedstuffs on gastrointestinal mucin production and on threonine requirement in the growing pig. MSc thesis, University of Alberta, Edmonton, Alberta, Canada.

Myrie, S.B., Bertolo, R.F., Sauer, W.C. and Ball, R.O. (2008) Effect of common antinutritional factors and fibrous feedstuffs in pig diets on amino acid digestibilities with special emphases on threonine. *Journal of Animal Science* 86, 609–619.

National Research Council (1998) *Nutrient Requirements for Swine*, 10th edn. The National Academies Press, Washington, DC.

National Research Council (2012) National Research Council, *Nutrient Requirements for Swine*, 11th edn. The National Academies Press, Washington, DC.

Nyachoti, C.M., de Lange, C.F.M., MacBride, B.W. and Schulze, H. (1997) Significance of endogenous gut nitrogen losses in the nutrition of growing pigs: a review. *Canadian Journal of Animal Science* 77, 149–163.

Opstvedt, J., Miller, R., Hardy, R.W. and Spinelli, J. (1984) Heat-induced changes in sulfhydryl groups and disulfide bonds in fish protein and their effect on protein and amino acid digestibility in rainbow trout. *Journal of Agricultural and Food Chemistry* 32, 929–935.

Prolla, I.R.D., Rafii, M., Courtney-Martin, G., Elango, R., da Silva, L.P., Ball, R.O. and Pencharz, P.B. (2013) Lysine from cooked white rice consumed by healthy young men is highly metabolically available when assessed using the indicator amino acid oxidation technique *Journal of Nutrition* 143, 302–306.

Reeds, P.J. and Beckett, P.R. (1996) Protein and amino acids. In: Zeigler, E.E. and Filer, L.J. (eds) *Present Knowledge in Nutrition*, 7th edn. ILSI (International Life Sciences Institute) Press, Washington, DC, pp. 67–86.

Rowan A.M., Moughan, P.J., Wilson, M.N., Maher, K. and Tasma-Jones, C. (1994) Comparison of the ileal and faecal digestibility of dietary amino acids in humans and evaluation of the pig as a model animal for digestion studies in man. *British Journal of Nutrition* 71, 29–42.

Rutherfurd, S.M. and Moughan, P.J. (1997) Application of a new method for determining digestible reactive lysine to variably heated protein sources. *Journal of Agricultural and Food Chemistry* 45, 1582–1586.

Sauer, W.C. and Ozimek, L. (1986) Digestibility of amino acids in swine: results and their practical applications. A review. *Livestock Production Science* 15, 367–388.

Schmit, T.M., Verteramo, L. and Tomek, W.G. (2008) Implications of growing biofuels demands on Northeast livestock feed costs. Paper from: *Proceedings of the NCCC-134 Conference on Applied Commodity Price Analysis, Forecasting, and Market Risk Management, St Louis, Missouri, April 21–22, 2008*. Available at: http://www.farmdoc.illinois.edu/nccc134/conf_2008/pdf/confp01-08.pdf (accessed 12 September 2014).

Seiquer, J., Diaz-Aquacil, J., Delgado-Andrada, C., López-Fríaz, M., Hoyos, A.M., Galdó, G. and Navarroa, M.P. (2006) Diets rich in Maillard reaction products affect protein digestibility in adolescent males ages 11–14 yr. *The American Journal of Clinical Nutrition* 83, 1082–1088.

Souffrant, W.B. (1991) Endogenous nitrogen losses during digestion in pigs. In: Verstegen, M.W.A., Huisman, J. and den Hartog, L.A. (eds) *Digestive Physiology in Pigs. Proceedings 5th International Symposium on Digestive Physiology in Pigs, Wageningen (Doorwerth), Netherlands, 24–26 April 1991*. EAAP (European Association for Animal Production) Publication 54. Wageningen Academic Publishers, Wageningen, the Netherlands, pp. 147–166.

Stein, H.H., Sève, B., Fuller, M.F. and de Lange, C.F.M. (2007) Invited review: amino acid bioavailability and digestibility in pig feed ingredients: terminology and application. *Journal of Animal Science* 85, 172–180.

USDA ARS (2006) Composition of Foods: Raw, Processed, Prepared. *USDA National Nutrient Database for Standard Reference, Release 19, August 2006*. US Department of Agriculture, Agricultural Research Service, Beltsville, Maryland. Available at: http://www.ars.usda.gov/Services/docs.htm?docid=8964 (accessed 12 September 2014).

Van Barneveld, R.J., Batterham, E.S. and Norton, B.W. (1994) The effect of heat on amino acids for growing pigs. 3. The availability of lysine for heat-treated field peas (*Pisum sativum* cultivar Dundale) determined using the slope-ratio assay. *British Journal of Nutrition* 72, 257–275.

Zebrowska, T., Low, A.G. and Zebrowska, H. (1983) Studies on gastric digestion of protein and carbohydrate, gastric secretion and exocrine pancreatic secretion in the growing pig. *British Journal of Nutrition* 40, 401–410.

Zello, G.A., Pencharz, P.B. and Ball, R.O. (1993) Dietary lysine requirement of young adult males determined by oxidation of L-[1-^{13}C] phenylalanine. *American Journal of Physiology* 259, E835–E843.

27 Toxicology of Non-Protein Amino Acids

J.P.F. D'Mello*
Formerly of SAC (Scottish Agricultural College), University of Edinburgh King's Buildings Campus,Edinburgh, UK

27.1 Abstract

The appearance of a wide range of non-protein amino acids in foliage, fruits, seeds and root exudates is an important expression of secondary metabolism in higher plants. The ubiquitous distribution of these amino acids is exemplified by their presence in food and forage plants as well as in temperate and tropical species. Canavanine occurs widely in legumes such as *Canavalia ensiformis*, *Medicago sativa*, *Gliricidia sepium*, *Dioclea megacarpa* and *Robinia pseudocacia*. Significant concentrations of the selenoamino acids and of *S*-methylcysteine sulfoxide are present in *Brassica oleracea*, including broccoli, cabbage, Brussels sprouts and kale. In contrast, the signalling agonists *β*-*N*-oxalylamino-L-alanine and *β*-*N*-methylamino-L-alanine occur in unrelated species, namely *Lathyrus sativus* and *Cycas circinalis*, respectively, while *γ*-aminobutyrate accumulates in plants in response to biotic and environmental stimuli. The aromatic amino acid, mimosine, occurs primarily in *Leucaena leucocephala*, a tropical legume yielding timber and prolific quantities of palatable forage. Of all the non-protein amino acids cited in this chapter, hypoglycin is unique in its natural distribution; two forms, hypoglycin A and its glutamyl derivative, hypoglycin B, occur in the arils of the unripe fruit of *Blighia sapida*.

Representative examples of non-protein amino acids have been selected for review to emphasize the relationship between allelochemical activity and toxicity in diverse living organisms. Canavanine and its metabolites in the canaline-urea cycle share striking structural analogy with corresponding amino acids in the ornithine-urea cycle. Additive and synergistic phytotoxic effects of canaline-urea cycle amino acids are reversed by their respective analogues in the ornithine-urea cycle. A similar canavanine–arginine antagonism has been demonstrated in insects and higher animals. Equally noteworthy are the structural analogues of the sulfur amino acids. On replacement of the sulfur atom with selenium, a wide range of congeners are formed, notably selenomethionine, *Se*-methylselenocysteine and *Se*-allylselenocysteine. It is universally recognized that the selenoamino acids are toxic to both plants and animals, their effects being characterized by inability to thrive. Selenoamino acids interfere with cellular biochemical reactions, and methionine status decreases with the accumulation of tissue selenomethionine in plants. Another analogue, *S*-methylcysteine sulfoxide, is associated with haemolytic anaemia in ruminant animals. Both *β*-*N*-methylamino-L-alanine and *γ*-aminobutyrate affect plant development; however, in humans, the former amino acid, together with *β*-*N*-oxalylamino-L-alanine, have been implicated, respectively, in amyotrophic lateral sclerosis and Parkinsonism-dementia (or Guam dementia) and in neurolathyrism. The phytotoxicity of mimosine is now unequivocal and with wide-ranging effects, including inhibition of seed germination, root growth, lignification and key enzymes. Mimosine is also detrimental to animals, but the effects in ruminants are modulated by differences in microbial ecology. In contrast, hypoglycin is specifically associated with the seasonal incidence of a human disorder known as 'vomiting sickness'.

*E-mail address: jpfdmello@hotmail.co.uk

The adverse effects of non-protein amino acids are mediated via diverse mechanisms generally consistent with unifying principles of allelochemical activity and metabolic mimicry. Canavanine competes with arginine for cellular transport and in the modulation of nitric oxide synthesis. In addition, canavanine may replace arginine in protein synthesis, enhance arginase activity and depress polyamine synthesis. Selenium analogues of the sulfur amino acids act by competing with methionine and cysteine at the levels of transport and protein synthesis. Both *β-N*-methylamino-L-alanine and *β-N*-oxalylamino-L-alanine operate as potent glutamate receptor agonists. The allelochemical action of mimosine has been linked with inhibition of phenylalanine ammonia-lyase and peroxidase. This diversity of mechanisms has been compared with the complex mode of action of the xenobiotic non-protein amino acid, *β*-aminobutyric acid.

The phytotoxic effects of canavanine and associated metabolites are reversed by respective analogues in the ornithine-urea cycle. In animals, deleterious effects may be substantially reduced through controlled feeding of forages containing toxic non-protein amino acids or by provision of appropriate nutritional supplements. Under commercial conditions, substantial mitigation has been achieved by the manipulation of rumen function to metabolize mimosine to innocuous residues.

The emerging consensus confirms that non-protein amino acids and their associated derivatives display considerable potential as plant protectants and as models for the development of novel pharmaceuticals. However, it is imperative that food safety and medicinal efficacy are not compromised in the quest for diversification. It is also essential that any ecological risks associated with the commercial exploitation of these compounds are minimized.

27.2 Introduction

An inevitable consequence of secondary metabolism in higher plants is the synthesis of a diverse array of heat-stable, non-protein amino acids which are differentially expressed in foliage, fruits, seeds and root exudates (D'Mello, 1991a). Variations in the structural features and chemistry of these non-protein amino acids are as diverse as their distribution in plants. The most striking aspect relates to structural analogy with the 'canonical' (natural or standard) amino acids that participate in protein synthesis or in other physiological functions such as signal transduction or the disposal of excess metabolic nitrogen. In particular, the principal non-protein amino acids occur as congeners or antagonists of urea cycle intermediates, sulfur amino acids, signalling amino acids and aromatic amino acids. There are also other examples of non-protein amino acids that are not compatible with conventional classification systems but which are of relevance here due to their specialized functional characteristics.

With a few exceptions, the non-protein amino acids of higher plants occur as free molecules and are thus in readily accessible forms for assimilation or participation in biochemical and physical reactions. This unconjugated status may be contrasted with the involvement of the glucose moiety in other secondary compounds, such as cyanogens and glucosinolates. However, the issue of availability depends upon the source of the non-protein amino acid and the type of matrix under consideration. In general, non-protein amino acids in seeds are more easily assimilated than those in dried leaf products derived from the same plant. Seeds also contain higher concentrations of non-protein amino acids than leaves or stems. The diversity and concentrations of non-protein amino acids are greater in legumes than in other plant species. Among leguminous plants, the range and levels of these amino acids are greater in tropical than in temperate species (D'Mello, 1995). A further factor of note is the co-occurrence of at least two non-protein amino acids in the same plant. The ubiquitous distribution of non-protein amino acids, generally in association with other secondary compounds of diverse origin, continues to undermine efforts to fully exploit the potential of species such as *Canavalia ensiformis*, *Indigofera spicata*, *Brassica oleracea*, *Lathyrus sativus* and *Leucaena leucocephala* as sources of nutrients for humans and domesticated animals.

Non-Protein amino acids are metabolized in living organisms, but the extent and efficacy of this depend on the species and the amino acid in question. Symbiotic microbes are capable of extensive degradation of non-protein amino acids, sometimes to innocuous residues, but in at least one instance, microorganisms may instigate toxicity.

Adverse effects of non-protein amino acids occur in a wide range of organisms but most consistently and decisively in insects and higher animals, including humans and domesticated livestock. As might be predicted, chemistry and animal species determine the extent and severity of the manifestations observed following the consumption of non-protein amino acids. Deleterious effects include alteration of feeding behaviour, growth inhibition, organ and cellular damage, reduced substrate utilization and reproductive cycle aberrations (D'Mello, 1991a, 2003, 2012).

A major question arises as to the role of non-protein amino acids in plants. Hitherto, there has been only tentative reference to the protective functions of these secondary compounds. Judgement was reserved owing to natural caution on the part of authors, given the paucity of satisfactory data. The emerging and more forthright consensus is that a number of non-protein amino acids may act to mitigate stress induced by biotic and environmental factors and may also exert some additional roles in primary metabolism and chemical ecology. Stress biology is a complex issue outwith the scope of this chapter, but the concept of oxidative stress amelioration in selenium-deficient seedlings by a non-protein amino acid has been advanced in at least two studies and might well provide the momentum for further research (Lalitha and Kulothungan, 2007a,b). Furthermore, the induction of systemic resistance to plant pathogens and environmental stress by xenobiotic non-protein amino acids is an emerging issue that exemplifies complex interactions with phytohormones (Cohen, 2002).

The purpose of this chapter is to consider in more detail the main themes highlighted in the foregoing analysis and to integrate these within a multidisciplinary review of the emerging evidence. Research continues on a wide range of issues, including the synthesis, metabolism and toxicology of non-protein amino acids. Interactions imposed by chemistry and species of organisms are also of relevance, as are the ecological implications. Research is also driven by the quest for novel protein polymer design. Connor and Tirrell (2007) suggested that the 20 canonical amino acids are somewhat restrictive for macromolecular design. Powerful methodologies are now available to insert artificial (non-canonical) amino acids into polypeptides and proteins to enable the incorporation of a wide range of functional groups for novel applications. The unique features and diversity of plant non-protein amino acids may well contribute to this effort.

A broad interpretation of the term 'toxicology' is envisaged here, notwithstanding the considerable pharmaceutical potential of a number of non-protein amino acids. Due to the volume of data, particular emphasis will be placed on urea cycle analogues and aromatic amino acid congeners, significance in economically important plants and the multidisciplinary implications of recent evidence relating to these secondary compounds.

27.3 Classification

A variety of methods may be employed in the taxonomy of non-protein amino acids, none of which is entirely satisfactory. For example, classification has been expressed solely in terms of toxicological criteria to include cytotoxicity or specific adverse effects relating to organ dysfunction. Thus, several non-protein amino acids are hepatotoxic, whereas others are nephrotoxic. Moreover, excitotoxicity is an emerging issue with respect to specific groups of non-protein amino acids. Alternatively, along conventional lines, structural chemistry may constitute the primary basis of classification. Thus, aliphatic, aromatic, acidic, neutral and basic non-protein amino acids may be recognized. The occurrence of sulfur- and selenium-containing non-protein amino acids is another distinguishing feature of diverse species of plants. Yoganathan and Vederas (2010) employed side-chain characteristics to distinguish eight groups. One category included non-protein amino acids with alkyl, cycloalkyl or haloalkyl side chains; amino acids with alkenyl or alkynyl side chains were listed in a second group. The remaining six categories comprised non-protein amino acids with aryl, amino, hydroxyl, carbonyl, heterocyclic or other functional groups as part of the side chain. However, this classification, while useful for cataloguing, contains anomalies and, more importantly, lacks unifying themes for toxicological or mechanistic inference. In addition, the sixth category contains non-protein amino acids with contrasting toxicological properties.

The classification adopted in this chapter has been designed to exemplify diverse antimetabolite features of non-protein amino acids (Table 27.1). The aim is to emphasize alternative structural or functional features that contribute to specific antagonisms and toxic manifestations. A number of non-protein amino acids are structural analogues of some of the 20 canonical amino acids. In addition, structural similarities may be apparent between non-protein amino acids and intermediates arising during the metabolism of the canonical amino acids. Further, the classification adopted here includes certain non-protein amino acids on the basis of their particular physiological properties as agonists of glutamate signalling.

The amino acids in the first group shown in Table 27.1, canavanine, homoarginine and indospicine, have long been recognized as structural congeners of arginine, a key intermediate in the ornithine-urea cycle. In addition, lysine acts as a potent antagonist of arginine and, consequently, is capable of modulating the adverse effects of certain urea cycle analogues. As will be apparent shortly, structural congeners also exist for all the other intermediates of the urea cycle. The unusual non-protein amino acid, $N(\delta)$-acetylornithine, emerged in studies aimed at identifying those *Arabidopsis thaliana* metabolites that are strongly induced by the phytohormone methyl jasmonate (JA). Following JA elicitation, arginine, proline and glutamate are metabolized to ornithine which, in turn, is acetylated by NATA1 (*N*-acetyltransferase activity1) to yield $N(\delta)$-acetylornithine (Adio *et al.*, 2011).

Table 27.1. Classification of non-protein amino acids on the basis of structural and functional criteria.

Group	Non-Protein amino acid
Arginine analogues	Canavanine
	Homoarginine
	Indospicine
Analogues of sulfur amino acids	Selenomethionine
	Se-Methylselenocysteine
	Selenocystathionine
	Selenocystine
	S-Methylcysteine sulfoxide (SMCO)
	Djenkolic acid
Signalling amino acids	β-*N*-Oxalylamino-L-alanine (BOAA)
	β-Cyanoalanine (BCNA)
	α, γ-Diaminobutyric acid (DABA)
	β-*N*-Methylamino-L-alanine (BMAA)
	γ-Aminobutyrate (GABA)
	β-Aminobutyric acid (BABA) (xenobiotic)
Aromatic amino acids	Mimosine
	3,4-Dihydroxyphenylalanine (DOPA)
Other	Hypoglycin A
	Hypoglycin B

Striking structural analogues exist for methionine, cysteine, cystine and cystathionine. When selenium replaces the sulfur atom of these amino acids (Table 27.1), a number of congeners are formed. Those non-protein amino acids of primary interest here are selenomethionine, *Se*-methylselenocysteine and selenocystine. Other analogues of the sulfur-containing amino acids include *S*-methylcysteine sulfoxide (SMCO) and djenkolic acid (D'Mello, 1991a).

D'Mello (1991a) has alluded to structural homology among the signalling non-protein amino acids. Although these amino acids have not been linked, in structural terms, with the amino acids found in proteins, it is worth noting that β-*N*-oxalylamino-L-alanine (BOAA) and β-*N*-methylamino-L-alanine (BMAA) are potent glutamate receptor agonists in animals. The non-protein amino acids γ-aminobutyrate (GABA) and β-aminobutyrate (BABA) are worthy of consideration at this juncture. Of particular interest here is GABA, normally associated with neurotransmission in animals, but now widely linked with stress responses and other functions in higher plants (Kinnersley and Turano, 2000; Shelp *et al.*, 2012). Equally, BABA, a xenobiotic amino acid and a glycine receptor agonist in animals (Schmieden and Betz, 1995), is effective in priming plants against pathogens and osmotic stress, while significantly enhancing acquired thermotolerance (Slaughter *et al.*, 2012).

The aromatic non-protein amino acids of some significance include mimosine and 3,4-dihydroxyphenylalanine (DOPA). Mimosine is generally perceived as a structural analogue of tyrosine and its neurotransmitter derivatives (dopamine and noradrenaline) found in the brain of animals. The initial degradation products of mimosine may also act as structural antagonists of tyrosine-related compounds such as the thyroid hormones. It is well established that DOPA, a natural component of the seeds of some legumes, also functions as a precursor of these neurotransmitters.

The final category of non-protein amino acids presented in Table 27.1 includes hypoglycin A, a compound that is not structurally associated with any of the canonical amino acids participating in the primary metabolism of living organisms.

It will be apparent that the arrangement shown in Table 27.1 includes examples that are not consistent with the stringent definition of antagonisms in the classical sense. However, this scheme has been retained in order to integrate a wider range of the non-protein amino acids into a framework that would allow a comprehensive exposition of the diverse distribution, toxicology and mechanisms of action of these amino acids. Furthermore, at least two of these amino acids may be regarded as potent glutamate receptor agonists and their activities are, therefore, of relevance here not least due to their metabolic and physiological impacts in plants and animals.

27.4 Distribution

Higher plants synthesize an almost infinite array and number of non-protein amino acids, depositing these in the leaves, stems, fruits, pods, seeds and root exudates. The non-protein amino acids that are found in plants of economic or ecological significance are listed in Table 27.2. The seeds of legumes are, arguably, the most concentrated sources of non-protein amino acids, and include examples of aliphatic, sulfur-bearing and aromatic amino acids.

It is clear that canavanine is ubiquitous in both grain and tree legumes and that in the seeds of

Table 27.2. Distribution and concentrations of principal non-protein amino acids in plant products. (Data from D'Mello, 1991a, enlarged and updated as indicated in the footnotes.)

Amino acid	Species	Product	Concentration ($g\ kg^{-1}$ dry weight, except where stated otherwise)
Canavanine	*Canavalia ensiformis*	Seed	51
	Gliricidia sepium	Seed	40
	Robinia pseudocacia	Seed	98
	Dioclea megacarpa	Seed	127
	Indigofera spicata	Seed	8.9
	Sesbania sesban	Seed	6.2
	Medicago sativa[a]	Seed	8-20
	Medicago sativa[b]	Seed exudate	1.04 $\mu mol\ ml^{-1}$
	Sutherlandia frutescens[c]	Leaf extract	0.08–0.55 $mg\ g^{-1}$ extract
Homoarginine	*Lathyrus cicera*	Seed	12
Indospicine	*Indigofera spicata*	Seed	20
S-Methylcysteine sulfoxide	*Brassica oleracea*	Forage	40–60
Djenkolic acid	*Leucaena esculenta*	Seed	2.2
	Pithecolobium lobatum[d]	Seed	20
β-N-Oxalylamino-L-alanine	*Lathyrus sativus*	Seed	25
β-Cyanoalanine	*Vicia sativa*	Seed	1.5
α,γ-Diaminobutyric acid	*Lathyrus sylvestris*	Seed	16
Mimosine	*Leucaena leucocephala*	Seed	145
	Leucaena leucocephala	Leaf meal	10–25
	Pithecolobium ondulatum[e]	Seed	8.4
3,4-Dihydroxyphenylalanine	*Mucuna pruriens*[f]	Seed	45.8–50.5

[a]Natelson (1985).
[b]Keshavan *et al.* (2005).
[c]Shaik *et al.* (2011).
[d]Current name *Archidendron jiringa*.
[e]*Pithecolobium umbellatum*, current name *Cathormion umbellatum*
[f]Thakur *et al.* (2011).

trailing or creeping indigo (*I. spicata*) it also co-occurs with indospicine. As is well documented, legumes contain a wide variety of non-protein amino acids and Table 27.2 provides confirmation of the exact nature and diversity of these compounds.

A number of plants accumulate analogues of the sulfur-containing amino acids in which the sulfur atom is replaced by selenium. *Se*-Methylselenocysteine occurs in *Astragalus* and *Brassica* spp. (Table 27.2). A diverse array of selenoamino acids, including selenomethionine and selenocystathionine, have been identified in other plants (see D'Mello, 1991a). Various species of *Brassica* are widely grown in temperate countries, but the occurrence of the non-protein amino acid SMCO restricts the utilization of these crops as fodder for cattle and sheep. The highest concentrations are found in the young leaves and growing shoots of kale (*Brassica oleracea*) and increases also occur as the plant matures.

Several lathyrogenic and neurotoxic non-protein amino acids arise during transformations of a key metabolite, β-cyanoalanine (BCNA) (Chapters 6 and 12) to BOAA, α,γ-diaminobutyric acid (DABA), γ-*N*-oxalyl-α,γ-diaminobutyric acid and β-aminopropionitrile. Glutamyl derivatives of BCNA and β-aminopropionitrile may also be synthesized. The distribution of these amino acids varies according to the type of legume, with the parent compound, BCNA, and its glutamyl dipeptide occurring in *Vicia sativa* and the remaining amino acids in at least five *Lathyrus* species (D'Mello, 1991a). Another neurotoxic amino acid, BMAA, occurs in the seed of *Cycas circinalis*.

Ipil ipil or koa haole (*L. leucocephala*) is widely distributed in the tropics, yielding timber and palatable forage for ruminant livestock, but the presence of mimosine limits its utilization as animal feed.

The fruit of the ackee tree (*B. sapida*) is the source of hypoglycin A and its glutamyl derivative, hypoglycin B (Table 27.1). The ackee tree is indigenous to West Africa but was introduced into the West Indies where the arils of the ripe fruit are used for human consumption.

27.5 Metabolic Fate

The rapid and extensive metabolism of canavanine, SMCO and mimosine is entirely consistent with the occurrence of these non-protein amino acids in readily available forms. Metabolism may imply detoxification, but this is not always the case, as the examples below are designed to demonstrate.

27.5.1 Canavanine

It has been established that *C. ensiformis*, and presumably certain other leguminous plants, normally contain the full complement of the ornithine-urea cycle enzymes that are required for the synthesis and metabolism of canavanine (Fig. 27.1). Notable exceptions are developing soybean embryos in which the enzyme appears to be inoperative (Goldraij and Polacco, 1999). Although arginase has been almost exclusively associated with the mammalian production of urea as a means of disposing of excess amino-N (Caldwell *et al.*, 2012), efforts to evaluate the general significance of this enzyme in the N metabolism of plants are now gathering momentum. A specific arginase-based assay for canavanine in leguminous plants has been available since 1982 (Korpela *et al.*, 1982). However, arginase is of interest for other reasons. For example, arginase induction may exert a defensive role by repressing gall development during clubroot infection in *Arabidopsis* (Gravot *et al.*, 2012). Conversely, arginase suppression may increase the availability of arginine for NO signalling in root development (Flores *et al.*, 2008). Previously it had been stated that stress-induced arginase may perform similar roles in diverse biological systems. Chen *et al.* (2004) observed profound similarities in the regulation of arginase in plants and animals. In summary, it may be stated that plants generally possess the potential to metabolize canavanine via the action of arginase. Following this metabolism in *C. ensiformis*, canaline, *O*-ureidohomoserine and canavaninosuccinate are synthesized in reactions that are analogous to those for the synthesis of ornithine, citrulline and argininosuccinate, respectively, in the urea cycle (Fig. 27.1).

The tobacco budworm, (*Heliothis virescens*), displays resistance to canavanine toxicity by employing an alternative pathway of metabolism (Melangeli *et al.*, 1997). This process involves the mediation of a constitutive enzyme of the larval gut to catalyse the irreversible hydrolysis of canavanine to homoserine and hydroxyguanidine. The enzyme has been tentatively named as canavanine hydrolase.

The extent to which higher animals metabolize canavanine depends upon the mode of N elimination. The ornithine-urea cycle is the primary means for the disposal of waste N in mammals, and arginine synthesis also occurs by this pathway (Fig. 27.1). It follows that the mammalian metabolism of canavanine should involve the production of canaline, *O*-ureidohomoserine and canavaninosuccinate. The experimental evidence confirms that the major route for canavanine degradation involves arginase-mediated hydrolysis to canaline and urea (see D'Mello, 1991a). In addition, alternative, albeit minor, pathways have been proposed for the metabolism of canavaninosuccinate to guanidinosuccinate and homoserine, and for the transamidination of canavanine with glycine to generate canaline and guanidinoacetate.

In avian species, the urea cycle is non-functional, although residual arginase expression occurs in the kidney and liver. Consequently, birds have no means of synthesizing arginine and readily succumb to the effects of arginine antagonists such as lysine and canavanine. Therefore, arginase-mediated metabolism of canavanine to urea and canaline is likely in birds. In addition, canavanine may undergo transamidination in avian as in mammalian species, but the synthesis of *O*-ureidohomoserine from canaline is not possible owing to the absence of carbamoyl phosphate synthase in avian tissues. It follows that the synthesis of canavaninosuccinate, guanidinosuccinate and homoserine is unlikely to occur in birds.

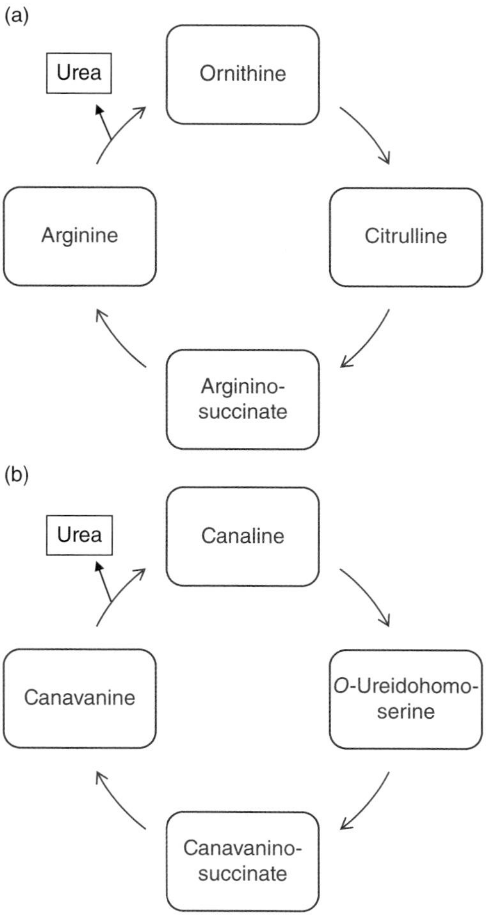

Fig. 27.1. Outline of (a) arginine and (b) canavanine metabolism in higher organisms.

27.5.2 Analogues of sulfur amino acids

The metabolism of selenomethionine and SMCO are well established (D'Mello, 2012). With respect to selenomethionine, Nakamuro *et al.* (2000) maintain that disturbances in the methylation reactions of the detoxification pathway may precede manifestations of hepatotoxicity in mammalian systems.

In the plant, SMCO co-occurs with its metabolite, methyl methane thiosulfinate. The deleterious effects of SMCO in ruminant animals fed mainly or solely on *Brassica* forage are well recognized (see D'Mello, 1991a). Adverse effects are precipitated after the synthesis of a reactive derivative (dimethyl disulfide) during fermentation by rumen microorganisms. Thus, in this particular instance, the symbiotic rumen microorganisms enhance reactivity instead of protecting the host animal as in the case of mimosine.

27.5.3 Mimosine

The microbial breakdown of mimosine is now well established particularly in symbiotic relationships with legumes and certain species of animals. In *Rhizobium* spp. (strain TAL 1145), it has been suggested that the *midK*-encoded pyruvate carboxylase is required for the efficient conversion of mimosine to 3-hydroxy-4(1*H*)-pyridone (3,4-DHP; Awaya *et al.*, 2008).

The metabolic fate of mimosine in ruminants (for example cattle, sheep and goats) is largely determined by the rate and extent of degradation by the rumen microorganisms. During this breakdown, 3,4-DHP is produced. However, this metabolite may also arise in the leaf of *Leucaena* itself following postharvest enzymatic action (see D'Mello, 1991a). Ruminants consuming the legume macerate the leaves and add alkaline saliva, thereby providing optimum conditions for the autolysis of mimosine to the extent that a substantial proportion of the non-protein amino acid may be converted to 3,4-DHP before even reaching the rumen. Additional quantities of 3,4-DHP arise during the rumen fermentation of mimosine and, moreover, another isomer, 2,3-DHP may also be produced. Some rumen microbes are capable of degrading the two forms of DHP to as yet unidentified compounds. Despite these transformations, considerable quantities of mimosine and 3,4-DHP may escape degradation to emerge in the faeces. Furthermore, conjugated forms of 3,4-DHP may also appear in the faeces and urine, while the parent amino acid may itself undergo decarboxylation within the tissues of the ruminant to yield mimosinamine which is then excreted in the urine. As will be pointed out later, all of these transformations are subject to differences in the microbial ecology of ruminants at various geographical locations. What is more, the ability to degrade DHP may decline or disappear entirely on removing these animals from *Leucaena* pasture.

It would appear that 3,4-DHP is a common initial intermediate of mimosine metabolism in diverse living organisms. In fact, there is an enzyme in extracts of *Leucaena* seedlings that is also endowed with the capacity to degrade mimosine, yielding 3,4-DHP, pyruvate and ammonia in the process (Smith and Fowden, 1966).

27.6 Adverse Effects

The underlying principles of toxicokinetics indicate the importance of several factors in the response of living organisms to non-protein amino acids. These include dose, absorption/uptake/assimilation, partition between organs and tissues, cellular metabolism (microbial/endogenous) and excretion. Other components of relevance include duration of exposure, nature (structure and reactivity) of the non-protein amino acid in question and species of target organism. These principles will form the basis of the following review.

It is salutary to note at the outset that virtually all amino acids, irrespective of classification, possess the potential to precipitate adverse effects in living organisms (D'Mello, 2003). Thus, of 21 amino acids tested, the L-isomers of phenylalanine, histidine, tyrosine and homoserine, as well as both isomers of threonine, inhibited growth of *Methylococcus capsulatus* (Eccleston and Kelly, 1972). Furthermore, the bactericidal activity of cysteine has also been demonstrated in strains of *Streptococcus mutans* (Cowman *et al.*, 1983). On the basis of nutritional assays, adverse effects in animals have been organized within three categories: imbalance, antagonism and toxicity. This model constitutes a satisfactory basis for classifying the deleterious effects of amino acids in human subjects. However, two subdivisions, namely 'clinical amino acid imbalance' and 'excitotoxicity', have been introduced to more accurately incorporate current interpretations of harmful reactions observed in humans (D'Mello, 2012). The category of antagonism is of particular significance because it includes the classical examples of the lysine–arginine interaction established with rats (Jones *et al.*, 1966) and avian species (D'Mello and Lewis, 1970a,b), as well as allelopathy induced by non-protein amino acids (D'Mello, 1991a, 1994, 2003, 2012).

Non-Protein amino acids are associated with deleterious effects in a wide range of living organisms. This is entirely within expectations as, in many instances, these compounds bear structural analogy with the metabolically important amino acids or their derivatives. Moreover, the initial degradation products of certain non-protein amino acids may also exert deleterious effects towards key intermediates through potent structural antagonisms. At the cellular level, alteration in enzyme activity is invariably the first indicator of the anti-metabolic effects of non-protein amino acids (Prasad and Subhashini, 1994; Andrade *et al.*, 2008), progressing to and culminating in gross effects of complex aetiology (D'Mello, 1991a, 1994, 2003, 2012; Rosenthal, 2001).

27.6.1 Anti-microbial activity

The effects of non-protein amino acids on microbial ecology are widely recognized and documented. Selected evidence is presented here to highlight observations that have been made with canavanine, selenomethionine, BMAA, DABA and mimosine.

Comparative physiological effects of incorporated amino acid analogues in *Escherichia coli* have been published (Pine, 1978). In general, such incorporations were no more than bacteriostatic, with canavanine being lethal in minimal media. The effects of canavanine are probably, at least in part, mediated via repression of arginine biosynthesis in *E. coli* (Faanes and Rogers, 1972). Also, and consistent with other evidence, canavanine synthesis in lucerne (*Medicago sativa*) interferes with quorum sensing in *Sinorhizobium meliloti*, which is capable of establishing N-fixing symbiosis with its legume host (Keshavan *et al.*, 2005). The population biology of *Bacillus cereus* is also adversely influenced by canavanine, effects that may be reversed by one of the basic amino acids, including arginine, histidine or lysine (Emmert *et al.*, 1998). It should also be noted that indospicine acts as another antagonist of arginine in *E. coli* and *Pseudomonas aeruginosa* (Lesinger *et al.*, 1972). In contrast, selenomethionine supports indefinite growth of *E. coli* following supplementation with many non-essential metabolites (Pine, 1978).

Exogenous supplements of BMAA caused marked and concentration-dependent decreases in the growth of *Synechocystis* PCC 6803, accompanied by substantial loss of photosynthetic pigmentation. Similar but attenuated responses were observed with DABA (Downing *et al.*, 2012).

Mimosine is present in the root nodules and exudates of *Leucaena*. Roots and germinating seeds of this legume secrete mimosine, which adversely affects the growth of microorganisms in the rhizosphere. However, some *Rhizobium* strains isolated from nodules of *Leucaena* are able to degrade mimosine, utilizing the amino acid as a source of carbon and nitrogen (Ali *et al.*, 2010). It has been established that the *mid* genes of *Rhizobium* spp. are required for degradation of mimosine to 3,4-DHP and are inducible by the amino acid itself (Borthakur *et al.*, 2003). It is also clear that the *pyd* genes of *Rhizobium* are required for completing the breakdown of 3,4-DHP (Awaya *et al.*, 2005). Nevertheless, mimosine is capable of depressing the growth of some root nodule bacteria, as measured by colony forming units (Ali *et al.*, 2010). It is suggested that this effect is mediated via the capacity of both mimosine and 3,4-DHP to chelate iron, thereby inhibiting the activity of critical enzymes dependent upon this element. In view of the economic and ecological implications, screening of *Rhizobium* isolates for mimosine degradation continues to attract attention (Ali *et al.*, 2010).

The xenobiotic non-protein amino acid, BABA, exerts its effects indirectly by inducing resistance in plants to a wide range of plant pathogens, including *Pseudomonas syringae*, *Alternaria brassicicola* (Flors *et al.*, 2008), *Phytophthora infestans* (Eschen-Lippold *et al.*, 2010) and *Plasmopara viticola* (Dubreuil-Maurizi *et al.*, 2010). Further details are presented later in this chapter to exemplify the diversity of mechanisms involved in the protective action of BABA.

27.6.2 Phytotoxicity

Limited evidence (Table 27.3) has indicated that canavanine and canaline were inhibitory towards organogenesis in leafy spurge (*Euphorbia esula*), specifically with respect to root and shoot growth (Davis, 1997). However, indole-3-acetic acid (IAA) partially reversed the inhibition of root formation, a remarkable feature related to the partial efficacy of arginine in restoring root development. Canavanine inhibited elongation of the second leaf sheath of rice seedlings, and a disruption of arginine metabolism was implied due to its unusual accumulation in rice shoots treated with canavanine (Nakajima *et al.*, 2001).

It is consistently maintained that the selenoamino acids are toxic to plants that are deemed 'non-accumulators' with reference to the Se component. Brown and Shrift (1982) indicated that selenoamino acids interfere with cellular biochemical reactions, while Wu *et al.* (1997) demonstrated that tissue methionine status in lucerne was negatively affected by increases in tissue selenomethionine concentrations. The overall effect is that sensitive plants are unable to thrive in soils burdened with Se.

Application of BMAA to *A. thaliana* seedlings increased hypocotyl elongation but inhibited cotyledon opening (Brenner *et al.*, 2000). The effect on hypocotyl elongation was light specific.

Table 27.3. Phytotoxicity and associated effects of non-protein amino acids.

Amino acid	Observations	Reference
Canavanine	Inhibition of root and shoot growth in leafy spurge (*Euphorbia esula*); Indole-3-acetic acid (IAA) partially reversed inhibition of root formation; partial efficacy of arginine in restoring root development	Davis (1997)
	Suppression of second leaf sheath elongation in rice seedlings; unusual accumulation of arginine in rice shoots	Nakajima *et al.* (2001)
Selenoamino acids	Interference with cellular biochemical reactions	Brown and Shrift (1982)
	Methionine status in lucerne (*Medicago sativa*) negatively correlated with tissue selenomethionine concentrations; sensitive plants unable to thrive in high-Se soils	Wu *et al.* (1997)
β-N-Methylamino-L-alanine	Application to thale cress (mouse-ear cress; *Arabidopsis thaliana*) seedlings increases hypocotyl elongation but inhibits cotyledon opening; effect on hypocotyl elongation is light specific	Brenner *et al.* (2000)
γ-Aminobutyrate	Role in stress mitigation in higher plants	Kinnersley and Turano (2000)
	Accumulation causes cell elongation defects and a decrease in expression of genes encoding secreted and cell wall-associated proteins in *Arabidopsis*	Renault *et al.* (2011)
Mimosine	Mimosine and 3-hydroxy-4(1*H*)-pyridone (3,4-DHP) depress growth of mung bean seedlings; inhibition reversed by tyrosine, although nicotinic acid, pyridoxal phosphate or ferrous ions also effective	Smith and Fowden (1966)
	Inhibition of seed germination, seedling growth and enzymes in rice (*Oryza sativa*); enzymes so affected include nitrate reductase, peroxidase and its isoenzymes, catalase and IAA oxidase; ecophysiological implications of mimosine phytotoxicity suggested	Prasad and Subhashini (1994)
	Reduced germination in wheat (*Triticum aestivum*) and sickle pod (*Senna obtusifolia*); maximum growth inhibition occurred after mimosine was provided via seed imbibition; mimosine considerably more phytotoxic than albizzine in terms of chlorophyll development	Williams and Hoagland (2007)
	Several indices of germination in perennial ryegrass (*Lolium perenne*) and cocksfoot (*Dactylis glomerata*) impaired by mimosine	Iftikhar *et al.* (2008)
	Inhibition of root growth, phenylalanine ammonia lyase (PAL) and peroxidase in soybean (*Glycine max*)	Andrade *et al.* (2008)
	Emerging consensus that mimosine may serve as a model or parent compound for the development of potent bio-herbicides	Xuan *et al.* (2006)
3,4-Dihydroxy-phenylalanine	Inhibition of seed germination and root growth in several plant species	Soares *et al.* (2011)

It is now well established that GABA exerts an important role in stress mitigation in higher plants (Kinnersley and Turano, 2000). However, Renault *et al.* (2011) observed that GABA accumulation causes cell elongation defects and a decrease in expression of genes encoding secreted and cell wall-associated proteins in *A. thaliana*.

There is convincing experimental evidence to demonstrate that mimosine is toxic to crop plants and weeds (Table 27.3). Initial studies indicated that mimosine and 3,4-DHP depressed the growth of mung bean seedlings. Of considerable significance was the observation that this inhibition was reversed by tyrosine, although nicotinic acid, pyridoxal phosphate or ferrous ions were also effective (Smith and Fowden, 1966). Mimosine inhibited seed germination, seedling growth and various enzymes of rice (*Oryza sativa* L.). The enzymes so affected included nitrate reductase, peroxidase and its isoenzymes, catalase and IAA oxidase. Ecophysiological impacts of mimosine phytotoxicity were implied in the light of these observations (Prasad and Subhashini, 1994). Mimosine also reduced germination in wheat (*Triticum aestivum*) and sickle pod (*Senna obtusifolia*). Maximum growth inhibition occurred after mimosine was provided via seed imbibition (Williams and Hoagland, 2007). Furthermore, in comparison with another non-protein amino acid (albizzine), mimosine was considerably more phytotoxic, specifically with respect to the inhibition of chlorophyll development. Several indices of germination in grasses, including *Lolium perenne* and *Dactylis glomerata*, were also impaired by mimosine (Iftikhar *et al.*, 2008). Of particular interest is the inhibition by mimosine of root growth, phenylalanine ammonia lyase (PAL) and peroxidase (POD) in soybean (*Glycine max*) (Andrade *et al.*, 2008). Consistent with these and other observations is the emerging consensus that mimosine may serve as a model or parent compound for the development of potent bio-herbicides (Xuan *et al.*, 2006).

The roots of the velvet bean (*Mucuna pruriens*) release DOPA into the rhizosphere which can then be absorbed by other plants with potentially detrimental effects. For example, Soares *et al.* (2011) indicated that DOPA was absorbed by soybean roots and metabolized to melanin while reactive oxygen species (ROS) was reduced.

27.6.3 Insecticidal activity

Marked insecticidal activity of non-protein amino acids has been recorded over several decades. A selection of the evidence is presented here to support emerging proposals that these amino acids may serve directly or as parent molecules for the development of novel bio-insecticides. Mimosine inhibits growth and development of *Tribolium castaneum* larvae, causing dose-dependent suppression of trehalase, invertase and amylase activities (Ishaaya *et al.*, 1991). Efficacy of this amino acid towards *Heliothis punctiger* has also been examined in terms of the growth and development of larvae on artificial diets (Elliot *et al.*, 1984). However, these results should be balanced against the rapid spread of the *Leucaena* psyllid (*Heteropsylla cubana*) westwards during the 1980s from Central America to Australia and Africa, despite the presence of mimosine in the legume.

A more detailed and consistent analysis of insecticidal potential relates to canavanine as a result of the pioneering work of Rosenthal and other investigators. Rosenthal (2001) attributed the insecticidal properties of canavanine to its mimicry of arginine resulting in the formation of aberrant dysfunctional canavanyl peptides and proteins. In *Manduca sexta*, the biological penalties of such products persisted throughout the life cycle of canavanine-treated larvae. Other derivatives, such as homocanavanine and the methyl and ethyl esters of canavanine, possess variable activities with respect to the inhibition of larval growth and the incidence of pupal deformities (Rosenthal *et al.*, 1995). Recent studies suggest that canavanine may also act as a repellent, judging by observations with *Drosophila*. It is suggested that receptors enable these flies to sense food containing canavanine (Mitri *et al.*, 2009). Lee *et al.* (2012) identified the gustatory receptor that is associated with specificity for the avoidance of canavanine.

Another dimension to the efficacy of canavanine concerns its effect on the pathogenicity of *Bacillus thuringiensis* (Bt), a microbial insecticide, in *M. sexta*, as reported by Felton and Dahlman in 1984. At that time, the results represented the first indication of a plant allelochemical enhancing the efficiency of Bt *in vivo*. It was suggested that canavanine increased the effect of Bt on gut permeability and active transport.

Notwithstanding these results, canavanine tolerance has been observed in the tobacco budworm (Melangeli *et al.*, 1997). It appears that a constitutive enzyme in the larval gut, canavanine hydrolase, catalyses an irreversible hydrolysis of canavanine to homoserine and hydroxyguanidine.

The defensive function of *N*(δ)-acetylornithine has been postulated on the basis of transient expression of the acetylating enzyme, NATA1, and the addition of *N*(δ)-acetylornithine to an artificial diet. Both procedures reduced reproduction of green peach aphid (*Myzus persicae*), suggesting a direct toxic or deterrent effect (Adio *et al.*, 2011).

The xenobiotic non-protein amino acid, BABA, is another potential bio-insecticide (Hodge *et al.*, 2005). Application as a root drench to legumes inhibited growth and reproduction in the pea aphid (*Acyrthosiphon pisum*).

27.6.4 Manifestations in higher animals

Adverse effects of non-protein amino acids are observed most consistently and decisively in mammals and avian species (D'Mello, 1991a, 2003), presenting clear implications for human health (D'Mello, 2012).

Limited acute toxicity data are available and these are tabulated by D'Mello (1991a). For canavanine, LD_{50} values for rats were in the range 5.0–5.9 g kg^{-1} body weight, following administration in a single subcutaneous injection. In contrast, BCNA LD_{50} values for rats were between 135 to 225 mg kg^{-1} body weight and for the chick just 70 mg kg^{-1}. It is clear that BCNA is markedly more potent in terms of lethality than canavanine and this assessment is also borne out by other criteria of toxicity.

Chronic exposure to non-protein amino acids in feeding trials is the conventional procedure for assessing potential risk to animal health and welfare. In general, the plant is used as the source of the amino acid under investigation, but interpretation of results may be confounded by the presence of co-occurring non-protein amino acids and/or other anti-nutritional factors. Manifestations of deleterious effects range from loss of appetite and reduced nutrient utilization to profound neurological disorders and even death.

Canavanine contributes significantly to the toxicity of the jack bean (*C. ensiformis*) for young chicks. Adverse effects may also arise through the formation of canaline, a structural analogue of ornithine, by the action of arginase on canavanine. The mammalian metabolism of canavanine corresponds with that of arginine in the urea cycle (Fig. 27.1). As this pathway is essentially non-functional in avian species, they are unable to regenerate arginine and, consequently, readily succumb to the adverse effects of canavanine in jack beans (D'Mello *et al.*, 1989). These effects included reduced growth rates and efficiency of food and N utilization in chicks fed diets containing jack beans previously heat treated to denature potent lectins. Canavanine appeared in the serum of chicks fed diets containing jack beans and, in addition, circulating levels of urea exceeded control values. Lysine supplementation aggravated the effects on growth and on the efficiencies of food and N utilization. In contrast, arginine fortification of jack bean diets enhanced growth and food utilization efficiency. These results were interpreted to indicate the existence of a canavanine–arginine interaction, similar to that between lysine and arginine. Thus, in both antagonisms, arginine requirements and urea excretion are enhanced, although the relative proportions of this additional urea arising from canavanine and arginine remain to be established. The failure of supplementary arginine to substantially reduce circulating levels of the respective antagonists is a feature common to both interactions. Creatine fortification markedly improved the efficiencies of food and N utilization in chicks fed jack bean diets (D'Mello *et al.*, 1990). It is noteworthy that Austic and Nesheim (1972) also reported improvements in food utilization efficiency with creatine supplementation in the lysine–arginine antagonism. However, differences between the two interactions are also apparent in that arginine enhanced the food and N utilization efficiencies in chicks fed excess lysine, but not in those fed diets containing canavanine. It is possible that there is a residual effect of canaline formation depressing polyamine synthesis in chicks fed jack bean diets. Homoarginine exacerbated the effects of canavanine-induced toxicity in chicks fed jack bean diets, illustrating the complexity of the interactions among the analogues and antagonists of arginine (D'Mello, unpublished).

It is widely acknowledged that indospicine is a powerful teratogen, inducing cleavage of the secondary palate and general somatic dwarfism in rats given a single oral dose of an extract of *I. spicata* seeds. Furthermore, subcutaneous injections of indospicine precipitate fat deposition and cytological abnormalities in the liver of mice. The fat accretion can be prevented by simultaneous injections of arginine but not of canavanine (see D'Mello, 1991a).

Striking allelochemicals of the sulfur amino acids exist naturally in plants, particularly in those species where the sulfur atom is replaced by selenium. The debilitating disorders associated with these selenoamino acids have been attributed to selenium poisoning (D'Mello, 1991a). The selenoamino acids are undoubtedly toxic to animals. For example, selenomethionine is as harmful as selenite or selenate in acute tests (Schrauzer, 2000), precipitating pancreatic acinar cell necrosis in rats (Hietaranta *et al.*, 1990) and histopathological aberrations in the liver and kidneys of juvenile white sturgeon (Tashjian *et al.*, 2006). Selenomethionine is readily incorporated into tissue proteins of animals (Schrauzer, 2000), but the balance between synthetic and detoxification pathways is dependent upon methionine status. At suboptimal intakes of methionine, selenomethionine is directed towards tissue protein incorporation. With adequate intakes of methionine, the Se analogue follows the pathway resulting in the formation of the trimethylselenonium ion. According to Schrauzer (2000), release of selenomethionine from body proteins by catabolic processes during illness should not result in toxicity as a steady state is established that prevents the uncontrolled accumulation of Se.

Another sulfur amino acid analogue, SMCO, constitutes a significant deterrent to the exploitation of forage and root *Brassica* crops as fodder for ruminant animals. The adverse effects of SMCO develop after its metabolism by rumen bacteria to dimethyl disulfide (Smith, 1980). A severe haemolytic anaemia appears within 1–3 weeks in animals fed mainly or exclusively on *Brassica* forage. Initial overt indications of the disorder include loss of appetite and reduced milk production, whereas internal changes are reflected in the appearance of refractile, stainable granules (Heinz-Ehrlich bodies) within the erythrocytes, accompanied by reduced blood haemoglobin concentrations. Extensive organ damage is another feature of this condition, with the liver becoming swollen, pale and necrotic. Spontaneous but incomplete recovery may be observed in animals continuing to graze the crop, with further fluctuations in blood haemoglobin status. Withdrawal of the forage usually results in the restoration of normal blood composition within 3–4 weeks (Smith, 1980).

With BOAA, BCNA, DABA and BMAA, there are common manifestations of toxicity, including tremors, convulsions and death following administration of the pure forms of these non-protein amino acids. There is overwhelming evidence that BOAA is one of the most potent neurotoxic amino acids found in plants. Neurological and biochemical lesions in animal models include tremors, convulsive seizures and increased brain concentrations of ammonia and glutamine (D'Mello, 1991a). The precipitation of adverse effects is influenced by age and by physiological factors such as the induction of an acidotic state, but not by animal species. Other studies confirm that BOAA is a powerful glutamate receptor agonist (Andersson *et al.*, 1997).

The neurotoxicity of BCNA is well established following systematic work with animal models (see D'Mello, 1991a). Single doses of this amino acid are sufficient to provoke severe metabolic and neurological manifestations of toxicity. Effects include cystathioninuria, hyperactivity, tremors, convulsions, rigidity and death. Pyridoxal hydrochloride administration delays the onset of the neurological lesions and increases the dose required to induce adverse reactions.

Toxicological investigations with DABA reveal increased susceptibility of mammals relative to avian species (D'Mello, 1991a). Effects in the rat include upper extremity tremors, convulsions and death. Biochemical changes may also occur, including increased ammonia concentrations in the blood and brain, elevated glutamine levels in the brain and the inhibition of ornithine carbamoyltransferase, with reduced urea production. There has been some interest in the use of *L. sativus* and *V. sativa* grains as feed for poultry. However, the occurrence of the lathyrogenic amino acids in the seeds may represent a primary deterrent. Ruminants may be able to degrade neurolathyrogens, as shown by the absence of toxicity in wether lambs on feeding *Lathyrus sylvestris* hay with a relatively high DABA concentration (Forster *et al.*, 1991).

Mimosine, present in *L. leucocephala* (Table 27.2) is widely regarded as a structural analogue of tyrosine and its neurotransmitter derivatives, dopamine and norepinephrine. The biochemical or physiological evidence for such an assumption has yet to emerge in animal studies. The extensive deleterious consequences of mimosine intake appear to be unrelated to any structural properties and include disruption of reproductive processes, teratogenic effects, loss of hair and wool, and even death. Similar reactions may be observed on feeding *Leucaena* to ruminant livestock. Hence, the defleecing effects in sheep may be precipitated by administration of pure mimosine or by feeding *Leucaena* forage. The toxicity of *Leucaena* is determined by geographical differences in rumen microbial ecology and is critically dependent upon the rate and extent of bacterial degradation of mimosine. The synthesis of 3,4-DHP is significant as it is capable of causing loss of appetite, goitre and reductions in circulating concentrations of thyroxine (Jones, 1985). Another isomer and goitrogen, 2,3-DHP, may also arise during ruminal degradation. Thus, the association of mimosine with tyrosine metabolism appears to be mediated via the two forms of DHP. Some rumen bacteria are capable of detoxifying both isomers of DHP. Despite these transformations, considerable quantities of mimosine and 3,4-DHP may escape degradation in the rumen, and other derivatives may also be formed and subsequently excreted. Ruminants in Australia, the USA and Kenya lack the requisite bacteria involved in the detoxification of the two DHP isomers and, consequently, succumb to their goitrogenic effects if relatively high intakes of *Leucaena* are maintained over protracted periods of time. Updated evidence suggests that mimosine and DHP toxicity in cattle grazing *Leucaena*–grass pastures is a continuing issue in Australia (Dalzell *et al.*, 2012). Then again, in certain other regions where *Leucaena* is indigenous (Central America) or is naturalized (Hawaii and Indonesia), ruminants possess the full complement of bacteria that are required for DHP degradation, which accounts for the absence of *Leucaena* toxicity in these countries (Jones, 1985).

In animal models, hypoglycin A induces fetal malformations and inhibits gluconeogenesis. Research continues on the toxicology of hypoglycin A. For example, Blake *et al.* (2006) suggested that the form in which the amino acid is administered to rats could affect outcome in dose-response investigations, and concluded that for the purpose of risk assessment, hypoglycin A should be incorporated within the matrix of the fruit (ackee) in which it is found.

27.6.5 Human health risks

Although the data on the adverse effects of amino acids are derived primarily from studies with rodent models, it is appropriate and instructive to consider the implications for human health.

Canavanine occurs in lucerne (alfalfa) seeds, which are regularly recommended as herbal supplements. In monkeys, feeding lucerne sprouts induced haematological and serological abnormalities similar to those observed in human systemic lupus erythematosus (SLE). Subsequent feeding of canavanine reactivated this condition in these monkeys (see Spencer and Berman, 2003). Notwithstanding the absence of direct definitive data, it is assumed here that canavanine might precipitate similar effects in humans consuming lucerne supplements (Table 27.2).

The primary source of homoarginine is the seed of *Lathyrus cicera* (Table 27.2). In rodent models, homoarginine has been shown to induce hypersensitivity and mortality as well as a depression in brain concentrations of lysine (O'Kane *et al.*, 2006; Zinnanti *et al.*, 2007), ornithine and arginine (D'Mello, 1991a). Kakoki *et al.* (2006) showed that homoarginine can influence NO production in endothelial cells by altering cellular arginine transport mechanisms. Homoarginine exacerbates the canavanine–arginine interaction in avian species (D'Mello, 2003), serving to illustrate the complexity of interactions among the analogues and antagonists of arginine. As *L. cicera* is not a normal constituent of human diets, the toxicity of homoarginine is of minor relevance as regards health issues. Nevertheless, there is continuing research interest in the activities of the pure form of this amino acid as a substrate for the modulation of NO production in infection and immunity (Degnan *et al.*, 1998), in human pregnancy (Valtonen *et al.*, 2008) and in the functional metabolism of the renal medulla (Kakoki *et al.*, 2004). Furthermore, Henningsson and Lundquist (1998) reported that homoarginine was less potent as an insulin secretagogue than arginine, but equally effective for inducing glucagon release.

In experimental models, indospicine has been shown to inhibit both constitutive and inducible forms of nitric oxide synthase (NOS) (Pass *et al.*, 1996) as well as arginase (Hey *et al.*, 1997). As *I. spicata* is not a normal component of human diets, it appears that interest in indospicine primarily resides in its ability to alter the partition of arginine metabolism between urea and NO production (Hey *et al.*, 1997).

With the selenoamino acids, a cautious approach is recommended to circumvent the potential cytotoxic effects of Se. In the plant, SMCO co-occurs with its metabolite, methyl methane thiosulfinate. At doses of 0.5 and 1.0 mmol kg^{-1}, this metabolite was observed to induce severe toxicity in mice (Marks *et al.*, 1993). There is no evidence that SMCO or methyl methane thiosulfinate exerts any adverse effects in humans consuming high levels of *Brassica* vegetables.

Intake of BOAA via the grain of *L. sativus*, has been positively associated with the incidence of neurolathyrism in humans living in India and in parts of China. This condition is characterized by muscular rigidity, weakness and paralysis of the leg muscles. Neurolathyrism affects all ages and several members of the same family may be similarly affected (Spencer and Berman, 2003). The role of BCNA in human neurolathyrism has been discounted as *V. sativa* generally only occurs as a contaminant of other legume grains. Human exposure to DABA is limited as *L. sylvestris* is not a normal dietary component.

Following extensive research, BMAA has been designated as a potent glutamate receptor agonist (Andersson *et al.*, 1997) and, controversially, has been implicated in a specific neurodegenerative disorder in humans. In a mouse model, BMAA induced neuronal cell death *in vivo* (Santucci *et al.*, 2009), while in dissociated rat brain cells, the amino acid caused an elevation in intracellular Ca levels (Brownson *et al.*, 2002). Using dissociated mixed spinal cord cultures, Rao *et al.* (2006) observed that pure BMAA selectively induced motor neuron loss via activation of the AMPA/kainate receptor (an ionotropic receptor that responds to the neurotransmitter glutamate), which was replicated with the use of cycad seed extracts. Spencer and Berman (2003) discounted any link between cycad-derived BMAA and the incidence of the amyotrophic lateral sclerosis/Parkinsonism-dementia complex (ALS/PDC) in the Chamorro communities of Guam. Instead, the co-occurring glycoside, cycasin in its aglycone form, was implicated in this condition. Others had previously concluded that processed cycad flour, as prepared by the Chamorros, contained extremely low levels of BMAA, which would not account for the occurrence of ALS/PDC (Duncan *et al.*, 1990). However, contemporary views are that BMAA may well be the causative agent. Murch *et al.* (2004) found the amino acid in free form in 83% of ALS/PDC patients and in protein-bound form in 100% of these individuals. Both forms were also found in two Canadians who had died of progressive neurodegenerative disease. BMAA is now thought to be more ubiquitous, being a product of all cyanobacteria, which are universally distributed in terrestrial, freshwater and marine environments (see Santucci *et al.*, 2009). Thus, the amino acid can be biomagnified within the ecosystem and accumulate in the food chain. Furthermore, BMAA concentrations in the protein fraction of cycad seeds have hitherto been underestimated. In addition, evidence demonstrating a similar efficacy of cycad seed extracts to that of the pure form of BMAA in precipitating motor neuron injury is consistent with the hypothesis that BMAA exerts a major role in the aetiology of ALS/PDC (Rao *et al.*, 2006). Nevertheless, it is still possible that BMAA and the aglycone of cycasin may act additively, synergistically or in a potentiating mode to precipitate ALS/PDC in Chamorro communities.

The leaves, pods and seeds of *Leucaena* are widely used in Central America, Indonesia and Thailand to prepare a soup. Anecdotal evidence indicates that loss of hair is a frequent outcome among those individuals who regularly consume this broth. The adverse effects are attributed to ingestion of mimosine, but the potential goitrogenic effects of the two isomers of DHP have yet to be explored. As will be elaborated later, current interest in mimosine has shifted to its role as an inhibitor of DNA replication and cell cycle progression (Dong and Zhang, 2003; Perry *et al.*, 2005).

In the West Indies, consumption of the arils and seed of unripe fruits of ackee is associated with violent vomiting, convulsions, coma and mortality (Spencer and Berman, 2003). Incidence of this 'vomiting sickness' tends to be familial, affecting undernourished children during periods of scarcity of mature fruits.

27.7 Mechanisms

The foregoing account indicates considerable diversity in the number, structure and reactivity of non-protein amino acids. This degree of complexity implies an equally diverse range of biochemical activity in respect of mechanisms involved in the expression of toxicity, host–parasite interactions and the response to abiotic factors. The emerging data appear to support this assumption. In developing the following review, phytochemical evidence as well as data obtained with animals will be used to demonstrate common underlying mechanisms implicating transport, misincorporation of amino acids, enzyme inhibition/enhancement and interactions with other bioactive/signalling molecules. Metabolic mimicry is often invoked as an underlying feature of the activity of non-protein amino acids and this theme is worthy of further consideration here (Rosenthal, 2001; D'Mello, 2012).

27.7.1 Biochemical basis of toxicity

The original hypothesis that an individual non-protein amino acid elicits its toxic effects via a single mechanism is no longer tenable (D'Mello, 1991a). On the contrary, there is overwhelming evidence to indicate that these adverse effects are mediated via diverse mechanisms (D'Mello, 2012). This multifactorial mode of action is observed individually and collectively for most of the significant non-protein amino acids present in plants. In addition, a number of structural analogues share common mechanisms. The following account represents an updated version of comprehensive reviews by D'Mello (1991a), Spencer and Berman (2003) and D'Mello (2012). Even though much of the evidence relates to work with mammalian and avian models, it is instructive to adopt a comparative view, using data obtained from other living organisms.

Limited evidence indicates that canavanine inhibits arginine metabolism in plants. This conclusion was based on the unusual accumulation of arginine in rice shoots treated with canavanine (Nakajima *et al.*, 2001). Whether this reflects competition for transport or misincorporation of canavanine in preference to arginine in protein synthesis remains unresolved. However, it should be recalled that Rosenthal *et al.* (1976) demonstrated that the ornithine-urea cycle amino acids effectively counteracted both the additive and synergistic growth-inhibiting effects of canaline-urea analogues in the common or lesser duckweed (*Lemna minor*).

The major route of canavanine metabolism in mammals follows the urea cycle, in which the amino acid undergoes arginase-induced hydrolysis to canaline and urea (Fig. 27.1). Canavanine administration elicits striking increases in serum and urinary concentrations of ornithine. This pathway should, therefore, provide a mechanism for at least partial detoxification of canavanine, although canaline forms a covalent complex with pyridoxal phosphate, thereby inhibiting the activities of enzymes dependent upon this cofactor. One such enzyme is ornithine decarboxylase, and its inhibition by canaline would account for the substantial accumulation of ornithine in body fluids. Ornithine decarboxylase is a key enzyme in the synthesis of specific polyamines required for the regulation of cell growth and differentiation. Inhibition of polyamine synthesis by canaline then provides an important focal point for the action of canavanine (Bence *et al.*, 2002). Canavanine may also act by competing with lysine and arginine for transport across cell membranes. In this context, it is significant to note that canavanine stimulated the efficacy of the microbial insecticide Bt in *M. sexta*, which was attributed to effects on gut permeability and active transport (Felton and Dahlman, 1984). At that time, the authors claimed this report to be the first demonstrating an enhancing effect of a plant allelochemical on the pathogenicity of Bt *in vivo*. Canavanine–arginine antagonism may also affect the population biology of *B. cereus* exposed to lucerne seed exudates (Emmert *et al.*, 1998). Previously, Faanes and Rogers (1972) had observed the repression of arginine biosynthesis by canavanine in mutants of *E. coli*. Other studies suggest that canavanine may act, like lysine, by inhibiting transamidinase activity, thereby reducing creatine synthesis. It is widely recognized that canavanine can replace arginine during protein synthesis, resulting in the formation of aberrant macromolecules with modified functional properties (Rosenthal, 2001; Igloi and Schiefermayr, 2009). Thus, in rats, canavanine is incorporated into liver enzymes and into brain components, yielding proteins of reduced activity; however, other studies with rats have shown negligible rates of canavanine incorporation into tissue proteins. Hartman *et al.* (2007) demonstrated that of four analogues of arginine, canavanine was utilized most efficiently in the synthesis of unnatural peptides in a series of peptide translation assays. D'Mello (1991a) suggested that canavanyl proteins may be degraded as rapidly as they are formed owing to their reduced biological activity. Any cycle of synthesis and degradation of these proteins might well result in increased protein turnover rates. Other studies have pointed to the incorporation of chlorinated analogues of valine and leucine into enzyme proteins of *E. coli* without loss of catalytic activity (Stigers *et al.*, 2011). Based on studies with canavanine, Yan *et al.* (2000) indicated that ubiquitin hydrolases in *A. thaliana* provide the metabolic machinery for degrading short-lived and abnormal proteins. In contrast, Rosenthal (2001) suggests that a key element of the insecticidal activity of canavanine is its mimicry of arginine, which facilitates the synthesis of dysfunctional canavanyl proteins that persist throughout the life cycle of *M. sexta* larvae. Mammalian–invertebrate differences in metabolic responses to canavanine may be associated with ureotelism and uricotelism, respectively. This would be a simplistic interpretation though, as other evidence indicates that the tobacco budworm is resistant to canavanine due to the presence of a constitutive enzyme in the larval gut that catalyses the irreversible hydrolysis of canavanine to homoserine and hydroxyguanidine. Arguably, the most interesting biochemical feature to emerge is the inhibition of constitutive and inducible forms of NOS by canavanine (Pass *et al.*, 1996; Rios-Santos *et al.*, 2007). The consequences of this on essential physiological processes have yet to be elucidated, but will probably show an adverse impact of canavanine on signalling, immunocompetence and cardiovascular function. Finally, canavanine may act in glial cells by decreasing the levels of reduced glutathione, an effect not related to its inhibition of NOS (Riganti *et al.*, 2003).

Current indications are that homoarginine competes with lysine for transport across the blood–brain barrier (O'Kane *et al.*, 2006). In the work of Zinnanti *et al.* (2007), for example, brain lysine concentrations were reduced by 50%. In the perfused rat kidney, it has been shown that homoarginine can modulate NO production by altering the transport of cellular arginine via the y^+ and y^+ L transport mechanisms (Kakoki *et al.*, 2006).

Indospicine operates via at least four mechanisms. It acts as a competitive inhibitor of arginase (Hey *et al.*, 1997), a property which accounts for the precipitation of hepatotoxicity in mammals owing to the critical role of this enzyme in the elimination of nitrogen in the urea cycle. Furthermore, indospicine strongly inhibits the aminoacylation of arginine, whereas the charging of leucine remains unaffected. The incorporation of arginine into proteins is thereby impaired. It has long been established that indospicine acts as an arginine antagonist in *E. coli* and *P. aeruginosa* (Lesinger *et al.*, 1972). Other evidence suggests that DNA synthesis is inhibited by indospicine and that this effect may be substantially reversed by increasing arginine supply. Indospicine is another inhibitor of constitutive and inducible forms of NOS, a property that it shares with canavanine (Pass *et al.*, 1996). Thus, indospicine has the capacity to alter the pathway of arginine utilization depending upon the cellular status of inducible NOS (Hey *et al.*, 1997).

Three types of reactions have been associated with the selenoamino acids: replacement of sulfur amino acids in protein synthesis; competition with sulfur amino acids for transport; and suppression of selenium methylation. Schrauzer (2000) suggested that substitution of methionine by its Se analogue does not significantly alter protein structure, but the functional properties may be affected if the replacement occurs near an active site. The release of Se stored in body proteins during illness is not considered to result in adverse effects, as no mechanism exists for the selective release of Se-methionine during degradative processes (Schrauzer, 2000). However, disturbances in the methylation pathway may lead to manifestations of selenosis (Nakamuro *et al.*, 2000).

In ruminant animals, the toxicity of SMCO arises after its metabolism to a reactive metabolite, dimethyl disulfide. In *Brassica* vegetables, SMCO coexists with its metabolite, methyl methane thiosulfinate. At relatively high doses (0.5 and 1.0 mmol kg^{-1} body weight), this metabolite precipitated severe acute toxicity in mice (Marks *et al.*, 1993). SMCO itself is considered to be safe for humans; indeed, it may confer beneficial effects.

Of the four neurotoxic amino acids listed in Table 27.1, two (BOAA and BMAA) act as potent glutamate receptor agonists (Andersson *et al.*, 1997). Brownson *et al.* (2002) suggest the involvement of a product of BMAA and CO_2 that mimics the structure of other excitatory amino acids, such as glutamate, in its mechanism of toxicity. Rao *et al.* (2006) demonstrated the capacity of BMAA to selectively damage motor neurons via AMPA/kainate receptor activation. It is significant that the effects of BMAA on the early morphogenesis of *A. thaliana* seedlings can be reversed by the simultaneous application of glutamate (Brenner *et al.*, 2000). In the case of *β*-cyanoalanine, complex formation with pyridoxal phosphate appears to be the major mechanism of action. With DABA, the inhibition of ornithine carbamoyltransferase induces primary liver dysfunction accompanied by secondary brain lesions in mammals, while avian models are unaffected by this neurotoxin (D'Mello, 1991a).

The two features of mimosine commonly associated with its adverse effects are complex formation with pyridoxal phosphate and the inhibition of DNA replication. Lin *et al.* (1996) noted that mimosine is an extremely effective inhibitor of DNA replication and specifically indicated that it targets serine hydroxymethyltransferase, which is involved in the penultimate step of thymidylate biosynthesis. Perry *et al.* (2005) concluded that mimosine attenuates serine hydroxymethyltransferase transcription by chelating Zn and considered the implications for the inhibition of DNA replication. Other relevant studies on mimosine include its effect in cell cycle progression (Kulp and Vulliet, 1996; Dong and Zhang, 2003) and its role in the induction of apoptosis (Hallak *et al.*, 2008). The reversal of mimosine toxicity by tyrosine in mung bean seedlings is suggestive of metabolic mimicry, but the relative efficacy of nicotinic acid, pyridoxal phosphate and ferrous ions adds a more complex dimension to the elucidation of underlying mechanisms (Smith and Fowden, 1966). In rice, mimosine inhibited the activities of key enzymes, including nitrate reductase, peroxidase and its isoenzymes, and catalase (Prasad and Subhashini, 1994). In soybean, mimosine suppressed PAL and peroxidase, implying allelochemical activity as an underlying mechanism (Andrade *et al.*, 2008). The role of PAL in the production of salicylic acid precursors has been proposed as a major function in the resistance of *Arabidopsis* towards downy mildew (*Peronospora parasitica*; Mauch-Mani and Slusarenko, 1996). Subsequently, Lois *et al.* (1989) indicated that PAL exerted a key role in phenylpropanoid metabolism, contributing to the synthesis of potentially protective compounds such as flavonoids, furanocoumarins, phytoalexins and cell-wall components. The role of PAL in plant responses to environmental factors and stress has been a continuing theme (Camm and Towers, 1973; Hyun *et al.*, 2011). Class III plant peroxidases are well recognized as part of the active defence-related proteins inducing both structural and metabolic responses to stress (Almagro *et al.*, 2009). Mhamdi *et al.* (2010) reviewed the evidence implicating catalase function in plant responses to changes in environmental stimuli and biotic challenge. Thus, despite limited evidence, it appears that mimosine undermines both signal transduction pathways and the synthesis of pathogenesis-related proteins in precipitating its phytotoxic action.

In *T. castaneum* larvae, mimosine caused dose-dependent suppression of the activities of trehalase, invertase and amylase (Ishaaya *et al.*, 1991). The accompanying growth inhibition may be attributed, at least in part, to enzyme inhibition at the fundamental level of energy acquisition.

According to Spencer and Berman (2003), hypoglycin A metabolism results in the inhibition of fatty acid transport, acyl-CoA dehydrogenases and neoglucogenesis. Tanaka (2005) showed that glucose-6-phosphatase activity is also impaired. The resulting hypoglycaemia is a characteristic feature of hypoglycin A toxicity in humans.

A number of general rules may be formulated in the light of the evidence presented above. Urea cycle enzymes appear to be an important focal point for the action of at least three non-protein amino acids. Canavanine enhances and indospicine inhibits arginase activity. In mammals, the effect of canavanine is observed after synthesis of canaline, which then precipitates the adverse effect, but the action of indospicine is more direct due to the competitive inhibition of arginase, which is required for the proper functioning of the urea cycle. The inhibition of ornithine carbamoyltransferase by DABA is also of significance in mammals due to disruption of the urea cycle.

Another common theme relates to the modulation of NO production by canavanine, homoarginine and indospicine, while competition during cellular amino acid transport is a feature of the individual actions of canavanine, homoarginine and the selenoamino acids.

Despite their diverse nature, a number of non-protein amino acids, including canaline, mimosine and BCNA form complexes with pyridoxal phosphate and so are associated with the inhibition of certain aminotransferases and decarboxylases. The association between mimosine toxicity and pyridoxal phosphate in mung beans has long been recognized (Smith and Fowden, 1966). These three amino acids also inhibit cystathionase, a feature which accounts for the cystathioninuria observed in rats administered with mimosine or with BCNA.

27.7.2 Stress-resistance mechanisms

Stress resistance is relevant here because it reflects further the diversity of mechanisms available in plants to counteract external pressures. The metabolism of specific non-protein amino acids may be affected by or may modulate the response to stress and may well be integrated into a general model for regulation. Food quality and safety may also be enhanced through an improved understanding of innate defence mechanisms directed at these factors. In addition, the potentiation or 'priming' of these processes by exogenous means is currently under investigation, as discussed below and elsewhere in this volume. It should be noted, however, that other organic molecules may also be affected by or contribute to stress mitigation, for example catecholamines (Świędrych *et al.*, 2004), arginase (Brauc *et al.*, 2012) and vitamin B_1 (Ahn *et al.*, 2005).

Biotic factors of particular importance include fungal infection and insect predation. The biochemical basis of host–pathogen interactions underlying resistance or susceptibility to disease is the subject of much speculation (Staskawicz, 2001). Elucidating the common mechanisms that might involve the non-protein amino acids in responses to pathogens (or insect predators) is likely to be one of the challenging issues in the foreseeable future. Abiotic factors include a variety of environmental stressors such as temperature, water deficit and salinity.

It is widely reported that GABA accumulates rapidly in plants in response to biotic and abiotic stress (Kinnersley and Turano, 2000; Bouche *et al.*, 2003; Roberts, 2007). Specifically, Bouche *et al.* (2003) suggested a role for the GABA shunt in preventing the accumulation of ROS and cell death. The possible existence of GABA receptors in plants that are similar to those in animals has been proposed. Metabolic interactions between GABA and plant hormones have also been identified. Suggested mechanisms might involve a role for GABA in N metabolism, in signalling to engage in host–parasite interactions and as an intracellular mediator. It has long been recognized that GABA accumulation occurs in response to anaerobic stress in radish (*Raphanus sativus*; Streeter and Thompson, 1972). The responses of *Nicotiana* plants to water/salt stress have been attributed to the antioxidant (Liu *et al.*, 2011a) and osmotic and pH regulation functions of GABA (Akçay *et al.*, 2012). Depending upon the nature of the environmental stress, GABA accumulation may occur as a result of pH- or calmodulin-dependent activation of glutamic acid decarboxylase (Kinnersley and Turano, 2000). Furthermore, extracellular GABA may, at least in part, mediate communication during competitive ecological pressure (Shelp *et al.*, 2006). In other studies, evidence has been presented to indicate that exogenous treatment with GABA induces chilling tolerance in peach fruits after long-term cold storage (Shang *et al.*, 2011). What is more, it has been suggested that GABA accumulation may be perceived as a rapidly initiated, local protective mechanism intended as a preliminary reaction to inhibit insect predation (Brown *et al.*, 2006). The role of GABA as a mediator in taste signal transduction has also been advanced (Ueno *et al.*, 2012).

Considerable attention is currently being given to the role and effects, in plants, of the xenobiotic amino acid analogue BABA (Cohen *et al.*, 1994;

Zimmerli *et al.*, 2000; Ton *et al.*, 2005; Flors *et al.*, 2008; Slaughter *et al.*, 2012). Its efficacy in potentiating stress resistance mechanisms in plants is emerging as a prominent feature of recent research. There is substantial evidence, not just in *Arabidopsis* mutants, to suggest that BABA is capable of exerting significant 'priming' activity towards disease organisms. For example, Cohen *et al.* (1994) observed that BABA induced the accumulation of pathogenesis-related proteins in tomato (*Lycopersicon esculentum* [now designated *Solanum lycopersicum*]) plants, thereby conferring resistance to late blight infection caused by *P. infestans*. It was further suggested that the mechanism of action did not involve ethylene. Dubreuil-Maurizi *et al.* (2010) indicated that another mechanism, the priming of NADPH oxidase-dependent ROS, also contributed to BABA-induced resistance in the *Vitis–Plasmopara* pathosystem. According to Cohen (2002), BABA functions in diverse defence mechanisms, including both physical and biochemical modes of action. BABA enhances deposition of callose-rich structures under the appressoria of plant pathogenic fungi (Ton *et al.*, 2005). Specific metabolic effects of BABA include the induction of salicylic acid (SA) synthesis, phytoalexin accumulation and production of pathogenesis-related proteins. The potential role of leucine-rich repeat motifs in specific proteins as part of this mechanism deserves consideration (Staskawicz, 2001). The interactions involving JA, SA and abscisic acid (ABA) signalling during basal and induced resistance towards *P. syringae* and *A. brassicicola* have been examined by Flors *et al.* (2008). The results suggest that BABA prevents pathogen-induced inhibition of ABA accumulation, sensitizes host tissues to ABA and potentiates the deposition of callose. However, a recurrent theme concerns the indispensability of a functional SA pathway for BABA-induced systemic resistance against *P. infestans* in potato, whereas oxylipins are irrelevant in this respect (Eschen-Lippold *et al.*, 2010).

Other factors and potential mechanisms should also be considered. For example, foliar applications of spermidine and arsenic appear to be as effective as tobacco mosaic virus (TMV) and BABA in stimulating defence mechanisms to subsequent challenge with TMV (Lazzarato *et al.*, 2009). In addition, glutamine suppresses BABA-induced priming and stress resistance in *Arabidopsis* (Wu *et al.*, 2010).

In a significant development, Slaughter *et al.* (2012) demonstrated that the progeny of *Arabidopsis* plants primed with BABA exhibited greater resistance to pathogenic infection than the descendants of untreated plants. This effect occurred without additional treatment with BABA, with the descendants of primed plants showing a faster and higher accumulation of the transcripts of defence-associated genes in the SA signalling system, accompanied by enhanced disease resistance, following challenge with a virulent isolate of *P. syringae*. The disease resistance in the progeny of primed plants was increased even further by additional treatment with BABA. Comparisons with the immune responses in animals are unavoidable. Following initial challenge with a novel antigen, there is usually a lag phase after which time antibodies appear in concentrations that increase with time up to a maximal point before receding. Presentation of the antigen on a subsequent occasion elicits a secondary response with a reduced lag phase, followed by a significantly enhanced rate of antibody synthesis compared with that in the primary response. Moreover, the duration of antibody production is more persistent in the secondary phase (D'Mello, 1991b).

Due to its synergistic interaction with fungicides, BABA may well constitute the basis of plant disease management systems under field conditions, although any implementation of this approach would be subject to satisfactory environmental and toxicological evaluation (Cohen, 2002). It should be noted that repeated treatment with high doses of BABA induces female sterility in *Arabidopsis* (Jakab *et al.*, 2001). The observation that BABA is an agonist at the glycine receptor is also of relevance here and warrants further investigation in any assessment of product safety (Schmieden and Betz, 1995). Nevertheless, the role of BABA in plant protection continues along distinct lines outlined above in a scenario that is still evolving. Thus, the application of BABA as a root drench to legumes inhibits the growth and fecundity of the pea aphid, indicating the potential for invertebrate control in plant protection (Hodge *et al.*, 2005).

Of particular significance has been the concept that BABA may enhance the response of plants to environmental stress. As such, there is accumulating recognition of the role of BABA in a general role of potentiating the stress resistance of plants. For example, Zimmerli *et al.* (2008) noted that BABA increased thermotolerance in *Arabidopsis*, implicating cross-talk with ABA signalling cascades as part of the mechanism. In addition, proteomic analysis indicated large-scale changes in primary metabolism in response to BABA application and on exposure of *Arabidopsis* to simulated acid rain. Liu *et al.* (2011b) identified 175 proteins responding to BABA in the absence and presence of simulated acid rain. It appears that such changes are complementary to the activation of antioxidant systems and SA, JA and ABA signalling mechanisms.

27.8 Detoxification

The diverse mechanisms underlying the toxicity of non-protein amino acids imply limited scope for detoxification by affected organisms. Complete mitigation of adverse effects is only possible if this complexity of mechanisms is matched by an equally versatile metabolic defence system in organisms exposed to non-protein amino acids.

In *Arabidopsis*, the ubiquitin-specific proteases are considered to be an important family of hydrolases that degrade short-lived and abnormal proteins synthesized during the misincorporation of amino acid analogues. Yan *et al.* (2000) demonstrated that the *AtUBP 1* and *2* subfamilies of these proteases confer resistance to canavanine and are required for the abnormal protein turnover in *Arabidopsis*. It is relevant to recall that the synthesis of aberrant proteins represents a metabolic burden that persists throughout the remaining life cycle of *M. sexta* larvae treated with canavanine (Rosenthal, 2001).

The tropical bruchid beetle (*Caryedes brasiliensis*) embodies the principle of metabolic versatility in its detoxification of canavanine. In addition to possessing enzymes capable of distinguishing between canavanine and arginine during protein synthesis, this insect has abandoned the normal uricotelic mode of nitrogen elimination used by other invertebrates in favour of ammoniotelism and ureotelism. Uricotelism inevitably confers dependence upon dietary arginine and, consequently, increased sensitivity to its structural analogues (Rosenthal, 1982). It follows that mammals also possess a significant capacity to detoxify canavanine by virtue of their ureotelism, whereas avian species, being uricotelic, succumb readily to relatively low dietary concentrations of canavanine. However, considerable mitigation of the adverse effects of canavanine in young chicks may be achieved by dietary additions of arginine or creatine, while lysine exacerbates the toxicity (D'Mello *et al.*, 1989; D'Mello, 1991a).

Rumen microbial fermentation precipitates SMCO toxicity in sheep through the synthesis of a highly reactive metabolite, but manipulation of rumen function by dietary means may be used to modulate SMCO metabolism. For example, administration of oral supplements of methionine to lambs grazing kale enhanced the rumen pool and plasma concentrations of SMCO, implying reduced ruminal degradation of SMCO. Accompanying these changes was the development of a slightly less severe manifestation of haemolytic anaemia in methionine-supplemented lambs (see D'Mello, 1991a).

In contrast to this approach, rumen microbes display the requisite features of biochemical versatility in the complete detoxification of mimosine, as outlined earlier in this chapter, though this capacity is subject to geographical differences in rumen ecology. In many parts of the tropics, cattle and goats harbour microorganisms with the ability to degrade mimosine to innocuous residues, whereas in Australia only partial breakdown to DHP is naturally observed in ruminants. None the less, recent biotechnological advances have enabled the transfer of DHP-degrading bacteria from animals in Hawaii to Australian cattle with complete success. Inoculated cattle grazing *Leucaena* showed markedly higher weight gains and serum thyroxine concentrations than untreated controls (Jones and Megarrity, 1986; Quirk *et al.*, 1988). Dosed cattle rapidly reduced DHP excretion in their urine, despite a doubling of *Leucaena* intake. The isolation of active DHP-degrading bacteria from the faeces of inoculated cattle confirmed that treatment of just a few animals was sufficient to avoid *Leucaena* toxicity in the entire herd (Quirk *et al.*, 1988). Notwithstanding these initial results in controlled trials, Dalzell *et al.* (2012) maintain that subclinical DHP toxicity remains a problem under commercial farm conditions in Queensland. Nevertheless, geographical variations in *I. spicata* toxicity have also been observed, implying that any differences in rumen ecology may be exploited in the detoxification of its constituent non-protein amino acid, indospicine (see D'Mello, 1991a).

Considerable mitigation of adverse effects is possible by controlled feeding of forages containing toxic non-protein amino acids. For example, current advice for the prevention of SMCO toxicity in sheep involves some control of *Brassica* intake, particularly in late winter when SMCO concentrations in the forage are maximal. In addition, further control of SMCO intake may be achieved by feeding *Brassica* forage grown on low-sulfur soils. Such crops contain markedly lower SMCO concentrations than those grown on soils with a normal sulfur status (reviewed by D'Mello, 1991a). Similarly, when sheep are conditioned to *Leucaena* by gradually increasing intake of the legume, mimosine toxicity may be averted.

Because toxic non-protein amino acids occur in unconjugated forms, their removal by extraction of plant products with aqueous solutions has been attempted (D'Mello and Walker, 1991). As an example, if whole seeds of *C. ensiformis* are extracted with dilute $KHCO_3$ solution at 80 °C, concentrations of canavanine decline to negligible levels within 48 h. Following heat treatment, the extracted seeds may be incorporated into broiler chick diets at concentrations of up to 280g kg^{-1} without deleterious effects. The relative ease with which canavanine is extracted from intact beans is striking and may constitute the basis for the detoxification of other crop plants containing non-protein amino acids. Cost-benefit analysis would be the final determinant of the efficacy of any extraction procedure.

The use of plant breeding programmes to reduce or eliminate the toxicity of products containing non-protein amino acids may be justified. It should be recognized though that these very toxic amino acids may confer resistance to insects (Ishaaya *et al.*, 1991; Rosenthal, 2001) and fungal pathogens, and removal of these compounds might result in agronomic disadvantages. In this respect, it is salutary to recall that infestation by psyllid insects has all but decimated the *Leucaena* crop in some countries (Beardsley 1986) and any attempt to develop low-mimosine lines are likely to exacerbate this problem. In addition, the exact identity of the principal toxic agent of plants must be ascertained if plant breeding methods are to succeed. Thus, *Leucaena* and *Lathyrus* species contain at least two toxic amino acids (Table 27.2) and, furthermore, *Leucaena* contains other anti-nutrients that may also contribute to its toxicity (D'Mello, 1991a).

27.9 Potential Applications

Despite the overall theme of negativity in this chapter, it is important to emphasize the potential benefits that may emerge in future research and development programmes. This is arguably a major driving force in ensuring that investigations with non-protein amino acids continue. The potential role of BABA in enhancing plant defence mechanisms towards biotic and environmental stress has been reviewed above. There are also other examples of the positive attributes of non-protein amino acids that deserve attention.

Mimosine displays strong phytotoxicity, a feature which may be exploited in the development of bio-herbicides (Xuan *et al.*, 2006). Mimosine also mitigates oxidative stress in selenium-deficient seedlings of *Vigna radiata* (Lalitha and Kulothungan, 2007a,b). Both mimosine and canavanine are now increasingly perceived as potential insecticides in their own right or as models for the development of more effective compounds (Ishaaya *et al.*, 1991; Rosenthal, 2001). However, earlier observations with derivatives of canavanine indicated limited potential for the D-isomer or the 1-methyl or 1-ethyl esters (Rosenthal *et al.*, 1995). Synergistic effects of canavanine and the microbial insecticide Bt have been reported in tests with *M. sexta*. It is postulated that canavanine enhanced the effect of Bt on gut permeability and active transport (Felton and Dahlman, 1984).

There is increasing evidence that non-protein amino acids may be used for clinical benefit in humans. The potential scope for the use of specific non-protein amino acids in cancer therapy appears to be more promising than for other conditions. Canavanine, *Se*-methylselenocysteine, *Se*-allylselenocysteine, selenomethionine, SMCO, mimosine and selected congeners have been tested in animal and *in vitro* models with varying degrees of efficacy. For example, NaPhuket *et al.* (1998) reported that several ester derivatives of canavanine were markedly more effective than the parent compound in suppressing the growth of cultured pancreatic carcinoma cells, whereas analogues based on modification of the carbon chain length or of the terminal functional group were less successful. More substantive evidence is available for the selenoamino acids. Thus, Whanger (2002) concluded that *Se*-methylselenocysteine was most effective against mammary tumours, whereas Rustum *et al.* (2004) reported that this amino acid was highly effective in potentiating the efficiency of anti-cancer drugs and in protecting against drug-induced toxicity. Indeed, Azrak *et al.* (2007) suggested that the efficacy of methylselenocysteine in increasing the therapeutic index of the anti-cancer drug irinotecan was dependent upon the dose of the amino acid. While there is good evidence of the therapeutic potential of methylselenocysteine (see also El-Bayoumy and Sinha, 2004; and Rayman, 2005), it should be recalled that in the studies of Ip *et al.* (1999), *Se*-allylselenocysteine was more effective than a number of other selenoamino acids in the chemoprevention of mammary cancer in a rat methylnitrosourea model. Limited work points to the potential therapeutic use of selenomethionine (Kumar *et al.*, 2005), but Wu *et al.* (2009) recommended a cautious approach to minimize cytotoxic effects of the amino acid as a supplement. Marks *et al.* (1993) examined the efficacy of SMCO and its metabolite, methyl methane thiosulfinate, on mouse genotoxicity and concluded that these two organosulfur compounds may contribute to the anti-carcinogenic properties of *Brassica* vegetables. However, epidemiological evidence only supports the case for prevention of gastric and lung cancers through consumption of these vegetables (Kim and Park, 2009). Finally, interest in mimosine as a potential anti-cancer agent is typified by three investigations. It is well known that mimosine blocks cell cycle progression, and the report of Kulp and Vulliet (1996) with asynchronous human breast cancer cells is consistent with this particular property. The work of Dong and Zhang (2003) provides data illustrating part of the mechanism for this effect. Mimosine attenuates the transcription of serine hydroxymethyltransferase by chelating Zn, and Perry *et al.* (2005) consider the implications of this for the inhibition of DNA replication, a well-recognized effect of this non-protein amino acid (Lin *et al.*, 1996). The third investigation relates to mimosine-induced apoptosis. A molecular mechanism has been proposed involving oxidative stress and mitochondrial activation, with both factors exerting functional roles in the induction of cell death (Hallak *et al.*, 2008).

Other pharmaceutical applications may emerge with increasing interest in the incorporation of an expanded set of amino acid analogues for the synthesis of unnatural peptides (Hartman *et al.*, 2007). Connor and Tirrell (2007) suggested that, for the protein chemist, the 20 canonical amino acids are much too restrictive for macromolecular design. Effective technologies are now available to insert artificial (non-canonical) amino acids into polymers to enable the incorporation of a wide range of functional groups for novel applications. The unique features of amino acid analogues may well be of relevance to this effort.

27.10 Conclusions

A wide range of potentially toxic non-protein amino acids occur naturally in higher plants, primarily but not exclusively as structural or functional analogues of the canonical amino acids. The seed is generally the most concentrated source of non-protein amino acids. As a general rule, legumes contain a more diverse array and higher concentrations of these secondary compounds than other plant species. An emerging issue concerns the co-occurrence of non-protein amino acids in different tropical legumes. For example, mimosine and djenkolic acid occur in certain species of *Leucaena*, while canavanine and indospicine are present in *I. spicata*. Furthermore, in some *Lathyrus* species, homoarginine may occur with BOAA. So it is now apparent that the distribution of toxic non-protein amino acids is not restricted to exotic plants. Arguably, the most striking structural analogues are those of the sulfur amino acids. When selenium replaces the sulfur atom of these amino acids in certain *Brassica* species, a number of congeners are formed. Those of primary interest are selenomethionine, *Se*-methylselenocysteine and selenocystine. Other analogues of the sulfur-containing amino acids include SMCO, also found in *Brassica* plants.

It is universally accepted that non-protein amino acids are associated with adverse effects in a wide range of living organisms. This is entirely predictable because, in virtually all instances, these compounds bear a structural analogy to the canonical amino acids or their immediate derivatives. Further, the initial degradation products of certain non-protein amino acids may also exert deleterious effects towards key intermediates through potent structural antagonisms. At the cellular level, alteration in enzyme activity is invariably the first indicator of the anti-metabolic effects of non-protein amino acids, culminating in gross effects of complex aetiology. Non-Protein amino acids are undoubtedly toxic to plants, precipitating both direct and indirect effects. Thus, canavanine inhibits the growth and development of plants, Its presence in root exudates may also be detrimental to N-fixing bacteria through interference with quorum sensing, nodulation and the synthesis of the symbiotically important exopolysaccharide II. Inhibitory effects of canavanine and its metabolites in the urea cycle are quantitatively reversed by the corresponding amino acids in the ornithine-urea pathway.

It is consistently stated that the selenoamino acids are toxic to plants that are deemed 'non-accumulators' with reference to the Se component. Selenoamino acids interfere with cellular biochemical reactions and methionine status correlates negatively with increases in tissue selenomethionine concentrations. The overall effect is that sensitive plants are unable to thrive in soils burdened with Se.

There is convincing experimental evidence to demonstrate that mimosine is toxic to a variety of higher plants, inhibiting seed germination, seedling growth and key enzymes. The enzymes so affected include nitrate reductase, PAL, peroxidase and its isoenzymes, catalase and IAA oxidase. What is more, in comparison with another non-protein amino acid (albizzine), mimosine is considerably more phytotoxic, specifically with respect to the inhibition of chlorophyll development. In root exudates, mimosine may also operate indirectly by suppressing the growth and metabolism of symbiotic bacteria in legumes.

In contrast, the novel xenobiotic non-protein amino acid BABA enhances protective mechanisms in plants, inducing resistance to pathogens and environmental stress. Similarly, the accumulation of GABA in plants is considered to be part of a rapid-response local resistance system directed against insect herbivory and other stress factors.

Potent insecticidal activity of non-protein amino acids has been reported over several decades, particularly with respect to canavanine and mimosine, but the adverse effects of non-protein amino acids are observed most consistently and decisively in higher animals, including humans. The manifestations of their deleterious effects range from loss of appetite and reduced nutrient utilization to profound neurological disorders and even death.

The adverse effects of non-protein amino acids are mediated via diverse mechanisms that incorporate the principles of allelochemical activity and metabolic mimicry. Thus, competition with the canonical amino acids for transport, signalling and protein synthesis represents the primary mode of action, together with the inhibition of key enzymes. This diversity of mechanisms is similar to the complex mode of action exhibited by BABA.

Despite the overall theme of negativity in this chapter it is important not to overlook a number of potential applications as novel biopesticides and pharmaceuticals. Canavanine and mimosine have emerged as prime candidates for herbicidal and insecticidal preparations. Furthermore, canavanine may enhance the efficacy of the microbial insecticide Bt. In addition, BABA acts synergistically with fungicides to stimulate activity. Moreover, the potential scope for non-protein amino acids in cancer therapy appears to be reasonably promising. Varying degrees of efficacy have been reported with canavanine, *Se*-methylselenocysteine, *Se*-allylselenocysteine, selenomethionine, SMCO and mimosine, and selected congeners and metabolites, in studies involving animal and *in vitro* models. Of these amino acids, *Se*-methylselenocysteine was most effective against mammary tumours, with the capacity of potentiating the efficacy of anti-cancer agents and protecting against drug-induced toxicity.

Adverse effects in animals may be significantly reduced through the controlled feeding of forages containing toxic non-protein amino acids or by the provision of appropriate nutritional supplements. However, substantial mitigation under commercial conditions has been achieved by manipulation of rumen function to metabolize mimosine to innocuous residues.

The evidence reviewed above supports an emerging consensus that non-protein amino acids may either serve directly or as parent molecules for the development of bio-pesticides and pharmaceuticals with unique potentiating properties. However, food safety, medicinal efficacy and ecological implications should be considered before the commercial exploitation of any novel product.

References

Adio, A.M., Casteel, C.L., Kim, J.H. and Jander, G. (2011) Biosynthesis and defensive function of *N*δ-acetylornithine, a jasmonate-induced *Arabidopsis* metabolite. *The Plant Cell* 23, 3303–3318.

Ahn, I-P., Kim, S. and Y.-H. Lee (2005) Vitamin B_1 functions as an activator of plant disease resistance. *Plant Physiology* 138, 1505–1515.

Akçay, N., Bor, M., Karabudak, T., Özdemir, F. and Türkan, I. (2012) Contribution of gamma-aminobutyric acid (GABA) to salt stress responses of *Nicotiana sylvestris* CMS II mutant and wild type plants. *Journal of Plant Physiology* 169, 452–458.

Ali, S.F., Rawat, L.S., Meghvansi, M.K. and Mahna, S.K. (2010) Screening of rhizobial isolates of *Leucaena leucocephala* Lam. for mimosine degradation ability. *Journal of Phytology* 2, 1–6.

Almagro, L., Gómez Ros, L.V., Belchi-Navarro, S., Bru, R., Ros Barceló and Pedreño, M.A. (2009) Class III peroxidases in plant defence reactions. *Journal of Experimental Botany*, 60, 377–390.

Andersson, H., Lindqvist, E. and Olson, L. (1997) Plant-derived amino acids increase hippocampus BDNF, NGF, *c-fos* and hsp 70 mRNAs. *Neuroreport* 8, 1813–1817.

Andrade, A.B., Ferrarese, M.L.L., Teixeira, A.F. and Ferraresse-Filho, O. (2008) Mimosine inhibits soybean (*Glycine max*) root growth, lignification and related enzymes. *Allelopathy Journal* 21, 145–154.

Austic, R.E. and Nesheim, M.C. (1972) Arginine and creatine interrelationships in the chick. *Poultry Science* 51, 1098–1105.

Awaya, J.D., Fox, P.M. and Borthakur, D. (2005) *pyd* Genes of *Rhizobium* sp. strain TAL 1145 are required for degradation of 3-hydroxy-4-pyridone, an aromatic intermediate of mimosine metabolism. *Journal of Bacteriology* 187, 4480–4487.

Awaya, J.D., Tittabutr, P., Li, Q.X. and Borthakur, D. (2008) Pyruvate carboxylase is involved in metabolism of mimosine by *Rhizobium* sp. strain TAL 1145. *Archives of Microbiology* 190, 409–415.

Azrak, R.G., Cao, S., Durrani, F.A. and Rustum, Y.M. (2007) Efficacy of increasing the therapeutic index of irinotecan, plasma and tissue selenium concentrations is methylselenocysteine dose dependent. *Biochemical Pharmacology* 73, 1280–1287.

Beardsley, J.W. (1986) Psyllidae or jumping plant lice: notes on biology and control. *Leucaena Research Reports* 7, 2–5.

Bence, A.K., Worthen, D.R., Adams, V.R. and Crooks, P.A. (2002) The antiproliferative and immunotoxic effects of L-canavanine and L-canaline *in vitro*. *Anticancer Drugs* 13, 313–320.

Blake, O.A., Bennink, M.R. and Jackson, J.C. (2006) Ackee (*Blighia sapida*) hypoglycin A toxicity: dose response assessment in laboratory rats. *Food and Chemical Toxicology* 44, 207–213.

Borthakur, D., Soedarjo, M., Fox, P.M. and Webb, D.T. (2003) The *mid* genes of *Rhizobium* sp. strain TAL 1145 are required for the degradation of mimosine into 3-hydroxy-4-pyridone and are inducible by mimosine. *Microbiology* 149, 537–546.

Bouche, N., Fait, A., Bouchez, D., Møller, S.G. and Fromm, H. (2003) Mitochondrial succinic-semialdehyde dehydrogenase of the γ-aminobutyrate shunt is required to restrict levels of reactive oxygen intermediates in plants. *Proceedings of the National Academy of Sciences of the United States of America* 100, 6843–6848.

Brauc, S., De Vooght, E., Claeys, M., Höfte, M. and Angenon, G. (2012) Over-expression of arginase in *Arabidopsis thaliana* influences defence responses against *Botrytis cinerea*. *Plant Biology (Stuttgart, Germany)* 1, 39–45.

Brenner, E.D., Clark, A.P., Liang, Q.S., Stevenson, D.W. and Coruzzi, G.M. (2000) *Arabidopsis* mutants resistant to *S*(+)-β-methyl-α,β-diaminopropionic acid, a cycad-derived glutamate receptor agonist. *Plant Physiology* 124, 1615–1624.

Brown, A.W., Macgregor, K.B. and Shelp, B.J. (2006) Gamma-aminobutyrate: defense against invertebrate pests? *Trends in Plant Science* 11, 424–427.

Brown, T.A. and Shrift, A. (1982) Selenium toxicity and tolerance in higher plants. *Biological Reviews* 57, 59–84.
Brownson, D.M., Mabry, T.J. and Leslie, S.W. (2002) The cycad neurotoxic amino acid, β-*N*-methylamino-L-alanine (BMAA), elevates intracellular calcium levels in dissociated rat brain cells. *Journal of Ethnopharmacology* 82, 159–167.
Caldwell, R.W., Chandra, S. and Caldwell, R.B. (2012) Arginase. In: D'Mello, J.P.F. (ed.) *Amino Acids in Human Nutrition and Health*. CAB International, Wallingford, UK, pp.51–71.
Camm, E.L. and Towers, G.H.N. (1973) Phenylalanine ammonia lyase. *Phytochemistry* 12, 961–973.
Chen, H., McCaig, B.C., Melotto, H., He, S.Y. and Howe, G.A. (2004) Regulation of plant arginase by wounding, jasmonate and the phytotoxin coronatine. *The Journal of Biological Chemistry* 279, 45998–46007.
Cohen, Y.R. (2002) β-Aminobutyric acid-induced resistance against plant pathogens. *Plant Disease* 86, 448–457.
Cohen, Y.R., Niderman, T., Mosinger, E. and Fluhl, R. (1994) β-Aminobutyric acid induces the accumulation of pathogenesis-related proteins in tomato (*Lycopersicon esculentum* L.) plants and resistance to late blight infection caused by *Phytophthora infestans*. *Plant Physiology* 104, 59–66.
Connor, R.E. and Tirrell, D.A. (2007) Non-canonical amino acids in protein polymer design. *Polymer Reviews* 47, 9–28.
Cowman, R.A., Baron, S.S. and Fitzgerald, R.J. (1983) Cysteine toxicity for oral streptococci and effect of branched-chain amino acids. *Infection and Immunity* 39, 1107–1115.
Dalzell, S.A., Burnett, D.J., Dowsett, J.E., Forbes, V.E. and Shelton, H.M. (2012) Prevalence of mimosine and DHP toxicity in cattle grazing *Leucaena leucocephala* pastures in Queensland, Australia. *Animal Production Science* 52, 365–372.
Davis, D.G. (1997) Polyamines, auxins and organogenesis in leafy spurge (*Euphorbia esula* L.). *Journal of Plant Physiology* 151, 603–609.
Degnan, B.A., Palmer, J.M., Robson, T., Jones, C.E.D., Mellor, G.D., Diamond, A.G., Kehoe, M.A. and Goodacre, J.A. (1998) Inhibition of human peripheral blood mononuclear cell proliferation by *Streptococcus pyogenes* cell extract is associated with arginine deaminase activity. *Infection and Immunity* 66, 3050–3058.
D'Mello, J.P.F. (1991a) Toxic amino acids. In: D'Mello, J.P.F., Duffus, C.M. and Duffus, J.H. (eds) *Toxic Substances in Crop Plants*. The Royal Society of Chemistry, Cambridge, UK, pp. 21–48.
D'Mello, J.P.F. (1991b) Antigenic proteins. In: D'Mello, J.P.F., Duffus, C.M. and Duffus, J.H. (eds) *Toxic Substances in Crop Plants*. The Royal Society of Chemistry, Cambridge, UK, pp. 107–125.
D'Mello, J.P.F. (1994) Amino acid imbalances, antagonisms and toxicities. In: D'Mello, J.P.F. (ed.) *Amino Acids in Farm Animal Nutrition*. CAB International, Wallingford, UK, pp. 63–97.
D'Mello, J.P.F. (1995) Anti-nutritional substances in legume seeds. In: D'Mello, J.P.F. (ed.) *Tropical Legumes in Animal Nutrition*. CAB International, Wallingford, UK, pp. 135–172.
D'Mello, J.P.F. (2003) Adverse effects of amino acids. In: D'Mello, J.P.F. (ed.) *Amino Acids in Animal Nutrition*. CAB International, Wallingford, UK, pp. 125–142.
D'Mello, J.P.F. (2012) Adverse effects. In: D'Mello, J.P.F. (ed.) *Amino Acids in Human Nutrition and Health*. CAB International, Wallingford, UK, pp. 322–352.
D'Mello, J.P.F. and Lewis, D. (1970a) Amino acid interactions in chick nutrition. 1. The interrelationship between lysine and arginine. *British Poultry Science* 11, 299–311.
D'Mello, J.P.F. and Lewis, D. (1970b) Amino acid interactions in chick nutrition. 3. Interdependence in amino acid requirements. *British Poultry Science* 11, 367–385.
D'Mello, J.P.F. and Walker, A.G. (1991) Detoxification of jack beans (*Canavalia ensiformis*): studies with young chicks. *Animal Feed Science and Technology* 33, 117–127.
D'Mello, J.P.F., Acamovic, T. and Walker, A.G. (1989) Nutritive value of jack beans (*Canavalia ensiformis*) (L.) (DC) for young chicks: effects of amino acid supplementation. *Tropical Agriculture (Trinidad)* 66, 201–205.
D'Mello, J.P.F., Walker, A.G. and Noble, E. (1990) Effects of dietary supplements on the nutritive value of jack beans (*Canavalia ensiformis*) for the young chick. *British Poultry Science* 31, 759–768.
Dong, Z. and Zhang, J-T. (2003) E1F3p170, a mediator of the mimosine effect on protein synthesis and cell cycle progression. *Molecular Biology of the Cell* 14, 3942–3951.
Downing, S., van de Venter, M. and Downing, T.G. (2012) The effect of exogenous β-*N*-methylamino-L-alanine on the growth of *Synechocystis* PCC 6803. *Microbial Ecology* 63, 149–156.
Dubreuil-Maurizi, C., Trouvelot, S., Pugin, A., Wendehenne, D. and Poinssot, B. (2010) β-Aminobutyric acid primes an NADPH oxidase-dependent reactive oxygen species production during grapevine-triggered immunity. *Molecular Plant–Microbe Interactions* 23, 1012–1021.

Duncan, M.W., Steele, J.C., Kopin, I.J. and Markey, S.P. (1990) 2-Amino-3-(methylamino)-propanoic acid (BMAA) in cycad flour. An unlikely cause of amyotrophic lateral sclerosis and Parkinsonism-dementia of Guam. *Neurology* 40, 767.

Eccleston, M. and Kelly, D.P. (1972) Assimilation and toxicity of exogenous amino acids in the methane-oxidizing bacterium *Methylococcus capsulatus*. *Microbiology* 71, 541–554.

El-Bayoumy, K. and Sinha, R. (2004) Mechanisms of mammary cancer chemoprevention by organoselenium compounds. *Mutation Research* 551, 181–197.

Elliot, R., Milton, J.T.B., Gordon, G.H. and Blamey, F.P.C. (1984) Effect of L-mimosine in artificial diet on the growth and development of *Heliothis punctiger* larvae. *The Journal of Agricultural Science* 103, 477–479.

Emmert, E.A.B., Milner, J.L., Lee, J.C. and Handelsman, J. (1998) Effect of canavanine from alfalfa seeds on the population biology of *Bacillus cereus*. *Applied and Environmental Microbiology* 64, 4683–4688.

Eschen-Lippold, L., Altmann, S. and Rosahl, S. (2010) DL-β-aminobutyric acid-induced resistance of potato against *Phytophthora infestans* requires salicylic acid but not oxylipins. *Molecular Plant–Microbe Interactions* 23, 585–592.

Faanes, R. and Rogers, P. (1972) Repression of enzymes of arginine biosynthesis by L-canavanine in arginyl-transfer ribonucleic acid synthetase mutants of *Escherichia coli*. *Journal of Bacteriology* 112, 102–113.

Felton, G.W. and Dahlman, D.L. (1984) Allelochemical-induced stress: effects of L-canavanine on the pathogenicity of *Bacillus thuringiensis* in *Manduca sexta*. *Journal of Invertebrate Pathology* 44, 187–191.

Flores, T., Todd, C.D., Dhanoa, P.K., Mullen, R.T., Lamattina, L. and Polacco, J.C. (2008) Arginase-negative mutants of *Arabidopsis* exhibit increased nitric oxide signalling in root development. *Plant Physiology* 147, 1936–1946.

Flors, V., Ton, J., Van Doorn, R., Jakab, G., García-Agustin, A. and Mauch-Mani, B. (2008) Interplay between JA, SA and ABA signalling during basal and induced resistance against *Pseudomonas syringae* and *Alternaria brassicicola*. *The Plant Journal* 54, 81–92.

Forster, L.A., Fontenot, J.P., Perry, H.D., Foster, J.G. and Allen, V.G. (1991) Apparent digestibility and nutrient balance in lambs fed different levels of flatpea hay. *Journal of Animal Science* 69, 1719–1725.

Goldraij, A. and Polacco, J.C. (1999) Arginase is inoperative in developing soybean embryos. *Plant Physiology* 119, 297–304.

Gravot, A., Debu, C., Wagner, G., Lariagon, C., Lugan, R., Todd, C.D., Wendehenne, D., Delourme, R., Bouchereau, A. and Manzanares-Dauleux, M.J. (2012) Arginase induction represses gall development during clubroot infection in *Arabidopsis*. *Plant and Cell Physiology* 53, 901–911.

Hallak, M., Vazana, L., Shpilberg, O., Levy, I., Mazar, J. and Nathan, I. (2008) A molecular mechanism for mimosine-induced apoptosis involving oxidative stress and mitochondrial activation. *Apoptosis* 13, 147–155.

Hartman, M.C.T., Josephson, K., Lin, C.-W. and Szostak, J.W. (2007) An expanded set of amino acid analogues for the ribosomal translation of unnatural peptides. *PLoS ONE* 2(10): e972.

Henningsson, R. and Lundquist, I. (1998) Arginine-induced insulin release is decreased and glucagon increased in parallel with islet NO production. *American Journal of Physiology – Endocrinology and Metabolism* 275, E500–E506.

Hey, C., Boucher, J.-L., Goff, V.-L., Ketterer, G., Wessler, I. and Racke, K. (1997) Inhibition of arginase in rat and rabbit alveolar macrophages by N_{ω}-hydroxy-D,L-indospicine: effects on L-arginine utilization by NO synthase. *British Journal of Pharmacology* 121, 395–400.

Hietaranta, A.J., Nevalainen, T.J., Aho, H.J., Hamalainen, O.M. and Suortamo, S.H. (1990) Pancreatic acinar cell necrosis with intact storage of digestive enzymes in selenomethionine-treated rats. *Virchows Archives of Cell Pathology* 58, 397–403.

Hodge, S., Thompson, G.A. and Powell, G. (2005) Application of DL-beta-aminobutyric acid (BABA) as a root drench to legumes inhibits the growth and reproduction of the pea aphid *Acyrthosiphon pisum* (Hemiptera: Aphididae). *Bulletin of Entomological Research* 95, 449–455.

Hyun, M.W., Yun, Y.H., Kim, J.Y. and Kim, S.H. (2011) Fungal and plant phenylalanine ammonia-lyase. *Mycobiology* 39, 257–265.

Iftikhar, H.M., Gonzalez-Rodriguez, L. and Reigosa, M.J. (2008) Germination and growth response of four plant species to different allelochemicals and herbicides. *Allelopathy Journal* 22, 101–110.

Igloi, G.L. and Schiefermayr, E. (2009) Amino acid discrimination by arginyl-tRNA synthetases as revealed by an examination of natural specificity variants. *The FEBS Journal* 276, 1307–1318.

Ip, C., Zhu, Z., Thompson H.J., Lisk, D. and Ganther, H.E. (1999) Chemoprevention of mammary cancer with *Se*-allylselenocysteine and other selenoamino acids in the rat. *Anticancer Research* 19, 2875–2880.

Ishaaya, I., Hirashima, A., Yablonski, S., Tawata, S. and Eto, M. (1991) Mimosine, a nonprotein amino acid, inhibits growth and enzyme systems in *Tribolium castaneum*. *Pesticide Biochemistry and Physiology* 39, 35–42.
Jakab, G., Cottier, V., Toquin, V., Rigoli, G., Zimmerli, L., Métraux, J.-P. and Mauch-Mani, B. (2001) β-Aminobutyric acid-induced resistance in plants. *European Journal of Plant Pathology* 107, 29–37.
Jones, J.D., Wolters, R. and Burnett, P.C. (1966) Lysine-arginine-electrolyte relationships in the rat. *Journal of Nutrition* 89, 171–188.
Jones, R.J. (1985) *Leucaena* toxicity and the ruminal degradation of mimosine. In: Seawright, A.A., Hegarty, M.P., James, L.F. and Keeler, R.F. (eds) *Plant Toxicology*. Queensland Poisonous Plants Committee, Yeerongpilly, Queensland, Australia, pp. 111–119.
Jones, R.J. and Megarrity, R.G. (1986) Successful transfer of DHP-degrading bacteria from Hawaiian goats to Australian ruminants to overcome the toxicity of *Leucaena*. *Australian Veterinary Journal* 63, 259–262.
Kakoki, M., Kim, H.-S., Arendshorst, W.J. and Mattson, D.L. (2004) L-Arginine uptake affects nitric oxide production and blood flow in the renal medulla. *American Journal of Physiology – Regulators, Integrative and Comparative Physiology* 287, R1478–R1485.
Kakoki, M., Kim, H-S., Edgell, C.J., Maeda, N., Smithies, O. and Mattson, D.L. (2006) Amino acids as modulators of endothelium-derived nitric oxide. *American Journal of Physiology – Renal Physiology* 291, F297–F304.
Keshavan, N.D., Chowdhary, P.K., Haines, D.C. and Gonzales, J.E. (2005) L-Canavanine made by *Medicago sativa* interferes with quorum sensing in *Sinorhizobium meliloti*. *Journal of Bacteriology* 187, 8427–8436.
Kim, M.K. and Park, J.H. (2009) Cruciferous vegetable intake and the risk of human cancer: epidemiological evidence. *Proceedings of the Nutrition Society* 68, 103–110.
Kinnersley, A.M. and Turano, F.J. (2000) Gamma aminobutyric acid (GABA) and plant responses to stress. *Critical Reviews in Plant Sciences* 19, 479–509.
Korpela, T.K., Lorenz, H. and Laakso, S. (1982) A specific arginase-based assay of L-canavanine in leguminous plants. *Journal of Biochemical and Biophysical Methods* 7, 67–70.
Kulp, K.S. and Vulliet, P.R. (1996) Mimosine blocks cell cycle progression by chelating iron in asynchronous human breast cancer cells. *Toxicology and Applied Pharmacology* 139, 356–364.
Kumar, B., Nahreini, P., Hanson, A.J., Prasad, J.E. and Prasad, K.N. (2005) Selenomethionine prevents degeneration induced by over expression of wild-type human α-synuclein during differentiation of neuroblastoma cells. *Journal of the American College of Nutrition* 24, 516–523.
Lalitha, K. and Kulothungan, S.R. (2007a) Mimosine mitigates oxidative stress in selenium-deficient seedlings of *Vigna radiata*. Part I: Restoration of mitochondrial function. *Biological Trace Element Research* 118, 84–96.
Lalitha, K. and Kulothungan, S.R. (2007b) Mimosine mitigates oxidative stress in selenium-deficient seedlings of *Vigna radiata*. Part II: Mitochondrial uptake of ^{75}Se and mimosine. *Biological Trace Element Research* 118, 269–286.
Lazzarato, L., Trebbi, G., Pagnucco, C., Franchin, C., Torrigiani, P. and Betti, L. (2009) Exogenous spermidine, arsenic and β-aminobutyric acid modulate tobacco resistance to tobacco mosaic virus and affect local and systemic glucosylsalicylic acid levels and arginine decarboxylase gene expression in tobacco leaves. *Journal of Plant Physiology* 166, 90–100.
Lee, Y., Kang, M.J., Shim, J., Cheong, C.U., Moon, S.J. and Montell, C. (2012) Gustatory receptors required for avoiding the insecticide L-canavanine. *The Journal of Neuroscience* 32, 1429–1435.
Lesinger, T., Haas, D. and Hegarty, M.P. (1972) Indospicine as an arginine antagonist in *Escherichia coli* and *Pseudomonas aeruginosa*. *Biochimica et Biophysica Acta* 262, 214–219.
Lin, H.-B., Falchetto, R., Mosca, P.J., Hunt, D.F. and Hamlin, J.L. (1996) Mimosine targets serine hydroxymethyltransferase. *The Journal of Biological Chemistry* 271, 2548–2556.
Liu, C., Zhao, L. and Yu, G. (2011a) The dominant glutamic acid metabolic flux to produce γ-aminobutyric acid over proline in *Nicotiana tabacum* leaves under water stress relates to its significant role in antioxidant activity. *Journal of Integrative Plant Biology* 53, 608–618.
Liu, T., Jiang, X., Shi, W., Chen, J., Pei, Z. and Zheng, H. (2011b) Comparative proteomic analysis of differentially expressed proteins in β-aminobutyric acid-enhanced *Arabidopsis thaliana* tolerance to simulated acid rain. *Proteomics* 11, 2079–2094.
Lois, R., Dietrich, A., Hahlbrock, K. and Schulz, W. (1989) A phenylalanine ammonia-lyase gene from parsley: structure, regulation and identification of elicitor and light-responsive *cis*-acting elements. *The EMBO Journal* 8, 1641–1648.

Marks, H.S., Anderson, J.A. and Stoewsand, G.S. (1993) Effect of *S*-methyl cysteine sulphoxide and its metabolite methyl methane thiosulphinate, both occurring naturally in brassica vegetables, on mouse genotoxicity. *Food and Chemical Toxicology* 31, 491–495.
Mauch-Mani, B. and Slusarenko, A.J. (1996) Production of salicylic acid precursors is a major function of phenylalanine ammonia-lyase in the resistance of *Arabidopsis* to *Peronospora parasitica*. *The Plant Cell* 8, 203–212.
Melangeli, C., Rosenthal, G.A. and Dahlman, D.L. (1997) The biochemical basis for L-canavanine tolerance by the tobacco budworm *Heliothis virescens* (Noctuidae). *Proceedings of the National Academy of Sciences of the United States of America* 94, 2255–2260.
Mhamdi, A., Queval, G., Chaouch, S., Vanderauwera, S. and Noctor, G. (2010) Catalase function in plants: a focus on *Arabidopsis* mutants as stress-mimic models. *Journal of Experimental Botany* 61, 4197–4220.
Mitri, C., Soustelle, L., Framery, B., Parmentier, M.L. and Grau, Y. (2009) Plant insecticide L-canavanine repels *Drosophila* via the insect orphan GPCR DmX. *PLoS Biology*, 7(6): e1000147.
Murch, S.J., Cox, P.A., Banack, S.A., Steele, J.C. and Sacks, O.W. (2004) Occurrence of β-*N*-methylamino-L-alanine (BMAA) in ALS/PDC patients from Guam. *Acta Neurologica Scandinavica* 110, 267–269.
Nakajima, N., Hiradate, S. and Fujii, Y. (2001) Plant growth inhibitory activity of L-canavanine and its mode of action. *Journal of Chemical Ecology* 27, 19–31.
Nakamuro, K., Okuno, T. and Hasegawa, T. (2000) Metabolism of selenoamino acids and contribution of selenium methylation to their toxicity. *Journal of Health Science* 46, 418–421.
NaPhuket, S.R., Trifonov, L.S., Crooks, P.A., Rosenthal, G.A., Freeman, J.W. and Strodel, W.E. (1998) Synthesis and structure–activity studies of some antitumor congeners of L-canavanine. *Drug Development Research* 40, 325–332.
Natelson, S. (1985) Canavanine in alfalfa (*Medicago sativa*). *Experientia* 41, 257–259.
O'Kane, R.L., Viña, J.R., Simpson, I., Zaragozá, R., Mokashi, A. and Hawkins, R.A. (2006) Cationic amino acid transport across the blood–brain barrier is mediated exclusively by system y^+. *American Journal of Physiology – Endocrinology and Metabolism* 291, E412–E419.
Pass, M.A., Arab, H., Pollitt, S. and Hegarty, M.P. (1996) Effects of the naturally occurring arginine analogues indospicine and canavanine on NO mediated functions in aortic endothelium and peritoneal macrophages. *Natural Toxins* 4, 135–140.
Perry, C., Sastry, R., Nasrallah, I.M. and Stover, P.J. (2005) Mimosine attenuates serine hydroxymethyltransferase transcription by chelating zinc. Implications for inhibition of DNA replication. *The Journal of Biological Chemistry* 280, 396–400.
Pine, M.J. (1978) Comparative physiological effects of incorporated amino acid analogs in *Escherichia coli*. *Antimicrobial Agents and Chemotherapy* 13, 676–685.
Prasad, M.N.V. and Subhashini, P. (1994) Mimosine-inhibited seed germination, seedling growth and enzymes of *Oryza sativa* L. *Journal of Chemical Ecology* 20, 1689–1696.
Quirk, M.F., Bushell, J.J., Jones, R.J., Megarrity, R.G. and Butler, K.L. (1988) Liveweight gains on *Leucaena* and native grass pastures after dosing cattle with rumen bacteria capable of degrading DHP, a ruminal metabolite from *Leucaena*. *The Journal of Agricultural Science* 11, 165–170.
Rao, S.D., Banack, S.A., Cox, P.A. and Weiss, J.H. (2006) BMAA selectively injures motor neurons via AMPA/kainate receptor activation. *Experimental Neurology* 201, 244–252.
Rayman, M.P. (2005) Selenium in cancer prevention: a review of the evidence and mechanism of action. *Proceedings of the Nutrition Society* 64, 527–542.
Renault, H., El Amrani, A., Palanivela, R., Renou, J.P., Preuss, O. and Deleu, C. (2011) GABA accumulation causes cell elongation defects and a decrease in expression of genes encoding secreted and cell wall-related proteins in *Arabidopsis thaliana*. *Plant and Cell Physiology* 52, 894–908.
Riganti, C., Aldieri, E., Bergandi, L., Miraglia, E. and Ghigo, D. (2003) Nitroarginine methyl ester and canavanine lower intracellular reduced glutathione. *Free Radical Biology and Medicine* 35, 1210–1216.
Rios-Santos, F., Alves-Filho, J.C., Spiller, F., Freitas, A., Soares, M.B.P., dos Santos, R.R., Teixeira, M.M. and de Queiroz Cunha, F. (2007) Down-regulation in severe sepsis is mediated by inducible nitric oxide synthase-derived nitric oxide. *American Journal of Respiratory and Critical Care Medicine* 175, 490–497.
Roberts, R.M. (2007) Does GABA act as a signal in plants?: hints from molecular studies. *Plant Signaling and Behaviour* 2, 407–408.
Rosenthal, G.A. (1982) *Plant Nonprotein Amino and Imino Acids*. Academic Press, New York.
Rosenthal, G.A. (2001) L-Canavanine: a higher plant insecticidal allelochemical. *Amino Acids* 21, 319–330.

Rosenthal, G.A., Gulati, D.K. and Sabharwal, P.S. (1976) Additive and synergistic growth-inhibiting properties of the canaline-urea cycle amino acids. *Plant Physiology* 57, 493–496.
Rosenthal, G.A., Dahlman, D.L., Crooks, P.A., Na Phuket, S. and Trifonov, L.S. (1995). Insecticidal properties of some derivatives of L-canavanine. *Journal of Agricultural and Food Chemistry* 43, 2728–2734.
Rustum, Y.M., Cao, S., Durrani, F.A. and Fakih, M. (2004) *Se*-(methyl)selenocysteine (MSC) potentiates the antitumor activity of irinotecan against human xenografts and protects against drug induced toxicity. *Journal of Clinical Oncology* 22, 2068.
Santucci, S., Zsurger, N. and Chabry, J. (2009) β-*N*-methylamino-L-alanine induced *in vivo* retinal cell death. *Journal of Neurochemistry* 109, 819–825.
Schmieden, V. and Betz, H. (1995) Pharmacology of the inhibitory glycine receptor: agonist and antagonist actions of amino acids and piperidine carboxylic acid compounds. *Molecular Pharmacology* 48, 919–927.
Schrauzer, G.N. (2000) Selenomethionine: a review of its nutritional significance, metabolism and toxicity. *Journal of Nutrition* 130, 1653–1656.
Shaik, S., Singh, N. and Nicholas, A. (2011) HPLC and GC analysis of *in vitro*-grown leaves of the cancer bush *Lessertia (Sutherlandia) frutescens* L. reveal higher yields of bioactive compounds. *Plant Cell, Tissue and Organ Culture* 105, 431–438.
Shang, H., Cao, S., Yang, Z., Cai, Y. and Zheng, Y. (2011) Effect of exogenous γ-aminobutyric acid treatment on proline accumulation and chilling injury in peach fruit after long-term cold storage. *Journal of Agricultural and Food Chemistry* 59, 1264–1268.
Shelp, B.J., Brown, A.W. and Faure, D. (2006) Extracellular γ-aminobutyrate mediates communication between plants and other organisms. *Plant Physiology* 142, 1350–1352.
Shelp, B.J., Mullen, R.T. and Waller, J.C. (2012) Compartmentation of GABA metabolism raises intriguing questions. *Trends in Plant Science* 17, 57–59.
Slaughter, A., Daniel, X., Flors, V., Hohn, B. and Mauch-Mani, B. (2012) Descendants of primed *Arabidopsis* plants exhibit resistance to biotic stress. *Plant Physiology* 158, 835–843.
Smith, I.K. and Fowden, L. (1966) A study of mimosine toxicity in plants. *Journal of Experimental Botany* 17, 750–761.
Smith, R.H. (1980) Kale poisoning: the brassica anaemia factor. *Veterinary Record* 107, 12–15.
Soares, A.R., de Lourdes Lucio Ferrarese, M., de Cássia Siqueira-Soares, R., Marchiosi, R., Finger-Teixeira, A. and Ferrarese-Filho, O. (2011) The allelochemical L-DOPA increases melanin production and reduces reactive oxygen species in soybean roots. *Journal of Chemical Ecology* 37, 891–898.
Spencer, P.S. and Berman, F. (2003) Plant toxins and human health. In: D'Mello, J.P.F. (ed.) *Food Safety. Contaminants and Toxins*. CAB International, Wallingford, UK, pp. 1–23.
Staskawicz, B.J. (2001) Genetics of plant–pathogen interactions specifying plant disease resistance. *Plant Physiology* 125, 73–76.
Stigers, D.J., Watts, Z.I., Taylor, M.C., Keillor, J.W., Dixon, N.E. and Easton, C.J. (2011) Incorporation of chlorinated analogues of aliphatic amino acids during cell-free protein synthesis. *Chemical Communications* 47, 1839–1841.
Streeter, J.G. and Thompson, J.F. (1972) Anaerobic accumulation of γ-aminobutyric acid and alanine in radish leaves (*Raphanus sativus* L.). *Plant Physiology* 49, 572–578.
Świędrych, A., Lorenc-Kukuła Skirycz, A. and Szopa, J. (2004) The catecholamine biosynthesis route in potato is affected by stress. *Plant Physiology and Biochemistry* 42, 593–600.
Tanaka, K. (2005) Inhibition of gluconeogenesis by hypoglycin: alternate interpretations. *Hepatology* 7, 1377–1379.
Tashjian, D.H., The, S.J., Sogomonyan, A. and Hung, S.S.O. (2006) Bioaccumulation and chronic toxicity of dietary L-selenomethionine in juvenile white sturgeon (*Acipenser transmontanus*). *Aquatic Toxicology* 79, 401–409.
Thakur, N.S., Verma, K.S. and Rana, R.C. (2011) Effect of tree–crop combination and nitrogen levels in velvet bean (*Mucuna pruriens* L.) grown in agrihorti-silvipasture systems of agroforestry in mid-Himalayan region. *Journal of Non-Timber Forest Products* 18, 303–312.
Ton, J., Jakab, G., Toquin, V., Flors, V. and Mauch-Mani, B. (2005) Dissecting the β-aminobutyric acid-induced priming phenomenon in *Arabidopsis*. *The Plant Cell* 17, 987–999.
Ueno, H., Inoue, Y., Matsukawa, S. and Nakamura, Y. (2012) Glutamate decarboxylase. In: D'Mello, J.P.F. (ed.) *Amino Acids in Human Nutrition and Health*. CAB International, Wallingford, UK, pp. 103–121.
Valtonen, P., Laitinen, T., Heiskanen, N., Vanninen, E. and Heinonen, S. (2008) Serum L-homoarginine concentration is elevated during normal pregnancy and is related to flow-mediated vasodilation. *Circulation Journal* 72, 1879–1884.

Whanger, P.D. (2002) Selenocompounds in plants and animals and their biological significance. *Journal of the American College of Nutrition* 21, 223–232.

Williams, R.D. and Hoagland, R.E. (2007) Phytotoxicity of mimosine and albizzine on seed germination and seedling growth of crops and weeds. *Allelopathy Journal* 19, 423–430.

Wu, C.-C., Singh, P., Chen, M.-C. and Zimmerli, L. (2010) L-Glutamine inhibits beta-aminobutyric acid-induced stress resistance and priming in *Arabidopsis*. *Journal of Experimental Botany* 61, 995–1002.

Wu, J., Lyons, G.H., Graham, R.D. and Fenech, M.F. (2009) The effect of selenium, as selenomethionine, on genome stability and cytotoxicity in human lymphocytes. *Mutagenesis* 24, 225–232.

Wu, L., Guo, X. and Banuelos, G.S. (1997) Accumulation of seleno-amino acids in legume and grass plant species grown in selenium-laden soils. *Environmental Toxicology and Chemistry* 16, 491–497.

Xuan, T.D., Elzaawely, A.A., Deba, F., Fukuta, M. and Tawata, S. (2006) Mimosine in *Leucaena* as a potent bio-herbicide. *Agronomy for Sustainable Development* 26, 89–97.

Yan, N., Doelling, J.H., Flabel, T.G., Durski, A.M. and Vierstra, R.D. (2000) The ubiquitin-specific protease family from *Arabidopsis*. AtUBP 1 and 2 are required for the resistance to the amino acid analog canavanine. *Plant Physiology* 124, 1828–1843.

Yoganathan, S. and Vederas, J.C. (2010) Nonprotein-L- amino acids. In: Vederas, J.C. (ed.) *Comprehensive Natural Products II: Chemistry and Biology, Volume 5: Amino Acids, Peptides and Proteins*. Elsevier, Kidlington, UK, pp. 5–69.

Zimmerli, L., Jakab, G., Metraux, J.-P and Mauch-Mani, B. (2000) Potentiation of pathogen-specific defense mechanisms in *Arabidopsis* by β-aminobutyric acid. *Proceedings of the National Academy of Sciences of the United States of America* 97, 22920–22925.

Zimmerli, L., Hou, B.-H., Tsai, C.-H., Jakab, G., Mauch-Mani, B. and Somerville, S. (2008) The xenobiotic β-aminobutyric acid enhances *Arabidopsis* thermotolerance. *The Plant Journal* 53, 144–156.

Zinnanti, W.J., Housman, C., O'Callaghan, J.P., Simpson, I., Goodman, S.I., Jacobs, R.E. and Cheng, K.C. (2007) Mechanism of age-dependent susceptibility and novel treatment strategy in glutaric acidemia type I. *Journal of Clinical Investigation* 117, 3258–3270.

28 Delivering Innovative Solutions and Paradigms for a Changing Environment

J.P.F. D'Mello*

Formerly of SAC (Scottish Agricultural College), University of Edinburgh King's Buildings Campus, Edinburgh, UK

28.1 Abstract

Recent advances in the fundamental disciplines of biochemistry, molecular biology, genetics and ecological sciences have provided the impetus for the publication of *Amino Acids in Higher Plants*. The choice of contributors to this volume underlines the diversity of research interests in a rapidly evolving field. The purpose here is to highlight the salient issues raised in the foregoing chapters with a view to providing an integrated model of amino acid metabolism in higher plants. A review of the main conclusions should serve as a salutary reflection of achievements so far, while also providing the basis for unlocking the full potential of emerging developments.

The quantitative and obligatory demands for amino acids in primary metabolism have long been recognized and continue to be the focus of current research. However, considerable attention has recently been devoted to the relationships of amino acids with signal transduction and defence metabolism in plants subjected to stress. Certain amino acids operate directly as signalling molecules or perform defence and adaptive functions in response to internal and external stimuli. Glutamate is the ideal example, with transduction functions in both plants and animals. Of equal significance is the indirect role of amino acids as precursors, activators or structural components of key plant hormones. For example, arginine is the source of nitric oxide, methionine is the precursor of ethylene and isoleucine is an activator of jasmonic acid. Among the aromatic amino acids, phenylalanine is, at least in part, the source of salicylic acid and tryptophan is the precursor of the auxin indole acetic acid. Interactions with other networks involving abscisic acid and the role of leucine zipper proteins, particularly under stress conditions, add to the complexity of signal transduction in plants. Leucine-rich repeat receptor kinases constitute a substantive subfamily of transmembrane receptor-like proteins in plants. The majority of receptor-like kinases are classified as serine/threonine kinases. Leucine-rich repeat motifs provide a versatile structural framework for facilitating interactions between proteins. Nitration and phosphorylation of tyrosine residues in certain proteins are, respectively, central to peroxynitrite and brassinosteroid signal transduction. Although the pivotal role of plant hormones is unequivocally established, other complex molecules such as the oligopeptide systemin also contribute to defence pathways during adverse conditions.

A diverse array of defence molecules are also synthesized during the course of the secondary metabolism of amino acids. A number of these compounds represent the constitutive framework for plant protection, but others are produced in response to specific biotic or environmental challenges involving the upregulation of metabolic pathways. Thus, the catecholamine biosynthesis route has been implicated in plant responses to stress and similar comments apply to the secondary metabolism of phenylalanine and tryptophan. The intersection between primary and secondary metabolism provides the potential for regulation at the gene and the enzyme level, and the opportunity for metabolic intervention. In a number of instances, amino acids serve as precursors of both signal transduction molecules and secondary compounds. Hence, methionine is the precursor of ethylene and

*E-mail address: jpfdmello@hotmail.co.uk

aliphatic glucosinolates, while tryptophan metabolism results in the synthesis of indole acetic acid, camalexin and indole glucosinolates. In addition, phenylalanine is the substrate for salicylic acid as well as flavonoids, coumarins and stilbenes.

A model for defence metabolism in plants has been presented in this chapter as a first approximation and with a view to stimulating additional refinement of the sequential processes. The pathways involved in plant responses to stress are highly complex, entailing initial perception, signal transduction and defence gene expression, culminating in the establishment of systemic acquired resistance. The contribution of secondary metabolites and pathogenesis-related proteins to this process are accepted, although the evidence is fragmentary at best. In any advanced model, recognition will need to be given to the issue of the wound response with respect to the role of complex interactions associated with signal transduction cascades, expression of inducible genes and the deployment of specific defence compounds. Furthermore, the potential significance of cross-talk between biotic and abiotic defence pathways is only just unfolding, but needs to be incorporated into a comprehensive model. Plant–herbivore and plant–pathogen interactions are in a dynamic state of flux, and adaptation to stress source and defence molecules is to be expected. For example, in response to aphid feeding, *Arabidopsis* plants increase defensive capability by interconversions of indole glucosinolates. A co-evolutionary process of plant responses and herbivore counter-adaptation is currently viewed as the driving force in the 'escalation of phytochemical diversity' and versatility. This concept has been expanded further with the proposition that some herbivorous insects are able to sequester and even reutilize plant secondary compounds for their own protection.

The comparative metabolism of amino acids has long been recognized as an important issue, particularly with respect to the reactions associated with glutamate dehydrogenase, aminotransferases and polyamine biosynthesis, but there is now an emerging consensus of a much greater degree of metabolic compatibility, specifically that between plants and animals. Thus, the synthesis of γ-aminobutyric acid, dopamine, serotonin and melatonin all require the action of appropriate decarboxylases in plants, as in animals. In addition, signal transduction, whether effected directly or via the associated metabolites of amino acids, represents another significant example transcending the traditional concepts of plant and mammalian biochemistry. The concept of molecular communication is a significant unifying feature of comparative metabolism, justifying separate consideration in this chapter.

Exploitation of botanical pharmaceuticals is likely to add value in the continuing diversification of crop production systems. The notion that plants should be grown for medicinal purposes rather than as sources of nutrients and fibre is rapidly gaining momentum in view of current interest in 4-hydroxyisoleucine occurring in fenugreek, indole phytoalexins and glucosinolates from crucifers, and the salidroside and tyrosol present in *Rhodiola crenulata*. In quantitative terms though, deficiencies and metabolic availability of essential amino acids in plant products used for human and livestock consumption remain outstanding issues at the present time.

Despite significant developments in recent decades, immense challenges lie ahead, not least at the fundamental level in discerning the genes encoding enzymes of the major pathways of metabolism, particularly insofar as they impact on metabolic regulation and the defence responses of plants. An issue requiring resolution is the role of amino acid conjugates of salicylic acid and indole acetic acid in gene expression and disease resistance. Concurrently, the complexities and significance of additive, synergistic and antagonistic interactions of signal transduction in plant immunity require elucidation in the context of biotic and environmental pressures. Nevertheless, there is currently the potential to drive forward a comprehensive and dynamic agenda aimed at delivering innovative solutions and paradigms for the future.

28.2 Background

This chapter title may appear to be a socio-economic statement and there are undoubtedly strong arguments for developing such a theme. Considerable progress has been made in elucidating novel roles of amino acids and their associated derivatives in the metabolism of higher plants against stringent economic constraints and competing demands for diminishing financial resources. However, the primary purpose of this chapter is to collate the evidence presented in the preceding chapters of this volume with a view to discerning common underlying themes and highlighting promising advances. Hitherto, attention has focused on enhancing the protein quality of foods destined for human and animal nutrition. It is intended that the chapters on enzymes, metabolism and transport in this volume will underpin changing research emphasis towards defence metabolism and environmental protection. The chemical ecology and involvement of amino acids in defence and stress responses will be summarized in unifying models that will attempt to accommodate the complex interactions that have been demonstrated.

Another emerging theme relates to the issue of food quality and safety, as plant products serve as a major source of nutrients for humans and animals. Where possible, and to underpin future developments, unique aspects of amino acid metabolism in plants will also be highlighted. All the above comments will be set against an overall context of comparative metabolism. Finally, the outlook for future advances in research will be presented to complete this chapter and the volume.

28.3 Approach

In a wide and rapidly expanding field of research such as that covered by *Amino Acids in Higher Plants*, it is important and instructive to highlight salient themes that are likely to motivate the next generation of biochemists and molecular scientists. Even a perfunctory examination of recent progress should demonstrate both the diversity and complexity of amino acid metabolism in higher plants. Current evidence confirms unique individual as well as collective effects. The approach in this chapter is to review emerging knowledge on the roles of individual amino acids with regard to the elucidation of unique or characteristic functions in higher plants. Amino acids are involved collectively in primary metabolism as components of structural and catalytic proteins. Even though the underlying biochemistry relating to protein synthesis is well documented, a perpetuating theme in this volume is the characterization of the molecular genetics of the plethora of enzymes catalysing obligatory reactions in plant metabolism. In regulatory proteins, leucine confers critical biophysical and interactive attributes. Leucine-rich repeat receptor kinases (LRR-RKs) constitute a substantive subfamily of transmembrane receptor-like proteins in plants associated with brassinosteroid (BR) perception and signal transduction. LRR motifs provide a versatile structural framework for facilitating interactions between proteins. Another significant aspect is the interaction between regulatory proteins and DNA, as exemplified by leucine zipper transcription factors.

At the individual level, amino acids are associated with diverse but, none the less, critical secondary functions, which constitute the basis of several chapters in *Amino Acids in Higher Plants*. The approach adopted in this chapter is to set these functions within the context of unifying themes. One important function relates to the direct actions of particular amino acids. Thus, glutamate and aspartate form or contribute to the basic structures of entire families of other amino acids (Jander and Joshi, 2010). The biosynthesis of amino acids in general is an important issue due to enduring interest in enhancing the protein quality of foods. The development of improved herbicides is another factor driving efforts to unravel regulatory networks and identify potential targets for metabolic manipulation. In addition, certain amino acids operate directly in signal transmission or perform defence and adaptive functions in response to internal and external stimuli.

Just as importantly, amino acids exert indirect activities, serving as precursors of certain key hormones or as activators of other signalling compounds. For example, arginine is the source of nitric oxide (NO), methionine is the precursor of ethylene (ethene, ET) and isoleucine is an activator of jasmonic acid (JA). Among the aromatic amino acids, phenylalanine is, at least in part, the source of salicylic acid (SA) and tryptophan is the precursor of the auxin indole acetic acid (IAA). Interactions with other signal transduction networks involving abscisic acid (ABA) and the role of leucine zipper proteins particularly, under stress conditions, add to the complexity of adaptive responses in plants. Although the pivotal role of these hormones is unequivocally established, other complex molecules, such as the oligopeptide systemin, also contribute to signalling pathways during adverse conditions.

In addition, a wide range of defence molecules are derived during the course of the secondary metabolism of amino acids. While a number of these compounds represent the constitutive framework for plant protection, others are produced in response to specific biotic or environmental challenges. The pathways involved in plant responses to stress are worthy of consideration to evaluate the role of the complex interactions associated with signal transduction cascades, the expression of inducible genes and the deployment of specific defence compounds.

The following review is based primarily on the outcomes established in the foregoing chapters of *Amino Acids in Higher Plants*. However, it is appropriate to include additional evidence to provide a more comprehensive perspective of recent developments and future prospects. Nevertheless, limitations of space mean that other issues, such as the role of pathogenesis-related proteins in defence pathways, cannot be considered in detail.

28.4 Glutamate

All living organisms synthesize and utilize glutamate in diverse metabolic reactions. Although glutamate is a major constituent of proteins, its ubiquitous distribution as an unconjugated molecule and its signal transduction roles place it in a unique position in the biochemistry of amino acids. Indeed, glutamate has been referred to as an amino acid of 'particular distinction' (Young and Ajami, 2000). Its metabolic fate is determined by the action of several key enzymes, including glutamate dehydrogenase (GDH; Chapter 1), various aminotransferases (Chapters 2–4), glutamine synthetase (Chapter 5) and glutamate decarboxylase (Chapter 7). The relative activities of these enzymes provide the basis of glutamate homeostasis in plants. Forde and Lea (2007) reviewed the evidence indicating that diurnal glutamate concentrations tend to remain remarkably constant irrespective of growth conditions, whereas glutamine concentrations fluctuate considerably. Thus, glutamate status appears to be a relatively stable entity in a dynamic metabolic system.

Current interest relates to putative signalling roles of glutamate in plants in a manner that may be analogous to that in animals (Brenner *et al.*, 2006), and Forde and Lea (2007) implied that glutamate may be considered to be a phylogenetically conserved signal amino acid. Plant glutamate-like receptors (GLRs) are considered to be homologous with the ionotropic glutamate receptors present in animals, and attributed with the potential to serve as amino acid sensors in plants (Price *et al.*, 2012). Numerous GLR genes have been identified in plant genomes, as reviewed by Davenport (2002). Subsequent investigations have centred on phylogeny and evolutionary issues, as well as expression and transcriptional responses, and primary metabolism, including C to N partitioning. The prevailing consensus is that plant GLRs participate in various biochemical and physiological processes in addition to N sensing. Glutamate applied exogenously elicits particular responses in the root. The effects are specific to glutamate and entail both suppression of primary root growth and stimulation of root branching. Glutamate appears to be sensed directly in the apical region, where it depresses meristematic activity. Li *et al.* (2006) indicated that in rice a GLR gene is critical for the division and survival of individual cells in the root apical meristem at the early seedling stage. The involvement of GLR genes has been implied (Walch-Liu and Forde, 2007). Contemporary evidence based on molecular and genetic studies reflect a distinctly more confident view with respect to structural and functional characterization (Miller *et al.*, 2010). In animals, fluxes in cytosolic free Ca concentrations represent a primary transduction feature (Flores *et al.*, 2012). Michard *et al.* (2011) showed in studies with mutants that tobacco and *Arabidopsis* GLR genes facilitate Ca^{2+} influx across the plasma membrane, modulating apical Ca^{2+} gradients and pollen tube growth and morphogenesis. Furthermore, the regulation of GLR activity by pistil D-serine implies an amino acid-mediated communication normally associated with animal neurobiology (Hines, 2011; Smith *et al.*, 2012). Vincill *et al.* (2012) are convinced that the family of plant GLRs operate as amino acid-gated, Ca^{2+} permeable channels, capable of eliciting a cytoplasmic Ca^{2+} signal. However, plant GLRs are characterized by a structurally diverse agonist profile compared with the corresponding mammalian glutamate receptors. Earlier the same group (Stephens *et al.*, 2008) had identified three classes of channels relating to activation and desensitization by glutamate only, or by alanine, cysteine, glutamate or glycine, or by asparagine, alanine, cysteine, glutamate, glycine and serine. Forde *et al.* (2013) concluded that there is now genetic evidence for the existence of a glutamate signal transmission pathway in plants similar to that operating in animals.

28.5 Proline

As a derivative of glutamate, proline is ubiquitous in living organisms and is now emerging as a key multifunctional amino acid in primary metabolism and in the response of plants to environmental stress (Sharma and Dietz, 2006). Proline is associated with specific attributes that include the regulation of cellular homeostasis, acting as a signal transduction molecule that modulates mitochondrial function and with a role in gene expression, particularly under adverse conditions (Szabados and Savouré, 2010). In addition, Lehmann *et al.* (2010) emphasized the physical properties of proline that arise from its cyclic structure, which restricts conformational flexibility, as well as its activity as a compatible solute in the protection of cellular components. It has been suggested that efforts to manipulate proline metabolism might assist in unlocking further potential to enhance stress tolerance in plants.

28.6 Arginine

As in other organisms, arginine is metabolized by arginase (arginine amidohydrolase, ARGAH) to ornithine and urea (Fig. 27.1). In a second pathway, arginine is also the substrate for nitric oxide synthase (NOS), the reaction resulting in the production of NO and citrulline. A third pathway involves the action of arginine decarboxylase leading to the synthesis of putrescine.

The urea cycle, its enzymes and substrates are normally associated with the disposal of excess N in mammals. However, the key enzyme, arginase, is widely distributed in nature, even occurring in diverse species of higher plants. For example, it has been detected in the ovaries and young fruits of tomato (*Lycopersicon esculentum* [now designated *Solanum lycopersicum*]; Alabadi *et al.*, 1996). In studies on the regulation of plant arginase by wounding, jasmonate and the phytotoxin coronatine, Chen *et al.* (2004) reported the characterization of two genes, *LeARG1* and *LeARG2*, encoding arginase in tomato. Marked similarities in the regulation of arginase in plants and animals were noted. In particular, *LeARG2* expression and arginase activity were induced by stress factors. The overexpression of the two genes (designated *ARGAH1* and *ARGAH2*) in *Arabidopsis thaliana* has also been reported (Brauc *et al.*, 2012). Previously, Brownfield *et al.* (2008) had proposed multiple functions for arginase genes in *Arabidopsis*. The modulation of arginase activity is a significant emerging theme emphasized in studies with chilling tolerance in tomato fruits (Zhang *et al.*, 2010) and in abiotic stress (Shi *et al.*, 2013) and clubroot infection in *Arabidopsis* (Gravot *et al.*, 2012). Notwithstanding the foregoing evidence, it is worth noting that arginase is inoperative in developing embryos (Goldraij and Polacco, 1999).

With respect to the second pathway, it has been proposed that arginine or an associated derivative is a potential substrate for NO synthesis (Flores *et al.*, 2008). In arginase-negative mutants of *Arabidopsis*, this pathway is enhanced with increased production of NO, thereby potentiating the action of auxin in roots.

Arginine synthesis also occurs via the urea cycle (Fig. 27.1), and argininosuccinate synthetase (ASS) plays an important role in this pathway in plants as in other organisms. The importance of histidine at the active site of ASS has been established in soybean (Shargool, 1985).

28.7 Ornithine

Ornithine-δ-aminotransferase (OAT) catalyses the transamination of ornithine to glutamate-γ-semialdehyde, but the potential significance of OAT may reside in stress-induced proline synthesis (Stránská *et al.*, 2008, 2010). The concept that proline is toxic to plants under non-stress conditions was alluded to by Kalamaki *et al.* (2009a), with the suggestion that overproduction of its precursor, ornithine, may offer the mechanism to provide plants with an amino acid that can readily be used for proline production under stress conditions. It was proposed that the genetic manipulation of plants to accumulate a non-toxic metabolite that can rapidly be transformed into a protective molecule might present an alternative method for inducing stress tolerance. The efficacy of ornithine is further underlined in exogenous feeding studies with ornithine and glutamate in salt-stressed cashew leaves; under these conditions, proline concentrations were several times higher in ornithine-fed plants than those in glutamate or salt-treated plants (da Rocha *et al.*, 2012). A significant contribution of the OAT pathway was implied.

28.8 Citrulline

Yokota *et al.* (2002) observed that under drought conditions in the presence of strong sunlight, high concentrations of citrulline, arginine and glutamate may accumulate in the leaves of wild watermelon. It was suggested that citrulline and arginine synthesis may be related to the induction of DRIP-1, a homologue of ArgE (*N*-acetylornithinedeacetylase, which catalyses hydrolysis of the *N*-acetyl group of *N*-acetylornithine to give ornithine and acetate) in *Escherichia coli*, in which it functions to incorporate the C-skeleton of glutamate into the urea cycle. Salt treatment of wild melon increased the concentrations of γ-aminobutyrate (GABA), glutamine and alanine, in addition to smaller quantities of citrulline. The relationship between citrulline accumulation and salt tolerance during vegetative growth of melon has been examined by Dasgan *et al.* (2009), who concluded that citrulline may protect green tissues from secondary oxidative stress. Alterations in the molar ratios of citrulline to free proline suggested that salt tolerance might be linked with high values for this ratio and the converse for sensitivity. The potential use of citrulline as a biochemical marker for salt tolerance in melon genotypes is now an emerging issue.

28.9 Glycine

Another amino acid with diverse roles in living organisms is glycine. For example, Dubos *et al.* (2003) suggested the involvement of glycine in the gating of plant *N*-methyl-D-aspartate (NMDA)-like receptors. Glycine is the primary inhibitory neurotransmitter in the spinal cord, brainstem and retina of animals. Expression of glycine-gated chloride channels in the neurons of the central nervous system has been confirmed (D'Mello, 2012).

Glycine is a major substrate available for light-induced oxidation in plant leaf mitochondria (Oliver *et al.*, 1990). The supporting metabolic system includes the glycine/serine transporter, optimum levels of the glycine decarboxylase multienzyme complex and serine hydroxymethyl transferase, as well as an efficient oxaloacetate transporter enabling the export of NADH produced by glycine utilization. In C_3 species, the enzyme complex is expressed in both mesophyll and bundle-sheath cells, whereas in C_3–C_4 intermediate and C_4 species, the enzyme is present only in the bundle-sheath cells (Hylton *et al.*, 1988). Douce *et al.* (2001) presented the glycine decarboxylase system as an intriguing complex comprising four components designated the P-, H-, T- and L-proteins, with the pivotal role being attributed to the lipoate-dependent H-protein. The latter moiety successively interacts with the other proteins in the catalytic and transfer cycle. The emerging consensus is that glycine decarboxylase controls the flow of carbon through both photosynthesis and photorespiration (Timm *et al.*, 2012). Chandok *et al.* (2003) suggested that the pathogen-inducible NOS in plants is a variant of the P-protein component of the glycine decarboxylase complex. It was concluded that, in contrast to animal NOS, the variant P-protein of glycine decarboxylase uses very different chemistry for the synthesis of NO by plants. Inhibition of the H-protein component in isolated pea leaf mitochondria has been linked to the cytotoxic effects of lipid peroxidation resulting in reduced glycine-dependent respiration. Furthermore, *in vivo* lipid peroxidation induced by paraquat and by environmental stresses such as chilling and water deficit inhibit the mitochondrial oxidation of glycine (Taylor *et al.*, 2002). Inhibition of glycine decarboxylase by the host-selective toxin, victorin, has been reported, and it has been suggested that the interaction between victorin and glycine decarboxylase is an underlying feature of this inhibition (Navarre and Wolpert, 1995). Structural abnormality of the H-protein has been observed in a patient with non-ketotic hyperglycinemia (Kikuchi and Hiraga, 1982), while overexpression of glycine decarboxylase activity appears to be associated with tumorigenesis in lung cancer (Zhang *et al.*, (2012). Thus, the glycine cleavage system constitutes a significant pathway in plants and mammals in normal and stress conditions. Moreover, glycine has long been linked with the alleviation of stress caused by methionine toxicity in animals (Hafez and Chavez, 1978).

Glycine and choline are precursors of glycine betaine (GB), a low molecular mass molecule which may confer protection to plants in stress conditions (Holmström *et al.*, 2000; Sakamoto and Murata (2002). GB is considered to act as a compatible solute, exerting a role in response to abiotic stresses such as high salinity and low environmental temperature. Initial work emphasized the role of GB in the maintenance of cellular osmotic potential in plants, but current studies demonstrate the protective functions of GB to a variety of abiotic stress factors. At the concentrations present, the effects of GB are attributed more significantly to stress tolerance than to the maintenance of cellular osmotic potential (Chen and Murata, 2011). GB accumulation is more effective in chloroplasts than in the cytosol for protection against abiotic stress, judging by the responses of transgenic tomato plants (Park *et al.*, 2007). In addition, GB accumulation varies, being relatively high in the chloroplasts of spinach and barley compared with other species such as *Arabidopsis* and tobacco. Transgenic tobacco lines accumulating GB show increased tolerance to salinity and low temperatures (Holmström *et al.*, 2000). It may be argued that GB accumulation is a product of stress and not an adaptive response. However, exogenous applications of GB may elicit significant increases in growth and crop yield under environmental stress conditions, although species differences in responses are apparent (Ashraf and Foolad, 2007).

28.10 Sulfur Amino Acids

In its role as a dietary essential for animals, methionine occupies a central position in the sulfur amino acid group. For different reasons, methionine status is also of critical importance in plants, particularly as a substrate for a diverse range of pathways (Fig. 28.1). In plants, the synthesis of methionine proceeds along a well-established pathway commencing with the precursors cysteine and *O*-phosphohomoserine which, via the action of cystathionine γ-synthase, yield cystathionine. The latter is converted to homocysteine by cystathionine γ-lyase. In the final step, the action of methionine synthase results in the formation of methionine (Ravanel *et al.*, 1998). The biosynthesis of methionine is under complex regulation brought about by competition between cystathionine γ-synthase and threonine synthase for their common substrate, *O*-phosphohomoserine (Hacham *et al.*, 2002). Indeed, Amir *et al.* (2002) concluded that methionine

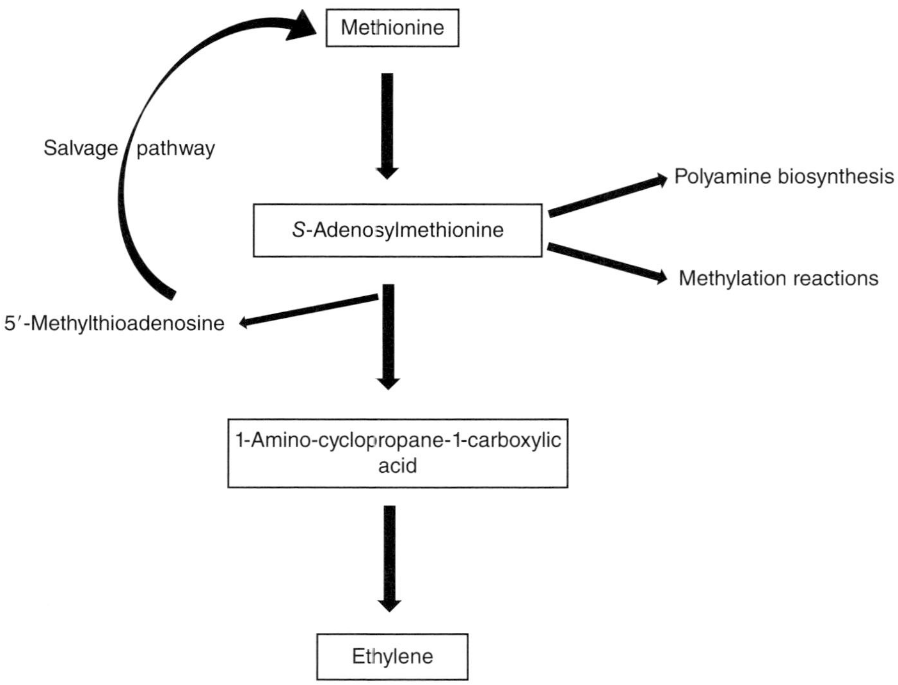

Fig. 28.1. Outline pathway for the biosynthesis of ethylene in plants. It has been established that a series of recycling reactions (incorporated within the Yang cycle) ensures the salvage of sulfur and the regeneration of methionine. The central position of *S*-adenosylmethionine in polyamine biosynthesis and in methyl transfer is emphasized. Key enzymes associated with these pathways are identified in the text.

synthesis and degradation were regulated by novel post-transcriptional and post-translational processes that are coordinated within the N-terminal part of cystathionine γ-synthase. In a second mechanism, methionine exerts feedback control of cystathionine synthesis, possibly through suppression of cystathionine γ-synthase gene expression.

Research interest in methionine continues primarily owing to its role in *S*-adenosylmethionine (SAM; AdoMet) biosynthesis (Fig. 28.1). It is established that SAM is essential as a methyl donor and as an activator of threonine synthase. In addition, SAM contributes to the biosynthesis of polyamines (Chapter 10) and ET, a significant signal transduction molecule in plants (Ravanel *et al.*, 1998). Current investigations focus on the synthesis of the intermediate 1-amino-cyclopropane-1-carboxylic acid (ACC) and the associated enzyme ACC synthase (Huang *et al.*, 2013). This enzyme is encoded by a number of genes, some of which are inducible by hypoxia, while others are involved in *Botrytis cinerea*-induced ET production in *Arabidopsis*. Although ET was initially associated with fruit ripening, it is now also implicated in seed germination, root development, abscission and flower senescence (Wang *et al.*, 2002); its interactions with other hormones, particularly under stress conditions, ensure that ET will continue to be the focus of considerable attention. For instance, antagonism between ABA and JA-ET transduction pathways in the modulation of defence gene expression has been reported (Anderson *et al.*, 2004). ET antagonizes ABA signalling during seed dormancy/development (Beaudoin *et al.*, 2000; Ghassemian *et al.*, 2000); it also renders the JA response pathway insensitive to subsequent suppression by SA (Leon-Reyes *et al.*, 2010).

The complexity of partition dynamics has been described elsewhere in this volume (Chapters 5, 12, 13 and 16) and also in relation to key enzymes associated with the metabolism of amino acids in animals (D'Mello, 2012). A similar level of intricacy is emerging with respect to the localization of methionine metabolism in plants. For example, Ravanel *et al.* (2004) demonstrated that chloroplasts are autonomous for the *de novo* synthesis of methionine by virtue of the occurrence of a particular isoform of methionine synthase located in these organelles. Other functional isoforms exist in the cytosol, where SAM synthesis occurs exclusively. Ravanel *et al.* (2004) further demonstrated that SAM is transported into the chloroplasts by a carrier-mediated diffusion process. In the biosynthesis of ET, the sulfur in methionine has to be recycled via a complex salvage pathway known as the Yang cycle. The data of Pommerrenig *et al.* (2011) show that this pathway in higher plants operates primarily in the phloem.

It has long been known that choline is the precursor of the osmolyte GB. BeGora *et al.* (2010) commented on the dearth of information concerning the enzymes or factors regulating the biosynthesis of choline and phosphatidylcholine. However, evidence was presented to demonstrate that the enzyme phosphomethylethanolamine *N*-methyltransferase catalyses three sequential *N*-methylations of phosphoethanolamine to produce phosphocholine, with SAM providing the methyl groups. In addition, it was claimed that the activity of this enzyme was not the rate-limiting factor and that choline transport into chloroplasts and insufficiency of other precursors should be considered as potential constraints in the supply of GB. This issue is relevant in the light of evidence indicating that for abiotic stress protection in plants, GB accumulation is more effective in the chloroplasts than in the cytosol (Park *et al.*, 2007).

Cysteine is associated with diverse patterns of metabolism in all living organisms, providing important intermediates for biosynthetic reactions (Chapters 11 and 12). Vitamins, cofactors, antioxidants and several defence compounds are all derived from cysteine. The modulation of cysteine biosynthesis is effected by an enzyme complex that is sensitive to the sulfur status of the plant.

Striking structural analogues exist for methionine, cysteine, cystine and cystathionine. When selenium replaces the sulfur atom of these amino acids (Table 27.1), a number of congeners are formed. Those non-protein amino acids of primary interest here are selenomethionine, *Se*-methylselenocysteine and selenocystine. Other analogues of the sulfur-containing amino acids include *S*-methylcysteine sulfoxide (SMCO) and djenkolic acid (Chapter 27).

28.11 Branched-Chain Amino Acids

The branched-chain amino acids (BCAAs) evoke enduring research interest that is consistent with the structural antagonisms among individual members of this group. Miflin (1969) demonstrated that the growth suppression of barley seedlings induced by valine could be reversed by isoleucine and that both isoleucine and valine were required to alleviate the inhibition precipitated by leucine. The leucine–valine interaction was subsequently confirmed by Borstlap (1981) in growth studies with *Spirodela polyrhiza* (duckweed). Other responses included antagonism by valine + isoleucine with respect to the growth inhibition caused by leucine. These results broadly reflect the BCAA antagonisms observed with certain bacteria (De Felice *et al.*, 1979) and with mammalian and avian animals (D'Mello, 2003a). However, in animals, the adverse effects of leucine dominate the antagonisms with isoleucine and valine, and valine appears to be highly sensitive to antagonisms imposed by leucine and isoleucine.

In studies with legumes, Prell *et al.* (2009) demonstrated the regulation of *Rhizobium* bacteroid development and persistence through the supply of BCAAs. It appears that this effect is specific and is reflected in a major reduction in the transcription and activity of enzymes involved in the biosynthesis of BCAAs. It was implied that the bacteroids assume the status of symbiotic auxotrophs with respect to the supply and utilization of host BCAAs.

Another intriguing feature of interdependence relates to BCAA metabolism and glucosinolate biosynthesis in plants (Chapter 22). At least in *Arabidopsis*, leucine biosynthesis and degradation and the methionine chain-elongation process in aliphatic glucosinolate production are catalysed by branched-chain aminotransferase (BCAT). BCAT4 has been identified as the enzyme involved in the initial stages of chain elongation (Schuster *et al.*, 2006), whereas BCAT3 appears to be associated with the terminal steps in this process (Knill *et al.*, 2008). It would be of much interest to determine the extent to which BCAA antagonisms might affect glucosinolate biosynthesis and the chain-elongation process in particular.

Of the three BCAAs, leucine continues to be the focus of considerable attention as a result of its biophysical and interactive attributes. LRR-RKs constitute a substantive subfamily of transmembrane receptor-like proteins in plants. The majority of receptor-like kinases are classified as Ser/Thr kinases (Friedrichsen *et al.*, 2000). LRR motifs provide a versatile structural framework for facilitating interactions between proteins (Kobe and Kajava, 2001). It is widely acknowledged that LRR-RK-regulated activities include cell proliferation and hormone perception, as well as responses to phytopathogens and wounding (Torii, 2004). Interactions with leucine-rich repeat receptor-like proteins (LRR-RP) are another feature of relevance. Xu *et al.* (2008) reported that two LRR-RKs mediate signal transduction, linking cell wall biosynthesis and ACC synthase. This enzyme converts adenosylmethionine to ACC in the biosynthesis of ET in plants. LRR-RKs have also been implicated in BR perception and signal transduction (Li and Chory, 1997). Both kinases autophosphorylate on tyrosine residues and are thus dual-specificity kinases (Oh *et al.*, 2009). It had previously been concluded that brassinosteroid-insensitive-1 is ubiquitously expressed and exerts a role in steroid hormone signalling through Ser/Thr phosphorylation (Friedrichsen *et al.*, 2000). BRs interact with other signalling molecules in responses to biotic and environmental challenges, in this way modulating the growth and development of the plant (Li *et al.*, 1996; Clouse, 2011). For example, Belkhadir *et al.* (2012) concluded that BRs modulate the efficiency of immune function specifically to microbe-associated molecular patterns.

Another significant aspect of leucine relates to the interaction between regulatory proteins and DNA, as exemplified by leucine zipper transcription factors. The orientation of leucine residues along the segments of the protein double helix promotes binding to DNA. Certain members of this class of proteins are attributed with mediating the response of plants to a range of processes, including ABA-dependent signal transduction under drought and high-salinity conditions (Uno *et al.*, 2000). ABA exerts a pivotal role in stress responses in plants, and Kang *et al.* (2002) emphasized that leucine zipper proteins were integrated into this process. Others have demonstrated that leucine zipper transcription factors regulate chloroplast number and photosynthetic capacity (Hymus *et al.*, 2013).

Although the pathways of biosynthesis and metabolism of isoleucine are well established, other important functional features have recently emerged (Yu *et al.*, 2013). In an *Arabidopsis* mutant with low isoleucine biosynthesis, root development was impaired but transcript levels of the genes encoding the major enzymes involved in BCAA catabolism were reduced. In addition, a number of genes promoting the biosynthesis of methionine-derived glucosinolates were upregulated. It was suggested that isoleucine may exert a signal transduction role in root development (Yu *et al.*, 2013).

A highly significant development in the function of isoleucine has recently emerged, partly as a result of metabolic and gene expression studies on plant responses to wounding and insect herbivory (Howe, 2004). The oxylipin signal JA has been implicated in a variety of plant defence responses. It is now known that activation of JA requires prior conjugation to isoleucine (Ile) to yield a JA-Ile complex (Staswick and Tiryaki, 2004). In 2004, activation by conjugation to amino acids was a relatively novel concept. It is well established that isoleucine is synthesized from threonine via the α-ketobutyrate pathway. Silencing threonine deaminase activity and the activation process impairs the efficacy of JA-Ile-mediated defences of plants against insect herbivory (Kang *et al.*, 2006). Investigations continue in an effort to elucidate the metabolic fate of the JA-Ile conjugate, including its catabolism and inactivation by cytochrome P450 (Koo *et al.*, 2011), turnover via the action of amidohydrolases (Widemann *et al.*, 2013) and the effects of light (Svyatyna *et al.*, 2014).

28.12 Aromatic Amino Acids

Despite decades of research, there are still significant issues concerning the biosynthesis and subsequent metabolism of phenylalanine, tyrosine and tryptophan. Animals are unable to synthesize phenylalanine or tryptophan, but can utilize phenylalanine to produce tyrosine. While the shikimate pathway (Hu *et al.*, 2009) has long been recognized as the mechanism for the synthesis of aromatic amino acids in plants (Fig. 28.2), some uncertainty is apparent over the terminal phases of this process. Furthermore, recognition that important signal transduction and defence compounds arise during the course of the metabolism of aromatic amino acids has provided new perspectives, particularly on regulatory networks.

Arguments prevail as to the relative role of arogenate and phenylpyruvate as immediate precursors of phenylalanine. A partial resolution of this issue was attempted by Cho *et al.* (2007), who proposed the existence of six genes and their corresponding proteins as arogenate dehydratases in *Arabidopsis*, indicative of the importance of this route in phenylalanine biosynthesis. This pathway was confirmed recently by Corea *et al.* (2012), who concluded that arogenate dehydratase isoenzymes profoundly and differentially modulated carbon flow into lignins. However, Yoo *et al.* (2013) concluded that plants also use phenylpyruvate to synthesize phenylalanine and that flux through this pathway is increased when the arogenate supply chain is restricted. Unexpectedly, Yoo *et al.* (2013) observed that the phenylpyruvate route utilized a cytosolic aminotransferase associated with the synchronized degradation of tyrosine as the source of the amino group.

Opinion is also divided over the mechanisms of SA biosynthesis in higher plants (Fig. 28.2). Initial observations indicated that the production of SA precursors was a major function of phenylalanine ammonia-lyase (PAL) in the resistance of *Arabidopsis* to *Peronospora parasitica* (Mauch-Mani and Slusarenko, 1996). Taken together with lack of correlation between expression patterns of chorismate synthase and isochorismate synthase genes with SA levels (Pál *et al.*, 2013), it might be concluded that the PAL pathway is the primary mode of SA biosynthesis. Others claim that the utilization of phenylalanine or chorismate as precursors for SA synthesis depends upon plant species, developmental stage and growth conditions (Tounekti *et al.*, 2013). More definitively, Yamasaki *et al.* (2013) stated that the isochorismate pathway exerted a central role in immunity-related SA biosynthesis in *Arabidopsis*, and implicated the PAL mechanism for SA production in *Nicotiana tabacum*. The critical role of chloroplasts in the biosynthesis of SA and its subsequent transport into the cytoplasm and epidermal cells were also emphasized (Yamasaki *et al.*, 2013).

It is widely accepted that SA is a critical signal transduction molecule for the induction of immune responses in plants towards fungal pathogens and particularly in the establishment of localized and systemic acquired resistance (SAR; Yalpani *et al.*, 1993; Mauch-Mani and Slusarenko, 1996; Achuo *et al.*, 2004; Wang *et al.*, 2007; Hu *et al.*, 2009; Moreau *et al.*, 2012). Indeed, the SA-dependent defence mechanism is effective against a wide range of pathogens,

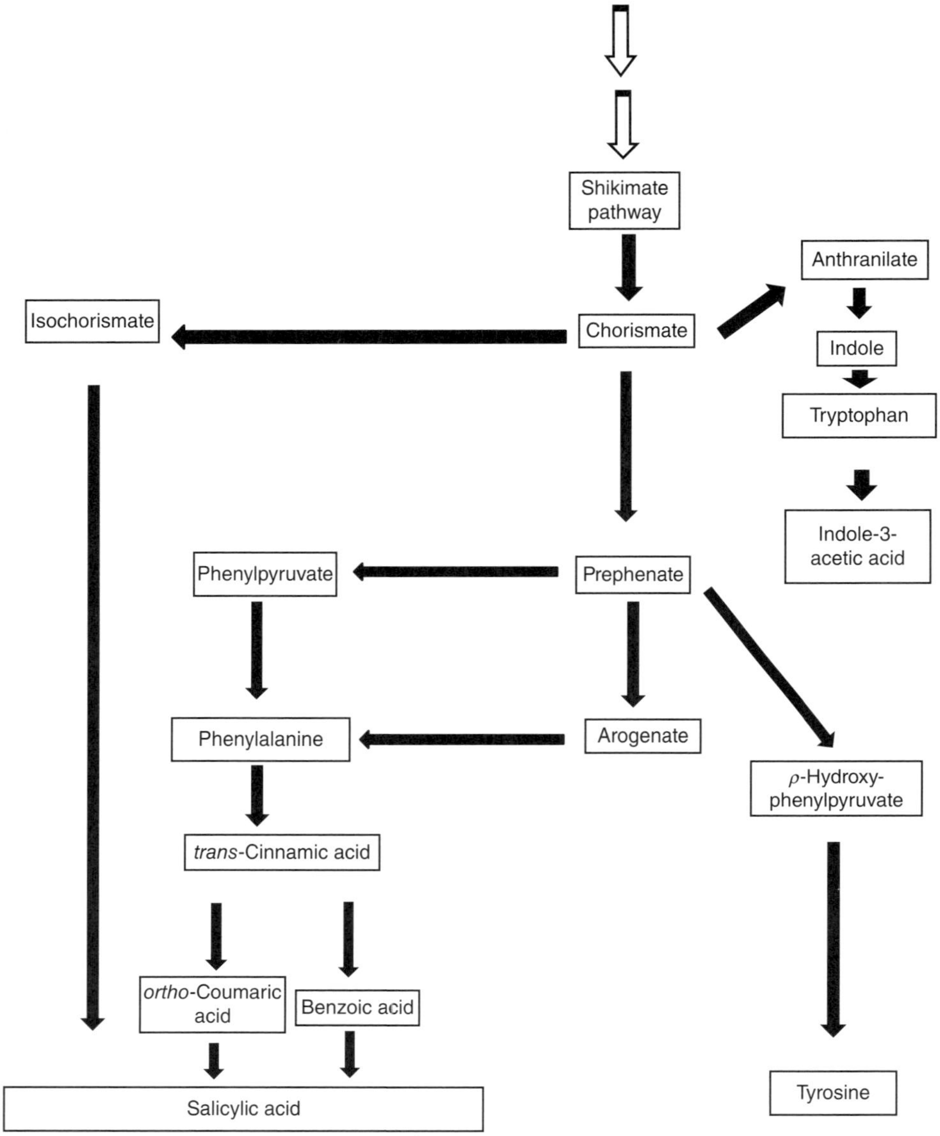

Fig. 28.2. Simplified scheme depicting the biosynthesis of aromatic amino acids, salicylic acid and indole-3-acetic acid (IAA) in higher plants. The central position of chorismate has long been established, although its functional contribution to plant–microbe interactions is now emerging (Hu *et al.*, 2009). The immediate source of salicylic acid (SA) depends upon plant species, developmental stage and growth conditions (Tounekti *et al.*, 2013), but the debate continues (Pál *et al.*, 2013). It should be noted that four primary pathways have been proposed for the biosynthesis of IAA (auxin) in plants, possibly operating in parallel but subject to endogenous and exogenous factors (Chapter 19). Key enzymes associated with these pathways are identified in the text and elsewhere in *Amino Acids in Higher Plants*.

including viruses (Yalpani *et al.*, 1993), *P. parasitica* (Mauch-Mani and Slusarenko, 1996), *B. cinerea* (Achuo *et al.*, 2004) and *Pseudomonas syringae* (Chen *et al.*, 2013). In addition, SA has been linked with plant responses to abiotic stress by the modulation of terpenoid and flavonoid metabolism (Tounekti *et al.*, 2013). Another aspect relates to the occurrence and activity of the conjugate SA-Asp (aspartate), which functions as a mobile molecule, possibly in the induction of pathogenesis-related gene expression (Chen *et al.*, 2013). It is then clear that SA exerts its defence functions in two ways: through the induction of SAR (Maier *et al.*, 2011; Moreau *et al.*, 2012) and by stimulating the synthesis of key secondary metabolites such as terpenoids and flavonoids (Tounekti *et al.*, 2013).

The pathway for tyrosine biosynthesis is presented in Fig. 28.2, showing prephenate as the key intermediate. The metabolic characterization of tyrosine decarboxylase and its role in the synthesis of tyrosol and salidroside has been updated (Lan *et al.*, 2013). Health benefits of the plant *Rhodiola crenulata* have been attributed to these products of tyrosine metabolism.

The chorismate-anthranilate pathway (Fig. 28.2) is a well-recognized route for the synthesis of tryptophan, occurring exclusively in plants (Zhao and Last, 1996) and microorganisms. The evidence obtained with specific auxotrophs of *Neurospora* and *E. coli* was presented by Greenberg (1961). Tryptophan is a rich source of physiologically active metabolites in both plants and animals. The syntheses of IAA (Chapter 19), serotonin (Chapter 20), melatonin (Chapter 21) and indole glucosinolates (Chapter 22) illustrate the diversity of pathways of tryptophan metabolism, which continue to attract attention to the present day. For example, the question of the role of IAA conjugates has recently emerged as a factor in the regulation and homeostasis of tryptophan metabolism (Ludwig-Müller, 2011). IAA normally exists in free form, but amide conjugates with alanine, leucine, phenylalanine, aspartate, glutamate and tryptophan (Trp) also occur naturally in plants. IAA-Trp is thought to be an inhibitor of auxin activity. Although auxin is a recognized plant hormone regulating various pathways in plant growth and development, a paradox is now emerging relating to the activity of IAA-Asp in the promotion of disease. González-Lamothe *et al.* (2012) claim that fungal and bacterial pathogens divert auxin metabolism in *A. thaliana* towards conjugation with aspartate, and that this conjugate increases pathogen progression in the plant by modulating the transcription of virulence genes.

It is also now clear that the induction of SA and defence molecules such as indole alkaloids, flavanones, isoflavanones and phytoalexins is accompanied by upregulation of the associated biosynthetic enzymes. Thus, Hu *et al.* (2009) elucidated the functional contribution of chorismate synthase, anthranilate synthase and chorismate mutase to penetration resistance in barley–powdery mildew interactions. Huang *et al.* (2010) examined the role of the *Arabidopsis PAL* gene family on the response to abiotic stress, while Ishihara *et al.* (2008) implicated the tryptophan pathway in defence responses to pathogens via serotonin production.

28.13 Secondary Metabolism

It is now acknowledged that the term 'secondary' in this particular context is a misnomer as it implies functional passivity in the biochemical response of plants to endogenous and environmental stimuli. However, there is almost universal agreement and ample evidence (see below) that secondary compounds actively support ecological competitiveness, survival and growth of plants. Amino acids serve as sources of a diverse range of secondary compounds, including signal transduction molecules and defence agents.

Conventional accounts of secondary metabolism tend to focus on the biosynthesis of low-molecular weight phytochemicals, particularly in response to stress challenges such as biotic or physical pressures. It is generally assumed that these compounds support the constitutive defences in the plant, but it may be argued that the production of more complex molecules, such as enzymes encoded by duplicate genes and proteinase inhibitors that are specifically aimed at insect herbivores and plant pathogens should also be considered as an integral feature of secondary metabolism. In the case of threonine dehydratase, two genes are recognized, one fulfilling housekeeping functions and the other providing insect deterrence (Gonzales-Vigil *et al.*, 2011). In addition, the appearance of the oligopeptide systemin has been associated with stress responses in plants (Léon *et al.*, 2001). It is also important to recognize that pathogenesis-related proteins contribute markedly to resistance to fungal infection in diverse species of plants (El-kereamy *et al.*, 2011). It has previously been established that ET and JA are synergistically involved in the induction of pathogenesis-related defence genes (Xu *et al.*, 1994).

Amino acids in higher plants are a prolific source of secondary compounds, including auxins (Chapter 19), serotonin (Chapter 20), melatonin (Chapter 21), glucosinolates (Chapter 22), natural phytotoxins (Chapter 23), alkaloids (Petterson *et al.*, 1991), condensed tannins (Griffiths, 1991), cyanogens (Davis, 1991), indole phytoalexins (Glawischnig, 2007) and flavonoids (Huang *et al.*, 2010). Plants also contain a diverse array of non-protein amino acids (Chapter 27). Although a number of secondary compounds form part of the constitutive defence system, others are synthesized as a result of upregulation of specific biosynthetic genes and pathways in response to biotic or environmental elicitors. In addition, defence versatility in plants may be enhanced via metabolic transformations, as for example in the conversion of one indole glucosinolate into another following insect attack (Kim and Jander, 2007). However, insect herbivores may not always succumb to the adverse effects of secondary metabolites and, in certain cases, may even sequester these compounds for use in counter-defence (Fürstenberg-Hägg *et al.*, 2013).

28.14 Comparative Metabolism

The universal features of amino acid metabolism have long been recognized, particularly with respect to the reactions catalysed by GDH (Chapter 1), aminotransferases (Chapters 2–4) and enzymes associated with polyamine biosynthesis (Chapter 10). There is now an emerging consensus of a much greater degree of metabolic compatibility, specifically between plants and animals, though significant differences are also apparent. Selected data are presented in Table 28.1 to illustrate the major features in the comparative metabolism of amino acids. While GDH is expressed in all living organisms, Smith and Stanley (2012) state that the animal enzyme is allosterically regulated by a wide range of metabolites, whereas it is mainly controlled at the transcriptional level in other kingdoms. Even though GDH is the pre-eminent exponent of its class, it should be recognized that other dehydrogenase enzymes participate actively in the N transactions and stress responses of plants. For instance, two proline dehydrogenase isoforms have been identified in *Arabidopsis* that differ in spatial and developmental expression and in their expression during the regulation of salinity stress (Funck *et al.*, 2010). The induction of betaine aldehyde dehydrogenase in response to osmotic stress in barley has been reported by Ishitani *et al.* (1995). Distinctive effects have also been observed in the activity of isovaleryl-coenzyme A dehydrogenase that appear to suggest a novel linkage in the leucine and valine catabolism in plants, but such dual functions are absent in animals (Daschner *et al.*, 2001).

Within the aminotransferases, there is appreciable diversity in terms of metabolic and functional integration. Comparability with mammalian biochemistry is clearly evident in the examples presented in Table 28.2. In photorespiration, peroxisomal glutamate:glyoxylate aminotransferase (GGAT) promotes the reaction between glutamate and glyoxylate to yield α-ketoglutarate and glycine. In combination, both GGAT and serine:glyoxylate aminotransferase contribute significantly to the synthesis of other amino acids (Igarashi *et al.*, 2006). In mammals, GGAT is considered to be identical to alanine-2-oxoglutarate aminotransferase (Noguchi *et al.*, 1977). In humans, primary hyperoxaluria type 1 is a severe kidney disorder caused by mutations in alanine:glyoxylate aminotransferase (Hopper *et al.*, 2008).

The functions of decarboxylases reflect another feature of biochemical compatibility between plants and animals. Thus, the synthesis of GABA (Chapter 7), polyamines, dopamine and serotonin (Chapter 20) all require the action of an appropriate decarboxylase in plants (Table 28.3), as in animals (D'Mello, 2003b; Sahin and Kirik, 2012). Furthermore, in the production of melatonin (Chapter 21), the role of tryptophan decarboxylase is considered in two branches of the biosynthetic pathway. The glycine–serine interconversion is well established in both plants and animals, providing one-carbon units for a multitude of biochemical reactions (Bauwe and Kolukisaoglu, 2003). However, glycine decarboxylase is also attributed with unique photorespiratory activity that modulates photosynthesis and plant growth (Hylton *et al.*, 1988; Timm *et al.*, 2012). In addition, this enzyme complex appears to be the target of inhibition by the host-specific toxin victorin produced by *Cochliobolus victoriae*, the causal agent of victoria blight in oats (Navarre and Wolpert, 1995). The mitochondrial glycine

Table 28.1. Comparative aspects of amino acid metabolism in plants and mammals: relative roles of selected enzymes, pathways and bioactive compounds. The evidence presented is designed to highlight common as well as unique features.

Enzyme	Plants	Mammals
Glutamate dehydrogenase (GDH)	Catalyses the reversible oxidative deamination of glutamate (Chapter 1)	Catalyses the reversible oxidative deamination of glutamate (Smith and Stanley, 2012)
Aminotransferases	Transfer of amino groups (Chapters 2–4; see also Table 28.2)	Transfer of amino groups (Conway, 2012)
Decarboxylases	Removal of CO_2 accompanied by the synthesis of critical bioactive molecules (Chapters 7 and 10; see also Table 28.3)	Removal of CO_2 accompanied by the synthesis of critical bioactive molecules (Ueno *et al*., 2012)
Glutamine synthetase (GS)	Condensation of ammonia with glutamate to form glutamine (Chapter 5)	Synthesis of glutamine (Mok and Hankard, 2012)
Arginase	Limited synthesis of urea; synthesis of NO (Chapter 8); provides ornithine for polyamine biosynthesis (Chapters 9 and 10)	Synthesis of urea in a fully functional ornithine cycle; primary mechanism for the disposal of waste N; synthesis of NO; provides precursor for polyamine biosynthesis (Caldwell *et al*., 2012)
Serine racemase	Converts L-serine to D-isomer in plants (Hepler *et al*., 2012); glutamate receptor-like genes regulated by D-serine (Michard *et al*., 2011)	D-Serine is the endogenous ligand for the 'glycine' binding site of the *N*-methyl-D-aspartate subtype of glutamate receptor in the brain and retina (Smith *et al*., 2012)
S-Adenosylmethionine synthetase	Enables synthesis of ethylene (ET) (Wang *et al*., 2002)	Enzyme well conserved through evolution and as a component of the methionine cycle facilitates transmethylation reactions (Mato *et al*., 2012)
Shikimate pathway	Synthesis of tryptophan, tyrosine and phenylalanine (Tan *et al*., 2006)	Tryptophan and phenylalanine dietary essentials (Pillai and Kurpad, 2012)
Phenylalanine-ammonia lyase (PAL)	Facilitates synthesis of salicylic acid (SA) (Klessig *et al*., 2000)	Enzyme absent in humans, but recombinant form of PAL has been proposed for enzyme-substitution therapy in patients with phenylketonuria (PKU) (Koch and Moseley, 2012) SA derivative, aspirin, used as an anti-inflammatory and analgesic drug (Deng *et al*., 2001)

cleavage system constitutes a major route for the catabolism of glycine and serine in mammals. In a study of a patient with non-ketotic hyperglycinemia, adverse effects were attributed to structural abnormality of the H-protein in this enzyme complex (Kikuchi and Hiraga, 1982). In addition, overexpression of glycine decarboxylase has been implicated in tumorigenesis (Zhang *et al*., 2012). It remains to be established whether the partial alleviation of methionine toxicity in animals by glycine supplementation is another facet of the activity of this enzyme or instead a factor involving one-carbon metabolism (Katz and Baker, 1975).

In addition, signal transduction, whether effected directly or via associated metabolites of amino acids, represents another significant example transcending the traditional concepts of

Table 28.2. Role of principal aminotransferase enzymes in plants.

Enzyme	Role
Glutamate-glyoxylate aminotransferase	Catalyses the reaction between glutamate and glyoxylate to produce α-ketoglutarate and glycine (Igarashi *et al.*, 2006)
Alanine aminotransferase	Biosynthesis of alanine from pyruvate (Chapter 2)
Aspartate aminotransferase	Biosynthesis of aspartate from oxaloacetate (Chapter 3)
Tyrosine aminotransferase	Transamination of tyrosine to ρ-hydroxyphenylpyruvate (Sandorf and Hollander-Czytko, 2002; Chapter 4)
Ornithine-δ-aminotransferase	Essential for arginine catabolism but not for proline biosynthesis (Funck *et al.*, 2008)
Diaminopimelate aminotransferase	Lysine biosynthesis (Sobolev *et al.*, 2013; Chapter 13)
Histidinol-phosphate aminotransferase	Histidine biosynthesis (Petersen *et al.*, 2010; Chapter 14)
Tryptophan aminotransferase	Conversion of tryptophan to indole-3-pyruvate (Won *et al.*, 2011; Chapter 19)
Prephenate aminotransferase	Promotes phenylalanine biosynthesis via arogenate (Maeda *et al.*, 2011)
Branched-chain aminotransferase	Involved in the metabolism of leucine, isoleucine and valine; branched-chain aminotransferase 4 implicated in chain elongation in biosynthesis of methionine-derived glucosinolates (Schuster *et al.*, 2006; Chapter 22)

Table 28.3. Enzymes catalysing the decarboxylation of amino acids in higher plants. In the removal of CO_2, important, and, in certain instances, identical bioactive molecules are produced in both plants and animals.

Enzymes	Significance
Glutamate decarboxylase	Synthesis of γ-aminobutyrate (GABA) (Chapter 7)
Lysine decarboxylase	Synthesis of quinolizidine alkaloids (Petterson *et al.*, 1991; Bunsupa *et al.*, 2012)
Arginine decarboxylase	Required for polyamine and GABA signalling (Kim *et al.*, 2013)
Ornithine decarboxylase	Synthesis of polyamines (Chapters 9 and 10)
Glycine decarboxylase	Controls flow of carbon through both photosynthesis and photorespiration (Timm *et al.*, 2012)
Serine decarboxylase	Direct synthesis of ethanolamine (Rontein *et al.*, 2001)
S-Adenosylmethionine decarboxylase	Synthesis of spermidine and spermine (Torrigiani *et al.*, 2005; Chapter 10)
Histidine decarboxylase-like mRNA	Implicated in tomato fruit ripening (Picton *et al.*, 1993)
Tyrosine decarboxylase	Synthesis of tyramine (Park *et al.*, 2012)
3,4-Dihydroxyphenylalanine (DOPA) decarboxylase	Synthesis of dopamine, a precursor of isoquinoline alkaloids (Facchini and DeLuca, 1995); enzyme activated by drought stress (Świędrych *et al.*, 2004)
Phenylacetaldehyde synthase	Bifunctional homotetrameric enzyme catalysing the decarboxylation and oxidation of phenylalanine (Kaminaga *et al.*, 2006)
Tryptophan decarboxylase	Serotonin biosynthesis (Kang *et al.*, 2007; Chapter 20)

plant and mammalian biochemistry. Ca^{2+} conduction by an amino acid-gated ion channel has been attributed to the existence of glutamate-like receptors regulated by D-serine. The concept of signal transduction is a significant unifying feature of comparative biochemistry, justifying separate consideration as set out in Section 28.15.

An intriguing observation relates to the occurrence and activity of serine racemase (Table 28.1), which is responsible for converting L-serine to its D-isomer in both plants (Gogami *et al.*, 2009; Hepler *et al.*, 2012) and animals (Smith *et al.*, 2012). Specifically, the occurrence of D-serine and the characterization of its racemase have been established in *Arabidopsis* (Fujitani *et al.*, 2006), barley (Fujitani *et al.*, 2007) and rice (Gogami *et al.*, 2009). In plants, glutamate receptor-like genes regulated by D-serine have been reported leading to the conclusion of a novel plant signal transduction cascade similar to the amino acid-mediated neural communication normally associated with animal systems (Michard *et al.*, 2011). D-Serine is the endogenous ligand for the 'glycine' binding site of the *N*-methyl-D-aspartate subtype of the glutamate receptor in the brain and retina (Smith *et al.*, 2012).

The presence of alanine racemase in plants has also been confirmed by Ono *et al.* (2006) in studies with lucerne seedlings. The enzyme resembles bacterial alanine racemase in terms of catalytic requirements and activity. Recent evidence indicates tolerance of transgenic plants to exogenous D-alanine but inhibition by D-valine. Compartmentation of the deamination product 3-methyl-2-oxobutanoate within the plastids failed to prevent D-valine-induced suppression (Gisby *et al.*, 2012). Exogenously supplied D-amino acids are readily transported into plants and further metabolized to the corresponding L isomers, so soil-borne D-amino acids may be actively absorbed and utilized via central metabolic hubs (Gördes *et al.*, 2011). Animals are capable of utilizing certain D isomers with variable efficacy (D'Mello, 2003b). Thus, D-methionine utilization ranges from 75 to 100% efficiency, whereas D-lysine and D-threonine are not utilized at all. However, in the presence of racemic mixtures of other amino acids, D-methionine efficacy is reduced in comparison with the L form.

Comparative aspects are also apparent in plant–insect interactions, as for example in the occurrence and activity of conjugates of plant hormones. The oxylipin JA is now confirmed as a significant factor in the initiation of plant defence responses to biotic and environmental challenges. It is known that activation of JA requires prior conjugation to isoleucine to yield a JA-Ile phytohormone (Staswick and Tiryaki, 2004). In addition, the conjugate SA-Asp has been implicated in the induction of pathogenesis-related gene expression and disease resistance in *Arabidopsis* (Chen *et al.*, 2013). At this level, molecular interactions between the specialist herbivore *Manduca sexta* and its natural host plant *Nicotiana attenuata* have also been reported. Fatty acid-amino acid conjugates in the insect oral secretions appear to be necessary and sufficient to initiate herbivore-specific plant defence cascades (Halitschke *et al.*, 2001).

28.15 Signal Transduction

Although conventionally associated with mammalian neurobiology, signalling has recently emerged as a concept of considerable relevance and debate in the context of plant growth and development, justifying the publication of dedicated journals such as *Plant Signaling and Behavior*. In addition, *Plant Physiology*, *The Plant Cell* and other well-established journals publish an increasing number of papers on plant signalling, particularly at the molecular level and in reference to modulating factors.

Amino acids may act directly in signal transduction, as has been demonstrated in numerous studies with mammalian models, including human subjects affected by neurological and neurodegenerative disorders (Flores *et al.*, 2012). The distinction between excitatory (glutamate and aspartate) and inhibitory (glycine) amino acids in synaptic transmission has emerged as a dominant feature in neurobiology. Amino acids may also act indirectly as precursors of signalling molecules or as components of a more complex order of molecular communication involving phosphorylation cascades and post-translational modifications. Furthermore, in addition to acting as precursors, amino acids may participate in conjugation reactions with derivative hormone signals.

The direct mode of action also exists in plants in a comparable manner to that in mammals (Table 28.4). For example, Ca^{2+} conduction by an amino acid-gated ion channel has been attributed to the existence of GLR homologues of mammalian ionotropic glutamate receptors (iGluRs); in contrast to iGluRs, these are activated by at least three structurally diverse amino acids (Vincill *et al.*, 2012), with methionine being highly effective as an agonist of plant GLR (Tapken *et al.*, 2013). GLR genes are regulated by D-serine in a novel plant signalling mechanism between male gametophyte and pistil tissue (Michard *et al.*, 2011). The involvement of D-serine in glutamate neurotransmission in mammals has been reviewed by Smith *et al.* (2012).

In higher plants, the indirect mode of action of amino acids is exemplified in several ways (Table 28.4). For instance, GABA may act as a signal mediating plant–animal and plant–microbe interactions (Roberts, 2007); in mammals, GABA also participates in signalling networks, exerting an important function as an inhibitory neurotransmitter (Ueno *et al.*, 2012).

Current evidence and opinion implicate a diverse array of phytohormones and other bioactive compounds in complex signal transduction networks that contribute significantly to defence reactions following biotic or environmental challenge. Examples frequently cited are ABA, SA, JA, ET (Fujita *et al.*, 2006), NO and IAA. Hormone signals are relevant here and elsewhere in this volume for several reasons. Importantly, in their respective biosynthetic pathways, ET (Fig. 28.1), SA (Fig. 28.2), NO (Chapter 8) and IAA (Chapter 19) can all be traced back to specific precursor amino acids. Subsequent signalling activity may be modulated by conjugation with various amino acids, as in the cases of SA, JA and IAA. Thus, SA-Asp is the endogenous conjugate that accumulates after pathogen infection (Chen *et al.*, 2013). SA has long been implicated in both localized and systemic disease resistance in plants (Yalpani *et al.*, 1993). JA-Ile is the classical example of a conjugate required for optimal signalling in *Arabidopsis* (Staswick and Tiryaki, 2004; Koo *et al.*, 2011; Widemann *et al.*, 2013) and which is modulated by antagonistic interactions with SA (Kazan and Manners, 2008). Uniquely, IAA-Trp is generally implicated in the regulatory inhibition of auxin activity (Ludwig-Müller, 2011). Other examples of the indirect role of amino acids include LRR-RKs in BR signalling, effective through serine/threonine phosphorylation (Friedrichsen *et al.*, 2000), and the nitration of protein tyrosine residues in the peroxynitrite mediation of defence responses (Vandelle and Delledonne, 2011).

As implied above, several of the proteins facilitating or promoting hormone signalling in plants have now been elucidated, including specific receptors (Santner and Estelle, 2009). The receptor kinase involved in BR signalling (Oh *et al.*, 2009) and the LRR-RKs mediating ET biosynthesis (Xu *et al.*, 2008) are worthy of note in this respect. Furthermore, hormone activity interacts at different levels during plant growth and development (Santner and Estelle, 2009) and in response to a plethora of endogenous and environmental factors.

The context for signal transduction is generally associated with stress and defence metabolism and the initiation of induced resistance, though signalling compounds rarely operate in isolation. The emerging evidence confirms complex interactions or cross-talk in communication networks of plants exposed to biotic and environmental challenges (Fujita *et al.*, 2006). Koornneef and Pieterse (2008) extended the concept of cross-talk to specific signalling compounds. It is generally accepted that SA signals mediate defences against biotrophic pathogens, whereas JA activity is effective against insect herbivores and necrotrophic pathogens. However, SA/JA cross-talk occurs in both directions, and Koornneef and Pieterse (2008) recommend additional work by molecular biologists and ecologists to unravel the ramifying interactions in a collaborative effort. It is hoped that the publication of *Amino Acids in Higher Plants* will assist in this activity.

28.16 Molecular Interactions

As stated by D'Mello (2012), interactions comprise the very essence of amino acid metabolism, exerting profound effects in all living organisms. Nevertheless, the full extent of these interrelationships is often underestimated or, in certain cases, yet to be determined. The potentiating actions of NO and auxin in root development in arginase-negative mutants of *Arabidopsis* is one result of an interaction (Flores *et al.*, 2008). The oxylipin signal jasmonic acid is activated by an enzyme that conjugates it to isoleucine for optimal efficacy, as demonstrated in *Arabidopsis* (Staswick and Tiryaki, 2004). Another significant example concerns tyrosine phosphorylation of the specific receptor kinase as a component of BR signal transduction (Oh *et al.*, 2009). An advanced order of complexity often prevails in higher organisms, including plants, and greater elucidation of these interactions should emerge with increasing momentum in genetics and molecular biology.

It is instructive to consider the complexity of interactions in further detail, recognizing that the evidence generally conforms within the

Table 28.4. Amino acids implicated in signal transduction in higher plants. Participation may be direct, as in the case of glutamate and proline. Of equal importance is the indirect role of amino acids via: (i) associated enzymes; (ii) the provision of specific precursors for synthesis of plant hormones; (iii) the activation of hormones through conjugation; or (iv) post-translational interactions with receptors or other structures. Homology with mammalian metabolism is implied and developed further in this edition of *Amino Acids in Higher Plants*.

Amino acid	Role in signal transduction
Glutamate	Ca^{2+} conduction by an amino acid-gated ion channel attributed to the existence of glutamate-like receptor (GLR) homologues of mammalian ionotropic glutamate receptors (iGluRs); activated by at least three structurally diverse amino acids, in contrast to iGluRs (Vincill *et al.*, 2012); methionine highly effective as an agonist of plant GLR (Tapken *et al.*, 2013); GLR genes regulated by D-serine in a novel plant signalling mechanism between male gametophyte and pistil tissue (Michard *et al.*, 2011)
	γ-Aminobutyrate (GABA) implicated as a signal that mediates plant–animal and plant–microbe interactions (Roberts, 2007)
Proline	Signalling molecule modulating mitochondrial function (Szabados and Savouré, 2010)
Hydroxyproline	Systemic signalling mediated by hydroxyproline-rich glycopeptides to activate gene expression (Pearce and Ryan, 2003)
Aspartate	Aspartate aminotransferase implicated in signal transduction (Chapter 3)
Arginine	Source of NO, a long-established signal transduction molecule in mammals. Attributed with diverse functions in plants (Chapter 8); increasing evidence that NO operates as a signalling molecule in plants (Chandok *et al.*, 2003); arginase-negative mutants display increased NO signalling in root development (Flores *et al.*, 2008)
Ornithine	Precursor of polyamines in plants and other organisms. Polyamines may function as 'surrogate messengers' facilitating other signalling molecules, such as plant hormones and NO (Mattoo and Handa, 2008; Chapters 9 and 10)
Glycine	Dubos *et al.* (2003) identified a synergistic action of glutamate and glycine at *N*-methyl-D-aspartate (NMDA)-like receptors, predating the divergence of plants and animals
Methionine	Precursor of ethylene (ET) involving the initial synthesis of *S*-adenosylmethionine (Wang *et al.*, 2002)
Leucine	Leucine-rich repeat receptor kinases (LRR-RKs) mediate signalling in cell wall biosynthesis (Xu *et al.*, 2008)
	Leucine zipper transcription factors involved in abscisic acid (ABA)-dependent signal transduction (Uno *et al.*, 2000)
Isoleucine	Activates oxylipin signal, jasmonic acid (JA) (Staswick and Tiryaki, 2004)
	Signalling role in the regulation of root development (Yu *et al.*, 2013)
Phenylalanine	Phenylalanine ammonia-lyase (PAL) yields cinnamic acid, a key intermediate in the synthesis of phenylpropanoids, which are precursors of flavonoids, isoflavonoids, coumarins and stilbenes. These compounds function as regulatory molecules in signal transduction and communication with other organisms (Huang *et al.*, 2010)
	Phenylalanine-derived salicylic acid (SA) is attributed with critical signalling functions (Klessig *et al.*, 2000), with the SA-Asp (aspartate) conjugate operating as a 'mobile' molecule activating gene expression (Chen *et al.*, 2013)
Tyrosine	Tyrosine phosphorylation of receptor kinase activates brassinosteroid (BR) signal transduction (Oh *et al.*, 2009); nitration of tyrosine residues is central to peroxynitrite signal (Vandelle and Delledonne, 2011)
Tryptophan	Auxins comprise a group of tryptophan-derived signal molecules (Pelagio-Flores *et al.*, 2011; Won *et al.*, 2011; Chapter 19)
	Source of serotonin (Chapter 20), a signal molecule conserved in plants and animals; assumed to act as a natural auxin inhibitor (Pelagio-Flores *et al.*, 2011)
	Tryptophan-derived melatonin attributed with signalling/antioxidant functions in plants in stress situations (Chapter 21)

widely accepted system of classification of the metabolic effects of bioactive compounds in living organisms (Table 28.5). Thus, examples of additive, synergistic, potentiating and antagonistic effects have been recorded. The evidence presented indicates that complex interactions occur between structurally similar amino acids in plants as they do in animals (D'Mello, 2003a), although most of the observations relate to the effects of signal transduction compounds derived from or activated by amino acids after conjugation or post-translational interactions. Predictably, the majority of examples listed in Table 28.5 are associated with synergistic and antagonistic effects in the modulation of plant responses during growth and development or their responses to biotic and environmental stress factors.

28.16.1 Synergistic effects

The distinctions between effects are not consistently apparent. For example, ET and ABA signalling cascades act synergistically in inhibiting root growth but antagonistically in the control of seed dormancy (Beaudoin *et al.*, 2000). Glutamate and glycine act synergistically at NMDA-like receptors (Dubos *et al.*, 2003), while upregulated synthesis and reduced catabolism synergistically increase seed lysine content (Zhu and Galili, 2003).

SA exerts a critical role in the activation of defence mechanisms (Klessig *et al.*, 2000), whereas the activation of PAL and accumulation of SA is light dependent (Zeier *et al.*, 2004). Synergism has been implicated in the effect of kinetin on auxin-induced ET synthesis (Lau and Yang, 1973), and in the induction of defence genes by MeJA and ET (Xu *et al.*, 1994). Complex synergistic interactions have been attributed to the effects of volicitin, JA and ET in mediating insect-induced volatile emissions (Schmelz *et al.*, 2003), and in the mode of action of JA and glucose in the modulation of glucosinolate production (Guo *et al.*, 2013).

Synergism is further exemplified by interactions between NO and reactive oxygen species (NO-ROS) modulating both the activation of disease resistance mechanisms (Delledonne *et al.*, 2001) and stomatal closure in response to abiotic stress (Neill *et al.*, 2008). Arginine decarboxylase activates polyamine and GABA synthesis and triggers NO and ROS bursts, ultimately resulting in cell death and defence reactions (Kim *et al.*, 2013).

Higher orders of synergism emerge in BR-mediated stress tolerance modulated by interactions with the ABA, ET and SA pathways (Divi *et al.*, 2010). In addition, BRs act synergistically with light, gibberellic acid and auxin signalling pathways (Ryu and Hwang, 2013).

28.16.2 Antagonisms

As indicated in Table 28.5, antagonisms are widely documented at the level of structurally related amino acids. Included here is the canavanine–arginine antagonism, operational in both plants and animals (Chapter 27), branched-chain amino acid antagonisms affecting seedling growth (Miflin, 1969) and the inhibition of PAL and lignin biosynthesis by mimosine (Andrade *et al.*, 2008).

Antagonisms have also been reported to occur among signal transduction molecules derived from amino acids and other substrates, particularly in plants subjected to biotic and environmental stresses. Hence, the expression of pathogenesis-related protein genes is affected by antagonistic modulation of SA and JA activities (Niki *et al.*, 1998). In addition, SA inhibits the JA-induced resistance of *Arabidopsis* to *Spodoptera exigua* (Cipollini *et al.*, 2004), while increased SA synthesis antagonizes JA and JA-dependent defences in plants after herbivory (Rayapuram and Baldwin, 2007). Other reports indicate that SA inhibits pathogen growth by repressing auxin signalling (Wang *et al.*, 2007), and that mutually antagonistic interactions operate between the SA and JA pathways (Kazan and Manners, 2008). The concept that SA engages in antagonistic or inhibitory effects is emerging as a consistent underlying feature of activity.

Examples of interactions between ABA and other signal transduction compounds relate to both abiotic and biotic stress. For example, salinity responses are modulated by ABA–JA antagonisms (Moons *et al.*, 1997), with a similar type of interaction between SAR and the ABA-mediated abiotic stress responses (Yasuda *et al.*, 2008). In biotic stress, an antagonism between ABA and JA-ET transduction pathways modulates defence

Table 28.5. Interactions involving amino acids or specific plant hormones: classification of effects as reported in the literature. Hormones selected on the basis of biosynthetic relationships or other indirect association with amino acids, involving conjugation or underpinning structural or functional support for optimal biochemical activity. A selection of the evidence is presented here to illustrate the diversity of effects in higher plants. In general, the examples below fall within well-established criteria for the classification of biochemical responses in living organisms exposed to stress.

Classification	Effects
Additive	Additive growth-inhibiting properties of the canaline-urea cycle amino acids (Rosenthal *et al.*, 1976)
	Simultaneous activation of salicylic acid (SA)-dependent systemic acquired resistance and jasmonic acid (JA)-dependent induced systemic resistance pathways resulted in an additive effect on induced disease protection (van Wees *et al.*, 2000)
	Nitric oxide and reactive oxygen species (NO-ROS) interactions may induce additive responses in stomatal closure and abiotic stress (Neill *et al.*, 2008)
	Additive effect in ethylene (ET) synthesis by combined mutants (Wang *et al.*, 2002); exogenous H_2O_2 treatment induced an additive effect on the alternative respiratory pathway in salt stress in comparison with ET (Wang *et al.*, 2010)
Synergistic/activating/ mediating	Synergistic effects of metabolically related amino acids on plant growth (Dunham and Bryan, 1971); synergistic growth-inhibiting properties of the canaline-urea cycle amino acids (Rosenthal *et al.*, 1976); glutamate and glycine act synergistically at *N*-methyl-D-aspartate (NMDA)-like receptors (Dubos *et al.*, 2003); upregulated synthesis and reduced catabolism synergistically increase seed lysine content (Zhu and Galili, 2003)
	SA exerts critical role in activation of defence mechanisms (Klessig *et al.*, 2000); activation of PAL and accumulation of SA is light dependent (Zeier *et al.*, 2004)
	Synergistic effect of kinetin on auxin-induced ET synthesis (Lau and Yang, 1973); synergistic induction of defence genes by methyl jasmonate (MeJA) and ET (Xu *et al.*, 1994); synergy between MeJA and ET pathways required for induction of defensin gene (Penninckx *et al.*, 1998); synergistic interactions between volicitin, JA and ET mediate insect-induced volatile emissions (Schmelz *et al.*, 2003); JA and glucose synergistically modulate accumulation of glucosinolates (Guo *et al.*, 2013)
	NO and ROS exert key roles in activation of disease resistance mechanisms (Delledonne *et al.*, 2001); NO-ROS interactions may induce synergistic responses in stomatal closure and abiotic stress (Neill *et al.*, 2008); arginine decarboxylase activates polyamine and γ-aminobutyrate (GABA) synthesis and triggers NO and ROS bursts, ultimately resulting in cell death and defence reactions (Kim *et al.*, 2013)
	Synergism between Ca ions and ABA in prevention of stomatal opening (De Silva *et al.*, 1985); ABA and ET cascades synergistically inhibit root growth (Beaudoin *et al.*, 2000)
	Brassinosteroid (BR)-mediated stress tolerance modulated by interactions with ABA, ET and SA pathways (Divi *et al.*, 2010); BRs act synergistically with light, gibberellic acid and auxin signalling pathways (Ryu and Hwang, 2013)
	Synergistic derepression of GA transduction by removal of negative regulators (Dill and Sun, 2001)
Potentiating	Potentiation of pathogen-specific defence mechanisms by BABA, a xenobiotic amino acid (Zimmerli *et al.*, 2000)
	Pathogen-induced accumulation of JA and camalexin enhanced further in the dark (Zeier *et al.*, 2004)

Continued

Table 28.5. Continued.

Classification	Effects
Antagonistic	
Structural (competitive) inhibition	Canavanine–arginine antagonism demonstrated in plants and animals (Chapter 27)
	Complex branched-chain amino acid antagonisms affect seedling growth (Miflin, 1969)
	Inhibition of phenylalanine ammonia-lyase (PAL) and lignin biosynthesis by mimosine (Andrade *et al.*, 2008)
Functional (non-competitive) inhibition	Glutamine inhibits β-aminobutyrate (BABA)-induced stress resistance and priming (Wu *et al.*, 2010)
	Herbicidal inhibition of amino acid biosynthesis in crop plants widely reported (Tan *et al.*, 2006)
	Expression of pathogenesis-related protein genes is affected by antagonistic effects of SA and JA (Niki *et al.*, 1998); SA inhibits JA-induced resistance of *Arabidopsis* to *Spodoptera exigua* (Cipollini *et al.*, 2004); increased SA synthesis antagonizes JA and JA-dependent defences in plants after herbivory (Rayapuram and Baldwin, 2007); SA inhibits pathogen growth by repressing auxin signalling (Wang *et al.*, 2007); mutually antagonistic interactions between SA and JA pathways (Kazan and Manners, 2008)
	Salt stress modulated by abscisic acid (ABA)–JA antagonism (Moons *et al.*, 1997); antagonism between ABA and JA-ET signalling pathways modulates defence gene expression (Anderson *et al.*, 2004); antagonistic interaction between SAR and ABA-mediated abiotic stress responses (Yasuda *et al.*, 2008); antagonism between SA and ABA reflects incipient host–pathogen conflict and affects plant defence responses (de Torres Zabala *et al.*, 2009)
	ET antagonizes ABA signalling during seed dormancy/development (Beaudoin *et al.*, 2000; Ghassemian *et al.*, 2000); ET renders the JA response pathway insensitive to subsequent suppression by SA (Leon-Reyes *et al.*, 2010)
	Amide-linked tryptophan conjugates of indole acetic acid (IAA) and JA operate as auxin antagonists (Staswick, 2009a,b); auxin–cytokinin antagonism modulates plant–pathogen interactions (Naseem and Dandekar, 2012)
Toxicity	Threonine accumulation in a mutant of *Arabidopsis* suppressed growth (Sarrobert *et al.*, 2000)
	Proline toxicity in an *Arabidopsis* mutant deficient in proline dehydrogenase (Nanjo *et al.*, 2003)
	Phytotoxicity of 3,4-dihydroxyphenylalanine (DOPA) attributed to oxidative damage caused by ROS and free radical species generated during the biosynthesis of melanin (possible structural antagonism) (Hachinohe and Matsumoto, 2007)

gene expression (Anderson *et al.*, 2004), whereas the antagonism between SA and ABA reflects incipient host-pathogen conflict which affects plant defence responses (de Torres Zabala *et al.*, 2009). ET antagonizes ABA signalling during seed dormancy/development (Beaudoin *et al.*, 2000; Ghassemian *et al.*, 2000). In addition, ET renders the JA response pathway insensitive to subsequent suppression by SA (Leon-Reyes *et al.*, 2010).

Uniquely, amide-linked Trp conjugates of IAA and JA operate as auxin antagonists (Staswick, 2009a,b) and an auxin–cytokinin antagonism modulates plant–pathogen interactions (Naseem and Dandekar, 2012).

28.16.3 Integration

It is accepted that the summary of interactions as presented in Table 28.5 at best represents an approximation of processes that prevail in the molecular communication of plants subjected to stress. In other words, analysis of two- or three-way interactions is unlikely to result in reliable conclusions. Thus, BR-mediated stress tolerance involves interactions with the ET, SA and ABA pathways (Divi *et al.*, 2010). The work of Leon-Reyes *et al.* (2010) illustrates this point further with reference to prioritization of the ET/JA pathway over the SA network during multifactorial stress challenges. The final resolution should emerge with additional investigations designed to elucidate the full extent and metabolic significance of cross-talk and integration into complex models. The work of Felton *et al.* (1999) illustrates the potential difficulties that lie ahead. Tobacco plants with reduced SAR (by silencing *PAL*) were more resistant to insect herbivory than plants with elevated levels of phenylpropanoids. The observation that the auxin conjugate IAA-Asp promotes pathogen progression in plants by regulating the transcription of virulence genes adds to the perception of a significantly greater complexity of molecular communication and defence networks in higher plants (González-Lamothe *et al.*, 2012). For instance, Zeier (2013) proposed that in addition to defence cascades regulated by the classical stress hormones, distinct amino acid metabolic pathways also constitute integral components of the immune system in plants.

Reconciliation of certain aspects of the evidence presented in Table 28.5 may be another issue of relevance. For example, on the one hand, simultaneous activation of SA-dependent systemic acquired resistance and JA-dependent induced systemic resistance pathways resulted in an additive effect on induced disease protection (van Wees *et al.*, 2000). On the other hand, the expression of pathogenesis-related protein genes is affected by the antagonistic effects of SA and JA (Niki *et al.*, 1998), while SA inhibits the JA-induced resistance of *Arabidopsis* to *S. exigua* (Cipollini *et al.*, 2004). In addition, increased SA synthesis antagonizes JA and JA-dependent defences in plants after herbivory (Rayapuram and Baldwin, 2007); mutually antagonistic interactions between SA and JA pathways have also been reported (Kazan and Manners, 2008). It is possible that the plant species and the source of the biotic stress may determine the type of signal transduction interaction that prevails in any particular instance. Other evidence indicates that ET and ABA signalling cascades operate synergistically in repression of root growth and antagonistically in the control of seed dormancy (Beaudoin *et al.*, 2000). Thus, the organ and developmental stage add to the complexity of unravelling signal transduction networks in plants. Another issue worthy of consideration is the ability of pathogens to modulate the ABA signalling pathway to suppress defence responses in the plant, thereby promoting disease (de Torres-Zabala *et al.*, 2007).

This review would be incomplete without reference to the symbiotic relationships existing in legumes. N fixation is generally attributed to the action of *Rhizobium* bacteroids residing within the root nodules of leguminous plants. The traditional view has been that these bacteroids obtain carbon and energy from the host in the form of dicarboxylic acids. However, Lodwig *et al.* (2003) claimed that a complex exchange of amino acids occurs in a manner that might, on initial interpretation, appear to resemble a futile activity. Nevertheless, it is suggested that the plant supplies the bacteroids with amino acids, permitting them to curtail ammonium assimilation. The bacteroids then return these amino acids to the plant for asparagine biosynthesis (see Chapter 6).

28.17 Biotic and Environmental Stress Responses

Biotic stress responses in plants are induced after infection with pathogenic microorganisms or damage by insect herbivores. Following initial microbial infection, plants generally exhibit enhanced resistance to subsequent challenge by that particular pathogen. This response (SAR) is widely recognized and constitutes a major feature of defence strategies in plants (Conrath, 2006). Signal transduction and synthesis of pathogenesis-related proteins and defence compounds (including phytoalexins) are major manifestations of the establishment and efficacy of SAR. Insect herbivory activates several endogenous signals from the wounded tissues, resulting in the deployment of plant defence compounds. Collectively, these processes comprise the 'wound response' of plants to insect damage (Léon *et al.*, 2001; Howe, 2004; Fürstenberg-Hägg *et al.*, 2013). Genetic analysis of the wound response suggests that systemin and prosystemin are upstream components of a JA-dependent cascade, though a JA-independent pathway is also possible (Howe, 2004). Others implicate oligosaccharides, ET and ABA as additional contributors to the wound response (Léon *et al.*, 2001). In plants exposed to abiotic stress, similar generic cascades may occur that include the involvement of receptors, phosphorylation, transcription factors, gene expression and signalling pathways (Xiong *et al.*, 2002).

That the metabolism of amino acids in higher plants is associated with, or modulated, by defence responses to biotic and environmental stresses is now unquestionable (Table 28.6). Equally, it is important to recognize that several amino acids probably do not participate in the innate immunology of plants. In such instances, changes in the metabolism or disposition of amino acids merely reflect the incidental result of external influences. In order to establish definitively any defence functions of amino acids, it is necessary to employ sophisticated methodologies primarily based on molecular genetics. Although the evidence presented above in support of signal transduction interactions alluded to defence responses, it is instructive to consider the effects of specific elicitors, whether of biotic or environmental origin. In the case of biotic stress, it is relevant to consider the effects of specific pathogens and insect herbivores.

The indirect role of glutamate in plant defences is an evolving issue. The evidence as presented in Table 28.6 indicates several lines of investigation including: the modulation of GDH after infection by viruses and bacteria (Pageau *et al.*, 2006); GDH transcription enhancement in tissues infected by *B. cinerea* (Dulermo *et al.*, 2009); and the involvement of a putative glutamate receptor in response to cryptogein, a well-recognized elicitor of defence responses (Vatsa *et al.*, 2011). It was on the basis of genetic evidence that Manzoor *et al.* (2013) concluded that the glutamate receptor AtGLR3.3 is a specific component of defence signalling and resistance of *Arabidopsis* to pathogenic fungi.

It is well documented that proline accumulates in plants exposed to abiotic stress (Sharma and Dietz, 2006; Lehmann *et al.*, 2010; Szabados and Savouré, 2010; Chapter 15). In heavy metal toxicity, proline is attributed with at least three properties, including osmoregulation and redox regulation, chelation of mineral ions and the scavenging of free radicals (Sharma and Dietz, 2006). Proline may also confer structural integrity to cellular components by virtue of its cyclic configuration. In addition, it may act as a compatible solute, protecting cellular architecture (Lehmann *et al.*, 2010), as a signalling molecule and as a modulator of gene expression (Szabados and Savouré, 2010).

A diverse range of genes may be induced in plants in response to changing environmental conditions and to other internal and external factors. In soybean, a gene encoding a proline-rich protein (PRP) and its transcription are regulated by salinity and drought stresses and the expression of PRP mRNA occurred in response to SA and virus infection (He *et al.*, 2002).

The indirect involvement of arginine in defence responses include: overexpression of plant arginase to degrade arginine in the herbivore midgut (Chen *et al.*, 2005); upregulation of arginase in *Arabidopsis* in response to *B. cinerea* (Brauc *et al.*, 2012); arginase induction to suppress gall development during clubroot infection in *Arabidopsis* (Gravot *et al.*, 2012) and arginine-derived NO (Chapter 8) accumulation in the cytosol following salinity stress in *Arabidopsis* plants (Corpas *et al.*, 2009). Furthermore, arginine decarboxylase has been implicated in key defence and cell death regulation in *Xanthomonas campestris*-challenged *Capsicum annuum* (Kim *et al.*, 2013), although contrasting observations have been noted. Accordingly, the upregulation of arginase may exert a defensive role by reducing the availability of arginine as a nutrient for phytophagous insects (Gravot *et al.*, 2012), but may also negatively affect tolerances of plants to abiotic factors, including water deficit and salinity stress (Shi *et al.*, 2013).

As with proline, GB is recognized as an osmolyte/compatible solute, accumulating in a number of plant species in response to stresses such as high salinity and low temperatures (Sakamoto and Murata, 2002). The synthesis of GB varies according to plant species, with relatively high accumulation in the chloroplasts of spinach and barley in comparison with *Arabidopsis* and tobacco. The mode of action of GB is attributed to activity towards the photosystem II complex and Rubisco and to the maintenance of cellular structure (Sakamoto and Murata, 2002). It may be argued that GB accumulation is merely a manifestation of adverse effects of stress and

Table 28.6. Amino acids/enzymes implicated in, or modulated by, stress and defence responses in higher plants. Mediation by amino acids may be direct at physiologically relevant concentrations or at excess accumulations. In addition, modulation may be indirect, via specific enzymes, hormones or other metabolites and structures associated with amino acids.

Amino acid/ enzyme	Observations
Free amino acids	Contribution of physiologically relevant concentrations of free amino acids to salinity stress regulation by maintaining optimal K^+/Na^+ ratios (Cuin and Shabala, 2006)
Glutamate/glutamate dehydrogenase (GDH)	Modulation of GDH after infection of *Nicotiana tabacum* by viruses and bacteria (Pageau *et al*., 2006); GDH transcription enhanced in tissues infected by *Botrytis cinerea* (Dulermo *et al*., 2009); involvement of a putative glutamate receptor in plant defence in response to a recognized elicitor (Vatsa *et al*., 2011); Genetic evidence that a specific glutamate receptor participates in elicitor/pathogen-mediated defence pathways in the resistance of *Arabidopsis thaliana* to *Hyaloperonospora arabidopsidis* (Manzoor *et al*., 2013)
Glutamine/glutamine synthetase (GS)	Modulation of cytosolic GS after infection of *N. tabacum* by specified phytopathogenic viruses and bacteria (Pageau *et al*., 2006); influence on the pathogenicity of the fungus *Colletotrichum lindemuthianum* to *Phaseolus vulgaris*, with glutamine accumulation correlated with increases in cytosolic GS mRNA (Tavernier *et al*., 2007)
Proline	Proline accumulates in response to environmental stress (Szabados and Savouré, 2010); proline dehydrogenase 2 upregulated in salt stress (Funck *et al*., 2010); regulation of proline metabolism by NO in salinity stress (Fan and Du, 2012)
	Oxidative cross-linking of a proline-rich plant cell wall protein precedes the expression of transcription-dependent defences (Bradley *et al*., 1992)
Aspartate/aspartate aminotransferase (AAT)	Overexpression of AAT influences amino acid metabolism and defence responses against *B. cinerea* infection in *A. thaliana* (Brauc *et al*., 2011)
	Aspartate oxidase modulates stomatal immunity (Macho *et al*., 2012)
Lysine	Catabolic intermediate pipecolic acid regulates systemic acquired resistance (SAR), defence priming and local resistance to bacterial pathogens (Zeier, 2013)
Arginine/arginase	Overexpression of plant arginase degrades arginine in the herbivore midgut (Chen *et al*., 2005); upregulation of arginase in *A. thaliana* influences defence responses towards *B. cinerea* (Brauc *et al*., 2012); arginase induction represses gall development during clubroot infection in *A. thaliana* (Gravot *et al*., 2012)
	Arginine-derived NO (Chapter 8) accumulates in the cytosol following salinity stress in *Arabidopsis* plants (Corpas *et al*., 2009)
	Arginine decarboxylase implicated in key defence and cell death regulation in *Xanthomonas campestris*-challenged *Capsicum annuum* plants (Kim *et al*., 2013)
Ornithine	Accumulation of ornithine suggested as an alternative mode of increasing abiotic stress tolerance directly (Kalamaki *et al*., 2009a), and specifically to salinity and drought (Kalamaki *et al*., 2009b), or via ornithine δ-aminotransferase activity (Stránská *et al*., 2008; da Rocha *et al*., 2012)
N (δ)-Acetylornithine	Insect predation and *Pseudomonas syringae* infection increase accumulation of this non-protein amino acid in *Arabidopsis* (Adio *et al*., 2011)
Glycine	Glycine and choline are precursors of glycine betaine (GB), a low molecular mass molecule which may enhance tolerance in plants to stress induced by salinity, photoinhibition and low temperature (Holmström *et al*., 2000; Sakamoto and Murata (2002); GB is considered to act as a compatible solute exerting a role in response to abiotic stresses such as high salinity and low environmental temperature
	Glycine-rich proteins accumulate in vascular tissues as structural components of a defence mechanism (Gómez *et al*., 1988; Mousavi and Hotta, 2005)

Continued

Table 28.6. Continued.

Amino acid/ enzyme	Observations
Cysteine	Cytosolic homeostasis modulates the plant response to pathogens (Chapter 12)
Threonine/threonine deaminase (TD)	One of the genes encoding TD (*TD2*) is attributed with insect defence functions; silencing *TD2* impairs JA-Ile-mediated defences towards *Manduca sexta* (Kang *et al.*, 2006)
Histidine	Functions as a ligand for Ni; hyperaccumulation of Ni implicated in plant defence towards predation and pathogens (Chapter 14)
Isoleucine/branched-chain amino acids (BCAAs)	Jasmonic acid-isoleucine (JA-Ile) conjugate (Staswick and Tiryaki, 2004) participates in defence processes directed at insect herbivory (Kang *et al.*, 2006)
	Catabolism of BCAAs affect plant resistance to distinct pathogen classes by modulating defences regulated by salicylic acid (SA)-JA interactions (Zeier, 2013)
Phenylalanine	Regulates localized and systemic disease resistance in virus-inoculated tobacco via pathway to SA biosynthesis (Yalpani *et al.*, 1993); production of SA precursors is a major function of phenylalanine ammonia-lyase (PAL) in resistance of *Arabidopsis* to *Peronospora parasitica* (downy mildew; Mauch-Mani and Slusarenko, 1996); SA-dependent defence pathway is effective against *Oidium neolycopersici* and *B. cinerea* (Achuo *et al.*, 2004)
Tyrosine	Tyrosine decarboxylase increased by high salt; 3,4-dihydroxyphenylalanine (DOPA) decarboxylase and tyrosine hydroxylase activated by drought (Świędrych *et al.*, 2004); tyrosine decarboxylase cDNA substantially induced by wounding (Lehmann and Pollmann, 2009)
Tryptophan	Source of auxin that differentially affects resistance to separate pathogen groups (Kazan and Manners, 2009; Chapter 19); indole acetic acid-aspartate (IAA-Asp) conjugate promotes plant disease development (González-Lamothe *et al.*, 2012)
	Involvement of the tryptophan pathway in the defence responses against pathogenic infection via serotonin synthesis (Ishihara *et al.*, 2008; see also Chapter 20)
	Overexpression of tryptophan synthase beta1 confers tolerance to cadmium stress (Sanjaya *et al.*, 2008)
	Precursor of camalexin implicated in defence towards a wide range of pathogens including *B. cinerea* and *Alternaria brassicicola* (Glawischnig, 2007; Møldrup *et al.*, 2013)
	Source of melatonin, implicated in its antioxidant role in plant stress defence mechanisms (Chapter 21)
	Substrate for synthesis of deterrent indole glucosinolates (Chapter 22); conversion of one indole glucosinolate to another enhances defensive benefit towards insect predation (Kim and Jander, 2007)
	Indole (and other) alkaloids implicated in plant defence against insect predation and fungal infection (Petterson *et al.*, 1991)
β-Aminobutyric acid (BABA)	Potentiates pathogen-specific defence mechanisms in plants (Chapter 27)
γ-Aminobutyric acid (GABA)	Accumulation in response to environmental stress and invertebrate herbivory (Chapter 27)

not part of the mechanism to restore normality (Ashraf and Foolad, 2007). However, Holmström *et al.* (2000) observed enhanced tolerance to salinity and low temperature stresses in transgenic tobacco lines that accumulated GB. The advantage of GB accretion is that it occurs at all stages of the life cycle and includes enlarged fruits and flowers and increased seed number (Chen and Murata, 2011). What is more, exogenous or transgenically derived sources of GB can induce the expression of specific stress-response genes, including those for ROS-scavenging enzymes (Chen and Murata, 2011). The site of accumulation may also be significant according to Park *et al.* (2007), who concluded that chloroplast GB was more effective for cellular protection than cytosolic accretion.

Evidence for the indirect effects of phenylalanine include: the regulation of localized and systemic disease resistance in virus-inoculated tobacco via the pathway to SA biosynthesis (Yalpani *et al.*, 1993); the production of SA precursors as a major function of PAL in the resistance of *Arabidopsis* to *P. parasitica* (downy mildew; Mauch-Mani and Slusarenko, 1996); and the efficacy of SA-dependent defence pathways against different pathogens (*Oidium neolycopersici* and *B. cinerea*) in tomato and tobacco (Achuo *et al.*, 2004). It has been demonstrated that SA-mediated disease resistance in plants operates by suppressing the auxin signalling network (Wang *et al.*, 2007). On the basis of extensive evidence, it is now recognized that SA is one of the prerequisites for the establishment of SAR and the activating genes associated with this priming function (Conrath, 2006).

The association of tyrosine with stress/defence responses in plants is exemplified by effects on its decarboxylase. Thus, tyrosine decarboxylase is increased by high salt and activated by drought (Świędrych *et al.*, 2004), while tyrosine decarboxylase cDNA is substantially induced by wounding (Lehmann and Pollmann, 2009). The role of tyrosine nitration is another emerging feature that has been implicated in NO-mediated signalling following pathogen infection (Vandelle and Delledonne, 2011).

Tryptophan is a significant source of metabolites with protective potential (Table 28.6). Indeed, the pathways and enzymes involved in its biosynthesis and metabolism are influenced by or provide tolerance/resistance to stresses induced by herbicides, ozone and biotic elicitors (Sanjaya *et al.*, 2008). Tryptophan is the source of auxin (IAA), which differentially affects resistance to separate pathogen groups (Kazan and Manners, 2009; Chapter 19). In addition, the involvement of the tryptophan pathway in the defence responses against pathogenic infection via serotonin synthesis is an emerging and novel issue (Ishihara *et al.*, 2008; see also Chapter 20). Tryptophan is also the precursor of camalexin, which is implicated in the defence mechanisms directed towards a wide range of pathogens, including *B. cinerea* and *Alternaria brassicicola* (Moldrup *et al.*, 2013). Of particular interest is the role of tryptophan in the synthesis of deterrent indole glucosinolates (Chapter 22). In other studies, *Verticillium* suppression has been correlated with the glucosinolate status of *Arabidopsis* (Witzel *et al.*, 2013). Conversion of one indole glucosinolate to another enhances the defensive benefit towards insect predation (Kim and Jander, 2007). Probably uniquely, the IAA-Asp conjugate promotes plant disease progression by regulating the transcription of virulence genes (González-Lamothe *et al.*, 2012).

Considerable attention is currently being given to the role and effects, in plants, of the xenobiotic amino acid analogue, β-aminobutyrate (BABA) (Chapter 27). Its efficacy in potentiating stress resistance mechanisms in plants is emerging as a prominent feature of recent research. There is substantial evidence, not just in *Arabidopsis* mutants, to suggest that BABA is capable of exerting significant 'priming' activity towards disease organisms and other stress factors (Zimmerli *et al.*, 2008).

Although signalling compounds have been considered separately as derivatives of specific amino acids, the concept of cross-talk in hormonal activity is gaining momentum. Robert-Seilaniantz *et al.* (2011) indicated that such interactions were 'more than just JA–SA antagonism' and that other components such as ET, auxin, ABA and BR are important regulators of plant disease and defence. Therefore, elements of both synergism and antagonism contribute to the eventual outcome of biotic and abiotic elicitors.

28.17.1 A general model

A general model for stress metabolism in higher plants may now be considered (Fig. 28.3). It is readily admitted that this scheme is too simplistic to accommodate the complex chain of molecular events that eventually result in the expression of defence responses or the incidence of disease or insect damage. Nevertheless, Fig. 28.3 should be perceived as an instructive initial model that can be developed further as the full scale of biochemical interactions are established for particular stress challenges.

There is an emerging consensus that the primary stress responses in plants are broadly analogous to those observed in animals in terms of passive and active immunity and the mediation of signalling and other regulatory molecules. In addition, local and systemic defence processes are also recognized in both plants and animals. Furthermore, the question of host–parasite relationships is a common issue relating in particular to pathogen adaptation to defence compounds produced *in vivo* or administered exogenously. The biochemical impact of stress in plants, as outlined in Fig. 28.3, is generally assumed to commence with signal transduction from injured or infected tissues. The signal molecules generated there are translocated via the phloem to undamaged tissues, where they are assumed to be detected. In response, a diverse array of low molecular weight secondary compounds and proteins are synthesized and released to augment pre-existing constitutive defences against predators or phytopathogens. The net effects include intoxication, deterrence or resistance to the biotic challenge, depending upon the source of injury or infection.

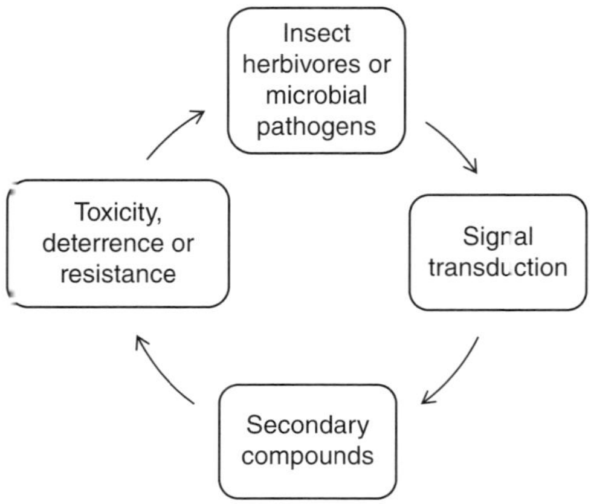

Fig. 28.3. A general model for biotic stress dynamics in higher plants, as exemplified by responses to insect herbivores or to microbial pathogens. The responses to abiotic (environmental) stress would also be consistent with the above pathway, but with particular exceptions as elaborated in the text. The relevance to amino acid metabolism in higher plants and to the dynamics of host–parasite relationships, as portrayed here, are based on hypotheses and evidence presented in the text of this chapter. In essence, certain amino acids, including methionine, phenylalanine and tryptophan, serve as precursors of respectively ethylene (ET), salicylic acid (SA) and indole acetic acid (IAA) as well as parent molecules for a diverse array of secondary compounds that are implicated in plant defences. The nitration of tyrosine residues in certain proteins is central to peroxynitrite signalling following pathogen infection. In addition, isoleucine activates jasmonic acid (JA) and, separately, leucine-rich repeat receptor kinases (LRR-RKs) and leucine zipper transcription factors are involved in signal transduction, as detailed in the text. There is evidence of insect or phytopathogen adaptation to secondary defence compounds, which is indicative of a dynamic flux in host–parasite relationships. Certain secondary volatile compounds may attract beneficial insects to augment endogenous protection strategies.

Amino acids may act directly as signalling molecules or may yield other compounds with similar functions (Table 28.4). In addition, amino acids may act indirectly by facilitating the functional capacity of other signalling compounds such as SA, BRs and JA. In response to signals, defence molecules may be deployed to reduce or neutralize the physiological impact of the predator, pathogen or abiotic challenge. A wide range of secondary and polymeric compounds are attributed with defence functions including, among many others, amino acid-derived polyphenolics, serotonin (Chapter 20), phytoalexins, indole glucosinolates (Chapter 22) and GABA (Chapters 7 and 27), as well as specific proteins with catalytic or even inhibitory activities.

The induction of defence compounds is a complex process involving the interaction of a diverse range of signal transduction networks. Thus, camalexin deployment in response to fungal infection, requires ROS and SA signal transduction, but glutathione status may also be an important component of the pathway. The targets of this signalling cascade include the tryptophan and camalexin biosynthetic genes which are transcriptionally upregulated in affected tissues. Although particular pathogens are effectively inhibited by camalexin, other fungi are resistant owing to their active degradation of this indole compound (Glawischnig, 2007).

28.17.2 Specific examples

The impact of stress on amino acid metabolism in plants is elegantly exemplified in the effects of insect herbivory (Gonzales-Vigil *et al.*, 2011). Following initiation of damage to tissues, a cascade of molecular events ensue that vary according to the species of plant and insect predator.

Several signals emanate from injured tissues, including calcium ion fluxes, phosphorylation reactions and JA signal transduction (Fürstenberg-Hägg *et al.*, 2013). In plants, threonine dehydratase catalyses the dehydration/deamination of threonine. In the tomato plant in particular, this enzyme is encoded by duplicate genes (Gonzales-Vigil *et al.*, 2011). One acts in the housekeeping mode, allowing the biosynthesis of isoleucine. The other (*TD2*), is attributed with insect defence functions. On insect-provoked injury, JA signal transduction induces *TD2* and the resulting enzyme causes the depletion of threonine in the insect gut, thereby depriving the predator of an essential amino acid. The enhanced survival and growth rates of lepidopteran larvae on transgenic tomato lines silenced for *TD2* expression support the hypothesis that threonine dehydratase serves in defence towards insects (Gonzales-Vigil, personal communication). In this example, the product of secondary metabolism is a protein with catalytic activity, as primary functions in the plant can be achieved via the housekeeping form of threonine deaminase. The concept that the catabolism of amino acids in the insect midgut by host enzymes exerts a protective function had previously been advanced with respect to overexpression of arginase (Chen *et al.*, 2005). In tobacco plants, there is only one copy of *TD* and a different sequence of events following insect damage, in which JA activates the wound response, which involves the expression of multiple defence genes (Kang *et al.*, 2006).

That plant–herbivore and plant–pathogen interactions are in a dynamic state of flux and adaptation to stress source and defence molecules is to be expected. For example, in response to aphid feeding, *Arabidopsis* plants increase their defensive capability by converting one indole glucosinolate to another (Kim and Jander, 2007). Gonzales-Vigil *et al.* (2011) maintains that a co-evolutionary process of plant responses and herbivore counter-adaptation is the driving mechanism in the 'escalation of phytochemical diversity' and versatility. Fürstenberg-Hägg *et al.* (2013) developed this concept further with the suggestion that some herbivorous insects are able to sequester and even reutilize plant secondary compounds for their own defence. Also, certain secondary volatile compounds may attract beneficial insects to augment endogenous protection strategies (Howe and Jander, 2008). Other observations indicate that the ABA signalling pathway in *Arabidopsis* can be manipulated by *P. syringae* as a virulence strategy to cause disease (de Torres-Zabala *et al.*, 2007). In addition, an inverse relationship between the systemic resistance of plants to microbes and to insect herbivory has been reported by Felton *et al.* (1999). Thus, resistance mechanisms differ according to the type of biotic stress, which is consistent with the evidence for signalling responses on exposure to pathogens (Chen *et al.*, 2013) and to insect predation (Fürstenberg-Hägg *et al.*, 2013), which involve, respectively, the SA and JA pathways for the deployment of defensive compounds.

28.18 Plant Products

Plant products are the primary sources of nutrients and beneficial compounds for humans and livestock. The major attributes of these products reside in their contributions of energy, protein, fibre, minerals and vitamins, as well as secondary compounds. Quantitatively, virtually all of the carbohydrate and lipid energy and a substantial quantity of protein consumed by humans and livestock emanate from plant products.

A wide range of secondary compounds have been associated with beneficial properties in humans. For instance, the consumption of the non-protein amino acid SMCO, which is present in *Brassica* plants, may confer health benefits (D'Mello, 2012; Chapter 27). There is also increasing awareness of the potential value of 4-hydroxyisoleucine through its occurrence in fenugreek (*Trigonella foenum-graecum*). On the basis of favourable metabolic results, Jette *et al.* (2009) concluded that this amino acid should be considered as a novel candidate compound for the treatment of type-2 diabetes, with obesity and dyslipidemia representing key components of the 'metabolic syndrome' in human subjects.

Flavonoids are associated with diverse protective properties towards adverse effects induced by allergens, viruses and inflammation, but the primary focus is now directed at their antioxidant potential. Flavonoids are generally credited with the ability to suppress and scavenge free radicals (Pietta, 2000). Glucosinolates (Chapter 22), possibly in combination with SMCO, have been implicated in the protective attributes of *Brassica* vegetables (Stoewsand, 1995). The accepted consensus is that the breakdown products of the glucosinolates, particularly isothiocyanates (ITCs), confer the primary benefits. The reactivity of these products is currently of relevance in a wider context. For example, Hanschen *et al.* (2012) showed in a model system that the preferred site of activity of ITCs was the thiol group of cysteine rather than the amino group and that even endogenously released ITCs were capable of reacting relatively quickly with cysteine in broccoli sprouts. There may well be implications here in the elucidation of plant defence mechanisms against fungal pathogens (Witzel *et al.*, 2013). There is also interest in the health benefits of *Rhodiola crenulata*, which have been attributed to its content of salidroside and tyrosol, both derived from tyrosine. The metabolic characterization of tyrosine decarboxylase is the focus of current attempts to enhance concentrations of these secondary compounds (Lan *et al.*, 2013).

It has long been established that plant proteins in general are deficient in certain essential amino acids, including lysine (Chapter 13), methionine, threonine and tryptophan, and efforts continue to increase amino acid content by traditional and unconventional means. Zeh *et al.* (2001) demonstrated that antisense inhibition of threonine synthase enhanced the methionine content of potato tubers without compromising threonine status. In the feeding of livestock, it has been suggested that limitations in protein quality may be rectified by increasing dietary protein concentrations proportionately to sufficiently high levels to ensure provision of essential amino acids at the recommended nutritional standards. It is envisaged that the quantity of protein used should compensate for the inferior quality of plant proteins. This proposal was tested with young chicks fed diets based on cereals and groundnut meal in different combinations to provide crude protein (CP) levels ranging from 120 to 420 g kg^{-1} diet (Wethli *et al.*, 1975). Here, the assumed minimal amino acid needs of the young chick were considered to be satisfied at high inclusion rates of the main ingredients of plant origin. As expected, the growth performance of chicks improved with successive increments of CP up to 360 g kg^{-1} diet, but nevertheless failed to match the growth of chicks fed lower levels of CP derived from an animal protein source (herring meal) in a separate series of diets. Of several hypotheses examined, Wethli *et al.* (1975) concluded that the amino acids supplied by low-quality plant protein sources were in such disproportion relative to the needs of the chick as to impair utilization of the first-limiting amino acid. It was suggested that amino acid imbalance is a primary factor limiting the efficacy of plant protein sources for the growth of young chicks, though a more plausible explanation for these results may be associated with subsequent reports indicating the formation of advanced glycation end products (AGEs) in high-protein dietary regimes (Uribarri and Tuttle, 2006). In such circumstances, post-translational modification of amino acid residues of proteins may occur, resulting in the appearance of AGEs via Maillard-type reactions. The residue regularly implicated in AGE formation is lysine, the first-limiting amino acid in virtually all plant products. The synthesis of AGE under normal physiological conditions would be accompanied by a depletion of the metabolic pools of lysine (and possibly other amino acids), leading to the reported reductions in the utilization of these nutrients. In addition, AGEs might induce nephrotoxicity and other physiological aberrations, such as intestinal inflammation and endothelial dysfunction. So the reduced metabolic availability (Chapter 26) of critical amino acids and AGE-induced physiological stress may well account for the results with high-protein diets based on low-quality plant products (Wethli *et al.* (1975). Meanwhile, the role of AGEs in the biomedical context is emerging, justifying a full review of the human health implications (Bengmark, 2012).

28.19 Summary

28.19.1 Enlightenment and debate in equal measure

A number of significant developments serve to define the achievements of the past few decades, as new initiatives emerge to reshape research strategies for the future. The notion that amino acid pathways in higher plants should be investigated to satisfy academic curiosity has been replaced with a highly focused approach based on well-defined objectives and the development of novel supporting technologies.

The critical role of amino acids in the primary metabolism of higher plants remains undisputed throughout several decades of sustained research. Thus, the contribution of amino acids to the structural and catalytic functions of proteins and other macromolecules is well recognized and continues to offer new opportunities for potential applications. However, other aspects have recently emerged to add to the diversity of objectives; for example, signal transduction and the secondary metabolism of amino acids are increasingly integrated within plant defence networks (Table 28.7).

In a number of instances, amino acids serve as sources of both the signal transduction molecules and secondary compounds that are regularly implicated in plant defence systems. Methionine is the precursor of ET and aliphatic glucosinolates, while tryptophan metabolism results in the synthesis of auxin, camalexin and indole glucosinolates. In addition, phenylalanine is the substrate for SA as well as flavonoids, coumarins and stilbenes (Huang *et al.*, 2010).

The indirect role of amino acids in plants (as in other living organisms) is exemplified by complex structural–functional relationships which are evident in virtually all aspects of metabolism (Chapters 1–16, 27). For instance, basic leucine zipper proteins mediate ABA-dependent signal transduction in drought and salinity stress, reflecting protein–DNA interactions (Uno *et al.*, 2000; Kang *et al.*, 2002). In BR signalling, tyrosine phosphorylation at specific residues enables the regulation of developmental processes and disease resistance in plants (Friedrichsen *et al.*, 2000; Oh *et al.*, 2009). In addition, the selective nitration of tyrosine residues in certain proteins is central to peroxynitrite signalling functions (Vandelle and Delledonne, 2011). LRR-RKs are transmembrane structures also regulating a diverse range of developmental and defence-related processes, but via protein–protein interactions (Kobe and Kajava, 2001).

Complex interactions occur between amino acids based on structural analogy, as for example in the arginine–canavanine antagonism in plants and in animals. The effect of mimosine on PAL activity is another type of antagonism observed in plants, though other interactions involving amino acid-derived signalling hormones, including NO, ET, SA and IAA, represent an elaborate order of complexity, particularly in relation to the activities of the JA-Ile conjugate and ABA. The antagonistic and inhibitory effects of SA and ET are notable features of these interactions (Table 28.5). However, any resolution of the overall physiological or immunological impact in higher plants would probably require the development of sophisticated software for simulation modelling.

Historically, interest in plant hormones has been driven by the need to elucidate their complex effects on plant growth and development. Thus, the effects of ET on seed germination, root

Table 28.7. Stress/defence/environmental issues specifically addressed in the current edition of *Amino Acids in Higher Plants*.

Chapter	Details
2	Alanine aminotransferase: role in stress biology; hypoxia and other abiotic and biotic stress
6	Asparagine synthetase: environmental stress and carbohydrate depletion
7	The γ-aminobutyrate (GABA) shunt pathway and stress
8	Modulation of L-arginine-dependent nitric oxide synthase (NOS) activity during plant development and under stress conditions
10	Polyamines in plants: stress-response roles
15	Proline accumulation and adaptive responses to stress; GABA accumulation in responses to plant stress; alanine accumulation, a universal phenomenon under stress; aspartate-derived amino acids in response to stress
19	Environmental regulation of auxin synthesis
20	Tryptophan pathway-derived secondary metabolites in defence responses
21	Melatonin: searching for specific roles in plants; accumulation in abiotic stress conditions
22	Glucosinolates in defensive, offensive, pre-emptive or proactive actions
27	The xenobiotic non-protein amino acid β-aminobutyrate (BABA) in enhancing protective mechanisms in plants, including resistance to pathogens and environmental stress

nodulation, flower senescence, abscission and fruit ripening were recognized as important functions (Wang *et al.*, 2002). Similarly, IAA, cytokinin and bioactive gibberellin production have been associated with fruit set and early fruit growth in tomato (Mariotti *et al.*, 2011). The potential role of plant hormones in defence is likely to remain the enduring theme in the future (Table 28.6). Resistance mechanisms differ according to the type of biotic stress; this is consistent with the evidence for signalling responses on exposure to pathogens (Chen *et al.*, 2013) and to predation by insects (Fürstenberg-Hägg *et al.*, 2013), which, respectively, involve the SA and JA pathways for the deployment of defensive compounds. Probably uniquely, the IAA-Asp conjugate promotes plant disease progression by regulating the transcription of virulence genes (González-Lamothe *et al.*, 2012). Fujita *et al.* (2006) stressed that cross-talk between biotic and abiotic challenges involves signalling pathways regulated by several of the major plant hormones. In addition, it is important to recognize that pathogenesis-related proteins contribute markedly to disease resistance in diverse species of plants (El-kereamy *et al.*, 2011). Reviewing data relating to the impact of osmotic stress and the effects of acid rain, Horváth *et al.* (2007) concluded that tolerance to abiotic stress may be induced by SA. Yasuda *et al.* (2008) extended this role with observations of antagonistic cross-talk at multiple steps between SA and ABA following NaCl treatment of *Arabidopsis*. So SA appears to exert a major influence in both biotic and environmental stresses and, in addition, is involved in the induction of low molecular weight defence secondary compounds (Glawischnig, 2007; Tounekti *et al.*, 2013), as well as in pathogenesis-related gene expression (van Wees *et al.*, 2000; Chen *et al.*, 2013). However, the role of JA and other signalling compounds in defence reactions and the implications of complex phytohormone interactions should not be underestimated (Niki *et al.*, 1998; Howe, 2004; Adio *et al.*, 2011; Fürstenberg-Hägg *et al.*, 2013). For instance, synergy between ET and JA is essential for the induction of the osmotin gene (Xu *et al.*, 1994) and the *Arabidopsis* defensin gene (*PDF 1.2*; Penninckx *et al.*, 1998). It will be recalled that ET has also been associated with the biosynthesis of defence compounds, particularly phytoalexins (Matsukawa *et al.*, 2013).

Light regulates diverse developmental and morphological changes in plants, and so effects on amino acid metabolism are only to be expected (Chapters 6 and 21). A role for BR in the light-dependent development of higher plants has been proposed (Li *et al.*, 1996). In addition, evidence for the involvement of glutamate receptors in light-signal transduction has been advanced by Lam *et al.* (1998). Another theme relates to the functional association of glycine decarboxylase with the control of carbon flow through both photosynthesis and photorespiration (Hylton *et al.*, 1988; Timm *et al.*, 2012). Several defence reactions, including upregulation of PAL, SA synthesis and the expression of pathogenesis-related protein PR-1 appear to be light dependent, whereas pathogen-induced production of JA, camalexin and myrosinase are enhanced in the dark (Zeier *et al.*, 2004). Of particular note is the effect of light on the promotion of JA-Ile conjugation via two distinct mechanisms (Svyatyna *et al.*, 2014).

An inescapable feature now emerging is homology between plants and other living organisms in the metabolism of amino acids. Comparative aspects relate to a diverse range of enzymes (Tables 28.1–28.4) and secondary metabolites (Chapters 10, 20 and 21). Increasingly, hormone relationships and activity are also integrated within this context. Thus, SA, which is an endogenous signalling compound of plants is endowed with anti-inflammatory and anti-neoplastic activities in humans; its derivative, aspirin, is a common analgesic drug with diverse pharmacological properties (Deng *et al.*, 2001). In addition, leucine-rich repeats involved in BR signal transduction in plants are thought to be analogous to similar motifs found in animal hormone receptors (Li and Chory, 1997). Clouse (2011) comments on structural similarities between BRs and animal steroid hormones. Furthermore, both plants and animals use pattern recognition receptors to sense conserved microbial-associated molecular profiles. In *Arabidopsis*, for example, these pattern recognition receptors are mainly LRR-RKs (Belkhadir *et al.*, 2012).

Recent developments also focus on chemical ecology, food safety and environmental protection. It is important not to overlook a number of potential applications of amino acids as novel bio-pesticides and pharmaceuticals. Canavanine and mimosine have emerged as prime candidates for herbicidal and insecticidal preparations (Chapter 27). What is more, canavanine may enhance the efficacy of the microbial insecticide, *Bacillus thuringiensis*. In addition, β-aminobutyric acid (BABA) acts synergistically with fungicides to stimulate efficacy. Moreover, the potential scope for non-protein amino acids in cancer therapy appears to be reasonably promising. Varying degrees of efficacy have been reported with canavanine, *Se*-methylselenocysteine, *Se*-allylselenocysteine, selenomethionine, SMCO, mimosine and selected congeners and metabolites in studies involving animal and *in vitro* models. Of these amino acids, *Se*-methylselenocysteine was most effective against mammary tumours, with the capacity of potentiating the efficacy of anti-cancer agents and protecting against drug-induced toxicity. Several indole phytoalexins have also been associated with significant anti-proliferative activity towards various cancer cells and with chemopreventive properties in models of mammary and skin carcinogenesis (Mezencov *et al.*, 2003). In addition, results of preclinical efficacy studies prompted an analysis of the biochemical and molecular mechanisms underlying the anti-proliferative properties of phytoalexins (Romagnolo *et al.*, 2012).

Despite recent advances reviewed in this volume and elsewhere relating to the complements of genes and enzymes, debate continues across diverse aspects of amino acid metabolism in plants (Yasuor *et al.*, 2006), while other issues remain largely unexplored (Tzin and Galili, 2010). Conflicting evidence has fuelled debate over the significance of particular pathways and mechanisms of regulation. The data of Yasuor *et al.* (2006) are pertinent in this respect in that the expected reduction of auxin through inhibition of the shikimic pathway by glyphosate did not occur in a consistent manner. Cho *et al.* (2007) commented on uncertainty regarding the relative roles of arogenate and phenylpyruvate as obligatory precursors in the biosynthesis of phenylalanine, identifying six genes and the corresponding proteins attributed with arogenate dehydratase activity. However, Yoo *et al.* (2013) confirmed the existence of an alternative pathway via cytosolic tyrosine:phenylpyruvate aminotransferase.

Controversy also prevails over the pathways of SA biosynthesis in higher plants. Initial observations indicated that the production of SA precursors was a major function of PAL in disease resistance (Mauch-Mani and Slusarenko, 1996). In view of the lack of correlation between the expression patterns of chorismate synthase and isochorismate synthase genes with SA levels (Pál *et al.*, 2013), it might be assumed that the PAL pathway is the primary mode of SA biosynthesis. Others claim that utilization of phenylalanine or chorismate as precursors for SA synthesis depends upon plant species, developmental stage and growth conditions (Tounekti *et al.*, 2013). More definitively, Yamasaki *et al.* (2013) stated that the isochorismate pathway exerted a central role in immunity-related SA biosynthesis in *Arabidopsis*, and implicated the PAL mechanism for SA production in *N. tabacum*. In addition, PAL promotes the deamination of phenylalanine and, in this way, provides the first reaction for the phenylpropanoid pathway that leads to the synthesis of a wide range of defence compounds, including flavonoids, isoflavonoids, coumarins and stilbenes (Huang *et al.*, 2010). Of particular significance is the role of PAL at the juncture of primary and secondary metabolism, where it represents an important regulatory point.

28.19.2 Amino acids of 'particular distinction'

It is universally agreed that all 20 canonical amino acids are equally critical in the ribosomal biosynthesis of proteins and in other features of primary metabolism. This has been, and will continue to be, the enduring and immutable tenet that will challenge those biochemists seeking to expand the 'genetic lexicon' (Young and Schultz, 2010). In addition, glutamate has been accorded with the title of an amino acid of 'particular distinction' in the context of mammalian metabolism (Young and Ajami, 2000). Similarly, the diversity of functions ensures an important position for glutamate in the biochemistry of higher plants, although the statement by Young and Ajami (2000) inevitably invites contradiction. There is much to support the concept that methionine and the aromatic amino acids are strong contenders for this title as well, by virtue of their roles in the provision of signalling hormones and critical secondary compounds deployed in adaptive and defence responses to biotic and environmental stress. Thus, methionine is the precursor of SAM, which provides for the synthesis of ET and participates in transmethylation reactions, polyamine biosynthesis and as an allosteric activator of threonine synthase (Ravanel *et al.*, 1998). In addition, methionine contributes to the biosynthesis of glucosinolates. Leucine biosynthesis and degradation and the methionine chain-elongation process in aliphatic glucosinolate production are catalysed by branched-chain aminotransferase. BCAT4 has been identified as the enzyme involved in the initial stages of chain elongation (Schuster *et al.*, 2006), whereas BCAT3 appears to be associated with the terminal steps in this process (Knill *et al.*, 2008).

As a group or individually, the aromatic amino acids are also endowed with characteristics of particular distinction. Phenylalanine metabolism results in the synthesis of SA as well as a host of defence molecules, including flavanones, isoflavanones, phenylpropanoid phytoalexins and lignin (Hu *et al.*, 2009). Tryptophan is exceptional in that during its synthesis and subsequent metabolism, IAA (Chapter 19), indole phytoalexins, alkaloids and glucosinolates (Chapter 22) are synthesized in addition to serotonin (Chapter 20).

28.19.3 Innovation

Virtually all aspects of amino acid metabolism reviewed in *Amino Acids in Higher Plants* are still evolving and provide the basis for continuing research. The interrelationships between key pathways in the biosynthesis and metabolism of amino acids and their modulation by plant hormones in response to diverse elicitors are complex issues destined to test the imagination and skills of future generations of researcher–workers.

Progress in the future will undoubtedly be determined by advances in our knowledge of the genes encoding enzymes of the major pathways of metabolism, particularly insofar as they have an impact on the defence responses of plants. Genetic evidence is accumulating to establish the role of glutamate receptors in plant defence signalling and resistance to specific diseases but, clearly, other initiatives are required (Manzoor *et al.*, 2013). In the case of ACC synthase (Fig. 28.1), eight genes encoding active enzyme proteins and one encoding an inactive enzyme have been identified in *Arabidopsis*. Some of these genes are inducible by hypoxia while others are inducible by *B. cinerea*-induced ET biosynthesis (Huang *et al.*, 2013). It would be of considerable relevance to ascertain the genes involved in ET stimulation of secondary metabolism and the synthesis of phytoalexins for defence purposes (Matsukawa *et al.*, 2013). From a different viewpoint, the regulatory components or mediators controlling gene expression in SA- and JA-responsive pathways within a global perspective require elucidation (Mathugala *et al.*, 2012). The assignment of gene function in biosynthetic pathways has also been contemplated with respect to the secondary metabolism of tryptophan and the formation of camalexin. This issue is of particular significance owing to the established defence functions of this phytoalexin (Møldrup *et al.*, 2013).

The agenda for the biosynthesis and metabolism of aromatic amino acids (Fig. 28.2) has been effectively outlined by Tzin and Galili (2010) and is of relevance to the present time. Thus, the requirement to identify the genes encoding prephenate aminotransferase is emphasized. In addition, overlap in the functions of isoenzymes associated with the biosynthesis of phenylalanine deserves examination as well as various aspects of tyrosine formation. Tzin and Galili (2010) maintain that transcription factors regulate different steps in the biosynthesis of aromatic amino acids and their associated secondary derivatives, but these do not represent the complete set. Furthermore, some transcription factors are considered to regulate genes encoding both the primary and secondary metabolism of aromatic amino acids. Huang *et al.* (2010) commented on the role of PAL at the divergence of primary and secondary metabolism, indicating a potentially important regulatory point for the modulation of phenylalanine utilization. The existence of an alternative pathway via cytosolic tyrosine:phenylpyruvate aminotransferase has recently been proposed for phenylalanine biosynthesis (Yoo *et al.*, 2013), and it was concluded that their results revealed new targets for metabolic regulation and genetic manipulation.

The literature abounds with instances of structural and non-specific antagonisms between amino acids and among their signalling derivatives (Table 28.5; Kazan and Manners, 2008). A dominant feature is the SA antagonism of other signalling pathways in both the biotic and abiotic responses of plants (Table 28.5). In particular, the SA–JA antagonism in the expression of pathogenesis-related genes (Niki *et al.*, 1998) and resistance to insect herbivores (Cipollini *et al.*, 2004; Rayapuram and Baldwin, 2007), and the SA–JA–ET interactions in response to multiple elicitors (Leon-Reyes *et al.*, 2010), require further elucidation. The SA–ABA antagonism in host–pathogen relationships adds to the difficulties in unravelling the molecular mechanisms so involved (de Torres-Zabala *et al.*, 2009). Although IAA has long been associated with plant defence signalling, the molecular mechanisms and interactions remain essentially unexplored (Kazan and Manners, 2009). Intriguingly, the IAA-Asp conjugate promotes plant disease progression by regulating the transcription of virulence genes, adding to the complexity of hormone–pathogen interactions (González-Lamothe *et al.*, 2012). In abiotic stress responses, antagonisms regulated by ABA may also provide the basis for more investigation (Moons *et al.*, 1997; Yasuda *et al.*, 2008). An integrated view of the complex interactions involving phytohormones, mediators and transcription factors has been attempted, but considerable uncertainty still exists (Kazan and Manners, 2008).

A major feature concerns the precise processes integrating the contribution of signal transduction hormones, secondary metabolism and defence mechanisms in responses to plant pathogens and insect herbivores (Fig. 28.3). For example, on insect-provoked injury, JA signalling induces *TD2* and the associated enzyme causes the depletion of threonine in the insect gut, thereby depriving the predator of an essential amino acid. The enhanced survival and growth rates of lepidopteran larvae on transgenic tomato lines silenced for *TD2* expression support the hypothesis that threonine dehydratase serves in defence towards insects (Gonzales-Vigil *et al.*, 2011). This type of evidence has emerged as a result of innovative developments in transgenic methodologies (Kang *et al.*, 2006). The upregulation of genes and associated enzymes involved in secondary metabolism of amino acids needs to be more actively pursued. Świędrych *et al.* (2004) and Ishihara *et al.* (2008) both place emphasis on pathways as well as on individual secondary compounds in defence responses to stress. In addition, the precise role of secondary compounds with defence attributes remains largely elusive. Camalexin is induced by a diverse array of plant pathogens, but its efficacy as a growth inhibitor towards fungi tends to be variable (Glawischnig, 2007); similar comments may apply to glucosinolates, isoflavones, anthocyanins and other putative defence compounds. There is scope here for elucidating the underlying factors affecting the efficacy of these compounds towards different fungi and insect herbivores. Proline and citrulline accumulate in plants exposed to abiotic stress (Sharma and Dietz, 2006; Dasgan et al., 2009; Lehmann *et al.*, 2010; Szabados and Savouré, 2010), but the sequence of metabolic events from signalling to protective functions remains obscure, despite extensive speculation as to the mechanisms involved in osmolyte functions. Szabados and Savouré (2010) invoked effects on homeostasis, mitochondrial function, cellular proliferation, apoptosis and gene expression as potential areas for research. In other abiotic stress studies, ABA (Gómez *et al.*, 1988) and SA signalling (Horváth *et al.*, 2007) have been implicated. Specifically, the role of SA in modulating terpenoid and flavonoid metabolism in plant responses to abiotic stress has been investigated (Tounekti *et al.*, 2013).

Further innovation is required in order to elucidate the biochemical basis of plant–pathogen and plant–herbivore interactions. The molecular mechanisms whereby a bacterial phytopathogen can manipulate ABA signalling pathways to suppress defence responses awaits elucidation (de Torres-Zabala *et al.*, 2007). The priming of defence pathways follows perception of the molecular patterns of pathogens or symbionts. However, priming can also be induced by natural or synthetic compounds (Chapter 27; Conrath, 2011). Interactions are generally in a dynamic state of flux, and adaptation to stress source and defence molecules is inevitable. For example, Gonzales-Vigil *et al.* (2011) indicated the potential role of plant–herbivore interactions in the 'escalation of phytochemical diversity' and versatility. Fürstenberg-Hägg *et al.* (2013) developed this concept with the proposal of sequestration and reutilization of plant secondary compounds by insect herbivores for their own defence. Other secondary compounds may attract beneficial insects to augment endogenous protection strategies (Howe and Jander, 2008). In addition, fatty acid–amino acid conjugates in the oral secretions of *M. sexta* have been implicated in herbivore-specific responses in its natural host, *N. attenuata* (Halitschke *et al.*, 2001).

So there is ample justification for the determination of the molecular basis of plant interrelationships with pathogens and insect herbivores.

It is universally accepted that plant proteins are deficient in certain essential amino acids, including lysine (Chapter 13), methionine, threonine and tryptophan. Efforts therefore continue in increasing amino acid content by traditional and innovative technologies. Thus, Zeh *et al.* (2001) demonstrated that antisense inhibition of threonine synthase enhanced the methionine content of potato tubers without compromising threonine status. The molecular and structural factors determining the metabolic availability of amino acids in dietary proteins derived from plants is an outstanding issue affecting nutrient utilization and health in both livestock and humans (Chapter 26). The biochemical and physical interactions associated with advanced glycation end-products in plant proteins should form the basis of new initiatives.

28.20 Outlook

Although significant advances have been reported over recent decades, there is increasing awareness of the need to redouble efforts to address emerging issues in diverse aspects of amino acid metabolism in plants. Further progress is anticipated in view of the major innovations in molecular biology, which should provide the underpinning methodologies to increase the scope of the requisite research.

Progress in the future will undoubtedly be determined by developments in our knowledge of the genes encoding enzymes of the major pathways of metabolism, particularly insofar as they provide new insights into regulatory mechanisms and defence responses in plants. Furthermore, some transcription factors are considered to regulate the genes encoding both primary and secondary metabolism of aromatic amino acids, and this feature may also apply to arginine, glycine, methionine and threonine. The role of key enzymes at the confluence of primary and secondary metabolism indicates a potentially important regulatory point for the modulation of amino acid utilization by plants subjected to a variety of biotic and environmental stresses.

Despite considerable research over several decades, there is still uncertainty about the relative significance of the pathways of amino acid metabolism in plants. The terminal phases of these pathways are generally associated with ambiguity, for instance in the synthesis of secondary compounds. This is a particular feature with the aromatic amino acids and with the synchronized action of certain enzymes in determining the flow of carbon. In addition, the partition that exists between the cytosol and organelles provides yet more challenges in the exploration of amino acid fluxes and dynamics in plants.

In the course of recent research, complex interactions have emerged with respect to signal transduction involving, in particular, phytohormones derived from amino acids or those in structural associations with isoleucine, leucine or tryptophan. The identification of additive, synergistic and antagonistic effects, and the influence of biotic and abiotic stresses on the outcomes of these interactions, underline the need for continuing investigations (Table 28.5). In response to adverse conditions, signal transduction mechanisms initiate SAR, resulting in the upregulation of key enzymes and the deployment of defence compounds. A simplified model is presented here (Fig. 28.3), but the role of pathogenesis-related proteins (El-kereamy *et al.*, 2011) remains to be incorporated within an enlarged multifactorial model for defence pathways. In addition, the wound response is often invoked in a restricted context of insect herbivory, but there may also be implications in other defence responses. The task is daunting, given the complexity of interactions (Table 28.5) and the persistent evidence of cross-talk between biotic and abiotic pathways (Divi *et al.*, 2010).

The production of botanical pharmaceuticals is likely to add value in the emerging diversification of arable farming systems (Raskin *et al.*, 2002). The notion that crop plants should be grown for medicinal purposes as well as for the provision of nutrients and fibre is rapidly gaining momentum based on sound fundamentals. However, the issues of amino acid deficiencies and their metabolic availability in plant products intended for human and livestock nutrition remain intractable problems at the present time.

There is, therefore, a rich diversity of objectives and sufficient impetus to underpin future research. In particular, the relationship between amino acids, signal transduction and defence metabolism in higher plants requires concerted action. The prospects for further advances will be critically dependent upon research funding as worldwide austerity measures take effect in the years ahead. The crucial factor will be the extent to which the commercial sector can make up the inevitable shortfall and react to academic aspirations for strategic research.

References

Achuo, E.A., Audenaert, K., Maziane, H. and Hofte, M. (2004) The salicylic acid-dependent defence pathway is effective against different pathogens in tomato and tobacco. *Plant Pathology* 53, 65–72.

Adio, A.M., Casteel, C.L., Kim, J.H. and Jander, G. (2011) Biosynthesis and defensive function of *N*δ-acetylornithine, a jasmonate-induced *Arabidopsis* metabolite. *The Plant Cell* 23, 3303–3318.

Alabadi, D., Aguero, M.S., Perez-Amador, M.A. and Carbonell, J. (1996) Arginase, arginine decarboxylase, ornithine decarboxylase and polyamines in tomato ovaries. Changes in unpollinated ovaries and parthenocarpic fruits induced by auxin or gibberellin. *Plant Physiology* 112, 1237–1244.

Amir, R., Hacham, Y. and Galili, G. (2002) Cystathionine γ-synthase and threonine synthase operate in concert to regulate carbon flow towards methionine in plants. *Trends in Plant Science* 7, 153–156.

Anderson, J.P., Badruzsaufari, E., Schenk, P.M. and Kazan, K. (2004) Antagonistic interaction between abscisic acid and jasmonate-ethylene signalling pathways modulates defense gene expression and disease resistance in *Arabidopsis*. *The Plant Cell* 16, 3460–3479.

Andrade, A.B., Ferrarese, M.L.L., Teixeira, A.F. and Ferraresse-Filho, O. (2008) Mimosine inhibits soybean (*Glycine max*) root growth, lignification and related enzymes. *Allelopathy Journal* 21, 145–154.

Ashraf, M. and Foolad, M.R. (2007) Roles of glycine betaine and proline in improving plant abiotic stress resistance. *Environmental and Experimental Botany* 59, 206–216.

Bauwe, H. and Kolukisaoglu, U. (2003) Genetic manipulation of glycine decarboxylation. *Journal of Experimental Botany* 54, 1523–1535.

Beaudoin, N., Serizet, C., Gosti, F. and Giraudat, J. (2000) Interactions between abscisic acid and ethylene signalling cascades. *The Plant Cell* 12, 1103–1115.

BeGora, M.D., Macleod, M.J.R., Summers, P.S. and Weretilnyk, E.A. (2010) Identification of phosphomethylethanolamine *N*-methyltransferase from *Arabidopsis* and its role in choline and phospholipid metabolism. *The Journal of Biological Chemistry* 285, 29147–29155.

Belkhadir, Y., Jaillais, Y., Epple, P., Dangl, J.L. and Chory, J. (2012) Brassinosteroids modulate the efficiency of plant immune responses to microbe-associated molecular patterns. *Proceedings of the National Academy of Sciences of the United States of America* 109, 297–302.

Bengmark, S. (2012) Modified amino acid-based molecules: accumulation and health implications. In: D'Mello, J.P.F. (ed.) *Amino Acids in Human Nutrition and Health*. CAB International, Wallingford, UK, pp. 382–405.

Borstlap, A.C. (1981) Interactions between the branched-chain amino acids in the growth of *Spirodela polyrhiza*. *Planta* 151, 314–319.

Bradley, D.J., Kjellbom, P. and Lamb, C.J. (1992) Elicitor- and wound-induced oxidative cross-linking of a proline-rich plant cell wall protein: a novel, rapid defense response. *Cell* 70, 21–30.

Brauc, S., De Vooght, E., Claeys, M., Höfte, M. and Angenon, G. (2011) Influence of over-expression of cytosolic aspartate aminotransferase on amino acid metabolism and defence responses against *Botrytis cinerea* infection in *Arabidopsis thaliana*. *Journal of Plant Physiology* 168, 1813–1819.

Brauc, S., De Vooght, E., Claeys, M., Geuns, J.M.C., Höfte, M. and Angenon, G. (2012) Overexpression of arginase in *Arabidopsis thaliana* influences defence responses against *Botrytis cinerea*. *Plant Biology (Stuttgart, Germany)* 1, 39–45.

Brenner, E.D., Stahlberg, R., Mancuso, S., Vivanco, J. and Volkenburgh, E.V. (2006) Plant neurobiology: an integrated view of plant signaling. *Trends in Plant Science* 11, 413–419.

Brownfield, D.L., Todd, C.D. and Deyholos, M.K. (2008) Analysis of *Arabidopsis* arginase gene transcription patterns indicates specific biological functions for recently diverged paralogs. *Plant Molecular Biology* 67, 429–440.

Bunsupa, S., Katayama, K., Toyooka, K., Saito, K. and Yamazaki, M. (2012) Lysine decarboxylase catalyzes the first step of quinolizidine alkaloid biosynthesis and coevolved with alkaloid production in *Leguminosae*. *The Plant Cell* 24, 1202–1216.

Caldwell, R.W., Chandra, S. and Caldwell, R.B. (2012) Arginase. In: D'Mello, J.P.F. (ed.) *Amino Acids in Human Nutrition and Health*. CAB International, Wallingford, UK, pp. 51–71.

Chandok, M.R., Ytterberg, A.J. and Klessig, D.F. (2003) The pathogen-inducible nitric oxide synthase (NOS) in plants is a variant of the P protein of the glycine decarboxylase complex. *Cell* 113, 469–482.

Chen, H., McCaig, B.C., Melotto, H., He, S.Y. and Howe, G.A. (2004) Regulation of plant arginase by wounding, jasmonate and the phytotoxin coronatine. *The Journal of Biological Chemistry* 279, 45998–46007.

Chen, H., Wilkerson, C.G., Phinney, B.S. and Howe, G.A. (2005) Jasmonate-induced plant enzymes degrade essential amino acids in the herbivore midgut. *Proceedings of the National Academy of Sciences of the United States of America* 102, 19237–19242.

Chen, T.H.H. and Murata, N. (2011) Glycine betaine protects plants against abiotic stress: mechanisms and biotechnological applications. *Plant, Cell and Environment* 34, 1–20.

Chen, Y., Shen, H., Wang, M. and He Z. (2013) Salicyloyl-aspartate synthesized by the acetyl-amido synthetase GH 3.5 is a potential activator of plant immunity in *Arabidopsis*. *Acta Biochimica et Biophysica Sinica* 45, 827–836.

Cho, M.-H., Corea, O.R.A., Yang, H. and Lewis, N.G. (2007) Phenylalanine biosynthesis in *Arabidopsis thaliana*: identification and characterization of arogenate dehydratases. *The Journal of Biological Chemistry* 282, 30827–30835.

Cipollini, D., Enright, S., Traw, M.B. and Bergelson, J. (2004) Salicylic acid inhibits jasmonic acid-induced resistance of *Arabidopsis thaliana* to *Spodoptera exigua*. *Molecular Ecology* 13, 1643–1653.

Clouse, S.D. (2011) Brassinosteroid signal transduction: from receptor kinase to transcriptional networks regulating plant development. *The Plant Cell* 23, 1219–1230.

Conrath, U. (2006) Systemic acquired resistance. *Plant Signaling and Behavior* 1, 179–184.

Conrath, U. (2011) Molecular aspects of defence priming. *Trends in Plant Science* 16, 524–531.

Conway, M.E. (2012) Aminotransferases. In: D'Mello, J.P.F. (ed.) *Amino Acids in Human Nutrition and Health*. CAB International, Wallingford, UK, pp. 24–50.

Corea, O.R.A., Cardenas, C.L., Brewer, S.E. and Lewis, N.G. (2012) Arogenate dehydratase isoenzymes profoundly and differentially modulate carbon flow into lignins. *The Journal of Biological Chemistry* 287, 11446–11459.

Corpas, F.J., Hayashi, M., Mano, S., Nishimura, M. and Barroso, J.B. (2009) Peroxisomes are required for *in vivo* nitric oxide accumulation in the cytosol following salinity stress of *Arabidopsis* plants. *Plant Physiology* 15, 2083–2094.

Cuin, T.A. and Shabala, S. (2006) Amino acids regulate salinity-induced potassium efflux in barley root epidermis. *Planta* 225, 753–761.

da Rocha, I.M., Vitorello, V.A., Silva, J.S. and Silveira, J.A. (2012) Exogenous ornithine is an effective precursor and the δ-ornithine aminotransferase pathway contributes to proline accumulation under high N recycling in salt-stressed cashew leaves. *Journal of Plant Physiology* 169, 41–49.

Daschner, K., Couee, I. and Binder, S. (2001) The mitochondrial isovaleryl-coenzyme A dehydrogenase of *Arabidopsis* oxidizes intermediates of leucine and valine catabolism. *Plant Physiology* 126, 601–612.

Dasgan, H.Y., Kusvuran, S., Abak, K. and Bouchereau, A. (2009) The relationship between citrulline accumulation and salt tolerance during the vegetative growth of melon (*Cucumis melo* L.). *Plant, Soil and Environment* 55, 51–57.

Davenport, R. (2002) Glutamate receptors in plants. *Annals of Botany* 90, 549–557.

Davis, R.H. (1991) Cyanogens. In: D'Mello, J.P.F., Duffus, C.M. and Duffus, J.H. (eds) *Toxic Substances in Crop Plants*. The Royal Society of Chemistry, Cambridge, UK, pp. 202–225.

De Felice, M., Levinthal, M., Iaccarino, M. and Guardiola, J. (1979) Growth inhibition as a consequence of antagonism between related amino acids: effect of valine in *Escherichia coli* K-12. *Microbiological Reviews* 43, 42–58.

Delledonne, M., Zeier, J., Marocco, A. and Lamb, C. (2001) Signal interactions between nitric oxide and reactive oxygen intermediates in the plant hypersensitive disease resistance response. *Proceedings of the National Academy of Sciences of the United States of America* 98, 13454–13459.

Deng, W.-G., Ruan, K.-H., Du, M., Saunders, M.A. and Wu, K.K. (2001) Aspirin and salicylate bind to immunoglobulin heavy chain binding protein (BiP) and inhibit its ATPase activity in human fibroblasts. *The FASEB Journal* 15, 2463–2470.

De Silva, D.L.R., Hetherington, A.M. and Mansfield, T.A. (1985) Synergism between calcium ions and abscisic acid in preventing stomatal opening. *New Phytologist* 100, 473–482.

de Torres-Zabala, M., Truman, W., Bennett, M.H., Mansfield, J.W. and Grant, M. (2007) *Pseudomonas syringae*. pv. *tomato* hijacks the *Arabidopsis* abscisic acid signaling pathway to cause disease. *The EMBO Journal* 26, 1434–1443.

de Torres-Zabala, M., Bennett, M.H., Truman, W.H. and Grant, M.R. (2009) Antagonism between salicylic and abscisic acid reflects early host–pathogen conflict and moulds plant defence responses. *The Plant Journal* 59, 375–386.

Dill, A. and Sun, T.-p. (2001) Synergistic derepression of gibberellin signaling by removing RGA and GAI function in *Arabidopsis thaliana*. *Genetics* 159, 777–785.

Divi, U.K., Rahman, T. and Krishna, P. (2010) Brassinosteroid-mediated stress tolerance in *Arabidopsis* shows interactions with abscisic acid, ethylene and salicylic acid pathways. *BMC Plant Biology* 10:151.

D'Mello, J.P.F. (2003a) Adverse effects of amino acids. In: D'Mello, J.P.F. (ed.) *Amino Acids in Animal Nutrition*. CAB International, Wallingford, UK, pp. 125–142.

D'Mello, J.P.F. (2003b) Amino acids as multifunctional molecules. In: D'Mello, J.P.F. (ed.) *Amino Acids in Animal Nutrition*. CAB International, Wallingford, UK, pp. 1–14.

D'Mello, J.P.F. (2012) Emergence of a new momentum. In: D'Mello, J.P.F. (ed.) *Amino Acids in Human Nutrition and Health*. CAB International, Wallingford, UK, pp. 479–524.

Douce, R., Bourguignon, M., Neuburger, M. and Rebeille, F. (2001) The glycine decarboxylase system: a fascinating complex. *Trends in Plant Science* 6, 167–176.

Dubos, C., Huggins, D. Grant, G.H. and Campbell, M.M. (2003) A role for glycine in the gating of plant NMDA-like receptors. *The Plant Journal* 35, 800–810.

Dulermo, T., Bligny, R., Gout, E. and Cotton, P. (2009) Amino acid changes during sunflower infection by the necrotrophic fungus, *B. cinerea*. *Plant Signaling and Behavior* 4, 859–861.

Dunham, V.L. and Bryan, J.K. (1971) Synergistic effects of metabolically related amino acids on the growth of a multicellular plant. *Plant Physiology* 47, 91–97.

El-kereamy, A., El-sharkawy, I., Ramamoorthy, R., Taheri, A., Errampalli, D., Kumar, P. and Jayasankar, S. (2011) *Prunus domestica* pathogenesis-related protein-5 activates the defense response pathway and enhances the resistance to fungal infection. *PLoS ONE* 6(3): e17973.

Facchini, P.J. and DeLuca, V. (1995) Phloem-specific expression of tyrosine/Dopa decarboxylase genes and the biosynthesis of isoquinoline alkaloids in opium poppy. *The Plant Cell* 7, 1811–1821.

Fan, H.-F. and Du, C.-X. (2012) Effect of nitric oxide on proline metabolism in cucumber seedlings under salinity stress. *Journal of the American Society for Horticultural Science* 137, 127–133.

Felton, G.W., Korth, K.L., Matthews, M.C., Lamb, C. and Dixon, R.A. (1999) Inverse relationship between systemic resistance of plants to microorganisms and to insect herbivory. *Current Biology* 9, 317–320.

Flores, G., Negrete-Díaz, J.V., Carrión, M., Andrade-Talavera, Y., Bello, S.A., Sihra, T.S. and Rodríguez-Moreno, A. (2012) Excitatory amino acids in neurological and neurodegenerative disorders. In: D'Mello, J.P.F. (ed.) *Amino Acids in Human Nutrition and Health*. CAB International, Wallingford, UK, pp. 427–453.

Flores, T., Todd, C.D., Dhanoa, P.K., Mullen, R.T., Lamattina, L. and Polacco, J.C. (2008) Arginase-negative mutants of *Arabidopsis* exhibit increased nitric oxide signalling in root development. *Plant Physiology* 147, 1936–1946.

Forde, B.G. and Lea, P.J. (2007) Glutamate in plants: metabolism, regulation and signaling. *Journal of Experimental Botany* 58, 2339–2358.

Forde, B.G., Cutler, S.R., Zaman, N. and Krysan, P.J. (2013) Glutamate signaling via a MEKK 1 kinase-dependent pathway induces changes in *Arabidopsis* root architecture. *The Plant Journal* 75, 1–10.

Friedrichsen, D.M., Joazeiro, C.A.P., Hunter, T. and Chory, J. (2000) Brassinosteroid-insensitive-1 is a ubiquitously expressed leucine-rich repeat receptor serine/threonine kinase. *Plant Physiology* 123, 1247–1256.

Fujita, M., Fujita, Y., Narusaka, Y. and Shinozaki, K. (2006) Crosstalk between abiotic and biotic stress responses: a current view from the points of convergence in the stress signaling networks. *Current Opinion in Plant Biology* 9, 436–442.

Fujitani, Y., Nakajima, N., Ishihara, K. and Sugimoto, M. (2006) Molecular and biochemical characterization of a serine racemase from *Arabidopsis thaliana*. *Phytochemistry* 67, 668–674.

Fujitani, Y., Horiuchi, T., Ito, K. and Sugimoto, M. (2007) Serine racemase from barley, *Hordeum vulgare* L., and other plant species represent a distinct eukaryotic group: gene cloning and recombinant protein characterization. *Phytochemistry* 68, 1530–1536.

Funck, D., Stadelhofer, B. and Koch, W. (2008) Ornithine-δ-aminotransferase is essential for arginine catabolism but not for proline biosynthesis. *BMC Plant Biology* 8:40.

Funck, D., Eckard, S. and Müller, G. (2010) Non-redundant functions of two proline dehydrogenase isoforms in *Arabidopsis*. *BMC Plant Biology* 10:70.

Fürstenberg-Hägg, J., Zagrobeiny, M. and Bak, S. (2013) Plant defense against insect herbivores. *International Journal of Molecular Sciences* 14, 10242–10297.
Ghassemian, M., Cutler, S., Kawaide, H. and McCourt, P. (2000) Regulation of abscisic acid signaling by the ethylene response pathway in *Arabidopsis*. *The Plant Cell* 12, 1117–1126.
Gisby, M.F., Mudd, E.A. and Day, A. (2012) Growth of transplastomic cells expressing D-amino acid oxidase in chloroplasts is tolerant to D-alanine and inhibited by D-valine. *Plant Physiology* 160, 2219–2226.
Glawischnig, E. (2007) Camalexin. *Phytochemistry* 68, 401–406.
Gogami, Y., Ito, K., Kamitani, Y., Matsushima, Y. and Oikawa, T. (2009) Occurrence of D-serine in rice and characterization of rice serine racemase. *Phytochemistry* 70, 380–387.
Goldraij, A. and Polacco, J.C. (1999) Arginase is inoperative in developing soybean embryos. *Plant Physiology* 119, 297–304.
Gómez, J., Sánchez-Martínez, M., Stiefel, V., Rigau, J., Puigdomènech, P. and Pages, M. (1988) A gene induced by the plant hormone abscisic acid in response to water stress encodes a glycine-rich protein. *Nature* 334, 262–264.
González-Lamothe, R., El Oirdi, M., Brisson, N. and Bouarab, K. (2012) The conjugated auxin indole-3-acetic acid-aspartic acid promotes plant disease development. *The Plant Cell* 24, 762–777.
Gonzalez-Vigil, E., Bianchetti, C.M., Phillips, G.N. Jr and Howe, G.A. (2011) Adaptive evolution of threonine deaminase in plant defense against insect herbivores. *Proceedings of the National Academy of Sciences of the United States of America* 108, 5897–5902.
Gördes, D., Kolukisaoglu, Ü. and Thurow, K. (2011) Uptake and conversion of D-amino acids in *Arabidopsis thaliana*. *Amino Acids* 40, 553–563.
Gravot, A., Debu, C., Wagner, G., Lariagon, C., Lugan, R., Todd, C.D., Wendehenne, D., Delourme, R., Bouchereau, A. and Manzanares-Dauleux, M.J. (2012) Arginase induction represses gall development during clubroot infection in *Arabidopsis*. *Plant and Cell Physiology* 53, 901–911.
Greenberg, D.M. (1961) Biosynthesis of amino acids and related compounds. In: Greenberg, D.M. (ed.) *Metabolic Pathways: Volume II*. Academic Press, New York and London, pp.173–235.
Griffiths, D.W. (1991) Condensed tannins. In: D'Mello, J.P.F., Duffus, C.M. and Duffus, J.H. (eds) *Toxic Substances in Crop Plants*. The Royal Society of Chemistry, Cambridge, UK, pp.180–201.
Guo, R., Shen, W., Qian, H., Zhang, M., Liu, L. and Wang, Q. (2013) Jasmonic acid and glucose synergistically modulate the accumulation of glucosinolates in *Arabidopsis thaliana*. *Journal of Experimental Botany* 64, 5707–5719.
Hacham, S.Y., Avraham, T. and Amir, R. (2002) The *N*-terminal region of *Arabidopsis* cystathionine γ-synthase plays an important regulatory role in methionine metabolism. *Plant Physiology* 128, 454–462.
Hachinohe, M. and Matsumoto, H. (2007) Mechanism of selective phytotoxicity of L-3,4-dihydroxyphenylalanine (L-DOPA). *Journal of Chemical Ecology* 33, 1919–1926.
Hafez, Y.S.M. and Chavez, E. (1978) Methionine toxicity in chicks and poults. *Poultry Science* 57, 699–703.
Halitschke, R., Schittko, U., Pohnert, G., Boland, W. and Baldwin, I.T. (2001) Molecular interactions between the specialist herbivore *Manduca sexta* (Lepidoptera, Sphingidae) and its natural host *Nicotiana attenuata*. III. Fatty acid-amino acid conjugates in herbivore oral secretions are necessary and sufficient for herbivore-specific plant responses. *Plant Physiology* 125, 711–717.
Hanschen, F.S., Brüggemann, N., Brodehl, A., Mewis, I., Schreiner, M., Rohn, S. and Kroh, L.W. (2012) Characterization of products from the reaction of glucosinolate-derived isothiocyanates with cysteine and lysine derivatives formed in either model systems or broccoli sprouts. *Journal of Agricultural and Food Chemistry* 60, 7735–7745.
He, C.Y., Zhang, J-S. and Chen, S.Y. (2002) A soybean gene encoding a proline-rich protein is regulated by salicylic acid, endogenous circadian rhythm and by various stresses. *Theoretical and Applied Genetics* 104, 1125–1131.
Hepler, P.K., Kunkel, J.G., Rounds, C.M. and Winship, L.J. (2012) Calcium entry into pollen tubes. *Trends in Plant Science* 17, 32–38.
Hines, P.J. (2011) Glutamate receptors in pollen. *Science Signaling* 4(170), ec 116.
Holmström, K.-O., Somersalo, S., Mandal, A., Palva, T.E. and Welin, B. (2000) Improved tolerance to salinity and low temperature in transgenic tobacco producing glycine betaine. *Journal of Experimental Botany* 51, 177–185.
Hopper, E.D., Pittman, A.M.C., Fitzgerald, M.C. and Tucker, C.L. (2008) *In vivo* and *in vitro* examination of stability of primary hyperoxaluria-associated human alanine:glyoxylate aminotransferase. *The Journal of Biological Chemistry* 283, 30493–30502.

Horváth, E., Szalai, G. and Janda, T. (2007) Induction of abiotic stress tolerance by salicylic acid signaling. *Journal of Plant Growth Regulation* 26, 290–300.

Howe, G.A. (2004) Jasmonates as signals in the wound response. *Journal of Plant Growth Regulation* 23, 223–237.

Howe, G.A. and Jander, G. (2008) Plant immunity to insect herbivores. *Annual Review of Plant Biology* 59, 41–66.

Hu, P., Meng, Y. and Wise, R.P. (2009) Functional contribution of chorismate synthase, anthranilate synthase and chorismate mutase to penetration resistance in barley-powdery mildew interactions. *Molecular Plant–Microbe Interactions* 22, 311–320.

Huang, J., Gu, M., Lai, Z., Zhou, Y.-H. and Chen, Z. (2010) Functional analysis of the *Arabidopsis PAL* gene family in plant growth, development and response to environmental stress. *Plant Physiology* 153, 1526–1538.

Huang, S.-J., Chang, C.-L., Wang, P.-H. and Chang, I.-F. (2013) A type III ACC synthase, ACS7, is involved in root gravitropism in *Arabidopsis thaliana*. *Journal of Experimental Botany* 64, 4343–4360.

Hylton, C.M., Rawsthorne, S., Smith, A.M., Jones, D.A. and Woolhouse, H.W. (1988) Glycine decarboxylase is confined to the bundle-sheath cells of leaves of C_3–C_4 intermediate species. *Planta* 175, 452–459.

Hymus, G.J., Cai, S., Kohl, E.A., Ratcliffe, O.J., Reuber, T.L. and Creelman, R.A. (2013) Application of HB17, an *Arabidopsis* class II homeodomain-leucine zipper transcription factor, to regulate chloroplast number and photosynthetic capacity. *Journal of Experimental Botany* 64, 4479–4490.

Igarashi, D., Tsuchida, H. and Ohsumi, C. (2006) Glutamate-glyoxylate aminotransferase modulates amino acid content during photorespiration. *Plant Physiology* 142, 901–910.

Ishihara, A., Hashimoto, Y., Tanaka, C. and Wakasa, K. (2008) The tryptophan pathway is involved in the defense responses of rice against pathogenic infection via serotonin production. *The Plant Journal* 54, 481–495.

Ishitani, M., Nakamura, T., Han, S.Y. and Takabe, T. (1995) Expression of the betaine aldehyde dehydrogenase gene in barley in response to osmotic stress and abscisic acid. *Plant Molecular Biology* 27, 307–315.

Jander, G. and Joshi, V. (2010) Recent progress in deciphering the biosynthesis of aspartate-derived amino acids in plants. *Molecular Plant* 3, 54–65.

Jette, L., Harvey, L., Eugeni, K. and Levens, N. (2009) 4-Hydroxyisoleucine: a plant-derived treatment for metabolic syndrome. *Current Opinion in Investigational Drugs* 10, 353–358.

Kalamaki, M.S., Merkouropoulos, G. and Kanellis, A.K. (2009a) Can ornithine accumulation modulate stress tolerance in *Arabidopsis*? *Plant Signaling and Behavior* 4, 1099–1101.

Kalamaki, M.S., Alexandrou, D.A., Lazari, D., Pateraki, I. and Kanellis, A.K. (2009b) Over-expression of a tomato *N*-acetyl-L-glutamate synthase gene (*SlNAG S1*) in *Arabidopsis thaliana* results in high ornithine levels and increased tolerance in salt and drought stresses. *Journal of Experimental Botany* 60, 1859–1871.

Kaminaga, Y., Schnepp, J., Peel, G., Kish, C.M., Schloss, J.V. and Dudareva, N. (2006) Plant phenylacetaldehyde synthase is a bifunctional homotetrameric enzyme that catalyzes phenylalanine decarboxylation and oxidation. *The Journal of Biological Chemistry* 281, 23357–23366.

Kang, J.-H., Wang, L., Giri, A. and Baldwin, I.T. (2006) Silencing threonine deaminase and *JAR4* in *Nicotiana attenuata* impairs jasmonic acid-isoleucine-mediated defenses against *Manduca sexta*. *The Plant Cell* 18, 3303–3320.

Kang, J.-Y., Choi, H., Im, M. and Kim, S.Y. (2002) *Arabidopsis* basic leucine zipper proteins that mediate stress-responsive abscisic acid signaling. *The Plant Cell* 14, 343–357.

Kang, S., Kang, K., Lee, K. and Back, K. (2007) Characterization of rice tryptophan decarboxylases and their direct involvement is serotonin biosynthesis in transgenic rice. *Planta* 227, 263–272.

Katz, R.S. and Baker, D.H. (1975) Methionine toxicity in the chick: nutritional and metabolic implications. *Journal of Nutrition* 105, 1168–1175.

Kazan, K. and Manners, J.M. (2008) Jasmonate signaling: towards an integrated view. *Plant Physiology* 146, 1459–1468.

Kazan, K. and Manners, J.M. (2009) Linking development to defense: auxin in plant–pathogen interactions. *Trends in Plant Science* 14, 373–382.

Kikuchi, G. and Hiraga, K. (1982) The mitochondrial glycine cleavage system. *Molecular and Cellular Biochemistry* 45, 137–149.

Kim, J.H. and Jander, G. (2007) *Myzus persicae* (green peach aphid) feeding on *Arabidopsis* induces the formation of a deterrent indole glucosinolate. *The Plant Journal* 49, 1008–1019.

Kim, N.H., Kim, B.S. and Hwang, B.K. (2013) Pepper arginine decarboxylase is required for polyamine and γ-aminobutyric acid signaling in cell death and defense response. *Plant Physiology* 162, 2067–2083.

Klessig, D.F., Noad, R., Navarre, D.A., Kumar, D., Lam, E. and Silva, H. (2000) Nitric oxide and salicylic acid signaling in plant defense. *Proceedings of the National Academy of Sciences of the United States of America* 97, 8849–8855.

Knill, T., Schuster, J., Gershenzon, J. and Binder, S. (2008) *Arabidopsis* branched-chain aminotransferase 3 functions in both amino acid and glucosinolate biosynthesis. *Plant Physiology* 146, 1028–1039.

Kobe, B. and Kajava, A.V. (2001) The leucine-rich repeat as a protein recognition motif. *Current Opinion in Structural Biology* 11, 725–732.

Koch, R. and Moseley, K. (2012) Phenylketonuria: newborn identification through to adulthood. In: D'Mello, J.P.F. (ed.) *Amino Acids in Human Nutrition and Health*. CAB International, Wallingford, UK, pp. 406–417.

Koo, A.J.K., Cooke, T.F. and Howe, G.A. (2011) Cytochrome P450 CYP94B3 mediates catabolism and inactivation of the plant hormone jasmonoyl-L-isoleucine. *Proceedings of the National Academy of Sciences of the United States of America* 108, 9298–9303.

Koornneef, A. and Pieterse, C.M.J. (2008) Cross talk in defense signaling. *Plant Physiology* 146, 839–844.

Lam, H.-M., Chiu, J., Hsieh, M.-H., Oliveira, I.C. and Coruzzi, G. (1998) Glutamate receptor genes in plants. *Nature* 396, 125–126.

Lan, X., Chang, K., Zeng, L., Liu, X., Qiu, F., Zheng, W., Quan, H., Liao, Z., Chen, M. Huang, W., Liu, W. and Wang, Q. (2013) Engineering salidroside biosynthetic pathway in hairy root cultures of *Rhodiola crenulata* based on metabolic characterization of tyrosine decarboxylase. *PLoS ONE* 8(10): e754590.

Lau, O.-L. and Yang, S.F. (1973) Mechanism of a synergistic effect of kinetin on auxin-induced ethylene production. *Plant Physiology* 51, 1011–1014.

Lehmann, S., Funck, D., Szabados, L. and Rentsch, D. (2010) Proline metabolism and transport in plant development. *Amino Acids* 39, 949–962.

Lehmann, T. and Pollmann, S. (2009) Gene expression and characterization of a stress-induced tyrosine decarboxylase from *Arabidopsis thaliana*. *FEBS Letters* 583, 1895–1900.

León, J., Rojo, E. and Sánchez-Serrano, J.J. (2001) Wound signalling in plants. *Journal of Experimental Botany* 52, 1–9.

Leon-Reyes, A., Du, Y., Koornneef, A., Proietti, S., Körbes, A.P., Memerlink, J., Pieterse, C.M.J. and Ritsema, T. (2010) Ethylene signaling renders the jasmonate response of *Arabidopsis* insensitive to future suppression by salicylic acid. *Molecular Plant–Microbe Interactions* 23, 187–197.

Li J. and Chory, J. (1997) A putative leucine-rich repeat receptor kinase involved in brassinosteroid signal transduction. *Cell* 90, 929–938.

Li J., Nagpal, P., Vitart, V., McMorris, T.C. and Chory, J. (1996) A role for brassinosteroids in light-dependent development of *Arabidopsis*. *Science* 272, 398–401.

Li J., Zhu, S., Song, X., Shen, Y., Chen, H. and Deng, X.W. (2006) A rice glutamate receptor-like gene is critical for the division and survival of individual cells in the root apical meristem. *The Plant Cell* 18, 340–349.

Lodwig, E.M., Hosie, A.H., Bourdès, A., Allaway, D., Karunakaran, R., Findlay, K., Downie, J.A. and Poole, P.S. (2003) Amino-acid cycling drives nitrogen fixation in the legume–*Rhizobium* symbiosis. *Nature* 422, 722–726.

Ludwig-Müller, J. (2011) Auxin conjugates: their role for plant development and in the evolution of land plants. *Journal of Experimental Botany* 62, 1757–1773.

Macho, A.P., Boutrot, F., Rathjen, J.P. and Zipfel, C. (2012) Aspartate oxidase plays an important role in *Arabidopsis* stomatal immunity. *Plant Physiology* 159, 1845–1856.

Maeda, H., Yoo, H. and Dudareva, N. (2011) Prephenate aminotransferase directs plant phenylalanine biosynthesis via arogenate. *Nature Chemical Biology* 7, 19–21.

Maier, F., Zwicker, S., Funck, J. and Pfitzner, U.M. (2011) Nonexpressor of pathogenesis-related proteins 1 (NPR1) and some NPR1-related proteins are sensitive to salicylic acid. *Molecular Plant Pathology* 12, 73–91.

Manzoor, H., Kelloniemi, J., Chiltz, A., Wendehenne, D., Pugin, A., Poinssot, B. and Garcia-Brugger, A. (2013) Involvement of the glutamate receptor AtGLR3.3 in plant defense signaling and resistance to *Hyaloperonospora arabidopsidis*. *The Plant Journal* 76, 466–480.

Mariotti, L., Piciarelli, P., Lombardi, L. and Ceccarelli, N. (2011) Fruit-set and early fruit growth in tomato are associated with increases in indoleacetic acid, cytokinin and bioactive gibberellin contents. *Journal of Plant Growth Regulation* 30, 405–415.

Mathugala, D.L., Hemsley, P.A., Moffat, C.S. and Knight, H. (2012) The mediator subunit SFR6/MED 16 controls defence gene expression mediated by salicylic acid and jasmonate responsive pathways. *New Phytologist* 195, 217–230.

Mato, J.M., Martínez-Chantar, M.L. and Lu, S.C. (2012) Methionine metabolism. In: D'Mello, J.P.F. (ed.) *Amino Acids in Human Nutrition and Health*. CAB International, Wallingford, UK, pp. 173–188.

Matsukawa, M., Shibata, Y., Ohtsu, M., Mizutani, A., Mori, H., Wang, P., Ojika, M., Kawakita, K. and Takemoto, D. (2013) *Nicotiana benthamiana* calreticulin 3a is required for the ethylene-mediated production of phytoalexins and disease resistance against oomycete pathogen *Phytophthora infestans*. *Molecular Plant–Microbe Interactions* 26, 880–892.

Mattoo, A.K. and Handa, A.K. (2008) Higher polyamines restore and enhance metabolic memory in ripening fruit. *Plant Science* 174, 386–393.

Mauch-Mani, B. and Slusarenko, A.J. (1996) Production of salicylic acid precursors is a major function of phenylalanine ammonia-lyase in the resistance of *Arabidopsis* to *Peronospora parasitica*. *The Plant Cell* 8, 203–212.

Mezencov, R., Mojzis, J., Pilatova, M. and Kutschy, P. (2003) Antiproliferative and cancer chemopreventive activity of phytoalexins: focus on indole phytoalexins from crucifers. *Neoplasma* 50, 239–245.

Michard, E., Lima, P.T., Borges, F., Silva, A.C., Portes, M.T., Cavalho, J.E., Gilliham, M., Liu, L.-H., Obermeyer, G. and Feijó, J.A. (2011) Glutamate receptor-like genes form Ca^{2+} channels in pollen tubes and are regulated by pistil D-serine. *Science* 332, 434–437.

Miflin, B.J. (1969) The inhibitory effects of various amino acids on the growth of barley seedlings. *Journal of Experimental Botany* 20, 810–819.

Miller, N.D., Durham, T.L. Assadi, A.H. and Spalding, E.P. (2010) Detection of a gravitropism phenotype in *glutamate receptor-like 3.3* mutants *of Arabidopsis thaliana* using machine vision and computation. *Genetics* 186, 585–593.

Mok, E. and Hankard, R. (2012) Amino acids in clinical and nutritional support: glutamine in Duchenne muscular dystrophy. In: D'Mello, J.P.F. (ed.) *Amino Acids in Human Nutrition and Health*. CAB International, Wallingford, UK, pp. 312–321.

Møldrup, M.E., Geu-Flores, F. and Halkier, B.A. (2013) Assigning gene function in biosynthetic pathways: camalexin and beyond. *The Plant Cell* 25, 360–367.

Moons, A., Prinsen, E., Bauw, G. and Van Montagu, M. (1997) Antagonistic effects of abscisic acid and jasmonates on salt stress-inducible transcripts in rice shoots. *The Plant Cell* 9, 2243–2259.

Moreau, M., Tian, M. and Klessig, D.F. (2012) Salicylic acid binds NPR3 and NPR4 to regulate NPR1-dependent defense responses. *Cell Research* 22, 1631–1633.

Mousavi, A. and Hotta, Y. (2005) Glycine-rich proteins: a class of novel proteins. *Applied Biochemistry and Biotechnology* 120, 169–174.

Nanjo, T., Fujita, M., Seki, M., Kato, T., Tabata, S. and Shinozaki, K. (2003) Toxicity of free proline in an *Arabidopsis* T-DNA-tagged mutant deficient in proline dehydrogenase. *Plant and Cell Physiology* 44, 541–548.

Naseem, M. and Dandekar, T. (2012) The role of auxin-cytokinin antagonism in plant–pathogen interactions. *PloS Pathogens* 8(11): e10003026.

Navarre, D.A. and Wolpert, T.J. (1995) Inhibition of the glycine decarboxylase multienzyme complex by the host-selective toxin victorin. *The Plant Cell* 7, 463–471.

Neill, S., Barros, R., Bright, J., Hancock, J. and Wilson, I. (2008) Nitric oxide, stomatal closure and abiotic stress. *Journal of Experimental Botany* 59, 165–176.

Niki, T., Mitsuhara, I., Seo, S. and Ohashi, Y. (1998) Antagonistic effect of salicylic acid and jasmonic acid on the expression of pathogenesis (PR) protein genes in wounded mature tobacco leaves. *Plant and Cell Physiology* 39, 500–507.

Noguchi, T., Takada, Y. and Kido, R. (1977) Glutamate-glyoxylate aminotransferase in rat liver cytosol: purification, properties and identity with alanine-2-oxoglutarate aminotransferase. *Hoppe-Seyler's Zeitschrift für Physiologische Chemie* 358, 1533–1542.

Oh, M.-H., Wang, X., Kota, U., Clouse, S.D. and Huber, S.C. (2009) Tyrosine phosphorylation of the BRI1 receptor kinase emerges as a component of brassinosteroid signaling in *Arabidopsis*. *Proceedings of the National Academy of Sciences of the United States of America* 106, 658–663.

Oliver, D.J., Neuburger, M., Bourguignon, J. and Douce, R. (1990) Glycine metabolism by plant mitochondria. *Physiologia Plantarum* 80, 487–491.

Ono, K., Yanagida, K., Oikawa, T., Ogawa, T. and Soda, K. (2006) Alanine racemase of alfalfa seedlings (*Medicago sativa* L.): first evidence for the presence of an amino acid racemase in plants. *Phytochemistry* 67, 856–860.

Pageau, K., Reisdorf-Cren, M., Morot-Gaudry, J.-F. and Masclaux-Daubresse, C. (2006) The two senescence-related markers, GS1 (cytosolic glutamine synthetase) and GDH (glutamate dehydrogenase), involved in nitrogen mobilization, are differentially regulated during pathogen attack and by stress hormones and reactive oxygen species in *Nicotiana tabacum* L. leaves. *Journal of Experimental Botany* 57, 547–557.

Pál, M., Kovács, V., Szalai, G., Soós, V., Ma, X., Liu, H., Mei, H. and Janda, T. (2013) Salicylic acid and abiotic stress responses in rice. *Journal of Agronomy and Crop Science* 200, 1–11.

Park, E.-J., Jeknić, Z., Pino, M.-T., Murata, N. and Chen, T.H.-H. (2007) Glycinebetaine accumulation is more effective in chloroplasts than in the cytosol for protecting transgenic tomato plants against abiotic stress. *Plant, Cell and Environment* 30, 994–1005.

Park, S., Lee, K., Kim, Y.S., Chi, Y.-T., Shin, J.S. and Back, K. (2012) Induced tyramine overproduction in transgenic rice plants expressing a rice tyrosine decarboxylase under the control of methanol-inducible rice tryptophan decarboxylase promoter. *Bioprocess and Biosystems Engineering* 35, 205–210.

Pearce, G. and Ryan, C.A. (2003) Systemic signaling in tomato plants for defense against herbivores. Isolation and characterization of three novel defense-signaling glycopeptide hormones coded in a single precursor gene. *The Journal of Biological Chemistry* 278, 30044–30050.

Pelagio-Flores, R., Ortíz-Castro, R., Méndez-Bravo, A., Macías-Rodríguez, L. and López-Bucio, J. (2011) Serotonin, a tryptophan-derived signal conserved in plants and animals, regulates root system architecture probably acting as a natural auxin inhibitor in *Arabidopsis thaliana*. *Plant and Cell Physiology* 52, 490–508.

Penninckx, I.A.M.A., Thomma, B.P.H.J., Buchala, A. and Broekaert, W.F. (1998) Concomitant activation of jasmonate and ethylene response pathways is required for induction of a plant defensin gene in *Arabidopsis*. *The Plant Cell* 10, 2103–2113.

Petersen, L.N., Marineo, S., Mandala, S., Davids, F., Sewell, B.T. and Ingle, R.A. (2010) The missing link in plant histidine biosynthesis: *Arabidopsis myoinositol monophosphatase-like 2* encodes a functional histidinol-phosphate phosphatase. *Plant Physiology* 152, 1186–1196.

Petterson, D.S., Harris, D.J. and Allen, D. (1991) Alkaloids. In: D'Mello, J.P.F., Duffus, C.M. and Duffus, J.H. (eds) *Toxic Substances in Crop Plants*. The Royal Society of Chemistry, Cambridge, UK, pp.148–179.

Picton, S., Gray, J.E., Payton, S. and Grierson, D. (1993) A histidine decarboxylase-like mRNA is involved in tomato fruit ripening. *Plant Molecular Biology* 23, 627–631.

Pietta, P.G. (2000) Flavonoids as antioxidants. *Journal of Natural Products* 63, 1035–1042.

Pillai, R.R. and Kurpad, A.V. (2012) Amino acid requirements: quantitative estimates. In: D'Mello, J.P.F. (ed.) *Amino Acids in Human Nutrition and Health*. CAB International, Wallingford, UK, pp. 267–290.

Pommerrenig, B., Feussner, K., Zierer, W., Feussner, I. and Sauer, N. (2011) Phloem-specific expression of Yang cycle genes and identification of novel Yang cycle enzymes in *Plantago* and *Arabidopsis*. *The Plant Cell* 23, 1904–1919.

Prell, J., White, J.P., Bourdes, A., Bunnewell, S., Bongaerts, R.J. and Poole, P.S. (2009) Legumes regulate *Rhizobium* bacteroid development and persistence by the supply of branched-chain amino acids. *Proceedings of the National Academy of Sciences of the United States of America* 106, 12477–12482.

Price, M.B., Jelesko, J. and Okumoto, S. (2012) Glutamate receptor homologs in plants: functions and evolutionary origins. *Frontiers in Plant Science* 3, 235–245.

Raskin, I., Ribnicky, D.M., Moreno, D.A., O'Neal, J.M. and Fridlender, B. (2002) Plants and human health in the twenty-first century. *Trends in Biotechnology* 20, 522–531.

Ravanel, S., Gakière, B., Job, D. and Douce, R. (1998) The specific features of methionine biosynthesis and metabolism in plants. *Proceedings of the National Academy of Sciences of the United States of America* 95, 7805–7812.

Ravanel, S., Block, M.A., Rippert, P., Curien, G. and Douce, R. (2004) Methionine metabolism in plants: chloroplasts are autonomous for *de novo* methionine synthesis and can import *S*-adenosylmethionine from the cytosol. *The Journal of Biological Chemistry* 279, 22548–22557.

Rayapuram, C. and Baldwin, I.T. (2007) Increased SA in NPR1-silenced plants antagonizes JA and JA-dependent direct and indirect defenses in herbivore-attacked *Nicotiana attenuata* in nature. *The Plant Journal* 52, 700–715.

Robert-Seilaniantz, A., Grant, M. and Jones, J.D.G. (2011) Hormone crosstalk in plant disease and defense: more than just jasmonate–salicylate antagonism. *Annual Review of Phytopathology* 49, 317–343.

Roberts, R.M. (2007) Does GABA act as a signal in plants? Hints from molecular studies. *Plant Signaling and Behavior* 2, 407–408.

Romagnolo, D.F., Davis, C.D. and Milner, J.A. (2012) Phytoalexins in cancer prevention. *Frontiers in Bioscience (Landmark edition)* 17, 2035–2058.

Rontein, D., Nishida, I., Basset, G. and Hanson, A.D. (2001) Plants synthesize ethanolamine by direct decarboxylation of serine using a pyridoxal phosphate enzyme. *The Journal of Biological Chemistry* 276, 35523–35529.

Rosenthal, G.A., Gulati, D.K. and Sabharwal, P.S. (1976) Additive and synergistic growth-inhibiting properties of the canaline-urea cycle amino acids. *Plant Physiology* 57, 493–496.
Ryu, H. and Hwang, I. (2013) Brassinosteroids in plant developmental signaling networks. *Journal of Plant Biology* 56, 267–273.
Sahin, G. and Kirik, D. (2012) Efficacy of L-DOPA therapy in Parkinson's disease. In: D'Mello, J.P.F. (ed.) *Amino Acids in Human Nutrition and Health*. CAB International, Wallingford, UK, pp. 454–463.
Sakamoto, A. and Murata, N. (2002) The role of glycine betaine in the protection of plants from stress: clues from transgenic plants. *Plant, Cell and Environment* 25, 163–171.
Sandorf, I. and Hollander-Czytko, H. (2002) Jasmonate is involved in the induction of tyrosine aminotransferase and tocopherol biosynthesis in *Arabidopsis thaliana*. *Planta* 216, 173–179.
Sanjaya, H.P.-Y., Su, R.-C., Ko, S.-S., Tong, C.-G. and Chan, M.-J. (2008) Overexpression of *Arabidopsis thaliana* tryptophan synthase beta1 (AtTSB1) in *Arabidopsis* and tomato confers tolerance to cadmium stress. *Plant, Cell and Environment* 31, 1074–1085.
Santner, A. and Estelle, M. (2009) Recent advances and emerging trends in plant hormone signalling. *Nature* 459, 1071–1078.
Sarrobert, C., Thibaud, M.-C., Contard-David, P., Gineste, S., Bechtold, N., Robaglia, C. and Nussaume, L. (2000) Identification of an *Arabidopsis thaliana* mutant accumulating threonine resulting from mutation in a new dihydrodipicolinate synthase gene. *The Plant Journal* 24, 357–368.
Schmelz, E.A., Albom, H.T. and Tumlinson, J.H. (2003) Synergistic interactions between volicitin, jasmonic acid and ethylene mediate insect-induced volatile emission in *Zea mays*. *Physiologia Plantarum* 117, 403–412.
Schuster, J., Knill, T., Gershenzon, J. and Binder, S. (2006) Branched-chain aminotransferase 4 is part of the chain elongation pathway in the biosynthesis of methionine-derived glucosinolates in *Arabidopsis*. *The Plant Cell* 18, 2664–2279.
Shargool, P.D. (1985) Evidence for the importance of histidine at the active site of argininosuccinate synthetase from soyabean. *Plant Cell Reports* 4, 206–209.
Sharma, S.S. and Dietz, K.-J. (2006) The significance of amino acids and amino acid-derived molecules in plant response and adaptation to heavy metal stress. *Journal of Experimental Botany* 57, 711–726.
Shi, H., Ye, T., Chen, F., Cheng, Z., Wang, Y., Yang, P., Zhang, Y. and Chan, Z. (2013) Manipulation of arginase expression modulates abiotic stress tolerance in *Arabidopsis*: effect on arginine metabolism and ROS accumulation. *Journal of Experimental Botany* 64, 1367–1379.
Smith, S.B., Ha, Y. and Ganapathy, V. (2012) D-serine and serine racemase in the retina. In: D'Mello, J.P.F. (ed.) *Amino Acids in Human Nutrition and Health*. CAB International, Wallingford, UK, pp. 137–149.
Smith, T.J. and Stanley, C.A. (2012) Glutamate dehydrogenase. In: D'Mello, J.P.F. (ed.) *Amino Acids in Human Nutrition and Health*. CAB International, Wallingford, UK, pp. 3–23.
Sobolev, V., Edelman, M., Unger, T., Albeck, S. and Galili, G. (2013) Structure of ALD1, a plant-specific homologue of the universal diaminopimelate aminotransferase enzyme of lysine biosynthesis. *Acta Crystallographica, Section F, Structural Biology and Crystallization Communications* 69, 84–89.
Staswick, P.E. (2009a) Plant hormone conjugation: a signal decision. *Plant Signaling and Behavior* 4, 757–759.
Staswick, P.E. (2009b) The tryptophan conjugates of jasmonic and indole-3-acetic acids are endogenous auxin inhibitors. *Plant Physiology* 150, 1310–1321.
Staswick, P.E. and Tiryaki, I. (2004) The oxylipin signal jasmonic acid is activated by an enzyme that conjugates it to isoleucine in *Arabidopsis*. *The Plant Cell* 16, 2117–2127.
Stephens, N.R., Qi, Z. and Spalding, E.P. (2008) Glutamate receptor subtypes evidenced by differences in desensitization and dependence on the *GLR 3.3* and *GLR 3.4* genes. *Plant Physiology* 146, 529–538.
Stoewsand, G.S. (1995) Bioactive organosulfur phytochemicals in *Brassica oleracea* vegetables: a review. *Food and Chemical Toxicology* 33, 537–543.
Stránská, J., Kopečný, D., Tylichová, M., Snégaroff, J. and Šebela, M. (2008) Ornithine δ-aminotransferase: an enzyme implicated in salt tolerance in higher plants. *Plant Signaling and Behavior* 3, 929–935.
Stránská, J., Tylichová, M., Kopečný, D., Snégaroff, J. and Šebela, M. (2010) Biochemical characterization of pea ornithine-δ-aminotransferase: substrate specificity and inhibition by di- and polyamines. *Biochimie* 92, 940–948.
Svyatyna, K., Jikumaru, Y., Brendel, R., Reichelt, M., Mithöfer, A., Takano, M., Kamiya, Y., Nick, P. and Riemann, M. (2014) Light induces jasmonate-isoleucine conjugation via OsJAR 1-dependent and -independent pathways in rice. *Plant, Cell and Environment* 37, 827–834.
Świędrych, A., Lorenc-Kukuła, K., Skirycz, A. and Szopa, J. (2004) The catecholamine biosynthesis route in potato is affected by stress. *Plant Physiology and Biochemistry* 42, 593–600.
Szabados, L. and Savouré, A. (2010) Proline: a multifunctional amino acid. *Trends in Plant Science* 15, 89–97.

Tan, S., Evans, R. and Singh, B. (2006) Herbicidal inhibitors of amino acid biosynthesis and herbicide-tolerant crops. *Amino Acids* 30, 195–204.

Tapken, D., Anschütz, U., Liu, L.-H., Huelsken, T., Seebohm, G., Becker, D. and Hollmann, M. (2013) A plant homolog of animal glutamate receptors is an ion channel gated by multiple hydrophobic amino acids. *Science Signaling* 6 (279), ra47.

Tavernier, V., Pageau, K., Langin, T. and Masclaux-Daubresse, C. (2007) The plant nitrogen mobilization promoted by *Colletotrichum lindemuthianum* in *Phaseolus* leaves depends on fungus pathogenicity. *Journal of Experimental Botany* 58, 3351–3360.

Taylor, N.L., Day, D.A. and Millar, A.H. (2002) Environmental stress causes oxidative damage to plant mitochondria leading to inhibition of glycine decarboxylase. *The Journal of Biological Chemistry* 277, 42663–42668.

Timm, S., Florian, A., Arrivault, S., Stitt, M., Fernie, A.R. and Bauwe, H. (2012) Glycine decarboxylase controls photosynthesis and plant growth. *FEBS Letters* 586, 3692–3697.

Torrigiani, P., Scaramagli, S., Mayer, M. and Biondi, S. (2005) Expression of an antisense *Datura stramonium* *S*-adenosylmethionine decarboxylase cDNA in tobacco: changes in enzyme activity, putrescine-spermidine ratio, rhizogenic potential and response to methyl jasmonate. *Journal of Plant Physiology* 162, 559–571.

Torii, K.U. (2004) Leucine-rich repeat receptor kinases in plants: structure, function and signal transduction pathways. *International Review of Cytology* 234, 1–46.

Tounekti, T., Hernández, I. and Munné-Bosch, S. (2013) Salicylic acid biosynthesis and role in modulating terpenoid and flavonoid metabolism in plant responses to abiotic stress. In: Hayat, S., Ahmad, A. and Alyemeni, M.N. (eds) *Salicylic Acid*. Springer, Dordrecht, The Netherlands, pp. 141–160.

Tzin, V. and Galili, G. (2010) The biosynthetic pathways for shikimate and aromatic amino acids in *Arabidopsis thaliana*. *The Arabidopsis Book* 8: e0132.

Ueno, H., Inoue, Y., Matsukawa, S. and Nakamura, Y. (2012) Glutamate decarboxylase. In: D'Mello, J.P.F. (ed.) *Amino Acids in Human Nutrition and Health*. CAB International, Wallingford, UK, pp. 103–121.

Uno, Y., Furihata, T., Abe, H., Yoshida, R. and Yamaguchi-Shinozaki, K. (2000) *Arabidopsis* basic leucine zipper transcription factors involved in an abscisic acid-dependent signal transduction pathway under drought and high-salinity conditions. *Proceedings of the National Academy of Sciences of the United States of America* 97, 11632–11637.

Uribarri, J. and Tuttle, K.R. (2006) Advanced glycation end products and nephrotoxicity of high-protein diets. *Clinical Journal of the American Society of Nephrology* 1, 1293–1299.

Vandelle, E. and Delledonne, M. (2011) Peroxynitrite formation and function in plants. *Plant Science* 181, 534–539.

van Wees, S.C.M., de Swart, E.A.M., van Pelt, J.A. and Pieterse, C.M.J. (2000) Enhancement of induced disease resistance by simultaneous activation of salicylate- and jasmonate-dependent defense pathways in *Arabidopsis thaliana*. *Proceedings of the National Academy of Sciences of the United States of America* 97, 8711–8716.

Vatsa, P., Chiltz, A., Bourque, S., Wendehenne, D., Garcia-Brugger, A. and Pugin, A. (2011) Involvement of putative glutamate receptors in plant defence signaling and NO production. *Biochimie* 93, 2095–2101.

Vincill, E.D., Bieck, A.M. and Spalding, E.P. (2012) Ca^{2+} conduction by an amino acid-gated ion channel related to glutamate receptors. *Plant Physiology* 159, 40–46.

Walch-Liu, P. and Forde, B.G. (2007) L-Glutamate as a novel modifier of root growth and branching: what's the sensor? *Plant Signaling and Behavior* 2, 284–286.

Wang, D., Hendrickson, A. and Dong, X. (2007) Salicylic acid inhibits pathogen growth in plants through repression of the auxin signaling pathway. *Current Biology* 17, 1784–1790.

Wang, H., Liang, X., Huang, J., Liu, Z. and Bi, Y. (2010) Involvement of ethylene and hydrogen peroxide in induction of alternative respiratory pathway in salt-treated *Arabidopsis* calluses. *Plant and Cell Physiology* 51, 1754–1765.

Wang, K.L.-C., Li, H. and Ecker, J.R. (2002) Ethylene biosynthesis and signaling networks. *The Plant Cell* 14, S131–S151.

Wethli, E., Morris, T.R. and Shresta, T.P. (1975) The effect of feeding high levels of low-quality proteins to growing chickens. *British Journal of Nutrition* 34, 363–373.

Widemann, E., Miesch, L., Lugan, R., Holder, E., Heinrich, C., Aubert, Y., Miesch, M., Pinot, F. and Heitz, T. (2013) The amidohydrolases IAR3 and ILL6 contribute to jasmonyl-isoleucine hormone turnover and generate 12-hydroxyjasmonic acid upon wounding in *Arabidopsis* leaves. *The Journal of Biological Chemistry* 288, 31701–31714.

Witzel, K., Hanschen, F.S., Krumbein, A., Ruppel, S. and Grosch, R. (2013) *Verticillium* suppression is associated with the glucosinolate composition of *Arabidopsis thaliana* leaves. *PloS ONE* 8(9): e71877.

Won, C., Shen, X., Zheng, Z., Chory, J. and Zhao, Y. (2011) Conversion of tryptophan to indole-3-acetic acid by tryptophan aminotransferases of *Arabidopsis* and YUCCAs in *Arabidopsis*. *Proceedings of the National Academy of Sciences of the United States of America* 108, 18518–18523.

Wu, C.-C., Singh, P., Chen, M.-C. and Zimmerli, L. (2010) L-Glutamine inhibits beta-aminobutyric acid-induced stress resistance and priming in *Arabidopsis*. *Journal of Experimental Botany* 61, 995–1002.

Xiong, L., Schumaker, K.S. and Zhu, J.-K. (2002) Cell signaling during cold, drought and salt stress. *The Plant Cell* 14, S165–S183.

Xu, S.-L., Rahman, A., Baskin, T.I. and Kieber, J.J. (2008) Two leucine-rich repeat receptor kinases mediate signaling, linking cell wall biosynthesis and ACC synthase in *Arabidopsis*. *The Plant Cell* 20, 3065–3079.

Xu, Y., Chang, D., Liu, M.L., Narasimhan, K.G., Raghothama, P.M., Hasegawa, P.M. and Bressan, R.A. (1994) Plant defense genes are synergistically induced by ethylene and methyl jasmonate. *The Plant Cell* 6, 1077–1085.

Yalpani, N., Léon, J., Lawton, M.A. and Raskin, I. (1993) Pathway of salicylic acid biosynthesis in healthy and virus-inoculated tobacco. *Plant Physiology* 103, 315–321.

Yamasaki, K., Motomura, Y., Yagi, Y., Nomura, H., Kikuchi, S., Nakai, M. and Shiina, T. (2013) Chloroplast envelope localization of EDS5, an essential factor for salicylic acid biosynthesis in *Arabidopsis thaliana*. *Plant Signaling and Behavior* 8(4):e23603.

Yasuda, M., Ishikawa, A., Jikumaru, Y., Seki, M., Umezawa, T., Asami, T., Maruyama-Nakashita, A., Kudo, T., Shinozaki, K., Yoshida, S. and Nakashita, H. (2008) Antagonistic interaction between systemic acquired resistance and the abscisic acid-mediated abiotic stress response in *Arabidopsis*. *The Plant Cell* 20, 1678–1692.

Yasuor, H., Belausov, E., Sadot, E. and Rubin, B. (2006) Glyphosate-induced anther indehiscence in cotton is partially temperature dependent and involves cytoskeleton and secondary wall modification and auxin accumulation. *Plant Physiology* 141, 1306–1315.

Yokota, A., Kawasaki, M., Iwano, M. and Akashi, K. (2002) Citrulline and DRIP-1 protein (ArgE homologue) in drought tolerance of wild watermelon. *Annals of Botany* 89, 825–832.

Yoo, H., Widhalm, J.R., Qian, Y., Maeda, H., Cooper, B.R., Jannasch, A.S., Gonda, I., Lewinsohn, E., Rhodes, D. and Dudareva, N. (2013) An alternative pathway contributes to phenylalanine biosynthesis via a cytosolic tyrosine-phenylpyruvate aminotransferase. *Nature Communications* 4, Article no.: 3833.

Young, T.S. and Schultz, P.G. (2010) Beyond the canonical 20 amino acids: expanding the genetic lexicon. *The Journal of Biological Chemistry* 285, 11039–11044.

Young, V.R. and Ajami, A.M (2000) Glutamate: an amino acid of particular distinction. *Journal of Nutrition* 130, 892S–900S.

Yu, H., Zhang, F., Wang, G., Liu, Y. and Liu, D. (2013) Partial deficiency of isoleucine impairs root development and alters transcript levels of the genes involved in branched-chain amino acid and glucosinolate metabolism in *Arabidopsis*. *Journal of Experimental Botany* 64, 599–612.

Zeh, M., Casazza, A.P., Kreft, O., Hoefgen, R. and Hesse, H. (2001) Antisense inhibition of threonine synthase leads to high methionine content in transgenic potato plants. *Plant Physiology* 127, 792–802.

Zeier, J. (2013) New insights into the regulation of plant immunity by amino acid metabolic pathways. *Plant, Cell and Environment* 36, 2085–2103.

Zeier, J., Pink, B., Mueller, M.J. and Berger, S. (2004) Light conditions influence specific defence responses in incompatible plant–pathogen interactions: uncoupling systemic resistance from salicylic acid and PR-1 accumulation. *Planta* 219, 673–683.

Zhang, W.C., Yang, H., Rai, A., Robson, P. and Lim, B. (2012) Glycine decarboxylase activity drives non-small cell lung cancer tumor-initiating cells and tumorigenesis. *Cell* 148, 259–272.

Zhang, X., Shen, L., Li, F., Zhang, Y., Meng, D. and Sheng, J. (2010) Up-regulating arginase contributes to amelioration of chilling stress and the antioxidant system in cherry tomato fruits. *Journal of the Science of Food and Agriculture* 90, 2195–2202.

Zhao, J. and Last, R.L. (1996) Coordinate regulation of the tryptophan biosynthetic pathway and indolic phytoalexin accumulation in *Arabidopsis*. *The Plant Cell* 8, 2235–2244.

Zhu, X. and Galili, G. (2003) Increased lysine synthesis coupled with a knockout of its catabolism synergistically boosts lysine content and also transregulates the metabolism of other amino acids in *Arabidopsis* seeds. *The Plant Cell* 15, 845–853.

Zimmerli, L., Jakab, G., Metraux, J.-P and Mauch-Mani, B. (2000) Potentiation of pathogen-specific defense mechanisms in *Arabidopsis* by β-aminobutyric acid. *Proceedings of the National Academy of Sciences of the United States of America* 97, 22920–22925.

Zimmerli, L., Hou, B.H., Tsai, C.H., Jakab, G., Mauch-Mani, B. and Somerville, S. (2008) The xenobiotic β-aminobutyric acid enhances *Arabidopsis* thermotolerance. *The Plant Journal* 54, 144–151.

Index